Research on the Properties of Polysaccharides, Starch, Protein, Pectin, and Fibre in Food Processing

Research on the Properties of Polysaccharides, Starch, Protein, Pectin, and Fibre in Food Processing

Editors

Jianhua Xie
Yanjun Zhang
Hansong Yu

MDPI

Basel • Beijing • Wuhan • Barcelona • Belgrade • Novi Sad • Cluj • Manchester

Editors
Jianhua Xie
Nanchang University
Nanchang, China

Yanjun Zhang
Chinese Academy of Tropical Agricultural Sciences
Wanning, China

Hansong Yu
Jilin Agricultural University
Changchun, China

Editorial Office
MDPI
St. Alban-Anlage 66
4052 Basel, Switzerland

This is a reprint of articles from the Special Issue published online in the open access journal *Foods* (ISSN 2304-8158) (available at: https://www.mdpi.com/journal/foods/special_issues/processing_protein).

For citation purposes, cite each article independently as indicated on the article page online and as indicated below:

Lastname, A.A.; Lastname, B.B. Article Title. *Journal Name* **Year**, *Volume Number*, Page Range.

ISBN 978-3-0365-9674-7 (Hbk)
ISBN 978-3-0365-9675-4 (PDF)
doi.org/10.3390/books978-3-0365-9675-4

Contents

About the Editors

Jianhua Xie

Jianhua Xie is a Professor of Nanchang University. He received his B.S., M.S., and Ph.D. degrees in Food Science and Engineering, Food Science and Nutrition from Nanchang University in July 2005, January 2008, and January 2015, respectively. He joined the department of Food Science, Purdue University as a Postdoctoral Fellow from 2016 to 2017 with a government scholarship from the China Scholarship Council. He is the Executive Council Member of the Society of Processing and Storage of Agro-products of the China Agricultural Society, a Council Member of the Food Science and Technology of Jiangxi Province, a Member of the Editorial Board of Journal of Food Nutrition and Dietetics, a Member of Institute of Food Technologist, a Member of Institute of Food Technologist, and a Member of American Chemical Society. His research mainly focuses on the structure and bioactivities of food carbohydrate polymers (dietary fiber, polysaccharide), and the structure–function relationships of polysaccharides. He has published more than 200 research papers, and has played a leading role in many scientific and technological projects sponsored by the National Natural Science Foundation of China, Science and Technology Support Project sponsored by the Ministry of Science and Technology, Jiangxi Province, and so on.

Yanjun Zhang

Yanjun Zhang is a Professor at the Spice and Beverage Research Institute, Chinese Academy of Tropical Agricultural Sciences. He graduated from Nanchang University with a major in Food Science and Engineering in 2012. He is engaged in scientific research on starch. Taking tropical grain crops and jackfruit starch as representative material, research has been carried out on the basic theory and application of technology with ideas including the relationship between starch functional characteristics and molecular structure, the pasting and retrogradation properties of starch controlled by the molecular weight of amylopectin, and starch digestion improved through innovative methods. The special functional properties and molecular structure of jackfruit starch have been clarified. The control theory of starch pasting and retrogradation has been enriched based on the cognition of starch structure and function. An efficient preparation method has also been established for improving the digestibility and utilization rate of jackfruit starch. The previous research led to 64 publications, 7 national invention patents, and the Hainan Provincial Science and Technology Progress Award in 2017. The applicant was selected from national-level talents: Shennong Talents, Jiangxi Province Double Thousand Plan leading talents, Hainan Province South Sea famous, 515 talent project second level, hot science excellent youth, and other candidates.

Hansong Yu

Hansong Yu is a Professor and Doctoral Supervisor at the College of Food Science and Engineering, Jilin Agricultural University. Professor Yu graduated from Jilin University's Department of Biochemistry and Molecular Biology, receiving his doctorate degree in 2008. From 2013 to 2014, he collaborated on the intensive processing of soybean products as a Fellow of the American Academy of Food Sciences (SAM K.C. CHANG, IFT Fellow, ACS Fellow) at North Dakota State University and Mississippi State University in the United States. He has presided over 13 projects of the National Agricultural Industry Technology System, the National Natural Science Foundation of China, the Major National Science and Technology Support Project, National Ministry of Agriculture 948 Project, and other national and Jilin major projects. He won the first prize of the Jilin Provincial Science and Technology Progress Award, the first prize of the National Harvest Award for agriculture, animal husbandry, and fishery, two first prizes of the Jilin Provincial Natural Science Achievement Award, four second prizes of the Science and Technology Progress Award, and one third prize. He has published more than 300 papers (90 papers included in SCI and EI) and two monographs and textbooks. Moreover, he has applied for and obtained 25 patents, and participated in the formulation of four national and industry standards.

Preface

Polysaccharides, starch, protein, pectin, and fibre are naturally occurring polymers with efficient, safe, and green characteristics. These food components show different physico-chemical properties, functional properties, and biological activity, and these natural components have broad application prospects in food processing. Therefore, research on the modification and application of polysaccharides, starch, protein, pectin, and fibre is needed. Our Special Issue, "Research on the Properties of Polysaccharides, Starch, Protein, Pectin, and Fibre in Food Processing", with four reviews and thirty articles, is an attempt to provide a broad perspective on the recent in-depth study of these components. The Guest Editors are very grateful to all the authors for their contributions, and the entire MDPI team for their professional support.

Jianhua Xie, Yanjun Zhang, and Hansong Yu
Editors

foods

Editorial

Research on the Properties of Polysaccharides, Starch, Protein, Pectin, and Fibre in Food Processing

Xin Qi [1], Yanjun Zhang [2], Hansong Yu [3] and Jianhua Xie [1,*]

1 State Key Laboratory of Food Science and Technology, Nanchang University, Nanchang 330047, China
2 Spice and Beverage Research Institute, Chinese Academy of Tropical Agricultural Sciences, Wanning 571533, China
3 College of Food Science and Engineering, Jilin Agricultural University, Changchun 130118, China
* Correspondence: jhxie@ncu.edu.cn; Tel./Fax: +86-791-8830-4347

As food components, polysaccharides, starch, protein, pectin, and fibre are often used in the food industry due to their particular functional properties, as well as their efficient, safe, and green characteristics. As shown in Table 1, they have been used to improve food texture, increase freshness and shelf-life, produce food packaging materials and health care products, deliver food active ingredients, and so on. However, these native components still have some drawbacks, such as lower solubility and instability under certain conditions, which limit their application in food processing. Therefore, the modification of natural food components by physical, chemical, and enzymatic means seeks to obtain components with superior functional properties and to expand their applications in the food field. In this sense, a Special Issue entitled "Research on the Properties of Polysaccharides, Starch, Protein, Pectin, and Fibre in Food Processing" was launched in *Foods* (MDPI) to provide a forum for researchers to communicate their latest research findings on polysaccharides, starch, protein, and other food components. A total of 63 manuscripts were submitted to this Special Issue from various countries, of which 34 were accepted for publication in the peer-review process, including 4 reviews and 30 research articles.

Table 1. The properties and applications of food components in food.

Food Components	Properties	Applications in Food	Refs.
Polysaccharides	Antibacteria, immunomodulatory, anti-tumour, hypoglycaemic, hypolipidemic, anti-oxidation effects, thickening, stabilizing, gelling and emulsifying properties, etc.	Packaging materials, preservative, health care products, food active ingredient carriers, texture modifiers, etc.	[1–3]
Starch	Thickening, stabilizing, gelling, emulsifying, film-forming properties, water holding capacity, etc.	Ingredient of gelled foodstuff, texture modifiers, biodegradable packaging materials, food active ingredient carriers, etc.	[4,5]
Protein	Gelation, water holding and binding capacities, foaming ability and stability, solubility, emulsification, etc.	Texture modifiers, nutrient supplements, plant-based meats, fat replacement, food active ingredient carriers, edible packaging film, etc.	[6,7]
Pectin	Thickening, stabilizing, gelling, emulsifying, film-forming properties, water and oil holding capacities, reduce blood lipid, cholesterol, radiation resistance, adsorption of heavy metal ions capacities, etc.	Texture modifiers, edible packaging film, antioxidant, fat replacement, health care products, food active ingredient carriers, etc.	[8,9]

Citation: Qi, X.; Zhang, Y.; Yu, H.; Xie, J. Research on the Properties of Polysaccharides, Starch, Protein, Pectin, and Fibre in Food Processing. *Foods* **2023**, *12*, 249. https://doi.org/10.3390/foods12020249

Received: 21 December 2022
Revised: 27 December 2022
Accepted: 3 January 2023
Published: 5 January 2023

Table 1. *Cont.*

Food Components	Properties	Applications in Food	Refs.
Fibre	Water holding, water swelling, water retention, oil holding, glucose retardation index, stabilizing, gel-forming, antioxidant capacities, water solubility, etc.	Texture modifiers, antioxidant, fat replacement, product freshness and shelf-life fortifier, food packaging, etc.	[10,11]

Among these published papers, there were plenty of studies on polysaccharide biological activities. Li et al. [12] isolated and purified polysaccharides from yam. They found the molecular weight of yam polysaccharides was 20.89 kDa, which primarily comprised galactose, glucose, and galacturonic acid with a ratio of 28.57:11.28:37.59. Moreover, pretreatment with yam polysaccharides attenuated oxidative damage in IEC-6 cells by H_2O_2 through the mitogen-activated protein kinase (MAPK) signalling pathway, improved cell viability, superoxide dismutase (SOD) activity, and reduced intracellular reactive oxygen species (ROS) production and malondialdehyde (MDA) content after H_2O_2 injury. Therefore, yam polysaccharides possess a good protective effect against oxidative damage, which could provide a reference for the development of functional foods and clinical therapeutics against the damage resulting from oxidative stress. Wang et al. [13] revealed that *Ganoderma atrum* (*G. atrum*) polysaccharide (PSG-1) activated the mammalian target of the rapamycin (mTOR) signalling pathway, and also alleviated the acrolein-induced impairment of viability, autophagy, and apoptosis of IEC-6 cells. Moreover, another study reported that short-chain fatty acids in the faeces of mice treated with high *Holothuria leucospilota* polysaccharides (HLP) doses were significantly higher than those treated with lower doses and the normal group. After oral administration of HLP, the glutathione peroxidase and SOD activities increased, and malondialdehyde contents in the mouse livers decreased, which revealed the good performance of HLP with respect to liver antioxidants [14]. Natural polysaccharides come from a wide range of sources; in addition to higher plants, microorganisms (bacteria and fungi) and algae are also rich in polysaccharides. Sun et al. [15] found that a polysaccharide produced by *Chaetomium globosum* CGMCC 6882 had antibacterial activity, and it enriched the abundance of gut microbiota and the firmicutes/bacteroidetes ratio was increased from 0.52 to 0.72. Chemical selenylation enhanced the immuno-modulatory effect of *Portulaca oleracea* L. polysaccharides (PSPO) on the murine splenocytes and RAW 264.7 macrophages [16]. In recent years, various natural polysaccharides, including *Dictyophora indusiata* polysaccharides, *Sarcodon aspratus* polysaccharides, *Flammuliana velutipes* polysaccharides, and *Pleurotus eryngii* polysaccharides, have shown great potential in attenuating systemic inflammation activities, such as inflammatory bowel disease (IBD). Wang et al. [17] summarized the main mechanisms of polysaccharides on IBD, including immune regulation, anti-oxidation, and regulation of probiotics in the intestine. Moreover, some polysaccharides were found to help control type 2 diabetes with minimal side effects. Wan et al. [18] reported that potential mechanisms for the antidiabetic function of β-glucan may include the retardation of macronutrient absorption and the inhibition of digestive enzymes. In addition to having a variety of biological activities, polysaccharides are also a good delivery carrier. Ferron et al. [19] used a polysaccharide fraction isolated from *Camelina sativa* L. Krantz (CCP) as a carrier for purple corn cob extract (MCE), and they found CCP could supply a protective barrier for MCE, effectively increasing storage stability and bioaccessibility. At the same time, polysaccharide-based materials have advantages such as edibility and environmentally friendly performance; as such, polysaccharides as a new kind of packaging material have been proposed. The most commonly used polysaccharide as a packaging material is chitosan. Chitosan has good biological activities, including antibacterial and antioxidant activity and enzyme activity inhibition, and thus possesses great potential in the field of food packaging and preservation [20]. Additionally, cellulose, hemicellulose and starch are also important natural sources of packaging materials. Liu et al. [20] and Zhao et al. [21] reviewed the latest advances in polysaccharide-based materials in food packaging and provided references for the development of modern edible packaging and novel preservation technologies.

Starch is widely used in the food industry, primarily in bakery products, noodles, instant foods, and snacks. The physicochemical properties of starches, such as pasting, rheological and retrogradation behaviours, could determine the qualities of their end products. However, many native starches have some drawbacks, such as poor solubility, stability against heat and shear during pasting, undesired paste consistency, and easy retrogradation. These drawbacks have limited the application of starch in the food industry. Therefore, many modification methods, such as chemical, enzymatic, and physical methods, or some combination of these, have been used to enhance the physicochemical properties of native starch to meet the demands of consumers. Xu et al. [22] modified potato starch by a pre-gelatinization method and built correlations between starch physicochemical properties and the degree of starch gelatinization (DSG). They found that the endothermic enthalpy, gelatinization range, and short-range ordered structure of starch were negatively correlated with DSG, whereas onset gelatinization temperature, apparent viscosity, and water-binding capacity were positively correlated. The viscoelasticity of starch gels was negatively correlated with the DSG after full gelatinization (DSG > 39.41%). Starch granules gradually lost their typical shape, and less birefringence could be observed with increasing DSG. Dai et al. [23] adjusted the physicochemical properties of rice, potato, and pea starches with betanin. Betanin decreased the peak, trough, and final viscosity of rice and potato starches, but increased those of pea starch. The poor short-range molecular order, low crystallinity, and low retrogradation enthalpy of starches were induced by betanin during retrogradation, suggesting that betanin could inhibit the retrogradation of starches. Similarly, Xu et al. [24] investigated the effect of different degrees of polymerization (DP) of polymeric proanthocyanidins (PPC) on the physicochemical characteristics and in vitro starch digestibility of potato starch. PPC enhanced the thermal stability, delayed starch gelatinization of potato starch, and affected the elastic properties, instead of the viscous properties, of potato starch pasting. PPC with a lower DP more evidently influenced starch gelling performance, whereas larger PPC molecules exhibited a greater impact on starch long-term retrogradation and digestive properties. In another study, native starch and fermented starch featured significantly higher hygroscopicity than starch granulate [25]. Resistant starch is the sum of starch and starch degradation products not absorbed by the small intestine of a healthy individual, which could improve insulin resistance and glucose homeostasis, maintain colon health, control body weight, elevate large-bowel short-chain fatty acids, and especially lower blood lipid. Zhang et al. [26] evaluated the effects of jackfruit seed sourced resistant starch (JSRS) on mice gut microbes and hyperlipidaemia. JSRS promoted the growth of *Bifidobacterium pseudolongum*, and *Bifidobacterium pseudolongum* interacted with JSRS to significantly reduce bodyweight and serum lipid levels, having a therapeutic effect on hepatic steatosis in mice. Their findings will help in the development of "synbiotics" for the treatment of hyperlipidaemia.

Dietary fibre is a polysaccharide that can neither be digested or absorbed by the gastrointestinal tract, and it is recognized as a seventh category nutrient by the nutritional supplement community. Therefore, improving the quality and functional properties of insoluble dietary fibre through physical, chemical, and biological approaches is a prerequisite for its wide application in the food sector. He et al. [27] used a microfluidizer system to improve the structures and functionalities of pea fibre. Cheng et al. [28] investigated the effects of γ-irradiation on the structure and functional properties of pea fibre. Their results showed that, when the γ-irradiation dose was 2 kGy, the highest oil-holding capacity, swelling capacity and water-holding capacity of pea fibre were 8.12 ± 0.12 g/g, 19.75 ± 0.37 mL/g, and 8.35 ± 0.18 g/g, respectively. γ-irradiation technology as an effective method significantly improved the functional properties of pea fibre. Lyu et al. [29] explored the properties of high-purity insoluble dietary fibre from *Okara* (HPIDF); after fermentation, it showed a higher water-holding capacity, water-swelling capacity, heavy-metals-adsorption capacity, and harmful-substances-adsorption capacity than HPIDF due to the changes in structure caused by fermentation. Based on the above studies, it is clear that dietary fibre changed its own properties after fermentation. Does the presence of dietary fibre affect the fermentation of other substances in the intestine? Yang et al. [30] explored the effect of four dietary fibres (fructose-oligosaccharides, pectin, β-glucan and arabinoxylan) on the modulation of cyanidin-3-O-glucoside (C3G) fermentation patterns. The results showed that the C3G metabolites after in vitro fermentation mainly included cyanidin, protocatechuic acid, 2,4,6-trihydroxybenzoic acid, and 2,4,6-trihydroxybenzaldehyde. Dietary fibres changed the proportions of C3G metabolites, which were dependent on the dietary fibres structures. It is well known that dietary fibre is an important component for promoting human health and managing calorie consumption. Fan et al. [31] showed that the *Okara* insoluble dietary fibre of three particle sizes could improve the elevation of blood lipids and the disturbance of gut microbiota caused by a high-fat diet; this finding provides valuable information regarding its future application in functional foods. In addition, dietary fibre is also a good carrier for the protection of bioactive substances. He et al. [32] found that soybean insoluble dietary fibre effectively improved the stability of malvidin-3-O-glucoside under different pH values (1.0~7.0), high temperatures, and sunlight conditions.

Protein is widely used in all kinds of foods, such as candy, pastry, dairy products, and so on. It not only enhances nutrition but can improve the quality and texture of various foods due to gelation, water holding and binding capacities, foaming ability, and stability, solubility, emulsification, etc. However, the functional properties of many proteins do not meet expectations, so multiple methods (e.g., physical, chemical, and enzymatic modifications) have been used to improve the functional properties of proteins. Pang et al. [33] prepared (−)-epigallocatechin gallate (EGCG)-hemp (*Cannabis sativa* L.) seed protein (HPI) conjugates to modify HPI. The solubility and emulsifying properties of the EGCG-HPI conjugates were improved compared with native HPI. Enzymatic hydrolysis was considered as a promising method for the improvement in functional properties of protein. Shuai et al. [34] found enzymatic modification treatment distinctly boosted the solubility of pea protein (PP), while the foaming and emulsification of PP were improved after enzymatic treatment. Zhang et al. [35] reported that the foaming property and water/oil holding capacity of walnut protein isolates were improved after oxidation by 2,2′-azobis (2-amidinopropane) dihydrochloride. Isoelectric solubilization/precipitation (co-precipitation) was used to produce protein mixtures from pea (PPI) and grass carp (CPI) [36]. The solubility and emulsifying properties of co-precipitates (PPI-CPI) were greater than those of PPI, CPI, and isolated protein blends (BL), and the foaming activity was higher than that of BL and CPI. Zhou et al. [37] mixed these four protein systems (PPI-CPI, PPI, CPI, BL) with soybean oil and water to form a pre-emulsification system to replace animal fat and prepare fish sausage. Their results showed that the overall quality of PPI-CPI fish sausage, as a whole, was significantly superior to that of PPI and BL, which was equivalent to CPI fish sausage. PPI-CPI fish sausage had a significantly higher water-holding capacity and hardness ($p < 0.05$). Generally, protein has gelling properties, and protein-based gelled products have become an increasingly popular food, such as surimi-based products. Heat treatment and the addition of some exogenous substances determine the quality of the gel matrix. Jiang et al. [38] reported that starch addition had a positive effect on the weak gel matrix by direct heating. Zhu et al. [39] found that carboxymethyl chitosan oligosaccharide stabilized the microstructure of the surimi gels associated with improved gel strength, viscoelasticity, water-holding capacities, and whiteness during frozen storage. Their studies are a route by which to widen the development of protein-based gelled products, and they provide useful information for food cryoprotectants. From the above studies, it is clear that proteins and polysaccharides are often added to the same system to better nullify the disadvantages of protein and polysaccharide alone, such as poor stability, gelling properties, etc. Li et al. [40] prepared zein nanoparticles/flaxseed gum complexes to stabilize Pickering emulsion. The emulsions remained stable without a serum phase for 14 days and exhibited improved stability at low-temperature storage; this provides guidance for the preparation of protein–polysaccharide complexes and Pickering emulsions. In the complex food system, food constituents (pectin, starch, sucrose, and protein) interact with each other and thus change their properties. Liu et al. [41] showed food constituents in the food system decreased the activity of polyphenol oxidase but increased the thermostability of polyphenol oxidase, and pectin exhibited the strongest protective effect against thermal inactivation. Therefore, it is meaningful to study the interactions between food macromolecules because they could affect molecular properties and food texture. Zhang et al. [42] studied the interaction between gelatin and *Mulberry* leaf polysaccharides obtained by four different extraction conditions (hot buffer = HBSS; chelating agent = CHSS; dilute alkali = DASS; concentrated alkali = CASS). They found that the addition of gelatin reduced the apparent viscosity of the four polysaccharide solutions. Gelatin-polysaccharides obtained by CHSS were highly stable, and gelatin-polysaccharides obtained by CASS were more suitable as stabilizers in the freezing process. Therefore, different extraction methods affect the properties of polysaccharides. He et al. [43] optimized the extraction process of quinoa (*Chenopodium quinoa* Willd.) protein isolate (QPI), and found the hydrolysis degree and total amino acid content at the end of digestion were significantly lower than those untreated at 90 °C and 121 °C for 30 min. Therefore, heat treatment could affect the vitro digestion characteristics of QPI. Among the many food proteins, there are some proteins, such as egg protein and peanut protein, which can cause allergic reactions and give consumers an unpleasant experience. Therefore, various studies have focused on exploring effective and safe technologies that might reduce the allergenicity of proteins. Wen et al. [44] used an emerging and efficient technology, superheated steam (SS), to change the allergenicity and molecular structure of ovomucoid (OVM). The IgG-binding ability of OVM decreased and the releases of β-hexosaminidase and TNF-γ were inhibited after SS treatment. The natural OVM structure was disrupted, leading to increased protein digestibility. Moreover, peptide Phe-Thr-Gly-Met-Leu (FTGML) inhibited the tyrosinase activity and melanin synthesis of B16F10 melanoma cells. FTGML reduced melanogenesis in B16F10 cells by downregulating the cAMP-PI3K/Akt and MAPK pathways (p38 and JNK) [45].

In short, the articles in this Special Issue cover various fields such as improvements in the physical properties of major food components, biological activity studies, and applications in food preservation. The pursuit of food texture and taste and the stringent requirements for safe, green, and efficient food materials are the main driving forces for continued research in this field.

Author Contributions: X.Q. and J.X. conceived and wrote this editorial; writing—review and editing, X.Q., Y.Z., H.Y. and J.X. All authors have read and agreed to the published version of the manuscript.

Conflicts of Interest: The authors declare no conflict of interest.

References

1. Dedhia, N.; Marathe, S.J.; Singhal, R.S. Food polysaccharides: A review on emerging microbial sources, bioactivities, nanoformulations and safety considerations. *Carbohydr. Polym.* **2022**, *287*, 119355. [CrossRef]
2. Chen, X.; Yang, J.; Shen, M.; Chen, Y.; Yu, Q.; Xie, J. Structure, function and advance application of microwave-treated polysaccharide: A review. *Trends Food Sci. Technol.* **2022**, *123*, 198–209. [CrossRef]
3. Yang, J.; Shen, M.; Luo, Y.; Wu, T.; Chen, X.; Wang, Y.; Xie, J. Advanced applications of chitosan-based hydrogels: From biosensors to intelligent food packaging system. *Trends Food Sci. Technol.* **2021**, *110*, 822–832. [CrossRef]
4. Rong, L.; Shen, M.; Wen, H.; Xiao, W.; Li, J.; Xie, J. Effects of xanthan, guar and *Mesona chinensis* Benth gums on the pasting, rheological, texture properties and microstructure of pea starch gels. *Food Hydrocoll.* **2022**, *125*, 107391. [CrossRef]
5. Cheng, H.; Chen, L.; McClements, D.J.; Yang, T.; Zhang, Z.; Ren, F.; Miao, M.; Tian, Y.; Jin, Z. Starch-based biodegradable packaging materials: A review of their preparation, characterization and diverse applications in the food industry. *Trends Food Sci. Technol.* **2021**, *114*, 70–82. [CrossRef]
6. Razi, S.M.; Fahim, H.; Amirabadi, S.; Rashidinejad, A. An overview of the functional properties of egg white proteins and their application in the food industry. *Food Hydrocoll.* **2023**, *135*, 108183. [CrossRef]
7. Munialo, C.D.; Stewart, D.; Campbell, L.; Euston, S.R. Extraction, characterisation and functional applications of sustainable alternative protein sources for future foods: A review. *Future Foods* **2022**, *6*, 100152. [CrossRef]
8. Singhal, S.; Swami Hulle, N.R. Citrus pectins: Structural properties, extraction methods, modifications and applications in food systems—A review. *Appl. Food Res.* **2022**, *2*, 100215. [CrossRef]

9. Munoz-Almagro, N.; Montilla, A.; Villamiel, M. Role of pectin in the current trends towards low-glycaemic food consumption. *Food Res. Int.* **2021**, *140*, 109851. [CrossRef]
10. Elleuch, M.; Bedigian, D.; Roiseux, O.; Besbes, S.; Blecker, C.; Attia, H. Dietary fibre and fibre-rich by-products of food processing: Characterisation, technological functionality and commercial applications: A review. *Food Chem.* **2011**, *124*, 411–421. [CrossRef]
11. Pathania, S.; Kaur, N. Utilization of fruits and vegetable by-products for isolation of dietary fibres and its potential application as functional ingredients. *Bioact. Carbohydr. Diet. Fibre* **2022**, *27*, 100295. [CrossRef]
12. Li, Z.; Xiao, W.; Xie, J.; Chen, Y.; Yu, Q.; Zhang, W.; Shen, M. Isolation, Characterization and Antioxidant Activity of Yam Polysaccharides. *Foods* **2022**, *11*, 800. [CrossRef] [PubMed]
13. Wang, Y.; Chang, X.; Zheng, B.; Chen, Y.; Xie, J.; Shan, J.; Hu, X.; Ding, X.; Hu, X.; Yu, Q. Protective Effect of *Ganoderma atrum* Polysaccharide on Acrolein-Induced Apoptosis and Autophagic Flux in IEC-6 Cells. *Foods* **2022**, *11*, 240. [CrossRef] [PubMed]
14. Wang, W.; Yuan, Y.; Cao, J.; Shen, X.; Li, C. Beneficial Effects of *Holothuria leucospilota* Polysaccharides on Fermentability In Vivo and In Vitro. *Foods* **2021**, *10*, 1884. [CrossRef] [PubMed]
15. Sun, X.; Wang, Z.; Hu, X.; Zhao, C.; Zhang, X.; Zhang, H. Effect of an Antibacterial Polysaccharide Produced by *Chaetomium globosum* CGMCC 6882 on the Gut Microbiota of Mice. *Foods* **2021**, *10*, 1084. [CrossRef]
16. Lin, Y.; Guan, Q.; Li, L.; Tang, Z.; Zhang, Q.; Zhao, X. In Vitro Immuno-Modulatory Potentials of Purslane (*Portulaca oleracea* L.) Polysaccharides with a Chemical Selenylation. *Foods* **2022**, *11*, 14. [CrossRef]
17. Wang, Y.; Zhu, H.; Wang, X.; Yu, Y.; Xie, J. Natural Food Polysaccharides Ameliorate Inflammatory Bowel Disease and Its Mechanisms. *Foods* **2021**, *10*, 1288. [CrossRef]
18. Wan, Y.; Xu, X.; Gilbert, R.G.; Sullivan, M.A. A Review on the Structure and Anti-Diabetic (Type 2) Functions of beta-Glucans. *Foods* **2021**, *11*, 57. [CrossRef]
19. Ferron, L.; Milanese, C.; Colombo, R.; Pugliese, R.; Papetti, A. A New Polysaccharide Carrier Isolated from *Camelina* Cake: Structural Characterization, Rheological Behavior, and Its Influence on Purple Corn Cob Extract's Bioaccessibility. *Foods* **2022**, *11*, 1736. [CrossRef]
20. Liu, T.; Li, J.; Tang, Q.; Qiu, P.; Gou, D.; Zhao, J. Chitosan-Based Materials: An Overview of Potential Applications in Food Packaging. *Foods* **2022**, *11*, 1490. [CrossRef]
21. Zhao, Y.; Li, B.; Li, C.; Xu, Y.; Luo, Y.; Liang, D.; Huang, C. Comprehensive Review of Polysaccharide-Based Materials in Edible Packaging: A Sustainable Approach. *Foods* **2021**, *10*, 1845. [CrossRef]
22. Xu, F.; Zhang, L.; Liu, W.; Liu, Q.; Wang, F.; Zhang, H.; Hu, H.; Blecker, C. Physicochemical and Structural Characterization of Potato Starch with Different Degrees of Gelatinization. *Foods* **2021**, *10*, 1104. [CrossRef] [PubMed]
23. Dai, T.; He, X.; Xu, J.; Geng, Q.; Li, C.; Sun, J.; Liu, C.; Chen, J.; He, X. Effects of Betanin on Pasting, Rheology and Retrogradation Properties of Different Starches. *Foods* **2022**, *11*, 1600. [CrossRef] [PubMed]
24. Xu, J.; Dai, T.; Chen, J.; He, X.; Shuai, X.; Liu, C.; Li, T. Effects of Three Types of Polymeric Proanthocyanidins on Physicochemical and In Vitro Digestive Properties of Potato Starch. *Foods* **2021**, *10*, 1394. [CrossRef]
25. Ocieczek, A.; Mesinger, D.; Toczek, H. Hygroscopic Properties of Three Cassava (*Manihot esculenta* Crantz) Starch Products: Application of BET and GAB Models. *Foods* **2022**, *11*, 1966. [CrossRef] [PubMed]
26. Zhang, Z.; Wang, Y.; Zhang, Y.; Chen, K.; Chang, H.; Ma, C.; Jiang, S.; Huo, D.; Liu, W.; Jha, R.; et al. Synergistic Effects of the Jackfruit Seed Sourced Resistant Starch and *Bifidobacterium pseudolongum* subsp. *globosum* on Suppression of Hyperlipidemia in Mice. *Foods* **2021**, *10*, 1431. [CrossRef]
27. He, X.; Dai, T.; Sun, J.; Liang, R.; Liu, W.; Chen, M.; Chen, J.; Liu, C. Disintegrating the Structure and Improving the Functionalities of Pea Fiber by Industry-Scale Microfluidizer System. *Foods* **2022**, *11*, 418. [CrossRef]
28. Cheng, T.; Liu, C.; Hu, Z.; Wang, Z.; Guo, Z. Effects of gamma-Irradiation on Structure and Functional Properties of Pea Fiber. *Foods* **2022**, *11*, 1433. [CrossRef]
29. Lyu, B.; Wang, Y.; Zhang, X.; Chen, Y.; Fu, H.; Liu, T.; Hao, J.; Li, Y.; Yu, H.; Jiang, L. Changes of High-Purity Insoluble Fiber from Soybean Dregs (*Okara*) after Being Fermented by Colonic Flora and Its Adsorption Capacity. *Foods* **2021**, *10*, 2485. [CrossRef]
30. Yang, Z.; Huang, T.; Li, P.; Ai, J.; Liu, J.; Bai, W.; Tian, L. Dietary Fiber Modulates the Fermentation Patterns of Cyanidin-3-O-Glucoside in a Fiber-Type Dependent Manner. *Foods* **2021**, *10*, 1386. [CrossRef]
31. Fan, H.; Zhang, Y.; Swallah, M.S.; Wang, S.; Zhang, J.; Fang, J.; Lu, J.; Yu, H. Structural Characteristics of Insoluble Dietary Fiber from *Okara* with Different Particle Sizes and Their Prebiotic Effects in Rats Fed High-Fat Diet. *Foods* **2022**, *11*, 1298. [CrossRef] [PubMed]
32. He, Y.; Chen, D.; Liu, Y.; Sun, X.; Guo, W.; An, L.; Shi, Z.; Wen, L.; Wang, Z.; Yu, H. Protective Effect and Mechanism of Soybean Insoluble Dietary Fiber on the Color Stability of Malvidin-3-O-glucoside. *Foods* **2022**, *11*, 1474. [CrossRef] [PubMed]
33. Pang, X.H.; Yang, Y.; Bian, X.; Wang, B.; Ren, L.K.; Liu, L.L.; Yu, D.H.; Yang, J.; Guo, J.C.; Wang, L.; et al. Hemp (*Cannabis sativa* L.) Seed Protein-EGCG Conjugates: Covalent Bonding and Functional Research. *Foods* **2021**, *10*, 1618. [CrossRef] [PubMed]
34. Shuai, X.; Gao, L.; Geng, Q.; Li, T.; He, X.; Chen, J.; Liu, C.; Dai, T. Effects of Moderate Enzymatic Hydrolysis on Structure and Functional Properties of Pea Protein. *Foods* **2022**, *11*, 2368. [CrossRef] [PubMed]
35. Zhang, X.; Yang, X.; Li, Y.; Wang, Z.; He, X.; Sun, J. Effect of Peroxyl Radical-Induced Oxidation on Functional and Structural Characteristics of Walnut Protein Isolates Revealed by High-Resolution Mass Spectrometry. *Foods* **2022**, *11*, 385. [CrossRef]
36. Zhou, X.; Zhang, C.; Cao, W.; Zhou, C.; Zheng, H.; Zhao, L. A Comparative Functional Analysis of Pea Protein and Grass Carp Protein Mixture via Blending and Co-Precipitation. *Foods* **2021**, *10*, 3037. [CrossRef]

37. Zhou, X.; Zhang, C.; Zhao, L.; Zhou, X.; Cao, W.; Zhou, C. Effect of Pre-Emulsion of Pea-Grass Carp Co-Precipitation Dual Protein on the Gel Quality of Fish Sausag. *Foods* **2022**, *11*, 3192. [CrossRef]
38. Jiang, X.; Chen, Q.; Xiao, N.; Du, Y.; Feng, Q.; Shi, W. Changes in Gel Structure and Chemical Interactions of *Hypophthalmichthys molitrix* Surimi Gels: Effect of Setting Process and Different Starch Addition. *Foods* **2021**, *11*, 9. [CrossRef]
39. Zhu, X.; Zhu, M.; He, D.; Li, X.; Shi, L.; Wang, L.; Xu, J.; Zheng, Y.; Yin, T. Cryoprotective Roles of Carboxymethyl Chitosan during the Frozen Storage of Surimi: Protein Structures, Gel Behaviors and Edible Qualities. *Foods* **2022**, *11*, 356. [CrossRef]
40. Li, Y.; Xu, G.; Li, W.; Lv, L.; Zhang, Q. The Role of Ultrasound in the Preparation of Zein Nanoparticles/Flaxseed Gum Complexes for the Stabilization of Pickering Emulsion. *Foods* **2021**, *10*, 1990. [CrossRef]
41. Liu, J.; Zhang, J.; Liao, T.; Zhou, L.; Zou, L.; Liu, Y.; Zhang, L.; Liu, W. Thermal Inactivation Kinetics of Kudzu (*Pueraria lobata*) Polyphenol Oxidase and the Influence of Food Constituents. *Foods* **2021**, *10*, 1320. [CrossRef] [PubMed]
42. Zhang, X.X.; Liao, B.Y.; Guan, Z.J.; Thakur, K.; Khan, M.R.; Busquets, R.; Zhang, J.G.; Wei, Z.J. Interaction between Gelatin and *Mulberry* Leaf Polysaccharides in Miscible System: Physicochemical Characteristics and Rheological Behavior. *Foods* **2022**, *11*, 1571. [CrossRef]
43. He, X.; Wang, B.; Zhao, B.; Yang, F. Ultrasonic Assisted Extraction of Quinoa (*Chenopodium quinoa* Willd.) Protein and Effect of Heat Treatment on Its In Vitro Digestion Characteristics. *Foods* **2022**, *11*, 771. [CrossRef] [PubMed]
44. Wen, P.W.; Tu, Z.C.; Hu, Y.M.; Wang, H. Effects of Superheated Steam Treatment on the Allergenicity and Structure of Chicken Egg Ovomucoid. *Foods* **2022**, *11*, 238. [CrossRef]
45. Hu, Z.; Sha, X.; Zhang, L.; Huang, S.; Tu, Z. Effect of Grass Carp Scale Collagen Peptide FTGML on cAMP-PI3K/Akt and MAPK Signaling Pathways in B16F10 Melanoma Cells and Correlation between Anti-Melanin and Antioxidant Properties. *Foods* **2022**, *11*, 391. [CrossRef] [PubMed]

foods

MDPI

Article

Effect of Pre-Emulsion of Pea-Grass Carp Co-Precipitation Dual Protein on the Gel Quality of Fish Sausage

Xiaohu Zhou [1,2,3,4], Chaohua Zhang [1,3,5,*], Liangzhong Zhao [2,4,*], Xiaojie Zhou [2,4], Wenhong Cao [1,3,5] and Chunxia Zhou [1,3,5]

1 College of Food Science and Technology, Guangdong Ocean University, Zhanjiang 524088, China
2 College of Food and Chemical Engineering, Shaoyang University, Shaoyang 422000, China
3 Guangdong Provincial Key Laboratory of Aquatic Products Processing and Safety, Zhanjiang 524088, China
4 Hunan Provincial Key Laboratory of Soybean Products Processing and Safety Control, Shaoyang 422000, China
5 Collaborative Innovation Center of Seafood Deep Processing, Dalian Polytechnic University, Dalian 116034, China
* Correspondence: zhangch@gdou.edu.cn (C.Z.); sys169@163.com (L.Z.)

Abstract: Currently, the processing method of introducing plant protein into meat products has attracted great attention. However, the direct addition of plant protein often leads to a decline in meat product quality. This paper aims to provide an efficient method for incorporating plant protein into fish sausage. Pea protein isolate (PPI), grass carp protein isolate (CPI) and pea-grass carp coprecipitated dual protein (Co) were derived from pea and grass carp by an isoelectric solubilisation/precipitation method. At the same time, the blended dual protein (BL) was obtained by blending PPI with CPI, and the plant and animal protein content of Co and BL was both controlled to be the same. The four proteins were combined with soybean oil and water to form a three-phase pre-emulsification system of protein-oil-water, which was added to grass carp meat as a replacement for animal fat to prepare fish sausage. The gelation properties of the four fish sausages and those without protein were analysed. The results showed that the gel quality of PPI fish sausage is poor, while the overall quality of Co fish sausage as a whole was significantly superior to that of PPI and BL, which was equivalent to CPI fish sausage. The sensory score of the Co fish sausage was slightly lower than that of CPI, but it had significantly higher water-holding capacity and hardness ($p < 0.05$). The Co fish sausage showed the synergistic effect of heterologous proteins, while BL had some antagonistic effects. This study shows that Co pre-emulsion is an effective strategy to introduce plant protein, so it has a good application prospect in the meat industry.

Keywords: pea; dual protein; plant protein; co-precipitation dual protein; blended dual protein; fish sausage; pre-emulsion

Citation: Zhou, X.; Zhang, C.; Zhao, L.; Zhou, X.; Cao, W.; Zhou, C. Effect of Pre-Emulsion of Pea-Grass Carp Co-Precipitation Dual Protein on the Gel Quality of Fish Sausage. *Foods* **2022**, *11*, 3192. https://doi.org/10.3390/foods11203192

Academic Editors: Jianhua Xie, Yanjun Zhang and Hansong Yu

Received: 31 May 2022
Accepted: 10 October 2022
Published: 13 October 2022

1. Introduction

Fish represent a high-quality animal protein with high protein content and bioavailability [1]. China has abundant fish resources, with the total production of fish resources reaching 35.21 million tonnes in 2020 [2]. Making surimi products from fish, such as fish sausage, is popular among consumers because of its distinct flavour and low price. However, traditional fish sausages have a high fat content, which has some defects, such as a rough gel structure and health risks [3]. At present, there is a processing trend to use plant-based components instead of animal components [4,5].

Pre-emulsification is the emulsification of oil with food protein itself as a natural emulsifier, which can form a crosslink with continuous-phase protein gel matrix by covalent or non-covalent bond [6]. Liu [7] found that soy protein pre-emulsified plant oil significantly increased the whiteness and water-holding capacity of surimi. However, the above study showed that when the proportion of plant protein increased, the quality deteriorated obviously. With the rise of plant-based meat, pea protein is favoured by researchers. A series of studies on pea protein added to pork [8], beef [9], and chicken [10] gels have been conducted. It was demonstrated that pea protein could increase the thermal denaturation temperature of myosin head structure, thus improving gel-forming ability and gel quality significantly. Due to the limitation of using a single protein, practical processing is more likely to use two or more proteins. Recent research shows that the addition of dual proteins can significantly improve the hardness, springiness, cooking loss, and water-holding capacity of surimi gel, especially when the ratio of plant and animal protein is 1:1 [11]. It can be seen that dual protein is a feasible way to introduce plant protein into meat products. However, there have been few reports on the application of pea protein and its compound dual protein in surimi.

Dual proteins can be divided into blended dual protein (BL) and coprecipitated dual protein (Co) according to their preparation methods [12]. BL refers to the protein produced by the direct mixing of protein isolate. Studies have shown that BL from peas and cods has superior emulsification properties than a single protein, reflecting the synergistic effect [13]. However, studies have confirmed that as the content of plant protein increases due to different species of BL proteins, the incompatibility of the molecular configurations of heterologous proteins causes steric hindrance and damage to gel structures (such as globulin and myofibril), resulting in antagonistic effects on gel properties [14,15]. Co was prepared using the isoelectric solubilisation/precipitation (ISP) method. Driven by pH, heterogeneous proteins are dissolved and precipitated in the same dispersion system to promote interactions between heterogeneous proteins, such as the formation of disulphide bonds and changes in protein subunit composition, surface charge, solubility, and surface hydrophobicity, thus effectively improving functional properties [16]. In our previous study, it was proved that Co has superior emulsification [17] and gel [18] properties than BL and a single protein. Additionally, we found that pre-emulsifying soybean oil can improve the gel quality of fish sausage [19]. However, no literature has reported the actual performance of the emulsification and gel properties of BL and Co in fish sausage.

In this study, a "protein-oil-water" three-phase pre-emulsion system was constructed, and pre-emulsion of PPI, CPI, and BL and without protein were used as controls to explore the effect of Co on the characteristics (such as water-holding capacity and TPA) of fish-sausage gel. The aim of this study was to find a more ideal way to introduce plant protein into fish sausage than the direct addition method so as to improve the gel quality of fish sausage and expand the application scope of Co in food processing. This study can provide a theoretical basis for increasing the application of pea protein in animal-based food.

2. Materials and Methods

2.1. Materials

Split peas were purchased from Foshan Jinnuoyi Processing Factory of Agricultural Products (Foshan, China); the protein content was 21.12% (6.25 of N, wet weight). The grass carp was purchased from the local fish market in Shaoyang City, Hunan Province. The gills, heads, tails, bones, and blood were all removed, leaving only the white muscles on the back, which were retained and frozen at a low temperature. The fish muscles were brought back to the laboratory and stored at −20 °C; the protein content was 18.27% (6.25 of N, wet weight). Natural-flavoured rice wine was purchased from Shandong Luhua Group Co., Ltd. (Yantai, China). Edible salt was purchased from China National Salt Corporation (Beijing, China). Edible-grade sodium nitrite and edible-grade sodium tripolyphosphate were purchased from Sichuan Jinshan Pharmaceutical Co., Ltd. (Meishan, China) and Hubei Xingfa Co., Ltd. (Yichang, China), respectively. Five-spice powder and pepper powder were purchased from the supermarket in Shaoyang City, Hunan Province. Sheep casings were purchased from the Xuchang Shenyuan casing company (Xuchang, China). Other analytical grade reagents were ordered from the Guangzhou Chemical Reagent Factory (Guangzhou, China).

2.2. Preparation of PPI, CPI, Co, and BL

PPI, CPI, and Co are prepared using the ISP method as described in reference [17], where Co is prepared by dissolving and precipitating pea powder and fish muscle in the same container simultaneously. Peas were powdered and sieved through a 100-mesh sieve (0.15 mm). The pea powder was mixed with deionised water at 4 °C in a ratio of 1:9 (*w*/*v*). The pH was adjusted to 10.0 by dissolving 1 M NaOH for 30 min, followed by centrifugation at 10,000× *g* and 4 °C for 20 min. The precipitate was removed, and 1 M HCl was slowly added to the supernatant. The mixture was fully stirred, and the pH was adjusted to 5.0. After washing the precipitation with deionised water and adjusting the pH to 7.0, dialysis was performed. PPI was obtained after 48 h of freeze drying. The fish muscles were defrosted and chopped, and CPI was prepared according to the steps indicated above for PPI. The pea powder was mixed with fish muscle, and Co was prepared following the steps described above. BL was obtained by directly mixing powdered PPI and CPI. To ensure a similar comparison, the protein ratio of pea and grass carp in BL and Co was kept constant at 1:1 (W:W) in this study.

2.3. Non-Reduced SDS-PAGE

The original protein compositions of BL and Co were characterised by non-reduced sodium dodecyl sulphate-polyacrylamide gel electrophoresis (non-reduced SDS-PAGE) as follows. Solutions of 1% protein concentration were prepared. The volume fraction of concentrated gel (upper gel) was 5%, while the volume fraction of separated gel (lower gel) was 12%. No reducing agent dithiothreitol (DTT) was added; the protein marker was 16–270 kDa. Protein band abundance was expressed by relative optical density (%) measured using Image LabTM V4.0 software (Bio-Rad Laboratories, Hercules, CA, USA).

2.4. Determination of Chemical Composition of PPI, CPI, Co, and BL

Chemical compositions, including moisture (AOAC 935.29), ash (AOAC 938.08), protein (AOAC 2001.11), and fat (AOAC 920.39), were determined according to the methods recommended by AOAC [20]. Carbohydrates were calculated using the following formula: 100 − (moisture + ash + protein + fat).

2.5. Preparation of Pre-Emulsion

The protein powder of PPI, CPI, BL, and Co was used as an emulsifier to prepare pre-emulsified soybean oil [21]. The material was weighed using a mass ratio of m (protein powder):m (soybean oil):m (water) = 12:44:44. The protein was mixed with water, slowly stirred for two hours, and then mixed with soybean oil. The mixture was evenly dispersed with a T18 homogeneous shear machine (IKA Laboratory Equipment, Staufen, GER) at 11,000 rpm for one min in an ice bath, with a five-minute interval, and repeated five times and kept at 4 °C. At the same time, the blank control group without protein was established, and the homogenisation method was the same as that of M (protein powder):m (soybean oil):m (water) = 0:44:44.

2.6. Preparation of Fish Sausage

Formula: Fish sausage (g/100 g) was determined as 64.0 g fish meat, 30.0 g pre-emulsion, 3.0 g rice wine, 2.0 g edible salt, 0.5 g five-spice powder, 0.3 g tripolyphosphate, and 0.2 g pepper. In addition, 0.01 g sodium nitrite was added to prevent *Clostridium botulinum*. The fish-sausage emulsion in the blank control group was 26.4 g.

Marinating: Mince the fish into small pieces, add the marinade (soybean oil, edible salt, tripolyphosphate, five-spice powder, pepper, sodium nitrite) into it, and marinate it at 4 °C for 48 h.

Chopping: Mix the pickled lean meat with the pre-emulsion, then slowly chop and mix at a low temperature (10 °C) for five minutes.

Pouring: After eliminating the air pockets, use a small enema device to pour into the sausage casing, tie knots every 10 cm, count by weight, and control the diameter at about 2.5 cm.

Cooking: Fish sausage was placed in a 90 °C constant temperature water bath for 45 min, cooled in an ice-water bath for 15 min, drained of water on the surface of the casings, weighed and counted, and stored in 4 °C for 8 h.

2.7. Cooking Loss

The cooking loss (CL) of the gel was determined using the Youssef method [22]. The gel was cut into a cylinder of about 15 × 15 × 5 mm and weighed (W_1), then put into a cooking bag and sealed. After cooking for 20 min in a 90 °C water bath, it was stored at 4 °C in a refrigerator for 24 h. Again, the gel was weighted (W_2) after carefully blotting its surface with filter paper. The CL was calculated as follows:

$$CL(\%) = \frac{W_1 - W_2}{W_1} \times 100 \quad (1)$$

where W_1 is the mass before cooking (g), and W_2 is the mass after cooking (g).

2.8. Water Holding Capacity

Next, we wrapped 8 g gel samples in double filter paper and placed them in a centrifugal tube before weighing the total weight. After centrifuging at 5000× g at 4 °C for 10 min, the gel was taken out and weighed. Formula 2 is used to calculate WHC, where M is the mass of centrifugal tube (g); M_1 is the mass of gel and centrifugal tube before centrifugation (g); M_2 is the mass of gel and centrifugal tube after centrifugation (g).

$$WHC(\%) = (M_2 - M)/(M_1 - M) \times 100 \quad (2)$$

2.9. Folding Test

According to Kudo [23], the fish sausage was cut to a thickness of 3 mm and extruded between the thumb and index finger. The folding ability of the fish sausages was divided into five levels: (1) Sausage slices were broken into two pieces on the first folding; (2) sausage slices were cracked but not broken on the first folding; (3) sausage slices did not show any cracks on the first folding, but they were broken into two pieces on the second folding; (4) sausages slices did not show cracks on the first folding but did show cracks on the second folding; and (5) sausage slices did not show cracks even after the second folding.

2.10. TPA

Texture profile analysis were measured by a P35 cylindrical flat bottom probe with a texture analyser (LS5, Ametek, Berwyn, PA, USA). Parameter setting: sample surface flatness height of 20 mm; front, middle, and final test speed of 4.0 mm/s, 3.0 mm/s, and 4.0 mm/s; descent distance of 40%; compression time of 5 s; trigger force of 5 g.

2.11. Colour

The colourimeter was used to measure the colourimetry of fish sausage, which was represented by the L*, a*, and b* values. Whiteness is calculated as follows:

$$\text{Whiteness} = 100 - \sqrt{(100 - L^*)^2 + a^{*2} + b^{*2}} \quad (3)$$

2.12. LF-NMR

Low-field nuclear magnetic resonance (MQC-23, Oxford, UK) was used to determine the water distribution of the fish sausage [24]. The fish sausage was packed into a 20 cm NMR tube, sealed with plastic wrap, and the transverse relaxation time (T) of the protein gel was measured by Carr–Purcell–Meiboom–Gill (CPMG) pulse sequence. The parameters were set as: P90 = 6, P180 = 12, SFI = 23.4 MHz, DW = 1.0, SW = 1,000,000 Hz, SI = 1,

NECH = 256, and NS = 16; WinDXP software was used to invert CPMG and save its data. The relaxation time diagram was made by drawing software. The peak area in the atlas was accumulated, and the peak area represented the percentage content of water in the group.

2.13. SEM

The samples were cut into 3 mm × 2 mm and placed in a covered container. They were soaked in 2.5% pH 7.2 phosphate buffer at 4 °C for more than eight hours and then washed three times with 0.1 mol/mL pH 7.2 phosphate buffer for 10 min each. Gradient elution was performed with 50%, 70%, 80%, and 90% ethanol solutions for 10 min each. Elution was repeated three times with 100% anhydrous ethanol for 10 min each. The dehydrated samples were defatted with the proper amount of trichloromethane for one hour, then replaced with a mixture of tert-butyl alcohol and ethanol with a volume ratio of 1:1 for 15 min, and after that replaced with 100% tert-butyl alcohol for 20 min and freeze-dried for 48 h. The surface morphology of the sample was observed by a scanning electron microscope (Gemini300, Zeiss, Oberkochen, Germany) under a 5000× field of vision.

2.14. Sensory Evaluation

The sensory evaluation team of sausages comprised ten teachers and postgraduates (five males and five females, aged 24–35) from the major of food Science and Engineering at Shaoyang University. All of these people have experience with food sensory evaluation and were familiar with the experimental process and operating norms. Prior to the analysis, all participants (volunteers) in this experiment were informed of the aim of the sensory evaluation, and they provided informed consent to perform it. The group members' scores ranged from 0 (poor sensorial property) to 10 (excellent sensorial property), indicating that their expectations for appearance, smell, taste, texture, and overall acceptability ranged from very low to very high. The scoring criteria are shown in Table 1. Sausages were re-heated in water for two minutes and kept in an oven at 40 °C before being served to the panel randomly on white porcelain plates under natural light at room temperature. A glass of water was also provided to cleanse the palate.

Table 1. Scoring criteria for sensory evaluation of grass carp sausage.

Project	Features	Score
Appearance (10)	Dark or grey, and there are numerous holes in the section	0.0–3.9
	Grey-white, with a few holes in the section	4.0–6.9
	Light yellow or fish white, bright colour; a few holes in section in the cross-section	7.0–10.0
Smell (10)	Fishy or bean fishy, unpleasant	0.0–3.9
	Light aroma or a slight characteristic fishy smell, without peculiar smell	4.0–6.9
	Ideal fish sausage has a strong aroma and a pleasant smell	7.0–10.0
Taste (10)	Oil tastes heavier or beans bitter taste heavier, difficult to swallow	0.0–3.9
	Mild astringency, umami, and saltiness are mild or heavy and can be tasted	4.0–6.9
	It has the taste of fish sausage, umami, and saltiness, worth tasting	7.0–10.0
Texture (10)	Too hard or too soft, too dry or too much water or oil, difficult to chew	0.0–3.9
	Normal, slightly uncomfortable to chew, but acceptable	4.0–6.9
	Good taste and suitable hardness, toughness, and elasticity	7.0–10.0
Overall acceptability (10)	Hard to accept ~ Good to accept	0.0–10.0

2.15. Statistical Analysis

SPSS 25.0 software of IBM company was used for statistical analysis. The statistical differences between the two groups were analysed by one-way analysis of variance

(ANOVA) and Duncan's multiple range test ($p < 0.05$). The original Pro 2021 software was used for drawing.

3. Results and Discussion

3.1. Protein Composition Analysis

Non-reducing electrophoresis cannot disrupt the disulphide bonds. Therefore, it reflects the original composition of the various proteins in a mixed protein (Figure 1 and Table 2), and both BL and Co contained myosin heavy chain (MHC, ~200 kDa), convicilin (~70 kDa), legumin (~62 kDa), vicilin (~50 kDa), actins (AC, ~42 kDa), legumin α (~38 kDa), tropomyosin (TM, ~34 kDa), and legumin β (~22 kDa), all aggregates consistent with the literature [25,26]. Compared with legumin α + β, vicilin is more flexible and exhibits higher interfacial activity [27]. Many studies have confirmed that the vicilin/legumin a + B ratio is positively related to a protein's functional properties, including solubility and foaming and emulsifying properties [28,29]. Thus, the ratio of vicilin to legumin a + β is very important. As shown in Table 2, the proportion of Co (203.99%) is 2.82-fold larger than that of BL (72.36%). Therefore, the analysis of protein composition indicates that the functional properties of Co might be superior to those of BL.

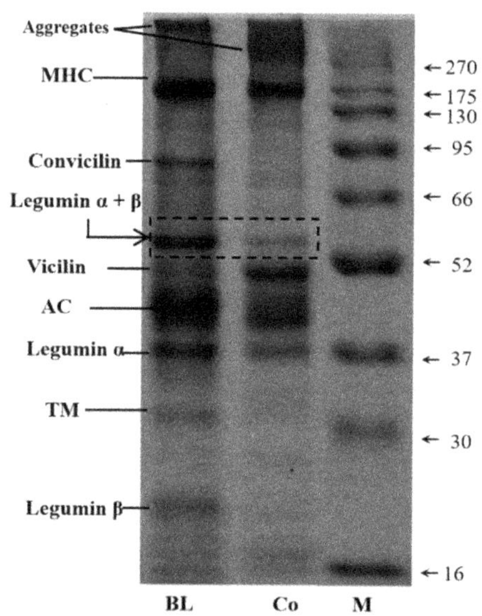

Figure 1. Non-reducing electrophoretograms of BL and Co.

Table 2. Relative band optical density (%) of BL and Co (non-reducing electrophoretogram).

Samples	Aggregates	MHC	Convicilin	Leg α + β	Vicilin	AC	Leg α	TM	Leg β	Others	Vicilin/Leg α + β
BL	11.61	15.90	4.60	9.98	7.22	19.31	10.63	5.31	9.98	5.45	0.72
Co	19.62	15.27	0.11	6.55	13.36	16.46	8.06	5.36	4.63	10.57	2.04

3.2. Chemical Composition of PPI, CPI, Co and BL

The chemical composition analysis of PPI, CPI, BL, and Co is shown in Table 3. The protein content of the four protein samples was about 85%, which belongs to high-quality protein powder, indicating that the ISP method is an effective way to concentrate pea protein and grass carp protein. Co was the highest ($p < 0.05$) at 87.91%, indicating that Co has the potential to be used as the raw material of high-protein food. Compared with BL, Co contains less fat. When Co is added to fish sausage, the damage to the gel structure and nutritional value caused by fat oxidation can be reduced, and the shelf life of the product can be prolonged [30]. We previously speculated that the carbohydrate content of Co should be between PPI and CPI. Interestingly, experimental data showed that the carbohydrate content of Co was only 0.16% lower than CPI ($p < 0.05$), suggesting that co-precipitation enhanced the removal rate of starch or muscle glycogen. It is speculated that this is because the interaction of heterologous proteins reduces protein–carbohydrate (glycoprotein) binding during co-dissolution or precipitation. In addition, Co has low fat and low carbohydrate characteristics, which is in line with the consumption trend. Therefore, Co, as an isolated protein powder, has greater application potential in meat products than BL.

Table 3. Chemical composition of PPI, CPI, Co, and BL.

Sample	Moisture	Protein	Fat	Carbohydrate	Ash
PPI	6.71 ± 0.14^{a}	84.84 ± 0.03^{c}	0.35 ± 0.02^{d}	5.89 ± 0.19^{a}	2.21 ± 0.16^{a}
CPI	5.30 ± 0.07^{b}	85.30 ± 0.13^{b}	6.15 ± 0.05^{a}	1.68 ± 0.13^{c}	1.57 ± 0.11^{c}
BL	4.01 ± 0.05^{c}	85.07 ± 0.09^{b}	5.24 ± 0.03^{b}	3.79 ± 0.04^{b}	1.89 ± 0.09^{b}
Co	5.28 ± 0.11^{b}	87.91 ± 0.02^{a}	4.84 ± 0.08^{c}	0.16 ± 0.08^{d}	1.81 ± 0.01^{b}

Note: Different letters in the same column indicate a significant difference between samples ($p < 0.05$).

3.3. Cooking Loss

The effect of adding different protein pre-emulsified soybean oil on the CL of fish sausage is shown in Figure 2. In this study, due to different emulsifiers, the introduction of soybean oil in the form of protein pre-emulsification will affect CL. CPI fish sausage had the lowest cooking loss (5.24%, $p < 0.05$), indicating that adding fish protein homologous to the fish-sausage matrix in the pre-emulsification emulsion can reduce cooking loss [31]. Compared with the control group, the cooking loss of PPI fish sausage was higher ($p < 0.05$). This is similar to the finding that replacing pig fat with pre-emulsified olive oil stabilised by soy protein isolate increased cooking loss. This is because the plant protein in fish sausage is not tightly bound to the fish matrix, resulting in a loose structure of the sausage that makes it difficult to hold water [32]. However, there was no significant difference in the CL of BL and Co when compared with the control group ($p > 0.05$). It can be seen that the introduction of plant protein in the form of dual proteins in fish sausage can significantly reduce CL.

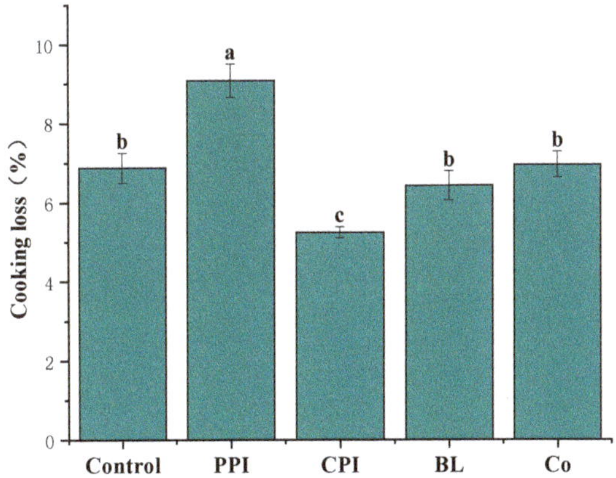

Figure 2. Effects of different protein pre-emulsified soybean oils on the cooking loss of fish sausage. Note: Different letters indicate that there is a significant difference between samples ($p < 0.05$). The error bars represent the standard deviation ($n = 3$).

3.4. Water-Holding Capacity

The WHC can reflect the cross-linking degree of the blend gel of pre-emulsified protein and fish meat. The effect of soybean oil pre-emulsified with four kinds of proteins on the water-holding capacity of fish sausage is shown in Figure 3. The results showed that the WHC of the five kinds of fish sausage was greater than 93%, with the control group having the lowest (93.20% $p < 0.05$), indicating that increased protein concentration results in a stronger gel network. This is consistent with the results reported in the literature [33]. The WHC of Co-, PPI-, and CPI fish sausage was higher than 96.00%. Under centrifugation, the water in the capillary of the gel permeates the surface and is removed by filter paper. Therefore, it can be considered that the addition of Co, PPI, and CPI reduces the flow of free water [34]. In addition, the WHC of Co fish sausage was the highest, while that of BL fish sausage was the lowest. Combined with the previous results [18], we speculate that this is due to the antagonistic effect of PPI and CPI in BL on heat-induced gelation, which leads to a loose structure and poor WHC of fish sausage. It can be seen that Co is more beneficial to the water retention of fish-sausage gels than BL.

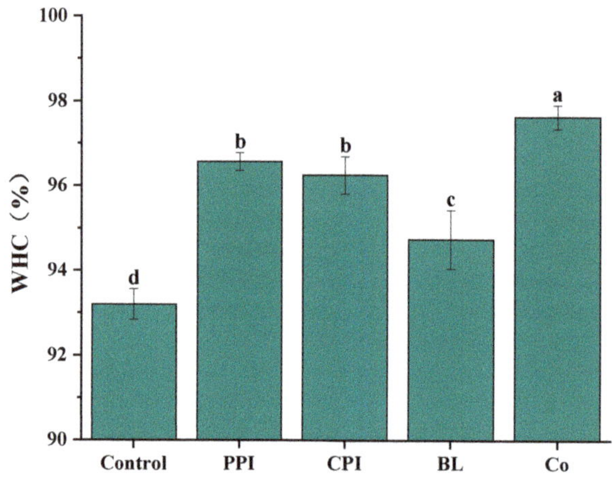

Figure 3. Effect of soybean oil with different protein pre-emulsification on the water retention of fish sausage. Note: Different letters indicate a significant difference between samples ($p < 0.05$). The error bars represent the standard deviation ($n = 3$).

3.5. Folding Test

The folding test can directly reflect the gel strength of fish sausage and allow researchers to quickly understand its texture characteristics [35]. The folding test results of the five kinds of fish sausages are shown in Figure 4, which confirms the results of CL and WHC. The folding ability of the control group was lower than that of PPI-, CPI-, and Co fish sausage, which was attributed to the weak gel network formed by the lower protein content [36]. However, the lowest folding ability of BL fish sausage was only level 3 ($p < 0.05$). This is due to the antagonism of PPI and CPI in BL, leading to the gel structure loosening.

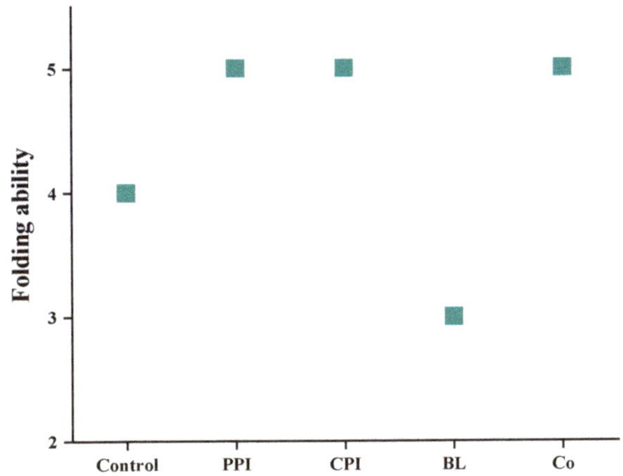

Figure 4. Effects of different protein pre-emulsified soybean oil on the folding capacity of fish sausage.

3.6. TPA

The TPA results of the five sausage samples are shown in Table 4. The hardness, cohesiveness, and chewiness of fish sausage added with Co were significantly higher than those of PPI and CPI ($p < 0.05$), which indicated the synergistic effect of Co. According to previous studies [18], it is speculated that disulphide bonds account for the highest proportion of intermolecular forces of Co gel, which makes the hardness and cohesion of Co fish sausage stronger and thus necessitating greater energy consumption during chewing. However, the thermal denaturation temperature of PPI (82 °C) [37] and CPI (56 °C) [38] contained in BL is obviously different, which leads to the loose structure of heat-induced gel formed by BL, which leads to the lowest hardness, cohesion, and chewiness of BL fish sausage ($p < 0.05$), showing an obvious antagonistic effect. The hardness, cohesion, and chewiness of the blank control group were the lowest, but the springiness was the highest ($p < 0.05$). This is similar to the research results of improving the texture characteristics of fish sausage by adding pea protein [39]. Adding protein can enhance the interaction between protein molecules in surimi gel, thus improving the texture characteristics of the gel.

Table 4. Effects of different protein pre-emulsified soybean oil on the fish-sausage texture.

Sample	Hardness (g)	Springiness (mm)	Cohesiveness	Chewiness (N·mm)
Control	354.581 ± 5.252 ^e^	0.885 ± 0.003 ^a^	0.575 ± 0.004 ^c^	170.868 ± 4.234 ^d^
PPI	562.330 ± 9.359 ^c^	0.812 ± 0.002 ^e^	0.597 ± 0.009 ^b^	272.554 ± 7.860 ^b^
CPI	586.078 ± 13.727 ^b^	0.832 ± 0.003 ^d^	0.580 ± 0.008 ^c^	282.826 ± 7.474 ^b^
BL	424.238 ± 7.348 ^d^	0.867 ± 0.006 ^b^	0.554 ± 0.005 ^d^	207.450 ± 6.173 ^c^
Co	659.296 ± 12.823 ^a^	0.854 ± 0.005 ^c^	0.622 ± 0.006 ^a^	350.302 ± 9.637 ^a^

Note: Different letters in the same column indicate a significant difference between samples ($p < 0.05$).

3.7. Colour

Colour impacts its application in food, and colour tests can objectively evaluate the colour of protein samples. L* represents sample brightness; the positive value of a* denotes redness, and the negative value represents greenness. A positive value of b* is yellowness, and a negative value is blueness. The effect of different protein pre-emulsified soybean oil on the colour of fish sausage is shown in Figure 5 and Table 5. The brightness (L*) and whiteness of PPI fish sausage were lower than those of the other groups ($p < 0.05$), and yellowness (b*) was higher than those in other groups ($p < 0.05$). This is related to the yellowish colour of PPI, which is attributed to the natural yellow plant pigments in peas that are thermally stable. The brightness (L*) of CPI fish sausage was the highest, reaching 86.37 ($p < 0.05$), and the yellowness (b*) was the lowest ($p < 0.05$), but the whiteness was significantly higher than that of other groups ($p < 0.05$) [40]. The redness/greenness (a*) of the fish sausage was lower in the five groups, which had little influence on the colour. The four colour indices of the fish sausage in the control group, i.e., BL and Co, were all between PPI and CPI, indicating that the relative white of CPI had a masking effect on the relative yellow of PPI [41]. Therefore, adding plant protein in the form of dual proteins can reduce the adverse effect of yellow pigments in plants on fish sausage, and the colour quality of BL is superior than that of Co.

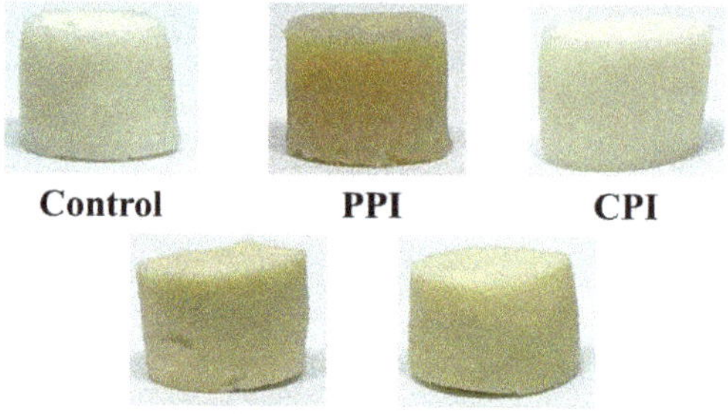

Figure 5. Appearance of soybean oil fish sausage pre-emulsified with different proteins.

Table 5. Effect of soybean oil with different protein pre-emulsification on the colour of fish sausage.

Sample	L*	a*	b*	Whiteness
Control	83.51 ± 0.03^{b}	-1.42 ± 0.10^{c}	17.93 ± 0.22^{d}	75.60 ± 0.15^{b}
PPI	76.42 ± 0.11^{e}	0.64 ± 0.05^{a}	23.69 ± 0.03^{a}	66.57 ± 0.07^{e}
CPI	86.37 ± 0.10^{a}	-2.37 ± 0.01^{d}	8.89 ± 0.10^{e}	83.56 ± 0.03^{a}
BL	82.37 ± 0.15^{c}	-1.14 ± 0.05^{b}	18.31 ± 0.03^{c}	74.56 ± 0.08^{c}
Co	80.75 ± 0.04^{d}	-1.15 ± 0.03^{b}	19.37 ± 0.03^{b}	72.67 ± 0.01^{d}

Note: Different letters in the same column indicate a significant difference between samples ($p < 0.05$).

3.8. Water Distribution Analysis

Low-field nuclear magnetic resonance (LF-NMR) is a useful method to study the content, state, distribution, migration, diffusion, and the interaction of water molecules with other molecules. These properties of water are the key factors in determining the quality of fish-sausage gel [42]. Figure 6 shows the effect of soybean oil pre-emulsified with different proteins on water distribution in fish sausage. The relaxation time T of hydrogen protons in fish sausage reflects their degree of binding or freedom. The higher the T, the lower the binding force of hydrogen protons, leading to decreased stability and increased fluidity. The peak area of each stage represents water content with a different binding degree in fish sausage. T1 represents the bound water tightly adsorbed with macromolecules such as protein; T2 represents the water locked in the gel network of fish sausage, which has a certain fluidity and is classified as not easily allowing the flow of water; and T3 represents the water outside the gel network structure of fish sausage: free water with high fluidity [43]. It can be seen from Figure 6 that the T1 peak area of fish sausage containing pea protein components, such as PPI-, BL-, and Co fish sausage, is more significant. This may be due to the existence of pea protein in the gel in the form of filling after emulsification, and the degree of interaction with the gel matrix is weak. In contrast, the adsorption capacity of plant protein (PPI) to water is considerably stronger than that of fish protein, resulting in a higher volume of bound water (T1) [44]. Only the BL fish sausage had a pronounced T3 peak, indicating that it had a high free-flowing water content and a low gel water-holding capacity. However, the difference between T1 and T2 peaks of

CPI fish sausage was not noticeable, and the peak distribution was similar to that of the control group, indicating that the CPI bonded well with the fish gel matrix during gelation, leaving appropriate gaps to lock the adsorbed water in the gel network [45]. Generally speaking, the water distribution of Co fish sausage was between that of CPI- and PPI fish sausage, with no discernible T3 peak. Again, BL fish sausage showed an antagonistic effect.

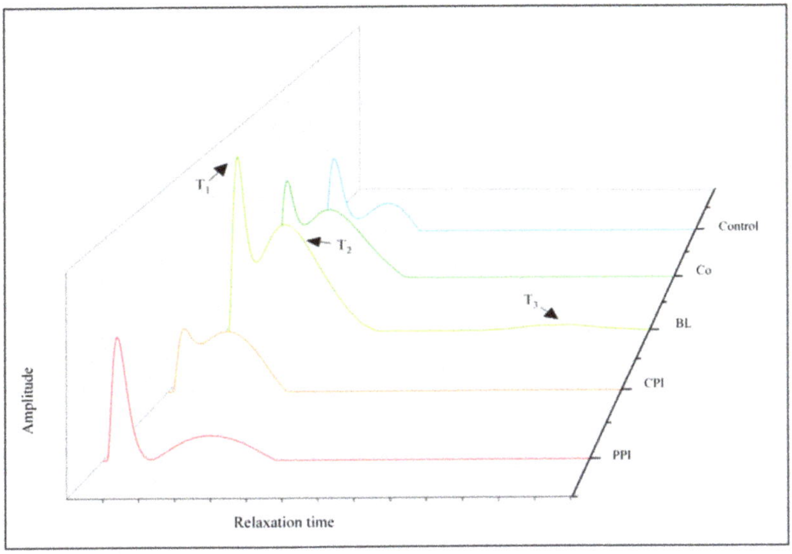

Figure 6. Effects of different protein pre-emulsified soybean oil on water distribution of fish sausage.

3.9. Microstructure

To accurately understand the characteristics of the protein emulsion and fish matrix blend gel, the microstructure of the protein pre-emulsified fish-sausage gel was observed by scanning electron microscopy at 40,000×, as shown in Figure 7. Compared with the microstructure of protein gel in the previous report [18], the gel network structure of fish sausage in the five groups was more compact and even, which could be attributed to the high content of TG enzyme in fresh surimi and forms a higher proportion of covalent cross-linking [46]. The microstructure of the control group and CPI fish sausage was compact and consistent with the results reported in other literature, which indicated that CPI participated in the formation of fish-sausage gel through cross-linking [41]. However, CPI fish sausages may show to an uneven distribution of TGase [47].

The presence of pea globulin particles in PPI-, BL-, and Co sausages confirmed that plant protein components filled the gel network. One possible explanation is that CPI acts as filler for the holes in the fish gel structure, allowing the two separate structures to interact and become intertwined to form a homogeneous structure [39,48,49]. The PPI was aggregation-filled on the gel's surface, which was different from the fish matrix gel. However, there are many large holes in BL fish sausage, which are consistent with the above research results and are attributed to the antagonistic effect of PPI and CPI. It is worth noting that Co fish sausage also has a few holes, but the overall structure of the gel is dense. Co is "embedded" in the gel matrix, which is beneficial to lock water and improve gel strength, confirming that the common fish sausage has the highest WHC and hardness.

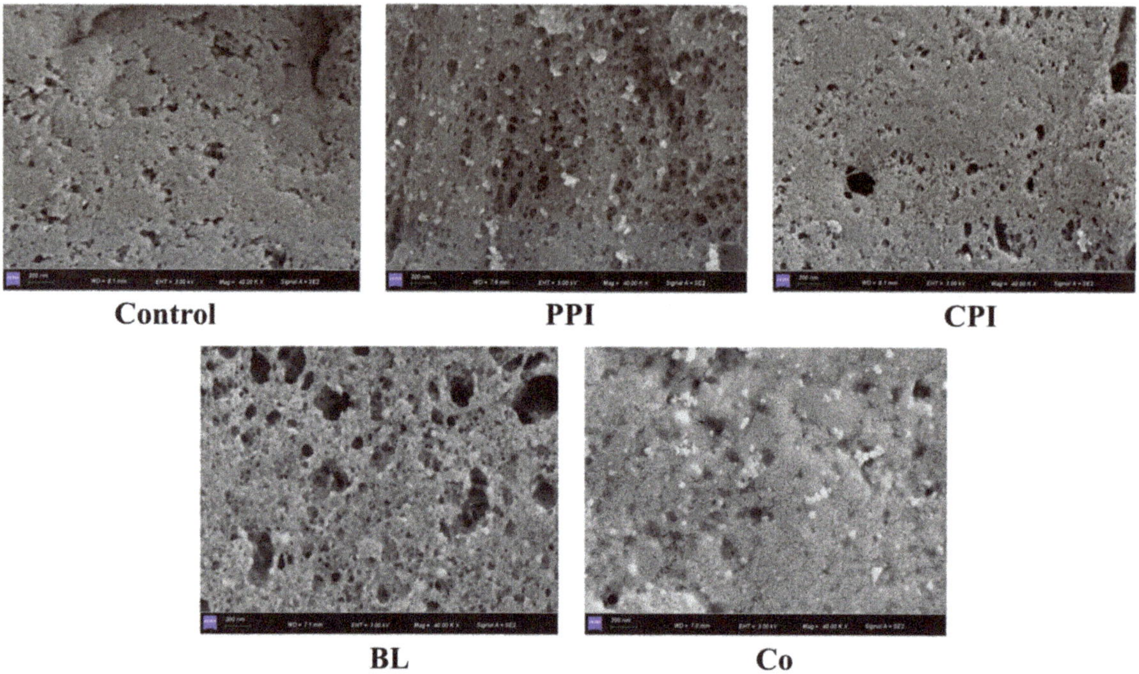

Figure 7. Effects of different protein pre-emulsified soybean oil on the microstructure of fish sausage (×40,000).

3.10. Sensory Evaluation

Sausages were tested for sensory attributes, such as appearance, texture, taste, smell, and overall acceptability. It can be seen in Figure 8 that the control group had the lowest taste score, indicating that protein pre-emulsification improves the taste of fish sausage. CPI fish sausages had the highest scores except for texture ($p < 0.05$). The scores of appearance and overall acceptability were 7.22 and 7.56, respectively. This is due to the whiteness, smell, and taste of CPI, which are all similar to that of fish. Similar observations were made by Santana, who reported that the addition of fish protein into sausages did not cause changes in their smell, oiliness, and colour [50]. PPI fish sausage scored the lowest in appearance because of its yellow colour.

Texture was the main attribute that affected consumer preferences [51]. The texture score of Co fish sausage was higher than CPI ($p < 0.05$), and the overall acceptability was second only to CPI, indicating that the colour and smell of pea components would affect the sensory quality of the fish sausage. It is noteworthy that the overall acceptability, texture, and smell of Co fish sausages are significantly higher than those of PPI- and BL fish sausages. The results were consistent with the folding test level, WHC, and texture properties, which again indicated that Co addition was a more ideal method for introducing plant protein compared to PPI and BL.

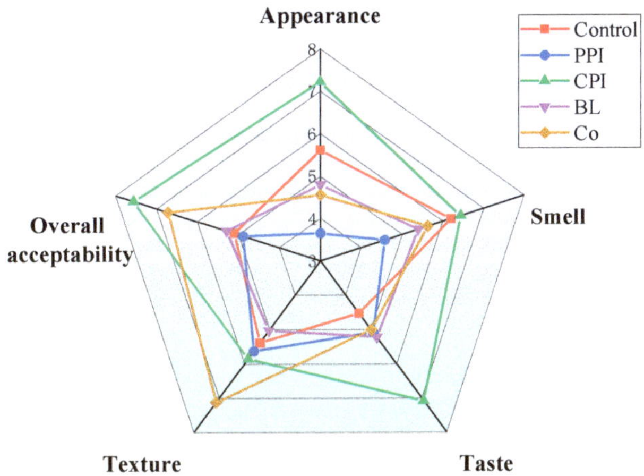

Figure 8. Sensory intensity radar diagram of fish sausage made with soybean oil and various pre-emulsified proteins.

4. Conclusions

On the basis of the previous research on the emulsion and gel properties of dual proteins (BL and Co), the quality of fish-sausage gel formed by pre-emulsification of PPI, CPI, BL, and Co was discussed in this paper. Comprehensive analysis showed that adding protein pre-emulsified soybean oil to fish sausage could effectively improve the gel network and edible quality of fish sausage. Because CPI and fish-sausage matrix gel are homologous, fish-sausage gel prepared using the CPI pre-emulsion method has a higher quality. The sensory score of Co-fish-sausage gel was slightly lower than that of CPI. However, because Co has the functional characteristics of both grass carp protein and pea protein, it has significant advantages in terms of WHC, hardness, and microstructure. This shows that the overall quality of Co-fish-sausage gel is comparable to or even slightly higher than CPI. In addition, the quality of PPI- and BL-fish-sausage gel is poor, with BL having a particularly pronounced antagonistic effect. Therefore, co-precipitation dual protein is an effective method for introducing plant protein into fish sausage. This study provides a theoretical reference for the application of Co in meat products. We will further explore the influence of the combination and ratio of other plant and animal proteins on the food gel system.

Author Contributions: Conceptualization, C.Z. (Chaohua Zhang); methodology, X.Z. (Xiaojie Zhou) and C.Z. (Chunxia Zhou); investigation, X.Z. (Xiaohu Zhou) and L.Z.; resources, C.Z. (Chaohua Zhang) and C.Z. (Chunxia Zhou); supervision, C.Z. (Chaohua Zhang); writing—original draft, X.Z. (Xiaohu Zhou); writing—review and editing, W.C. and C.Z. (Chaohua Zhang). All authors have read and agreed to the published version of the manuscript.

Funding: This research was funded by the Guangdong Higher Education Institutions Processing and Utilization Innovation Team of High Value of Aquatic Products (GDOU2016030503), the Science and Technology Innovation Program of Hunan Province (2019TP1028, 2019SK2122, 2019NK4229, S2021GCZDYF0535), the Natural Science Foundation of Hunan Province (2022JJ50240), and the Open Project of Hunan Provincial Key Laboratory of Soybean Products Processing and Safety Control (DZPJG202009).

Data Availability Statement: Research data are not shared.

Conflicts of Interest: The authors declare no conflict of interest.

References

1. Kondjoyan, A.; Sicard, J.; Cucci, P.; Audonnet, F.; Elhayel, H.; Lebert, A.; Scislowski, V. Predicting the Oxidative Degradation of Raw Beef Meat during Cold Storage Using Numerical Simulations and Sensors—Prospects for Meat and Fish Foods. *Foods* **2022**, *11*, 1139. [CrossRef]
2. Department of Fisheries and Fisheries Administration, Ministry of Agriculture and Rural Affairs. *China Fishery Statistics Yearbook*; China Agriculture Press: Beijing, China, 2021.
3. Wang, Y.; Wang, Y.N.; Shang, Y.B. The effect and mechanism of fat preemulsification on the gel properties of myofibrin-fat complex system. *Food Ferment. Ind.* **2018**, *44*, 76–82.
4. Wang, H.; Zhang, W.; Liu, X.L.; Qiao, M.; Yi, S.M.; Li, X.P.; Li, J.R. Effects of chickpea and peanut protein isolates on the gelling properties of hairtail (*Trichiurus haumela*) myosin. *Lwt* **2022**, *163*, 113562. [CrossRef]
5. Xiong, F.J.; Ma, L.Z.; Wang, Y. Effects of different types and proportions of fat on quality and characteristics of fish tofu. *Meat Res.* **2018**, *32*, 1–6.
6. Kang, Z.L.; Chen, F.S.; Ma, H.J. Effect of pre-emulsified soy oil with soy protein isolate in frankfurters: A physical-chemical and Raman spectroscopy study. *LWT* **2016**, *74*, 465–471. [CrossRef]
7. Liu, X.Y.; Ji, L.; Zhang, T.; Xue, Y.; Xue, C.H. Effects of pre-emulsification by three food-grade emulsifiers on the properties of emulsified surimi sausage. *J. Food Eng.* **2019**, *247*, 30–37. [CrossRef]
8. Ji, H.F.; Li, S.S.; Zhang, L.W.; Wang, X.F.; Chen, F.S.; Ma, H.J. Effect of Pea Protein on Physicochemical Properties of Salt—Soluble Protein of Pork. *Sci. Technol. Food Ind.* **2019**, *40*, 31–36.
9. Ji, H.F.; Li, S.S.; Zhang, L.W.; Wang, X.F.; Chen, F.S.; Ma, H.J. Effects of pea protein on co-gel properties of salt-soluble proteins in beef. *Food Ferment. Ind.* **2019**, *45*, 89–95.
10. Ji, H.F.; Li, S.S.; Zhang, L.W.; Wang, X.F.; Kang, Z.L.; Chen, F.S.; Ma, H.J. Effect of Pea Protein on Quality Properties and Microstructure of Heat-Induced Gel of Chicken Batter. *Food Sci.* **2020**, *41*, 74–79.
11. Yang, L.; Jiang, Y.; Tu, Y.; Bi, Y.W.; Liu, J.; Yang, Y.J.; Xu, M.S. Effect of Ovalbumin and Soybean Protein Isolate Mixtures on Gel Properties and Microstructure of Minced Meat Products. *Food Sci.* **2021**, *42*, 22–27.
12. Wu, C.; Wang, T.; Ren, C.; Ma, W.; Wu, D.; Xu, X.; Wang, L.S.; Du, M. Advancement of food-derived mixed protein systems: Interactions, aggregations, and functional properties. *Compr. Rev. Food Sci. Food Saf.* **2021**, *20*, 627–651. [CrossRef]
13. Tomé, A.S.; Pires, C.; Batista, I.; Sousa, I.; Raymundo, A. Protein gels and emulsions from mixtures of Cape Hake and pea proteins. *J. Sci. Food Agric.* **2015**, *95*, 289–298. [CrossRef]
14. Nicolai, T. Gelation of food protein-protein mixtures. *Adv. Colloid Interface Sci.* **2019**, *270*, 147–164. [CrossRef] [PubMed]
15. Wu, C.; Yan, X.Y.; Wang, T.; Ma, W.C.; Xu, X.B.; Du, M. A self-sorted gel network formed by heating a mixture of soy and cod proteins. *Food Funct.* **2019**, *10*, 5140–5151. [CrossRef] [PubMed]
16. Kristensen, H.T.; Christensen, M.; Hansen, M.S.; Hammershøj, M.; Dalsgaard, T.K. Protein-protein interactions of a whey-pea protein co-precipitate. *Int. J. Food Sci. Tech.* **2021**, *56*, 5777–5790. [CrossRef]
17. Zhou, X.H.; Zhang, C.H.; Cao, W.H.; Zhou, C.X.; Zheng, H.N.; Zhao, L.Z. A Comparative Functional Analysis of Pea Protein and Grass Carp Protein Mixture via Blending and Co-Precipitation. *Foods* **2021**, *10*, 3037. [CrossRef]
18. Zhou, X.H.; Zhang, C.H.; Zhao, L.Z.; Zhou, X.J.; Huang, Z.R.; Zhou, C.X.; Cao, W.H.; Zheng, H.N. Comparative Analysis of Heat-induced Gel Properties of Pea-grass Carp Dual-protein Prepared by Different Methods. *J. Guangdong Ocean. Univ.* **2022**, *42*, 79–86.
19. Song, C.; Lin, Y.; Hong, P.; Liu, H.; Zhou, C. Low-Content Pre-Emulsified Safflower Seed Oil Enhances the Quality and Flavor of the Nemipterus Virgatus Surimi Gel. *Gels* **2022**, *8*, 106. [CrossRef]
20. AOAC. *Official Methods of Analysis of the Association of Official Analytical Chemists International*, 18th ed.; AOAC: Rockville, MD, USA, 2005.
21. Gani, A.; Benjakul, S. Impact of virgin coconut oil nanoemulsion on properties of croaker surimi gel. *Food Hydrocoll.* **2018**, *82*, 34–44. [CrossRef]
22. Youssef, M.K.; Barbut, S. Fat reduction in comminuted meat products-effects of beef fat, regular and pre-emulsified canola oil. *Meat Sci.* **2011**, *87*, 356–360. [CrossRef]
23. Kudo, G.; Okada, M.; Miyauchi, D. Gel-forming capacity of washed and unwashed flesh of some Pacific coast species of fish. *Mar. Fish. Rev.* **1973**, *35*, 10.
24. Chin, K.B.; Go, M.Y.; Xiong, Y.L. Konjac flour improved textural and water retention properties of transglutaminase-mediated, heat-induced porcine myofibrillar protein gel: Effect of salt level and transglutaminase incubation. *Meat Sci.* **2009**, *81*, 565–572. [CrossRef] [PubMed]
25. Chen, M.; Lu, J.; Fei, L.; Nsor-Atindana, J.; Fang, Z. Study on the emulsifying stability and interfacial adsorption of pea proteins. *Food Hydrocoll.* **2019**, *88*, 247–255. [CrossRef]
26. Xu, Y.; Wang, R.; Zhao, H.L.; Zhao, J.M.; Li, X.P.; Yi, S.M.; Li, J.R.; Sun, X.T. Binding of aldehydes to myofibrillar proteins as affected by two-step heat treatments. *J. Sci. Food Agric.* **2020**, *100*, 1195–1203. [CrossRef]
27. Lam, A.C.Y.; Can Karaca, A.; Tyler, R.T.; Nickerson, M.T. Pea protein isolates: Structure, extraction, and functionality. *Food Rev. Int.* **2018**, *34*, 126–147. [CrossRef]
28. Lu, B.Y.; Quillien, L.; Popineau, Y. Foaming and emulsifying properties of pea albumin fractions and partial characterisation of surface-active components. *J. Sci. Food Agric.* **2000**, *80*, 1964–1972. [CrossRef]

29. Kimura, A.; Fukuda, T.; Zhang, M.; Motoyama, S.; Maruyama, N.; Utsumi, S. Comparison of physicochemical properties of 7S and 11S globulins from pea, fava bean, cowpea, and French bean with those of soybean-French bean 7S globulin exhibits excellent properties. *J. Agric. Food Chem.* **2008**, *56*, 10273–10279. [CrossRef] [PubMed]
30. Lima, J.L.; Assis, B.B.; Olegario, L.S.; Galvão, M.D.S.; Soares, Á.J.; Arcanjo, N.M.; González-Mohino, A.; Bezerra, T.K.; Madruga, M.S. Effect of adding byproducts of chicken slaughter on the quality of sausage over storage. *Poult. Sci.* **2021**, *100*, 101178. [CrossRef] [PubMed]
31. Utrera, M.; Morcuende, D.; Estévez, M. Fat content has a significant impact on protein oxidation occurred during frozen storage of beef patties. *LWT* **2014**, *56*, 62–68. [CrossRef]
32. Bloukas, J.G.; Paneras, E.D.; Fournitzis, G.C. Effect of replacing pork backfat with olive oil on processing and quality characteristics of fermented sausages. *Meat Sci.* **1997**, *45*, 133–144. [CrossRef]
33. Surasani, V.K.R.; Raju, C.V.; Singh, A.; Joshi, S. Quality changes in fish sausages supplemented with pangas protein isolates as affected by frozen storage and casing material. *J. Food Sci. Technol.* **2022**, *59*, 2127–2140. [CrossRef] [PubMed]
34. Yang, H.; Khan, M.A.; Yu, X.; Zheng, H.; Han, M.; Xu, X.; Zhou, G. Changes in protein structures to improve the rheology and texture of reduced-fat sausages using high pressure processing. *Meat Sci.* **2016**, *121*, 79–87. [CrossRef] [PubMed]
35. Surasani, V.; Raju, C.; Sofi, F.; Shafiq, U. Utilization of protein isolates from rohu (*Labeo rohita*) processing waste through incorporation into fish sausages; quality evaluation of the resultant paste and end product. *J. Sci. Food Agric.* **2022**, *102*, 1263–1270. [CrossRef] [PubMed]
36. Surasani, V.K.R.; Raju, C.V.; Shafiq, U.; Chandra, M.V.; Lakshmisha, I.P. Influence of protein isolates from Pangas processing waste on physico-chemical, textural, rheological and sensory quality characteristics of fish sausages. *LWT* **2020**, *117*, 108662. [CrossRef]
37. Liang, H.N.; Tang, C.H. pH-dependent emulsifying properties of pea [*Pisum sativum* (L.)] proteins. *Food Hydrocoll.* **2013**, *33*, 309–319. [CrossRef]
38. Xiao, J.H.; Zhong, Q.X. Suppression of retrogradation of gelatinized rice starch by anti-listerial grass carp protein hydrolysate. *Food Hydrocoll.* **2017**, *72*, 338–345. [CrossRef]
39. Borderías, A.J.; Tovar, C.A.; Domínguez-Timón, F.; Díaz, M.T.; Pedrosa, M.M.; Moreno, H.M. Characterization of healthier mixed surimi gels obtained through partial substitution of myofibrillar proteins by pea protein isolates. *Food Hydrocoll.* **2020**, *107*, 105976. [CrossRef]
40. Broucke, K.; Van Poucke, C.; Duquenne, B.; De Witte, B.; Baune, M.C.; Lammers, V.; Terjung, N.; Ebert, S.; Gibis, M.; Weiss, J.; et al. Ability of (*extruded*) pea protein products to partially replace pork meat in emulsified cooked sausages. *Innov. Food Sci. Emerg.* **2022**, *78*, 102992. [CrossRef]
41. Zakaria, N.A.; Sarbon, N.M. Physicochemical properties and oxidative stability of fish emulsion sausage as influenced by snakehead (*Channa striata*) protein hydrolysate. *Lwt* **2018**, *94*, 13–19. [CrossRef]
42. Bertram, H.C.; Ersen, H.J. Applications of NMR in Meat Science. *Annu. Rep. NMR Spectrosc.* **2004**, *53*, 157–202.
43. Sun, Q.X.; Sun, F.D.; Xia, X.F.; Xu, H.H.; Kong, B.H. The comparison of ultrasound-assisted immersion freezing, air freezing and immersion freezing on the muscle quality and physicochemical properties of common carp (*Cyprinus carpio*) during freezing storage. *Ultrason. Sonochemistry* **2019**, *51*, 281–291. [CrossRef] [PubMed]
44. Halim, N.R.A.; Yusof, H.M.; Sarbon, N.M. Functional and bioactive properties of fish protein hydolysates and peptides: A comprehensive review. *Trends Food Sci. Technol.* **2016**, *51*, 24–33. [CrossRef]
45. Zhang, L.L.; Xue, Y.; Xu, J.; Li, Z.J.; Xue, C.H. Effects of high-temperature treatment ($\geq$100 °C) on Alaska Pollock (*Theragra chalcogramma*) surimi gels. *J. Food Eng.* **2013**, *115*, 115–120. [CrossRef]
46. Yi, S.M.; Qiao, C.P.; Li, X.P.; Zhu, W.H.; Li, J.R.; Li, Y.J. Effect of Fish Bone Powder on Sea Bass Myofibrillar Protein Gel Properties. *J. Chin. Inst. Food Sci. Technol.* **2018**, *18*, 77–85.
47. Xu, Y.; Xia, W.; Yang, F.; Nie, X. Physical and chemical changes of silver carp sausages during fermentation with Pediococcus pentosaceus. *Food Chem.* **2010**, *122*, 633–637. [CrossRef]
48. Lin, D.; Lu, W.; Kelly, A.L.; Zhang, L.; Zheng, B.; Miao, S. Interactions of vegetable proteins with other polymers: Structure-function relationships and applications in the food industry. *Trends Food Sci. Tech.* **2017**, *68*, 130–144. [CrossRef]
49. Sun, J.X.; Wu, Z.; Xu, X.L.; Li, P. Effect of peanut protein isolate on functional properties of chicken salt-soluble proteins from breast and thigh muscles during heat-induced gelation. *Meat Sci.* **2012**, *91*, 88–92. [CrossRef] [PubMed]
50. Santana, P.; Huda, N.; Yang, T.A. Physicochemical properties and sensory characteristics of sausage formulated with surimi powder. *J. Food Sci. Technol.* **2015**, *52*, 1507–1515. [CrossRef]
51. Cavenaghi-Altemio, A.D.; Alcade, L.B.; Fonseca, G.G. Low-fat frankfurters from protein concentrates of tilapia viscera and mechanically separated tilapia meat. *Food Sci. Nutr.* **2013**, *1*, 445–451. [CrossRef]

foods

MDPI

Article

Effects of Moderate Enzymatic Hydrolysis on Structure and Functional Properties of Pea Protein

Xixiang Shuai [1], Lizhi Gao [1], Qin Geng [1], Ti Li [1,*], Xuemei He [2], Jun Chen [1], Chengmei Liu [1] and Taotao Dai [2,*]

[1] State Key Laboratory of Food Science and Technology, Nanchang University, Nanchang 330047, China
[2] Guangxi Academy of Agricultural Sciences, Nanning 530007, China
* Correspondence: liti@ncu.edu.cn (T.L.); daitaotao@gxaas.net (T.D.)

Abstract: Pea protein (PP) was moderately hydrolyzed using four proteolytic enzymes including flavourzyme, neutrase, alcalase, and trypsin to investigate the influence of the degree of hydrolysis (*DH*) with 2%, 4%, 6%, and 8% on the structural and functional properties of PP. Enzymatic modification treatment distinctly boosted the solubility of PP. The solubility of PP treated by trypsin was increased from 10.23% to 58.14% at the 8% *DH*. The results of SDS-PAGE indicated the protease broke disulfide bonds, degraded protein into small molecular peptides, and transformed insoluble protein into soluble fractions with the increased *DH*. After enzymatic treatment, a bathochromic shift and increased intrinsic fluorescence were observed for PP. Furthermore, the total sulfhydryl group contents and surface hydrophobicity were reduced, suggesting that the unfolding of PP occurred. Meanwhile, the foaming and emulsification of PP were improved after enzymatic treatment, and the most remarkable effect was observed under 6% *DH*. Moreover, under the same *DH*, the influence on the structure and functional properties of PP from large to small are trypsin, alcalase, neutrase and flavourzyme. This result will facilitate the formulation and production of natural plant-protein-based products using PP.

Keywords: pea protein; moderate enzymatic hydrolysis; structure properties; functional properties

Citation: Shuai, X.; Gao, L.; Geng, Q.; Li, T.; He, X.; Chen, J.; Liu, C.; Dai, T. Effects of Moderate Enzymatic Hydrolysis on Structure and Functional Properties of Pea Protein. *Foods* **2022**, *11*, 2368. https://doi.org/10.3390/foods11152368

Academic Editor: Christine Scaman

Received: 30 June 2022
Accepted: 5 August 2022
Published: 7 August 2022

1. Introduction

In recent years, the use of plant proteins as an alternative to animal proteins has attracted considerable consumer interest due to their wide availability, affordability, hypoallergenicity, cholesterol-lowering effects, and ease of digestion [1]. Pea protein (PP) is a valuable plant protein with the advantages of being hypoallergenic, non-GMO, amino-acid-balanced and rich in lysine [2,3]. The amino acid ratio of PP is balanced, the content of the other seven kinds of human essential amino acids except methionine is close to the recommended model value of FAO and WHO, and it is easy to digest and absorb. Meanwhile, the lysine content is higher than other plant proteins.

Although PP possesses high nutritional value, its application in formula food is limited due to its poor water solubility and limited functional properties [3,4]. In order to overcome these drawbacks, physical modification, chemical modification, and enzymatic modification have been applied to improve the properties of PP. Physical modification, including heating, magnetic field, freezing, ultrasound, and microfluidization, has been reported to improve the structural, functional, and nutritional value of proteins [5,6]. Although physical modification does not involve the addition of exogenous substances, it is not widely used because the modification effect is limited [7]. Chemical modification was considered as an effective method to improve the functional properties of protein by introducing new organic groups into the protein molecule to break or polymerize the amino, carboxyl, sulfhydryl, or carbonyl groups on the main or side chains of the protein, resulting in changes in the structural and physicochemical properties of the protein [8]. However, chemical modification was also not widely used in the food industry because of the addition of chemicals that may be harmful to human health, such as toxicity.

Enzymatic hydrolysis is a promising technology in the protein field. Compared to physical and chemical modification, enzymatic modification is widely used because of the advantages of low by-products, high specificity, and easy control. It can convert native protein into peptides of various sizes and free amino acids [9]. Enzymatic hydrolysis was considered as one of the most promising methods for the modification of tailor-made protein preparations [10,11]. Arteaga et al. [12] used 11 enzymes to hydrolyze PP under different enzymolysis time, altering the molecular weight distribution and, thus, improving the technical functionality and sensory properties of PP. Klost et al. [13] investigated the ability of trypsin treatment to improve the poor solubility and interfacial properties of PP, and the potential to improve the overall stability of PP emulsions. Tamm et al. [14] found that trypsin treatment improved various properties of PP emulsions through characterization on molecular weight distribution, interfacial activity, expansive characterization, and emulsion properties. Klost et al. [15] discovered that the treatment of PP with trypsin improved the gel properties of fermented PP, which helped to develop PP yogurt alternatives for tailoring such gels. However, most studies mainly focused on single enzyme treatment and comparison of different enzyme treatments at the same enzymatic hydrolysis time. The effects of different enzyme treatments on PP at the same degree of enzymolysis (*DH*) were not clear. In addition, mild enzymatic hydrolysis modification does not improve the protein properties significantly, and excessive enzymatic hydrolysis modification would produce strong bitterness, thus affecting the organoleptic properties of PP [10,12]. Therefore, moderate enzymatic hydrolysis is required to improve the protein properties and produce less bitterness at the same time. Furthermore, systematic studies and comparative information on the effects of different *DH* on PP are limited.

Based on the above analysis, the present study focused on investigating the effects of different enzymes (flavourzyme, neutrase, alcalase, and trypsin) on the structural and functional properties of PP at the same *DH* under moderate enzymatic hydrolysis conditions.

2. Materials and Methods

2.1. Material

Pea protein (PP, purity $\geq$ 80%) was provided by the Shuangta Food Co., Ltd. (Qingdao, China). Enzymes including Alcalase 2.4 L FG (2.4 AU/Ml, endo protease), Neutrase 0.8 L (0.8 AU/g, endo protease), Trypsin (4000 U/g, acts on carboxyl side cleavage of lysine and arginine residues in polypeptide chains), and Flavourzyme 1000 L (1000 LAPU/g, mixture of endopeptidase and exopeptidase) were purchased from Novozymes (Bagsvaerd, Denmark). The SDS-PAGE kit and Micro Total Mercapto Assay Kit were purchased from Solarbio Science and Technology (Beijing, China). 1-anilino-8-naphthalene-sulfonate (ANS), sodium dodecyl sulfate, and ammonium peroxodisulfate were purchased from Sigma company (St. Louis, MO, USA). All other chemicals used were of analytical grade. Double-distilled water was used throughout the research.

2.2. Enzymatic Hydrolysis of PP

The hydrolysates of PP with different enzymatic hydrolysis degrees (*DH*) were prepared based on the study by Phoon et al. with some modification [16]. The specific steps were as follows: 10 g PP was dispersed in distilled water according to the ratio of protein to distilled water 1:15 (*w*/*w*), and stirred overnight at room temperature to make the protein fully hydrated. The pH of the aqueous solution of protein was adjusted to the optimum pH of the enzyme (alcalase and trypsin were pH 8; neutrase and flavourzyme were pH 7), and kept in a water bath (50 °C) for 15 min. Then, 1% enzyme (the ratio of enzyme to substrate was 1:100) was added for enzymolysis. In the process of enzymatic hydrolysis, 1 mol/L NaOH solution was added continuously to maintain the pH. After the enzymatic hydrolysis, the enzyme suspension was quickly moved into a water bath at 95 °C and heated for 15 min to inactivate the used enzyme. After cooling to ambient temperature, the pH adjusted to neutral.

2.3. Structure Properties

2.3.1. Degree of Enzymolysis (*DH*)

The degree of enzymolysis (*DH*) is defined as the number of hydrolyzed peptide bonds relative to the number of peptide bonds per unit weight expressed as a percentage. The *DH* of PP was calculated using the pH-stat method [17]. In the enzymatic hydrolysis process, the *DH* of PP was calculated in real time by representing the volume of NaOH consumed by the sample. When the *DH* reached 2, 4, 6, and 8%, the enzymatic hydrolysis process was stopped and samples for subsequent experiments were obtained, Equation (1) is as follows:

$$DH = \frac{B \times N_b}{\alpha \times h_{hot} \times M_p} \times 100\% \quad (1)$$

where B represents the volume of NaOH consumed by the sample; N_b refers to the concentration of the calibrated NaOH; α is the degree of amino dissociation, which has different values under different enzymolysis conditions; Mp refers to the net protein content of PP; and h_{hot} is the number of millimoles of peptide bonds per gram of protein, which is usually an empirical value. The h_{hot} of PP is 7.55 mep/g.

2.3.2. Intrinsic Fluorescence Spectroscopy

The intrinsic fluorescence spectra of PP and PP with different *DH* samples were determined according to previous papers by using an fluorescence spectrophotometer (F-7000, Hitachi, Tokyo, Japan) equipped with a 10 mm square quartz cell [18]. The emission spectra of PP samples were collected to investigate the effect of hydrolysis on the intrinsic fluorescence of PP. The critical parameters used in fluorescence were: excitation wavelength 295 nm, excitation slit 2.5 nm, emission slit 2.5 nm, and scan rate 1200 nm/min. Then, the emission spectra of PP samples were collected to investigate the effect of the enzyme on the intrinsic fluorescence of PP.

2.3.3. Surface Hydrophobicity Measurements

The surface hydrophobicity of PP and PP with different *DH* samples was determined according to Li et al. [19]. ANS was used as a fluorescence probe; the excitation wavelength was set at 380 nm and emission spectra were collected from 390 to 660 nm. Both excitation and emission slits widths were fixed at 5 nm. Briefly, 20 μL 8 mM ANS was added into 3 mL samples with different protein concentrations (0.004%, 0.008%, 0.012%, 0.016%, and 0.020%). Under conditions of fluorescent probes present in excess, protein relative fluorescence intensity F was plotted against protein concentration C, and the slope of the line was represented as Equation (2):

$$S_0 = \frac{\Delta F}{\Delta C} \quad (2)$$

2.3.4. Circular Dichroism (CD) Spectroscopy Studies

The secondary structure of PP and PP with different *DH* samples were measured utilizing a MOS-450 CD spectrometer (Claix, France) as in a previous study described by Li et al. [20]. The spectra of samples were collected from 180 to 260 nm. The PBS was measured as a background to correct the CD spectra. The secondary structure contents including α-helix, β-sheet, β-turn, and random coil of PP samples were calculated by the analytical approach of previous research using online CONTIN program [19].

2.3.5. Morphology Observation

The microstructure of PP and PP with different *DH* samples was imaged utilizing a scanning electron microscope (Quanta-200, FEI Company, Eindhoven, The Netherlands). The freeze-dried protein powders were stuck onto one side of double adhesive tape attached to a circular specimen stub and then sputter-coated with a thin film of gold. Then, the microstructural images of the samples were captured under an accelerating voltage of 5.0 kV with magnification at 300-fold.

2.3.6. Sodium Dodecyl Sulfate Polyacrylamide Gel Electrophoresis (SDS-PAGE)

SDS-PAGE under nonreducing conditions was performed to determine the protein fractions patterns of PP and PP with different *DH* samples according to He et al. [5]. Briefly, samples were dispersed in 10 mM phosphate buffer (pH 7) to obtain a 1 mg/mL protein sample. A 60 μL protein solution was mixed with 20 μL 4× loading buffer, followed by heating at 95 °C for 8 min. An aliquot containing 10 μg protein and the Thermo Scientific Page Ruler Prestained Protein Ladder (ranging from 11–245 kDa) were loaded to the specific cell. Electrophoresis was run for 15 min at 80 V for stacking gel (5%) and 55 min at 120 V for separating gel (12%). After electrophoresis, the gel was stained using 0.25% Coomassie Brilliant Blue in 50% methanol and 10% acetic acid for at least 50 min. Then, the destaining processing was performed using a water solution of 5% methanol and 7.5% acetic acid.

2.3.7. Determination of Sulfhydryl Groups (SH)

The total sulfhydryl content of PP and PP with different *DH* samples was detected using a micro total mercapto assay kit (Sigma-Aldrich Co., Ltd., Shanghai, China) according to the method reported by Yang et al. [21]. Briefly, the sample of 0.1 g was weighed, and 1 mL of extract was added to prepare 10% homogenate. The sample was centrifuged at room temperature for 10 min at 8000× *g*, and the supernatant was taken to be measured. A certain amount of the samples and reagents was added according to the requirements of the kit, and the absorbance value was measured at a 412 nm.

2.3.8. Amino Acid Composition Analysis

Amino acid composition was analyzed according to Xie et al. [22]. Briefly, 1.0 g samples, 10 mL hydrochloric acid (6 mol/L), and 1.0 g phenol were added to a sealed tube and the solution was treated with nitrogen for 15 min. The above samples were hydrolyzed for 24 h at 110 °C. The hydrolyzed samples were transferred to a 50 mL volumetric flask for constant volume and were then filtrated. The filtrated samples (1.0 mL) were evaporated to dryness at 60 °C in a water bath. Subsequently, the dried samples were diluted by sample diluent (3–5 mL, pH 2.2, 0.02 mol/L Sodium citrate buffer solution). The mixed samples were filtrated by a 0.22 μm filter membrane and then analyzed utilizing an automatic amino acid analyzer (S-433D, Sykam, Munich, German). The amino acid compositions for PP with different *DH* were presented as g/100 g protein.

2.3.9. Hydrophobicity Analysis Based on Amino Acid Composition

The hydrophobicity of PP with different *DH* samples was determined according to previous research [17], Equations (3) and (4) are as follows:

$$Q = \sum \Delta Qi \tag{3}$$

$$\Delta Q = [(AAi \ / \ Mi\ )/(\sum AAi/\ Mi\ )] \times \Delta f \tag{4}$$

where AAi refers to the amount of various amino acids in 100 mL of protein; Mi represents the molar mass of various amino acids; $\sum AAi/Mi$ is the total number of moles of amino acids; Δf refers to the free energy values of amino acid side chains; and ΔQ is the free energy for the transfer of an amino acid side chain from ethanol to water.

2.3.10. Sensory Evaluation

The color, taste, beany flavor, and bitterness of PP and PP with different *DH* samples were evaluated by a slightly modified method described by Garcia-Arteaga et al. [12] In brief, 10 mL sample solution was used in this test and served at random to the panelists. The samples were tested at 25 °C, in a uniformly illuminated room, by a five-member panel selected from a pool of students and staff members of our research team. The evaluation criteria were as follows: color (light to dark), taste (rough to fine), and beany flavor (serious to slight). Those sensory properties' intensity was estimated on a five-point scale, and bitterness (light to heavy) was also evaluated on a ten-point scale. Water was provided for rinsing between samples.

2.4. Functional Properties

2.4.1. Solubility

The solubility of PP and PP with different *DH* samples was determined by the Lowry method [23]. The specific methods are as follows: the lyophilized samples were dissolved in distilled water; then, pH was adjusted to neutral. After that, the suspension was centrifuged at 4500× *g* for 15 min. The protein content of the supernatant was determined. Bovine serum albumin (BSA) was used as the standard. Solubility was calculated according to Equation (5).

$$\text{Solubility } (\%) = \frac{\text{protein content in supernatant}}{\text{total protein content}} \times 100\% \tag{5}$$

2.4.2. Foaming Performance

The foaming properties of PP and PP with different *DH* samples were characterized at room temperature using a method described previously with some modification [24]. A fixed volume (18 mL) of protein solution was subjected to mechanical shearing (10,000 r/min, 60 s) using a high-shear mixer to generate foam (Ultra TURRAX homogenizer, T18digital, IKA, Staufen, Germany). The foamability and foam stability of each test sample were then calculated using the following Equations (6) and (7):

$$\text{Foamability } (\%) = 100 \times \frac{V_0}{18} \tag{6}$$

$$\text{Foam stability } (\%) = 100 \times \frac{V_{20}}{V_0} \tag{7}$$

where 18 is the volume of the test sample before shearing (18 mL), V_0 refers to the volume of the foam (mL) immediately after shearing, and V_{20} is the volume of the foam (mL) at 20 min after shearing.

2.4.3. Emulsifying Performance

The emulsifying ability index (*EAI*) and emulsifying stability index (*ESI*) of PP and PP with different *DH* samples were measured according to previous papers with some modifications [25]. Then, 16 mL of protein solution was mixed with 4 mL of soybean oil, and followed by stirring at 10,000 rpm for 1 min using the high-shear mixer mentioned above. Then, 50 μL of emulsion was dispersed into 5 mL of 0.1% SDS solution at 0.5 cm at the bottom of the container, vortexed, and mixed. The absorbance values were measured at 500 nm for 0 and 10 min. *EAI* and *ESI* were calculated using the following Equations (8) and (9):

$$EAI\ (\text{m}^2/\text{g}) = \frac{2 \times 2.303 \times A_0 \times DF}{C \times (1-\varphi) \times \theta \times 10{,}000} \tag{8}$$

$$ESI\ (\text{min}) = \frac{A_0 \times \Delta t}{A_0 - A_{10}} \tag{9}$$

where A_0 refers to the absorbance at 500 nm of the sample solution, *DF* represents the dilution factor, *C* is the protein concentration, φ is the oil volume fraction, and θ refers to the path length. In Equation (9), Δt is the time difference (10 min) and A_{10} represents the absorbance at 500 nm of sample solution after 10 min.

2.5. Statistical Analyses

The experiments were conducted in triplicate. The statistical analyses were performed by SPSS 26.0 (IBM Inc., Chicago, IL, USA). Significant differences ($p < 0.05$) between the means of parameters were analyzed with one-way ANOVA test followed by Duncan's LSD test.

3. Results and Discussion

3.1. Structure Properties

3.1.1. Degree of Enzymolysis (*DH*)

In this study, pea protein (PP) was hydrolyzed by four enzymes (flavourzyme, neutrase, alcalase, and trypsin) for 5 h. As shown in Figure 1, the enzymolysis degree (*DH*) of PP increased firstly and then tended to be flat with the increase in enzymolysis time. The *DH* of PP was most intense during the first 1 h, its *DH* increased rapidly, and the *DH* of PP increased slowly after 1 h. This was consistent with Avramenko et al. [26], who pointed out that the *DH* of the lentil isolate hydrolyzed by trypsin was most rapid in the first 40 min, and then the *DH* increased slowly. Figure 1 indicated that the *DH* of PP treated by four enzymes reached 19.88% (trypsin), 16.78% (alcalase), 8.97% (neutrase), and 7% (flavourzyme) at 5 h. In the same enzymatic hydrolysis time, the enzymatic hydrolysis degree after four kinds of enzyme treatment from large to small is trypsin, alcalase, neutrase, and flavourzyme. This was closely related to the activity unit and specific restriction site of the enzyme. A similar result had been reported for pea protein isolate treated with 11 proteolytic enzymes at different enzymatic hydrolysis times [12]. Compared to previous studies, it was found that excessive enzymatic hydrolysis ($DH \geq 10\%$) showed adverse effects on the properties of PP. In addition, it was worth noting that the enzymatic hydrolysis limit of flavourzyme was 7%. Therefore, in this study, the enzymatic hydrolysates with the *DH* of 2%, 4%, 6%, and 8% (flavourzyme is 7%) of the four enzymes were selected for subsequent studies in order to clarify the effects of different enzymes on the physical and chemical properties of PP at the same *DH*.

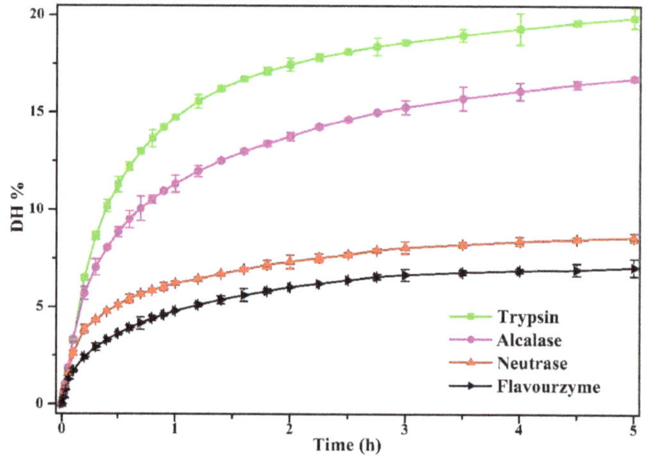

Figure 1. The *DH* of PP enzymolyzed by four enzymes (alcalase, neutrase, flavourzyme, and trypsin) using pH-stat method for 5 h.

3.1.2. Intrinsic Fluorescence Spectroscopy

The intrinsic fluorescence of aromatic amino acid residues (Trp, Tyr, and Phe) is very sensitive to the microenvironment, and the emission fluorescence spectra of proteins are used to investigate the changes in their tertiary structure [27]. Therefore, the maximum fluorescence intensity and the wavelength (λmax) at the maximum fluorescence intensity are effective indicators to observe the structure and conformational changes in proteins [28]. Figure 2A–D show the fluorescence intensity of PP treated with flavourzyme, neutrase, alcalase, and trypsin, respectively. The intensities of the emission fluorescence of PP exhibited a maximum absorption at 337 nm. The λmax was affected by enzymatic treatment, and a shift in λmax to longer wavelengths (bathochromic shift) was observed for PP treated by flavourzyme enzymolysis (345.8, 345.4, 350.4, and 350.4 nm for 2%, 4%, 6%, and 7%, respectively), neutrase enzymolysis (346.8, 346.4, 350.2, and 350.4 nm for 2%, 4%, 6%, and 8%, respectively), alcalase enzymolysis (349, 349.6, 350.2, and 349.2 nm for 2%, 4%, 6%, and 8%, respectively), and trypsin enzymolysis (349.2, 350.4, 350.6, and 350.4 nm for 2%, 4%, 6%, and 8%, respectively). These results suggested that the microenvironment of chromophores in PP became more polar and hydrophilic after enzymatic treatment owing to the increased contact between the fluorophore and the aqueous medium [29]. In addition, the fluorescence intensity of PP after enzymatic hydrolysis increased significantly; the fluorescence intensity of all the four enzymes changed most significantly when *DH* increased from 0 to 2%. This indicated that the enzymatic hydrolysis process made more aromatic groups be exposed to solvent and more available to emit fluorescence [30,31]. With the increase in the *DH*, this process gradually slowed down. Among the four enzymes, trypsin-enzymolyzed PP had the highest fluorescence absorption peak. This indicated that trypsin changed the structure of PP most obviously.

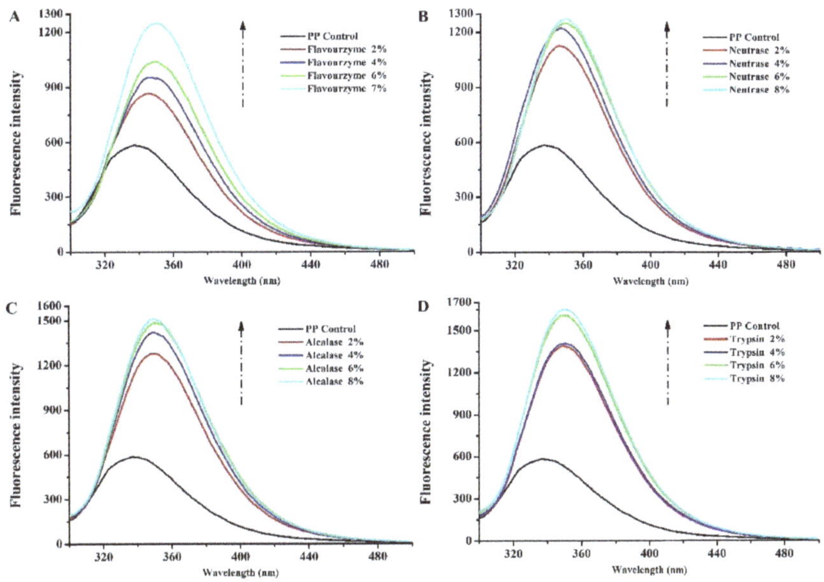

Figure 2. Intrinsic fluorescence of PP and PP with different *DH* samples ((**A–D**) are flavourzyme, neutrase, alcalase and trypsin treatment, respectively).

3.1.3. Surface Hydrophobicity

The surface hydrophobicity largely determines the protein structure and properties [32]. The changes in surface hydrophobicity of proteins after enzymatic hydrolysis were related to the type of enzyme and protein and time of enzymatic hydrolysis [33]. The surface hydrophobicity of PP after enzymatic hydrolysis by four enzymes is shown in Figure 3. Compared with PP (16,102 ± 1136), flavourzyme and neutrase showed a gradual decrease in H_0 with increasing enzymatic hydrolysis. The H_0 of flavourzyme 2%, 4%, 6%, and 7% was 10,418 ± 784, 9010 ± 527, 8804 ± 485, and 8693 ± 153, respectively. The H_0 of neutrase 2%, 4%, 6%, and 8% was 19,191 ± 1234, 17,037 ± 478, 14,353 ± 222, and 11,761 ± 563, respectively. Speculatively, the above two enzymes caused the PP molecules to stretch and be partially hydrolyzed into smaller fragments, and at the same time, these small fragments would reassemble under the action of hydrophobic and disulfide bonds, resulting in a decrease in surface hydrophobic groups [34]. The H_0 of alcalase 2%, 4%, 6%, and 8% was 19,191 ± 1234, 17,037 ± 478, 14,353 ± 222, and 11,761 ± 564, respectively. The H_0 of trypsin 2%, 4%, 6%, and 8% was 21,274 ± 554, 18,869 ± 937, 14,387 ± 78, and 13,219 ± 987, respectively. The surface hydrophobicity of the PP treated by alcalase and trypsin tended to increase and then decrease with the increase in the *DH*, and the surface hydrophobicity of PP was greatest at a *DH* of 2%. A similar result has been reported for rice glutelin treated with trypsin [18]. The increase in hydrophobicity was due to the partial enzymatic hydrolysis that fully stretched the protein structure, thus exposing the hydrophobic sites wrapped inside the protein molecules [35]. With further increases in enzymatic hydrolysis, the hydrophobicity of the protein decreased. Enzymatic hydrolysis broke down hydrophobic regions or led to protein–protein aggregation, thereby reducing the number or surface area of hydrophobic groups exposed to water [26,34]. In general, trypsin enzymatic hydrolysis has the greatest effect on the surface hydrophobicity of PP.

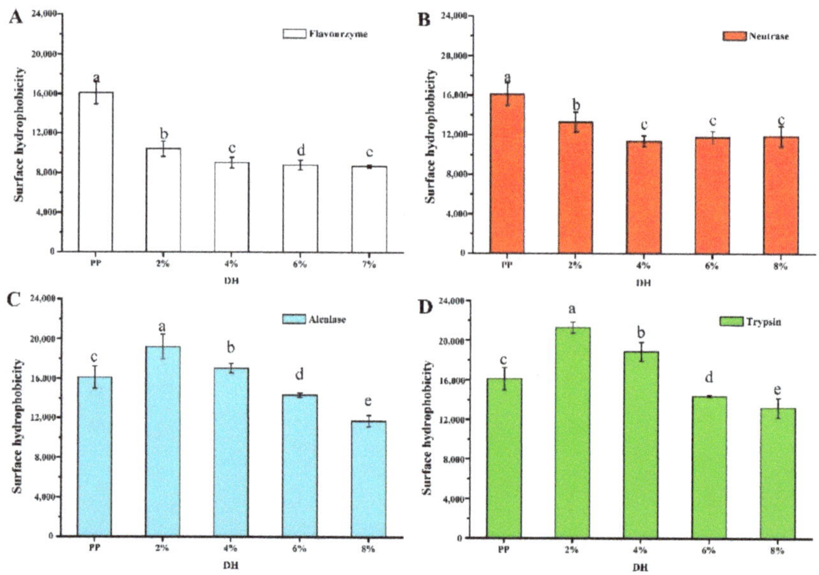

Figure 3. Surface hydrophobicity of PP and PP with different *DH*. The values reported represent means (n = 3) ± SDs and different superscript letters indicate significant difference ($p < 0.05$). ((**A–D**) are flavourzyme, neutrase, alcalase and trypsin treatment, respectively).

3.1.4. Circular Dichroism (CD) Spectroscopy Studies

As shown in Table 1, the second structure compositions of PP were 15.0% α-helix, 28.8% β-sheet, 22.1% β-turn, and 34.2% random coils. After enzymatic hydrolysis with four enzymes, the β-sheet content of enzymatic hydrolysis samples decreased with the increase in *DH*, the α-helix content increased significantly, and there was no significant change in β-turn angle and random coil. When the *DH* was 8%, the β-sheet content of PP treated by flavourzyme, neutrase, alcalase, and trypsin decreased from 28.8% to 21.7%, 20.7%, 18.5%, and 14.1%, respectively. Correspondingly, the α-helix content increased from 15.0% to 22.2%, 24.6%, 26.9%, and 32.1%, respectively. These results indicated that enzymatic hydrolysis caused perturbations of the PP secondary structure and, thus, may have an impact on the function properties of PP. Among them, the structure of β-sheet was more compact and the structure of α-helix was looser, so it can be deduced that the ordered β-sheet in the enzymatic hydrolysis products was disrupted to a more flexible and spreading α-helix structure after enzyme treatment [36]. According to the results, under the same *DH*

(2%, 4%, 6%, and 8%), trypsin showed the most obvious effect on the secondary structure of PP, followed by alcalase, neutrase, and flavourzyme. The changes in the secondary structure of PP can correspond to the results of the intrinsic fluorescence and surface hydrophobicity.

Table 1. Secondary structure content of PP and PP with different *DH* samples determined by circular dichroism.

Protein	Flavourzyme				Protein	Neutrase			
	α-Helix (%)	β-Sheet (%)	β-Turn (%)	Random (%)		α-Helix (%)	β-Sheet (%)	β-Turn (%)	Random (%)
PP	15.0	28.8	22.1	34.2	PP	15.0	28.8	22.1	34.2
2%	20.1	24.4	19.6	35.9	2%	21.2	23.5	20.4	34.9
4%	20.5	23.8	19.8	35.9	4%	22.5	22.8	19.0	35.7
6%	21.6	22.4	19.8	36.2	6%	23.0	22.2	18.2	36.6
7%	22.2	21.7	19.7	36.4	8%	21.2	23.5	20.4	34.9
Protein	**Alcalase**				**Protein**	**Trypsin**			
	α-Helix (%)	β-Sheet (%)	β-Turn (%)	Random (%)		α-Helix (%)	β-Sheet (%)	β-Turn (%)	Random (%)
PP	15.0	28.8	22.1	34.2	PP	15.0	28.8	22.1	34.2
2%	24.5	21.7	19.5	34.3	2%	25.9	20.8	19.1	34.2
4%	24.7	21.4	19.5	34.4	4%	27.9	19.3	19.0	33.8
6%	25.6	20.3	19.0	35.1	6%	29.2	16.9	19.9	34.0
8%	26.9	18.5	18.3	36.3	8%	32.1	14.1	19.0	34.8

3.1.5. Morphology Observation

To gain insight into the effect of enzymatic hydrolysis on the change in the microstructure of PP, SEM micrographs were imaged for PP and PP with different *DH* samples. As shown in Figure 4, the natural PP is irregularly spherical with a smooth surface and a tight structure. After being treated with flavourzyme, neutrase, alcalase, and trypsin, the spherical structure of PP was broken into fragments. In addition, the microstructure of PP showed that the particles gradually became smaller with the increase in *DH*. The smaller particle size of PP made it have more chance to come into contact with water molecules during the dispersion process, thus changing the physicochemical properties of PP [29]. Furthermore, at the same *DH*, the protein particles treated by the four enzymes from small to large were trypsin, alcalase, neutrase, and flavourzyme, respectively. This result can correspond to the above result. The disruption of the protein microstructure may affect the amino acid groups embedded in the molecule and change the surface hydrophobicity, interaction forces, and secondary structure [37].

3.1.6. SDS-PAGE

The molecular weight distribution was analyzed to investigate the effect of enzymatic hydrolysis on the PP. Figure 5 shows the SDS-PAGE electrophoresis of PP and PP with different *DH* samples. The electrophoresis pattern of PP showed multiple cleaned bands. The band at 17 kDa, 20–22 kDa, ~40 kDa, ~18 and ~50 kDa, ~60 kDa, ~75 kDa, and ~100 kDa could be attributed to 2S albumin polypeptides and/or γ-vicilin, legumin β, legumin α, vicilin, legumin, convicilin, and lipoxygenase, respectively [5]. After being treated with flavourzyme, neutrase, alcalase, and trypsin, the protein bands became progressively smaller as *DH* increased. Additionally, after enzymatic hydrolysis by flavourzyme and neutrase, the bands were clear at low *DH* (2% and 4%), but with the increase in *DH*, the protein bands were aggregated downward and the bands were gradually indistinguishable. After trypsin and alcalase treatment, the bands gradually moved toward the small molecules with the increase in *DH*, and the color of bands gradually became lighter, indicating that they kept aggregating toward the small molecule fragments after enzymatic hydrolysis.

This result was consistent with the results of microstructure. It can also be shown that with the increase in *DH*, the molecular weight of PP gradually becomes smaller, which may affect the functional properties of PP. In addition, the effect of flavourzyme and neutrase on molecular weight of PP was lower than trypsin and alcalase under the same *DH*.

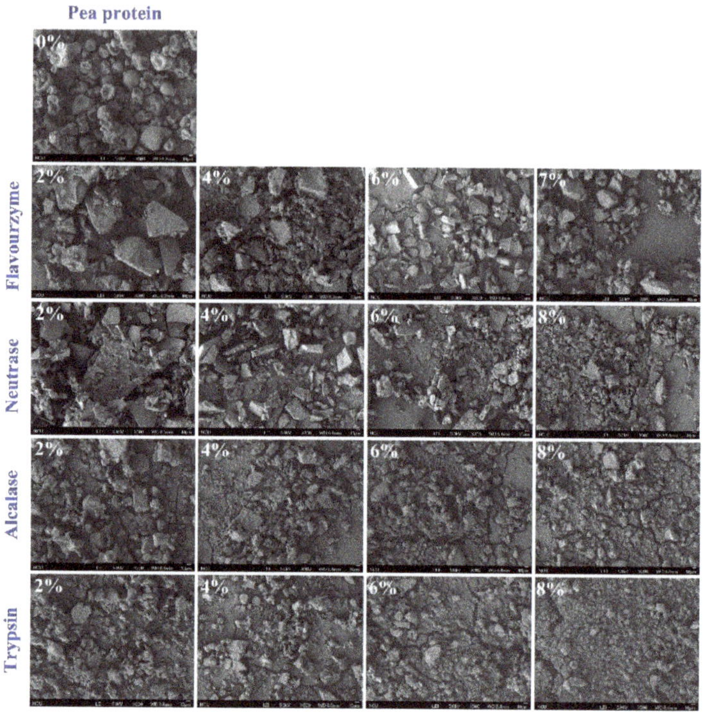

Figure 4. Morphology of PP and PP with different *DH* samples.

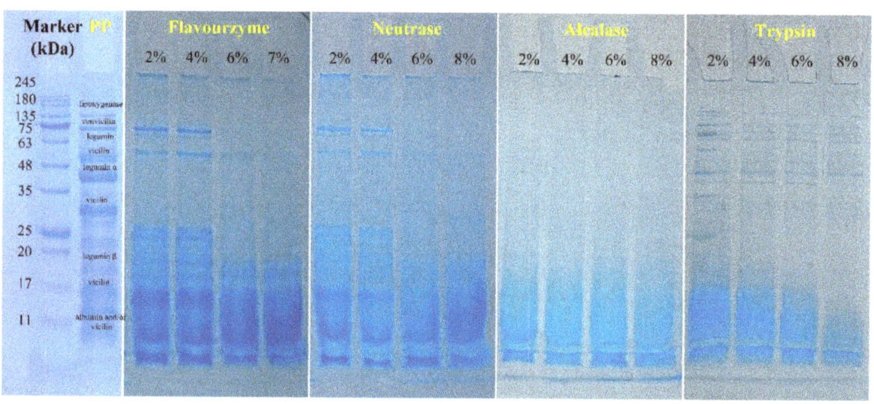

Figure 5. SDS-PAGE profiles of PP and PP with different *DH* samples under nonreducing conditions.

3.1.7. Determination of Total Sulfhydryl Groups

The content of sulfhydryl groups (SH) and disulfide bonds in proteins determines their rigid structure and has a significant impact on their functional properties. Figure 6 shows the variation in the total SH of samples. Compared to PP (3.94 μmol/g), the enzymatic hydrolysis samples resulted in a lower total SH. The decrease in SH content may be due to the destruction of SH within the protein molecule by enzymatic hydrolysis. As known, the spatial structure will unfold after enzymolysis; the more SH is exposed, the SH will be destroyed during enzymatic hydrolysis. As reported by Lepedda et al. [38], the SH in protein can be oxidized by enzymes, resulting the decrease in total SH. When the *DH* was 2% to 6%, the reduced SH content was the same, but when the *DH* was increased to 8%, the total SH of the samples treated by flavourzyme and alcalase further decreased significantly, which may be due to the enhanced oxidation of the SH during the enzymatic hydrolysis process. Moreover, the total SH content was reduced by trypsin > alcalase > neutrase > flavourzyme after treatment with the four enzymes, indicating that the effect of the four enzymes on the SH content and structure of PP was also trypsin > alcalase > neutrase > flavourzyme.

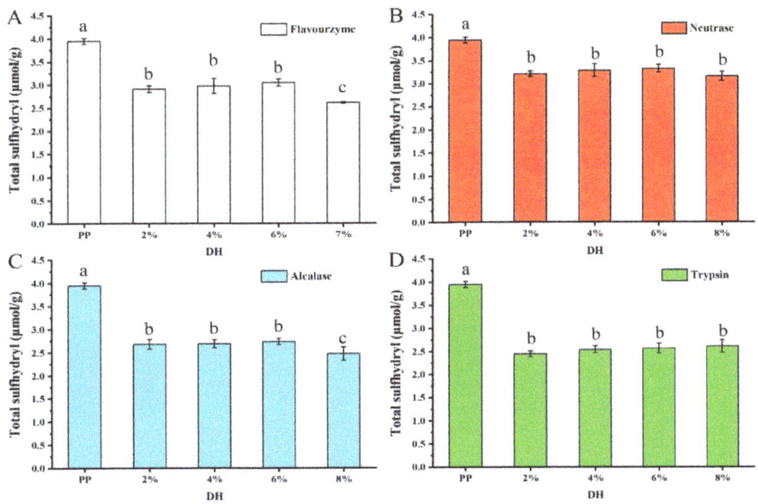

Figure 6. Total sulfhydryl content of PP and PP with different *DH* samples. The values reported represent means ($n = 3$) ± standard deviations and different superscript letters indicate significant difference ($p < 0.05$). ((**A–D**) are flavourzyme, neutrase, alcalase and trypsin treatment, respectively).

3.1.8. Amino Acid Composition and Average Hydrophobicity Analysis

PP was considered a high-quality protein because its balanced amino acid ratio can fulfil FAO/WHO recommendations [39]. Table 2 shows the results of the amino acid composition of PP with different *DH* samples. Due to the difference in the specific cutting sites of different proteases, different amino acid compositions of peptides can be produced by enzymatic hydrolysis. As summarized in Table 2, 16 amino acids were identified and quantified among all the hydrolysates, and GLU had the highest content (it ranged from 15.70 to 18.99 g/100 g), which was similar to the amino acid composition of PP [3]. It was worth noting that the hydrophobic amino acid (HAA) (including ALA, VAL, MET, ILE, LEU, TYR, PHE, and PRO) content of PP treated by flavourzyme, neutrase, alcalase, and trypsin increased with increasing *DH* (from 2% to 8%). Meanwhile, proteolytic enzyme treatment of food protein can improve the functional and nutritional properties, but also introduce undesirable attributes [40]. Among these, bitterness is considered to be one of the main disadvantages in utilizing protein hydrolysates in food applications. In addition, sensory evaluation showed the color gradually deepened, and the roughness of the taste and the beany flavor gradually decreased with the increase in *DH* after enzymatic hydrolysis with the four enzymes. Meanwhile, the bitterness assessed by tasting was also enhanced with the increase in *DH* (data are displayed in Supplementary Materials Figure S1), and at the same *DH*, the bitterness intensity of PP treated with different enzymes was in the following order: trypsin > alcalase > neutrase > flavourzyme.

Table 2. Amino acid composition of PP with different *DH* samples treated with flavourzyme, neutrase, alcalase, and trypsin (g/100 g). HAA: hydrophobic amino acids.

Amino Acid	Flavourzyme				Neutrase				Alcalase				Trypsin			
	2%	4%	6%	7%	2%	4%	6%	8%	2%	4%	6%	8%	2%	4%	6%	8%
ASP	7.79 ± 0.01 a	8.66 ± 0.02 c	8.09 ± 0.00 b	8.61 ± 0.02 c	9.09 ± 0.02 a	9.17 ± 0.01 b	9.58 ± 0.03 c	9.70 ± 0.01 d	10.24 ± 0.03 c	10.02 ± 0.01 a	10.13 ± 0.02 b	10.18 ± 0.03 b	9.58 ± 0.01 a	9.70 ± 0.02 b	9.58 ± 0.04 a	9.73 ± 0.02 b
THR	2.05 ± 0.01 a	2.25 ± 0.00 d	2.18 ± 0.01 c	2.07 ± 0.02 b	2.76 ± 0.02 a	2.89 ± 0.00 b	3.06 ± 0.01 c	3.10 ± 0.01 d	3.05 ± 0.01 a	3.05 ± 0.03 a	3.04 ± 0.01 a	3.24 ± 0.02 b	3.11 ± 0.00 b	3.06 ± 0.02 a	3.13 ± 0.01 b	3.18 ± 0.03 c
SER	3.36 ± 0.00 c	3.38 ± 0.01 c	2.98 ± 0.02 b	2.67 ± 0.00 a	3.85 ± 0.01 a	3.83 ± 0.02 a	3.93 ± 0.01 b	4.04 ± 0.02 c	4.01 ± 0.01 c	3.86 ± 0.02 a	3.82 ± 0.01 a	3.95 ± 0.02 b	4.05 ± 0.03 c	3.92 ± 0.01 a	3.93 ± 0.01 a	3.98 ± 0.01 b
GLU	18.59 ± 0.05 c	18.99 ± 0.03 d	18.05 ± 0.06 b	17.39 ± 0.03 a	16.73 ± 0.03 d	15.89 ± 0.02 a	16.25 ± 0.02 b	16.62 ± 0.01 c	18.52 ± 0.02 d	17.52 ± 0.03 c	16.49 ± 0.01 a	16.83 ± 0.04 b	16.99 ± 0.01 d	15.90 ± 0.02 c	15.70 ± 0.01 a	15.78 ± 0.02 b
PRO	2.68 ± 0.02 a	3.00 ± 0.01 b	3.13 ± 0.01 c	3.18 ± 0.01 d	3.19 ± 0.02 a	3.25 ± 0.01 b	3.36 ± 0.01 c	3.40 ± 0.00 d	3.44 ± 0.01 a	3.45 ± 0.02 a	3.50 ± 0.00 b	3.63 ± 0.01 c	3.42 ± 0.02 a	3.42 ± 0.01 a	3.43 ± 0.00 a	3.44 ± 0.01 a
GLY	2.75 ± 0.01 a	2.90 ± 0.00 b	2.89 ± 0.02 b	2.92 ± 0.01 b	2.95 ± 0.02 a	2.99 ± 0.00 b	3.08 ± 0.01 c	3.12 ± 0.01 d	3.22 ± 0.00 c	3.17 ± 0.02 b	3.09 ± 0.01 a	3.31 ± 0.01 d	3.15 ± 0.00 b	3.09 ± 0.01 a	3.13 ± 0.02 b	3.16 ± 0.01 b
ALA	2.55 ± 0.00 a	2.82 ± 0.02 b	2.89 ± 0.00 c	2.98 ± 0.01 d	3.08 ± 0.01 a	3.15 ± 0.01 b	3.29 ± 0.01 c	3.31 ± 0.01 c	3.30 ± 0.01 a	3.33 ± 0.01 a	3.39 ± 0.01 b	3.45 ± 0.01 c	3.28 ± 0.01 a	3.35 ± 0.01 b	3.36 ± 0.01 b	3.39 ± 0.02 b
VAL	2.94 ± 0.01 a	3.24 ± 0.01 b	3.33 ± 0.03 c	3.54 ± 0.02 d	3.45 ± 0.01 a	3.56 ± 0.01 b	3.67 ± 0.02 c	3.81 ± 0.02 d	3.68 ± 0.02 a	3.71 ± 0.01 a	3.72 ± 0.01 a	3.84 ± 0.02 b	3.64 ± 0.01 a	3.70 ± 0.02 b	3.74 ± 0.00 c	3.77 ± 0.01 d
MET	0.05 ± 0.00 a	0.19 ± 0.01 b	0.22 ± 0.01 b	0.25 ± 0.00 c	0.33 ± 0.01 a	0.40 ± 0.01 b	0.42 ± 0.00 b	0.45 ± 0.01 c	0.34 ± 0.01 a	0.36 ± 0.00 a	0.43 ± 0.00 b	0.49 ± 0.01 c	0.42 ± 0.01 a	0.45 ± 0.01 a	0.46 ± 0.01 a	0.54 ± 0.00 b
ILE	2.76 ± 0.02 a	3.03 ± 0.03 b	3.16 ± 0.01 c	3.33 ± 0.01 d	3.19 ± 0.01 a	3.32 ± 0.02 b	3.42 ± 0.01 c	3.46 ± 0.02 c	3.40 ± 0.01 a	3.45 ± 0.02 b	3.46 ± 0.01 b	3.61 ± 0.02 c	3.31 ± 0.02 a	3.33 ± 0.01 a	3.36 ± 0.01 b	3.40 ± 0.01 c
LEU	4.72 ± 0.01 a	5.43 ± 0.01 b	5.63 ± 0.03 c	5.81 ± 0.01 d	5.75 ± 0.02 a	5.89 ± 0.01 b	6.09 ± 0.02 c	6.17 ± 0.01 d	6.15 ± 0.01 a	6.19 ± 0.02 a	6.25 ± 0.02 b	6.34 ± 0.02 c	6.14 ± 0.01 a	6.15 ± 0.03 a	6.31 ± 0.02 b	6.44 ± 0.01 c
TYR	1.32 ± 0.01 a	1.56 ± 0.00 b	1.69 ± 0.01 c	1.87 ± 0.00 d	1.83 ± 0.00 a	2.03 ± 0.01 b	2.17 ± 0.01 c	2.19 ± 0.01 c	2.17 ± 0.00 a	2.34 ± 0.01 b	2.36 ± 0.00 b	2.54 ± 0.01 c	2.37 ± 0.01 a	2.41 ± 0.01 b	2.57 ± 0.01 c	2.61 ± 0.02 d
PHE	2.91 ± 0.01 a	3.80 ± 0.02 b	3.90 ± 0.00 c	4.03 ± 0.03 d	3.92 ± 0.01 a	4.03 ± 0.01 b	4.18 ± 0.01 c	4.20 ± 0.01 c	4.08 ± 0.01 a	4.15 ± 0.01 b	4.16 ± 0.01 b	4.30 ± 0.01 c	4.10 ± 0.00 a	4.13 ± 0.01 b	4.20 ± 0.01 c	4.25 ± 0.01 d
LYS	5.61 ± 0.03 b	5.79 ± 0.01 c	5.63 ± 0.03 b	5.06 ± 0.02 a	5.99 ± 0.02 a	5.95 ± 0.01 a	6.21 ± 0.02 b	6.21 ± 0.01 b	6.42 ± 0.02 c	6.29 ± 0.01 b	5.99 ± 0.02 a	6.26 ± 0.03 b	6.08 ± 0.02 c	5.83 ± 0.03 a	5.83 ± 0.01 a	5.94 ± 0.00 b
HIS	1.01 ± 0.00 b	1.16 ± 0.01 c	0.92 ± 0.00 a	0.90 ± 0.02 a	1.13 ± 0.00 b	1.12 ± 0.01 b	1.13 ± 0.00 b	1.07 ± 0.00 a	1.07 ± 0.00 b	1.09 ± 0.01 b	1.04 ± 0.01 a	1.12 ± 0.01 b	1.07 ± 0.00 a	1.09 ± 0.01 b	1.06 ± 0.00 a	1.10 ± 0.01 b
ARG	6.13 ± 0.03 c	6.45 ± 0.03 d	5.44 ± 0.01 b	4.98 ± 0.02 a	7.26 ± 0.02 b	7.16 ± 0.02 a	7.39 ± 0.01 c	7.49 ± 0.01 d	7.65 ± 0.02 c	7.42 ± 0.03 b	7.09 ± 0.01 a	6.41 ± 0.02 b	7.47 ± 0.01 c	7.11 ± 0.00 b	7.07 ± 0.01 a	7.11 ± 0.03 ab
HAA	19.92 ± 0.03 a	23.08 ± 0.02 b	23.94 ± 0.06 c	24.99 ± 0.02 d	24.73 ± 0.05 a	25.62 ± 0.03 b	26.59 ± 0.01 c	26.99 ± 0.02 d	26.58 ± 0.05 a	26.99 ± 0.05 b	27.61 ± 0.01 c	28.18 ± 0.03 d	26.50 ± 0.03 a	27.13 ± 0.02 b	27.43 ± 0.03 c	27.83 ± 0.05 d
Q value	1039.94 ± 1.21 a	1074.08 ± 0.86 b	1108.16 ± 1.11 c	1126.82 ± 2.01 d	1089.13 ± 1.01 a	1101.133 ± 2.11 b	1104.35 ± 1.33 b	1108.58 ± 0.56 c	1083.81 ± 1.51 a	1098.21 ± 2.11 b	1104.91 ± 0.96 c	1111.09 ± 1.71 d	1094.78 ± 1.44 a	1105.858 ± 1.66 b	1110.83 ± 1.13 c	1112.40 ± 0.56 c

The same enzyme and different letters on the same row indicate significant differences ($p < 0.05$).

It is well known that bitter peptides are spontaneously produced during the proteolytic process and more obvious at high *DH* [41]. The degree of hydrophobicity is considered the most important predictor of peptide bitterness. So far, the Q rule proposed by Ney is still used for the relationship between hydrophobicity and bitterness of polypeptide chains [42]. As expected, the Q values tended to increase with the increase in enzymatic hydrolysis by the four enzymes, which meant that the bitterness gradually increased with the enzymatic hydrolysis. This can be explained by the results of HAA content as mentioned above. However, the Q rule found that the bitter taste was presented when the Q value was greater than 1400 cal/mol, and vice versa [41]. In this study, the Q value of all samples was below the threshold, which was not consistent with the actual bitterness of the products through tasting. Therefore, the hydrophobicity data calculated based on amino acid composition from this study do not support Ney's Q rule as a predictor of bitterness of PP hydrolysates.

In general, the secondary structure of PP changed from ordered β-sheet to α-helix and the molecular weight of PP gradually decreased with the increase in *DH* after treatment with the four enzymes. Furthermore, the total sulfhydryl content and surface hydrophobicity of PP were changed by enzymatic hydrolysis. In addition, the bitterness of PP after enzymatic hydrolysis increased with the increase in *DH*.

3.2. Functional Properties

3.2.1. Solubility

Solubility plays a crucial role in the functionality of the protein and protein-based systems including gels, emulsions, and foams [43]. Especially for plant protein, boosted properties can benefit from the increase in solubility [44]. The solubility of PP and PP with different *DH* samples is shown in Figure 7. Compared with PP (10.23%), the solubility of PP gradually increased with the increase in *DH*. The solubility of flavourzyme 2%, 4%, 6%, and 7% was 17.54%, 23.68%, 25.46%, and 29.44%, respectively. The solubility of neutrase 2%, 4%, 6%, and 8% was 19.18%, 24.78%, 27.62%, and 30.47%, respectively. The solubility of alcalase 2%, 4%, 6%, and 8% was 40.55%, 41.52%, 46.37%, and 54.97%, respectively. The solubility of trypsin 2%, 4%, 6%, and 8% was 41.91%, 43.49%, 54.22%, and 58.14%, respectively. The increased solubility may be due to the decreased molecular size of PP and smaller peptides when the PP was hydrolyzed. This was consistent with the results of SDS-PAGE. In addition, the effect of different enzymes on PP at the same *DH* was different. From the experimental results, the degree of change induced by the four enzymes on PP solubility was: trypsin > alcalase > neutrase > flavourzyme. This meant that trypsin had the best enzymatic hydrolysis effect on PP of all the enzymes.

3.2.2. Foaming Performance

The foaming capacity and foaming stability of PP and PP with different *DH* samples are shown in Figure 8. The foaming ability of the enzymatically digested PP was significantly higher than the native PP, which was consistent with the solubility results. Solubility is a prerequisite for protein-foaming ability, and the protein should first dissolve in the aqueous phase and then rapidly stretch to form a dense layer of protein molecules around the air and foaming [45]. The foaming ability and foaming stability of protein was also related to the *DH*. After the four enzymes' treatment, the foaming ability of PP treated with flavourzyme, alcalase, and trypsin showed an increasing trend and then decreased, reaching the best levels at 6% (101.48%), 6% (164.44%), and 4% (168.88%) of *DH*, respectively. The foaming ability of PP treated by neutrase increased gradually. Speculatively, limited enzymatic hydrolysis produced peptides with balanced hydrophilic/hydrophobic groups, and the amphiphilic parts of proteins and peptides have a stronger capacity to decrease the surface tension at the air–water interface. However, high *DH* usually had an adverse effect on foaming properties. On the one hand, the increase in *DH* increased the protein solubility, but the best structure of protein spheres was destroyed, thus resulting in a reduction in the foaming ability. On the other hand, excessive enzymolysis made the hydrolysates more hydrophilic, disturbing the hydrophilic/hydrophobic balance, which led to the decreased foam formation ability [46]. In addition, after the four enzymes (flavourzyme, neutrase, alcalase, and trypsin) treatment, the foaming stability of PP was decreased compared with native PP.

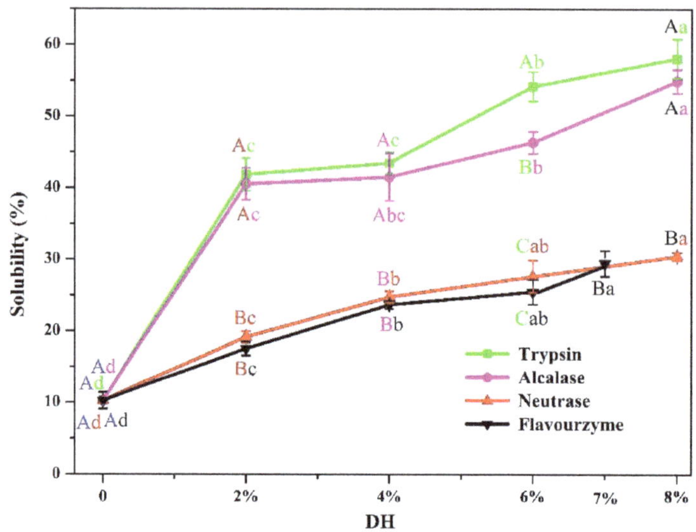

Figure 7. Solubility of PP and PP with different *DH* samples. Different lowercase letters are used to indicate the significant differences ($p < 0.05$) of the same enzyme under different *DH*, while different uppercase letters are used to indicate the significant differences ($p < 0.05$) of different enzymes under the same *DH*.

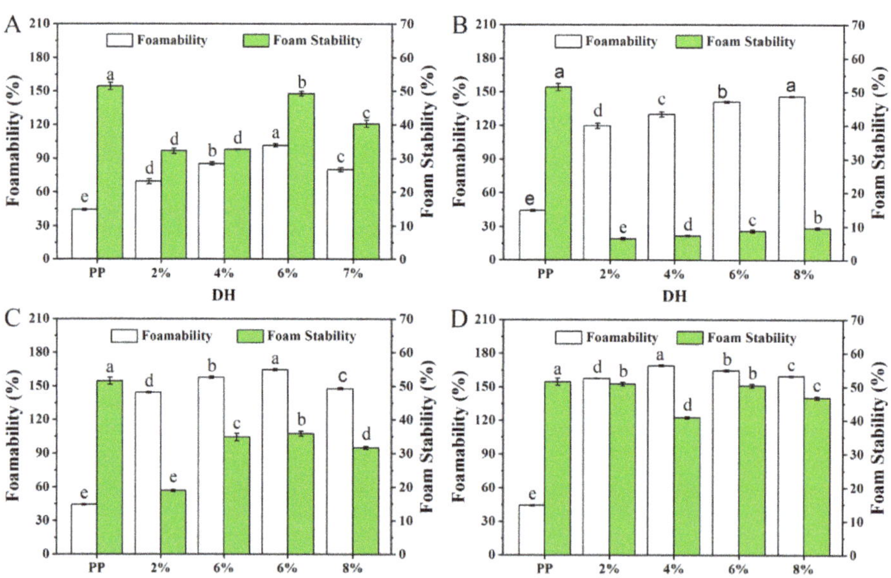

Figure 8. The foamability and foam stability of PP and PP with different *DH* samples ((**A–D**) are flavourzyme, neutrase, alcalase, and trypsin treatment, respectively). The values reported represent means ($n = 3$) ± SDs and different superscript letters indicate significant difference ($p < 0.05$).

3.2.3. Emulsifying Performance

Proteins are a kind of natural emulsifier that are widely used in food emulsion preparation due to their amphiphilic nature. The emulsifying property of protein represents the ability of protein molecules in an emulsion to adsorb to the oil–water interface and is usually characterized by emulsifying activity and emulsifying stability [47]. Usually, the stronger the emulsification, the more stable the emulsion which forms and the less likely it is to form an aggregation. Figure 9 shows the changes in the emulsifying properties and emulsion stability of PP and PP with different *DH* samples. As shown in Figure 9, the emulsification ability and emulsion stability of PP treated with the four enzymes were significantly improved, which may be related to the increased solubility and peptide chain flexibility of the PP enzymatic hydrolysis products. All four enzymes showed an increasing trend and then a decrease in emulsifying properties as the *DH* increased, and all of them were highest at a *DH* of 6%. When the *DH* was 6%, the *EAI* of PP treated by flavourzyme, neutrase, alcalase, and trypsin increased from 24.55 to 42.36, 52.22, 49.55, and 52.16 m^2/g, respectively, while the *ESI* increased from 30.02 to 87.54, 89.23, 77.93, and 84.88%, respectively. This may be due to the fact that excessive enzymolysis further reduces the molecular weight of the protein and reduces the amphiphilicity of the peptide, thus inhibiting the interaction between protein molecules at the oil–water interface and reducing the viscoelasticity of the interface membrane [47,48]. In addition, it may be due to the reduced charge repulsion between low-molecular-weight peptides preventing the proteins from either stretching or rearranging at the interface [49].

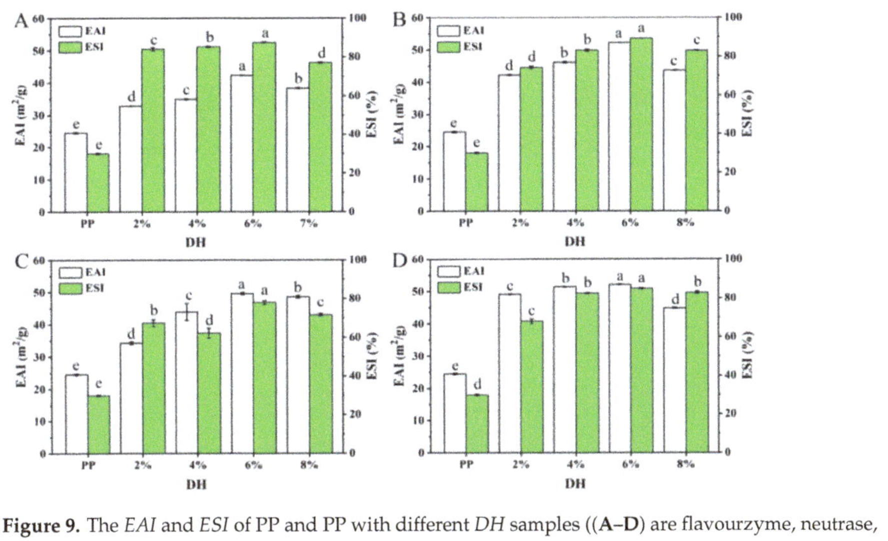

Figure 9. The *EAI* and *ESI* of PP and PP with different *DH* samples ((**A–D**) are flavourzyme, neutrase, alcalase, and trypsin treatment, respectively). The different superscript letters of same index indicate significant difference ($p < 0.05$).

Based on the above results and analysis, the solubility of PP gradually increased with the increase in *DH* after treatment with the four enzymes. Furthermore, the foaming capacity of PP treated with four enzymes was significantly improved, and the best foaming capacity was observed for flavourzyme, alcalase, trypsin, and neutrase with *DH*s of 6%, 6%, 4%, and 8%, respectively. In addition, the emulsification ability and emulsion stability of PP treated with the four enzymes were significantly improved, and all of them were the highest at a *DH* of 6%. Overall, the functional properties of PP treated with the four enzymes were the best when the *DH* was 6%.

4. Conclusions

The present study aimed to investigate the effects of four different enzymes on the structure and functional properties of PP at the same *DH*. At the same *DH*, trypsin had the strongest effect on improving the properties of pea protein, followed by alcalase, neutrase, and flavourzyme. Trypsin-treated PP had smaller surrounding particles, smaller molecular weight, less total sulfhydryl content, higher solubility, and higher fluorescence intensity. With the increase in *DH*, the improvement effect on the foaming and emulsification of PP tended to increase and then decrease and reached the best value at 6% *DH*. This study provided important information about moderate enzymatic hydrolysis, which can broaden the application of pea protein in plant-based proteins.

Supplementary Materials: The following supporting information can be downloaded at: https://www.mdpi.com/article/10.3390/foods11152368/s1, Figure S1: Sensory evaluation results of PP and PP with different *DH* samples.

Author Contributions: Conceptualization, X.S., L.G., Q.G., C.L. and T.D.; methodology, T.D.; data curation, X.H., L.G. and T.D.; writing—original draft preparation, X.S., L.G. and T.D.; writing—review and editing, T.L., T.D., C.L. and J.C.; visualization, Q.G., C.L., J.C. and T.L.; supervision, X.H., L.G., Q.G. and T.L. All authors have read and agreed to the published version of the manuscript.

Funding: This work was supported by National Natural Science Foundation of China [Grant No. 32101948 and 32060524], China Postdoctoral Science Foundation [Grant No. 2020M6832211], Postdoctoral Foundation of Guangxi Academy of Agricultural Sciences (Grant No. 2020037), and Guangxi Natural Science Foundation (Grant No. 2022GXNSFBA035522).

Institutional Review Board Statement: Not applicable.

Informed Consent Statement: Not applicable.

Data Availability Statement: Data are contained within the article.

Conflicts of Interest: The authors declare no conflict of interest.

References

1. Burger, T.G.; Zhang, Y. Recent progress in the utilization of pea protein as an emulsifier for food applications. *Trends Food Sci. Technol.* **2019**, *86*, 25–33. [CrossRef]
2. Dai, T.; Li, T.; Li, R.; Zhou, H.; Liu, C.; Chen, J.; McClements, D.J. Utilization of plant-based protein-polyphenol complexes to form and stabilize emulsions: Pea proteins and grape seed proanthocyanidins. *Food Chem.* **2020**, *329*, 127219. [CrossRef] [PubMed]
3. Ge, J.; Sun, C.X.; Corke, H.; Gul, K.; Gan, R.Y.; Fang, Y. The health benefits, functional properties, modifications, and applications of pea (Pisum sativum L.) protein: Current status, challenges, and perspectives. *Compr. Rev. Food Sci. Food Saf.* **2020**, *19*, 1835–1876. [CrossRef] [PubMed]
4. Lan, Y.; Xu, M.; Ohm, J.B.; Chen, B.; Rao, J. Solid dispersion-based spray-drying improves solubility and mitigates beany flavour of pea protein isolate. *Food Chem.* **2019**, *278*, 665–673. [CrossRef]
5. He, X.; Chen, J.; He, X.; Feng, Z.; Li, C.; Liu, W.; Dai, T.; Liu, C. Industry-scale microfluidization as a potential technique to improve solubility and modify structure of pea protein. *Innov. Food Sci. Emerg. Technol.* **2020**, *67*, 102582. [CrossRef]
6. Mahdavian Mehr, H.; Koocheki, A. Effect of atmospheric cold plasma on structure, interfacial and emulsifying properties of Grass pea (Lathyrus sativus L.) protein isolate. *Food Hydrocoll.* **2020**, *106*, 105899. [CrossRef]
7. Wu, D.; Wu, C.; Ma, W.; Wang, Z.; Yu, C.; Du, M. Effects of ultrasound treatment on the physicochemical and emulsifying properties of proteins from scallops (Chlamys farreri). *Food Hydrocoll.* **2019**, *89*, 707–714. [CrossRef]
8. Jiang, S.; Ding, J.; Andrade, J.; Rababah, T.M.; Almajwal, A.; Abulmeaty, M.M.; Feng, H. Modifying the physicochemical properties of pea protein by pH-shifting and ultrasound combined treatments. *Ultrason.-Sonochem.* **2017**, *38*, 835–842. [CrossRef]
9. Wouters, A.G.B.; Rombouts, I.; Fierens, E.; Brijs, K.; Delcour, J.A. Relevance of the Functional Properties of Enzymatic Plant Protein Hydrolysates in Food Systems. *Compr. Rev. Food Sci. Food Saf.* **2016**, *15*, 786–800. [CrossRef]
10. Meinlschmidt, P.; Sussmann, D.; Schweiggert-Weisz, U.; Eisner, P. Enzymatic treatment of soy protein isolates: Effects on the potential allergenicity, technofunctionality, and sensory properties. *Food Sci. Nutr.* **2016**, *4*, 11–23. [CrossRef]
11. Schlegel, K.; Sontheimer, K.; Hickisch, A.; Wani, A.A.; Eisner, P.; Schweiggert-Weisz, U. Enzymatic hydrolysis of lupin protein isolates-Changes in the molecular weight distribution, technofunctional characteristics, and sensory attributes. *Food Sci. Nutr.* **2019**, *7*, 2747–2759. [CrossRef] [PubMed]

12. García Arteaga, V.; Apéstegui Guardia, M.; Muranyi, I.; Eisner, P.; Schweiggert-Weisz, U. Effect of enzymatic hydrolysis on molecular weight distribution, techno-functional properties and sensory perception of pea protein isolates. *Innov. Food Sci. Emerg. Technol.* **2020**, *65*, 102449. [CrossRef]
13. Klost, M.; Drusch, S. Functionalisation of pea protein by tryptic hydrolysis–Characterisation of interfacial and functional properties. *Food Hydrocoll.* **2019**, *86*, 134–140. [CrossRef]
14. Tamm, F.; Herbst, S.; Brodkorb, A.; Drusch, S. Functional properties of pea protein hydrolysates in emulsions and spray-dried microcapsules. *Food Hydrocoll.* **2016**, *58*, 204–214. [CrossRef]
15. Klost, M.; Giménez-Ribes, G.; Drusch, S. Enzymatic hydrolysis of pea protein: Interactions and protein fractions involved in fermentation induced gels and their influence on rheological properties. *Food Hydrocoll.* **2020**, *105*, 105793. [CrossRef]
16. Phoon, P.Y.; San Martin-Gonzalez, M.F.; Narsimhan, G. Effect of hydrolysis of soy β-conglycinin on the oxidative stability of O/W emulsions. *Food Hydrocoll.* **2014**, *35*, 429–443. [CrossRef]
17. Adler-Nissen, J. *Enzymic Hydrolysis of Food Proteins*; Elsevier Applied Science Publishers: New York, NY, USA, 1986; pp. 2210–2230. [CrossRef]
18. Xu, X.; Liu, W.; Liu, C.; Luo, L.; Chen, J.; Luo, S.; McClements, D.J.; Wu, L. Effect of limited enzymatic hydrolysis on structure and emulsifying properties of rice glutelin. *Food Hydrocoll.* **2016**, *61*, 251–260. [CrossRef]
19. Li, T.; Hu, P.; Dai, T.; Li, P.; Ye, X.; Chen, J.; Liu, C. Comparing the binding interaction between beta-lactoglobulin and flavonoids with different structure by multi-spectroscopy analysis and molecular docking. *Spectrochim. Acta Part A Mol. Biomol. Spectrosc.* **2018**, *201*, 197–206. [CrossRef]
20. Li, X.; Dai, T.; Hu, P.; Zhang, C.; Chen, J.; Liu, C.; Li, T. Characterization the non-covalent interactions between beta lactoglobulin and selected phenolic acids. *Food Hydrocoll.* **2020**, *105*, 105761. [CrossRef]
21. Yang, X.; Xu, J.; Fu, C.; Jia, Z.; Yao, K.; Chen, X. The cataract-related S39C variant increases gammaS-crystallin sensitivity to environmental stress by destroying the intermolecular disulfide cross-links. *Biochem. Biophys. Res. Commun.* **2020**, *526*, 459–465. [CrossRef]
22. Xie, H.; Huang, J.; Woo, M.W.; Hu, J.; Xiong, H.; Zhao, Q. Effect of cold and hot enzyme deactivation on the structural and functional properties of rice dreg protein hydrolysates. *Food Chem.* **2021**, *345*, 128784. [CrossRef] [PubMed]
23. Lowry, O.; Rosebrough, N.; Farr, A.L.; Randall, R. Protein Measurement with the Folin Phenol Reagent. *J. Biol. Chem.* **1951**, *193*, 265–275. [CrossRef]
24. Fathollahy, I.; Farmani, J.; Kasaai, M.R.; Hamishehkar, H. Characteristics and functional properties of Persian lime (Citrus latifolia) seed protein isolate and enzymatic *Hydrolysates. LWT-Food Sci. Technol.* **2021**, *140*, 110765. [CrossRef]
25. Hou, F.; Ding, W.; Qu, W.; Oladejo, A.O.; Xiong, F.; Zhang, W.; He, R.; Ma, H. Alkali solution extraction of rice residue protein isolates: Influence of alkali concentration on protein functional, structural properties and lysinoalanine formation. *Food Chem.* **2017**, *218*, 207–215. [CrossRef] [PubMed]
26. Avramenko, N.A.; Low, N.H.; Nickerson, M.T. The effects of limited enzymatic hydrolysis on the physicochemical and emulsifying properties of a lentil protein isolate. *Food Res. Int.* **2013**, *51*, 162–169. [CrossRef]
27. Roger, P.H. Determining the fluorescence spectrum of a protein. *Curr. Protoc. Protein Sci.* **2004**, *7*, 7.7.1. [CrossRef]
28. Vera, A.; Valenzuela, M.A.; Yazdani-Pedram, M.; Tapia, C.; Abugoch, L. Conformational and physicochemical properties of quinoa proteins affected by different conditions of high-intensity ultrasound treatments. *Ultrason. Sonochem.* **2019**, *51*, 186–196. [CrossRef]
29. Ling, B.; Ouyang, S.; Wang, S. Effect of radio frequency treatment on functional, structural and thermal behaviors of protein isolates in rice bran. *Food Chem.* **2019**, *289*, 537–544. [CrossRef]
30. Burstein, E.A.; Vedenkina, N.S.; Ivkova, M.N. Fluorescence and the location of tryptophan residues in protein molecules. *Photochem. Photobiol.* **1973**, *18*, 263–279. [CrossRef]
31. Sponton, O.E.; Perez, A.A.; Carrara, C.; Santiago, L.G. Effect of limited enzymatic hydrolysis on linoleic acid binding properties of beta-lactoglobulin. *Food Chem.* **2014**, *146*, 577–582. [CrossRef]
32. Kang, D.C.; Zou, Y.H.; Cheng, Y.P.; Xing, L.J.; Zhou, G.H.; Zhang, W.G. Effects of power ultrasound on oxidation and structure of beef proteins during curing processing. *Ultrason. Sonochem.* **2016**, *33*, 47–53. [CrossRef] [PubMed]
33. Zhao, Q.; Selomulya, C.; Wang, S.; Xiong, H.; Chen, X.D.; Li, W.; Peng, H.; Xie, J.; Sun, W.; Zhou, Q. Enhancing the oxidative stability of food emulsions with rice dreg protein hydrolysate. *Food Res. Int.* **2012**, *48*, 876–884. [CrossRef]
34. Jung, S.; Murphy, P.A.; Johnson, L.A. Physicochemical and functional properties of soy protein substrates modified by low levels of protease hydrolysis. *J. Food Sci.* **2005**, *70*, 180–187. [CrossRef]
35. Guan, X.; Yao, H.; Chen, Z.; Shan, L.; Zhang, M. Some functional properties of oat bran protein concentrate modified by trypsin. *Food Chem.* **2007**, *101*, 163–170. [CrossRef]
36. Yong, Y.H.; Yamaguchi, S.; Matsumura, Y. Effects of enzymatic deamidation by protein-glutaminase on structure and functional properties of wheat gluten. *J. Agric. Food Chem.* **2006**, *54*, 6034–6040. [CrossRef] [PubMed]
37. Zou, Y.; Shi, H.; Chen, X.; Xu, P.; Jiang, D.; Xu, W.; Wang, D. Modifying the structure, emulsifying and rheological properties of water-soluble protein from chicken liver by low-frequency ultrasound treatment. *Int. J. Biol Macromol.* **2019**, *139*, 810–817. [CrossRef]

38. Lepedda, A.J.; Zinellu, A.; Nieddu, G.; Zinellu, E.; Carru, C.; Spirito, R.; Guarino, A.; De Muro, P.; Formato, M. Protein sulfhydryl group oxidation and mixed-disulfide modifications in stable and unstable human carotid plaques. *Oxidative Med. Cell. Longev.* **2013**, *2013*, 66–67. [CrossRef]
39. Gorissen, S.H.M.; Crombag, J.J.R.; Senden, J.M.G.; Waterval, W.A.H.; Bierau, J.; Verdijk, L.B.; van Loon, L.J.C. Protein content and amino acid composition of commercially available plant-based protein isolates. *Amino Acids* **2018**, *50*, 1685–1695. [CrossRef]
40. Guo, S.G.; Zhao, M.; Wang, S.; Cui, C. Proteolytic degradation and amino acid liberation during extensive hydrolysis of porcine blood hemoglobin by protease admixture. *J. Food Process Eng.* **2007**, *30*, 640–659. [CrossRef]
41. Cho, M.J.; Unklesbay, N.; Hsieh, F.H.; Clarke, A.D. Hydrophobicity of bitter peptides from soy protein hydrolysates. *J. Agric. Chem.* **2004**, *52*, 5895–5901. [CrossRef]
42. Ney, K.H. Bitterness of peptides: Amino acid composition and chain length. *Acs Symp. Ser.* **1979**, *115*, 149–173. [CrossRef]
43. Noman, A.; Xu, Y.; Al-Bukhaiti, W.Q.; Abed, S.M.; Ali, A.H.; Ramadhan, A.H.; Xia, W. Influence of enzymatic hydrolysis conditions on the degree of hydrolysis and functional properties of protein hydrolysate obtained from Chinese sturgeon (Acipenser sinensis) by using papain enzyme. *Process. Biochem.* **2018**, *67*, 19–28. [CrossRef]
44. Malik, M.A.; Saini, C.S. Polyphenol removal from sunflower seed and kernel: Effect on functional and rheological properties of protein isolates. *Food Hydrocoll.* **2017**, *63*, 705–715. [CrossRef]
45. Tang, S.; Hettiarachchy, N.S.; Horax, R.; Eswaranandam, S. Physicochemical properties and functionality of rice bran protein hydrolyzate prepared from heat-stabilized defatted rice bran with the aid of enzymes. *J. Food Sci.* **2003**, *68*, 152–157. [CrossRef]
46. Hettiarachchy, N.S.; Ziegler, G.R. Ift Basic Symposium Series. In *Protein Functionality in Food Systems*; Marcel Dekker, Inc.: New York, NY, USA, 1994.
47. Mokni Ghribi, A.; Maklouf Gafsi, I.; Sila, A.; Blecker, C.; Danthine, S.; Attia, H.; Bougatef, A.; Besbes, S. Effects of enzymatic hydrolysis on conformational and functional properties of chickpea protein isolate. *Food Chem.* **2015**, *187*, 322–330. [CrossRef]
48. Xia, Y.; Bamdad, F.; Ganzle, M.; Chen, L. Fractionation and characterization of antioxidant peptides derived from barley glutelin by enzymatic hydrolysis. *Food Chem.* **2012**, *134*, 1509–1518. [CrossRef]
49. Severin, S.; Xia, W.S. Enzymatic hydrolysis of whey proteins by two different proteases and their effect on the functional properties of resulting protein hydrolysates. *J. Food Biochem.* **2006**, *30*, 77–97. [CrossRef]

foods

MDPI

Article

Hygroscopic Properties of Three Cassava (*Manihot esculenta* Crantz) Starch Products: Application of BET and GAB Models

Aneta Ocieczek [1,*], Dominika Mesinger [1] and Henryk Toczek [2]

[1] Faculty of Management and Quality Science, Gdynia Maritime University, 81-225 Gdynia, Poland; d.mesinger@sd.umg.edu.pl
[2] Faculty of Marine Engineering, Gdynia Maritime University, 81-225 Gdynia, Poland; h.toczek@wm.umg.edu.pl
* Correspondence: a.ocieczek@wznj.umg.edu.pl

Abstract: This study aimed to compare hygroscopicity properties of three cassava (*Manihot esculenta* Crantz) products: native starch powder (NS), fermented starch powder (FS), and starch granulate (SG). The analyzed properties were compared based on the statistical evaluation of differences in the course of sorption isotherms and the identification and comparison of parameters in two theoretical models of sorption. Empirical data were generated by means of the static-desiccator method. Measurements were made using AquaLab apparatus. The size, shape, and number of tapioca particles were characterized using a Morphology automatic particle analyzer. The study demonstrated that in-depth exploration of empirical data describing hygroscopicity of samples with the use of mathematical tools allows evaluating their physical parameters. The results obtained were analyzed in terms of correlations between physical and physicochemical properties determining utility traits of cassava starch. The NS and SS featured significantly higher hygroscopicity than SG, as evidenced by the values of all parameters analyzed in this study. The study results provided new information related to the management of the production process, safety, and stability of these products.

Keywords: tapioca; hygroscopicity; *BET* and *GAB* models; particle size characteristics; storage stability

check for updates

Citation: Ocieczek, A.; Mesinger, D.; Toczek, H. Hygroscopic Properties of Three Cassava (*Manihot esculenta* Crantz) Starch Products: Application of BET and GAB Models. *Foods* **2022**, *11*, 1966. https://doi.org/10.3390/foods11131966

Academic Editors: Jianhua Xie, Yanjun Zhang and Hansong Yu

Received: 29 May 2022
Accepted: 27 June 2022
Published: 2 July 2022

1. Introduction

Cassava (*Manihot esculenta* Crantz), belonging to the *Euphorbiaceae* family, is a highly popular crop in the tropical zones of the globe, likewise, e.g., wheat in European countries [1]. It is one of the most often cultivated crops worldwide, providing starch for foodstuff and feedstuff production [2,3]. It is of particular importance to the economies of such countries as Nigeria, Thailand, Ghana, Brazil, and Indonesia, as it plays the role of a strategic raw material [4–6]. In the years 2000–2018, its production increased by approximately 100 million tons only in the countries claimed to be its main producers [7], which points to the growing demand for this crop.

Cassava (*Manihot esculenta* Crantz) roots are used to manufacture various starch products with a different fraction size composition, the common feature of which is a high dehydration degree. Water content reduction enables a radical decrease in water activity, and by this means, achieves enzymatic and microbiological stability, which, in turn, ensures adequate storage stability [8,9]. This form of preservation and storage of cassava products is driven by its higher susceptibility to spoilage compared to other tuber roots (potato, sweet potato, yam) [10]. Consequently, the cassava root, which is a storage organ, like other tuberous organs, does not fall into dormancy even under favorable conditions [11]. Granulated products obtained from cassava starch in the form of uneven, polyhedral, or spherical granules are called tapioca [12,13].

Tapioca is becoming more and more popular and is also used in European countries. There are three basic forms of this starch available on the market: starch powder, starch granulate, and fermented starch powder produced in the lactic fermentation process. Fermented starch is used as an additive to some baked products, both sweet and dry ones. In Ghana, fermented starch is used to produce six groups of products traditionally consumed in African countries, including: fermented pulps—fufu and akpu; fermented flours—lafun, kanyanga, and luku; smoked fermented balls—kumkum and pupuru; steamed and fermented chips—chikwangue and ntuka; baked grits—gari, agbelima, kapok, and popogari; steamed grits—attieke.

Lactic acid fermentation of edible cassava roots is extremely important as it removes cyanogenic glycosides, linamarin, and lotaustralin, which are present in the plant in various proportions. Cyanogenic glycosides are toxic because their hydrolytic degradation releases hydrogen cyanide. Therefore, the fermentation of cassava is strongly recommended [14].

The food industry extensively uses cassava starch (tapioca) in each of the three mentioned forms. However, starch powder and granules are the most common in the European market, whereas fermented starch cannot be commonly purchased in Poland yet. Tapioca starch (non-fermented) is used as, e.g., a stabilizer or a thickener due to its favorable technological properties and very neutral or utterly neutral taste. Tapioca-based food products are recommended to persons susceptible to allergens, especially those allergic to wheat and/or maize proteins. In addition, the starch of this type is increasingly often applied in food products intended for children, owing to its beneficial physical properties, texture, stability, and neutral taste. However, the most popular food products available on the European market are Asian-style noodles, Asian-style crispy bread, gluten-free bread, or the very popular molecular caviar (used in the production of Bubbletea, i.e., tapioca starch in the form of balls soaked in a fruit-flavored syrup and added in this form to lemonade or tea). Tapioca can also be used to prepare, e.g., pancakes, cakes, or puddings [15–17].

Edible cassava starch, i.e., tapioca, is also used in pharmaceutical products as an excipient and a filling substance. In the native form, it is a constituent of many pharmaceutical formulations—a binder, disintegrant, and a diluent of the active substance. In turn, the modified tapioca starch has an even more comprehensive range of applications, depending on the type of modification and intended use, etc. [18].

Carbohydrates are the most important and, at the same time, the dominant component of each starch, determining its technological usefulness as well as the stability of functional properties and consumer safety. In particular, attention is paid to the ratio of amylose to amylopectin, which in tapioca is at the level of 17–24% amylose and 76–83% amylopectin [19]. The ratio of these fractions may be necessary for shaping the sorption, including hygroscopicity. The crystalline form of granules of the dominant component of the cassava root causes the water present in it to be characterized by a high level of its activity, amounting to approximately 0.85 [20]. At the same time, it should be emphasized that the degree of its processing affects both the water content and its activity. At the same time, there are reports showing that the natural variety of cassava and the degree of its processing generally have a negligible or unclear effect on sorption behavior [21]. This statement had to be verified in the light of the growing interest in cassava products on the European market.

This study aimed to compare the sorption properties, in terms of hygroscopicity, of three cassava (*Manihot esculenta* Crantz) starch products (native, fermented, granulated), by comparing their sorption isotherms and the parameters of selected mathematical models used to describe these isotherms. The use of the *BET* (Brunauer, Emmett, and Teller) and *GAB* (Guggenheim, Anderson, and De Boer) models was conditioned by their theoretical nature, allowing for a physical interpretation of the parameters of these models. Moreover, its goal was to identify selected parameters describing the microstructure of the tested products. The study also assessed the differentiation of particle size and shape distributions of the studied starch samples. It was assumed that the analysis of the differentiation of the hygroscopic properties of starch, taking into account the variability of selected physical parameters of these starch particles, will be a source of new valuable information for the management of the production process, safety, and stability of these products.

2. Material and Methods

2.1. Materials

Three products made of cassava (*Manihot esculenta* Crantz) starch, also called tapioca, were studied: native starch powder (NS), starch granulate (SG), and fermented starch powder (FS). All three starches were produced under industrial conditions. Native tapioca powder (NS) and tapioca granulate (SG) were purchased in Poland, whereas fermented starch (FS) was imported from Brazil. Native tapioca powder was produced in Thailand and distributed in Poland by De Care Group Sp. z o.o. i wsp. Sp. Kom. Tapioca granulate was produced in Thailand by THAI WORLD IMPORT & EXPORT Co., Ltd. (Bangkok, Thailand) and distributed in Poland by KK Polska Sp. z o.o. Fermented tapioca was produced in Brazil and distributed by SUCOS DO BRASIL—PRODUCTOS LATINO GMB (Neuss, Germany) under the trademark YOKI™. When opened, the experimental material was kept in a tightly closed package, at room temperature, according to producers' recommendations.

The chemical reagents used to maintain appropriate relative humidity conditions in the desiccators were high-quality, pure, analytical compounds. The water used to prepare the saturated solutions of these reagents was distilled water.

2.2. Methods

Differences in the physical properties of the analyzed starches were established by comparing value distributions of parameters characterizing the size and shape of their particles determined using a Morphology G3 automatic particle analyzer (Malvern Instruments, Malvern, UK). The analyzer enables determination of the size distribution of solid particles, with sizes ranging from 0.5 to 10,000 μm. Estimations were made for value distributions of such parameters as: diameter, circularity, convexity, elongation, shape coefficient, and solidity [22].

Water content was determined according to the Polish Standard (PN-ISO 712:2002) in the AquaLab apparatus series 4 model TE (Decagon Devices, Inc., Pullman, WA, USA), exact to ± 0.003, at a temperature of 20 ± 1 °C.

Sorption isotherms were plotted using the static-desiccator method at ambient temperature of 20 ± 1 °C, in a water activity (a_w) range from 0.07 to 0.98, and the time (30 days) required to reach the dynamic equilibrium between the analyzed samples and saturated solutions of respective substances was: $NaOH \cdot H_2O$ (0.0698); $LiCl \cdot H_2O$ (0.1114); $KC_2H_3O_7 \cdot 1.5H_2O$ (0.231); $MgCl_2 \cdot 6H_2O$ (0.3303); $K_2CO_3 \cdot 2H_2O$ (0.440); $Na_2Cr_2O_7 \cdot 2H_2O$ (0.548); KJ (0.6986); NaCl (0.7542); KCl (0.8513); KNO_3 (0.932); and $K_2Cr_2O_7$ (0.9793). Equilibrium water contents were determined based on the initial masses of the analyzed samples with established water content, and then their changes triggered by the incubation process in desiccators. The values achieved allowed plotting of the isotherms of water vapor sorption in the tested range of water activities. Each point on each of the plotted isotherms was the arithmetic mean from three parallel determinations. Differences in the course of the sorption isotherms in the entire a_w range were analyzed statistically using the Student's *t*-test of differences between mean values for dependent variables. Differences were considered statistically significant at $p < 0.05$.

The course of sorption isotherms was analyzed mathematically using the *BET* Equation (1):

$$v = \frac{v_m \cdot c_{BET} \cdot a_w}{(1 - a_w) \cdot [1 + (c_{BET} - 1) \cdot a_w]} \quad (1)$$

where:

v—adsorption, g H_2O/100 g d.m.;

v_m—maximal adsorption value corresponding to the complete coverage of the surface with a monomolecular layer of the adsorbate, g H_2O/100 g d.m.;

c_{BET}—energy constant *BET*, describing the difference between the chemical potential of crude adsorbate molecules and those in the first adsorption layer, kJ/mol;

a_w—water activity at the adsorption temperature [23].

The analysis also used the *GAB* Equation (2):

$$v = \frac{v_m \cdot c_{GAB} \cdot k \cdot a_w}{(1 - k \cdot a_w) \cdot (1 - k \cdot a_w + c_{GAB} \cdot k \cdot a_w)} \quad (2)$$

where:

v—adsorption, g H_2O/100 g d.m.;
v_m—maximal adsorption value corresponding to the complete coverage of the surface with a monomolecular layer of the adsorbate, g H_2O/100 g d.m.;
c_{GAB}—energy constant *GAB*, describing the difference between the chemical potential of adsorbate molecules in the first adsorption layer and higher layers, kJ/mol;
k—constant correcting properties of multilayer molecules compared to the liquid phase;
a_w—water activity at the adsorption temperature [23].

The characterization of the sorption properties, in terms of hygroscopicity, using the *BET* and *GAB* models consisted of determining the maximal adsorption value corresponding to the complete coverage of the surface with a monomolecular layer of the adsorbate, called the monolayer (v_m). Estimation of the monolayer allowed computing of the specific adsorption area (a_{sp}) using Equation (3):

$$a_{sp} = \omega \cdot \frac{v_m}{M} \cdot N \quad (3)$$

where:

a_{sp}—specific sorption area, m^2/g;
N—Avogadro number, 6.023×10^{23} molecules/mol;
M—molecular weight of water, 18 g/mol;
ω—water cross-section area, 1.05×10^{-19} m^2/molecule [24].

In addition, energy constants c_{BET} and c_{GAB} were determined, and the k constant in the *GAB* equation was estimated.

Sizes and volumes of capillaries of the examined starch samples were determined for the area of capillary condensation using Kelvin's Equation (4), assuming the cylindrical shape of the capillaries.

$$lna_w = -\frac{2 \cdot \sigma \cdot V}{r_c \cdot R \cdot T} \quad (4)$$

where:

V—molar volume of the liquid, g/mol;
σ—surface tension of the liquid, N/m;
R—universal gas constant, J/(mol K);
T—temperature, K;
r_c—capillary radius, nm [24].

The *BET* equation parameters were identified based on empirical data in the water activity range of $0.07 < a_w < 0.50$ [25]. In turn, *GAB* equation parameters were identified based on empirical data from the entire a_w range studied [26].

2.3. Statistical Analysis

The statistical significance of differences between the mean water content and water activity in the tested samples was measured using the ANOVA test and the Tukey post hoc test.

Parameters of *BET* and *GAB* equations were determined using a numerical algorithm based on non-linear regression and a Monte Carlo algorithm. Minimizing the residual sum of squares (*RSS*) was adopted as the target function. Standard errors of the equations' determined parameters were estimated by a numerical algorithm using the SolverAid macro command, which reflects estimates of uncertainty of values of parameters obtained from Solver [24]. Finally, the fit of empirical data to both equations was characterized based on the evaluation of the residual sum of squares (*RSS*) Equation (5):

$$RSS = \sum (v_e - v_0)^2 \tag{5}$$

and root mean square (*RMS*) error Equation (6) expressed in % [27,28]:

$$RMS = \sqrt{\frac{\sum\left(\frac{v_e - v_0}{v_e}\right)^2}{N}} \cdot 100\% \tag{6}$$

where:

N—number of data;
v_e—experimental equilibrium water content, g H_2O/100 g d.m.;
v_o—predicted equilibrium water content, g H_2O/100 g d.m. [28,29].

All computations were made in Excel 2019.

3. Results and Discussion

The first stage of the study involved a comparative analysis of the characteristics differentiating the parameters of cassava starch particles. This study stage aimed to identify the most effective applications of the starch products tested. In this case, the knowledge of physical properties is the second indicator, after the characterization of biological and chemical properties, enabling the rational management of the production process involving cassava starch, e.g., in food or pharmaceutical industries.

Hygroscopic properties represent a cumulative effect of many factors determining the affinity to water molecules [23]. These factors include, for instance, particle morphology and chemical composition [30]. The analyzed starches differed in their physical properties described by means of such parameters as: diameter, circularity, convexity, elongation, shape coefficient, and solidity (Table 1).

Table 1. Selected physical characteristics of the analyzed cassava starches.

Characteristics of Value Distribution	Parameter					
	Min.	Max.	Mean ± SD	D[*n*, 0.1]	D[*n*, 0.5]	D[*n*, 0.9]
Native starch (NS) (particles counted *n* = 214,733) (optic used: 10×)						
diameter, μm	1.1	145.3	12.7 ± 7.8	4.1	11.5	21.8
circularity	0.036	1.000	0.871 ± 0.147	0.666	0.922	0.990
convexity	0.338	1.000	0.983 ± 0.041	0.917	0.986	0.997
elongation	0.000	0.968	0.196 ± 0.153	0.025	0.162	0.416
shape coefficient	0.032	1.000	0.804 ± 0.153	0.582	0.834	0.971
solidity	0.089	1.000	0.974 ± 0.051	0.881	0.975	0.998
Starch granulate (SG) (particles counted *n* = 86) (optic used: 2.5×)						
diameter, μm	1297.6	3330.9	2465.6 ± 330.4	1283.0	1847.7	1862.4
circularity	0.417	0.965	0.801 ± 0.138	0.578	0.842	0.942
convexity	0.774	0.990	0.929 ± 0.057	0.843	0.946	0.982
elongation	0.001	0.515	0.131 ± 0.128	0.028	0.079	0.356
shape coefficient	0.485	0.999	0.869 ± 0.128	0.642	0.920	0.970
solidity	0.810	0.998	0.963 ± 0.038	0.904	0.976	0.992
Fermented starch (FS) (particles counted *n* = 317,432) (optic used: 10×)						
diameter, μm	1.1	380.8	13.3 ± 6.9	6.9	12.3	19.9
circularity	0.028	1.000	0.895 ± 0.145	0.697	0.959	0.993
convexity	0.329	1.000	0.984 ± 0.040	0.916	0.986	0.997
elongation	0.000	0.977	0.159 ± 0.145	0.019	0.104	0.376
shape coefficient	0.023	1.000	0.841 ± 0.145	0.622	0.893	0.978
solidity	0.177	1.000	0.976 ± 0.052	0.883	0.978	0.998

D[*n*, 0.1]—10% of the particles are smaller than this diameter. D[*n*, 0.5]—half of the particles are smaller than this diameter, and half are longer. D[*n*, 0.9]—90% of the particles are smaller than this diameter.

In terms of physical properties, the particles of the analyzed starches differed mainly in size, which was compared based on diameter. The fine-powdered starches (NS and FS) were polydisperse systems differing significantly in their particle diameters but with normal distributions in the values of this parameter. The diameter of fermented starch (FS) particles was greater than that of native starch (NS). This observation was also confirmed by higher values of solidity parameter recorded in the case of FS starch. In turn, the particles of starch granulate (SG) had significantly greater diameters than the two other starches and featured a bimodal distribution of the values of this parameter. The particles of the fine-powdered starches (NS and FS) showed greater variability in their shape regularity. Although the mean values of the circularity parameter pointed to their more remarkable similarity to a square (0.871 ÷ 0.895) compared to starch granulate (SG) particles (0.801), the comparison of the range of circularity values indicated the opposite. Among the particles of the fine-grained powders, there were also perfectly circular particles (1.000) as well as those not circular at all (0.028 ÷ 0.036). In contrast, there were no perfectly circular particles (0.965) of the starch granulate, but even the least circular particles of this starch type were more regular (0.417) than those of the fine-powdered starches (NS and FS). Analogous observations were made for the convexity values of the cassava starch particles. The particles of fine-powdered starches (NS and FS) were similar in terms of distribution of convexity value. In contrast, the starch granulate (SG) particles were characterized by greater clustering of their values manifested in a narrower range of the extreme values. By analogy, the values of the parameter described as elongation fitted within a substantially broader range in the case of SG particles compared to the NS and SS particles. As a consequence, the shape coefficient values of SG particles also fell in a significantly narrower range than those determined for the particles of the two other starch types. It may be concluded that the starch particles in the form of granules (SG) differed significantly from those of the fine-powdered starches (NS and FS), which also differed significantly in their sizes (diameter and solidity). The starch granulate (SG) differed significantly in terms of its particle characteristics from the two other starches (NS and FS). It may also be concluded that the fermentation process applied to produce fermented starch modified the size of its particles, including their diameter and solidity values. In contrast, it did not modify the shape, circularity, convexity, or elongation of cassava starch particles (Table 1).

A comparison of native cassava starch (NS) to other starches shows that, in terms of the diameter of its particles, it is very similar to native potato starch. However, its particles are more solid (massive) than those of potato starch [31].

The hygroscopic properties represent a cumulative effect of: interactions between body surface and water, water vapor condensation in capillaries, concentration and type of water-soluble substances, and water content of a product [32]. Water significantly affects the physical, chemical, and biochemical properties and microbiological safety of a biomatrix. In addition, it determines its susceptibility to degradation [31]. Hence, the water content and activity of the analyzed cassava starches were determined to compare their hygroscopicity (Table 2).

Table 2. Water content and water activity of the analyzed cassava starch samples.

Product	Water Content [g H_2O/100 g d.m.]	SD	Water Activity [−]	SD
Native starch (NS)	14.347	0.552	0.492	0.005
Starch granulate (SG)	10.108	0.081	0.380	0.001
Fermented starch (FS)	13.524	0.232	0.457	0.001

Data collated in Table 2 show that the starch granulate (SG) had the lowest content and, consequently, the lowest water activity, probably due to its preparation method (loose starch compression). On the other hand, more minor differences in water content and activity were found between the fermented starch (FS) and native starch powder (NS). In addition, the water-content-to-water-activity ratio enables conclusion of a greater affinity of water to fermented starch (FS).

The statistical assessment of the significance of differences between the mean values of water content and activity performed with the ANOVA test and the Tukey post hoc test (T) showed that only native and fermented starch did not differ significantly in terms of water content ($T_{crit.} = 0.868$; $T_{NS/SG} = 4.2417$; $T_{NS/FS} = 0.8223$; $T_{SG/FS} = 3.4193$). On the other hand, the assessment of the differentiation in the water activity level showed statistically significant differences between all the samples ($T_{crit.} = 0.0081$; $T_{NS/SG} = 0.1125$; $T_{NS/FS} = 0.0357$; $T_{SG/FS} = 0.0768$).

Despite significant water content differences, each starch sample's water activity was low enough to ensure their microbiological stability. Microorganisms with the most minor demands in terms of water availability require water activity over 0.6 to initiate their vital functions [33].

Valuable information on the state of water in the material is provided by sorption isotherms, which enable establishment of product sensitivity to water in the form of vapor and the degree of water absorption by this product, as well as predicting changes in the material during storage affected by water availability [34]. In addition, the shape of an adsorption isotherm enables identification of a characteristic water-binding mechanism in a given material [35,36]. Figure 1 presents sorption isotherms of the analyzed cassava starch products.

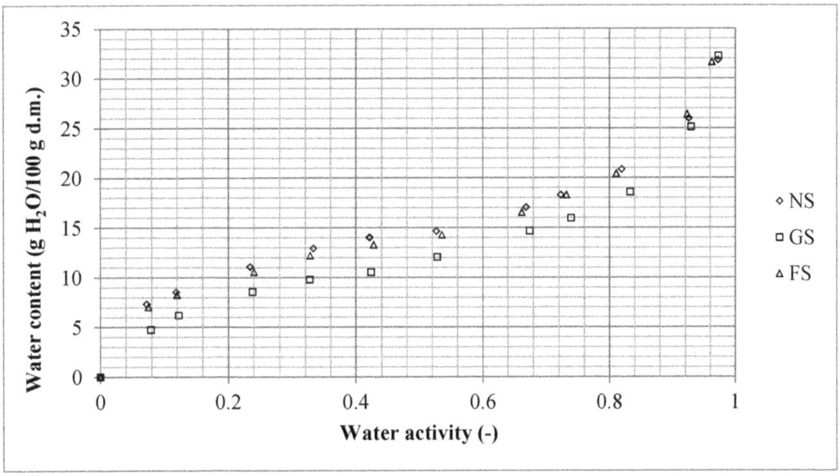

Figure 1. Comparison of water vapor sorption isotherms of native starch (NS), starch granulate (SG), and fermented starch (FS) at a temp. of 20 °C. Source: results of the present study.

Sorption isotherms of the analyzed starches had a sigmoidal shape and continuous course across the entire range of water activities tested. This indicates that water adsorption by starch caused no changes in the structure of starch granules, and, in particular, it did not lead to the crystallization of components [35].

The sorption isotherm of the starch granulate (SG) differed significantly from the other starches in terms of its position in the reference frame. Such a position of the isotherm indicates that, during storage of the analyzed starches in the environment with a specified humidity level, starch granulate (SG) will adsorb significantly less water reaching the same level of its activity as the other two fine-powdered starches (NS and FS). In addition, if all analyzed starches have the same water content, the starch granulate (SG) will feature the highest water activity. This, in turn, will make it the least stable, primarily regarding microbiological safety. The most likely reason behind the significantly different hygroscopic properties of SG is its smaller specific surface area compared to the developed surface of fine-powdered starches.

The second inflexion on the sorption isotherms of all starch samples was observed at $a_w \approx 0.7$. Suriyatem and Rachtanapun [37] reported similar findings. This characteristic point indicates the intensification of the sorption process due to the initiated phenomenon of capillary condensation [38]. This phenomenon consists of the filling of capillaries on the surface of a solid body with water molecules, leading to the modification of its properties to resemble those of free water [31].

The course of sorption isotherms in the entire a_w range was also compared in the statistical analysis using the Student's t-test for differences between mean values for dependent pairs [22,31]. The isotherms of all analyzed starches differed significantly in their course ($t_{SN/FS} = 3.3096$, $t_{SN/SG} = 6.7958$, $t_{FS/SG} = 6.9045$, $t_{crit.} = 2.228$). The determined statistical values confirm a huge difference between the hygroscopic properties of starch granulate (SG) and those of the other two starch samples. In addition, they show that the fine-powdered starches (NS and FS) also differed in their hygroscopicity. The above results allow concluding that both starch fermentation and compression into compact granules (granulation) lead to significant differences in its hygroscopic properties. This is an important finding, as certain works demonstrate a paucity of data that would confirm the correlation between the physical properties (size and shape) and hygroscopic properties of powders or small-sized solid bodies [39]. The results of the research conducted by Ikhu-Omoregbe [40] showed that the fermentation of cassava starch causes a significant change in sorption properties. Furthermore, the results of many other studies have shown that starch fermentation changes the sorption properties of cassava starch. At the same time, it does not cause changes in the differentiation in the course of adsorption and desorption (hysteresis loop) [21].

The empirically determined sorption isotherms were described using two theoretical mathematical models: *BET* and *GAB* [36]. The *BET* model is most frequently employed to describe the structure of a product and sorption phenomena. It assumes that the shape of a sorption isotherm is due to the complex character of the sorption process on porous bodies [39]. Using the simplifying assumptions described in the literature, this model estimates the water-bound content in the so-called monolayer [37]. When making these estimates, caution should be exercised regarding the model's limitation resulting from a tendency for significant overestimation of the predicted results in the range of high a_w values [39]. Table 3 presents the parameters of the *BET* equation of the analyzed cassava starches.

Table 3. Parameters of *BET* equation determined for the analyzed cassava starches.

Parameter	Native Starch (NS)		Starch Granulate (SG)		Fermented Starch (FS)	
	Value	Error	Value	Error	Value	Error
c_{BET}	0.8620	0.2610	1.1832	0.3646	0.9694	0.2741
v_m	9.3989	3.7904	4.6637	2.0328	7.5257	2.9108
RMS	10.11		13.70		8.88	
RSS	6.6325	1.4869	4.7435	1.2574	5.7693	1.3868

The goodness of fit of empirical data to the results generated using the *BET* model was evaluated based on the residual sum of squares (*RSS*) and root mean square (*RMS*) errors. The comparison of the *RSS* values shows that the *BET* model described empirical data of all starches with similar accuracy. In turn, the *RMS* value determined for SG exceeded the threshold of 10%, indicating a good fit of the model to experimental data [41]. This means that the *BET* model better describes the hygroscopicity of fine-grained powders than granules; hence, the results characterizing the fine-grained powders can be claimed to be more reliable.

The monolayer (v_m) describes the sorption capacity of the adsorbent and the availability of polar sites to the water vapor [36]. Karel [42] demonstrated that the monolayer of various natural products ranged from 4 to 11 kg H_2O per 100 kg dry matter. The v_m values estimated for cassava starches fitted within this range. The v_m values estimated for individual starches differed, indicating that both cassava starch fermentation and granulation affected its water-binding capability. The fermentation process modified the physical parameters of native, which means a crystalline form of starch, having sparse amorphous regions due to peripheral damage of starch granules [35]. As a result, starch particles became more solid, and the monolayer capacity decreased. In turn, upon the granulation process, the physical properties of starch particles changed radically (Table 1), as did the monolayer capacity.

Knowledge of the monolayer is essential for designing processes of food drying and storage, and also for ensuring appropriate conditions during its transport [43]. Furthermore, the maximum permissible water content in the product is determined in practice based on the monolayer estimation [44]. Given the above, it may be concluded that the native cassava starch (NS) is less sensitive to water content changes in the environment compared to the fermented (FS) or granulated (SG) starches. When the monolayer capacity is low, the lesser amount of water adsorbed from the environment will result in the critical humidity. The latter, in turn, will trigger undesirable physical modifications, i.e., product caking and hardening, and especially hazardous microbiological changes [45].

The energy constant c_{BET} informed about the energy released during sorption. Its values determined for all analyzed starch were low, pointing to the similar course of the sorption phenomenon and its physical nature. The enthalpy value approximating 20 kJ·mol^{-1} usually does not affect the nature of physically adsorbed molecules [46]. Notably, the highest load of energy was released during the adsorption of water molecules on the surface of starch granulate.

Ocieczek and Mesinger [31] conducted an analogous study with the numerical method of estimating *BET* model parameters for gluten-free wheat, maize, and potato starches. A comparison of the present study results with the findings from their study shows that cassava starch has a significantly larger monolayer. In turn, the sorption process in all these starches entails similar energetic transformations.

The second model tested in the study was the *GAB* model. Table 4 presents the parameters of the *GAB* equation of the analyzed cassava starches. Its use to study the sorption properties of dry products spurs a growing interest, especially among food technologists [36]. The *GAB* model was based on the *BET* theory by taking into account the modified properties of an adsorbent in terms of multi-layer adsorption [47,48]. It enables description of sorption isotherms in almost the entire a_w range (from 0.05 to 0.93) and extrapolates data obtained at different temperatures [49]. Due to the above, this model may be applied in calculations used in product logistics management.

Table 4. Parameters of *GAB* equation determined for the analyzed cassava starches.

Parameter	Native Starch (NS)		Starch Granulate (SG)		Fermented Starch (FS)	
	Value	Error	Value	Error	Value	Error
c_{GAB}	38.2050	4.6761	21.6854	1.8444	38.3936	4.0099
k	0.5732	0.0129	0.6535	0.0079	0.6182	0.0089
v_m	11.1898	0.2784	8.7150	0.1690	10.3235	0.1913
RMS	7.71		9.64		7.23	
RSS	0.9842	0.3508	0.3867	0.2199	0.5681	0.2665

Lewicki demonstrated that maintaining the calculation error of the monolayer estimated based on the *GAB* equation at ±15.5% required the k constant to fit the range of 0.24 ÷ 1, and the c_{GAB} constant to be higher than 5.67 [50]. These conditions were met in each of the analyzed cases. In addition, the estimated *RSS* and *RMS* values indicate that the *GAB* model proved very good in describing water vapor sorption by the analyzed starch samples. Model parameters were calculated for each starch type with the same accuracy. In turn, a comparison of *RMS* values shows that the *GAB* model proved better than the *BET* model in describing the sorptive properties of cassava starches [36].

The monolayer values estimated using the *GAB* equation were higher than those determined with the *BET* equation; however, the distribution of *GAB* model results was analogous to that achieved with the *BET* model. Native starch (NS) had the largest monolayer (11.2), whereas starch granulate (SG) had the smallest monolayer (8.7), which was most likely due to the modifications in the physical structure of starch granules [51] caused by fermentation or granulation.

The energy constant c_{GAB} was defined as a difference between the enthalpy of adsorbate molecules in the first adsorption layer and the higher layers [24]. In turn, the strong exothermal interactions between the solid body matrix and water molecules were ascribed to a reduced process temperature and increased c_{GAB} value [52]. Given the above considerations, it may be speculated that water sorption by the analyzed starch samples was physical in nature.

The k parameter also describes the deviation between desorption enthalpy and liquid adsorbent evaporation enthalpy and corrects the properties of molecules building the monolayer compared to the liquid phase. Such a deviation does not occur only when k equals 1 [49]. In addition, as Caurie [53] claims, the k value enables distinction between monolayer ($k \leq 0.5$) and multilayer ($k > 0.5$) adsorption. For this reason, the determined values of k make the *GAB* model a reliable tool for describing the sorption properties of the analyzed starches. Furthermore, the k constant values determined for individual starch types were similar, which explicitly confirms the similarity of the examined material (cassava starch), which has earlier been pinpointed by Chirife and Iglesias [54]. Therefore, it may be concluded that the energy status of the molecules building multilayer systems on particular starch samples was very similar [26].

The last stage of the study aimed to determine selected parameters describing the microstructure of the surface of the analyzed starches. Table 5 presents the parameters describing the microstructure of the surface of the cassava starches. The first determined parameter was the specific sorption area, which is a derivative of the monolayer. Hence, the results obtained were analogous to those describing the monolayer. The second parameter was the total volume of capillaries, describing the total volume occupied by water as a result of filling the micro-, meso-, and macrocapillaries. Finally, the third parameter was the capillary radius that was filled as a result of initiating capillary condensation [55].

Table 5. Microstructural characteristics of the surface of the analyzed cassava starches.

Product	Specific Sorption Area (m^2/g d.m.)	Total Volume of Capillaries (mm^3/100 g d.m.)	Capillary Radius Filled at a_w = 0.7 (nm)
Native starch (NS)	393.1	122.4	2.18
Starch granulate (SG)	306.2	109.3	1.84
Fermented starch (FS)	362.7	121.6	2.13

The results obtained indicate that the native (NS) starch, and therefore semicrystalline starch, was characterized by the most developed specific sorption surface area. As a result of the fermentation process and granulation, the specific surface of the starch decreased. This change, however, cannot be associated with an increase in the ordering of starch particles, but only with their physical modifications. Starch fermentation resulted in the increased solidity of its particles. However, as a result of granulation, all parameters describing the physical state of starch particles changed radically (the diameter, solidity, circularity, convexity, and shape coefficient increased radically, whereas elongation decreased) (Table 1). Moreover, the granulation of cassava starch contributed to the reduction in the diameter and total volume of capillaries filled during capillary condensation. Fermentation did not cause significant changes in the surface microstructure parameters.

Cassava starch-based products can complement the assortment of starch products available on the Polish market. Native cassava starch is similar to native potato starch in terms of particle characteristics but significantly different in this aspect from native wheat gluten-free starch and native maize starch [31].

Edible cassava starches, regardless of their type (native, fermented, granulated), showed strong hygroscopic properties [49], which were described and compared graphically using sorption isotherms and also mathematically based on the parameters of sorption models.

Sorption isotherms plotted for all tested cassava starches (NS, SG, FS) were sigmoidal and continuous over the entire range of water activities. This finding justified using *BET* and *GAB* models to identify the parameters of hygroscopicity characteristics.

The mathematical sorption models (*BET* and *GAB*) used to explore experimental data were based on well-established theoretical foundations [39]. These models described the experimental data very well, which was confirmed by the low *RSS* and *RMS* values, and by errors with which the parameters of both equations were estimated. However, taking into account the recommendations of the European COST 90 Project [56], a broader range of experimental data used to determine the parameters of the *GAB* model, the possibility of transferring the results obtained using the *GAB* model to different temperatures, and slightly better results of this model fitting to the original data, it can be concluded that the *GAB* model should be recommended for the description and comparison of the hygroscopic properties of all tested cassava starches.

4. Conclusions

The analysis of the results obtained using both models indicates that:

1. Native cassava starch significantly differs from potato starch in terms of hygroscopicity described by the parameters of the *BET* model, despite similar particle size characteristics.
2. Modification of cassava starch, both through fermentation and granulation, significantly modifies its physical and hygroscopic parameters.
3. Modification of physical/chemical properties, and consequently hygroscopic properties of cassava starches, opens avenues for its more comprehensive and targeted use in both the food and pharmaceutical industries.
4. Modification of cassava starch may contribute to the rational management of the food or pharmaceutical production processes, considering the diverse needs of these industries and consumers.

Author Contributions: Conceptualization A.O. and D.M.; methodology A.O.; software, D.M. and H.T.; formal analysis A.O. and D.M.; investigation, D.M. and H.T.; resources, D.M.; data curation, A.O.; writing—original draft preparation, A.O. and D.M.; writing—review and editing A.O., D.M. and H.T.; visualization, A.O. and D.M.; supervision A.O.; project administration, A.O. and D.M. All authors have read and agreed to the published version of the manuscript.

Funding: This research and the APC were funded by Gdynia Maritime University, grant number WZNJ/2022/PZ/05, WM/2022/PZ/06, and GMU Doctoral Fund.

Institutional Review Board Statement: Not applicable.

Informed Consent Statement: Not applicable.

Data Availability Statement: Data is contained within the article.

Conflicts of Interest: The authors declare no conflict of interest.

References

1. Aryee, F.N.A.; Oduro, I.; Ellis, W.O.; Afuakwa, J.J. The physicochemical properties of flour samples from the roots of 31 varieties of cassava. *Food Control* **2006**, *17*, 916–922. [CrossRef]
2. Hongbété, F.; Mestres, C.; Akissoé, N.; Pons, B.; Hounhouigan, D.J.; Cornet, D.; Nago, C.M. Effects of cultivar and harvesting conditions (age, season) on the texture and taste of boiled cassava roots. *Food Chem.* **2011**, *126*, 127–133. [CrossRef]

3. Nwokocha, L.; Aviara, N.A.; Senan, C.; Williams, P.A. A comparative study of some properties of cassava (*Manihot esculenta*, Crantz) and cocoyam (*Colocasia esculenta*, Linn) starches. *Carbohydr. Polym.* **2009**, *76*, 362–367. [CrossRef]
4. Léotard, G.; Duputié, A.; Kjellberg, F.; Douzery, E.J.; Debain, C.; de Granville, J.J.; McKey, D. Phylogeography and the origin of cassava: New insights from the northern rim of the Amazonian basin. *Mol. Phylogenet. Evol.* **2009**, *53*, 329–334. [CrossRef] [PubMed]
5. Zanetti, E.G.B.; Cardoso, E.M.G.; Dourado, D.P.; Reina, E.; Muraishi, C.T. Performance of two cassava varieties submitted to different spacings, grown in the Cerrado region. *Appl. Res. Agrotechnol.* **2014**, *7*, 39–46. [CrossRef]
6. CONAB (Companhia Nacional de Abastecimento). *Perspectiva Para a Agropecuária*; Companhia Nacional de Abastecimento: Brasília, Brazil, 2017; Volume 2, pp. 1–155.
7. FAOSTAT—Food and Agriculture Data. Available online: http://www.fao.org/faostat/en/ (accessed on 8 July 2020).
8. Dias, L.T.; Leonel, M. Phisico-chemical characteristics of cassava flours from different regions of Brazil. *Ciência Agrotecnologia* **2006**, *30*, 692–700. [CrossRef]
9. Falade, K.O.; Akingbala, J.O. Utilization of Cassava for Food. *Food Rev. Int.* **2010**, *27*, 51–83. [CrossRef]
10. An, D.; Yang, J.; Zhang, P. Transcriptome profiling of low temperature-treated cassava apical shoots showed dynamic responses of tropical plant to cold stress. *BMC Genom.* **2012**, *13*, 64. [CrossRef]
11. Cooke, R.D.; Rickard, J.E.; Thompson, A.K. The Storage of Tropical Root and Tuber Cropscassava, Yam and Edible Aroids. *Exp. Agric.* **1988**, *24*, 457–470. [CrossRef]
12. Adebowale, A.A.; Sanni, L.O.; Onitilo, M.O. Chemical composition and pasting properties of tapioca grits from different cassava varieties and roasting methods. *Afr. J. Agric. Food Secur.* **2019**, *7*, 1–6.
13. Chisté, R.C.; Silva, P.A.; Lopes, A.S.; da Silva Pena, R. Sorption isotherms of tapioca flour. *Int. J. Food Sci. Technol.* **2012**, *47*, 870–874. [CrossRef]
14. Iwu, M.M. *Food as Medicine: Functional Food Plants of Africa*; Taylor & Francis: Abingdon, UK, 2017; Chapter 4; pp. 63–85. [CrossRef]
15. Breuninger, W.F.; Piyachomkwan, K.; Sriroth, K. Tapioca/Cassava Starch: Production and Use. In *Starch: Chemistry and Technology*, 3rd ed.; Elsevier: Amsterdam, The Netherlands, 2009. [CrossRef]
16. Domaradzki, M.; Korpal, W. Diety eliminacyjne w alergiach pokarmowych. *Inżynieria Przetwórstwa Spożywczego* **2017**, *121*, 5–8.
17. Kolawole, O.P.; Agbetoye, L.A.S.; Ogunlowo, A.S. Evaluation of cassava mash dewatering methods. *J. Bioinform. Seq. Anal.* **2011**, *3*, 23–30. [CrossRef]
18. Charoenthai, N.; Sanga-ngam, T.; Puttipipatkhachorn, S. Use of modified tapioca starches as pharmaceutical excipients. *Pharm. Sci. Asia* **2018**, *45*, 195–204. [CrossRef]
19. Charles, A.L.; Chang, Y.H.; Ko, W.C.; Sriroth, K.; Huang, T.C. Influence of amylopectin structure and amylose content on the gelling properties of five cultivars of cassava starches. *J. Agric. Food Chem.* **2005**, *53*, 2717–2725. [CrossRef] [PubMed]
20. Oriol, E.; Raimbault, M.; Roussos, S.; Viniegra-Gonzales, G. Water and water activity in the solid state fermentation of cassava starch by Aspergillus niger. *Appl. Microbiol. Biotechnol.* **1988**, *27*, 498–503. [CrossRef]
21. Van der Werf, L.; Chapuis, A.; Courtois, F. A global sorption equation for raw and processed cassava based on a review and compilation of sorption properties. *Dry. Technol.* **2022**, 1–14. [CrossRef]
22. Ocieczek, A.; Skotnicka, M.; Baranowska, K. Sorptive properties of modified maize starch as indicators of their quality. *Int. Agrophys.* **2017**, *31*, 383–392. [CrossRef]
23. Figura, L.O.; Teixeira, A.A. *Food Physics. Physical Properties—Measurement and Applications*, 1st ed.; Springer: Berlin/Heidelberg, Germany, 2007.
24. Paderewski, M. *Procesy Adsorpcyjne w Inżynierii Chemicznej*, 1st ed.; WNT: Warsaw, Poland, 1999.
25. Rizvi, S.S.H. Thermodynamic Properties of Food in Dehydration. In *Engineering Properties of Foods*, 3rd ed.; Rao, M.A., Rizvi, S.S.H., Eds.; Taylor & Francis: Boca Raton, FL, USA, 2005; pp. 223–309. [CrossRef]
26. Andrade, R.D.; Lemus, R.M.; Pérez, C.C. Models of sorption isotherms for food: Uses and limitations. *Vitae* **2011**, *18*, 325–334.
27. Ocieczek, A.; Schur, J. Ocena wpływu wybranych dodatków na właściwości sorpcyjne miękiszu pieczywa pszennego. *ŻYWNOŚĆ Nauka Technol. Jakość* **2015**, *1*, 143–154. [CrossRef]
28. Lewicki, P.P. A three-parameter equation for food moisture sorption isotherms. *J. Food Process Eng.* **1998**, *21*, 127–144. [CrossRef]
29. Pałacha, Z.; Sas, A. Właściwości sorpcyjne wybranych gatunków ryżu. *Acta Agrophys.* **2016**, *23*, 681–694.
30. Erbas, M.; Ertugay, M.F.; Certel, M. Moisture adsorption behaviour of semolina and farina. *J. Food Eng.* **2005**, *69*, 191–198. [CrossRef]
31. Ocieczek, A.; Mesinger, D. Porównawcza charakterystyka właściwości sorpcyjnych popularnych rodzajów skrobi z zastosowaniem modelu BET w rozwinięciu analitycznym i numerycznym. *Przemysł Chem.* **2020**, *99*, 1000–1004. [CrossRef]
32. Czapski, J. Wybrane Własności Produktów Żywnościowych. In *Opakowania Żywności*; Czerniawski, B., Michniewicz, J., Eds.; Agro Food Technology: Czeladź, Poland, 1998.
33. Sandle, T. The Important of Water Activity for Risk Assessing Pharmaceutical Products. *J. Pharm. Microbiol.* **2016**, *2*, 1.
34. Al-Muhtaseb, A.H.; McMinn, W.A.M.; Magee, T.R.A. Moisture sorption isotherm characteristics of food products: A review. *Food Bioprod. Process.* **2002**, *80*, 118–128. Available online: https://www.sciencedirect.com/science/article/abs/pii/S0960308502703052 (accessed on 7 February 2008). [CrossRef]

35. Al-Muhtaseb, A.H.; McMinn, W.A.M.; Magee, T.R.A. Water sorption isotherms of starch powders Part 1: Mathematical description of experimental data. *J. Food Eng.* **2004**, *61*, 297–307. [CrossRef]
36. Furmaniak, S.; Terzyk, A.P.; Gauden, P.A. The general mechanism of water sorption of foodstuffs—Importance of the multitemperature fitting of data and the hierarchy of models. *J. Food Eng.* **2007**, *82*, 528–535. [CrossRef]
37. Suriyatem, R.; Rachtanapun, P. Prediction modeling for moisture sorption isotherms of rice starch/carboxymethyl cellulose from durion rind blend films. *Appl. Mech. Mater.* **2013**, *431*, 32–36. [CrossRef]
38. Newman, A.; Zografi, G. An examination of water vapor sorption by multicomponent crystalline and amorphous solids and its effects on their solid-state properties. *J. Pharm. Sci.* **2019**, *108*, 1061–1680. [CrossRef]
39. Peleg, M. Models of sigmoid equilibrium moisture sorption isotherms with and without the monolayer hypothesis. *Food Eng. Rev.* **2020**, *21*, 1–13. [CrossRef]
40. Ikhu-Omoregbe, D.I.O. Comparison of the sorption isotherm characteristics of two cassava products. *Int. J. Food Prop.* **2006**, *9*, 167–177. [CrossRef]
41. Stępień, A.; Witczak, M.; Witczak, T. Moisture sorption characteristics of food powders containing freeze dried avocado, maltodextrin and inulin. *Int. J. Biol. Macromol.* **2020**, *149*, 256–261. [CrossRef] [PubMed]
42. Karel, M. Water Activity and Food Preservation. In *Physical Principles of Food Preservation. Principles of Food Science. Part 2*; Karel, M., Fennema, O.R., Lund, D.B., Eds.; Marcel Dekker Inc.: New York, NY, USA, 1975.
43. Górnicki, K.; Kaleta, A.; Trajer, J. Modelling of dried apple rehydration indices using ANN. *Int. Agrophys.* **2019**, *33*, 285–296. [CrossRef]
44. Pérez-Alonso, C.; Fabela-Morón, M.F.; Guadarrama-Lezama, A.Y.; Barrera-Pichardo, J.F.; Alamilla-Beltrán, L.; Rodríguez-Huezo, M.E. Interrelationship between the structural features and rehydration properties of spray dried mano chilli sauce microcapsules. *Rev. Mex. Ing. Química* **2009**, *8*, 187–196.
45. Cassini, A.S.; Marczak, L.D.F.; Noreña, C.P.Z. Water adsorption isotherms of texturized soy protein. *J. Food Eng.* **2006**, *77*, 194–199. [CrossRef]
46. Atkins, P.W. *Chemia Fizyczna*; Wydawnictwo Naukowe PWN: Warsaw, Poland, 2003.
47. Timmermann, E.O.; Chirife, J.; Iglesias, H.A. Water sorption isotherms of food and foodstuffs: BET Or GAB parameters? *J. Food Eng.* **2001**, *48*, 19–31. [CrossRef]
48. Timmermann, E.O. Multilayer sorption parameters: BET or GAB values? *Colloids Surf. A Physicochem. Eng. Asp.* **2003**, *220*, 235–260. [CrossRef]
49. Dooshima, I.B.; Mbanengen, S.F.; Julius, I.; Julius, A. Moisture Adsorption Studies on Soy—Mumu Supplemented with Moringa Leaf Powder. *Adv. Biosci. Bioeng.* **2016**, *4*, 67–73. [CrossRef]
50. Lewicki, P.P. The applicabillity of the GAB model to food water sorption isotherms. *Int. J. Food Sci. Technol.* **1997**, *32*, 553–557. [CrossRef]
51. Stępniewska, S.; Cacak-Pietrzak, G. Skrobia żytnia—Budowa, właściwości i metody badań. *Przegląd Zbożowo-Młynarski* **2018**, *62*, 25–29.
52. Diosady, L.L.; Rizvi, S.S.H.; Cai, W.; Jagdeo, D.J. Moisture sorption isotherms of canola meals, and applications to packing. *J. Food Sci.* **1996**, *61*, 204–208. [CrossRef]
53. Caurie, M. The derivation of the GAB adsorption equation from the BDDT adsorption theory. *Int. J. Food Sci. Technol.* **2006**, *41*, 173–179. [CrossRef]
54. Chirife, J.; Iglesias, H.A. Estimation of precision of isosteric heat of sorption determined from the temperature dependence of food isotherms. *LWT* **1992**, *25*, 83–84.
55. Ocieczek, A.; Mesinger, D.; Kamińska, D. Porównanie Właściwości Sorpcyjnych Wybranych Typów Kaszek Dla Dzieci w Kontekście Szacowania Ich Stabilności Przechowalniczej. In *Żywność w XXI Wieku. Od Producent do Konsumenta*; Gałkowska, D., Kowalski, S., Zięć, G., Eds.; Polskie Towarzystwo Technologów Żywności: Warsaw, Poland, 2021.
56. Barbosa-Cánovas, G.V.; Fontana, A.J., Jr.; Schmidt, S.J.; Labuza, T.P. *Water Activity in Foods. Fundamentals and Applications*; Blackwell Publishing and the Institute of Food Technologists: Hoboken, NJ, USA, 2007.

foods

Article

A New Polysaccharide Carrier Isolated from Camelina Cake: Structural Characterization, Rheological Behavior, and Its Influence on Purple Corn Cob Extract's Bioaccessibility

Lucia Ferron [1,2], Chiara Milanese [3], Raffaella Colombo [1], Raffaele Pugliese [4] and Adele Papetti [1,*]

1 Department of Drug Sciences, University of Pavia, 27100 Pavia, Italy; lucia.ferron01@universitadipavia.it (L.F.); raffaella.colombo@unipv.it (R.C.)
2 FlaNat Research Italia Srl, 20017 Rho, Italy
3 C.S.G.I. & Department of Chemistry, Physical Chemistry Section, University of Pavia, 27100 Pavia, Italy; chiara.milanese@unpv.it
4 NeMO Lab, ASST Grande Ospedale Metropolitano Niguarda, 20162 Milan, Italy; raffaele.pugliese@nemolab.it
* Correspondence: adele.papetti@unipv.it; Tel.: +39-0382-98-7863

Abstract: A polysaccharide fraction obtained from camelina cake (CCP), selected as a carrier to encapsulate purple corn cob extract (MCE), was investigated. A wide population of carbohydrate polymers (with a polydispersivity index of 3.26 ± 0.07 and an average molecular weight of about $139.749 \times 103 \pm 4.392 \times 10^3$ g/mol) with a gel-like behavior and a thixotropic feature characterized the fraction. MCE-CCP combinations (50–50 and 25–75, *w*/*w*), selected based on CCP encapsulation efficiency, were tested for their stability and MCE polyphenols' bioaccessibility during digestion (monitored using an in vitro static procedure). During the oral and gastric phases of the digestion process, CCP gradually swelled and totally released MCE polyphenols. MCE-CCP50 had the fastest release. Moreover, anthocyanins were still detectable during the duodenal phase, in both MCE-CCP ingredients. Furthermore, CCP (5 mg/mL) exerted in vitro potential hypocholesterolemic activity via bile salts binding during digestion.

Keywords: camelina cake polysaccharide; carrier; stabilizer; purple corn cob; by-product; anthocyanin bioaccessibility

Citation: Ferron, L.; Milanese, C.; Colombo, R.; Pugliese, R.; Papetti, A. A New Polysaccharide Carrier Isolated from Camelina Cake: Structural Characterization, Rheological Behavior, and Its Influence on Purple Corn Cob Extract's Bioaccessibility. *Foods* **2022**, *11*, 1736. https://doi.org/10.3390/foods11121736

Academic Editors: Jianhua Xie, Yanjun Zhang and Hansong Yu

Received: 23 May 2022
Accepted: 10 June 2022
Published: 14 June 2022

1. Introduction

Nowadays, the agro-food industry generates a huge amount of waste, which constitutes one of the main environmental problems due to the high organic charge [1], but it could represent a potential source of valuable bioactive compounds, suitable for food, pharmaceutical, and textile industries [1,2]. Recently, the search for new and renewable sources of bioactive compounds has been encouraged by the European Commission, which targeted the achievement of a fully circular economy by 2050, also through the farm to fork strategy. This program represents the heart of the European green deal, which aims to achieve healthier and more environmentally friendly food systems [3].

Therefore, great effort has been made by scientists and companies in order to investigate and characterize the potential applications of agro-food by-product bioactive compounds or metabolites in food, pharmaceutical, and cosmetic industries [1,2]. These natural compounds can be used as both nutraceutical ingredients and additives thanks to their strong antioxidant capacities, and are thus potentially useful for the development of products with enhanced nutritional value, potential health benefits, and long shelf-lives [1].

Among agro-food wastes, corn cobs are the main by-products generated during corn processing, and purple corn cobs are still lacking added-value applications [4]. Purple corns (*Zea mays* L.) are pigmented corn varieties from South America, mainly Peru and Bolivia, which have strong antioxidant properties thanks to their high anthocyanin content [5].

Several studies have investigated the anthocyanin contents in purple corn extracts obtained from different tissues, i.e., kernel, cob, husk, and silk, and tested their bioprotective effects [6,7]. Husk and cob extracts are richer in anthocyanins than kernel extracts. Contents range from 0.49 to 4.6% and from 1.25 to 13.18% (*w*/*w* dry material) for cob and husk, respectively [7]; moreover, the husk and cob differ from the kernel by the presence of perlagonidin in their phytocomplex [8]. Moradyn is a new Italian purple corn variety obtained from a Peruvian corn line (Morado), and it is characterized by highly pigmented cobs. The extract obtained from Moradyn cob (MCE) is characterized by a phytocomplex rich in antioxidant compounds, such as anthocyanins, quercetin, and kaempferol derivatives, but a preliminary study highlighted that the bioaccessibility of these MCE polyphenols was markedly affected by digestion, leading to a decrease in bioactivity [8], as already reported for other natural extracts [9]; this behavior limits polyphenols bioavailability and their subsequent efficacy.

During the last five years, there has been growing interest in the application of encapsulation technology in order to entrap antioxidant compounds, protect them against environmental conditions during storage, extend their shelf lives, and enhance their bioavailability [10,11]. Different carriers have been employed, depending on the type and nature of core materials and on the targeted applications of the microencapsulated ingredients [10]. Several authors reported that vegetable polysaccharides such as Arabic gum, maltodextrins, inulin, and other purified native gums are carriers more stable, biocompatible, biodegradable, and versatile than proteins (which might be melted or denatured), since they better resist high temperatures (>40 °C) and pH changes [12]. In fact, their functional groups make them extremely versatile, as they can interact with a wide range of both hydrophilic and hydrophobic bioactive compounds. Examples are anthocyanin and quercetin derivatives which are consistent with water-based gel formulations, including gum, maltodextrin, and starch [13]. Moreover, polysaccharides are efficient at entrapping bioactives due to their high molecular weights and the presence of numerous functional groups in their structures [12]. In addition, among natural polysaccharides, gums have remarkable and specific rheological features, which have been strictly related to their biological properties and technological applications [5,14].

In our previous work, the polysaccharide fraction isolated from camelina cake (*Camelina sativa* L. Krantz) was selected as potential carrier for MCE, based on the high encapsulation efficiency values registered at 1:1 and 1:3 core/wall material ratios. Camelina sativa is an important seed oil crop belonging to the Brassicaceae family. It is cultivated worldwide, and its oil is a valuable and well-characterized source of unsaturated fatty acids, mainly linolenic and α-linoleic acids [15].

To the best of our knowledge, the potential application in the food industry of Camelina sativa's by-product has never been investigated, so the use of camelina cake polysaccharide fraction (CCP) as a carrier for MCE could represent a valuable technological innovation. Preliminary spectroscopic analysis performed both on CCP and on MCE encapsulated with CCP indicated that CCP is mainly characterized by mannose, arabinose, and rhanmose residues, and by the presence of a protein component with a random coil conformation, suggesting a gum-like composition [16] maintained after MCE addition. CCP also increased MCE shelf-life when used as a carrier at 50 and 75% *w*/*w* [17].

Therefore, considering these preliminary results, the aim of the present work was to determine the CCP's average molecular weight and rheological properties, since these physico-chemical and structural parameters are strictly related to the biological behavior of natural polysaccharides [16,18,19]. The relationship between the CPP's physico-chemical features and its stabilizing properties was evaluated by submitting a MCE-based ingredient to simulated gastrointestinal conditions (applying a slightly modified INFOGEST protocol) [20] and monitoring the MCE bioaccessibility index for thirteen selected markers. CCP hypocholesterolemic activity was assessed by testing its bile salt binding activity [21].

2. Materials and Methods

2.1. Chemicals

Ethanol (96% *v/v*), methanol, HPLC-grade acetonitrile, phosphoric acid (85% *w/v*), and sodium chloride were obtained from Carlo Erba (Milan, Italy). HPLC-grade formic acid, hydrochloric acid (37% *w/v*), Type VI-porcine pancreatic α-amylase, pepsin from porcine gastric mucosa (≥400 U mg^{-1}), bile extract porcine, pancreatin (8 × USP) from porcine pancreas, sodium taurocholate (NaTC), sodium glycocholate (NaGC), sodium taurodeoxycholate (NaTCDC), sodium glycochenoxycholate (NaGCDC), protease from Streptomyces griseus type XIV (≥3.5 U mg^{-1}), viscozyme L cellulolytic enzyme mixture, sodium azide, and cholesterolamine were provided by Merck KGaA (Darmstadt, Germany).

Water was obtained from a Millipore Direct-QTM system (Merck-Millipore, Milan, Italy).

Pullulan gel filtration chromatography standard kit was purchased from Waters Corporation (Milford, MA, USA).

2.2. Camelina Cake Polysaccharide (CCP) Extraction

Camelina cake was kindly provided by FlaNat Research Italia S.r.l. (Milan, Italy).

A CCP dried extract was prepared following the procedure previously optimized [17]. Briefly, the cake obtained from cold-press oil production was immediately soaked with water (1:10 solvent/raw material ratio) and extracted at 115 °C for 15 min. After filtration of the supernatant, CCP was precipitated by adding a 96% ethanol solution (1:2.5, *v/v*) at 4 °C overnight. Finally, the polysaccharide fraction was separated by centrifugation at 5000 rpm (Neya 8 ZFKN-39276, Remi Eletrotechnik LTD, Mumbai, India) for 10 min at 25 °C, freeze dried (Modulyo freeze-drier s/n 5101, 5 Pascal, Trezzano sul Naviglio, Italy), and then used in the experiments.

2.3. Moradyn Cob Extract (MCE) Preparation

Moradyn chopped cobs were kindly provided by FlaNat Research Italia S.r.l. (Milan, Italy) and extracted with 50% aqueous ethanol for 3 h at 50 °C. The extract (MCE) was filtered through 0.45 μm membrane filters (Merck-Millipore, Milan, Italy) and the organic solvent removed under reduced pressure at 40 °C (Buchi R-II, Büchi Labortechnik AG, Flawil, Switzerland) [8]. Finally, MCE was freeze-dried (Modulyo freeze-drier s/n 5101, 5 Pascal, Trezzano sul Naviglio, Italy) and used in the experiments.

2.4. Preparation of MCE-CCP Ingredients

MCE was encapsulated in 50 or 75% CCP (*w/w*) to obtain MCE-CCP ingredients as follows: freeze-dried CCP was dissolved in water at 10 mg/mL final concentration and hydrated at 4 °C overnight. The carrier solution and MCE (core) were mixed at 1:1 or 1:3 ratio (*w/w*), and the obtained ingredients were dried under vacuum in a drying oven (42 °C, 8 mbar) for 48 h (Goldbrunn 1400, Expondo, Berlin, Germany) [17].

2.5. Mechanical Testing

The CCP rheological properties were investigated using a Kinexus DSR Rheometer (Netzsch, Selb, Germany) equipped with a parallel-plate geometry (acrylic diameter 20 mm; gap 34 μm). CCP viscosity was measured using a flow step program, at increasing shear rate (0.001–1000 s^{-1}), to investigate their non-Newtonian behavior. To evaluate the storage (G′) and loss (G″) moduli of CCP, frequency sweep experiment results were recorded as a function of angular frequency (0.1–100 Hz) at 0.5% fixed strain. To test CCP thixotropy, shear-thinning tests were performed by a series of peak hold tests in which shear rates were kept constant, as previously reported by Pugliese et al., (2021) [22]. Briefly, a shear rate of 0.01 s^{-1} and a shear rate of 5.3 s^{-1} were applied in sequence for 60 and 20 s, respectively. Subsequently, a high shear rate of 1000 s^{-1} for 20 s followed by a shear rate of 5.3 s^{-1} for 20 s were applied. Lastly, a shear rate of 0.01 s^{-1} was used to simulate the shear condition of CCP at rest. Each experiment was performed in triplicate, and data were processed using OriginLabTM 8 software.

2.6. CCP Average Molecular Mass Determination

The CCP average molecular weight (Mw) and polydispersivity index (Pi) were determined by size exclusion chromatography (SEC) using an Agilent 1200 chromatographic system coupled with a G7162A Refractive Index Detector (RID) (Agilent Technologies, Santa Clara, CA, USA). Narrow pullulan standards in the $5-642 \times 10^3$ g/mol range were used for the calibration curve (Waters Corporation, Milford, MA, USA). An Ultrahydrogel™ 2000 column (12 μm, 7.8 mm × 300 mm) and an Ultrahydrogel™ 250 column (6 μm, 7.8 mm × 300 mm) (Waters Corporation, Milford, MA, USA) were coupled in series and operated at a constant flow rate of 0.8 mL/min. The mobile phase consisted of a 0.1 M NaCl solution (*w*/*v*) containing 0.02% NaN_3 (*w*/*v*), filtered through a 0.45 μm PTFE membrane (Merck-Millipore, Milan, Italy). Columns and detector were maintained at 40 °C.

Samples were prepared by dissolving CCP or pullulan standards in 0.1 M NaCl (1 mg/mL final concentration) and filtered (cellulose microporous membrane filter, 0.45 mm, Merck-Millipore, Milan, Italy) before injection (injection volume 50 μL).

2.7. Determination of Total Polyphenol Content in CCP-MCE Ingredients by RP-HPLC-WVD Analysis

The encapsulated polyphenols were extracted following a procedure described by Norkaew et al. (2019) [23] with slight modifications: 5 mg of dried ingredients was dissolved in 1 mL deionized water, mixed, and then sonicated for 20 min; a methanol/acetonitrile/formic acid (60:35:5, *v*/*v*/*v*) mixture was added to obtain a 10 mL final volume. Samples were concentrated up to 0.25 mL under reduced pressure at 40 °C (Buchi R-II, Büchi Labortechnik AG, Flawil, Switzerland) and then diluted to a 5 mL final volume by means of a binary mixture consisting of 0.1% formic acid aqueous solution and 0.01% formic acid in acetonitrile 80:20 (*v*/*v*) before HPLC analysis.

Analyses were carried out using a 1260 Infinity II technology series system (Agilent Technologies, Santa Clara, CA, USA), equipped with a quaternary gradient pump, a vial sampler, a degasser, a thermostatted column system set at 25.0 ± 0.5 °C, and a variable wavelength detector (VWD). The HPLC-VWD system was controlled using Agilent OpenLab CDS ChemStation software—Windows 10. The chromatographic separation was carried out on a Gemini® C18 analytical column (150 × 2.0 mm i.d., 5 μm, Phenomenex, Torrance, CA, USA) operating at 0.3 mL/min constant flow rate (injection volume 20 μL), using the mobile phase and the gradient elution program already applied and validated by Ferron et al. (2021) [24]. Chromatograms were recorded at 520 and 370 nm. The selected marker compounds' identification was based on co-cromatography with analytical standards (when commercially available) and comparison with literature data [8]. The peak area (mAU) registered for the undigested compound was used to calculate the bioaccessibility index.

2.8. In Vitro Digestion Procedure

MCE-CCP 50% or MCE-CCP 75% water solutions (1 mg/mL) were submitted to an in vitro gastrointestinal simulation digestion process following the INFOGEST protocol; briefly, simulated salivary (SSF), gastric (SGF), and intestinal (SIF) fluids were prepared using proper mixtures of electrolytes, bile salts, water, and enzymes. Pepsin and pancreatin were directly added to SGF and SIF, respectively [20]. Changes occurring in phytocomplex composition were monitored by collecting 2.5 mL of sample directly from the flask at different time points during each digestion step (after 2 min for the oral phase; after 15, 30, 60, and 120 min for the gastric phase; and after 30 and 120 min for the intestinal phase). At the end of each monitored time, enzymes were inactivated (90 °C, 5 min) and samples were centrifuged (30 min, 4 °C, 5000 rpm) (Centrifuge 5804 R Eppendorf, Hamburg, Germany).

Supernatants were freeze-dried (Modulyo freeze-drier s/n 5101, 5 Pascal, Trezzano sul Naviglio, Italy) and stored at −20 °C until analyses.

2.9. Bioaccessibility Evaluation

The percentage of soluble polyphenols in each collected digested sample represented the bioaccessible MCE fraction available for absorption.

The samples collected during the digestion process were dissolved in 2.5 mL of 0.1% formic acid aqueous solution—0.01% formic acid in acetonitrile (80:20, *v*/*v*)—and filtered through 0.2 μm nylon syringe filters (Phenomenex, Torrance, CA, USA) before HPLC analysis.

The bioaccessibility index for each monitored polyphenolic compound was calculated as:

$$\text{Bioaccessibility index } (\%) = A_{dig} / A_{tot} \times 100 \quad (1)$$

where A_{dig} corresponds to the peak area (mAU) of the marker compound in the ingredients after digestion, and A_{tot} represents the peak area (mAU) of the marker in the undigested sample, considered as 100%.

2.10. CCP Bile Salts Binding Capacity

The CCP bile salts' binding capacity was evaluated following the protocol reported by Lin et al. (2020) [21], with some modifications. The procedure basically involved following three steps: CCP in vitro digestion in the presence of bile salts (BS), collection of free BS by centrifugation, and their quantification by RP-HPLC.

Five different CCP concentration levels (0.5, 0.75, 2.5, 5, 10 mg/mL) were tested, and cholestyramine (10 mg/mL) was used as a positive control [25]. Water was used as blank in the digestion process.

The in vitro digestion procedure was carried out following the standardized INFOGEST protocol, but reproducing the intestinal phase conditions with a 10 mM BS mixture (35% NaTC, 35% NaGC, 15% NaTCDC and 15% NaGCDC) instead of commercial bile, in order to mimic the BS composition and concentration typically present in an adult intestine under the fed condition. Samples collected at the end of the intestinal phase were centrifuged at 14,000 rpm for 30 min at 4 °C (Centrifuge 5804 R Eppendorf, Hamburg, Germany); subsequently the supernatants were filtered (0.2 μm nylon syringe filters, Phenomenex, Torrance, CA, USA) and immediately submitted to RP-HPLC analysis.

BS separation and quantification were performed by HPLC-WVD (1260 Infinity II system Agilent Technologies, Santa Clara, CA, USA), using a Zorbax SB-C18 column (150 mm × 4.6 i.d., 5μm, Agilent Technologies, Santa Clara, CA, USA) operating at 0.8 mL/min constant flow rate (injection volume 100 mL). The mobile phase consisted of 0.3 M phosphoric acid (solvent A) and acetonitrile (solvent B) with the following gradient table: 0–1 min, 25% B; 1–10 min, 25–43% B; 10–12 min, 43–44% B; 12–22 min, 44–90% B; 22–24 min, 90–25% B, and 10 min column reconditioning. Chromatograms were recorded at 200 nm.

A BS 10 mM mixture prepared dissolving NaTC, NaGC, NaTCDC, and NaGCDC in SIF was used to assess the separation efficiency of the HPLC method. To quantify the un-bound BS, a five-point standard curve was prepared for each BS in the concentration range 1–10 mM.

The binding activity was calculated according to Equation (2):

$$\text{Binding Activity } (\%) = [(BS\ blank - BS\ unbound)/BS\ blank] \times 100 \quad (2)$$

where *BS blank* is BS total concentration (expressed in mM) registered after the water digestion process and *BS unbound* is BS concentration (expressed in mM) detected in supernatants after the CCP intestinal digestion phase.

2.11. Statistical Analysis

Statistical analysis of the data was performed using Microsoft Excel (version 365). The significant differences ($p < 0.05$) were evaluated by variance analysis (ANOVA). Experiments were performed at least in three replicates.

3. Results and Discussion

3.1. CCP Molecular Parameters

Considering that the average molecular weight and the molecular weight distribution affect the polysaccharide rheological and functional features [26], these parameters were extrapolated for camelina cake polysaccharides following the slice method reported by Garcia-Lopera et al. (2005) [27], based on data obtained from a SEC-RID system calibrated with pullulans (external standard method).

The registered chromatogram clearly indicated the presence of a varied population of carbohydrate polymers (number average molecular weight, Mn: 3.234×10^3 g/mol) eluting in the range 17–23 min and representing about 100% of the eluted material. CCP molecular distribution weight was extrapolated from the pullulan standard calibration curve, and it was in the range $7.224 \times 103 - 698.297 \times 10^3$ g/mol. The average Mw was about $139.749 \times 10^3 \pm 4.392 \times 10^3$ g/mol; the Pi (Mn/Mw) was 3.26 ± 0.066.

These data and the results previously obtained by Fourier-Transform Infrared Spectroscopy (FT-IR) analysis [17] indicated that CCP Mw and Mn values were far lower than those reported for flaxseed (*Linum usitatis-simum* L.) mucilage, which had a composition and rheological behavior close to those of camelina seed mucilage [15]; however, CCP features were very similar to those obtained for a neutral polysaccharide fraction isolated from flaxseed mucilage. Therefore, we could hypothesize that CCP represented the neutral polysaccharide fraction of camelina mucilage, whose structural features might also be due to a partial degradation of polysaccharide chains, which occurs during the extraction process at temperatures higher than 100 °C [5,28].

3.2. CCP Rheological Properties

The viscoelastic properties of CCP were evaluated by using shear-rate rheology. Firstly, we performed viscosity tests in order to assess the Newtonian/non-Newtonian flow behavior of CCP (Figure 1a). CCP showed non-Newtonian behavior through a decrease in viscosity as the shear-rate increased. The high viscosity of the sample (5.9 Pa·s) at low shear rates ($0.001\ s^{-1}$) provided insights for its long-term stability in the application of hydrocolloidal systems and could be attributed to the intermolecular interactions among protein and polysaccharide molecules, which would result in the formation of entangled networks, similar to those observed for Camelina seed gum by Li et al. [29]. On the other hand, the reduction in viscosity at a high shear rate ($1000\ s^{-1}$) highlighted its shear-thinning propensity, which typical of hydrogel-like materials and was also observed in other food-derived materials (i.e., soybean and lupine peptides) [30].

CCP mechanical properties were evaluated by measuring the storage (G′) and loss (G″) moduli using oscillatory shear rheological experiments (Figure 1b). G′ reflects the stiffness, and G″ represents the energy dissipated during the oscillatory test and correlated with the liquid-like response of the sample. The ratio between G′ and G″ provided insights into the viscoelastic profile, i.e., whether a material behaved as an elastic solid ($G' > G''$) or a viscous liquid ($G' < G''$) [31]. In Figure 1b, G′ (in blue) and G″ (in red) moduli trends of CCP showed typical hydrogel-like profiles, featuring predominant solid-elastic behavior (G′) as compared to the viscous component (G″). Throughout the tested frequency range (0.1–100 Hz) at a fixed strain (0.5%), CCP displayed G′ and G″ mean values of 9.8 Pa and 0.6 Pa, respectively (Figure 1c), in agreement with the data obtained for camelina gum fibers by Li et al., (2016) [29].

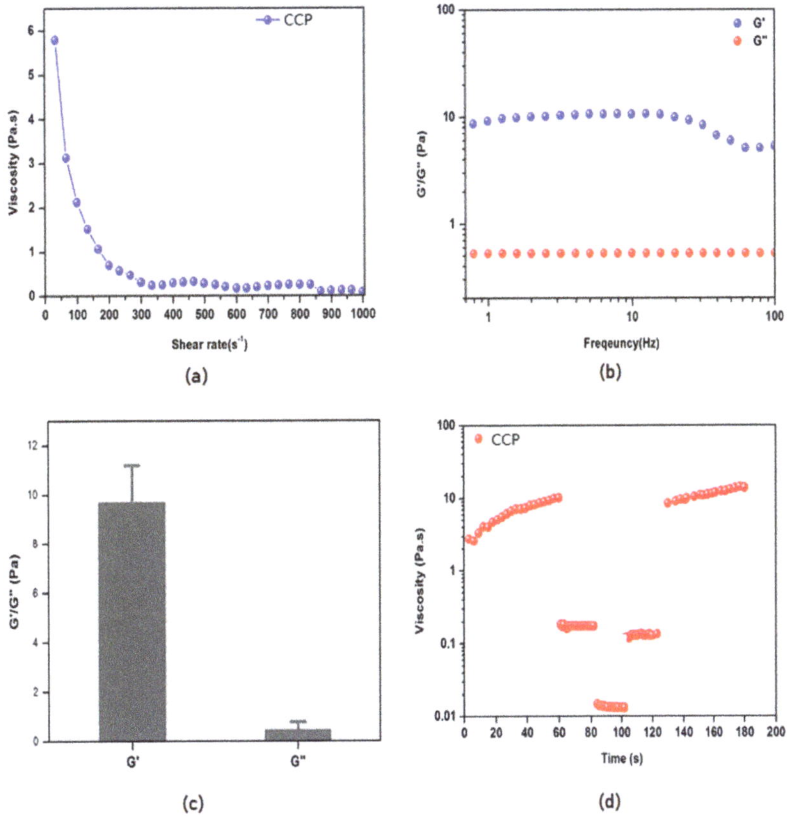

Figure 1. Rheological studies on the viscoelastic properties of CCP. (**a**) Viscosity measurements for increasing the shear rate of CCP. (**b**) Frequency-dependent oscillatory rheology (0.1–100 Hz) of CCP featuring a predominant solid-elastic behavior (G′) as compared to the viscous component (G″), (**c**) Average values of storage (G′—blue color) and loss (G″—red color) moduli obtained from frequency-sweep tests. (**d**) Thixotropy test of CCP solution showing its space-filling propensity.

Lastly, CCP thixotropy (i.e., its propensity to recover the initial viscosity after shear-rate changing, by simulating an injection) was evaluated. CCP exhibited fast recovery after injection simulation through a series of constant shear rate tests (see Materials and Methods for further details) (Figure 1d). This fast viscosity recovery hinted to its space-filling propensity, which allowed CCP gel-like structure to break and then to recover its structure when the stress was removed.

3.3. CCP Stabilizing Effect on Bioaccessibility of Encapsulated MCE Polyphenols

Considering the above results and the effect of CCP in prolonging MCE shelf-life [17], the improvement of MCE phytocomplex bioaccessibility in the ingredients obtained by the encapsulation of MCE with CCP 50% (MCE-CCP50) and CCP 75% (MCE-CCP75) was investigated.

Thirteen different marker compounds selected in MCE were monitored in MCE-CCP50 and MCE-CCP75 at different time intervals during the in vitro simulated digestion process, as reported in Tables 1 and 2, respectively.

Table 1. Bioaccessibility index registered for each selected Moradyn Cob Extract (MCE) polyphenol during in vitro digestion of MCE-CCP50.

	Bioaccessibility Index (%)						
Compound	**Oral Phase**	**Gastric Phase (15′)**	**Gastric Phase (30′)**	**Gastric Phase (1 h)**	**Gastric Phase (2 h)**	**Duodenal Phase (30′)**	**Duodenal Phase (2 h)**
Expected	50	25	25	25	25	12.5	12.5
Cyanidin-3-*O*-glucoside	35.36 ± 0.06	22.18 ± 1.81	22.33 ± 2.59	20.59 ± 1.56	23.12 ± 0.48	4.58 ± 1.25	2.72 ± 0.01
Perlagonidin-3-*O*-glucoside	39.26 ± 2.11	24.35 ± 1.37	23.49 ± 2.54	22.53 ± 1.63	24.12 ± 0.78	6.08 ± 0.62	5.18 ± 0.03
Peonidin-3-*O*-glucoside	48.96 ± 0.48	26.94 ± 1.41	27.45 ± 3.18	26.31 ± 2.26	28.04 ± 0.67	5.53 ± 1.69	2.60 ± 0.26
Ferulic acid derivative	44.87 ± 1.95	21.36 ± 0.13	22.48 ± 1.79	25.38 ± 2.12	23.41 ± 2.29	n.d.	n.d.
Myricetin-7-*O*-hexoside	33.25 ± 2.85	21.05 ± 0.23	24.55 ± 0.64	23.28 ± 0.28	24.46 ± 3.94	12.14 ± 1.13	n.d.
Isorhamnetin-3,7-di-*O*-hexoside	41.82 ± 2.84	23.29 ± 0.01	23.37 ± 0.07	26.78 ± 1.25	24.22 ± 3.34	11.45 ± 1.99	3.66 ± 1.04
Quercetin-7-*O*-*p*-cumaroylhexoside	93.07 ± 1.95	70.28 ± 7.67	79.11 ± 0.91	90.27 ± 3.62	91.34 ± 10.61	33.49 ± 7.23	35.42 ± 4.88
Quercetin-7-*O*-glucoside	64.09 ± 0.66	36.02 ± 2.17	35.97 ± 1.12	39.46 ± 4.00	35.88 ± 1.15	13.80 ± 2.96	10.30 ± 1.86
Kaempferol-7-*O*-(6″-*O*-malonyl)-hexoside	62.69 ± 0.69	32.02 ± 3.80	34.40 ± 1.56	33.68 ± 1.56	33.74 ± 0.40	12.96 ± 3.55	10.48 ± 1.68
Isorhamnetin-7-*O*-rutinoside	59.98 ± 0.46	30.23 ± 0.17	31.81 ± 0.05	51.04 ± 10.65	39.32 ± 6.49	13.11 ± 2.75	14.63 ± 1.87
Isorhamnetin-3-*O*-hexoside	57.57 ± 1.47	25.80 ± 0.82	27.55 ± 1.00	27.86 ± 3.70	27.36 ± 1.53	11.63 ± 3.02	11.64 ± 3.02
Luteolin-7-*O*-glucoside	66.52 ± 1.56	21.48 ± 0.54	23.24 ± 0.19	27.02 ± 0.47	25.86 ± 2.72	10.20 ± 2.69	11.83 ± 1.76
Kaempferol-3-*O*-hexosyl-7-*O*-glucuronilhexoside	66.34 ± 1.85	36.71 ± 1.62	34.68 ± 0.74	35.85 ± 2.17	35.22 ± 2.57	20.18 ± 3.67	19.17 ± 3.45

Note: Following INFOGEST protocol, samples were diluted 1:1 (*v*/*v*) at the beginning of each digestion phase; thus, the expected bioaccessibility index was based on this dilution factor. n.d.: not determined.

Table 2. Bioaccessibility index registered for each selected Moradyn Cob Extract (MCE) polyphenol during in vitro digestion of MCE-CCP75.

Compound	Bioaccessibility Index (%)						
	Oral Phase	Gastric Phase (15′)	Gastric Phase (30′)	Gastric Phase (1 h)	Gastric Phase (2 h)	Duodenal Phase (30′)	Duodenal Phase (2 h)
Expected	50	25	25	25	25	12.5	12.5
Cyanidin-3-*O*-glucoside	56.09 ± 1.77	30.28 ± 4.40	27.75 ± 4.34	24.83 ± 1.49	28.83 ± 0.96	5.35 ± 0.08	2.78 ± 0.07
Perlagonidin-3-*O*-glucoside	56.77 ± 0.13	32.11 ± 3.10	29.19 ± 3.86	26.83 ± 0.03	30.33 ± 2.10	6.34 ± 0.53	2.24 ± 0.19
Peonidin-3-*O*-glucoside	75.24 ± 0.07	38.32 ± 4.80	35.03 ± 5.10	31.98 ± 1.65	36.37 ± 1.82	6.59 ± 0.75	n.d.
Ferulic acid derivative	63.59 ± 0.32	32.02 ± 2.67	24.14 ± 4.41	18.12 ± 1.02	20.06 ± 0.01	n.d.	n.d.
Myricetin-7-*O*-hexoside	46.49 ± 1.89	30.51 ± 1.30	30.76 ± 1.46	23.83 ± 1.04	24.51 ± 0.18	13.06 ± 0.33	2.14 ± 0.82
Isorhamnetin-3,7-di-*O*-hexoside	63.54 ± 1.54	29.06 ± 1.34	30.94 ± 6.35	23.87 ± 1.92	25.05 ± 1.28	16.27 ± 3.38	3.51 ± 0.31
Quercetin-7-*O*-p-cumaroylhexoside	89.89 ± 0.62	57.80 ± 4.00	60.58 ± 2.20	49.51 ± 0.29	53.40 ± 3.28	26.03 ± 5.83	23.50 ± 2.63
Quercetin-7-*O*-glucoside	n.d.	71.07 ± 2.74	66.77 ± 5.37	61.71 ± 0.43	67.24 ± 0.32	24.95 ± 1.86	7.06 ± 2.99
Kaempferol-7-*O*-(6″-*O*-malonyl)-hexoside	70.85 ± 1.43	39.86 ± 4.29	37.27 ± 8.36	30.89 ± 1.02	33.97 ± 1.40	13.27 ± 1.33	6.85 ± 1.72
Isorhamnetin-7-*O*-rutinoside	67.80 ± 1.94	45.17 ± 5.16	35.40 ± 0.35	32.84 ± 0.61	44.30 ± 9.83	26.23 ± 4.00	20.00 ± 2.97
Isorhamnetin-3-*O*-hexoside	64.63 ± 1.32	30.41 ± 3.53	27.60 ± 4.89	23.92 ± 1.05	26.81 ± 1.15	12.22 ± 0.90	10.51 ± 0.28
Luteolin-7-*O*-glucoside	55.98 ± 1.82	26.90 ± 2.16	29.86 ± 10.74	20.93 ± 1.70	25.72 ± 2.94	12.45 ± 1.52	10.62 ± 0.64
Kaempferol-3-*O*-hexosyl-7-*O*-glucuronilhexoside	81.84 ± 2.97	40.21 ± 3.39	31.82 ± 12.33	29.30 ± 2.68	33.13 ± 2.32	27.88 ± 2.57	24.96 ± 1.72

Note: Following the INFOGEST protocol, samples were diluted 1:1 (*v*/*v*) at the beginning of each digestion phase; thus, the expected bioaccessibility index was based on this dilution factor. n. d.: not determined.

The bioaccessibility index was calculated for each of anthocyanin (cyanidin-3-*O*-glucoside, perlagonidin-3-*O*-glucoside, and peonidin-3-*O*-glucoside), flavonol (myricetin-7-*O*-hexoside, isorhamnetin-3,7-di-*O*-hexoside, quere-cetin-7-*O*-*p*-cumaroylhexoside, quercetin-7-*O*-glucoside, kaempferol-7-*O*-(6″-*O*-malonyl)-hexoside, isorhamnetin-7-*O*-rutinoside, isorhamnetin-3-*O*-hexoside, luteolin-7-*O*-glucoside, kaempferol-3-*O*-hexosyl-7-*O*-glucuronilhexoside), and hydroxycinnamic acid (ferulic acid derivative) as the percentage of soluble compound detected in the collected digestive fractions in comparison with that expected following the gradual dilution occurring during the static in vitro digestion procedure [8].

Regarding MCE-CCP50 (Table 1), during the oral and gastric phases, the bioaccessibility indexes for kaempferol, myricetin, and quercetin derivatives were higher than expected, and the following observed reductions during the monitoring period were probably attributable to the dilution occurring during the digestion. The same trend was observed in the MCE-CCP75 fraction (Table 2), suggesting that the release of polyphenols from both the ingredients was very fast during the oral phase and at the beginning of the gastric phase, as showed by the highest bioaccessibility index registered after 15 min of digestion. Conversely, during the intestinal phase, all the registered bioaccessibility indexes were lower than those expected on dilution basis for both the digested ingredients, but still detectable, differently from what previously observed for MCE [8]; therefore, this behavior could suggest that CCP effectively improved MCE bioaccessibility.

The flavonols' release trend registered for CCP-based ingredients agreed with that reported for polysaccharide-based hydrogels [32], probably due to CCP matrix gradual swelling and erosion during the digestion process.

The only exception was registered for quercetin-7-*O*-*p*-cumaroylhexoside: during MCE-CCP50 digestion, its bioaccessibility index was 93.97% after the oral phase, it decreased to 70.28% at the beginning of the gastric phase, and then was 91.34% after 2 h, suggesting that the release rate of this compound was higher than those registered for the other markers and balanced the gradual dilution occurring during the digestion process.

Note: Following the INFOGEST protocol, samples were diluted 1:1 (v/v) at the beginning of each digestion phase; thus, the expected bioaccessibility index was based on this dilution factor.

Conversely, the quercetin-7-*O*-*p*-cumaroylhexoside release trend from MCE-CCP75 was slower during the gastric phase, and its bioaccessibility index was still 53.4% after 2 h under gastric conditions, and it did not balance the dilution factor. Among MCE polyphenols, anthocyanins were the most abundant and representative compounds known for their easy degradation [8]. Various delivery systems were tested to improve their bioaccessibility [9,33–36]. The use of CCP strongly improved the bioaccessibility index for all the anthocyanins present in MCE (Figure 2a–c); in fact, in undigested samples, only the unbound fraction was soluble and bioaccessible, and the registered bioaccessibility index was different according to the anthocyanin and the ingredient content, ranging from 40.68% to 47.93% in MCE-CCP50 and from 3.95% to 9.07% in MCE-CCP75.

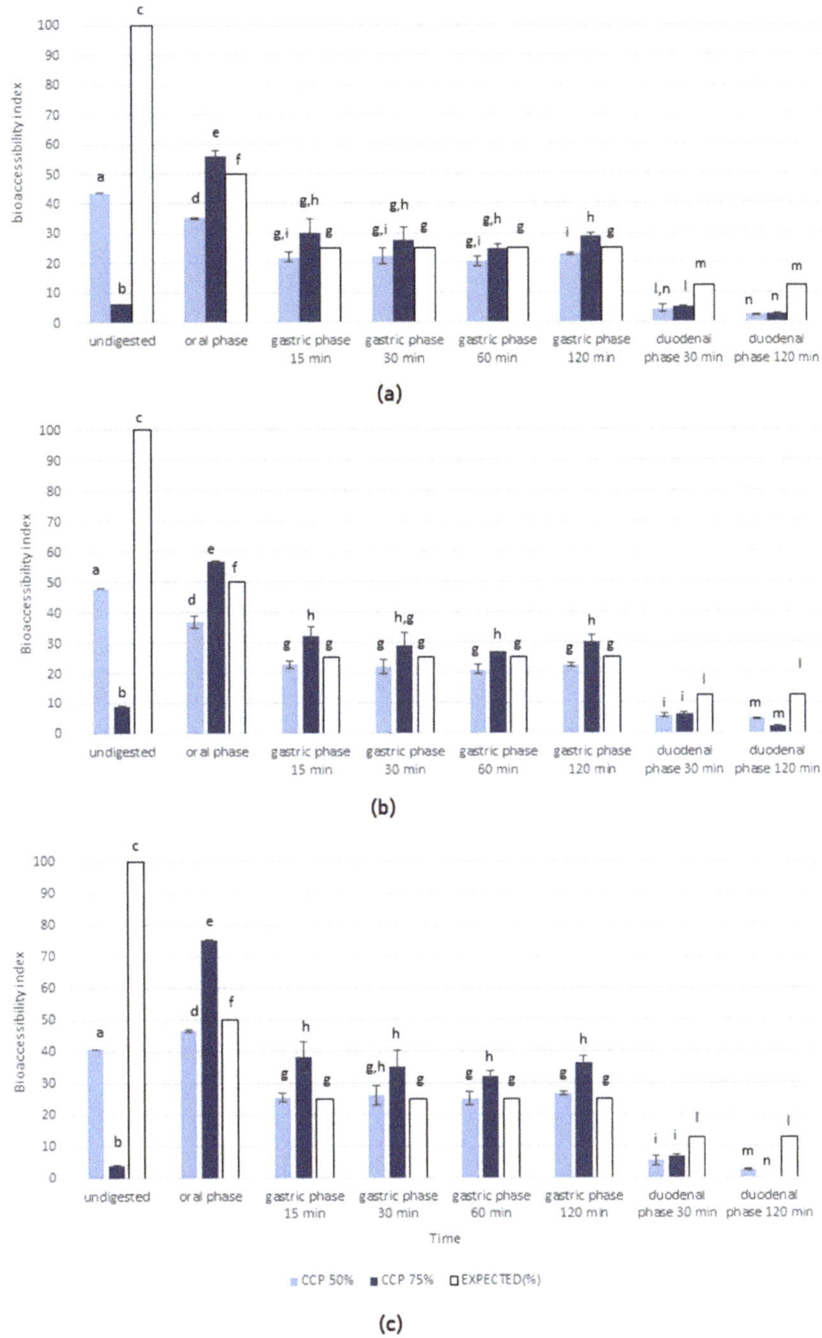

Figure 2. Experimental bioaccessibility index. Values registered for (**a**) cyanidin-3-*O*-glucoside, (**b**) perlagonidin-3-*O*-glucoside, and (**c**) peonidin-3-*O*-glucoside in MCE-CCP50 (light blue bars) and MCE-CCP75 (dark blue bars) at different time points. White bars represented the expected bioaccessibility index on a gradual dilution basis (occurring during the in vitro digestion process). Different lowercase letters indicate significant differences for each compound ($p < 0.05$).

After the oral phase, the three anthocyanins were quickly released and the bioaccessibility index was strongly improved; in particular, in MCE-CCP75, cyanidin-3-*O*-glucoside, perlagonid-in-3-*O*-glucoside, and peonidin-3-*O*-glucoside, increased from 6.34%, 9.07%, and 3.95% to 56.09%, 56.77%, and 75.24% respectively. During the gastric phase, almost all the anthocyanin content was already bioaccessible after 15 min in both the fractions. Under gastric acidic environment, anthocyanins were in flavylium cation form and thus relatively stable, as evident from the constant bioaccessible amount until the end of gastric phase. This behavior was already reported in literature by Norkaew et al. [23] and Huang and Zhou [37], who investigated the bioaccessibility of the anthocyanin fraction from black rice encapsulated in two different Arabic gum-based delivery systems. The release rates of these two ingredients were higher during oral and gastric phases due to the rapid degradation of the carrier by α-amylase and pepsin, as probably occurred for CCP, whose released mechanism was based on swelling and erosion according to our tests. Differently from the Arabic gum-based delivery systems, for which during the intestinal phase, the color turned to a dark blue–green in a few minutes due to the pH value change (responsible for anthocyanins degradation to colorless carbinol pseudobase, which in turn degraded to protocatechuic acid and phloroglucinaldehyde) [23,37], the anthocyanins released from MCE-CCP ingredients were still detectable and quantifiable, and the intestinal simulated fluid was pale pink until the end of digestion process. This could suggest that at the intestinal level, CCP was present in a fully hydrated form, which created a viscous matrix able to affect the polyphenols' diffusion and thus prevent their degradation [12].

Overall, the effects of vegetable carrier agents such as polysaccharides or gums on polyphenols during digestion have been widely investigated [22,38]; and by preserving these compounds from degradation and enhancing their solubility in the intestinal fluids, their bioprotective effects could be maintained, or sometimes enhanced [39].

The results of CCP BS-binding activity at the different tested concentrations are summarized in Table 3. Cholestyramine (10 mg/mL) totally bound 10 mM bile salts, in agreement with literature data [25,40]. Conversely, when CCP was tested at the concentrations used in the ingredients (0.5 mg and 0.75 mg/mL), CCP was not able to trap bile salts, and therefore, the purified polysaccharide fraction could not exert any hypocholesterolemic activity; conversely, at the highest concentrations of 2.5–10 mg/mL, its activity quickly increased, reaching 100%. No significant difference was detected among the four primary bile salts (tested at the same concentration used in the mixture) and their mixture ($p < 0.05$).

Table 3. Camelina Cake Polysaccharide (CCP) binding capacity (%) calculated for each primary bile salt (BS) or for the BS mixture (35% NaTC, 35% NaGC, 15% NaTCDC and 15% NaGCDC).

	Bile Acid Binding Activity (%)				
CCP (mg/mL)	**NaTC**	**NaGC**	**NaTCDC**	**NaGCDC**	**BS Mixture**
0.50	N.A.	N.A.	N.A.	N.A.	N.A.
0.75	N.A.	N.A.	N.A.	N.A.	N.A.
2.50	38.21 ± 2.29	40.26 ± 2.25	41.27 ± 0.04	41.27 ± 2.13	39.88 ± 0.72
5.00	100.00 ± 0.13	100.00 ± 0.08	100.00 ± 0.01	100.00 ± 0.01	100.00 ± 2.97
10.00	100.00 ± 0.01	100.00 ± 0.07	100.00 ± 0.01	100.00 ± 0.01	100.00 ± 1.28

N.A. indicates no activity.

Considering that the experimental setup required a centrifugation step to isolate unbound BS [39], the high activity registered for CCP in the concentration range 2.5–10 mg/mL could be attributed to BS absorption, and these results are similar to those obtained by Gomez, Singh, Acharya, Jayaprakasha, and Patil [41], who investigated the BS binding activity of 3 g/mL of fresh garnet stem leaf powder under the same conditions. However, based on CCP extraction yield [17], 5 mg/mL of the purified polysaccharide fraction corresponded to 50 mg of crude cake; therefore, its binding activity should be considered far higher than the activity reported in literature for other matrices, which were usually crude flours or dietary-fiber-enriched food ingredients [38,39,41].

3.4. Bile Salts Binding Capacity of CCP

Another goal of this study was to investigate a putative functional property of the CCP-based ingredient by assessing the relationships between the CCP's structural and rheological properties and its capacity to retain BS during in vitro simulated digestion.

Soluble and insoluble polysaccharides were reported to interact with bile salts, preventing their reabsorption and promoting their transit to the colon, thereby increasing the hepatic synthesis of primary bile acids and reducing serum cholesterol synthesis [39].

CCP BS-binding ability was tested at five different concentrations in the range 0.5–10 mg/mL and compared with that of cholesteryramine, a known synthetic bile acid sequestrant [25,40]. An HPLC method that had been previously setup and validated was applied, and the quantification was performed using the external standard method [20]. The calibration curve constructed for NaTC, NaGC, NaTCDC, and NaGCDC had $R^2 > 0.990$, and in Figure 3, the profile of the standard mixture is reported.

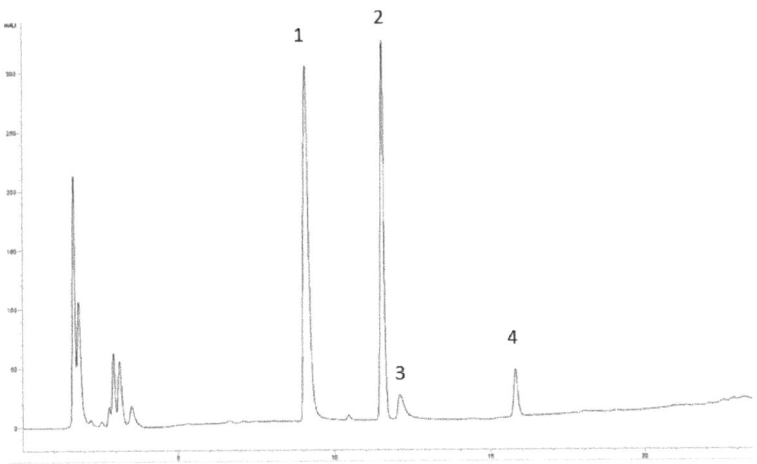

Figure 3. Chromatographic profile of BS mixture in SIF tested at 10 mM final concentration, registered at 208 nm. **(1)** NaTC, **(2)** NaGC, **(3)** NaTCDC, **(4)** NaGCDC.

4. Conclusions

CCP was the alcohol-insoluble polysaccharide fraction previously isolated from camelina cake, the main by-product generated from camelina oil production, able to pro-long the shelf life of an MCE-based ingredient when used at 50% and 75% (*w*/*w*). The structural and rheological properties of CCP were deeply investigated in this work, confirming that its carbohydrate fraction is mainly composed by neutral polysaccharides with a broad molecular weight distribution around 139.749×103 g/mol. Moreover, CCP showed a typical hydrogel-like profile, resulting in the formation of an entangled network of proteins and polysaccharides with a thixotropic feature. This property could gain it attention in the food or food supplement field, as it could allow the controlled delivery of bioactive compounds.

Furthermore, these rheological properties could justify the results obtained from in vitro digestion of the two MCE-CCP50 and MCE-CCP75 ingredients. In fact, the bioaccessibility index registered for selected polyphenols in MCE highlighted that CCP, at both 50% and 75%, allowed the gradual delivery of such compounds during oral and gastric phases by a swelling mechanism, which was already known of by a composite gel obtained from the interaction of lotus root extract and whey proteins.

Considering CCP's structural and rheological features and the results obtained from solid-state stability and bioaccessibility studies performed on MCE-CCP50 and MCE-CCP75, it could be concluded that CCP can supply a protective barrier for MCE polyphenols, increasing their storage stability and bioaccessibility. Moreover, CCP, by improving anthocyanins' stability and keeping constant their concentrations during the digestion process, is supposed to preserve the antioxidant potency of MCE until the intestine.

Therefore, since the ingredient containing CCP 75% represents a valuable solution for stabilizing MCE polyphenols and enhancing their bioaccessibility, hypoglycemic and anti-glycative in vitro tests will be carried out after digestion in order to fully characterize the potential efficacy of this final ingredient in the prevention of chronic and age-related disease risk factors.

Author Contributions: Conceptualization, A.P., C.M. and L.F.; methodology, C.M., R.P. and L.F.; software, R.P. and L.F.; validation, L.F., C.M. and A.P.; formal analysis, C.M., R.P. and L.F.; investigation, L.F., C.M., R.P., R.C. and A.P.; data curation, C.M., R.P. and L.F.; writing—original draft preparation, L.F., R.P., C.M. and A.P.; writing—review and editing, A.P. and C.M.; supervision, A.P. All authors have read and agreed to the published version of the manuscript.

Funding: This research received no external funding.

Institutional Review Board Statement: Not applicable.

Informed Consent Statement: Not applicable.

Data Availability Statement: The data presented in this study are available in the present article.

Conflicts of Interest: The authors declare no conflict of interest. Flanat Research only host in its lab PhD student in order to perform some tests. There is no potential conflict of interest.

References

1. Nayak, A.; Bhushan, B. An overview of the recent trends on the waste valorization techniques for food wastes. *J. Environ. Manag.* **2019**, *233*, 352–370. [CrossRef] [PubMed]
2. Rao, P.; Rathod, V. Valorization of food and agricultural waste: A step towards greener future. *Chem. Rec.* **2018**, *18*, 1858–1871. [CrossRef] [PubMed]
3. European Commission. A new Circular Economy Action Plan For a Cleaner and More Competitive Europe. Communication from the Commission to the European Parliament, the Council, the European Economic and Social Committee and the Committee of the Regions. 2020. Available online: https://eur-lex.europa.eu/legal-content/EN/TXT/?uri=COM:2020:98:FIN&WT.mc_id=Twitter (accessed on 29 April 2022).
4. Imman, S.; Laosiripojana, N.; Champreda, V. Effects of liquid hot water pretreatment on enzymatic hydrolysis and physicochemical changes of corncobs. *Appl. Biochem. Biotechnol.* **2018**, *184*, 432–443. [CrossRef] [PubMed]
5. Zhang, Q.; de Mejia, E.G.; Luna-Vital, D.; Tao, T.; Chandrasekaran, S.; Chatham, L.; Juvik, J.; Singh, V.; Kumar, D. Relationship of phenolic composition of selected purple maize (*Zea mays* L.) genotypes with their anti-inflammatory, anti-adipogenic and anti-diabetic potential. *Food Chem.* **2019**, *289*, 739–750. [CrossRef] [PubMed]
6. Yang, Z.; Zhai, W. Identification and antioxidant activity of anthocyanins extracted from seed and cob of purple corn (*Zea mays* L.). *Innov. Food Sci. Emerg. Technol.* **2010**, *11*, 169–176. [CrossRef]
7. Li, C.Y.; Kim, H.W.; Won, S.R.; Min, H.; Park, K.J.; Park, J.Y.; Ahn, M.S.; Rhee, H.I. Corn husk as a potential source of anthocyanins. *J. Agric. Food Chem.* **2008**, *56*, 11413–11416. [CrossRef]
8. Ferron, L.; Colombo, R.; Mannucci, B.; Papetti, A. A new Italian purple corn variety (Moradyn) byproduct ex-tract: Antiglycative and hypoglycemic in vitro activities and preliminary bioaccessibility studies. *Molecules* **2020**, *25*, 1958. [CrossRef]
9. Costantin, O.E.; Stănciuc, N.; Yan, Y.; Ghinea, I.O.; Ungureanu, C.; Cîrciumaru, A.; Wang, D.; Poklar Ulrihe, N.; Râpeanua, G. Polymers and protein-associated vesicles for the microencapsulation of anthocyanins from grape skins used for food applications. *J. Sci. Food Agric.* **2020**, *101*, 2676–2686. [CrossRef]
10. Jafari, S.M.; Mahdavi-Khazaei, K.; Hemmati-Kakhki, A. Microencapsulation of saffron petal anthocyanins with cress seed gum compared with Arabic gum through freeze drying. *Carbohydr. Polym.* **2016**, *140*, 20–25. [CrossRef]
11. Pieczykolan, E.; Kurek, M.A. Use of guar gum, gum arabic, pectin, beta-glucan and inulin for microencapsulation of anthocyanins from chokeberry. *Int. J. Biol. Macromol.* **2019**, *129*, 665–671. [CrossRef]
12. Fathi, M.; Martin, A.; McClements, D.J. Nanoencapsulation of food ingredients using carbohydrate based delivery systems. *Trends Food Sci. Technol.* **2014**, *39*, 18–39. [CrossRef]
13. Mahdavi, S.A.; Jafari, S.M.; Assadpoor, E.; Dehnad, D. Microencapsulation optimization of natural anthocyanins with maltodextrin, gum Arabic and gelatin. *Int. J. Biol. Macromol.* **2016**, *85*, 379–385. [CrossRef] [PubMed]

14. Taheri, A.; Jafari, S.M. Gum-based nanocarriers for the protection and delivery of food bioactive compounds. *Adv. Colloid Interface Sci.* **2019**, *269*, 277–295. [CrossRef]
15. Sarv, V.; Trass, O.; Diosady, L.L. Preparation and characterization of Camelina sativa protein isolates and mucilage. *J. Am. Oil Chem. Soc.* **2017**, *94*, 1279–1285. [CrossRef]
16. Ding, H.H.; Cui, S.W.; Goff, H.D.; Wang, Q.; Chen, J.; Han, N.F. Soluble polysaccharides from flaxseed kernel as a new source of dietary fibers: Extraction and physicochemical characterization. *Food Res. Int.* **2014**, *56*, 166–173. [CrossRef]
17. Ferron, L.; Milanese, C.; Colombo, R.; Pugliese, R.; Papetti, A. Selection, structural characterization and solid-state stability studies of an innovative alcohol insoluble polysaccharide fraction stabilizing purple corn cob extract. *Antioxidants* **2022**, *11*, 916. [CrossRef]
18. Maric, M.; Grassino, A.N.; Zhu, Z.; Barba, F.J.; Brnčić, M.; Brnčić, S.R. An overview of the traditional and inno-vative approaches for pectin extraction from plant food wastes and by-products: Ultrasound-, microwaves-, and enzyme-assisted extraction. *Trends Food Sci. Technol.* **2018**, *76*, 28–37. [CrossRef]
19. Naupane, S.; Bittkau, K.S.; Alban, S. Size distribution and chain conformation of six different fucoidans using size-exclusion chromatography with multiple detection. *J. Chromatogr. A* **2020**, *1612*, 460658–460669. [CrossRef]
20. Minekus, M.; Alminger, M.; Alvito, P.I.S.; Bohn, T.; Bourlieu, C.; Carrière, F.; Boutrou, R.; Corredig, M.; Dupont, D.; Dufour, C.; et al. A standardised static in vitro digestion method suitable for food—An international consensus. *Food Funct.* **2014**, *5*, 1113–1124. [CrossRef]
21. Lin, T.; O'Keefe, S.; Duncan, S.; Fernàndez-Fraguas, C. Retention of primary bile salts by dry beans (*Phaseolus vulgaris* L.) during in vitro digestion: Role of bean components and effect of food processing. *Food Res. Int.* **2020**, *137*, 109337–109351. [CrossRef]
22. Pugliese, R.; Arnoldi, A.; Lammi, C. Nanostructure, Self-Assembly, mechanical properties, and antioxidant activity of a lupin-derived peptide hydrogel. *Biomedicines* **2021**, *13*, 294. [CrossRef] [PubMed]
23. Norkaew, O.; Thitisut, P.; Mahatheeranont, S.; Pawin, B.; Sookwong, P.; Yodpitak, S.; Lungkaphin, A. Effect of wall materials on some physicochemical properties and release characteristics of encapsulated black rice anthocyanin microcapsules. *Food Chem.* **2019**, *294*, 493–502. [CrossRef] [PubMed]
24. Ferron, L.; Milanese, C.; Colombo, R.; Papetti, A. Development of an accelerated stability model to estimate pur-ple corn cob extract powder (Moradyn) shelf-life. *Foods* **2021**, *10*, 1617. [CrossRef] [PubMed]
25. Hamauzu, Y.; Suwannachot, J. Non-extractable polyphenols and in vitro bile acid-binding capacity of dried persimmon (*Diospyros kaki*) fruit. *Food Chem.* **2019**, *293*, 127–133. [CrossRef]
26. Chouana, T.; Pierre, G.; Vial, C.; Gardarin, C.; Wadouachi, A.; Cailleuc, D.; Le Cerf, D.; Bouala, Z.; Ould El Hadja, M.D.; Michaud, P.; et al. Structural characterization and rheological properties of a galactomannan from *Astragalus gombo* Bunge seeds harvested in Algerian Sahara. *Carbohydr. Polym.* **2017**, *175*, 387–394. [CrossRef]
27. Garcia-Lopera, R.; Codoner, A.; Bano, M.C.; Abad, C.; Campos, A. Size exclusion chromatographyc determination of polymer molar mass averages using a fractal calibration. *J. Chromatogr. Sci.* **2005**, *43*, 226–234. [CrossRef]
28. Elboutachfaiti, R.; Delattre, C.; Quéro, A.; Roulard, R.; Duchêne, J.; Mesnard, F.; Petit, E. Fractionation and structural characterization of six purified rhamnogalacturonans type I from flaxseed mucilage. *Food Hydrocholl.* **2017**, *62*, 273–279. [CrossRef]
29. Li, N.; Qi, G.; Sun, S.X.; Wang, D. Characterization of gum isolated from Camelina seed. *Ind. Crops Prod.* **2016**, *83*, 268–274. [CrossRef]
30. Pugliese, R.; Bollati, C.; Gelain, F.; Arnoldi, A.; Lammi, C. A supramolecular approach to develop new soybean and lupin peptide nanogels with enhanced dipeptidyl peptidase IV (DPP-IV) inhibitory activity. *J. Agric. Food Chem.* **2019**, *67*, 3615–3623. [CrossRef]
31. Pugliese, R.; Beltrami, B.; Regondi, S.; Lunetta, C. Polymeric biomaterials for 3D printing in medicine: An over-view. *Ann. 3D Print. Med.* **2021**, *2*, 100011–100021. [CrossRef]
32. Bajaj, S.R.; Marathe, S.J.; Singhal, R.S. Co-encapsulation of vitamins B12 and D3 using spray drying: Wall material optimization, product characterization, and release kinetics. *Food Chem.* **2021**, *335*, 127642–127650. [CrossRef] [PubMed]
33. Liu, K.; Kong, X.; Li, Q.; Zhang, H.; Zha, X. Stability and bioavailability of vitamin D3 encapsulated in composite gels of whey protein isolate and lotus root amylopectin. *Carbohydr. Polym.* **2020**, *227*, 115337–115346. [CrossRef] [PubMed]
34. Fredes, C.; Osorio, M.J.; Parada, J.; Robert, P. Stability and bioaccessibility of anthocyanins from maqui (*Aristotelia chilensis* [Mol.] Stuntz) juice microparticles. *LWT—Food Sci. Technol.* **2018**, *91*, 549–556. [CrossRef]
35. Moser, P.; Nicoletti Telis, V.R.; De Andrade Neves, N.; Garcìa-Romero, E.; Gòmez-Alonso, S.; Hermosìn-Gutiérrez, I. Storage stability of phenolic compounds in powdered BRS Violeta grape juice microencapsulated with protein and maltodextrin blends. *Food Chem.* **2017**, *214*, 308–318. [CrossRef]
36. Xiong, J.; Hsuan Chan, Y.; Rathinasabapathy, T.; Gracea, M.H.; Komarnytskya, S.; Lila, M.A. Enhanced stability of berry pomace polyphenols delivered in protein polyphenol aggregate particles to an in vitro gastrointestinal digestion model. *Food Chem.* **2020**, *331*, 127279–127287. [CrossRef]
37. Huang, Y.; Zhou, W. Microencapsulation of anthocyanins through two-step emulsification and release characteristics during in vitro digestion. *Food Chem.* **2019**, *278*, 357–363. [CrossRef]
38. Guo, R.; Chang, X.; Guo, X.; Brennan, C.S.; Li, T.; Fu, X.; Liu, R.H. Phenolic compounds, antioxidant activity, antiproliferative activity and bioaccessibility of sea buckthorn (*Hippophanaë rhamnoides* L.) berries as affected by in vitro digestion. *Food Funct.* **2017**, *8*, 4229–4240. [CrossRef]

39. Naumann, S.; Schweiggert-Weisz, U.; Bader-Mittermaier, S.; Haller, D.; Eisner, P. Differentiation of adsorptive and viscous effects of dietary fibers on bile acid release by means of In vitro digestion and dialysis. *Int. J. Mol. Sci.* **2018**, *19*, 2193. [CrossRef]
40. Khalon, T.S.; Milczarek, R.R.; Chiu, M.M. In vitro bile acid binding of mustard greens, kale, broccoli, cabbage and green bell pepper improves with sautéing compared with raw or other methods of preparation. *Food Sci. Nutr.* **2012**, *3*, 951–958. [CrossRef]
41. Gomez, M.K.; Singh, J.; Acharya, P.; Jayaprakasha, G.K.; Patil, S.B. Identification and quantification of phytochemicals, antioxidant activity, and bile acid-binding capacity of garnet stem dandelion (*Taraxacum officinale*). *J. Food Sci.* **2018**, *83*, 1569–1578. [CrossRef]

foods

Article

Effects of Betanin on Pasting, Rheology and Retrogradation Properties of Different Starches

Taotao Dai [1,2], Xiaohong He [3], Jiahui Xu [3], Qin Geng [3], Changhong Li [3], Jian Sun [1,2], Chengmei Liu [3], Jun Chen [3,*] and Xuemei He [1,2,*]

1 Guangxi Academy of Agricultural Sciences, Nanning 530007, China; ncubamboo@163.com (T.D.); jiansun@gxaas.net (J.S.)
2 Guangxi Key Laboratory of Fruits and Vegetables Storage-Processing Technology, Nanning 530007, China
3 State Key Laboratory of Food Science and Technology, Nanchang University, Nanchang 330047, China; hexiaohongmaimai@163.com (X.H.); 412314919061@email.ecu.edu.cn (J.X.); 357900210011@email.ecu.edu.cn (Q.G.); lichanghong@163.com (C.L.); liuchengmei@ncu.edu.cn (C.L.)
* Correspondence: chenjun@ncu.edu.cn (J.C.); xuemeihe1981@126.com (X.H.)

Abstract: As a natural pigment with high antioxidative activity, betanin is underutilized owing to less attention. This study aimed to investigate the impact of betanin on pasting, rheology and retrogradation properties of rice, potato and pea starches. Betanin decreased the peak, trough and final viscosity of rice and potato starches, but increased those of pea starch. Rheology measurements implied that betanin had the greatest effect on the hysteresis loops and dynamic modulus of potato starch. Betanin endowed starch pastes with a vivid red appearance and maintained the color of the starch pastes during storage. XRD analysis indicated that betanin weakened the diffraction intensities and reduced the crystallinity of the retrograded starches. Meanwhile, betanin reduced the short-range ordered structure of the retrograde starches. The results of DSC analysis found that betanin significantly depressed the retrogradation enthalpy and retrogradation rate, implying that the long-term retrogradation of starches was delayed. Furthermore, the changed morphology of the retrograded starches was observed. These results suggested that betanin could be applied as an excellent colorant and inhibitor of retrogradation in foods such as bread and pastry products.

Keywords: betanin; starch; gelatinization; rheology; retrogradation

Citation: Dai, T.; He, X.; Xu, J.; Geng, Q.; Li, C.; Sun, J.; Liu, C.; Chen, J.; He, X. Effects of Betanin on Pasting, Rheology and Retrogradation Properties of Different Starches. *Foods* **2022**, *11*, 1600. https://doi.org/10.3390/foods11111600

Academic Editor: Alessandra Marti

Received: 11 May 2022
Accepted: 26 May 2022
Published: 29 May 2022

1. Introduction

Natural pigments have attracted more and more attention owing to the rejection by consumers of synthetic colorants and their adverse effects [1]. Betalains are a type of natural pigments which are commercially obtained from red beet root. In addition, red dragon fruit and pitaya are also abundant sources of betalains, which can be explored as viable alternatives [2]. Betanin is a major betalain and a soluble pigment that is also present in the pitaya fruit [3]. Meanwhile, betanin is a nontoxic betalain approved for use in foods [4] that can be used as a food colorant. In addition, betalains, including betanin, have antioxidant properties [5]. However, previous research about applying betalains mainly concentrated on their stability, due to their sensitivity to thermal and photochemical decomposition. For example, betalains were applied in dairy products, such as cow milk [6] and ice-cream [7], because these products are commonly stored under chilled conditions and could maintain a higher pigment stability. In order to expand the application of this excellent resource, other functions of betalains remain to be explored in addition to being used as a colorant.

Starches serve as an important ingredient, which have been applied in many processed foods such as bakery products, noodles, instant foods and snacks [8]. The physicochemical properties of starches, such as pasting, rheological and retrogradation behaviors, are the main parameters that determine their technological properties and the qualities of their end products. Nevertheless, native starches have some disadvantages, such as weak shear resistance, undesired paste consistency and easy retrogradation. These insufficient characteristics of native starches limited their practical applications in the food industry to a certain extent [9]. Briefly, the retrogradation of starches profoundly affects the textural attributes, shelf life and acceptability of starch-based products [10]. In order to suit specific applications, modification methods, such as chemical, enzymatic and physical methods or some combination of these, were reported to regulate the properties of starches [11]. Recently, plant bioactive substances such as phenolics and plant extracts have been increasingly incorporated to adjust the physicochemical properties of starches. Yu et al. [10] investigated the influences of the selected phenolic acids (cinnamic acid, caffeic acid and ferulic acid) on the retrogradation of corn starch, and found that phenolic acids inhibited the retrogradation of corn starch via the interactions of hydrogen bonding or hydrophobic interaction. Wu et al. [12] suggested that green tea polyphenols reduced pasting attributes, gelatinization enthalpy and the retrogradation degree, as well as improved the freeze–thaw stability of rice starch. Wang et al. [13] reported on the inhibition effect of three common proanthocyanidins (grape seed proanthocyanidins, peanut skin proanthocyanidins and pine bark proanthocyanidins) on the retrogradation properties of maize starch, which was reflected by lowering the melting enthalpy and degree of relative crystallinity. Nevertheless, to our best knowledge, there is little information about the impacts of betanin on the physicochemical properties of starches. Therefore, the aim of this study was to investigate the effects of betanin on pasting, rheology and retrogradation behaviors of rice, potato and pea starches, and the co-gelatinization of betanin and starches was selected as the treatment condition. The results of the study probably provide some useful knowledge for improving the quality of starchy foods, and open up new ideas for the potential utilization of betanin.

2. Materials and Methods

2.1. Material

Betanin was purchased from Yuanye Bio-Technology Co., Ltd. (Shanghai, China). Rice starch (RS, 10.64% moisture, 0.92% fat, 22.49% amylose, *w*/*w*) was obtained from Sigma-Aldrich Inc. (St. Louis., MO, USA). Potato starch (PoS, 9.78% moisture, fat not detected, 26.54% amylose, *w*/*w*) and pea starch (PeS, 10.47% moisture, 0.49% fat, 38.67% amylose, *w*/*w*) were purchased from Rogate Starch Company (Jiangsu, China). All other chemical reagents were of analytical grade and supplied by Aladdin Chemical Company (Shanghai, China). Distilled water was used throughout the experiments.

2.2. Rapid Viscosity Analysis (RVA)

The pasting properties of starches (RS, PoS and PeS) and starch–betanin mixtures were investigated by using an RVA (D8-ADVANCE, Perten, Sweden) according to a previous study [14]. Briefly, the starch sample (2 g) was mixed with betanin at a dosage of 0% or 5% (*w*/*w*) and added to 20 mL of distilled water. The "standard 2" thermal program offered by the supplier was used to determine pasting properties. The samples were held at 50 °C for 60 s, then heated to 95 °C within 225 s and maintained at 95 °C for 150 s, cooled to 50 °C at the same rate as heating, and held at 50 °C for 120 s to develop the final paste viscosity. The rotational speed was set at 960 rpm for the initial 10 s, then changed to 160 rpm. The pasting curves were obtained and parameters were recorded, including peak viscosity (PV), trough viscosity (TV), final viscosity (FV), breakdown (BD) and setback (SB). Some of the gelatinized samples were transferred for rheological and chromaticity value analysis, and others were reserved at 4 °C for 7 days to prepare the retrograded samples. In addition, the same mass ratio of starch to betanin (20:1) was used for DSC analysis of starch–betanin samples.

2.3. Rheological Measurements

The rheological characteristics of starches (RS, PoS and PeS) and starch–betanin pastes were analyzed by an MCR 302 rheometer (Anton-Paar, Graz, Austria) with a parallel-plate measuring system, according to He et al. [15]. In brief, the gels of starch and starch–betanin pastes obtained from Section 2.2 were transferred to a rheometer plate with a probe type of PP50 and a gap of 1 mm. The pastes were equilibrated at ambient temperature for 5 min before measurement.

2.3.1. Steady Shear Analysis

The changes in shear stress of the samples were measured within the range of increasing shear rate from 0.01 to 1000 s^{-1} and then decreasing shear rate from 1000 to 0.01 s^{-1} by referencing the method of Zhu et al. [16]. The total area of the hysteresis loops, referring to the region of shuttle between the up and down curve of the fluid properties, was integrated using the Origin software (Version 8.0, Microcal Inc., Northampton, MA, USA). The obtained curve was fitted with the power law model for fitting:

$$\sigma = K \cdot \gamma^n$$

where σ is the shear stress (Pa), γ is the shear rate (s^{-1}), K is the consistency coefficient ($Pa \cdot s^n$), and n is the flow behavior index ($n < 1$ for a shear-thinning fluid and $n = 1$ for a Newtonian fluid).

2.3.2. Dynamic Rheological Analysis

Firstly, deformation sweep tests were carried out to determine the maximum deformation attainable by all samples, and the strain γ ranged from 0.01% to 100% at a constant frequency of 1 Hz. The linear viscoelastic region for all samples was in the strain range of 0.1~1.6%. An oscillatory frequency sweep measurement was conducted at room temperature with 1% strain (within the linear viscoelastic (LVE) region of all samples), and a frequency range of 0.1~20.8 Hz was selected according to the previous method [17,18]. The storage modulus (G′) and loss modulus (G″) were obtained, and the loss factor tanδ (G″/G′) was calculated according to the modulus.

2.4. The Chromaticity Value Analysis

The chromaticity value of the samples was measured by a colorimeter (CM-5, Ke Sheng Instrument Co., Ltd., Shanghai). Before measurement, the instrument was calibrated with the black and the white board. During measurement, the 10 g sample obtained from Section 2.2 was moved into the sample plate to record the L^* value (brightness), a^* value (red-green) and b^* value (yellowish-blue).

2.5. X-ray Diffraction (XRD)

The crystalline properties of the native starches, retrograded starches and starch–betanin samples were determined by a diffractometer (D8 ADVANCEX, BRUKER, Germany) according to a previous study [19]. The samples were scanned from 5 to 40° (2θ) at a scan step size of 0.02° (2θ). The relative crystallinity was calculated as the ratio of the area of the crystal region to total area using the Origin software with the following equation:

$$Relative\ crystallinity\ (\%) = \frac{A_c}{A_c + A_a} \times 100$$

where A_c and A_a represent the crystalline and amorphous areas.

2.6. Fourier Transform-Infrared (FT-IR) Spectroscopy

Spectroscopic properties of starches, betanin and starch–betanin samples were characterized by using an FT-IR spectrometer (Thermo Nicolet-5700, Nicolet, Rhinelander, WI, USA) according to the study by Li et al. [20]. The retrograded starch or starch–betanin samples (1–3 mg) were ground with KBr (140 mg) with an agate mortar, and then compressed into disk-shaped pellets. The FT-IR spectra were recorded over the range of 4000 to 400 cm^{-1}. Raw spectra were deconvoluted by using Omnic 8.0 software to obtain 1047/1022 cm^{-1} and 995/1022 cm^{-1} values.

2.7. Differential Scanning Calorimetry (DSC)

The thermal properties of starches (RS, PoS and PeS) and starch–betanin samples were analyzed by using DSC (7000X, HITACHI, Japan) based on a previous study [21]. Samples (2–3 mg) were accurately weighed and placed in an aluminum pan. Distilled water was added to the pan, and the mass ratio of sample/water was 1:2. The samples were heated from 40 to 100 °C at a rate of 10 °C/min, and the empty pan was used as the reference. The onset (T_o), peak (T_p), conclusion (T_c) temperature and gelatinization enthalpy (ΔH_g) were obtained from the DSC curve. The retrograded samples obtained from Section 2.2 were lyophilized and milled. Then the retrograded starch–betanin samples were reheated in the same conditions to determine the retrogradation enthalpy (ΔH_r). Finally, the degree of retrogradation rate R was calculated according to the ratio of ΔH_r and ΔH_g.

2.8. Scanning Electron Microscopy (SEM)

Morphological properties of the retrograded starches and starch–betanin samples were observed using SEM (JSM 6701F, JEOL, Japan). The longitudinal section of each freeze-dried sample was fixed on a metal sample holder using double-backed cellophane tape, and then sprayed with a layer of gold to a level of 250–500 nm at an operating voltage of 5 kV. The magnification was 100 times.

2.9. Statistical Analyses

The results were expressed as means ± standard deviation (SD) of triplicate analyses for each sample. Data analysis adopted Duncan's test using SPSS 24.0 statistical software, and differences were considered to be significant at $p < 0.05$.

3. Results and Discussion

3.1. Pasting Properties of Starches

The pasting properties of different starches and starch–betanin samples were shown in Figure 1 and Table 1. RS and PoS exhibited typical pasting curves with significant convex and concave peaks. Upon heating, the starches swelled early and leached out amylose, which resulted in an increase in viscosity and displaying PV. The PV value of RS and PoS was 1032 and 3839 mPa·s, respectively. When constantly shearing at 95 °C, starch granules were disintegrated, and the BD value represented the degree of disintegration. The BD value of RS and PoS was 124 and 2353 mPa·s, respectively. Subsequently, cooling promoted the rearrangement of amylose molecules, and short-term retrogradation occurred. However, in the pasting patterns of PeS, the concave peak was extremely weak (Figure 1) and the TV value was close to that of the PV, accompanied by a very small BD value. Generally, pea starches were characterized by a high amylose content [22]. As reported by Han et al. [23,24], pasting properties of starches are related to their amylose content. The discrepancy of pasting patterns between RS, PoS and PeS may be affected by their amylose content, resources and other inherent characteristics of starch.

For starch–betanin samples, the combination with betanin affected pasting properties in varying degrees, and the change in pasting parameters for PoS was the biggest. Meanwhile, betanin lowered the PV, TV, BD and FV of RS and PoS. On the contrary, these parameters of PeS were slightly increased. The carboxylate groups in betanin interacted with phosphate groups in PoS, which impeded the absorption of water, swelling of PoS, and leaching out of amylose, thus significantly reducing the viscosity parameters of PoS. In comparison with PeS, RS, with a relatively high content of lipids, was prone to form a complex with betanin; this situation was also not conducive for RS to absorb water and gelatinize. Nevertheless, PeS had a high amylose content and steric hindrance owing to the presence of betanin was beneficial for the interactions between amylose molecules during gelatinization. Despite the increasing or decreasing of the viscosity parameters of starches, the combination with betanin just slightly elevated the SB value of the three starches, indicating the weak ability of betanin to inhibit the short-term retrogradation of starches.

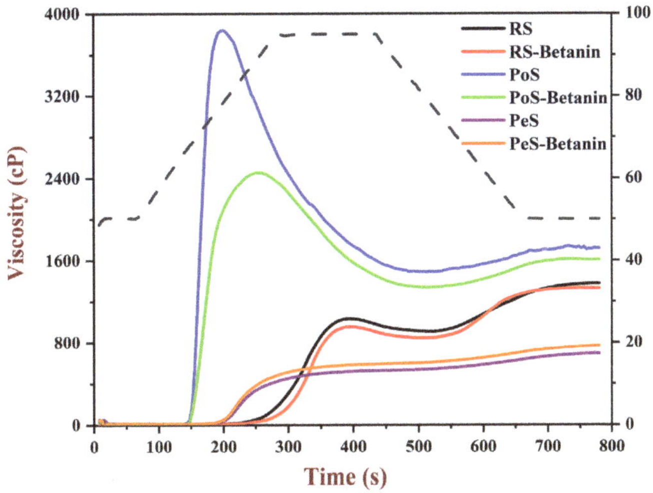

Figure 1. RVA pasting profiles of starches and starch–betanin samples. RS: rice starch; PoS: potato starch; PeS: pea starch.

Table 1. Pasting properties of starches and starch–betanin samples.

Samples	Peak Viscosity (mPa·s)	Trough Viscosity (mPa·s)	Breakdown (mPa·s)	Final Viscosity (mPa·s)	Setback (mPa·s)
RS	1032 ± 4 c	908 ± 8 c	124 ± 7 c	1375 ± 8 c	467 ± 1 b
RS–betanin	953 ± 8 d	845 ± 6 d	108 ± 10 d	1328 ± 6 d	483 ± 5 a
PoS	3839 ± 18 a	1486 ± 21 a	2353 ± 10 a	1716 ± 15 a	230 ± 12 d
PoS–betanin	2490 ± 11 b	1339 ± 2 b	1151 ± 6 b	1611 ± 9 b	272 ± 9 c
PeS	522 ± 8 f	506 ± 6 f	16 ± 6 e	692 ± 6 f	186 ± 5 f
PeS–betanin	586 ± 6 e	568 ± 2 e	18 ± 4 e	767 ± 3 e	199 ± 5 e

Differences between values, indicated by different letters in the same columns, are significant at 0.05 level of confidence. RS: rice starch; PoS: potato starch; PeS: pea starch.

3.2. Rheological Properties of Starches

3.2.1. Steady Shear Rheological Properties

The steady shear rheology curves of starch and starch–betanin pastes are shown in Figure 2A. RS and PoS exhibited typical upward and downward curves as a function of shear rate, and hysteresis loops were observed. However, PeS showed an irregular profile of shear stress, and it was difficult to obtain a hysteresis loop. It is likely that PeS pastes easily and rapidly formed a fragile gel during measurement due to their high content of amylose. The shear stress of RS–betanin and PoS–betanin pastes were significantly lower than that of RS and PoS at the shear rate ranges of 0.01~1000 s^{-1} and 1000~0.01 s^{-1}; that is, the corresponding apparent viscosity of RS and PoS was lowered when combined with betanin. As for PeS–betanin, typical curves of shear stress and a small hysteresis loop (8943.9 $Pa{\cdot}s^{-1}$) were present, indicating that betanin improved the shear stability of PeS. Additionally, the hysteresis loop area of RS–betanin (30,553.2 $Pa{\cdot}s^{-1}$) and PoS–betanin (25,944.7 $Pa{\cdot}s^{-1}$) was smaller than RS (31,483.2 $Pa{\cdot}s^{-1}$) and PoS (52,972.6 $Pa{\cdot}s^{-1}$), respectively (Table 2). Betanin had the biggest effect on the hysteresis loop area of PoS, as the hysteresis loop area was reduced by 40.6%, which indicated that betanin reduced the thixotropy and enhanced the shear stability of starch paste, especially for PoS. The fitting results of the shear stress curves to the power law model are shown in Table 2. Owing to an irregular shear stress curve, PeS was unable to fit the model. Except for PeS, the R^2 of shear stress curves for other samples was greater than 0.97, indicating the high fitting accuracy to the power law model. The fluid indexes *n* of the fitted starches and starch–betanin pastes were all less than 1, indicating that they were typical pseudoplastic fluids with shear thinning behavior. Betanin seemed to have no effect on the *n* of RS and PoS, but reduced their consistency coefficient *K*. It was implied that betanin weakened the thickening property and enhanced the pseudoplasticity of RS and PoS. Meanwhile, betanin endowed PeS with shear thinning behavior, and the biggest *K* and the smallest *n* were presented for PeS–betanin.

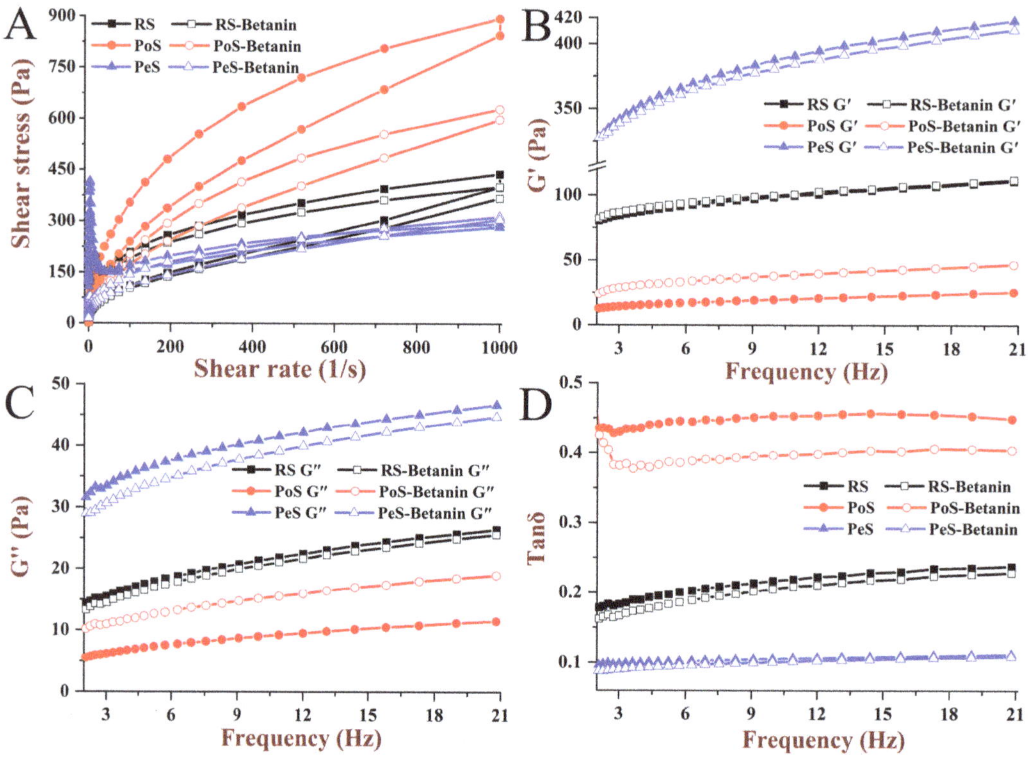

Figure 2. Shear rheology curves (**A**), storage moduli (G′) (**B**), loss moduli (G″) (**C**) and loss tangents (tanδ) (**D**) of starches and starch–betanin samples determined by rheological measurements. RS: rice starch; PoS: potato starch; PeS: pea starch.

Table 2. The steady shear rheological parameters of starches and starch–betanin samples.

Samples	Hysteresis Loop Area ($Pa \cdot s^{-1}$)	Up Curve			Down Curve		
		$K/Pa \cdot s^n$	n	R^2	$K/Pa \cdot s^n$	n	R^2
RS	31483.2	54.67	0.30	0.98	12.98	0.48	0.98
RS–betanin	30553.2	53.98	0.28	0.99	11.84	0.48	0.98
PoS	52972.6	52.76	0.50	0.98	22.63	0.52	0.98
PoS–betanin	25944.7	21.01	0.41	0.97	16.06	0.52	0.98
PeS	-	-	-	-	-	-	-
PeS–betanin	8943.9	56.01	0.23	0.97	24.92	0.35	0.97

RS: rice starch; PoS: potato starch; PeS: pea starch; *K*: the consistency coefficient; *n*: the flow behavior index.

3.2.2. Dynamic Rheological Properties

A dynamic frequency sweep range from 0.1 to 20.8 Hz was employed to investigate the viscoelastic properties of starch and starch–betanin pastes, and their storage modulus (G′), loss modulus (G″) and loss tangent (tan δ) as functions of frequency were depicted in Figure 2B–D, respectively. Owing to the unstable state in the measurements of initial small frequency, the results were recorded from 0.6~20.8 Hz. G′ and G″ represent the elasticity and viscosity of the tested samples, and tanδ represents the ratio of G″ and G′, which can be used to explain the viscoelastic behavior [25]. G′ and G″ of all samples were increased along with the frequency, and G′ was bigger than G″, indicating that starch and starch–betanin were typical weak gels. Compared to RS and PoS, PeS had the highest moduli, with the reason possibly being that the high content of amylose contributed to promoting the crosslinking of gel networks. The combination with betanin affected the dynamic rheological properties of starches, and the change in the degree of PoS was the most significant, which was similar to that of the pasting properties. Furthermore, the effect of betanin on G′ was stronger than that of G″; thus, betanin performed a more effective influence on the elasticity than on the viscous properties of RS, PoS and PeS. The tanδ values of RS, PoS and PeS were less than 1 over the whole frequency range, and PeS exhibited the lowest tanδ values, showing the strongest elastic behavior. The addition of betanin decreased the tanδ value of starches; that is, starch–betanin exhibited lower tanδ values than that of the corresponding starch. The effect of betanin on the tanδ of PoS was the biggest, and the smallest effect was observed on the tanδ value of PeS. The decreased tanδ values implied that the structure of the gel network was enhanced, which was probably attributed to increasing the junctions or crosslinking between amylose and swollen fragments caused by the presence of betanin. The result demonstrated that betanin changed the network structure of PoS pastes to a greater extent compared to RS and PeS, which was consistent with the result of the steady shear rheological properties.

3.3. Color Observation

In addition to its high antioxidant activity and many health-beneficial effects, betanin is a natural food colorant [26,27]. Therefore, observing the changes in the color of pastes could help foresee the visual quality of pastes during storage, which plays an important role in determining the acceptability and desirability of the products for consumers. Figure 3 displays the appearance of starch and starch–betanin pastes during retrogradation, and the lightness (L^*), redness (a^*) and yellowness (b^*) of these samples are listed in Table 3. The fresh PoS paste was transparent, and RS and PeS pastes tended to be white. The L^* value for PoS (29.67) was the lowest, and those of RS and PeS pastes were 40.17 and 57.85, respectively. Nevertheless, fresh starch–betanin pastes exhibited a vivid and attractive red color, which indicated that betanin could endow starch paste with a good visual image. As shown in Table 3, the values of a^* for fresh starch–betanin pastes were also significantly higher than that of the corresponding fresh starch pastes. After retrogradation for 7 days, L^* values of pure starch pastes were higher than that of the corresponding fresh starch paste. Starch molecules were rearranged to form crosslinking during the retrogradation, resulting in the color to gradually become cloudy. The transparency was decreased and the whiteness was increased, giving rise to the increasing of lightness. In regard to the starch–betanin pastes, similar alterations as occurred to the pure starch pastes after retrogradation were found. Starch–betanin pastes after retrogradation for 7 days presented higher L^* and a^* values than fresh starch–betanin pastes. In light of the analysis of retrogradation properties in the next sections, betanin inhibited the long-term retrogradation of all starches. In other words, betanin delayed the transformation to cloudy; thus, the redness a^* of starch–betanin pastes was increased after storage. According to the appearance of the pastes, the retrograded starch–betanin pastes seemed to slightly fade, which was mainly related to the phenomenon of becoming cloudy during retrogradation. Overall, the retrograded starch–betanin pastes maintained a bright red color, which indicates that betanin can also be applied as a good colorant during the storage of starches.

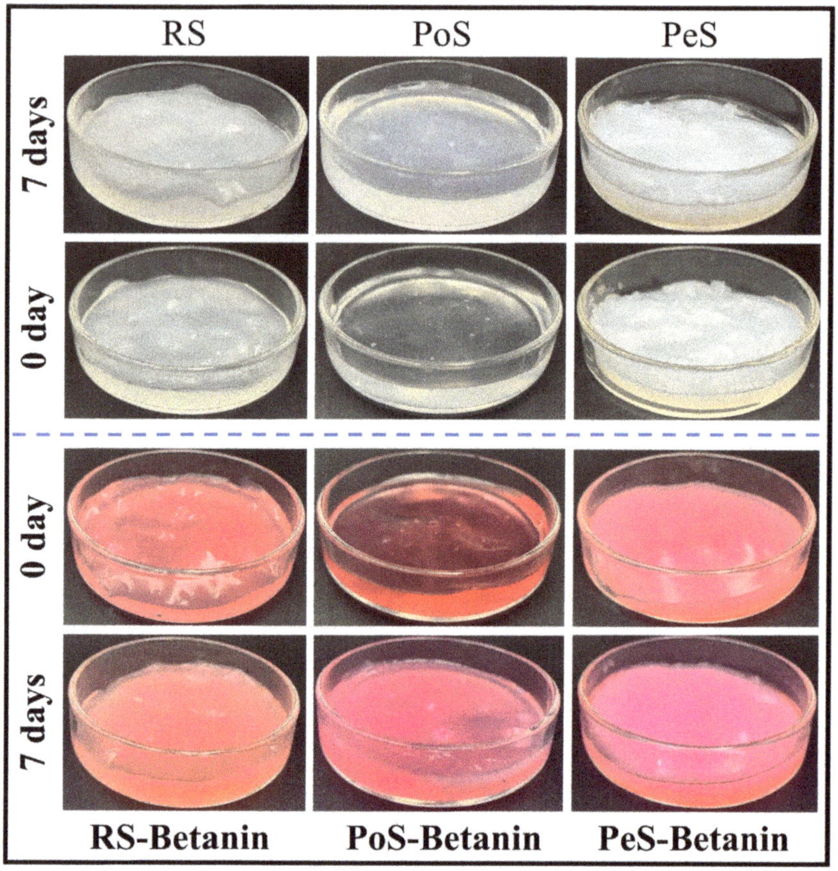

Figure 3. The appearance of starches and starch–betanin samples after retrogradation for 0 days and 7 days. RS: rice starch; PoS: potato starch; PeS: pea starch.

Table 3. The color of starches and starch–betanin samples after retrogradation for 0 days and 7 days.

Samples	Storage Time	L^*	a^*	b^*
RS	0 days	40.17 ± 0.02 [g]	−0.96 ± 0.00 [i]	−8.57 ± 0.01 [h]
RS	7 days	48.73 ± 0.02 [d]	−1.74 ± 0.00 [j]	−7.81 ± 0.02 [f]
RS–betanin	0 days	35.30 ± 0.01 [j]	9.30 ± 0.04 [e]	−5.55 ± 0.01 [d]
RS–betanin	7 days	41.77 ± 0.01 [f]	10.26 ± 0.04 [d]	−2.52 ± 0.02 [c]
PoS	0 days	29.67 ± 0.02 [k]	−0.40 ± 0.01 [g]	−1.39 ± 0.00 [a]
PoS	7 days	37.18 ± 0.01 [h]	−0.64 ± 0.01 [h]	−8.07 ± 0.01 [g]
PoS–betanin	0 days	27.41 ± 0.07 [l]	3.96 ± 0.03 [f]	−1.49 ± 0.01 [b]
PoS–betanin	7 days	35.56 ± 0.01 [i]	10.69 ± 0.01 [c]	−8.69 ± 0.01 [i]
PeS	0 days	57.85 ± 0.02 [b]	−2.19 ± 0.00 [k]	−9.49 ± 0.01 [k]
PeS	7 days	66.03 ± 0.02 [a]	−2.57 ± 0.00 [l]	−8.96 ± 0.01 [j]
PeS–betanin	0 days	41.97 ± 0.01 [e]	16.36 ± 0.03 [b]	−7.58 ± 0.01 [e]
PeS–betanin	7 days	50.77 ± 0.00 [c]	20.92 ± 0.01 [a]	−10.54 ± 0.01 [l]

Differences between values, indicated by different letters in the same columns, are significant at 0.05 level of confidence. RS: rice starch; PoS: potato starch; PeS: pea starch; L^*: brightness value; a^*: red-green value; b^*: yellowish-blue value.

3.4. XRD Analysis

The X-ray diffraction patterns of the retrograded starch and starch–betanin samples were displayed in Figure 4. Native rice starch (NRS), native potato starch (NPoS) and native pea starch (NPeS) exhibited typical A-, B- and C-type diffraction patterns, respectively. After retrogradation for 7 days, RS with an A-type diffraction pattern was transformed to a V-type pattern, and B-type patterns for the retrograded PoS and PeS were shown. The relative crystallinity of the retrograded RS, PoS and PeS was 16.8%, 10.9% and 9.5%. The lowest crystallinity of the retrograded PeS may be related to its low amylopectin content, because the crystallinity of the processed starch was considered to be associated with retrogradation of amylopectin [28]. In regard to all the retrograded starch–betanin samples, their diffraction intensities were observably weakened in comparison with the retrograded starch. For example, the diffraction peak at 19.6° for the retrograded RS–betanin almost disappeared, and peaks at 17.3° for the retrograded PoS–betanin and PeS–betanin narrowed significantly. The relative crystallinity was reduced by 14.2%, 8.8% and 4.4% for the retrograded RS–betanin, PoS–betanin and PeS–betanin, respectively. These results imply that betanin impeded the rearrangement of starch molecules during long-term retrogradation. The presence of betanin may alter the distribution and rearrangement of starch molecules, and retrogradation was restrained by steric hindrance. The degree of inhibiting crystallinity by betanin was the highest for PeS, and it is likely that betanin and the high amylose content of PeS occupied the space to interfere with the association of amylopectin.

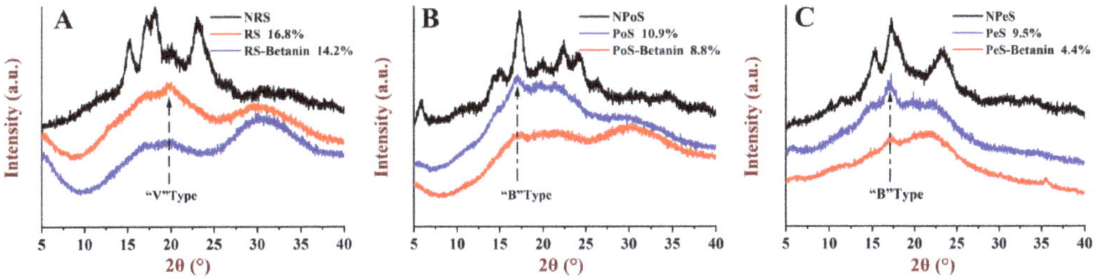

Figure 4. XRD diagrams of starches and starch–betanin samples after retrogradation; (**A**) rice starch, (**B**) potato starch, and (**C**) pea starch. NRS: native rice starch; NPoS: native potato starch; NPeS: native pea starch. RS: retrograded rice starch; PoS: retrograded potato starch; PeS: retrograded pea starch.

3.5. Short-Range Ordered Structure Analysis

Internal structural changes of starches after retrogradation could also be determined by a FTIR spectrometer that was used to analyze the short-range ordered structure, and the FITR spectra of starch and starch–betanin after retrogradation for 7 days were displayed in Figure 5A. Compared to the retrograded starches, no new characteristic absorption peaks appeared in the spectra of the retrograded starch–betanin samples, and just some of the wavenumbers of peaks were shifted, which implied that the primary structure of the retrograded starches was maintained and no covalent bond was formed. As demonstrated by Sevenou et al. [29], the absorbance at approximately 995 cm^{-1}, 1047 cm^{-1} and 1022 cm^{-1} could be associated with the structural order of starch chains near the granule surface. Meanwhile, the ratio of 1047/1022 cm^{-1} and 995/1022 cm^{-1} could be used to determine a degree of short-range order for the retrograded starches [30]. As displayed in Figure 5B, the retrograded starch–betanin behaved with lower ratio values of 1047/1022 cm^{-1} and 995/1022 cm^{-1} as compared to the retrograded starch. For example, the ratios of 995/1022 cm^{-1} for RS, PoS and PeS were 0.974, 0.967 and 0.972, respectively. However, the ratios of RS–betanin, PoS–betanin and PeS–betanin were decreased to 0.908, 0.937 and 0.946, respectively. These results suggest that betanin reduced the short-range order structure of the retrograded starches, which was in accord with the effect of betanin on crystallinity in the XRD analysis.

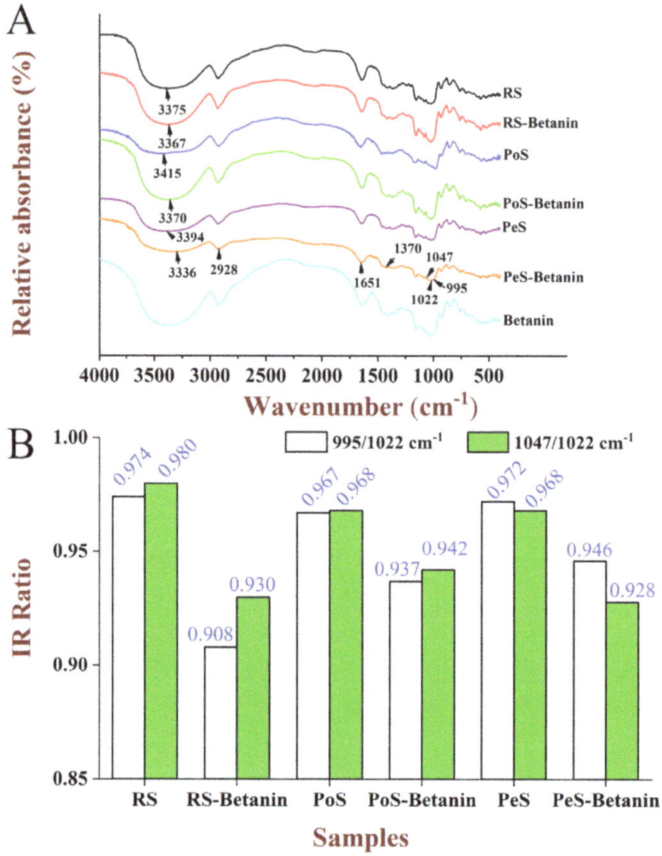

Figure 5. FTIR spectra (**A**) and 1047 cm^{-1}/1022 cm^{-1} ratios from deconvoluted FTIR spectra (**B**) of starches and starch–betanin samples after retrogradation for 7 days. RS: rice starch; PoS: potato starch; PeS: pea starch.

3.6. Thermal Properties Analysis

The thermal and retrogradation properties of starch and starch–betanin determined by DSC are presented in Table 4. The presence of betanin increased the gelatinization transition temperatures (T_o, T_p and T_c) of all starches, and it was implied that starch gelatinization was delayed by betanin, which might be related to the strong hydrophilicity of betanin and its ability to hinder starch from absorbing water. ΔH_g represents the required thermal energy for melting the double-helix structure and destroying crystallinity within the starch granule [31]. In comparison to pure starches, a lower ΔH_g for starch–betanin was found. Similar results were reported in our previous paper [14], in which the addition of polymeric proanthocyanidin decreased the gelatinization enthalpy of RS, PoS and PeS. Simultaneously, betanin significantly delayed the long-term retrogradation of the three starches, resulting in a reduction in the retrogradation enthalpy (ΔH_r) and retrogradation rate (R). After incorporating betanin, the R of RS, PoS and PeS declined from 34.13%, 42.95% and 31.82% to 23.24%, 31.48% and 25.80, respectively. It was indicated that betanin affected the formation and weakened the order degree of crystallinity during the long-term retrogradation of starches, which conformed to the results of the XRD and FITR analyses. In light of these results, it is known that betanin could be used as an excellent inhibitor of retrogradation in products such as bread and pastry.

Table 4. Thermal properties of starches and starch–betanin samples after retrogradation for 7 days.

Samples	T_o (°C)	T_p (°C)	T_c (°C)	ΔH_g (J/g)	ΔH_r (J/g)	*R* (%)
RS	63.65 ± 0.10^c	66.16 ± 0.03^c	70.54 ± 0.47^c	8.41 ± 0.39^d	2.87 ± 0.12^c	34.13
RS–betanin	64.76 ± 0.11^b	68.00 ± 0.16^b	73.49 ± 0.82^a	6.11 ± 0.30^e	1.42 ± 0.11^d	23.24
PoS	58.91 ± 0.41^e	63.23 ± 0.21^e	66.54 ± 0.26^d	12.69 ± 0.41^a	5.45 ± 0.03^a	42.95
PoS–betanin	60.75 ± 0.07^d	65.93 ± 0.51^d	69.74 ± 0.98^c	9.69 ± 0.21^c	3.05 ± 0.07^c	31.48
PeS	60.44 ± 0.35^d	67.73 ± 0.38^b	71.87 ± 0.24^b	12.41 ± 0.13^a	3.95 ± 0.26^b	31.82
PeS–betanin	70.82 ± 0.26^a	71.80 ± 0.41^a	74.21 ± 0.13^a	11.55 ± 0.34^b	2.98 ± 0.03^c	25.80

Differences between values, indicated by different letters in the same columns, are significant at 0.05 level of confidence. RS: rice starch; PoS: potato starch; PeS: pea starch. T_o: onset temperature; T_p: peak temperature; T_c: conclusion temperature; ΔH_g: gelatinization enthalpy; ΔH_r: retrogradation enthalpy; *R*: the degree of retrogradation $(\Delta H_r/\Delta H_g) * 100$.

3.7. Morphology Analysis

Microstructures of the retrograded starches and starch–betanin samples were examined through SEM to observe the changes from adding betanin, and images are depicted in Figure 6. The retrograded RS displayed a dense surface with a uniformly distributed cavity, and the retrograded PoS and PeS presented relatively rough network-like structures with some fragments. Nevertheless, the incorporation of betanin altered the formation of the gel network of the retrograded starches, and the retrograded starch–betanin samples displayed more porous and loose structures. In comparison with the morphology of the retrograded starches, RS–betanin and PoS–betanin showed an increased cavity volume, and porous spongy-like morphology was observed for PeS–betanin. This phenomenon was similar to the report by Xu et al. [32], who found *Vaccinium bracteatum Thunb. leaf* pigment loosened matrices of rice starch gels. The microstructural changes reflected that the addition of betanin affected the gel network of the retrograded starches. That is, betanin might inhibit the long-term retrogradation of starches, which coincides with the results of the aforementioned investigations.

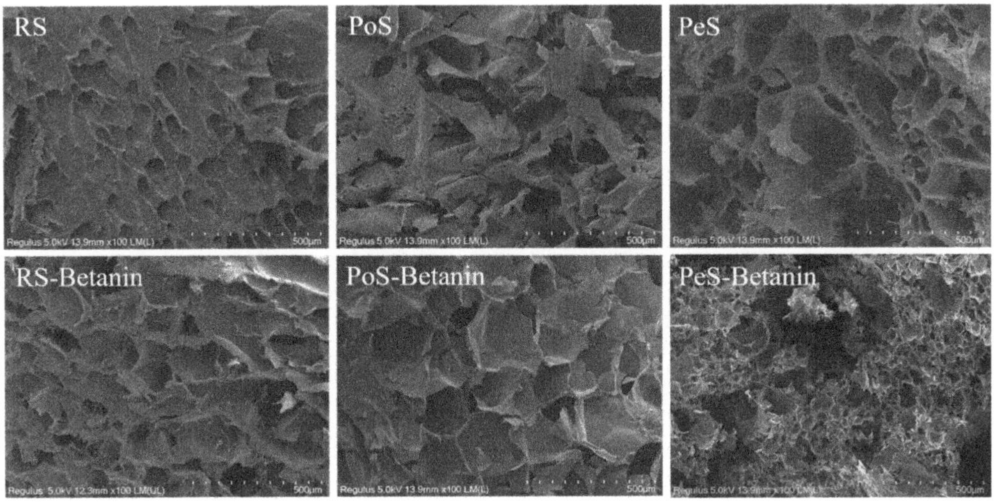

Figure 6. SEM photographs of starches and starch–betanin samples after retrogradation for 7 days. Magnification was 100. RS: rice starch; PoS: potato starch; PeS: pea starch.

4. Conclusions

The pasting, rheology and retrogradation properties of rice, potato and pea starches were changed by the presence of betanin. Betanin decreased the peak, trough and final viscosity of rice and potato starches, but increased these of pea starch. Rheological properties including thixotropy, the dynamic modulus and the loss factor of the three starches were varied by different degrees after incorporating betanin. Betanin endowed starch pastes with a vivid red appearance and maintained color during storage. Furthermore, the poor short-range molecular order, low crystallinity and low retrogradation enthalpy of starches were induced by betanin during retrogradation, suggesting that betanin could inhibit the retrogradation of starches. The micromorphology of the retrograded starches was also altered by betanin. These findings provide guidance for the application of betanin in developing foods that require colorant and an inhibitor of starch retrogradation.

Author Contributions: Conceptualization, T.D., J.C. and C.L. (Changhong Li); methodology, X.H. (Xiaohong He), J.X. and T.D.; investigation, X.H. (Xuemei He), Q.G. and X.H. (Xiaohong He); data curation, X.H. (Xuemei He) and T.D.; writing—original draft preparation, X.H. (Xiaohong He), T.D. and C.L. (Changhong Li); writing—review and editing, J.S., C.L. (Chengmei Liu) and J.C.; visualization, J.S. and X.H. (Xuemei He); supervision, T.D. and C.L. (Chengmei Liu); funding acquisition, J.S. and T.D. All authors have read and agreed to the published version of the manuscript.

Funding: This work was supported by the China Postdoctoral Science Foundation [Grant No. 2020M6832211], Postdoctoral Foundation of Guangxi Academy of Agricultural Sciences (Grant No. 2020037), Guangxi Natural Science Foundation (Grant No. 2022GXNSFBA035522), Guangxi Key Laboratory of Fruits and Vegetables Storage-processing Technology Project (20-065-68), National Natural Science Foundation of China (Grant No. 32101948), China Agriculture Research System (CARS-31-12), and Special Fund for Guangxi Bagui Scholars (Grant No. (2016)21).

Data Availability Statement: The data presented in this study are available in this article.

Acknowledgments: The authors would like to thank the Centre of Analysis and Testing of Nanchang University and State Key Laboratory of Food Science and Technology for their expert technical assistance.

Conflicts of Interest: The authors declare no conflict of interest.

References

1. McCann, D.; Barrett, A.; Cooper, A.; Crumpler, D.; Dalen, L.; Grimshaw, K.; Kitchin, E.; Lok, K.; Porteous, L.; Prince, E.; et al. Food additives and hyperactive behaviour in 3-year-old and 8/9-year-old children in the community: A randomised, double-blinded, placebo-controlled trial. *Lancet* **2007**, *370*, 1560–1567. [CrossRef]
2. Stintzing, F.C.; Schieber, A.; Carle, R. Betacyanins in fruits from red-purple pitaya, Hylocereus polyrhizus (Weber) Britton & Rose. *Food Chem.* **2002**, *77*, 101–106. [CrossRef]
3. Shaaruddin, S.; Ghazali, H.M.; Mirhosseini, S.H.; Muhammad, K. Stability of betanin in pitaya powder and confection as affected by resistant maltodextrin. *LWT-Food Sci. Technol.* **2017**, *84*, 129–134. [CrossRef]
4. Pagano, A.P.E.; Khalid, N.; Kobayashi, I.; Nakajima, M.; Neves, M.A.; Bastos, E.L. Microencapsulation of betanin in monodisperse W/O/W emulsions. *Food Res. Int.* **2018**, *109*, 489–496. [CrossRef] [PubMed]
5. Goncalves, L.C.P.; Trassi, M.A.D.; Lopes, N.B.; Dorr, F.A.; dos Santos, M.T.; Baader, W.J.; Oliveira, V.X.; Bastos, E.L. A comparative study of the purification of betanin. *Food Chem.* **2012**, *131*, 231–238. [CrossRef]
6. Guneser, O. Pigment and color stability of beetroot betalains in cow milk during thermal treatment. *Food Chem.* **2016**, *196*, 220–227. [CrossRef] [PubMed]
7. Kumar, S.S.; Manoj, P.; Shetty, N.P.; Prakash, M.; Giridhar, P. Characterization of major betalain pigments -gomphrenin, betanin and isobetanin from Basella rubra L. fruit and evaluation of efficacy as a natural colourant in product (ice cream) development. *J. Food Sci. Technol.-Mysore* **2015**, *52*, 4994–5002. [CrossRef]
8. Fonseca, L.M.; El Halal, S.L.M.; Dias, A.R.G.; Zavareze, E.D. Physical modification of starch by heat-moisture treatment and annealing and their applications: A review. *Carbohydr. Polym.* **2021**, *274*, 118665. [CrossRef]
9. Chen, S.; Qin, L.; Chen, T.; Yu, Q.; Chen, Y.; Xiao, W.H.; Ji, X.Y.; Xie, J.H. Modification of starch by polysaccharides in pasting, rheology, texture and in vitro digestion: A review. *Int. J. Biol. Macromol.* **2022**, *207*, 81–89. [CrossRef]
10. Yu, M.H.; Zhu, S.; Zhong, F.; Zhang, S.H.; Du, C.D.; Li, Y. Insight into the multi-scale structure changes and mechanism of corn starch modulated by different structural phenolic acids during retrogradation. *Food Hydrocoll.* **2022**, *128*, 107581. [CrossRef]
11. Molavi, H.; Razavi, S.M.A.; Farhoosh, R. Impact of hydrothermal modifications on the physicochemical, morphology, crystallinity, pasting and thermal properties of acorn starch. *Food Chem.* **2018**, *245*, 385–393. [CrossRef] [PubMed]
12. Wu, Y.; Niu, M.; Xu, H.L. Pasting behaviors, gel rheological properties, and freeze-thaw stability of rice flour and starch modified by green tea polyphenols. *LWT-Food Sci. Technol.* **2020**, *118*, 108796. [CrossRef]
13. Wang, M.T.; Chen, J.C.; Chen, S.G.; Ye, X.Q.; Liu, D.H. Inhibition effect of three common proanthocyanidins from grape seeds, peanut skins and pine barks on maize starch retrogradation. *Carbohydr. Polym.* **2021**, *252*, 117172. [CrossRef]

14. Xu, J.H.; Li, X.; Chen, J.; Dai, T.T.; Liu, C.M.; Li, T. Effect of polymeric proanthocyanidin on the physicochemical and in vitro digestive properties of different starches. *LWT-Food Sci. Technol.* **2021**, *148*, 111713. [CrossRef]
15. He, X.H.; Xia, W.; Chen, R.Y.; Dai, T.T.; Luo, S.J.; Chen, J.; Liu, C.M. A new pre-gelatinized starch preparing by gelatinization and spray drying of rice starch with hydrocolloids. *Carbohydr. Polym.* **2020**, *229*, 115485. [CrossRef] [PubMed]
16. Zhu, F.; Wang, Y.J. Rheological and thermal properties of rice starch and rutin mixtures. *Food Res. Int.* **2012**, *49*, 757–762. [CrossRef]
17. Liu, S.C.; Shen, M.Y.; Xiao, Y.H.; Luo, Y.; Xie, J.H. Effect of maize, potato, and pea starches with Mesona chinensis polysaccharide on pasting, gelatinization properties, granular morphology and digestion. *Food Hydrocoll.* **2020**, *108*, 106047. [CrossRef]
18. Wang, R.S.; Wan, J.; Liu, C.M.; Xia, X.; Ding, Y.P. Pasting, thermal, and rheological properties of rice starch partially replaced by inulin with different degrees of polymerization. *Food Hydrocoll.* **2019**, *92*, 228–232. [CrossRef]
19. Guo, X.; He, X.; Dai, T.; Liu, W.; Liang, R.; Chen, J.; Liu, C. The physicochemical and pasting properties of purple corn flour ground by a novel low temperature impact mill. *Innov. Food Sci. Emerg. Technol.* **2021**, *74*, 102825. [CrossRef]
20. Li, Y.T.; Wang, R.S.; Liang, R.H.; Chen, J.; He, X.H.; Chen, R.Y.; Liu, W.; Liu, C.M. Dynamic high-pressure microfluidization assisting octenyl succinic anhydride modification of rice starch. *Carbohydr. Polym.* **2018**, *193*, 336–342. [CrossRef]
21. Xu, J.H.; Dai, T.T.; Chen, J.; He, X.M.; Shuai, X.X.; Liu, C.M.; Li, T. Effects of three types of polymeric proanthocyanidin on physicochemical and in vitro digestive properties of potato starch. *Foods* **2021**, *10*, 1394. [CrossRef] [PubMed]
22. Ratnayake, W.S.; Hoover, R.; Warkentin, T. Pea starch: Composition, structure and properties-A review. *Starch-Starke* **2002**, *54*, 217–234. [CrossRef]
23. Han, L.H.; Cao, S.P.; Yu, Y.T.; Xu, X.C.; Cao, X.H.; Chen, W.J. Modification in physicochemical, structural and digestive properties of pea starch during heat-moisture process assisted by pre- and post-treatment of ultrasound. *Food Chem.* **2021**, *360*, 129929. [CrossRef] [PubMed]
24. Zavareze, E.D.; Dias, A.R.G. Impact of heat-moisture treatment and annealing in starches A review. *Carbohydr. Polym.* **2011**, *83*, 317–328. [CrossRef]
25. Chen, X.J.; Liu, Y.; Xu, Z.K.; Zhang, C.C.; Liu, X.X.; Sui, Z.Q.; Corke, H. Microwave irradiation alters the rheological properties and molecular structure of hull-less barley starch. *Food Hydrocoll.* **2021**, *120*, 106821. [CrossRef]
26. Amjadi, S.; Almasi, H.; Hamishehkar, H.; Khaledabad, M.A.; Lim, L.T. Coating of betanin and carvone Co-loaded nanoliposomes with synthesized cationic inulin: A strategy for enhancing the stability and bioavailability. *Food Chem.* **2022**, *373*, 131403. [CrossRef]
27. Amjadi, S.; Ghorbani, M.; Hamishehkar, H.; Roufegarinejad, L. Improvement in the stability of betanin by liposomal nanocarriers: Its application in gummy candy as a food model. *Food Chem.* **2018**, *256*, 156–162. [CrossRef]
28. Wang, S.J.; Li, C.L.; Copeland, L.; Niu, Q.; Wang, S. Starch Retrogradation: A Comprehensive Review. *Compr. Rev. Food Sci. Food Saf.* **2015**, *14*, 568–585. [CrossRef]
29. Sevenou, O.; Hill, S.E.; Farhat, I.A.; Mitchell, J.R. Organisation of the external region of the starch granule as determined by infrared spectroscopy. *Int. J. Biol. Macromol.* **2002**, *31*, 79–85. [CrossRef]
30. Flores-Morales, A.; Jimenez-Estrada, M.; Mora-Escobedo, R. Determination of the structural changes by FT-IR, Raman, and CP/MAS C-13 NMR spectroscopy on retrograded starch of maize tortillas. *Carbohydr. Polym.* **2012**, *87*, 61–68. [CrossRef]
31. He, X.H.; Luo, S.J.; Chen, M.S.; Xia, W.; Chen, J.; Liu, C.M. Effect of industry-scale microfluidization on structural and physicochemical properties of potato starch. *Innov. Food Sci. Emerg. Technol.* **2020**, *60*, 102278. [CrossRef]
32. Xu, Y.; Fan, M.C.; Zhou, S.M.; Wang, L.; Qian, H.F.; Zhang, H.; Qi, X.G. Effect of *Vaccinium bracteatum* Thunb. leaf pigment on the thermal, pasting, and textural properties and microstructure characterization of rice starch. *Food Chem.* **2017**, *228*, 435–440. [CrossRef] [PubMed]

foods

Article

Interaction between Gelatin and Mulberry Leaf Polysaccharides in Miscible System: Physicochemical Characteristics and Rheological Behavior

Xiu-Xiu Zhang [1], Bu-Yan Liao [1], Zi-Jing Guan [1], Kiran Thakur [1,2], Mohammad Rizwan Khan [3], Rosa Busquets [4], Jian-Guo Zhang [1,2] and Zhao-Jun Wei [1,2,*]

1 School of Food Science and Biological Engineering, Hefei University of Technology, Hefei 230009, China; 2020111406@mail.hfut.edu.cn (X.-X.Z.); liaobuyan@afc.edu.cn (B.-Y.L.); guanzj@mail.hfut.edu.cn (Z.-J.G.); kumarikiran@hfut.edu.cn (K.T.); zhangjianguo@hfut.edu.cn (J.-G.Z.)
2 Ningxia Key Laboratory for the Development and Application of Microbial Resources in Extreme Environments, School of Biological Science and Engineering, North Minzu University, Yinchuan 750021, China
3 Department of Chemistry, College of Science, King Saud University, Riyadh 11451, Saudi Arabia; mrkhan@ksu.edu.sa
4 School of Life Sciences, Pharmacy and Chemistry, Kingston University London, Kingston upon Thames KT1 2EE, UK; r.busquets@kingston.ac.uk
* Correspondence: zjwei@hfut.edu.cn; Tel./Fax: +86-551-6290-2227

Abstract: In this study, the miscible system was formed by mixing gelatin (G) with mulberry leaf polysaccharides (MLPs) continuously extracted with a hot buffer (HBSS), a chelating agent (CHSS), a dilute alkali (DASS), and a concentrated alkali (CASS), and the zeta potential, turbidity, particle size, distribution, and rheological properties of the miscible systems were evaluated. Under acidic conditions, the miscible systems of four polysaccharides and gelatin were in a clear state; under alkaline conditions, G-HBSS and G-CHSS were clarified, and G-DASS and G-CASS changed from clarification to turbidity. The zeta potential changed from positive to negative with the increase in pH. When the pH was at 7, it increased with the increase in polysaccharide concentration but was still negative. The four miscible systems all showed polydispersity. The particle sizes of G-HBSS and G-CHSS decreased with the increase in pH, while the particle sizes of G-DASS and G-CASS were increased. The four miscible systems showed "shear thinning" behavior, and the addition of gelatin reduced the apparent viscosity of the four polysaccharide solutions. G-CHSS was highly stable, and G-CASS was more suitable as a stabilizer in the freezing process.

Keywords: gelatin; mulberry leaf polysaccharides; miscible system; zeta potential; rheological properties

Citation: Zhang, X.-X.; Liao, B.-Y.; Guan, Z.-J.; Thakur, K.; Khan, M.R.; Busquets, R.; Zhang, J.-G.; Wei, Z.-J. Interaction between Gelatin and Mulberry Leaf Polysaccharides in Miscible System: Physicochemical Characteristics and Rheological Behavior. *Foods* **2022**, *11*, 1571. https://doi.org/10.3390/foods11111571

Academic Editors: Jianhua Xie, Yanjun Zhang and Hansong Yu

Received: 22 April 2022
Accepted: 24 May 2022
Published: 26 May 2022

1. Introduction

Mulberry, belonging to the mulberry genus of the mulberry family, is widely distributed throughout the world [1]. In particular, the planting area of mulberry trees in China ranks first in the world, and the yield of mulberry leaves is rich [2]. Mulberry leaves are mainly used for silkworm breeding, and a few are used for the preparation of tea and fruit juice [3,4]. In China, mulberry leaf is also a medicinal resource which is widely used in traditional Chinese medicine [5]. Mulberry leaves contain a variety of functional components, such as alkaloids, polyphenols and flavonoids, proteins, amino acids, and carbohydrates [6–9]. A variety of functional components endow mulberry leaves with different biological activities, such as hypoglycemic, anti-atherosclerotic, antioxidant, and antibacterial activities [10–13]. Though there are many studies on the active components of alkaloids, flavonoids, and polyphenols in mulberry leaves, in recent years, more studies have emerged on the activity of polysaccharides.

Plant polysaccharides have attracted great attention in recent years because of their various biological activities, such as antitumor, antiviral, hypoglycemic, and antioxidant activities [14–17]. Because of their rheological properties, polysaccharides can also be used as adhesives and gelling agents in the food and cosmetics industries [18]. Previous studies have shown that plant polysaccharides can be used as a source of natural antioxidants in the food, pharmaceutical, and other industries [19]. At the same time, because of their rheological properties, plant polysaccharides can be used as thickeners and adhesives in natural materials and food industries [20].

The food system is a complex system composed of compounds with different properties, such as polysaccharides, proteins, and minerals. The current research focus is on regulating and controlling the interaction of biological macromolecules in the process of food processing to change their characteristics or structure [21]. Polysaccharides and proteins are the main nutrients in food formulas and can be miscible under certain conditions and change the system by forming the electrostatic complex [22]. This change affects the food structure and rheological properties, which is of great significance in food processing. As a kind of protein macromolecule, gelatin has been widely used in food processing [23].

In our previous study, four mulberry leaf polysaccharides (MLPs) were extracted continuously with a thermal buffer, a chelating agent, dilute alkalis, and concentrated alkali extractant to obtain thermal buffer soluble solids (HBSSs), chelating agent soluble solids (CHSSs), dilute alkali soluble solids (DASSs), and concentrated alkali soluble solids (CASSs), respectively [19,20]. Four MLPs had a shear-thinning characteristic and the properties of non-Newtonian fluid [20], and exhibited significant differences in the physicochemical and antioxidant activities. The CASS displayed higher thermal properties and suitability as a supplement in hot processed foods [19]; The DASS showed the highest apparent viscosity at different tested concentrations [20]. However, there are no reports on the interaction between mulberry leaf polysaccharides and gelatin in the miscible system. In this study, the effects of gelatin on zeta potential, turbidity, particle size, distribution, and rheological properties of gelatin–polysaccharide miscible systems were studied.

2. Materials and Methods

2.1. Materials and Chemicals

The mulberry leaves (*Morus alba* L.) used in the experiment were provided by Anhui Academy of Agricultural Sciences (Anhui Province, Hefei, China). Four types of mulberry leaf polysaccharides (HBSS, CHSS, DASS, and CASS) were obtained by continuous extraction with four solvents [19,20]. All chemicals were analytical-grade.

2.2. Preparation of Gelatin–Polysaccharide Miscible Solution

A total of 0.5 g gelatin was dissolved in 100 mL water at 40 °C, 1 h, then stored at 4 °C for 24 h, and soaked in water at 25 °C for 18 h to prepare a stable gelatin solution with a mass concentration of 5 mg/mL [24]. Then, 0.1 g of mulberry leaf polysaccharides was added to 100 mL of 5 mg/mL gelatin solution and stirred until it was completely dissolved to obtain gelatin–polysaccharide miscible solutions (G-MLPs) with a concentration of 1 mg/mL for the phase diagram experiment, in which 2.3. 5 mg/mL gelatin solution was used as the solvent for the preparation of gelatin–polysaccharide solutions with different concentrations.

2.3. Phase Diagram

The gelatin–polysaccharide miscible solutions prepared in Section 2.2. were used as a mother liquor to prepare gelatin–polysaccharide miscible systems with different pHs (3–10, interval 0.5). At first, the miscible system was placed at room temperature for 1 h, and then the solution state was observed. The state was divided into the following three types: clear solution, cloudy solution, precipitation, and cloudy solution [25]. The phase diagram was drawn according to the change in pH.

2.4. Measurement of Zeta Potential, Particle Size, and Distribution

The zeta potential, particle size, and distribution of gelatin–polysaccharide miscible solution at different pHs (3.5, 7, 10) with a concentration of 20 mg/mL and different concentrations (5, 10, 15, and 20 mg/mL) at pH 7 were determined by Malvin particle size analyzer (Zetasizer Nano-ZSE, Malvern Instruments, Worcestershire, UK) [26]. The calculations of zeta potential and particle size were as follows:

$$\zeta\left(\times 10^{3}\right) = \frac{3\,\eta\,U_{E}}{2_{\varepsilon f}\,(k\alpha)} \tag{1}$$

ε (F/M) is numerically equal to εrε0; εr is the relative permittivity of the mixture (25 °C water is about 78.2); ε0 is the vacuum relative permittivity (about 8.854×10^{-12} F/m); η (Pa·s) is the viscosity of the dispersion system (25 °C water is about 8.937×10^{-4} Pa·s); f (κα) is the ratio of the particle radius α to Debye's length in Henry's equation and approximately 1.5 in the Smoluchowski model.

$$d\left(\times 10^{9}\right) = \frac{K_{B}T}{3\pi\,\eta\,D} \tag{2}$$

η (Pa·s) is the viscosity of the dispersion system (25 °C water is about 8.937×10^{-4} Pa·s); KB is Boltzmann's constant, and T is absolute temperature (k); the refractive index was set as follows: water 1.33, G 1.45, G-MLPs 1.59; the absorbance of the material was 0.01.

2.5. Determination of Rheological Properties

2.5.1. Effect of Concentration, pH, Na^+ Concentration, and Temperature on Apparent Viscosity of G-MLPs

The rheological properties of G-MLPs were determined using the previously described method [20]. The rheological properties of the gelatin–polysaccharide miscible system were mainly studied through the effects of concentration, pH, Na^+ concentration, and temperature on the apparent viscosity of G-MLPs. In short, gelatin–polysaccharide aqueous solutions with concentration gradients of 5, 10, 15, and 20 mg/mL were prepared to study the effect of concentration on apparent viscosity, and 5 mg/mL gelatin solution was used as control. The pH value of 10 mg/mL gelatin–polysaccharide aqueous solution was adjusted to 3.5, 7.0, and 10.0 with HCl and NaOH solution to study the effect of pH on apparent viscosity. NaCl solutions with concentrations of 0, 0.1, 0.2, 0.4, and 0.8 mol/L were prepared, respectively, and then gelatin was dissolved and prepared in 5 mg/mL gelatin–NaCl solution, and finally polysaccharides were added to the solution to make the concentration of polysaccharides 10 mg/mL. When studying the effect of temperature on the apparent viscosity of the miscible system, we selected −20, 25, and 100 °C for cold and heat treatment of 10 mg/mL gelatin–polysaccharide miscible systems. Rheometer (DHR-3, TA instruments, New Castle, DE, USA) conditions were as follows: steady-state shear mode; clamp: diameter 40 mm; cone plate: cone angle 2°; shear speed range: 0.01–1000 s^{-1}; temperature: 20 °C.

2.5.2. Effect of G-MLPs on Viscoelasticity

The changes in apparent viscosity of gelatin–polysaccharide aqueous solution with mass concentrations of 5, 10, 15, and 20 mg/mL were measured under different shear oscillation frequencies. Test conditions were as follows: DHR-3: clamp: diameter 40 mm; cone plate: cone angle 2°; shear speed range: 0.01–100 Hz; temperature: 20 °C [20].

2.6. Statistical Analysis

All the tests reported were repeated three times. One-way ANOVA through SPSS 21.0 was used for the data analysis at $p < 0.05$, and origin 8.0 software was used for graphs.

3. Results and Discussion

3.1. Phase Diagram of Gelatin–Mulberry Leaf Polysaccharides (G-MLPs) Miscible System

Table 1 shows the effect of pH on the phase behavior of the gelatin–mulberry leaf polysaccharide miscible system. The electrical properties of proteins were reported to vary according to the pH of their environment. When it was in an electronegative state, proteins co-dissolved with anionic polysaccharides due to electrostatic repulsion, and under electropositive conditions, proteins aggregated due to the electrostatic binding reaction with anionic polysaccharides, resulting in a cloudy and precipitated solution [27]. At pH < 7, the dispersions of the four miscible systems were clear. When pH was 7–8.5, G-HBSS and G-CHSS were still clear, indicating that the reaction between G-HBSS and G-CHSS affected the intermolecular aggregation of polysaccharides [22]. G-DASS and G-CASS were the alkaline extracts; therefore, their resulting systems changed from clear to cloudy. When the pH was close to alkaline conditions, the repulsion force between G-DASS and G-CASS molecules decreased, resulting in aggregation, and the miscible system changed from clear to cloudy [25]. With the continuous increase in pH value, both the G-HBSS and G-CHSS miscible systems changed from clear to cloudy, while G-DASS and G-CASS changed to cloudy and precipitated systems. At pH 9, the four samples were unstable, indicating that the system environment with high pH was conducive to the formation of an insoluble gelatin–polysaccharide complex, which ultimately increased the instability of the system [25].

Table 1. State diagrams of G-MLPs' mixed solutions (1 mg/mL) as a function of pH.

Samples	pH													
	3	3.5	4	4.5	5	5.5	6	7	7.5	8	8.5	9	9.5	10
G-HBSS	▲	▲	▲	▲	▲	▲	▲	▲	▲	▲	▲	■	■	■
G-CHSS	▲	▲	▲	▲	▲	▲	▲	▲	▲	▲	▲	■	■	■
G-DASS	▲	▲	▲	▲	▲	▲	▲	▲	■	■	■	•	•	•
G-CASS	▲	▲	▲	▲	▲	▲	▲	▲	■	■	■	•	•	•

Note: The solubility or insolubility was evaluated by visual observation. (▲: clear solution; ■: cloudy solution; •: precipitation and cloudy solution).

3.2. Zeta Potential Analysis of Gelatin–Mulberry Leaf Polysaccharides (G-MLPs) Miscible System

The change in the zeta potential of gelatin and mulberry leaf polysaccharide miscible systems with respect to pH is shown in Figure 1a. With the increase in pH value, the zeta potential of the four miscible systems G-HBSS, G-CHSS, G-DASS, and G-CASS changed from positive to negative. Moreover, the zeta potential of the miscible system also decreased with the increase in pH. The zeta potential of the miscible system was negatively charged with the increase in pH, which may have been due to the electrostatic binding reaction between gelatin and mulberry polysaccharides, and the deprotonation reaction of amino and carboxyl functional groups on gelatin molecules and carboxyl groups in the mulberry polysaccharide structure with the increase in pH [28]. At pH 7.0, with the increase in mulberry leaf polysaccharide concentration, the zeta potential of the miscible system gradually increased, but it was still negative, as shown in Figure 1b. This may be due to the fact that the combination of mulberry leaf polysaccharides and gelatin resulted in the movement of the net charge of the miscible system to the positive direction [29,30].

3.3. Particle Size Analysis of Gelatin–Mulberry Leaf Polysaccharides (G-MLPs) Miscible System

The particle size and distribution of the gelatin solution and four mulberry leaf polysaccharide solutions at the same concentration (10 mg/mL) are shown in Figure 2. The particle size of the HBSS, CHSS, DASS, and CASS had two, one, three, and two peaks, respectively. Unlike the DASS, the other three mulberry leaf polysaccharides were monodisperse, indicating that the composition of the HBSS, CHSS, and CASS was relatively uniform and may have formed electrostatic complexes with the aqueous solution. This result is consistent with the particle size distribution of water-soluble polysaccharides extracted from kidney beans [31]. The particle size of gelatin had two peaks, and the particle size was larger, which may have been due to the high pH value of the solution, which led to its positive or negative charge and increased the interaction between gelatin and water molecules [32].

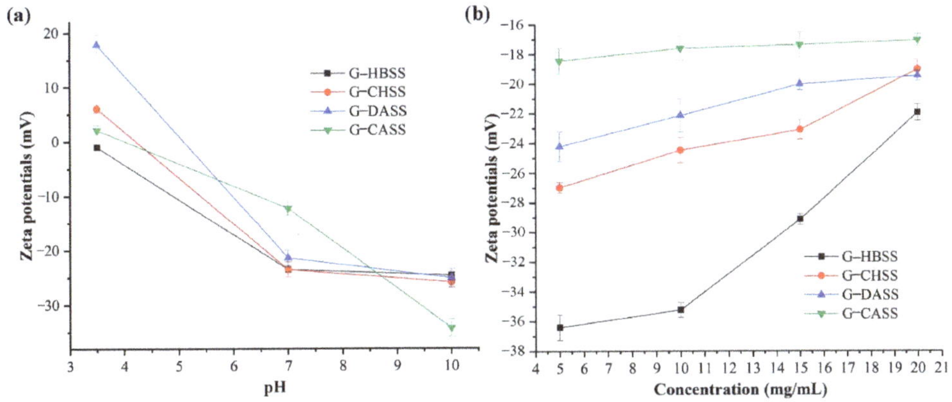

Figure 1. Zeta potential of G-MLPs at 20 mg/mL as a function of pH (**a**); zeta potential of G-MLPs as a function of concentration at pH 7 (**b**).

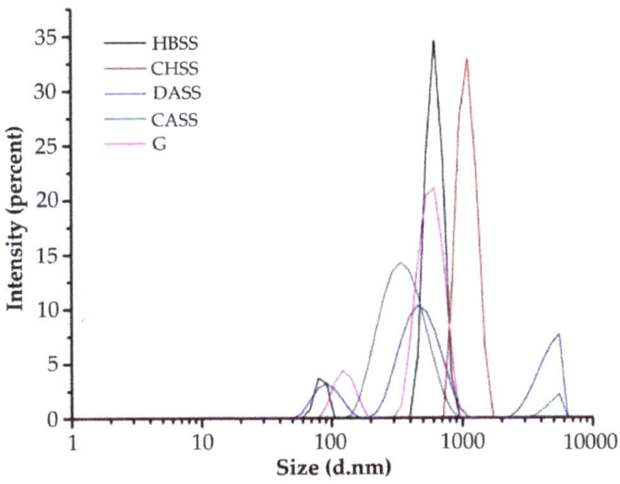

Figure 2. Particle size distribution of MLPs and gelatin (G).

Figure 3 shows the particle size and distribution of four miscible systems at pH 3.5, 7.0, and 10.0. At pH 3.5, G-HBSS, G-CHSS, G-DASS, and G-CASS formed two, one, two, and three main peaks, respectively. When the pH was 7, there were two, one, three, and two peaks, respectively. When the pH was 10, there were three, two, three, and two peaks, respectively. Compared with the particle size of gelatin and four polysaccharides, the particle size distribution of the four miscible systems of gelatin and mulberry leaf polysaccharides changed significantly. Under different pH conditions, the G-CHSS solution still maintained high monodispersity, which may have been caused by the electrostatic complex formed after the miscibility of the CHSS and gelatin [30]. The three miscible systems of G-HBSS, G-DASS, and G-CASS showed polydispersity, which may have been due to the charge repulsion between gelatin and these three mulberry leaf polysaccharides [33]. With the increase in pH, the change trend of particle size of the four miscible systems was also different. The particle size of G-HBSS and G-CHSS miscible systems decreased with the increase in pH, while the particle size of G-DASS and G-CASS increased with the increase in pH. These phenomena may have been caused by the different extractants of the HBSS, CHSS, DASS, and CASS. Specifically, the reason for the change trend of G-HBSS and G-CHSS may be that with the increase in pH, gelatin and two polysaccharides produced repulsion, which reduced the particle size of the complex [30], while the reason for the change of G-DASS and G-CASS may be that both were alkaline extracts. Under alkaline conditions, the electrostatic complexation reaction intensity between gelatin and polysaccharides increased, and the formed complex was easy to aggregate, resulting in the increase in particle size, which was the same as the change of the solution state reflected in the above phase diagram [33].

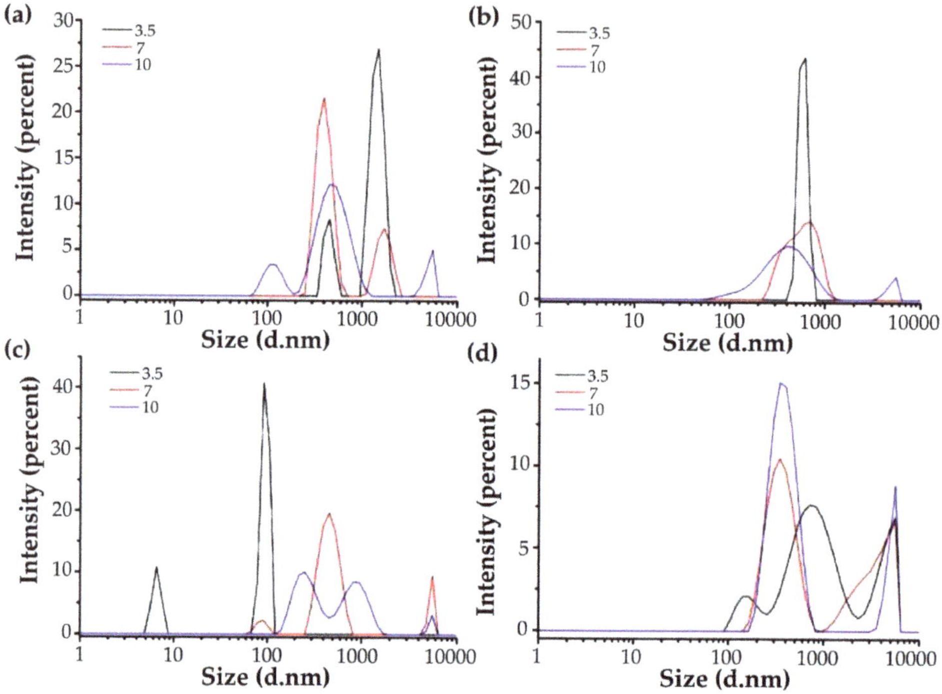

Figure 3. Particle size distribution of G-MLPs at different pHs. (**a**) G-HBSS; (**b**) G-CHSS; (**c**) G-DASS; (**d**) G-CASS.

The particle size distribution and size of the four miscible systems at different concentrations are shown in Figure 4. It can be seen from the figure that the miscible system showed polydispersity, but the particle size did not change much with the change in concentration, indicating that the change in concentration only affected the distribution of particle size, and had little effect on its size. This may be because with the increase in the polysaccharide concentration, it had little effect on the intermolecular force of the miscible system of gelatin and polysaccharides [32].

3.4. Rheological Property of Gelatin–Mulberry Leaf Polysaccharide (G-MLPs) Miscible System

3.4.1. Effect of Concentration on Apparent Viscosity of G-MLPs

The change in and influence of apparent viscosity of four miscible systems with different concentrations and shear rates and the comparison with the apparent viscosity of the gelatin solution are shown in Figure 5. Under the condition of a low shear rate, the viscosity of the four miscible systems decreased with the increase in shear rate, and they showed the "shear thinning" behavior [20]. The main reason for the "shear thinning" behavior was that gelatin and polysaccharide were colloidal particles of macromolecules. Under the condition of a static or low shear rate, they entangled with each other and had a higher viscosity, while with the increase in shear rate, the shear stress between the flow layers increased, which made the scattered gelatin–polysaccharide chain particles roll, rotate, and shrink into clusters, resulting in reducing the hooking effect between the chain molecules and the viscosity of gelatin–polysaccharide solutions. On the other hand, with the increase in flow rate, the molecular force between miscible systems decreased, the flow direction of gelatin–polysaccharide macromolecules changed from disorder to ordered, the flow direction appeared consistent, and the viscosity decreased [34]. This was a typical non-Newtonian fluid behavior, indicating that the miscible systems were a non-Newtonian fluid like pure polysaccharide solutions. As mentioned above, "Shear thinning" behavior is important in food processing industries for the development of various food products with desirable properties [20].

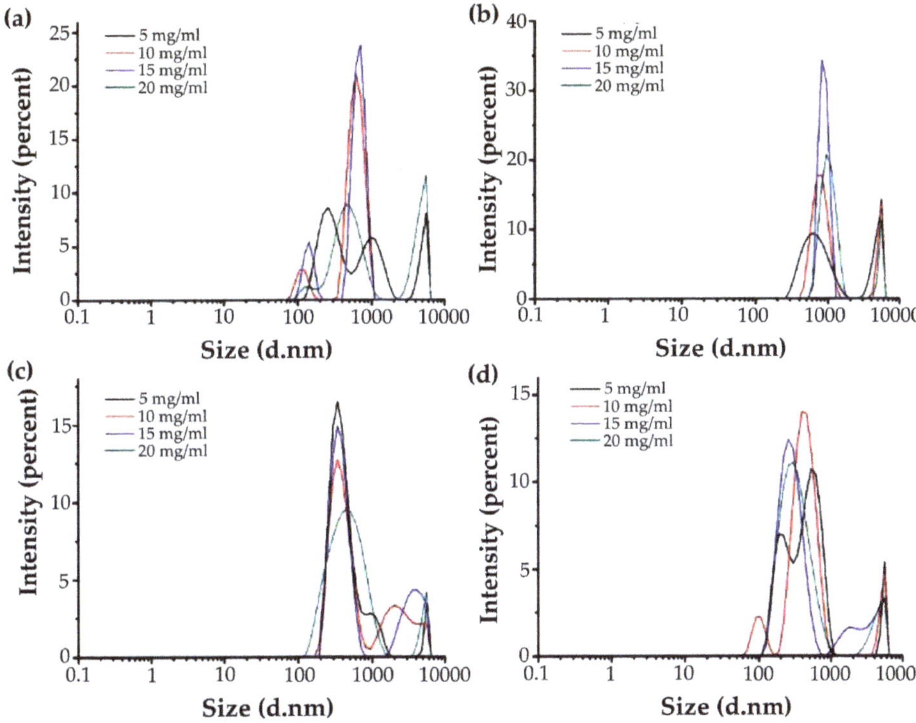

Figure 4. Particle size distribution of G-MLPs at different concentrations. (**a**) G-HBSS; (**b**) G-CHSS; (**c**) G-DASS; (**d**) G-CASS.

The apparent viscosity of the four miscible systems varied with the increase in mass concentration (Figure 5). Except for the apparent viscosity of the G-CHSS miscible system at the highest concentration, which was higher than that of the gelatin solution, the apparent viscosity of other miscible systems was lower than that of the gelatin solution, which indicated that the addition of polysaccharides reduced the apparent viscosity of the gelatin solution. Specifically, after the four mulberry leaf polysaccharide solutions were mixed with the gelatin solution, the apparent viscosity of G-HBSS, G-DASS, and G-CASS miscible systems decreased with the increase in mass concentration, while the apparent viscosity of the G-CHSS miscible system increased with the increase in mass concentration, which showed relatively appreciable thickening behavior of G-CHSS. One study reported that the apparent viscosity of polysaccharides increased with their concentration, while their apparent viscosity decreased after the addition of gelatin, which may be related to the fact that the particle size of the miscible system in the particle size experiment did not increase with the increase in solution concentration [35]. The previous studies [18,20] showed that the apparent viscosity increased with the increase in mass concentration, mainly because the fluid viscosity results from intermolecular internal friction. In this study, the particle size in the miscible system did not increase, resulting in the decrease in apparent viscosity. Within the studied concentration range, the fluid types of gelatin and mulberry leaf polysaccharide miscible systems did not change due to the change in concentration.

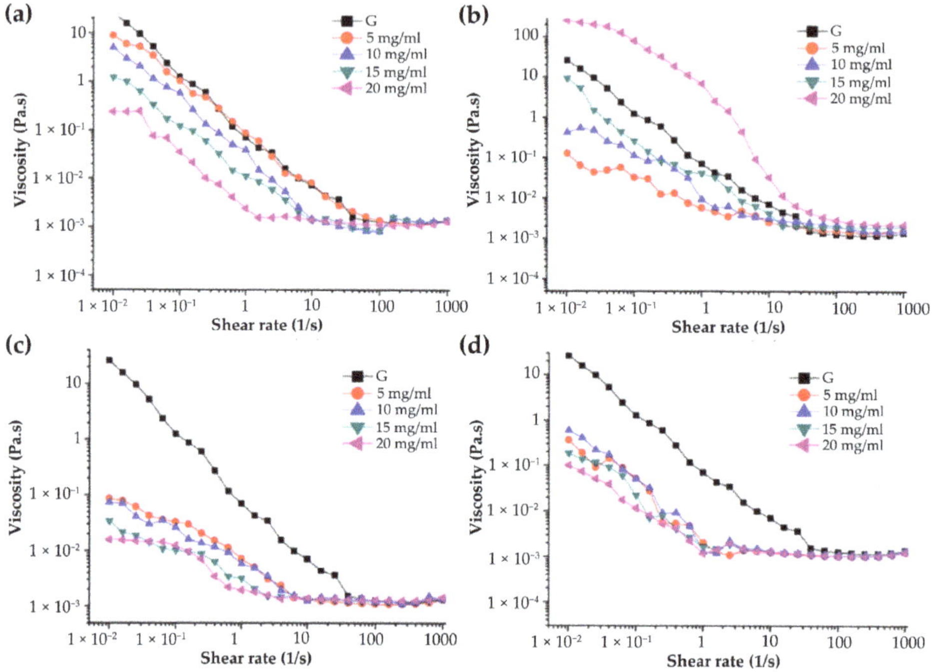

Figure 5. Effect of different concentrations on the viscosity of G-MLPs. (**a**) G-HBSS; (**b**) G-CHSS; (**c**) G-DASS; (**d**) G-CASS.

It can also be seen from Figure 5 that the rheological properties of the solutions of the miscible systems with different concentrations were the same as those of mulberry leaf polysaccharides. With the increase in concentration, the pseudoplastic flow was dominant. When the shear rate was 0.01–10 s^{-1}, the viscosity of the four miscible systems of G-HBSS, G-CHSS, G-DASS, and G-CASS decreased rapidly with the increase in shear rate, and then became stable as the shear rate increased. With the increase in shear rate, the miscible system showed the properties of Newtonian fluid, which is characterized by the gradual stabilization of apparent viscosity. Compared with pure mulberry leaf polysaccharides, G-CHSS had the highest apparent viscosity, and G-DASS had the lowest apparent viscosity among the four miscible systems. The above results show that G-CHSS can be used as a food thickener, gelling agent, and binder in food, medicine, and cosmetics industries [18].

3.4.2. Effect of pH on Apparent Viscosity of G-MLPs

The four miscible systems showed different apparent viscosity changes in low- and high-pH solutions (Figure 6). G-CHSS and G-DASS showed a decrease in apparent viscosity under acidic and alkaline conditions, which may be because the intermolecular force of the miscible system was weakened in the acidic or alkaline environment, resulting in the decomposition of the polymer of the polysaccharide and gelatin miscible system and a decrease in molecular weight and apparent viscosity [36]. G-HBSS and G-CASS had the highest apparent viscosity under acidic conditions, and the apparent viscosity decreased with the increase in pH. The reason for this may be that the two polysaccharide miscible systems belonged to an acidic polysaccharide miscible system, and their structure was much more stable under acidic conditions, while in the alkaline environment, the solvation effect of water increased with the increase in OH- in the solution, which may have broken the intramolecular and intermolecular hydrogen bonds of the molecules and reduced the interactions, resulting in making the structure of the miscible system loosen and decreasing the viscosity of G-MLPs [37]. Among the four miscible systems, the CHSS had the highest stability, and it can be used as an important ingredient in acidic or alkaline beverages and other products.

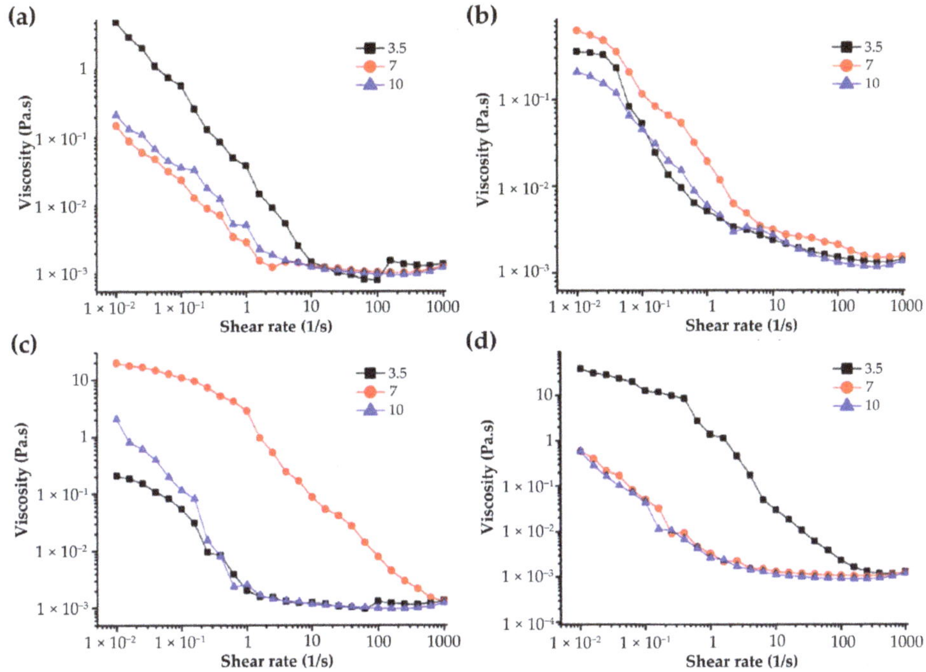

Figure 6. Effect of pH on the viscosity of G-MLPs. (**a**) G-HBSS; (**b**) G-CHSS; (**c**) G-DASS; (**d**) G-CASS.

3.4.3. Effects of Na^+ Concentration on Apparent Viscosity of G-MLPs

Figure 7 shows the effect of Na^+ concentration on the apparent viscosity of 10 mg/mL gelatin–polysaccharide miscible systems. It can be seen from the figure that the apparent viscosity of G-CASS (Figure 7b) increased with the increase in Na^+ concentration. The reason for this phenomenon may be that the addition of Na^+ affected the molecular conformation of the miscible system and improved the structural entanglement, resulting in the increase in apparent viscosity [38,39]. The apparent viscosity of G-CHSS decreased with the increase in Na^+ concentration (Figure 7b). This may be due to the salting out of the polysaccharide solution of the miscible system due to the addition of Na^+, which would reduce the concentration of the polysaccharide solution and the apparent viscosity. The higher the NaCl concentration, the more serious the salting out phenomenon and the greater the viscosity drop [18,20]. The apparent viscosity of G-HBSS and G-DASS did not change regularly with the concentration of Na^+, which may have been caused by the particularity of the structure of the HBSS and DASS. The action law of Na^+ on their miscible systems was different from that of G-CHSS and G-CASS. In other words, the low concentration of Na^+ increased the structural entanglement between molecules, resulting in the increase in apparent viscosity, while the high concentration of Na^+ salted out the system, resulting in the decrease in apparent viscosity [18]. The inconsistency of the rheological behavior of samples may also have been due to the different intensity of the interaction between gelatin and the HBSS, CHSS, DASS, and CASS in different concentrations of sodium chloride solution. Therefore, during the food processing, the content of salt ions needs to be strictly controlled to ensure the desired products.

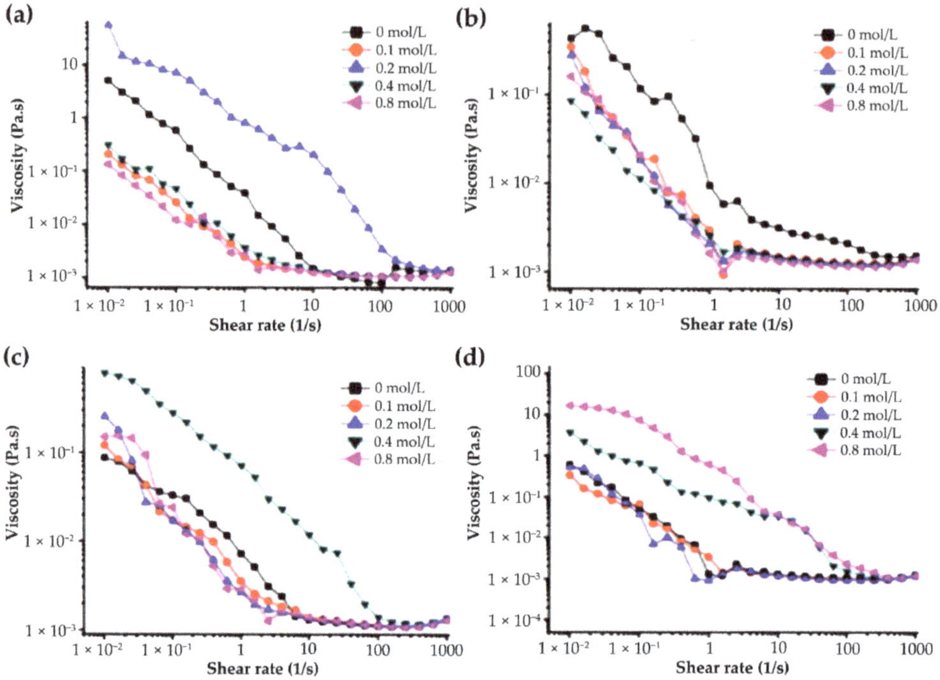

Figure 7. Effect of Na^+ on the viscosity of G-MLPs (10 mg/mL). (**a**) G-HBSS; (**b**) G-CHSS; (**c**) G-DASS; (**d**) G-CASS.

3.4.4. Effect of System Temperature on Apparent Viscosity of G-MLPs

The different effects of cold treatment and heat treatment on the apparent viscosity of the four miscible systems are shown in Figure 8. Gelatin is thermo-reversible, meaning that the viscosity of a gelatin solution would increase with decreasing temperature to form a gel, and vice versa. Thus, heating and freezing followed by reversing the system temperature to room temperature will have no significant effect on gelatin viscosity [39]. The rheological properties of the four miscible systems after cold and heat treatment were different. After cold treatment, the apparent viscosity of G-HBSS decreased, while the apparent viscosity of G-CHSS, G-DASS, and G-CASS was increased. After heat treatment, the apparent viscosity of G-HBSS was decreased. The increase in the apparent viscosity of G-DASS was higher than that of cold treatment, while the increase in the apparent viscosity of G-CASS was lower than that of cold treatment, and the apparent viscosity of G-CASS after freezing treatment was the highest among the four miscible systems, which indicated that G-CASS was more suitable as a stabilizer in the freezing process. After heat treatment, the apparent viscosity of the G-CHSS sample generally showed an upward trend with the increase in shear rate and changed from "shear thinning" to "shear thickening" expansive fluid [40]. The reason for this phenomenon may be that the interaction between gelatin and the CHSS was strengthened during the heat treatment, resulting in the very stable miscible system. When the temperature returned to room temperature, it was still in an orderly distribution. Therefore, with the increase in shear rate, the order in the system was disturbed and transformed into disorder, and the apparent viscosity increased with the increase in shear force. The above results show that the four miscible systems, except G-HBSS, had good freeze–thaw stability and could be applied to the products that need to be stabilized after freezing, while the "shear thickening" characteristic of G-CHSS marks its suitability as a thickener in heat processing.

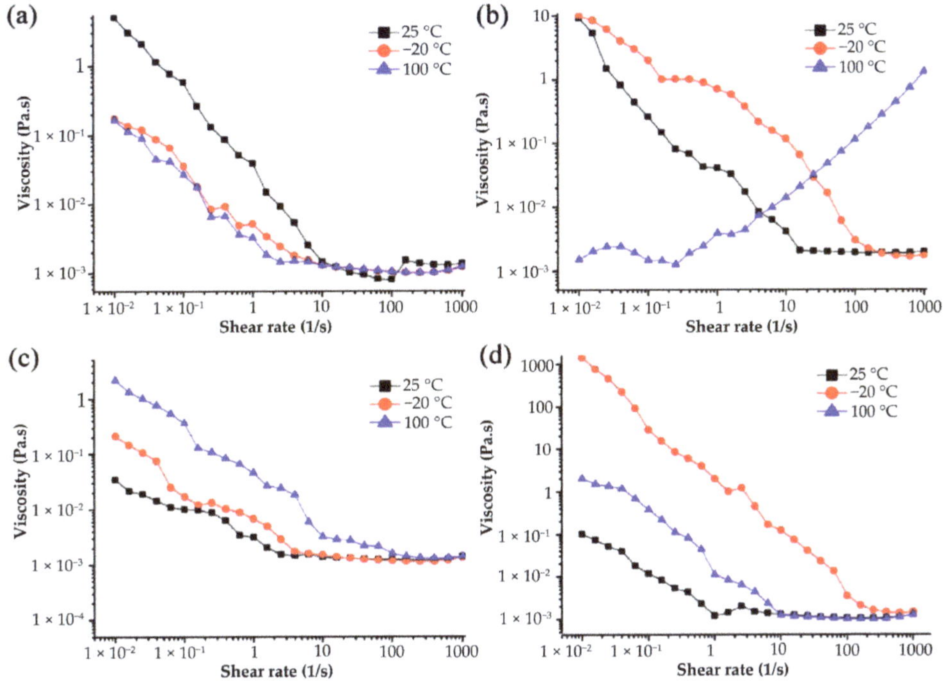

Figure 8. Effect of freezing and heating on the viscosity of G-MLPs. (**a**) G-HBSS; (**b**) G-CHSS; (**c**) G-DASS; (**d**) G-CASS.

3.4.5. Effect of G-MLPs on Viscoelasticity

At 25 °C, the storage modulus G′ and loss modulus G″ of gelatin and mulberry leaf polysaccharide miscible systems changed with frequency, as shown in Figure 9. By dynamically measuring the G′ and G″ of the samples, the advantages of solid elasticity or liquid viscosity of the sample could be quantified [41]. Within the range of shear oscillation frequency measured in this study, the storage modulus G′ and loss modulus G″ of the four miscible systems had a certain dependence on the shear oscillation frequency, which showed that they continued to increase with the increase in shear oscillation frequency and indicated that gelatin and mulberry leaf polysaccharides solutions were a viscoelastic material [18,20]. The storage modulus G′ and loss modulus G″ of G-HBSS were different from those of the other three miscible systems. At 5 mg/mL and 10 mg/mL, G′ was greater than G″ under the low shear vibration frequency of G-HBSS, and solution mainly showed the elastic properties of a solid. With the increase in frequency, G′ was less than G″, and the main property of the solution changed to the viscosity of the liquid. The frequency of shear oscillation further increased until the G′ of the solution was greater than G″. The elastic properties of the solid were dominant [42]. The changes between the storage modulus G′ and loss modulus G″ of G-HBSS at 15 mg/mL and 20 mg/mL were the same as those of the other three miscible systems. At low frequency, the storage modulus G′ was less than the loss modulus G″, and the solution showed the viscosity of the liquid. With the increase in frequency, a crossover value is reported [43]. At this time, the loss modulus G′ curve and the storage modulus G″ curve was close to or intersected, which also meant that the crossover value was negatively correlated with the elastic contribution at the beginning of the elastic behavior [44]. With the further increase in oscillation frequency, G′ began to exceed G″, which indicated that the elastic properties were dominant. Additionally, the miscible system mainly showed the elastic properties of solids. The crossover values of four gelatin and mulberry leaf polysaccharide miscible systems varied with different concentrations. At 5 mg/mL and 15 mg/mL, the crossover value of G-CHSS was lower than that of the other three, while at 10 mg/mL and 20 mg/mL, the crossover value of G-DASS was lowest.

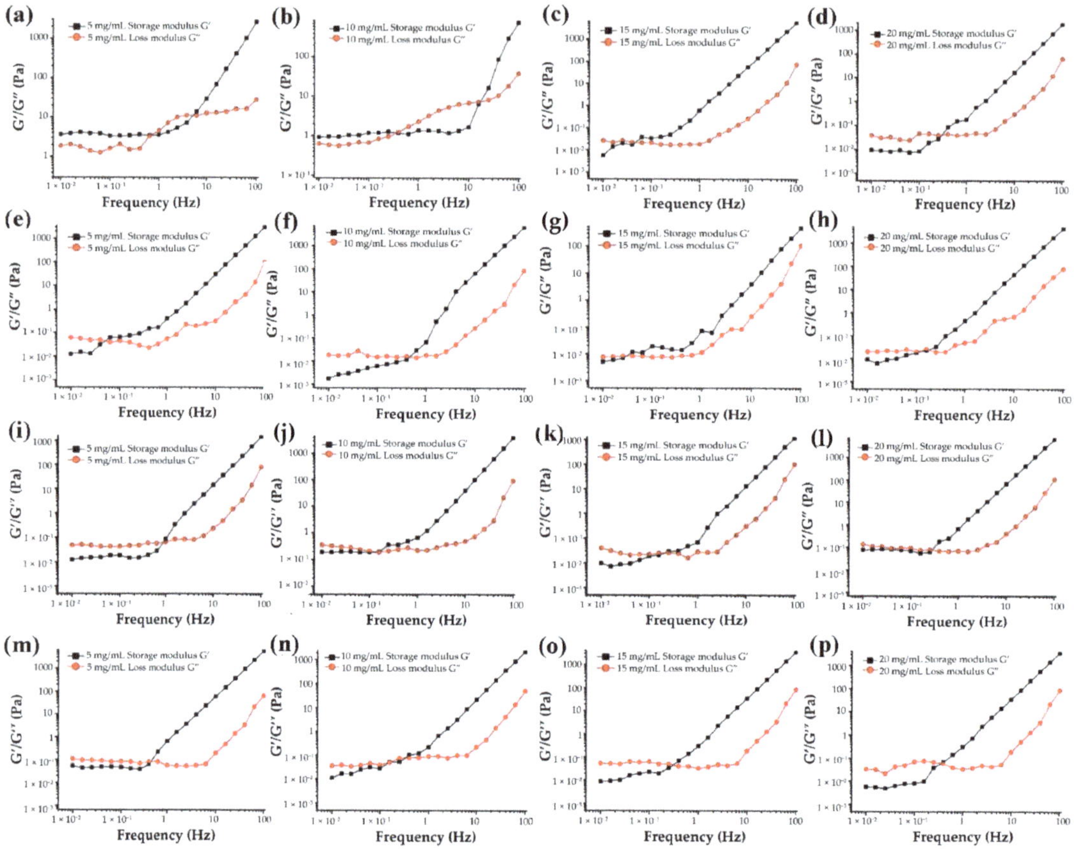

Figure 9. Frequency dependence of storage (G′) and loss (G″) modulus of G-HBSS (**a**–**d**), G-CHSS (**e**–**h**), G-DASS (**i**–**l**), and G-CASS (**m**–**p**) at different concentrations.

4. Conclusions

The zeta potential, turbidity, rheological properties, and particle size of gelatin–mulberry leaf polysaccharide miscible systems were evaluated. Under acidic conditions, the four miscible systems were in a clear state. Under alkaline conditions, G-HBSS and G-CHSS were in a clear state, while G-DASS and G-CASS changed from a clear to cloudy state. With the increase in pH, the potential of the four miscible systems changed from positive to negative and decreased gradually. At pH 7.0, the zeta potential of the four miscible systems increased with the increase in MLP concentration, but it was still negative. The particle size distribution of the four miscible systems showed polydispersity at different concentrations, but the particle size did not change much. The four miscible systems showed "shear thinning" behavior, and the addition of gelatin reduced the apparent viscosity of four polysaccharides solutions. The apparent viscosity of G-HBSS, G-DASS, and G-CASS decreased with the increase in mass concentration, while the apparent viscosity of the G-CHSS miscible system increased with the increase in mass concentration, and it had relatively appreciable thickening. The effect of pH value on the stability of the miscible system was in the order of G-CHSS > G-HBSS > G-CASS > G-DASS. Na^+ had different effects on the apparent viscosity of the solutions of the four miscible systems. With the increase in Na^+ concentration, the apparent viscosity of G-CASS increased, while G-HBSS, G-CHSS, and G-DASS decreased. After cold and heat treatment, G-CHSS changed from a "shear thinning" to "shear thickening" expansive fluid. Dynamic oscillatory shear data revealed that the miscible systems exhibited viscous properties with the increase in oscillation frequency.

Author Contributions: J.-G.Z. and Z.-J.W. proposed and designed the experiment; X.-X.Z., B.-Y.L. and Z.-J.G. experimented; X.-X.Z., B.-Y.L. and K.T. wrote the paper; M.R.K. and R.B. processed the data; R.B., J.-G.Z. and Z.-J.W. corrected the language of the paper. All authors have read and agreed to the published version of the manuscript.

Funding: This work was supported by the key research and development projects in Ningxia Province (2021BEF02013), the National Natural Science Foundation of Ningxia Province (2021AAC02019), the youth talent cultivation project of North Minzu University (2021KYQD27, FWNX14), the Researchers Supporting Project No. (RSP-2021/138) King Saud University (Riyadh, Saudi Arabia), and the Major Projects of Science and Technology in Anhui Province (201903a06020021, 202004a06020042, 202004a06020052).

Institutional Review Board Statement: Not applicable.

Informed Consent Statement: All related data and methods are presented in this paper. Additional inquiries should be addressed to the corresponding author.

Data Availability Statement: Not applicable.

Conflicts of Interest: The authors declare no conflict of interest.

References

1. Fan, W.; Xia, Z.; Liu, C.; Ma, S.; Liu, S.; Wu, Y.; Zhu, B.; Xu, C.; Zhao, A. Ionomics, transcriptomics and untargeted metabolomics analyses provide new insights into the Cd response and accumulation mechanisms of mulberry. *Environ. Exp. Bot.* **2022**, *196*, 104821. [CrossRef]
2. Ma, G.Q.; Chai, X.Y.; Hou, G.G.; Zhao, F.L.; Meng, Q.G. Phytochemistry, bioactivities and future prospects of mulberry leaves: A review. *Food Chem.* **2022**, *372*, 131335. [CrossRef] [PubMed]
3. Chen, X.; Li, Y.; Zeng, J.; Zheng, L.; Zhou, Y.; Xu, K.; Politova, E.D.; Li, G. Effect of rhombohedral-orthorhombic phase transition on structure and electrical properties of KNN-based lead-free piezoceramics. *Ferroelectrics* **2016**, *505*, 10–16. [CrossRef]
4. Eruygur, N.; Dural, E. Determination of 1-Deoxynojirimycin by a developed and validated HPLC-FLD method and assessment of In-vitro antioxidant, α-Amylase and α-Glucosidase inhibitory activity in mulberry varieties from Turkey. *Phytomedicine* **2019**, *53*, 234–242. [CrossRef]
5. Joh, B.; Jeon, E.S.; Lim, S.H.; Park, Y.L.; Park, W.; Chae, H. Intercultural Usage of Mori Folium: Comparison Review from a Korean Medical Perspective. *Evid. Based Complementary Altern. Med.* **2015**, *2015*, 379268. [CrossRef]
6. Chan, E.W.C.; Wong, S.K.; Tangah, J.; Inoue, T.; Chan, H.T. Phenolic constituents and anticancer properties of *Morus alba* (white mulberry) leaves. *J. Integr. Med.* **2020**, *18*, 189–195. [CrossRef]
7. Qadir, R.; Anwar, F.; Gilani, M.A.; Zahoor, S.; Misbah ur Rehman, M.; Mustaqeem, M. RSM/ANN based optimized recovery of phenolics from mulberry leaves by enzyme-assisted extraction. *Czech. J. Food Sci.* **2019**, *37*, 99–105. [CrossRef]
8. Hu, X.Q.; Jiang, L.; Zhang, J.G.; Deng, W.; Wang, H.L.; Wei, Z.J. Quantitative determination of 1-deoxynojirimycin in mulberry leaves from 132 varieties. *Ind. Crops Prod.* **2013**, *49*, 782–784. [CrossRef]
9. Li, R.L.; Xue, Z.H.; Jia, Y.N.; Wang, Y.J.; Li, S.Q.; Zhou, J.N.; Liu, J.Y.; Zhang, M.; He, C.W.; Chen, H.X. Polysaccharides from mulberry (*Morus alba* L.) leaf prevents obesity by inhibiting pancreatic lipase in high-fat diet induced mice. *Int. J. Biol. Macromol.* **2021**, *92*, 452–460. [CrossRef]
10. Liao, B.Y.; Hu, H.M.; Thakur, K.; Chen, G.H.; Li, L.; Wei, Z.J. Hypoglycemic activity and the composition analysis of the polysaccharide extracted from the fruit of *Mori multicaulis*. *Curr. Top. Nutraceut. Res.* **2018**, *16*, 1–8.
11. Hu, X.Q.; Thakur, K.; Chen, G.H.; Hu, F.; Zhang, J.G.; Zhang, H.B.; Wei, Z.J. Metabolic Effect of 1 Deoxynojirimycin from Mulberry Leaves on db/db Diabetic Mice Using Liquid Chromatography−Mass Spectrometry Based Metabolomics. *J. Agric. Food Chem.* **2017**, *65*, 4658–4667. [CrossRef]
12. Zhang, Q.; Zhang, F.; Thakur, K.; Wang, J.; Wang, H.; Hu, F.; Zhang, J.G.; Wei, Z.J. Molecular mechanism of anti-cancerous potential of Morin extracted from mulberry in Hela cells. *Food Chem. Toxicol.* **2018**, *112*, 466–475. [CrossRef] [PubMed]
13. Thakur, K.; Zhang, Y.Y.; Mocan, A.; Zhang, F.; Zhang, J.G.; Wei, Z.J. 1-Deoxynojirimycin, its potential for management of non-communicable metabolic diseases. *Trends Food Sci. Technol.* **2019**, *89*, 88–99. [CrossRef]

14. Li, L.; Thakur, K.; Cao, Y.Y.; Liao, B.Y.; Zhang, J.G.; Wei, Z.J. Anticancerous potential of polysaccharides sequential extracted from Polygonatum cyrtonema Hua in Human cervical cancer Hela cells. *Int. J. Biol. Macromol.* **2020**, *148*, 843–850. [CrossRef] [PubMed]
15. Kalinina, T.S.; Zlenko, D.V.; Kiselev, A.V.; Litvin, A.A.; Stovbun, S.V. Antiviral activity of the high-molecular-weight plant polysaccharides (Panavir). *Int. J. Biol. Macromol.* **2020**, *161*, 936–938. [CrossRef]
16. Mirzadeh, M.; Keshavarz Lelekami, A.; Khedmat, L. Plant/algal polysaccharides extracted by microwave: A review on hypoglycemic, hypolipidemic, prebiotic, and immune-stimulatory effect. *Carbohydr. Polym.* **2021**, *266*, 118134. [CrossRef] [PubMed]
17. Ji, Y.H.; Liao, A.M.; Huang, J.H.; Thakur, K.; Li, X.L.; Wei, Z.J. Physicochemical and antioxidant potential of polysaccharides sequentially extracted from Amana edulis. *Int. J. Biol. Macromol.* **2019**, *131*, 453–460. [CrossRef]
18. Dimopoulou, M.; Alba, K.; Sims, I.M.; Kontogiorgos, V. Structure and rheology of pectic polysaccharides from baobab fruit and leaves. *Carbohydr. Polym.* **2021**, *273*, 118540. [CrossRef]
19. Liao, B.Y.; Zhu, D.Y.; Thakur, K.; Li, L.; Zhang, J.G.; Wei, Z.J. Thermal and antioxidant properties of polysaccharides sequentially extracted from mulberry leaves (*Morus alba* L.). *Molecules* **2017**, *22*, 22. [CrossRef]
20. Liao, B.Y.; Li, L.; Tanase, C.; Thakur, K.; Zhu, D.Y.; Zhang, J.G.; Wei, Z.J. The rheological behavior of polysaccharides from mulberry leaves (*Morus alba* L.). *Agronomy* **2020**, *9*, 10. [CrossRef]
21. Guo, X.; Chen, M.; Li, Y.; Dai, T.; Shuai, X.; Chen, J.; Liu, C. Modification of food macromolecules using dynamic high pressure microfluidization: A review. *Trends Food Sci. Technol.* **2020**, *100*, 223–234. [CrossRef]
22. Zhang, Z.; Hao, G.; Liu, C.; Fu, J.; Hu, D.; Rong, J.; Yang, X. Recent progress in the preparation, chemical interactions and applications of biocompatible polysaccharide-protein nanogel carriers. *Food Res. Int.* **2021**, *147*, 110564. [CrossRef] [PubMed]
23. Luo, Q.; Hossen, M.A.; Zeng, Y.; Dai, J.; Li, S.; Qin, W.; Liu, Y. Gelatin-based composite films and their application in food packaging: A review. *J. Food Eng.* **2022**, *313*, 110762. [CrossRef]
24. Pei, Y.; Zheng, Y.; Li, Z.; Liu, J.; Zheng, X.; Tang, K.; Kaplan, D.L. Ethanol-induced coacervation in aqueous gelatin solution for constructing nanospheres and networks: Morphology, dynamics and thermal sensitivity. *J. Colloid Interface Sci.* **2021**, *582*, 610–618. [CrossRef]
25. Azarikia, F.; Abbasi, S. Mechanism of soluble complex formation of milk proteins with native gums (tragacanth and Persian gum). *Food Hydrocoll.* **2016**, *59*, 35–44. [CrossRef]
26. Esteban, P.P.; Jenkins, A.T.; Arnot, T.C. Elucidation of the mechanisms of action of Bacteriophage K/nano-emulsion formulations against S. aureus via measurement of particle size and zeta potential. *Colloid Surf. B* **2016**, *139*, 87–94. [CrossRef] [PubMed]
27. Li, H.; Wang, T.; Hu, Y.; Wu, J.; Van der Meeren, P. Designing delivery systems for functional ingredients by protein/polysaccharide interactions. *Trends Food Sci. Technol.* **2022**, *119*, 272–287. [CrossRef]
28. Wu, Y.; Ding, W.; Jia, L.; He, Q. The rheological properties of tara gum (*Caesalpinia spinosa*). *Food Chem.* **2015**, *168*, 366–371. [CrossRef]
29. Zhang, J.; Wang, G.; Liang, Q.; Cai, W.; Zhang, Q. Rheological and microstructural properties of gelatin B/tara gum hydrogels: Effect of protein/polysaccharide ratio, pH and salt addition. *LWT-Food Sci. Technol.* **2019**, *103*, 108–115. [CrossRef]
30. Zhang, T.; Xu, X.; Ji, L.; Li, Z.; Wang, Y.; Xue, Y.; Xue, C. Phase behaviors involved in surimi gel system: Effects of phase separation on gelation of myofibrillar protein and kappa-carrageenan. *Food Res. Int.* **2017**, *100*, 361–368. [CrossRef]
31. Nakamura, A.; Ohboshi, H.; Sakai, M.; Nomura, K.; Nishiyama, S.; Ashida, H. Extraction of water-soluble polysaccharides from kidney beans and examination of their protein dispersion and stabilization properties under acidic conditions. *Food Res. Int.* **2021**, *144*, 110357. [CrossRef] [PubMed]
32. Ahsan, S.M.; Rao, C.M. The role of surface charge in the desolvation process of gelatin: Implications in nanoparticle synthesis and modulation of drug release. *Int. J. Nanomed.* **2017**, *12*, 795–808. [CrossRef] [PubMed]
33. Hosseini, S.M.H.; Emam-Djomeh, Z.; Sabatino, P.; Van der Meeren, P. Nanocomplexes arising from protein-polysaccharide electrostatic interaction as a promising carrier for nutraceutical compounds. *Food Hydrocoll.* **2015**, *50*, 16–26. [CrossRef]
34. Ji, X.; Yin, M.; Hao, L.; Shi, M.; Liu, H.; Liu, Y. Effect of inulin on pasting, thermal, rheological properties and in vitro digestibility of pea starch gel. *Int. J. Biol. Macromol.* **2021**, *193*, 1669–1675. [CrossRef] [PubMed]
35. Xie, F.; Zhang, H.; Xia, Y.; Ai, L. Effects of tamarind seed polysaccharide on gelatinization, rheological, and structural properties of corn starch with different amylose/amylopectin ratios. *Food Hydrocoll.* **2020**, *105*, 105854. [CrossRef]
36. Yang, X.; Nisar, T.; Liang, D.; Hou, Y.; Sun, L.; Guo, Y. Low methoxyl pectin gelation under alkaline conditions and its rheological properties: Using NaOH as a pH regulator. *Food Hydrocoll.* **2018**, *79*, 560–571. [CrossRef]
37. Han, W.Y.; Meng, Y.H.; Hu, C.Y.; Dong, G.R.; Qu, Y.L.; Deng, H.; Guo, Y.R. Mathematical model of Ca2+ concentration, pH, pectin concentration and soluble solids (sucrose) on the gelation of low methoxyl pectin. *Food Hydrocoll.* **2017**, *66*, 36–48. [CrossRef]
38. Pan, M.K.; Zhou, F.F.; Liu, Y.; Wang, J.-H. Na^+-induced gelation of a low-methoxyl pectin extracted from *Premna microphylla* Turcz. *Food Hydrocoll.* **2021**, *110*, 106153. [CrossRef]
39. Wang, R.; Hartel, R.W. Confectionery gels: Gelling behavior and gel properties of gelatin in concentrated sugar solutions. *Food Hydrocoll.* **2022**, *124*, 107132. [CrossRef]
40. Ghosh, A.; Majumdar, A.; Singh Butola, B. Modulating the rheological response of shear thickening fluids by variation in molecular weight of carrier fluid and its correlation with impact resistance of treated p-aramid fabrics. *Polym. Test.* **2020**, *91*, 106830. [CrossRef]
41. Ali Razavi, S.M.; Alghooneh, A.; Behrouzian, F.; Cui, S.W. Investigation of the interaction between sage seed gum and guar gum: Steady and dynamic shear rheology. *Food Hydrocoll.* **2016**, *60*, 67–76. [CrossRef]

42. Chen, L.; Dai, Y.; Hou, H.; Wang, W.; Ding, X.; Zhang, H.; Li, X.; Dong, H. Effect of high pressure microfluidization on the morphology, structure and rheology of sweet potato starch. *Food Hydrocoll.* **2021**, *115*, 106606. [CrossRef]
43. Gharanjig, H.; Gharanjig, K.; Hosseinnezhad, M.; Jafari, S.M. Development and optimization of complex coacervates based on zedo gum, cress seed gum and gelatin. *Int. J. Biol. Macromol.* **2020**, *148*, 31–40. [CrossRef]
44. Cui, T.; Wu, Y.; Ni, C.; Sun, Y.; Cheng, J. Rheology and texture analysis of gelatin/dialdehyde starch hydrogel carriers for curcumin controlled release. *Carbohydr. Polym.* **2022**, *283*, 119154. [CrossRef] [PubMed]

foods

MDPI

Review

Chitosan-Based Materials: An Overview of Potential Applications in Food Packaging

Tong Liu, Junbo Li, Qilong Tang, Peng Qiu, Dongxia Gou and Jun Zhao *

College of Food Science and Engineering, Changchun University, Changchun 130022, China; liut@ccu.edu.cn (T.L.); lijunbo0111@163.com (J.L.); 200301046@mails.ccu.edu.cn (Q.T.); a17835093337@163.com (P.Q.); goudx@ccu.edu.cn (D.G.)
* Correspondence: zhaoj70@ccu.edu.cn; Tel.: +86-0431-8511-5756 (ext. 5751)

Abstract: Chitosan is a multifunctional biopolymer that is widely used in the food and medical fields because of its good antibacterial, antioxidant, and enzyme inhibiting activity and its degradability. The biological activity of chitosan as a new food preservation material has gradually become a hot research topic. This paper reviews recent research on the bioactive mechanism of chitosan and introduces strategies for modifying and applying chitosan for food preservation and different preservation techniques to explore the potential application value of active chitosan-based food packaging. Finally, issues and perspectives on the role of chitosan in enhancing the freshness of food products are presented to provide a theoretical basis and scientific reference for subsequent research.

Keywords: chitosan; biological activity; modification; packaging; preservation

Citation: Liu, T.; Li, J.; Tang, Q.; Qiu, P.; Gou, D.; Zhao, J. Chitosan-Based Materials: An Overview of Potential Applications in Food Packaging. *Foods* **2022**, *11*, 1490. https://doi.org/10.3390/foods11101490

Academic Editors: Jianhua Xie, Yanjun Zhang and Hansong Yu

Received: 22 April 2022
Accepted: 18 May 2022
Published: 20 May 2022

1. Introduction

Chitosan, with the scientific composition of (1,4)-2-amino-2-deoxy-D-glucose, is the only natural cationic alkaline polysaccharide. Chitin is mainly found in shrimp, crab [1], and insect shells [2]. Chitosan can be produced by the deacetylation of chitin with NaOH, as shown in Figure 1, and by fermentation with some microbial cultures [3]. As a renewable green bio-resource from a wide range of sources, chitosan has superior bacterial inhibition and antioxidant properties and enzyme inhibition activity, and it is edible, nontoxic, and biodegradable [4], so there is a great deal of research being carried out on it in the food, medicine, environmental protection, chemical, cosmetics, and textile fields, among others [5]. It has good prospects in food preservation and green packaging because of its antibacterial [6–9], antioxidant [10–14], and enzyme inhibiting [15–18] activities and biodegradability [19–21]. However, the tensile strength, structural strength, and physical properties, such as oxygen and water vapor transmission, after being made into film cannot fully meet the requirements of food packaging, so the performance of chitosan as a food packaging material must be improved by employing derivatization, preparing nanocomposites, and compounding it with biological extracts. Many studies have shown that the future direction of food preservation is to make it green, safe, and environmentally friendly preservation [22–26]. Therefore, the use of chitosan as a base material for food preservation has become a research hotspot [27,28]. Even so, research on chitosan to date has not actually resulted in a commercially available packaging product. This paper reviews reports on the relevant biological activities of chitosan in food preservation and discusses the potential applications of chitosan derivatization, chitosan-based nanocomposites, and chitosan–extract composites, as shown in Figure 2, to provide a reference for the application of chitosan in food preservation and the development of novel preservation technologies.

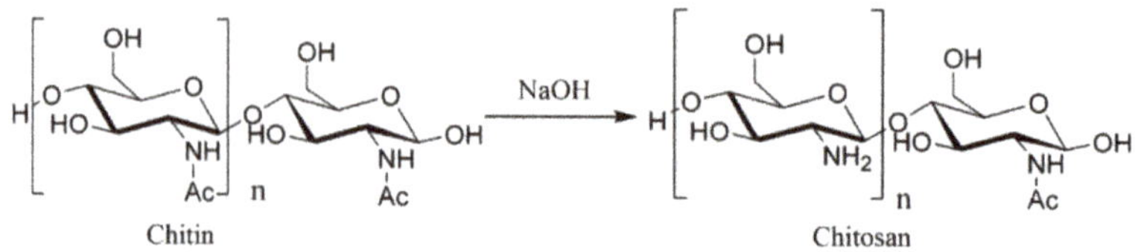

Figure 1. Preparation of chitosan from chitin by deacetylation.

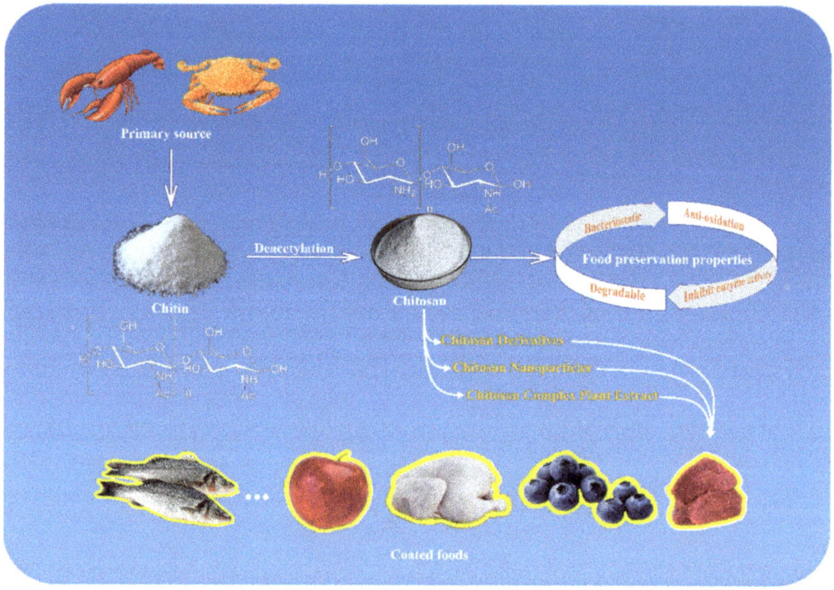

Figure 2. Sources, properties, and applications of chitosan.

2. Research Methods

In this review, information on the biological activity of chitosan and food packaging based on chitosan was consulted through various literature databases and search engines on the Internet, including but not limited to X-MOL, Google Scholar, PubMed, CNKI, AbleSci, Scopus. Through the integration of the obtained information, the hypothesis of the biological activity mechanism of chitosan in the field of food preservation is proposed, and the advantages and disadvantages of chitosan in the preservation of fresh products by coating or making packaging films are evaluated.

3. Biological Activity of Chitosan

3.1. Bacteriostatic Properties

The broad-spectrum bacterial inhibitory properties of chitosan have been confirmed by in vitro experiments and complex food antimicrobial assays [29–32]. In the process of food preservation, the use of chitosan coating or film covering the surface of food can, to a certain extent, resist microbial attack, slow down food spoilage, and thus extend the shelf life of food. For example, Wei et al. [33] studied the effects of different concentrations of chitosan on the color and surface microbial populations of frozen duck skin. The results showed that chitosan could form oxygenation competition with color-presenting substances, and 3% chitosan could effectively reduce the total number of surface colonies of duck skin during storage. In addition to the effect of chitosan concentration on the inhibition effect, molecular weight, degree of deacetylation, and bacterial species have also been shown to have an effect. Benhabiles et al. [6] studied the effect of molecular weight and degree of deacetylation on the inhibition effect, comparing the inhibition performance of chitosan on four Gram-positive and seven Gram-negative bacteria. The results showed that chitosan with low molecular weight and a high degree of deacetylation had a better inhibition effect, and among the tested species, chitosan had an inhibition effect on all bacteria except *Salmonella typhimurium*. The pH of the solution and the presence of metal ions also have an effect on the bacterial inhibitory effect of chitosan. Jing et al. [34] used ANOVA to test the data and concluded that the inhibitory activity of chitosan increased with decreasing pH in the range of >4 and <6. Comparing the effects of metal ions on the inhibitory activity of chitosan at a pH of 4, the results showed that Ca^{2+} and Mg^{2+} significantly reduced the inhibitory ability against Gram-negative bacteria. It can be seen that the inhibitory activity of chitosan is related to a variety of factors, with interlocking effects among them, and the best inhibition results can be obtained only by comprehensively analyzing all of the factors.

Although the information on the antimicrobial activity of chitosan is controversial, it is generally accepted that yeasts and mycobacteria are the most sensitive organisms to chitosan, followed by Gram-positive and Gram-negative bacteria [35]. The antifungal activity of chitosan is mainly attributed to the inhibition of mycelial growth and spore germination [36]. Early studies showed that chitosan inhibits spore germination, budding tube elongation, and radial growth by sequestering metals, minerals, trace elements, or nutrients required for fungal growth [37]. The inhibition of pathogenic fungi increases with increased chitosan concentration [38], but its application is hindered by low solubility at physiological pH due to its low charge density. Therefore, increasing the positive charge density of chitosan may be the most effective way to improve its solubility and antifungal activity [39]. The antifungal activity of chitosan also depends on its molecular weight [40], which affects the morphogenesis of the fungal cell wall [41]. It has also been shown that chitosan's inhibition of pathogenic fungi in fruits may be related to its ability to induce increased production of, e.g., polyphenol oxidase and peroxidase [42].

As research progressed, hypotheses on the mechanism of chitosan inhibition were put forward, but the exact mechanism of inhibition has still not been determined. Feng et al. [43] measured the OD of cell contents under UV light at a wavelength of 260 nm and showed an increase in OD. Using transmission electron microscopy to observe changes in the ultrastructure of *E. coli* and *S. aureus* before and after the action of chitosan, it was observed that the cell membrane was broken, accompanied by a large amount of content spillage (DNA and mRNA) at the periphery and bacterial apoptosis. Similarly, Hui et al. [44] evaluated the antibacterial activity of chitosan acetate solution against *E. coli* and *S. aureus*. The morphology of bacteria after chitosan treatment was observed by transmission electron microscopy. The integrity of the cell membranes of both bacteria and the permeability of the inner and outer membranes of *E. coli* were investigated by measuring the absorption values at 260 nm UV, fluorescence changes in cells treated with the fluorescent probe 1-N-phenylnaphthylamine, and the release of cytoplasmic h-galactosidase activity. The results showed that chitosan ultimately damaged the bacteria by increasing the permeability of the inner and outer cell membranes, releasing the cell contents. This damage is most likely caused by electrostatic interactions between the NH^{3+} group in the chitosan acetate solution and the phosphoryl group of the phospholipid fraction of the cell membrane. Imelda et al. [29] confirmed the disruptive effect of chitosan on protein synthesis by β-galactosidase expression assay.

The above studies showed that chitosan can inhibit bacteria by disrupting bacterial cell membranes and protein synthesis. Ming et al. [45] pointed out that chitosan has high chelating ability for various metal ions (including Ni^{2+}, Zn^{2+}, Co^{2+}, Fe^{2+}, Mg^{2+}, and Cu^{2+}) under acidic conditions. As a chelating agent, chitosan selectively chelates metal ions, thus playing a key role in microbial growth by inhibiting the growth and reproduction of microorganisms. Gooday et al. [46] pointed out that when chitosan crosses the cell wall of fungal pathogens with plant hydrolases as host, it penetrates the nucleus of the fungus, and the positively charged chitosan interacts with the negatively charged DNA, affecting RNA transcription and protein synthesis, thus achieving fungal inhibition.

In summary, there are three generally accepted theories about the mechanism of chitosan inhibition, the first of which is inter-charge interactions. As shown in Figure 3 [17], the electrostatic interaction between $R\text{-}NH^{3+}$ of chitosan and negatively charged groups on the bacterial surface destabilizes the structure of the cell membrane, and the leakage of intracellular material leads to microbial death. The second theory concerns the chelating property of chitosan with metal ions. Chitosan chelates metal ions on the surface of bacteria, forming a microbially unavailable chelate, thus inhibiting microbial growth. The third theory holds that chitosan enters the cell and binds to DNA, and affects RNA transcription and protein expression, thus achieving the effect of bacterial inhibition. Although the above views are generally accepted, there are still many scholars who believe that the molecular weight and degree of deacetylation, and the differences between bacterial species, should also be considered when investigating the inhibition mechanism of chitosan.

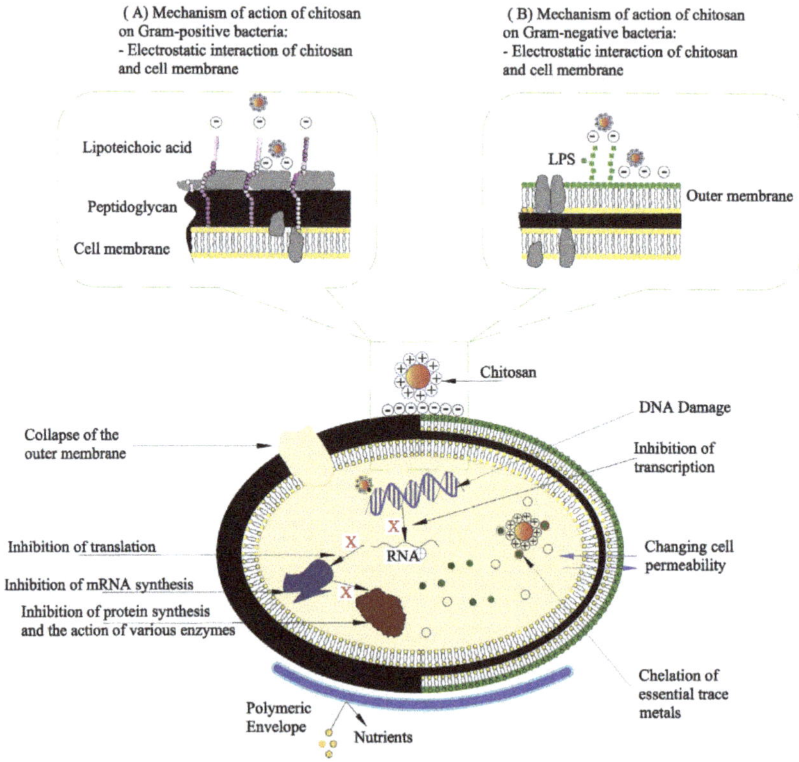

Figure 3. Mechanisms used to explain antibacterial activity of chitosan.

3.2. Antioxidant Properties

The antioxidant properties of chitosan have been demonstrated [10,47,48]. It is well known that oxidation is one of the most important issues affecting the quality of meat, and heterocyclic amines, which are generated during the high-temperature processing of foods containing proteins, are carcinogenic substances. Many factors influence the formation of heterocyclic amines, mainly cooking methods, processing conditions (time, temperature), and the presence of antioxidant substances. Therefore, reducing heterocyclic amines by natural compounds and reducing or inhibiting the formation of these carcinogens in cooked meat by plant extracts containing antioxidants have become hot research topics [49].

When chitosan is added to meat products as a food additive, chitosan concentration and temperature are important factors that affect the experimental results. Fatih et al. [50] investigated the effect of adding different concentrations of chitosan (0.25, 0.50, 0.75, 1% *w*/*w*) to meatballs cooked at different temperatures on the formation of heterocyclic aromatic amines and the quality of the meatballs. The results showed that the heterocyclic amine content of the meatballs increased with increasing temperature (150, 200, and 250 °C), and the heterocyclic amine content showed a decreasing trend with increased chitosan concentration at the same temperature. Similarly, a study by Hojat et al. [51] showed that the addition of chitosan to huso fillets for cooking was effective in reducing heterocyclic amine production. Given that the production of heterocyclic amines usually requires high-temperature conditions, chitosan coatings or films have been less studied for reducing the production of heterocyclic amines. Kader et al. [52] studied four types of chitosan coating with different degrees of deacetylation and molecular weights for cherry preservation and evaluated the changes in total phenolic content, antioxidant capacity, total anthocyanin content, ascorbic acid, total pectin content, hardness, and color of cherries. The results showed that the antioxidant capacity of chitosan and the ascorbic acid content in cherries had a trend of increasing and then decreasing with the degree of deacetylation and molecular weight, and the chitosan with 81.22% deacetylation and 273 kDa molecular weight at 4 °C showed the best antioxidant capacity.

Relatively stable DPPH radicals have been widely used to test the ability of compounds to act as free radical scavengers or hydrogen donors [53]. The IC_{50} value is the concentration at which 50% free radical scavenging is obtained, and the lower the IC_{50}, the more active the sample is as an antioxidant compound and the greater its ability to absorb free radicals. Rainy et al. [54] investigated the antioxidant properties of edible chitosan–galactose complex coating by compounding chitosan and galactose (0, 0.5, 1, and 1.5 g), and performed an in vitro test to evaluate the coating and analyze the parameters of antioxidant activity (DPPH method). The results showed a decreasing trend in the IC_{50} values, followed by an increasing trend. The lowest IC_{50} value of 43.20 ppm was recorded for the combination of chitosan and 1 g of galactose, which was the best antioxidant in the tested treatments. Similarly, Zhang et al. [55] reported the results of free radical scavenging experiments with different molecular weight chitosan, showing that high-molecular-weight chitosan had a high scavenging effect on hydroxyl radicals, and low-molecular-weight chitosan had a better scavenging effect on superoxide anion radicals and DPPH.

Giuseppina et al. [10] studied the effect of chitosan-based coatings on the freshness of figs by assessing the activities of enzymes, such as catalase, ascorbate peroxidase, polyphenol oxidase, and peroxidase. The results showed that the addition of chitosan coating significantly increased the total polyphenol, anthocyanin, and flavonoid contents and the antioxidant activity of stored figs, reduced oxidative stress, and prevented browning reactions compared with the untreated group. Miriam et al. [56] studied the effect of nano-chitosan/propolis coating on the shelf life and antioxidant capacity of strawberries, and the results showed higher total phenol and flavonoid content and antioxidant capacity of strawberries in the nano-chitosan/propolis coated group than in the untreated group at the end of the storage period. Chitosan delays the ripening and aging process of food, which extends the shelf life; largely maintains all sensory qualities of food; reduces enzymatic browning; decreases water loss; maintains the bright color, taste, and texture of food; and makes the aroma more durable, thus effectively maintaining the commercial value of food. There are many factors involved in the antioxidant performance of chitosan, and the main research concerns are concentration, degree of deacetylation, molecular weight, and antioxidant enzymes.

As the antioxidant mechanism of chitosan has been investigated, hypotheses have been proposed. Chitosan films in combination with other plant-based flavonoids have antioxidant properties as coatings on food surfaces [57]. Yuntao et al. [13] showed that high-molecular-weight chitosan films have a denser structure and better antibacterial properties. Braber et al. [58] concluded that chitosan biopolymers scavenge free radicals or chelate metal ions through a hydrogen or lone pair electron donor mechanism. The -OH and -NH_2 groups in chitosan are the key functional groups for its antioxidant activity [59]. Zhang et al. [60] analyzed the barrier properties of chitosan–cyanidin films and concluded that the increased hydrogen bonding between cyanidin and chitosan molecules leads to a tighter arrangement between molecules inside the film, which improves its gas barrier properties.

The comprehensive analysis of the above studies led to three conjectures regarding chitosan's antioxidant properties. The first is the barrier effect: chitosan coating can act as a barrier on the fruit surface, changing the internal gas atmosphere, reducing water loss, and delaying ripening. Second, many oxidation processes are carried out with the participation of metal ions, which play a role in transferring electrons during valence changes, and chitosan removes metal ions through chelation, thereby inhibiting oxidation reactions. Third, high-molecular-weight chitosan has denser intramolecular hydrogen bonds, which further prevents oxidation reactions from occurring when food comes into contact with air.

3.3. Enzyme Activity Inhibition

Enzymatic reactions play a non-negligible role in the softening and browning of foods. Tissue browning is inevitable in some fruits and vegetables with damaged surfaces due to the action of polyphenol oxidase. He et al. [18] studied the effect of clove oil–chitosan coating on the quality of fresh-cut lemons at four temperatures of 0, 4, 7, and 10 °C. By analyzing the changes in peroxidase, polyphenol oxidase, and phenolic acids, it was concluded that the chitosan coating had a stronger inhibitory effect on enzyme activity as the temperature decreased. Seafood is also highly susceptible to deterioration due to enzymatic reactions, which is the main reason for its quality decline [17].

Chitosan concentration is another important factor influencing enzyme activity. A-Dan et al. [61] investigated the effect of chitosan on papain and showed that with increased chitosan concentration, there was an inhibitory effect on papain when the concentration was greater than about 8.0288 g/L. Arisa et al. [62] investigated the effect of chitosan coating with different molecular weights at a concentration of 1% on the physicochemical properties of red bananas. The results showed that chitosan coating was able to reduce fruit respiration rate, ethylene production, and pectin hardness by inhibiting the activity of cell wall degrading enzymes (polygalacturonase and pectin lyase) that are important in pectin degradation, and with increasing molecular weight, chitosan coating showed better inhibition of enzymatic activity, thus delaying banana spoilage and deterioration. Although the mechanism by which chitosan inhibits the browning of fruits and vegetables and softening of meat tissues has not been elaborated, it must be related to the inhibition of enzymes associated with the occurrence of enzymatic reactions in foods.

With the study of the mechanism of chitosan's enzymatic activity inhibition, related hypotheses have been proposed. Kurita et al. [63] indicated that polycations may compete with divalent metals, such as Mg^{2+} and Ca^{2+}, present in the cell wall, thus disrupting the integrity of the cell wall or affecting the activity of degradative enzymes. To resist local browning in fruits and vegetables, chitosan binds suspended polyphenol oxidase molecules through electrostatic interactions to inactivate the enzyme, and the inactivated polyphenol oxidase is unable to transfer electrons directly to molecular oxygen, thus acting as an anti-browning agent [56]. Riaz et al. [64] studied the effect of chitosan-based apple peel polyphenol composite coating on improving the storage quality of strawberries. The antioxidant content of the coated fruits was relatively stable compared to uncoated fruits. The reduced antioxidant activity during storage may lead to decay and senescence. The ability of chitosan-based apple peel polyphenol composite coating to reduce decay, decrease enzymatic activity, and retain the quality attributes of fruits, leading to the degradation of antioxidant compounds, is closely related to the presence of efficient oxygen radical scavengers [65]. Yage et al. [66] used chitosan/nano-titanium dioxide coating to induce film formation on the surface of mangoes, creating a microenvironment that reduced decay and water loss and delayed respiratory peaks to preserve fruit flavor.

In summary, the hypotheses on the mechanism of action of chitosan's enzyme activity inhibition include the following: First, the chelation of metal ions disrupts the enzyme activation pathway, and calpain and matrix metalloproteinases are Ca^{2+} dependent endogenous enzymes. Second, chitosan molecules penetrate the sarcoplasm and bind directly to the enzyme, thereby affecting enzyme activity. Third, chitosan may bind to structural proteins, which can affect the degradation and dissociation of enzyme proteins. Fourth, the chitosan coating alters the microenvironment of the sample, thus synthetically modulating the biochemical characteristics of the preserved food. Unfortunately, there are not enough relevant studies, and more research is needed to develop systematic interpretations of these hypotheses.

3.4. Biodegradability

Biodegradable polymer materials can be degraded by microorganisms or their secretions under the action of enzymes or chemical decomposition with a certain time and under certain conditions [67]. The degradation of biodegradable materials is generally accompanied by changes in their chemical and physical properties, reduced molecular weight, and the production of low-molecular-weight substances (CO_2, H_2O, CH_4). Ruchir et al. [20] reported that chitosan coatings are environmentally friendly, biodegradable, and in most cases edible. Aris et al. [19] pointed out that although polymers synthesized with chitosan cannot be completely degraded, mixing chitosan with synthetic polymers can improve the decomposition rate of plastics that are more difficult to decompose. Polyvinyl alcohol (PVA), which is nontoxic and water-soluble, is one of the most commonly used synthetic polymers for chitosan. Yueming Li et al. [68] studied the effect of biodegradable, antibacterial chitosan starch composite film on the freshness of red grapes. The results showed that the biodegradable film had good water retention and antimicrobial efficacy and could be applied to grape preservation.

Zhang et al. [69] observed a link between degradation rate and molecular weight, degree of deacetylation, and distribution of N-acetyl D-glucosamine residues. Chitosan is a semi-crystalline polymer, and the relationship between chitosan biodegradability and the degree of deacetylation also depends on the degree of crystallinity, which reaches a maximum at a deacetylation degree of 0% (in the form of titin) or 100% (fully deacetylated chitosan) and decreases at intermediate values, with the rate of biodegradation increasing as the degree of crystallinity decreases. Similarly, Croisier et al. [70] found that acetyl residues distributed along the chitosan chain also affect the crystallinity of chitosan and, thus, the biodegradation rate. It can be concluded that smaller chitosan chains biodegrade more efficiently than higher-molecular-weight chitosan. In vivo, chitosan can be degraded by several enzymes, including lysozyme, a nonspecific enzyme present in all mammalian tissues. The degradation products are nontoxic oligosaccharides, and in vitro degradation of chitosan occurs by oxidative, chemical, or enzymatic hydrolysis. These methods are commonly used for low-molecular-weight chitosan prepared under controlled conditions.

In addition to being a polymer with an amino group, chitosan is also a polysaccharide and therefore contains readily breakable glycosidic bonds. Chitosan appears to be degraded in vivo by nonspecific enzymes, but mainly lysozyme has been reported to have this property. The biodegradation of chitosan leads to the formation of nontoxic oligosaccharides. These oligosaccharides, with variable lengths, can potentially be incorporated into metabolic pathways or excreted from the body [71,72]. Therefore, it is clear that chitosan is a strong natural candidate to replace nondegradable plastics in the future and its position as a natural alternative is established. Based on the current research, the relative application of chitosan in food packaging will be developed and commercialized shortly.

4. Status of Research on Chitosan Preservation

4.1. Enhanced Freshness Preservation Performance

Despite the many biological activities of chitosan described above, its antimicrobial and antioxidant capacities are still lacking when it is used for food preservation [73,74]. To address these issues and expand its application, many methods have been investigated to enhance its preservation performance, mainly including derivatization, chitosan nanoparticles, and combinations with natural extracts, essential oils, natural polymers, synthetic polymers, etc.

4.1.1. Chitosan Derivatization

The functional groups of chitosan contain an amino group and two hydroxyl groups, which can easily react with a variety of chemical groups. Introducing new groups by modifying chitosan molecules to produce derivatives with excellent physical and chemical properties is important to expand the applications of chitosan and its derivatives [75]. Previous and current studies have shown that chitosan derivatives have positive effects in terms of antioxidant activity. Xiao et al. [76] studied the use of gallic acid–chitosan derivatives to preserve cherry tomatoes and evaluated the antioxidant and endogenous enzyme activities. The results showed that the enzyme-grafted gallic acid–chitosan derivatives had excellent antioxidant capacity in scavenging DPPH, hydroxyl, and superoxide anion radicals. This resulted in prolonged fruit ripening, reduced weight loss, high hardness, and little change in epidermal color in the treated group, and aromatic compounds remained relatively constant throughout storage due to delayed postharvest senescence.

Similarly, Neslihan et al. [77] synthesized two novel chitosan–Schiff base derivatives by condensation reactions of high- and low-molecular-weight chitosan with cotton phenol and studied their scavenging activity against 1,1-diphenyl-2-picrylhydrazyl (DPPH) radicals. They concluded that both chitosan–cotton phenol derivatives had better scavenging ability against DPPH radicals than unmodified chitosan. The modified chitosan had significantly improved antibacterial activity against foodborne bacterial spoilage. Similarly, chitosan derivatives have good film-forming and antimicrobial properties, and the use of film coating treatment has significant effects on inhibiting browning, reducing enzyme activity, maintaining fruit and vegetable quality, and prolonging freshness. Meriem et al. [78] prepared active nanocomposite films of cellulose nanocrystals reinforced with styrene-based quinoxaline-grafted chitosan by the solvent casting method and studied their antibacterial activity against five common bacteria. The films demonstrated good antibacterial activity against *Pseudomonas aeruginosa*.

Furthermore, films prepared from chitosan derivatives have different physical properties depending on the type and degree of substitution of the grafting groups. Bingnan et al. [79] found that the mechanical properties of chitosan products could be improved with a small amount of differently structured formyl saccharides by establishing intermolecular formyl–sugar bonds. When the amount of sucralfate and cottonseed aldehyde was 0.5%, the breaking strength and strain of chitosan films increased from 28 to 38 MPa and 19% to 48%, respectively, and the wet stability and toughness were improved at the same time. Currently, various chitosan derivatives have been successfully designed and applied to food preservation, as shown in Table 1. There are not many research cases of chitosan derivatives for food preservation due to the concern for their potential toxicity. Therefore, future research should not only explore more chitosan derivatives with specific and enhanced functions, but also develop green chemical processes to avoid toxic residues.

Table 1. Related studies on chitosan derivatives for food preservation.

Product	Derivative	Reference
Beef	Chitosan/lauric	[80]
Tomato	N, O-carboxymethyl chitosan	[81]
Cucumber	Chitosan-g-salicylic	[82]
Strawberry	N-succinyl chitosan	[83]
Mulberry	Chitosan-g-caffeic	[84]
Blueberry	Carboxymethyl chitosan-peptide	[85]
Button mushroom	Gallic acid-grafted chitosan	[86]
Pork	Ferulic acid-grafted chitosan	[87]

4.1.2. Chitosan Nanoparticles

Due to its large particle size, chitosan does not come into contact with food spoilage factors when used for food preservation, thus limiting the biological activity of chitosan alone when used for this purpose [88]. Studying chitosan nanoparticles is another efficient option to enhance the utilization of chitosan. Melo et al. [89] developed composite pectin films supplemented with copper algae mud chitosan nanoparticles and evaluated their physical and mechanical properties. The mechanical analysis of maximum stress and elongation showed that nanoparticles as fillers increased the toughness of the pectin films. Water vapor permeability tests showed that nanostructured films containing copper sulfur had better barrier properties. More studies on using chitosan nanoparticles to enhance the physical and mechanical properties of chitosan films are presented in Table 2.

Ran et al. [90] studied the use of chitosan nanoparticles for fish preservation and found that the storage life of fish fillets wrapped in composite nanofilms could be extended by 6–8 days through sensory evaluation, microbiological analysis, pH, total volatile alkaline nitrogen value, thiobarbituric acid value, color, texture, and other storage quality indicators. The plant essential oil anthocyanin composite film had the best effect on fish fillet preservation, and the anthocyanin chitosan composite nanoparticle film had the best effect on protecting fish fillet appearance. In another study, modified magnetic chitosan nanomaterials obtained by combining modified chitosan with magnetic materials retained the properties of chitosan with the strong chelating ability of metal ions and had good regenerative properties, greatly expanding the application prospects of chitosan [91].

There are several methods to prepare chitosan nanoparticles, including the ionic gel [92], microemulsion [93], emulsification solvent diffusion [94], polyelectrolyte complex [95], and reversed-phase micellar [96] methods. Chitosan nanoparticles have both the biological properties of chitosan and the small size of nanoparticles with a large contact area. The final performance of nanoparticles also depends on several other parameters, such as the composition of secondary materials in the matrix, the material concentration/ratio, and the reaction time [97]. In addition, studies have reported using different types of chitosan (varying by molecular weight, degree of deacetylation, etc.) [98,99]. The applications of chitosan-based nanoparticles are gradually increasing, and it is expected that chitosan-based nanoparticle formulations will enter the market soon.

Table 2. Correlation of mechanical properties of chitosan nanoparticles after film formation.

Property	Nanomaterials	Reference
Barrier	Chitosan/tripolyphosphate	[100]
Water sorption	Chitosan/starch	[101]
Stretchability	Chitosan/chlorogenic acid	[102]
Water vapor	Chitosan/TiO_2 and Ag	[103]
Transparency	Chitosan/curcumin	[104]
Breathability	Chitosan/polyoxyethylene	[105]

4.1.3. Chitosan Plant Extract Compound

Plant extracts are often rich in low-molecular-weight bioactive ingredients, such as polyphenols, terpenoids, and terpenoids, which are considered to be powerful antibacterial and antioxidant agents [106]. Plant extracts are known not only to act as free radical scavengers in vitro but also to protect the body from free radical activity. The respiration and transpiration of fruits and vegetables during storage can cause water loss, and when water loss reaches 5%, wilting and withering can occur, which can seriously affect the edible value [107]. Table 3 summarizes some recent literature on the use of plant extracts in combination with chitosan-based coatings/films for different food products.

The use of chitosan-based composite coating and preservation solution can reduce water loss, inhibit cellular action, and prolong the storage time of fruits and vegetables. Rambabu et al. [108] investigated the application of mango leaf extract incorporated into chitosan film to enhance the antioxidant activity and characterized the tensile strength. They found that the mango leaf extract–chitosan composite film had better tensile strength (maximum tensile strength of 23.06 ± 0.19 MPa) and less elongation compared to the pure chitosan film (18.14 ± 0.72 MPa). They also evaluated the antioxidant activity of chitosan films in terms of total phenolic content, DPPH radical scavenging ability, iron ion reducing ability, and ABTS radical scavenging ability, and the results showed that the antioxidant activity of increased with the addition of mango leaf extract. Yang et al. [109] conducted similar experiments, in which blueberry leaf extract was added to chitosan coating to maintain the postharvest quality of fresh blueberries. The composite film synthesized with chitosan and plant extracts also had UV-blocking and antioxidant abilities [110]. Wanli et al. [111] studied the process of preparing chitosan–banana peel extract composite film. Different contents of banana peel extract (4%, 8%, and 12%) were added to chitosan membranes using various characterization methods, and the experimental results showed that the chitosan–banana peel extract composite membranes had good antioxidant activity in different food samples.

Plant extracts, when synergized with chitosan, usually have an inhibitory effect on bacterial growth, which is related to metabolic disorders caused in bacteria through disruption of cell membranes, enzyme systems, or genetic material. Ali et al. [112] investigated chicken meat coated with chitosan and containing 1% essential oil of oregano and 1% or 2% grape seed extract stored in the refrigerator. On the 20th day of refrigeration, the minimum viable count of each treatment organism (Enterobacteriaceae, *Pseudomonas* spp., *Lactobacillus*, and *Saccharomyces* (yeast-mold)) was 3.54–4.51 log CFU/mL. Compared with the untreated group, the results indicated that the compound coating was effective in retarding microbial growth and oxidative activity. At present, most of the plant extracts compounded with chitosan are from easily available and nontoxic substances, such as herbs and spices, and they are used in food preservation by obtaining their essential oils, which have active antioxidant and antibacterial active properties when compounded with chitosan. However, there are still problems, such as the complex extraction process, low efficiency, and the effect on the sensory quality of food. Therefore, the extraction and bio-preservation ability of plant extracts still need to be explored in the future in order to provide better technical support for the development of food preservation.

Table 3. Studies on chitosan–plant extract complexes for food coatings.

Commodity	Plant Extract	Reference
Blueberry	Blueberry leaf	[109]
Banana	Gum arabic	[113]
Sierra fish fillets	Tomato	[114]
Strawberry	Peony	[115]
Mango	Pomegranate peel	[116]
Apple	Banana peels	[117]
Sweet basil leaves	Thyme volatile oil	[118]
Le Conte pear	Beeswax–pollen grains	[119]

4.2. Freshness Preservation Technology

In recent years, the preparation of chitosan-based preservation films has become increasingly standardized, and the growing demand for food preservation has greatly contributed to the development of chitosan biofilm preservation technology. In terms of operation, traditional coating and film-making methods for food preservation have been widely used and have shown reliable results. The thickness of chitosan films is usually in the range of 6–80 μm, with an average of 25 μm, as reported by Elena et al. [120]. In addition, emerging technologies, such as film production from plant extracts, by lamination with plastic films, and by laminated layer self-assembly (LbL), are emerging. Both direct coating and bioactive packaging film preservation methods are systematically discussed at a technical level below.

4.2.1. Coating Preservation

Coating preservation refers to covering the food surface with a layer of chitosan solution, which can usually be applied by dip coating, electrostatic spraying, or brushing. Due to its outstanding advantages of convenience and operability, it is often used in food preservation research. For example, Yong et al. [121,122] used electrostatic spraying of chitosan solution to preserve strawberries, which prolonged their shelf life and slowed down the degradation of quality. Rui et al. [123] investigated the potential application of chitosan as a natural growth regulator for bean sprouts by dipping bean sprouts in different concentrations of chitosan solution. The results showed a significant increase in the hypocotyl length and fresh weight of bean sprouts compared to the blank control. In addition, chitosan solution combined with plant extracts have been used as dip coatings to retard the quality deterioration of fruits, vegetables, meat, etc. Mehran et al. [124] applied chitosan solution mixed with Berberis extract and peppermint essential oil to turkey breast meat, and microbial counts and oxidation levels were significantly reduced under refrigerated conditions.

Usually, chitosan-based dip-coating treatment consists of the following steps: First, a pre-determined concentration of chitosan solution is prepared, and then the food is immersed in the solution for a pre-determined time, depending on the characteristics of the food. Finally, the food is removed, the excess solution adhering to the surface is drained, and the product is dried in a specific airflow environment. Regarding other coating methods, all steps are similar to dip coating except for the intermediate steps. Sometimes, repeated coating is applied to enhance preservation during storage [125].

Compared to other coating methods, it is obvious that dip coating is easy to operate under simple conditions. However, it is also prone to problems with dilution and contamination in practical applications. In addition, gravity tends to cause uneven film thickness on the food surface and reduces productivity to some extent [126]. The preservation substrate determines the thickness of the coating, the equipment requirements, and the drying technology used. In addition, self-healing coatings are intelligent materials that can repair coating damage and restore its properties, reducing or eliminating the adverse effects of damage. As a method of fabricating multilayer coatings, laminated layer self-assembly (LbL) is often used to construct self-healing coatings. Yu et al. [127] investigated a self-healing coating by assembling chitosan (CS) with sodium alginate (SA) layer-by-layer, and the damaged area could be completely repaired after three assembly cycles. The self-healing schematic is shown in Figure 4. The mechanical properties, water barrier, and oxygen barrier of the repaired coating were 97%, 95%, and 63% of those of the intact coating, respectively, which basically restored the barrier properties of the intact coating, although the permeability increased. In addition, the coating reduced the effect of coating damage by restoring the barrier properties of the coating and extending the shelf life of the strawberries. Thus, it can be seen that chitosan coating not only slows down food quality degradation, but also demonstrates efficient antioxidant and antibacterial effects. However, the uniformity of covering food and rough sensory effects need to be further improved. Therefore, the application of chitosan coating for freshness is worth further study, as it has the potential to become a low-cost alternative technology for fresh food preservation.

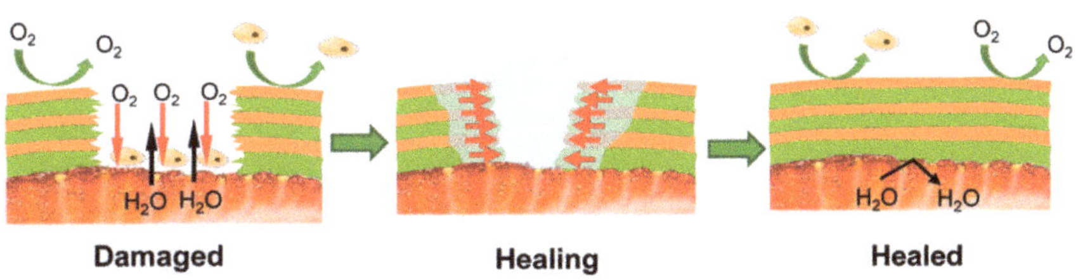

Figure 4. Diagram of self-healing behavior of chitosan (CS)/sodium alginate (SA) coating.

4.2.2. Film-Making and Preservation

Bioactive chitosan packaging films are made by co-blending chitosan solution with bioactive materials to make cling film in which to wrap food and provide freshness. Cling films can be produced in advance in large quantities without the condition of preserved products. They can be broadly divided into two types according to the composite materials: packaging film based on chitosan solution and hybrid film prepared in combination with food-grade plastic film.

Chitosan-based biofilms are often prepared by co-blending with extracts derived from natural plants, which are of increasing interest to the food industry because of their eco-safety and rich nutritional value. The addition of active ingredients not only enhanced the inherent bioactivity of the films, but also improved the final mechanical properties and applicability of the films. For example, Zhang et al. [128] improved the mechanical properties of chitosan films by vanillin modification and showed that the film stretch of vanillin/chitosan composite films increased by 53.3% and moisture permeability decreased by 36.5% compared to pure chitosan films. Ting et al. [129] added spice extracts to chitosan films to extend the shelf life of frozen pork. The data showed that chitosan could interact intermolecularly with the polyphenols in the spices and also caused the formation of covalent bonds between the hydrogen and water molecules of the polyphenols, increasing the hydrophilicity of the composite films. In addition, plant essential oil, an aromatic oily liquid extracted from plant tissues and organs, can not only scavenge free radicals in the body and perform antioxidant functions, but also has the functions of regulating the balance of intestinal flora, killing pathogenic bacteria, and promoting the secretion of digestive juices, which is a plant extract with more applications at present. When the plant essential oil is compounded with a chitosan solution in a certain ratio, the prepared chitosan film will also have the relevant nutritional properties of plant extracts, and the plant extracts can function at low concentrations. The selective gas (CO_2 and O_2) permeability and good mechanical properties of chitosan make it an excellent film-forming material. However, the low water barrier properties of chitosan films limit their application, as effective control of water transfer is a desirable property for most food packaging, especially in humid environments [130]. Several studies have focused on the moisture permeability of chitosan films containing essential oils or other plant extracts. López et al. [131] found that the addition of carvacrol (0.5, 1.0, and 1.5% v/v) significantly reduced the moisture permeability of chitosan films. When essential oils or other plant extracts (tea tree, carvacrol, cinnamon, turmeric, etc.) were added to chitosan films, the moisture permeability was significantly reduced, which may be due to the hydrophobicity of the essential oil particles. It can be seen that various properties of the film or coating can be modified by adding essential oils, and chitosan with essential oils has been shown to increase hydrophobicity, reduce water vapor permeability, and improve the antimicrobial activity of the film.

For the preparation of multilayer or hybrid structured packaging films in combination with food-grade plastic films, synthetic plastics and biodegradable polymers provide sufficient mechanical strength to meet the packaging requirements and are widely used matrix materials [132]. With regard to structured packaging films, the physical and mechanical properties depend heavily on the mixing state and compatibility between the constituent components. Shun et al. [133] prepared chitosan–polylactic acid plastic films by an extrusion method and used them to preserve grouper fillets, then analyzed their physical properties and antibacterial activity. The results showed that the water vapor transmission rate and water content of the plastic films was increased, and inhibition of *Escherichia coli*, *Pseudomonas fluorescens*, and *Staphylococcus aureus* reached over 95%. For hybrid structured packaging films, the thermal stability of the active component is a key factor to consider. Sudharsan et al. [133] prepared a compatible PLA/chitosan composite film with a thickness of 0.25 mm by the solvent casting method, and it demonstrated good thermal stability, low oxygen permeability, and high mechanical properties. Although all the above studies are still at the experimental stage and still some distance from commercialization, it is believed that there will be a positive development trend of bioactive chitosan packaging films through the continuous development of film making technology.

5. Conclusions and Outlook

In summary, chitosan has good biological activities, including antibacterial and antioxidant activity and enzyme activity inhibition, and has demonstrated great potential in the field of food preservation. Thus, it has attracted widespread attention. Although there are inconsistent views among domestic and foreign scholars on the mechanism behind chitosan's preservation of food, the generally accepted views are as follows: (1) inter-charge interaction, (2) chelation of metal ions, (3) disruption of cell membranes to enter bacteria, bind DNA, and affect protein expression, (4) dense intramolecular hydrogen bonding of chitosan, and (5) binding to enzymes or structural proteins, which affects enzyme activity. In addition, chitosan preservation methods are constantly changing and will continue to be integrated.

This review is a discussion and summary of the current status of chitosan in the field of food preservation, to help the reader understand its importance as a sustainable material in this field and related strategies to improve its properties to enlighten future developments. While much work has been done to establish methods and elucidate the positive performance of chitosan in food preservation, there are still many difficulties and theoretical shortcomings in practical application. Therefore, future developments should focus on the following aspects: in-depth research on the mechanism behind chitosan's food preservation properties, and technical optimization and enrichment of strategies to further approach practical applications. In addition, studies analyzing the synergistic effects between components and methods as well as potential toxicity are also needed. Finally, the advantages of various technologies and composite components should be summarized regularly to help achieve effective the targeted inhibition of food spoilage sources. The future of chitosan as a sustainable material in the food preservation industry is bright, and expanding its application will be an inevitable trend as the technology continues to be developed.

Author Contributions: T.L. performed the conceptual design and supervised the work. J.L. was responsible for writing—original manuscript. Q.T. is responsible for data collection and literature search. P.Q. and D.G. were responsible for the article review. Project management and funding acquisition were handled by J.Z. All authors have read and agreed to the published version of the manuscript.

Funding: This study was funded by the Science and Technology Research Project of the Jilin Provincial Education Department (JJKH20210618KJ) and the Science and Technology Program of the Jilin Provincial Science and Technology Department (20210203135SF).

Acknowledgments: We thank the Science and Technology Research Project of Jilin Provincial Department of Education and the Science and Technology Program of Jilin Provincial Department of Science and Technology for financial support, and the laboratory of Changchun University College of Food for technical support.

Conflicts of Interest: The authors declare no conflict of interest. The funders had no role in the design of the study; in the collection, analyses, or interpretation of data; in the writing of the manuscript, or in the decision to publish the results.

References

1. Mulyani, R.; Mulyadi, D.; Yusuf, N. Chitosan Membrane from Shrimp Shells (*Panaeus Modonon*) as an Antibacterial Food. *Proc. J. Phys. Conf. Ser.* **2020**, *147*, 072006. [CrossRef]
2. Mohan, K.; Ganesan, A.R.; Muralisankar, T.; Jayakumar, R.; Sathishkumar, P.; Uthayakumar, V.; Chandirasekar, R.; Revathi, N.J. Recent insights into the extraction, characterization, and bioactivities of chitin and chitosan from insects. *Trends Food Sci. Technol.* **2020**, *105*, 17–42. [CrossRef] [PubMed]
3. Tan, Y.N.; Lee, P.P.; Chen, W.N.J.F. Dual extraction of crustacean and fungal chitosan from a single Mucor circinelloides fermentation. *Fermentation* **2020**, *6*, 40. [CrossRef]

4. Wang, W.; Xue, C.; Mao, X. Chitosan: Structural modification, biological activity and application. *J. Biol. Macromol.* **2020**, *164*, 4532–4546. [CrossRef] [PubMed]
5. Kou, S.G.; Peters, L.M.; Mucalo, M.R. Chitosan: A review of sources and preparation methods. *Int. J. Biol. Macromol.* **2021**, *169*, 85–94. [CrossRef]
6. Benhabiles, M.S.; Salah, R.; Lounici, H.; Drouiche, N.; Goosen, M.F.A.; Mameri, N. Antibacterial activity of chitin, chitosan and its oligomers prepared from shrimp shell waste. *Food Hydrocoll.* **2012**, *29*, 48–56. [CrossRef]
7. Cheah, W.Y.; Show, P.-L.; Ng, I.-S.; Lin, G.-Y.; Chiu, C.-Y.; Chang, Y.K. Antibacterial activity of quaternized chitosan modified nanofiber membrane. *Int. J. Biol. Macromol.* **2019**, *126*, 569–577. [CrossRef]
8. Hajji, S.; Chaker, A.; Jridi, M.; Maalej, H.; Jellouli, K.; Boufi, S.; Nasri, M.J.E.S.; Research, P. Structural analysis, and antioxidant and antibacterial properties of chitosan-poly (vinyl alcohol) biodegradable films. *Environ. Sci. Pollut. Res.* **2016**, *23*, 15310–15320. [CrossRef]
9. Youssef, A.M.; Abou-Yousef, H.; El-Sayed, S.M.; Kamel, S.J.I. Mechanical and antibacterial properties of novel high performance chitosan/nanocomposite films. *Int. J. Biol. Macromol.* **2015**, *76*, 25–32. [CrossRef]
10. Adiletta, G.; Zampella, L.; Coletta, C.; Petriccione, M. Chitosan Coating to Preserve the Qualitative Traits and Improve Antioxidant System in Fresh Figs (*Ficus carica* L.). *Agriculture* **2019**, *9*, 84. [CrossRef]
11. Hadidi, M.; Pouramin, S.; Adinepour, F.; Haghani, S.; Jafari, S.M. Chitosan nanoparticles loaded with clove essential oil: Characterization, antioxidant and antibacterial activities. *Carbohydr. Polym.* **2020**, *236*, 116075. [CrossRef] [PubMed]
12. Huang, Y.L.; Tsai, Y.H. Extraction of chitosan from squid pen waste by high hydrostatic pressure: Effects on physicochemical properties and antioxidant activities of chitosan. *Int. J. Biol. Macromol.* **2020**, *160*, 677–687. [CrossRef] [PubMed]
13. Liu, Y.; Yuan, Y.; Duan, S.; Li, C.; Hu, B.; Liu, A.; Wu, D.; Cui, H.; Lin, L.; He, J.; et al. Preparation and characterization of chitosan films with three kinds of molecular weight for food packaging. *Int. J. Biol. Macromol.* **2020**, *155*, 249–259. [CrossRef]
14. Maringgal, B.; Hashim, N.; Mohamed Amin Tawakkal, I.S.; Muda Mohamed, M.T. Recent advance in edible coating and its effect on fresh/fresh-cut fruits quality. *Trends Food Sci. Technol.* **2020**, *96*, 253–267. [CrossRef]
15. Duran, A.; Kahve, H.I. The effect of chitosan coating and vacuum packaging on the microbiological and chemical properties of beef. *Meat Sci.* **2020**, *162*, 107961. [CrossRef] [PubMed]
16. Hamad, A.F.; Han, J.H.; Kim, B.C.; Rather, I.A. The intertwine of nanotechnology with the food industry. *Saudi J. Biol. Sci.* **2018**, *25*, 27–30. [CrossRef]
17. Inanli, A.G.; Tümerkan, E.T.A.; Abed, N.E.; Regenstein, J.M.; Özogul, F. The impact of chitosan on seafood quality and human health: A review. *Trends Food Sci. Technol.* **2020**, *97*, 404–416. [CrossRef]
18. Li, H.; Shui, Y.; Li, S.; Xing, Y.; Xu, Q.; Li, X.; Lin, H.; Wang, Q.; Yang, H.; Li, W.; et al. Quality of fresh cut lemon during different temperature as affected by chitosan coating with clove oil. *Int. J. Food Prop.* **2020**, *23*, 1214–1230. [CrossRef]
19. Giannakas, A.; Vlacha, M.; Salmas, C.; Leontiou, A.; Katapodis, P.; Stamatis, H.; Barkoula, N.M.; Ladavos, A. Preparation, characterization, mechanical, barrier and antimicrobial properties of chitosan/PVOH/clay nanocomposites. *Carbohydr. Polym.* **2016**, *140*, 408–415. [CrossRef]
20. Priyadarshi, R.; Rhim, J.-W. Chitosan-based biodegradable functional films for food packaging applications. *Innov. Food Sci. Emerg. Technol.* **2020**, *62*, 102346. [CrossRef]
21. Matica, A.; Menghiu, G.; Ostafe, V.J.N.F.C. Biodegradability of chitosan based products. *New Front. Chem.* **2017**, *26*, 75–86.
22. Ribeiro-Santos, R.; Andrade, M.; de Melo, N.R. Use of essential oils in active food packaging: Recent advances and future trends. *Trends Food Sci. Technol.* **2017**, *61*, 132–140. [CrossRef]
23. Juan, L.; Ying, F.; Nuerguli, M. Application of different compounding techniques of tea polyphenol complexes in food preservation. *J. Food Saf. Guide* **2021**, 44–45. [CrossRef]
24. Zhen, N.; Wang, X.; Li, X.; Xue, J.; Zhao, Y.; Wu, M.; Zhou, D.; Liu, J.; Guo, J.; Zhang, H.J.M.B. Protein-based natural antibacterial materials and their applications in food preservation. *Microb. Biotechnol.* **2022**, *15*, 1324–1338. [CrossRef]
25. Bajpai, V.K.; Kamle, M.; Shukla, S.; Mahato, D.K.; Chandra, P.; Hwang, S.K.; Kumar, P.; Huh, Y.S.; Han, Y.-K. Prospects of using nanotechnology for food preservation, safety, and security. *J. Food Drug Anal.* **2018**, *26*, 1201–1214. [CrossRef]
26. Zhao, Y.; Li, B.; Li, C.; Xu, Y.; Luo, Y.; Liang, D.; Huang, C.J.F. Comprehensive Review of Polysaccharide-Based Materials in Edible Packaging: A Sustainable Approach. *Foods* **2021**, *10*, 1845. [CrossRef]
27. Dutta, P.; Tripathi, S.; Mehrotra, G.; Dutta, J.J.F. Perspectives for chitosan based antimicrobial films in food applications. *Food Chem.* **2009**, *114*, 1173–1182. [CrossRef]
28. Ansorena, M.R.; Marcovich, N.E.; Pereda, M.J.H.E.; Springer: Berlin, G. *Food Biopackaging Based on Chitosan*; Springer: Berlin/Heidelberg, Germany, 2019; pp. 2057–2083.
29. Galvan Marquez, I.; Akuaku, J.; Cruz, I.; Cheetham, J.; Golshani, A.; Smith, M.L. Disruption of protein synthesis as antifungal mode of action by chitosan. *Int. J. Food Microbiol.* **2013**, *164*, 108–112. [CrossRef]
30. Liu, N.; Chen, X.-G.; Park, H.-J.; Liu, C.-G.; Liu, C.-S.; Meng, X.-H.; Yu, L.-J. Effect of MW and concentration of chitosan on antibacterial activity of *Escherichia coli*. *Carbohydr. Polym.* **2006**, *64*, 60–65. [CrossRef]
31. Yilmaz Atay, H. Antibacterial activity of chitosan-based systems. In *Functional Chitosan*; Springer: Berlin/Heidelberg, Germany, 2019; pp. 457–489.
32. Sun, M.W.; Tong, J.J.; Jiang, L.S.; Xiong, B.H. Factors influencing the antimicrobial activity of chitosan and its mechanism of action. *J. Acta Zool. Sin.* **2018**, *30*, 4327–4333.

33. Wei, C.; Liao, X.; Huang, Y.; Zhou, F.; Liao, Y.; Zhou, Z. Effects of different levels of chitosan and licorice antioxidants on skin color and surface microbial counts of frozen duck meat. *J. Food Sci. Technol.* **2020**, *45*, 280–285. [CrossRef]
34. Yingjun, J.; Youjin, H.; Hui, Q.; Ying, S.; Desen, L.; Rongqian, D. Analysis of the antibacterial activity of chitosan and its antibacterial mechanism. *Chin. J. Antibiot.* **2006**, *19*, 361–365.
35. Aider, M.J.L. Chitosan application for active bio-based films production and potential in the food industry. *LWT* **2010**, *43*, 837–842. [CrossRef]
36. Arceo-Martinez, M.T.; Jiménez-Mejía, R.; Salgado-Garciglia, R.; Santoyo, G.; Lopez-Meza, J.E.; Loeza-Lara, P.D.J.A. In vitro and in vivo anti-fungal effect of chitosan on post-harvest strawberry pathogens. *Agrociencia* **2019**, *53*, 1297–1311.
37. Jackson, S.; Heath, I.J.M.R. Roles of calcium ions in hyphal tip growth. *Microbiol. Rev.* **1993**, *57*, 367–382. [CrossRef] [PubMed]
38. Kanawi, M.A.; AL Haydar, M.; Radhi, W.N.J.E.J.B. Effect of Chitin and Chitosan in Improvement of Plant Growth and Anti-Fungal Activity. *Egypt. J. Bot.* **2021**, *61*, 513–519. [CrossRef]
39. Qin, Y.; Li, P.; Guo, Z.J.C.P. Cationic chitosan derivatives as potential antifungals: A review of structural optimization and applications. *Carbohydr. Polym.* **2020**, *236*, 116002. [CrossRef]
40. Palma-Guerrero, J.; Jansson, H.B.; Salinas, J.; Lopez-Llorca, L.J.J.M. Effect of chitosan on hyphal growth and spore germination of plant pathogenic and biocontrol fungi. *J. Appl. Microbiol.* **2008**, *104*, 541–553. [CrossRef]
41. El Ghaouth, A.; Arul, J.; Grenier, J.; Asselin, A.J.P. Antifungal activity of chitosan on two postharvest pathogens of strawberry fruits. *Phytopathology* **1992**, *82*, 398–402. [CrossRef]
42. Pastor, C.; Sánchez-González, L.; Marcilla, A.; Chiralt, A.; Cháfer, M.; González-Martínez, C. Quality and safety of table grapes coated with hydroxypropylmethyl cellulose edible coatings containing propolis extract. *Postharvest Biol. Technol.* **2011**, *60*, 64–70. [CrossRef]
43. Feng, X.; Li, X.; Yang, S.; Wang, T.; Su, Z. Effect of chitosan on bacterial cell membranes. *Food Sci.* **2009**, *30*, 63–67.
44. Liu, H.; Du, Y.; Wang, X.; Sun, L.J.I. Chitosan kills bacteria through cell membrane damage. *Int. J. Food Microbiol.* **2004**, *95*, 147–155. [CrossRef] [PubMed]
45. Kong, M.; Chen, X.G.; Xing, K.; Park, H.J. Antimicrobial properties of chitosan and mode of action: A state of the art review. *Int. J. Food Microbiol.* **2010**, *144*, 51–63. [CrossRef] [PubMed]
46. Gooday, G.W.; Jeuniaux, C.; Muzzarelli, R. *Chitin in Nature and Technology*; Springer Science & Business Media: Berlin/Heidelberg, Germany, 2012.
47. Yen, M.-T.; Yang, J.-H.; Mau, J.-L.J.C. Antioxidant properties of chitosan from crab shells. *Carbohydr. Polym.* **2008**, *74*, 840–844. [CrossRef]
48. Vinsova, J.; Vavrikova, E.J.C. Chitosan derivatives with antimicrobial, antitumour and antioxidant activities—A review. *Curr. Pharm. Des.* **2011**, *17*, 3596–3607.
49. Nadeem, H.R.; Akhtar, S.; Ismail, T.; Sestili, P.; Lorenzo, J.M.; Ranjha, M.M.A.N.; Jooste, L.; Hano, C.; Aadil, R.M.J.F. Heterocyclic aromatic amines in meat: Formation, isolation, risk assessment, and inhibitory effect of plant extracts. *Foods* **2021**, *10*, 1466. [CrossRef] [PubMed]
50. Oz, F.; Zaman, A.; Kaya, M. Effect of chitosan on the formation of heterocyclic aromatic amines and some quality properties of meatball. *J. Food Process. Preserv.* **2017**, *41*, e13065. [CrossRef]
51. Mirsadeghi, H.; Alishahi, A.; Ojagh, M.; Pourashouri, P.J. Preservation. The effect of different kinds of chitosans and cooking methods on the formation of heterocyclic aromatic amines in huso (*Huso huso*) fillet. *J. Food Process. Preserv.* **2019**, *43*, e14253. [CrossRef]
52. Tokatli, K.; Demirdoven, A. Influences of chitosan coatings on functional compounds of sweet cherries. *J. Food Sci. Technol.* **2021**, *58*, 1808–1818. [CrossRef]
53. Kanatt, S.R.; Chander, R.; Sharma, A.J.F.C. Chitosan glucose complex—A novel food preservative. *Food Chem.* **2008**, *106*, 521–528. [CrossRef]
54. Sulistijowati, R.; Husain, R.; Datau, M.C. Antioxidant, antibacterial and antifungal activity of edible coating chitosan-galactose complex. In Proceedings of the IOP Conference Series: Earth and Environmental Science, Makassar, Indonesia, 22 June 2019; p. 012032.
55. Zhang, Z.T.; Zhong, Z.M. Study on antioxidant properties of chitosan with different molecular weights. *Chem. Manag.* **2019**, *344*, 1690–1696. [CrossRef]
56. Martínez-González, M.C.; Bautista-Baños, S.; Correa-Pacheco, Z.N.; Corona-Rangel, M.L.; Ventura-Aguilar, R.I.; Del Río-García, J.C.; Ramos-García, M.L. Effect of Nanostructured Chitosan/Propolis Coatings on the Quality and Antioxidant Capacity of Strawberries During Storage. *Coatings* **2020**, *10*, 90. [CrossRef]
57. Kumar, S.; Mukherjee, A.; Dutta, J. Chitosan based nanocomposite films and coatings: Emerging antimicrobial food packaging alternatives. *Trends Food Sci. Technol.* **2020**, *97*, 196–209. [CrossRef]
58. Braber, N.L.V.; Paredes, A.J.; Rossi, Y.E.; Porporatto, C.; Allemandi, D.A.; Borsarelli, C.D.; Correa, S.G.; Montenegro, M.A.J.I. Controlled release and antioxidant activity of chitosan or its glucosamine water-soluble derivative microcapsules loaded with quercetin. *Int. J. Biol. Macromol.* **2018**, *112*, 399–404. [CrossRef] [PubMed]
59. Kosaraju, S.L.; Weerakkody, R.; Augustin, M.A.J.J. Chitosan−glucose conjugates: Influence of extent of Maillard reaction on antioxidant properties. *J. Agric. Food Chem.* **2010**, *58*, 12449–12455. [CrossRef]

60. Zhang, L.; Chen, Y.; Teng, Y.; Zhang, Z. Study of chitosan-proanthocyanidin high antioxidant antibacterial films. *Packag. Eng.* **2020**, *41*, 103–109. [CrossRef]
61. Pan, A.D.; Zeng, H.Y.; Foua, G.B.; Alain, C.; Li, Y.Q. Enzymolysis of chitosan by papain and its kinetics. *Carbohydr. Polym.* **2016**, *135*, 199–206. [CrossRef]
62. Wantat, A.; Rojsitthisak, P.; Seraypheap, K. Inhibitory effects of high molecular weight chitosan coating on 'Hom Thong' banana fruit softening. *Food Packag. Shelf Life* **2021**, *29*, 100731. [CrossRef]
63. Kurita, K.J.P.D. Chemistry and application of chitin and chitosan. *Polym. Degrad. Stab.* **1998**, *59*, 117–120. [CrossRef]
64. Riaz, A.; Aadil, R.M.; Amoussa, A.M.O.; Bashari, M.; Abid, M.; Hashim, M.M. Preservation. Application of chitosan-based apple peel polyphenols edible coating on the preservation of strawberry (*Fragaria ananassa* cv. *Hongyan*) fruit. *J. Food Process. Preserv.* **2021**, *45*, e15018. [CrossRef]
65. Sogvar, O.B.; Saba, M.K.; Emamifar, A. Aloe vera and ascorbic acid coatings maintain postharvest quality and reduce microbial load of strawberry fruit. *Postharvest Biol. Technol.* **2016**, *114*, 29–35. [CrossRef]
66. Xing, Y.; Yang, H.; Guo, X.; Bi, X.; Liu, X.; Xu, Q.; Wang, Q.; Li, W.; Li, X.; Shui, Y.J.S.H. Effect of chitosan/Nano-TiO_2 composite coatings on the postharvest quality and physicochemical characteristics of mango fruits. *Sci. Hortic.* **2020**, *263*, 109135. [CrossRef]
67. Kumar Tiwari, A.; Gautam, M.; Maurya, H.K. Recent Development of Biodegradation Techniques of Polymer. *Int. J. Res. Granthaalayah* **2018**, *6*, 414–452. [CrossRef]
68. Li, Y.-M.; Zhang, H.; Zhou, S.-J.; Sun, Q.-S. Study on the effect of biodegradable chitosan starch antibacterial composite film on the preservation of red grapes. *J. Food Saf. Qual. Insp.* **2017**, *8*, 1579–1584. [CrossRef]
69. Zhang, H.; Neau, S.H.J.B. In vitro degradation of chitosan by a commercial enzyme preparation: Effect of molecular weight and degree of deacetylation. *Biomaterials* **2001**, *22*, 1653–1658. [CrossRef]
70. Croisier, F.; Jérôme, C.J.E. Chitosan-based biomaterials for tissue engineering. *Eur. Polym. J.* **2013**, *49*, 780–792. [CrossRef]
71. Kean, T.; Thanou, M.J.R. Chapter 10—Chitin and chitosan: Sources, production and medical applications. In *Renewable Resources for Functional Polymers and Biomaterials*; The Royal Society of Chemistry: Cambridge, UK, 2011; Volume 12, pp. 292–318.
72. Dash, M.; Chiellini, F.; Ottenbrite, R.M.; Chiellini, E.J.P. Chitosan—A versatile semi-synthetic polymer in biomedical applications. *Prog. Polym. Sci.* **2011**, *36*, 981–1014. [CrossRef]
73. Liu, J.; Liu, S.; Wu, Q.; Gu, Y.; Kan, J.; Jin, C.J.F.H. Effect of protocatechuic acid incorporation on the physical, mechanical, structural and antioxidant properties of chitosan film. *Food Hydrocoll.* **2017**, *73*, 90–100. [CrossRef]
74. Narasagoudr, S.S.; Hegde, V.G.; Vanjeri, V.N.; Chougale, R.B.; Masti, S.P.J.C.P. Ethyl vanillin incorporated chitosan/poly(vinyl alcohol) active films for food packaging applications. *Carbohydr. Polym.* **2020**, *236*, 116049. [CrossRef]
75. Huang, G.H. Application of chitosan and its derivatives in the food industry. *Food Res. Dev.* **2015**, *36*, 131–134.
76. Zhang, X.; Wu, H.; Zhang, L.; Sun, Q. Horseradish peroxidase-mediated synthesis of an antioxidant gallic acid-g-chitosan derivative and its preservation application in cherry tomatoes. *RSC Adv.* **2018**, *8*, 20363–20371. [CrossRef] [PubMed]
77. Beyazit, N.; Cakran, H.S.; Cabir, A.; Akiscan, Y.; Demetgul, C. Synthesis, characterization and antioxidant activity of chitosan Schiff base derivatives bearing (-)-gossypol. *Carbohydr. Polym.* **2020**, *240*, 116333. [CrossRef] [PubMed]
78. Fardioui, M.; Meftah Kadmiri, I.; Qaiss, A.E.K.; Bouhfid, R. Bio-active nanocomposite films based on nanocrystalline cellulose reinforced styrylquinoxalin-grafted-chitosan: Antibacterial and mechanical properties. *Int. J. Biol. Macromol.* **2018**, *114*, 733–740. [CrossRef] [PubMed]
79. Mu, B.; Wu, Q.; Xu, L.; Yang, Y. A sustainable approach to synchronous improvement of wet-stability and toughness of chitosan films. *Food Hydrocoll.* **2022**, *123*, 107138. [CrossRef]
80. Hoa, V.-B.; Song, D.-H.; Seol, K.-H.; Kang, S.-M.; Kim, H.-W.; Kim, J.-H.; Cho, S.H. Coating with chitosan containing lauric acid (C12: 0) significantly extends the shelf-life of aerobically–Packaged beef steaks during refrigerated storage. *Meat Sci.* **2022**, *184*, 108696. [CrossRef]
81. Benhabiles, M.; Tazdait, D.; Abdi, N.; Lounici, H.; Drouiche, N.; Goosen, M.; Mameri, N. Assessment of coating tomato fruit with shrimp shell chitosan and N, O-carboxymethyl chitosan on postharvest preservation. *J. Food Meas. Charact.* **2013**, *7*, 66–74. [CrossRef]
82. Zhang, Y.; Zhang, M.; Yang, H.J.F. Postharvest chitosan-g-salicylic acid application alleviates chilling injury and preserves cucumber fruit quality during cold storage. *Food Chem.* **2015**, *174*, 558–563. [CrossRef]
83. Niu, X.; Zhu, L.; Xi, L.; Guo, L.; Wang, H.J.F.C. An antimicrobial agent prepared by N-succinyl chitosan immobilized lysozyme and its application in strawberry preservation. *Food Control* **2020**, *108*, 106829. [CrossRef]
84. Yang, C.; Han, B.; Zheng, Y.; Liu, L.; Li, C.; Sheng, S.; Zhang, J.; Wang, J.; Wu, F.J. The quality changes of postharvest mulberry fruit treated by chitosan-g-caffeic acid during cold storage. *J. Food Sci.* **2016**, *81*, C881–C888. [CrossRef]
85. Liu, X.; Xue, F.; Li, C.; Adhikari, B.J.I.J.B.M. Physicochemical properties of films produced using nanoemulsions stabilized by carboxymethyl chitosan-peptide conjugates and application in blueberry preservation. *Int. J. Biol. Macromol.* **2022**, *202*, 26–36. [CrossRef]
86. Liu, J.; Liu, S.; Zhang, X.; Kan, J.; Jin, C.J. Effect of gallic acid grafted chitosan film packaging on the postharvest quality of white button mushroom (*Agaricus bisporus*). *Postharvest Biol. Technol.* **2019**, *147*, 39–47. [CrossRef]
87. Wang, G.; Liu, Y.; Yong, H.; Zong, S.; Jin, C.; Liu, J.J.F. Effect of Ferulic Acid-Grafted-Chitosan Coating on the Quality of Pork during Refrigerated Storage. *Foods* **2021**, *10*, 1374. [CrossRef] [PubMed]

88. Xiao, T.; Ting-Ting, Q.; Ruonan, L.; Lisa, Z.; Yan-Yun, Z.; Shun-Sheng, C.; Hong-Cai, Z. Preparation of chitosan nanoparticles and progress of research in food antibacterial. *J. Food Sci.* **2020**, *41*, 347–353.
89. Melo, P.T.S.; Nunes, J.C.; Otoni, C.G.; Aouada, F.A.; de Moura, M.R. Combining Cupuassu (*Theobroma grandiflorum*) Puree, Pectin, and Chitosan Nanoparticles into Novel Edible Films for Food Packaging Applications. *J. Food Sci.* **2019**, *84*, 2228–2233. [CrossRef] [PubMed]
90. Zhao, R.; Guan, W.; Zheng, P.; Tian, F.; Zhang, Z.; Sun, Z.; Cai, L. Development of edible composite film based on chitosan nanoparticles and their application in packaging of fresh red sea bream fillets. *Food Control* **2022**, *132*, 108545. [CrossRef]
91. Tang, Q.; Fu, J.; Dou, X.; He, S.; Zhao, J. Progress of modified chitosan magnetic nanomaterials. *J. Compos.* **2022**, *39*, 1017–1025. [CrossRef]
92. Agarwal, M.; Agarwal, M.K.; Shrivastav, N.; Pandey, S.; Das, R.; Gaur, P. Preparation of Chitosan Nanoparticles and their In-vitro Characterization. *Int. J. Life Sci. Sci. Res.* **2018**, *4*, 1713–1720. [CrossRef]
93. Asab, G.; Zereffa, E.A.; Abdo Seghne, T. Synthesis of Silica-Coated Fe3O4 Nanoparticles by Microemulsion Method: Characterization and Evaluation of Antimicrobial Activity. *Int. J. Biomater.* **2020**, *2020*, 4783612. [CrossRef]
94. Grenha, A. Chitosan nanoparticles: A survey of preparation methods. *J. Drug Target* **2012**, *20*, 291–300. [CrossRef]
95. Kumar, A.; Ahuja, M. Carboxymethyl gum kondagogu-chitosan polyelectrolyte complex nanoparticles: Preparation and characterization. *Int. J. Biol. Macromol.* **2013**, *62*, 80–84. [CrossRef]
96. Kafshgari, M.H.; Khorram, M.; Mansouri, M.; Samimi, A.; Osfouri, S. Preparation of alginate and chitosan nanoparticles using a new reverse micellar system. *Iran. Polym. J.* **2012**, *21*, 99–107. [CrossRef]
97. Zhang, S.; He, Z.; Xu, F.; Cheng, Y.; Waterhouse, G.I.; Sun-Waterhouse, D.; Wu, P.J.F.H. Enhancing the performance of konjac glucomannan films through incorporating zein–pectin nanoparticle-stabilized oregano essential oil Pickering emulsions. *Food Hydrocoll.* **2022**, *124*, 107222. [CrossRef]
98. Babii, O.; Wang, Z.; Liu, G.; Martinez, E.C.; Chen, L. Low molecular weight chitosan nanoparticles for CpG oligodeoxynucleotides delivery: Impact of molecular weight, degree of deacetylation, and mannosylation on intracellular uptake and cytokine induction. *Int. J. Biol. Macromol.* **2020**, *159*, 46–56. [CrossRef] [PubMed]
99. Gomes, L.; Paschoalin, V.; Del Aguila, E.J.R.V.Q. Chitosan nanoparticles: Production, physicochemical characteristics and nutraceutical applications. *Rev. Virtual Quim.* **2017**, *9*, 387–409. [CrossRef]
100. De Moura, M.R.; Aouada, F.A.; Avena-Bustillos, R.J.; McHugh, T.H.; Krochta, J.M.; Mattoso, L.H.J.J.F.E. Improved barrier and mechanical properties of novel hydroxypropyl methylcellulose edible films with chitosan/tripolyphosphate nanoparticles. *J. Food Eng.* **2009**, *92*, 448–453. [CrossRef]
101. Othman, S.H.; Kechik, N.R.; Shapi'i, R.A.; Talib, R.A.; Tawakkal, I.S.J.J.N. Water sorption and mechanical properties of starch/chitosan nanoparticle films. *J. Nanomater.* **2019**, *2019*, 3843949. [CrossRef]
102. Nallamuthu, I.; Devi, A.; Khanum, F.J.A.J.P.S. Chlorogenic acid loaded chitosan nanoparticles with sustained release property, retained antioxidant activity and enhanced bioavailability. *Asian J. Pharm. Sci.* **2015**, *10*, 203–211. [CrossRef]
103. Amin, K.A.M.; Panhuis, M.I.H. Reinforced materials based on chitosan, TiO_2 and Ag composites. *Polymers* **2012**, *4*, 590–599. [CrossRef]
104. Xin, S.; Xiao, L.; Dong, X.; Li, X.; Wang, Y.; Hu, X.; Sameen, D.E.; Qin, W.; Zhu, B.J. Preparation of chitosan/curcumin nanoparticles based zein and potato starch composite films for *Schizothorax prenati* fillet preservation. *Int. J. Biol. Macromol.* **2020**, *164*, 211–221. [CrossRef]
105. Yue, T.-T.; Li, X.; Wang, X.-X.; Yan, X.; Yu, M.; Ma, J.-W.; Zhou, Y.; Ramakrishna, S.; Long, Y.-Z.J.N. Electrospinning of carboxymethyl chitosan/polyoxyethylene oxide nanofibers for fruit fresh-keeping. *Nanoscale Res. Lett.* **2018**, *13*, 1–8. [CrossRef]
106. Zhang, Q.; Chunhong, W. Research progress on the application of chitosan and plant extracts in food preservation. *Mod. Food* **2021**, *78–81*, 94. [CrossRef]
107. Zhibing, J.; Shuang, W.; Xu, Y.; Yinmei, C.; Zhenyi, F. Application of edible chitosan coating film in food preservation. *J. Jilin Med. Coll.* **2018**, *39*, 42–44. [CrossRef]
108. Rambabu, K.; Bharath, G.; Banat, F.; Show, P.L.; Cocoletzi, H.H. Mango leaf extract incorporated chitosan antioxidant film for active food packaging. *Int. J. Biol. Macromol.* **2019**, *126*, 1234–1243. [CrossRef]
109. Yang, G.; Yue, J.; Gong, X.; Qian, B.; Wang, H.; Deng, Y.; Zhao, Y.J. Blueberry leaf extracts incorporated chitosan coatings for preserving postharvest quality of fresh blueberries. *Postharvest Biol. Technol.* **2014**, *92*, 46–53. [CrossRef]
110. Sun, M.; Liu, N.; Ni, S.; Bian, H.; Fu, Y.; Chen, X. Poplar Hot Water Extract Enhances Barrier and Antioxidant Properties of Chitosan/Bentonite Composite Film for Packaging Applications. *Polymers* **2019**, *11*, 1614. [CrossRef]
111. Zhang, W.; Li, X.; Jiang, W. Development of antioxidant chitosan film with banana peels extract and its application as coating in maintaining the storage quality of apple. *Int. J. Biol. Macromol.* **2020**, *154*, 1205–1214. [CrossRef]
112. Mojaddar Langroodi, A.; Nematollahi, A.; Sayadi, M. Chitosan coating incorporated with grape seed extract and Origanum vulgare essential oil: An active packaging for turkey meat preservation. *J. Food Meas. Charact.* **2021**, *15*, 2790–2804. [CrossRef]
113. Le, K.H.; La, D.D.; Nguyen, P.T.M.; Nguyen, M.D.-B.; Vo, A.T.K.; Nguyen, M.T.H.; Tran, D.L.; Chang, S.W.; Nguyen, X.H.; Nguyen, D.D.J.P.O.C. Fabrication of Cleistocalyx operculatus extracts/chitosan/gum arabic composite as an edible coating for preservation of banana. *Prog. Org. Coat.* **2021**, *161*, 106550. [CrossRef]

114. Ramírez-Guerra, H.; Castillo-Yañez, F.; Montaño-Cota, E.; Ruíz-Cruz, S.; Márquez-Ríos, E.; Canizales-Rodríguez, D.; Torres-Arreola, W.; Montoya-Camacho, N.; Ocaño-Higuera, V.J.J.C. Protective effect of an edible tomato plant extract/chitosan coating on the quality and shelf life of sierra fish fillets. *J. Chem.* **2018**, *2018*, 2436045. [CrossRef]
115. Pagliarulo, C.; Sansone, F.; Moccia, S.; Russo, G.L.; Aquino, R.P.; Salvatore, P.; Di Stasio, M.; Volpe, M.G. Preservation of strawberries with an antifungal edible coating using peony extracts in chitosan. *Food Bioprocess Technol.* **2016**, *9*, 1951–1960. [CrossRef]
116. Kumar, N.; Petkoska, A.T.; AL-Hilifi, S.A.; Fawole, O.A.J.C. Effect of Chitosan–Pullulan Composite Edible Coating Functionalized with Pomegranate Peel Extract on the Shelf Life of Mango (*Mangifera indica*). *Coatings* **2021**, *11*, 764. [CrossRef]
117. Hassan, F.; Ali, E.; Mostafa, N.; Mazrou, R.J.I.J.B.M. Shelf-life extension of sweet basil leaves by edible coating with thyme volatile oil encapsulated chitosan nanoparticles. *Int. J. Biol. Macromol.* **2021**, *177*, 517–525. [CrossRef]
118. Sultan, M.; Hafez, O.M.; Saleh, M.A.; Youssef, A.M.J.R.A. Smart edible coating films based on chitosan and beeswax–pollen grains for the postharvest preservation of Le Conte pear. *RSC Adv.* **2021**, *11*, 9572–9585. [CrossRef] [PubMed]
119. Poverenov, E.; Arnon-Rips, H.; Zaitsev, Y.; Bar, V.; Danay, O.; Horev, B.; Bilbao-Sainz, C.; McHugh, T.; Rodov, V.J.F. Potential of chitosan from mushroom waste to enhance quality and storability of fresh-cut melons. *Food Chem.* **2018**, *268*, 233–241. [CrossRef] [PubMed]
120. Jiang, Y.; Yu, L.; Hu, Y.; Zhu, Z.; Zhuang, C.; Zhao, Y.; Zhong, Y. Electrostatic spraying of chitosan coating with different deacetylation degree for strawberry preservation. *Int. J. Biol. Macromol.* **2019**, *139*, 1232–1238. [CrossRef] [PubMed]
121. Jiang, Y.; Yu, L.; Hu, Y.; Zhu, Z.; Zhuang, C.; Zhao, Y.; Zhong, Y. The preservation performance of chitosan coating with different molecular weight on strawberry using electrostatic spraying technique. *Int. J. Biol. Macromol.* **2020**, *151*, 278–285. [CrossRef]
122. Yang, R.; Jiang, Y.; Xiu, L.; Huang, J. Effect of chitosan pre-soaking on the growth and quality of yellow soybean sprouts. *J. Sci. Food Agric.* **2019**, *99*, 1596–1603. [CrossRef]
123. Sayadi, M.; Langroodi, A.M.; Pourmohammadi, K. Combined effects of chitosan coating incorporated with Berberis vulgaris extract and Mentha pulegium essential oil and MAP in the shelf life of turkey meat. *J. Food Meas. Charact.* **2021**, *15*, 5159–5169. [CrossRef]
124. Yuceer, M.; Caner, C. Antimicrobial lysozyme-chitosan coatings affect functional properties and shelf life of chicken eggs during storage. *J. Sci. Food Agric.* **2014**, *94*, 153–162. [CrossRef]
125. Andrade, R.D.; Skurtys, O.; Osorio, F.A. Atomizing Spray Systems for Application of Edible Coatings. *Compr. Rev. Food Sci. Food Saf.* **2012**, *11*, 323–337. [CrossRef]
126. Du, Y.; Yang, F.; Yu, H.; Cheng, Y.; Guo, Y.; Yao, W.; Xie, Y. Fabrication of novel self-healing edible coating for fruits preservation and its performance maintenance mechanism. *Food Chem.* **2021**, *351*, 129284. [CrossRef] [PubMed]
127. Zhang, Z.-H.; Han, Z.; Zeng, X.-A.; Xiong, X.-Y.; Liu, Y.-J. Enhancing mechanical properties of chitosan films via modification with vanillin. *Int. J. Biol. Macromol.* **2015**, *81*, 638–643. [CrossRef] [PubMed]
128. Liu, T.; Liu, L.; Gong, X.; Chi, F.; Ma, Z. Fabrication and comparison of active films from chitosan incorporating different spice extracts for shelf life extension of refrigerated pork. *Lwt* **2021**, *135*, 110181. [CrossRef]
129. Sawaya, W.N. Impact of food losses and waste on food security. In *Water, Energy & Food Sustainability in the Middle East*; Springer: Berlin/Heidelberg, Germany, 2017; pp. 361–388.
130. López-Mata, M.A.; Ruiz-Cruz, S.; Silva-Beltrán, N.P.; Ornelas-Paz, J.D.J.; Zamudio-Flores, P.B.; Burruel-Ibarra, S.E.J.M. Physicochemical, antimicrobial and antioxidant properties of chitosan films incorporated with carvacrol. *Molecules* **2013**, *18*, 13735–13753. [CrossRef]
131. Cazón, P.; Vázquez, M. Mechanical and barrier properties of chitosan combined with other components as food packaging film. *Environ. Chem. Lett.* **2020**, *18*, 257–267. [CrossRef]
132. Chang, S.H.; Chen, Y.J.; Tseng, H.J.; Hsiao, H.I.; Chai, H.J.; Shang, K.C.; Pan, C.L.; Tsai, G.J. Antibacterial Activity of Chitosan-Polylactate Fabricated Plastic Film and Its Application on the Preservation of Fish Fillet. *Polymers* **2021**, *13*, 696. [CrossRef] [PubMed]
133. Kasirajan, S.; Umapathy, D.; Chandrasekar, C.; Aafrin, V.; Jenitapeter, M.; Udhyasooriyan, L.; Packirisamy, A.S.B.; Muthusamy, S.J.J. Bioengineering. Preparation of poly (lactic acid) from *Prosopis juliflora* and incorporation of chitosan for packaging applications. *J. Biosci. Bioeng.* **2019**, *128*, 323–331. [CrossRef]

foods

MDPI

Article

Protective Effect and Mechanism of Soybean Insoluble Dietary Fiber on the Color Stability of Malvidin-3-*O*-glucoside

Yang He [1,†], Dongxia Chen [1,†], Yuheng Liu [1], Xiaozhen Sun [1], Wenrui Guo [1], Lingyu An [1], Zhenming Shi [1], Liankui Wen [1], Zhitong Wang [1,*] and Hansong Yu [1,2,*]

1 College of Food Science and Engineering, Jilin Agricultural University, Changchun 130118, China; heyang200704@jlau.edu.cn (Y.H.); chendongxia1218@163.com (D.C.); lyhh5689@163.com (Y.L.); sunxiaozhen825@163.com (X.S.); gwr13694434337@163.com (W.G.); aly18104373339@163.com (L.A.); szm17767897978@163.com (Z.S.); wenliankui@jlau.edu.cn (L.W.)
2 Division of Soybean Processing, Soybean Research & Development Center, Chinese Agriculture Research System, Changchun 130118, China
* Correspondence: wangzhitong03@163.com (Z.W.); yuhansong@jlau.edu.cn (H.Y.)
† These authors contributed equally to this work.

Abstract: Anthocyanins have great health benefits, especially malvidin. *Vitis amurensis Rupr* are rich in malvidin, and malvidin-3-*O*-glucoside (Mv3G) monomer is the most abundant. However, natural anthocyanins are unstable, which limits their wide application in the food field. Soybean insoluble dietary fiber (SIDF) has high stability, and it can be used as an inert substrate to construct a stable system, which may improve the stability of anthocyanins. The optimal condition to construct a stable system of SIDF and Mv3G at pH 3.0 was determined by an orthogonal experiment. The results indicated that SIDF effectively improved the stability of Mv3G under different pH values (1.0~7.0), high temperature (100 °C for 100 min), and sunlight (20 ± 2 °C for 30 d) conditions. The absorption peak intensity of the UV–VIS spectrum of SIDF-Mv3G was enhanced, which indicated that there was interaction between SIDF and Mv3G. Fourier transform infrared spectroscopy analyses revealed that the -OH stretching vibration peak of SIDF-Mv3G was changed, which indicated that the interaction between SIDF and Mv3G was due to hydrogen bonding. X-ray diffraction analysis showed that the crystalline morphology of SIDF was opened, which was combined with Mv3G, and SIDF made Mv3G change to a more stable state. Scanning electron microscope analysis showed that SIDF and Mv3G were closely combined to form an inclusion complex. Overall, this study provides valuable information for enhancing the color stability of anthocyanins, which will further expand the application of anthocyanins in the food field.

Keywords: anthocyanins; insoluble dietary fiber; stable system; inclusion complex; hydrogen bonding

Citation: He, Y.; Chen, D.; Liu, Y.; Sun, X.; Guo, W.; An, L.; Shi, Z.; Wen, L.; Wang, Z.; Yu, H. Protective Effect and Mechanism of Soybean Insoluble Dietary Fiber on the Color Stability of Malvidin-3-*O*-glucoside. *Foods* **2022**, *11*, 1474. https://doi.org/10.3390/foods11101474

Academic Editor: Lavanya Reddivari

Received: 25 April 2022
Accepted: 17 May 2022
Published: 19 May 2022

1. Introduction

Anthocyanins are gorgeous in color and have many human health benefits, including antioxidant, antiobesity, and antidiabetes [1–3]. In recent years, people's awareness of the relationship between diet and health has gradually increased, resulting in an increase in consumer demand for foods containing natural anthocyanins. Therefore, anthocyanins are widely used in the fields of food, natural pigments, cosmetics, and medicine [4]. The distribution of the six most common anthocyanins in nature is cyanidin, delphinidin, pelargonidin, peonidin, petunidin, and malvidin, accounting for 50%, 12%, 12%, 12%, and 7%, 7% of total anthocyanins, respectively [5]. Malvidin has high biological activity, but the overall distribution is less in nature [5]. It is worth noting that the high-yielding *Vitis amurensis Rupr* in the Changbai Mountains of China are rich in anthocyanins (the average content is above 150 mg/100 g·FW, and individual varieties are as high as 400 mg/100 g·FW) [6,7]. Malvidin accounts for 55~65% of total anthocyanins from *Vitis amurensis Rupr*, among which malvidin-3-O-glucoside (Mv3G) is the most abundant monomer [8]. There are many phenolic hydroxyl groups on the carbon skeleton of anthocyanins, so pH, light, heat, oxygen, metal ions, and other environmental factors can affect the color stability of anthocyanins [6,7]. The instability of anthocyanins greatly affects their application in food. Therefore, enhancing the color stability of anthocyanins is an important problem that needs to be solved urgently.

During food processing, polyphenols can interact with starch, protein, and cellulose to form a complex, which can improve the stability of polyphenols [9–12]. In particular, Quan et al. [13] found that adding soybean protein to purple-fleshed sweet potato anthocyanins could effectively improve the color stability at 100 °C. Sun et al. [14] reported that linear dextrin could be used as a food-grade carrier of curcumin. UV–VIS spectrophotometry (UV–VIS) and Fourier transform infrared spectroscopy (FT-IR) analyses showed that linear dextrin and curcumin formed an inclusion complex (a mixture in which the molecules of one component are contained in the crystal lattice of another component) by hydrogen bonding, which improved the stability of curcumin. Li et al. [15] also reported the interaction between soluble dietary fiber and polyphenols through conjugation and hydrogen bonding improved the stability of the polyphenols–soluble dietary fiber complex. Zhang et al. [16] microencapsulated ethylcellulose and polyphenols to diminish the instability of polyphenols as water-soluble compounds, particularly under harsh processing and storage conditions. Thus, the protective effect of macromolecules on polyphenols contributes to their application in more complex food systems.

Okara is the main byproduct during the processing of traditional soybean products such as tofu, soybean milk, and soybean oil. Okara is abundant in nutrients, especially soybean insoluble dietary fiber (SIDF), which accounts for 45~55% of the dry matter of okara [17,18]. SIDF has high stability and is not easy to react with other food ingredients, so it can be used as an inert substrate to build a stable system [19]. SIDF has a loose structure and rough surface, which can promote the interaction with polyphenols by adsorption or embedding [20]. Zhao et al. [21] reported that the hydration of insoluble dietary fiber (IDF) improved as its particle size decreased because of the greater surface area, increased number of polar groups, and exposure to other water-binding sites of IDF to the surrounding water. IDF has a high denaturation temperature, so the thermal characteristics are stable [22]. There is little research on the interaction between IDF and anthocyanins. It is concluded that SIDF has the potential to promote the stability of anthocyanins. In order to further explore the interaction between SIDF and Mv3G, this study explained the stabilization mechanism of SIDF and Mv3G, which will be more conducive to anthocyanins as stable pigments or functional components added to food.

This study aimed to optimize the stable system of SDIF and Mv3G and explain the stabilization mechanism. In this paper, the color stabilities of the stable system of SIDF-Mv3G (hereinafter referred to as SIDF-Mv3G) at different pH values, thermal conditions, or sunlight conditions were investigated. Furthermore, the interaction mechanism of SIDF and Mv3G was investigated through UV–VIS, FT-IR, X-ray diffraction (XRD), and scanning electron microscopy (SEM).

2. Materials and Methods

2.1. Material

The purity of Mv3G was ≥95%, which was extracted in our lab according to a previous study [8]. The purity of SIDF was ≥90%, which was extracted in our lab according to a previous study [23]. All chemicals used in this work were all analytical or HPLC grade.

2.2. Optimal Design of SIDF-Mv3G Stable System

The appropriate amount of SIDF and Mv3G was weighed and mixed in a certain proportion with PBS (citric acid–disodium hydrogen phosphate) at pH 3.0 to 100 mL (the final Mv3G concentration was 0.12 mg/100 mL). Under a certain temperature range, the stable system was prepared by stirring and emulsifying at 500 rpm/min.

According to the results of the pre-experiment, the SIDF–Mv3G ratio A (20:1, 30:1, 40:1; final sample concentration of Mv3G was 0.12 mg/mL), SIDF particle size B (200-, 300-, 400-mesh sieve), emulsification temperature C (20, 30, 40 °C), and emulsification time D (20, 30, 40 min) were optimized using an L_9 (4^3) orthogonal design. The stable effect was evaluated by measuring the absorbance at 521 nm and sedimentation eccentricity (S, %). Sedimentation eccentricity was calculated by Equation (1).

$$S\ (\%) = \frac{W_2 - W_1}{W_3 - W_1} \times 100\% \qquad (1)$$

where W_1 is the weight of the centrifuge pipe, W_2 is the weight of sediment added to the centrifuge pipe after centrifugation, and W_3 is the weight of the centrifuge pipe plus sample.

The optimal conditions of the orthogonal experiment were selected to construct the stable system of SIDF-Mv3G. Mv3G solution and SIDF solution under the same conditions were used as the control. SIDF-Mv3G and Mv3G solutions were used for sunlight stability, thermal stability, and UV–VIS spectrum analyses. SIDF-Mv3G, SIDF, and Mv3G were lyophilized and used for pH stability, FT-IR, XRD, and SEM analyses.

2.3. *Stability of SIDF-Mv3G2.3.1. pH Stability*

2.3.1. pH Stability

SIDF-Mv3G and Mv3G were dispersed in PBS solution with a pH of 1.0~7.0, respectively. Color changes of the SIDF-Mv3G and Mv3G solutions were measured using a Color Flex EZ colorimeter (HunterLab, American) with a white standard color plate as the background. Hunter color values (L*, a*, and b*) of the SIDF-Mv3G and Mv3G solutions were obtained, and the browning degree (ΔE) was calculated as follows:

$$\Delta E = \sqrt{(L_t - L^*)^2 + (a_t - a^*)^2 + (b_t - b^*)^2} \qquad (2)$$

where L_t (51.85), a_t (−26.26), and b_t (14.32) are the L*, a*, and b* values of the standard white color plate.

2.3.2. Thermal Stability

The samples were prepared in 5 mL glass tubes wrapped in tin foil, which were heated in a water bath at 100 °C for 100 min. Every 20 min, the samples were quickly cooled in ice. After 5 min, the absorbance of the upper supernatant was measured at 521 nm by UV–VIS spectrophotometry (The turbidity problem of an appropriate amount of SIDF is not serious, and the influ-ence is deducted by using the same amount of SIDF as the instrument benchmark in the experimental process). The retention rate (R, %) of anthocyanins was used as the evaluation index to evaluate the thermal stability.

$$R\ (\%) = \left(\frac{A_1}{A}\right) \times 100 \qquad (3)$$

where A is the absorbance before sunlight or thermal stability treatment, and A_1 is the absorbance after sunlight or thermal stability treatment.

2.3.3. Sunlight Stability

The SIDF-Mv3G and Mv3G solutions were sealed in vials and irradiated with sunlight at room temperature (20 ± 2 °C; samples were placed on the indoor windowsill, and the daily insolation time was not less than 6 h). Samples were taken out after 0, 5, 10, 15, 20, 25, and 30 d. The absorbance of the upper supernatant was measured at 521 nm by UV–VIS spectrophotometry (The turbidity problem of an appropriate amount of SIDF is not serious, and the influence is deducted by using the same amount of SIDF as the instrument benchmark in the experimental process), and the retention rate of anthocyanins was calculated according to Equation (3).

2.4. Structural Characterization of SIDF-Mv3G

2.4.1. UV–VIS Spectroscopy

The UV–VIS absorption spectra of the SIDF-Mv3G and Mv3G solutions were measured using a T6 UV-Vis spectrophotometer (Beijing Universal Instrument Co., Ltd., Beijing, China) in the wavelength range of 400~700 nm

2.4.2. FT-IR Spectroscopy

The FT-IR spectra of the SIDF-Mv3G, SIDF and Mv3G were recorded using an IR Prestige-21 spectrometer (Shimadzu, Japan). The samples were blended with KBr at a ratio of 1:100 (*w*/*w*) and pressed into tablets for further measurement. The FT-IR spectra were measured in absorbance mode between 4000 and 500 cm^{-1}.

2.4.3. XRD

The XRD diffraction patterns of SIDF-Mv3G, SIDF, and Mv3G were collected using an X-ray diffractometer (D8 ADVANCE, Germany), equipped with copper Cu Kα radiation (λ = 0.154 nm) as the X-ray source (2θ range of 5~50°).

2.5. Statistical Analysis

All tests were repeated three times, and the results were expressed as the average value (±SD). SPSS was used for statistical analysis (ANOVA), and when $p \leq 0.05$, the difference between the two groups was evaluated.

3. Results

3.1. Optimal Design of SIDF-Mv3G Stable System

The variation coefficient method, which assigns different weights to the indexes based on their degree of variation [24], is typically employed in the comprehensive evaluation of multiple indicators. The variation coefficient V_i of the *i*th characteristic parameter value was calculated according to Equation (4). The weight value W_i of the *i*th characteristic parameter was calculated according to Equation (5).

$$V_i = \frac{S_i}{X_i} \quad (4)$$

where V_i is the variation coefficient, S_i is the standard deviation, and X_i is the arithmetic mean.

$$W_i = \frac{V_i}{\sum_{i=1}^{n} V_i} \quad (5)$$

where W_i is the weight, and $\sum_{i=1}^{n} V_i$ is the sum of *n* indexes.

According to Table 1, the weight of absorbance and sedimentation rate were 0.411 and 0.589, respectively, and the results showed that the sedimentation rate had a greater influence on the stable system of SIDF-Mv3G than absorbance. The optimal result of the test was $A_2B_1C_2D_3$ (1.246).

According to the *K* value, the order of the effect of the four factors on the stable system were C (emulsification temperature) > B (SIDF particle size) > A (SIDF–Mv3G ratio) > D (emulsification time), and the optimal condition was $A_2B_2C_2D_3$.

The variance analysis showed that the effects of the single factors of emulsification temperature were significant ($p < 0.01$) on the stable system, and the SIDF particle size and SIDF–Mv3G ratio exhibited were significant on the stable system ($p < 0.05$), whereas emulsification time was not significant on the stable system. In addition, micronization reduced the interfacial tension and could expose more polar groups, surface area, and water-binding sites to the surrounding water, which could be closely combined with Mv3G. For the above reasons, the optimal combination was adjusted to $A_2B_2C_2D_3$: SIDF–Mv3G ratio was 30:1 (*w/w*), SIDF particle size was a 300-mesh sieve, emulsification temperature was 30 °C, and emulsification time was 50 min. This result was confirmed by three parallel validations, and the actual score for the optimal condition was 1.387 ± 0.034. The relative standard deviation RSD = 0.024%, which showed that the results of the orthogonal design had good reproducibility.

Table 1. Results of orthogonal experiment.

Samples	A: SIDF [1]–Mv3G [2] Ratio	B: SIDF Particle Size (Mesh Sieve)	C: Emulsification Temperature (°C)	D: Emulsification Time (min)	Absorbance	Precipitation Eccentricity (%)	Composite Scores
1	1 (1:20)	1 (200)	1 (20)	1 (30)	0.807	0.855	−1.128
2	1	2 (300)	2 (30)	2 (40)	0.880	0.786	1.2
3	1	3 (400)	3 (40)	3 (50)	0.855	0.800	0.195
4	2 (1:30)	1	2	3	0.885	0.781	1.246
5	2	2	3	1	0.850	0.786	0.719
6	2	3	1	2	0.830	0.857	−0.793
7	3 (1:40)	1	3	2	0.819	0.889	−1.507
8	3	2	1	3	0.828	0.829	−0.355
9	3	3	2	1	0.864	0.818	0.406
K_1	0.27	−1.39	−2.28	0.00			
K_2	1.17	1.56	2.85	−1.10			
K_3	−1.46	−0.19	−0.59	1.09			
R	0.88	0.98	1.71	0.73			
	Index		Average value	Standard deviation		Variable coefficient	Weight
	Absorbance		0.846	0.026		0.030	0.411
	Precipitation eccentricity		0.825	0.035		0.043	0.589
				Standardized index			
Samples		Absorbance				Precipitation eccentricity	
1		−1.520				−0.855	
2		1.325				1.112	
3		0.351				0.086	
4		1.520				1.054	
5		0.156				1.112	
6		−0.623				−0.912	
7		−1.052				−1.825	
8		−0.701				−0.114	
9		0.701				0.200	
	Projects	Sums of squared deviations	Freedom	Mean square		F value	Significance
	SIDF–Mv3G ratio	1.19	2	0.59		4.48	*
	SIDF particle size	1.47	2	0.74		5.54	*
	Emulsification temperature	4.56	2	2.28		17.16	**
	Emulsification time	0.80	2	0.40		3.00	
	Error	0.80	6	0.13			

SIDF [1]: soybean insoluble dietary fiber; Mv3G [2]: malvidin-3-*O*-glucoside; *: the difference was significant ($p < 0.05$); **: the difference was significant ($p < 0.01$).

3.2. Stability of SIDF-Mv3G

3.2.1. pH Stability

Figure 1 shows the color changes of the SIDF-Mv3G and Mv3G solutions at different pH values (1.0~7.0). Specifically, both SIDF-Mv3G and Mv3G solutions changed from bright red to pink, and finally to purple. The color diversity of the solutions was caused by the structural changes of anthocyanins [6]. With the increase in pH value, the structures of the flavylium cation were destroyed and transformed into purple quinone base structures [25].

Color is one of the most important parameters of food and affects the application of anthocyanins. The subtle color changes of anthocyanins are difficult to judge by visual inspection. Therefore, the color of the SIDF-Mv3G and Mv3G solutions was described by the color measurement. As shown in Table 2, with the increase in pH value, the L*, a*, and b* values gradually decreased, which indicated that the anthocyanins had faded [26]. With the gradual increase in ΔE, the browning caused by pH was aggravated, and brown substances were produced due to the existence of oxygen. Compared with Mv3G, the L*, a*, and b* values of SIDF-Mv3G were higher, and ΔE was lower. The results showed that SIDF can effectively prevent Mv3G degradation and protect the color stability of Mv3G.

Generally, when the pH value of the anthocyanin solution is 3.0, the structure of anthocyanins can be transformed into a stable flavylium cation. However, when the pH value is lower than 3.0, some biological macromolecules cannot protect against anthocyanin degradation [10]. At the same time, a lower pH value could cause the taste quality of food to decrease. Koh et al. [27] reported that when the pH value was 2.6, the flavonoid cations of anthocyanins were more dominant in the electrostatic effect; however, the charge of pectin was minimal, which affected their interaction; when the pH value was 3.0, the interaction between them was enhanced. Most fruit juice drinks on the market have a pH value of about 3.0. If SIDF-Mv3G is applied to beverages, it will ensure the bright color of beverages, prolong the stability of anthocyanins, and have potential benefits to human health.

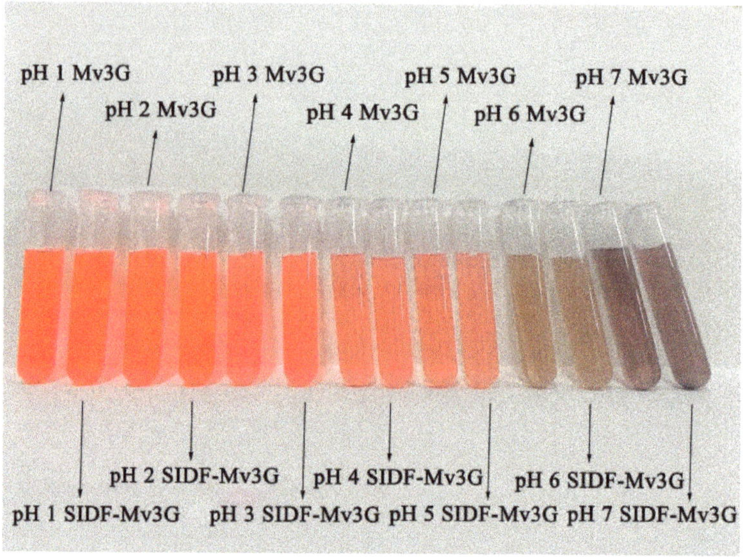

Figure 1. Color change of SIDF-Mv3G and Mv3G. SIDF-Mv3G: the stable system solution of soybean insoluble dietary fiber and malvidin-3-*O*-glucoside; Mv3G: malvidin-3-*O*-glucoside solution.

Table 2. Color measurement analysis.

	L*	a*	b*	ΔE
pH 1 Mv3G solution	43.02 [b]	9.34 [a]	2.11 [bc]	38.88 [gh]
pH 1 Mv3G-SIDF [1]	49.06 [a]	10.14 [a]	3.37 [a]	38.31 [h]
pH 2 Mv3G solution	43.18 [b]	7.62 [b]	1.43 [c]	37.51 [hi]
pH 2 Mv3G-SIDF	46.28 [a]	9.15 [a]	2.63 [ab]	37.81 [hi]
pH 3 Mv3G solution	42.88 [b]	6.02 [d]	−2.48 [d]	37.69 [hi]
pH 3 Mv3G-SIDF	43.84 [b]	7.21 [bc]	2.56 [abc]	36.57 [i]
pH 4 Mv3G solution	37.94 [c]	5.73 [d]	−7.20 [f]	41.26 [ef]
pH 4 Mv3G-SIDF	41.26 [b]	7.38 [bc]	−3.99 [e]	39.95 [fg]
pH 5 Mv3G solution	33.76 [de]	3.96 [e]	−8.43 [g]	42.15 [e]
pH 5 Mv3G-SIDF	35.45 [cd]	6.61 [cd]	−6.85 [f]	42.62 [de]
pH 6 Mv3G solution	31.69 [ef]	2.79 [f]	−12.32 [h]	44.49 [c]
pH 6 Mv3G-SIDF	32.76 [ef]	6.07 [d]	−7.81 [fg]	43.80 [cd]
pH 7 Mv3G solution	26.70 [g]	2.64 [f]	−14.83 [i]	48.36 [a]
pH 7 Mv3G-SIDF	30.59 [f]	4.18 [e]	−12.91 [h]	46.27 [b]

SIDF-Mv3G [1]: the stability system solution of soybean insoluble dietary fiber and malvidin-3-*O*-glucoside. Continuous different letters indicate significant difference ($p < 0.05$).

3.2.2. Thermal Stability

The protective effect of SIDF on Mv3G at 100 °C is shown in Figure 2. Anthocyanins degraded rapidly at high temperature, and the structure of anthocyanins was opened and converted into chalcone, resulting in unpleasant brown substances [28]. After heating for 100 min, the retention rates of the SIDF-Mv3G and Mv3G solutions were 81.54% and 65.55%, respectively. The results showed that the retention rate of SIDF-Mv3G was significantly higher than the Mv3G solution, which indicated that SIDF-Mv3G had high thermal stability.

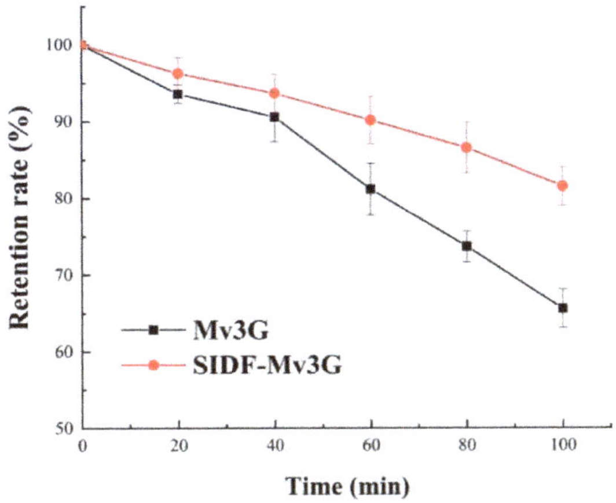

Figure 2. Retention rate of heat treatment at 100 °C for 100 min. SIDF-Mv3G: the stability system solution of soybean insoluble dietary fiber and malvidin-3-*O*-glucoside; Mv3G: malvidin-3-*O*-glucoside solution.

3.2.3. Sunlight Stability

During food processing or shelf storage, food containing anthocyanins can fade and degrade under sunlight. Therefore, the retention rate of the SIDF-Mv3G and Mv3G solutions stored in natural light (20 ± 5 °C) for 30 d was measured to evaluate the sunlight stability of Mv3G. Figure 3 shows that the retention rates of the SIDF-Mv3G and Mv3G solutions in sunlight were 75.16% and 60.54%, respectively. The results showed that SIDF could prolong the storage time of Mv3G in sunlight.

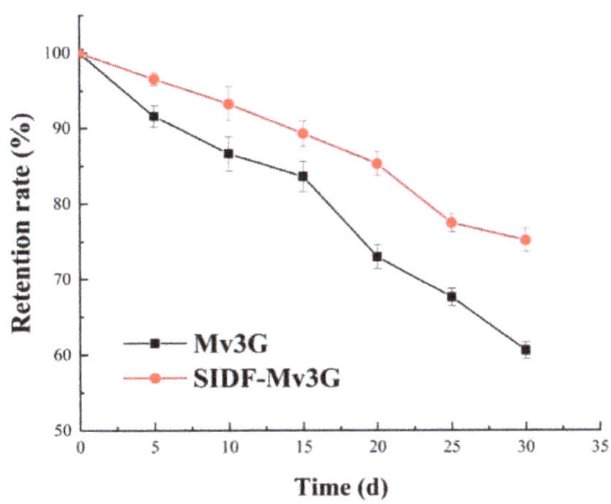

Figure 3. Retention rate of sunlight treatment at room temperature. SIDF-Mv3G: the stability system solution of soybean insoluble dietary fiber and malvidin-3-*O*-glucoside; Mv3G: malvidin-3-*O*-glucoside solution.

3.3. Structural Characterization of SIDF-Mv3G

3.3.1. UV–VIS Spectroscopy

Figure 4 shows the UV–VIS spectra of the SIDF-Mv3G and Mv3G solutions. The figure shows that the absorption peak intensity of SIDF-Mv3G was significantly higher than the Mv3G solution, which is due to the copigmentation between SIDF and Mv3G, leading to Mv3G moving to a stable flavylium cation structure [29]. In addition, the increasing absorbance might be due to the fact that polysaccharides provide nucleation sites for anthocyanins, and then Mv3G is adsorbed on SIDF [9].

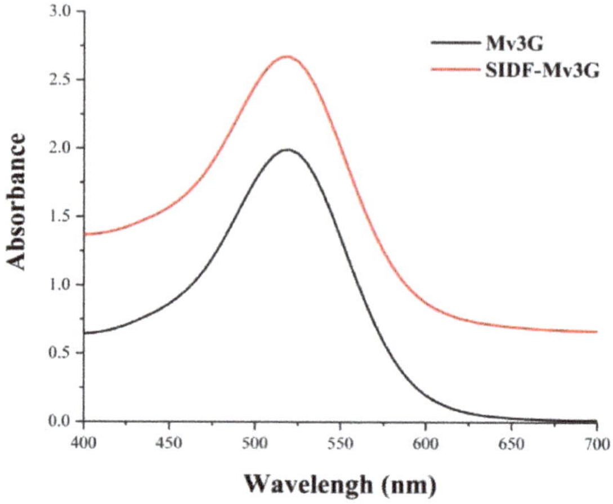

Figure 4. Ultraviolet–visible spectra. SIDF-Mv3G: the stability system solution of soybean insoluble dietary fiber and malvidin-3-*O*-glucoside; Mv3G: malvidin-3-*O*-glucoside solution.

3.3.2. FT-IR Spectroscopy

If the FT-IR spectra shift obviously after adding other components to the matrix, it indicates that there is an obvious interaction between the components [30]. Figure 5 shows the FT-IR spectra of SIDF-Mv3G, SIDF, and Mv3G. Compared with SIDF and Mv3G, the -OH stretching vibration peak of SIDF-Mv3G moved from 3397 cm^{-1} and 3389 cm^{-1} to 3442 cm^{-1}, indicating that the hydrogen bond between SIDF and Mv3G was increased [4]. The C-H stretching vibration of SIDF-Mv3G at 2806~2723 cm^{-1} was obviously weakened, indicating that the hydrophobic groups in SIDF were decreased [31]. The characteristic peak of the C-O angle deformation of phenols compounds remained at 1332 cm^{-1}, which indicated that anthocyanins were combined into SIDF [4]. FT-IR spectra synthesis showed that SIDF was combined with Mv3G through hydrogen bonding as the main driving force.

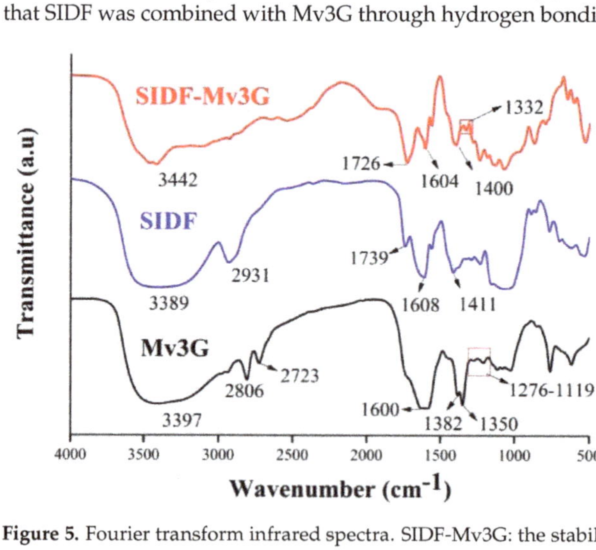

Figure 5. Fourier transform infrared spectra. SIDF-Mv3G: the stability system of soybean insoluble dietary fiber and malvidin-3-*O*-glucoside; Mv3G: malvidin-3-*O*-glucoside; SIDF: soybean insoluble dietary fiber.

3.3.3. XRD

XRD is an important technique for structure detection [32]. Figure 6 shows the XRD patterns of SIDF-Mv3G, SIDF, and Mv3G. According to the XRD pattern of SIDF, the diffraction peaks at 14.67°, 15.11°, and 24.23° were typical cellulose diffraction peaks, which indicated that SIDF has a crystalline structure [33]. According to the XRD pattern of Mv3G, a broad peak appeared at 20.83°, which indicated that Mv3G has an amorphous structure [26]. Qin et al. [34] also found a similar amorphous structure in blueberry anthocyanins. In the XRD pattern of SIDF-Mv3G, the intensity of diffraction peaks around 14.67°, 15.11°, and 24.23° was weakened, and the absorption peak at 20.83° was weakened and widened, which indicated that the crystalline state of SIDF was opened and combined with Mv3G after mixing with SIDF [35,36]. Finally, SIDF made Mv3G change to a more stable state.

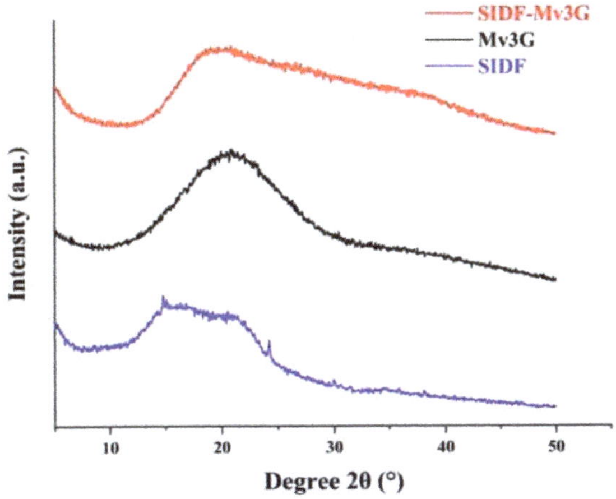

Figure 6. X-ray diffraction patterns. SIDF-Mv3G: the stability system of soybean insoluble dietary fiber and malvidin-3-*O*-glucoside; Mv3G: malvidin-3-*O*-glucoside; SIDF: soybean insoluble dietary fiber.

3.3.4. SEM

Figure 7 shows the SEM results of SIDF-Mv3G, SIDF, and Mv3G. After the small particles of Mv3G were magnified, the surface was lamellar and irregular, and the specific surface area was large. SIDF had a loose and stratiform network structure, which might be due to the degradation of starch and protein caused by enzymatic hydrolysis in the extraction process [37]. The folding structure of SIDF was opened, and it may be better combined with anthocyanins. SIDF-Mv3G was an inclusion complex, which showed that the loose structure of SIDF was closely combined with the layered structure of Mv3G. SIDF wrapped Mv3G, thus protecting Mv3G from environmental factors such as different pH values, light, and high temperature.

Based on the above conclusion, there are two main reasons why SIDF improves the anythocyanin stability of Mv3G. The first reason why biological macromolecules improve the stability of polyphenols may be the interaction between biological macromolecules and polyphenols through hydrogen bonds, van der Waals forces, electrostatic action, and hydrophobic action. Gao et al. [38] reported that hydrogen bonding was the main driving force between (-)-epigallocatechin gallate and oat *β*-glucan, which was similar to our study. Guan et al. [39] reported that gum arabic and anthocyanins formed a complex through hydrophobic interaction, which improved the thermal stability of anthocyanins. This result was different from our research. This is because of the different structure of biological macromolecules, and the binding sites of macromolecules with anthocyanins are different, which result in different forces. Fernandes et al. [9] found in the beverage model that the complex between grape pectic polysaccharides and malvidin-3-*O*-β-D-glucoside was formed by electrostatic and hydrophobic interactions, which protected the chromophore of anthocyanins, thus possibly shielding the highly electrophilic C2-position of the flavylium cation. The C2 position was easily attacked by water due to chemical degradation, so protecting the C2 position can improve the stability of anthocyanins.

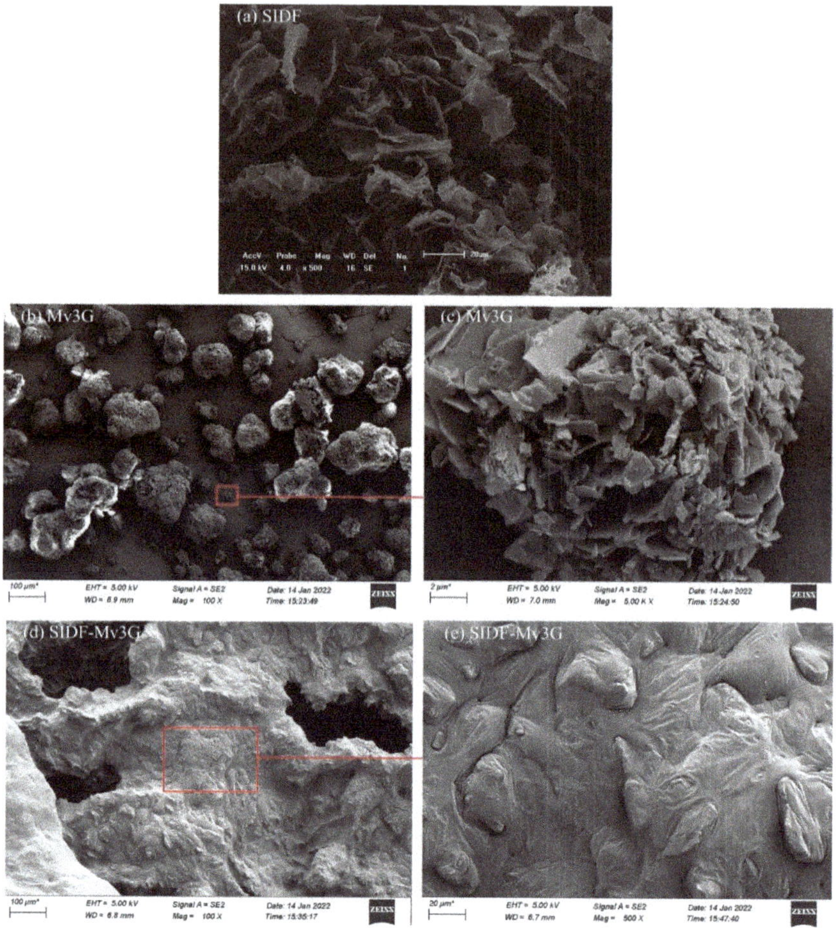

Figure 7. Scanning electron microscopy. SIDF-Mv3G: the stability system of soybean insoluble dietary fiber and malvidin-3-*O*-glucoside; Mv3G: malvidin-3-*O*-glucoside; SIDF: soybean insoluble dietary fiber. (**a**) Scanning electron microscope image of SIDF ($\times$500); (**b**) scanning electron microscope image of Mv3G ($\times$100); (**c**) scanning electron microscope image of Mv3G ($\times$5000); (**d**) scanning electron microscope image of SIDF-Mv3G ($\times$100); (**e**) scanning electron microscope image of SIDF-Mv3G ($\times$500).

The second reason is that when SIDF is exposed to an aqueous solution, it can spontaneously combine with polyphenols in order to reduce SIDF contact with water. Meanwhile, SIDF has a network structure, which can adsorb and form an inclusion complex with anthocyanins. A study by Cai et al. [4] showed that the thermal stability of blueberry anthocyanins was improved by encapsulation with carboxymethyl starch and xanthan gum. Pradhan et al. [40] reported that the half-life of berberis lycium anthocyanins increased by 1.74 (60 °C), 1.28 (75 °C), and 1.44 (90 °C) times when anthocyanins were encapsulated by the hydrophobic cavity of β-cyclodextrin, thus improving the stability of anthocyanins.

4. Conclusions

The best stability conditions of SIDF and Mv3G were as follows: SIDF–Mv3G ratio was 30:1 (w/w), SIDF particle size was a 300-mesh sieve, emulsification temperature was 30 °C, and emulsification time was 50 min. When the pH value gradually increased from 1.0 to 7.0, the color of the SIDF-Mv3G and Mv3G solutions changed from rose-red to purple, which indicated that the structure of anthocyanins changed from flavylium cation to purple quinone base structures. ΔE increased with increasing pH, indicating that pH leads to the aggravation of anthocyanin browning, but SIDF could effectively prevent anthocyanins from browning. Meanwhile, the L*, a*, and b* values of SIDF-Mv3G were higher than the Mv3G solution, which indicated that SIDF could stabilize the color of anthocyanins. After the SIDF-Mv3G and Mv3G solutions were heated at 100 °C for 100 min, the retention rate of SIDF-Mv3G increased by 15.99% compared with the Mv3G solution. When the SDIF-Mv3G and Mv3G solutions were stored for 30 d at 20 $\pm$ 2 °C under sunlight, the retention rate of SIDF-Mv3G was 14.62% higher than the Mv3G solution, which indicated that SIDF could protect the color stability of anthocyanins under sunlight. The stable system of SIDF-Mv3G was mainly combined by hydrogen bonding to form an inclusion complex. This inclusion complex structure was relatively stable, which protected Mv3G from environmental factors, thus improving the stability of anthocyanins.

Considering the health benefits of anthocyanins and insoluble dietary fiber to the human body, stable SIDF-Mv3G can be applied to natural pigments and also be added to the food matrix as a functional factor. This study provides a new idea for the application of anthocyanins in the food industry.

Author Contributions: Conceptualization, L.W.; methodology, Y.H.; software, D.C.; validation, Z.W.; formal analysis, Z.W. and L.W.; investigation, Y.L., X.S., W.G., L.A. and Z.S.; resources, H.Y.; writing—original draft preparation, Y.H. and D.C.; writing—review and editing, Y.H., D.C., Y.L., X.S., W.G., L.A. and Z.S.; project administration, Z.W. and H.Y.; funding acquisition, Y.H. and Z.W. All authors have read and agreed to the published version of the manuscript.

Funding: This research was funded by Jilin Provincial Science and Technology Department, grant number: 20210202105NC.

Institutional Review Board Statement: No applicable.

Informed Consent Statement: Not applicable.

Data Availability Statement: Data are contained within the article.

Acknowledgments: We thank the Jilin Agricultural University Agricultural Products Processing and Storage Laboratory for support, and also thank the China Agriculture Research System of MOF and MARA (Project No. CARS-04) for providing soybean insoluble dietary fiber materials.

Conflicts of Interest: The authors declare no conflict of interest.

References

1. Maciel, L.G.; do Carmo, M.A.V.; Azevedo, L.; Daguer, H.; Molognoni, L.; de Almeida, M.M.; Granato, D.; Rosso, N.D. Hibiscus sabdariffa anthocyanins-rich extract: Chemical stability, in vitro antioxidant and antiproliferative activities. *Food Chem. Toxicol.* **2018**, *113*, 187–197. [CrossRef] [PubMed]
2. Hamed, Y. Antioxidant and anti-inflammatory activities of target anthocyanins di-glucosides isolated from syzygium cumini pulp by high speed counter-current chromatography. *J. Food Biochem.* **2020**, *44*, 1050–1062.
3. Han, F.; Yang, P.; Wang, H.; Fernandes, I.; Mateus, N.; Liu, Y. Digestion and absorption of red grape and wine anthocyanins through the gastrointestinal tract. *Trends Food Sci. Technol.* **2019**, *83*, 211–224. [CrossRef]
4. Cai, X.; Du, X.; Cui, D.; Wang, X.; Yang, Z.; Zhu, G. Improvement of stability of blueberry anthocyanins by carboxymethyl starch/xanthan gum combinations microencapsulation. *Food Hydrocoll.* **2019**, *91*, 238–245. [CrossRef]
5. Khoo, H.E.; Azlan, A.; Tang, S.T.; Lim, S.M. Anthocyanidins and anthocyanins: Colored pigments as food, pharmaceutical ingredients, and the potential health benefits. *Food Nutr. Res.* **2017**, *61*, 1361779. [CrossRef]
6. Castañeda-Ovando, A.; Pacheco-Hernández, M.d.L.; Páez-Hernández, M.E.; Rodríguez, J.A.; Galán-Vidal, C.A. Chemical studies of anthocyanins: A review. *Food Chem.* **2009**, *113*, 859–871. [CrossRef]

7. He, Y.; Wen, L.; Yu, H.; Zheng, F.; Wang, Z.; Xu, X.; Zhang, H.; Cao, Y.; Wang, B.; Chu, B.; et al. Effects of high hydrostatic pressure-assisted organic acids on the copigmentation of vitis amurensis rupr anthocyanins. *Food Chem.* **2018**, *268*, 15–26. [CrossRef]
8. He, Y.; Wen, L.; Yu, H.; Cao, Y.; Nan, H.; Gou, M.; Xie, C.; Xue, H. Isolation and structural identification of the main anthocyanin monomer in vitis amurensis rupr. *Nat. Prod. Res.* **2018**, *32*, 867–870. [CrossRef]
9. Fernandes, A.; Brandão, E.; Raposo, F.; Maricato, É.; Oliveira, J.; Mateus, N.; Coimbra, M.A.; de Freitas, V. Impact of grape pectic polysaccharides on anthocyanins thermostability. *Carbohydr. Polym.* **2020**, *239*, 116240. [CrossRef]
10. Cortez, R.; Luna-Vital, D.A.; Margulis, D.; Gonzalez de Mejia, E. Natural pigments: Stabilization methods of anthocyanins for food applications. *Compr. Rev. Food Sci. Food Saf.* **2017**, *16*, 180–198. [CrossRef]
11. Chung, C.; Rojanasasithara, T.; Mutilangi, W.; McClements, D.J. Enhancement of colour stability of anthocyanins in model beverages by gum arabic addition. *Food Chem.* **2016**, *201*, 14–22. [CrossRef] [PubMed]
12. Zhu, F.; Sakulnak, R.; Wang, S. Effect of black tea on antioxidant, textural, and sensory properties of chinese steamed bread. *Food Chem.* **2016**, *194*, 1217–1223. [CrossRef] [PubMed]
13. Quan, W.; He, W.; Qie, X.; Chen, Y.; Zeng, M.; Qin, F.; Chen, J.; He, Z. Effects of β-cyclodextrin, whey protein, and soy protein on the thermal and storage stability of anthocyanins obtained from purple-fleshed sweet potatoes. *Food Chem.* **2020**, *320*, 126655. [CrossRef] [PubMed]
14. Sun, B.; Tian, Y.; Chen, L.; Jin, Z. Linear dextrin as curcumin delivery system: Effect of degree of polymerization on the functional stability of curcumin. *Food Hydrocoll.* **2018**, *77*, 911–920. [CrossRef]
15. Li, S.; Li, J.; Zhu, Z.; Cheng, S.; He, J.; Lamikanra, O. Soluble dietary fiber and polyphenol complex in lotus root: Preparation, interaction and identification. *Food Chem.* **2020**, *314*, 126219. [CrossRef] [PubMed]
16. Zheng, L.; Ding, Z.; Zhang, M.; Sun, J. Microencapsulation of bayberry polyphenols by ethyl cellulose: Preparation and characterization. *J. Food Eng.* **2011**, *104*, 89–95. [CrossRef]
17. Préstamo, G.; Rupérez, P.; Espinosa-Martos, I.; Villanueva, M.J.; Lasunción, M.A. The effects of okara on rat growth, cecal fermentation, and serum lipids. *Eur. Food Res. Technol.* **2007**, *225*, 925–928. [CrossRef]
18. Kim, H.-S.; Yu, O.-K.; Byun, M.-S.; Cha, Y.-S. Okara, a soybean by-product, prevents high fat diet-induced obesity and improves serum lipid profiles in c57bl/6j mice. *Food Sci. Biotechnol.* **2016**, *25*, 607–613. [CrossRef]
19. Zhang, Y.; Sun, Y.; Zhang, H.; Mai, Q.; Zhang, B.; Li, H.; Deng, Z. The degradation rules of anthocyanins from eggplant peel and antioxidant capacity in fortified model food system during the thermal treatments. *Food Biosci.* **2020**, *38*, 100701. [CrossRef]
20. Zheng, Y.; Wang, Q.; Huang, J.; Fang, D.; Zhuang, W.; Luo, X.; Zou, X.; Zheng, B.; Cao, H. Hypoglycemic effect of dietary fibers from bamboo shoot shell: An in vitro and in vivo study. *Food Chem. Toxicol.* **2019**, *127*, 120–126. [CrossRef]
21. Zhao, G.; Zhang, R.; Dong, L.; Huang, F.; Tang, X.; Wei, Z.; Zhang, M. Particle size of insoluble dietary fiber from rice bran affects its phenolic profile, bioaccessibility and functional properties. *LWT—Food Sci. Technol.* **2018**, *87*, 450–456. [CrossRef]
22. Wang, X.; Zhang, Y.; Li, Y.; Yu, H.; Wang, Y.; Piao, C. Insoluble dietary fibre from okara (soybean residue) modified by yeast kluyveromyces marxianus. *LWT—Food Sci. Technol.* **2020**, *134*, 110252. [CrossRef]
23. Wang, B.; Yu, H.; He, Y.; Wen, L.; Gu, J.; Wang, X.; Miao, X.; Qiu, G.; Wang, H. Effect of soybean insoluble dietary fiber on prevention of obesity in high-fat diet fed mice via regulation of the gut microbiota. *Food Funct.* **2021**, *12*, 7923–7937. [CrossRef] [PubMed]
24. Wang, L.; Li, Y.; Wang, Y.; Guo, J.; Xia, Q.; Tu, Y.; Nie, P. Compensation benefits allocation and stability evaluation of cascade hydropower stations based on variation coefficient-shapley value method. *J. Hydrol.* **2021**, *599*, 126277. [CrossRef]
25. Wang, Y.; Zhang, J.; Zhang, L. An active and ph-responsive film developed by sodium carboxymethyl cellulose/polyvinyl alcohol doped with rose anthocyanin extracts. *Food Chem.* **2022**, *373*, 131367. [CrossRef]
26. Fan, L.; Wang, Y.; Xie, P.; Zhang, L.; Li, Y.; Zhou, J. Copigmentation effects of phenolics on color enhancement and stability of blackberry wine residue anthocyanins: Chromaticity, kinetics and structural simulation. *Food Chem.* **2019**, *275*, 299–308. [CrossRef]
27. Koh, J.; Xu, Z.; Wicker, L. Binding kinetics of blueberry pectin-anthocyanins and stabilization by non-covalent interactions. *Food Hydrocoll.* **2020**, *99*, 105354. [CrossRef]
28. Patras, A.; Brunton, N.P.; O'Donnell, C.; Tiwari, B.K. Effect of thermal processing on anthocyanin stability in foods; mechanisms and kinetics of degradation. *Trends Food Sci. Technol.* **2010**, *21*, 3–11. [CrossRef]
29. Tang, B.; He, Y.; Liu, J.; Zhang, J.; Li, J.; Zhou, J.; Ye, Y.; Wang, J.; Wang, X. Kinetic investigation into ph-dependent color of anthocyanin and its sensing performance. *Dye. Pigment.* **2019**, *170*, 107643. [CrossRef]
30. Wu, Y.; Geng, F.; Chang, P.R.; Yu, J.; Ma, X. Effect of agar on the microstructure and performance of potato starch film. *Carbohydr. Polym.* **2009**, *76*, 299–304. [CrossRef]
31. Zhao, X.; Chen, J.; Chen, F.; Wang, X.; Zhu, Q.; Ao, Q. Surface characterization of corn stalk superfine powder studied by ftir and xrd. *Colloids Surf. B Biointerfaces* **2013**, *104*, 207–212. [CrossRef] [PubMed]
32. Le Bourvellec, C.; Guyot, S.; Renard, C.M.G.C. Non-covalent interaction between procyanidins and apple cell wall material: Part i. Effect of some environmental parameters. *Biochim. Biophys. Acta (BBA)—Gen. Subj.* **2004**, *1672*, 192–202. [CrossRef] [PubMed]
33. Ullah, I.; Yin, T.; Xiong, S.; Huang, Q.; Ziaud, D.; Zhang, J.; Javaid, A.B. Effects of thermal pre-treatment on physicochemical properties of nano-sized okara (soybean residue) insoluble dietary fiber prepared by wet media milling. *J. Food Eng.* **2018**, *237*, 18–26. [CrossRef]

34. Qin, Y.; Liu, Y.; Yong, H.; Liu, J.; Zhang, X.; Liu, J. Preparation and characterization of active and intelligent packaging films based on cassava starch and anthocyanins from lycium ruthenicum murr. *Int. J. Biol. Macromol.* **2019**, *134*, 80–90. [CrossRef] [PubMed]
35. Zhao, L.; Pan, F.; Mehmood, A.; Zhang, Y.; Hao, S.; Rehman, A.U.; Li, J.; Wang, C.; Wang, Y. Protective effect and mechanism of action of xanthan gum on the color stability of black rice anthocyanins in model beverage systems. *Int. J. Biol. Macromol.* **2020**, *164*, 3800–3807. [CrossRef] [PubMed]
36. Mansour, M.; Salah, M.; Xu, X. Effect of microencapsulation using soy protein isolate and gum arabic as wall material on red raspberry anthocyanin stability, characterization, and simulated gastrointestinal conditions. *Ultrason. Sonochem.* **2020**, *63*, 104927. [CrossRef]
37. Wang, S.; Sun, W.; Swallah, M.S.; Amin, K.; Lyu, B.; Fan, H.; Zhang, Z.; Yu, H. Preparation and characterization of soybean insoluble dietary fiber and its prebiotic effect on dyslipidemia and hepatic steatosis in high fat-fed c57bl/6j mice. *Food Funct.* **2021**, *12*, 8760–8773. [CrossRef]
38. Gao, R.; Liu, H.; Peng, Z.; Wu, Z.; Wang, Y.; Zhao, G. Adsorption of (−)-epigallocatechin-3-gallate (egcg) onto oat β-glucan. *Food Chem.* **2012**, *132*, 1936–1943. [CrossRef]
39. Guan, Y.; Zhong, Q. The improved thermal stability of anthocyanins at ph 5.0 by gum arabic. *LWT—Food Sci. Technol.* **2015**, *64*, 706–712. [CrossRef]
40. Pradhan, P.C.; Mandal, A.; Dutta, A.; Sarkar, R.; Kundu, A.; Saha, S. Delineating the behavior of berberis anthocyanin/β-cyclodextrin inclusion complex in vitro: A molecular dynamics approach. *LWT—Food Sci. Technol.* **2022**, *157*, 113090. [CrossRef]

foods

MDPI

Article

Effects of γ-Irradiation on Structure and Functional Properties of Pea Fiber

Tianfu Cheng, Caihua Liu, Zhaodong Hu, Zhongjiang Wang and Zengwang Guo *

College of Food Science, Northeast Agricultural University, Harbin 150030, China; ctf2303@163.com (T.C.); 17686963949@163.com (C.L.); 2021nodanger@163.com (Z.H.); wzjname@126.com (Z.W.)
* Correspondence: gzwname@163.com; Tel.: +86-182-4955-4700

Abstract: In this study, pea residue reserve insoluble diet fiber (hereinafter referred to as pea fiber) was used as a raw material. The effects of γ-irradiation doses (0, 0.5, 1, 2, 3, and 5 kGy) on the structural properties (main composition, particle size and specific surface area, scanning electron microscope (SEM) microstructure, Fourier transform infrared spectroscopy, and X-ray diffraction) and functional properties (oil-holding capacity, swelling and water-holding capacity, and adsorption properties) of pea fiber were explored. The results show that, when the γ-irradiation dose was 2 kGy, compared with the untreated sample, the contents of cellulose, hemicellulose and lignin in pea fiber decreased by 1.34 ± 0.42%, 2.56 ± 0.03% and 2.02 ± 0.05%, respectively, and the volume particle size of pea fiber decreased by 17.43 ± 2.35 μm. The specific surface area increased by 23.70 ± 2.24 m^2/kg and the crystallinity decreased by 7.65%. Pore and irregular particles appeared on the microstructure surface of the pea fiber treated with γ-irradiation. The results of the infrared spectrum showed that the hemicellulose and lignin in pea fiber were destroyed by γ-irradiation. These results indicate that γ-irradiation can significantly affect the structural properties of pea fiber. When the γ-irradiation dose was 2 kGy, the highest oil-holding capacity, swelling capacity and water-holding capacity of pea fiber were 8.12 ± 0.12 g/g, 19.75 ± 0.37 mL/g and 8.35 ± 0.18 g/g, respectively, and the adsorption capacities of sodium nitre, cholesterol and glucose were also the strongest. These results indicate that the functional properties of pea fiber are improved by γ-irradiation. In this study, γ-irradiation technology was used as pretreatment to provide a theoretical basis for the application of pea fiber in food processing.

Keywords: γ-irradiation; pea fiber; structural properties; functional properties

Citation: Cheng, T.; Liu, C.; Hu, Z.; Wang, Z.; Guo, Z. Effects of γ-Irradiation on Structure and Functional Properties of Pea Fiber. *Foods* **2022**, *11*, 1433. https://doi.org/10.3390/foods11101433

Academic Editors: Jianhua Xie, Yanjun Zhang and Hansong Yu

Received: 29 March 2022
Accepted: 11 May 2022
Published: 16 May 2022

1. Introduction

Pea is a leguminous plant, which is widely planted all over the world because of its strong adaptability. According to the United Nations, the area of pea harvested in 2018 was behind only soybean, common bean, chickpea and cow pea, while the production of pea ranked fourth, behind soybean, common bean, and chickpea [1]. Pea is one of the most important edible beans in China. In recent years, the pea deep-processing industry has developed rapidly in China. The main products are pea starch, pea protein, pea protein peptides, and pea protein artificial meat. Most of the by-products of pea processing are crushed and used as animal feed, with low added value. Pea dregs are one of the main by-products of pea processing. It is rich in dietary fiber, with a total dietary fiber content of 14–26%, insoluble dietary fiber content of 10–15%, and soluble dietary fiber content of 2–9% [2]. Dietary fiber has the effect of reducing the cholesterol content in blood, preventing heart disease, controlling blood sugar, preventing diabetes, promoting gastrointestinal peristalsis, preventing constipation, and removing harmful toxins from the human body [3]. However, after the unmodified insoluble dietary fiber is directly incorporated into baked food, meat products, beverage products, jams and other fields, it will affect the sensory quality and physical and chemical properties of food and cannot be directly applied in large quantities [4]. As a result, a new process has been employed to improve the application quality and processing characteristics of insoluble dietary fiber in pea.

γ-irradiation is a cold sterilization technology with simple operation, short processing time, large processing capacity, easy control, no chemical requirements and no pollution to the environment. It has broad application prospects in industrial production [5]. Studies show that irradiated food is safe when the dose is less than 10 kGy. The WTO announced that when the irradiation dose is greater than 10 kGy, there is no harm to food [6]. At present, irradiation treatment in the food field is mainly used in food sterilization, protein modification, enzyme elimination, promote fiber hydrolysis and other applications. It was found that γ-irradiation alters the structure of the polymer by forming several intermediates that follow several rapid reaction pathways and form new bonds within the polymer chain [7]. Studies by Fei [8] and Li [9] found an increase in irradiation dose; depolymerization of hemicellulose, cellulose, and lignin; and the association of a cracking structure with a change in the dietary fiber. They also found that, as the free group increased, dietary fiber in the fiber form of crystalline and amorphous forms was destroyed, and more of the crystalline form was broken into an amorphous form. This reduced crystallinity, degree of polymerization, and thermal stability. However, the enzymatic hydrolysis yield of dietary fiber increased. Zhu [10] found that the combination of γ-irradiation and micropulverization has the best degradation effect on soybean dietary fiber, and can improve the physical and chemical properties of soybean dietary fiber. Therefore, it is regarded as an ideal method to improve the quality of soybean dietary fiber. These studies indicate that γ-irradiation treatment is a promising technology and has a prospective application in changing fiber structure and improving fiber function.

Studies have proved that γ-irradiation technology has the effect of changing fiber structure, as well as its physical and chemical properties; it is mainly used in food component modification. However, there are few reports on the effects of irradiation on the physicochemical and functional properties of pea residue dietary fiber. Therefore, in this study, pea residue insoluble dietary fiber (hereinafter referred to as pea fiber) was used as a raw material to explore the effects of γ-irradiation doses (0, 0.5, 1, 2, 3, and 5 kGy) on the physicochemical and functional properties of pea fiber. This study provides a theoretical basis for the wide industrial application of pea fiber in the future.

2. Materials and Methods

2.1. Materials and Reagents

KBr (Beijing Chemical Reagent Factory, Beijing, China), P-aminobenzene sulfonic acid (Nanjing Chemical Reagent Company, Nanjing, China), Naphthalene hydrochloride (Tianjin Dongli Tianda Chemical Reagent Factory, Tianjin, China), Dinitrosalicylate (DNS), Nitrite and Glucose (Sigma Chemical Company, St. Louis, MO; USA).

2.2. Sample Prepare

The irradiation test was carried out at No.2 cobalt source facility of Heilongjiang Institute of Atomic Energy (activity, 200 kCi; temperature, 20 ± 1 °C; average dose rate, 10 Gy/h; inhomogeneity < 5%). The dried pea fiber samples were crushed by ST-G200 high-speed mill (Beijing Xuxinshengke Co., LTD., Beijing, China), and passed through a 60-mesh sieve. The irradiation experiments were carried out after the samples were packed in Ziplock bags (polyethylene). The irradiation doses were 0.5, 1, 2, 3 and 5 kGy, respectively. The irradiation dose of the samples was tracked by a silver dichromate chemical dosimeter (RadEye GF-10, Thermo Fisher Scientific, Waltham, MA, USA), and the absorbed dose was measured by a UV–visible spectrophotometer (SPECORD®210 PLUS, Analytik Jena AG, Jena, GER). After irradiation, samples were stored in sealed polyethylene bags at room temperature.

2.3. Determination of Structural Properties of Pea Fiber

2.3.1. Determination of Main Components

Referring to the research method of Guo et al. [11], cellulose and hemicellulose were prepared by alkali method (24% KOH for 2 h and 10% KOH for 16 h), and lignin was prepared using the acid method (72% H_2SO_4 for 1 h). The separated cellulose and lignin were dried at 105 °C for 2 h and hemicellulose was dried at 60 °C for 24 h for further analysis.

2.3.2. Determination of Particle Size and Specific Surface Area

The 3 g sample was dissolved in absolute ethanol, and ultrasonically dispersed for 3 min using an MS2000 Masterizer particle-size analyzer (Malvern company, Malvern, UK) to determine the volume-average particle size and specific surface area of pea fibers.

2.3.3. Determination of Microstructure

The method of determination of microstructure is modified according to Li et al. [12]. The microstructure of pea fiber was determined by SU8010 field emission scanning electron microscope (Hitachi, Japan). The sample particles of about 1 mg were placed on the tape of the circular aluminum sample sub and coated with palladium for 90 s at a current of 15 mA. The specimen stubs were then placed in the observation room. These samples were observed at an accelerator potential of 5 kV with a 6000x increase.

2.3.4. Determination of Fourier Transform Infrared (FTIR) Spectroscopy

According to the method of Guo et al. [13], the pea fiber and KBr (1:250 w/w) were fully mixed and pressed into a disk, and then Scimitar 2000 FTIR spectrometer (Agilent, Santa Clara, CA, USA) was used to obtain the infrared spectrum of the sample with a wave number of 400–4000 cm^{-1}. Each sample was scanned 32 times.

2.3.5. Determination of X-ray Diffraction

The crystal structure determination of pea fiber was slightly modified with reference to the research method of Yang et al. [14]. Scanning from 10 to 40° (2θ°) at 1 π/min was performed with an XPert Powder Multifunctional Powder X-ray diffractometer (Dandong Tonda Technology Co., Ltd., Dandong, China). The generator voltage was 40 kV, and the incident current was 40 mA. The formula of relative crystallinity (RC) of pea fiber is as follows:

$$\mathrm{RC}(\%) = \frac{\mathrm{A_c}}{\mathrm{A_c}+\mathrm{A_a}} \times 100 \tag{1}$$

where A_c is the area of the crystalline region; Aa is the area of the amorphous region.

2.4. Determination of Functional Properties of Pea Fiber

2.4.1. Determination of Oil-Holding Capacity

Referring to the oil-holding capacity determination method of Liu et al. [15], 0.5 g (M0) pea fiber was mixed with 5 mL soybean oil. After standing in a centrifuge tube for 24 h, the samples were centrifuged for 20 min at 4000 rpm in a refrigerated high-speed centrifuge (X1R, Thermo Fisher Scientific, Waltham, MA, USA). The supernatant was removed, and the weight of the residue was recorded as M1. The oil-holding capacity (OHC) of pea fiber is calculated as follows:

$$\mathrm{OHC}(\mathrm{g/g}) = \frac{M_1 - M_0}{M_0} \tag{2}$$

where, M_0 and M_1 are the weights of pea dietary fiber before and after oil absorption (g).

2.4.2. Determination of Swelling and Water-Holding Capacity

According to the determination method of Wang et al. [16], the pea fiber suspension was hydrated for 24 h after diluting 0.3 g (m_0) of pea fiber to 15 mL (V_0) with deionized water; the volume was recorded as V1. Then, the pea fiber suspension was centrifuged at 6000 rpm for 20 min. The supernatant was removed and wet fiber weight, m1, was recorded. Swelling (SC) and water-holding capacity (WHC) were calculated using the following formula:

$$SC(mL/g) = \frac{V_1 - V_0}{m_0} \tag{3}$$

where V_0 is the dilution volume of pea fiber 15 mL; V1 is the volume of the suspension after pea fiber hydration for 24 h (mL).

$$WHC(g/g) = \frac{m_1 - m_0}{m_0} \tag{4}$$

where m_0 and m_1 are the weight of pea fiber before and after water absorption (g).

2.4.3. Determination of Adsorption Properties

Nitrite

According to Luo et al. [17], the adsorption capacity of pea fiber to $NaNO_2$ was measured. A 0.1 g sample was added to a 5 mL 20 μg/mL $NaNO_2$ solution; the environments of the small intestine and stomach were simulated at pH = 7.0 and pH = 2.0, respectively. Then, the obtained mixture was left to stand at room temperature for 2 h, centrifuged at 4800 rpm for 10 min, and then the 0.5 mL of supernatant was placed in a glass tube. According to Gan et al. [18], $NaNO_2$ levels in the supernatant were determined using p-aminobenzene sulfonic acid and naphthalenediamide hydrochloride. Deionized water was added to the tube until the volume of the mixture was 2 mL; then, 2 mL p-aminobenzene sulfonic acid (4 μg/mL) and 1 mL naphthalene hydrochloride (2 μg/mL) were added to the mixture. The solution was left in the dark to react for 30 min, at 538 nm. This enabled us to measure the concentration of $NaNO_2$ and obtain the standard curve value (y = 1.9832x + 0.0467, R2 = 0.9981, y is the absorbance value, x is the concentration of $NaNO_2$). The adsorption capacity of pea fiber to $NaNO_2$ (NIAC) was calculated as follows:

$$NIAC(\mu g/g) = \frac{C_1 - C_2}{W} \times V \tag{5}$$

where C_1 and C_2 are, respectively, the concentration of $NaNO_2$ in the supernatant before and after adsorption (μg/L), W is the weight of pea fiber (g), and V is the volume of $NaNO_2$ solution (mL).

Cholesterol

Cholesterol was added using the method used by Benitez et al. [19] with slight modifications. Fresh egg yolks were diluted with 9 times the weight of distilled water, then beaten until completely emulsified. Next, 0.5 g of pea fiber was added to 25 mL of the egg yolk emulsion and stirred to combine; then, the solution was shaken at 37 °C for 2, 5, 10, 15, 25, 40, 60, 90 and 120 min, respectively. After centrifugation at 4000 rpm for 15 min, the cholesterol content in the supernatant was determined at a wavelength of 550 nm using the phthalaldehyde method and a UV-2700 spectrophotometer (Shimadzu Company, Kyoto, Japan) (Gan et al., 2020). The standard curve was obtained (y = 1.6578x + 0.0254, R2 = 0.9975, y is the absorbance value, x is the cholesterol concentration). The cholesterol adsorption capacity (CAC) of pea fiber was calculated using the following equation:

$$CAC(mg/g) = \frac{C_1 - C_2}{W} \tag{6}$$

where C_1 and C_2 are the weight of cholesterol before and after adsorption (mg), and W is the weight of pea fiber (g).

Glucose

According to the method of Ma et al. [20] and Chen et al. [21], the glucose adsorption capacity of pea fiber samples was determined. A 0.5 g sample of pea fiber and 100 mL glucose solutions with concentrations of 10, 50, 100 and 200 mmol/L were prepared; they were then shaken at 37 °C for 6 h, centrifuged at 4000 rpm for 15 min, and retained in the supernatant. The glucose content in the supernatant was determined by dinitro salicylate (DNS) chromogenic reagent [18]. Next, 0.5 mL of supernatant was added into a glass tube, wherein deionized water was added until the volume reached 3 mL. This was then mixed with 2 mL dinitrosalicylate (DNS) chromogenic reagent. The mixture was continuously shaken in a water bath at 100 °C for 6 min. After the solution was cooled to room temperature, the glucose concentration was measured at 520 nm to generate a standard curve (y = 14674x + 0.0543, R2 = 0.9983, y is the absorbance value, x is the glucose concentration). The calculation formula of glucose adsorption capacity (GAC) of pea fiber is as follows:

$$GAC(mmol/g) = \frac{G_1 - G_2}{W} \times V \quad (7)$$

where G_1 and G_2 are glucose concentration before and after adsorption, respectively (mmol/g), W is weight of pea fiber (g), and V is volume of supernatant (mL).

2.5. Statistical Analysis

Three pea fiber samples were prepared, and all samples were tested for three times in parallel. The data were analyzed by one-way ANOVA with SPSS 22.0 software. The data result was mean ± SD, and the difference was significant with $p < 0.05$. Origin 9.0 software was used for data analysis, fitting and standard curve drawing. The areas of crystalline and amorphous regions of X-ray diffraction were calculated by Peakfit version 4.12 software.

3. Results and Discussion

3.1. Analysis of the Effect of Electron Beam γ-Irradiation on the Physicochemical Properties of Pea Fiber

3.1.1. Analysis of Content of Main Components

According to Table 1, compared with the untreated sample, the contents of cellulose, hemicellulose and lignin in pea fiber significantly decreased ($p < 0.05$) following γ-irradiation. The contents of cellulose, hemicellulose and lignin significantly decreased ($p < 0.05$) with the increase in γ-irradiation dose. This may be a lignin with strong covalent bonds, hydrogen bonds and hemicellulose to form a stable compound composed of cellulose and wrapped up, but it is difficult to predict for certain in outside conditions decomposition [22]. Additionally, γ-irradiation of strong energy rays can promote the depolymerization of hemicellulose and cellulose; lignin association cracking occurred, and the part of dietary fiber component was transformed into oligosaccharides; then, the composition of insoluble dietary fiber in pea dregs changed [23]. The energy absorbed in the irradiation process destroyed the monomeric units of hemicellulose and lignin, cracking into small molecular components or other free radicals, thereby reducing their content [11]. With the increase in γ-irradiation dose, the structure of pea fiber was opened, the exposed cellulose was destroyed and degraded, and the content of pea fiber decreased [24]. The results showed that the binding sites and structures of cellulose, hemicellulose and lignin in pea fiber were destroyed by γ-irradiation, which changed the content of pea fiber.

Table 1. Effects of γ-irradiation on the content of main components of pea fiber.

Treatment Dose (kGy)	Cellulose (%)	Hemicellulose (%)	Lignin (%)
0	37.44 ± 0.18 a	15.10 ± 0.04 a	14.27 ± 0.05 a
0.5	37.09 ± 0.23 b	13.31 ± 0.05 b	13.81 ± 0.06 b
1	36.23 ± 0.15 c	12.97 ± 0.06 c	12.52 ± 0.07 c
2	36.10 ± 0.23 d	12.54 ± 0.07 d	12.25 ± 0.05 d
3	36.25 ± 0.18 c	12.10 ± 0.06 e	12.11 ± 0.06 e
5	36.31 ± 0.23 d	11.50 ± 0.06 f	12.02 ± 0.06 f

Note: Different lowercase letters in the same column indicate significant difference ($p < 0.05$)—same below.

3.1.2. Analysis of Particle Size and Specific Surface Area

According to Table 2, compared with the untreated sample, with the increase in γ-irradiation dose, the average particle size of pea fiber volume decreased significantly, and the specific surface area increased significantly ($p < 0.05$). Combined with the results in Section 3.1.1, γ-irradiation causes the depolymerization of lignin and hemicellulose in pea fibers by breaking molecular chains, reducing the average particle size and enhancing the specific surface area. Guo [11] and Al-Sheraji et al. [25] also showed that irradiation would destroy the glycosidic bonds between cellulose and reduce cellulose, resulting in a decrease in the particle size of dietary fiber and an increase in the specific surface area of insoluble dietary fiber. This indicates that γ-irradiation can affect the structure and physicochemical properties of pea fibers by breaking glycosidic bonds and changing fiber components.

Table 2. Effects of γ-irradiation on particle size and specific surface area of okra pea fiber.

Treatment Dose (kGy)	D10 (μm)	D50 (μm)	D90 (μm)	Volumetric Mean Particle Size (μm)	Specific Surface Area (m^2/kg)
0	46.15 ± 2.03 a	75.57 ± 2.33 a	165.47 ± 3.01 a	135.62 ± 3.14 a	78.68 ± 1.42 d
0.5	42.35 ± 2.43 a	73.67 ± 2.06 ab	163.98 ± 3.14 a	133.82 ± 2.61 a	85.36 ± 1.77 c
1	39.98 ± 1.50 b	70.79 ± 1.99 c	155.29 ± 2.67 b	129.38 ± 2.81 ab	94.78 ± 1.91 ab
2	36.16 ± 1.29 c	65.47 ± 2.30 d	153.09 ± 2.93 bc	118.19 ± 2.01 c	102.38 ± 2.54 a
3	37.26 ± 1.43 c	67.19 ± 2.19 cd	154.48 ± 3.01 b	124.29 ± 2.14 b	100.94 ± 2.63 a
5	41.23 ± 1.89 b	70.65 ± 2.31 c	154.98 ± 2.60 b	127.21 ± 1.54 ab	95.79 ± 1.95 b

3.1.3. Analysis of SEM

According to Figure 1, the surface microstructure of untreated samples shows fiber strip structure, and cracks and pores appear in samples treated by γ-irradiation. When the dose of γ-irradiation increased from 0 to 2 kGy, the pore size of the samples gradually increased. When the irradiation dose was 2 kGy, the largest pores appeared in the fiber structure of the sample. When the irradiation dose increased from 2 kGy to 5 kGy, the pore size of the sample showed a downward trend, and when the irradiation dose reached 5 kGy, the surface of pea fiber breaks and forms a lamellar microstructure. This is consistent with the specific surface area results in Section 3.1.2. This may be due to the fact that the increase in γ-irradiation intensity can break the glycosidic bond in the molecular chain, reduce the molecular weight and weaken the interaction between molecules, promoting the formation of carbonyl and double bonds, and leading to the relaxation of the pea fiber's structure, which demonstrated a honeycomb structure and had larger pores [26,27]. The decrease in particle size and changes in the microstructure of pea fibers may generate capillary action and form a larger specific surface area, which may be important for its absorption capacities with some other compounds [17]. Jiang et al. [28] also demonstrated that high-intensity energy field treatment can change the surface structure and affect the specific surface area by destroying the intermolecular crosslinking of fibers. When the irradiation dose is too high, layered substances appear on the surface of pea fiber. This may be due to the accumulation of residual protein on the surface of the fiber and degraded fiber fragments caused by excessive dose irradiation. The research shows that the hydration properties and glucose absorption capacity of dietary fiber are mainly related to the porosity of the fiber structure [29,30]. This indicates that the appropriate dose γ-irradiation treatment can improve the functional activity by adjusting the microstructure and porosity of pea fiber.

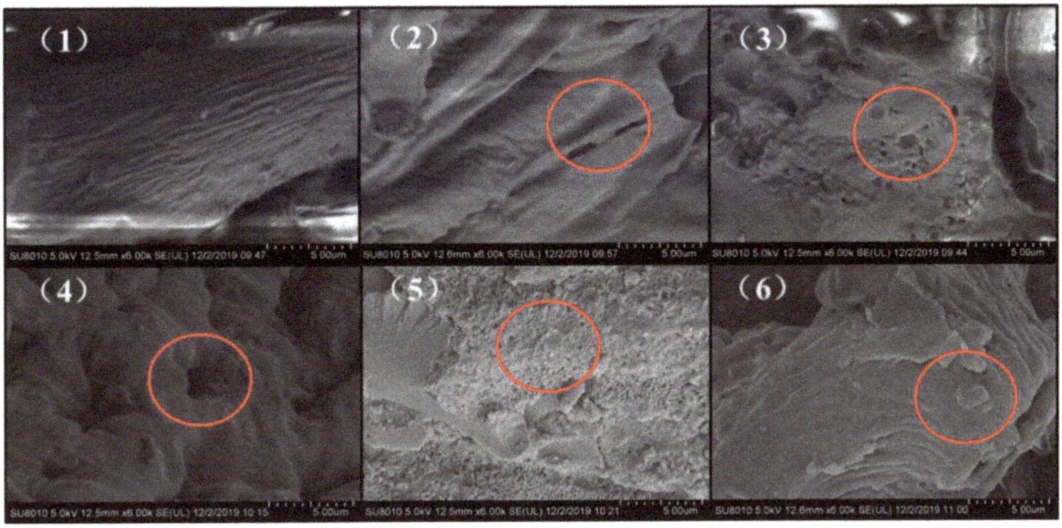

Figure 1. Scanning electron microscope (SEM) images of pea fiber after γ-irradiation. Note: (**1**) as the control, the irradiation doses of (**2**)–(**6**) were 0.5, 1, 2, 3, and 5 kGy, respectively, and the magnification was 6000×. Note: The red circles represent cracks, pores, and layered structures present in the γ-irradiation treatment samples.

3.1.4. Analysis of Fourier Transform Infrared Spectroscopy

Fourier transform infrared spectroscopy is an infrared absorption spectrum formed according to the vibration of molecules at different wavelengths, which is used to detect changes in molecular groups and chemical bonds of pea fibers. The position and strength of the absorption peak are mainly affected by the types of chemical bonds or molecular groups, and the position of the absorption peak changes with induction, conjugation or steric hindrance [31]. According to Figure 2, all the pea fiber samples showed similar spectral curves while retaining characteristic bands specific to each procedure. The absorption bands of all pea fibers in the range of 3000–3700 cm^{-1} are due to O-H bond tensile vibration, and these absorption bands also indicate the presence of pectin and hemicellulose in soybean residue fibers [32]. The absorption peaks at 2853 cm^{-1} and 2925 cm^{-1} were attributed to the asymmetric and symmetrical C-H vibrational bands in the polysaccharide compound methylene [33]. The absorption peak was observed at 1235 cm^{-1}, indicating the presence of a crystalline region. With the increase in γ-irradiation dose, the intensity of these absorption bands weakened, probably due to the destruction of intramolecular hydrogen bonds in cellulose and hemicellulose compared with untreated samples [34]. The intensity of the absorption peak at 1725 cm^{-1} weakened, indicating that the adsorption of water on the fiber matrix became weak [35]. The decrease in absorption bands near 1000 cm^{-1} was due to the stretching of C=O and aromatic skeleton of aldehyde/ester groups of hemicellulose and lignin, indicating that the hemicellulose and lignin in pea fiber were destroyed by γ-irradiation [36]. The reactive groups play an important role in the physicochemical and functional properties of dietary fibers, such as hydration, adsorption, cation exchange capacity, and metal chelation [37]. This indicates that γ-irradiation treatment can modulate physicochemical and functional properties by altering the reactive groups of pea fibers.

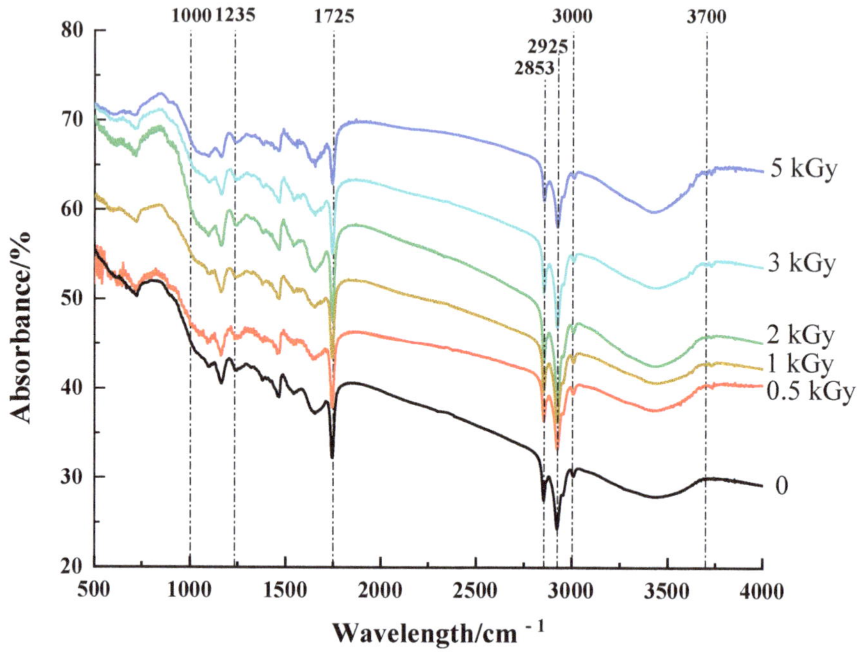

Figure 2. Effect of γ-irradiation on infrared spectrum of pea fiber.

3.1.5. Analysis of X-ray Diffraction

The effect of γ-irradiation on the crystalline properties of pea fiber was studied by X-ray diffraction. The overall peak shape represents the crystal type of the sample. The increase in diffraction peak intensity indicates that the crystallinity at the diffraction angle increases. It can be seen from Figure 3 that the irradiation treatment does not cause the peak shape to change, and each sample has obvious absorption peaks at the scanning angles (2θ) of 19.96° and 34.56° (except 2 kGy, which is 35.05°). This indicates that the crystals of the six groups of fiber samples are all of the type I cellulose type, which consists of ordered crystalline cellulose regions and disordered cellulose and hemicellulose regions, and irradiation treatment does not change the cellulose type of the samples. According to Figures 2 and 3, the highest crystallinity of untreated sample is 34.22%. With the increase in irradiation dose, the crystallinity of pea fiber decreased first and then increased. When the irradiation dose was 2 kGy, the crystallinity was the lowest with a value of 26.57%, which might be because the crystalline form and the amorphous form of pea fiber were destroyed by γ-irradiation, and the destroyed crystalline form was decomposed into the amorphous form, thus reducing the crystallinity of pea fiber [9]. When the irradiation dose was greater than 2 kGy, the γ-irradiation destroyed the monomer units of hemicellulose and lignin in the amorphous region, but the structural damage degree of cellulose monomer units in the crystallization region is small, so the destruction degree of the amorphous region is greater than that of the crystallization region, resulting in the increase in crystallinity [38]. The results showed that γ-irradiation changed the crystal structure of pea fibers by changing the contents of cellulose, hemicellulose and lignin in pea fibers.

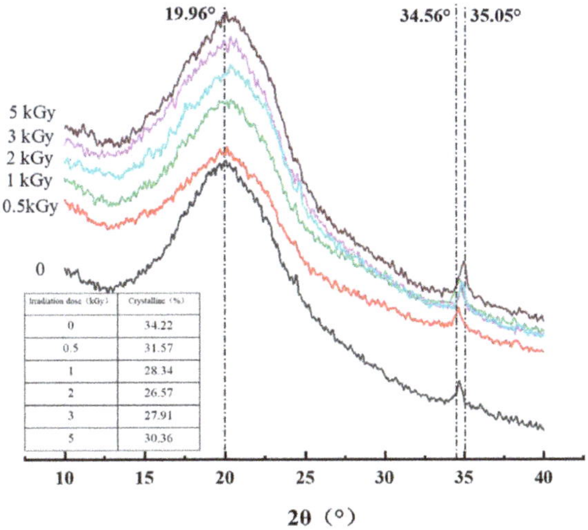

Figure 3. Effects of γ-irradiation on X-ray diffraction of pea fiber.

3.2. *Analysis of the Effect of γ-Irradiation on the Functional Properties of Pea Fiber*

3.2.1. Analysis of Oil-Holding Capacity

The oil-holding capacity of dietary fiber can significantly improve the sensory characteristics of food, and help to prolong the shelf life of food [12]. Meanwhile, the high oil-holding capacity of dietary fiber can reduce the absorption of lipids in the intestinal tract [39]. According to Figure 4, the oil-holding capacity of pea fiber increased first and then decreased with the increase in γ-irradiation dose. When the γ-irradiation dose was 2 kGy, the oil-holding capacity of pea fiber was the highest. This may be because the γ-irradiation treatment can increase the porosity and specific surface area of pea fibers by destroying the microstructure. This is helpful for powering more oil droplets to contact and embed in pea fibers, thereby improving the oil-holding capacity of pea dietary fibers [25]. However, when the irradiation dose was greater than 2 kGy, the pores on the microstructure surface of pea fiber decreased, and the specific surface area decreased; therefore, that the oil-holding capacity decreased.

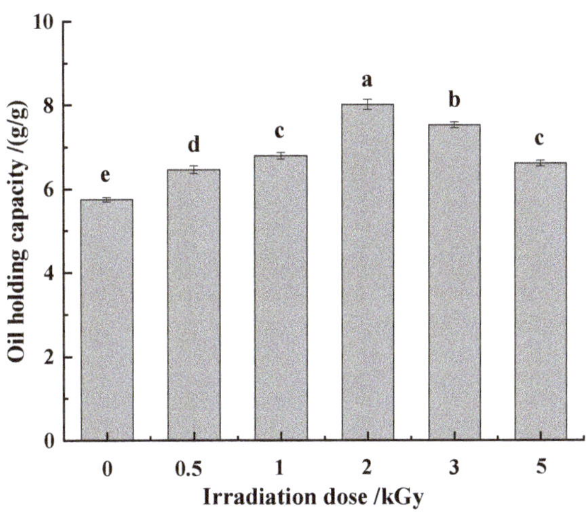

Figure 4. Effect of γ-irradiation on oil-holding capacity of pea fiber. Note: Different lowercase letters indicate significant differences in oil-holding capacity ($p < 0.05$).

3.2.2. Analysis of Swelling and Water-Holding Capacity

The water swelling capacity is a functional characteristic in food processing, which is used to evaluate the hydration properties of dietary fiber [12]. The swelling properties and water-holding capacity of dietary fiber are beneficial to intestinal function by increasing chyme bulk and enhancing peristalsis. According to Figure 5, compared with the untreated sample, the swelling capacity and water-holding capacity of pea fiber were significantly increased ($p < 0.05$). With the increase in γ-irradiation dose, the swelling capacity and water-holding capacity increased first and then decreased. Additionally, when the irradiation dose was 2 kGy, the swelling capacity and water-holding capacity were highest. It shows that the appropriate irradiation dose can significantly improve the swelling capacity and water-holding capacity of dietary fiber. Alam reported in his study that particle size can affect the swelling capacity of dietary fiber, and the smaller the particle size, the higher the swelling capacity of dietary fiber [40]. This is consistent with the results in Section 3.1.2. Furthermore, the moderate irradiation could break some fibers, thereby loosening the dense network structure of dietary fibers and increasing the specific surface area. This is also one of the reasons for the increased swelling capacity of dietary fiber. Irradiation treatment can break the glycosidic bond of dietary fiber, and the microstructure of fiber has obvious pores and honeycomb structure at 2 kGy. This provides more space for the storage of water molecules, thereby enhancing the water-holding capacity of the pea fibers [41,42]. However, when the irradiation dose was too high, the swelling and water-holding capacity of pea fiber decreased significantly ($p < 0.05$). This may be because the swelling capacity is related to the crystallinity. Studies have shown that the crystallinity of γ-irradiated potato starch is reduced, which leads to a reduction in its swelling capacity [43]. At the same time, high-dose irradiation will lead to the destruction of the network structure of the fibers, resulting in a decrease in water-holding capacity [12].

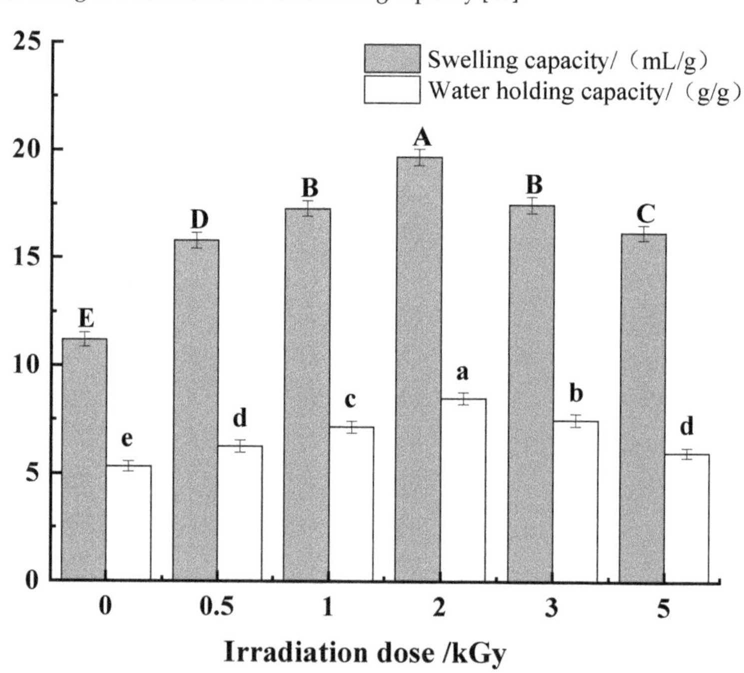

Figure 5. Effects of γ-irradiation on the swelling and water-holding capacity of pea fiber. Note: Different uppercase letters indicate significant differences in swelling capacity ($p < 0.05$), and different lowercase letters indicate significant differences in water-holding capacity ($p < 0.05$).

3.2.3. Analysis of Adsorption Properties

Nitrite

Nitrite is a relatively toxic compound and can form carcinogenic compounds during human digestion [28]. Some studies have shown that nitro compounds with carcinogenicity to animals can also enter the fetus through the placenta and have teratogenic effects on the fetus [44,45]. As an important indicator, pH has a great influence on the ability of dietary fiber to adsorb nitrite ions. Therefore, the adsorption capacity of each pea fiber sample for nitrite ions under simulated gastric environment (pH = 2) and intestinal environment (pH = 7). Figure 6 shows that the nitrite adsorption capacity of pea fiber is affected by pH value and γ-irradiation dose. When pH = 2, the nitrite adsorption capacity was much higher than that of pH = 7. This may be because, in acidic conditions, NO_2^- reacts with H^+ to produce HNO2 and then forms nitrogen oxides, including the strong electron affinity compound N_2O_3, which can combine with the negatively charged oxygen atoms of phenolic acid groups in dietary fiber and cause adsorption [46]. Another explanation is that the active groups such as uronic acid, amino acid, and especially phenolic acid contained in the structure of dietary fiber have a strong adsorption effect on nitroso groups under acidic conditions; this is why the nitrite ion adsorption capacity of pea fiber sample at pH = 2 is significantly higher than that at pH = 7 [47]. With the increase in irradiation dose, the nitrite adsorption capacity of pea fiber increases first and then decreases at two pH values. When the irradiation dose is 2 kGy, the nitrite adsorption capacity of pea fiber was the highest. The γ-irradiation changed the structure of pea fibers, the surface appeared loose and porous, and the specific surface area became larger, which exposed more adsorption sites for nitrite ions to a certain extent. The capillary effect formed by these pores accelerates the entry of nitrite ions into the interior of the fiber structure, so the adsorption capacity of pea fibers for nitrite ions after moderate irradiation treatment is improved in the gastrointestinal environment.

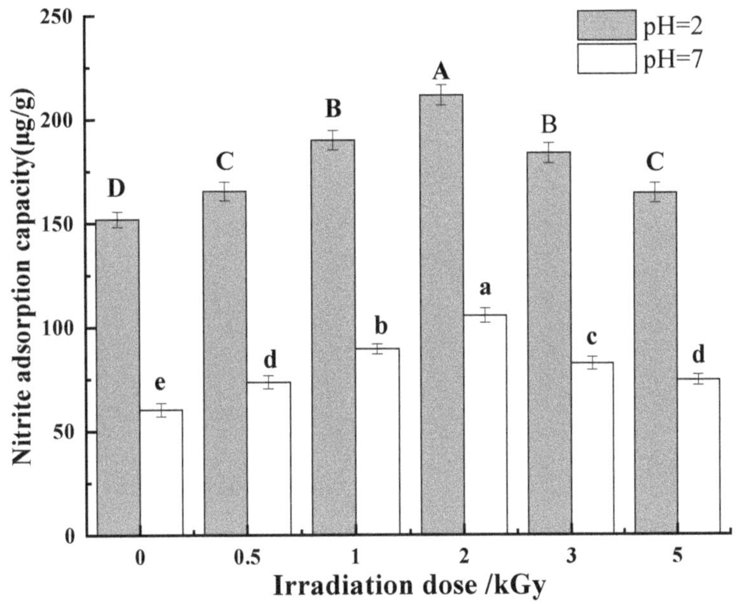

Figure 6. Effect of γ-irradiation on nitrite adsorption of pea fiber. Note: Different uppercase letters indicate significant difference in nitrite adsorption capacity under simulated gastric environment (pH = 2) ($p < 0.05$), and different lowercase letters indicate significant difference in nitrite adsorption capacity under simulated intestinal environment (pH = 7) ($p < 0.05$).

Cholesterol

The occurrence of coronary heart disease is directly related to the content of cholesterol in the blood. There is evidence that dietary fibers reduce the risk of cardiovascular disease and serum cholesterol levels through the absorption of cholesterol [48,49]. Thus, it is of great significance to improve the cholesterol adsorption capacity of dietary fiber. According to Figure 7, the cholesterol adsorption capacity of pea fiber was affected by pH value and γ-irradiation dose. With the increase in γ-irradiation dose, the cholesterol adsorption capacity of pea fiber increased first and then decreased at both pH values. When the γ-irradiation dose was 2 kGy, the cholesterol adsorption capacity of pH = 2 and pH = 7 was the largest, and the cholesterol adsorption capacity of pH = 7 was greater than that of pH = 2. This may be due to the repulsion between H+ and some positive charges in dietary fiber and cholesterol under acidic conditions, which affects the binding of dietary fiber and cholesterol, thus reducing the adsorption amount of cholesterol so that the absorption of cholesterol by pea fiber in intestinal environment is more effective than that in stomach environment. When the irradiation dose was lower than 2 kGy, the γ-irradiation increases the specific surface area of the insoluble dietary fiber of pea dregs and changes the structure of hemicellulose and lignin to enhance the capillary action, thereby increasing the adsorption capacity of cholesterol [18,47,50]. However, when the irradiation dose is too high, the surface pores and surface area of pea fiber decrease, which weakens the effect of capillaries, thus weakening the adsorption effect of pea fiber on cholesterol and reducing the adsorption amount of cholesterol [51].

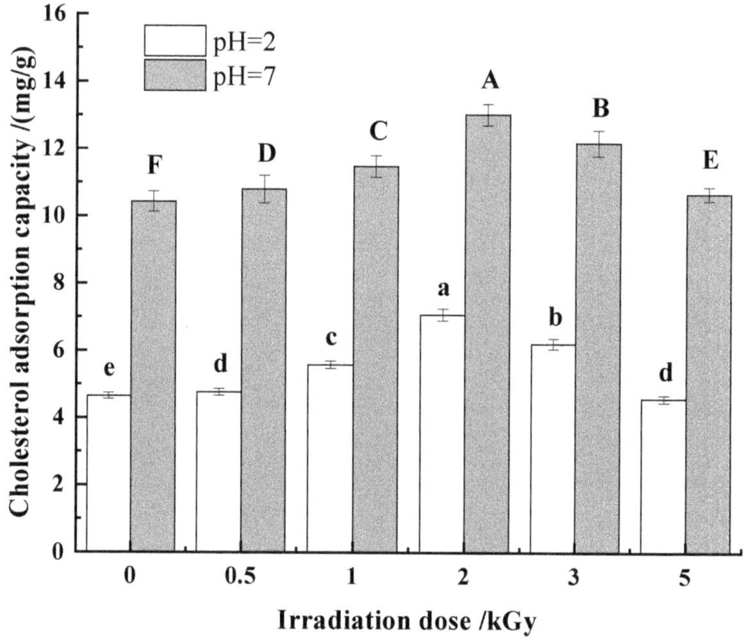

Figure 7. Effect of γ-irradiation on cholesterol adsorption of pea fiber. Note: Different uppercase letters indicate significant difference in cholesterol adsorption capacity under simulated gastric environment (pH = 2) ($p < 0.05$), and different lowercase letters indicate significant difference in cholesterol adsorption capacity under simulated intestinal environment (pH = 7) ($p < 0.05$).

Glucose

The absorption of glucose by dietary fiber can delay or reduce the digestion and absorption of glucose in the gastrointestinal tract, thus playing a role in reducing blood glucose, which is also an important functional characteristic of dietary fiber [12]. It can be seen from Figure 8 that the adsorption capacity of pea fiber on glucose is affected by glucose concentration and γ-irradiation dose. With the increase in γ-irradiation dose, the adsorption capacity of pea fiber to different concentrations of glucose first increased and then decreased. When the dose of γ-irradiation was 2 kGy, the adsorption capacity of pea fibers to different concentrations of glucose was the highest, which indicated that γ-irradiation could improve the glucose adsorption capacity of pea fibers, possibly due to the increase in specific surface area and porosity caused by moderate irradiation, making it easier for glucose to enter the interior of the fiber and bind more tightly to the interior of the dietary fiber. Studies have shown that the increased hydration of dietary fiber makes it easier for glucose to bond to the fiber's surface [52]; this is confirmed by the findings in Section 3.2.2 of this paper. Furthermore, the adsorption capacity positively correlated with glucose concentration. This may be due to the increased contact probability between glucose and fiber, which improves the adsorption capacity of pea fiber. However, when the irradiation dose was too high, the damage to polar and non-polar groups in pea fiber structure weakens the interaction with glucose molecules [53], thus reducing the adsorption amount of glucose.

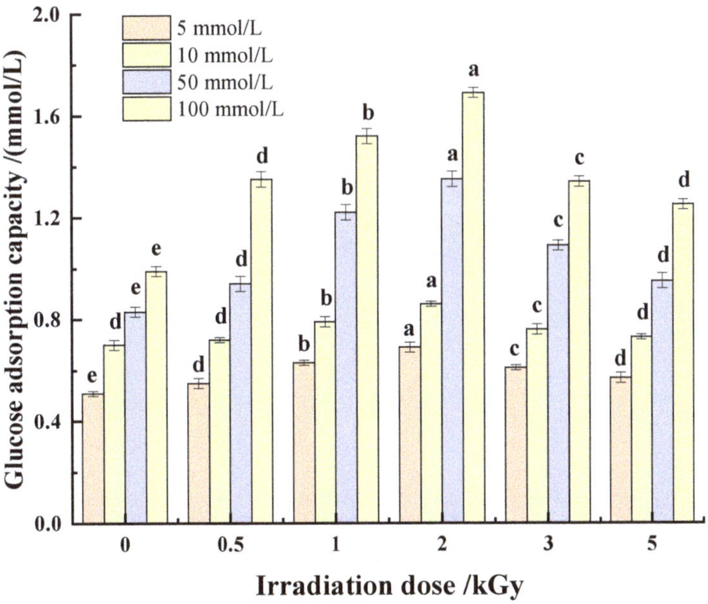

Figure 8. Effect of γ-irradiation on glucose adsorption of pea fiber. Note: Different small letters at the same concentration indicate significant differences in glucose adsorption capacity ($p < 0.05$).

4. Conclusions

Pea fiber was irradiated with different doses (0, 0.5, 1, 2, 3, and 5 kGy) by γ-irradiation technology to investigate the effects of irradiation dose on the structure and functional characteristics of pea fiber. According to the structural characteristics of pea fiber, when the γ-irradiation dose was 2 kGy, the contents of cellulose, hemicellulose and lignin in pea fiber decreased by 1.34 ± 0.42%, 2.56 ± 0.03% and 2.02 ± 0.05%, respectively, and the crystallinity of pea fiber decreased by 7.65%. The pore and irregular particles appeared on the microstructure surface of pea fiber treated by γ-irradiation. The results of infrared spectroscopy showed that the hemicellulose and lignin in pea fiber were destroyed by γ-irradiation. The results of the functional characteristics of pea fiber showed that, when the γ-irradiation dose was 2 kGy, the highest oil-holding capacity, swelling capacity and water-holding capacity of pea fiber were 8.12 ± 0.12 g/g, 19.75 ± 0.37 mL/g and 8.35 ± 0.18 g/g, respectively. Additionally, the adsorption capacity of sodium nitre, cholesterol and glucose were also the strongest in these conditions. These results suggest that the functional properties of pea fiber can be improved by γ-irradiation, changing the structural properties of pea fiber. In this study, γ-irradiation technology was used as pretreatment to provide a theoretical basis for the application of pea fiber in food processing.

Author Contributions: T.C.: conceptualization, software, writing—original draft; C.L.: visualization, software; Z.H.: methodology, investigation, methodology; Z.W.: writing—review and editing, validation, funding acquisition, project administration; Z.G.: investigation, supervision, funding acquisition, project administration. All authors have read and agreed to the published version of the manuscript.

Funding: This research was funded by Major industrial key projects for the transformation of new and old kinetic energy in Shandong Province; Key R&D plan of Shandong Province (major sci-entific and technological innovation project), grant number 2022CXGC010603; Major scientific and technological special projects of "millions" project in Heilongjiang Province, grant number 2021ZX12B02; Heilongjiang Postdoctoral Scientific Research Developmental Fund, grant number LBH-Q20008.

Institutional Review Board Statement: Not applicable.

Informed Consent Statement: Not applicable.

Data Availability Statement: Data is contained within the article.

Conflicts of Interest: The authors declare no conflict of interest.

References

1. FAOSTAT. *FAOSTAT Data Base*; Food and Agriculture Organization of the United Nations—Statistic Division: New York, NY, USA, 2018.
2. Che, L.; Chen, H.; Yu, B.; He, J.; Zheng, P.; Mao, X.; Yu, J.; Huang, Z.; Chen, D. Long-Term Intake of Pea Fiber Affects Colonic Barrier Function, Bacterial and Transcriptional Profile in Pig Model. *Nutr. Cancer* **2014**, *66*, 388–399. [CrossRef] [PubMed]
3. Queiroz-Monici, K.D.S.; Costa, G.E.; da Silva, N.; Reis, S.M.; de Oliveira, A.C. Bifidogenic effect of dietary fiber and resistant starch from leguminous on the intestinal microbiota of rats. *Nutrition* **2005**, *21*, 602–608. [CrossRef] [PubMed]
4. Lee, J.Y.; Lim, T.; Kim, J.; Hwang, K.T. Physicochemical characteristics and sensory acceptability of crackers containing red ginseng marc. *J. Food Sci. Technol.* **2022**, *59*, 212–219. [CrossRef] [PubMed]
5. Kuan, Y.H.; Bhat, R.; Patras, A.; Karim, A.A. Radiation processing of food proteins—A review on the recent developments. *Trends Food Sci. Technol.* **2013**, *30*, 105–120. [CrossRef]
6. Skovgaard, N. High-dose Irradiation: Wholesomeness of Food Irradiated with Doses above 10 kGy.: Report of a Joint FAO/IAEA/WHO Study Group. Technical report series, No. 890, 1999, vi+197 pages (available in English; French and Spanish in preparation). Paper back; WHO Marketing and Dissemination, 1211 Geneva 27, Switzerland; order no. ISBN 92 4 120890 2; Sw.fr. 42.-/US $37.In developing countries: Sw.fr. 29.20. *Int. J. Food Microbiol.* **2000**, *58*, 130–131. [CrossRef]
7. Chmielewski, A.G.; Haji-Saeid, M.; Ahmed, S. Progress in radiation processing of polymers. *Nucl. Instrum. Methods Phys. Res.* **2005**, *236*, 44–54. [CrossRef]
8. Fei, X.; Jia, W.; Wang, J.; Chen, T.; Ling, Y. Study on enzymatic hydrolysis efficiency and physicochemical properties of cellulose and lignocellulose after pretreatment with electron beam irradiation. *Int. J. Biol. Macromol.* **2020**, *145*, 733–739. [CrossRef]
9. Li, T.; Wang, L.; Chen, Z.; Li, C.; Li, X.; Sun, D. Structural changes and enzymatic hydrolysis yield of rice bran fiber under electron beam irradiation. *Food Bioprod. Process.* **2020**, *122*, 62–71. [CrossRef]
10. Zhu, L.; Yu, B.; Chen, H.; Yu, J.; Yan, H.; Luo, Y.; He, J.; Huang, Z.; Zheng, P.; Mao, X.; et al. Comparisons of the micronization, steam explosion, and gamma irradiation treatment on chemical composition, structure, physicochemical properties, and in vitro digestibility of dietary fiber from soybean hulls. *Food Chem.* **2021**, *366*, 130618. [CrossRef]
11. Guo, X.; Zhang, T.; Shu, S.; Zheng, W.; Gao, M. Compositional and structural changes of corn cob pretreated by electron beam irradiation. *ACS Sustain. Chem. Eng.* **2017**, *5*, 420–425. [CrossRef]
12. Li, X.; Wang, B.; Hu, W.; Chen, H.; Sheng, Z.; Yang, B.; Yu, L. Effect of γ-irradiation on structure, physicochemical property and bioactivity of soluble dietary fiber in navel orange peel. *Food Chem. X* **2022**, *14*, 100274. [CrossRef] [PubMed]
13. Guo, Z.; Teng, F.; Huang, Z.; Lv, B.; Lv, X.; Babich, O.; Yu, W.; Li, Y.; Wang, Z.; Jiang, L. Effects of material characteristics on the structural characteristics and flavor substances retention of meat analogs. *Food Hydrocoll.* **2020**, *105*, 105752. [CrossRef]
14. Yang, W.; Kong, X.; Zheng, Y.; Sun, W.; Chen, S.; Liu, D.; Zhang, H.; Fang, H.; Tian, J.; Ye, X. Controlled ultrasound treatments modify the morphology and physical properties of rice starch rather than the fine structure. *Ultrason. Sonochem.* **2019**, *59*, 104709. [CrossRef] [PubMed]
15. Liu, Y.; Zhang, H.; Yi, C.; Quan, K.; Lin, B. Chemical composition, structure, physicochemical and functional properties of rice bran dietary fiber modified by cellulase treatment. *Food Chem.* **2021**, *342*, 128352. [CrossRef] [PubMed]

16. Wang, C.; Li, L.; Sun, X.; Qin, W.; Wu, D.; Hu, B.; Raheem, D.; Dong, H.; Vasanthan, T.; Zhang, Q.; et al. High-speed shearing of soybean flour suspension disintegrates the component cell layers and modifies the hydration properties of okara fibers. *LWT* **2019**, *116*, 108505. [CrossRef]
17. Luo, X.; Wang, Q.; Fang, D.; Zhuang, W.; Chen, C.; Jiang, W.; Zheng, Y. Modification of insoluble dietary fibers from bamboo shoot shell: Structural characterization and functional properties. *Int. J. Biol. Macromol.* **2018**, *120*, 1461–1467. [CrossRef]
18. Gan, J.; Huang, Z.; Yu, Q.; Peng, G.; Chen, Y.; Xie, J.; Nie, S.; Xie, M. Microwave assisted extraction with three modifications on structural and functional properties of soluble dietary fibers from grapefruit peel. *Food Hydrocoll.* **2020**, *101*, 105549. [CrossRef]
19. Benitez, V.; Rebollo-Hernanz, M.; Hernanz, S.; Chantres, S.; Aguilera, Y.; Martin-Cabrejas, M.A. Coffee parchment as a new dietary fiber ingredient: Functional and physiological characterization. *Food Res. Int.* **2019**, *122*, 105–113. [CrossRef]
20. Ma, M.; Mu, T.; Sun, H.; Zhang, M.; Chen, J.; Yan, Z. Optimization of extraction efficiency by shear emulsifying assisted enzymatic hydrolysis and functional properties of dietary fiber from deoiled cumin (*Cuminum cyminum* L.). *Food Chem.* **2015**, *179*, 270–277. [CrossRef]
21. Chen, J.; Zhao, Q.; Wang, L.; Zha, S.; Zhang, L.; Zhao, B. Physicochemical and functional properties of dietary fiber from maca (*Lepidium meyenii* Walp.) liquor residue. *Carbohydr. Polym.* **2015**, *132*, 509–512. [CrossRef]
22. Sun, S.-N.; Li, M.-F.; Yuan, T.-Q.; Xu, F.; Sun, R.-C. Effect of Ionic Liquid Pretreatment on the Structure of Hemicelluloses from Corncob. *J. Agric. Food Chem.* **2012**, *60*, 11120–11127. [CrossRef] [PubMed]
23. Guo, X.; Shu, S.; Zhang, W.; Wang, E.; Hao, J. Synergetic Degradation of Corn Cob with Inorganic Salt (or Hydrogen Peroxide) and Electron Beam Irradiation. *ACS Sustain. Chem. Eng.* **2016**, *4*, 1099–1105. [CrossRef]
24. Huang, J.-Y.; Liao, J.-S.; Qi, J.-R.; Jiang, W.-X.; Yang, X.-Q. Structural and physicochemical properties of pectin-rich dietary fiber prepared from citrus peel. *Food Hydrocoll.* **2020**, *110*, 106140. [CrossRef]
25. Al-Sheraji, S.H.; Ismail, A.; Manap, M.Y.; Mustafa, S.; Yusof, R.M.; Hassan, F.A. Purification, characterization and antioxidant activity of polysaccharides extracted from the fibrous pulp of Mangifera pajang fruits. *LWT* **2012**, *48*, 291–296. [CrossRef]
26. Fan, X.T. Radiation chemistry of food components. In *Food Irradiation Research and Technology*, 2nd ed.; Wiley-Blackwell: Hoboken, NJ, USA, 2012; pp. 75–97.
27. Zhang, Y.; Qi, J.; Zeng, W.; Huang, Y.; Yang, X. Properties of dietary fiber from citrus obtained through alkaline hydrogen peroxide treatment and homogenization treatment. *Food Chem.* **2020**, *311*, 125873. [CrossRef] [PubMed]
28. Jiang, Y.; Yin, H.; Zheng, Y.; Wang, D.; Liu, Z.; Deng, Y.; Zhao, Y. Structure, physicochemical and bioactive properties of dietary fibers from *Akebia trifoliata* (Thunb.) Koidz. seeds using ultrasonication/shear emulsifying/microwave-assisted enzymatic extraction. *Food Res. Int.* **2020**, *136*, 109348. [CrossRef]
29. Gu, M.; Fang, H.; Gao, Y.; Su, T.; Niu, Y.; Yu, L. Characterization of enzymatic modified soluble dietary fiber from tomato peels with high release of lycopene. *Food Hydrocoll.* **2020**, *99*, 105321. [CrossRef]
30. Dong, W.; Wang, D.; Hu, R.; Long, Y.; Lv, L. Chemical composition, structural and functional properties of soluble dietary fiber obtained from coffee peel using different extraction methods. *Food Res. Int.* **2020**, *136*, 109497. [CrossRef]
31. Uruakpa, F.; Arntfield, S. Surface hydrophobicity of commercial canola proteins mixed with κ-carrageenan or guar gum. *Food Chem.* **2006**, *95*, 255–263. [CrossRef]
32. Chylińska, M.; Szymańska-Chargot, M.; Kruk, B.; Zdunek, A. Study on dietary fibre by Fourier transform-infrared spectroscopy and chemometric methods. *Food Chem.* **2016**, *196*, 114–122. [CrossRef]
33. Yan, X.; Ye, R.; Chen, Y. Blasting extrusion processing: The increase of soluble dietary fiber content and extraction of soluble-fiber polysaccharides from wheat bran. *Food Chem.* **2015**, *180*, 106–115. [CrossRef] [PubMed]
34. Wu, C.; Teng, F.; McClements, D.; Zhang, S.; Li, Y.; Wang, Z. Effect of cavitation jet processing on the physicochemical properties and structural characteristics of okara dietary fiber. *Food Res. Int.* **2020**, *134*, 109251. [CrossRef] [PubMed]
35. Wen, Y.; Niu, M.; Zhang, B.; Zhao, S.; Xiong, S. Structural characteristics and functional properties of rice bran dietary fiber modified by enzymatic and enzyme-micronization treatments. *LWT* **2017**, *75*, 344–351. [CrossRef]
36. Chen, B.; Cai, Y.; Liu, T.; Huang, L.; Deng, X.; Zhao, Q.; Zhao, M. Improvements in physicochemical and emulsifying properties of insoluble soybean fiber by physical-chemical treatments. *Food Hydrocoll.* **2019**, *93*, 167–175. [CrossRef]
37. Jiang, Y.; Yu, L.; Hu, Y.; Zhu, Z.; Zhuang, C.; Zhao, Y.; Zhong, Y. Electrostatic spraying of chitosan coating with different deacetylation degree for strawberry preservation. *Int. J. Biol. Macromol.* **2019**, *139*, 1232–1238. [CrossRef]
38. Karthika, K.; Arun, A.; Rekha, P. Enzymatic hydrolysis and characterization of lignocellulosic biomass exposed to electron beam irradiation. *Carbohydr. Polym.* **2012**, *90*, 1038–1045. [CrossRef]
39. Elleuch, M.; Bedigian, D.; Roiseux, O.; Besbes, S.; Blecker, C.; Attia, H. Dietary fibre and fibre-rich by-products of food processing: Characterisation, technological functionality and commercial applications: A review. *Food Chem.* **2011**, *124*, 411–421. [CrossRef]
40. Alam, S.A.; Järvinen, J.; Kirjoranta, S.; Jouppila, K.; Poutanen, K.; Sozer, N. Influence of Particle Size Reduction on Structural and Mechanical Properties of Extruded Rye Bran. *Food Bioprocess Technol.* **2013**, *7*, 2121–2133. [CrossRef]
41. Xia, Q.; Gu, M.; Liu, J.; Niu, Y.; Yu, L. Novel composite gels of gelatin and soluble dietary fiber from black bean coats with interpenetrating polymer networks. *Food Hydrocoll.* **2018**, *83*, 72–78. [CrossRef]
42. Fan, X.; Chang, H.; Lin, Y.; Zhao, X.; Zhang, A.; Li, S.; Feng, Z.; Chen, X. Effects of ultrasound-assisted enzyme hy-drolysis on the microstructure and physicochemical properties of okara fibers. *Ultrason. Sonochem.* **2020**, *69*, 105247. [CrossRef]
43. Singh, S.; Singh, N.; Ezekiel, R.; Kaur, A. Effects of gamma-irradiation on the morphological, structural, thermal and rheological properties of potato starches —ScienceDirect. *Carbohydr. Polym.* **2011**, *83*, 1521–1528. [CrossRef]

44. Quist, A.; Inoue-Choi, M.; Weyer, P.J.; Anderson, K.E.; Cantor, K.P.; Krasner, S.; Freeman, L.; Ward, M.H.; Jones, R.R. Ingested nitrate and nitrite, disinfection by-products, and pancreatic cancer risk in postmenopausal women. *Int. J. Cancer* **2018**, *142*, 251–261. [CrossRef] [PubMed]
45. Zheng, Y.; Xu, B.; Shi, P.; Tian, H.; Li, Y.; Wang, X.; Wu, S.; Liang, P. The influences of acetylation, hydroxypropylation, enzymatic hydrolysis and crosslinking on improved adsorption capacities and in vitro hypoglycemic properties of millet bran dietary fibre. *Food Chem.* **2022**, *368*, 130883. [CrossRef] [PubMed]
46. Lyu, B.; Wang, H.; Swallah, M.S.; Fu, H.; Shen, Y.; Guo, Z.; Feng, Z.; Jiang, L. Structure, properties and potential bioactivities of high-purity insoluble fibre from soybean dregs (Okara). *Food Chem.* **2021**, *364*, 130402. [CrossRef]
47. Chu, J.; Zhao, H.; Lu, Z.; Lu, F.; Bie, X.; Zhang, C. Improved physicochemical and functional properties of dietary fiber from millet bran fermented by Bacillus natto. *Food Chem.* **2019**, *294*, 79–86. [CrossRef]
48. Grunberger, G.; Jen, K.L.C.; Artiss, J.D. The benefits of early intervention in obese diabetic patients with FBCx™—A new dietary fibre. *Diabetes/Metab. Res. Rev.* **2007**, *23*, 56–62. [CrossRef]
49. Nsor-Atindana, J.; Zhong, F.; Mothibe, K.J. In vitro hypoglycemic and cholesterol lowering effects of dietary fiber pre-pared from cocoa (*Theobroma cacao* L.) shells. *Food Funct.* **2012**, *3*, 1044–1050. [CrossRef]
50. Chen, Y.; Ye, R.; Yin, L.; Zhang, N. Novel blasting extrusion processing improved the physicochemical properties of soluble dietary fiber from soybean residue and in vivo evaluation. *J. Food Eng.* **2014**, *120*, 1–8. [CrossRef]
51. Zhu, Y.; Chu, J.; Lu, Z.; Lv, F.; Bie, X.; Zhang, C.; Zhao, H. Physicochemical and functional properties of dietary fiber from foxtail millet (*Setaria italic*) bran. *J. Cereal Sci.* **2018**, *79*, 456–461. [CrossRef]
52. Zhang, Z.S.; Li, C.Y.; Liu, Y.L. Effect of Mildew and γ-Irradiation on the Physicochemical Property and Structure of Soybean Proteins. *Mod. Food Sci. Technol.* **2015**, *31*, 191–196, 282.
53. Saikia, S.; Mahanta, C.L. In vitro physicochemical, phytochemical and functional properties of fiber rich fractions de-rived from by-products of six fruits. *J. Food Sci. Technol.* **2015**, *53*, 1496–1504. [CrossRef] [PubMed]

foods MDPI

Article

Structural Characteristics of Insoluble Dietary Fiber from Okara with Different Particle Sizes and Their Prebiotic Effects in Rats Fed High-Fat Diet

Hongliang Fan [1,2], Ying Zhang [1,2], Mohammed Sharif Swallah [3], Sainan Wang [1,2], Jiarui Zhang [1,2], Jiaqi Fang [1,2], Jiahong Lu [1,2] and Hansong Yu [1,2,*]

1 College of Food Science and Engineering, Jilin Agricultural University, Changchun 130118, China; f_hongliang@163.com (H.F.); zhang577726644@163.com (Y.Z.); 18644941109@163.com (S.W.); jiarui197@163.com (J.Z.); fjq2825516331@163.com (J.F.); lujiahong1222@163.com (J.L.)
2 National Soybean Industry Technology System Processing Laboratory, Changchun 130118, China
3 Science Island Branch of Graduate School, University of Science and Technology of China, Hefei 230026, China; m.s.swallah@gmail.com
* Correspondence: yuhansong@jlau.edu.cn; Tel./Fax: +86-0431-84533104

Citation: Fan, H.; Zhang, Y.; Swallah, M.S.; Wang, S.; Zhang, J.; Fang, J.; Lu, J.; Yu, H. Structural Characteristics of Insoluble Dietary Fiber from Okara with Different Particle Sizes and Their Prebiotic Effects in Rats Fed High-Fat Diet. *Foods* **2022**, *11*, 1298. https://doi.org/10.3390/foods11091298

Academic Editor: Antonella Pasqualone

Received: 1 April 2022
Accepted: 26 April 2022
Published: 29 April 2022

Abstract: Dietary fiber, which is utilized to make functional meals, is an important component for promoting human health and managing calorie consumption. In this study, three different particle sizes of OIDF (Okara insoluble dietary fiber) were characterized. Their lipid-lowering effects and the impacts on gut microbiota were determined by OIDF intervention in high-fat diet rats. Scanning electron microscopy (SEM) results showed that the three particle sizes of OIDF have different morphologies. Fourier transform infrared spectroscopy (FT-IR) results showed that the three sources of IDF samples have similar active groups, but the thermogravimetric analysis/differential scanning calorimetry (TGA/DSC) and X-ray diffraction (XRD) showed that three different particle sizes of OIDF have different retention and crystallinity. Among the three OIDFs, OIDF-10 exhibited the stronger WSC, OHC, CAC, and SCAC. The results after the feeding showed that the OIDF of three particle sizes could improve the elevation of blood lipids and the disturbance of gut microbiota caused by the high-fat diet. Therefore, this study demonstrated the functional significance of the three particle sizes of OIDF and provided a reference for its application in functional food processing, aiming at maintaining healthy blood lipid and intestinal flora levels.

Keywords: okara; insoluble dietary fiber; prebiotic; gut microbiota

1. Introduction

Okara is a by-product of soybean processing. Over 20 million tons of wet soybean dregs (okara) are produced each year in China [1], which is one of the world-leading producers and consumers of soybean. This huge sum of okara produced during soybean processing is evidenced to be packed with a significant amount of nutritional and non-nutritional constituents. However, these are usually used as animal feeds, fertilizers, landfills, or discarded as waste due to their high susceptibility to spoilage, which exerts serious economic and socio-environmental problems [2]. Hence, its valorization will be important to help utilize the untapped nutrients and minimize the environmental problems caused by this waste disposal. Okara is high in nutrients, with insoluble fiber fraction accounting for 70% of the total dietary fiber [1]. Besides, the functional characteristics of insoluble dietary fiber have recently been discovered and gained recognition. Dietary fiber is tagged as "The Seventh Nutrient" [3], and is necessary for a balanced diet [4]. Hence, can be utilized in shaping the gut microbiota [5]. The consumption of a high dietary fiber diet is evidenced by decreases in the bioavailability of some important nutritional components, such as some vitamins and minerals [6], and may also impact the rate of digestion of food substances as well as energy metabolism. The intake of a high-fiber diet can also improve the activities of the gut microbial composition [7].

The gut ecosystem is a home for over 10^{14} microbial cells, and collectively comprise of at least 150 times more genes than their host [8]. In both humans and animals, the gastrointestinal tract is host to a complex community of varied microorganisms, whose activities considerably impact host nutrition and health [9]. The composition of the gut microbiota is said to vary based on genetic characteristics, sex, age, and diet [10]. The association between the host gut-microbiota are dynamic and are highly susceptible to several environmental conditions, mainly diet [8], and the interactive relationship between prebiotics and dietary fiber is reliant on variations in the gut microflora as well as their colonic degradation [7]. Numerous studies indicates that the gut microorganisms can directly impact the physiological conditions of the host organism by encouraging the immune system through stimulating the defensive mechanisms against pathogens and inflammatory bowel diseases, as well as improving the roles of the intestinal barrier, regulating autoimmunity, producing biological metabolites, destroying cancer cells, regulating diabetes and preventing obesity. The host gut-microbiota interactions are highly influenced by several environmental conditions, buy mainly by diet [3,10]. Dietary fiber intake is reported to impact gastrointestinal health via encouraging the gastrointestinal barrier function, nutrient absorption, as well as selectively stimulating the composition and activities of the gut microbes, that can confer health benefits to the host, including lowering the risk of cardiovascular disease, cancer, diabetes, and other ailments [11,12]. For instance, dietary fiber alters the gastrointestinal barrier function through increasing cells and the mucins that produce the "goblet cells" [11]. Mucins are large glycoproteins that, along with antibodies, bacteria, lipids, proteins, ions, antimicrobial peptides, and water, form what is known as mucus. Mucus shields the gut epithelium from mechanical stress, to inhibit the translocation of toxic substances and to lubricate the intestine as well as encourage smooth transportation of digested material [3,12]. In addition, dietary fiber may absorb and excrete toxic chemicals in the intestines, promote intestinal flora, and offer energy and nutrients for probiotic growth [13]. The majority of studies on okara's insoluble dietary fiber focus on changing their fundamental physical and chemical characteristics [14,15], and only a few studies have been conducted on the functional qualities of okara due to their unpleasant taste [16]. The prolonged intake of a high-fat diet poses a major hazard to human/animal health. Emerging evidence has indicated that high-fat-diet can alter the composition of the microbiota, by inducing dysbiosis or imbalance of the gut microbial ecosystem. Dysbiosis is associated with diseases and metabolic health ranging from insulin resistance, glucose intolerance and to metabolic syndrome [17]. Furthermore, the gut microbiota can influence lipid metabolism via short-chain fatty acid (SCFA) metabolites [18]. SCFAs are primarily generated in the large intestine by anaerobic bacteria fermenting indigestible carbohydrates. They provide the material foundation for the gut flora's "conversation" with the host [13]. They have the ability to stimulate cell development, enhance intestinal function, and have an impact on cardiovascular metabolism, as well as anti-inflammatory, anti-tumor, and immunological regulatory activities [19].

In this work, high-pressure microfluidic technology was employed to extract three insoluble dietary fibers (IDFs) with varying particle sizes from okara (purity > 90%) to investigate the effects of insoluble dietary fiber from okara (OIDF) on blood lipid levels and intestinal flora in rats fed a high-fat diet. Okara can obtain high-purity OIDF and increase its value addition.

2. Materials and Methods

2.1. Preparation of OIDFs with Three Different Particle Sizes

Okara, a by-product of producing soybean protein isolate with a protein level of 10–15% (Liaocheng, China), was generously provided by Shandong Jiahua Health Care Products Co., Ltd. The IDF from okara was further processed into three different particle sizes using High Pressure Microfluidic Technology to obtain the OIDFs. The steps comprise of α-amylase treatment at 95 °C for 35 min, being starch glucosidase treated at 60 °C for 30 min, being neutral protease treated 60 °C for 30 min, 70 °C water precipitation for 1.5 h, centrifugation at 3500 rpm for 10 min, alcohol precipitation until colorless, and being freeze-dried for standby. According to the Chinese national standard GB 5009.88-2014, OIDF has a purity of 90.50 percent [20]. The OIDF (purity higher than 90.50%) extracted by biological enzymatic method was processed by dynamic high pressure microfluidization (DHPM). OIDF weighing 20 g was mixed with 800 mL of deionized water, and treated at 10 MPa, 50 MPa, and 150 MPa for 2 min, respectively. Three distinct particle sizes of OIDFs were named OIDF-10 (μm), OIDF-50 (μm), and OIDF-100 (μm).

2.2. *Determination of the Basic Components of the Three OIDFs*

The protein content was determine according to the Kjeldahl's method in GB5009.5-2016 "Determination of Protein in Food"; Ash content was determine based on GB5009.4-2016 "Determination of ash in food; Moisture content was determined by MB35 Ohaus moisture analyzer; SDF (soluble dietary fiber), IDF and TDF content were determined using GB5009.88-2014 "Determination of Dietary Fiber in Food".

2.3. *Structure of the Three Different Granularity of OIDFs*

2.3.1. Particle Size Determination

In total, a 0.1% (*m*/*v*) OIDF-10, OIDF-50 and OIDF-100 suspension was prepared. The particle size and specific surface area were analyzed using a BT-9300HT laser particle sizer (Bettersize Instruments Ltd., Dandong, China).

2.3.2. Scanning Electron Microscopy (SEM)

After spraying with gold-palladium alloy, the microscopic appearance of OIDF-10, OIDF-50 and OIDF-100 was studied using a scanning electron microscope (SEM; Zeiss Supra55VP, Carl Zeiss, Jena, Germany). The scanning pictures were taken at magnifications ranging from 1000 (scale bar 20 μm) to 8000 (scale bar 2 μm) at accelerating voltages of 20 kV.

2.3.3. Fourier Infrared Spectrum (FT-IR)

The three distinct granularities of OIDF were evenly mixed with KBr at a 1/100 (*w*/*w*) ratio and pressed into tablets. Each sample was scanned using a Fourier-infrared spectrometer (IRTracer-100, Shimadzu, Kyoto, Japan) throughout a wave range of 4000 cm^{-1} to 400 cm^{-1} with a resolution of 4 cm^{-1}.

2.3.4. Determination of Thermogravimetry (TG)

Thermogravimetry (TG) is a technical method to measure the relationship between the mass of a substance and temperature under the condition of a program-controlled specific temperature. Thermogravimetric analysis of OIDF-10, OIDF-50, and OIDF-100 was carried out by the HCT-3 microcomputer differential thermal balance. A sample of weighed 3–4 mg was placed into a crucible, placed on a sling balance, and the parameters were set as follows: initial temperature: 25 °C; heating rate: 10 °C/min; termination temperature: 600 °C; nitrogen flow: 50 mL/min. Mass and temperature/time were recorded continuously to obtain thermogravimetric curves.

2.3.5. Determination of X-ray Diffraction (XRD)

The structures of OIDF-10, OIDF-50 and OIDF-100 were determined using an X-ray diffractometer model XRD-7000. Using Cu-Kα radiation (λ = 0.15418 nm), the working voltage of 40.0 kV, and a working current of 30.0 mA, we took an appropriate amount of sample and placed it on the surface of the sample plate, flattened it with a flat and smooth glass sheet, put it into the diffractometer, and set the parameters as follows: measuring angle: 5°~90°; step size: 0.02°; scanning speed: 4°/min. The specific crystallinity (degree of crystallinity) calculation formula is as follows:

$$Dc(\%) = \frac{I_{002} - I_{am}}{I_{002}} \times 100 \tag{1}$$

where Dc—degree of crystallinity; I_{002}—diffraction intensity of crystalline region (2θ = 20.78°); I_{am}—diffraction intensity of amorphous region (2θ = 18.0°).

2.4. Physicochemical Properties of the Three OIDFs

2.4.1. Water-Holding Capacity (WHC), Oil-Holding Capacity (OHC) and Water-Swelling Capacity (WSC)

Referring to the method of Zhang et al. [20] with a slight modification, we accurately weighed 0.5 g of OIDF-10, OIDF-50 and OIDF-100 samples into 50 mL centrifuge tubes, added 30 mL of distilled water, shook them evenly and placed them at room temperature for 24 h. They were then centrifuged at 4000 r/min for 20 min, supernatant was then removed, and the mass of the precipitate was weighed. The calculation formula is as follows:

$$\text{WHC}(g/g) = \frac{W_2 - W_1}{W_1} \tag{2}$$

where W_1—sample mass (g); W_2—precipitate mass after centrifugation (g).

We accurately weighed 0.5 g of OIDF-10, OIDF-50 and OIDF-100 samples in three separate 50 mL centrifuge tubes, added 30 mL of soybean oil, and evenly shook them. They were then centrifuged at 4000 r/min for 20 min, the upper layer oil was then removed, and the mass of the sediment was weighed. The calculation formula is as follows:

$$\text{OHC}(g/g) = \frac{W_2 - W_1}{W_1} \tag{3}$$

where W_1—sample mass (g); W_2—precipitate mass after centrifugation (g).

We accurately weighed 0.5 g of OIDF-10, OIDF-50 and OIDF-100 samples in triplicate and placed them in a 25 mL graduated cylinder, and read the dry product volume V1. Then, we added 20 mL of distilled water to the graduated cylinder and mixed well. After standing at room temperature for 24 h, we recorded the volume of the fully swollen sample. The calculation formula is as follows:

$$\text{SW}(mL/g) = \frac{V_2 - V_1}{W} \tag{4}$$

where V_1—sample volume (mL); V_2—swollen volume (mL); W—sample mass (g).

2.4.2. Adsorption Capacity of the Three Particle Sizes of OIDF

Cholesterol-Adsorption Capacity (CAC)

A total of 1 mg/mL cholesterol in ethanol was made. Then, 0.2 g OIDF-10/OIDF-50/OIDF-100 was mixed with 10 mL cholesterol solution, adjusted to pH = 2/7 (adding water phase would precipitate cholesterol), incubated at 37 °C for 2 h (pH = 2/4) or 4 h (pH = 7), centrifuged at 4000 rpm for 20 min, and supernatant obtained to test cholesterol concentration using the OPA technique [21].

Sodium Cholate-Adsorption Capacity (SCAC)

A total of 0.1 g OIDF-10/OIDF-50/OIDF-100 was mixed with 10 mL sodium cholate standard solution (0.2 g sodium cholate + 15 mmol/L NaCl aq 100 mL), adjusted to pH = 2, 4, 7, incubated at 37 °C for 2 h (pH = 2, 4) or 4 h (pH = 7), centrifuged at 4000 rpm for 20 min, and the permeate was taken to determine the exact concentration of sodium cholate using this method [22].

2.5. Animals and Experimental Diets

The experiment was conducted in compliance with Jilin Agricultural University's Laboratory Animals Guidelines and was authorized by Jilin Agricultural University's Laboratory Animal Welfare and Ethics Committee (no. 20210422001). SD rats were kept in a temperature-controlled (20–25 °C) environment with a 12-h light/dark cycle. SD rats were randomly separated into five groups (n = 10) after a week of adaption, with the average weight in each group being comparable. The diet for each of the five groups of SD rats was divided into a normal diet (NC), HD, HD-OIDF-10, HD-OIDF-50, HD-OIDf-100. NC: Normal diet rats; HD: high-fat diet rats; HD-OIDF-10: High-fat diet + OIDF-10 (1000 mg/kg); HD-OIDF-50: High-fat diet + OIDF-50 (1000 mg/kg); HD-OIDF-100: High-fat diet + OIDF-100 (1000 mg/kg); After eight weeks of feeding, all rats were starved for 12 h and then anesthetized at the end of the experiment. For the serum preparation, blood samples were obtained and kept at −80 °C for future use. The liver and cecal contents were kept at a temperature of −80 °C. The composition of diets for each group is shown in Table 1.

Table 1. Composition of experimental diets.

Ingredient	NC		HFD/OIDF-10/OIDF-50/OIDF-100	
	gm	kcal	gm	kcal
Casein, 80 Mesh	200	800	200	800
L-Cystine	3	12	3	12
Corn Starch	315	1260	0	0
Maltodextrin 10	35	140	125	500
Sucrose	350	1400	68.8	275.2
Cellulose, BW200	50	0	50	0
Soybean Oil	25	225	25	225
Lard	20	180	245	2205
Mineral Mix	10	0	10	0
DiCalcium Phosphate	13	0	13	0
Calcium Carbonate	5.5	0	5.5	0
Potassium Citrate,1H_2O	16.5	0	16.5	0
Vitamin Mix	10	40	10	40
Choline Bitartrate	2	0	2	0
FD&C Blue Dye#1	0.05	0	0.05	0
Total	1055.05	4057	773.85	4057
	Composition			
	gm%	kcal%	gm%	kcal%
Protein	19.2	20	26.2	20
Carbohydrate	67.3	70	26.3	20
Fat	4.3	10	34.9	60
Total		100		100
kcal/gm	3.85		5.24	

2.6. Lipid Analysis

The blood concentrations of TC, TG, LDL-C, HDL-C, were measured using commercial test kits employing an enzymatic technique. A microplate reader was used to measure all of the data.

2.7. Histological Analysis

Liver tissues were fixed in four percent paraformaldehyde, embedded in paraffin for tissue slice preparation (5 m thickness), and stained with hematoxylin and eosin for histological investigation (H&E).

2.8. *RNA Extraction and Quantitative Real-Time PCR Analysis*

Total RNA was extracted from the frozen liver using Trizol reagent according to the manufacturer's instructions for quantitative real-time PCR analysis. A spectrophotometer (Thermo Fisher Scientific, Waltham, MA, USA) was used to determine the RNA content and purity at 260 and 280 nm. The cDNA Synthesis Kit was used to reverse-transcribe the total RNA (1 g) into cDNA. Then, using a Biometra TProfessional PCR, mRNA expression was measured using quantitative real-time PCR with the SYBR Premix Ex Taq TM mix (Takara, Shiga, Japan). The manufacturers' websites include the sequences of the primers that were used to amplify the target genes. The subsequent PCR amplification protocol was used: 95 °C for 30 s, followed by 40 cycles of 95 °C for 10 s, 56 °C for 30 s, and 72 °C for 30 s. The 2-ΔΔCT technique was used to calculate relative quantification.

2.9. *Western Blot Analyses*

RIPA lysis buffer combined with a protease inhibitor cocktail was used to lyse liver samples (Beyotime Biotechnology, Shanghai, China). The lysates were centrifuged for 20 min at 12,000 rpm at 4 °C, the supernatants were collected, and the protein concentration was measured using the BCA protein assay kit (Beyotime Biotechnology, Shanghai, China). SDS-PAGE was used to separate the protein lysates, which were then transferred to PVDF membranes. The blotted membranes were blocked for 2 h with 5% skimmed milk, then washed and incubated overnight at 4 °C with antibodies against CYP7A1, HMG-CoA, LDL-R. The bands were seen using an enhanced chemiluminescence reagent and assessed in an iBright CL1000 imaging system after incubation with the relevant primary and secondary antibodies (Invitrogen, Singapore). The expression of β-actin was used as a control for normalizing protein expressions.

2.10. *Gut Microbiota Sequencing Analysis*

For the sequencing of cecal contents for microflora sequencing, the QIAamp® DNA Stool Mini Kit (QIAGEN, Hilden, Germany) was used to extract DNA from feces according to the manufacturer's instructions. The quality of extracted DNAs was assessed using an agarose gel electrophoresis and subsequently a NanoDrop NC2000 spectrophotometer to precisely measure the DNA concentration (Thermo Fisher, Waltham, MA, USA).

Each sample's pure DNA was used to amplify the V3–V4 regions of the 16S rRNA gene. With forward primers, 338F (5′-ACTCCTACGGGAGGCAGCA-3′) and the reverse primer 806R (5′-GGACTACHVGGGTWTCTAAT-3′) were utilized in the PCR amplification. The Data science PicoGreen dsDNA Assay Kit was used to determine the concentrations of PCR products (Invitrogen, Carlsbad, CA, USA). Following that, all of the PCR results were pooled in equal proportions, and paired-end sequencing was conducted using the Illumina MiSeq platform, which is run by Shanghai Personal Biotechnology Co., Ltd. (Shanghai, China).

2.11. *Statistical Analysis*

The trials were conducted three times, and the findings were presented as mean, standard deviations (SD). GraphPad Prism 6 was used for statistical analysis (GraphPad Software, Inc., California, CA, USA). Using IBM SPSS 25.0 (SPSS Inc., Chicago, IL, USA), were evaluated using one-way analysis of variance (ANOVA), and $p < 0.05$ was considered to be statistically significant.

3. Results

3.1. *Determination of the Basic Components of the Three OIDFs*

As shown in Table 2, the protein level of OIDF reduced from 1.85% to 1.63% after it was digested by DHPM. The TDF content of OIDF-10 increased, as did the TDF and SDF (soluble dietary fiber) content, which could be attributed to the fact that after high-pressure treatment, some insoluble hemicellulose, cellulose, and pectin, etc., link bonds are broken, the structure is destroyed, and the molecules are converted into soluble short-chain small molecules. This could be owing to the DF's intricate structure. DF and protein particles will agglomerate to form a whole in most cases [23]. They are subjected to various strong effects such as pressure, shear force, and friction force during the high-pressure microfluidic process, resulting in part of the protein being converted into a free state and released, so it gradually increases with the increase in pressure. During the particle size reduction process, the gliadin protein is lost, and the protein content is reduced.

Table 2. Analysis of basic components of OIDF with different particle sizes (%).

OIDF	Protein (%)	Ash (%)	Moisture (%)	TDF (%)	IDF (%)	SDF (%)
OIDF-10	1.63 ± 0.28^{a}	2.95 ± 0.11^{a}	6.06 ± 0.56^{a}	90.06 ± 0.12^{a}	89.08 ± 0.04^{a}	0.98 ± 0.09^{a}
OIDF-50	1.66 ± 0.35^{a}	3.05 ± 0.23^{a}	5.87 ± 0.68^{b}	89.42 ± 0.08^{b}	88.63 ± 0.16^{a}	0.79 ± 0.13^{b}
OIDF-100	1.85 ± 0.24^{b}	2.98 ± 0.18^{a}	5.36 ± 0.84^{c}	89.81 ± 0.06^{b}	89.18 ± 0.03^{a}	0.63 ± 0.11^{c}

Different letters in the same column (a, b, c) are significantly different ($p < 0.05$). The results are expressed as mean $\pm$ SD ($n = 3$).

3.2. Structure of the Three Different Granularity of OIDFs

From Figure 1a, with regard to particle size analysis: the particle size distribution curves of the three OIDFs were all unimodal, and the particle size distribution of OIDF-10 is relatively concentrated, with a higher peak and a normal distribution. When the average particle size of OIDF is smaller and more uniform, the difference in its physical and chemical properties is noticeable.

From Figure 1b, as shown in the SEM figure, the surface of OIDF-100 processed by 10 MPa pressure processing is relatively smooth, the structure is compact and complete, but the particle size is not uniform, and there are large flake particles. The surface of OIDF-50 after high-pressure treatment at 50 MPa is rough, curled in many places, and the structure is partially damaged, which is related to the uneven force during grinding [24], The surface of the sample OIDF-10 is very different from OIDF-50 and OIDF-100, the particle size is significantly reduced, the surface is flocculent, becomes loose and porous, the internal structure is exposed, and there is a uniform accumulation phenomenon.

From Figure 1c, the FT-IR spectrum shows that the three particle sizes of OIDF have similar characteristic curve peaks, the broad peak at 3420 cm^{-1} is the stretching vibration peak of hydroxyl (-OH), and hydrogen bonds in cellulose and hemicellulose [25] are a characteristic band of all cellulose. The absorption peak at 2927 cm^{-1} is the stretching vibration peak on the carboxymethyl group and methylene group (-CH) in hemicellulose [26]. Characteristic absorption peaks at 1739 cm^{-1} and 1630 cm^{-1} are due to the asymmetric stretching vibrations of C=O of acetyl groups or esters, including methyl esterification or free carboxyl groups [27]. As the particle size decreases, the intensity of the absorption peak gradually weakens, which may be due to the destruction of some ester structures during the grinding process. There is also literature showing that the weakening of the absorption peak here will lead to a poor water holding capacity [28], which provides a theory for water holding research. The absorption peak at 1531 cm^{-1} is the characteristic absorption peak of C=O in the imide bond; the absorption peak at 1245 cm^{-1} is the vibration of the OH or -CO group of hemicellulose, and 1053 cm^{-1} is the C-O-C stretching vibration peak. The absorption peak at 890 cm^{-1} is the characteristic absorption peak of the β-configuration glycosidic bond [29]. To sum up, compared with the characteristic peaks of OIDF-50 and OIDF-100, the peak shape, position and number of OIDF-10 did not change significantly, indicating that the main components and structure of OIDF did not change substantially after DHPM treatment.

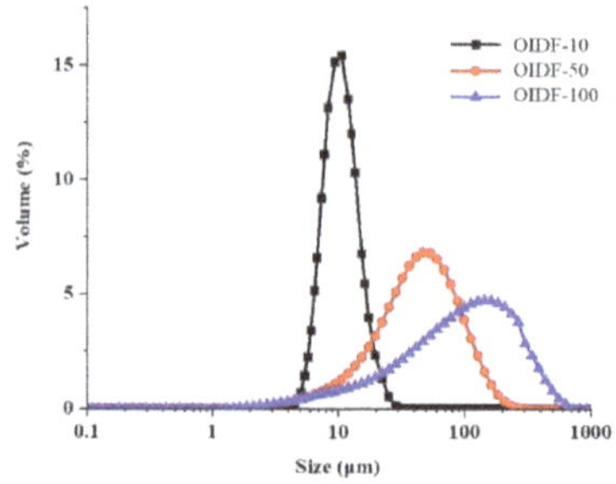

(a)

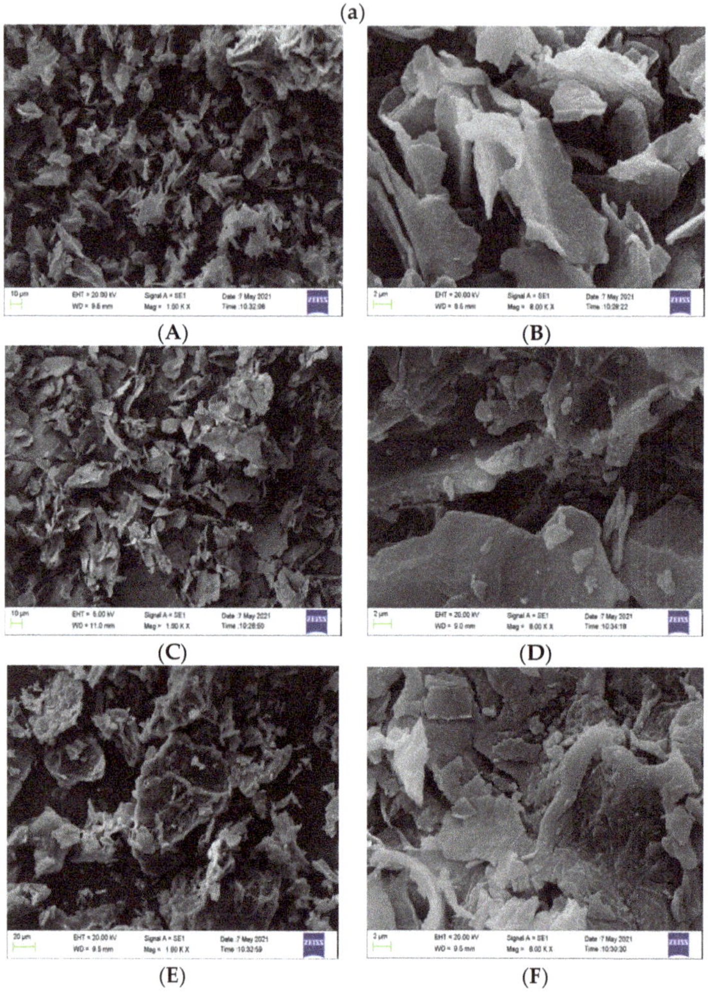

(b)

Figure 1. *Cont.*

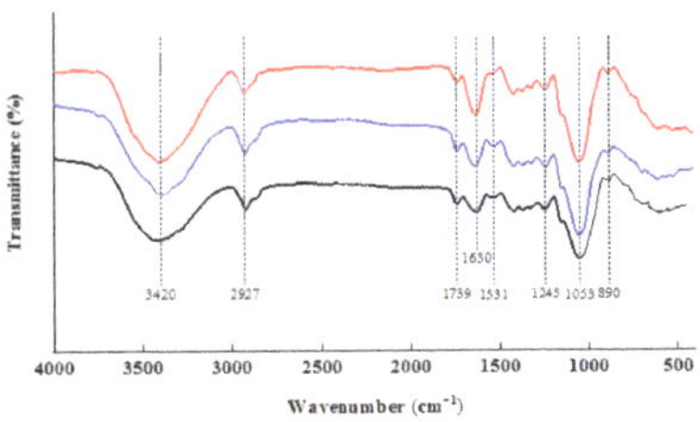

(c)

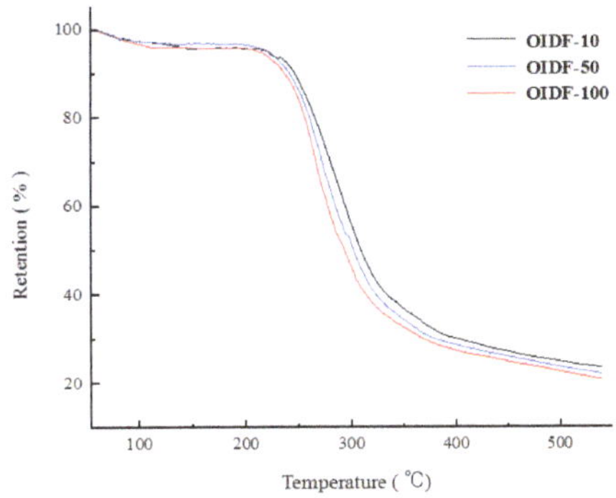

(d)

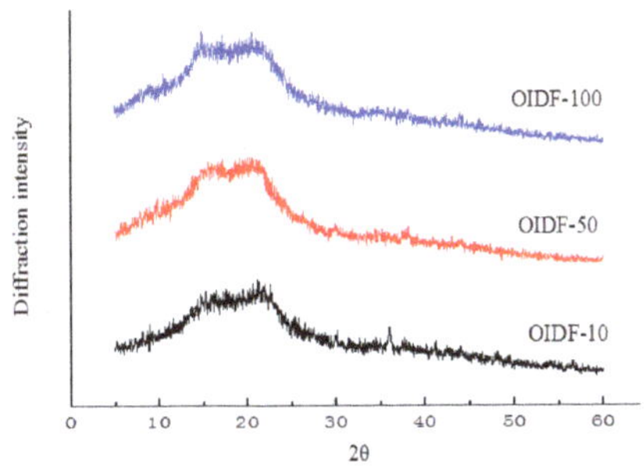

(e)

Figure 1. Structural characterization of OIDF with three particle sizes: (**a**) Particle size characterization; (**b**) SEM characterization; (**c**) FI-IR characterization; (**d**) TG characterization; (**e**) XRD characterization). Figure 1b (**A**: OIDF-10, 1000×; **B**: OIDF-10, 8000×; **C**: OIDF-50, 1000×; **D**: OIDF-50, 8000×; **E**: OIDF-100, 1000×; **F**: OIDF-100, 8000×).

From Figure 1d, the OIDF thermogravimetric curves of the three particle sizes showed the same trend, mainly composed of four stages. The first stage is the drying stage. From the initial temperature of 25 °C to 110.25 °C, the thermogravimetric curves of the three particle sizes of OIDF all show a weight loss of about 3.9%, which is due to the evaporation of free water and crystal water inside the fiber. The second stage is the pre-carbonization stage, the temperature range is 110.25–208 °C, and the weight loss is 0.73%, which belongs to the process of slow decomposition of macromolecules; the third stage occurs at about 208.38–355.55 °C, which belongs to the carbonization stage. The decrease was significant, with weight loss as high as 62%, which may be due to thermal decomposition or degradation of cellulose and hemicellulose [30]. It is consistent with the conclusion obtained by Fourier transform infrared spectroscopic analysis of OIDF with different particle sizes. The final stage is the combustion stage. When the temperature exceeds 355 °C, the thermogravimetric curve trend is stable due to the gradual decomposition of the residual sample into carbon and ash. The retention rate is the percentage of the final residue of the sample to the original mass. The retention rate of OIDF-10 is 22.89%, the retention rate of OIDF-50 is 21.77%, and the retention rate of OIDF-100 is 20.35%.

As shown in Figure 1e, any crystalline substance has its own unique X-ray diffraction pattern, and the position and shape of the characteristic diffraction peaks can be used to qualitatively and quantitatively analyze the substance, and to determine the crystal structure and degree of crystallinity. The results are shown in Figure 1e. The diffraction angles (2θ angles) are 15.42°, 20.78°, and 36.08°, which are typical characteristic diffraction peaks of cellulose, indicating that the crystal types of the three OIDFs belong to the cellulose

I type. It is a state in which two phases coexist in a crystalline region and an amorphous region [31]. According to the calculation, the crystallinity of OIDF-100 is 33.65%, the crystallinity of OIDF-50 is 30.81%, and the crystallinity of OIDF-10 is 27.08%, indicating that the crystallinity of cellulose in OIDF-10 is significantly reduced after high-pressure microfluidization. Ullah et al. [32] also found that insoluble dietary fiber in soybean dregs was treated with high-energy wet media, and the crystallinity of cellulose gradually decreased with the prolongation of treatment time. In addition, Kang and Lu et al. [33,34] also came to the same conclusion.

3.3. Physicochemical Properties of the Three OIDFs

The shear force is increased in the high-pressure micro-jet process, as illustrated in Figure 2, due to the mechanical effect of ultra-high pressure. Water holding capacity drops, oil holding capacity decreases, and swelling property rises when the OIDF particle size reduces, which might be related to an increase in the specific surface area as the particle size lowers. Ultra-high pressure damages the fiber structure, causing more hydroxyl groups of hydrophilic groups to be destroyed, resulting in a weakening of the fiber's hydrophilic ability and a loss in water holding capacity and moisture content [35]. For example, Jasim et al. [36] and Zheng et al. [37] discovered that following ultrafine pulverization, the water-holding and oil-holding capabilities of insoluble dietary fiber declined as the particle size dropped. This might be due to certain hydroxyl and ester structures being destroyed during processing, which is consistent with the findings of this study.

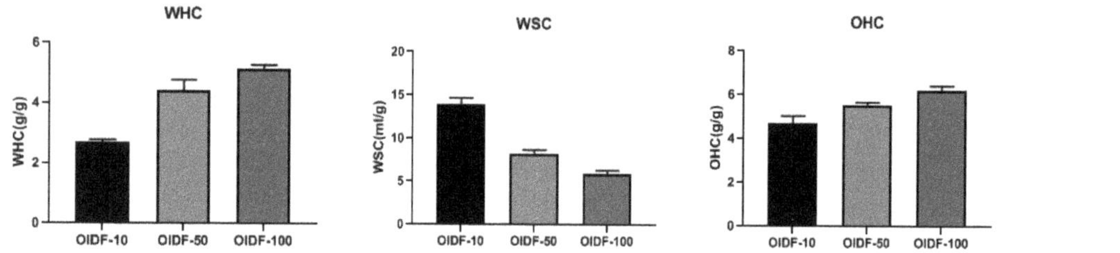

Figure 2. Effects of different particle sizes of OIDF on the water holding capacity (WHC), swelling capacity (WSC) and oil holding capacity (OHC).

As shown in Figure 3, the three particle sizes of dietary fibers have varied adsorption capabilities for cholesterol in the three pH conditions. The smaller the particle size of dietary fibers in the same pH environment, the greater the adsorption capability. The adsorption ability of three types of dietary fiber to cholesterol steadily declined as pH increased. This finding suggests that dietary fiber has good cholesterol adsorption capabilities in a stomach acid environment. Similarly, in three different acidic circumstances, dietary fiber exhibited a similar adsorption rule to sodium cholate and adsorbed cholesterol, indicating that dietary fiber had high adsorption qualities to cholate in the stomach acid environment.

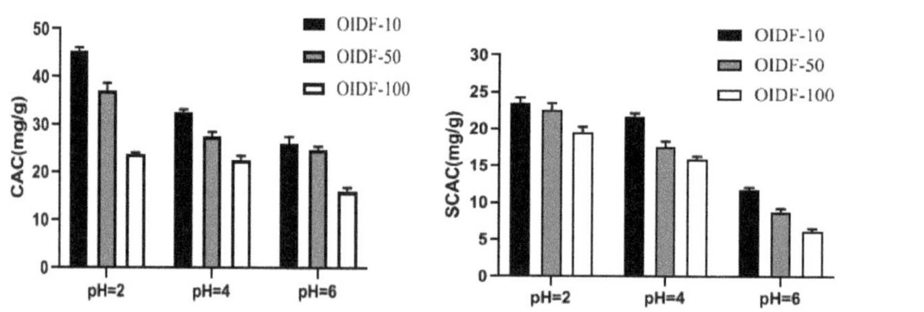

Figure 3. The effect of different particle sizes of OIDF on the adsorption of cholesterol (CAC) and sodium cholate (SCAC).

3.4. Effects of Different Particle Sizes of OIDF on Blood Lipid Levels in High-Fat Diet Rats

As shown in Figure 4, a high-fat diet can significantly increase the content of TC and TG in blood. After OIDF intervention, there is a relief, and the group fed with OIDF-10 is the most significant. OIDF-50 and OIDF-100 groups have significant differences in reducing TC levels but no significant difference in reducing TG levels. This may be due to the large particle size, which leads to low intestinal utilization of OIDF, and thus, the reducing TG levels are not apparent. Regarding HDL-C, the OIDF-10 group significantly increased. HDL-C is mainly synthesized in the liver, which can promote the reverse transport of cholesterol, thereby achieving a lipid-lowering effect. Under a high-quality diet, the content of LDL-C in serum is higher, indicating that it is positively correlated with high blood lipids. OIDF-10 can significantly reduce LDL-C content, while the intervention effect of OIDF-50 and OIDF-100 is not significant. This is consistent with the findings of Zhou [38] and Hoang [39], but the exact reason is not clear.

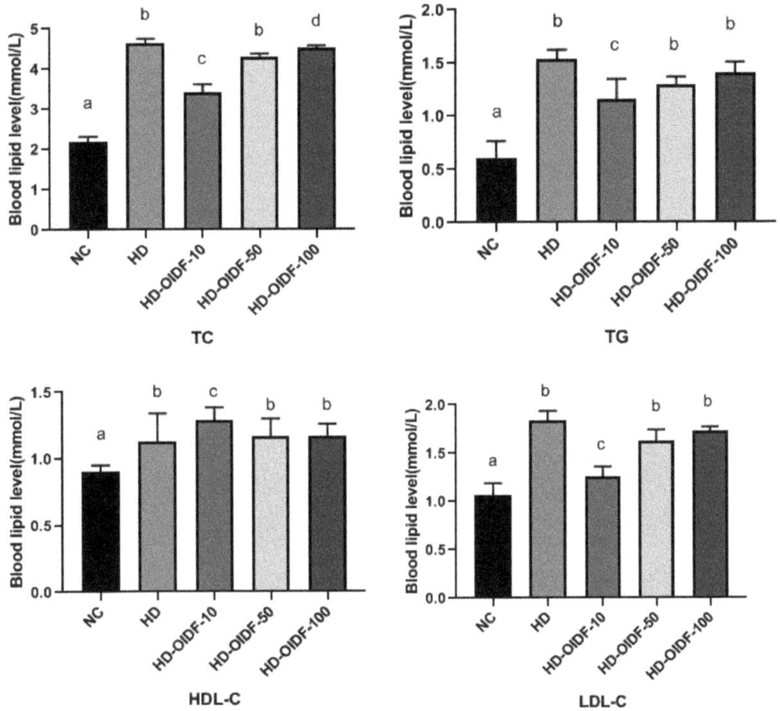

Figure 4. Effects of different particle sizes of OIDF on blood lipid levels in rats (TC, TG, HDL-C, LDL-C). Different letters in the same column (a, b, c and d) are significantly different ($p < 0.05$). The results are expressed as mean ± SD ($n = 3$).

3.5. Effects of Different Particle Sizes of OIDF on Hepatic Steatosis in High-Fat Diet Rats

After H&E staining, the morphological changes of hepatocytes and fat droplets can be visually observed by microscope observation. The morphological observation results of the liver tissue of rats in each group are shown in Figure 5A. It can be seen from the figure that the size of hepatocytes in the NC group was normal throughout the eight weeks of feeding, the cells had complete cytoplasm and obvious nuclei, and the cell boundaries were clear and neatly arranged, while in the other groups, different degrees of accumulation of fat droplets were observed. In histological section, the white circles indicate fat droplets, which may be due to the accumulation of lipids in the body due to a long-term high-fat diet and the lack of exercise in cage feeding. After circulation in the liver, aggregation occurred. Different degrees of fat droplets were observed in other groups. Fat droplets accumulate. From the figure, we can clearly observe that in the HD group, after eight weeks of high-fat feeding, a large number of fat droplets appeared in the liver slices, and the liver was severely fatty. After eight weeks of intervention, compared with the HD group, the steatosis of liver cells in the HD-OIDF-50 and HD-OIDF-100 groups was slightly reduced, and the degree of lipid accumulation was not as fast as that in the HD group, but the morphological changes were different from those in the HD group, although not by much as the effect of inhibiting liver adipose is not as significant as the OIDF-10 group.

As shown in Figure 5B, RT-PCR was performed on the mRNA levels of three key genes of 3-hydroxy-3-methylglutaryl coenzyme (HMG-CoAr), Cholesterol 7α-hydroxylase (CYP7A1), and Low-density lipoprotein receptor (LDL-R) in rat liver. HMG-CoAr is a key rate-limiting enzyme for cholesterol synthesis. From the mRNA expression level of the gene, the relative expression of HMG-CoAr gene in the HD group was up-regulated in the NC group, indicating that a high-fat diet would increase the synthesis of cholesterol. This result was consistent with the results of lipid expression in serum and liver of HD group. As shown in Figure 5B,C, the effect of PCR was not significant. Combined with the analysis of the translation level Western Blot data, high-fat feeding in the HD group led to the high expression of HMG-CoAr gene, and OIDF intervention could reduce its expression, of which OIDF-10 had the most reduction effect. LDL-R can mediate the endocytosis of LDL and reduce the synthesis of hepatic cholesterol by increasing the reabsorption of LDL [40,41]. Therefore, the up-regulation of LDL-R in the OIDF-10 group may be due to the decrease in the blood lipid level of LDL-C. After eight weeks of OIDF dietary intervention, the expression level of CYP7A1 gene in the HD-OIDF-10 group was significantly higher than that in the HD group. Some researchers believe that FXR can activate bile acid synthesis by inducing the expression of CYP7A1 [42]. The high expression of the gene may promote the synthesis of bile acids, thereby promoting the metabolism of cholesterol.

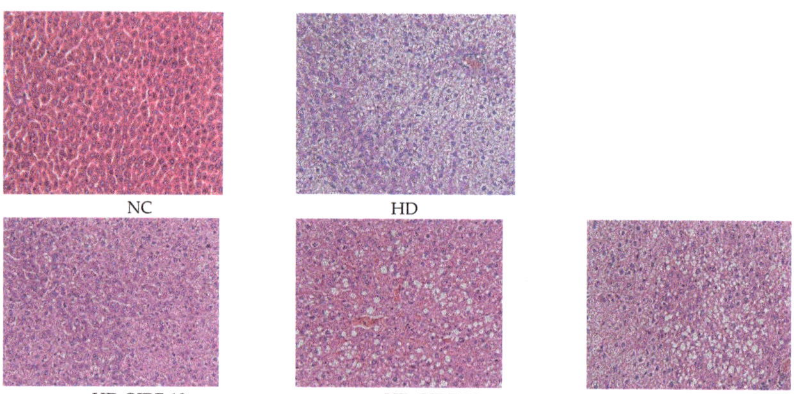

(A)

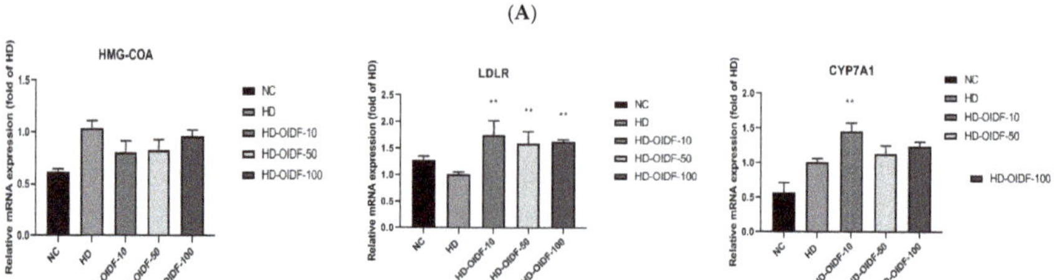

(B)

Figure 5. *Cont.*

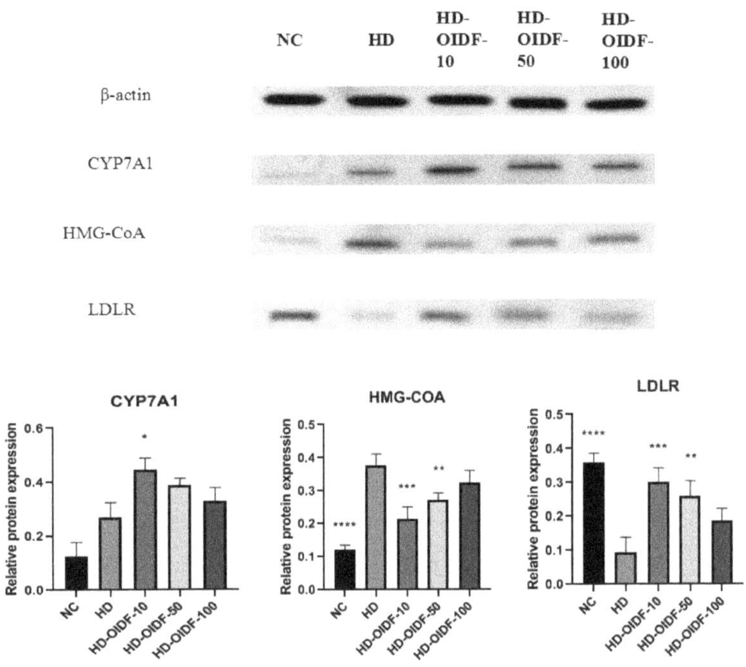

(C)

Figure 5. Effects of different particle sizes of OIDF on diet-induced hepatic steatosis in mice. Histological changes in liver sections were measured by H&E staining at 400× magnification: (**A**) Relative mRNA expression levels of CYP7A1, HMG-CoAr, LDL-R were determined by qRT-PCR; (**B**) The protein expressions of CYP7A1, HMG-CoAr and LDL-R in liver were determined by Western blotting; (**C**) Values are expressed as mean ± SD ($n = 10$). * represents $p < 0.05$, ** represents $p < 0.01$, and *** represents $p < 0.001$, and **** represents $p < 0.0001$, compared to the HD group. NC, normal diet-fed group; HD, high-fat diet-fed group; HD-OIDF-10, HD-OIDF-50, HD-OIDF-100, representing three distinct particle sizes of OIDF.

The above three genes were confirmed by Western Blot, as shown in Figure 5C, and their protein level expression and mRNA expression trend were the same.

3.6. Bacterial Composition Analysis between Different Populations

By statistic on the ASV/OTU table after leveling, the specific composition table of the microbial community in each sample at each classification level can be obtained. Through this table, the number of taxa contained in different samples at each taxonomic level can be calculated first. Analysis software: QIIME2 (2019.4); Personal company self-compiled perl script. Analysis step: According to the results of the sequence species taxonomic annotation and the selected samples, we can count the number of taxa contained in each of the seven taxonomic levels of domain, phylum, class, order, family, genus, and species in the species annotation results of these samples. Some studies have used Firmicutes/Bacteroidetes ratios to explore the degree of obesity [43]. Compared with the HD group, the HD-OIDF-10 group was able to significantly reduce the ratio of Firmicutes/Bacteroidetes. As shown in Figure 6A and Table 3 (percentages shown in Table 3 are the mean value), at the phylum level, Firmicutes NC group accounted for 80.32%, HD group accounted for 80.67%, HD-OIDF-10 group accounted for 88.25%, HD-OIDF-50 group accounted for 93.04%, HD-OIDF-100 group accounted for 90.64%. Firmicutes are the dominant beneficial bacteria in the intestinal tract, indicating that OIDF with different particle sizes can increase the abundance of the Firmicutes after the intervention of a high-fat diet. Regarding Bacteroidetes, the NC group accounted for 1.27%, the HD group 0.03%, the HD-OIDF-10 group 4.44%, the HD-OIDF-50 group 2.34%, and the HD-OIDF-100 group 2.06%. Bacteroidetes are engaged in colonic metabolism, including glucose fermentation and bile acid and steroid biotransformation, and their numbers have grown. In the HD group, a high-fat diet resulted in a significant reduction in Bacteroidetes. The proportion of Bacteroidetes rose after OIDF intervention, notably in the HD-OIDF-10 group. The OIDF-10 intervention might dramatically reduce the rise in *Actinobacteria* and *Proteobacteria* produced by a high-fat diet when compared to the NC group. At the genus level, as shown in Figure 6B, Oscillospira showed the largest quantitative difference among the groups, and Oscillospira was able to produce short-chain fatty acids (SCFAs) such as butyrate, suggesting that it may play an essential role in various aspects of human function and health [44]. At the same time, studies have shown that Oscillospira can ferment complex plant carbohydrates [44]. After the intervention of three particle sizes of OIDF, the content of Oscillospira in the HD-OIDF-10 group was the highest at 20.25%, which was significantly higher than that in the HD group by 2.4 times. It shows that the bacteria in the gut can obtain carbon sources by decomposing the plant polysaccharide OIDF. As shown in Figure 6C, the elements of the phylogenetic tree graph mainly include: the phylogenetic tree graph, coloring ASV characteristic sequences or OTU representative sequences (tips in the graph) according to the taxonomic level. The first column is the sequence ID, and the second column is the relative abundance of the sequence in each sample of the grouping scheme. From left to right are the HD group, the NC group, the OIDF-10 group, the OIDF-50 group, and the OIDF-100 group, with six samples in each group. As shown in Table 3, at the genus level, OIDF was also able to modulate gut microbiota composition, enabling the development of gut microbiota towards a healthy state, reducing the abundance in *Allobaculum* bacteria in the gut of rats on a high-fat diet. *Ruminococcus*, ruminant digestive tract flora, is fermented to produce lactic acid, and the OIDF was able to increase *Ruminococcus* abundance, of which OIDF-10 had the most significant effect, possibly through the production of lactate, which is produced by gut bacteria, making a valuable contribution to colon health. It helps promote gut health and the bacteria that produce it protect against disease.

Table 3. Ratio of microbiota (top10) at the phylum and genus levels (%) (n = 6).

Taxonomy		NC	HD	HD-OIDF-10	HD-OIDF-50	HD-OIDF-100
phylum	*Firmicutes*	80.32%	80.67%	88.25%	93.04%	90.64%
	Actinobacteria	10.75%	17.04%	5.65%	2.94%	2.37%
	Proteobacteria	1.69%	2.06%	1.40%	1.15%	3.89%
	Bacteroidetes	1.27%	0.03%	4.44%	2.34%	2.06%
	Verrucomicrobia	5.40%	0.09%	0.10%	0.01%	0.01%
	Deferribacteres	0.01%	0.01%	0.05%	0.31%	0.81%
	Tenericutes	0.23%	0.02%	0.08%	0.12%	0.08%
	Cyanobacteria	0.04%	0.00%	0.00%	0.03%	0.01%
	TM7	0.02%	0.00%	0.01%	0.00%	0.02%
	Spirochaetes	0.01%	0.00%	0.00%	0.00%	0.00%
	Others	0.26%	0.08%	0.03%	0.07%	0.12%
Genus	*Oscillospira*	1.38%	8.40%	20.25%	5.71%	12.58%
	Staphylococcaceae	21.00%	0.47%	1.68%	7.87%	7.68%
	Corynebacterium	15.81%	7.72%	3.10%	2.08%	1.05%
	Lactobacillus	7.75%	4.22%	0.10%	1.77%	0.51%
	Clostridiaceae	2.51%	2.21%	1.27%	3.63%	3.43%
	Jeotgalicoccus	4.18%	1.67%	1.06%	0.95%	1.86%
	Ruminococcus	0.11%	1.65%	3.62%	0.91%	1.62%
	Allobaculum	0.58%	5.88%	0.26%	0.06%	0.37%
	Desulfovibrio	1.41%	0.85%	0.33%	0.64%	3.18%
	Coprococcus	0.74%	1.01%	2.69%	0.82%	0.90%
	Others	44.53%	65.91%	65.64%	75.57%	66.82%

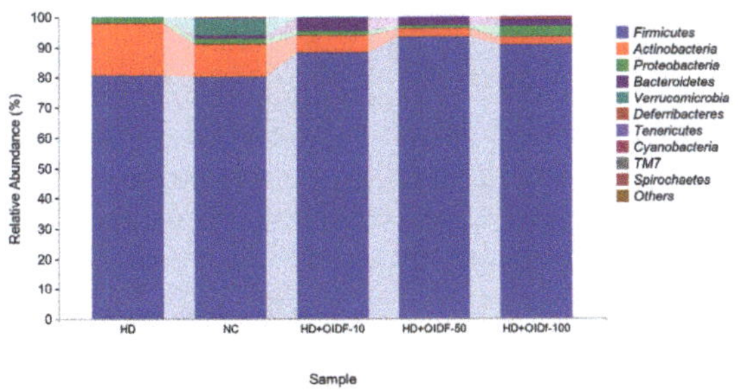

(**A**)

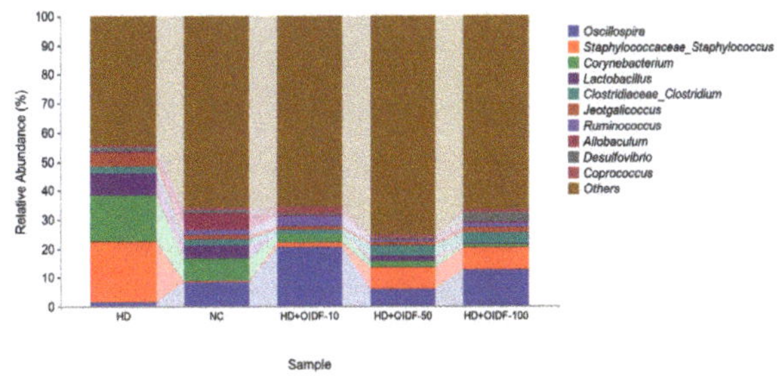

(**B**)

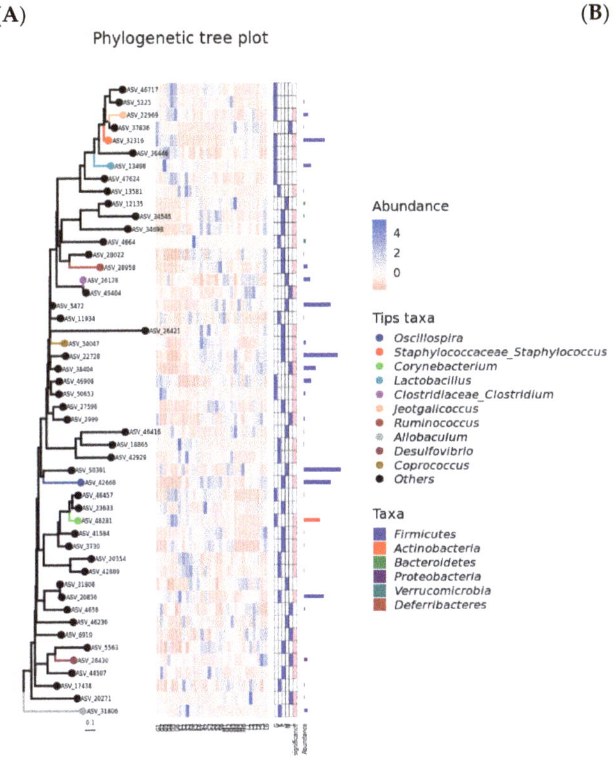

(**C**)

Figure 6. Bacterial composition analysis between different populations: (**A**) t phylum level; (**B**) genus level; (**C**) Species composition analysis.

3.7. Alpha Diversity Index and Beta diversity PCoA Analysis

As can be seen from Figure 7, the chao1 index of intestinal flora in the HD group was the smallest, which is a measure of its richness. The emphasis on rare species was the lowest, which was significantly different from the HD-OIDF-10 group. The Shannon index of intestinal flora in the HD group was significantly lower than that in the NC and HD-OIDF-10 groups ($p < 0.01$), that is, the intestinal microecological diversity of the HD group was significantly lower than that of the NC and OIDF intervention groups. The reason may be due to the high-fat diet changes the intestinal ecology, and the excessive reproduction of harmful bacteria that reduces the diversity of intestinal flora [40]. The richness and diversity of the intestinal flora of the HD-OIDF-10 group were closer to those of the NC group, and the results of other types of the analysis showed the same trend, indicating that the species richness between HD and NC, OIDF-10, and OIDF-100 and diversity are significantly different.

As the picture shows, β-diversity analysis can be used to study the similarity and difference of the bacterial community structure in different samples. Commonly used analysis methods include sample hierarchical cluster analysis, principal coordinate analysis, and multi-dimensional scaling analysis [45]. This study focuses on principal coordinate analysis (PCoA). As shown in Figure 7, the HD group and the NC group were clustered separately. The difference was obvious, that is, the microflora structure of the high-fat diet rats was different from that of the normal mice. In contrast, the coordinates of the OIDF dietary intervention group were shifted to the NC group, indicating that OIDF was able to modulate gut microbiota composition in rats fed a high-fat diet. Figure 7 shows that HD-OIDF-10 intervention in high-fat diet rats has a significant difference with HD group rats ($p < 0.05$), which indicates that OIDF-10 can change the intestinal flora structure of high-fat diet rats. These results suggest that OIDF can modulate the gut microbiota structure in rats fed a high-fat diet.

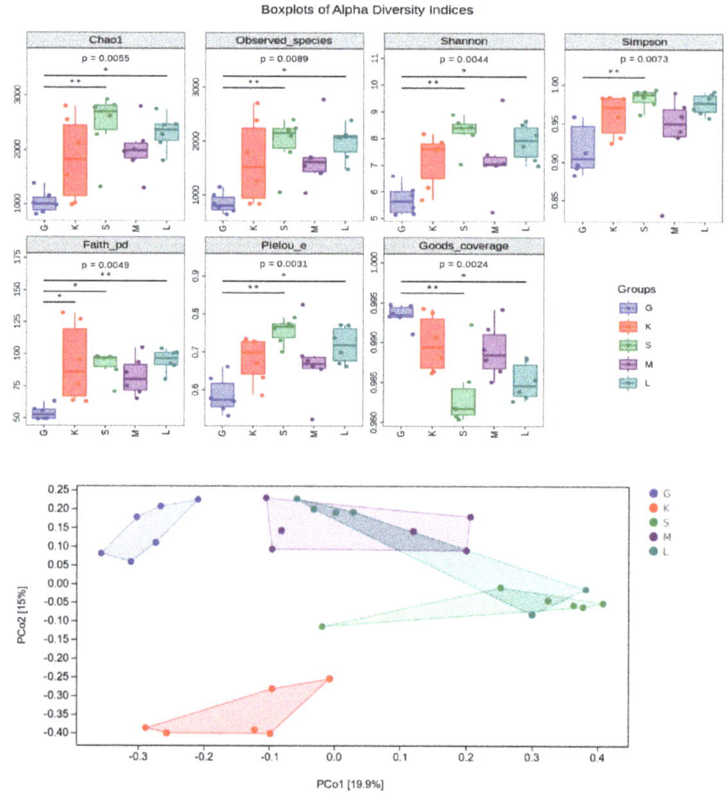

Figure 7. Alpha Diversity Index and Beta diversity PCoA analysis (G: HD group; K: NC group; S: HD-OIDF-10 group; M: HD-OIDF-50 group; L: HD-OIDF-100 group. * represents $p < 0.05$, ** represents $p < 0.01$).

3.8. Alpha Diversity Index and Beta Diversity PCoA Analysis

As shown in Figure 8A, cluster analysis was performed between each group, and the top 20 bacterial species were obtained, which were *Ruminococcus*, *Clostridium*, *Desulfovibrio*, *Brevibacterium*, *Oscillospira*, *Ruminococcus*, *Coprococcus*, *Bacteroides*, *Yaniella*, *Blautia*, *Sporosarcina*, *Jeotgalicoccus*, *Corynebacterium*, *Lactobacillus*, *Dorea*, *Adlercreutzia*, *Allobaculum*, *Turicibacter*, *Akkermansia*, *Facklamia*. Compared with the HD group, HD-OIDF-10 group increased the abundance of *Brevibacterium, Oscillospira, Ruminococcus, Coprococcus, Bacteroides, Yaniella*, and HD-OIDF-50 increased the abundance of *Ruminococcus, Clostridium*, HD-OIDF-100 compared with HD group increased the abundance of *Ruminococcus, Clostridium, Desulfovibrio, Oscillospira, Ruminococcus, Bacteroides* compared with the HD group, indicating that the OIDF can improve the dysbiosis caused by a high-fat diet. As shown in Figure 8B, in the PCA analysis, PC1 = 52.7%, PC2 = 22.2%, and the differences in PCA analysis between groups were significant. The HD-OIDF-10 group and the HD group with extremely significant differences were selected for MetagenomeSeq analysis, and the occurrence frequency of ASV was greater than or equal to 0.3 as the condition, as shown in Figure 8C. The sample groups were compared using the metagenomeSeq method. This method avoids the influence of data sparse (Rarefaction) process on the accuracy of results, and is especially suitable for sparse microbial composition data. The dotted line separated the significant difference (above) and the insignificant ASV, significant difference. The points are marked by colored dots or circles, gray circles indicate the insignificant ones and the significantly up-regulated ones in this group are indicated by colored solid circles. The color of the dots identifies its phylum level name and is marked at the bottom of the figure; for the top 10 genera with significantly up-regulated points, a grayscale background is added, and the final analysis shows that the difference ASV number is 62, and after classification, it is concentrated in *Actinobacteria, Bacteroidetes, Firmicutes, Proteobacteria*.

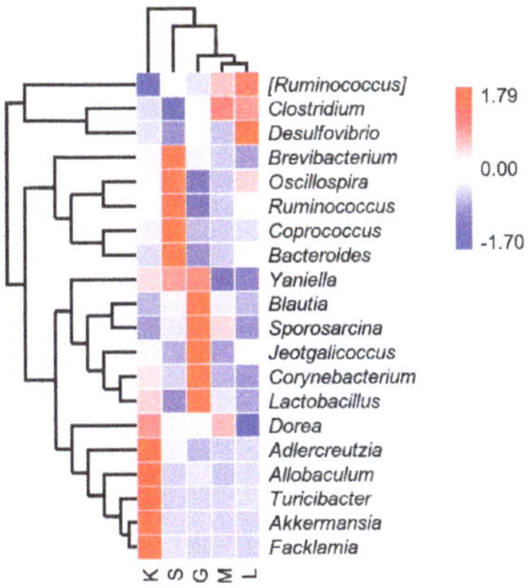

(A)

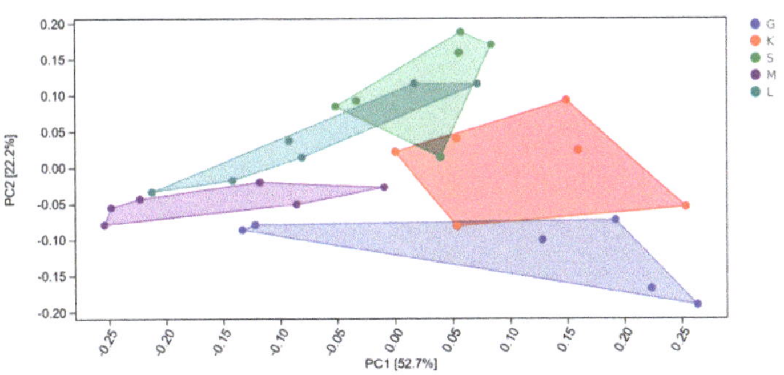

(B)

Figure 8. *Cont.*

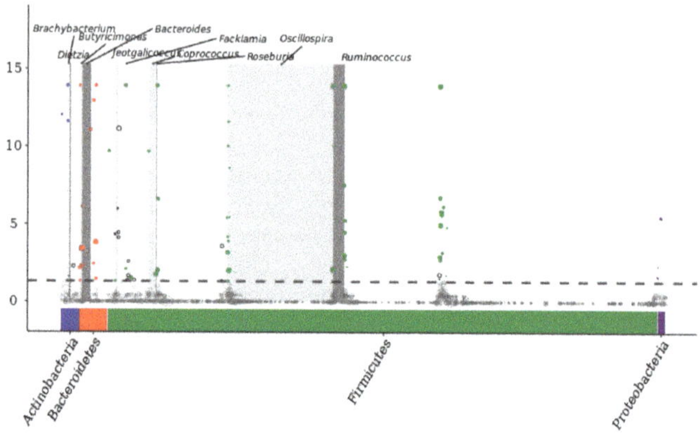

(**C**) (between HD-OIDF-10 and HD)

Figure 8. Species differences and marker analysis: (**A**) Species composition heatmap; (**B**): PCA analysis; (**C**): MetagenomeSeq analysis. (G: HD group; K: NC group; S: HD-OIDF-10 group; M: HD-OIDF-50 group; L: HD-OIDF-100 group).

4. Conclusions

In this study, three distinct particle sizes of OIDF were prepared from high-purity okara insoluble dietary fiber, and their structures and physicochemical properties were characterized, such as OIDF-10 having the higher WSC, OHC, CAC and SCAC. Using three distinct particle sizes of OIDF to interfere with the lipid-lowering effect of high-fat diet rats, we observed that OIDF supplementation could reduce blood lipid levels, alleviate hepatic steatosis, and regulate mRNA expression and protein of genes related to fat metabolism in the liver. The OIDF supplementation may provide a theoretical basis for OIDF with different particle sizes as functional food or dietary supplements by improving lipid metabolism and intestinal flora function in rats induced by a high-fat diet.

Author Contributions: H.Y. contributed to the conceptualization and supervision. Y.Z. contributed to the resources. H.F., S.W., M.S.S., J.F., J.L. and J.Z. contributed to the investigation, formal analysis, and writing. All the authors critically reviewed and contributed to the editing. All authors have read and agreed to the published version of the manuscript.

Funding: This research was funded by China Agriculture Research System of MOF and MARA (CARS-04); Young and Middle-Aged Technological Innovation Outstanding Talent (Team) Project (Innovation), grant number (20210509015RQ).

Institutional Review Board Statement: The study was conducted according to the guidelines of the Declaration of Helsinki, and approved by the Ethics Committee of Laboratory Animal Welfare and Ethics Committee of Jilin Agricultural University (no. 20210422001, 22 April 2021).

Informed Consent Statement: Not applicable.

Data Availability Statement: Raw data can be provided by the corresponding author on request.

Acknowledgments: The authors acknowledge Shanghai Personal Biotechnology Co., Ltd., (Shanghai, China). For their support to the 16S rDNA Sequencing.

Conflicts of Interest: The authors declare no conflict of interest.

References

1. Wang, S.N.; Yu, H.S.; Gu, C.; He, Y.; Zhang, T.; Xu, X. Preventive effect of soybean insoluble dietary fiber on high fat diet induced obesity in mice. *Sci. Technol. Food Ind.* **2020**, *41*, 295–301, 314.
2. Swallah, M.S.; Fan, H.; Wang, S.; Yu, H.; Piao, C. Prebiotic Impacts of Soybean Residue (Okara) on Eubiosis/Dysbiosis Condition of the Gut and the Possible Effects on Liver and Kidney Functions. *Molecules* **2021**, *26*, 326. [CrossRef] [PubMed]

3. Lyu, B.; Wang, Y.; Fu, H.; Li, J.; Yang, X.; Shen, Y.; Swallah, M.S.; Yu, Z.; Li, Y.; Wang, H.; et al. Intake of high-purity insoluble dietary fiber from Okara for the amelioration of colonic environment disturbance caused by acute ulcerative colitis. *Food Funct.* **2022**, *13*, 213–226. [CrossRef] [PubMed]
4. Ussar, S.; Griffin, N.W.; Bezy, O.; Fujisaka, S.; Vienberg, S.; Softic, S.; Deng, L.; Bry, L.; Gordon, J.I.; Kahn, C.R. Interactions between gut microbiota, host genetics and diet modulate the predisposition to obesity and metabolic syndrome. *Cell Metab.* **2015**, *22*, 516–530. [CrossRef] [PubMed]
5. Cantu-Jungles, T.M.; Hamaker, B.R. New view on dietary fiber selection for predictable shifts in gut microbiota. *MBio* **2020**, *11*, e02179-19. [CrossRef]
6. Adams, S.; Che, D.; Hailong, J.; Han, R.; Qin, G.; Danquah, K. Dietary supplementation of pulverised Astragalus membranaceus improved performance, immunity and diarrhoea incidence in weaned piglets. *Indian J. Anim. Res.* **2018**, B-936. [CrossRef]
7. Adams, S.; Sello, C.T.; Qin, G.X.; Che, D.; Han, R. Does dietary fiber affect the levels of nutritional components after feed formulation? *Fibers* **2018**, *6*, 29. [CrossRef]
8. Adams, S.; Che, D.; Qin, G.; Rui, H.; Sello, C.T.; Hailong, J. Interactions of Dietary Fibre with Nutritional Components on Gut Microbial Composition, Function and Health in Monogastrics. *Curr. Protein Pept. Sci.* **2018**, *19*, 1011–1023. [CrossRef]
9. Abedin, A.; Evera, E.; Qingzhen, Z.; Zewei, S. Gastrointestinal Interaction between Dietary Amino Acids and Gut Microbiota: With Special Emphasis on Host Nutrition. *Curr. Protein Pept. Sci.* **2020**, *21*, 785–798.
10. Adams, S.; Xiangjie, K.; Hailong, J.; Guixin, Q.; Sossah, F.L.; Dongsheng, C. Prebiotic effects of alfalfa (Medicago sativa) fiber on cecal bacterial composition, short-chain fatty acids, and diarrhea incidence in weaning piglets. *RSC Adv.* **2019**, *9*, 13586–13599. [CrossRef]
11. Piel, C.; Montagne, L.; Sève, B.; Lalleès, J.-P. Increasing digesta viscosity using carboxymethylcellulose in weaned piglets stimulates ileal goblet cell numbers and maturation. *J. Nutr.* **2005**, *135*, 86–91. [CrossRef] [PubMed]
12. Lamont, J.T. Mucus: The front line of intestinal mucosal defense. *Ann. N. Y. Acad. Sci.* **1992**, *664*, 190–201. [CrossRef]
13. Li, L.; Ma, L.; Fu, P. Gut microbiota–derived short-chain fatty acids and kidney diseases. *Drug Des. Dev. Ther.* **2017**, *11*, 3531. [CrossRef] [PubMed]
14. Marques, F.Z.; Nelson, E.; Chu, P.Y.; Horlock, D.; Fiedler, A.; Ziemann, M.; Tan, J.K.; Kuruppu, S.; Rajapakse, N.W.; El-Osta, A.; et al. High-fiber diet and acetate supplementation change the gut microbiota and prevent the development of hypertension and heart failure in hypertensive mice. *Circulation* **2017**, *135*, 964–977. [CrossRef] [PubMed]
15. Lee DP, S.; Gan, A.X.; Kim, J.E. Incorporation of biovalorised okara in biscuits: Improvements of nutritional, antioxidant, physical, and sensory properties. *LWT* **2020**, *134*, 109902.
16. Chan, L.Y.; Takahashi, M.; Lim, P.J.; Aoyama, S.; Makino, S.; Ferdinandus, F.; Ng, S.Y.; Arai, S.; Fujita, H.; Tan, H.C.; et al. Eurotium Cristatum Fermented Okara as a Potential Food Ingredient to Combat Diabetes. *Sci. Rep.* **2019**, *9*, 17536. [CrossRef]
17. Swallah, M.S.; Yu, H.; Piao, C.; Fu, H.; Yakubu, Z.; Sossah, F.L. Synergistic two-way interactions of dietary polyphenols and dietary components on the gut microbial composition: Is there a positive, negative, or neutralizing effect in the prevention and management of metabolic diseases? *Curr. Protein Pept. Sci.* **2021**, *22*, 313–327. [CrossRef]
18. Xu, T.; Wu, X.; Liu, J.; Sun, J.; Wang, X.; Fan, G.; Meng, X.; Zhang, J.; Zhang, Y. The regulatory roles of dietary fibers on host health via gut microbiota-derived short chain fatty acids. *Curr. Opin. Pharmacol.* **2022**, *62*, 36–42. [CrossRef]
19. Rodríguez, R.; Jimenez, A.; Fernández-Bolanos, J.; Guillen, R.; Heredia, A. Dietary fibre from vegetable products as source of functional ingredients. *Trends Food Sci. Technol.* **2006**, *17*, 3–15. [CrossRef]
20. Zhang, H.; Troise, A.D.; Qi, Y.; Wu, G.; Zhang, H.; Fogliano, V. Insoluble dietary fibre scavenges reactive carbonyl species under simulated physiological conditions: The key role of fibre-bound polyphenols. *Food Chem.* **2021**, *349*, 129018. [CrossRef]
21. Lyu, B.; Wang, H.; Swallah, M.S.; Fu, H.; Shen, Y.; Guo, Z.; Tong, X.; Li, Y.; Yu, H.; Jiang, L. Structure, Properties and Potential Bioactivities of High-purity Insoluble Fibre from Soybean Dregs (Okara). *Food Chem.* **2021**, *364*, 130402. [CrossRef] [PubMed]
22. Xie, F.; Wang, Y.; Wu, J.; Wang, Z. Functional Properties and Morphological Characters of Soluble Dietary Fibers in Different Edible Parts of Angelica Keiskei. *J. Food Sci.* **2016**, *81*, C2189. [CrossRef] [PubMed]
23. Wu, C.; Teng, F.; McClements, D.J.; Zhang, S.; Li, Y.; Wang, Z. Effect of cavitation jet processing on the physicochemical properties and structural characteristics of okara dietary fiber. *Food Res. Int.* **2020**, *134*, 109251. [CrossRef] [PubMed]
24. Liu, C.; Fan, L.; Yang, Y.; Jiang, Q.; Xu, Y.; Xia, W. Characterization of surimi particles stabilized novel Pickering emulsions: Effect of particles concentration, pH and NaCl levels. *Food Hydrocoll.* **2021**, *117*, 106731. [CrossRef]
25. Ma, M.; Mu, T. Modification of deoiled cumin dietary fiber with laccase and cellulase under high hydrostatic pressure. *Carbohydr. Polym.* **2016**, *136*, 87–94. [CrossRef]
26. Liu, C.; Liu, Q.; Sun, J.; Jiang, B.; Yan, J. Extraction of water-soluble polysaccharide and the antioxidant activity from Semen cassiae. *J. Food Drug Anal.* **2014**, *22*, 492–499. [CrossRef]
27. Alba, K.; Macnaughtan, W.; Laws, A.P.; Foster, T.J.; Campbell, G.M.; Kontogiorgos, V. Fractionation and characterisation of dietary fibre from blackcurrant pomace. *Food Hydrocoll.* **2018**, *81*, 398–408. [CrossRef]
28. Wen, Y.; Niu, M.; Zhang, B.; Zhao, S.; Xiong, S. Structural characteristics and functional properties of rice bran dietary fiber modified by enzymatic and enzyme-micronization treatments. *LWT-Food Sci. Technol.* **2017**, *75*, 344–351. [CrossRef]
29. Wang, Q.H.; Shu, Z.P.; Xu, B.Q.; Xing, N.; Jiao, W.J.; Yang, B.Y.; Kuang, H.X. Structural characterization and antioxidant activities of polysaccharides from *Citrus aurantium* L. *Int. J. Biol. Macromol.* **2014**, *67*, 112–123. [CrossRef]

30. Tanpichai, S.; Biswas, S.K.; Witayakran, S.; Yano, H. Water Hyacinth: A Sustainable Lignin-Poor Cellulose Source for the Production of Cellulose Nanofibers. *ACS Sustain. Chem. Eng.* **2019**, *7*, 18884–18893. [CrossRef]
31. Zhang, Z.; Song, H.; Peng, Z.; Luo, Q.; Ming, J.; Zhao, G. Characterization of stipe and cap powders of mushroom (Lentinus edodes) prepared by different grinding methods. *J. Food Eng.* **2012**, *109*, 406–413. [CrossRef]
32. Jiang, L.; Xu, Q.X.; Qiao, M.; Ma, F.F.; Thakur, K.; Wei, Z.J. Effect of superfine grinding on properties of Vaccinium bracteatum Thunb leaves powder. *Food Sci. Biotechnol.* **2017**, *26*, 1571–1578. [CrossRef] [PubMed]
33. Kang, X.; Kuga, S.; Wang, C.; Zhao, Y.; Wu, M.; Huang, Y. Green preparation of cellulose nanocrystal and its application. *ACS Sustain. Chem. Eng.* **2018**, *6*, 2954–2960. [CrossRef]
34. Lu, X.; Zhang, H.; Li, Y.; Huang, Q. Fabrication of milled cellulose particles-stabilized Pickering emulsions. *Food Hydrocoll.* **2018**, *77*, 427–435. [CrossRef]
35. Park, J.; Choi, I.; Kim, Y. Cookies formulated from fresh okara using starch, soy flour and hydroxypropyl methylcellulose have high quality and nutritional value. *LWT-Food Sci. Technol.* **2015**, *63*, 660–666. [CrossRef]
36. Ahmed, J.; Al-Attar, H.; Arfat, Y.A. Effect of particle size on compositional, functional, pasting and rheological properties of commercial water chestnut flour. *Food Hydrocoll.* **2016**, *52*, 888–895. [CrossRef]
37. Zheng, Y.; Li, Y. Physicochemical and functional properties of coconut (*Cocos nucifera* L.) cake dietary fibres: Effects of cellulase hydrolysis, acid treatment and particle size distribution—ScienceDirect. *Food Chem.* **2018**, *257*, 135–142. [CrossRef]
38. Zhou, A.L.; Hergert, N.; Rompato, G.; Lefevre, M. Whole Grain Oats Improve Insulin Sensitivity and Plasma Cholesterol Profile and Modify Gut Microbiota Composition in C57BL/6J Mice. *J. Nutr.* **2015**, *145*, 222–230. [CrossRef]
39. Hoang, M.H.; Houng, S.J.; Jun, H.J.; Lee, J.H.; Choi, J.W.; Kim, S.H.; Kim, Y.R.; Lee, S.J. Barley Intake Induces Bile Acid Excretion by Reduced Expression of Intestinal ASBT and NPC1L1 in C57BL/6J Mice. *J. Agric. Food Chem.* **2011**, *59*, 6798–6805. [CrossRef]
40. Kemper, M.F.; Srivastava, S.; Todd King, M.; Clarke, K.; Veech, R.L.; Pawlosky, R.J. An Ester of β-Hydroxybutyrate Regulates Cholesterol Biosynthesis in Rats and a Cholesterol Biomarker in Humans. *Lipids* **2015**, *50*, 1185–1193. [CrossRef]
41. Levinson, S.S.; Wagner, S.G. Implications of reverse cholesterol transport: Recent studies. *Clin. Chim. Acta* **2015**, *439*, 154–161. [CrossRef] [PubMed]
42. Musso, G.; Gambino, R.; Cassader, M. Cholesterol metabolism and the pathogenesis of non-alcoholic steatohepatitis. *Prog. Lipid Res.* **2013**, *52*, 175–191. [CrossRef] [PubMed]
43. Huang, J.; Lin, X.; Xue, B.; Luo, J.; Gao, L.; Wang, Y.; Ou, S.; Peng, X. Impact of polyphenols combined with high-fat diet on rats' gut microbiota. *J. Funct. Foods* **2016**, *26*, 763–771. [CrossRef]
44. Konikoff, T.; Gophna, U. Oscillospira: A Central, Enigmatic Component of the Human Gut Microbiota. *Trends Microbiol.* **2016**, *24*, 523–524. [CrossRef]
45. Xie, X.; Zhang, L.; Yuan, S.; Li, H.; Zheng, C.; Xie, S.; Sun, Y.; Zhang, C.; Wang, R.; Jin, Y. Val-Val-Tyr-Pro protects against non-alcoholic steatohepatitis in mice by modulating the gut microbiota and gut-liver axis activation. *J. Cell. Mol. Med.* **2021**, *25*, 1439–1455. [CrossRef]

foods

MDPI

Article

Isolation, Characterization and Antioxidant Activity of Yam Polysaccharides

Zhedong Li, Wenhao Xiao, Jianhua Xie, Yi Chen, Qiang Yu, Weidong Zhang and Mingyue Shen *

State Key Laboratory of Food Science and Technology, Nanchang University, Nanchang 330047, China; 15639052109@163.com (Z.L.); xwh970726@163.com (W.X.); jhxie@ncu.edu.cn (J.X.); chenyi15@ncu.edu.cn (Y.C.); yuqiang8612@163.com (Q.Y.); 15502618197@163.com (W.Z.)
* Correspondence: shenmingyue1107@163.com; Tel./Fax: +86-791-8830-4347

Abstract: This study aimed to characterize the structure of Chinese yam (*Dioscoreae Rhizoma*) polysaccharide (CYP) and to investigate its protective effect against H_2O_2-induced oxidative damage in IEC-6 cells. The chemical composition and structural characteristics of the samples were analyzed by chemical and instrumental methods, including high-performance gel permeation chromatography, high-performance anion-exchange chromatography (HPAEC), Fourier transformed infrared (FT-IR), ultraviolet (UV), and scanning electron microscopy (SEM). Antioxidant activity was evaluated by establishing a cellular model of oxidative damage. The molecular weight of CYP was 20.89 kDa. Analysis of the monosaccharide composition revealed that CYP was primarily comprised of galactose (Gal), glucose (Glu), and galacturonic acid (GalA), and the ratio between them was 28.57:11.28:37.59. Pretreatment with CYP was able to improve cell viability, superoxide dismutase (SOD) activity, and reduce intracellular reactive oxygen species (ROS) production and malondialdehyde (MDA) content after H_2O_2 injury. CYP also attenuated oxidative damage in cells through the mitogen-activated protein kinase (MAPK) signaling pathway. This study showed that CYP was an acidic heteropolysaccharide with a good protective effect against oxidative damage, and it thus has good prospects in food and biopharmaceutical industries.

Keywords: Chinese yam; polysaccharide; structural characterization; oxidative stress; antioxidant activity

Citation: Li, Z.; Xiao, W.; Xie, J.; Chen, Y.; Yu, Q.; Zhang, W.; Shen, M. Isolation, Characterization and Antioxidant Activity of Yam Polysaccharides. *Foods* **2022**, *11*, 800. https://doi.org/10.3390/foods11060800

Academic Editors: Philippe Michaud and Lovedeep Kaur

Received: 23 December 2021
Accepted: 8 March 2022
Published: 10 March 2022

1. Introduction

Cellular and tissue oxidative stress is a response to stimulation by increased reactive oxygen species (ROS) and free radicals in the body, resulting in dysregulation of the intracellular oxidative antioxidant system [1]. Cells can catalyze the superoxide anion produced in oxidative stress into H_2O_2, which is broken down into water and oxygen in one step by inducing superoxide dismutase (SOD), catalase, and glutathione peroxidase [2,3]. Oxidative stress could also disrupt intestinal homeostasis, which is a critical element in the development of intestinal damage. Stress can induce intestinal cells to produce large amounts of ROS metabolites, which affect the stability of intracellular nucleic acids, proteins, and lipids, increase apoptosis, inflict intestinal mucosal damage, and induce inflammatory bowel disease [4,5]. Many small intestinal epithelial cells exist on the surface of the intestine, and as a mediator of the internal and external environment of the intestine, the small intestinal epithelial cells are also part of the immune barrier that can defend the body against pathogens and other harmful substances [6–9]. Studies have shown that polysaccharides can scavenge ROS and enhance the antioxidant system to improve intestinal inflammation, therefore, it is relevant to find an antioxidant to protect the intestine from oxidative damage [10–13].

Yam is a plant from the *Dioscoreaceae* family. It grows in warm, low-altitude environments and is distributed throughout Asia, primarily in Korea, Japan, and China [14]. It contains polysaccharides, amino acids, fatty acids, trace elements, starch, protein, and other components [15]. Polysaccharides extracted from yam have a variety of activities such as antioxidant, antitumor, immunomodulatory, and hypoglycemic. The monosaccharide composition of yam polysaccharides is complicated and caused by factors such as the growth conditions and different extraction methods. However, ribose is not found in the monosaccharide composition of most yam polysaccharides. Polysaccharides obtained from Chinese yams generally show good antioxidant activity in vitro, whereas glucose accounts for a large proportion of their monosaccharide composition. Based on the report of Liu et al., the molar ratios of glucose in the polysaccharides they obtained from Chinese yam all exceed 80% [16–20].

Over the past decade, antioxidant studies on yam polysaccharides have primarily focused on the free-radical scavenging ability in vitro. According to Zhu et al. [21], CYP has a strong antioxidant capacity and could effectively scavenge DPPH, ABTS+ and ·OH radicals, especially for DPPH radicals, whose scavenging ability is comparable to that of ascorbic acid (Vc). However, few studies have been conducted on the antioxidant activity of CYP at the cellular level. Despite salvia glycoproteins can increase the activity of antioxidant enzymes and reduce malondialdehyde (MDA) levels, the potential mechanism remains unknown. Therefore, it is meaningful to study this mechanism at the cellular level [22].

Despite the multiple biological activities of CYP, its structural features remain unclear and incomplete [23–25]. The structures of yams grown in different environments at different times are not quite the same, and their biological activities might still be different [26]. In the present study, we established a model of oxidative loss damage of cells by H_2O_2 and explored the protective effects of CYP on the oxidative damage of IEC-6 through cell-viability experiments and the levels of SOD and MDA levels.

2. Materials and Methods

2.1. Materials and Reagents

Fresh yam (*Dioscoreae Rhizoma*), which was plucked in November and purchased from Jiujiang, Jiangxi, China, was powdered with a high-speed disintegrator (XL-20B, Zhengzhou Fengli Crushing Equipment Co., Ltd., Zhengzhou, China). Fetal bovine serum (FBS) and Dulbecco's modified eagle medium (DMEM) were purchased from Solarbio Science and Technology Co., Ltd. (Beijing, China). The cell counting kit-8 (CCK-8) was purchased from Dojindo Molecular Technologies, Inc. (Kyushu Island, Japan). The reactive oxygen species kit, superoxidase dismutase kit, and malondialdehyde kit were purchased from Beyotime Biotechnology (Shanghai, China). Hydrogen peroxide (H_2O_2) was purchased from Sigma (Sigma, Burlington, MA, USA). The IEC-6 cell line was purchased from the Cell Bank of the Chinese Academy of Sciences (Shanghai, China). Antibodies used in this study were purchased from Cell Signalingas (Bever, MA, USA). All the other chemicals used in the present study were analytical reagent grade.

2.2. Extraction and Purification of Polysaccharides

The yam was cleaned and dried for 12 h and then peeled and sliced the next day. The sliced yam (thickness = 3 mm) was placed in a tray, dried in an oven (VYJG-9420, Hangzhou Yijie Technology Co., Hangzhou, China) at 55 °C for 48 h, taken out and ground into powder (60 mesh) with a high-speed disintegrator, after soaking in ethanol (95%) overnight to remove fat, color, and small molecules, the precipitate was collected the next day and heated in a water bath in a ventilated area to remove the ethanol and dry completely. The obtained powder was placed in a self-sealing bag and prepared for later experiments. A total of 300 g of yam powder was weighed, and 6 L of ultrapure water was added at a ratio of 1:20. After extraction at 80 °C for 3 h, the above process was repeated twice, and the extracts were combined, centrifuged, and concentrated to 1/10 of the original volume. After adding 95% ethanol (concentrate:ethanol = 1:5.33) overnight precipitate at 4 °C. The next day, the precipitate was taken and re-solubilized, then the solution pH was adjusted to 4.5 with hydrochloric acid, and glycosylase was added at 55 °C for 1 h, followed by pH adjustment to 6.5 with sodium hydroxide, amylase was added at 95 °C for 2 h, finally papain was added at 65 °C for 1.5 h, high temperature (100 °C) inactivated for 30 min, and proteins were removed 3 times by Sevage method [27], (polysaccharide liquid:chloroform:n-Butyl alcohol = 25:4:1). Then the solution was centrifuged, and the upper layer was gathered to obtain the polysaccharide solution with proteins removed, and then the solution was subjected to dialysis (Mw:8000, tap water for 24 h, distilled water for 24 h, ultrapure water for 24 h) to remove minor molecules and the inorganic ions. The yam polysaccharide solution was concentrated under reduced pressure with a vacuum pump, and four times the volume of anhydrous alcohol was added for 24 h. After centrifugation, the precipitate was placed in a lyophilizer to obtain refined yam polysaccharide.

2.3. *Structure Characterization*

2.3.1. Determination of the Content of Total Carbohydrate, Uronic Acid and Protein

Carbohydrate content was determined by phenol sulfate method according to the method of Huang et al. [28]. In short, prepare 0.1 mg/mL standard glucose solution was prepared, 0.1, 0.2, 0.4, 0.6, 0.8 and 1.0 mL was pipetted in a test tube, and then add distilled water to 1 mL, after that, 1 mL of 3% phenol was added, gradually add 4 mL of concentrated sulfuric acid, shake while adding, measuring the absorbance at 490 nm after reaction for 30 min at room temperature.

Content of the uronic acid was determined by Carbazole-sulfate method [29]. In brief, galacturonic acid standard solution was configured in the same way, then 6 mL of superior pure sulfuric acid was added in an ice bath, shaking while adding, water bath at 85 °C for 25 min, taken out and cooled to room temperature, 0.2 mL of 0.1% carbazole-ethanol (25 mg carbazole dissolved in 25 mL ethanol) was added, and the reaction was carried out for 2 h. The absorbance was measured at 530 nm.

Content of protein was measured with Coomassie brilliant blue method [30]. Configure 0.1 mg/mL of BSA standard solution, and Komas brilliant blue solution (100 mg of Komas brilliant blue dissolved in 50 mL of 90% ethanol, add 100 mL of 85% phosphoric acid, distilled water volume to 1000 mL), 5 mL of Komas brilliant blue was added to the standard solution and CYP solution, plug the cap and invert 3–5 times, measure the absorbance at 595 nm.

2.3.2. Determination of Molecular Weight (Mw)

Molecular weight (Mw) was evaluated on a high-performance liquid chromatography system with an ultraviolet absorption and refractive index detector.

The chromatographic column of UltrahydrogelTM-1000 was eluted with ultrapure water at a flow rate of 0.6 mL/min. The column and RI detector were maintained at 35 °C. Glucose and dextrans (T-10, T-25, T-40, T-50, T-70, and T-500) were used as standards. We prepared 1 mg/mL CYP solution and added the polysaccharide solution to the injection vial by using a syringe combined with a 0.22 μm aqueous-phase needle filter for determination.

2.3.3. Monosaccharide Composition

The monosaccharide composition of CYP was determined based on the method of Xu et al. [31] with few changes. CYP (5 ± 0.05 mg) was put into a tube, and 0.5 mL of 12 M H_2SO_4 was added in an ice bath with magnetic stirring for 0.5 h to dissolve it completely. After 2 mL of ultrapure water was added, tubes were placed in an oil bath at 105 °C for 4 h. After heating, polysaccharide solution was fixed the volume with a 250 mL volume volumetric flask, 1 mL of sample was taken out and was filtered through 0.22 μm membrane for HPAEC analysis. Water and sodium hydroxide were used as the eluent and at a flow rate of 0.25 mL/min.

2.3.4. Scanning Electron Microscope (SEM) Analysis

CYP micromorphology was observed by SEM. In a typical procedure, polysaccharide samples were placed on a conductive plate and held in place. After CYP was sprayed with gold powder, SEM was used to record images at different magnifications (500–2000 times) at an acceleration voltage of 5 kV [32].

2.3.5. Spectra Analysis

The UV spectrum of CYP (0.5 mg/mL) was obtained on a spectrophotometer (TU-1900, Pgenenal, Beijing, China) within the scanning range of 190–400 nm. The FT-IR spectrum of CYP was measured with a Nicolet 5700 FT-IR spectrometer (Thermo Electron, Madison, WI, USA) within the 4000–400 cm^{-1} wavenumber region [33].

2.4. Cytoprotective Activity

2.4.1. Cell Viability Assay

In this study, a Cell Counting Kit-8 (CCK8) kit was conducted to assess the viability of the cells. IEC-6 cells were seeded into a 96-well plate, 10^4 cells/well. After incubation in 5% CO_2 and 37 °C for 24 h, CYP (200,400, and 800 μg/mL) was inserted into the well for co-culturing 24 h, CCK-8 solution was added and incubated with IEC-6 cells for 30 min, followed by measuring the absorbance at 450 nm using a microplate reader.

Different concentrations (100–500 μM) of H_2O_2 were used to stimulate IEC-6 cells for 4 h in 96-well plates, respectively. After adding CCK-8 solution for 4 h, the absorbance was measured at 450 nm to screen the damage concentration of H_2O_2.

After incubating IEC-6 cells in 5% CO_2 at 37 °C for 24 h in 96-well plates, CYP at 200, 400, and 800 μg/mL was incubated in the wells together for 24 h, respectively. IEC-6 cells were stimulated with H_2O_2 for 4 h, and then CCK-8 assay was performed.

2.4.2. Measurement of SOD and MDA Levels

The cells (2×10^5 cells/well) were seeded onto a six-well plate and incubated for 24 h to adhere to the bottom of the well plates. After CYP (200, 400, and 800 μg/mL, respectively) were added for co-culturing 24 h, cells were exposed to a final concentration of 300 μM of H_2O_2 for 4 h. Finally, the content of MDA and SOD contents were measured according to the manufacturer's instructions of the assay kits.

2.4.3. Measurement of Intracellular ROS Level

Intracellular ROS was tested by using 2,7-dichlorofuorescin diacetate (DCFH-DA). IEC-6 cells (2×10^5 cells/well, 2 mL/well) were seeded onto a six-well plate. After incubation for 24 h, cells were exposed to H_2O_2 (300 μM) for 4 h before different concentrations of CYP pre-treated cells for 4 h. Fluorescent dye was added to bind to the cells for 30 min, then washed with PBS and 1640 medium, respectively. Finally, the fluorescence intensity was detected by fluorescence spectrophotometry.

2.4.4. Western Blot Analysis

Aiming to further evaluate the expression of proteins in the mitogen-activated protein kinase (MAPK) pathway associated with cellular oxidative stress, pre-treat IEC-6 cells with different concentrations of CYP, after adding H_2O_2 to stimulate for 4 h, the cells were washed twice with PBS and then lysed the cells with RIPA buffer. After centrifugation, the samples buffer was added to the supernatant and heated at 95 °C for 5 min to obtain the total protein. Immediately prior to electrophoresis with 10% SDS-PAGE, protein samples were transferred onto PVDF membranes (Millipore Co., Belford, MA, US). After blocking with 5% bovine serum albumin (BSA), the membranes were incubated with primary antibody overnight at 4 °C and secondary antibody at room temperature for 1 h. Following incubation with developer, the fluorescent signal was detected using a Molecular Imager ChemiDoc™ XRS Imaging System (Bio-Rad Laboratories, Hercules, CA, USA).

2.5. Statistical Analysis

The results have been represented as the mean ± standard deviation (SD). Significance between the two groups was analyzed by Student's *t*-test using the SPSS 20.0 (SPSS Inc.,

Chicago, IL, USA) software package, and p-values less than 0.05 were considered to be statistically different.

3. Results and Discussion

3.1. Physicochemical Properties and Molecular Weight

In Table 1, the yield of CYP obtained from Chinese yam was 0.21 ± 0.01%, and the contents of carbohydrate, uronic acid, protein were 33.62 ± 0.08%, 34.95 ± 0.21%, and 5.21 ± 0.26%, respectively. The finding indicated that CYP was an acidic polysaccharide. The molecular weight of CYP was 20.89 kDa based on T-series dextran as the standard.

Table 1. Chemical composition of CYP.

Sample	CYP
Yield (%)	0.21 ± 0.01
Carbohydrate (%)	33.62 ± 0.08
Uronic acid (%)	34.95 ±0.21
Protein (%)	5.21 ± 0.26
Mw (kDa)	20.89
Monosaccharide compositions	Ratio (%)
Arabinose	4.51
Galactose	28.57
Glucose	11.28
Mannose	6.77
Xylose	2.26
Rhamnose	4.89
Galacturonic acid	37.59
Glucuronic acid	4.14

Data are expressed as the mean ± SD.

3.2. Monosaccharide Compositions

Monosaccharide compositions were determined by ion chromatography. The results demonstrated that CYP was composed of arabinose (Ara), galactose (Gal), glucose (Glu), mannose (Man), xylose (Xyl), rhamnose (Rha), galacturonic acid (GalA), and glucuronic acid (GluA) (Table 1). CYP mainly consisted of Gal (28.57%), GluA (11.28%), and GalA (37.59%), which indicated that Gal, Glu, and GalA might form the backbone structure of CYP. However, the result was different from the polysaccharides prepared by Huang et al. [34], which could be due to raw materials produced in different periods and different extraction methods.

3.3. UV and FT-IR Spectrum Analysis

UV spectrum analysis of CYP showed that there was no absorption at 260 and 280 nm (Figure 1A), which indicated that CYP barely contained nucleic acids and proteins. This finding was consistent with the previous physicochemical-property analysis [35]. As shown in Figure 1B, a broad and strong absorption peak at 3410 cm^{-1} was attributed to the O–H stretching vibration, and a relatively weak absorption peak at 2910 cm^{-1} was assigned to the C–H stretching vibration [36]. The absorption peak at 1646 cm^{-1} might be caused by the O–H bending vibration [37]. A weak absorption band at 1420 cm^{-1} could originate from the stretching vibration of the carboxyl symmetry [38], which proved the presence of uronic acid. It could be confirmed by the results of the physicochemical properties and monosaccharide composition of CYP. The absorption peak at 1250 and 1070 might be caused by the stretching vibration of C–O [39], indicating the presence of the pyranose ring [40]. Combined with the above analysis, CYP had a clear polysaccharide characteristic.

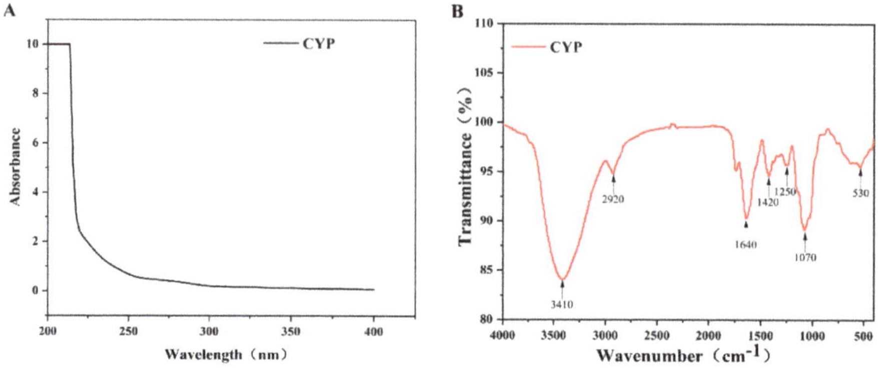

Figure 1. (**A**) UV-vis and FT-IR spectra of CYP (**B**) UV-vis spectra were recorded in the range of 200–400 nm. FT-IR spectra were recorded with a Nicolet 5700 FT-IR spectrometer between 400 and 4000 cm^{-1} using the KBr-disk method.

3.4. SEM Analysis

Scanning electron microscopy (Figure 2) can observe the surface morphology of polysaccharides, providing direct information on their microscopic appearance. Different from most polysaccharides that have a granular shape [32], CYP consists of irregular pieces that were observed at low magnification. After the magnification of 2000 times, many polymerized polygonal pieces can be seen on the surface of turmeric polysaccharide [41], but the surface of CYP is relatively smooth. This finding may be due to the relatively small molecular weight of CYP.

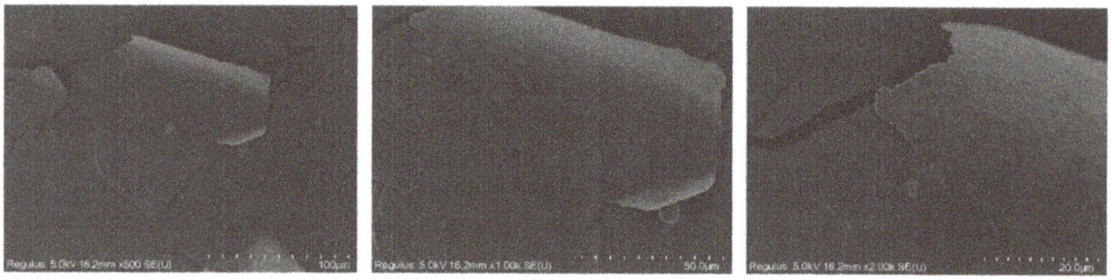

(**A**) (**B**) (**C**)

Figure 2. SEM micrographs showing surface microstructure of CYP ((**A–C**), 500×, 1000×, 2000× respectively).

3.5. Effect of CYP on IEC-6 Cells Viability

The cell viability of different concentrations of CYP cultured with IEC-6 cells is shown in Figure 3A. Compared with the control group, IEC-6 cells co-cultured with three concentrations of CYP had comparable cell viability, indicating that CYP had no toxic effect on IEC-6 cells.

In constructing a model of cellular oxidative damage by stimulating IEC-6 cells with different concentrations of H_2O_2, all concentrations of H_2O_2 reduced the viability of IEC-6 cells in a dose-dependent manner (Figure 3B). The cell viability of the 300 μM group was 55.83% compared with the control group. Therefore, 300 μM H_2O_2 was used as the concentration of the model group for the subsequent experiments.

The cell viability of the model group was significantly reduced after H_2O_2 stimulation of the cells. The pre-protected polysaccharide group was able to effectively slow down this trend, where the cell viability of the high-concentration group was comparable to that of the control group (Figure 3C). Thus, CYP had a better protective ability to deal with the negative effects of H_2O_2, consistent with a previous report [42].

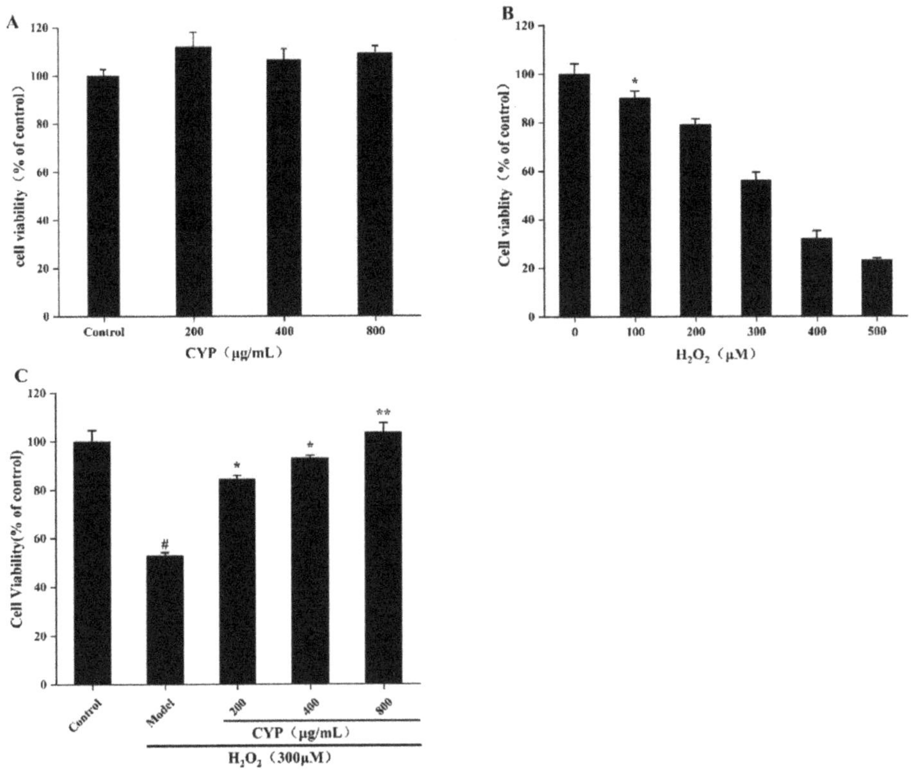

Figure 3. (**A**) Toxicity test of CYP on cell viability of IEC-6 cells (% of control), (**B**) Effects of H_2O_2 on cell viability of IEC-6 cells (% of control), (**C**) Effects of CYP on cell viability in H_2O_2-injured IEC-6 cells (% of control). Results shown are expressed as means ± SD ($n = 3$). # $p < 0.05$ compared with normal group, * $p < 0.05$ compared with H_2O_2 group alone ** $p < 0.01$ compared with H_2O_2 group alone.

3.6. Effects of CYP on MDA and SOD in H_2O_2-Stimulated IEC-6 Cells

As two important indexes for detecting the level of oxidative cell damage, SOD is an important antioxidant enzyme in organisms, which catalyzes the dismutation of superoxide anions to produce H_2O_2 and O_2 [43]. MDA is a natural product of lipid oxidation in living organisms; lipid oxidation occurs when oxidative stress occurs in animal or plant cells, some fatty acids are oxidized and gradually decompose into a complex series of compounds, including MDA. Incubation of IEC-6 cells with 200 μg/mL CYP before H_2O_2 damage was able to increase SOD activity, and CYP showed a dose-increasing trend to enhance the activity of SOD. However, the MDA content remained high (6.04 μM). With an increase in CYP concentration to 400 and 800 μg/mL, the level of MDA decreased to 4.77 and 2.54 μM compared to control (1.46 μM) (Figure 4). Pretreatment with CYP significantly inhibited H_2O_2-induced oxidative damage, reduced MDA levels, and increased antioxidant enzyme activity. These results suggested the potential antioxidant capacity of CYP [44].

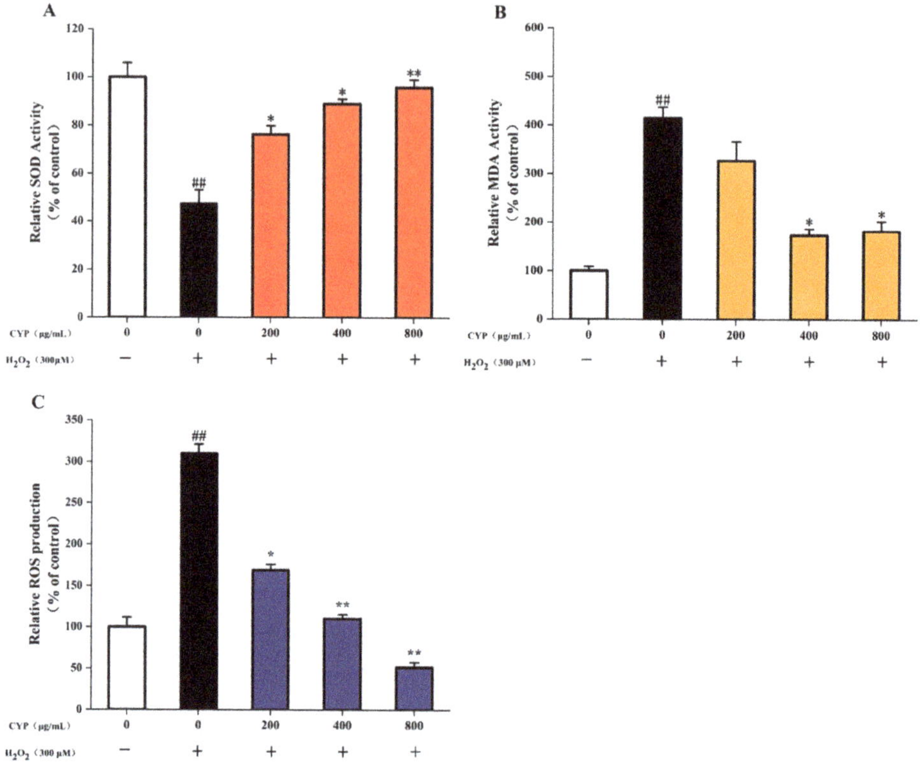

Figure 4. (**A**) Effects of CYP on levels of SOD in H_2O_2-injured IEC-6 cells, (**B**) Effects of CYP on levels of MDA in H_2O_2-injured IEC-6 cells, (**C**) Effects of CYP on the ROS production in H_2O_2-injured IEC-6 cells. Results shown are expressed as means ± SD (n = 3). ## $p < 0.01$ compared with normal group, * $p < 0.05$ compared with H_2O_2 group alone, ** $p < 0.01$ compared with H_2O_2 group alone.

3.7. ROS Production in H_2O_2-Induced IEC-6 Cells

Excessive ROS production can damage cells and lead to cellular oxidative stress, which can cause the development of various diseases [6]. H_2O_2 significantly increased the ROS level in IEC-6 cells to 3.1-fold (Figure 4C) compared with the control group. The high dose (800 μg/mL) of CYP significantly reduced the release of intracellular ROS. Interestingly, this effect was higher when compared to control. Moreover, pretreatment of IEC-6 cells with different CYP concentrations reduced the level of ROS in a significant dose-dependent manner from 200 μg/mL to 800 μg/mL. These results were similar to those of Liao et al. [7].

3.8. Effect of CYP on MAPK Pathway in Oxidative Stressed IEC-6 Cells

MAPK is associated with oxidative stress, and the excessive activation of the MAPK pathway can lead to oxidative damage in cells [8]. Meanwhile, ROS overproduction could trigger the MAPK pathway when oxidative stress occurs in cells [45]. As observed in Figure 5, the ratios of p-JNK to JNK, p-ERK to ERK and p-P38 to P38 were significantly increased after H_2O_2 treatment compared to the control group. It suggested that oxidative damage to cells might be related to the MAPK pathway.

The JNK family is a key molecule in cellular signaling in response to various stressors induced by stressors and is involved in cellular responses to radiation, osmotic stress, temperature changes, and other stressors. After stimulation by H_2O_2, JNK protein was phosphorylated, while the level of intracellular JNK phosphorylation was significantly decreased after pretreatment with CYP.

ERK is a protein involved in the regulation of cell proliferation and differentiation. Upon oxidative stimulation, ERK phosphorylation in cells significantly increased. This phenomenon was effectively inhibited by CYP at 400 and 800 mg/mL.

P38 mediates inflammation and apoptosis. In the model group, the phosphorylation level of P38 was significantly higher than that of the control group, and the phosphorylation level of the polysaccharide group decreased relative to that of the model group. Unlike the JNK and ERK proteins, the phosphorylation level of the polysaccharide group did not decrease to that of the control group, which may be due to the fact that as an important mediator of apoptosis, the cells underwent apoptosis after oxidative damage stimulated by H_2O_2.

The ratios of all three proteins and their phosphorylated proteins were significantly decreased after pretreatment with CYP, which indicated that CYP might inhibit the activation of the MAPK pathway and thus protect IEC-6 cells.

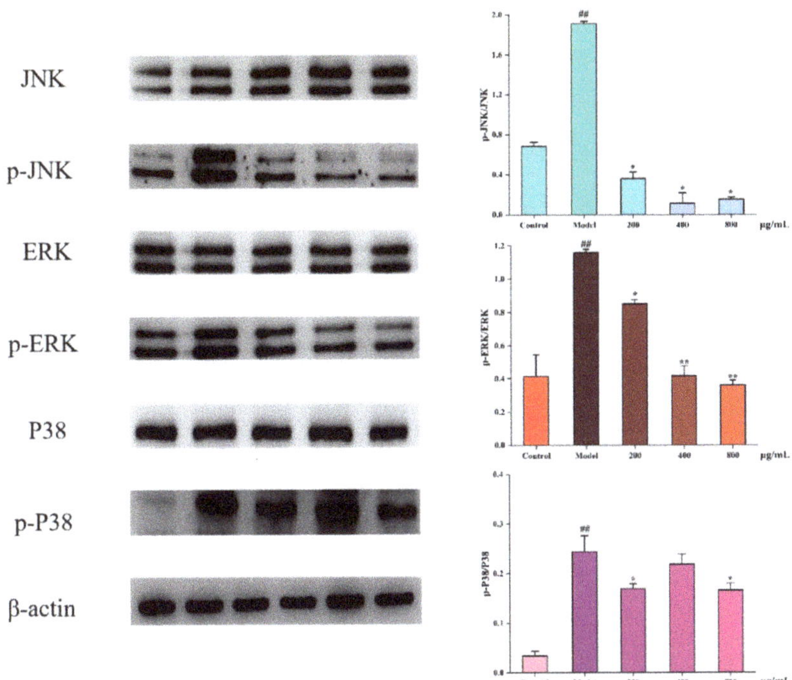

Figure 5. Effects of CYP on the H_2O_2 induced MAPK pathway of IEC-6 cells. Results shown are expressed as means ± SD ($n = 3$). ## $p < 0.01$ compared with normal group, * $p < 0.05$ compared with H_2O_2 group alone, ** $p < 0.01$ compared with H_2O_2 group alone.

4. Conclusions

In the present study, we isolated a polysaccharide and evaluated its protective effect against oxidative damage in cells. A water-soluble acidic polysaccharide (molecular weight = 20.89 kDa) was successfully extracted from Chinese yam. Analysis of monosaccharide composition showed that CYP primarily comprised galactose (28.57%), glucose (11.28%), and galacturonic acid (37.59%). CYP was able to increase SOD activity, inhibit MDA production, and reduce intracellular ROS production. Moreover, CYP exhibited better antioxidant activity, and it was able to alleviate the reduction of cell viability induced by oxidative stress due to H_2O_2 by inhibiting the activation of the MAPK pathway. It also provided a theoretical basis for the development of functional foods and clinical therapeutics against the damage caused by oxidative stress.

Author Contributions: Data curation, Z.L.; Formal analysis, Z.L.; Methodology, M.S.; Supervision, M.S.; Validation, W.Z.; Writing—original draft, Z.L.; Writing—review and editing, W.X., J.X., Q.Y., Y.C. and M.S. All authors have read and agreed to the published version of the manuscript.

Funding: This research work was financially supported by the National Natural Science Foundation of China (81960708).

Data Availability Statement: Not applicable.

Conflicts of Interest: The authors declare no conflict of interest.

References

1. Medini, F.; Bourgou, S.; Lalancette, K.G.; Snoussi, M.; Mkadmini, K.; Coté, I.; Abdelly, C.; Legault, J.; Ksouri, R. Phytochemical analysis, antioxidant, anti-inflammatory, and anticancer activities of the halophyte *Limonium densiflorum* extracts on human cell lines and murine macrophages. *S. Afr. J. Bot.* **2015**, *99*, 158–164. [CrossRef]
2. Yun, B.; King, M.; Draz, M.S.; Kline, T.; Alex, R.P. Oxidative reactivity across kingdoms in the gut: Host immunity, stressed microbiota and oxidized foods. *Free Radic. Biol. Med.* **2021**, *178*, 97–110. [CrossRef] [PubMed]
3. Valko, M.; Leibfritz, D.; Moncol, J.; Mark, T.D.C.; Mazur, M.; Telser, J. Free radicals and antioxidants in normal physiological functions and human disease. *Int. J. Biochem. Cell Biol.* **2007**, *39*, 44–84. [CrossRef] [PubMed]
4. Salinas, F.L.S.; Gonzalez, A.P.; Casique, A.A.; Ballote, A.I.; Diaz, A.; Treviño, S.; Murrieta, N.H.R.; Peña, L.M.P.; Maycotte, P. Reactive oxygen species: Role in carcinogenesis, cancer cell signaling and tumor progression. *Life Sci.* **2021**, *284*, 119942. [CrossRef] [PubMed]
5. Moloney, J.N.; Cotter, T.G. ROS signalling in the biology of cancer. *Semin. Cell Dev. Biol.* **2018**, *80*, 50–64. [CrossRef]
6. Xie, L.M.; Shen, M.Y.; Wen, P.W.; Hong, Y.Z.; Liu, X.; Xie, J.H. Preparation, characterization, antioxidant activity and protective effect against cellular oxidative stress of phosphorylated polysaccharide from *Cyclocarya paliurus*. *Food Chem. Toxicol.* **2020**, *145*, 111754. [CrossRef] [PubMed]
7. Liao, B.W.; Zhou, C.H.; Liu, T.T.; Dai, Y.Y.; Huang, H.H. A novel Hericium erinaceus polysaccharide: Structural characterization and prevention of H_2O_2-induced oxidative damage in GES-1 cells. *Int. J. Biol. Macromol.* **2019**, *154*, 1460–1470. [CrossRef]
8. Ghosh, J.; Das, J.; Manna, P.; Sil, P.C. Taurine prevents arsenic-induced cardiac oxidative stress and apoptotic damage: Role of NF-κB, p38 and JNK MAPK pathway. *Toxicol. Appl. Pharmacol.* **2009**, *240*, 73–87. [CrossRef]
9. Qiu, H.M.; Veeraperumal, S.; Lv, J.H.; Wu, T.C.; Zhang, Z.P.; Zeng, Q.K.; Liu, Y.; Chen, X.Q.; Aweya, J.J.; Cheong, K.L. Physicochemical properties and potential beneficial effects of porphyran from Porphyra haitanensis on intestinal epithelial cells. *Carbohydr. Polym.* **2020**, *246*, 116626. [CrossRef]
10. Yu, Y.; Zhu, H.; Shen, M.; Yu, Q.; Chen, Y.; Xie, J. Sulfation modification enhances the intestinal regulation of Cyclocarya paliurus polysaccharides in cyclophosphamide-treated mice via restoring intestinal mucosal barrier function and modulating gut microbiota. *Food Funct.* **2021**, *12*, 12278–12290. [CrossRef]
11. Li, G.Q.; Chen, P.F.; Zhao, Y.T.; Zeng, Q.H.; Ou, S.Y.; Zhang, Y.H.; Wang, P.C.; Chen, N.H.; Ou, J.Y. Isolation, structural characterization and anti-oxidant activity of a novel polysaccharide from garlic bolt. *Carbohydr. Polym.* **2021**, *267*, 118194. [CrossRef] [PubMed]
12. Chen, P.L.; Tong, M.Y.; Zeng, H.L.; Zheng, B.D.; Hu, X.H. Structural characterization and in vitro fermentation by rat intestinal microbiota of a polysaccharide from Porphyra haitanensis. *Food Res. Int.* **2021**, *147*, 110546. [CrossRef] [PubMed]
13. Dong, X.; Zhu, C.P.; Huang, G.Q.; Xiao, J.X. Fractionation and structural characterization of polysaccharides derived from red grape pomace. *Process Biochem.* **2021**, *109*, 37–45. [CrossRef]
14. Yang, W.Y.; Wang, Y.; Li, X.P.; Yu, P. Purification and structural characterization of Chinese yam polysaccharide and its activities. *Carbohydr. Polym.* **2015**, *117*, 1021–1027. [CrossRef] [PubMed]
15. Huang, R.; Xie, J.H.; Yu, Y.; Shen, M. Recent progress in the research of yam mucilage polysaccharides: Isolation, structure and bioactivities. *Int. J. Biol. Macromol.* **2019**, *155*, 1262–1269. [CrossRef] [PubMed]
16. Wang, X.; Huo, X.Z.; Liu, Z.; Yang, R.; Zeng, H.J. Investigations on the anti-aging activity of polysaccharides from Chinese yam and their regulation on klotho gene expression in mice. *J. Mol. Struct.* **2020**, *1208*, 127895. [CrossRef]
17. Zhou, S.Y.; Huang, G.L.; Chen, G.Y. Extraction, structural analysis, derivatization and antioxidant activity of polysaccharide from Chinese yam. *Food Chem.* **2021**, *361*, 130089. [CrossRef]
18. Luo, L.; Zheng, S.S.; Huang, Y.F.; Qin, T.; Xing, J.; Niu, Y.L.; Bo, R.N.; Liu, Z.G.; Huang, Y.; Hu, Y.L.; et al. Preparation and characterization of Chinese yam polysaccharide PLGA nanoparticles and their immunological activity. *Int. J. Pharm.* **2016**, *511*, 140–150. [CrossRef]
19. Xue, H.Y.; Li, J.R.; Liu, Y.G.; Gao, Q.; Wang, X.W.; Zhang, J.W.; Tanokura, M.; Xue, Y.L. Optimization of the ultrafiltration-assisted extraction of Chinese yam polysaccharide using response surface methodology and its biological activity. *Int. J. Biol. Macromol.* **2019**, *121*, 1186–1193. [CrossRef]

20. Liu, J.; Stefan, W.; Xu, C.L. A review of bioactive plant polysaccharides: Biological activities, functionalization, and biomedical applications. *Bioact. Carbohydr. Diet. Fibre* **2015**, *5*, 31–61. [CrossRef]
21. Zhu, Y.P.; Yang, L.; Zhang, C.N.; Tian, Y.L.; Zhang, F.F.; Li, X.T. Structural and functional analyses of three purified polysaccharides isolated from Chinese Huaishan-yams. *Int. J. Biol. Macromol.* **2018**, *120*, 693–701. [CrossRef] [PubMed]
22. Ji, H.Y.; Dai, K.Y.; Liu, C.; Yu, J.; Jia, X.Y.; Liu, A.J. Preparation, Antioxidant and Immunoregulatory Activities of a Macromolecular Glycoprotein from Salvia miltiorrhiza. *Foods* **2020**, *11*, 705. [CrossRef]
23. Li, P.; Xiao, N.; Zeng, L.P.; Xiao, J.; Huang, J.Z.; Xu, Y.N.; Chen, Y.L.; Ren, Y.H.; Du, B. Structural characteristics of a mannoglucan isolated from Chinese yam and its treatment effects against gut microbiota dysbiosis and DSS-induced colitis in mice. *Carbohydr. Polym.* **2020**, *250*, 116958. [CrossRef] [PubMed]
24. Li, M.; Chen, L.X.; Chen, S.R.; Deng, Y.; Zhao, J.; Wang, Y.; Li, S.P. Non-starch polysaccharide from Chinese yam activated RAW 264.7 macrophages through the Toll-like receptor 4 (TLR4)-NF-κB signaling pathway. *J. Funct. Food.* **2017**, *37*, 491–500. [CrossRef]
25. Liu, X.X.; Yan, Y.Y.; Liu, H.M.; Wang, X.D.; Qin, G.Y. Emulsifying and structural properties of polysaccharides extracted from Chinese yam by an enzyme-assisted method. *LWT-Food Sci. Technol.* **2019**, *111*, 242–251. [CrossRef]
26. Ju, Y.; Xue, Y.; Huang, J.L.; Zhai, Q.Z.; Wang, X.H. Antioxidant Chinese yam polysaccharides and its pro-proliferative effect on endometrial epithelial cells. *Int. J. Biol. Macromol.* **2014**, *66*, 81–85. [CrossRef]
27. Hao, B.H.; Yang, X.; Ma, Y. Study on deproteinization in extraction of polysaccharides from Patentillaunserina by Sevage. *Sci. Technol. Food Ind.* **2011**, *32*, 254–258.
28. Huang, L.X.; Shen, M.Y.; Zhang, X.W.; Jiang, L.; Song, Q.Q.; Xie, J.H. Effect of high-pressure treatment on the physicochemical properties and antioxidant activities of polysaccharide from Mesona chinensis Benth. *Carbohydr. Polym.* **2018**, *200*, 191–199. [CrossRef]
29. Liu, H.; Chen, F.S.; Yang, H.S.; Yao, Y.Z.; Gong, X.Z.; Xin, Y.; Ding, C.H. Effect of calcium treatment on nanostructure of chelate-soluble pectin and physicochemical and textural properties of apricot fruits. *Food Res. Int.* **2009**, *42*, 1131–1140. [CrossRef]
30. Guo, S.Y.; Li, L.; Wang, M.L.; Liang, G.J. The determination of protein content in polysaccharides from Stanuntonia Chinese with coomassie brilliant blue method. *Food Res. Dev.* **2008**, *29*, 115–116.
31. Xu, J.; Chen, Z.Y.; Liu, P.H.; Wei, Y.; Zhang, M.; Huang, X.D.; Peng, L.L.; Wei, X.L. Structural characterization of a pure polysaccharide from Bletilla striata tubers and its protective effect against H_2O_2-induced injury fibroblast cells. *Int. J. Biol. Macromol.* **2021**, *193*, 2281–2289. [CrossRef] [PubMed]
32. Wang, Z.J.; Xie, J.H.; Kan, L.J.; Wang, J.Q.; Shen, M.Y.; Li, W.J.; Nie, S.P.; Xie, M.Y. Sulfated polysaccharides from Cyclocarya paliurus reduce H_2O_2-induced oxidative stress in RAW264.7 cells. *Int. J. Biol. Macromol.* **2015**, *80*, 410–417. [CrossRef] [PubMed]
33. Shi, J.J.; Zhang, J.G.; Sun, Y.H.; Qu, J.; Li, L.; Prasad, C.; Wei, Z.J. Physicochemical properties and antioxidant activities of polysaccharides sequentially extracted from peony seed dreg. *Int. J. Biol. Macromol.* **2016**, *91*, 23–30. [CrossRef] [PubMed]
34. Huang, R.; Xie, J.H.; Liu, X.; Shen, M.Y. Sulfated modification enhances the modulatory effect of yam polysaccharide on gut microbiota in cyclophosphamide-treated mice. *Food Res. Int.* **2021**, *145*, 110393. [CrossRef] [PubMed]
35. Ji, X.L.; Zhang, F.; Zhang, R.; Liu, F.; Peng, Q.; Wang, M. An acidic polysaccharide from Ziziphus Jujuba cv. Muzao: Purification and structural characterization. *Food Chem.* **2019**, *274*, 494–499. [CrossRef]
36. Jing, L.; Sheng, J.W.; Jiang, J.R.; Wang, Y.; Shen, X.Y.; Liu, D.M.; Zhang, W.F.; Mao, S.M. Chemical characteristics and cytoprotective activities of polysaccharide fractions from Athyrium Multidentatum (Doll.) Ching. *Int. J. Biol. Macromol.* **2020**, *158*, 1227–1238. [CrossRef]
37. Wang, L.C.; Chen, L.Y.; Li, J.S.; Di, L.Q.; Wu, H. Structural elucidation and immune-enhancing activity of peculiar polysaccharides fractioned from marine clam *Meretrix meretrix* (Linnaeus). *Carbohydr. Polym.* **2018**, *201*, 500–513. [CrossRef]
38. Chai, Z.; Huang, W.Y.; Zhao, X.; Wu, H.; Zeng, X.X.; Li, C.Y. Preparation, characterization, antioxidant activity and protective effect against cellular oxidative stress of polysaccharide from Cynanchum auriculatum Royle ex Wight. *Int. J. Biol. Macromol.* **2018**, *119*, 1068–1076. [CrossRef]
39. Liao, W.Z.; Luo, Z.; Liu, D.; Ning, Z.X.; Yang, J.G.; Ren, J.Y. Structure Characterization of a Novel Polysaccharide from Dictyophora indusiata and Its Macrophage Immunomodulatory Activities. *J. Agric. Food Chem.* **2015**, *63*, 535–544. [CrossRef]
40. Tang, W.; Liu, C.C.; Liu, J.J.; Hu, L.Y.; Huang, Y.S.; Yuan, L.; Liu, F.W.; Pan, S.J.; Chen, S.P.; Bian, S.G.; et al. Purification of polysaccharide from Lentinus edodes water extract by membrane separation and its chemical composition and structure characterization. *Food Hydrocoll.* **2020**, *105*, 105851. [CrossRef]
41. Zong, P.Z.; Jiao, C.; Yi, C.; Yuntong, M.; Qing, S.Y.; Yun, Q.F.; Chao, M.F.; Boonjai, L.; Rui, L.; Wan, L. Extraction, structural characterization and antioxidant activity of turmeric polysaccharides. *LWT-Food Sci. Technol.* **2022**, *154*, 112805.
42. Wang, L.F.; Shi, Z.X.; Wang, X.Y.; Mu, S.; Xu, X.Y.; Shen, L.; Li, P. Protective effects of bovine milk exosomes against oxidative stress in IEC-6 cells. *Eur. J. Nutr.* **2021**, *60*, 317–327. [CrossRef] [PubMed]
43. Chen, L.; Hu, M.B.; Chen, Z.Y.; Wang, G.; Su, Q.; Liu, Y.J. Preparation, structural characterization and neuroprotective effects of polysaccharides from the pericarp of Zanthoxylum bungeanum Maxim against H_2O_2-induced oxidative damage in PC12 cells. *S. Afr. J. Bot.* **2021**, *142*, 165–174. [CrossRef]

44. Yu, Y.; Mo, S.R.; Shen, M.Y.; Chen, Y.; Yu, Q.; Li, Z.D.; Xie, J.H. Sulfated modification enhances the immunomodulatory effect of Cyclocarya paliurus polysaccharide on cyclophosphamide-induced immunosuppressed mice through MyD88-dependent MAPK/NF-κB and PI3K-Akt signaling pathways. *Food Res. Int.* **2021**, *150*, 110756. [CrossRef] [PubMed]
45. Meng, Y.; Feng, R.; Yang, Z.; Liu, T.T.; Huo, T.G.; Jiang, H. Oxidative stress induced by realgar in neurons: p38 MAPK and ERK1/2 perturb autophagy and induce the p62-Keap1-Nrf2 feedback loop to activate the Nrf2 signalling pathway. *J. Ethnopharmacol.* **2022**, *282*, 114582. [CrossRef]

foods

MDPI

Article

Ultrasonic Assisted Extraction of Quinoa (*Chenopodium quinoa* Willd.) Protein and Effect of Heat Treatment on Its In Vitro Digestion Characteristics

Xingfen He, Bin Wang, Baotang Zhao and Fumin Yang *

College of Food Science and Engineering, Gansu Agricultural University, Lanzhou 730070, China; hexingfen163@163.com (X.H.); wangbin_1519@163.com (B.W.); zhaobaotang@126.com (B.Z.)
* Correspondence: yfumin@gsau.edu.cn; Tel.: +86-13893337478

Abstract: To extract and utilise the protein in quinoa efficiently, we investigated the effect of rate of quinoa protein isolate (QPI) extraction by ultrasound-assisted alkaline extraction and traditional alkaline extraction methods using single-factor experiments and Box-Behnken design. The effect of different heat treatment temperature and time on QPI functional properties and in vitro digestion characteristics were also investigated. The results showed that the optimal conditions of ultrasound-assisted alkaline extraction process were: ultrasonic time 99 min, solid-liquid ratio 1:20 *w:v*, ultrasonic temperature 47 °C, and pH 10, and its extraction rate and purity were 74.67 ± 1.08% and 87.17 ± 0.58%, respectively. It was 10.18% and 5.49% higher than that of the alkali-soluble acid precipitation method, respectively. The isoelectric point (pI) of QPI obtained by this method was 4.5. The flexibility and turbidity of QPI had maximum values at 90 °C, 30 min, and 121 °C, 30 min, which were 0.42 and 0.94, respectively. In addition, heat treatment changed the 1.77–2.79 ppm protein characteristic region in QPI's nuclear magnetic resonance hydrogen spectroscopy (^{1}H NMR). After heating at 90 °C and 121 °C for 30 min, the hydrolysis degree and total amino acid content at the end of digestion (121 °C, 30 min) were significantly lower than those of untreated QPI by 20.64% and 27.85%. Our study provides basic data for the efficient extraction and utilisation of QPI.

Keywords: quinoa protein; ultrasounic-assisted extraction; in vitro digestion; amino acid content; nuclear magnetic resonance hydrogen spectroscopy

Citation: He, X.; Wang, B.; Zhao, B.; Yang, F. Ultrasonic Assisted Extraction of Quinoa (*Chenopodium quinoa* Willd.) Protein and Effect of Heat Treatment on Its In Vitro Digestion Characteristics. *Foods* **2022**, *11*, 771. https://doi.org/10.3390/foods11050771

Academic Editor: Filipa V. M. Silva

Received: 8 February 2022
Accepted: 3 March 2022
Published: 7 March 2022

1. Introduction

Quinoa (*Chenopodium quinoa* Willd.), also known as quinoa flour or grey rice. Its seeds are rich in high-quality protein and have high nutritional value [1]. In particular, the growing popularity of quinoa seeds is due to the rich content of proteins, dietary fibers, B vitamins, and dietary minerals as well as for being a gluten-free food [2]. It is a perfect "whole food" for humans and is listed as one of the top ten healthy nutritional foods globally [3]. Quinoa protein isolate (QPI) is widely used in the food industry for preparation of infant foods, edible films, beverages, sauces, and sausages due to its gluten-free nature [4]. In addition, quinoin (a type 1 ribosome-inactivating protein) is contained in quinoa seeds. Quinoin is considered a toxic protein present in both quinoa seeds and sprouts, which is resistant to both heat treatment and in vitro digestion [5]. Therefore, it is necessary to treat quinoa seeds by heating before consumption.

The traditional method of quinoa protein extraction mainly involves alkaline solubilisation and acid precipitation. However, due to disadvantages such as low extraction rate, long extraction time, and ability to change the structure and properties of the active material, the application of this traditional method for QPI extraction is limited. Ruiz et al. [6] demonstrated that the thermal stability of QPI worsened with an increase in pH at the time extraction. Guerreo-Ochoa et al. [7] reported that the yield of QPI obtained by the traditional alkaline extraction method was only 62.1%. Literature suggests that ultrasound, due to its cavitation properties, can shorten the extraction time and increase the extraction rate of proteins [8]. Additionally, improvement in solubility was reported for rice protein extracted by ultrasound-assisted alkaline extraction method [9]. The optimal ultrasound-assisted extraction conditions for rice bran protein were: amplitude 76%, extraction time 18 min, solid-liquid ratio 0.99 g/10 mL; and the protein yield was 4.73 ± 0.03% [10]. Moreover, in comparison with the traditional extraction method, ultrasound-assisted method increased the protein yield and protein content of duck liver by 67.7% and 4.6%, respectively [8].

The structural and functional properties of proteins play an important role in food processing and determine the application range of proteins. Therefore, research on the effects of different treatments on the structural and functional properties of food proteins is essential to better understand their roles in food systems. Various treatments such as heat [11], ultrasound [12], high pressure [13], and pH [14] significantly affect the structural and functional properties of proteins, among which the heat treatment has been one of the major methods to modify the characteristics of food ingredients and has a great impact on the structural and functional properties of proteins [15]. What's more, most food proteins undergo structural transformation during heat processing which could affect the digestion of proteins [16]. The heat treatment reduced the intermolecular forces of soluble proteins and increased flexibility [17]. Protein from Pacific oyster showed higher digestibility at relatively low temperature, but lower digestibility at relatively high temperature [18].

Although previous studies have documented the effects of heat treatment on the in vitro digestion characteristics of pea protein [19], whey protein [20], and soybean protein [21], there are few references about studies on the optimisation of ultrasounic-assisted alkaline extraction of QPI and the effect of heat treatment on the in vitro digestion characteristics of QPI. Therefore, this study aimed to (1) determine the optimum process conditions for ultrasound-assisted alkaline extraction of quinoa protein; (2) compare the optimum process conditions, extraction rate, and purity of QPI obtained by ultrasound-assisted alkaline extraction process and alkali-solution and acid-isolation extraction process; (3) investigate the effect of different heat treatment time and temperature on structural and functional properties of QPI; and (4) determine the effect of heat treatment on its in vitro digestion characteristics.

2. Materials and Methods

2.1. Materials

Quinoa was planted in the open field at Wanggeertang town (East longitude 102.847758, North latitude 35.216012, Altitude 2500 m), Xia he County, Gan nan Tibetan Autonomous Prefecture, China in May. Commercial mature quinoa was harvested in September of the same year. Within 3 h of mechanical shelling, quinoa was packed in woven bags and transported to the laboratory. Samples were stored at 4 °C for later use. Pepsin from porcine gastric mucosa (power, ≥250 units/mg solid) was purchased from Sigma-Aldrich (St. Louis, MO, USA). Pepsin (Potency: 1:3000) was purchased from Shanghai Yuanye Biological Technology Co., Ltd., (Shanghai, China). The rest of the chemicals used are of analytical grade.

2.2. Extraction of Quinoa Proteins

Extraction of quinoa proteins was carried out as described by Silventoinen, & Sozer [22]. The quinoa seeds was crushed, sifted using a 60-mesh sieve, and degreased with petroleum ether (Auto fat analyzer, Sex 406, Jinan Haineng Instrument Co., Ltd., Jinan, China). The degreased quinoa flour was air-dried followed by addition of distilled water to obtain a certain solid-liquid ratio. The pH was adjusted using 1 M NaOH followed by extraction of quinoa proteins at a set ultrasonic temperature. The extract was then centrifuged for 15 min at 6369× g (H−1850R, Changsha Xiangyi Centrifuge Co., Ltd., Changsha, China). And the supernatant was collected. The pH of the supernatant was adjusted to isoelectric point of QPI using 1 M HCl and incubated at 4 °C for 120 min. After that, the extract was centrifuged for 15 min at 6369× g to collect the precipitate. The precipitate was resuspended in distilled water and washed 3–5 times. The pH was adjusted to 7.0 using 1 M NaOH. QPI was obtained by vacuum freeze drying the resuspended precipitate in a vacuum freeze-drying machine (LyoQuest-85, Telstar Lab, Barcelona, Spain) and stored at −20 °C until use.

2.3. Determination of Extraction Rate and Purity of QPI

The extraction rate and purity of QPI were determined according to the method described by Ruiz et al. [1]. The protein content refered to the method of Miller et al. [23]

$$\text{The extraction rate of QPI } (\%) = \frac{\text{Isolate protein content } (\%) \times \text{ isolate weight } (\text{g})}{\text{flour protein content } (\%) \times \text{flour weight } (\text{g})} \times 100 \quad (1)$$

$$\text{The purity of QPI } (\%) = \frac{\text{Isolate protein content } (\%) \times \text{ isolate weight } (\text{g})}{\text{isolate weight } (\text{g})} \times 100 \quad (2)$$

2.4. Optimisation of Quinoa Protein Extraction Process

2.4.1. Single-Factor Test for Ultrasound-Assisted Alkaline Extraction

The basic conditions for extraction were: ultrasonic time 90 min, solid-liquid ratio 1:15, ultrasonic temperature 45 °C, and pH 10; for single-factor test, one of these parameters was changed at a time keeping the other three parameters constant. The different conditions tested were ultrasonic time (30, 60, 90, 120, 150 and 180 min), solid-liquid ratio (1:5, 1:10, 1:15, 1:20, 1:25 and 1:30; *w:v*), ultrasonic temperature (25, 35, 45, 55, 65 and 75 °C) and pH (8.0, 9.0, 10.0, 11.0, 12.0 and 13.0), with extraction rate of QPI as the index. The experiments were performed in triplicates.

2.4.2. Response Surface Optimisation Test

On the basis of the single-factor test, the four parameters for quinoa protein extraction including ultrasonic time (A), solid-liquid ratio (B), ultrasonic temperature (C) and pH(D) were optimised. The response surface test scheme was designed according to the Box-Behnken method in the Design-Expert 8.0.6 software, as shown in Tables S1 and S2.

2.4.3. Comparison between Ultrasound-Assisted and Traditional Alkaline Extraction Methods

The extraction rate and purity of QPI obtained by the ultrasound-assisted alkaline extraction method and the traditional alkaline extraction method were compared and the optimum technological conditions were determined.

2.5. Isoelectric Point

The isoelectric point of QPI was determined according to the method of Pedroche et al. [24]. Under the optimal extraction conditions of QPI, eight equal portions of protein extract supernatant were collected, and the pH was adjusted to 3.0, 3.5, 4.0, 4.5, 5.0, 5.5, 6.0, 6.5 with 1 M HCl, respectively. The optimum pH of acid precipitation was determined by measuring the extraction rate of the protein. The pH corresponding to the maximum protein extraction rate is the isoelectric point of QPI.

2.6. Heat Treatment

QPI was dissolved in phosphate buffer (0.2 M, pH 7.0), while stirring for 2 h at ambient temperature (25 °C) and stored it overnight at 4 °C to enhance hydration. The QPI solution with a concentration of 5% was treated at different temperatures of 60, 70, 80, 90, 100, and 121 °C for 5, 10, 20, and 30 min (Employed each temperature at different times), and then quickly cooled with ice water for 5 min. Samples were stored at 4 °C and generally analyzed within 2 d of treatment [25].

2.7. Flexibility and Turbidity Measurement

Trypsin solution (250 μL, 0.1%) was mixed with heat-treated QPI solution (14 mL, 0.1%), and after 5 min incubation at 38 °C, 4 mL of trichloroacetic acid solution (5%) was added. After centrifugation, the supernatant was taken, and its absorbance was measured at 280 nm, indicating flexibility [26]. The heat-treated QPI solution was diluted to 3 mg/mL, and the turbidity was measured by measuring its absorbance at 400 nm after shaking and mixing [27].

2.8. Samples Preparation for Proton Nuclear Magnetic Resonance (^1H NMR) Spectra Acquisition

Freeze-dried QPI (20.0 mg) was placed in a 2 mL sample tube with 1000 μL CD_4O aqueous solution (CD_4O:D_2O = 1:1, *v:v*; TMSP 0.5 mM). The mixed solution was placed in a tissue grinder for 300 s (70 Hz), repeated 2 times and then centrifuged (10 min, 12,738× *g*) 2 times. The supernatant (540 μL) was transferred into a nuclear magnetic tube for 1H NMR test on a 600 MHz nuclear magnetic resonance spectrometer (AVANCE III Brook (Beijing) Technology Co., Ltd., Beijing, China). The test conditions were: constant temperature of 298 K (25 °C), pulse interval D1 of 4.00 s, spectral width of 12,335.526 Hz, and 64 samples, using a ZGPR pulse sequence to suppress the water peak. The raw spectra were imported into MestReNova software for segment integration and normalization at 0.51 ppm [28].

2.9. In Vitro Simulation of Gastrointestinal Digestion of QPI

2.9.1. Preparation of Simulated Gastrointestinal Fluid

Simulated gastricfluid (SGF) and simulated intestinalfluid (SIF) were prepared following the harmonized protocol [29]. SGF: pepsin (0.3 g), NaCl (2.0 g) and HCl (12 M, 0.7 mL) were mixed and made up to 100 mL, and the pH of the solution was adjusted to 1.2. SIF: After KH_2PO_4 (0.68 g) was completely dissolved in deionized water (25 mL), NaOH solution (0.2 m, 19 mL), deionized water (40 mL) and trypsin (4.0 g) were successively added, and the volume was fixed to 100 mL. The pH of the solution is 7.5.

2.9.2. In Vitro Gastrointestinal Digestion

After the heat-treated QPI solution (Treatment at 90 °C and 121 °C for 30 min respectively), SGF and SIF were incubated at 37 °C for 10 min, equal volumes of QPI solution and SGF were taken into a centrifuge tube. The centrifuge tube was placed in a water bath shaker at 37 °C, and the digested products with reaction times of 0, 30, 60, 90 and 120 min were respectively taken into the test tube, and immediately boiled at 100 °C to kill the enzyme and cooled for later use. The pH of the remaining solution in the centrifuge tube was adjusted to 7.5, an equal volume of SIF was added, and the centrifuge tube was placed in a water bath shaker at 37 °C, and the digested products with reaction times of 150, 180, 210, 240 and 270 min were respectively taken into the test tube, and immediately boiled at 100 °C to kill the enzyme and cooled for later use [30].

2.9.3. Degree of Hydrolysis

The hydrolysis degree of QPI was determined by o-phthalaldehyde (OPA) method. The hydrolyzed supernatant (400 μL) was mixed with OPA (3 mL), and the absorbance was measured at 340 nm immediately after 2 min of reaction. Deionized water and serine standard solution (0.9516 mM) were used as blank and stand, respectively [31]. The degree of hydrolysis was calculated as follows:

$$SerineNH_2 = \frac{OD_{sample} - OD_{blank}}{OD_{stand} - OD_{blank}} \times 0.9516 \times \frac{N \times V}{X \times P} \tag{3}$$

where $SerineNH_2$ is the serine content per gram of protein, mmol/g, X is the quality of the sample, g, p is the protein content of the sample, %, N is dilution times, V is volume of supernatant, L.

$$DH\ (\%) = \frac{(SerineNH_2 - \beta)/\alpha}{h_{tot}} \times 100 \tag{4}$$

where β and α are constants 0.4 and 1, respectively, h_{tot} is number of peptide bonds of protein, mmol/g, the h_{tot} of quinoa protein is 7.4 mmol/g.

2.9.4. Total Amino Acid Content of In Vitro Digestion Products

The enzymolysis solution (2 mL) was mixed with 6 M HCl (8 mL) and placed in a digestive tube then filled with high-purity nitrogen for 3 min. The bottle plug was quickly sealed and digested at 110 °C for 24 h. It was taken out after cooling. All the hydrolysates were transferred to 25 mL volumetric flask. The digestive solution (2 mL) was dried in an oven at 60 °C, and a few solid or stains were left at the bottom. The residue was redissolved with 2 mL of ultrapure water, then dried again, repeated for 3 times, and diluted to 5 mL. Digestion solution (2 mL) was filtered with 0.22 µm aqueous membrane into sample bottles for automatic amino acid analyzer (LA8080 Hitachi High-Tech Science Co., Ltd. Naka Office, Tokyo, Japan) testing [32].

2.10. Data Analysis

Experimental data are presented as mean ± standard deviation (calculated by Excel 2007). Origin 2021 was used for plotting graphs. SPSS 17.0 was used for one-way ANOVA and multiple comparisons. The response surface results were analysed and graphs were plotted using Design-Expert 8.0. MestReNova was used to analyze the ^{1}H NMR spectrum.

3. Results and Discussion

3.1. Single-Factor Test Results

Quinoa protein extraction rate increased sharply with prolongation of ultrasonic time and reached a maximum value of 69.07 ± 0.89% at 90 min, and then decreased gradually (Figure 1A). When the ultrasonic time was between 30 and 90 min, cavitation and thermal effects of the ultrasonic wave caused the solvent molecules to rapidly enter the solid matter, resulting in quick dissolution of QPI and increase in extraction rate. When the ultrasonic time exceeded 90 min, the protein structure was destroyed and degraded due to long-term cavitation and thermal effects of the ultrasound. This resulted in protein folding and aggregation due to excessive increase in hydrophobicity, leading to a decline in the dissolution rate of QPI [33]. Therefore, the optimum ultrasonic time was 60–120 min.

With an increase in the amount of solvent, quinoa protein extraction rate rapidly increased initially, reached a maximum value of 71.44 ± 1.49% when the solid-liquid ratio was 1:15, and then decreased slowly (Figure 1B). This trend is observed because the quinoa particles do not fully collide with each other under the action of ultrasonic cavitation when the concentration of quinoa is low. Appropriate solid-liquid ratio enables the quinoa particles to fully accept the force of emptiness and collide strongly with each other, which increases the quinoa protein dissolution and extraction rate [34]. However, excessive quinoa concentration will affect the mass and heat transfer effect of the material, thereby affecting the cavitation effect and decreasing the quinoa protein extraction rate. Therefore, the optimum solid-liquid ratio was 1:10–1:20.

As the ultrasonic temperature increased, quinoa protein extraction rate gradually increased, reached a maximum value of 69.07 ± 0.89% at 45 °C, and then decreased rapidly (Figure 1C). This occurs because within a certain range, the increase in temperature is conducive to the acceleration of thermal motion inside the molecule, which increases the protein dissolution and extraction rate. However, very high temperatures damage the integrity of the protein structure and reduce the dissolution and extraction rate [35]. Therefore, 35–55 °C was considered as the optimum temperature range.

With increase in pH, quinoa protein extraction rate gradually increased, reached a maximum value of 71.10 ± 0.58% at pH 11.0, and then decreased rapidly (Figure 1D). It was observed that pH changes significantly affected the extraction rate of quinoa protein ($p < 0.05$). This is mostly likely due to the loosening of tight cell structure and destruction of the protein structure in an alkaline environment, thereby increasing the amount of protein eluted [36]. As the pH increased above 11.0, the extraction rate of quinoa protein decreased, which might be due to the destruction of secondary bonds in the protein by strong bases, resulting in denaturation of the protein secondary and tertiary structure [37]. However, previous studies have shown that the higher the pH when extracting QPI, the worse the thermal stability, and the purity of QPI is not high and the color is dark at this time [1]. Therefore, the pH value of the solvent was set to 10.

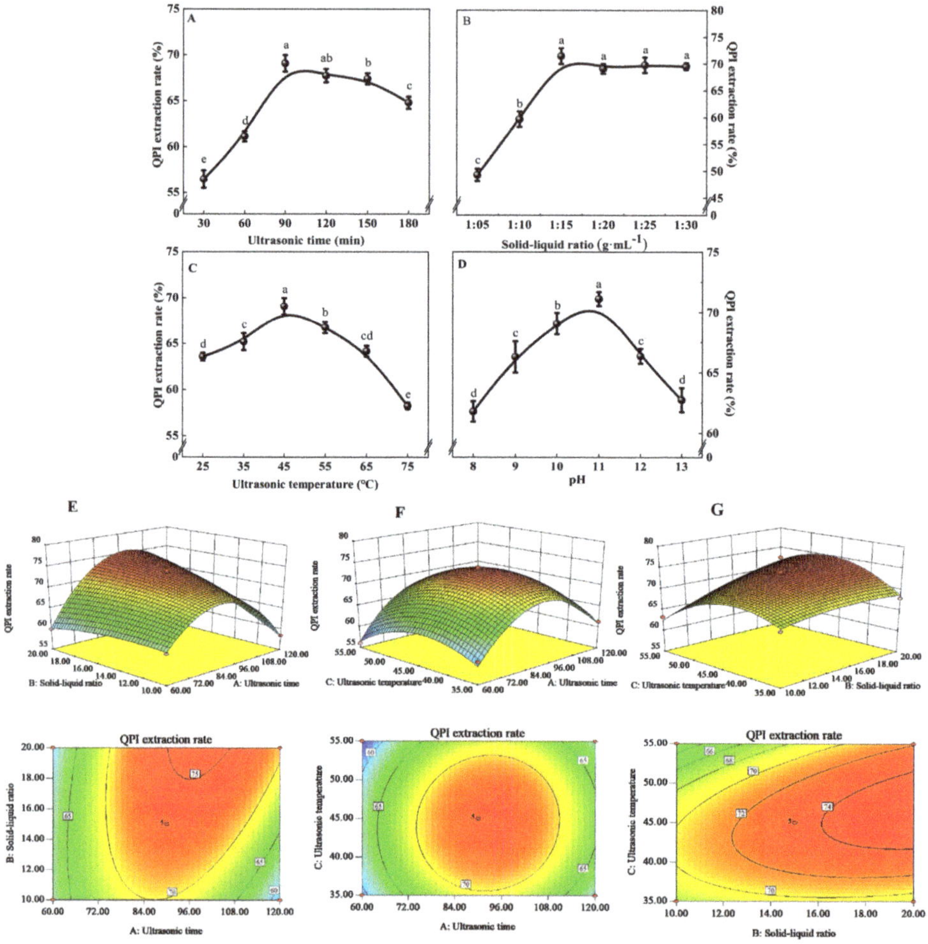

Figure 1. The effects of ultrasonic time (**A**), solid-liquid ratio (**B**), ultrasonic temperature (**C**), pH (**D**) and the response surface methodology and contour plots for the effects of various factors (**E**–**G**) on the extraction rate of QPI. Different letters (a–e) represent significant differences ($p < 0.05$).

3.2. *Response Surface Test Optimization Results*

3.2.1. Response Surface Test Design and Results

Response surface test design and results are shown in Table S2.

3.2.2. Regression Equation Fitting and Analysis of Variance (ANOVA)

A mathematical model with the regression equation was established by statistical analysis of the experimental data: extraction rate of QPI (%) = $73.48 + 1.61A + 2.55B - 0.16C + 3.66AB + 0.66AC + 2.09BC - 9.49A^2 - 1.05B^2 - 4.61C^2$

ANOVA was performed for the regression model (Table S3). $p < 0.01$ indicated that the model equation was highly significant and the model's lack of fit was insignificant.

Therefore, the selected model was appropriate. The correlation coefficients were $R^2 = 0.9895$ and $R_{Adj}{}^2 = 0.9759$, indicating that the equation fitted the test well, and can effectively reflect the relationship between each factor and its response value. Therefore, it was feasible to use this model equation to predict the process parameters of ultrasound-assisted quinoa protein extraction.

ANOVA of regression equation coefficients showed that the effect of ultrasonic time, the solid:liquid ratio on quinoa protein extraction rate, the interaction terms AB, and the interaction terms BC, were extremely significant; the effect of ultrasonic temperature on the quinoa protein extraction rate and the interaction terms AC was insignificant. The F value represents the degree of influence of ultrasonic time, solid-liquid ratio, and ultrasonic temperature on the extraction rate of QPI. A higher F value indicated greater influence. The order of influence of the three influencing factors on quinoa protein extraction rate, as inferred from the magnitude of F values, was solid-liquid ratio > ultrasonic time > ultrasonic temperature.

3.2.3. Response Surface Analysis of Interaction of Various Factors

Response surface diagram and contour diagram of the interaction among ultrasonic time (A), solid-liquid ratio (B), and ultrasonic temperature (C) were prepared according to the regression equation (Figure 1E–G). A very steep curve was obtained for interactions between ultrasonic time and solid-liquid ratio, and solid-liquid ratio and ultrasonic temperature, indicating that these interactions had extremely significant effects on the extraction rate of QPI. A gentle curve obtained for the interaction between ultrasonic time and temperature indicated that these interactions did not have significant effects on the protein extraction rate. This was consistent with the results of ANOVA of regression equation.

The optimal process parameters of ultrasound-assisted quinoa protein extraction, determined by the mathematical model, were as follows: ultrasonic time 98.51 min, solid-liquid ratio 1:20, ultrasonic temperature 46.82 °C, and pH 10; the predicted extraction rate of QPI under these conditions was 75.87%. For actual operation, the following conditions were considered: ultrasonic time 99 min, solid-liquid ratio 1:20 $g \cdot mL^{-1}$, ultrasonic temperature 47 °C, and pH 10; under these conditions, the extraction rate could reach 74.67 ± 1.08% and the purity of QPI obtained was 87.17 ± 0.58%. The time taken by the ultrasound-assisted method for quinoa protein extraction was significantly lower than the time taken by the traditional alkaline extraction method. This is probably because during the ultrasound-assisted extraction process, the rapidly formed cavitation bubbles have mechanical, chemical, and thermal effects on the medium, which significantly increases the extraction rate of QPI [38].

3.3. *Comparison of Ultrasound-Assisted Extraction and Traditional Alkaline Extraction of Quinoa Protein*

In an early phase of this study, single-factor and response surface test optimisation was carried out for the traditional alkaline extraction method using solid-liquid ratio, temperature, pH, and time as process parameters (Figure S1). The optimal conditions were as follows: time 139 min, solid-liquid ratio 1:29 *w:v*, temperature 56 °C, and pH 10; under these conditions, the protein extraction rate was 67.77 ± 1.50% and the purity of QPI was 82.63 ± 2.85% (Tables S4–S6). It was observed that the ultrasound-assisted extraction method significantly improved the extraction rate of QPI, i.e., by 10.18%, and increased the purity by 5.49% compared to the traditional alkaline extraction method. Moreover, using the ultrasound-assisted method, the extraction time and temperature were reduced by 40 min and 9 °C, respectively, thereby greatly improving the efficiency of quinoa protein extraction. Similar results were obtained by Li et al. [39] in the study of ultrasound-assisted extraction of capsicum seed protein. Thus, ultrasound-assisted quinoa protein extraction has a significant energy saving effect. In the pH range 3.0–6.5, the extraction rate of quinoa protein first increased, reached the maximum value at pH 4.5, and then decreased (Figure 2).

Thus, the pI of QPI was 4.5, which is in agreement with the results obtained by Mir et al. [40] and Elsohaimy et al. [36].

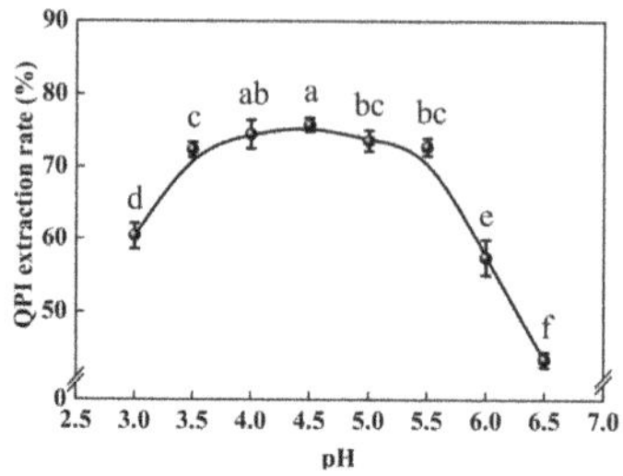

Figure 2. Isoelectric point of QPI. Different letters (a–f) represent significant differences ($p < 0.05$).

3.4. Functional Properties of QPI

3.4.1. Flexibility and Turbidity

The flexibility of a protein can be understood as the ability of its structure to change when the external environment of the protein changes, reflecting the sensitivity of the protein structure to environmental changes [41]. It has received increasing attention because of its key role in determining the functional properties of proteins, especially the interfacial properties. Studies have shown that bovine serum albumin with higher flexibility is more likely to form a better viscoelastic protein film at the interface, thereby showing better emulsifying properties [42]. In this study, it was found that when the heat treatment time was certain, QPI flexibility increased and then decreased with the increase of temperature, reaching a maximum value of 0.42 at 90 °C, 30 min, which was significantly higher than that of untreated 27.27%. In the range of 60–90 °C, the flexibility of QPI gradually increased with the prolongation of heat treatment time. Previous studies have shown that rigid heat-resistant proteins reach the flexibility of unstable proteins at higher temperatures [43]. This is probably due to the fact that moderate heat treatment will destroy the covalent or non-covalent forces that maintain the rigid structure of the protein, such as van der Waals forces, hydrogen bonds, electrostatic interactions, disulfide bonds, hydrophobic interactions, etc., thereby increasing the flexibility of the protein [44]. When the temperature was 100 °C and 121 °C, QPI flexibility then gradually decreased with time and fell to the lowest value of 0.37 at 121 °C, 30 min, which was still higher than the untreated 12.12% (Figure 3A). This is probably due to the excessive heating temperature that severely denatures the protein, forming a large number of insoluble agglomerates and decreasing solubility, which in turn leads to reduced flexibility [45].

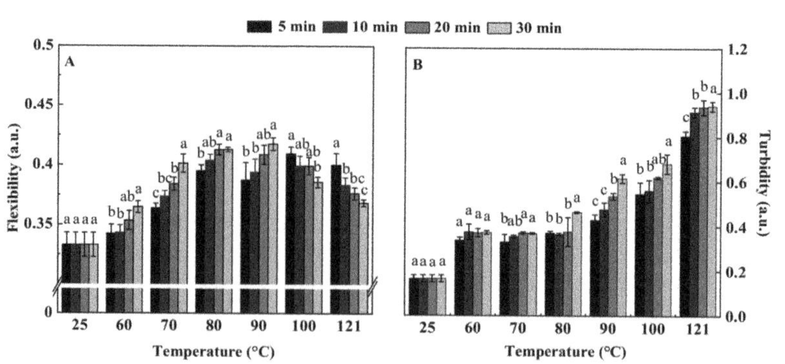

Figure 3. Effect of different heat treatment conditions on flexibility (**A**) and turbidity (**B**) of QPI. Different letters (a–c) indicate significant differences at same temperature ($p < 0.05$).

The measurement of turbidity can directly reflect the dispersion state, aggregation state and particle size of protein particles in the solution [46]. In this study, it was found that the turbidity of QPI showed a gradual increase with the increase of time and temperature. At lower temperatures (60 °C and 70 °C), the effect of heat treatment time on QPI turbidity was not significant, while at higher temperatures the effect was significant, and QPI turbidity increased continuously with time, reaching a maximum value of 0.94 at 121 °C and 30 min, which was 4.53 times higher than that of untreated (Figure 3B). This is consistent with the result that heat treatment increases the turbidity of whey protein [47]. Previous studies have shown that the increase in turbidity is related to the formation of protein molecular aggregates [48]. In addition, higher aggregation rates and higher turbidity increases were also observed for longer microwave heating times for grass carp sarcoplasmic and myogenic fiber proteins [49]. Therefore, we speculate that the increase in turbidity may be due to the unfolding of QPI molecules, when treated at higher temperature and for a long time, the intramolecular and intermolecular interaction forces of proteins are enhanced, resulting in substantial aggregation of proteins [50].

3.4.2. Nuclear Magnetic Resonance Hydrogen Spectroscopy (^{1}H NMR)

A variety of solvents were first screened and three were finally identified for comparison, namely D_2O, CD_4O and a mixture of $D_2O:CD_4O$ = 1:1. The best solution for QPI is $D_2O:CD_4O$ = 1:1 solvent mixture (Figure 4A). Further study of the effect of different heat treatment temperatures on the ^{1}H 600 MHz NMR spectra of QPI (Figure 4B). The peak at 0 ppm is the internal standard (sodium,2,2,3,3-tetradeuterio-3-trimethylsilylpropanoat -e), 4.50–4.80 ppm is the water peak, and 0.75–3.81 ppm is the amino characteristic range of protein [51,52]. The spectra of all QPIs in the range of 0.75 ppm to 3.81 ppm are very complete and clear, which is helpful for further data processing and analysis. In this study, the 1H 600 MHz NMR spectra of QPI were subjected to a series of operations such as phase correction, baseline correction, calibration, peak calibration, integration and normalization in the range of 0.75–3.81 ppm in units of 0.51 ppm to obtain the NMR integral spectra (Figure S2). The results showed that only the normalized relative percentages in the range of 3.30–3.81 ppm were dependent on the heat treatment temperature of QPI, which gradually decreased with increasing temperature in the range of 60–121 °C, and had a minimum value of 34.97 ± 0.82% at the heat treatment condition of 121 °C for 30 min, which was significantly lower than the untreated 8.46% ($p < 0.05$). Its variation in the interval from 0.75 to 1.26 ppm was not only dependent on temperature, but also at 121 °C, 30 min had a minimum value of 23.25 ± 0.34%, significantly lower than the untreated 2.47% ($p < 0.05$). Its variation in the interval of 1.77–2.28 ppm, 2.28–2.79 ppm and 2.79–3.30 ppm was not dependent on temperature and reached maximum values of 10.85 ± 0.07%, 2.26 ± 0.38% and 5.42 ± 0.96% at the heat treatment condition of 121 °C for 30 min, respectively, which were significantly higher than the untreated 24.57%, 4.51 times and 7.54% ($p < 0.05$) (Table 1). This indicates that 0.75–1.26 ppm, 3.30–3.81 ppm, 1.77–2.28 ppm, 2.28–2.79 ppm and 2.79–3.30 ppm are all susceptible to heat treatment.

Heat treatment temperature has a significant effect on the 1H NMR spectrum, turbidity and flexibility of QPI (Figure 4C). The ^{1}H NMR of 1.77–2.28 ppm and 2.28–2.79 ppm showed highly significant positive correlations ($p < 0.01$) with heat treatment temperature and turbidity, and 3.30–3.81 ppm showed highly significant negative correlations ($p < 0.01$) with turbidity and significant negative correlations ($p < 0.05$) with heat treatment temperature. In addition, 0.75–1.26 ppm also had a significant negative correlation with temperature ($p < 0.05$), while flexibility had a significant positive correlation with temperature ($p < 0.05$) and turbidity had a highly significant positive correlation with temperature ($p < 0.001$). This suggests that heat treatment affects the protein characteristic 1.77 to 2.79 ppm region of quinoa, which in turn alters the turbidity of the QPI. However, studies on the influence of heat treatment on 1H NMR spectra and microscopic substances are still in the initial stage, and the specific mechanisms and principles are not well understood, and further in-depth studies are needed.

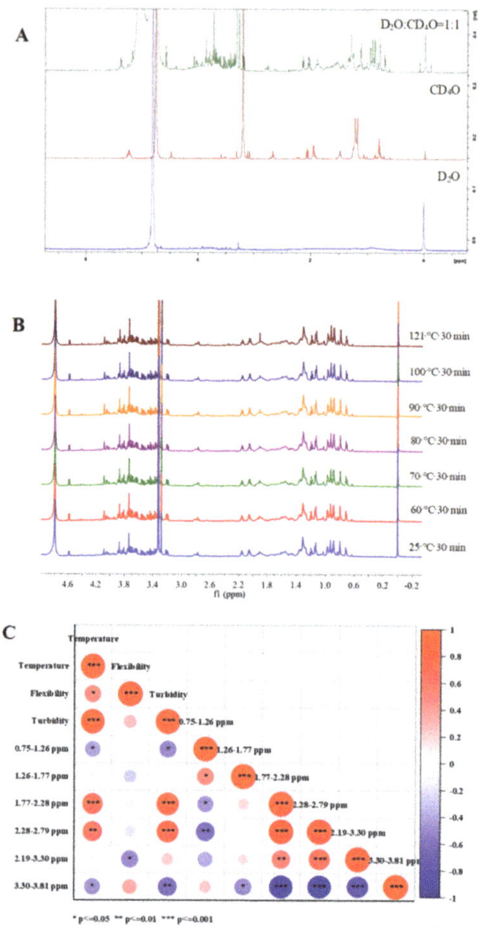

Figure 4. Effect of different solvents on dissolution of quinoa protein isolate (**A**), heat treatment temperatures on ^{1}H NMR spectra of QPI (**B**), correlation between the influence of heat treatment on QPI ^{1}H NMR spectrum and temperature, flexibility and turbidity (**C**).

Table 1. Effect of different heat treatment temperatures on the relative percentage content of QPI's ^{1}H NMR integrated spectra after normalization.

Temperature (°C)	Relative Percentage Content (%)					
	0.75–1.26 ppm	**1.26–1.77 ppm**	**1.77–2.28 ppm**	**2.28–2.79 ppm**	**2.79–3.30 ppm**	**3.30–3.81 ppm**
25	23.84 ± 0.19 ^{ab}	23.80 ± 0.16 ^{a}	8.71 ± 0.19 ^{c}	0.41 ± 0.07 ^{bc}	5.04 ± 0.29 ^{ab}	38.20 ± 0.07 ^{d}
60	23.30 ± 0.09 ^{c}	22.38 ± 0.06 ^{d}	8.64 ± 0.04 ^{c}	0.47 ± 0.13 ^{bc}	4.29 ± 0.29 ^{b}	40.93 ± 0.30 ^{a}
70	23.58 ± 0.14 ^{bc}	22.778 ± 0.04 ^{c}	8.758 ± 0.07 ^{c}	0.38 ± 0.16 ^{bc}	4.23 ± 0.25 ^{b}	40.30 ± 0.29 ^{ab}
80	23.96 ± 0.05 ^{a}	23.34 ± 0.12 ^{b}	8.84 ± 0.07 ^{c}	0.27 ± 0.08 ^{c}	4.12 ± 0.35 ^{b}	39.46 ± 0.25 ^{c}
90	23.47 ± 0.06 ^{bc}	22.91 ± 0.01 ^{c}	8.73 ± 0.04 ^{c}	0.52 ± 0.06 ^{bc}	4.38 ± 0.06 ^{b}	39.99 ± 0.08 ^{bc}
100	23.47 ± 0.17 ^{bc}	23.47 ± 0.05 ^{b}	9.27 ± 0.13 ^{b}	0.75 ± 0.13 ^{b}	4.40 ± 0.10 ^{b}	38.64 ± 0.11 ^{d}
121	23.25 ± 0.34 ^{c}	23.24 ± 0.23 ^{b}	10.85 ± 0.07 ^{a}	2.26 ± 0.38 ^{a}	5.42 ± 0.96 ^{a}	34.97 ± 0.82 ^{e}

Different letters ($^{a–e}$) represent significant differences in the same column ($p < 0.05$).

3.5. Hydrolysis Degree during In Vitro Digestion

Degree of hydrolysis (DH) is defined as the proportion of cleaved peptide bonds in the protein hydrolysate [53]. In this study, the trend of hydrolysis of QPI after heat treatment was similar to that of untreated QPI, and the hydrolysis of QPI after heat treatment at 90 and 121 °C for 30 min was lower than that of untreated QPI during the whole digestion process, among which the hydrolysis of QPI after treatment at 121 °C was the smallest, which was 25.53% after simulated in vitro digestion for 240 min, significantly lower than that of untreated QPI at 20.64% (Figure 5A). Previous studies found that the hydrolysis of soybean isolate protein (SPI) decreased under heat treatment conditions greater than 85 °C [54]. Reduction in protein digestibility of QPI after heating treatment has generally been attributed to the formation of cross-linked protein polymers, which are resistant to proteolysis [55]. Therefore, after heat treatment of the QPI suspension, pepsin became less effective in enzymatic digestion and its digestibility was lower than that of untreated protein [6]. This is consistent with the results of reduced digestibility of chicken breast proteins after heat treatment at 121 °C [56]. Previous studies have shown that heat treatment is an important process for the complete digestion of food proteins. These processes significantly affect the protein structure and thus its digestive resistance [57]. Therefore, we speculate that the QPIs treated at 90 °C and 121 °C may have structures that are difficult to degrade enzymatically and may inhibit the activity of digestive enzymes.

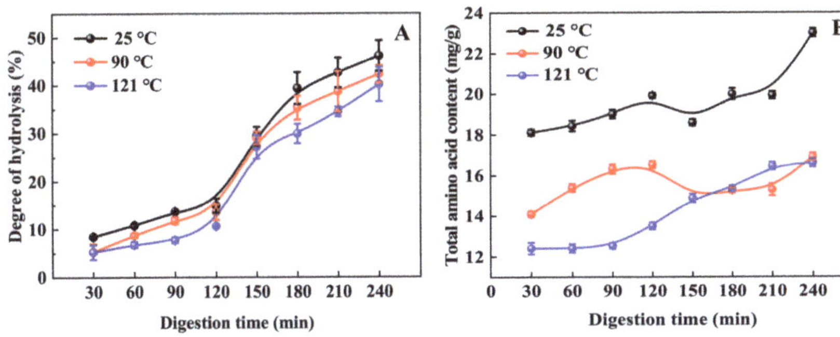

Figure 5. Effect of different heat treatment temperatures on the degree of hydrolysis (**A**) and total amino acid content (**B**) of QPI during in vitro digestion. Different letters indicate significant differences ($p < 0.05$).

3.6. Total Amino Acid Content during In Vitro Digestion

This study found that the content of each amino acid at the end point of pepsin digestion (120 min) was significantly lower than that of trypsin digestion end point (240 min) at the same temperature (Figure 5B). The amino acid content increased gradually with the prolongation of digestion time, and the amino acid content in the trypsin digestion stage was significantly higher than that in the pepsin digestion stage. The content of amino acids in the in vitro digestion products of untreated QPI was higher than that of the heat treated products. The content of amino acids in the in vitro digestion products of QPI after treatment at 121 °C was the lowest, and the content at the end of digestion was 16.60 mg/g (Table 2), which was significantly lower than that of untreated 27.85%. Previous studies have found that the amino acid content of heat-treated king oyster mushroom protein was significantly reduced after digestion in vitro [58]. Amino acid content of in vitro digested gluten protein was significantly reduced after microwave heating [59]. Furthermore, Giménez et al. [60] indicated that low levels of His, Pro and Ala in protein hydrolyzates were associated with a decrease in their antioxidant potential. Similar results were found in soy protein isolates [61]. In this study, the contents of His, Pro and Ala in the in vitro digestion products of QPI treated at 90 °C and 121 °C were significantly lower than those of untreated. Therefore, we speculate that the high heating temperature might cause serious damage to amino acids and irreversible decomposition, resulting in the reduction of content. This result confirms the reason why heat treatment leads to a decrease in the antioxidant activity of QPI in vitro.

Table 2. Total amino acid content in vitro digestion products of quinoa protein isolate after different heat treatment.

Type		Content ($mg \cdot g^{-1}$)					
		25 °C		90 °C		121 °C	
		120 min	240 min	120 min	240 min	120 min	240 min
Essential Amino Acids	Thr	0.82 ± 0.02	0.94 ± 0.04	0.68 ± 0.07	0.70 ± 0.02	0.54 ± 0.07	0.68 ± 0.03
	Met	0.48 ± 0.03	0.55 ± 0.01	0.39 ± 0.04	0.39 ± 0.04	0.31 ± 0.04	0.39 ± 0.03
	Val	1.12 ± 0.09	1.29 ± 0.18	0.92 ± 0.10	0.94 ± 0.05	0.74 ± 0.14	0.93 ± 0.05
	Lys	1.24 ± 0.12	1.44 ± 0.08	1.03 ± 0.09	1.05 ± 0.10	0.82 ± 0.04	1.03 ± 0.12
	Leu	1.76 ± 0.11	2.04 ± 0.14	1.45 ± 0.20	1.48 ± 0.08	1.16 ± 0.12	1.45 ± 0.08
	Ile	1.03 ± 0.04	1.19 ± 0.15	0.85 ± 0.07	0.87 ± 0.08	0.68 ± 0.06	0.85 ± 0.06
	Phe	1.02 ± 0.09	1.18 ± 0.12	0.84 ± 0.04	0.86 ± 0.08	0.67 ± 0.07	0.84 ± 0.06
Non-essential Amino Acids	Asp	1.90 ± 0.01	2.19 ± 0.07	1.58 ± 0.09	1.62 ± 0.06	1.27 ± 0.07	1.58 ± 0.02
	Tyr	0.80 ± 0.08	0.93 ± 0.05	0.66 ± 0.02	0.67 ± 0.09	0.53 ± 0.10	0.67 ± 0.06
	Ser	0.99 ± 0.01	1.15 ± 0.12	0.83 ± 0.11	0.85 ± 0.06	0.66 ± 0.09	0.83 ± 0.03
	Glu	3.10 ± 0.24	3.52 ± 0.19	2.62 ± 0.14	2.68 ± 0.12	2.11 ± 0.25	2.61 ± 0.19
	Gly	1.03 ± 0.12	1.20 ± 0.05	0.86 ± 0.04	0.88 ± 0.08	0.68 ± 0.07	0.86 ± 0.04
	Ala	1.02 ± 0.09	1.18 ± 0.16	0.84 ± 0.07	0.86 ± 0.07	0.67 ± 0.08	0.84 ± 0.04
	Cys	0.23 ± 0.01	0.28 ± 0.02	0.16 ± 0.02	0.18 ± 0.08	0.16 ± 0.05	0.20 ± 0.04
	Pro	0.64 ± 0.08	0.70 ± 0.03	0.52 ± 0.01	0.56 ± 0.08	0.45 ± 0.05	0.55 ± 0.04
	His	0.66 ± 0.05	0.77 ± 0.08	0.54 ± 0.04	0.55 ± 0.07	0.43 ± 0.05	0.55 ± 0.04
	Arg	2.10 ± 0.08	2.44 ± 0.07	1.74 ± 0.19	1.77 ± 0.09	1.39 ± 0.10	1.74 ± 0.11
	TAA	19.92 ± 0.08	23.00 ± 0.09	16.51 ± 0.08	16.90 ± 0.07	13.27 ± 0.09	16.60 ± 0.06

4. Conclusions

In this study, the effects of ultrasonic-assisted and traditional alkali-dissolving and acid-precipitation methods on the extraction of QPI were compared. The effect of heat treatment on the functional properties and in vitro digestion properties of QPI were studied. The results indicate that the factors influencing the extraction rate of QPI by ultrasound-assisted alkaline extraction method were solid-liquid ratio > ultrasonic time > ultrasonic temperature, in order of their influence. The optimal conditions for extraction were as follows: ultrasonic time 99 min, solid-liquid ratio 1:20 *w:v*, ultrasonic temperature 47 °C, and pH 10. Under these conditions, the extraction rate reached 74.67 ± 1.08% and the purity of QPI obtained was 87.17 ± 0.58%. In comparison with the traditional alkaline dissolution and acid precipitation method, the extraction rate and purity of QPI extracted by ultrasound-assisted method were increased by 10.18% and 5.49%, respectively. The pI of QPI is 4.5. Heat treatment had a significant effect on the ^{1}H NMR spectrum, turbidity and flexibility of QPI. Heat treatment changed the turbidity of QPI by affecting the 1.77 to 2.79 ppm region in the ^{1}H NMR spectrum of QPI. After heat treatment, the degree of hydrolysis and amino acid content of QPI in vitro digestion decreased. The results of this study provide a basis for processing and utilization of quinoa protein isolate.

Supplementary Materials: The following supporting information can be downloaded at: https://www.mdpi.com/article/10.3390/foods11050771/s1, Figure S1: The effects of time (A), solid-liquid ratio (B), temperature (C), pH (D) and the response surface methodology and contour plots for the effects of various factors (E, F, G) on the extraction rate of QPI. Different letters represent significant differences ($p < 0.05$). Figure S2: Effect of different heat treatment temperatures on 1H NMR integrated spectra of QPI.;Table S1: Codes and levels of factors for response surface methodology experiments. (Ultrasonic); Table S2: Central composite arrangement for independent variables A (Ultrasonic time, min), B (Solid-liquid ratio, g·mL^{-1}), C (Ultrasonic temperature, °C) and their response (QPI extraction rate, %); Table S3: Regression equation analysis of variance. (Ultrasonic); Table S4: Codes and levels of factors for response surface methodology experiments. (Alkali-solution and acid-isolation); Table S5: Central composite arrangement for independent variables A (Time, min), B (Solid-liquid ratio, g·mL^{-1}), C (Temperature, °C) and their response (QPI extraction rate, %); Table S6: Regression equation analysis of variance. (Alkali-solution and acid-isolation).

Author Contributions: Conceptualization, X.H.; methodology, X.H.; software, X.H.; validation, X.H.; formal analysis, B.W.; investigation, B.W.; resources, B.W.; data curation, X.H.; writing—original draft preparation, X.H.; writing—review and editing, B.W.; visualization, B.Z.; supervision, B.Z.; project administration, F.Y.; funding acquisition, F.Y. All authors have read and agreed to the published version of the manuscript.

Funding: This research was funded by Key Research and Development Project of Gansu Provincial Science and Technology Department grant number 18YF1NA076.

Institutional Review Board Statement: Not applicable.

Informed Consent Statement: Not applicable.

Data Availability Statement: No new data were created or analyzed in this study. Data sharing is not applicable to this article.

Conflicts of Interest: The authors declare no conflict of interest.

References

1. Ruiz, G.A.; Xiao, W.; Boekel, M.V.; Minor, M.; Stieger, M. Effect of extraction pH on heat-induced aggregation, gelation and microstructure of protein isolate from quinoa (*Chenopodium quinoa* Willd). *Food Chem.* **2016**, *209*, 203–210. [CrossRef]
2. Ceyhun, S.A.; Sanlier, N. A new generation plant for the conventional cuisine: Quinoa (*Chenopodium quinoa* Willd.). *Trends Food Sci. Technol.* **2019**, *86*, 51–58. [CrossRef]
3. Zhu, N.; Kikuzaki, H.; Vastano, B.; Nakatani, N.; Karwe, M.V.; Rosen, R.; Ho, C.T. Ecdysteroids of quinoa seeds (*Chenopodium quinoa* Willd.). *J. Agric. Food Chem.* **2001**, *49*, 2576–2578. [CrossRef]
4. Abugoch, L.E.; Tapia, C.; Villamán, M.C.; Yazdani-Pedram, M.; Díaz-Dosque, M. Characterization of quinoa protein–chitosan blend edible films. *Food Hydrocolloid.* **2011**, *25*, 879–886. [CrossRef]
5. Landi, N.; Roucco, M.R.; Ragucci, S.; Aliotta, F.; Maro, A.D. Quinoa as source of type 1 ribosome inactivating proteins: A novel knowledge for a revision of its consumption. *Food Chem.* **2020**, *342*, 128227. [CrossRef] [PubMed]
6. Ruiz, G.A.; Opazo-Navarrete, M.; Meurs, M.; Minor, M.; Sala, G.; Boekel, M.V.; Stieger, M.; Janssen, A.E.M. Denaturation and in vitro gastric digestion of heat-treated quinoa protein isolates obtained at various extraction pH. *Food Biophys.* **2016**, *11*, 184–197. [CrossRef] [PubMed]
7. Guerreo-Ochoa, M.R.; Pedreschi, R.; Chirinos, R. Optimised methodology for the extraction of protein from quinoa (*Chenopodium quinoa* Willd.). *Int. J. Food Sci. Technol.* **2015**, *50*, 1815–1822. [CrossRef]
8. Zou, Y.; Wang, L.; Li, P.; Cai, P.; Zhang, M.; Sun, Z.; Sun, C.; Geng, Z.; Xu, W.; Xu, X.; et al. Effects of ultrasound assisted extraction on the physiochemical, structural and functional characteristics of duck liver protein isolate. *Process Biochem.* **2016**, *52*, 174–182. [CrossRef]
9. Zhang, L.; Pan, Z.; Shen, K.; Cai, X.; Zheng, B.; Miao, S. Influence of ultrasound-assisted alkaline treatment on the structural properties and functionalities of rice protein. *J. Cereal. Sci.* **2018**, *79*, 204–209. [CrossRef]
10. Phongthai, S.; Lim, S.T.; Rawdkuen, S. Ultrasonic-assisted extraction of rice bran protein using response surface methodology. *J. Food Biochem.* **2017**, *41*, e12314. [CrossRef]
11. Li, F.; Kong, X.; Zhang, C.; Hua, Y. Effect of heat treatment on the properties of soy protein-stabilised emulsions. *Int. J. Food Sci. Technol.* **2011**, *46*, 1554–1560. [CrossRef]
12. Jiang, L.; Wang, J.; Li, Y.; Wang, Z.; Liang, J.; Wang, R.; Chen, Y.; Ma, W.; Qi, B.; Zhang, M. Effects of ultrasound on the structure and physical properties of black bean protein isolates. *Food Res. Int.* **2014**, *62*, 595–601. [CrossRef]
13. Cadesky, L.; Walkling-Ribeiro, M.; Kriner, K.T.; Karwe, M.V.; Moraru, C.I. Structural changes induced by high-pressure processing in micellar casein and milk protein concentrates. *J. Dairy. Sci.* **2017**, *100*, 7055–7070. [CrossRef]

14. Jiang, J.; Chen, J.; Xiong, Y.L. Structural and emulsifying properties of soy protein isolate subjected to acid and alkaline pH-shifting processes. *J. Agric. Food Chem.* **2009**, *57*, 7576–7583. [CrossRef]
15. Chen, J.; Mu, T.; Miao, Z.; Goffin, D. Effect of heat treatments on the structure and emulsifying properties of protein isolates from cumin seeds (cuminum cyminum). *Food Sci. Technol. Int.* **2018**, *24*, 108201321878875. [CrossRef]
16. Zhang, F.; Jiang, S.; Feng, X.; Wang, R.; Zeng, M.; Zhao, Y. Effect of heat treatment on physicochemical state and in vitro digestion of salt-soluble protein from pacific oyster (*Crassostrea gigas*). *LWT-Food Sci. Technol.* **2020**, *134*, 110126. [CrossRef]
17. Lertittikul, W.; Benjakul, S.; Tanaka, M. Characteristics and antioxidative activity of maillard reaction products from a porcine plasma protein-glucose model system as influenced by pH. *Food Chem.* **2007**, *100*, 669–677. [CrossRef]
18. Zhang, F.; Jiang, S.; Feng, X.; Wang, R.; Zhao, Y. Physicochemical state and in vitro digestibility of heat treated water-soluble protein from pacific oyster (*Crassostrea gigas*). *Food Biosci.* **2020**, *34*, 100528. [CrossRef]
19. Laguna, L.; Picouet, P.; Guàrdia, M.D.; Renard, C.; Sarkar, A.; Renard, C.M.G.C.; Sarkar, A. In vitro gastrointestinal digestion of pea protein isolate as a function of pH, food matrices, autoclaving, high-pressure and re-heat treatments. *LWT-Food Sci. Technol.* **2017**, *84*, 511–519. [CrossRef]
20. Ye, A.; Wang, X.; Lin, Q.; Han, J.; Singh, H. Dynamic gastric stability and in vitro lipid digestion of whey-protein-stabilised emulsions: Effect of heat treatment. *Food Chem.* **2020**, *318*, 126463. [CrossRef]
21. Rio, A.; Opazo-Navarrete, M.; Cepero-Betancourt, Y.; Tabilo-Munizaga, G.; Janssen, A. Heat-induced changes in microstructure of spray-dried plant protein isolates and its implications on in vitro gastric digestion. *LWT-Food Sci. Technol.* **2019**, *118*, 108795.
22. Silventoinen, P.; Sozer, N. Impact of ultrasound treatment and pH-shifting on physicochemical properties of protein-enriched barley fraction and barley protein isolate. *Foods* **2020**, *9*, 1055. [CrossRef]
23. Miller, E.L.; Bimbo, A.P.; Barlow, S.M.; SHeridan, B.; Burks, L.B.W.; Barrins, T.; Bassompierre, M.; Brodin, A.; Brunsgaard, G.; Henry, M.; et al. Repeatability and reproducibility of determination of the nitrogen content of fishmeal by the combustion (Dumas) method and comparison with the kjeldahl method: Interlaboratory study. *J Aoac Int.* **2007**, *15*, 90. [CrossRef]
24. Pedroche, J.; Yust, M.M.; Lqari, H.; Girón-Calle, J.; Alaiz, M.; Vioque, J.; Millán, F. Brassica carinata protein isolates: Chemical composition, protein characterization and improvement of functional properties by protein hydrolysis. *Food Chem.* **2004**, *88*, 337–346. [CrossRef]
25. Hall, A.E.; Moraru, C.I. Structure and function of pea, lentil and faba bean proteins treated by high pressure processing and heat treatment. *LWT-Food Sci. Technol.* **2021**, *152*, 112349. [CrossRef]
26. Kato, A.; Ibrahim, H.R.; Watanabe, H.; Honma, K.; Kobayashi, K. Structural and gelling properties of dry-heated egg white proteins. *J. Agric. Food Chem.* **1990**, *38*, 32–37. [CrossRef]
27. Kurganov, B.I. Kinetics of protein aggregation. quantitative estimation of the chaperone-like activity in test-systems based on suppression of protein aggregation. *Biochemistry* **2002**, *67*, 409–422.
28. Vidal, N.P.; Picone, G.; Goicoechea, E.; Laghi, L.; Manzanos, M.J.; Danesi, F.; Bordoni, A.; Capozzi, F.; Guillén, M.D. Metabolite release and protein hydrolysis during the in vitro digestion of cooked sea bass fillets. A study by ^{1}H NMR. *Food Res. Int.* **2016**, *88*, 293–301. [CrossRef]
29. Minekus, M.; Alminger, M.; Alvito, P.; Ballance, S.; Bohn, T. A standardised static in vitro digestion method suitable for food—An international consensus. *Food Funct.* **2014**, *5*, 1113–1124. [CrossRef]
30. Cho, S. Changes in the antioxidant properties of rice bran protein isolate upon simulated gastrointestinal digestion. *LWT-Food Sci. Technol.* **2020**, *126*, 109206. [CrossRef]
31. Nielsen, P.M.; Petersen, D.; Dambmann, C. Improved method for determining food protein degree of hydrolysis. *J. Food Sci.* **2001**, *66*, 642–646. [CrossRef]
32. Machado, S.; Costa, A.; Filipa, P.B.; Oliveira, M.; Alves, R.C. A study on the protein fraction of coffee silverskin: Protein/non-protein nitrogen and free and total amino acid profiles. *Food Chem.* **2020**, *326*, 126940. [CrossRef] [PubMed]
33. Wang, B.; Atungulu, G.G.; Khir, R.; Geng, J.J.; Ma, H.L.; Li, Y.L. Ultrasonic treatment effect on enzymolysis kinetics and activities of ACE-inhibitorry peptides from oat-isolated protein. *Food Biophys.* **2014**, *10*, 244–252. [CrossRef]
34. Sun, X.H.; Zhu, K.X.; Zhou, H.M. Optimization of a novel backward extraction of defatted wheat germ protein from reverse micelles. *Innov. Food Sci. Emerg.* **2009**, *10*, 328–333. [CrossRef]
35. Hammi, K.M.; Jdey, A.; Abdelly, C.; Majdoub, H.; Ksouri, R. Optimization of ultrasound-assisted extraction of antioxidant compounds from Tunisian *Zizyphus lotus* fruits using response surface methodology. *Food Chem.* **2015**, *184*, 80–89. [CrossRef]
36. Elsohaimya, S.A.; Refaaya, T.M.; Zaytounb, M.A.M. Physicochemical and functional properties of quinoa protein isolate. *Annal. Agric. Sci.* **2015**, *60*, 297–305. [CrossRef]
37. Chittapalo, T.; Noomhorm, A. Ultrasonic assisted alkaline extraction of protein from defatted rice bran and properties of the protein concentrates. *Int. J. Food Sci. Technol.* **2009**, *44*, 1843–1849. [CrossRef]
38. Li, H.; Chen, B.; Yao, S. Application of ultrasonic technique for extracting chlorogenic acid from *Eucommia ulmodies* Oliv. (*E. ulmodies*). *Ultrason. Sonochem.* **2005**, *12*, 295–300. [CrossRef]
39. Li, M.; Ni, Y.; Peng, Y.; Wen, X.; Wang, Y. Optimization of ultrasound-assisted extraction of capsicum seed protein isolate. *Trans. Chin. Soc. Agric. Eng.* **2016**, *32*, 309–314.
40. Mir, N.A.; Riar, C.S.; Singh, S. Effect of pH and holding time on the characteristics of protein isolates from *Chenopodium* seeds and study of their amino acid profile and scoring. *Food Chem.* **2019**, *272*, 165–173. [CrossRef]

41. Tang, C. Emulsifying properties of soy proteins: A critical review with emphasis on the role of conformational flexibility. *Crit. Rev. Food Sci. Nutr.* **2017**, *57*, 2636–2679. [CrossRef]
42. Tang, C.; Shen, L. Role of conformational flexibility in the emulsifying properties of bovine serum albumin. *J. Agric. Food Chem.* **2013**, *61*, 3097–3110. [CrossRef]
43. Liu, L.; Zeng, J.; Sun, B.; Zhang, N.; Zhu, X. Ultrasound-assisted mild heating treatment improves the emulsifying properties of 11s globulins. *Molecules* **2020**, *25*, 875. [CrossRef]
44. Freer, E.M.; Yim, K.S.; Fuller, G.; Radke, C.J. Interfacial rheology of globular and flexible proteins at the hexadecane/water interface: Comparison of shear and dilatation deformation. *J. Phys. Chem. B* **2004**, *108*, 3835–3844. [CrossRef]
45. Raikos, V.; Duthie, G.; Ranawana, V. Denaturation and oxidative stability of hemp seed (*Cannabis sativa* L.) protein isolate as affected by heat treatment. *Plant. Food Hum. Nutr.* **2015**, *70*, 304–309. [CrossRef]
46. Mohan, M.; Ramachandran, D.; Sankar, T.V.; Anandan, R. Physicochemical characterization of muscle proteins from different regions of mackerel (*Rastrelliger kanagurta*). *Food Chem.* **2008**, *106*, 451–457. [CrossRef]
47. Joyce, A.M.; Kelly, A.L.; O'Mahony, J.A. Controlling denaturation and aggregation of whey proteins during thermal processing by modifying temperature and calcium concentration. *Int. J. Dairy. Technol.* **2018**, *71*, 915. [CrossRef]
48. Cromwell, M.; Hilario, E.; Jacobson, F. Protein aggregation and bioprocessing. *Aaps. J.* **2006**, *8*, E572–E579. [CrossRef]
49. Cai, L.; Feng, J.; Cao, A.; Zhang, Y.; Lv, Y.; Li, J. Denaturation kinetics and aggregation mechanism of the sarcoplasmic and myofibril proteins from grass carp during microwave processing. *Food Bioprocess. Technol.* **2017**, *11*, 417–426. [CrossRef]
50. Li, Y.; Kong, B.; Xia, X.; Liu, Q.; Diao, X. Structural changes of the myofibrillar proteins in common carp (*Cyprinus carpio*) muscle exposed to a hydroxyl radical-generating system. *Process. Biochem.* **2013**, *48*, 863–870. [CrossRef]
51. Kruif, C.G.D.; Tuinier, R.; Holt, C.; Timmins, P.A.; Rollema, H.S. Physicochemical study of κ- and β-casein dispersions and the effect of cross-linking by transglutaminase. *Langmuir* **2002**, *18*, 4885–4891. [CrossRef]
52. Fulmer, G.R.; Miller, A.; Sherden, N.H.; Gottlieb, H.E.; Nudelman, A.; Stoltz, B.M.; Bercaw, J.E.; Goldberg, K.I. NMR chemical shifts of trace impurities: Common laboratory solvents, organics, and gases in deuterated solvents relevant to the organometallic chemist. *Organometallics* **2010**, *29*, 2176–2179. [CrossRef]
53. Franck, M.; Perreault, V.; Suwal, S.; Marciniak, A.; Bazinet, L.; Doyen, A. High hydrostatic pressure-assisted enzymatic hydrolysis improved protein digestion of flaxseed protein isolate and generation of peptides with antioxidant activity. *Food Res. Int.* **2019**, *115*, 467–473. [CrossRef]
54. Tian, T.; Teng, F.; Zhang, S.; Qi, B.; Wu, C.; Zhou, Y.; Li, L.; Wang, Z.; Li, Y. A study of structural change during in vitro digestion of heated soy protein isolates. *Foods* **2019**, *8*, 594. [CrossRef]
55. Zhao, R.; Bean, S.R.; Ioerger, B.R.; Wang, D.H.; Boyle, D.L. Impact of mashing on sorghum proteins and its relationship to ethanol fermentation. *J. Agric. Food Chem.* **2008**, *56*, 946–953. [CrossRef]
56. Sangsawad, P.; Kiatsongchai, R.; Chitsomboon, B.; Yongsawatdigul, J. Chemical and cellular antioxidant activities of chicken breast muscle subjected to various thermal treatments followed by simulated gastrointestinal digestion. *J. Food Sci.* **2016**, *81*, 2431–2438. [CrossRef]
57. Gámez, C.; Zafra, M.P.; Sanz, V.; Mazzeo, C.; Ibáñez, M.D.; Sastre, J.; Pozo, D.V. Simulated gastrointestinal digestion reduces the allergic reactivity of shrimp extract proteins and tropomyosin. *Food Chem.* **2015**, *173*, 475–481. [CrossRef]
58. Shi, R.; Chen, Z.; Fan, W.; Chang, M.; Meng, J.; Liu, J.; Feng, C. Research on the physicochemical and digestive properties of *Pleurotus eryngii* protein. *Int. J. Food Prop.* **2018**, *21*, 2785–2806. [CrossRef]
59. Xiang, S.; Zou, H.; Liu, Y.; Ruan, R. Effects of microwave heating on the protein structure, digestion properties and maillard products of gluten. *J. Food Sci. Technol.* **2020**, *57*, 2139–2149. [CrossRef]
60. Giménez, B.; Alemán., A.; Montero., P.; Gómez-Guillén., M.C. Antioxidant and functional properties of gelatin hydrolysates obtained from skin of sole and squid. *Food Chem.* **2009**, *114*, 976–983. [CrossRef]
61. Zhang, Y.; Zhou, F.; Zhao, M.; Ning, Z.; Sun-Waterhouse, D.; Sun, B. Soy peptide aggregates formed during hydrolysis reduced protein extraction without decreasing their nutritional value. *Food Funct.* **2017**, *8*, 4384–4395. [CrossRef] [PubMed]

foods MDPI

Article

Disintegrating the Structure and Improving the Functionalities of Pea Fiber by Industry-Scale Microfluidizer System

Xiaohong He [1], Taotao Dai [2], Jian Sun [2], Ruihong Liang [1], Wei Liu [1], Mingshun Chen [1], Jun Chen [1] and Chengmei Liu [1,*]

1 State Key Laboratory of Food Science and Technology, Nanchang University, Nanchang 330047, China; hexiaohongmaimai@163.com (X.H.); liangruihong@ncu.edu.cn (R.L.); liuwei@ncu.edu.cn (W.L.); chenshun1221@163.com (M.C.); chen-jun1986@hotmail.com (J.C.)
2 Agro-Food Science and Technology Research Institute, Guangxi Academy of Agricultural Sciences, Nanning 530007, China; ncubamboo@163.com (T.D.); jiansun@gxaas.net (J.S.)
* Correspondence: liuchengmei@ncu.edu.cn

Abstract: In the food industry, the most prominent and concerned points in the application of dietary fiber are hydration properties and oil absorption capacity. The target of this work was to investigate the impact of a novel industry-scale microfluidizer system (ISMS) on the changing structures and functionalities of pea fiber. Different ISMS treatment intensity (0–120 MPa for one pass and 120 MPa for two passes) was applied to treat pea fiber. ISMS treatment induced the reduction in particle size and the transformation of big compact blocks to loose flakes, and the destruction of the original ordered cellulose structure caused the decline of crystallinity. Meanwhile, the hydration properties of pea fiber were improved, and pre-pulverizer and industry-scale microfluidizer treatment together increased the swelling capacity and water retention capacity of fiber. The oil holding capacity of ISMS-treated fiber was increased to more than double the original one. The elevated functionalities of pea fiber by ISMS treatment could be attributed to loosening structure, exposing more surface area, and disordering the crystalline structure, which increased the sites of water binding and oil adsorption. These findings suggested that ISMS could be applied as an effective industrial technique to the disintegrate structure and improve the functionalities of pea fiber, so as to widen the application of pea fibers in foods.

Keywords: pea fiber; industry-scale microfluidizer; structure; functional properties

Citation: He, X.; Dai, T.; Sun, J.; Liang, R.; Liu, W.; Chen, M.; Chen, J.; Liu, C. Disintegrating the Structure and Improving the Functionalities of Pea Fiber by Industry-Scale Microfluidizer System. *Foods* **2022**, *11*, 418. https://doi.org/10.3390/foods11030418

Academic Editor: Gian Carlo Tenore

Received: 23 December 2021
Accepted: 29 January 2022
Published: 31 January 2022

1. Introduction

The appropriate consumption of dietary fibre has considered to be advantageous to health, such as by preserving gastrointestinal function, lowering blood lipids and cholesterol levels, and reducing the risk of cardiovascular disease [1], so we were encouraged to seek excellent fiber sources for the development of foods supplemented with dietary fibers. Pea fiber is a valuable and attractive food component which has been added to pasta for improving the nutritional value [2]. Commonly, the incorporation of raw pea fiber would negatively impact the sensory characteristics of the food matrix owing to its poor techno-functional properties. One of the areas of focus among researchers and in the field of functional food processing is the modification of fibers to increase their quality or functional properties by adopting physical, chemical and biological approaches [3]. In recent years, superfine pulverization techniques have gained much attention for their modifying properties of fiber. For instance, pulverization methods such as homogenization, microfluidization and ultrafine comminution have been reported to modify fibers from citrus [4,5], purple-fleshed potatoes [6], bamboo shoot shells [7] and carrot pomace [8]. Furthermore, microfluidization showed great potential in modifying fibers such as wheat and corn bran, peach and oat fiber, insoluble soybean fiber and hazelnut skin fiber, as reviewed by Guo, et al. [9] and Ozturk and Turasan [10]. Morales-Medina, et al. [11] also found a continuous defibrillation of pea hull fiber by microfluidization with the decreasing of particle size. Nevertheless, previous reports were mainly concentrated on treating fibers by small scale microfluidizers. As pointed out by Ozturk and Turasan [10], one of the biggest problems with this technology was the limitation of scale, and additional supporting equipment like a pretreatment miller was necessary if an industry-scale microfluidizer was provided. In our research group, an innovative industry-scale microfluidizer system (ISMS) was developed which combined a pre-pulverizer and an industry-scale microfluidizer (ISM). The schematic diagram and the main functions of the pre-pulverizer and industry-scale microfluidizer were demonstrated in our previous research in greater detail [12–14]. The industry-scale microfluidizer possesses a unique constructional microchannel with large orifice diameters, unusual impact modes and a high processing capacity which could continuously run to reach productivity of five tons per hour. ISMS has been successfully applied to produce stable whole soybean milk and improve the stability of whole corn slurry without filtering and removing any components. This indirectly reflected that the soybean and corn fibers in the whole component systems were modified. With regard to the low utilization of pea fiber, it was worth investigating the effect of ISMS treatment on its structural and functional properties. Ascertaining the efficiency and ability of ISMS to modify pea fiber can provide the possibility for its high-value utilization.

Therefore, the objective of the current work was to investigate the efficiency of ISMS modifying pea fiber. Pea fiber was treated by ISMS at a different intensity (60, 90, 120 MPa for one pass and 120 MPa for two passes after pre-pulverizer treatment). Subsequently, the structure and functionalities of ISMS-treated pea fiber (ISMS-treated PeaF) were determined. The structural properties were characterized by particle size distribution, confocal laser scanning microscopy (CLSM), scanning electron microscopy (SEM), and X-ray diffraction (XRD) analysis. The functionalities including swelling capacity (SC), water retention capacity (WRC) and oil holding capacity (OHC) of ISMS-treated PeaF were discussed. This study may identify a new technology that can be utilized to manufacture nutritious foods with high fiber content, thereby promoting the development of new food products.

2. Materials and Methods

2.1. Material

Pea fiber powder was provided by Shuangta Food Co., Ltd (Zhaoyuan, China), which contained 80.7% of total dietary fibre. The protein, starch, and ashes accounted for a much lower percentage (0.3, 3.6 and 3.4%, respectively). Calcofluor white was purchased from Aladdin Reagent Company (Shanghai, China). Corn oil was purchased from a local supermarket (Jinlongyu, Shanghai, China). All other chemical agents were of analytical grade.

2.2. ISMS Treatment of Pea Fiber

Native pea fiber (native PeaF) suspensions were prepared by mixing pea fiber powder in distilled water using a stirrer for overnight at approximately 500 rpm to completely hydrate, and a concentration of 1.2% (w/w) was selected according to the investigation of Bruno, et al. [5]. Completely hydrated pea fiber suspensions were added directly into the pre-pulverizer, then the pea fiber suspensions obtained from the pre-pulverizer were treated by ISM for one pass at 60, 90, and 120 MPa and two passes at 120 MPa in series. The temperature of the treated pea fiber suspensions (ISMS-treated PeaF) was less than 54 °C, and after ISM treatment, the ISMS-treated PeaF suspensions were immediately cooled to room temperature using an ice bath. Pea fiber treated by ISM at the corresponding intensity was designated as ISM-60 PeaF, ISM-90 PeaF, ISM-120 PeaF and ISM-120-T_2 PeaF, respectively. The fiber treated by pre-pulverizer but without ISM treatment was labeled as Pre-PeaF. Part of the suspensions were withdrawn to perform particle size distributions and CLSM analysis. Other suspension samples were freeze-dried and ground into powders. The suspensions were first frozen at −80 °C overnight, and then freeze-dried using the FreeZone® 4.5 L freeze-drier (Labconco, Kansas City, MO, USA) at low vacuum (≤10 bar) and temperature (~−40 °C). The freeze-dried PeaF was ground into power using a basic analytical mill (IKA® A11B S025, Merck KGaA, Darmstadt, Germany) by intermittent operating for dozens of seconds. The obtained powders were stored in a desiccator for further analysis.

2.3. Particle Size

The particle size distributions and mean diameters of pea fibers were determined (expressed in % volume) using a Malvern MasterSizer 3000 (Malvern Instrument, Ltd., Malvern, UK) by referencing the method of Chen, et al. [15] with slight modifications. This technique provides information on the equivalent sphere diameter of fiber particles with different geometrical shapes, and the term "diameter" will be used to substitute for "equivalent sphere diameter" for simplicity. The measurement was conducted with the refractive indices of 1.52 and 1.33 for pea fiber and water, respectively, and an absorption index of 0.1 was used. The mean diameters including volume weighted mean diameter $D_{[4,3]}$ (diameter of the sphere of equivalent volume to measured particles) and surface-weighted mean diameter $D_{[3,2]}$ (particle diameter that has the same specific surface as that of the full distribution) were evaluated. Cumulative percentiles of D_{10}, D_{50} and D_{90} were also calculated, which indicated that the size of 10%, 50% and 90% of the particles was below the specified diameter, respectively. The pan of particle size distributions was applied to characterize the width of the particle size distribution, which was calculated according to Equation (1):

$$\text{Span} = \frac{D_{90} - D_{10}}{D_{50}} \quad (1)$$

2.4. Confocal Laser Scanning Microscopy (CLSM) Analysis

The microstructure of pea fiber was observed using confocal laser scanning microscopy (Carl Zeiss LSM710, Jena, Germany). The preparation of samples were conducted by referencing the method of Huang, et al. [16]. Briefly, pea fibers were mixed with calcofluor white dye at a 50:1 volume ratio, and then the stained sample was added to a culture dish for observation. Images were acquired using a CLSM multiphoton system with a 40× objective lens. The excitation-emission wavelengths of 405–455 nm for calcofluor white and appropriate emission channels were used.

2.5. Scanning Electron Microscopy (SEM) Analysis

The morphology of pea fiber samples after freeze-drying and grinding was also examined using an environmental scanning electron microscope (Quanta-200, FEI Company, Eindhoven, The Netherlands). ISM-treated PeaF were prepared by sticking them onto double-sided adhesive tape attached to a circular specimen stub, following by sputtering a thin film of gold. The morphology of samples was observed at ×800 magnifications with an accelerating voltage of 5 kV voltage.

2.6. Bulk Density Analysis

1.8 g of pea fiber powder was accurately weighed and carefully added into a calibrated 25 mL graduated cylinder. Pressure was imposed manually to ensure no further decrease in sample volume. The bulk density was calculated as the volume of sample occupied by per gram dry weight (mL/g) [17].

2.7. X-ray Diffraction (XRD) Analysis

X-ray diffractograms of pea fiber powders after freeze-drying and grinding were obtained using an X-ray diffractometer (D8 Advance, Bruker, Berlin, Germany) operated at 40 kV and 40 mA with Cu Kα radiation. Before measurements, samples were stored in a desiccator where a saturated solution of NaCl maintained a constant humidity atmosphere (relative humidity = 75%) at 25 °C for five days. The scanning angle (2θ) of 5–50° with the interval of 0.02° was used to obtain XRD patterns. The crystallinity of samples was calculated based on the ratio of the area of the crystalline region to the total area in the XRD spectra using Origin software [18].

2.8. Measurement of Hydration Properties

Hydration properties including swelling capacity (SC) and water-retention capacity (WRC) were determined.

Swelling capacity was determined by referencing the method of Mateos-Aparicio, et al. [19]. Accurately weighed pea fiber (0.3 ± 0.001 g) was added into a 25 mL graduated cylinder containing 15 mL of distilled water. The sample was stirred gently, then left undisturbed at room temperature for 8 h to completely hydrate. The volume (mL) of the settled sample was recorded, and SC was expressed as the volume of the settled sample (mL) per gram of dry fiber.

Water retention capacity was measured by referencing the method of Morales-Medina, Dong, Schalow and Drusch [11], with some modifications. Accurately weighed pea fiber (0.3 ± 0.001 g) and 15 mL of distilled water were added in a 50 mL centrifuge tube, and the sample was stirred and allowed to hydrate at room temperature for 8 h. Subsequently, the fibre suspension was centrifuged at $1790\times\ g$ for 15 min, and the supernatant of each tube was carefully decanted. The excess of liquid was drained by turning the tubes upside down on a filter paper for several minutes. The weight of the hydrated sample (m_1) was recorded. Then the hydrated sample was freeze-dried, and the weight of the dried sample was labeled as m_2. WRC was calculated according to Equation (2):

$$\mathrm{WRC}(g/g) = \frac{m_1 - m_2}{m_1} \tag{2}$$

2.9. Measurement of Oil Holding Capacity

Oil holding capacity (OHC) was measured following a modified procedure by referencing the report of Jiang, et al. [20] and Meng, et al. [21]. Pea fiber powder (0.2 ± 0.001 g) and oil (10 g $\pm$ 0.01 g) were put into a 50 mL centrifuge tube and fully mixed by vortex mixer for several minutes. Then the centrifuge tube was kept at room temperature for 4 h. Subsequently, the fibre suspension was centrifuged at $1790\times\ g$ for 15 min, and the upper clear liquid was poured out gently. The weight of the pea fiber after absorbing oil was recorded. The OHC of pea fibers was calculated by Equation (3):

$$\mathrm{OHC}(g/g) = \frac{m_{oiled} - m}{m} \tag{3}$$

where m was the weight of the original ISMS-treated PeaF powder (0.2 g), and m_{oiled} was the weight of the ISMS-treated PeaF after absorbing oil (g).

2.10. Statistical Analysis

All experiments were carried out in triplicate using three samples, and then the mean and standard deviation were calculated by statistical analysis software (SPSS 25.0, SPSS Inc., Chicago, IL, United States). Significant differences between sample means ($p < 0.05$) were established according to Duncan's test using one-way analysis of variance (ANOVA).

3. Results and Discussion

3.1. Particle Size Characteristics

Initially, the influence of the ISMS treatment on the particle characteristics of pea fiber was examined. Particle size distributions for native PeaF and ISMS-treated PeaF aqueous suspensions are shown in Figure 1. Native PeaF aqueous suspensions exhibited a wide and asymmetric unimodal distribution (1.4~374.8 μm) with a shoulder around 8.1~37.7 μm. The pre-pulverizer treatment weakened the shape of the shoulder and slightly narrowed the particle size distribution. After ISM treatment, peaks of the distributions tended to be homogeneously distributed. Meanwhile, a progressive shift of the peaks to the left for ISMS-treated PeaF aqueous suspensions was observed as the treatment intensity increased, and the distribution gradually narrowed. Additionally, when ISM pressure was below 120 MPa, the peaks moved downward with the increasing of ISM pressure. Although the particle size distribution of ISM-120-T_2 PeaF was located at the leftmost, it was not significantly deviated from that of ISM-120 PeaF. These phenomena indicated that the increasing treatment pressure resulted in a decrease in the overall particle size, and ISMS treatment could grind the pea fiber to a micron size to a limited extent. As displayed in Table 1, all mean diameters analyzed decreased with increasing ISMS treatment intensity, and similar results had been found in another study of treating soybean insoluble dietary fiber by high-energy wet media milling [22]. For example, along with the treatment intensity, $D_{[4,3]}$ values decreased from 92.6 μm of native PeaF to 38.3 μm of ISM-120-T_2 PeaF. Furthermore, $D_{[3,2]}$ values of pea fiber significantly dropped from 34.5 μm to 19.3 μm, and those of ISM-60 PeaF, ISM-90 PeaF, and ISM-120 PeaF were 23.5, 20.4 and 19.4 μm, respectively. Bruno, et al. [5] reported that $D_{[3,2]}$ values of citrus fiber were approximately 26 μm and 20 μm after treatment by a M110P microfluidizer® at higher pressure for one pass (103.3 MPa and 172.2 MPa, respectively). The cumulative percentiles of $D_{(50)}$ and $D_{(90)}$ were also reduced to varying degrees. When pressures of ISM were 90 MPa and 120 MPa, $D_{(90)}$ the values of pea fiber were 93.2 μm and 82.5 μm, respectively. In the study of Morales-Medina, et al. [11], when $D_{(90)}$ values of pea hull fiber reached 100 μm and 80 μm, the conditions of treatment by the LM20 Microfluidizer® were predicted to be 109 MPa for two passes and 127 MPa for four passes, respectively. Meanwhile, $D_{(50)}$ values of pea hull fiber processed by the aforementioned conditions were slightly higher than these of ISM-90 PeaF and ISM-120 PeaF in this investigation. These phenomena indicated that ISM was a more powerful technique than conventional microfluidizers at disrupting fiber to a smaller particle size. Intriguingly, ISMS treatment did not cause a decrease in span, which was contrary to the observation of narrowing distributions from Figure 1. As depicted in Figure 1, the size range of native PeaF was actually larger than that of ISMS-treated PeaF, and native PeaF were mainly large size particles with concentrated distribution. Nevertheless, ISMS-treated PeaF possessed more homogeneous distribution with a high fraction of small sized particles and a slightly small size range. The concentrated distribution of large size particles and very low fraction of small size particles contributed to a lower span for native PeaF. As considered by Guo, et al. [12], the reduced particle size of pea fiber was possibly attributed to mechanical action initiated by ISMS, and higher processing strength provided a greater crushing effect. Meanwhile, the specific surface area of pea fiber was increased with the increasing of ISMS treatment intensity, which climbed from 173.6 m^2/kg to 309.4 m^2/kg when ISM pressure rose to 120 MPa (Table 1). The reduction in particle size and increase in particle specific surface area initiated by the crushing effect inevitably destroyed the structure of pea fiber. Therefore, the structure of ISMS-treated PeaF will be investigated in the next sections.

Table 1. Particle diameter size of ISMS-treated PeaF [1].

Samples	Specific Surface Area (m^2/kg)	$D_{[3,2]}$ (μm)	$D_{[4,3]}$ (μm)	$D_{(50)}$ (μm)	$D_{(90)}$ (μm)	Span
native PeaF	173.6 ± 1.7e	34.5 ± 0.4a	92.6 ± 0.5a	82.8 ± 1.2a	185.0 ± 1.4a	2.03 ± 0.05b
Pre-PeaF	197.3 ± 3.1d	30.4 ± 0.4b	72.7 ± 3.3b	64.2 ± 1.0b	145.0 ± 9.9b	2.03 ± 0.13b
ISM-60 PeaF	256.6 ± 0.1c	23.5 ± 0.1c	52.9 ± 0.4c	43.1 ± 0.1c	109.0 ± 1.4c	2.26 ± 0.03a
ISM-90 PeaF	293.4 ± 0.4b	20.4 ± 0.1d	45.0 ± 0.1d	36.1 ± 0.1d	93.2 ± 0.14d	2.31 ± 0.01a
ISM-120 PeaF	309.4 ± 1.3a	19.4 ± 0.1e	40.5 ± 0.4e	32.7 ± 0.1e	82.5 ± 1.0e	2.23 ± 0.02a
ISM-T_2-120 PeaF	310.0 ± 2.8a	19.3 ± 0.2e	38.3 ± 0.6e	32.1 ± 0.3e	76.2 ± 1.3e	2.08 ± 0.01b

[1] Reported results correspond to mean ± standard deviation. Different letters within the same column indicate significant differences ($p < 0.05$).

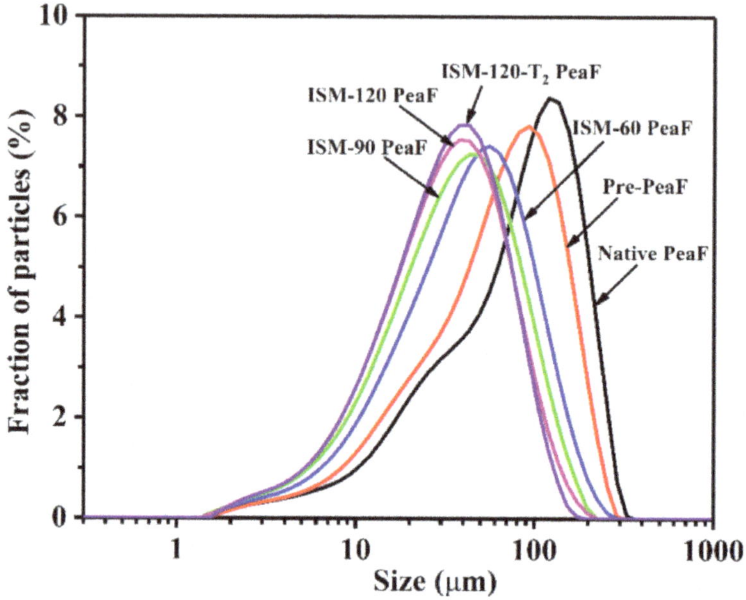

Figure 1. Particle size distributions of ISMS-treated PeaF.

3.2. CLSM

The microstructure of ISMS-treated PeaF suspensions was first visualized by means of CLSM to analyze the alterations of pea fiber induced by ISMS treatment, and blue fluorescence was observed, as shown in Figure 2A–F. Native PeaF revealed the predominance of a relatively big thick and compact structure, where the centre of some fibers presented weaker fluorescence than their well-defined edges, as shown by arrows in Figure 2A. In accordance with the results of particle size, pre-pulverizer treatment broke the fibers into relatively small structures with bright fluorescence (Figure 2B). Upon ISM treatment, the dense fiber seemed to be disrupted into loose fibers with the increasing of treatment intensity according to the significant change of blue fluorescence. ISM-60 PeaF showed diverse shapes of fluorescence with some strips as depicted in Figure 2C. When ISM pressure reached 90 MPa, the fluorescence intensity of fiber was weakened. ISM-90 PeaF presented a small flake-like structure with weak fluorescence intensity (Figure 2D). The further increasing of the treatment intensity led to a greater degree of tearing fiber, and ISM-120 PeaF (Figure 2E) and ISM-120-T$_2$ PeaF (Figure 2F) displayed more small fragments with faint fluorescence and multiform shape. Overall, as the intensity of the ISMS treatment increased, the fluorescence of ISMS-treated PeaF was weakened and blurred, and the area of fluorescence for most fibers significantly diminished. It was implied that the ISMS treatment disrupted the thick and compact block-like pea fibers into small loose fragments. As pictured in Figure 2G, ISMS-treated PeaF suspensions exhibited a larger sediment volume of fiber at the bottom of the centrifugal tube as the treatment intensity increased, since fibers with smaller and looser structure were more difficult to sink lower to the bottom. These results indicated that the ISMS treatment did cause serious damage to the pea fiber, which confirmed the observation of CLSM images.

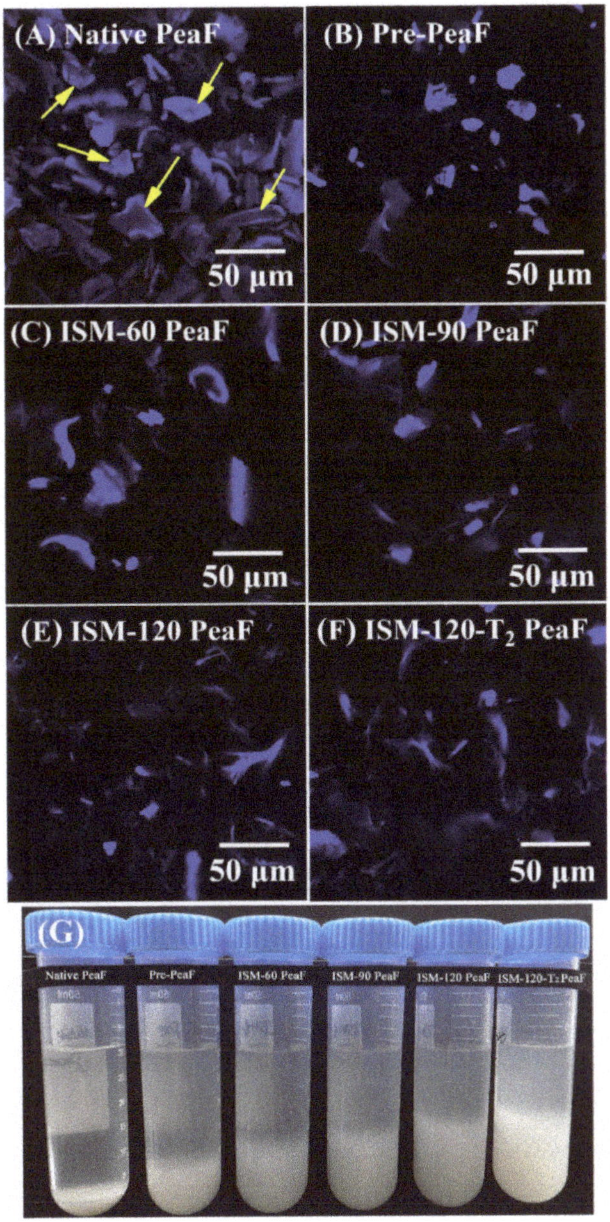

Figure 2. CLSM images and appearance of native PeaF and ISMS-treated PeaF. (**A**) Native PeaF; (**B**) Pre-PeaF; (**C**) ISM-60 PeaF; (**D**) ISM-90 PeaF; (**E**) ISM-120 PeaF; (**F**) ISM-120-T_2 PeaF; (**G**) Appearance of native peaF and ISMS-treated PeaF.

3.3. Morphology

SEM micrographs of pea fibers before and after ISMS treatments are displayed in Figure 3. Native PeaF displayed compact and thick blocks, and some fiber blocks with a smooth surface were distributed (Figure 3A). ISMS treatment could effectively alter the microstructure of the pea fiber, and compared to native PeaF, the structure of the ISMS-treated PeaF became looser and thinner. There was a flattened structure with some band-like sheets at the edge for Pre-PeaF (Figure 3B). After ISM treatment at low pressure, the fibers were highly distorted, and ISM-60 PeaF was tightly packed into a flat plate with a large surface (Figure 3C). When the treatment pressure reached 90 MPa, the fibers were torn into flat lamellas with interwoven filaments (Figure 3D). Furthermore, the appearance of a flimsy and crimped flake-like structure was observed for ISM-120 PeaF (Figure 3E) and ISM-120-T_2 PeaF (Figure 3F), implying that a greater degree of breaking pea fiber was induced by high treatment intensity. The change of SEM morphology was more clear than the observation of CLSM images to illustrate the effect of ISMS treatment on the pea fiber, further implying that mechanical effect initiated by ISM could destroy and break the architecture of the pea fiber. High-density energy could be produced by ISM treatment owing to powerful shear, turbulence, high-velocity impaction, high-frequency vibration, instantaneous pressure drop and cavitation forces which contributed to disrupt the pea fiber. As ISM treatment intensity increased, different degrees of damage and disruption forms of the fiber were observed. It was possible that the varied strength of mechanical forces exerted an effect on tearing the fiber when different treatment intensities were imposed. The appearance of a multi-branched flake-like structure most likely provided an explanation for the changes in the absorption properties of the fiber.

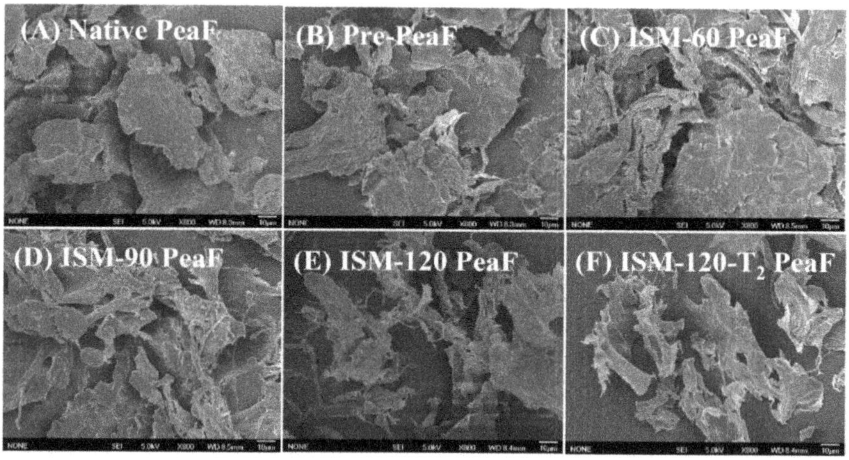

Figure 3. SEM micrographs. Magnification was 800×.

3.4. Bulk Density

The visual image of the packed native PeaF and ISMS-treated PeaF powers was shown in Figure 4A. ISMS treatment observably increased the packed volume of pea fiber in varying degrees in light of the treatment intensity. Figure 4B illustrated bulk density values of fibers versus the extent of ISMS treatment. The bulk density of native PeaF was 2.20 mL/g, and pre-pulverizer treatment increased the bulk density of fiber by about 3.07 fold. While the fiber was subjected to ISM treatment at 60 MPa, the bulk density of fiber rose, reaching 8.48 mL/g. Intriguingly, in comparison with ISM-60 PeaF, applying an ISM pressure of 90 MPa resulted in a smaller increase in bulk density (6.79 mL/g). Moreover, subsequent strengthening treatment induced a continuous increase in the bulk density, which was 7.59 and 8.21 mL/g for ISM-120 PeaF and ISM-120-T_2 PeaF, respectively. Indeed, the bulk density of fiber was related to their morphological characteristics, as mentioned by Wang, Sun, Zhou and Chen [17] and Wang, et al. [23]. It was universally acknowledged that the compact structure occupied a small space, thus native PeaF with compact and thick blocks had a low value of bulk density. As revealed in Figure 3, ISM-60 PeaF was a large dimensional flat plate in a distorted and stacked state, which produced its high bulk density. With regard to pea fibers treated by high intensity, there was more compressible space between fibers with loose flakes, thus ISM-90, ISM-120 PeaF and ISM-120-T_2 PeaF behaved at a lower value of bulk density than the ISM-60 PeaF.

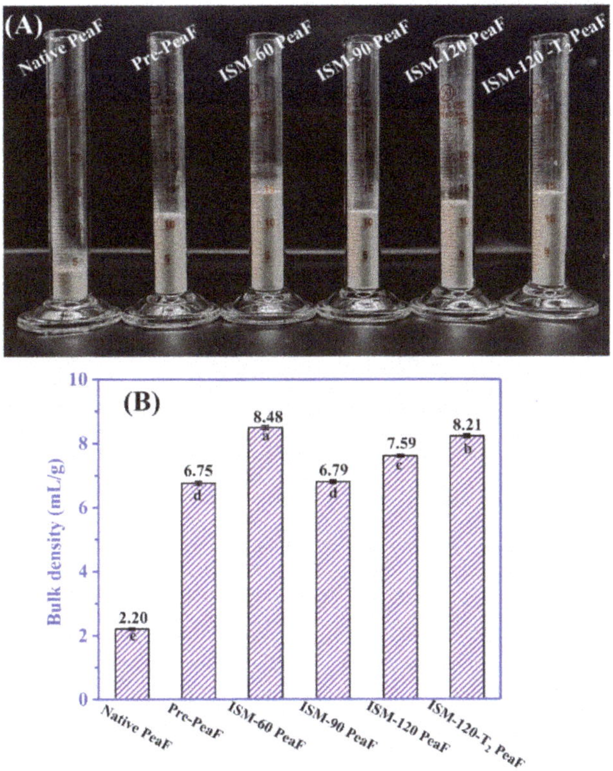

Figure 4. The graph of appearance volume (**A**) and bulk density (**B**) of ISMS-treated PeaF. Different letters in (**B**) indicated significant differences ($p < 0.05$) of bulk density between samples.

3.5. XRD Analysis

The fiber was mainly composed of cellulose, lignin and hemicellulose, and the crystalline structure was dominated by cellulose [24]. The X-ray patterns and crystallinity of native PeaF and ISMS-treated PeaF are determined and displayed in Figure 5. There was one strong diffraction peak near the 2θ diffraction angle of 22.40° and two weak ones at about 16.5° and 35.2° in the XRD pattern of native PeaF. It was indicated that pea fiber belonged to cellulose type I, where both crystalline and amorphous regions coexist [25]. No significant difference in the peak position between native PeaF and ISMS-treated PeaF was observed. Nevertheless, the intensity of the peak near 2θ at 16.5° was visibly attenuated, and the peak near 2θ at 35.2° even disappeared with the increasing of ISMS treatment intensity, implying that the crystalline structure of pea fiber was perturbed. The crystallinity of fiber was gradually decreased from 28.60% to 21.99% along with the treatment intensity, which also suggested that the fiber was decrystallized after ISMS treatment. The weakened crystalline structure was accompanied with a reduction in the crystallinity, and this is most likely because of the destruction of the original ordered cellulose structure as analyzed by Sun, et al. [26]. Decrystallization could activate the cellulosic fiber for functionalization [18]. Therefore, the functional properties of pea fiber after ISMS treatment would be altered.

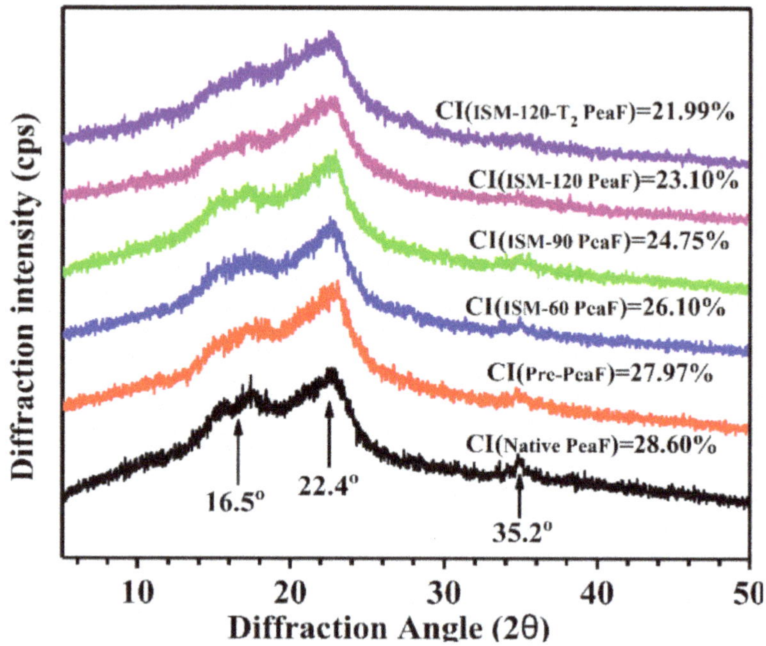

Figure 5. X-ray diffraction patterns of ISMS-treated PeaF.

3.6. Hydration Properties

The ISMS treatment-induced reduction in particle size, the increase in surface area, and changes in the microstructure and disruption of the crystalline structure could have an effect on the hydration properties of pea fiber, including swelling capacity (SC) and water retention capacity (WRC). SC and WRC of ISMS-treated PeaF is presented in Figure 6A,B, respectively. The value of SC and WRC for the native PeaF was 4.40 mL/g and 4.19 g/g, respectively. ISMS treatment significantly increased the SC and WRC of pea fiber. For instance, the SC of Pre-Pea, ISM-60 PeaF, ISM-90 PeaF, ISM-120 PeaF and ISM-120-T_2 PeaF was 9.86, 14.90, 14.21, 13.85 and 16.73 mL/g, respectively. The change trend of WRC was similar with that of SC as the treatment intensity strengthened. In particular, ISMS treatment at 120 MPa for two passes resulted in a 3.8 fold and 2.1 fold increase of SC and WRC, respectively, and this exhibited the largest values (16.73 mL/g and 8.73 g/g). ISMS-treated PeaF occupied a larger sediment volume in Figure 2G owing to its small, loose structure, which implied a greater tendency to absorb water. The transformation from dense to loose microstructure and reduction in size owing to the mechanical effect initiated by ISMS treatment endowed pea fiber with a larger surface area and more water binding sites (polar groups etc.) to the surrounding water [27], thus leading to a strengthened expansion in water and binding with water. Deleris and Wallecan [28] also pointed out that the WRC of fiber suspensions was affected by the crystalline characteristics of cellulose, since water molecules were not able to enter into the crystalline region of the cellulose. That is, disrupting the crystalline structure by ISMS treatment also favored pea fiber to binding water. Intriguingly, the values of WHC and SC were not gradually increased with the increasing of intensity during the ISMS treatment for one pass, which was not inconsistent with other studies [29,30] in which the hydration properties of insoluble dietary fiber were increased or decreased along with the reduction in particle size. The results from this study demonstrated that the hydration properties of pea fiber were dependent on several factors, such as the alteration of the microstructure and crystallinity, and particle size did not play a vital role.

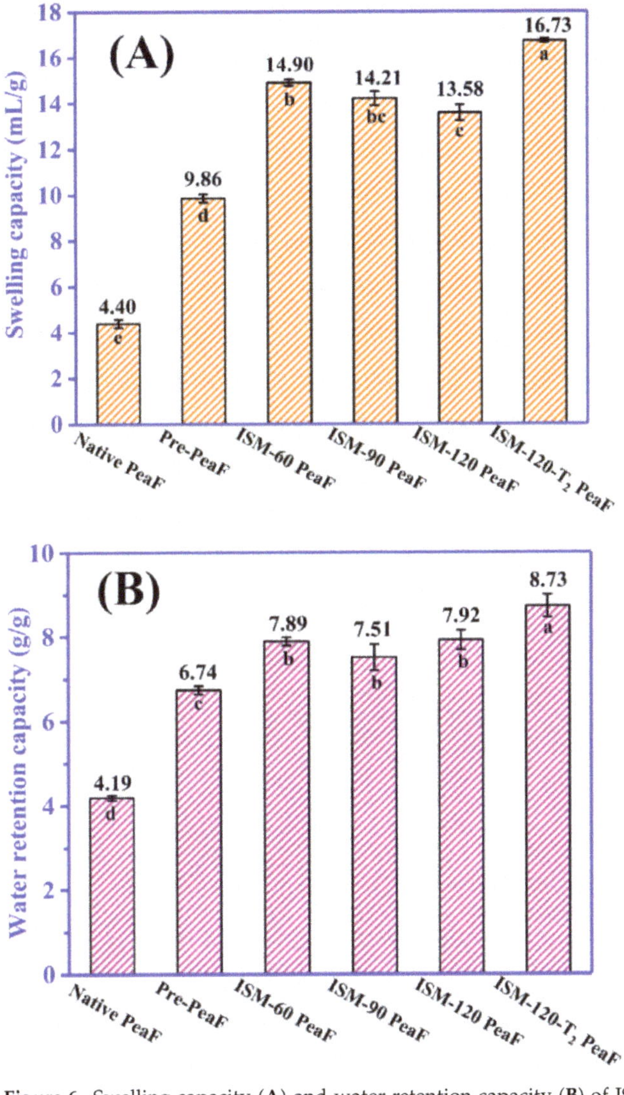

Figure 6. Swelling capacity (**A**) and water retention capacity (**B**) of ISMS-treated PeaF. Different letters in (**A**,**B**) indicated significant differences ($p < 0.05$) of hydration properties between samples.

3.7. Oil Holding Capacity

The intake of low-fat products using dietary fiber as a fat replacement could satisfy the requirement of lowering the amount of ingested fat and calories in the diet, so the capacity of fiber to retain oil is crucial for fibre-rich foodstuffs to exert an effect on cholesterol absorption and removing excess fat from the human body [31]. The OHC of ISMS-treated PeaF is shown in Figure 7. The OHC of native PeaF was 3.01 g/g. Similar to the effect of ISMS treatment on hydration properties, the OHC of fiber was significantly elevated. The OHC of all ISMS-treated PeaF was twice as high as that of native PeaF, and the change trend of OHC was similar to that of the bulk density with the increasing of treatment intensity. After ISMS treatment, an increase in bulk density, a more looser structure, more exposure of the fiber surface area and the disordering of the crystalline structure might increase the capillary attraction and adsorption sites of pea fibre, thus improving the oil-holding capacity. The increased OHC implied the potential of ISMS-treated PeaF to be used as an ingredient in meat products requiring oil absorption.

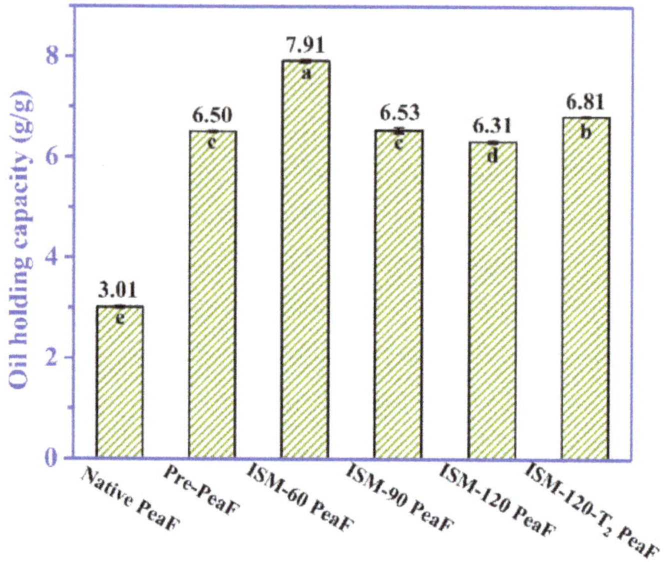

Figure 7. Oil holding capacity of ISMS-treated PeaF. Different letters in Figure 7 indicated significant differences ($p < 0.05$) of oil holding capacity between samples.

4. Conclusions

ISMS treatment significantly changed the structure of pea fiber. CLSM images revealed that fibers with a big and compact structure were disintegrated into slim and loose ones. The morphology of pea fiber was changed from compact thick blocks to flimsy crimped flakes. The crystalline structure was also destroyed, owing to the attacking of the original ordered cellulose, thus leading to the reduction of crystallinity. The alterations of the structure were accompanied with the narrowed particle size and the increased bulk density. In the meantime, the SC, WRC and OHC of pea fibers were evidently increased after the ISMS treatment. The improved hydration properties and the OHC of pea fiber was related to destroying the compact structure, providing more surface area and disrupting the crystalline structure by ISMS treatment, since more water binding and oil adsorption sites were exposed. These results suggested that the technology of ISMS facilitated the processing of pea fiber as the ingredient of foodstuffs such as meat products and jams at an industrial level.

Author Contributions: Conceptualization, T.D., J.C. and C.L.; methodology, X.H. and T.D.; investigation, X.H.; data curation, X.H., T.D. and M.C.; writing—original draft preparation, X.H., T.D. and R.L.; writing—review and editing, J.S., R.L. and J.C.; visualization, J.S. and M.C.; supervision, W.L. and C.L.; funding acquisition, J.C. and C.L. All authors have read and agreed to the published version of the manuscript.

Funding: This study was supported financially by the National Natural Science Foundation of China (32160572, 32101948), and the first batch of high-end talents (Youth) projects of science and technology innovation in Jiangxi Province (grant number jxsq2019201076), and the Research Program of State Key Laboratory of Food Science and Technology, Nanchang University (grant number SKLF-ZZA-201908), China Postdoctoral Science Foundation (grant number 2020M6832211), and Postdoctoral Foundation of Guangxi Academy of Agricultural Sciences (grant number 2020037).

Institutional Review Board Statement: Not applicable.

Informed Consent Statement: Not applicable.

Data Availability Statement: Data is contained within the article.

Acknowledgments: The authors would like to thank the Centre of Analysis and Testing of Nanchang University and State Key Laboratory of Food Science and Technology for their expert technical assistance.

Conflicts of Interest: The authors declare that they have no conflict of interest.

References

1. Gutohrlein, F.; Morales-Medina, R.; Boje, A.L.; Drusch, S.; Schalow, S. Modulating the hydration properties of pea hull fibre by its composition as affected by mechanical processing and various extraction procedures. *Food Hydrocoll.* **2020**, *107*, 105958. [CrossRef]
2. Muneer, F.; Johansson, E.; Hedenqvist, M.S.; Plivelic, T.S.; Markedal, K.E.; Petersen, I.L.; Sorensen, J.C.; Kuktaite, R. The impact of newly produced protein and dietary fiber rich fractions of yellow pea (*Pisum sativum* L.) on the structure and mechanical properties of pasta-like sheets. *Food Res. Int.* **2018**, *106*, 607–618. [CrossRef]
3. Gan, J.P.; Xie, L.; Peng, G.Y.; Xie, J.H.; Chen, Y.; Yu, Q. Systematic review on modification methods of dietary fiber. *Food Hydrocoll.* **2021**, *119*, 106872. [CrossRef]
4. Su, D.B.; Zhu, X.D.; Wang, Y.; Li, D.; Wang, L.J. Effect of high-pressure homogenization on rheological properties of citrus fiber. *LWT-Food Sci. Technol.* **2020**, *127*, 109366. [CrossRef]
5. Bruno, E.; Lupi, F.R.; Martin-Pinero, M.J.; Girimonte, R.; Baldino, N.; Munoz, J.; Gabriele, D. Influence of different dispersing systems on rheological and microstructural properties of citrus fiber suspensions. *LWT-Food Sci. Technol.* **2021**, *152*, 112270. [CrossRef]
6. Xie, F.; Li, M.; Lan, X.H.; Zhang, W.; Gong, S.X.; Wu, J.H.; Wang, Z.W. Modification of dietary fibers from purple-fleshed potatoes (*Heimeiren*) with high hydrostatic pressure and high pressure homogenization processing: A comparative study. *Innov. Food Sci. Emerg. Technol.* **2017**, *42*, 157–164. [CrossRef]
7. Luo, X.L.; Wang, Q.; Fang, D.Y.; Zhuang, W.J.; Chen, C.H.; Jiang, W.T.; Zheng, Y.F. Modification of insoluble dietary fibers from bamboo shoot shell: Structural characterization and functional properties. *Int. J. Biol. Macromol.* **2018**, *120*, 1461–1467. [CrossRef] [PubMed]
8. Yu, G.Y.; Bei, J.; Zhao, J.; Li, Q.H.; Cheng, C. Modification of carrot (*Daucus carota* Linn. var. Sativa Hoffm.) pomace insoluble dietary fiber with complex enzyme method, ultrafine comminution, and high hydrostatic pressure. *Food Chem.* **2018**, *257*, 333–340. [CrossRef]
9. Guo, X.J.; Chen, M.S.; Li, Y.T.; Dai, T.T.; Shuai, X.X.; Chen, J.; Liu, C.M. Modification of food macromolecules using dynamic high pressure microfluidization: A review. *Trends Food Sci. Technol.* **2020**, *100*, 223–234. [CrossRef]
10. Ozturk, O.K.; Turasan, H. Latest developments in the applications of microfluidization to modify the structure of macromolecules leading to improved physicochemical and functional properties. *Crit. Rev. Food Sci. Nutr.* **2021**, 1–23. [CrossRef]
11. Morales-Medina, R.; Dong, D.; Schalow, S.; Drusch, S. Impact of microfluidization on the microstructure and functional properties of pea hull fibre. *Food Hydrocoll.* **2020**, *103*, 105660. [CrossRef]
12. Guo, X.J.; McClements, D.J.; Chen, J.; He, X.M.; Liu, W.; Dai, T.T.; Liu, C.M. The nutritional and physicochemical properties of whole corn slurry prepared by a novel industry-scale microfluidizer system. *LWT-Food Sci. Technol.* **2021**, *144*, 111096. [CrossRef]
13. He, X.H.; Luo, S.J.; Chen, M.S.; Xia, W.; Chen, J.; Liu, C.M. Effect of industry-scale microfluidization on structural and physicochemical properties of potato starch. *Innov. Food Sci. Emerg. Technol.* **2020**, *60*, 102278. [CrossRef]
14. Li, Y.T.; Chen, M.S.; Deng, L.Z.; Liang, Y.Z.; Liu, Y.K.; Liu, W.; Chen, J.; Liu, C.M. Whole soybean milk produced by a novel industry-scale micofluidizer system without soaking and filtering. *J. Food Eng.* **2021**, *291*, 110228. [CrossRef]
15. Chen, J.L.; Gao, D.X.; Yang, L.T.; Gao, Y.X. Effect of microfluidization process on the functional properties of insoluble dietary fiber. *Food Res. Int.* **2013**, *54*, 1821–1827. [CrossRef]
16. Huang, K.; Zhang, S.R.; Guan, X.; Li, C.; Li, S.; Liu, Y.Y.; Shi, J.L. Effect of the oat beta-glucan on the development of functional quinoa (*Chenopodium quinoa* wild) milk. *Food Chem.* **2021**, *349*, 129201. [CrossRef] [PubMed]
17. Wang, T.; Sun, X.H.; Zhou, Z.X.; Chen, G.B. Effects of microfluidization process on physicochemical properties of wheat bran. *Food Res. Int.* **2012**, *48*, 742–747. [CrossRef]
18. Liu, Y.L.; Zhang, H.B.; Yi, C.P.; Quan, K.; Lin, B.P. Chemical composition, structure, physicochemical and functional properties of rice bran dietary fiber modified by cellulase treatment. *Food Chem.* **2021**, *342*, 128352. [CrossRef]
19. Mateos-Aparicio, I.; Mateos-Peinado, C.; Ruperez, P. High hydrostatic pressure improves the functionality of dietary fibre in okara by-product from soybean. *Innov. Food Sci. Emerg. Technol.* **2010**, *11*, 445–450. [CrossRef]
20. Jiang, G.H.; Bai, X.S.; Wu, Z.G.; Li, S.J.; Zhao, C.; Ramachandraiah, K. Modification of ginseng insoluble dietary fiber through alkaline hydrogen peroxide treatment and its impact on structure, physicochemical and functional properties. *LWT-Food Sci. Technol.* **2021**, *150*, 111956. [CrossRef]
21. Meng, X.M.; Liu, F.; Xiao, Y.; Cao, J.W.; Wang, M.; Duan, X.C. Alterations in physicochemical and functional properties of buckwheat straw insoluble dietary fiber by alkaline hydrogen peroxide treatment. *Food Chem. X* **2019**, *3*, 100029. [CrossRef] [PubMed]
22. Ullah, I.; Yin, T.; Xiong, S.B.; Zhang, J.; Din, Z.U.; Zhang, M.L. Structural characteristics and physicochemical properties of okara (soybean residue) insoluble dietary fiber modified by high-energy wet media milling. *LWT-Food Sci. Technol.* **2017**, *82*, 15–22. [CrossRef]

23. Wang, T.; Sun, X.H.; Raddatz, J.; Chen, G.B. Effects of microfluidization on microstructure and physicochemical properties of corn bran. *J. Cereal Sci.* **2013**, *58*, 355–361. [CrossRef]
24. Ma, M.M.; Mu, T.H. Effects of extraction methods and particle size distribution on the structural, physicochemical, and functional properties of dietary fiber from deoiled cumin. *Food Chem.* **2016**, *194*, 237–246. [CrossRef] [PubMed]
25. Chen, B.F.; Cai, Y.J.; Liu, T.X.; Huang, L.H.; Deng, X.L.; Zhao, Q.Z.; Zhao, M.M. Improvements in physicochemical and emulsifying properties of insoluble soybean fiber by physical-chemical treatments. *Food Hydrocoll.* **2019**, *93*, 167–175. [CrossRef]
26. Sun, C.C.; Wu, X.F.; Chen, X.J.; Li, X.J.; Zheng, Z.; Jiang, S.W. Production and characterization of okara dietary fiber produced by fermentation with *Monascus Anka*. *Food Chem.* **2020**, *316*, 126243. [CrossRef]
27. Chau, C.F.; Wen, Y.L.; Wang, Y.T. Effects of micronisation on the characteristics and physicochemical properties of insoluble fibres. *J. Sci. Food Agric.* **2006**, *86*, 2380–2386. [CrossRef]
28. Deleris, I.; Wallecan, J. Relationship between processing history and functionality recovery after rehydration of dried cellulose-based suspensions: A critical review. *Adv. Colloid Interface Sci.* **2017**, *246*, 1–12. [CrossRef]
29. Gupta, P.; Premavalli, K.S. Effect of particle size reduction on physicochemical properties of ashgourd (*Benincasa hispida*) and radish (*Raphanus sativus*) fibres. *Int. J. Food Sci. Nutr.* **2010**, *61*, 18–28. [CrossRef]
30. Raghavendra, S.N.; Swamy, S.R.R.; Rastogi, N.K.; Raghavarao, K.; Kumar, S.; Tharanathan, R.N. Grinding characteristics and hydration properties of coconut residue: A source of dietary fiber. *J. Food Eng.* **2006**, *72*, 281–286. [CrossRef]
31. Chen, J.J.; Zhao, Q.S.; Wang, L.W.; Zha, S.H.; Zhang, L.J.; Zhao, B. Physicochemical and functional properties of dietary fiber from maca (*Lepidium meyenii* Walp.) liquor residue. *Carbohydr. Polym.* **2015**, *132*, 509–512. [CrossRef] [PubMed]

foods MDPI

Article

Effect of Grass Carp Scale Collagen Peptide FTGML on cAMP-PI3K/Akt and MAPK Signaling Pathways in B16F10 Melanoma Cells and Correlation between Anti-Melanin and Antioxidant Properties

Zizi Hu [1], Xiaomei Sha [2], Lu Zhang [1], Sheng Huang [3] and Zongcai Tu [1,3,*]

1 National R&D Center for Freshwater Fish Processing, College of Chemistry and Chemical Engineering, Jiangxi Normal University, Nanchang 330022, China; Hzz13667949044@163.com (Z.H.); zhanglu00104@163.com (L.Z.)
2 College of Life Science, Jiangxi Normal University, Nanchang 330022, China; shaxiaomei1987@sina.com
3 State Key Laboratory of Food Science and Technology, Nanchang University, Nanchang 330047, China; hsheng21@126.com
* Correspondence: 004756@jxnu.edu.cn

Abstract: Peptide Phe-Thr-Gly-Met-Leu (FTGML) is a bioactive oligopeptide with tyrosinase inhibitory activity derived from gelatin hydrolysate of grass carp scales. Previous studies have shown that FTGML addition can effectively inhibit mushroom tyrosinase activity in vitro, and also has some effect on the inhibition of melanogenesis in zebrafish in vivo, but the underlying mechanism is not fully understood. In this study, we used FTGML to treat B16F10 melanoma cells, and found a significant inhibition of tyrosinase activity and melanin synthesis. Interestingly, the treatment showed a strong correlation between antioxidant activity and anti-melanin, which was associated with FTGML reducing the involvement of reactive oxygen species in melanin synthesis. Furthermore, FTGML reduced melanogenesis in B16F10 cells by downregulating the cAMP-PI3K/Akt and MAPK pathways (p38 and JNK). These results suggested that FTGML can reduce melanin production in mouse B16F10 melanoma cells through multiple pathways.

Keywords: collagen peptide; B16F10 cells; melanogenesis; tyrosinase; antioxidant

Citation: Hu, Z.; Sha, X.; Zhang, L.; Huang, S.; Tu, Z. Effect of Grass Carp Scale Collagen Peptide FTGML on cAMP-PI3K/Akt and MAPK Signaling Pathways in B16F10 Melanoma Cells and Correlation between Anti-Melanin and Antioxidant Properties. *Foods* **2022**, *11*, 391. https://doi.org/10.3390/foods11030391

Academic Editors: Rotimi Aluko, Jianhua Xie, Yanjun Zhang and Hansong Yu

Received: 8 December 2021
Accepted: 27 January 2022
Published: 29 January 2022

1. Introduction

Melanin is a macromolecule biological pigment, mainly composed of eumelanin and pheomelanin. Eumelanin is brown-black, and pheomelanin is red-brown. The different proportions of the two pigments cause the different colors of hair and skin [1]. Melanin can protect the skin from the harmful effects of environmental pollution factors, such as ultraviolet radiation and oxidative stress. However, excessive accumulation of melanin can lead to the appearance of many skin diseases, including melasma, freckles, age spots, and other pigmentation syndromes. The process of melanin synthesis is extremely complicated, mainly involving three enzymes in the tyrosinase gene family, namely tyrosinase (Tyrosinase, TYR), tyrosinase-related protein-1 (Tyrosinase-related protein-1, TRP-1), and tyrosinase-related protein-2 (Tyrosinase-related protein-2, TRP-2), in which tyrosinase is the key enzyme in this reaction [2].

It Is well known that melanocortin 1 receptor (MC1R)/α-melanocyte, stimulating hormone (α-MSH), phosphatidylinositol 3-kinase (PI3K)/activation of phosphorylation of protein kinase B (Akt), mitogen-activated protein kinase (MAPK), Wingless/Integrated (Wnt)/β-catenin, nitric oxide (NO), and other signaling pathways are involved in melanin production of melanocytes [3,4]. Microphthalmia-associated transcription factor (MITF) are the most important molecular targets in these pathways, and the change of MITF expression is directly related to abnormal skin and hair pigments [5]. MITF is a basic helix-loop-helix-leucine zipper transcription factor, which can share a highly conserved sequence with the m-box motif of the promoter region (TYR, TRP-1, TRP-2 in the promoter region; that is, 5′-AGTCATGTGCT-3′), and regulates the expression of TYR, TRP-1, and TRP-2; thus, regulating the production of melanin [6]. Previous studies have found that some peptides inhibit melanin synthesis mainly through ERK in the mitogen-activated protein kinase (MAKP) signaling pathway [7,8]. In the previous experiments of this study, we found a peptide Phe-Thr-Gly-Met-Leu (FTGML) could promote cell apoptosis, which was related to the other two pathways of MAKP (i.e., p38 and JNK) [9], but there was no relevant report that the peptide could inhibit melanin synthesis through these two pathways.

In recent years, people have tried to develop new natural-derived whitening cosmetics or skin bleaching agents, as natural products have the advantages of being less toxic and having fewer side effects compared to chemicals. Examples include tea [10], mung bean seed [2], etc. Fucoidan derived from hizikia fusiforme can significantly inhibit the expression pathway of tyrosinase and tyrosinase-related proteins in B16F10 cells by regulating the extracellular signal-regulated kinase mitogen-activated protein kinase (ERK-MAPK), and downregulating the MITF [11]. Tea catechins downregulate MITF expression by inhibiting cyclic adenosine monophosphate (cAMP), leading to subsequent phosphorylation of CREB, and a decrease in the levels of tyrosinase, TRP-1, and TRP-2, thereby reducing melanin synthesis [12]. However, at present, most research on natural products that inhibit melanin mostly focus on plant sources. Natural products from animal sources, especially fish, have less research in this area, and most of them focus on basic index evaluation. The previous research of our group found that grass carp fish scale gelatin hydrolysate has a good whitening effect, and a new peptide consisting of five amino acids (phenylalanine-threonine-glycine-methionine-leucine, FTGML) was obtained, with good in vitro tyrosinase inhibitory activity (IC_{50} value was 1.89 mM). However, the melanin inhibitory mechanism is still unclear.

In addition, there are reports that hydrogen peroxide (H_2O_2) and other ROS and RS are produced during melanogenesis, leading to advanced oxidative stress in melanocytes. Therefore, the use of free radical scavengers in this process can play a role in feedback regulation. Studies have shown that ROS scavengers and ROS generation inhibitors may inhibit ultraviolet-induced melanogenesis [13]. Antioxidants, such as ascorbic acid derivatives and reduced glutathione (GSH), have also been used as inhibitors of melanin production [14,15]. Therefore, it is necessary to pay attention to the antioxidant effect and melanin inhibitory effect of compounds or natural products at the same time, and explore the relationship between the two effects, but there are few studies on the relevant mechanism.

In terms of cell structure and melanin synthesis, the mouse melanoma cell line B16F10 is highly consistent with human melanoma cell. Furthermore, it is difficult to culture human melanoma cells when evaluating the efficacy of active whitening, due to its stricter requirements for cell culture and handling condition. The mouse melanoma cell line B16F10 has the advantages of multiple passages, rapid development, relatively simple culture conditions, high malignancy, and good tumorigenicity. As such, B16F10 cells are widely used as the effective cells for cell evaluation of whitening activity substances [16–18]. In this study, we aim to further analyze the anti-melanin effect and potential mechanism of FTGML based on the previous stage in the mouse melanoma cell line B16F10. It is also assumed that FTGML functionally causes cAMP-PI3K/Akt- and MAPK-mediated downregulation of MITF, which is the main cascade of melanin production, leading to the decrease of tyrosinase in melanocytes. In addition, we also studied the correlation between anti-melanin and anti-oxidation in melanocytes, and provided the possibility of predicting its anti-melanin potential through antioxidant activity in the future. Our research contributes to the development of new cosmetic ingredients, food supplements, and functional foods containing collagen peptides.

2. Materials and Methods

2.1. Experimental Materials

FTGML was obtained from the grass carp fish scale gelatin hydrolysate. In brief, grass carp fish scale gelatin was hydrolyzed by alcalase and gastrointestinal simulated digestion to obtain a mixture of peptides. The peptides with tyrosinase inhibitory activity were screened and identified by bioaffinity ultrafiltration-mass spectrometry, and peptide FTGML showed the best tyrosinase inhibitory activity. FTGML was synthesized in Shanghai Sangon Biotechnology Co., Ltd. (Shanghai, China). The purity of all synthetic peptides was above 95%. Mushroom tyrosinase (EC1.14.18.1, with specific activity 6680 U/mg) and L-DOPA were produced by Sigma-Aldrich (St. Louis, MO, USA). The murine melanoma B16F10 cells (CL0039) were acquired from Fenghui Biological Technology Co., Ltd. (Changsha, China). All other reagents were analytical grade.

2.2. Cell Culture

B16F10 cells was cultured by the medium of RPMI-1640, with 10% fetal bovine serum (FBS), and 1% penicillin/streptomycin at 37 °C in 5% CO_2.

2.3. Cell Viability Assay

Cell viability was assessed by Cell Counting Kit-8 (CCK-8) assay. The cells in the logarithmic growth phase were digested by trypsin, and counted under the microscope to make a cell suspension of 1~5 × 10^4 cells/mL. A 100 μL cell suspension was taken to a 96-well culture plate with 1~5 × 10^3 cells/well. Wells added with 100 μL medium were used as blank control, and wells added with different concentrations (0, 0.1, 0.2, 0.4, 0.8, 1.6 mg/mL) of FTGML solution and 0.75 mg/mL of kojic acid were used as the experimental groups and positive control groups, respectively. After 0 h, 24 h, and 48 h of incubation at 37 °C, the 1:10 volume ratio mixed CCK-8 and serum-free essential basic medium was added to the test well at 100 μL per well, and incubated for 1 h at 37 °C in a 5% CO_2 incubator. The absorbance at 450 nm was measured with a microplate reader.

2.4. Determination of Apoptosis Rate of B16F10 Cells

Cells in 2.3 were collected and performed according to the Annexin V-FITC kit (Thermo Fisher Scientific, Carlsbad, CA, USA). The apoptosis of B16F10 cells was analyzed by flow cytometry (Ex = 488 nm, FL1 Em = 525 ± 20 nm, FL2 Em = 585 ± 21 nm) and FlowJo software.

2.5. Assay of Tyrosinase Activity

Tyrosinase activity was measured using a Tyrosinase activity assay kit (BC4055, Beijing Solarbio Science & Technology Co., Ltd., Beijing, China). OD was measured at 475 nm.

2.6. Measurement of Melanin Synthesis

B16F10 cells were seeded in 6-well plates at a density of 2 × 10^6 cells/well. After incubation for 24 h, the cells were treated with FTGML (0–1.6 mg/mL) or kojic acid (0.75 mg/mL), and incubated at 37 °C in 5% CO_2 for 48 h. The cells were washed with PBS, digested with trypsin, and centrifuged at 1500 rpm for 10 min. To dissolve the cell precipitates, 1 mL 1 M NaOH containing 10% DMSO was used at 80 °C for 1 h. The optical density of each well at 405 nm was measured with a microplate reader, and the content of melanin in the cell was calculated [12]:

$$\text{Melanin content}/\% = \frac{(A_s - A_b)}{(A_c - A_b)} \times 100\% \tag{1}$$

where, A_s is the absorbance of experimental wells; A_c is the absorbance of control well; A_b is the absorbance of blank wells.

2.7. Assay of Antioxidant Activity

The centrifuge precipitate in Section 2.6 was dissolved with 1 mL 1 M NaOH containing 10% DMSO at 80 °C for 1 h were collected separately for antioxidant activity determination. The content of reduced glutathione (GSH), oxidized glutathione (GSSG), reactive oxygen species (ROS), and malondialdehyde (MDA), and the activity of superoxide dismutase (SOD), catalase (CAT), and glutathione peroxidase (GPX) were measured using different assay kits that were all from Beijing Solarbio Science & Technology Co., Ltd., Beijing, China, and the item numbers were as follows: GSH by BC1175; GSSG by BC1185; ROS by CA1410; MDA by BC0025; SOD by BC0175; CAT by BC0205; GPX by BC1195.

2.8. Measurement of Intracellular cAMP Concentration

B16F10 cells were inoculated in 6-well plates at a density of 2×10^5 cells/well. After 24 h incubation, the cells were treated with FTGML (0–1.6 mg/mL) or kojic acid (0.75 mg/mL), and incubated at 37 °C in 5% CO_2 for 48 h. Cells were washed with cold PBS, and dissolved on ice in a RIPA buffer containing protease and phosphatase inhibitors. The samples were centrifuged at 1000× *g* for 20 min to remove impurities and cell debris, and the supernatant was used for cAMP detection. A cAMP immunoassay kit (Jiancheng Bioengineering Institute, Nanjing, China) was used to measure intracellular cAMP concentration [12].

2.9. Western Blot Analysis

B16F10 cells were treated with FTGML (0–1.6 mg/mL) or kojic acid (0.75 mg/mL) for 48 h. Lysis buffer (150–250 μL) was added to each well to complete the lysis. Then, the lysed samples were centrifuged at 12,000× *g* for 15 minutes at 4 °C, and the supernatant was obtained for protein quantification (BCA assay kit, PICPI23223, Thermo Fisher Scientific, Carlsbad, CA, USA), and stored in a refrigerator at −80 °C for the next analysis.

An equal amount of protein (20 μg/sample) was bathed in boiling water for 10 min after being mixed with 5 × SDS loading buffer, separated by 12% SDS polyacrylamide gel electrophoresis, and transferred to a polyvinylidene fluoride (PVDF) membrane (Thermo Fisher Scientific, Carlsbad, CA, USA). The PVDF membrane was blocked in 5% skim milk in TBST buffer (PBS containing 0.05% Tween-20) for 1 h at room temperature. The membranes were incubated with several antibodies overnight at 4 °C after a short wash. These antibodies include anti-MITF (1:500), anti-GAPDH (1:2500), anti-p38 (1:1000), anti-p -p38 (1:1000), anti-p-JNK (1:200), anti-JNK (1:500), anti-p-PI3K (1:500), anti-PI3K (1:500), anti-p-Akt (1:2000), and anti-Akt (1:1000). After incubation, the membrane was washed thoroughly with TBST buffer, and further incubated with goat anti-rabbit HRP-labeled secondary antibody (1:1000, Beyotime Biotechnology Co., Ltd., Shanghai, China) at 37 °C for 1 h. The blots were visualized using enhanced chemiluminescence (ECL), and quantified by Image-pro plus software (Media Cybernetics, Rockville, MD, USA).

2.10. Statistical Analysis

All results were expressed as mean ± standard deviation. SPSS version 20 (SPSS Inc., Chicago, IL, USA) was used for analysis of variance ($p < 0.05$). The Duncan multi-range test was used for the comparison of means. Three replicates were used for each analysis. Principal component analysis (PCA) was used to analyze the correlation between the different properties of anti-melanin and anti-oxidation. The above indicators were used as active variables, and the concentration of FTGML was used as the observed value. Two-dimensional or three-dimensional graphs were drawn using Origin Pro 2018 software. Hierarchical cluster analysis (HCA) was used to visualize and emphasize the similarities between individuals. The difference is usually expressed by the distance between individuals [19]. Origin Pro 2018 software was used to determine the correlation between indicators by the Pearson correlation coefficient in binary linear correlation.

3. Results and Discussion

3.1. Effects of FTGML on B16F10 Cells Viability

B16F10 cells were treated with 0.1–1.6 mg/mL FTGML and 0.75 mg/mL kojic acid for 0 h, 24 h, and 48 h, respectively. The cell viability is expressed as a percentage relative to the cells in the blank control group (cells without any drugs), and the results are shown in Figure 1. FTGML and kojic acid showed low toxicity to B16F10 cells at all treatment concentrations (cell viability > 80%, [20]). Based on these results, the concentration of FTGML in the range of 0.1 mg/mL to 1.6 mg/mL can be used for subsequent experiments.

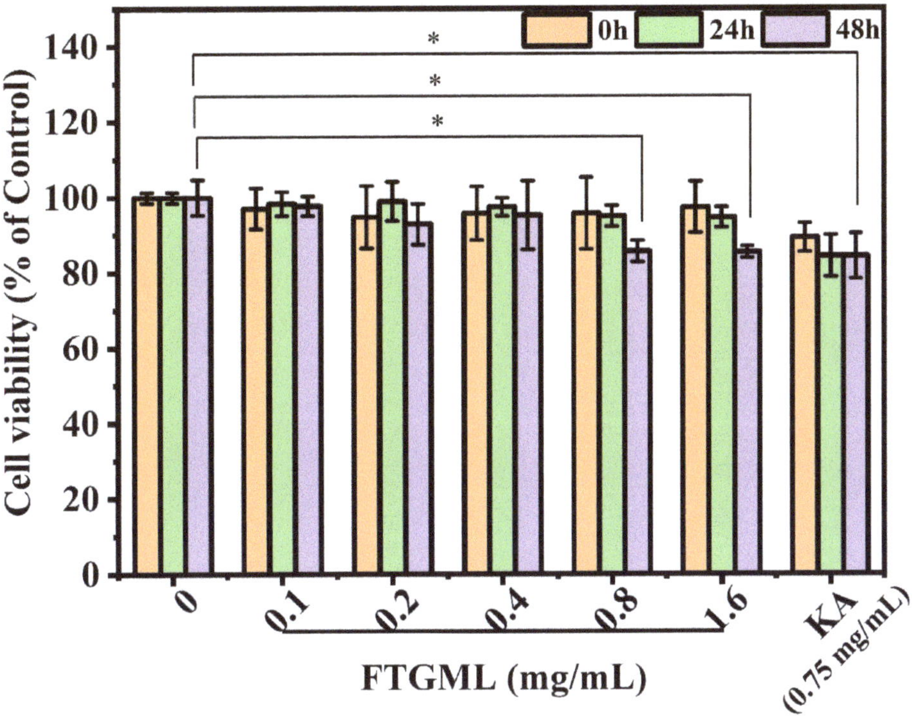

Figure 1. Effects of FTGML and kojic acid (KA) on cell viability. The CCK8 method was used to detect the effects of FTGML at 0, 0.1, 0.2, 0.4, 0.8, and 1.6 mg/mL; and kojic acid at 0.75 mg/mL on the activity of B16F10 cells (Note: *, $p < 0.05$).

3.2. Effects of FTGML on Apoptosis Rate of B16F10 Cells

Apoptosis is a basic biological phenomenon of cells, and plays an essential role in the removal of unwanted or abnormal cells by multicellular organisms. The disorder of apoptosis may be directly or indirectly related to the occurrence of many diseases, such as tumors and autoimmune diseases. Therefore, an apoptosis assay is often used to evaluate the development and application potential of active ingredients in food in the field of functional food [21].

The effect of 0.1~1.6 mg/mL FTGML treatment for 48 h on the apoptosis of B16F10 cells is shown in Figure 2. It can be seen from Figure 2A that as the concentration of FTGML increased, the proportion of normal living cells gradually decreased, and the proportion of early apoptotic cells, late apoptotic cells, and necrotic cells increased. Figure 2B shows the apoptosis rate of B16F10 cells (i.e., the total proportion of early apoptotic cells to late apoptotic cells). It can be seen from the figure that FTGML treatment promoted the apoptosis of cells, and the apoptosis rate was positively correlated with the concentration of

FTGML. There was no significant difference in apoptosis rate between B16F10 cells treated with kojic acid (0.75 mg/mL) and FTGML at high concentration (≥0.8 mg/mL) ($p > 0.05$).

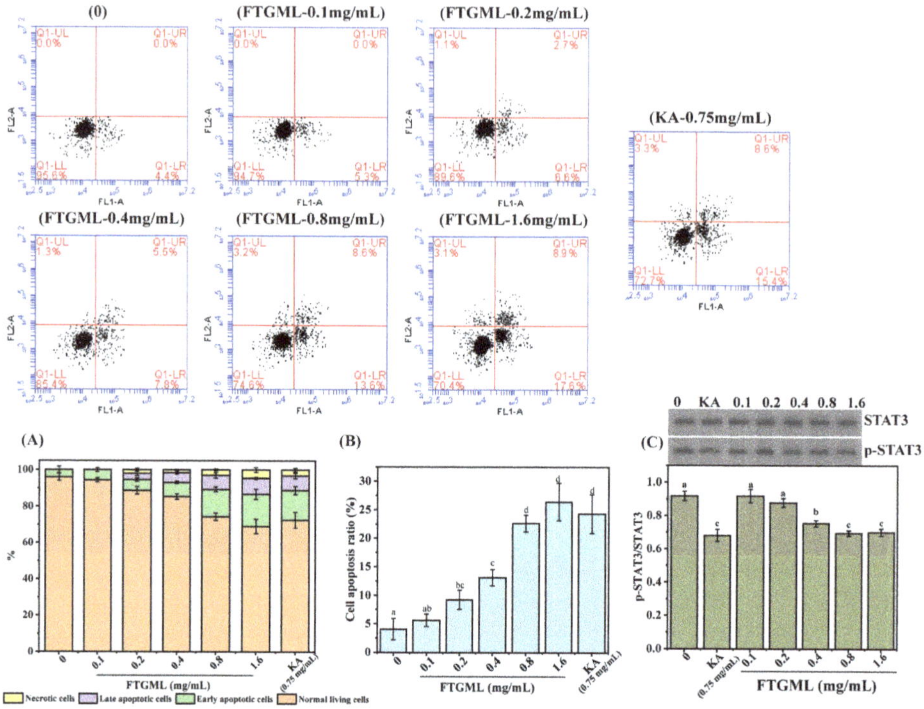

Figure 2. Effects of FTGML and kojic acid (KA) on cell apoptosis. (**A**) The proportion of different states (normal cells, early apoptotic cells, late apoptotic cells, and necrotic cells). (**B**) The apoptosis rates (the sum of the proportions of early and late apoptosis) of B16F10 cells at different treatments. (**C**) The effect of FTGML and kojic acid on the expression of STAT3. Means with different lowercase letters are significantly different ($p < 0.05$) among the different groups.

STAT3 (signal transducer and activator of transcription 3) is both a cytoplasmic signal molecule and a nuclear transcription factor, which is involved in cell proliferation, transformation, and migration. At present, STAT3 has been identified as a major oncogene in the development of melanoma [22]. There is also genetic evidence for a direct role of STAT3 in melanoma cell transformation [23]. As can be seen from Figure 2C, when the concentration of FTGML was greater than 0.4 mg/mL, the ratio of phosphorylated STAT3 (p-STAT3) to STAT3 was lower than 1, and significantly lower than that of the blank group ($p < 0.05$). This suggests that medium-high concentrations (≥0.4 mg/mL) of FTGML can reduce the activation/phosphorylation of STAT3. Some researchers have found that increasing the activation/phosphorylation level of STAT3 can promote the growth of melanoma, whereas silencing STAT3 can significantly inhibit the proliferation of melanoma cells, and promote cell apoptosis [22]. This is consistent with the results shown in Figure 2B.

3.3. Effects of FTGML on Intracellular Melanogenesis

The first step in studies of melanin production is usually to measure intracellular tyrosinase activity, since tyrosinase is the rate-limiting enzyme involved in melanin synthesis [12]. To test the inhibitory effect of FTGML on intracellular tyrosinase activity, B16F10 cells were exposed to FTGML solution in the concentration range of 0~1.6 mg/mL for 48 h. As shown in Figure 3A, FTGML significantly inhibited tyrosinase activity compared with untreated cells ($p < 0.05$). With the increase of FTGML concentration, intracellular tyrosinase activity decreased gradually. When the FTGML concentration was 1.6 mg/mL, the residual tyrosinase activity was 43.87 ± 5.89% of that of the control. This was not a statistically significant difference ($p > 0.05$) from kojic acid (positive control substance) at 0.75 mg/mL.

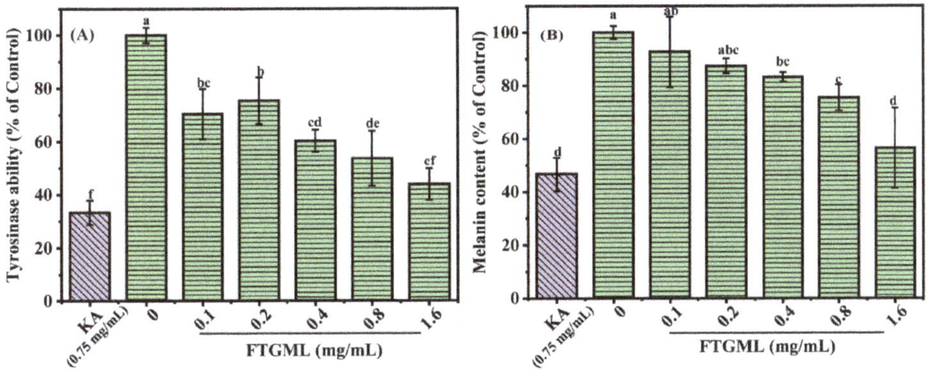

Figure 3. Effects of FTGML and kojic acid (KA) on melanin content and tyrosinase activity in B16F10 melanoma cells. Cells were treated with various concentrations for 48 h (0–1.6 mg/mL). (**A**) Tyrosinase activity. (**B**) Relative melanin content. Means with different lowercase letters are significantly different ($p < 0.05$) among the different groups.

To further explore the effect of FTGML on melanin production, we detected the residual melanin content in B16F10 cells after FTGML treatment. The results in Figure 3B show that FTGML significantly reduced melanin in B16F10 cells, especially when the FTGML concentration was higher than 0.8 mg/mL. When the concentration of FTGML was 1.6 mg/mL, the intracellular melanin content was 56.44 ± 15.05% of the control group. The content of residual melanin in cells after kojic acid (0.75 mg/mL) treatment was 46.67 ± 6.31% of that of the control group, which was not significantly different from that after FTGML (1.6 mg/mL) treatment ($p > 0.05$). These results suggest that FTGML can achieve whitening by inhibiting the synthesis of melanin in cells, and can achieve a similar effect to kojic acid to a certain extent. This is similar to the results of Han et al., who found that oyster hydrolysate significantly reduced intracellular melanin content [24].

3.4. *Effects of FTGML on Intracellular Antioxidant Activity*

In the production of melanin, tyrosine is required to produce dopamine in an oxidized environment and subsequently, to produce dopamine. Therefore, the existence of reactive oxygen species, such as superoxide anion and hydroxyl radical, is conducive to the synthesis of melanin [25]. Studies have shown that peptides with oxygen free radical scavenging ability can inhibit the biosynthesis of melanin in cells [8,26].

Glutathione (GSH) is an important intracellular regulatory metabolite, and its redox states (i.e., reduced glutathione (GSH) and oxidized glutathione (GSSG)) are important for many physiological processes. The effect of FTGML on GSH content in B16F10 cells is shown in Figure 4A. The content of GSH in cells is positively correlated with the concentration of added FTGML. When the concentration of FTGML is 1.6 mg/mL, the content of GSH in cells is 1.78 times higher than that in the blank group. GSH, as a small peptide substance with strong reducibility in cells, can scavenge intracellular free radicals, leading to its critical role in the production of melanin [27]. The effect of FTGML on GSSG content in B16F10 cells is shown in Figure 4B. The addition of FTGML reduced the content of GSSG in B16F10 cells. The GSG:GSSG ratios of B16F10 cells treated with different concentrations of FTGML (0, 0.1, 0.2, 0.4, 0.8, and 1.6 mg/mL) and 0.75 mg/mL kojic acid were 2.10 ± 0.23, 2.57 ± 0.09, 3.61 ± 0.10, 4.79 ± 0.25, 5.00 ± 0.45, 5.22 ± 0.29, 5.68 ± 0.22, respectively. These results indicate that FTGML can destroy the oxidative environment in cells. It can maintain the reductive power in cells, and regulate the level of GSH in cells, thus inhibiting the production of melanin.

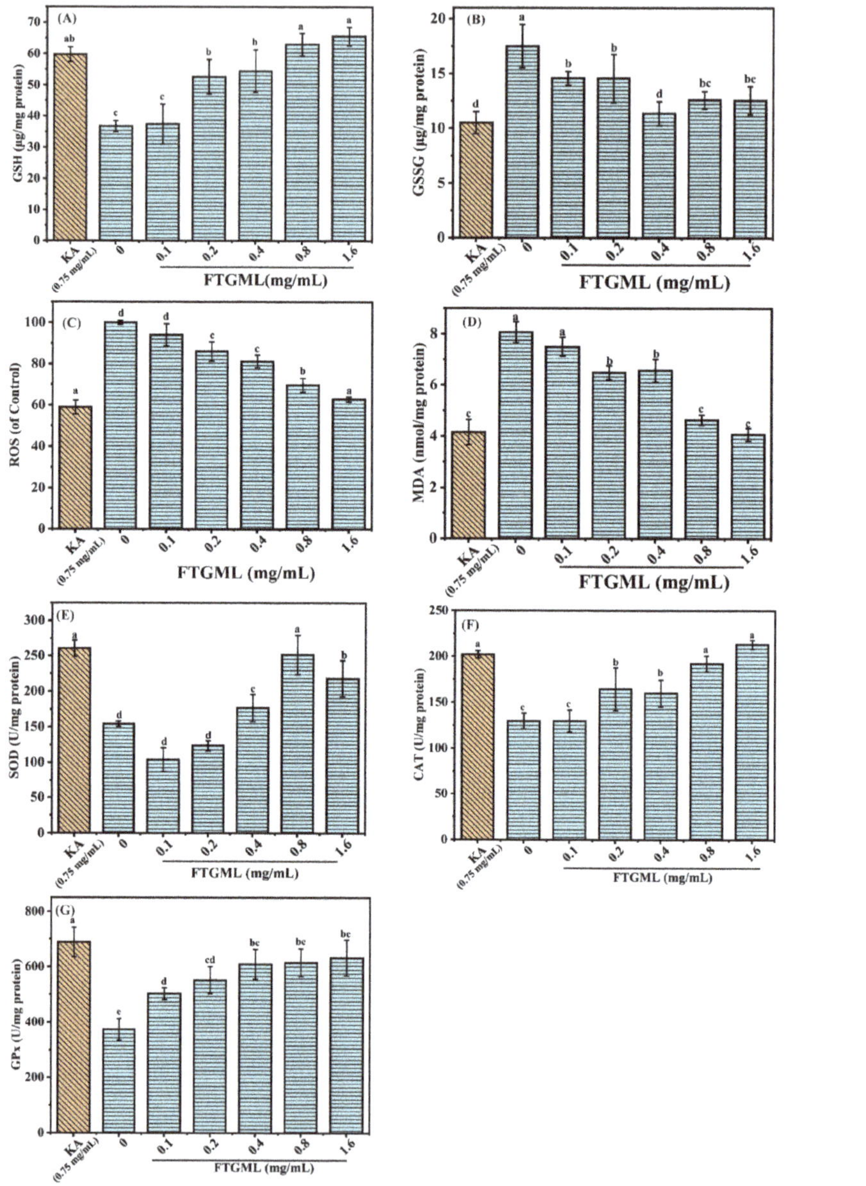

Figure 4. The effect of FTGML and kojic acid on intracellular antioxidant activity. The GSH contents (**A**), GSSG contents (**B**), ROS levels (**C**), MDA contents (**D**), SOD (**E**), CAT (**F**), and GPX (**G**) activities of B16F10 cells at different treatments. Means with different lowercase letters are significantly different ($p < 0.05$) among the different groups.

It has been shown that ROS play an important role in the regulation of melanogenesis and melanocyte proliferation [28]. ROS scavengers or formation inhibitors can reduce melanogenesis in melanocytes. Malondialdehyde (MDA) is one of the products of cellular membrane lipid peroxidation, and can be used to reflect the extent of oxidative stress damage to cells [29]. To elucidate the protective mechanism of FTGML against oxidative stress in B16F10 cells, the intracellular ROS level and MDA content of B16F10 cells were measured, as shown in Figure 4C,D, respectively. Intracellular ROS and MDA dropped to 62.81% and 50.46% of the untreated group, respectively, as the amount of FTGML was increased. After kojic acid treatment, intracellular ROS and MDA dropped to 58.94% and 51.49% of the untreated group, respectively. This means that after treatment with FTGML (1.6 mg/mL) and kojic acid (0.75 mg/mL), there was no significant difference ($p > 0.05$). This demonstrated that FTGML significantly inhibited intracellular ROS and MDA production, and protected B16F10 cells from oxidative damage. This is similar to the study by Huang et al., who found that [8]-gingerol inhibited melanogenesis in melanoma cells, and that the addition of 100 μM reduced intracellular ROS levels to 71.01 ± 1.45% of the blank group, indicating that it significantly depleted ROS levels in B16F10 cells [30].

SOD, CAT, and GPx are antioxidant enzymes that work together to decrease ROS, and protect cells from oxidative stress damage [31]. It can be seen from Figure 4E,G that the activities of SOD, CAT, and GPx increased by 63.10%, 64.53%, and 69.29%, respectively, after FTGML treatment of B16F10 cells, and increased by 68.51%, 56.00%, and 84.33%, respectively, after treatment with kojic acid. Compared with kojic acid, similar effects to FTGML were observed in SOD and CAT activities ($p > 0.05$). However, there is still a certain gap between the two in GPx activity ($p < 0.05$), and with the increase of FTGML concentration, there is no significant change in GPx activity ($p > 0.05$). The activity of antioxidant enzymes (SOD, CAT, and GPx) significantly affects the sensitivity of the skin to oxidative damage, including skin pigmentation problems [32]. Therefore, FTGML shows a protective effect on oxidation by increasing the activity of antioxidant enzymes in B16F10 cells, thereby avoiding melanin deposition. Some researchers' reports also indicate that peptides interfere with skin biochemical reactions by protecting cells against oxidative damage, for example, sorghum kafrins-derived peptide fraction [26], and this is mainly due to their ability to enhance the activity of antioxidant enzymes in the cell.

3.5. Effects of FTGML on the Melanogenesis-Related Signaling Pathway in B16F10 Melanoma Cells

In this study, we examined the expression of melanogenesis-related proteins, including MITF and cAMP, and melanogenesis-regulating molecules, including PI3K/AKT, p38, and JNK, to elucidate the potential mechanisms by which FTGML inhibits melanogenesis in B16F10 melanoma cells. The results are shown in Figure 5.

cAMP is a well-known intracellular second messenger, and its mediated signaling is the main cascade of melanin production, which is largely influenced by changes in intracellular cAMP levels [12]. As shown in Figure 5A, FTGML (1.6 mg/mL) and kojic acid (0.75 mg/mL) reduced intracellular cAMP levels in B16F10 cells by 48.77% and 45.38%, respectively, compared to untreated controls ($p < 0.05$). cAMP is an important mediator of intracellular signal-activated protein kinase A (PKA). In the absence of cAMP, PKA is inactive, and its two PKA catalytic subunits bind to the two PKA regulatory subunits. When cAMP is present, cAMP binds to the regulatory subunits of PKA, and induces dissociation of the PKA catalytic subunits from the PKA holoenzyme complex, and the released PKA catalytic subunits are activated, ultimately activating the cAMP response element binding protein (CREB), and thus promoting melanin synthesis [33]. Therefore, it can be concluded that FTGML treatment reduces the intracellular cAMP level, which is beneficial to the regulation of subsequent signaling pathways. Similar results were found by Han et al. Oyster hydrolysate exhibited anti-melanogenic activity by downregulating the cAMP signaling pathway, thereby reducing melanin synthesis [24]. Phosvitin-derived peptide Pt5 was shown to be involved in the cAMP pathway to inhibit melanogenesis, with no significant effect on the Wnt and MAPK signaling pathways [34].

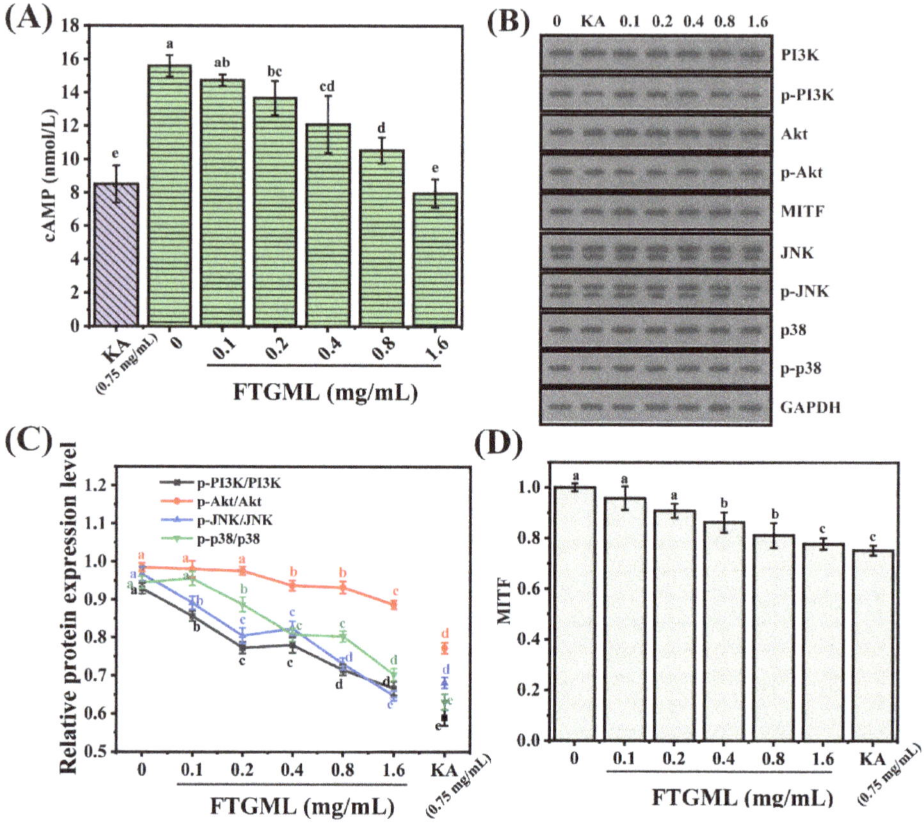

Figure 5. The effect of different concentrations of FTGML on the expression level of related proteins. 0.1, 0.2, 0.4, 0.8, and 1.6 represent the concentration of FTGML (mg/mL), KA represents kojic acid, and the concentration is 0.75 mg/mL. The cAMP contents (**A**), western blot analysis. (**B**), the relative protein expression was examined by western blot analysis (**C**), the relative protein expression of MITF was examined by western blot analysis (**D**). Means with different lowercase letters are significantly different ($p < 0.05$) among the different groups.

cAMP can also regulate melanogenesis through PKA non-dependent mechanisms, such as phosphatidylinositol-3-kinase (PI3K). One of the key effectors of PI3K is protein kinase B (Akt). The results shown in Figure 5C were obtained by analysis of Figure 5B. Both FTGML and kojic acid treatment decreased the expression of p-PI3K and p-Akt compared to untreated controls. p-PI3K/PI3K and p-Akt/Akt ratios of B16F10 cells were 0.67 and 0.89, respectively, after 1.6 mg/mL FTGML treatment. In addition, 0.75 mg/mL kojic acid treatment decreased the p-PI3K/PI3K and p-Akt/Akt ratios to 0.59 and 0.77, respectively. Under the stimulation of external signals and intracellular cAMP, activated Akt enhances the binding of MITF to the M-box by phosphorylating glycogen synthase kinase 3β (GSK3β), and promotes its loss of activity. The reduction in GSK3β activity enhances the binding of MITF to the M-box, and synergistically stimulates the tyrosinase promoter with MITF, enhancing its binding to the tyrosinase promoter, and thus, promoting melanogenesis [35]. The results showed that FTGML reduced the phosphorylation level of PI3K/Akt by downregulating cAMP. It has been shown that diosgenin has the effect of regulating the reduction of PI3K expression and phosphorylation of AKt, which affects the phosphorylation of downstream GSK-β, which, in turn, allows the downregulation of MITF expression, leading to a reduction in melanin synthesis [36].

In addition to the cAMP-mediated signaling pathway described above, there is also the mitogen-activated protein kinase (MAPK) signaling pathway, which is involved in the expression and activation of MITF [37]. Based on the apoptotic results in Section 3.2, we speculate that it may be FTGML that promotes apoptosis in B16F10 cells, resulting in reduced melanin. Several past studies have shown that JNK and p38 are the two most important factors in MAPK that affect apoptosis [9,38,39]. As shown in Figure 5C, both FTGML and kojic acid treatment reduced the expression of p-JNK and p-p38 compared to untreated controls. p-JNK/JNK and p-p38/p38 ratios of B16F10 cells after 1.6 mg/mL FTGML treatment were 0.65 and 0.70, respectively. In addition, 0.75 mg/mL kojic acid treatment reduced the p-JNK/JNK and p-p38/p38 ratios of B16F10 cells, which had p-JNK/JNK and p-p38/p38 ratios of 0.68 and 0.63, respectively. It has been reported that the inhibition of the JNK pathway reduces melanin production [40], and that phosphorylation of p38 activates MITF expression and upregulates melanogenesis-related proteins, which, in turn, affects melanin synthesis [41]. These results suggest that FTGML inhibits melanin synthesis through downregulation of the JNK and p38 signaling pathways.

MITF is located downstream of these signaling pathways and upstream of tyrosinase, and is involved in the activation of several melanocyte-related genes. It regulates both tyrosinase and its associated proteins on melanin synthesis, as well as the growth, development, and differentiation of melanocytes [42]. As shown in Figure 5D, the expression of MITF was significantly reduced after FTGML acted on melanocytes, and the inhibition was most evident at the highest concentration of 1.6 mg/mL of FTGML, which was consistent with our prediction, indicating that FTGML could reduce the expression of MITF. It was also verified that inhibition of the above signaling pathways (cAMP-PI3K/Akt and MAPK pathways) also significantly affected the expression of MITF. Based on these results, we hypothesize that FTGML downregulates the phosphorylation of PI3K/Akt in this signaling pathway by decreasing cAMP levels. It also downregulates the phosphorylation of JNK and p38 in the MAPK signaling pathway. Together, these two pathways are involved in FTGML's inhibition of MITF expression-mediated melanogenesis, with the PI3K signaling pathway being relatively more important. Of the available studies on tyrosinase inhibitory peptides, very little has been done on their signaling pathways. Peptides derived from the fermented microalga (*Pavlova lutheri*) [8] and oyster hydrolysate [24] have been demonstrated, but they have all been studied by selecting one of the signaling pathways. The findings of the present study suggest the possibility that multiple signaling pathways may still act together in melanin inhibition.

3.6. Principal Component (PCA), Cluster Analyses (HCA), and Correlation Analysis

Principal component analysis was performed on 10 indicators from 6 samples with different FTGML additions using SPSS 20.0 software. As shown in Table 1, a total of two principal components were extracted, and the cumulative value of the contribution of the principal components reached 94.836%, which can explain the vast majority of the original information. Principal component 1 is the most important, with a contribution of 87.144% of the variance, which can represent 87.144% of the total information. Principal component 2 has a variance contribution of 7.692%.

The corresponding eigenvectors and load matrices of the principal components are shown in Table 2. It can be concluded that PC1 is associated with 15% melanin, 17% cAMP, 15% GSH, 26% CAT, 23% MDA, 15% ROS, and 44% SOD, whereas PC2 is mainly associated with 34% tyrosinase, 44% GPx, and 57% GSSG. Using PC1 as the x-axis, and PC2 as the y-axis, a plot was drawn based on the corresponding load values, as shown in Figure 6A. Ten indicators were scattered in the first and third quadrants of the axes. The linear equation of the combined score of each principal component was derived from the eigenvector matrix of the two principal components, and the relative contribution of the variance corresponding to each principal component was used as the weight to build the comprehensive evaluation model: $F1 = 0.154 \times 1 + 0.170 \times 2 - 0.154 \times 3 - 0.146 \times 4 - 0.366 \times 5 - 0.259 \times 6 + 0.238 \times 7 + 0.225 \times 8 + 0.147 \times 9 - 0.442 \times 10$; $F2 = -0.011 \times 1 -$

0.024 × 2 + 0.342 × 3 − 0.001 × 4 + 0.567 × 5 + 0.130 × 6 − 0.436 × 7 − 0.089 × 8 + 0.005 × 9 + 0.372 × 10. Based on the 53.110% of principal component 1, and 41.727% of principal component 2, the composite score function can be derived as F = 0.53110 × F1 + 0.41727 × F2. The composite score of each sample was calculated and ranked by the above model (Table 3). The F1 and F2 scores for each sample are represented as coordinates in Figure 6A. Combined with the PCA composite evaluation model scores, the samples were further classified into four groups by hierarchical cluster analysis (HCA) (Figure 6B-1): control (0), low performing samples (0.1 and 0.2), mild samples (0.4), and high performing samples (0.8 and 1.6). The classification results indicated that FTGML had good activity at 0.8 mg/mL and 1.6 mg/mL addition, and was a suitable anti-melanin agent and antioxidant. The 10 indicators were systematically clustered, and the results are shown in Figure 6B-2, which can be divided into two major categories and four minor categories, which is consistent with what is shown in Figure 6A. Among the subcategories, the antioxidant indicators that were in the same category as the anti-melanin indicators were MDA and ROS.

Table 1. Characteristic values and contribution rates of principal components.

Items	Variance of Initial Eigenvalues			Extract Square Sum Load Variance		
	Eigenvalue	Variance Contribution Rate/%	Accumulative Contribution Rate/%	Eigenvalue	Variance Contribution Rate/%	Accumulative Contribution Rate/%
Principal component 1	8.714	87.144	87.144	5.311	53.110	53.110
Principal component 2	0.769	7.692	94.836	4.173	41.727	94.836
Principal component 3	0.325	3.246	98.083			
Principal component 4	0.140	1.401	99.484	–		
Principal component 5	0.052	0.516	100.000			
Principal component 6	2.834×10^{-16}	2.834×10^{-15}	100.000			
Principal component 7	4.313×10^{-17}	4.313×10^{-16}	100.000			
Principal component 8	-3.498×10^{-17}	-3.498×10^{-16}	100.000			
Principal component 9	-1.600×10^{-16}	-1.600×10^{-15}	100.000			
Principal component 10	-2.691×10^{-16}	-2.691×10^{-15}	100.000			

The correlation between anti-melanin properties and other antioxidant properties is shown in Figure 6C. Melanin content was positively correlated with cAMP and tyrosinase activity (both greater than 0.85), indicating a strong correlation between these three factors. GSH and ROS were negatively/positively correlated with melanin (−0.90 for GSH and 0.96 for ROS), indicating that an increase in GSH and a decrease in ROS favored a reduction in melanin content, which was related to the free radical scavenging capacity of GSH, and the fact that a decrease in ROS inhibited the melanin production process. MDA was positively correlated with melanin (0.95), suggesting that a reduction in the content of lipid peroxidation products in the cell membrane favors a reduction in melanin content. A strong correlation between MDA and ROS (0.99), and a negative correlation between CAT and melanin (−0.95), suggest that reactive oxygen species affecting the cell can directly regulate melanin production. Overall, GSH, CAT, MDA, and ROS are the main antioxidant factors influencing melanin production.

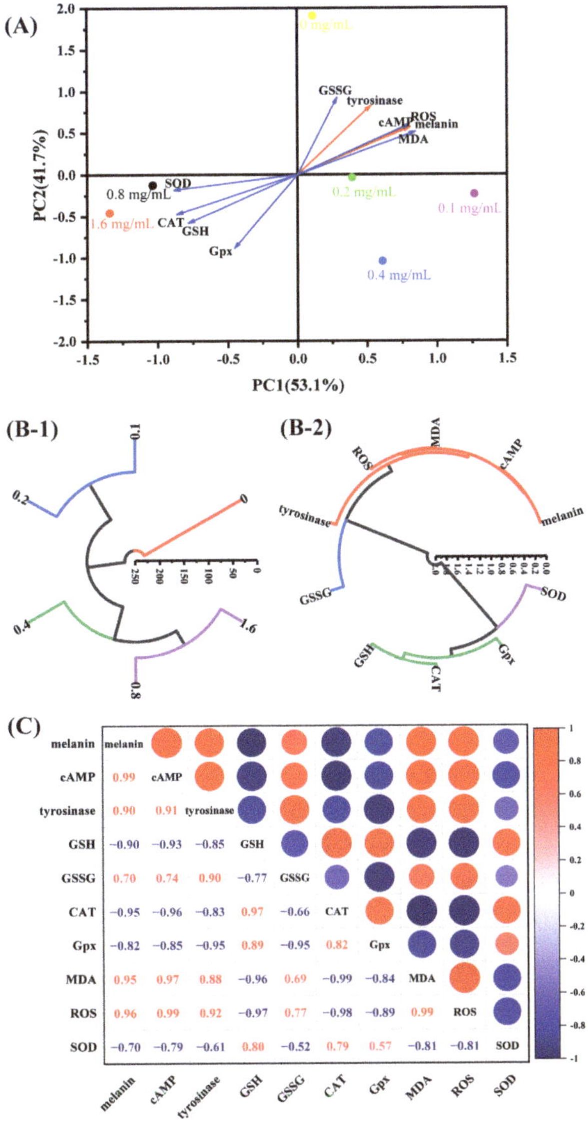

Figure 6. Principal component analysis (**A**) loading plot and score plot. (**B-1**) Hierarchical cluster analysis under different FTGML additions. (**B-2**) Hierarchical cluster analysis of various indicators under different FTGML additions. (**C**) Results of correlation analysis. melanin: mean melanin content (% of control); cAMP: mean cAMP content (nmol/L); tyrosinase: mean tyrosinase ability (% of control); GSH and GSSG: mean GSH and GSSG content (μg/mg protein), respectively; MDA: mean MDA content (nmol/mg protein); ROS: mean ROS content (IU/mL); CAT, GPx, and SOD: mean CAT, GPx, and SOD content (μg/mg protein), respectively.

Table 2. Eigenvectors and load matrices corresponding to principal components.

Items	Principal Component 1		Principal Component 2	
	Feature Vector	Load	Feature Vector	Load
melanin	0.154	0.778	−0.011	0.562
cAMP	0.170	0.807	−0.024	0.567
tyrosinase	−0.154	0.529	0.342	0.824
GSH	−0.146	−0.781	−0.001	−0.580
GSSG	−0.366	0.285	0.567	0.926
CAT	−0.259	−0.863	0.130	−0.475
Gpx	0.238	−0.451	−0.436	−0.884
MDA	0.225	0.845	−0.089	0.513
ROS	0.147	0.800	0.005	0.599
SOD	−0.442	−0.884	0.372	−0.185

Table 3. Principal component score, comprehensive score, and ranking.

Dosage of FTGML (mg/mL)	Principal Component 1 Score (F1)	Principal Component 2 Score (F2)	Comprehensive Score (F)	Ranking
0	0.10872	1.90461	85.25	6
0.1	1.26466	−0.23872	57.20	5
0.2	0.39347	−0.03859	19.29	4
0.4	0.60744	−1.04331	−11.27	3
0.8	−1.03487	−0.12622	−60.23	2
1.6	−1.33942	−0.45777	−90.24	1

4. Conclusions

This study investigated the effects of tyrosinase inhibitory peptide FTGML, derived from grass carp fish scale gelatin, on the melanin inhibitory signaling pathway and intracellular antioxidant activity in murine B16F10 melanoma cells. The results showed that FTGML significantly inhibited intracellular tyrosinase activity and melanin content, and had a positive effect on intracellular antioxidant activity. There was a strong correlation between MDA, ROS, GSH, and CAT among the antioxidant indicators for their ability to counteract melanin, which provides theoretical support for predicting the anti-melanin ability or antioxidant activity. In addition, two signaling pathways through which FTGML affects melanin synthesis were revealed. FTGML downregulates MITF expression by inhibiting the cAMP-PI3K/Akt signaling pathway, as well as p38 and JNK in the MAPK signaling pathway. On the other hand, p38 and JNK were found to be present for the first time in the peptide mediated melanin synthesis signaling pathway, whereas this signaling pathway and the STAT3 factors also promoted apoptosis in B16F10 cells. These results suggested that FTGML can reduce melanin production in mouse B16F10 melanoma cells.

Author Contributions: Conceptualization: Z.H., X.S. and Z.T.; Methodology: Z.H.; investigation: Z.H., L.Z. and S.H.; resources: Z.T.; supervision: X.S. and Z.T.; writing—original draft: Z.H.; writing—review and editing: X.S. and Z.T. All authors have read and agreed to the published version of the manuscript.

Funding: This study was supported by the National Key R&D Program of China (No. 2018YFD0901101), and the Key R&D Program of Jiangxi Province (No. 20192ACB60005), National Natural Science Foundation of China (No. 32160576, 31760445).

Institutional Review Board Statement: Not applicable.

Informed Consent Statement: Not applicable.

Data Availability Statement: The data presented in this study are available on request from the corresponding author.

Conflicts of Interest: The authors declare no conflict of interest.

References

1. Pillaiyar, T.; Manickam, M.; Jung, S.H. Recent development of signaling pathways inhibitors of melanogenesis. *Cell. Signal.* **2017**, *40*, 99–115. [CrossRef] [PubMed]
2. Chai, W.M.; Ou-Yang, C.; Huang, Q.; Lin, M.Z.; Wang, Y.X.; Xu, K.L.; Huang, W.Y.; Pang, D.D. Antityrosinase and antioxidant properties of mung bean seed proanthocyanidins: Novel insights into the inhibitory mechanism. *Food Chem.* **2018**, *260*, 27–36. [CrossRef] [PubMed]
3. Park, W.S.; Kwon, O.; Yoon, T.J.; Chung, J.H. Anti-graying effect of the extract of Pueraria thunbergiana via upregulation of cAMP/MITF-M signaling pathway. *J. Dermatol. Sci.* **2014**, *75*, 153–155. [CrossRef] [PubMed]
4. Kang, Y.G.; Choi, E.J.; Choi, Y.; Hwang, J.K. 5,7-Dimethoxyflavone induces melanogenesis in B16F10 melanoma cells through cAMP-dependent signalling. *Exp. Dermatol.* **2011**, *20*, 445–447. [CrossRef] [PubMed]
5. Shin, S.Y.; Gil, H.-N.; Choi, J.H.; Lim, Y.; Lee, Y.H. Agerarin inhibits α-MSH–induced TYR gene transcription via STAT3 suppression independent of CREB-MITF pathway. *J. Dermatol. Sci.* **2018**, *91*, 107–110. [CrossRef]
6. Hwang, Y.S.; Kim, Y.J.; Kim, M.O.; Kang, M.; Oh, S.W.; Nho, Y.H.; Park, S.H.; Lee, J. Cannabidiol upregulates melanogenesis through CB1 dependent pathway by activating p38 MAPK and p42/44 MAPK. *Chem.-Biol. Interact.* **2017**, *273*, 107–114. [CrossRef]
7. Huang, H.C.; Lin, H.; Huang, M.C. The lactoferricin B-derived peptide, LfB17-34, induces melanogenesis in B16F10 cells. *Int. J. Mol. Med.* **2017**, *39*, 595–602. [CrossRef]
8. Oh, G.W.; Ko, S.C.; Heo, S.Y.; Nguyen, V.T.; Kim, G.; Jang, C.H.; Park, W.S.; Choi, I.W.; Qian, Z.J.; Jung, W.K. A novel peptide purified from the fermented microalga Pavlova lutheri attenuates oxidative stress and melanogenesis in B16F10 melanoma cells. *Process Biochem.* **2015**, *50*, 1318–1326. [CrossRef]
9. Kim, E.K.; Choi, E.J. Pathological roles of MAPK signaling pathways in human diseases. *Biochim. Biophys. Acta-Mol. Basis Dis.* **2010**, *1802*, 396–405. [CrossRef]
10. Song, X.; Hu, X.; Zhang, Y.; Pan, J.; Gong, D.; Zhang, G. Inhibitory mechanism of epicatechin gallate on tyrosinase: Inhibitory interaction, conformational change and computational simulation. *Food Funct.* **2020**, *11*, 4892–4902. [CrossRef]
11. Wang, L.; Oh, J.Y.; Kim, Y.S.; Lee, H.G.; Lee, J.S.; Jeon, Y.J. Anti-Photoaging and Anti-Melanogenesis Effects of Fucoidan Isolated from Hizikia fusiforme and Its Underlying Mechanisms. *Mar. Drugs* **2020**, *18*, 427. [CrossRef] [PubMed]
12. Zhang, X.; Li, J.; Li, Y.; Liu, Z.; Lin, Y.; Huang, J.-A. Anti-melanogenic effects of epigallocatechin-3-gallate (EGCG), epicatechin-3-gallate (ECG) and gallocatechin-3-gallate (GCG) via down-regulation of cAMP/CREB /MITF signaling pathway in B16F10 melanoma cells. *Fitoterapia* **2020**, *145*, 104634. [CrossRef] [PubMed]
13. Yamakoshi, J.; Otsuka, F.; Sano, A.; Tokutake, S.; Saito, M.; Kikuchi, M.; Kubota, Y. Lightening Effect on Ultraviolet-Induced Pigmentation of Guinea Pig Skin by Oral Administration of a Proanthocyanidin-Rich Extract from Grape Seeds. *Pigment Cell Melanoma Res.* **2010**, *16*, 629–638. [CrossRef] [PubMed]
14. Imokawa, G. Analysis of initial melanogenesis including tyrosinase transfer and melanosome differentiation through interrupted melanization by glutathione. *J. Investig. Dermatol.* **1989**, *93*, 100–107. [CrossRef] [PubMed]
15. Kumano, Y.; Sakamoto, T.; Egawa, M.; Iwai, I.; Tanaka, M.; Yamamoto, I. In vitro and in vivo prolonged biological activities of novel vitamin C derivative, 2-O-alpha-D-glucopyranosyl-L-ascorbic acid (AA-2G), in cosmetic fields. *J. Nutr. Sci. Vitaminol.* **1998**, *44*, 345–359. [CrossRef] [PubMed]
16. Ohno, O.; Watabe, T.; Nakamura, K.; Kawagoshi, M.; Uotsu, N.; Chiba, T.; Yamada, M.; Yamaguchi, K.; Yamada, K.; Miyamoto, K. Inhibitory effects of bakuchiol, bavachin, and isobavachalcone isolated from Piper longum on melanin production in B16 mouse melanoma cells. *Biosci. Biotechnol. Biochem.* **2010**, *74*, 1504–1506. [CrossRef] [PubMed]
17. Kubo, I.; Nihei, K.; Tsujimoto, K. Methyl p-coumarate, a melanin formation inhibitor in B16 mouse melanoma cells. *Bioorg. Med. Chem.* **2004**, *12*, 5349–5354. [CrossRef]
18. Usuki, A.; Ohashi, A.; Sato, H.; Ochiai, Y.; Ichihashi, M.; Funasaka, Y. The inhibitory effect of glycolic acid and lactic acid on melanin synthesis in melanoma cells. *Exp. Dermatol.* **2003**, *12*, 43–50. [CrossRef] [PubMed]
19. Torres, E.A.F.D.; Garbelotti, M.L.; Neto, J.M.M. The application of hierarchical clusters analysis to the study of the composition of foods. *Food Chem.* **2006**, *99*, 622–629. [CrossRef]
20. Cha, J.Y.; Yang, H.J.; Moon, H.I.; Cho, Y.S. Inhibitory effect and mechanism on melanogenesis from fermented herbal composition for medical or food uses. *Food Res. Int.* **2012**, *45*, 225–231. [CrossRef]
21. Zhuang, Y.L.; Ma, Q.Y.; Guo, Y.; Sun, L.P. Protective effects of rambutan (Nephelium lappaceum) peel phenolics on H2O2-induced oxidative damages in HepG2 cells and D-galactose induced aging mice. *Food Chem. Toxicol.* **2017**, *108*, 554–562. [CrossRef] [PubMed]
22. Fu, X.Q.; Liu, B.; Wang, Y.P.; Li, J.K.; Zhu, P.L.; Li, T.; Tse, K.W.; Chou, J.Y.; Yin, C.L.; Bai, J.X.; et al. Activation of STAT3 is a key event in TLR4 signaling-mediated melanoma progression. *Cell Death Dis.* **2020**, *11*, 246. [CrossRef] [PubMed]

23. Niu, G.; Bowman, T.; Huang, M.; Shivers, S.; Reintgen, D.; Daud, A.; Chang, A.; Kraker, A.; Jove, R.; Yu, H. Roles of activated Src and Stat3 signaling in melanoma tumor cell growth. *Oncogene* **2002**, *21*, 7001–7010. [CrossRef] [PubMed]
24. Han, J.H.; Bang, J.S.; Choi, Y.J.; Choung, S.-Y. Anti-melanogenic effects of oyster hydrolysate in UVB-irradiated C57BL/6J mice and B16F10 melanoma cells via downregulation of cAMP signaling pathway. *J. Ethnopharmacol.* **2019**, *229*, 137–144. [CrossRef]
25. Yamamura, T.; Onishi, J.; Nishiyama, T. Antimelanogenic activity of hydrocoumarins in cultured normal human melanocytes by stimulating intracellular glutathione synthesis. *Arch. Dermatol. Res.* **2002**, *294*, 349–354. [CrossRef] [PubMed]
26. Montalvo-González, E.; Tovar-Pérez, E.; Alcantara, L. Anti-elastase, anti-tyrosinase and antioxidant properties of a peptide fraction obtained from sorghum (Sorghum bicolor L. Moench) grain. *Int. Food Res. J.* **2019**, *26*, 1813–1822.
27. Delmarmol, V.; Solano, F.; Sels, A.; Huez, G.; Libert, A.; Lejeune, F.; Ghanem, G. Glutathione Depletion Increases Tyrosinase Activity in Human-Melanoma Cells. *J. Investig. Dermatol.* **1993**, *101*, 871–874. [CrossRef]
28. Chou, T.H.; Ding, H.Y.; Hung, W.J.; Liang, C.H. Antioxidative characteristics and inhibition of α-melanocyte-stimulating hormone-stimulated melanogenesis of vanillin and vanillic acid from Origanum vulgare. *Exp. Dermatol.* **2010**, *19*, 742–750. [CrossRef]
29. Wu, J.H.; Sun, B.G.; Luo, X.L.; Zhao, M.M.; Zheng, F.P.; Sun, J.Y.; Li, H.H.; Sun, X.T.; Huang, M.Q. Cytoprotective effects of a tripeptide from Chinese Baijiu against AAPH-induced oxidative stress in HepG2 cells via Nrf2 signaling. *RSC Adv.* **2018**, *8*, 10898–10906. [CrossRef]
30. Huang, H.C.; Chou, Y.C.; Wu, C.Y.; Chang, T.M. [8]-Gingerol inhibits melanogenesis in murine melanoma cells through down-regulation of the MAPK and PKA signal pathways. *Biochem. Biophys. Res. Commun.* **2013**, *438*, 375–381. [CrossRef]
31. de Castro, R.J.S.; Sato, H.H. Biologically active peptides: Processes for their generation, purification and identification and applications as natural additives in the food and pharmaceutical industries. *Food Res. Int.* **2015**, *74*, 185–198. [CrossRef] [PubMed]
32. Lorencini, M.; Brohem, C.A.; Dieamant, G.C.; Zanchin, N.I.T.; Maibach, H.I. Active ingredients against human epidermal aging. *Ageing Res. Rev.* **2014**, *15*, 100–115. [CrossRef] [PubMed]
33. Azam, M.S.; Kwon, M.; Choi, J.; Kim, H.R. Sargaquinoic acid ameliorates hyperpigmentation through cAMP and ERK-mediated downregulation of MITF in α-MSH-stimulated B16F10 cells. *Biomed. Pharmacother.* **2018**, *104*, 582–589. [CrossRef]
34. Liu, Y.Y.; Su, X.R.; Liu, S.S.; Yang, S.S.; Jiang, C.Y.; Zhang, Y.; Zhang, S.C. Zebrafish phosvitin-derived peptide Pt5 inhibits melanogenesis via cAMP pathway. *Fish Physiol. Biochem.* **2017**, *43*, 517–525. [CrossRef] [PubMed]
35. Khaled, M.; Larribere, L.; Bille, K.; Aberdam, E.; Ortonne, J.-P.; Ballotti, R.; Bertolotto, C. Glycogen synthase kinase 3β is activated by cAMP and plays an active role in the regulation of melanogenesis. *J. Biol. Chem.* **2002**, *277*, 33690–33697. [CrossRef] [PubMed]
36. Hsieh, M.J.; Tsai, T.L.; Hsieh, Y.S.; Wang, C.J.; Chiou, H.L. Dioscin-induced autophagy mitigates cell apoptosis through modulation of PI3K/Akt and ERK and JNK signaling pathways in human lung cancer cell lines. *Arch. Toxicol.* **2013**, *87*, 1927–1937. [CrossRef]
37. Han, J.S.; Sung, J.H.; Lee, S.K. Antimelanogenesis Activity of Hydrolyzed Ginseng Extract (GINST) via Inhibition of JNK Mitogen-activated Protein Kinase in B16F10 Cells. *J. Food Sci.* **2016**, *81*, H2085–H2092. [CrossRef]
38. Chang, L.F.; Karin, M. Mammalian MAP kinase signalling cascades. *Nature* **2001**, *410*, 37–40. [CrossRef]
39. Krens, S.F.G.; Spaink, H.P.; Snaar-Jagalska, B.E. Functions of the MAPK family in vertebrate-development. *FEBS Lett.* **2006**, *580*, 4984–4990. [CrossRef]
40. Peng, H.Y.; Lin, C.C.; Wang, H.Y.; Shih, Y.; Chou, S.T. The Melanogenesis Alteration Effects of Achillea millefolium L. Essential Oil and Linalyl Acetate: Involvement of Oxidative Stress and the JNK and ERK Signaling Pathways in Melanoma Cells. *PLoS ONE* **2014**, *9*, e95186. [CrossRef]
41. Pillaiyar, T.; Manickam, M.; Jung, S.H. Downregulation of melanogenesis: Drug discovery and therapeutic options. *Drug Discov. Today* **2017**, *22*, 282–298. [CrossRef] [PubMed]
42. Levy, C.; Khaled, M.; Fisher, D.E. MITF: Master regulator of melanocyte development and melanoma oncogene. *Trends Mol. Med.* **2006**, *12*, 406–414. [CrossRef] [PubMed]

foods

MDPI

Article

Effect of Peroxyl Radical-Induced Oxidation on Functional and Structural Characteristics of Walnut Protein Isolates Revealed by High-Resolution Mass Spectrometry

Xuechun Zhang [1,2,3,4], Xi Yang [4], Yunqian Li [4], Zhenxing Wang [2], Xuemei He [1,*] and Jian Sun [3,*]

1 Guangxi Key Laboratory of Fruits and Vegetables Storage-Processing Technology, Nanning 530007, China; xuechun_zhang@163.com
2 Institute of Environmental Remediation and Human Health, Southwest Forestry University, Kunming 650224, China; wangzhenxingfood@163.com
3 Guangxi Academy of Agricultural Sciences, Nanning 530007, China
4 College of Life Sciences, Southwest Forestry University, Kunming 650224, China; xi_yang2076@163.com (X.Y.); 18388173746@163.com (Y.L.)
* Correspondence: xuemeihe1981@126.com (X.H.); jiansun@gxaas.net (J.S.); Tel.: +86-771-339-0388 (J.S.)

Abstract: The present study aims to investigate the structural and functional properties of oxidated walnut protein isolates (WPI) by 2,2′-azobis (2-amidinopropane) dihydrochloride (AAPH). The oxidation degree, changes in structural characteristics, processing properties, and protein modifications of WPI were measured. The results showed that oxidation significantly induced structural changes, mainly reflected by the increasing carbonyl content, and decreasing sulfhydryl and free amino groups. Moreover, the secondary structure of WPI was altered in response to oxidation, and large aggregates formed through disulfide cross-linking and hydrophobic interactions. Almost all the property indicators were significantly decreased by oxidation except the foaming property and water/oil holding capacity. Mass spectrometry analysis showed that 16 different modifications occurred in amino acid side chains, and most of the protein groups with higher numbers of modifications were found to be associated with allergies, which was further confirmed by the reduction in antigenicity of the major allergen (Jug r 1) in WPI. Meanwhile, we used oxidation-related proteins for gene ontology (GO) enrichment analyses, and the results indicated that 115, 204 and 59 GO terms were enriched in terms of biological process, molecular function, and cellular component, respectively. In conclusion, oxidation altered the groups and conformation of WPI, which in turn caused modification in the functional properties correspondingly. These findings might provide a reference for processing and storage of walnut protein foods.

Keywords: walnut protein isolates; oxidation; structural characteristics; functional properties; proteomics

Citation: Zhang, X.; Yang, X.; Li, Y.; Wang, Z.; He, X.; Sun, J. Effect of Peroxyl Radical-Induced Oxidation on Functional and Structural Characteristics of Walnut Protein Isolates Revealed by High-Resolution Mass Spectrometry. *Foods* **2022**, *11*, 385. https://doi.org/10.3390/foods11030385

Academic Editor: David Bongiorno

Received: 11 December 2021
Accepted: 24 January 2022
Published: 28 January 2022

1. Introduction

Walnut (*Juglans regia* L.) is one of the common tree nuts widely grown worldwide. The planted area and production of walnuts in China ranks first in the world [1]. It has been reported that walnut kernels are rich in unsaturated oils, protein, vitamins, minerals and some bioactive compounds which confer antioxidant, cholesterol reduction, arterial improving, immunomodulatory and other biological activities [2,3]. Therefore, walnut kernels are widely consumed in China (409,608 metric tons in 2019) and even worldwide (914,383 metric tons in 2019) for their health promotion and delicious taste [4].

In view of the high-cholesterol and saturated fatty acids content in animal-derived foods, plant proteins are gaining increasing interest for their specific advantages. Soy protein, peanut protein and walnut protein, etc. have been studied for their nutritional, processing and physiological properties. Among them, as one of the main by-products of the oil industry, walnut protein has great potential of high value utilization. Currently, there have been gradually increasing studies on walnut protein due to its large amount of consumption, such as classification and characterization [5], characteristics analysis and functional properties modification [3,6–9].

Oxidation is one of the three major chemical reactions. In contrast to other forms of modification, oxidation can occur during the harvesting, storage and processing of foods [10]. It has been reported that radical-induced oxidation could change the properties, conformation, digestibility and allergenicity of food proteins [11–14]. As we know, walnut kernels contain high amounts of unsaturated fatty acids, which could generate radical intermediates and secondary products through oxidative stress, and then induce the oxidized walnut protein, in turn, to undergo corresponding changes.

Peroxy radical (ROO·) is an important radical during the lipid oxidation process. However, the effect of peroxy radical on the properties and structure of walnut protein remains unclear. Thus, in this study, walnut protein isolates (WPI) were prepared and oxidized by 2,2′-azobis (2-amidinopropane) dihydrochloride (AAPH), a water-soluble azo-free radical initiator, and then, the effect of peroxy radical on the main groups, structure, functional properties, and antigenicity of WPI were investigated. Meanwhile, the oxidation site and modifications of WPI were analyzed by proteomics technology. This study hopes to provide advice on quality control and utilization of walnut protein.

2. Materials and Methods

2.1. Materials and Reagents

Walnut kernels were purchased from the local market in Baoshan, China, 2,2′-azobis (2-amidinopropane) dihydrochloride (AAPH), 2,4-dinitrophenylhydrazine (DNPH), 5,5′-Dithiobis-(2-nitrobenzoic acid) (DTNB) and 8-Anilino-1-naphthalenesulfonic acid (ANS) were bought from Sigma-Aldrich Co. LLC. (Shanghai, China), and double antibody sandwich ELISA Kits was bought from Shanghai Meilian Biotechnology Co., Ltd. (Shanghai, China). All other reagents used were analytically pure, and water was distilled water in this study.

2.2. Walnut Protein Isolates (WPI) Preparation

WPI was extracted by the alkali dissolution-acid precipitation method according to Qin et al. [9] with slight modifications. Walnut kernels were pulverized and degreased with six-fold volumes of hexane three times. After being dried in a fume hood at room temperature, 25-fold volumes of sodium hydroxide solution (pH 8.7) were added to the defatted walnut powder, stirring the extract for 1 h at 48 °C, followed by centrifugation at 10,000× *g* for 20 min. The supernatant was adjusted to pH 5, and centrifuged using the same conditions as described above. WPI was obtained by washing the precipitate with H_2O until neutral and freeze-drying it at −80 °C for 24 h. The WPI samples were stored frozen at −80 °C for spare use.

2.3. Oxidation of WPI

The oxidation of WPI was performed according to Duan et al. [15] with some modifications. Briefly, dried WPI was dispersed in phosphate buffer (PB, 0.01 mol/L, pH 7.4) to prepare 25 mg/mL protein suspension, then mixed with various concentrations of AAPH (0, 0.04, 0.20, 1.00, 3.00, 5.00 and 10.00 mmol/L), followed by incubation for 24 h at 37 °C under light protection. The reaction solutions were centrifuged for 30 min at 1000× *g* and 4 °C. The supernatants were dialyzed (molecular weight cut-off 7 kDa) against distilled water for 72 h at 4 °C. Next, the dialysates were freeze-dried at −80 °C for 24 h to obtain AAPH oxidized WPI (APPH-WPI) and were stored frozen at −80 °C for spare use.

2.4. Determination of Oxidation Degree of AAPH-Oxidized WPI (APPH-WPI)

2.4.1. Determination of Carbonyl Content

The carbonyl contents of WPI and APPH-WPI were measured by 2,4-dinitrophenylhydrazine (DNPH) assay. Briefly, 1.5 mL protein dispersion (10 mg/mL) was mixed with 1 mL DNPH solution (10 mmol/L, containing 2 mol/L HCl), after incubation in the dark for 1 h at room temperature, then 1 mL of trichloroacetic acid (TCA, 20%) was added and it was mixed well, followed by centrifugation at 8000× *g* for 15 min. The precipitate was washed with 1 mL of ethyl acetate-ethanol (1:1, *v/v*) three times, and then dissolved in 4 mL urea (6 mol/L) for 20 min at 37 °C. After centrifugation at 8000× *g* for 15 min, absorbance of the supernate was read at 370 nm by a microplate reader (Biotek, Winooski, VT, USA), and the carbonyl content of protein was expressed as nmol/mg protein using the molar extinction coefficient of 22,000 $L \cdot mol^{-1} \cdot cm^{-1}$ [16]. Measurement was repeated at least three times to calculate the carbonyl content of the samples.

2.4.2. Determination of Sulfhydryl Group Content

The sulfhydryl group content of WPI and APPH-WPI was measured in triplicate and repeated three times by using the DTNB assay. A 4 mg/mL DTNB solution was first prepared in 0.086 M Tris-Gly buffer (containing 5 mM EDTA, pH 8.0), followed by a suspension of protein samples in Tris-Gly buffer (containing 2% (*w/w*) sodium dodecyl sulfate (SDS)) to obtain a concentration of 10 mg/mL. Then, 1 mL of the sample solution was reacted with 10 μL of DTNB solution for 1 h at room temperature, followed by centrifugation at 10,000× *g* for 10 min, the absorbance of supernate was then measured at 412 nm against the blank (without DTNB and sample). The free sulfhydryl group contents of protein samples were calculated using the extinction coefficient of 13,600 $L \cdot mol^{-1} \cdot cm^{-1}$, while the total sulfhydryl content was measured as a similar procedure, except that 8 M urea was added to the sample buffer [17,18].

2.4.3. Determination of Free Amino Groups Content

The o-phthaldialdehyde (OPA) method was used to measure the free amino groups content of proteins samples. OPA was first dissolved in methanol solution (0.4 g/mL), followed by addition of 2.5 mL of SDS solution (200 g/L), 25 mL of boric acid solution (0.1 mol/L), and 100 μL of β-mercaptoethanol. The mixture was diluted to a certain volume with distilled water to obtain the OPA working solution for the subsequent analysis. Next, an amount of 200 μL of protein solution was mixed with 4 mL of OPA working solution, and then water bathed at 35 °C for 2 min. The absorbance at 340 nm was measured against distilled water as control. This trial was repeated three times to calculate the free amino group content [19].

2.5. Determination of Structural Characteristics of APPH-WPI

2.5.1. Particle Size Distributions

The particle size distribution of protein samples was determined by laser scattering particle size distribution analyzer (Partica, LA-960, Kyoto, Japan) with the relative refractive index and absorption set as 1.414 and 0.001, respectively. Briefly, the protein samples were suspended in distilled water to obtain a 5 mg/mL solution, and then the particle size distribution and the mean diameter of sample solutions were assessed and repeated at least three times, and then made into graphs [20].

2.5.2. Sodium Dodecyl Sulphate–Polyacrylamide Gel Electrophoresis (SDS-PAGE) Analysis

SDS-PAGE was performed under non-reducing and reducing conditions according to the previous reports [21,22]. Briefly, an amount of 15 μL of sample solution (1 mg/mL) was mixed with 5 μL of 4 × protein loading buffer, followed by heating in a 100 °C water bath for 5 min, and then immediately placed in an ice bath. Next, the sample solutions were centrifuged (10,000× *g*, 5 min) and a 10 μL portion of supernatant was taken for electrophoresis (Bio-rad, Powerpac Basic, Hercules, CA, USA) using 12% separating gel and 5% stacking gel, with standards of known molecular weight ranging from 10 to 200 kDa. The gels were decolored and photographed for further analysis. This experiment was repeated with three different batches of samples.

2.5.3. Intrinsic Fluorescence

The protein sample was dispersed in PB (0.01 mol/L, pH 7.9) to prepare a solution at 2.5 μg/mL, followed by centrifugation for 10 min at 10,000× *g*, 4 °C. The fluorescence intensity of supernate was scanned three times (Thermo Fisher Scientific, Lumina, Waltham, MA, USA) from 290 to 500 nm with excitation at 283 nm [23].

2.5.4. Surface Hydrophobicity (H_0)

The surface hydrophobicity (H_0) of WPI was determined by the 1-anilino-8-naphthalenesulfonic acid (ANS) fluorescence probe method [24]. Each protein sample was diluted to 0.005, 0.01, 0.02, 0.05, 0.1, 0.2 and 0.5 mg/mL in PB (10 mM), respectively, then a 2 mL portion of the aforementioned solution was mixed well with 20 μL ANS solution (in 10 mM PB) and allowed to stand for 3 min at room temperature. The fluorescence intensity was measured at 390 nm excitation wavelength and 470 nm emission wavelength. Next, a linearity curve was drawn with the protein concentration as the abscissa and fluorescence intensity as the vertical coordinate. The slope of the curve was calculated as surface hydrophobicity of protein sample. This experiment was repeated with three different batches of samples.

2.5.5. Fourier Transformed Infrared Spectroscopy (FT-IR)

Protein powder samples were mixed with potassium bromide and pressed into a thin tablet, followed by full-band scanning from 4000 to 400 cm^{-1} using a Fourier transform infrared spectrometer (Thermo Fisher Scientific, Nicolet IS50, Waltham, MA, USA), and scans were averaged three times for each spectrum. The data were analyzed and fitted with Peak fit 4.12 and OMNIC 8.2 software to obtain the proportions of the secondary structure (α-helix, β-sheet, β-turn, and random coil), respectively [15].

2.6. Determination of Physicochemical Properties of APPH-WPI

2.6.1. Solubility

A protein sample was suspended in distilled water to obtain a final solution at 0.02 mg/mL and stirred for 1 h at room temperature, followed by centrifugation (6500× *g*, 4 °C) for 20 min. The protein content of supernate was measured by the micro-Kjeldahl method, each replicate was determined three times, and the solubility was calculated as the percentage of the soluble protein (g) compared to total protein sample (g) [5].

2.6.2. Foaming ability (FA) and Foaming Stability (FS)

FA and FS of protein samples were measured according to Qian and Sun [8] with slight modification. A protein sample dispersion (1 mg/mL) was prepared by being suspended in PB (0.05 mol/L, pH 7.0), and after stirring for 1 h at room temperature, a 40 mL portion of the dispersion was homogenized at 10,000 rpm for 2 min. The volume of stirring solution at the first moment (0 min) and 60 min were recorded and calculated according to the follow equations to obtain FA and FS. This experiment was repeated with three different batches of samples.

$$FA(\%) = (V_1 - V)/V \quad (1)$$

$$FS(\%) = (V_2 - V)/(V_1 - V) \quad (2)$$

where V_1 is the volume of sample solution homogenized at the first moment, V is the initial volume of sample solution (40), and V_2 is the volume of sample solution kept for 30 min after being homogenized.

2.6.3. Water/Oil Holding Capacity (WHC/OHC)

An amount of 200 mg protein powder was mixed well with 10 mL water or soybean oil for 1 min in a weighted centrifuge tube using a vortex; afterward, the mixture was centrifuged at 3000× *g* for 20 min and the supernate was discarded, while the pellet was weighed. This trial was repeated three times and the WHC/OHC was expressed as the amount of absorbed water/oil (g) per gram protein sample [25].

2.7. Mass Spectrometry Analysis

Mass spectrometry analysis was carried out by nano liquid chromatography-Q Exactive (LC-QE) mass spectrum (Thermo Fisher Scientific, Q Exactive, Waltham, MA, USA). The control and oxidized WPI (10 mmol/L AAPH) samples were prepared by reduction and alkylation methods, followed by trypsin-hydrolyzed (1:50, *w/w*) at 37 °C for 20 h. The hydrolysates were lyophilized and re-dissolved in formic acid solution (FA, 0.1%, *w/w*) after desalting, and stored at −20 °C for further liquid chromatography (LC) separation. The LC conditions were as follows: Solution A, aqueous solution of 0.1% FA, solution B, 84% acetonitrile containing 0.1% FA. The hydrolysate was loaded into the trap column by automatic sampler after being balanced, then the full scan was performed and 20 fragment profiles were collected to obtain the mass charge ratios of peptides and peptide fragments. The mass spectrometry data were searched against the UniProtKB database by proteome discoverer software (Thermo Fisher Scientific, Version 1.4, Waltham, MA, USA).

2.8. Double Antibody Sandwich Enzyme-Linked Immunosorbent Assay (ELISA)

The allergenicity of WPI was evaluated by determining the content of Jug r 1, the major allergen of walnut, using double antibody sandwich ELISA Kits. WPI dispersions oxidized with 0, 5, 10, 15, and 30 mmol/L AAPH were prepared (3 mg/mL) and centrifuged at 5000× *g*, 4 °C for 30 min, and the supernatants were saved as sample solutions. A 50 µL portion of standard and mentioned sample solution were loaded onto the microplate wells, respectively, followed by being closed and incubated at 37 °C for 30 min, then the liquid was discarded and the microplates were washed 5 times by wash solution and patted dry. Next, enzyme-labeled reagent (50 µL) was added to each well except the blank well, after incubation and washing, and the chromogenic agents A (50 µL) and B (50 µL) were added to each well, mixed well and incubated in the dark at 37 °C for 10 min. Finally, the reaction was terminated by adding a stop solution (50 µL) to each well, and the absorbance at 450 nm was read (TECAN, F50, Männedorf, Switzerland). A standard curve was drawn with the standard concentration as abscissa and OD as the ordinate. This experiment was repeated with three different batches of samples, and the Jug r 1 content was calculated by the standard curve according to the corresponding OD.

2.9. Statistical Analysis

All experiments were repeated three times, and the results were expressed as average ± standard deviation (SD). The significance was determined by one-way ANOVA, and the graphs were drawn using origin 2021 software (Origin Lab, Northampton, MA, USA).

3. Results and Discussion

3.1. Oxidative Indicators Analysis

3.1.1. Carbonyl Groups

The process of peroxy radical oxidation of protein is complex and accompanied by corresponding products generation, among which carbonyl group is the most widely used indicator for preliminary evaluation of oxidation degree [10]. The effect of peroxy radical on carbonyl content of WPI is shown in Table 1. Carbonyl content of oxidized WPI was significantly ($p < 0.05$) increased with the concentration of AAPH, and a maximum (1.48 nmol/mg) at 10 mmol/L AAPH was observed. As reported by previous studies [26], peroxy radicals were generated linearly with the decomposition of AAPH, in turn attacking the main peptide chain or side chain of WPI, and thus carbonyl derivatives such as aldehydes and ketones were formed accordingly. These results were similar to that of AAPH oxidized casein [27] and ovalbumin [17].

Table 1. Carbonyl, free sulfhydryl, total sulfhydryl, and free amino groups of WPI oxidized with different concentration of AAPH.

AAPH (mmol/L)	Carbonyl (nmol/mg)	Free Sulfhydryl (nmol/mg)	Total Sulfhydryl (nmol/mg)	Free Amino (nmol/mg)
0.0	0.61 ± 0.08 c	6.66 ± 0.10 a	172.43 ± 16.17 ab	0.67 ± 0.07 a
0.2	0.85 ± 0.05 b	6.03 ± 0.21 b	173.09 ± 11.10 a	0.65 ± 0.05 ab
0.5	0.90 ± 0.10 b	5.11 ± 0.04 c	165.66 ± 4.59 ab	0.60 ± 0.03 b
1.0	0.93 ± 0.05 b	4.18 ± 0.13 d	155.70 ± 4.76 bc	0.54 ± 0.04 c
3.0	1.30 ± 0.15 a	3.16 ± 0.04 e	148.00 ± 3.69 cd	0.48 ± 0.03 d
5.0	1.42 ± 0.01 a	2.34 ± 0.04 f	135.64 ± 2.17 d	0.43 ± 0.02 e
10.0	1.48 ± 0.12 a	1.25 ± 0.09 g	114.26 ± 4.35 e	0.31 ± 0.04 f

WPI = walnut protein isolates, AAPH = 2, 2′-azobis (2-amidinopropane) dihydrochloride. Different letters within a column indicate significant differences ($p < 0.05$).

3.1.2. Sulfhydryl Groups

Cysteine residues are sensitive to oxidation, and are susceptible to corresponding modification, therefore, the degree of protein oxidation can be reflected by the content of sulfhydryl groups. As shown in Table 1, with the increase of AAPH concentration, the total and free sulfhydryl groups of WPI decreased significantly ($p < 0.05$), and the minimums were achieved (1.25 and 114.26 nmol/mg) respectively when AAPH concentration was 10 mmol/L. This result is similar to that of peroxyl radical oxidized chickpea protein reported by Zhu et al. [23], the sulfhydryl groups were oxidized to both disulfide bonds and irreversible sulfonic and sulfonic acid groups by radicals [28]; moreover, other covalent cross-linking might have occurred during oxidation [18].

3.1.3. Free Amino Groups

As depicted in Table 1, the free amino groups of oxidized WPI decreased with the AAPH concentration, and the minimum (0.31 nmol/mg) was discerned at a concentration of 10 mmol/L, which could be due to the oxidation of amino-containing side chain (threonine, proline, arginine and lysine residues) in WPI [29], and the formation of Schiff base with carbonyl groups by a covalent reaction [19]. This finding echoed that of the carbonyl result mentioned above and a similar result was obtained with oxidized *Coregonus peled* myofibrillar proteins [29].

3.2. Structural Characteristics Analysis

3.2.1. Endogenous Fluorescence

The endogenous fluorescence of proteins originates from the emission of tryptophan residues, and the polarity of the micro-environment in which can be reflected by the changes of fluorescence peak. Therefore, endogenous fluorescence was often used to indicate the transformation of protein conformation. The effects of peroxy radical on the endogenous fluorescence intensity and maximum fluorescence emission wavelength of WPI are shown in Figure 1a. With increasing AAPH concentration, the fluorescence intensity of WPI showed a downward trend. This might be attributed to the unfolding of the WPI tertiary structure, and the tryptophan residue is oxidized to kynurenine, thus inducing a reduction in the endogenous fluorescence intensity. In addition, a blue shift of maximum fluorescence peak due to oxidation was also observed in Figure 1a, which is similar to the report by Zhang et al. [18]; hence, it was concluded that the tryptophan residue of WPI was exposed to hydrophobic environment which in turn made hydrophobic interactions by oxidation.

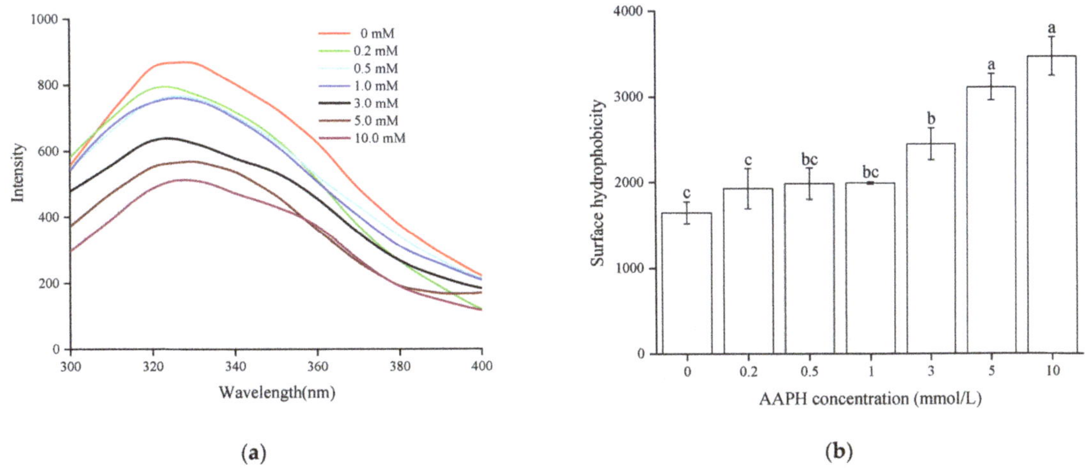

Figure 1. Endogenous fluorescence spectra (**a**) and surface hydrophobicity (**b**) of WPI oxidized with different concentration of AAPH. WPI = walnut protein isolates, AAPH = 2, 2′-azobis (2-amidinopropane) dihydrochloride. Different letters indicate significant differences between columns ($p < 0.05$).

3.2.2. Surface Hydrophobicity

Surface hydrophobicity reflects the distribution of hydrophobic amino acid residues on the protein's surface, and characterizes the changes of tertiary structure of protein [30]. The surface hydrophobicity of oxidized WPI increased significantly ($p < 0.05$) with the AAPH concentration, the maximum achieved at 10 mmol/L AAPH, which is 84% greater than the control (Figure 1b), indicating that the peptide chains were moderately unfolded and rearranged [16], and the surface hydrophobic groups were exposed to the polar environment. This result corresponded to those of endogenous fluorescence; meanwhile, oxidized myofibrillar proteins from *Culter alburnus* also showed similar trends [18].

3.2.3. Particle Size Distributions

Particle size is the common indicator to evaluate protein aggregation [31], and can be monitored by dynamic light scattering. Particle size distributions and average particle size of WPI oxidized with different concentrations of AAPH are shown in Figure 2. The particle size of control WPI was mainly distributed between 1–10 μm, while a large peak at 100 μm was observed after oxidation, illustrating formation of large aggregates through the interaction of protein peptides by oxidation [32]. Moreover, there is a rising trend on average particle size with higher AAPH concentration, with maximum values (119.2 μm) at 10 mmol/L AAPH, which is similar to the result of Fu et al. [16]. Overall, the above results suggested that the hydrophobic groups exposure (Figure 1b), as well as the disulfide and secondary bonds cross-linking (Table 1) might be responsible for the formation of larger protein aggregates.

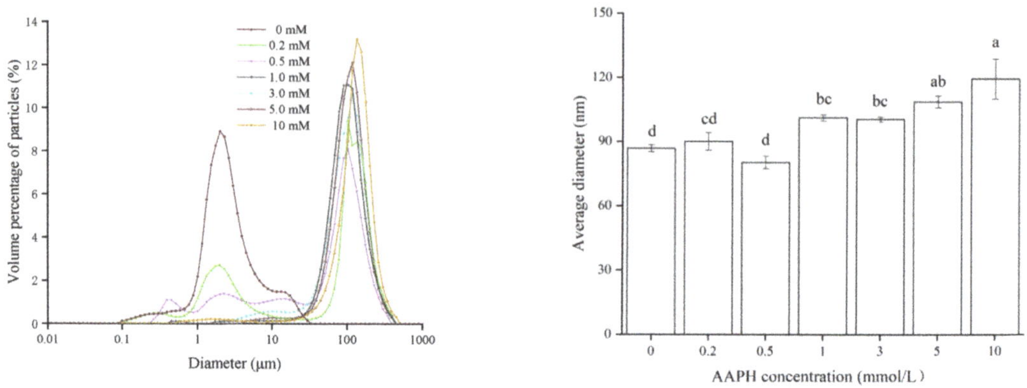

(a) (b)

Figure 2. Particle size distributions (**a**) and average diameter (**b**) of WPI oxidized with different concentrations of AAPH. WPI = walnut protein isolates, AAPH = 2, 2′-azobis (2-amidinopropane) dihydrochloride. Different letters indicate significant differences between columns ($p < 0.05$).

3.2.4. SDS-PAGE

SDS-PAGE has the advantages of being fast, of high resolution and high sensitivity, and is especially suitable for the analysis and identification of oligomers and their subunits, as well as molecular weight determination of protein. As shown in the non-reducing SDS-PAGE pattern (Figure 3b), the intensity of 7 S conglycinin bands decreased with increasing AAPH concentration; meanwhile, new large molecule weight bands were observed at the top of the gel, indicating that large polymers may be formed by stronger oxidation [19]. Moreover, most of the degraded bands caused by oxidation were recovered in the reducing SDS-PAGE pattern (Figure 3a), and thus we reasoned that disulfide bond polymerization is the predominant manner for aggregates formation, which is similar to the study of Fu et al. [16].

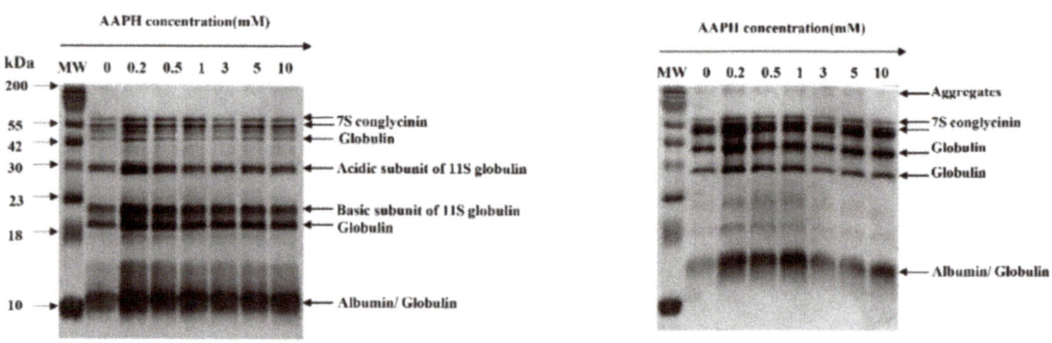

(a) (b)

Figure 3. Sodium dodecyl sulphate–polyacrylamide gel electrophoresis (SDS-PAGE) patterns of WPI oxidized with different AAPH concentration (**a**) Reducing (**b**) Non-reducing. WPI = walnut protein isolates, AAPH = 2, 2′-azobis (2-amidinopropane) dihydrochloride; MW= molecular weight.

3.2.5. Fourier Transform Infrared (FT-IR) Spectroscopy

The secondary structure of WPI could be reflected by the FT-IR spectrum, specifically, the amide 1 region was processed in de-convolution and second derivative, and the featured wavenumber of α-helix is 1648–1655 cm^{-1}, β-sheet is 1610–1629 cm^{-1}, β-turn is 1660–1681 cm^{-1}, and random coil is 1638 cm^{-1} [24]. As shown in Figure 4, the percentage of α-helix decreased whereas that of β-sheet and random coil increased with the AAPH concentration. Studies have shown that α-helix is associated with the weak hydrogen-bond interaction between amino groups and amidogen, whereas β-sheet is related to the hydrogen-bond interaction between peptide chains [33]. Thus we deemed that the hydrogen-bond interaction between amino bond of WPI was broken by oxidation, and the unfolding peptide chains further re-associated to form a new construction (β-sheet and random coil), which in turn decreased the flexibility of the protein structure [28]. This finding is in accordance with the oxidized chickpea protein isolates reported by Zhu et al. [23].

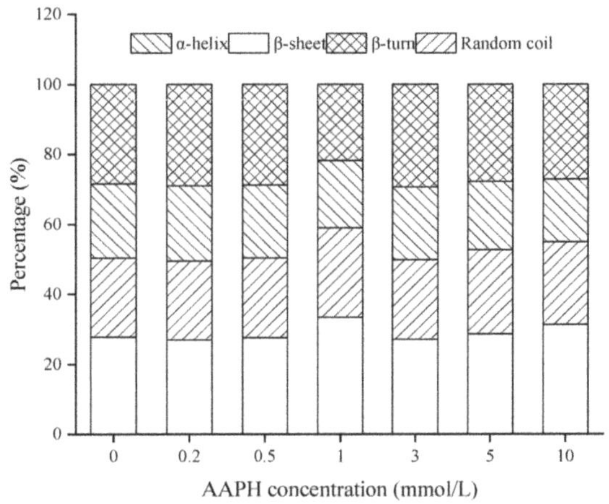

Figure 4. Secondary structure of WPI oxidized with different concentrations of AAPH. WPI = walnut protein isolates, AAPH = 2, 2′-azobis (2-amidinopropane) dihydrochloride.

3.3. *Physicochemical Properties Analysis*

3.3.1. Solubility

Solubility is one of the critical characteristics of food proteins, and is often used to represent the degree of cross-linking and aggregation to a certain extent. As shown in Table 2, with the increasing AAPH concentration, the solubility of WPI shows a significant downward trend ($p < 0.05$), and the lowest solubility (20.44%) was observed at 10 mmol/L AAPH. This may be due to the unfolding and destruction of WPI conformation by peroxy radical oxidation, then further polymerization occurs to form macromolecule insoluble aggregates by covalent or non-covalent interactions [34]. The result of solubility is similar to that of myofibrillar protein from *Culter alburnus* oxidized by hydroxyl radical [18], and also echoes the result of particle size distribution in this study.

Table 2. Solubility, WHC, OHC, and foaming property of WPI oxidized with different concentration of AAPH.

AAPH (mmol/L)	Solubility (%)	WHC (g/g)	OHC (g/g)	FA (%)	FS (%)
0.0	80.64 ± 3.41 a	3.34 ± 0.05 b	4.55 ± 0.07 d	22.66 ± 3.65 e	72.22 ± 6.80 c
0.2	69.25 ± 3.72 b	2.61 ± 0.03 d	4.63 ± 0.08 d	27.78 ± 2.72 cde	74.58 ± 8.72 cb
0.5	59.65 ± 2.96 c	2.98 ± 0.15 c	4.87 ± 0.07 bc	28.89 ± 3.44 bcd	78.50 ± 5.48 abc
1.0	46.87 ± 2.41 d	3.31 ± 0.12 bc	5.00 ± 0.04 b	36.67 ± 3.65 a	83.33 ± 3.65 a
3.0	39.71 ± 3.07 e	4.05 ± 0.17 a	5.84 ± 0.19 a	33.33 ± 4.22 ab	81.39 ± 5.00 ab
5.0	31.48 ± 3.79 g	3.86 ± 0.15 a	4.74 ± 0.03 cd	31.11 ± 3.44 bc	78.33 ± 2.58 abc
10.0	20.44 ± 2.49 h	3.35 ± 0.29 b	4.64 ± 0.07 d	25.33 ± 2.98 de	72.36 ± 6.46 c

WPI = walnut protein isolates, AAPH = 2, 2′-azobis (2-amidinopropane) dihydrochloride; WHC = water holding capacity; OHC = oil holding capacity; FA = foaming ability; FS = foaming stability. Different letters within a column indicate significant differences ($p < 0.05$).

3.3.2. Foaming Ability (FA) and Foaming Stability (FS)

Foaming property is an important interfacial property for protein processing, and is impacted by solubility, hydrophobicity, and flexibility of peptide chain. As indicated in Table 2, with the increase of AAPH concentration, the FA of WPI increased first and then decreased, and reached the maximum (36.67%) when the concentration of AAPH was 1 mmol/L. This is possibly because appropriate oxidation makes non-covalent bonds of WPI unfold, thereby the flexibility of the peptide chain increases. Consequently, more protein was adsorbed and formed a relatively stable gas–liquid interface. However, when WPI was over oxidized, the internal structure of the protein was largely unfolded and reconstructed, which led to the drop down of solubility markedly, and then a decline in FA. Meanwhile, FS also showed a similar trend by oxidation, and the maximum (83.33%) was identified at 1 mmol/L AAPH. The above research is similar to that on chickpea protein and rice bran globulin [23,24].

3.3.3. Water Holding Capacity (WHC) and Oil Holding Capacity (OHC)

WHC and OHC are known to be linked to the water/oil binding ability of protein, and can significantly affect the texture of protein foods. As presented in Table 2, the WHC and OHC of oxidized WPI exhibited an initial increment followed by a decreasing tendency with AAPH concentration, and both the WHC (4.05 g/g) and OHC (5.84 g/g) reached maximum at 3 mmol/L AAPH. This is probably because, on the one hand, the hydrophobic groups of WPI were partly unfolded by oxidation, and some charged amino acid exposed to the surface, thus made the net charge of WPI increased [12]; and these mentioned changes of protein would be an advantage for WHC. On the other hand, macromolecular protein aggregation was formed by intermolecular disulfide bond crosslinking in oxidation [18], which would be a disadvantage for WHC. Therefore, the positive effects appear to be stronger than the negative effects in mild oxidation, and the water molecules entered the internal protein with non-covalent bonds, that made the WHC increase correspondingly. However, when the concentration was higher than 3 mmol/L AAPH, the negative effect was stronger than positive effects, thus decreasing the WHC dramatically. To the OHC, the oil adsorption capacity of WPI was enhanced with the hydrophobicity in certain concentrations of AAPH (≤3 mmol/L), and the OHC increased accordingly. Nevertheless, in strong oxidative conditions (>3 mmol/L), WPI was crosslinked and aggregated, which reduced the OHC consequently.

3.4. *Proteomic Analysis of Oxidative Modified WPI*

In order to further investigate the impact of oxidation on the structure of WPI, a Nano LC-MS/MS peptide sequencing analysis was performed, and the total ion chromatograms (TIC) of unoxidized and oxidized protein were shown in Figure 5. Although the trend of the TIC plot of both samples seemed to be very similar, the differences in protein abundance still revealed changes.

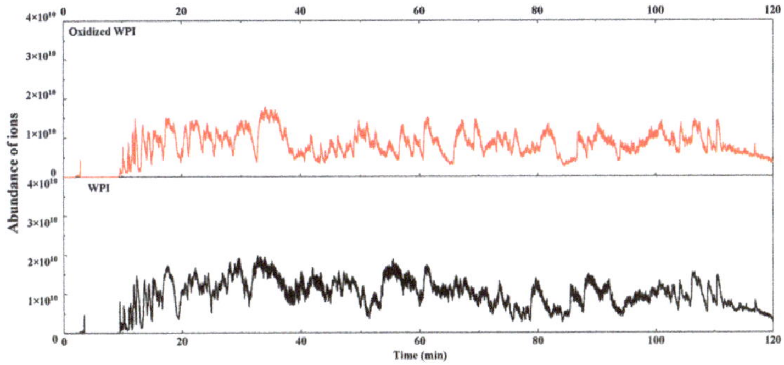

Figure 5. The total ion chromatogram (TIC) of WPI and oxidized WPI. WPI = walnut protein isolates.

By mass spectrometry analysis, a total of 3095 peptide spectra were detected, of which 646 protein groups were corresponding to 2396 unique peptide sequences (Supporting Information File S1, Table S1). According to their respective peptide sequences, the modifications in the functional group of the amino-acid side chains were analyzed and are listed in Table 3, including acetyl, carbamidomethyl, carboxyethylation, γ-glutamic semialdehydes (GGS), dehydrated, dehydroalanine, deoxidation, 4-hydroxy-2-nonenal (HNE), hydroxykynurenine, kynurenin, lactyllysine, malondialdehyde (MDA), oxidation, trioxidation, α-aminoadipic acid (AAA), and α-aminoadipic semialdehyde (AAS). For the different amino acids, cysteine, methionine, and tryptophan had the greatest number of side-chain modification types (5), followed by histidine, tyrosine and proline (4), and phenylalanine (3), whereas methionine and arginine created two types of modification. The above results are consistent with results previously reported on amino acids side-chains most prone to oxidative modification [35]. These modifications by oxidation may induce changes in the structure of the proteins, which in turn affected the function of proteins. For example, cysteine has unique reactivity which is associated with sulfhydryl reactive chemicals, hence, cysteine substitution has been a powerful tool to investigate the structure and function of proteins [36]. Methionine oxidation may influence the peptide hydrophobicity [37], where histidine residues in proteins are major targets for reaction with the lipid peroxidation product HNE [38], and so forth.

The UpSet plot and Venn diagram were constructed to further visualize the numbers and intersections of the identified peptide from different modifications. From Figure 6, the top five modifications ranked by modified number were carbamidomethyl, oxidation, HNE, MDA, and deoxidation. They were 243, 169, 64, 61, and 61 times, respectively. Of them, 124 times occurred alone for carbamidomethyl, and 119 times were the combination of carbamidomethyl with other modifications. The Wayne diagram in the upper part of the figure also showed the same results. It was generally believed that proteins from an animal source were more susceptible to undergoing oxidative reactions than plant proteins, while GGS and AAS were the significant protein oxidative indicators in meat and dairy products [39]. The results of this study indicated that numerous modifications also occurred in oxidized WPI, whereas carbamidomethyl and oxidation was the most significant indicator.

Table 3. The amino acid modified sites and types in the oxidation of WPI.

Amino Acid	Peptide Sequence	Modification Types												
		CAM	OX	DI	DHA	TRI	HNE	KYN	3-HK	GGS	CAE	ACE	AAS	AAA
Cysteine	**C**FDGSLFEY**C**AK	√												
Cysteine	GKLNFGRALE**C**FFLSSCSSPCFCMNSMESQDEFETR		√											
Cysteine	CFDGSLFEYCAK					√								
Cysteine	GLHGAAIPG**C**PETFQSESSSQFR			√										
Cysteine	M**CC**WNAPPCGFFYLNVDGAIFFYIHKAGVGAVVR						√							
Methionine	CQDE**M**R		√											
Methionine	QCCQQLSQ**M**DEQCQCEGLR			√										
Proline	CTNNAEKI**PP**GTVR			√										
Proline	CGDQVGC**P**CIGFDGLDVASECGPSCECGLECGNRSTQR		√											
Phenylalanine	**F**RPGTVALR		√											
Phenylalanine	NFYLAGNPDDE**F**RPQGQQEYEQHRR			√										
Phenylalanine	**F**MFWFACGVGGLEIICIFLVWCLLSR					√								
Histidine	**H**NLDTQTESDVFSR						√							
Histidine	ILRPVSVPG**H**FEAFHGSGGEDPESFYR			√										
Histidine	ILRPVSVPG**H**FEAFHGSGGEDPESFYR		√											
Histidine	QL**H**VILK					√								
Lysine	**K**EDIEMALTK						√							
Lysine	RVNALENVV**K**PR											√		
Lysine	**K**GSLGCIKYILK										√			
Lysine	VGESPSPATASS**K**PEQAR													√
Lysine	EAAYNLHLIY**K**												√	
Tyrosine	CDGCVRSIFPPF**Y**TCAQCGFFLHKSCVELSR		√											
Tyrosine	CLKLYCECFAAGI**Y**CVGSCACETCFNKPEYEDLVLDTR				√									
Tyrosine	ACGTCCARCDCVPPGTSGNYDACPC**Y**ANMTTHGGR			√										
Tyrosine	VVQCTEGERS**Y**HIFYQLCAGAPAALR					√								
Tryptophan	FVSVCYHNYEHVYCF**W**HRGMICVCCHT**W**PEILYYK		√											
Tryptophan	FLIGDDEHC**W**SENGVSNIEGGCYAK							√						
Tryptophan	FML**W**FACGVGGFEIICIFLVWCLLSR			√										
Tryptophan	ANRMG**W**FQLQR								√					
Tryptophan	GALYSDALYVPH**W**NLNAHSVVYALRGR					√								
Arginine	**R**ATGEGFEWVSFK									√				
Arginine	**R**QEEAEWEEEEAR						√							
Proline	ESFNLECGDVIRV**P**AGATVYVINQDSNERLEMVK			√										
Proline	ESFNLECGDVIRV**P**AGATVYVINQDSNERLEMVK	□	√	□	□	□	□	□	□	□	□	□	□	□

Note: Tick mark represents the corresponding modification was detected, the lower line and bold font represent the corresponding modification site. WPI = walnut protein isolates, CAM = carbamidomethyl; OX = oxidation; DI = dioxidation; TRI = trioxidation; DHA = dehydroalanine; HNE = 4-hydroxy-2-nonenal; KYN = kynurenin; 3-HK = 3-hydroxykynurenine; GGS = γ-glutamic semialdehydes; CAE = carboxyethylation; ACE = acetyl; AAS = α-aminoadipic semialdehyde; AAA = α-aminoadipic acid.

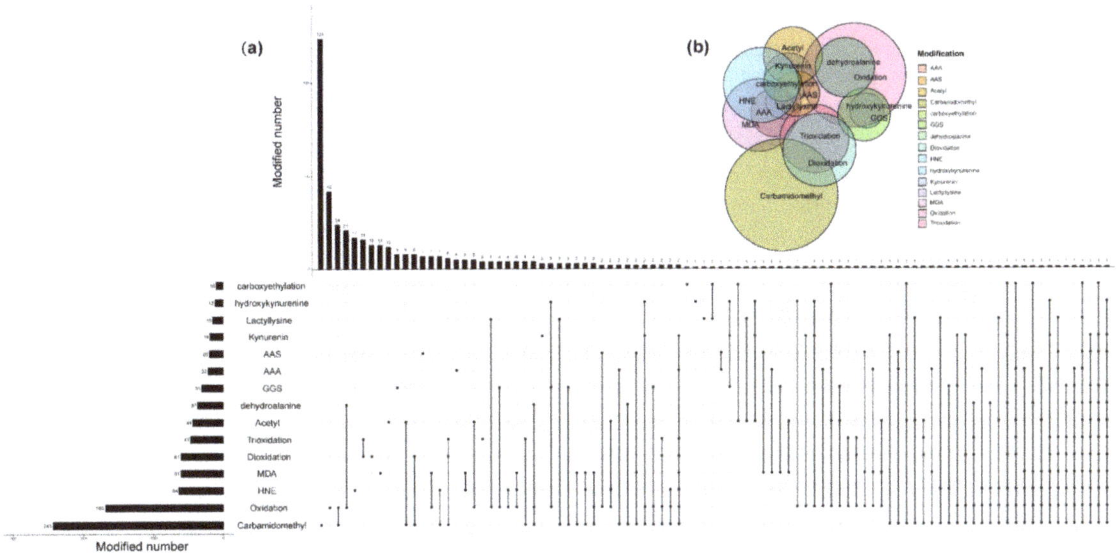

Figure 6. (**a**) UpSet plot of interactions between different modifications. (**b**) Venn diagram of interactions between different modifications. AAA = α-aminoadipic acid; AAS = α-aminoadipic semialdehyde; GGS = γ-glutamic semialdehydes; HNE = 4-hydroxy-2-nonenal; MDA = malondialdehyde.

To present the degree of modification on different protein groups more intuitively, a heatmap (Figure 7) was drawn in log10-transformed modification numbers (only total modification numbers equal to or greater than 5 are displayed). From Figure 7, oxidation, carbamidomethyl, deoxidation, HNE, and trioxidation were, in order, the most modified protein groups, and their modification numbers were greater than 5. This conclusion was basically consistent with the conclusion above. For different protein groups, the top 10 protein groups with the most modification numbers were P93 198 (190 times), A0A2I4F6R4 (127 times), A0A2I4DYF1 or Q9SEW4 (117 times), A0A2I4EG83 (108 times), Q2TPW5 (94 times), A0A2I4E5L6 (90 times), A0A2I4GEH1 (82 times), A0A2I4DYF1 (81 times), and Unknown (43 times). These proteins were mainly depicted as albumin seed storage protein, legumin-like protein, vicilin-like protein, and 11 S globulin seed storage protein. Interestingly, most of these proteins were found to be associated with allergies [40,41].

Mass spectrometry of these peptides detected the presence of 519 proteins, and detailed information of these proteins is presented in Supporting Information File S2: Table S2. To further understand the functions of these proteins, gene ontology (GO) annotation was performed (Supporting Information File S3: Table S3). Through analysis, 378 GO terms were annotated, including 115 terms in biological process (BP), 204 in molecular function (MF), and 59 in cellular component (CC), and the top 10 enriched GO terms of different classifications are displayed in Figure 8. It can be seen clearly that "cell redox homeostasis" was the most enriched (number = 31) in BP, while the most enriched (number = 46) in CC was "integral component of membrane". For MF, it was "ATP binding" with 33 enrichment numbers. Taking into consideration that protein modifications, especially post-translational modification, could regulate the structure and function of protein, including lifetime, assembly, localization, function, and degradation, it was therefore of great interest to further understand the main biological functions of these proteins [42].

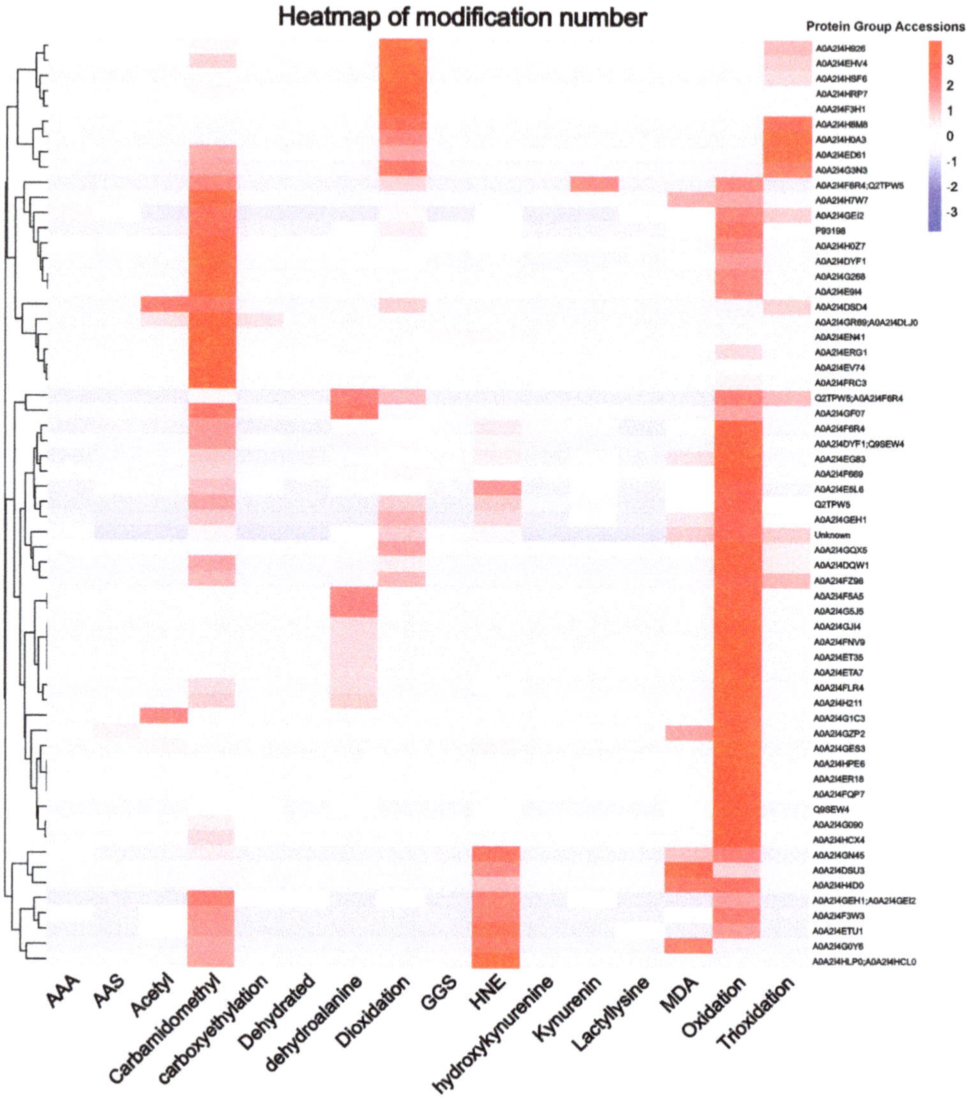

Figure 7. The heatmap of modification numbers. AAA = α-aminoadipic acid; AAS = α-aminoadipic semialdehyde; GGS = γ-glutamic semialdehydes; HNE = 4-hydroxy-2-nonenal; MDA = malondialdehyde.

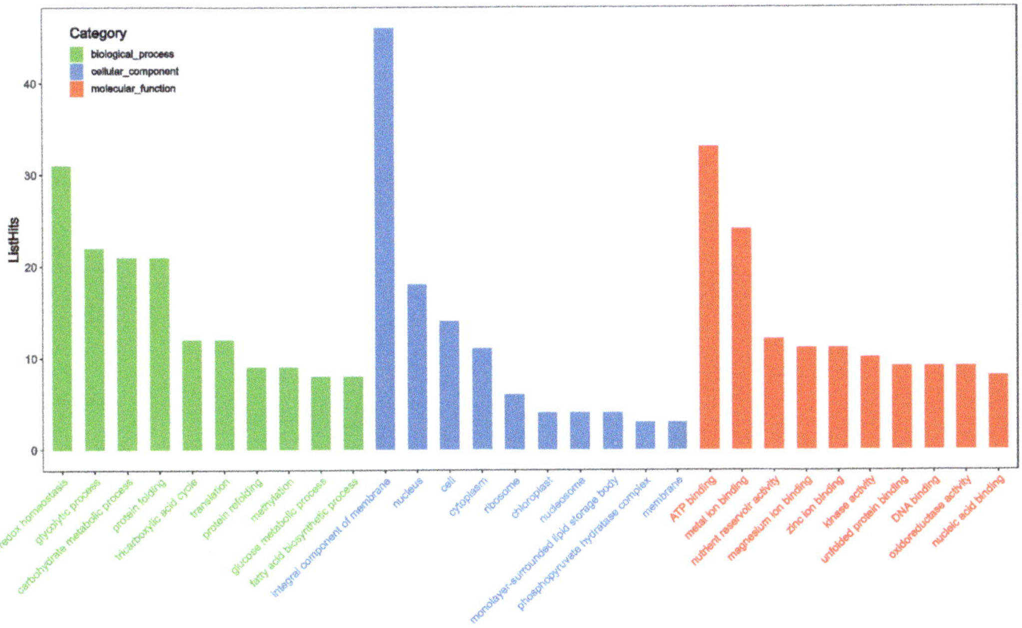

Figure 8. The top 10 functional enrichment analyses of gene ontology (GO).

3.5. Antigenicity of Jug r 1

Jug r 1, which is 16.4 kD long and contains 139 amino acids, is the major allergen protein in walnuts, and could reflect the allergenic properties of WPI. As hinted at by Figure 9, the effect of oxidation on the antigenicity of Jug r 1 was assessed. It can be observed that there was a gradual decrease of antigenicity with the AAPH concentration, and the lowest value (13.91 ng/mL) was obtained at 30 mmol/L, indicating a significant effect ($p < 0.05$) on antigenicity by radical oxidation. Combined with the mass spectrometry results, changes in oxidative indicators and secondary structure (Table 1 and Figure 4), the oxidative modification of WPI might be covalent, and the changes of protein side chains led to the destruction of allergen epitopes, which consequently decreased the antigenicity of Jug r 1 [43]. In addition, the unfolding of WPI blocked the epitopes and could not be identified by an antibody. Similar results were exhibited in acrolein-oxidized shrimp tropomyosin [44] and hydroxyl radical-treated β-conglycinin [45], whereas the specific reason for the reduction of antigenicity should be further researched.

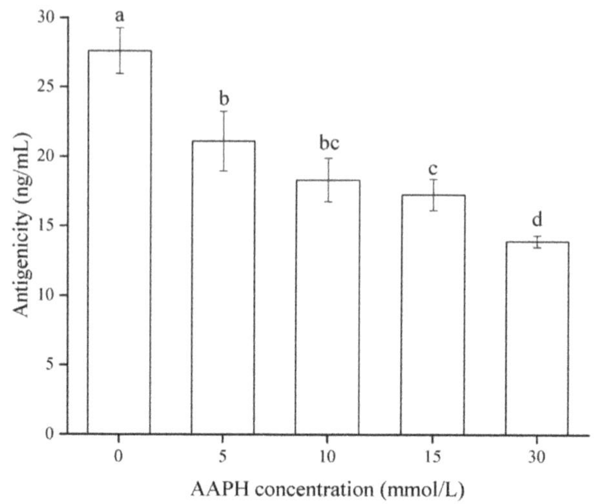

Figure 9. Effect of oxidation on antigenicity of Jug r 1. AAPH = 2, 2′-azobis (2-amidinopropane) dihydrochloride. Different letters within a column indicate significant differences ($p < 0.05$).

4. Conclusions

This paper explored the influence of peroxy radicals on structural and functional characters of WPI. The results showed that the carbonyl group of WPI was significantly increased with the concentration of AAPH, whereas the free amino and sulfhydryl groups decreased significantly ($p < 0.05$), which could be due to formation of disulfide bonds and irreversible sulfonic, sulfonic acid groups by radical oxidization. The reduction of fluorescence intensity and increasing of surface hydrophobicity illustrated the unfolding and rearranging of protein tertiary structure; meanwhile, FT-IR analysis also demonstrated alteration of the secondary structure. Besides, large aggregates were observed by particle size distributions and SDS-PAGE analysis, and thus unfolding and cross-linking of WPI were further confirmed. The aforementioned changes of conformation in turn affected the functional properties of WPI, and thus we saw a significant reduction in solubility, whereas different degrees of increase in foaming property and WHC/OHC were observed by mild oxidation. Nevertheless, all of the measured property indicators in this study were found to be significantly decreased after overoxidation. Mass spectrometry analysis revealed numerous modifications of the amino acid side chains, and these modifications are involved in many protein groups. Interestingly, these protein groups were found to be associated with allergies, which implied that oxidative modification could have an effect on the allergenic properties of WPI. The double antibody sandwich ELISA result indicated the gradual reduction of Jug r 1, the major allergen of WPI, and this further confirmed the mass spectrometry results. In conclusion, peroxy radical oxidation altered the conformation of WPI, which in turn caused modification in functional properties correspondingly, including allergenicity.

Proteins are one of the most important components of foods and are easily oxidized during their harvesting, storage, and processing. Although it is well known that protein structure strongly influences its functions, little is known about oxidative modifications and how it affects food protein structure and function. Generally, the preliminary properties, and secondary and tertiary structure of proteins can be investigated via multiple physical, chemical, and biological means, including a chromogenic assay, protein electrophoresis, spectroscopic studies, mass spectrometry, etc. However, they are often studied in isolation, and the results are incoherent. In this study, an integrated approach was used to produce results at multiple levels, seeking to provide a comprehensive understanding of the effect of oxidation induced by peroxy radical on the functional properties of WPI. Our results suggest that oxidation can lead to the modification of amino acids in proteins, subsequently altering their structure, which in turn changes their functional properties. Although the extent of protein modification is strongly increased with the increase of AAPH concentration, and it appears to perform adversely for the functional properties of the protein, appropriate oxidation is favorable for some properties, such as foaming property, WHC/OHC, and allergenicity. In conclusion, a potential methodology to limit oxidation, including antioxidants, should be taken into account. On the other hand, the above results also revealed the possibility of oxidation as a means to regulate protein functional properties in the future.

Supplementary Materials: The following supporting information can be downloaded at: https://www.mdpi.com/article/10.3390/foods11030385/s1, Table S1: peptide information; Table S2: protein information; Table S3: gene ontology (GO) annotation.

Author Contributions: Methodology, X.Z. and Z.W.; validation, X.Y. and Y.L.; formal analysis, X.H.; data curation, X.Y. and Y.L.; writing—original draft preparation, X.Z.; writing—review and editing, Z.W. and J.S.; supervision, X.H.; project administration, J.S.; funding acquisition, J.S. All authors have read and agreed to the published version of the manuscript.

Funding: This research was funded by the Postdoctoral Starting Foundation of Guangxi Academy of Agricultural Sciences (grant number 2018033), National Natural Science Foundation of China (grant number 31760440), and Yunnan Innovative Research Team (grant number 202005AE160017).

Institutional Review Board Statement: Not applicable.

Informed Consent Statement: Not applicable.

Data Availability Statement: The data presented in this study are available on request from the corresponding author.

Acknowledgments: The authors thank the support of the Applied Protein Technology Co., Ltd. for mass spectrometry support.

Conflicts of Interest: The authors declare no conflict of interest.

References

1. Mao, X.; Hua, Y. Composition, structure and functional properties of protein concentrates and isolates produced from walnut (*Juglans regia* L.). *Int. J. Mol. Sci.* **2012**, *13*, 1561–1581. [CrossRef] [PubMed]
2. Kong, X.Z.; Zhang, L.N.; Lu, X.; Zhang, C.M.; Hua, Y.F.; Chen, Y.M. Effect of high-speed shearing treatment on dehulled walnut proteins. *LWT Food Sci. Technol.* **2019**, *116*, 108500. [CrossRef]
3. Stanisic, J.; Ivkovic, T.; Romic, S.; Zec, M.; Culafic, T.; Stojiljkovic, M.; Koricanac, G. Beneficial effect of walnuts on vascular tone is associated with Akt signalling, voltage-dependent calcium channel LTCC and ATP-sensitive potassium channel Kv1.2. *Int. J. Food Sci. Nutr.* **2021**, *72*, 324–334. [CrossRef] [PubMed]
4. Nuts & Dried Fruits Statistical Yearbook 2020/2021. Available online: https://www.nutfruit.org/files/tech/1625230833_INC_Stats_2021.pdf (accessed on 19 November 2021).
5. Kong, X.; Zhang, L.; Song, W.; Zhang, C.; Hua, Y.; Chen, Y.; Li, X. Separation, identification and molecular binding mechanism of dipeptidyl peptidase IV inhibitory peptides derived from walnut (*Juglans regia* L.) protein. *Food Chem.* **2021**, *347*, 129062. [CrossRef] [PubMed]
6. Hu, H.; Fan, T.; Zhao, X.; Zhang, X.; Sun, Y.; Liu, H. Influence of pH and salt concentration on functional properties of walnut protein from different extraction methods. *J. Food Sci. Tech. Mys.* **2017**, *54*, 2833–2841. [CrossRef] [PubMed]
7. Moghadam, M.; Salami, M.; Mohammadian, M.; Emam-Djomeh, Z.; Jahanbani, R.; Moosavi-Movahedi, A.A. Physicochemical and bio-functional properties of walnut proteins as affected by trypsin-mediated hydrolysis. *Food Biosci.* **2020**, *36*, 100611. [CrossRef]
8. Sun, Q.; Ma, Z.F.; Zhang, H.X.; Ma, S.J.; Kong, L.M. Structural characteristics and functional properties of walnut glutelin as hydrolyzed: Effect of enzymatic modification. *Int. J. Food Prop.* **2019**, *22*, 265–279. [CrossRef]
9. Qin, Z.; Guo, X.; Lin, Y.; Chen, J.; Liao, X.; Hu, X.; Wu, J. Effects of high hydrostatic pressure on physicochemical and functional properties of walnut (*Juglans regia* L.) protein isolate. *J. Sci. Food Agric.* **2013**, *93*, 1105–1111. [CrossRef]
10. Hellwig, M. The chemistry of protein oxidation in food. *Angew. Chem. Int. Ed. Engl.* **2019**, *58*, 16742–16763. [CrossRef]
11. Li, X.; Liu, C.; Wang, J.; Zhou, K.; Yi, S.; Zhu, W.; Xu, Y.; Lin, H.; Li, J. Effect of hydroxyl radicals on biochemical and functional characteristics of myofibrillar protein from large yellow croaker (*Pseudosciaena crocea*). *J. Food Biochem.* **2020**, *44*, e13084. [CrossRef]
12. Nyaisaba, B.M.; Liu, X.X.; Zhu, S.C.; Fan, X.J.; Sun, L.L.; Hatab, S.; Miao, W.H.; Chen, M.L.; Deng, S.G. Effect of hydroxyl-radical on the biochemical properties and structure of myofibrillar protein from Alaska pollock (*Theragra chalcogramma*). *LWT Food Sci. Technol.* **2019**, *106*, 15–21. [CrossRef]

13. Chen, N.; Zhao, M.; Sun, W. Effect of protein oxidation on the in vitro digestibility of soy protein isolate. *Food Chem.* **2013**, *141*, 3224–3229. [CrossRef] [PubMed]
14. Lv, L.; Lin, H.; Li, Z.; Nayak, B.; Ahmed, I.; Tian, S.; Chen, G.; Lin, H.; Zhao, J. Structural changes of 2,2′-azobis(2-amidinopropane) dihydrochloride (AAPH) treated shrimp tropomyosin decrease allergenicity. *Food Chem.* **2019**, *274*, 547–557. [CrossRef] [PubMed]
15. Duan, X.; Li, M.; Shao, J.; Chen, H.; Xu, X.; Jin, Z.; Liu, X. Effect of oxidative modification on structural and foaming properties of egg white protein. *Food Hydrocoll.* **2017**, *75*, 223–228. [CrossRef]
16. Fu, Q.; Liu, R.; Wang, H.; Hua, C.; Song, S.; Zhou, G.; Zhang, W. Effects of Oxidation in Vitro on Structures and Functions of Myofibrillar Protein from Beef Muscles. *J. Agric. Food Chem.* **2019**, *67*, 5866–5873. [CrossRef] [PubMed]
17. Zhang, J.J.; Tu, Z.C.; Wang, H.; Hu, Y.M.; Du, P.C.; Yang, Y.P. Mechanism of the effect of 2, 2′-azobis (2-amidinopropane) dihydrochloride simulated lipid oxidation on the IgG/IgE binding ability of ovalbumin. *Food Chem.* **2020**, *327*, 127037. [CrossRef] [PubMed]
18. Zhang, Z.; Xiong, Z.; Lu, S.; Walayat, N.; Hu, C.; Xiong, H. Effects of oxidative modification on the functional, conformational and gelling properties of myofibrillar proteins from Culter alburnus. *Int. J. Biol. Macromol.* **2020**, *162*, 1442–1452. [CrossRef]
19. Guo, X.; Qiu, H.; Deng, X.; Mao, X.; Guo, X.; Xu, C.; Zhang, J. Effect of Chlorogenic Acid on the Physicochemical and Functional Properties of Coregonus Peled Myofibrillar Protein through Hydroxyl Radical Oxidation. *Molecules* **2019**, *24*, 3205. [CrossRef]
20. Shen, H.; Stephen Elmore, J.; Zhao, M.; Sun, W. Effect of oxidation on the gel properties of porcine myofibrillar proteins and their binding abilities with selected flavour compounds. *Food Chem.* **2020**, *329*, 127032. [CrossRef]
21. Trigui, I.; Zarai, Z.; Chevance, S.; Cheikh-Rouhou, S.; Attia, H.; Ayadi, M.A. Physicochemical properties, antioxidant activity and in vitro gastrointestinal digestion of purified proteins from black cumin seeds. *Int. J. Biol. Macromol.* **2019**, *126*, 454–465. [CrossRef]
22. Labuckas, D.; Maestri, D.; Lamarque, A. Effect of different oil extraction methods on proximate composition and protein characteristics of walnut (*Juglans regia* L.) flour. *LWT Food Sci. Technol.* **2014**, *59*, 794–799. [CrossRef]
23. Zhu, Z.; Mao, X.; Wu, Q.; Zhang, J.; Deng, X. Effects of oxidative modification of peroxyl radicals on the structure and foamability of chickpea protein isolates. *J. Food Sci.* **2021**, *86*, 824–833. [CrossRef] [PubMed]
24. Wu, W.; Li, F.; Wu, X. Effects of rice bran rancidity on oxidation, structural characteristics and interfacial properties of rice bran globulin. *Food Hydrocoll.* **2020**, *110*, 106123. [CrossRef]
25. Abbou, A.; Kadri, N.; Dahmoune, F.; Chergui, A.; Remini, H.; Berkani, F.; Adel, K.; Boukhalfa, F.; Madani, K. Optimising functional properties and chemical composition of Pinus halepensis Mill. Seeds protein concentrates. *Food Hydrocoll.* **2020**, *100*, 105416. [CrossRef]
26. Zhou, L.Y.; Zhang, Y.; Zhao, C.B.; Lin, H.J.; Wang, Z.J.; Wu, F. Structural and functional properties of rice bran protein oxidized by peroxyl radicals. *Int. J. Food Prop.* **2017**, *20*, 1456–1467. [CrossRef]
27. Wang, J.; Tan, Y.; Xu, H.; Niu, S.; Yu, J. Effect of 2,2-azobis (2-amidinopropane) dihydrochloride oxidized casein on the microstructure and microrheology properties of emulsions. *Food Sci. Biotechnol.* **2016**, *25*, 1283–1290. [CrossRef]
28. Li, F.; Wu, X.; Wu, W. Effects of oxidative modification by malondialdehyde on the in vitro digestion properties of rice bran protein. *J. Cereal Sci.* **2020**, *97*, 103158. [CrossRef]
29. Deng, X.; Lei, Y.; Liu, J.; Zhang, J.; Qin, J. Biochemical changes of Coregonus peled myofibrillar proteins isolates as affected by HRGS oxidation system. *J. Food Biochem.* **2019**, *43*, e12710. [CrossRef]
30. Wang, S.; Zhang, Y.; Chen, L.; Xu, X.; Zhou, G.; Li, Z.; Feng, X. Dose-dependent effects of rosmarinic acid on formation of oxidatively stressed myofibrillar protein emulsion gel at different NaCl concentrations. *Food Chem.* **2018**, *243*, 50–57. [CrossRef]
31. Li, X.P.; Liu, C.K.; Wang, J.X.; Li, W.X.; Lin, B.Y.; Zhu, W.H.; Xu, Y.X.; Yi, S.M.; Mi, H.B.; Li, J.R. Tea Polyphenols Affect Oxidative Modification and Solution Stability of Myofibrillar Protein from Grass Carp (Ctenopharyngodon idellus). *Food Biophys.* **2020**, *15*, 397–408. [CrossRef]
32. Pan, J.F.; Lian, H.L.; Jia, H.; Hao, R.Y.; Wang, Y.J.; Ju, H.P.; Li, S.J.; Dong, X.P. Dose affected the role of gallic acid on mediating gelling properties of oxidatively stressed Japanese seerfish myofibrillar protein. *LWT Food Sci. Technol.* **2020**, *118*, 108849. [CrossRef]
33. Zhu, Y.P.; Wang, Y.; Li, J.L.; Li, F.; Teng, C.; Li, X.T. Effects of Water-Extractable Arabinoxylan on the Physicochemical Properties and Structure of Wheat Gluten by Thermal Treatment. *J. Agric. Food Chem.* **2017**, *65*, 4728–4735. [CrossRef] [PubMed]
34. Zhang, D.; Li, H.; Wang, Z.; Emara, A.M.; Hu, Y.; He, Z. Effects of in vitro oxidation on myofibrillar protein charge, aggregation, and structural characteristics. *Food Chem.* **2020**, *332*, 127396. [CrossRef] [PubMed]
35. Heinonen, M.; Gürbüz, G.; Ertbjerg, P. Oxidation of proteins. In *Chemical Changes during Processing and Storage of Foods*; Rodriguez-Amaya, D.B., Amaya-Farfan, J., Eds.; Academic Press: Cambridge, MA, USA, 2021; pp. 85–123.
36. Akabas, M.H. Cysteine modification: Probing channel structure, function and conformational change. In *Novel Chemical Tools to Study Ion Channel Biology*; Ahern, C., Pless, S., Eds.; Springer New York: New York, NY, USA, 2015; pp. 25–54.
37. Lao, Y.W.; Gungormusler-Yilmaz, M.; Shuvo, S.; Verbeke, T.; Spicer, V.; Krokhin, O.V. Chromatographic behavior of peptides containing oxidized methionine residues in proteomic LC-MS experiments: Complex tale of a simple modification. *J. Proteom.* **2015**, *125*, 131–139. [CrossRef] [PubMed]
38. Uchida, K.; Stadtman, E.R. Modification of histidine residues in proteins by reaction with 4-hydroxynonenal. *Proc. Natl. Acad. Sci. USA* **1992**, *89*, 4544–4548. [CrossRef]

39. Estevez, M.; Ollilainen, V.; Heinonen, M. Analysis of protein oxidation markers alpha-aminoadipic and gamma-glutamic semialdehydes in food proteins using liquid chromatography (LC)-electrospray ionization (ESI)-multistage tandem mass spectrometry (MS). *J. Agric. Food Chem.* **2009**, *57*, 3901–3910. [CrossRef] [PubMed]
40. Teuber, S.S.; Dandekar, A.M.; Peterson, W.R.; Sellers, C.L. Cloning and sequencing of a gene encoding a 2S albumin seed storage protein precursor from English walnut (Juglans regia), a major food allergen. *J. Allergy Clin. Immunol.* **1998**, *101*, 807–814. [CrossRef]
41. Youle, R.J.; Huang, A.H. Albumin storage protein and allergens in cottonseeds. *J. Agric. Food Chem.* **1979**, *27*, 500–503. [CrossRef]
42. Lourenço dos Santos, S.; Petropoulos, I.; Friguet, B. The Oxidized Protein Repair Enzymes Methionine Sulfoxide Reductases and Their Roles in Protecting against Oxidative Stress, in Ageing and in Regulating Protein Function. *Antioxidants* **2018**, *7*, 191. [CrossRef]
43. Song, Y.; Li, Z.; Lin, H.; Du, S.; Hao, Z.; Lin, H.; Zhu, Z. Effect of malondialdehyde treatment on the IgE binding capacity and conformational structure of shrimp tropomyosin. *Food Chem.* **2015**, *175*, 374–380. [CrossRef]
44. Lv, L.; Lin, H.; Li, Z.; Wang, J.; Ahmed, I.; Chen, H. Changes of structure and IgE binding capacity of shrimp (Metapenaeus ensis) tropomyosin followed by acrolein treatment. *Food Funct.* **2017**, *8*, 1028–1036. [CrossRef] [PubMed]
45. Xu, J.; Chen, Z.J.; Han, D.; Li, Y.Y.; Sun, X.T.; Wang, Z.J.; Jin, H. Structural and Functional Properties Changes of β-Conglycinin Exposed to Hydroxyl Radical-Generating Systems. *Molecules* **2017**, *22*, 1893. [CrossRef] [PubMed]

foods

MDPI

Article

Cryoprotective Roles of Carboxymethyl Chitosan during the Frozen Storage of Surimi: Protein Structures, Gel Behaviors and Edible Qualities

Xiangwei Zhu [1,*], Minglang Zhu [1], Diheng He [1], Xueyin Li [1], Liu Shi [2], Lan Wang [2], Jianteng Xu [3], Yi Zheng [3] and Tao Yin [4,*]

1 National "111" Center for Cellular Regulation and Molecular Pharmaceutics, Key Laboratory of Fermentation Engineering (Ministry of Education), Hubei Key Laboratory of Industrial Microbiology, Hubei University of Technology, Wuhan 430068, China; datouzhu1216@163.com (M.Z.); hdiheng@163.com (D.H.); xiali12282021@163.com (X.L.)
2 Institute for Farm Products Processing and Nuclear-Agricultural Technology, Hubei Academy of Agricultural Science, Wuhan 430064, China; shiliu_hzau@163.com (L.S.); lilywang_2016@163.com (L.W.)
3 Department of Grain Science and Industry, Kansas State University, Manhattan, KS 66506, USA; jianteng@ksu.edu (J.X.); yzheng@ksu.edu (Y.Z.)
4 College of Food Science and Technology, Huazhong Agricultural University, Wuhan 430070, China
* Correspondence: xiangwei@ksu.edu (X.Z.); yintao@mail.hzau.edu.cn (T.Y.); Tel.: +86-182-7189-3897 (X.Z.)

Abstract: Carboxymethyl chitosan (CMCh) is an ampholytic chitosan derivative that manifests versatile applications in food industry, such as antibacterial ingredients and nutritional additives. However, its use as a cryoprotectant remains under-researched. In this study, the cryoprotective effect of CMCh oligosaccharide (CMCO) on frozen surimi (silver carp) was systematically investigated in terms of protein structures, gelling behaviors, and sensory qualities. CMCO (0.6%) was incorporated in the surimi before frozen storage (−18 °C for 60 days) while the commercial cryoprotectant (4% sucrose, 4% sorbitol) was used as a positive control. Results indicated that CMCO could inhibit the freezing-induced denaturation of myofibrillar protein, whose values of solubility, Ca^{2+}-ATPase and sulfhydryl content were 24.8%, 64.7%, and 17.1% higher than the nonprotected sample, respectively, while the surface hydrophobicity was 21.6% lower. Accordingly, CMCO stabilized microstructure of the surimi gels associated with improved gel strength, viscoelasticity, water-holding capacities, and whiteness. Moreover, the cryoprotective effect of CMCO with higher degree of carboxymethyl substitution (DS: 1.2) was more pronounced than that of low-DS-CMCO (DS: 0.8). Frozen surimi treated with high-DS-CMCO achieved competitive gelling properties and sensory acceptability to those with the commercial counterpart. This study provided scientific insights into the development of ampholytic oligosaccharides as food cryoprotectants.

Keywords: carboxymethyl chitosan; frozen surimi; myofibrillar protein; denaturation; gelling properties

Citation: Zhu, X.; Zhu, M.; He, D.; Li, X.; Shi, L.; Wang, L.; Xu, J.; Zheng, Y.; Yin, T. Cryoprotective Roles of Carboxymethyl Chitosan during the Frozen Storage of Surimi: Protein Structures, Gel Behaviors and Edible Qualities. *Foods* **2022**, *11*, 356. https://doi.org/10.3390/foods11030356

Academic Editors: Jianhua Xie, Yanjun Zhang and Hansong Yu

Received: 24 December 2021
Accepted: 24 January 2022
Published: 26 January 2022

1. Introduction

Surimi is a reconstituted food used for the production of various seafood dishes, such as fish balls, cakes, and croquettes [1,2]. Among all ingredients of the fish muscle, myofibrillar protein (MP), a salt-soluble protein, is mainly responsible for surimi functionalities [3]. Like other uncured meats, the shelf life of surimi is largely depending on frozen storage [3–5], which restrains intrinsic enzyme activities, microbial growth, and lipid oxidation of surimi during long-term preservation [6,7]. However, undesirable quality deteriorations still occur (e.g., water loss, structural weakening, and nutrition decay), and most of them are related to MP denaturation [8], which owes to the effect of ice crystallization, protein concentration and oxidation [9].

To address those problems, appropriate uses of food-grade cryoprotectants are necessary [10]. Given the unexpected taste or high caloric values of commercial cryoprotectants, such as sugars, polyols, and phosphates [11], research interest shifted to other natural additives derived from food proteins and saccharides. Among which, oligosaccharides, i.e., carrageenan oligomer [12], cellobiose [13], and konjac oligo-glucomannan [14] etc., have drawn particular attentions owing to their pronounced health benefits and consumer acceptability [15].

Oligosaccharides function as cryoprotectants through: (1) inhibiting ice recrystallization in frozen surimi and reducing physical damages to MP structures [9], (2) forming saccharide-MP binary complex to prevent MP from freezing-induced aggregation [16]. It is noteworthy that both ice-inhibition and complexation effects of saccharides were dependent on their strong polarities [9,17]. Therefore, some intrinsic structural characteristics, such as the charge properties, of exogenous oligosaccharides are believed to govern their cryoprotective effects. For example, Zhang et al. [18,19] reported that alginate oligosaccharide and trehalose could prevent the freezing denaturation of myosin. Though both saccharides exhibited the same degree of polymerization (DP = 2n), the charged alginate oligomer had stronger protecting effect than the noncharged trehalose.

Chitosan is an acid-soluble cationic polysaccharide that has been widely used as food additives [20,21]. To improve its solubility in water, carboxymethylation is mostly carried out [22]. The derived carboxymethyl chitosan (CMCh, Figure S1) is a typical ampholytic polysaccharide [23,24]. It demonstrates outstanding moisture-retention capacity [22], which is the key to regulate the water state and ice crystallization at subzero temperatures. Furthermore, the ampholytic structure of saccharide was also reported to be effective in modulating the protein aggregation [25]. On the basis of these observations, the ampholytic CMCh was a potential cryoprotectant to proteins. However, the development of CMCh-based additives remained less in frozen food when compared with other charged or neutral saccharides, such as chitosan (+), carrageenan (−), and konjac glucomannan (nonionic) [26,27]. Thus, the objective of this study is to: (1) investigate how the ampholytic structure of CMCh influences its cryoprotective effect to frozen surimi proteins, and (2) evaluate the edible properties of obtained surimi gels.

Since the functional properties of charged saccharides are highly related to their charge densities [25,26], CMCh with different degrees of carboxymethyl substitution (DS) were selected as cryoprotectants in this study, which were hydrolyzed to oligosaccharides and added to surimi before freezing. The storage stability of MP was thoroughly characterized, while the gelling behaviors and edible qualities of frozen surimi were also evaluated, including the rheological properties, microstructures, mechanical strength, water-holding capacities, sensory scores, and whiteness. This work aims at a paradigm shift for the development of ampholytic oligosaccharides as high-performance cryoprotectants for aquatic food preservations.

2. Materials and Methods

2.1. Materials

Alive farm raised-silver carp (Hypophthalmichthys molitrix, weighing 2.0 ± 0.1 kg) were purchased from a local market (Wuhan, China), transported to laboratory within 15 min, and subjected to percussive stunning. Carboxymethyl chitosan with two different DS, CMCh-A and -B, were kindly donated by Haobo biotech Corp (Henan, China). The DS values of CMCh-A and -B were characterized to be 0.8 and 1.2 by using the potentiometric titration [28]. Both CMCh exhibited similar molecular weight (M_w) at about 6.02×10^4 Da. Chitosanase, with an activity unit at 200,000 U/g, was purchased from Shengda biotech Corp (Henan, China). All other chemicals were purchased from Aladdin Bio-chem Technology Co., LTD (Shanghai, China). Ultrapure water was used in all experiments unless specified otherwise.

2.2. Preparation of CMCh Oligosaccharide

The CMCh oligosaccharide (CMCO) was prepared according to previous studies with some modifications [29,30]. Briefly, 500 mL of CMCh solution (10 mg/mL) was prepared at pH 5, and an aliquot of 60 mg chitosanase was added into the solution. The hydrolysis reaction was carried out at 50 °C for 5 h and terminated by heating the mixture at 100 °C for 10 min. The obtained solution was neutralized, filtered, and concentrated with a vacuum rotary evaporator (IkA-Works Inc., Staufen, Germany). Then, the hydrolysate was precipitated by mixing with 10 volumes of ethanol. After being redissolved in the water, the solution was filtered through a polymer membrane (Mw cut-off: 10 KDa) and the filtrates were lyophilized. The average DP for both CMCO-A and CMCO-B was characterized to be approximately 6.8, as reflected by the contents of reducing sugars.

2.3. Preparation of Frozen Surimi

The frozen surimi was prepared according to Zhangt et al. [31]. The fish was killed and its head, scales and internal organs were removed. Fresh fish meat was cut and washed three times by five volumes of distilled ice water. Then, the surimi was dispersed in 0.5% NaCl solution at the ratio of 1:2 (*w*/*v*) and dehydrated with nylon net to reach a final moisture content at 78%. The protein content in surimi was determined to be 15.2% through the Kjeldahl method. The obtained surimi was added with CMCO-A (0.6%), CMCO-B (0.6%) or commercial cryoprotectant (4% sucrose and 4% sorbitol), respectively, and packaged in polyethylene bags with the size of $20 \times 15 \times 2$ cm^3. Surimi without any cryoprotectant added was set as a blank control. All samples were frozen at −18 °C in air (with a freezing rate of 1.2 °C/min) and conditioned overnight to achieve uniform initial temperatures. Then, the surimi was transferred to a freezer and stored at −18 °C for 0, 10, 20, 30, 45 and 60 days. The frozen surimi was thawed at 4 °C for 20 h before further analysis.

2.4. Characterizations of MP in Frozen Surimi

2.4.1. Content of Salt-Soluble Protein

MP of the surimi was prepared by following the processes of Lin et al. with modifications [7]. Frozen surimi (5 g) was mixed with 45 mL of 0.6 M KCl solution (pH 7.0) and homogenized at 9000 rpm for 100 s (an operation of 20 s with an interval of 15 s). The homogenate was conditioned in an ice water bath for 30 s and centrifuged at 9000 rpm, 4 °C for 20 min. The supernatant was collected and mixed with three volumes of cold deionized water followed by centrifugation at 6000 rpm, 4 °C for 20 min. The precipitate was re-suspended in 20 mL of 0.6 M KCl solution and centrifuged again at 6000 rpm, 4 °C for 20 min. Finally, content of salt-soluble protein in the supernatant was determined by the Biuret method.

2.4.2. Ca^{2+}-ATPase Activity

Ca^{2+}-ATPase activity was determined according to Liu et al. [14]. Briefly, the Ca^{2+}-ATPase assay was performed at 37 °C for 10 min in a reaction medium that consisted of 0.5 M KCl, 5 mM $CaCl_2$, 25 mM Tris-maleate (pH 7.0), 6 mg/mL MP, and 1 mM ATP. The reaction was terminated by adding $HClO_4$ at 5% final concentration. The mixture was centrifuged at 9000 rpm for 10 min, and the released inorganic phosphate in the supernatant was measured by using phosphomolybdate method. The activity of Ca^{2+}-ATPase of MP was expressed as Pi amount per mg of MP per reaction time liberation min^{-1} (μmol/mg/min)$^{-1}$.

2.4.3. Surface Hydrophobicity

Surface hydrophobicity of MP from the frozen surimi was characterized by using 8-anilino-1-naphthalenesulfonic acid (ANS) as the fluorescence probe according to Lin et al. with modifications [7]. The MP solution was diluted with 0.6 M KCl-10 mM phosphate buffer (pH 6.0) to achieve a series of protein concentrations of 0.2, 0.3, 0.5 and 1.0 mg/mL, and 20 μL of 8 mM ANS-0.1 M phosphate buffer (pH 7.0) was added to 1 mL of the prepared protein solution. Fluorescence intensity (FI) of the mixture was measured immediately by using a fluorophotometer (F-4600, Hitiachi High Technologies Co., Tokyo, Japan) with excitation wavelength at 390 nm and emission wavelength at 420 nm. The recorded FI vs. the concentration of MP (mg/mL) was plotted, and the initial slope was used as the index of protein surface hydrophobicity.

2.4.4. Concentration of Sulfhydryl Group

The concentration of sulfhydryl (–SH) group was determined via DTNB (5,5′-Dithiobis-2-nitrobenzoic acid) assay as described by Jiang [32]. In brief, 1 mL of MP solution (4 mg/mL) was mixed with 9 mL buffer A (8 mol/L urea, 2% SDS and 10 mmol/L EDTA at pH 8.0), and 4 mL of the mixture was mixed with 0.5 mL buffer B (10 mM DTNB and 0.2 M Tris-HCl, pH 8.0). After incubation at 40 °C for 25 min, absorbance of the solution was determined at 412 nm. The concentration of –SH group was calculated as Equation (1):

$$x = \frac{A \times D}{C \times B} \tag{1}$$

where x represents the SH content (mol/g); D is the dilution factor; A is the absorbance of the mixture solution; C is the molar extinction (13,600 $mol^{-1} \cdot cm^{-1} \cdot L$); B is the protein concentration (mg/mL).

2.4.5. Intrinsic Fluorescence Intensity

Fluorescence intensity (FI) of MPs was measured with a fluorescence spectrophotometer (F-4600, Hitiachi High Technologies Co.) according to the method described by Walayat et al. [33]. Previously prepared MP solution (Section 2.3) was diluted to 0.05 mg/mL by using 0.6 M NaCl solution. The scan wavelength was in the range of 300 to 450 nm, with the scanning speed at 1000 nm/min, and the excitation wavelength was 295 nm with both excitation and emission slits at 10 nm.

2.5. Preparation of Surimi Gels

Surimi gel was prepared according to the method described by Tao et al. [8]. Frozen surimi was thawed at 4 °C, chopped for 5 min, mixed with 2.5% NaCl addition followed by chopping for another 5 min. The obtained surimi sol was then stuffed into a plastic polyvinylidene case and conditioned at 40 °C for 60 min followed by heating at 90 °C for 30 min (Figure S2). The obtained heat-set surimi gels were cooled in ice water for 30 min and stored at 4 °C overnight before further analysis.

2.6. Characterizations of Surimi Gels

2.6.1. Scanning Electron Microscopy

Scanning electron microscopy (SEM) was used to evaluate the microstructures of surimi gels. A cubic Surimi gel (3 × 3 × 3 mm^3) was conditioned in 2% glutaraldehyde at 4 °C for 18 h and dehydrated with a series of concentrations of ethanol (50, 70, 80, 90, and 100%), followed by lyophilization. Then, gel samples were mounted on an aluminum stub, sputter coated with gold, and observed by SEM (Hitachi S–3500 N, Hitachi, Japan) at an accelerating voltage of 10 kV.

2.6.2. Rheology Test

Small amplitude oscillatory strain (SAOS) tests were performed with a Physica controlled stress rheometer (MCR-301, Anton Paar, Graz, Austria). A small piece of surimi gel was loaded between two parallel plates (20 mm diameter and 1 mm gap). Disk-shaped samples were mounted on the lower plate of the rheometer at 20.0 ± 0.1 °C and allowed to rest for 15 min before analyses. Then, a frequency sweep test (0.1–100 Hz) was carried out to evaluate the rheological variations of the gels prepared from surimi stored for different periods of time. The strain was fixed at 0.5% which was within the viscoelastic region

(LVR), and the storage modulus (G′) and loss modulus (G″) were recorded. Values of tan δ were expressed as G″/G′ at the testing frequency of 1 Hz and the temperature of 20 °C.

2.6.3. Gel Strength

Gel strength was characterized on a TA. XT. plus Texture analyzer (SMS, Surrey, UK) according to the method of Liu et al. [14]. Surimi gels were cut into 25 mm high cylindrical shape slices. A slice was horizontally placed on the platform and penetrated by a spherical probe (type P/0.25). The testing speed of probe was 1 mm/s, the trigger point load was 0.1 N, and the depth of probe penetration was 10 mm. Gel strength (g/cm) was calculated by multiplying the breaking force (g) by penetration distance (mm).

2.6.4. Water-Holding Capacity (WHC)

WHC of surimi gels was measured according to the method described by Li et al. with minor modifications [4]. Surimi gels were cut into the thickness of 2.5 mm and centrifuged for 10 min at 1000 g and 4 °C, and the released moisture was absorbed by filter paper. The weight of the gels was recorded before (W1) and after centrifugation (W2). WHC was calculated as Equation (2):

$$\text{WHC}~(\%) = (\text{W1} - \text{W2})/\text{W1} \times 100\% \quad (2)$$

2.6.5. Whiteness

The whiteness of surimi gel was determined with a colorimeter (CR-400, Konica Minolta, Osaka, Japan) by measuring L* (lightness), a* (redness/greenness) and b* (yellowness/blueness) values. Whiteness was calculated as Equation (3):

$$\text{Whiteness} = 100 - 100 - \text{L}^*)^2 + \text{a}^{*2} + \text{b}^{*2}]^{1/2} \quad (3)$$

2.6.6. Sensory Assessment of Surimi Gels

Surimi gels (prepared in Section 2.5) were cut into square pieces (1 × 1 × 2 cm) for sensory assessment. The evaluation panelist was composed of 20 trained individuals, and the whole evaluation process was performed in a professional sensory evaluation laboratory at room temperature. A five-point hedonic score was recorded for each surimi gel, ranging from 5 (extremely like) to 1 (extremely dislike) in terms of the taste, smell, juiciness, color, texture and overall acceptability of the cooked surimi gel.

2.7. Statistical Analysis

All characterizations were repeated three times, and the reported results were shown as the mean ± standard deviation. Significant difference among treatments was analyzed by using SPSS (Version 13.0 SPSS, Chicago, IL, USA) through analysis of variance (ANOVA) with Duncan's test. Statistical significance was accepted at $p < 0.05$.

3. Results and Discussions

3.1. Effect of CMCO on Storage Stability of MP

3.1.1. Salt-Soluble Protein Content

MP is the dominant salt-soluble protein in the surimi, whose solubility variation is an important indicator to reflect its denaturation during frozen storage [34]. As shown in Figure 1a, the salt solubility of MP from all groups of surimi were about 86 mg/g before freezing and kept decreasing during the entire storage period. After 60 days, the solubility of the control group reached 58 mg/g. The decrease of protein solubility could be attributed to the freezing-induced denaturation/aggregation of MP in surimi as the storage time extended [12,35]. With the addition of cryoprotectants, the MP solubility of frozen surimi was significantly higher, whose values were 62, 70, and 72 mg/g, respectively, regarding to CMCO-A, -B and commercial group. Similar to most other saccharide cryoprotectants, CMCO could restrain the movement of water molecules and inhibit the growth of ice in frozen surimi, which alleviated the MP denaturation. Moreover, the MP solubility of CMCO-B group was higher than that of CMCO-A at the same storage time. It has been shown that the complexity of hydrogen bonding enhanced in the food matrix as the DS of charged polysaccharide increased [20,36]; thus, water molecules became more bounded to the protein networks, which led to a better cryoprotective effect to the MP, i.e., CMCO-B group > CMCO-A group, whose content of salt-soluble protein was 70 and 62 mg/g, respectively.

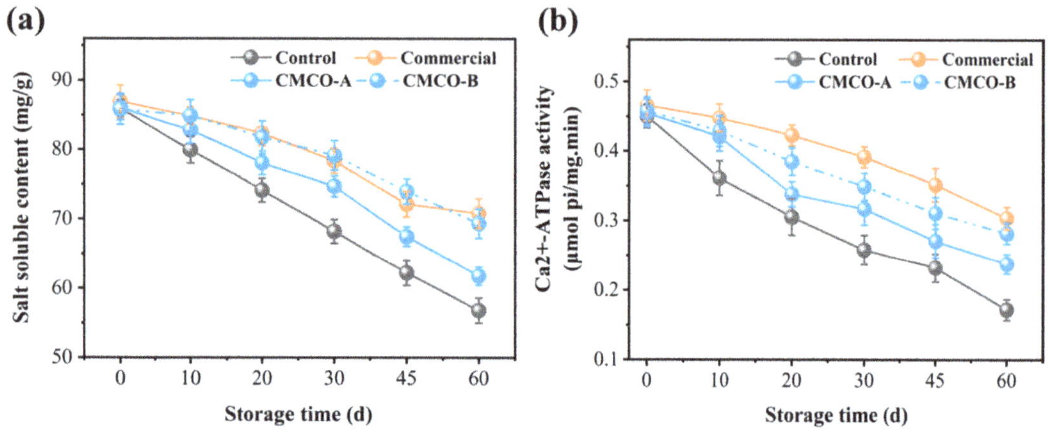

Figure 1. Effects of CMCO on salt-soluble protein content (**a**) and Ca^{2+}-ATPase activity (**b**) of MP at different storage time.

3.1.2. Ca^{2+}-ATPase Activity

Ca^{2+}-ATPase activity is an important indicator to measure the structural integrity of MP in fish muscle [8]. The protein conformations can be changed along with the breakdown of intermolecular noncovalent bonding during storage, resulting in the decrease of Ca^{2+}-ATPase activity [35]. Figure 1b showed that the initial Ca^{2+}-ATPase activity of MP of all frozen surimi samples were about at 0.45 μmol pi/mg.min before freezing and decreased to different degrees as storage time extended. The Ca^{2+}-ATPase activity of the control group experienced the most significant decrease (~60%) to 0.17 μmol pi/mg.min at day 60. In consistent with the results of MP solubility (Section 3.1), the use of cryoprotectants (i.e., CMCO-A/-B and the commercial additive) inhibited the protein denaturation. Specifically, the values of Ca^{2+}-ATPase activity of the CMCO-A (0.23 μmol pi/mg.min) and -B (0.28 μmol pi/mg.min) were significantly higher than that of the control group (0.17 μmol pi/mg.min) at day 60, even though they also decreased by 44.4% and 33.3%, respectively, during storage. Meanwhile, the CMCO-B group achieved comparable 60-day Ca^{2+}-ATPase activity (i.e., 0.28 vs. 0.30 μmol pi/mg.min) to the commercial group. A similar cryoprotective effect of pullulan was also reported by Jiang et al. [37].

3.1.3. Surface Hydrophobicity and Sulfhydryl Content

Variations of surface hydrophobicity reflect conformational changes of MP at different physical and chemical states. A reduced surface hydrophobicity suggests that the protein was less denatured (unfolded), and less hydrophobic groups were exposed, thus limiting the protein binding with the fluorescence probe [7]. As depicted in Figure 2a, surface hydrophobicity of the control group increased significantly from 3000 to 8800 after 60 days of storage. This phenomenon was reasonably due to the freezing-induced MP denaturation. Ice crystals that formed in surimi destroyed the hydration layer around the peptides, which resulted in more exposed hydrophobic amino acid residues and higher protein surface hydrophobicity [7]. In the presence of CMCO-A, -B and commercial additive, the surface hydrophobicity of MP was lower than that of the control group, suggesting their cryoprotective effects on protein structures. CMCO, an ampholytic saccharide with strong polarity, may interact with proteins and increase their hydrophilicities. Consequently, undesirable protein aggregation was inhibited as evidenced by the reduced hydrophobic interaction [38].

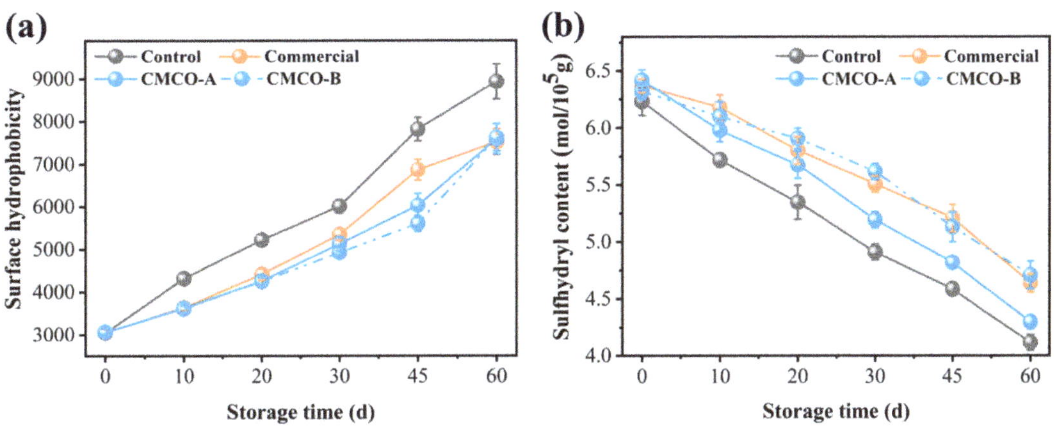

Figure 2. Effects of CMCO on surface hydrophobicity (**a**) and sulfhydryl content (**b**) of MP at different storage time.

To further elaborate the storage stability of MPs, the contents of their sulfhydryl groups were measured in the presence of different cryoprotectants (Figure 2b). During the 60 days of storage, the sulfhydryl content of the control group decreased significantly from 6.22 to 4.08 mol/10^5 g, while the addition of CMCO-A and B mitigated such decrease with the final sulfhydryl contents of 4.31 mol/10^5 g and 4.73 mol/10^5 g, respectively. Moreover, the CMCO-B group can achieve similar sulfhydryl contents of MP to the commercial group (4.69 mol/10^5 g). The sulfhydryl group is chemically reactive. It could be easily oxidized to form disulfide bonds under long-term frozen storage [33], leading to serious protein aggregation, denaturation, and deterioration of food quality [39]. Owing to the strong reducing power of oligosaccharide, CMCO could inhibit the oxidation and conversion of SH to disulfide bond, thus maintaining the storage stability of MP [14].

3.1.4. Intrinsic Fluorescence Intensity

The intrinsic fluorescence intensity (FI) of MP in frozen surimi was characterized by using fluorescence spectroscopy. It reflected the changes of chemical environment of tryptophan (Trp) residues, which suggested the variation of protein tertiary structures [40]. As shown in Figure 3, the FI of all MP samples was similar at Day 0, and decreased as far as the storage period elapsed, indicating the exposure of Trp residue towards solvent [41]. These results agreed with findings about the structural integrity of MP molecules (Ca^{2+}-ATPase activity in Figure 1b), which also experienced significant decreases as the tertiary structure of MP became collapsed during long-term frozen storage [41]. In the presence of different cryoprotectants, reduction of FI became less obvious compared to the control group, and relatively higher FI was observed for the CMCO-B group than the CMCO-A during the entire storage period. A similar charge-dependent cryoprotective behavior of ampholetic saccharides was also observed in our recent study [20], in which CMCh could stabilize the microstructures of wheat gluten in frozen dough, and the protective effect was more obvious as DS of CMCh increased.

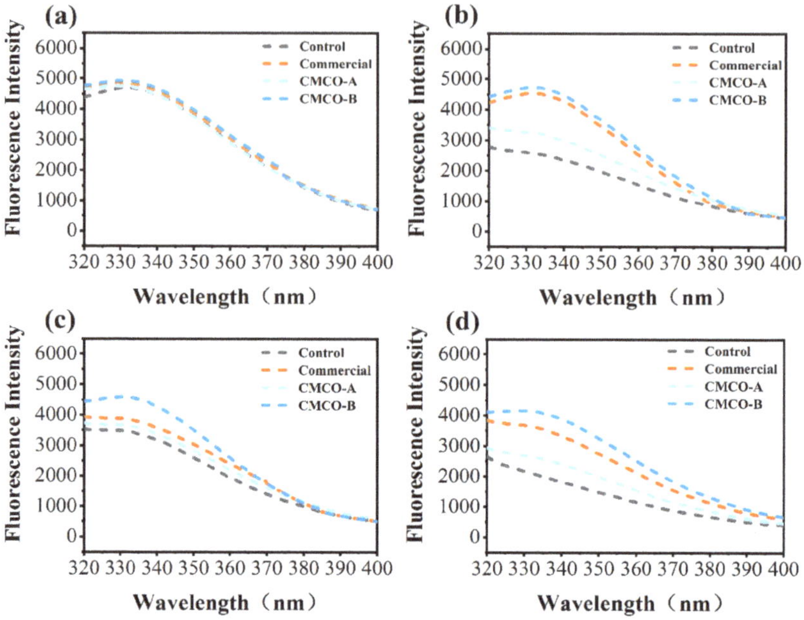

Figure 3. Effects of CMCO on fluorescence intensity of MP at different storage time: (**a**) day 0, (**b**) day 15, (**c**) day 30, and (**d**) day 60.

3.2. *Effect of CMCO on Gel Behaviors of Frozen Surimi*

3.2.1. Effect of CMCO on the Microstructure of Gels Prepared from Frozen Surimi

A porous protein network is formed during the two-stage thermal processing of surimi, which determines many important quality attributes of the gel products, such as texture, shelf life and digestion etc. [42,43]. Therefore, SEM was applied to observe the microstructure of gels prepared from frozen surimi. As shown in Figure 4, all gel samples of fresh surimi exhibited reticular and continuous protein network. After frozen storage, the gel matrix became loose for the control group, seen from the increase of heterogenic pore sizes. In contrast, the addition of cryoprotectants alleviated the breakdown of honeycomb-like network of surimi gels, which were beneficial to the mechanical strength of gel matrix. Besides, a more continuous and ordered protein architecture was observed for surimi gel of the CMCO-B group than that of the CMCO-A. A similar phenomenon was also reported by Tan et al. [13], who investigated the gel morphology of frozen surimi treated by cellulosic oligosaccharide.

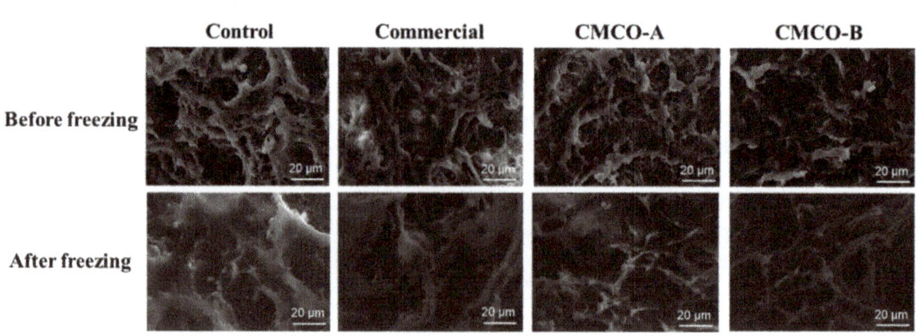

Figure 4. Microstructure of the gels prepared from frozen surimi with different cryoprotectants before and after storage for 60 days.

3.2.2. Effect of CMCO on the Rheological Properties of Surimi Gels

The rheological properties of surimi gels also changed significantly as the MP became denatured during frozen storage. As shown in Figure 5a,b, the elastic moduli (G′) of all surimi gels were higher than those of the viscous moduli (G″) at both day 0 and day 60, indicating their solidlike behaviors. To better illustrate the gel characteristics, loss tangents (tan δ) of surimi gels during frequency sweep were recorded (Figure 5c). The reduced elasticities were observed for all groups of surimi gels, as evidenced by their ever-increased tan δ (i.e., less elastic) after freezing storage. These results were correlated with the formed heterogeneous gel networks prepared from the stored surimi (Figure 4). In the cases of cryoprotected surimi, their gel structures were maintained, thus exhibiting lower tan δ compared to the control. After 60 days, the lowest tan δ was obtained for the CMCO-B group at 0.16. These observations were in consistent with the effect of antifreeze proteins on rheological properties of the frozen surimi [33,42].

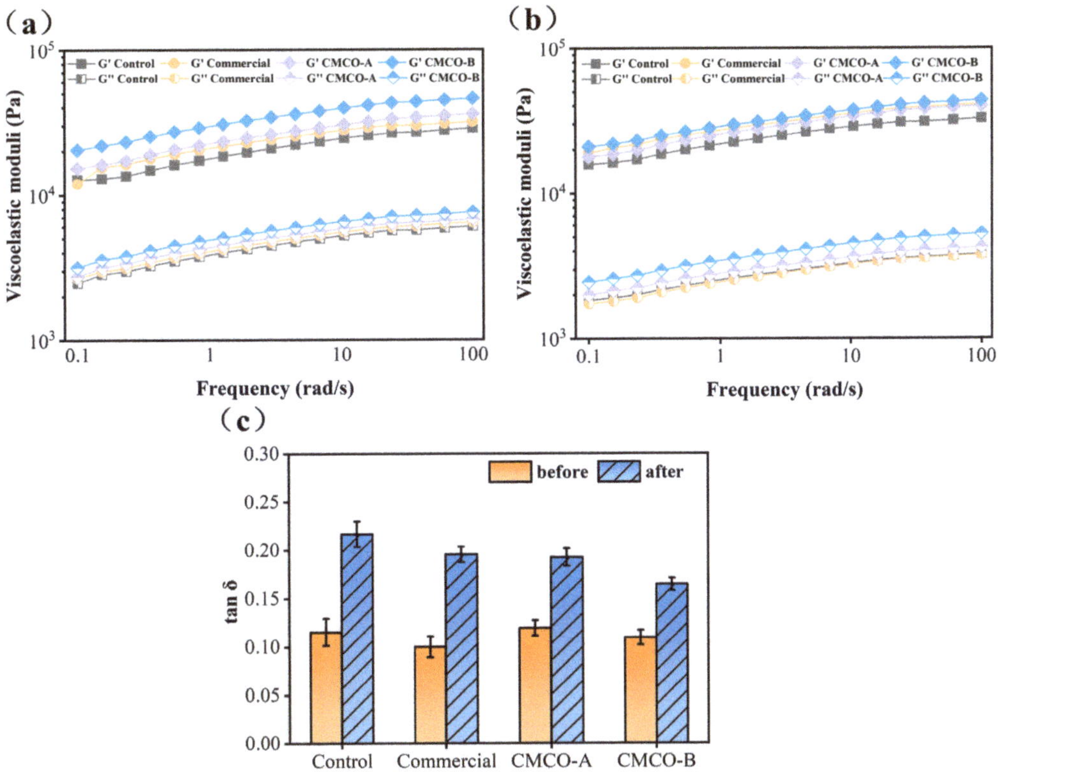

Figure 5. Dynamic rheological properties of the gels prepared from frozen surimi before (**a**) and after (**b**) frozen storage, and (**c**) tan δ of surimi gel within the linear viscoelastic region. Values with different uppercase letters indicate statistically significant difference among samples with different cryoprotectants at the same storage time ($p < 0.05$). Values with different lowercase letters indicate significant difference among samples with the same cryoprotectants at different storage time ($p < 0.05$).

3.2.3. Effect of CMCO on the Gel Strength and Water-Holding Capacity

Gel weakening is one of the most typical quality deteriorations for surimi products, which occurs during the whole process of cold chain transportation, storage and communications [44,45]. Thus, the gel strengths of frozen surimi with different cryoprotectants were characterized (Figure 6a). Few significant differences were observed among the samples at day 0, and the weakening effect was observed as storage time prolonged. After 60 days, gel strength of the control group exhibited the most significant decrease from 288 to 158 g·cm. The weakening effect was suppressed in the presence of cryoprotectants, and the highest gel strength of 219 g·cm was obtained for the CMCO-B group. As previously discussed (Section 3.1), introduction of CMCO to frozen surimi may preserve the MP integrity (solubility, Ca^{2+}-ATPase, sulphydryl, and intrinsic FI), which stabilized protein networks and enhanced mechanical strengths of surimi gels.

In addition to the gel strength, WHC is another important parameter to evaluate edible qualities of gel-type foods. As shown in Figure 6b, WHC of all surimi gels kept decreasing during storage, which could be owing to the muscle filament contraction and protein denaturation/ tertiary structural changes [5]. In the presence of different cryoprotectants, such reduction of WHC became less obvious. After 60 days, the CMCO-A, CMCO-B, and commercial groups exhibited significantly higher WHC (79.20%, 85.70%, and 86.50%, respectively) than the control group (68.50%). The improved WHC may be due to the cryoprotective effect of CMCO to the gel structures. The intrinsic ampholytic characteristics of CMCO may endow the saccharide superior hydrophilicity, which helped reduce the ice crystallization, improve the gelling property and finally enhance the WHC [12]. In consistent with the results of mechanical strength, as the DS of CMCO increased, the WHC of surimi gels were enhanced (CMCO-B > CMCO-A).

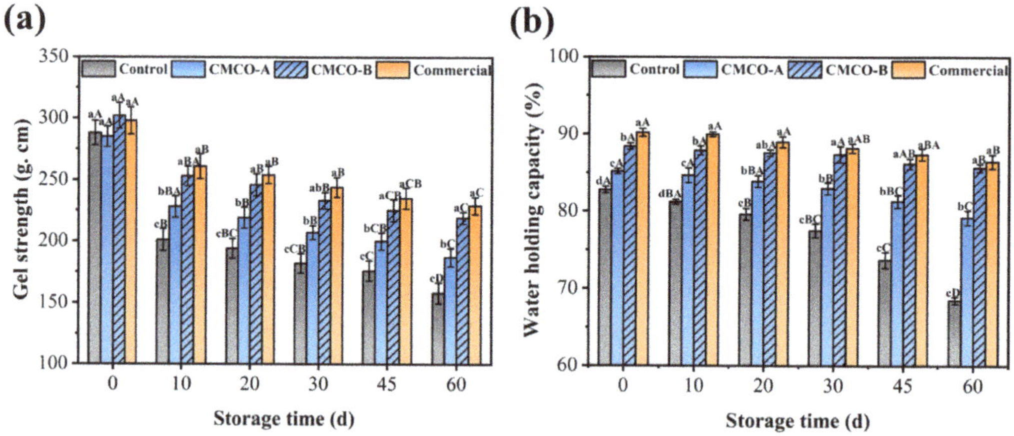

Figure 6. Effects of CMCO on the mechanical strength (**a**) and water-holding capacity (**b**) of the gels prepared from frozen surimi at different storage time. Values with different uppercase letters indicate statistically significant difference among samples with different cryoprotectants at the same storage time ($p < 0.05$). Values with different lowercase letters indicate significant difference among samples with the same cryoprotectants at different storage time ($p < 0.05$).

Collectively, the cryoprotective mechanism of CMCO can be described from molecular perspective (Figure 7). CMCO could interact with MP through both hydrogen bonding and electrostatic complexation, which replaces the water molecules at MP surface, and prevents the freezing-induced protein aggregation [41,42]. In addition, the ampholytic structure of oligosaccharide could entrap the water, modulate the growth of ice crystals, and stabilize MP structures [46]. Thus, after thermal processing, the gel of cryoprotected surimi exhibited more organized microstructures and improved mechanical properties. It is also noteworthy that the inhibitory effect of saccharide to ice crystallization may also account for the cryoprotective behaviors of CMCO to frozen surimi [47–49], and in most cases, both the protein-stabilization and ice-inhibition effects worked concurrently.

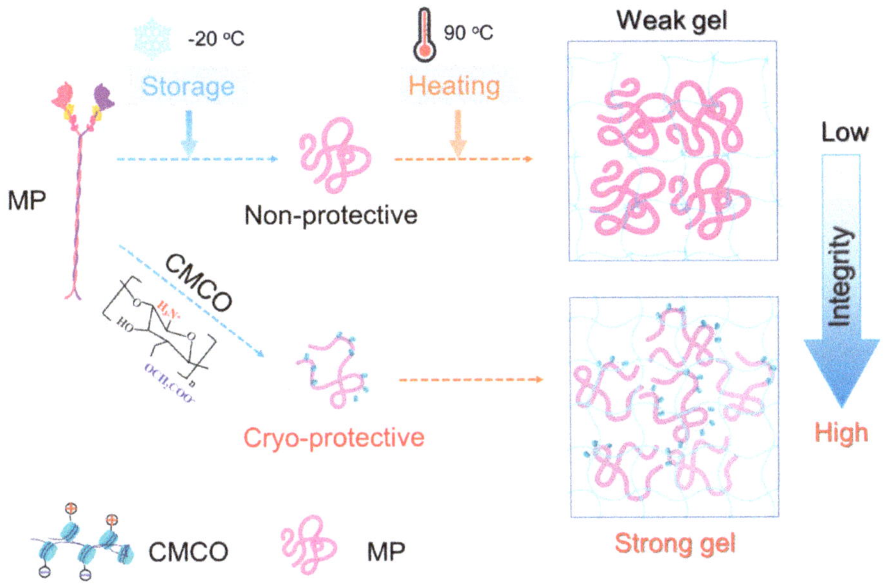

Figure 7. Schematic representation of the cryoprotective mechanism of CMCO for MP in frozen surimi.

3.3. *Effect of CMCO on the Whiteness and Sensory Quality of Frozen Surimi*

Decreased whiteness of gel products made from frozen surimi is a common problem impairing their appearance and edible qualities. As depicted in Figure 8, the whiteness value of the control group kept decreasing from 0 to 60 days. As expected, the addition of cryoprotectants could limit the loss of whiteness of surimi gels, whose values of CMCO-A, -B, and commercial group experienced no significant difference during the entire storage period. Similar phenomenon was also reported by Tao and Walayat [5,8] who attributed the decline of gel whiteness to the combined effects of nonenzymatic protein oxidation and crystallization-induced protein denaturation. The addition of oligosaccharides could prevent those protein deteriorations and thus maintain the whiteness of gels prepared from frozen surimi [50].

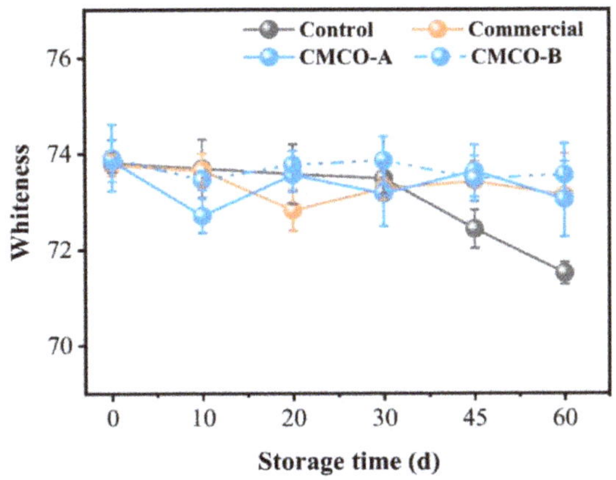

Figure 8. Effects of the CMCO on whiteness of the gels prepared from frozen surimi at different storage time.

Sensory evaluations of all surimi gels were performed by panelists (Table 1). The average sensory scores of surimi gels enhanced in terms of flavor, taste, juiciness, texture, and color through the addition of cryoprotectants. Moreover, gels of the CMCO-B group exhibited slightly higher overall acceptability than that of the CMCO-A and commercial group. In agreement with the results of gel strength, and whiteness (Figure 6), the CMCO-B could act as a high-performance cryoprotectant to improve the edible quality of the frozen stored surimi.

Table 1. Effect of CMCO on sensory attributes of the gels prepared from frozen surimi after 60 days of storage.

	Control	CMCO-A	CMCO-B	Commercial
Taste	3.78 ± 0.43 [b]	4.18 ± 0.31 [a]	4.34 ± 0.35 [a]	4.22 ± 0.27 [ab]
Smell	4.09 ± 0.45 [a]	4.38 ± 0.35 [a]	4.51 ± 0.38 [a]	4.46 ± 0.40 [a]
Texture	3.60 ± 0.25 [b]	3.93 ± 0.34 [ab]	4.17 ± 0.27 [a]	4.20 ± 0.32 [a]
Juiciness	3.52 ± 0.28 [b]	4.38 ± 0.31 [a]	4.51 ± 0.35 [a]	4.46 ± 0.30 [a]
Color	4.08 ± 0.24 [a]	4.12 ± 0.22 [a]	4.10 ± 0.25 [a]	4.15 ± 0.33 [a]
Overall acceptability	3.93 ± 0.35 [b]	4.38 ± 0.27 [ab]	4.67 ± 0.34 [a]	4.49 ± 0.31 [a]

Different letters indicate statistically significant differences among samples with different cryoprotectants ($p < 0.05$).

4. Conclusions

The present study designated the cryoprotective effects of ampholetic oligosaccharide (CMCO) on the storage stability of frozen surimi. An addition of 0.6% (*w*/*w*) CMCO can alleviate the denaturation of MP in the frozen surimi during 60 days of storage at −18 °C as indicated by their increased salt-protein solubility, Ca^{2+}-ATPase activity, and sulfhydryl content. The CMCO-protected MP experienced less conformation changes (reflected by the surface hydrophobicity and FI) than the nonprotected control. Accordingly, the obtained surimi gels demonstrated significantly improved elasticity, mechanical strength, and WHC, and had more uniform microstructure upon thermal processing. Moreover, the cryoprotective effect of CMCO-B (DS: 1.2) was more pronounced than that of CMCO-A (DS of 0.8). The gel strength and WHC of the CMCO-B group were comparable to those of the commercial counterpart (4% sucrose and 4% sorbitol). Results of sensory evaluation showed that CMCO addition also improved the taste, smell, texture, juiciness, whiteness, and overall acceptability of gels prepared from frozen surimi. Findings from this study deepen the scientific insights of ampholytic saccharides as high-performance cryoprotectants in food industry.

Supplementary Materials: The following supporting information can be downloaded at: https://www.mdpi.com/article/10.3390/foods11030356/s1, Figure S1: The chemical structure of carboxymethyl chitosan; Figure S2: A schematic demonstration to the processing of surimi.

Author Contributions: Conceptualization, X.Z. and M.Z.; methodology, M.Z., X.L. and D.H.; validation, L.W., T.Y. and Y.Z.; investigation, X.Z., X.L., L.S. and M.Z.; resources, T.Y. and L.S.; data curation, D.H., X.Z. and M.Z.; writing—original draft preparation, X.Z., D.H. and X.L. writing—review and editing, X.Z., T.Y., Y.Z. and J.X.; visualization, X.L. and M.Z.; project administration, X.Z. and T.Y.; funding acquisition: X.Z. All authors have read and agreed to the published version of the manuscript.

Funding: This research was funded by National Natural Science Foundation of China (31871870), Hubei Provincial Natural Science Foundation of China (Grant No. 2021CFB283) and Hubei University of Technology (BSQD-2020037, XBTK-2020001).

Institutional Review Board Statement: Not applicable.

Informed Consent Statement: Not applicable.

Data Availability Statement: Not applicable.

Acknowledgments: This work is financially supported by National Natural Science Foundation of China (31871870), Hubei Provincial Natural Science Foundation of China (Grant No. 2021CFB283), Doctoral start-up fund (BSQD-2020037) and Collaborative Grant-in-Aid of HBUT National "111" center for cellular Regulation and Molecular Pharmaceutics (XBTK-2020001) of Hubei University of Technology. The authors would like to thank Yunyun Zou from the shiyanjia lab (www.shiyanjia.com) for the fluorescence test.

Conflicts of Interest: The authors declare no conflict of interest.

References

1. Okita, A.; Takahashi, K.; Itakura, M.; Horio, A.; Yamamoto, R.; Nakamura, Y.; Osako, K. A novel soft surimi gel with functionality prepared using alcalase for people suffering from dysphagia. *Food Chem.* **2020**, *344*, 128641. [CrossRef]
2. Borderías, A.J.; Tovar, C.A.; Domínguez-Timón, F.; Díaz, M.T.; Pedrosa, M.M.; Moreno, H.M. Characterization of healthier mixed surimi gels obtained through partial substitution of myofibrillar proteins by pea protein isolates. *Food Hydrocoll.* **2020**, *107*, 105976. [CrossRef]
3. Dara, P.K.; Geetha, A.; Mohanty, U.; Raghavankutty, M.; Mathew, S.; Chandragiri Nagarajarao, R.; Rangasamy, A. Extraction and Characterization of Myofibrillar Proteins from Different Meat Sources: A Comparative Study. *J. Bioresour. Bioprod.* **2021**, *6*, 367–378. [CrossRef]
4. Li, J.; Munir, S.; Yu, X.; Yin, T.; You, J.; Liu, R.; Xiong, S.; Hu, Y. Double-crosslinked effect of TGase and EGCG on myofibrillar proteins gel based on physicochemical properties and molecular docking. *Food Chem.* **2021**, *345*, 128655. [CrossRef]
5. Walayat, N.; Xiong, Z.; Xiong, H.; Moreno, H.M.; Li, Q.; Nawaz, A.; Zhang, Z.; Wang, P.; Niaz, N. The effectiveness of egg white protein and β-cyclodextrin during frozen storage: Functional, rheological and structural changes in the myofibrillar proteins of Culter alburnus. *Food Hydrocoll.* **2020**, *105*, 105842. [CrossRef]
6. Singh, A.; Benjakul, S.; Prodpran, T.; Nuthong, P. Effect of Psyllium (*Plantago ovata* Forks) Husk on Characteristics, Rheological and Textural Properties of Threadfin Bream Surimi Gel. *Foods* **2021**, *10*, 1181. [CrossRef]
7. Lin, J.; Hong, H.; Zhang, L.; Zhang, C.; Luo, Y. Antioxidant and cryoprotective effects of hydrolysate from gill protein of bighead carp (*Hypophthalmichthys nobilis*) in preventing denaturation of frozen surimi. *Food Chem.* **2019**, *298*, 124868. [CrossRef]
8. Tao, L.; Tian, L.; Zhang, X.; Huang, X.; Long, H.; Chang, F.; Li, T.; Li, S. Effects of γ-polyglutamic acid on the physicochemical properties and microstructure of grass carp (Ctenopharyngodon idellus) surimi during frozen storage. *LWT* **2020**, *134*, 109960. [CrossRef]
9. Bao, Y.; Ertbjerg, P.; Estévez, M.; Yuan, L.; Gao, R. Freezing of meat and aquatic food: Underlying mechanisms and implications on protein oxidation. *Compr. Rev. Food Sci. Food Saf.* **2021**, *20*, 5548–5569. [CrossRef]
10. Tian, Y.; Zhu, Z.; Sun, D.W. Naturally sourced biosubstances for regulating freezing points in food researches: Fundamentals, current applications and future trends. *Trends Food Sci. Technol.* **2020**, *95*, 131–140. [CrossRef]
11. Huang, J.; Bakry, A.M.; Zeng, S.; Xiong, S.; Yin, T.; You, J.; Fan, M.; Huang, Q. Effect of phosphates on gelling characteristics and water mobility of myofibrillar protein from grass carp (*Ctenopharyngodon idellus*). *Food Chem.* **2019**, *272*, 84–92. [CrossRef]
12. Walayat, N.; Wang, X.; Liu, J.; Nawaz, A.; Zhang, Z.; Khalifa, I.; Rincón Cervera, M.Á.; Pateiro, M.; Lorenzo, J.M.; Nikoo, M.; et al. Kappa-carrageenan as an effective cryoprotectant on water mobility and functional properties of grass carp myofibrillar protein gel during frozen storage. *LWT* **2022**, *154*, 112675. [CrossRef]
13. Tan, M.; Ding, Z.; Mei, J.; Xie, J. Effect of cellobiose on the myofibrillar protein denaturation induced by pH changes during freeze-thaw cycles. *Food Chem.* **2021**, 131511. [CrossRef]
14. Liu, J.; Fang, C.; Luo, Y.; Ding, Y.; Liu, S. Effects of konjac oligo-glucomannan on the physicochemical properties of frozen surimi from red gurnard (*Aspitrigla cuculus*). *Food Hydrocoll.* **2019**, *89*, 668–673. [CrossRef]
15. Yosri, N.; Khalifa, S.A.M.; Guo, Z.; Xu, B.; Zou, X.; El-Seedi, H.R. Marine organisms: Pioneer natural sources of polysaccharides/proteins for green synthesis of nanoparticles and their potential applications. *Int. J. Biol. Macromol.* **2021**, *193*, 1767–1798. [CrossRef]
16. Tian, J.; Walayat, N.; Ding, Y.; Liu, J. The role of trifunctional cryoprotectants in the frozen storage of aquatic foods: Recent developments and future recommendations. *Compr. Rev. Food Sci. Food Saf.* **2021**, 321–339. [CrossRef]
17. Zhu, S.; Yu, J.; Chen, X.; Zhang, Q.; Cai, X.; Ding, Y.; Zhou, X.; Wang, S. Dual cryoprotective strategies for ice-binding and stabilizing of frozen seafood: A review. *Trends Food Sci. Technol.* **2021**, *111*, 223–232. [CrossRef]
18. Zhang, B.; Qi, X.E.; Mao, J.L.; Ying, X.G. Trehalose and alginate oligosaccharides affect the stability of myosin in whiteleg shrimp (*Litopenaeus vannamei*): The water-replacement mechanism confirmed by molecular dynamic simulation. *LWT* **2020**, *127*, 109393. [CrossRef]
19. Zhang, B.; Cao, H.J.; Lin, H.M.; Deng, S.G.; Wu, H. Insights into ice-growth inhibition by trehalose and alginate oligosaccharides in peeled Pacific white shrimp (*Litopenaeus vannamei*) during frozen storage. *Food Chem.* **2019**, *278*, 482–490. [CrossRef]
20. Zhu, X.; Yuan, P.; Zhang, T.; Wang, Z.; Cai, D.; Chen, X.; Shen, Y.; Xu, J.; Song, C.; Goff, D. Effect of carboxymethyl chitosan on the storage stability of frozen dough: State of water, protein structures and quality attributes. *Food Res. Int.* **2022**, *151*, 110863. [CrossRef]

21. Zhu, X.; Chen, J.; Hu, Y.; Zhang, N.; Fu, Y.; Chen, X. Tuning complexation of carboxymethyl cellulose/ cationic chitosan to stabilize Pickering emulsion for curcumin encapsulation. *Food Hydrocoll.* **2021**, *110*, 106135. [CrossRef]
22. Shariatinia, Z. Carboxymethyl chitosan: Properties and biomedical applications. *Int. J. Biol. Macromol.* **2018**, *120*, 1406–1419. [CrossRef]
23. Wang, H.; Gong, X.; Miao, Y.; Guo, X.; Liu, C.; Fan, Y.Y.; Zhang, J.; Niu, B.; Li, W. Preparation and characterization of multilayer films composed of chitosan, sodium alginate and carboxymethyl chitosan-ZnO nanoparticles. *Food Chem.* **2019**, *283*, 397–403. [CrossRef]
24. Liu, X.; Zhi, X.; Liu, Y.; Wu, B.; Sun, Z.; Shen, J. Effect of chitosan, O-carboxymethyl chitosan, and N-[(2-hydroxy-3-N, N-dimethylhexadecyl ammonium)propyl] chitosan chloride on overweight and insulin resistance in a murine diet-induced obesity. *J. Agric. Food Chem.* **2012**, *60*, 3471–3476. [CrossRef]
25. He, Y.; Wu, X.; Zhang, G.; Huang, X.; Zhang, W.; Tu, M.; Zeng, R. A zwitterionic serine modified chitosan derivative for improving protein stability and activity. *Int. J. Biol. Macromol.* **2020**, *163*, 1738–1746. [CrossRef]
26. Inanli, A.G.; Tümerkan, E.T.A.; El Abed, N.; Regenstein, J.M.; Özogul, F. The impact of chitosan on seafood quality and human health: A review. *Trends Food Sci. Technol.* **2020**, *97*, 404–416. [CrossRef]
27. Chen, J.; Deng, T.; Wang, C.; Mi, H.; Yi, S.; Li, X.; Li, J. Effect of hydrocolloids on gel properties and protein secondary structure of silver carp surimi. *J. Sci. Food Agric.* **2020**, *100*, 2252–2260. [CrossRef]
28. Ge, H.; Luo, D. Preparation of carboxymethyl chitosan in aqueous solution under microwave irradiation. *Carbohydr. Res.* **2005**, *340*, 1351–1356. [CrossRef]
29. Chang, C.; Lin, Y.; Lu, S.; Huang, C.; Wang, Y.; Chung, Y. Characterization of a Chitosanase from Jelly Fig (*Ficus awkeotsang* Makino) Latex and Its Application in the Production of Water-Soluble Low Molecular Weight Chitosans. *PLoS ONE* **2016**, *11*, e0150490. [CrossRef]
30. Wang, H.B.; Pan, S.K.; Wu, S.J. Chitooligosaccharides suppress the freeze-denaturation of actomyosin in Aristichthys nobilis surimi protein. *Int. J. Biol. Macromol.* **2014**, *63*, 104–106. [CrossRef]
31. Zhang, L.; Li, Q.; Hong, H.; Luo, Y. Prevention of protein oxidation and enhancement of gel properties of silver carp (*Hypophthalmichthys molitrix*) surimi by addition of protein hydrolysates derived from surimi processing by-products. *Food Chem.* **2020**, *316*, 126343. [CrossRef]
32. Jiang, X.; Chen, Q.; Xiao, N.; Du, Y.; Feng, Q.; Shi, W. Changes in Gel Structure and Chemical Interactions of Hypophthalmichthys molitrix Surimi Gels: Effect of Setting Process and Different Starch Addition. *Foods* **2021**, *11*, 9. [CrossRef]
33. Walayat, N.; Xiong, Z.; Xiong, H.; Moreno, H.M.; Niaz, N.; Ahmad, M.N.; Hassan, A.; Nawaz, A.; Ahmad, I.; Wang, P.K. Cryoprotective effect of egg white proteins and xylooligosaccharides mixture on oxidative and structural changes in myofibrillar proteins of Culter alburnus during frozen storage. *Int. J. Biol. Macromol.* **2020**, *158*, 865–874. [CrossRef]
34. Nikoo, M.; Benjakul, S.; Ahmadi Gavlighi, H.; Xu, X.; Regenstein, J.M. Hydrolysates from rainbow trout (*Oncorhynchus mykiss*) processing by-products: Properties when added to fish mince with different freeze-thaw cycles. *Food Biosci.* **2019**, *30*, 100418. [CrossRef]
35. Coombs, C.; Holman, B.; Collins, D.; Kerr, M.; Friend, M.; Hopkins, D. Effects of chilled-then-frozen storage (up to 52 weeks) on an indicator of protein oxidation and indices of protein degradation in lamb M. longissimus lumborum. *Meat Sci.* **2018**, *135*, 134–141. [CrossRef]
36. Ke, Y.; Wang, Y.; Ding, W.; Leng, Y.; Lv, Q.; Yang, H.; Wang, X.; Ding, B. Effects of inulin on protein in frozen dough during frozen storage. *Food Funct.* **2020**, *11*, 7775–7783. [CrossRef]
37. Jiang, L.; Wu, S. Pullulan suppresses the denaturation of myofibrillar protein of grass carp (*Ctenopharyngodon idella*) during frozen storage. *Int. J. Biol. Macromol.* **2018**, *112*, 1171–1174. [CrossRef]
38. Karnjanapratum, S.; Benjakul, S. Cryoprotective and antioxidative effects of gelatin hydrolysate from unicorn leatherjacket skin. *Int. J. Refrig.* **2015**, *49*, 69–78. [CrossRef]
39. Nawaz, A.; Xiong, Z.; Xiong, H.; Irshad, S.; Chen, L.; Wang, P.K.; Ahsan, H.M.; Walayat, N.; Qamar, S.H. The impact of hydrophilic emulsifiers on the physico-chemical properties, microstructure, water distribution and: In vitro digestibility of proteins in fried snacks based on fish meat. *Food Funct.* **2019**, *10*, 6927–6935. [CrossRef]
40. Yin, T.; He, Y.; Liu, L.; Shi, L.; Xiong, S.; You, J.; Hu, Y.; Huang, Q. Structural and biochemical properties of silver carp surimi as affected by comminution method. *Food Chem.* **2019**, *287*, 85–92. [CrossRef]
41. Shi, J.; Lei, Y.; Shen, H.; Hong, H.; Yu, X.; Zhu, B.; Luo, Y. Effect of glazing and rosemary (*Rosmarinus officinalis*) extract on preservation of mud shrimp (*Solenocera melantho*) during frozen storage. *Food Chem.* **2019**, *272*, 604–612. [CrossRef] [PubMed]
42. Walayat, N.; Xiong, Z.; Xiong, H.; Moreno, H.M.; Nawaz, A.; Niaz, N.; Hu, C.; Taj, M.I.; Mushtaq, B.S.; Khalifa, I. The effect of egg white protein and β-cyclodextrin mixture on structural and functional properties of silver carp myofibrillar proteins during frozen storage. *LWT* **2021**, *135*, 109975. [CrossRef]
43. Wang, W.; Shen, M.; Liu, S.; Jiang, L.; Song, Q.; Xie, J. Gel properties and interactions of Mesona blumes polysaccharide-soy protein isolates mixed gel: The effect of salt addition. *Carbohydr. Polym.* **2018**, *192*, 193–201. [CrossRef]
44. Chen, H.; Zou, Y.; Zhou, A.; Xiao, J.; Benjakul, S. Insight into the effect of ice addition on the gel properties of nemipterus virgatus surimi gel combined with water migration. *Foods* **2021**, *10*, 1815. [CrossRef] [PubMed]

45. Sharma, S.; Majumdar, R.K.; Siddhnath, K.; Mehta, N.K.; Saha, A.; Gupta, S. Effects of Partial and Complete Replacement of Synthetic Cryoprotectant with Carrot (*Daucus carota*) Concentrated Protein on Stability of Frozen Surimi. *J. Aquat. Food Prod. Technol.* **2019**, *28*, 808–820. [CrossRef]
46. Zhang, B.; Fang, C.D.; Hao, G.J.; Zhang, Y. yang Effect of kappa-carrageenan oligosaccharides on myofibrillar protein oxidation in peeled shrimp (*Litopenaeus vannamei*) during long-term frozen storage. *Food Chem.* **2018**, *245*, 254–261. [CrossRef] [PubMed]
47. Kamińska-Dwórznicka, A.; Łaba, S.; Jakubczyk, E. The effects of selected stabilizers addition on physical properties and changes in crystal structure of whey ice cream. *LWT* **2022**, *154*, 112841. [CrossRef]
48. Kamińska-Dwórznicka, A.; Skrzypczak, P.; Gondek, E. Modification of kappa carrageenan by β-galactosidase as a new method to inhibit recrystallization of ice. *Food Hydrocoll.* **2016**, *61*, 31–35. [CrossRef]
49. Chen, X.; Wu, J.; Li, X.; Yang, F.; Yu, L.; Li, X.; Huang, J.; Wang, S. Investigation of the cryoprotective mechanism and effect on quality characteristics of surimi during freezing storage by antifreeze peptides. *Food Chem.* **2022**, *371*, 131054. [CrossRef]
50. Shen, X.; Li, T.; Li, X.; Wang, F.; Liu, Y.; Wu, J. Dual cryoprotective and antioxidant effects of silver carp (*Hypophthalmichthys molitrix*) protein hydrolysates on unwashed surimi stored at conventional and ultra-low frozen temperatures. *LWT* **2022**, *153*, 112563. [CrossRef]

foods

MDPI

Article

Protective Effect of *Ganoderma atrum* Polysaccharide on Acrolein-Induced Apoptosis and Autophagic Flux in IEC-6 Cells

Yudan Wang, Xinxin Chang, Bing Zheng, Yi Chen, Jianhua Xie, Jialuo Shan, Xiaoyi Hu, Xiaomeng Ding, Xiaobo Hu and Qiang Yu *

State Key Laboratory of Food Science and Technology, China-Canada Joint Laboratory of Food Science and Technology (Nanchang), Key Laboratory of Bioactive Polysaccharides of Jiangxi Province, Nanchang University, 235 Nanjing East Road, Nanchang 330047, China; W15698331381@163.com (Y.W.); cxx13870698052@163.com (X.C.); ncuspyzhengbing@163.com (B.Z.); chenyi15@ncu.edu.cn (Y.C.); jhxie@ncu.edu.cn (J.X.); jialuoa1998@163.com (J.S.); huxiaoyi0101@126.com (X.H.); dxm19970128@163.com (X.D.); hxbxq2005@163.com (X.H.)

* Correspondence: yuqiang8612@ncu.edu.cn

Abstract: This study was designed to explore the beneficial effect and mechanism of *Ganoderma atrum* (G. atrum) polysaccharide (PSG-1) on acrolein-induced IEC-6 cells. Our results indicated that PSG-1 significantly reduced the impairment of acrolein on cell viability, decreased oxidative stress, and enabled normal expression of tight junction (TJ) proteins that were inhibited by acrolein in IEC-6 cells. Furthermore, PSG-1 attenuated the elevation of microtubule-associated proteins light chain 3 (LC3) and Beclin 1-like protein 1 (Beclin 1) and increased the protein levels of phospho-mTOR (p-mTOR) and phospho-akt (p-akt), indicating that PSG-1 activated the mammalian target of rapamycin (mTOR) signaling pathway and alleviated acrolein-induced autophagy in IEC-6 cells. Moreover, PSG-1 markedly attenuated the acrolein-induced apoptosis, as evidenced by the increase in mitochondrial membrane potential (MMP) and B-cell lymphoma 2 (Bcl-2) expression, and the decrease in cysteine aspartate lyase (caspase)-3 and caspase-9. In addition, autophagy the inhibitor inhibited acrolein-induced TJ and apoptosis of IEC-6 cells, while the apoptosis inhibitor also inhibited acrolein-induced TJ and autophagy, suggesting that autophagy and apoptosis were mutually regulated. Taken together, the present study proved that PSG-1 could protect IEC-6 cells from acrolein-induced oxidative stress and could repair TJ by inhibiting apoptosis and autophagic flux, where autophagy and apoptosis were mutually regulated.

Keywords: polysaccharide; acrolein; IEC-6 cells; autophagy; apoptosis

Citation: Wang, Y.; Chang, X.; Zheng, B.; Chen, Y.; Xie, J.; Shan, J.; Hu, X.; Ding, X.; Hu, X.; Yu, Q. Protective Effect of *Ganoderma atrum* Polysaccharide on Acrolein-Induced Apoptosis and Autophagic Flux in IEC-6 Cells. *Foods* **2022**, *11*, 240. https://doi.org/10.3390/foods11020240

Academic Editor: Antonio Cilla

Received: 3 December 2021
Accepted: 10 January 2022
Published: 17 January 2022

1. Introduction

Acrolein is an electrophilic, unsaturated aldehyde, a colorless, flammable, volatile, pure liquid, known for its pungent odor and strong irritation of mucous membranes [1], which comes from food, the environment, and the human body itself [2,3]. Acrolein in food is mainly produced by oil fumes or fried and baked foods, which produce acrolein mainly through the Maillard reaction and lipid oxidation [4]. Acrolein is neurotoxic, genotoxic, and potentially carcinogenic, and the question of how to mitigate the side effects of acrolein has generated scholars' attention [5]. Therefore, finding active substances that can reduce or antagonize the toxic effects of acrolein in food is important to protect humans from its harm.

Programmed cell death (PCD) is a self-regulatory mechanism of the organism, mainly including autophagy, apoptosis, and necrosis, the first two of which can play an indispensable role in the life activities of the organism by working closely together. Autophagy is a highly conserved intracellular degradation pathway, a protective mechanism for cells in response to various of hazardous stimuli, which maintains intracellular homeostasis by degrading damaged macromolecular proteins or organelles within the cell [6]. Autophagy is a multi-step regulated process with highly complex signaling. In this process, microtubule-associated proteins light chain 3 (LC3), Beclin 1-like protein 1 (Beclin 1), and so on, play remarkably important roles [7]. Apoptosis is an orderly and autogenous cell death that is regulated by the involvement of genes, acting in the extrinsic and intrinsic pathways [8]. Both pathways stimulate cysteine aspartate lyase (caspase), which is the basis of apoptosis [9]. The endogenous pathway surrounding mitochondria is mainly mediated by the proteins of B-cell lymphoma 2 (Bcl-2) family. Several structural features are altered during apoptosis: membrane blistering, cytoplasmic crumpling, DNA breaks, chromatin condensation, and DNA degradation of chromosomes [10].

The intestine is the body's largest immune system, which mainly relies on the intestinal mucosal barrier to resist the invasion of foreign toxic substances and prevent the invasion of adverse substances and displacement in the intestine [11]. Thus, the integrity of the intestinal mucosal screen is essential to maintain the normal physiological function of the intestine [12]. Many studies have found that autophagy and apoptosis act as vital players in the intestinal barrier and work together to sustain the complexity of the intestinal barrier [13].

Numerous studies have revealed that polysaccharides possessed great promises for food and health applications [14–16]. *Ganoderma atrum* (*G. atrum*) has been used as a medicine or diet with high nutritional value for thousands of years [17]. In the last decade, our laboratory has identified a polysaccharide from *G. atrum* (PSG-1), with purity > 99.8%. It is composed of glucose (Glc), mannose (Man), galactose (Gal), and galacturonic acid (GalA) in molar ratio of 4.91:1:1.28:0.71, and is rich in 17 proteins, including glutamic acid, asparagic acid, alanine, glycine, threonine, and serine; in addition, it is formed as a main chain of 1,3-linked and 1,6-linked β-Glcp residues, substituted at O-3 and O-6 position as the branch points, with the molecular weight of 1013 kDa. The residues of α-1,4-Galp, α-1,2-Manp, and α-1,4-Manp were also found in the backbone. Side chains were terminated by β-Glcp, with the composition of α-1,4-Galp, α-1,4-GalpA, β-1,3-Glcp, and β-1,6-Glcp [18]. PSG-1 have exhibited a broad range of beneficial health effects, including immunomodulatory, chemo-protective, antioxidant, and hypoglycemic activities [19–22]. In particular, latest studies have also proved that PSG-1 could modulate the mucosal immunity through epithelial cells [23,24]. However, there are few reports on acrolein-induced intestinal effects. Therefore, in this study, an acrolein-induced IEC-6 cells injury model was established, aiming to examine the protective characteristics of PSG-1 against acrolein-induced intestinal injury and its intrinsic mechanism.

2. Materials and Methods

2.1. Materials

Acrolein was purchased from Shandong Xiya Chemical Industry Co., Ltd. (Linyi, China). 3-Methyladenine (3-MA) (PI3K inhibitor) and Z-DEVD-FMK (caspase-3 inhibitor) were purchased from MedChem Express Co., Ltd. (Monmouth Junction, NJ, USA). Annexin V-FITC apoptosis detection kits were purchased from Beyotime Biotechnology (Nanjing, China). Mitochondrial membrane potential (JC-1) assay kit was obtained from Solarbio Science and Technology Co., Ltd. (Beijing, China). The Cell Counting Kit-8 (CCK-8) was from Dojindo Molecular Technologies, Inc. (Shanghai, China).

2.2. Cell Culture

IEC-6 cells (rat-derived intestinal epithelial cell line) were cultured in high glucose Dulbecco's Minimal Essential Medium (DMEM) (Solarbio Science and Technology Co., Ltd. Beijing, China), supplemented with 10% fetal bovine serum (FBS) (BI, 04-001-1ACS), 5% CO_2, temperature 37 °C and humidity saturation conditions. The passages 5–20 were used in this study.

2.3. Cell Viability

A 96-well microplate format was used. After 24 h of incubation, treating IEC-6 cells with acrolein (0–80 μM) for 20 h, a solution of CCK-8 was added to each well. Finally, the optical absorbance was read at 450 nm with a microplate reader (Thermo Fisher Scientific, Waltham, MA, USA). Under the same cultural conditions as above, PSG-1 at concentrations of 20, 40, 80, and 160 (μg/mL) were added to each group, while a blank control group was set up. The effect of PSG-1 on IEC-6 cells viability was detected.

2.4. Antioxidant Enzyme Assays

After cell culture, cells were washed once with phosphate-buffered saline (PBS), cells were digested with trypsin, and they were collected by centrifugation for 10 min at 1000× *g*. Protein concentrations were determined using the BCA protein assay kit, superoxide dismutase (SOD), malondialdehyde (MDA), and glutathione peroxidase (GPx) (Beyotime Biotechnology Nanjing, China) were estimated by referring to the instructions.

2.5. Apoptosis Rate Detection

The IEC-6 cells were collected after washing with PBS and digesting with trypsin, centrifuged, and 195 μL binding buffer was added to keep the cells in suspension; 5 μL annexin V-FITC and 10 μL PI solutions were added and mixed. The cells were then cultured at 20–25 °C for 10–20 min, protected from light, observed by flow cytometry, and detected within 1 h.

2.6. Assay for Mitochondrial Membrane Potential (MMP)

Cells in 6-well plates were washed once with PBS, then 1 mL of cell culture solution and 1 mL of JC-1 staining buffer were added and mixed thoroughly. Cells were cultured in an incubator at 37 °C for 20 min. Following staining, the cells were polished twice with JC-1 staining buffer and observed by fluorescence microscopy.

2.7. Western Blot Analysis

Protein samples were prepared by treating IEC-6 cells with acrolein and PSG-1 in 6-well plates, after 10% sodium dodecyl sulfate (SDS)-PAGE electrophoresis, and transported to a PVDF membrane (Millipore Co., Belford, MA, USA). The membranes were blocked with 5% bovine serum albumin (BSA) for 1 h. Membrane were incubated with primary antibodies (1:1000): ZO-1, claudin-1, occludin, LC3B, Beclin 1, phospho-mammalian target of rapamycin (p-mTOR), phospho-Protein Kinase B (p-akt), caspase-3, caspase-9, Bcl-2, and beta actin monoclonal antibody (β-actin) (CST, Boston, MA, USA) at 4 °C overnight under shaking and incubated with horseradish peroxidase-conjugated secondary antibodies (ZSGB Biotechnology, Beijing, China) for 1 h at room temperature. After incubation in ECL detection reagent, the fluorescent signal was detected using the Molecular Imager ChemiDoc™ XRS Imaging System (Bio-Rad Laboratories, Hercules, CA, USA).

2.8. Inhibitor Experiments

According to the instructions, cells induced by acrolein were pretreated with configured autophagy inhibitor 3-MA and the apoptosis inhibitor Z-DEVD-FMK, respectively. Cellular proteins were extracted after a certain period of time, and the proteins expression were examined by Western blot method to explore the interaction between TJ and autophagy, TJ and apoptosis, and the autophagy and apoptosis pathways.

2.9. Statistical Analysis

Statistical analysis was carried out using Graphpad Prism 6.01 software. Results were analyzed by a one-way ANOVA analysis of variance and expressed as the mean ± SD. $p < 0.05$ was considered statistically significant.

3. Results

3.1. PSG-1 Increased the IEC-6 Cells Viability Exposed to Acrolein

As shown in Figure 1A, acrolein inhibited the viability of IEC-6 cells in a dose-dependent manner. Besides, when acrolein concentration was 40 μM, the cell viability was about 50%. Thus, we chose an acrolein concentration of 40 μM for subsequent experiments to detect the potential cytoprotective capacity of PSG-1 ($p < 0.01$).

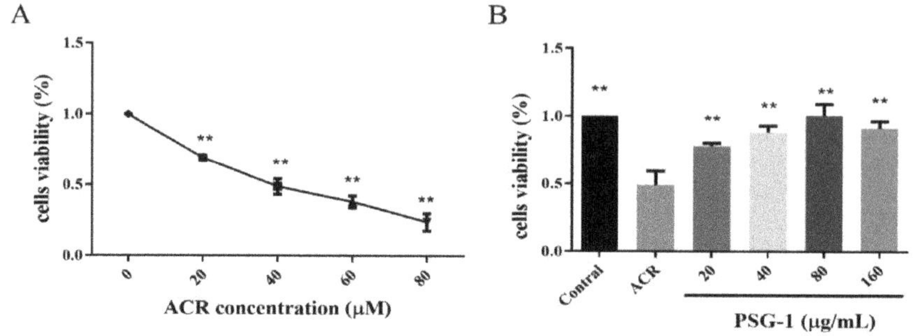

Figure 1. PSG-1 increased cell viability exposed to acrolein. (**A**) Acrolein effect on IEC-6 cells viability. (**B**) Effect of PSG-1 on acrolein-induced IEC-6 cells viability. Values are means ± SD (n = 3), ** $p < 0.01$ compared with the acrolein group.

Acrolein treatment caused a significant decrease in cell viability, while different concentrations of PSG-1 significantly inhibited the toxicity of acrolein. As can be seen in Figure 1B, all 4 concentration groups were effective, and cell viability tended to enhance and then diminish with increasing PSG-1 concentration, especially at a dose of 80 μg/mL, which almost restored cell viability to normal levels. The differences between the groups were not significant ($p < 0.01$).

3.2. PSG-1 Attenuated Acrolein-Induced Oxidative Damage of IEC-6 Cells

As shown in Figure 2A, the cellular SOD activity declined after acrolein treatment, the addition of PSG-1 restored the SOD activity to different degrees, and the PSG-1 dose group with 160 μg/mL had the highest effect on the increase in SOD activity.

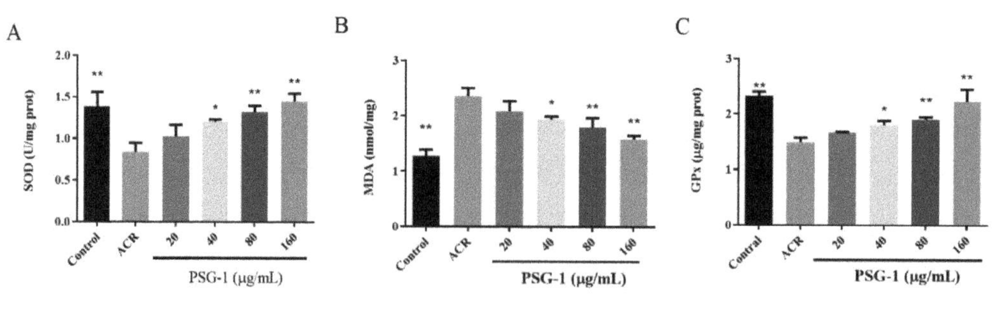

Figure 2. PSG-1 regulated acrolein-induced SOD, MDA, and Gpx. The oxidative state of IEC-6 cells, following acrolein (40 μM) and PSG-1 (20, 40, 80, and 160 μg/mL) treatment, was determined by measuring SOD (**A**), MDA (**B**) and GPx (**C**).Values are means ± SD (n = 3), * $p < 0.05$, ** $p < 0.01$ versus the acrolein group.

In Figure 2B, the MDA level significantly increased after the addition of acrolein, while PSG-1 treatment significantly decreased the MDA contents.

The significant decrease in GPx viability was caused by acrolein. Under the protection of PSG-1, GPx significantly increased especially in the 160 μg/mL dose group (Figure 2C), indicating that PSG-1 dramatically prevented the inhibition of acrolein-induced oxidative stress ($p < 0.05$).

3.3. Effects of PSG-1 on TJ Proteins Damaged by Acrolein in IEC-6 Cells

TJ mainly includes transmembrane proteins and intracellular plaque proteins, such as Zonula occludens (ZO), different claudin-1 family members, and occludin. As Figure 3 shows, acrolein-induced expression of ZO-1, claudin-1, and occludin were strikingly reduced, indicating that acrolein treatment significantly impaired intestinal barrier function, and PSG-1 treatment clearly promoted the expression of the three proteins ($p < 0.05$), suggesting that PSG-1 was able to tightly link the reduction in proteins and thus protect the intestinal barrier.

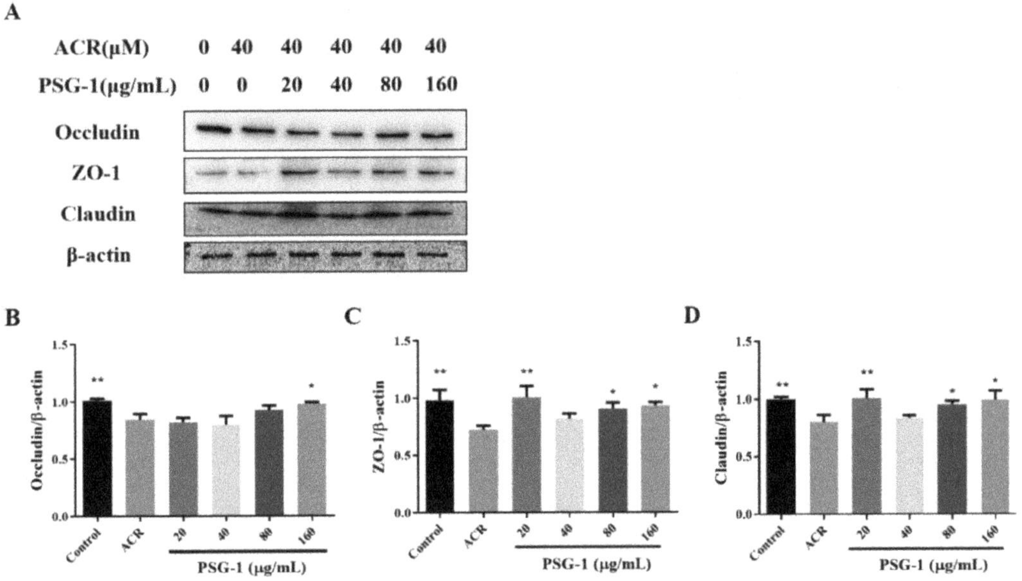

Figure 3. PSG-1 regulated the expression of TJ-related proteins in acrolein-induced IEC-6 cells. (**A**) The protein levels of occludin, claudin, and ZO-1. (**B–D**) Relative intensities of these proteins. Values are means ± SD ($n = 3$), * $p < 0.05$, ** $p < 0.01$ compared with the acrolein group.

3.4. Effects of PSG-1 on Autophagic Proteins Damaged by Acrolein in IEC-6 Cells

Figure 4B results revealed that the LC3-II/I ratio was significantly higher in acrolein-treated cells, while LC3 protein was significantly lower in PSG-1-treated cells than the model group. In addition, we examined key proteins of the PI3K/Akt/mTOR pathway and the Beclin 1 pathway, the phosphorylation levels of mTOR and akt were clearly lower and the level of Beclin 1 was significantly higher in acrolein-induced cells than in the control group. After treating IEC-6 cells with PSG-1, the levels of p-mTOR and p-akt were significantly enhanced, and the content of Beclin 1 was significantly lower than that of the acrolein group ($p < 0.05$) (Figure 4C–E).

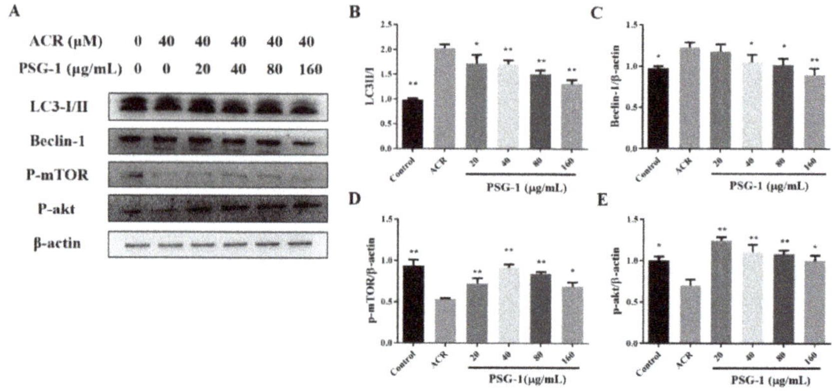

Figure 4. PSG-1-regulated, acrolein-induced autophagy-related proteins. (**A**) The protein levels of LC3, Beclin 1, p-mTOR, and p-akt. (**B–E**) Relative intensities of these proteins. Values are means ± SD (n = 3), * $p < 0.05$, ** $p < 0.01$ compared with the acrolein group.

3.5. Effect of PSG-1 on Acrolein-Induced Apoptosis

Double staining with annexin V and PI to detect apoptosis showed that the left lower quadrant had live cells and the right lower quadrant had apoptotic cells. Figure 5 showed that after acrolein induction, the apoptosis rate of cells increased significantly, reaching more than 70%, far exceeding that of normal controls. A decrease in the percentage of apoptotic cells could be clearly observed after PSG-1 treatment.

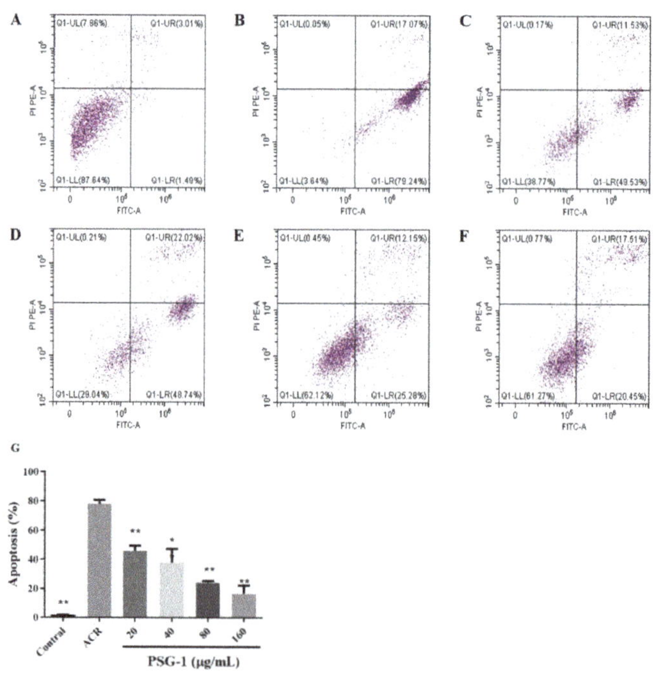

Figure 5. PSG-1 attenuated acrolein-induced apoptosis. Apoptosis measured by flow cytometry. (**A**) control cells; (**B**) the cells treated with acrolein, (**C–F**) the cells treated with acrolein and PSG-1 at concentrations of 20, 40, 80and 160μg/mL, (**G**) Effect of PSG-1 on acrolein-induced apoptosis. Values are means ± SD (n = 3), * $p < 0.05$, ** $p < 0.01$ compared with the acrolein group.

3.6. Effects of PSG-1 on MMP Induced by Acrolein in IEC-6 Cells

The process of apoptosis is often accompanied by a disruption of the MMP, which is thought to be one of the earliest events in the apoptotic process. The change of MMP was detected by JC-1 fluorescent probe. It can be seen that cells after acrolein induction have depolarized mitochondrial transmembrane potential and impaired mitochondrial permeability, while PSG-1 treatment restored cell morphology with enhanced red fluorescence intensity and weaker green fluorescence intensity, implying that PSG-1 attenuated acrolein-induced loss of MMP in IEC-6 cells, thereby preserving mitochondria (Figure 6).

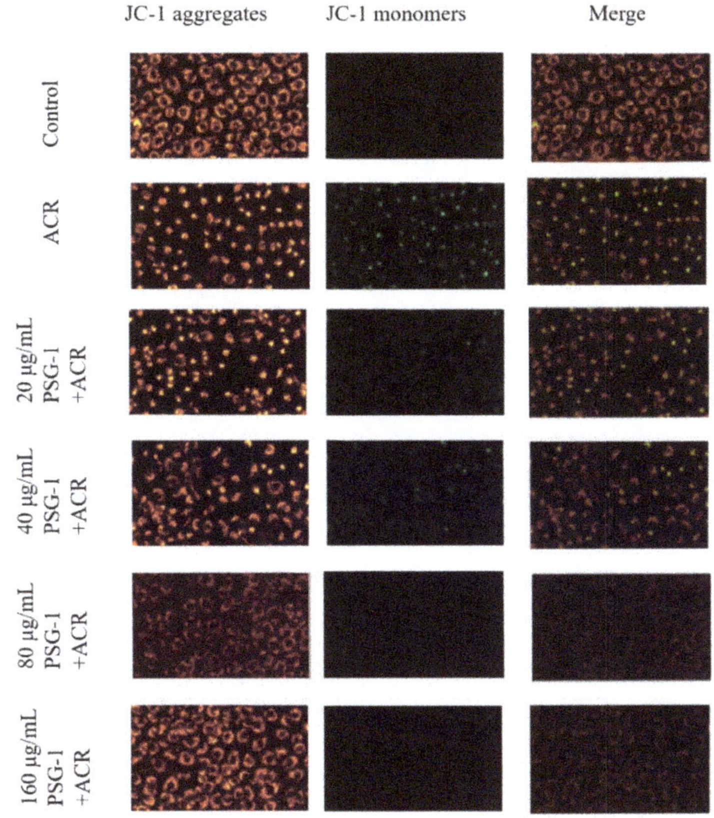

Figure 6. PSG-1 attenuated acrolein-induced MMP reduction. MMP was detected by fluorescence microscopy.

3.7. Effects of PSG-1 on Apoptotic Proteins Damaged by Acrolein in IEC-6 Cells

As depicted in Figure 7, acrolein significantly promoted the two positively related proteins, caspase-3 and caspase-9, and significantly declined the level of anti-apoptotic protein Bcl-2. The addition of PSG-1 decreased the expression of caspase-3 and caspase-9 proteins and increased the content of Bcl-2 protein ($p < 0.05$), which can indicate that PSG-1 can inhibit acrolein-induced apoptosis.

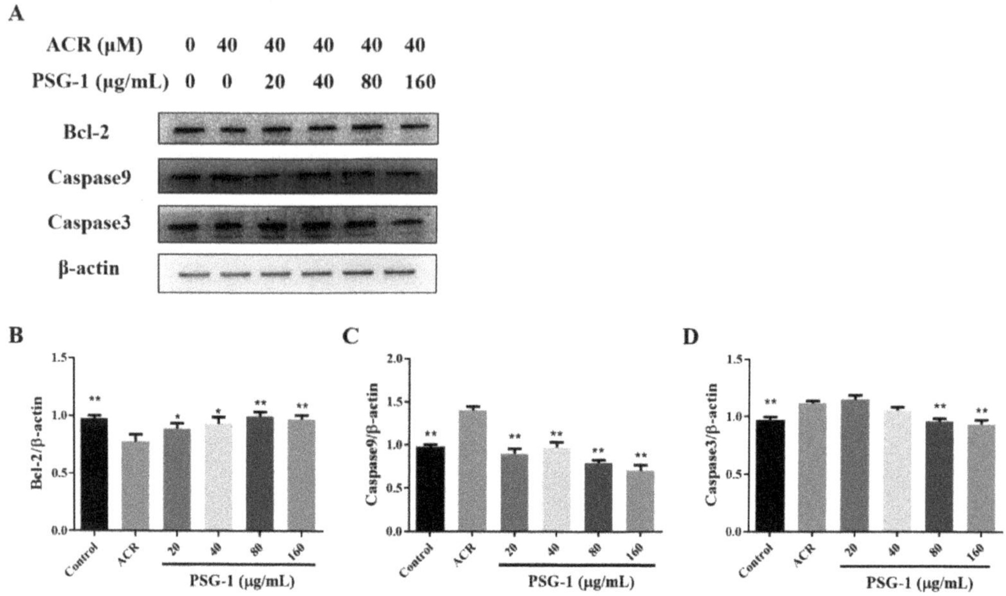

Figure 7. PSG-1 regulated the expression of apoptosis-related proteins in acrolein-induced IEC-6 cells. (**A**) The protein levels of caspase-9, caspase-3, and Bcl-2. (**B–D**) Relative intensities of these proteins. Values are means ± SD ($n = 3$), * $p < 0.05$, ** $p < 0.01$ compared with the acrolein group.

3.8. Effects of Autophagy Inhibitor and Apoptosis Inhibitor on the TJ in IEC-6 Cells

After treating cells with autophagy inhibitor and apoptosis inhibitor, we separately inspected the expression of three TJ proteins, and it was found that the expression of TJ protein was markedly increased in IEC-6 cells, where the inhibitor co-interacted with acrolein, compared with acrolein alone ($p < 0.05$) (Figure 8).

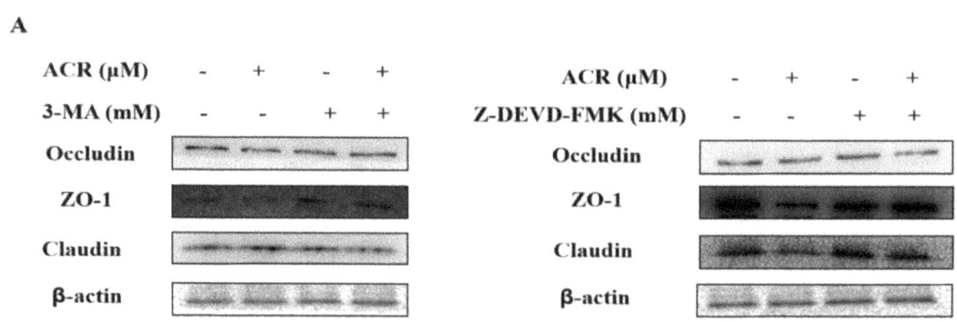

Figure 8. *Cont.*

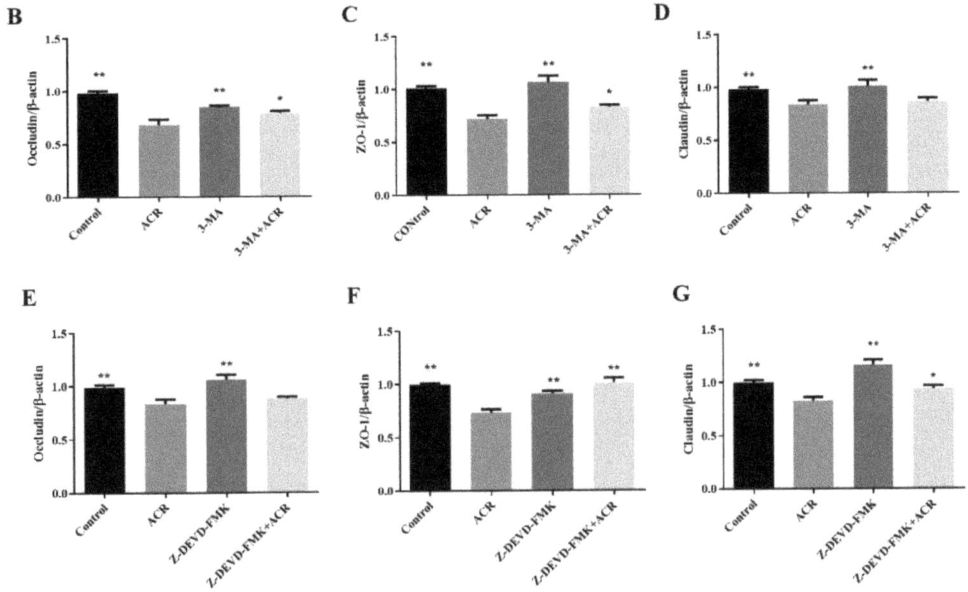

Figure 8. Autophagy and apoptosis inhibitor attenuated TJ. (**A**) The protein levels of occludin, claudin, and ZO-1, "+" indicates addition of acrolein or PSG-1 and "-" means not added. (**B–G**) Relative intensities of these proteins. Values are means ± SD (n = 3), * $p < 0.05$, ** $p < 0.01$ compared with the acrolein group.

3.9. The Effects of Autophagy Inhibitor on IEC-6 Cell Apoptosis

Autophagy inhibitor treatment of IEC-6 cells significantly decreased the increase in acrolein-induced autophagy-related protein LC3 (Figure 9B). After that, we examined the expression of apoptosis-related proteins Bcl-2, caspase-3, and caspase-9, and the results demonstrated that the autophagy inhibitor reduced the overexpression of acrolein-induced, apoptosis-related proteins caspase-3 and caspase-9 and increased the expression of anti-apoptotic protein Bcl-2 ($p < 0.05$) (Figure 9C–E).

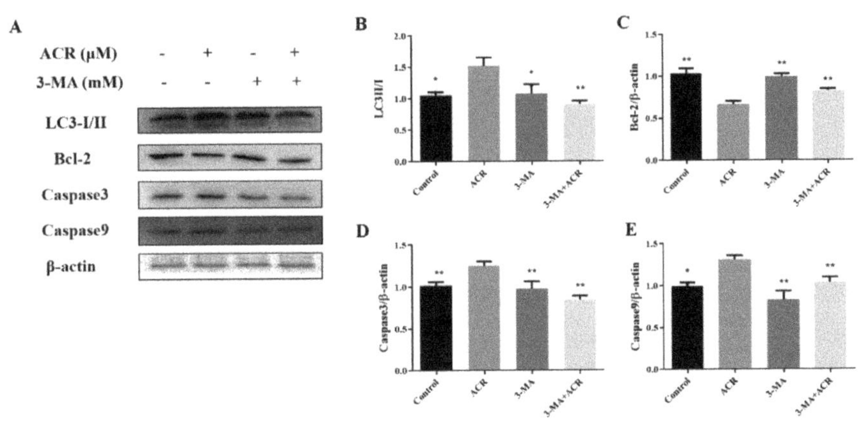

Figure 9. Autophagy inhibitor attenuated apoptosis. (**A**) The protein levels of LC3, caspase-9, caspase-3, and Bcl-2. (**B–E**) Relative intensities of these proteins. Values are means ± SD (n = 3), * $p < 0.05$, ** $p < 0.01$ compared with the acrolein group.

3.10. The Effects of Apoptosis Inhibitor on IEC-6 Cell Autophagy

In Figure 10B, treatment of IEC-6 cells with apoptosis inhibitors markedly inhibited the increase in acrolein-induced, apoptosis-related protein caspase-3. To verify the apoptosis inhibitor effect on cellular autophagy, we examined the expression of autophagy-related proteins Beclin 1, and the result showed that apoptosis inhibitor significantly reduced acrolein-induced overexpression of autophagy-related protein Beclin 1 ($p < 0.05$) (Figure 10C).

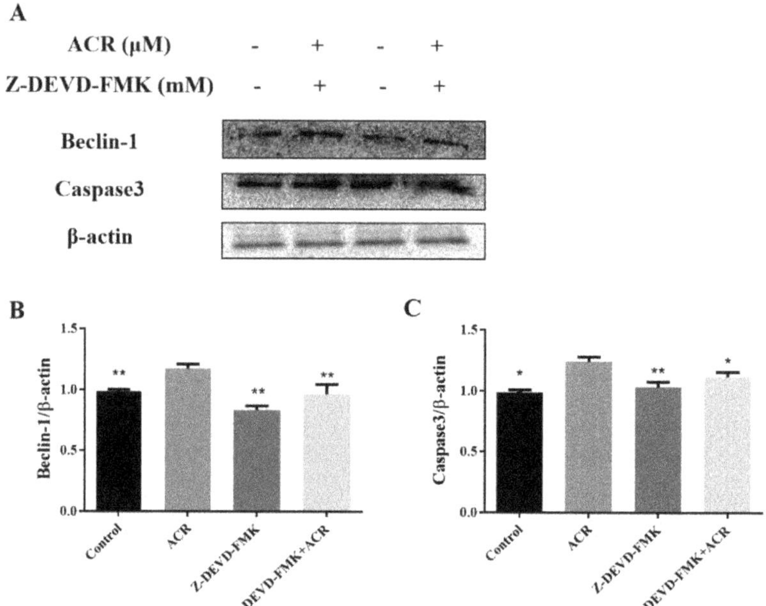

Figure 10. Apoptosis inhibitor attenuated autophagy. (**A**) The protein levels of caspase-3 and Beclin 1 proteins. (**B**,**C**) Relative intensities of these proteins. Values are means ± SD (n = 3), * $p < 0.05$, ** $p < 0.01$ compared to ACR group.

4. Discussion

The intestinal epithelial cells (IECs) have roles in the functions of digestion, absorption of nutrients [25], and the barrier function preventing toxic substances from entering the body, etc. [3,26]. They are the main component of the mechanical barrier of the intestinal mucosa and the first line of defense of the body against intestinal bacteria and toxins [27]. In this study, acrolein was found to cause a reduction in growth and viability of IEC-6 cells, and its cytotoxic action on IEC-6 cells was a concentration-dependent action, with a concentration of 40 μM significantly impairing the viability. Meanwhile, several concentrations of PSG-1 extracts were found to have cytoprotective effects against acrolein-induced damage in IEC-6 cells.

GPx is an essential peroxidase enzyme that is extensively present in living organisms [28]. MDA and SOD content is a vital indicator of the potential antioxidant capacity of the body [29,30]. Meanwhile, we found that acrolein induced oxidative stress in IEC-6 cells, decreased the SOD activity and GPx activity, and increased the intracellular lipid peroxide MDA content. PSG-1 prevented acrolein-induced oxidative stress by improving GPx and SOD levels and reducing MDA content. Therefore, we hypothesized that PSG-1 altered the antioxidant activity in IEC-6 cells under the effect of acrolein to enhance the antioxidant capacity of cells and prevented oxidative stress damage to cells, which indirectly achieved cell protection. The mechanism may be due to the PSG-1 combining with radicals to form a stable radical, or combining with the radical ions, which are necessary for radical chain reaction, and then the reaction is terminated [18].

The TJ of IECs is the main component connecting IECs to the intestinal mucosal barrier, which regulates the permeability of barrier and is the first line of defense of cell bypass [31]. The TJ determines the electrolyte and water balance in the intestine, which is the key factor of intestinal integrity and mucosal barrier function, and it consists mainly of occludin, claudins, and other protein families, which are composed of the linker complex proteins (ZO-1), and cytoskeletal structures which form the TJ complex [32]. Three common TJ proteins—occludin, claudin, and ZO-1—were assayed. Acrolein was able to decrease the level of TJ proteins, while the PSG-1 treatment markedly enhanced the expression of these proteins, which is beneficial for maintaining the integrity of the intestinal mucosa. This is consistent with the oxidative damage index data. The experiment suggests that PSG-1 can maintain intestinal barrier function by up-regulating the expression level of TJ proteins in IEC-6 cells.

Autophagy, known as type II programmed cell death, is a self-protective mechanism widely present in eukaryotic cells [33]. To confirm autophagy is engaged in the protection of acrolein-damaged IEC-6 cells by PSG-1, the expression of key proteins of autophagy was inspected. The expression ratios of LC3-II and LC3-I can be used as important indicators for autophagy experiments. The ratio of LC3-II/I was significantly higher in IEC-6 cells treated with acrolein, suggesting that autophagy was promoted by acrolein exposure.

Autophagy is a complex signaling process with multi-step regulations. Studies have revealed that autophagy mainly consists of the PI3K/Akt/mTOR pathway and the Beclin 1 pathway, which is a key regulation protein of autophagy [34]. The mTOR signaling pathway is a critical regulation of autophagy initiation, activated mTOR inhibits autophagy [35]. After acrolein treatment, the expression of Beclin 1 increased, and the p-mTOR and p-akt content significantly decreased. Moreover, PSG-1 therapy significantly reduced LC3-II/I and Beclin 1 expression and promoted p-mTOR and p-akt expression. Experiments suggested that PSG-1 was able to inhibit acrolein-stimulated excessive autophagy, which may be achieved by promoting the mTOR pathway.

Apoptosis is a fundamental and important biological phenomenon that plays an essential role in the clearance of abnormal cells [36]. It was found that acrolein caused apoptosis of IEC-6 cells at pathological concentrations. To investigate whether apoptosis has a function in acrolein-induced IEC-6 cells, we first looked at the rate of apoptotic cell death. The experiments suggested that the apoptosis rate of cells remarkably elevated after acrolein induction and decreased significantly after the addition of PSG-1, which suggested that acrolein can induce IEC-6 cells apoptosis and PSG-1 can protect IEC-6 cells by reducing the number of apoptotic cells. Various studies have shown that mitochondria are closely related to apoptosis and that apoptosis is irreversible once the transmembrane potential of the mitochondrial membrane collapses. Therefore, the change of MMP was detected, the result showed that IEC-6 cells treatment with acrolein caused MMP loss and PSG-1 treatment significantly improved this situation.

Afterwards, the expression of apoptosis-related proteins was also examined. Bcl-2 is located upstream of mitochondria and an important regulation of the altered permeability of mitochondrial membranes. Caspase-3 and caspase-9 play an irreplaceable role in apoptosis. Caspase-9 involves the initiation of apoptosis. Caspase-3 is the most critical apoptosis executor downstream of the caspase cascade reaction. We found that after acrolein induction of IEC-6 cells, the expression of Bcl-2 was significantly decreased, the expression of two apoptotic proteins, caspase3 and caspase-9, was significantly increased, and PSG-1 restored the expression of these three proteins to a certain extent. The above data suggested that acrolein-induced IEC-6 cells led to MMP loss, increased expression of caspase 3 and caspase 9 and decreased expression of the Bcl-2 protein, which indicated that PSG-1 effectively restored acrolein-induced apoptosis through the mitochondrial pathway. It had been established that the biological activity of mushroom polysaccharides were attributed to their chemical structure [14,37]. Thus, we speculated that the regulation of autophagy and apoptosis by PSG-1 was most likely related to the main structure β-(1 → 3, 1 → 6)-glucan, which was elucidated in our previous study [18,38].

Numerous studies have shown that autophagy and apoptosis play an invaluable function in the intestinal barrier and jointly maintain the integrity of the intestinal barrier. To explore the association between autophagy/apoptosis and TJ, we pretreated IEC-6 cells with specific inhibitors of autophagy and apoptosis pathways, respectively, examining the expression of occludin, claudin, and ZO-1 in each group. It was found that the expression of TJ proteins was increased after treatment of IEC-6 cells with two inhibitors, suggesting that acrolein-induced autophagy and apoptosis occur in IEC-6 cells could disrupt the integrity of the intestinal barrier.

It is well known that a complex interactive regulation exists between autophagy and apoptosis [39]. The relationship between autophagy and apoptosis in acrolein-induced IEC-6 cells by pathway inhibitors was explored in this experiment. It was revealed that the expression of autophagy-related proteins was reduced in autophagy inhibitor-treated IEC-6 cells, and that autophagy inhibitor inhibited the expression of key apoptotic proteins, suggesting that autophagy may be modulated by acrolein-induced apoptosis of IEC-6 cells. Apoptosis inhibitor-treated cells showed reduced expression of apoptosis-related proteins. In addition, the apoptosis inhibitor inhibited the expression of key autophagic proteins, suggesting that apoptosis may be modulated by acrolein-induced autophagy of IEC-6 cells.

5. Conclusions

This study proved that PSG-1 repaired oxidative stress in acrolein-induced IEC-6 cells, restored the expression of TJ proteins inhibited by acrolein, and attenuated acrolein-induced autophagy and apoptosis in IEC-6 cells. Furthermore, PSG-1 restored TJ by inhibiting apoptosis and autophagic flux, where autophagy and apoptosis were mutually regulated. The finding may provide a theoretical basis for the development and generation of mushroom β-(1 → 3, 1 → 6)-glucan as high value-added products.

Author Contributions: Conceptualization, Q.Y.; formal analysis, Y.W.; funding acquisition, Q.Y.; investigation, Y.W.; methodology, Y.W.; project administration, Y.C., X.H. (Xiaobo Hu) and Q.Y.; resources, B.Z.; software, Y.C. and X.H. (Xiaobo Hu); supervision, J.X. and J.S.; visualization, Y.C. and X.H. (Xiaobo Hu); writing—original draft preparation, X.C.; writing—review and editing, X.C., J.X., J.S., X.H. (Xiaoyi Hu), X.D. and Q.Y. All authors have read and agreed to the published version of the manuscript.

Funding: This research was funded by the National Natural Science Foundation of China (81760737).

Institutional Review Board Statement: Not applicable.

Informed Consent Statement: Not applicable.

Data Availability Statement: Not applicable.

Conflicts of Interest: The authors declare no conflict of interest. The funders had no role in the design of the study; in the collection, analyses, or interpretation of data; in the writing of the manuscript, or in the decision to publish the results.

References

1. Mohammad Reza Zirak, S.M.; Karimani, A.; Zeinali Majid, A.; Hayes, W.; Karimi, G. Mechanisms behind the atherothrombotic effects of acrolein, a review. *Food Chem. Toxicol.* **2019**, *129*, 38–53. [CrossRef] [PubMed]
2. Lijun Chen, X.W.; Zeb, F.; Huang, Y.; An, I.; Jiang, P.; Chen, A.; Xu, C.; Feng, G. Acrolein-induced apoptosis of smooth muscle cells through NEAT1Bmal1/Clock pathway and a protection from asparagus extrac. *Environ. Pollut.* **2020**, *258*, 113735. [CrossRef]
3. Jia Shi, X.-H.Z. Chemical features of the oligochitosan-glycated caseinate digest and its enhanced protection on barrier function of the acrylamide-injured IEC-6 cells. *Food Chem.* **2019**, *290*, 246–254.
4. Xiaoyue Wu, L.C.; Zeb, F.; Li, C.; Jiang, P.; Chen, A.; Xu, C.; Feng, Q. Clock-Bmal1 mediates MMP9 induction in acrolein-promoted atherosclerosis associated with gut microbiota regulation. *Environ. Pollut.* **2019**, *252*, 1455–1463.
5. Hou, K.Q.Y.; Hu, X.; Ding, X.; Hong, J.; Chen, Y.; Xie, J.; Nie, S.M.X. Protective effect of *Ganoderma atrum* polysaccharide on acrolein-induced macrophage injury via autophagy-dependent apoptosis pathway. *Food Chem. Toxicol.* **2019**, *133*, 110757. [CrossRef] [PubMed]

6. Mehmet Kandemir, F.S.Y.; Kucukler, S.; Caglayan, C.; Darendelioğlu, E.; Bahaeddin Dortbudak, M. Protective effects of morin against acrylamide-induced hepatotoxicity and nephrotoxicity: A multi-biomarker approach. *Food Chem. Toxicol.* **2020**, *138*, 111190. [CrossRef]
7. Isei Tanida, T.U.; Kominami, E. LC3 and autophagy. *Methods Mol. Biol.* **2008**, *445*, 77–88.
8. Dan Li, L.G.; Li, M.; Luo, Y.; Xie, Y.; Luo, T.; Su, L.; Tianqiao Yong, S.C.; Jiao, C.; Su, J.; Huang, S. Polysaccharide from spore of Ganoderma lucidum ameliorates paclitaxelinduced intestinal barrier injury: Apoptosis inhibition by reversing microtubule polymerization. *Biomed. Pharmacother.* **2020**, *130*, 110539.
9. Ueno, Y.; Kawamoto, Y.; Nakane, Y.; Natsume, R.; Miura, K.; Okumura, Y.; Murate, T.; Hattori, E.; Osawa, T. Oxidized perilla and linseed oils induce neuronal apoptosis by caspase-dependent and -independent pathways. *Foods* **2020**, *9*, 538. [CrossRef]
10. Tu, D.G.; Chyau, C.C.; Chen, S.Y.; Chu, H.L.; Wang, S.C.; Duh, P.D. Antiproliferative Effect and mediation of apoptosis in human hepatoma HepG2 cells induced by djulis husk and its bioactive compounds. *Foods* **2020**, *9*, 1514. [CrossRef]
11. Yao, F.D.J.; Cheng, F.; Yao, W.; Chen, P.; Guo, S.; Cao, Y.L.Z. Diterpene pekinenal from euphorbia pekinensis radix induced IEC-6 cells apoptosis mediated by mitochondria and death receptors. *Toxicol. Vitr.* **2019**, *57*, 1–8. [CrossRef] [PubMed]
12. Wang, X.-P.; Zhao, X.-H. Activity of the peptic-tryptic caseinate digest with caseinate oligochitosan-glycation in rat intestinal epithelial (IEC-6) cells via the Wnt/beta-catenin signaling pathway. *Chem. -Biol. Interact.* **2020**, *328*, 109201. [CrossRef]
13. Chen, W.-Y.; Zhang, J.; Barve, S.S.; McClain, C.J.; Swati, J.-B. Acrolein disrupts tight junction proteins and causes endoplasmic reticulum stress-mediated epithelial cell death leading to intestinal barrier dysfunction and permeability. *Am. J. Pathol.* **2017**, *187*, 2686–2697. [CrossRef] [PubMed]
14. Sheng, K.; Chen, B.; Kang, M.; Wang, M.; Liu, K.; Wang, M. Recent advances in polysaccharides from Lentinus edodes (Berk.): Isolation, structures and bioactivities. *Food Chem.* **2021**, *358*, 129883. [CrossRef]
15. Cheng, Z.Z.; Han, Y.; Wang, J.; Wang, Y.; Chen, X.; Shao, Y.; Cheng, Y.; Zhou, W.; Lu, X.; Wu, Z. A review on anti-cancer effect of green tea catechins. *J. Funct. Foods* **2020**, *74*, 104172. [CrossRef]
16. Che, H.Z.; Tian, Y.; Lai, P.F.H.; Xia, Y.; Wang, S.; Ai, L. Exopolysaccharide from Streptococcus thermophilus as stabilizer in fermented dairy: Binding kinetics and interactions with casein of milk. *Int. J. Biol. Macromol.* **2019**, *140*, 1018–1025. [CrossRef]
17. Chen, X.Y.H.; Meng, H.; Li, W.; Li, Q.; Luo, Y.; Wang, C.; Xie, J.; Wu, L.; Zhang, X.; Wu, Z.; et al. Characteristics of the emulsion stabilized by polysaccharide conjugates alkali-extracted from green tea residue and its protective effect on catechins. *Ind. Crops Prod.* **2019**, *140*, 111611. [CrossRef]
18. Zhang, H.; Nie, S.-P.; Chen, Y.; Wang, Y.-X.; Xie, M.-Y. Structural characterisation of a novel bioactive polysaccharide from *Ganoderma atrum*. *Carbohydr. Polym.* **2012**, *88*, 1047–1054. [CrossRef]
19. Zheng, B.; Xie, J.; Hong, J.; Liao, W.; Yu, Q.; Chen, Y.; Wang, Y.; Ding, X. A *Ganoderma atrum* polysaccharide alleviated DSS-induced ulcerative colitis by protecting the apoptosis/autophagy-regulated physical barrier and the DC-related immune barrier. *Food Funct.* **2020**, *11*, 10690–10699. [CrossRef]
20. Ding, X.; Yu, Q.; Hou, K.; Hu, X.; Wang, Y.; Chen, Y.; Xie, J.; Nie, S. Indirectly stimulation of DCs by *Ganoderma atrum* polysaccharide in intestinal-like Caco-2/DCs co-culture model based on RNA-seq. *J. Funct. Foods* **2020**, *67*, 103850. [CrossRef]
21. Wang, Y.; Chen, Y.; Xie, J.; Zheng, B.; Chang, X.; Liu, S.; Hu, X.; Yu, Q. "Dialogue" between Caco-2 and DCs regulated by *Ganoderma atrum* polysaccharide in intestinal-like Caco-2/DCs co-culture model. *Food Res. Int.* **2021**, *144*, 110310. [CrossRef] [PubMed]
22. Hu, X.; Hou, K.; Ding, X.; Chen, Y.; Xie, J.; Nie, S.; Xie, M. Regulatory effects of *Ganoderma atrum* polysaccharides on LPS-induced inflammatory macrophages model and intestinal-like Caco-2/macrophages co-culture inflammation model. *Food Chem. Toxicol.* **2020**, *140*, 111321. [CrossRef] [PubMed]
23. Yang, Y.; Jiang, G.; Lei, A.; Yu, Q.; Xie, J.; Chen, Y. Evaluation of the protective effects of *Ganoderma atrum* polysaccharide on acrylamideinduced injury in small intestine tissue of rats. *Food Funct.* **2019**, *10*, 5863–5872. [CrossRef]
24. Jiang, Q.; Wang, H.; Chen, Q.; Wu, X.; Yan, X.; Chen, Y.; Xie, M. Protective effects of a *Ganoderma atrum* polysaccharide against acrylamide induced oxidative damage via a mitochondria mediated intrinsic apoptotic pathway in IEC-6 cells. *Food Funct.* **2018**, *9*, 1133–1143. [CrossRef]
25. Barraza-Garza, G.; Castillo-Michel, H.; de la Rosa, L.A.; Martinez-Martinez, A.; Cotte, M.; Alvarez-Parrilla, E. Antioxidant effect of phenolic compounds (PC) at different concentrations in IEC-6 cells: A spectroscopic analysis. *Spectrochim. Acta Part A Mol. Biomol. Spectrosc.* **2020**, *227*, 117570. [CrossRef] [PubMed]
26. Qiu, H.-M.; Lv, J.-H.; Wu, T.-C.; Zhang, Z.-P.; Liu, Y.; Chen, X.-Q.; Aweya, J.J.; Cheong, K.-L. Physicochemical properties and potential beneficial effects of porphyran from Porphyra haitanensis on intestinal epithelial cells. *Carbohydr. Polym.* **2020**, *246*, 116626. [CrossRef]
27. Cai, B.; Chen, H.; Chen, D.; Chen, X.; Sun, H.; Pan, J. Composition characterization of oyster polysaccharides from Crassostrea hongkongensis and their protective effect against H2O2-induced oxidative damage in IEC-6 cells. *Int. J. Biol. Macromol.* **2019**, *124*, 246–254. [CrossRef]
28. Bettaiba, J.; Droguet, M.; Magné, C.; Boulaaba, M.; Giroux-Metges, M.-A.; Ksouri, R. Tamarix gallica phenolics protect IEC-6 cells against H2O2 induced stress by restricting oxidative injuries and MAPKs signaling pathways. *Biomed. Pharmacother.* **2017**, *89*, 490–498. [CrossRef]

29. Zhang, D.; Zhou, J.; Cai, W.; Qian, L. Milk Fat globule membrane ameliorates necrotizing enterocolitis in neonatal rats and suppresses lipopolysaccharide-induced inflammatory response in IEC-6 enterocytes. *J. Parenter. Enter. Nutr.* **2018**, *43*, 863–873. [CrossRef]
30. Ren, B.; Li, C.; Fu, X.; You, L.; Liu, R.H. Optimization of microwave-assisted extraction of Sargassum thunbergii polysaccharides and its antioxidant and hypoglycemic activities. *Carbohydr. Polym.* **2017**, *173*, 192–201. [CrossRef]
31. Xiao, J.; Wu, Y.; Wu, C.; Jia, X.; Dong, L.; Liu, L.; Chen, Y.; Bai, Y.; Zhang, M. Rice bran phenolic extract protects against alcoholic liver injury in mice by alleviating intestinal microbiota dysbiosis, barrier dysfunction and liver inflammation mediated by endotoxin-TLR4-NF-kB pathway. *J. Agric. Food Chem.* **2019**, *68*, 1237–1247. [CrossRef]
32. Díaz-Coránguez, M.; Liu, X.; Antonetti, D.A. Tight junctions in cell proliferation. *Int. J. Mol. Sci.* **2019**, *20*, 5972. [CrossRef]
33. Vukelić, I.; Batičić, L.; Potočnjak, I.; Domitrović, R. Luteolin ameliorates experimental colitis in mice through ERK-mediated suppression of inflammation, apoptosis and autophagy. *Food Chem. Toxicol.* **2020**, *145*, 111680. [CrossRef] [PubMed]
34. Huang, Y.; Wen, Q.; Huang, J.; Luo, M.; Xiao, Y.; Mo, R.; Wang, J. Manganese (II) chloride leads to dopaminergic neurotoxicity by promoting mitophagy through BNIP3-mediated oxidative stress in SH-SY5Y cells. *Cell. Mol. Biol. Lett.* **2021**, *26*, 23. [CrossRef] [PubMed]
35. Gao, X.; Jiang, Y.H.; Xu, Q.; Liu, F.; Pang, X.N.; Wang, M.J.; Li, Q.; Li, Z.C. 4-Hydroxyderricin promotes apoptosis and cell cycle arrest through regulating PI3K/AKT/mTOR pathway in hepatocellular cells. *Foods* **2021**, *10*, 2036. [CrossRef] [PubMed]
36. Hussain, S.S.; Zhang, F.; Zhang, Y.; Thakur, K.; Naudhani, M.; Cespedes-Acuña, C.L.; Wei, Z. Stevenleaf from Gynostemma Pentaphyllum inhibits human hepatoma cell (HepG2) through cell cycle arrest and apoptotic induction. *Food Sci. Hum. Wellness* **2020**, *9*, 295–303. [CrossRef]
37. Wu, Y.-j.; Zhang, F.-m.; Linhardt, R.J.; Sun, P.-l.; Zhang, A.-q. Structure, bioactivities and applications of the polysaccharides from Tremella fuciformis mushroom: A review. *Int. J. Biol. Macromol.* **2019**, *121*, 1005–1010. [CrossRef]
38. Chen, Y.; Nie, S.-P.; Li, C.; Wang, Y.-X. Purification, composition analysis and antioxidant activity of a polysaccharide from the fruiting bodies of *Ganoderma atrum*. *Food Chem.* **2008**, *107*, 231–241. [CrossRef]
39. Muniraj, N.; Nagalingam, A.; Walker, A.; Woo, J.; Győrffy, B.; Gabrielson, E.; Saxena, N.K.; Sharma, D. Withaferin A inhibits lysosomal activity to block autophagic flux and induces apoptosis via energetic impairment in breast cancer cells. *Carcinogenesis* **2019**, *40*, 1110–1120. [CrossRef] [PubMed]

foods

MDPI

Article

Effects of Superheated Steam Treatment on the Allergenicity and Structure of Chicken Egg Ovomucoid

Ping-Wei Wen [1,2,3], Zong-Cai Tu [1,2,3,*], Yue-Ming Hu [3] and Hui Wang [3]

1 National R&D Branch Center for Conventional Freshwater Fish Processing, College of Chemistry and Chemical Engineering, Jiangxi Normal University, Nanchang 330022, China; wenpingwei@ncu.edu.cn
2 Engineering Research Center of Freshwater Fish High-Value Utilization of Jiangxi Province, Jiangxi Normal University, Nanchang 330022, China
3 State Key Laboratory of Food Science and Technology, Nanchang University, Nanchang 330047, China; huyueming@ncu.edu.cn (Y.-M.H.); wanghui00072@ncu.edu.cn (H.W.)
* Correspondence: tuzc_mail@aliyun.com; Tel.: +86-791-8812-1868; Fax: +86-791-8830-5938

Abstract: The aim of this study was to explore the effects of an emerging and efficient heating technology, superheated steam (SS), on the allergenicity and molecular structure of ovomucoid (OVM). OVM was treated with 120–200 °C of SS for 2 to 10 min. The allergenicity (IgG/IgE binding abilities and cell degranulation assay) and molecular structure (main functional groups and amino acids modification) changes were investigated. The IgG-binding ability of OVM decreased and the releases of β-hex and TNF-γ were inhibited after SS treatment, indicating that the protein allergenicity was reduced. Significant increases in oxidation degree, free SH content and surface hydrophobicity were observed in SS-treated OVM. The protein dimer and trimer appeared after SS treatment. Meanwhile, obvious changes occurred in the primary structure. Specifically, serine can be readily modified by obtaining functional groups from other modification sites during SS treatment. Moreover, the natural OVM structure which showed resistance to trypsin digestion was disrupted, leading to increased protein digestibility. In conclusion, SS-induced OVM aggregation, functional groups and amino acids modifications as well as protein structure alteration led to reduced allergenicity and increased digestibility.

Keywords: ovomucoid; superheated steam; modification; allergenicity; mass spectrometry

Citation: Wen, P.-W.; Tu, Z.-C.; Hu, Y.-M.; Wang, H. Effects of Superheated Steam Treatment on the Allergenicity and Structure of Chicken Egg Ovomucoid. *Foods* **2022**, *11*, 238. https://doi.org/10.3390/foods11020238

Academic Editor: Jan Mei Soon

Received: 14 December 2021
Accepted: 12 January 2022
Published: 17 January 2022

1. Introduction

Egg is the main daily protein source for human beings, and also one of the major ingredients in the food industry [1]. However, egg allergy is common among infants and young children, and a high proportion of patients suffer the egg allergy throughout their lives [2]. Egg allergens are concentrated in egg white proteins, including ovomucoid (OVM), ovalbumin, ovotransferrin and lysozyme, among which, OVM is considered as the predominant component of egg allergens [3]. It is urgent to explore an effective and safe technology to reduce the allergenicity of eggs.

Previous studies have shown that heating, high pressure, glycation and hydrolysis can decrease the protein allergenicity [1,4,5]. Heat processing includes dry-heating and heat-moisture treatments. There are two drawbacks of traditional heating approaches for protein modification. One is that they often consume long periods of time, and the other is that the protein solubility is always decreased. Moreover, the traditional heating treatments are not drastic enough to alter specific sequences and conformation of allergenic epitopes and could not significantly decrease the protein allergenicity. Among the egg protein components, OVM showed the strongest heat resistance, and even 100 °C of heating for 60 min could not lead to the denaturation of OVM [6]. It is necessary to explore new high temperature heating methods to efficiently alter the OVM structure and modify/disrupt/bury the allergen epitopes of protein.

Superheated steam (SS) treatment is an emerging heating method and has attracted more and more attention for its advantages, including higher heat transfer efficiency, low-oxygen and pollution-free environment as well as good quality of processed products [7] During SS processing, a large amount of heat is transferred to food when steam condenses on food surfaces, which rapidly increases the food temperature. The efficiency of SS has recently been demonstrated in several different food processing fields by dramatically reducing the processing time from days and h to min or even s [8]. SS-treated vegetables, grains, meats, etc. showed better functional and textural properties as well as higher nutritional qualities than those treated by conventional heating methods [9,10]. SS was also proven to improve the physicochemical and nutritional properties of biological macromolecules, including starch and protein, etc. [11,12]. Hu, Wang, and Li [7] reported that SS treatment could promote the covalent and non-covalent interactions as well as alter the conformation of wheat gluten. However, to date, the application of SS on the modification of natural structure and nutrition of food components is still lacking. Particularly, the effects of SS processing on allergenicity and the molecular structure of egg proteins have rarely been reported.

The objective of the study is to find a new, easily operated, efficient and safe technology to reduce the allergenicity of proteins. Our preliminary experiments showed that SS is an effective approach to modifying egg proteins and reducing the allergenicity. It is necessary to perform experiments to prove those effects.

Previous research demonstrated that egg proteins allergies are mainly originated from immunoglobulin G (IgG) and immunoglobulin E (IgE)-mediated rapid allergic reaction. In addition, macrophages are also important effector cells that mediate immune response [2]. The mediators released from basophils may be responsible for the occurrence of Type I allergic reactions. Therefore, in this study, the allergenicity changes of SS-treated OVM were revealed by the following experiments: (1) IgG and IgE binding abilities analysis through enzyme-linked immunosorbent assay (ELISA) with rabbit polyclonal antibodies and sera from patients allergic to egg, respectively. (2) Histamine, β-hexosaminidase (β-hex), TNF-γ, and interleukin-6 (IL-6) releases determination using a basophil cell line (KU812) model.

The protein allergenicity changes are reported related to the alteration in molecular structure in terms of primary structure, main functional groups and degree of protein aggregation [1–5]. Therefore, in this study, the changes in the surface hydrophobicity, free sulfhydryl (SH) content, molecular weight and amino acids modification of OVM were investigated to preliminary interpret the mechanism of SS-induced protein allergenicity alteration.

2. Materials and Methods

2.1. Chemicals and Materials

OVM, D-glucose, goat antirabbit IgG-HRP conjugate, goat antihuman IgE-HRP conjugate and trypsin were purchased from Sigma-Aldrich (St. Louis, MO, USA). Dithiothreitol (DTT) was purchased from Thermo Fisher Scientific Inc. (Waltham, MA, USA). The KU812 cells were obtained from iCell Bioscience (Shanghai, China). All other reagents used were of analytical reagent grade. Ultrapure water from a water purification system (Millipore; Billerica, MA, USA) was used throughout this study.

The sera of six egg white allergy (EWA) patients were purchased from Plasma Lab International. The specific IgE levels of six groups of sera were 10.8, 26.3, 24.9, 13.5, 16.5 and 12.8 kU/L. The six groups of sera were mixed and served as human anti-OVM IgE serum and stored at −80 °C until used.

Preparation of rabbit anti-OVM IgG serum: Male Japanese rabbits (100 days) (permission number, SCXK (Gan) 2014–0005) were purchased from Longping (Nanchang, China). After acclimatizing in a breeding room for 10 days, the rabbits were intravenously injected with 1 mL OVM (0.8 mg/mL) emulsified with the Freund's complete adjuvant (*v/v*, 1:1). The booster immunization was taken after one, two and three weeks at a dose of 0.4 mL, respectively. After the rabbits were anesthetized, the rabbit anti-OVM IgG plasma was collected. After being centrifuged at 4 °C, the rabbit anti-OVM IgG serum was separated and stored at −80 °C until used.

2.2. SS Processing

An SS processing system developed by the Food Engineering Center of Nanchang University was used in this study. The schematic diagram was detailed and shown in the report of Wu et al. [13]. A total of 200 mg OVM powder was spread in a culture dish. When the temperature in the SS chamber was kept stable, the culture dish was inserted into the processing chamber. Processing of OVM samples was conducted at atmospheric pressure. The flow velocity of SS was 1.0 m/s. The treatment temperatures were 120, 140, 160, 180 and 200 °C. The processing times were 2, 4, 6, 8 and 10 min. After being treated for the set time, the OVM sample was taken out of the processing chamber and then cooled in an ice bath. Native OVM was served as the control. All the samples were stored at 4 °C until used.

2.3. IgG and IgE-Binding Abilities

The IgG and IgE-binding abilities of the OVM samples were estimated with indirect competitive ELISA according to our previous report [14]. Rabbit anti-OVM IgG serum and human anti-OVM IgE serum were used to determine IgG-binding ability and IgE-binding ability, respectively. A 96-well microplate was coated with 2 μg/mL of native OVM at 37 °C for 1 h. Then residual free binding sites were blocked with 1% fish gelatin solution and incubated at 37 °C for 1 h. Next, the diluted samples and the serum of rabbit or EWA patients were injected into the microplate and incubated at 37 °C for 1 h. After removing the solutions, the microplate was washed with PBST five times. Goat antirabbit IgG-HRP conjugate (100 μL, diluted at 1:2500) or goat antihuman IgE-HRP conjugate (100 μL, diluted at 1:700) was injected and then incubated at 37 °C for 1 h. Finally, tetramethylbenzidine (TMB, 100 μL) was immediately added to each well and purged with sulfuric acid. The absorbance was measured at 450 nm and the decline rate was calculated using the equation:

$$\text{Inhibition (\%)} = (1 - B/B_0) \times 100\% \quad (1)$$

where B_0 and B indicate the optical density (OD) value of a well with the control and SS-treated OVM, respectively.

2.4. Human Peripheral Blood Basophilic Leukemia Cells (KU812 Cells) Culture and Degranulation Assay

The basophil histamine release test was carried out according to the method of Wang et al. [15]. KU812 cells were cultured in RPMI-1640 medium with 20% fetal bovine serum (FBS) and 105 U/L penicillin/streptomycin for 24 h. A total of 5×10^5 cells was in each well. The culture condition was 37 °C with CO_2 (5%). The cells were then activated with human anti-OVM IgE serum for 24 h, and followed by, stimulated by 50 μg/well of samples for 4 h. Cells treated with PBS buffer were served as the negative control. The release of β-hex, TNF-γ, histamine and IL-6 were analyzed by ELISA kits, following the manufacturer's instructions [16,17].

2.5. Determination of Protein Carbonyl Content

The carbonyl content of the OVM samples was determined according to the method of Zhang et al. [18] with slight modifications. The measurements were performed by monitoring the reaction between 2,4-dinitrophenylhydrazine (DNPH) and the carbonyl group of the protein. An amount of 0.2 mL of OVM (4 mg/mL) was mixed with 0.4 mL of DNPH (10 mmol/L, in 2 mol/L HCl) and the mixture was incubated at 37 °C in the dark for 1 h. Followed by 0.5 mL of 20% trichloroacetic acid was added and the solution was incubated for 5 min. The sample was then centrifuged at 12,000 rpm and 4 °C for 15 min. The supernatant was removed, and the precipitate was retained. The sediment was washed using 1 mL of ethyl acetate/ethanol (1:1) mixture and then centrifuged with the method described above three times. The obtained protein was incubated with 1 mL of guanidine hydrochloride solution (6 mol/L), and the solution was centrifuged to remove precipitation. Absorption at 370 nm was determined using a U-2910 ultraviolet-visible spectrophotometer. The blank was applied with the same method except using HCl (2 mol/L) instead of DNPH. The carbonyl content was calculated with a molar extinction coefficient of 22,000 $mol^{-1} \cdot cm^{-1}$, and the result was expressed as nmol/mg protein.

2.6. Determination of Surface Hydrophobicity

The surface hydrophobicity of the OVM samples was determined by an 8-aniline-1-naphthalene sulfonic acid (ANS) fluorescence probe method [19]. The samples were diluted to 1, 0.5 and 0.25 mg/mL, and the fluorescence intensities were determined by mixing 4 mL of each sample with 20 μL of ANS solution (8 mmol/L). The measurement conditions were as follows: excitation wavelength of 390 nm; scanning emission wavelength range of 400–600 nm; slit width of 5 nm. A linear regression equation was prepared according to the fluorescence intensities of the samples at each concentration for curve fitting. The obtained slope was the surface hydrophobicity of the samples.

2.7. Determination of FREE SH Content

The free SH content of OVM samples was determined according to the method of Ellman [20] with slight modifications. Ellman's reagent was prepared according to the following procedure: first, 0.369 g of DTNB was dissolved in 50 mmol/L Na_2HPO_4 (pH 7.0); then, the volume was adjusted to 100 mL; the reagent was stored at 4 °C in the dark. The protein samples were diluted to 2 mg/mL with 0.1 mol/L phosphate buffer (pH 8.0) containing 0.086 mol/L Tris-HCl, 0.09 mol/L Gly, 4 mmol/L EDTA and 8 mol/L urea. The diluted protein solution (1 mL) was mixed with buffer solution (8 mL), and centrifuged at 8000 r/min for 20 min. Following this, 4.5 mL of supernatant was added to 0.5 mL of Ellman's reagent. The mixed solution was reacted at room temperature for 15 min after evenly oscillated. The absorption at 412 nm was measured. The free SH content was calculated according to the following formula:

$$SH\ (\mu mol/g) = 73.53 \times A_{412}/C \quad (2)$$

where $73.53 = 10^6/(1.36 \times 10^4)$; 1.36×10^4 is the molar absorption coefficient of DTNB; C is the protein concentration (mg/mL).

2.8. Matrix-Assisted Laser Desorption/Ionization Time of Flight Mass Spectrometry (MALDI TOF MS) Analysis

The molecule weight of OVM was determined by MALDI-TOF MS (4800 Plus MALDI-TOF/TOF Analyzer, AB Science, Framingham, MA, USA) according to the method of Liu et al. [21]. OVM samples were dissolved in distilled water at 1:100. The matrix solution was prepared by the mixing of sinapic acid (5 mg/mL) in 50% acetonitrile with 0.1% TFA. The OVM solution was mixed with the matrix solution at a ratio of 1:1. The mixtures (2.0 μL) were then dropped onto the MALDI target plate and allowed to dry at room temperature before analysis.

2.9. High Performance Liquid Chromatography Orbitrap Tandem Mass Spectrometry (HPLC Orbitrap MS/MS) Analysis

The OVM samples were digested by trypsin at 37 °C for 24 h. The digested sample solution was separated via a nanoliter flow HPLC system (UltiMate 3000RSLCnano, Thermo Fisher Scientific, Waltham, MA, USA). Solution A was an aqueous solution with 0.1% formic acid. Solution B was acetonitrile binary solution with 0.1% formic acid and 84% acetonitrile. The digested sample solution was injected into a RP-C18 column to remove insoluble or impure substances and then separated by another RP-C18 column at a flow speed of 300 μL/min. Gradient elution was then carried out.

After separation, the eluant was injected into an LTQ-Orbitrap Fusion Velos mass spectrometer (Thermo Fisher Scientific; Waltham, MA, USA) for analysis by tandem mass spectrometry (MS/MS) to identify protein modification forms and sites with a positive ion detection mode. The precursor ions were subjected to high-energy collisional dissociation (HCD) fragmentation to detect fragment ions. Twenty fragment maps (MS/MS^2 scans) showing the mass-to-charge ratio of the polypeptide and polypeptide fragments were collected at each full scan. Raw files were obtained from the corresponding database using Proteome Discoverer 1.4. Some parameters were set as follows: Enzymes, non-specified; Missed cleavage, 2; Modification, carbamidomethyl, oxidation, acetylation, phosphorylation, sulfonation, methylation, ubiquitination, nitro.

2.10. Statistical Analysis

The values were expressed as means ± standard deviation from three separate experiments. The analysis was performed using SPSS version 20.0 (SPSS Inc., Chicago, IL, USA). Statistical data were determined based on a two-tailed *t*-test using standard deviations.

3. Results and Discussion

3.1. IgG and IgE Binding Abilities

The IgG and IgE binding abilities of OVM were estimated through indirect competitive ELISA with sera from rabbit and EWA patients, respectively. As shown in Figure 1A, the IgG binding rate of SS-treated OVM samples significantly decreased with the increase of SS temperature and processing time. The IgG binding rate was markedly influenced by high temperatures and long periods of treatments, with a maximum decline to 28% when treated at 200 °C for 10 min. The reduction in the IgG-binding ability of OVM may be due to the structural changes derived from amino acid modification and denaturation [15]. The results implied that SS-created extreme high temperatures of up to 200 °C might influence the protein structure and destroy IgG allergenicity epitopes.

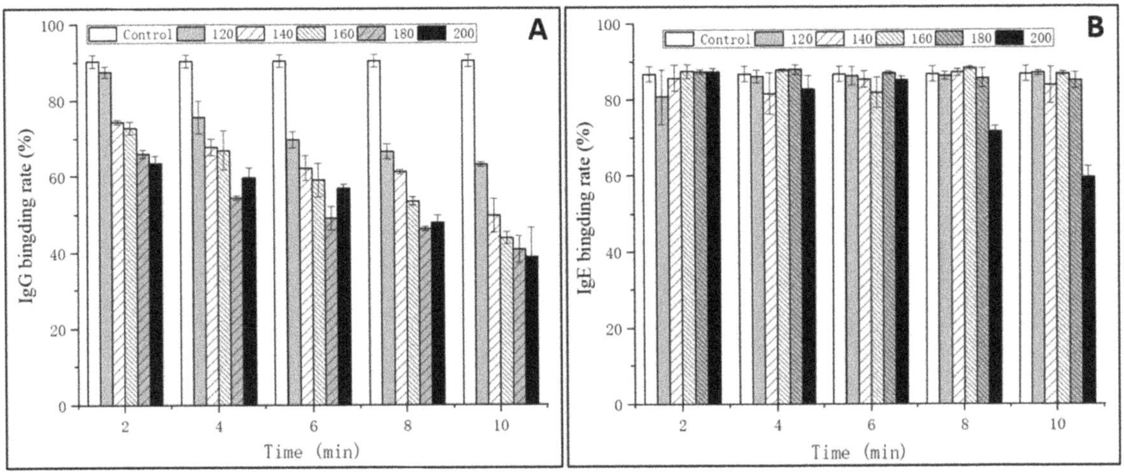

Figure 1. IgG (**A**) and IgE (**B**) binding rates of OVM treated by SS at different temperatures and times.

However, there were no obvious changes in the IgE binding rate of OVM samples treated by SS, except that at 200 °C for 8–10 min (Figure 1B). The results were in accordance with the research of Martos, Lopez-Exposito, Bencharitiwong, Berin, and Nowak-Wegrzyn [22], who reported that heat-treated OVM did not induce the symptoms of anaphylaxis in sensitized mice when administered orally. However, being heated by boiling water for 30 min did not completely destroy the IgE binding capacity of OVM. Native and SS-treated OVM samples at 120–180 °C for 2–10 min and 200 °C for 2–6 min showed high IgE binding rates, suggesting the persistence of linear epitopes recognized by IgE. This could be caused by the fact that OVM possesses high thermal stability and limited denaturation as the structure is made up of nine disulfide bonds and 25% carbohydrate [23]. Interestingly, when the heating temperature reached 200 °C and was maintained for 8–10 min, significant reductions of the IgE binding rate of OVM were observed, declining to 60% of that of native OVM when treated for 10 min. Therefore, it was confirmed that high heating temperatures could alter the OVM structure more and reduce the content of the IgE epitope in comparison with lower temperatures. In addition, macrophages are also important effector cells that mediate the immune response. The mediators released from basophils may be responsible for the occurrence of Type I allergic reactions. Therefore, the IgE-mediated allergic response was evaluated using a KU812 cell model in the subsequent experiment.

3.2. *Effects of OVM on the Viability and Degranulation of IgE Sensitized KU812 Cells*

To analyze the potential cytotoxic activities of the tested substances, KU812 cells were incubated for 24 h with increasing concentrations of OVM treated by SS at different temperatures and periods. As shown in Figure 2, compared with the control sample, the absorbance (OD_{450}) of cells increased after SS treatment, suggesting that SS heating can promote KU812 cells proliferation. Generally, the absorbance (OD_{450}) of cells increased with the increase of OVM concentration and SS processing period. These suggested that a higher OVM concentration and longer SS processing period promoted the KU812 cells proliferation.

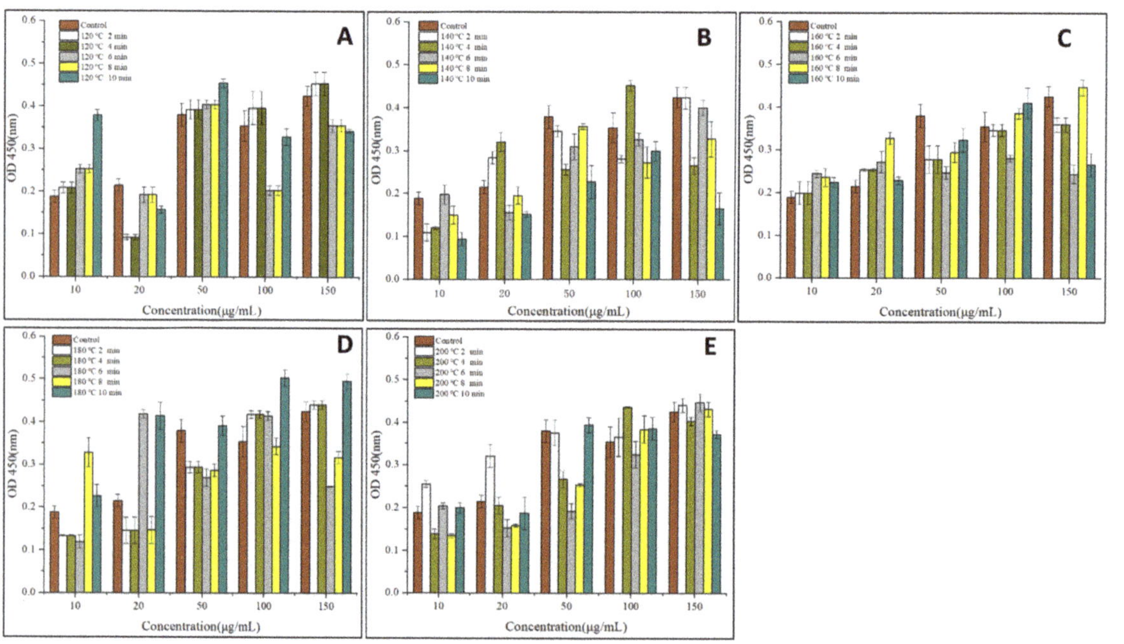

Figure 2. Effects of SS-treated OVM at different temperatures and times on the cell viability (**A**–**E**) of the KU812 cells sensitized with sera IgE from patients allergic to egg.

The degranulation rate of basophils and mast cells are key factors that determine the immunoreactivity of allergens. In this work, the effects of SS treatment on the release of β-hex, TNF-γ, histamine and IL-6 in KU812 cells are shown in Figure 3. The varying concentrations of 4 types of I cytokines indicated that SS treatment on OVM affected allergic response at the molecular level [24].

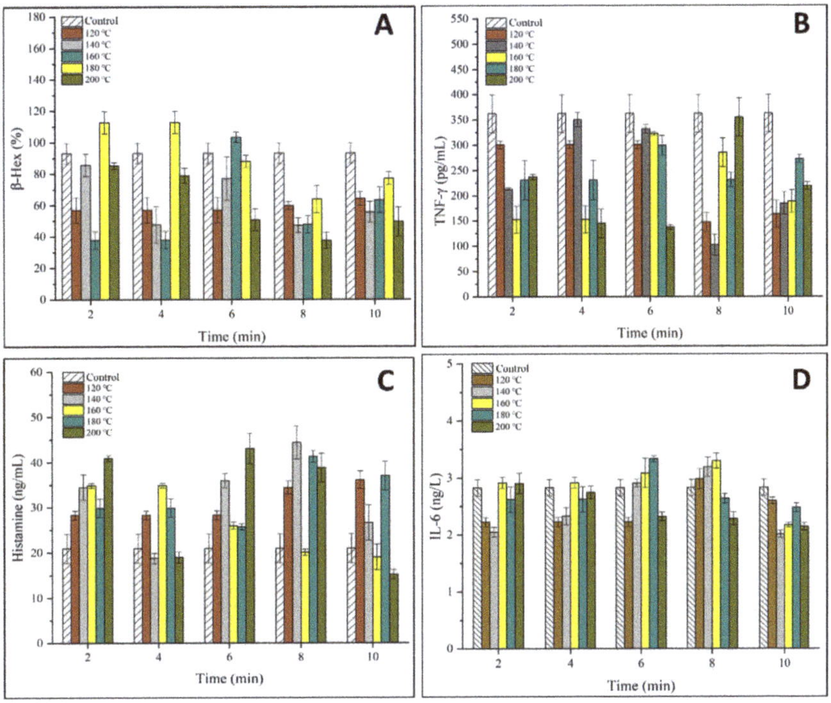

Figure 3. Effects of SS-treated OVM at different temperatures and times on the degranulation of KU812 cells sensitized with sera IgE from patients allergic to egg (**A**), the release of β-hex; (**B**), the release of TNF-γ; (**C**), the release of histamine; (**D**), the release of IL-6).

Figure 3A–C showed that SS treatment inhibited the release of β-hex and TNF-γ while promoting the release of histamine from KU812 cells generally. β-hex is an indicator of mast cell degranulation in regard to OVM allergenicity. The release of β-hex was decreased to a minimum of 36% for SS-treated OVM at 180–200 °C for 2–10 min, suggesting that SS treatment at high temperatures could lower the basophil degranulation. This might be due to the cross-linking and aggregation of OVM blocking the interaction between IgE and allergens by covering parts of the epitopes, leading to less degranulation [7]. TNF-γ is an important activity mediator of the immune response or allergenicity response produced or increased in inflammatory disease states [25]. The release of TNF-γ decreased by half when OVM was treated at 200 °C for 10 min, although there exists the experiment error.

Histamine is another key mediator found in KU812 cell granules which is released during antigen-specific IgE binding and plays an important role in the induction of adverse physiological symptoms of an allergic reaction [25]. Interestingly, compared with β-hex, a higher concentration of histamine was observed with OVM treated by SS at 120–200 °C for 2–10 min, except for several samples that showed negligible changes. Previous studies have reported similar findings on changes in the β-hex and histamine release, which suggested that histamine may not be a good indicator of mast cell degranulation in regard to OVM [24].

IL-6 is an important mediator produced in the process of mast cell degranulation. As shown in Figure 3D, no significant increase in the content of IL-6 was detected in KU812 cells, indicating that IL-6 was not stimulated by any antigen from OVM samples.

The above results indicated that the degranulation ability of the basophils was decreased, although there was little change in the IgE binding ability. In conclusion, SS treatment on OVM inhibited the anaphylactic reaction and reduced the release of β-hex and TNF-γ in the process of basophilic granulocyte degranulation. The allergenicity of SS-treated OVM decreased in the KU812 cell.

3.3. Analysis of Oxidation Degree

Generally, the content of carbonyl is an oxidative indicator of the protein [18]. In this study, the oxidation degree of OVM was analyzed through the determination of the contents of carbonyl and free SH. As shown in Figure 4A, a significant increase of the carbonyl content was observed with the rise of temperature and extension of processing time at 120–160 °C for 2–6 min and 200 °C for 2–4 min. No significant change was observed in OVM treated by SS at 200 °C for 6–10 min. In an environment of high temperature and high humidity, vaporized water molecules could react with oxygen ions, which will induce the protein backbone of amino acid residue to form peroxyl radicals. Subsequently, carbonyl derivatives were produced via an amidation reaction [26]. However, when the temperature was increased up to 200 °C and the processing time reached 6–10 min, large amounts of protein aggregation were generated, which influenced the monitoring of the carbony content.

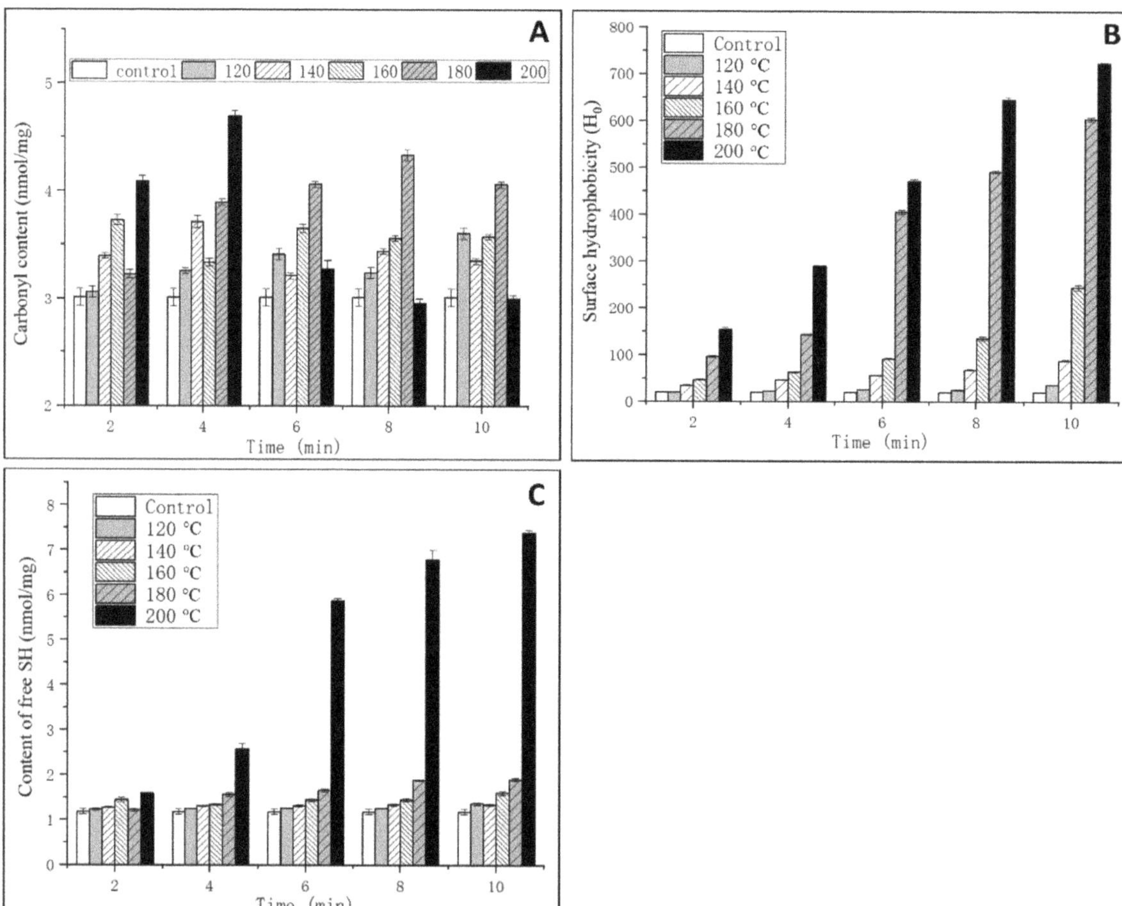

Figure 4. Effects of SS on the carbonyl content (**A**), surface hydrophobicity (**B**) and free SH content (**C**) of OVM.

3.4. Analysis of Surface Hydrophobicity

Surface hydrophobicity is an important index that indicates the binding ability between antibody and antigen. As one type of molecular force involved in immune reactions, surface hydrophobicity plays an important role in assisting in epitope recognition [27]. As shown in Figure 4B, the surface hydrophobicity significantly increased with the increase of temperature and time. The results indicated that SS treatment could induce the unfolding of protein structure, and the internal hydrophobic amino acids were exposed to a nonpolar microenvironment that enhanced the surface hydrophobicity [7]. High temperatures may cause a reduction in steric hindrance against some hydrophobic groups, and ANS is more likely to bind to amino acid residues with cationic groups, such as lysine and arginine. Therefore, it can be concluded that SS treatment could promote the combination of ANS and OVM. It was also concluded that the decrease in IgG binding ability may be related to the increased hydrophobic interactions, while this was not the cause for IgE binding ability.

3.5. Analysis of Free SH Content

OVM consists of three structurally independent tandem homologous domains and possesses nine intramolecular disulfide bridges but lacks interdomain disulfide bonds. This is the reason that OVM is resistant to heat denaturation and structural change. The surface hydrophobicity analysis suggested that an unfolding of OVM conformation could occur after SS treatment, which was related to the reduction of disulfide bridges. As shown in Figure 4C, the free SH content significantly increased with the increasing in SS temperature and processing time, and a maximum value was observed when the temperature and processing time reached 200 °C and 10 min. The native OVM has a free SH content of 1.1 nmol/mg. After SS treatment at 120–180 °C for 2–10 min, the free SH content increased to about 1.5 nmol/mg. It is worth noting that at a temperature of 200 °C, the free SH content drastically increased with the increase in processing time. Generally, high temperature treatment did not cleave the disulfide bridges. However, 200 °C SS treatment could catalyze the oxidation of the disulfide bridges to free SH, leading to the unfolding of the protein. Previous studies have reported that reducing the disulfide bonds in OVM can lower allergenicity in vitro [28], which was in accordance with the IgG-binding ability change in this study.

3.6. Molecular Weight Analysis

OVM has a molecular mass of 28.0 kDa, comprising 186 amino acids and 20–25% of carbohydrate [29]. Heat treatment could induce the oxidation or aggregation of OVM, increasing the molecular weight as well as forming dimer, trimer or polymer [21,30]. MALDI TOF MS was used to measure the molecular weights of SS-treated OVM samples. The dimer, trimer and tetramer molecule of OVM were observed after SS treatment at 120, 140, 160 and 180 °C for 10 min (Figure 5), indicating that covalent-binding reactions had occurred. However, there were no obvious changes in dimer among the native and SS-treated OVM samples at 120, 140, 160, 180 and 200 °C for 10 min. Interestingly, the single protein molecule and trimer disappeared in the MALDI TOF MS spectrometry of OVM treated at 200 °C for 10 min. It was presumed that SS treatment at higher temperatures for longer periods might markedly promote the aggregation of proteins, which was not detected by MALDI TOF MS.

3.7. Modification Sites of OVM

In order to identify the specific protein sites susceptible to SS treatment and explore the mechanism of the decreasing in OVM allergenicity, it is necessary to find out the precise modification sites of OVM. Previous studies have suggested that the antigenic epitopes in protein sequence mainly originated from hydrophobic amino acids and amino acids containing sulfhydryl groups [21,31]. Therefore, trypsin was used to digest native and SS-treated OVM at 120, 140 and 200 °C for 10 min. Table 1 showed the main modified peptides and sites of OVM. It is worth noting that the coverage of the control was 86.67%, much lower than SS-treated OVM, reaching 99%, 99% and 100%, respectively. This result indicated that although OVM was a trypsin inhibitor, it could also be hydrolyzed. Only a few structural domains were resistant to trypsin digestion. It also indicated that SS treatment could destruct the native structure of OVM and weaken the anti-digestion property.

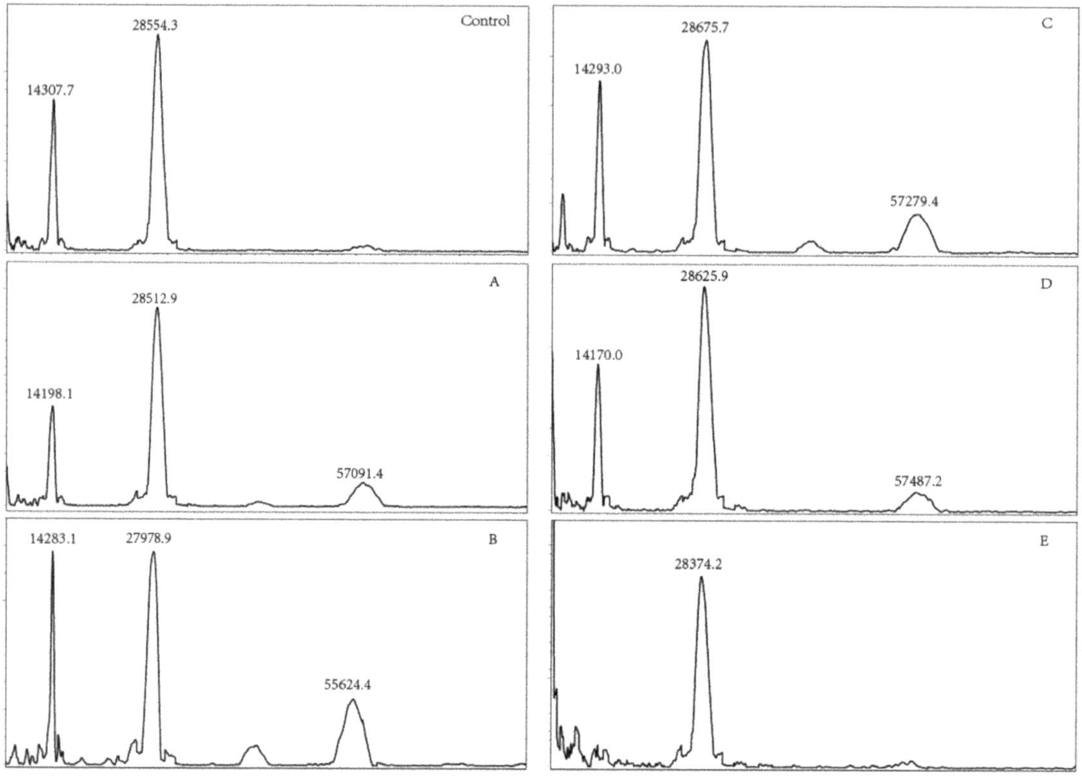

Figure 5. MALDI TOF MS of OVM treated by SS at different temperatures (**Control**, native OVM; (**A–E**), OVM treated with SS at 120, 140, 160, 180 and 200 °C, respectively, for 10 min).

OVM has a signal sequence of 24 amino acids and a protein sequence of 186 amino acids. Therefore, in this work, the SS-induced changes in 210 amino acids of OVM were explored and mainly concentrated in oxidation, nitro, phospho, carboxymethyl, glygly and sulfo modifications. Oxidative modification is typically associated with lower protein solubility, which can significantly hinder and confound the identification [21,32]. Therefore, only signal sequence M3 produced after treatment at 140 and 200 °C for 10 min and protein sequence M84 after treatment at 120, 140 and 200 °C for 10 min were analyzed. From the results, it can be concluded that SS could catalyze the oxidation of OVM (Table 1). It was also found that nitro modification occurred on some specific amino acids, such as Y73 and Y161. However, the mechanism of the reaction under SS treatment is not yet clear.

Interestingly, it was observed that the phospho modification of OVM occurred after treatment at 120, 140 and 200 °C for 10 min. There were two phospho sites S6 and T38 in native OVM, and three new phospho sites were observed in OVM treated by SS at 120, 140 and 200 °C for 10 min. It is worth noting that no phosphate groups participated in this reaction. Therefore, this result indicated that the phospho groups have shifted from phospho sites to non-phospho sites, which was similar to the observation in our previous research [33]. This finding provides a new method to promote the phospho modification of proteins using SS technology.

Table 1. Modified peptides and sites of the OVM treated by SS.

		Control	120 °C for 10 min	140 °C for 10 min	200 °C for 10 min
Coverage		**86.67%**	**99%**	**99%**	**100%**
		Signal sequence 1–24			
Modification	Oxidation			M3	M3
	Nitro	M3	M3		
	Phosphor	S11	S11	S11	S11
	Carboxymethyl	C15	C15	C15	C15
		Protein signal sequence 1–186			
Modification	Oxidation		M84	M84	M84
	Nitro	Y102, Y141	Y102, Y141	Y102, Y141	Y73, Y102, Y141, Y161
	Phosphor	S6, T38	S6, T12, T52, T160	S6, T12, T52, T160	S6, T12, T52, T160
	Carboxymethyl	C5, C22, C30, C41, C44, C87, C95, C106, C109, C138, C146, C154, C165, C168	C5, C22, C30, C41, C44, C70, C87, C95, C106, C109, C138, C146, C154, C165, C168	C5, C22, C30, C41, C44, C70, C87, C95, C106, C109, C138, C146, C154, C165, C168	C5, C22, C30, C41, C44, C70, C87, C95, C106, C109, C138, C146, C154, C165, C168
	Glygly	K14, K56, T32, S55, K112, K121	K14, K56, T32, S55, K112, K121, S71, S72, S78	K14, K56, T32, S55, K112, K121, S71, S72, S78	K14, K56, T32, S55, K112, K121, S71, S72, S78
	Sulfo	T36	T36	T36	T36, S6, S47, S156, S174

There was also an important finding in regard to glygly modification, which plays an important role in life activities. Similarly, three new glygly sites S71, S72 and S78 were observed in OVM treated at 120, 140 and 200 °C for 10 min, without the participation of the glygly group. The internal transfer might be the reason for the generation of new glygly modification sites in the OVM molecule. Finally, the sulfo modification sites among native and SS-treated OVM at 120, 140 and 200 °C for 10 min were also compared. There were four new sulfo sites (S6, S47, S156 and S174) observed in OVM treated at 200 °C for 10 min. The sulfuric groups may be originated from cysteine or other sulfo sites. Interestingly, serine could be readily modified with the acquisition of functional groups from other modification sites during SS treatment.

OVM molecule is organized into three well-separated domains which play an important role in inducing allergic reactions. It is important to explore the relationship between the sensitive modification sites and the three-dimensional (3D) structure of SS-treated OVM. The modification sites of OVM treated at 120 °C, 140 °C and 200 °C for 10 min are shown in Figure 6. OVM treated at 120 °C for 10 min exhibited similar modification sites with that treated at 140 °C for 10 min, and the 3D diagrams of three domains are presented in Figure 6A. The content of modification sites was higher in OVM treated at 200 °C for 10 min (Figure 6B). Interestingly, the modification sites including T52, T12, S71, S72, S78, M84 and T160 were all found in the β-sheet structure of SS-treated OVM at 120 °C and 140 °C for 10 min. However, for the SS-treated OVM at 200 °C for 10 min, S47 was found in the α-helix structure, and S156 was found in the structure. From these results, it suggested that the low temperature of SS always modifies the amino acids located on β-sheet which were relatively fragile. Meanwhile, the higher temperature treatment of 200 °C could also attack the amino acids located on α-helices and β-turn of OVM and promote the modification reaction.

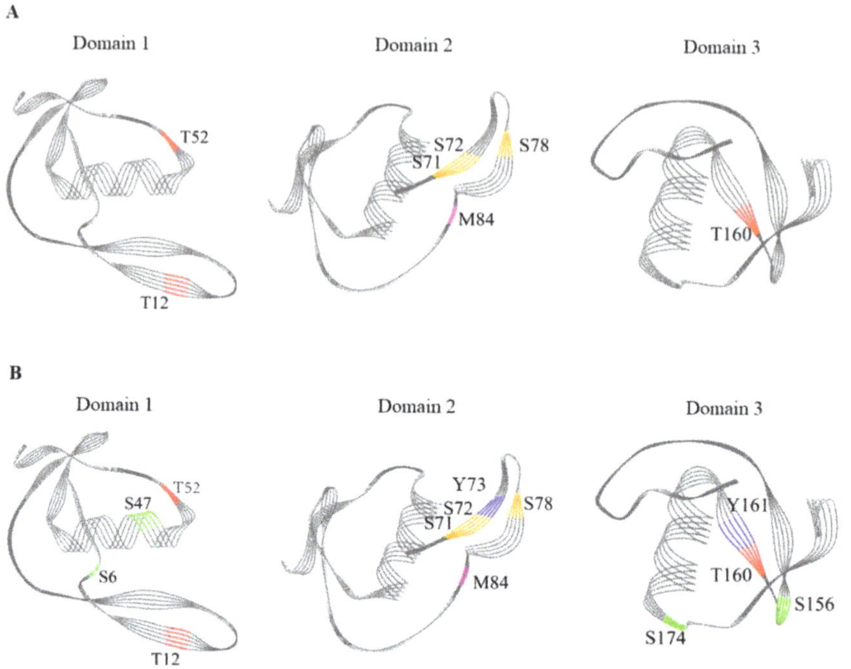

Figure 6. Modification sites in 3D diagrams of the three domains of SS-treated OVM. (**A**), 120 °C for 10 min and 140 °C for 10 min; (**B**), 200 °C for 10 min; purple, oxidation modification; blue, nitro modification; red, phosphorylation; orange, glygly modification; green, sulfo modification.

SS treatment, particularly at high temperatures and long processing times, could induce the OVM aggregation, increase surface hydrophobicity, modify the functional groups and amino acids in OVM. These phenomena disrupted and buried some allergen epitopes and made it difficult for the combination of antibodies and antigens [14,15]. Therefore, the allergenicity of OVM was decreased and the anaphylactic reaction in the KU812 cell was inhibited.

4. Conclusions

To our knowledge, this is the first study investigating the allergenicity and structure changes of OVM under SS treatment. A decrease in the IgG-binding ability was observed for SS-treated OVM with no significant changes in the IgE binding ability. SS treatment reduced the release of β-hex and TNF-γ, promoted the release of histamine, while having no significant effect on IL-6 release in KU812 cells. An obvious increase in the oxidation degree, free SH content and surface hydrophobicity were observed. Trimer and tetramer aggregations generated after SS treatment. In addition, there were obvious changes in the primary structure. Furthermore, some amino acids could be readily modified by obtaining functional groups from other modification sites during SS treatment. It can be inferred that SS-induced OVM aggregation, functional groups and amino acids modifications as well as protein structure alteration led to the reduction in allergenicity and increase in digestibility.

Author Contributions: P.-W.W.: Writing—original draft, Methodology, Validation, Formal analysis. Z.-C.T.: Funding acquisition, Validation, Resources. Y.-M.H.: Formal analysis, Supervision. H.W.: Supervision, Software, Visualization. All authors have read and agreed to the published version of the manuscript.

Funding: This research was funded by National Natural Science Foundation of China (No. 32101946) and Jiangxi Provincial Natural Science Foundation (20212BAB215037).

Data Availability Statement: The data presented in this study are available on request from the corresponding author. The data are not publicly available due to privacy or ethical restrictions.

Conflicts of Interest: The authors declare no competing financial interest.

Abbreviations

SS	superheated steam
IgE	immunoglobulin E
IgG	immunoglobulin G
KU812 cells	human peripheral blood basophilic leukemia cells
β-hex	β-hexosaminidase
IL-6	interleukin-6
DTT	dithiothreitol
ANS	8-aniline-1-naphthalene sulfonic acid
SH	sulfhydryl
ELISA	enzyme-linked immunosorbent assay
MALDI TOF MS	Matrix-assisted laser desorption/ionization time of flight mass spectrometry
HPLC Orbitrap MS/MS	high performance liquid chromatography Orbitrap tandem mass spectrometry
DNPH	2,4-dinitrophenylhydrazine

References

1. He, W.; Xu, H.; Lu, Y.; Zhang, T.; Li, S.; Lin, X.; Xu, B.; Wu, X. Function, digestibility and allergenicity assessment of ovalbumin-EGCG conjugates. *J. Funct. Foods* **2019**, *61*, 103490. [CrossRef]
2. Pablos-Tanarro, A.; Lozano-Ojalvo, D.; Molina, E.; López-Fandiño, R. Assessment of the allergenic potential of the main egg white proteins in BALB/c mice. *J. Agric. Food Chem.* **2018**, *11*, 2970–2976. [CrossRef] [PubMed]
3. Mine, Y.; Zhang, J.W. Comparative studies on antigenicity and allergenicity of native and denatured egg white proteins. *J. Agric. Food Chem.* **2002**, *50*, 2679–2683. [CrossRef] [PubMed]
4. Azdad, O.; Mejrhit, N.; Aarab, L. Reduction of the allergenicity of cow's milk α-lactalbumin under heat-treatment and enzymatic hydrolysis in Moroccan population. *Eur. Ann. Allergy Clin. Immunol.* **2018**, *50*, 177–183. [CrossRef]
5. Li, H.J.; Zhu, K.X.; Zhou, H.M.; Peng, W.; Guo, X.N. Comparative study of four physical approaches about allergenicity of soybean protein isolate for infant formula. *Food Agric. Immunol.* **2016**, *27*, 1–20. [CrossRef]
6. Fredericq, E.; Deutsch, H.F. Studies on ovomucoid. *J. Biol. Chem.* **1949**, *181*, 499–510. [CrossRef]
7. Hu, Y.; Wang, L.; Li, Z. Modification of protein structure and dough rheological properties of wheat flour through superheated steam treatment. *J. Cereal Sci.* **2017**, *76*, 222–228. [CrossRef]
8. Hu, Y.; Wang, L.; Hu, X.; Li, Z. Microbial decontamination of wheat grain with superheated steam. *Food Control* **2016**, *62*, 264–269. [CrossRef]
9. Sehrawat, R.; Nema, P.K.; Kaur, B.P. Effect of superheated steam drying on properties of foodstuffs and kinetic modeling. *Innov. Food Sci. Emerg.* **2016**, *34*, 285–301. [CrossRef]
10. Zhang, N.C.; Gao, Y.Q.; Tong, L.T.; Li, Z.G. Superheated steam processing improved the qualities of oats flour and noodles. *J. Cereal Sci.* **2018**, *83*, 96–100. [CrossRef]
11. Ceccanti, C.; Pellegrini, E.; Guidi, L. Effect of superheated steam and conventional steam roasting on nutraceutical quality of several vegetables. *LWT-Food Sci. Technol.* **2021**, *149*, 112014. [CrossRef]
12. Hu, X.; Guo, B.; Liu, C.; Yan, X.; Chen, J.; Luo, S.; Liu, Y.; Wang, H.; Yang, R.; Zhong, Y.; et al. Modification of potato starch by using superheated steam. *Carbohyd. Polym.* **2018**, *198*, 375–384. [CrossRef] [PubMed]
13. Wu, J.; Chen, J.; Liu, W.; Liu, C.; Zhong, Y.; Luo, D.; Li, Z.; Huang, Z. Selective peroxidase inactivation of lightly milled rice by superheated steam. *J. Cereal Sci.* **2014**, *60*, 623–630. [CrossRef]
14. Wang, X.M.; Tu, Z.C.; Ye, Y.H.; Liu, G.X.; Wang, H.; Hu, Y.M. Mechanism on the allergenicity changes of α-lactalbumin treated by sonication-assisted glycation during in vitro gastroduodenal digestion. *J. Agric. Food Chem.* **2021**, *69*, 6850–6859. [CrossRef]
15. Wang, X.M.; Ye, Y.H.; Tu, Z.C.; Hu, Y.M.; Wang, H.; Huang, T. Mechanism of the reduced IgG/IgE binding abilities of glycated β-lactoglobulin and its digests through high-resolution mass spectrometry. *J. Agric. Food Chem.* **2021**, *69*, 3741–3750. [CrossRef] [PubMed]
16. Kuehn, H.S.; Radinger, M.; Brown, J.M.; Ali, K.; Vanhaesebroeck, B.; Beaven, M.A.; Metcalfe, D.D.; Gilfillan, A.M. Btk-dependent Rac activation and actin rearrangement following FcεRI aggregation promotes enhanced chemotactic responses of mast cells. *J. Cell Sci.* **2010**, *123 Pt 15*, 2576–2585. [CrossRef]
17. Lv, L.T.; Lin, H.; Li, Z.X.; Nayak, B.; Ahmed, I.; Tian, S.L.; Chen, G.Z.; Lin, H.; Zhao, J.X. Structural changes of 2,2′-Azobis (2-amidinopropane) dihydrochloride (AAPH) treated shrimp tropomyosin decrease allergenicity. *Food Chem.* **2018**, *274*, 547–557. [CrossRef]

18. Zhang, J.J.; Tu, Z.C.; Wang, H.; Hu, Y.M.; Du, P.C.; Yang, Y.P. Mechanism of the effect of 2, 2′-azobis (2-amidinopropane) dihydrochloride simulated lipid oxidation on the IgG/IgE binding ability of ovalbumin. *Food Chem.* **2020**, *327*, 127037. [CrossRef]
19. Xi, C.; Kang, N.; Zhao, C.; Liu, Y.; Sun, Z.; Zhang, T. Effects of pH and different sugars on the structures and emulsification properties of whey protein isolate-sugar conjugates. *Food Biosci.* **2020**, *33*, 100507. [CrossRef]
20. Ellman, G.L. Tissues sulfhydryl groups. *Arch. Biochem. Biophys.* **1959**, *82*, 70–77. [CrossRef]
21. Liu, G.X.; Tu, Z.C.; Yang, W.H.; Wang, H.; Zhang, L.; Ma, D.; Huang, T.; Liu, J.; Li, X. Investigation into allergenicity reduction and glycation sites of glycated β-lactoglobulin with ultrasound pretreatment by high-resolution mass spectrometry. *Food Chem.* **2018**, *252*, 99–107. [CrossRef]
22. Martos, G.; Lopez-Exposito, I.; Bencharitiwong, R.; Berin, M.C.; Nowak-Węgrzyn, A. Mechanisms underlying differential food allergy response to heated egg. *J. Allergy Clin. Immun.* **2011**, *127*, 990–997. [CrossRef]
23. Djurtoft, R.; Pedersen, H.S.; Aabin, B.; Barkholt, V. Studies of food allergens: Soybean and egg proteins. *Adv. Exp. Med. Biol.* **1991**, *289*, 281–293. [PubMed]
24. Rupa, P.; Schnarr, L.; Mine, Y. Effect of heat denaturation of egg white proteins ovalbumin and ovomucoid on CD4+ T cell cytokine production and human mast cell histamine production. *J. Funct. Foods.* **2015**, *18*, 28–34. [CrossRef]
25. Cianferoni, A.; Muraro, A. Food-induced anaphylaxis. *Immunol. Allergy Clin. N. Am.* **2012**, *32*, 165–195. [CrossRef]
26. Zhou, F.B.; Zhao, M.M.; Zhao, H.F.; Sun, W.F.; Cui, C. Effects of oxidative modification on gel properties of isolated porcine myofibrillar protein by peroxyl radicals. *Meat Sci.* **2014**, *96*, 1432–1439. [CrossRef]
27. Chandrapala, J.; Zisu, B.; Palmer, M.; Kentish, S.; Ashokkumar, M. Effects of ultrasound on the thermal and structural characteristics of proteins in reconstituted whey protein concentrate. *Ultrason Sonochem.* **2011**, *18*, 951–957. [CrossRef] [PubMed]
28. Kovacs-Nolan, J.; Zhang, J.W.; Hayakawa, S.; Mine, Y. Immunochemical and structural analysis of pepsin-digested egg white ovomucoid. *J. Agric. Food Chem.* **2000**, *48*, 6261–6266. [CrossRef]
29. Mine, Y.; Yang, M. Recent advances in the understanding of egg allergens: Basic, industrial, and clinical perspectives. *J. Agric. Food Chem.* **2008**, *56*, 4874–4900. [CrossRef]
30. Ma, Y.; Zhang, H.; Xu, D.; Jin, Y.; Xu, X. Wheat flour superheated steam treatment induced changes in molecular rearrangement and polymerization behavior of gluten. *Food Hydrocolloid.* **2021**, *118*, 106769. [CrossRef]
31. Yang, Y.P.; Liu, G.X.; Wang, H. Investigation of the mechanism of conformational alteration in ovalbumin as induced by glycation with different monoses through conventional spectrometry and liquid chromatography high-resolution mass spectrometry. *J. Agric. Food Chem.* **2019**, *67*, 3096–3105. [CrossRef] [PubMed]
32. Dyer, J.M.; Clerens, S.; Thomas, A.; Callaghan, C.; Deb-Choudhury, S.; Haines, S. Photo-oxidation of whey proteins: Molecular markers of modification. *Int. Dairy J.* **2017**, *66*, 56–60. [CrossRef]
33. Wang, H.; Zong, B.Z.; Liu, G.X.; Zhang, L.; Chen, Y. Identification and quantification of the phosphorylated ovalbumin by high resolution mass spectrometry under dry-heating treatment. *Food Chem.* **2016**, *210*, 141–147. [CrossRef] [PubMed]

foods MDPI

Review

A Review on the Structure and Anti-Diabetic (Type 2) Functions of β-Glucans

Yujun Wan [1], Xiaojuan Xu [2], Robert G. Gilbert [1,3,*] and Mitchell A. Sullivan [4,*]

1 Centre for Nutrition and Food Sciences, Queensland Alliance for Agriculture and Food Innovation, The University of Queensland, Brisbane, QLD 4072, Australia; y.wan@uq.edu.au
2 College of Chemistry and Molecular Sciences, Wuhan University, Wuhan 430072, China; xuxj@whu.edu.cn
3 Joint International Research Laboratory of Agriculture and Agri-Product Safety, College of Agriculture, Yangzhou University, Yangzhou 225009, China
4 Glycation and Diabetes Group, Mater Research Institute, The University of Queensland, Translational Research Institute, Brisbane, QLD 4072, Australia
* Correspondence: b.gilbert@uq.edu.au (R.G.G.); mitchell.sullivan@mater.uq.edu.au (M.A.S.)

Abstract: Type 2 diabetes, a long-term chronic metabolic disease, causes severe and increasing economic and health problems globally. There is growing evidence that β-glucans can function as bioactive macromolecules that help control type 2 diabetes with minimal side effects. However, conflicting conclusions about the antidiabetic activities of β-glucans have been published, potentially resulting from incomplete understanding of their precise structural characteristics. This review aims to increase clarity on the structure–function relationships of β-glucans in treating type 2 diabetes by examining detailed structural and conformational features of naturally derived β-glucans, as well as both chemical and instrumental methods used in their characterization, and their underlying anti-diabetic mechanisms. This may help to uncover additional structure and function relationships and to expand applications of β-glucans.

Keywords: type 2 diabetes; β-glucan; molecular structure; anti-diabetic function; macronutrient absorption; enzyme inhibitor

Citation: Wan, Y.; Xu, X.; Gilbert, R.G.; Sullivan, M.A. A Review on the Structure and Anti-Diabetic (Type 2) Functions of β-Glucans. *Foods* **2022**, *11*, 57. https://doi.org/10.3390/foods11010057

Academic Editor: Diego A. Moreno

Received: 29 November 2021
Accepted: 24 December 2021
Published: 27 December 2021

1. Introduction

Type 2 diabetes (T2D), a disease associated with insulin resistance and poor blood glucose control, is a major public concern due to the potential severe complications and associated morbidity and its increasing rate of occurrence in many developed and developing countries. The major characteristic of this metabolic disease is chronic hyperglycemia, with severe complications leading to the long-term damage of various organs, including the eyes, kidneys, heart, blood vessels and nerves [1]. According to projections from the International Diabetes Federation, it has been estimated that in 2019 there were 463 million people with diabetes globally, and this is expected to reach 700 million by 2045, which will cost US$850 billion per year for diabetes healthcare [2,3]. Although several drugs have been used clinically to control T2D, such as biguanides [4], sulfonylureas [5], sodium-glucose co-transporter-2 (SGLT2) inhibitors [6], and thiazolidinediones [7], all of these drugs have some serious side-effects, especially resulting in gastrointestinal disorders [8,9] which affect both the drugs' efficacy and the patient's life. "Natural" medicine based on the concept of "food as medicine" has been proposed as an alternative strategy in the managing of metabolic diseases (such as T2D), due, among other things, to their safety [10]. Growing evidence has confirmed that certain bioactive nutrients in these foods, including polysaccharides [11,12], can help mitigate metabolic abnormalities [13].

Polysaccharides, one of the most important biomacromolecules for life, are polymers found in natural sustainable resources. From 1970, since the first discovery of the bioactive properties of lentinan, there has been a drive to research the biological functions of polysaccharides [14,15]. β-glucans are a type of naturally-derived polysaccharide which can be widely found in bacteria, alga, fungi, cereals, and higher plants. They are generally biopolymers with β-glucopyranosyl units, which normally contain a β-(1,3)-linked and/or β-(1,4)-linked backbone and may be branched with β-(1,6)-linked glucose.

β-glucans have been used in the food industry and in clinical practice because of their significant biological functions. For example, β-glucans with high molecular weights have been reported to enhance the binding capacity to the receptors involved in immune responses (e.g., dectin-1: a natural killer (NK)-cell-receptor-like C-type lectin involving in innate immune responses) and therefore improve their immunomodulatory activities, which can help to control some chronic diseases such as diabetes [16]. However, many studies have reported that degrading β-glucans to yield lower molecular weights can increase their anti-diabetic effects *in vitro*, as well as their antioxidant and antibacterial activities [17,18]. These differences in reported results may arise from differences in the detailed molecular structures of the substrates [19]. Therefore, this paper reviews structural and conformational features of naturally derived β-glucans, summarizing the potential underlying mechanisms of their anti-diabetic functions, for a better understanding of the structure–function relationships of β-glucans.

2. Structural Features of β-Glucans

It is commonly asserted that the functionalities of β-glucans are highly dependent on their molecular structure. The structural characteristics of β-glucans, including molecular weight distributions, glycosidic linkage patterns and branching degrees, vary with different sources and extraction methods. There are three main glycosidic linkage patterns identified in β-glucans (Figure 1): β-1,3-linked, β-1,4-linked and β-1,6-linked patterns. Normally, β-1,3-linked and β-1,4-linked patterns appear in the backbone of β-glucan, and β-1,6- linkages represent the branch points in the backbone.

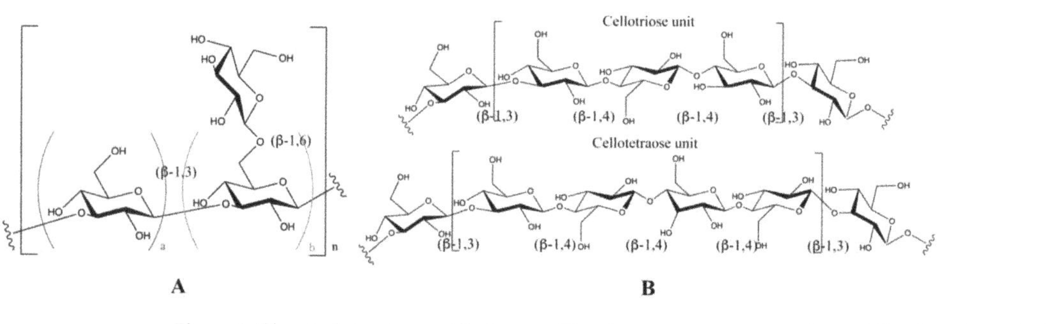

Figure 1. Chemical structures of β-glucans. The glucose monomers are shown following the symbol nomenclature for glycans. (**A**) The β-1,3-linked backbone of β-glucan with different branching degree of β-1,6-linked glucose. (**B**) The β-1,3-1,4-linked backbone of β-glucan, DP3: cellotriosyl, DP4: cellotetraosyl).

Branched or linear β-1,3-glucans (Figure 1A) are commonly isolated from fungi (e.g., mushroom [20]) and bacteria (e.g., yeast [20]). The occurrence of β-1,3-linkages together with β-1,4-linked glycosidic bonds (Figure 1B) are observed in the β-glucans from cereal grains (e.g., oat [21]). As shown in Figure 1B, there are two types of oligosaccharide subunits; one is three β-1,4-linked glucose monomers, termed cellotriosyl (DP3), and the other type is β-1,4-linked glucose monomers called cellotetraosyl (DP4). The molar ratio of DP3 and DP4 in β-glucans is specific for different cereals; this can be used as a tool to trace the origin of a given β-glucan structure [21].

The extraction methods of β-glucans vary from sample to sample. There are commonly four types of isolation method used in extracting β-glucans, such as water extraction [22–24], alkaline extraction [25,26], acidic extraction [27] and enzymatic extraction [28]. Some suggested structures of naturally derived β-glucans are shown in Table 1. These structures are generally inferred from a combination of results collected by both chemical tests and instrumental analysis. Interestingly, some similar repeating structural units exist in several β-glucans from different sources. For example, β-glucans from *Dictyophora indusiate*, *Hericium erinaceus*, *Grifola frondosa*, *Schizophyllan* and brown algae have the same repeating unit, viz., three β-1,3-Glc*p* backbone residues with a branch comprising one β-1,6-linked glucose residue. The branches of β-glucans, connected to the backbone via β-1,6 glycosidic bonds, play a major role in the solubility of the β-glucan. For example, curdlan, a linear β-glucan (i.e., without side chains) is insoluble in water [29], while the β-glucans with branched glucose residues, such as lentinan and Schizophyllan, are water-soluble [30,31]. However, these β-glucans exhibit some hydrophobicity due to hydrophobic carbon rings, resulting in limited water-solubility. Thus, β-glucans adopt different chain conformations to achieve stability. It is essential to consider their chain conformations in different solvents for the application of β-glucans in the food industry and medicine.

Table 1. Sources and deduced chemical structures of several β-glucans.

Name/Abbr.	Source	Extraction Solvent	Type of Glucan	Structure [a]	Ref.
Curdlan	*Alcaligenes faecalis* var.	NaOH	β-1,3 glucan	(A) a = 1, b = 0	[25,26]
APP	*Auricularia auricula*	NaCl	β-1,3 glucan	(A) a = 1, b = 2	[32–34]
DIP	*Dictyophora indusiata*	Water	β-1,3 glucan	(A) a = 2, b = 1	[22–24]
HEP	*Hericium erinaceus*	Water	β-1,3 glucan	(A) a = 2, b = 1	[35,36]
GFP	*Grifola frondosa*	Water	β-1,3 glucan	(A) a = 2, b = 1	[37–39]
Schizophyllan	*Schizophyllum*	Water	β-1,3 glucan	(A) a = 2, b = 1	[30,40]
Laminarin	Algae	Water	β-1,3 glucan	(A) a = 2, b = 1	[41,42]
Lentinan	*Lentinula edodes*	NaCl/NaOH	β-1,3 glucan	(A) a = 3, b = 2	[31,43]
GLP	*Ganoderma lucidum*	Water	β-1,3 glucan	(A) a = 5, b = 1	[44,45]
YBG	*Saccharomyces cerevisiae*	NaOH	β-1,3 glucan	(A) a = 5, b = 1	[46]
CSP	Wild *Cordyceps sinensis*	Water	β-1,3 glucan	(A) a = 5, b = 2	[47]
WBG	Wheat	Water	β-1,3-1,4 glucan	(B) DP3:DP4 = 3.0–4.5	[21,48]
BBG	Barley	Water/NaOH	β-1,3-1,4 glucan	(B) DP3:DP4 = 1.7–3.3	[48,49]
RBG	Rye	Water	β-1,3-1,4 glucan	(B) DP3:DP4 = 1.8–3.1	[50–52]
OBG	Oat	Water	β-1,3-1,4 glucan	(B) DP3:DP4 = 1.5–2.2	[48,53]

[a] The uppercase letters within this column represent the repeating units shown in Figure 1 (A. β-1,3-1,6 glucan; B. β-1,3-1,4 glucan). The lowercase letters indicate the molar ratio of each part in the repeating units.

3. Conformational Features of β-Glucans

Several chain conformations of β-glucans are found in different solutions (Figure 2), from a disordered conformation (e.g., random coil) to an ordered conformation (e.g., helix conformation). These more organized conformations can easily form a stable network (e.g., a triple helix), and the stabilization of this network arises from its inter- and intra-molecular hydrogen bonds. However, the dense triple helix conformation formed by the interaction of intramolecular polyhydroxy groups may result in its insolubility in aqueous solution [54].

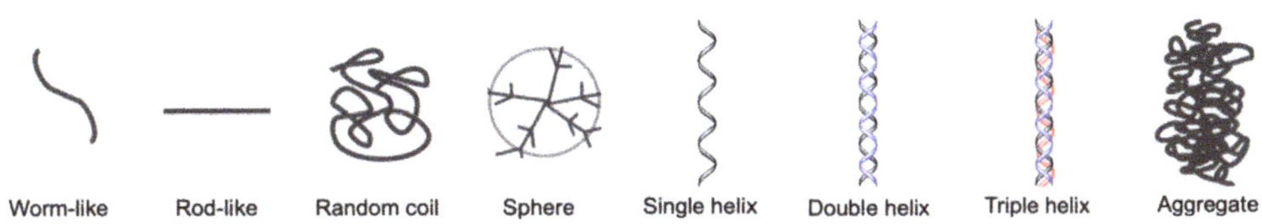

Figure 2. Various chain conformations of polysaccharides in different solvents.

There are many parameters that can affect conformational features of polysaccharides, including the molar mass per unit of contour length (M_L), the contour length of chains (L), the persistence length (q) and the chain diameter (d). The contour length of a polymer chain refers to its length at maximum physically possible extension, and the persistence length reflects the bending stiffness of a chain. Several reports indicate that β-glucans that exhibit strong biological functions show a triple-helix conformation, such as lentinan [55], curdlan [56] and yeast β-glucan [57]. Based on both theoretical and experimental results, the M_L and q values of a polysaccharide with a rigid triple-helix conformation usually range from 2000 to 2800 nm^{-1} and from 100 to 250 nm, respectively [20]. For example, lentinan, the first reported β-glucan with antitumor activities, exists as a triple-helix conformation in aqueous solution, with a reported M_L value of 2160 nm^{-1} and q value of 110 nm [58]. Schizophyllan, a widely studied β-glucan from *Schizophyllum*, has a reported M_L value of 2150 nm^{-1} and q value of 200 nm [59]. However, the triple-helix conformation of these β-glucans can be transferred to other conformations under special conditions, such as high temperature [60], high pH solvents [61] and strong polar solvents [62]. With disrupted conformations, their bioactivities and solubilities are also changed. Therefore, the structural and conformational features of β-glucan need to be well characterized for a correct understanding of their functionalities.

4. Characterization Methods for β-Glucan Structure and Conformation Analysis

The structural determination of polysaccharides is more complicated than for other biopolymers due to the diverse monosaccharide compositions and glycosidic linkage patterns. In addition to classical chemical characterization methods, many newly developed technologies for the characterization of polysaccharides have been, or could be, employed to help understand the structure-function relationship of bioactive polysaccharides. For example, enzymatic arrays [63] and matrix-assisted laser desorption ionization mass spectrometry [64] have been used to sequence polysaccharides, and ion mobility-mass spectrometry has been developed to analyze carbohydrate anomers [65]. Advanced microscopy techniques, such as atomic force microscopy [66] and confocal laser scanning microscopy [67], provide a new level of microstructure analysis. Recently, low-temperature scanning tunneling microscopy has been successfully applied to observe single glycans [68]. However, the exploration of polysaccharides has been much slower than that of polynucleotides and proteins because of limitations on the structural theories, the complexity of their structures and a poor understanding of the underlying mechanisms of their bioactivities.

The characterization of β-glucan structure therefore requires a combination of chemical and instrumental analyses (Figure 3). The structural information of β-glucans usually reported include its purity, molecular weight, monosaccharide composition, anomeric configuration, glycosidic linkage pattern and sequence of residues.

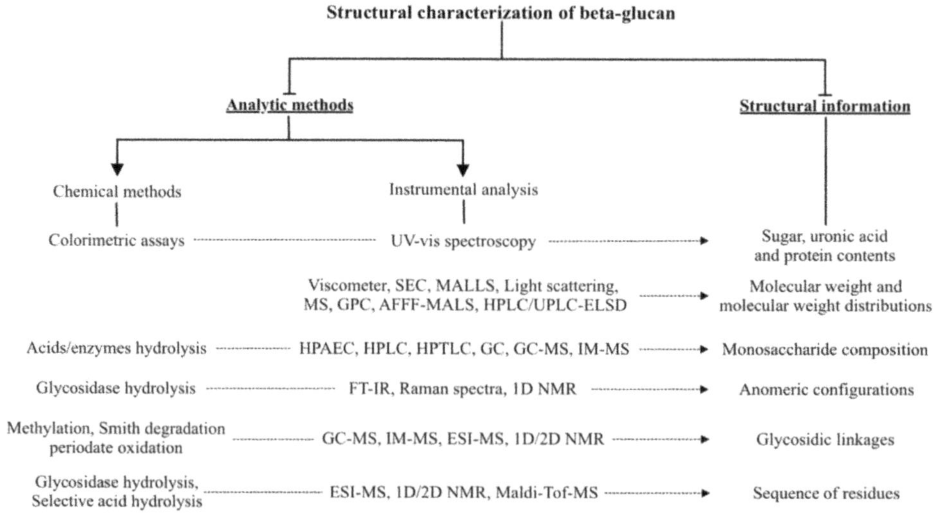

Figure 3. Chemical and instrumental methods used for β-glucan structure characterization.

To investigate detailed structure–function relationships of polysaccharides, the "purity" is one of the most important factors. Generally, measuring the purity of β-glucans includes obtaining several parameters, such as the sugar content and the molecular weight distribution. Colorimetric methods are commonly used as the first step to determine the purity of crude polysaccharides, including the measurement of sugar, uronic acid and protein contents. Then, the molecular weight distributions of polysaccharides can be analyzed using several instrumental methods, such as size exclusion chromatography (SEC). Sometimes, β-glucans coexists with other polysaccharides, such as arabinogalactans [69]. Therefore, it is essential to analyze its monosaccharide composition to identify the purity of a β-glucan. Hydrolysis with acids or enzymes is the first step to analyze the monosaccharide composition, after which the hydrolysate is characterized with various instruments. High-performance anion exchange chromatography (HPAEC) is considered one of the most effective instrumental analysis techniques to determine monosaccharide composition due to the high sensitivity and simple sample preparation [70]. A high-purity β-glucan should have a narrow molecular weight distribution and high glucose content.

Extraction of β-glucans from grains always results in some starch (an α-glucan) [54], and it is difficult to distinguish β-glucans from α-glucans through their sugar content or monosaccharide composition alone. However, a combination of glycosidase hydrolysis and instrumental analysis, such as Raman spectra, FT-IR and NMR, can easily identify the anomeric configuration of glucans. The sequence of residue and the branching degree of naturally derived β-glucans is highly dependent on its source, and can be characterized using NMR. For a comprehensive characterization of naturally derived β-glucans, it is necessary to use both chemical methods and instrumental analysis.

To identify the conformation of β-glucans, weight-average molecular weight (M_w), intrinsic viscosity ($[\eta]$), radius of gyration (R_g) and hydrodynamic radius (R_h) of β-glucan samples can be measured by static light scattering (SLS), dynamic light scattering (DLS) and viscometry, using the molecular-weight dependence of their solution properties. Several models, including the helical wormlike chains model and the Kratky-Porod model, can be used to deduce the four main parameters, M_L, L, q and d based on the results of measurements.

Conformation analysis can be performed using X-ray diffraction (XRD), e.g., for measuring the triple-helix β-glucan [71]. Atomic force microscopy (AFM) is a powerful tool to observe chain conformations in solution, including rod-, sphere- and fiber-like shapes. AFM can provide information on the chain length, chain diameter and even the M_L of a β-glucan sample [58]. In addition to these experimental techniques, molecular dynamics simulations have been used as a tool to explore the conformation of polymers, which can help illustrate the chain movements and conformations of polymers in different solutions [32], although the results always depend on the assumed model.

Although the characterization of β-glucans is complicated, an understanding of the structure/function relationships of these molecules is crucial if they are to advance further as a potential antidiabetic drug. Without this understanding the efficacy of a particular β-glucan is difficult to predict.

5. Amelioration of Type 2 Diabetes and Associated Mechanisms

T2D is the most frequent metabolic disorder which involves insulin resistance, followed by deficient insulin secretion by impaired pancreatic islet β-cells [72]. The two main factors that typically account for T2D are genetic factors and environmental factors. A genome-wide association study has confirmed that there are more than 400 gene variants associated with T2D, with most of them involving islet function [73]. The environmental factors that increase the risk of developing T2D include obesity, alcohol intake and smoking. The predominant factor accounting for T2D is the consumption of unhealthy diets, including those with energy-dense refined food [72].

5.1. Pharmacotherapy for T2D and Anti-Diabetic Mechanisms

There is a lack of effective drugs to treat T2D due to the complexity of pathogenesis [72,74]. However, several drugs are used in controlling T2D, and these drugs can be classified into seven main types based on their structures and mechanisms, including biguanides, sulfonylureas, thiazolidinediones, glucagon-like peptide (GLP-1), dipeptidyl peptidase (DPP-4) inhibitors, sodium-glucose co-transporter-2 (SGLT2) inhibitors and enzyme inhibitors. The underlying mechanisms and potential side-effects of these drugs are summarized in Table 2. As well as the injection of insulin being essential to control type 1 diabetes, this is also adopted to control T2D under certain conditions, such as functional failure of pancreatic islet β-cells due to the long-term suffering from T2D [75]. However, it should be noticed that all of these anti-diabetic drugs are companied by some severe side-effects (Table 2), such as gastrointestinal disorders, which have prolonged impact on the patient.

Table 2. Drugs used in amelioration of T2D.

Type	Drug Name	Mechanisms	Side-Effects	Ref.
Biguanides	Metformin, Phenformin	Lowering fasting plasma insulin concentration; enhancing insulin sensitivity; changing gut microbiota composition; promoting functional shifts in gut microbiome.	Gastrointestinal disorders; folate deficiency; increasing homocysteine levels	[8,76,77]
Sulfonylureas	Glibenclamide, Glipizide	As insulin secretagogues to stimulate insulin secretion.	Gastrointestinal disorders, headache	[9,78,79]
Thiazolidinediones	Rosiglitazone, Pioglitazone	Improving insulin sensitivity by up-regulation of adipokine.	Peripheral and pulmonary edema; fluid retention.	[80,81]
GLP-1	Liraglutide, Semaglutide	Suppressing glucagon release; delaying gastric emptying and increasing satiety.	Nausea, vomiting and diarrhoea	[82,83]
DPP-4 inhibitors	Vidagliptin, Saxagliptin	Enhancing incretin axis; improving meal-stimulated insulin secretion by sparing incretin hormones.	Nausea and gastrointestinal problems	[78,84–86]
SGLT2 inhibitors	Dapagliflozin, Cangliflozin	Inhibition of renal glucose reabsorption to lower plasma glucose levers.	Increasing the risk of developing diabetic ketoacidosis.	[6,87]
Enzyme inhibitors	α-amylase inhibitors, α-glucosidase inhibitors,	Reduction in the rate of glucose absorption in post-prandial blood	Lactic acidosis, diarrhoea, liver function disorders.	[88]

5.2. Glucans Used in Controlling T2D and Underlying Mechanisms

Naturally derived β-glucans have been promoted due to their various reported health-promoting activities and minimal side-effects. They have been widely adopted as health-improving ingredients to prevent some chronic diseases, especially for T2D. For example, a clinical trial showed that oat β-glucan can help manage glycemic index, carbohydrate metabolism and alter gut microbiota profile in T2D [89,90]. There are two main underlying mechanisms for the roles of β-glucans in controlling T2D, which can be explained through their detailed structural and/or conformational features.

5.2.1. Retardation of Macronutrient Absorption

Macronutrients within daily diets are necessary for life. However, ongoing quick absorption of these macronutrients in T2D can induce hyperglycemia and hyperlipidemia, and thereby cause metabolic disorders [91]. Hence, a way to help manage T2D is by preventing the absorption of macronutrients, resulting in a reduction in blood cholesterol levels and suppressing the postprandial increase of blood sugar levels [92]. This retardation effect of β-glucans has been shown to be highly dependent on their molecular weight and concentration. Wood et al. established the relationship of plasma glucose increment (DGpg) and structural features of oat β-glucan (concentration (c) and weight-average molecular weight (M_w)) as shown in the formula: $\text{DGpg} = A + B \times \log_{10}(c) + 0.72\text{B} \log_{10}(M_w)$ [93]. In addition, depolymerization of β-glucans (reducing molecular size) as a result of processing was reported to decrease its effect on decreasing the peak blood glucose response [17].

Additionally, β-glucans can play a role in increasing the viscosity of a meal during gastrointestinal digestion, limiting the absorption of macronutrients, slowing down gastric emptying, and entrapping bile acids and cholesterol throughout digestion. This lowers serum sugar and cholesterol levels in T2D [94]. These benefits are highly dependent on the structure and conformation of β-glucans, which can be explained by the Mark–Houwink equation for the intrinsic viscosity: $[\eta] \propto K\,M^{\alpha}$, where the values of the parameters K and α depend on the particular polymer solution system [95].

5.2.2. Inhibition of Digestive Enzyme

a-amylase and a-glucosidase are the two main enzymes necessary to hydrolyze carbohydrates in the digestive system. a-amylase can initiate carbohydrate hydrolysis by cleaving a-(1,4)-linked glycosidic bonds and yield smaller fractions, such as sucrose and maltose [96]. Then, a-glucosidase can hydrolyze these fractions into absorbable monosaccharides, such as glucose and fructose, during intestinal digestion [97]. Adequate free glucose can be generated after this digestive process, which may be excessively ingested into the bloodstream in T2D patients, leading to hyperglycemia. Therefore, inhibition of these enzymes to cause lower carbohydrate digestion can help control T2D. Ma et al. showed that a β-glucan adopting a triple helix conformation from *Hericium erinaceus* reduced wheat starch digestibility from 70% to less than 60% by inhibiting the digestive enzymes [98]. Qin et al. compared the glucose availability of the digestive system after adding different modified oat β-glucans, and all of these β-glucans decreased glucose availability, which indicated their potential hypoglycemic effect [17].

Lipases play an important role in hydrolyzing diet lipids; they convert lipids into cholesterol and fatty acids, which can be absorbed by enterocytes through lipid transporters [99]. Many studies have reported that β-glucans can inhibit lipase activity and therefore alleviate the hyperlipidemia seen in T2D. For example, barley β-glucans can slow lipolysis and reduce the release of free fatty acids; these inhibition effects were highly dependent on the molecular weights of the β-glucans [100].

There are some possible underlying mechanisms related to the enzyme inhibitor role of β-glucans, as shown in Figure 4. For example, these β-glucans, which can only be fermented in the large intestine, play a physical barrier role in the digestive system to inhibit the activity of these digestive enzymes [101]. At the same time, β-glucans can inhibit these digestive enzymes by mixed competitive and uncompetitive inhibition to suppress enzyme–substrate interactions and therefore reduce the digestion of substrates [102]. Both of these proposed mechanisms rely on the physiological conformations of the β-glucans *in vivo*. For example, the binding between β-glucans and enzymes follows the lock-and-key principle; therefore, the shapes (confirmations) of β-glucans under physiological conditions are a decisive factor for binding efficiency, thereby controlling the inhibition effect of these β-glucans [20,103].

It should be noted that other possible anti-diabetic mechanisms of β-glucans have been proposed. For example, the modulation effects of β-glucan on gut microbiota have been widely reported [89,104]. β-glucans can function as prebiotics, as they are mainly fermented in the large bowel and benefit the host-microbiota interactions in the whole gastrointestinal process. As a prebiotic, β-glucans alter the microbiota compositions by improving the number of beneficial bacteria, such as *Lactobacillus*, during large bowel fermentation [105,106], leading to an increase in short-chain fatty acids (SCFA), which can improve the colonic defense barrier in T2D [107]. In addition, β-glucans can regulate the superoxide dismutase and malondialdehyde levels in the livers of diabetic mice [108]. However, these reported effects need to be further explored. For example some particular structural features of β-glucans may be preferred by these beneficial bacteria. Again a better understanding of the structure and function relationship of β-glucans may help yield more potent health benefits.

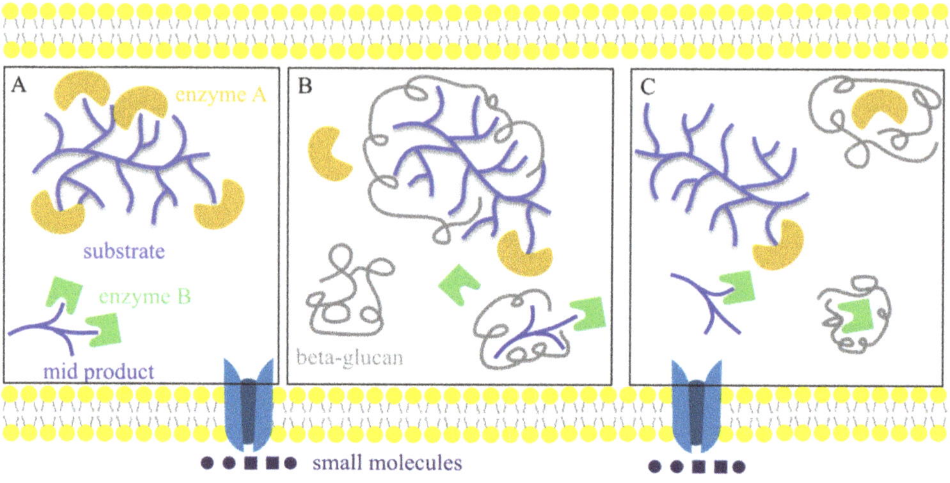

Figure 4. The proposed mechanism of enzyme inhibition. (**A**) The normal enzyme–substrate interactions during digestion. (**B**) β-glucans play as a physical barrier to inhibit enzyme–substrate interactions. (**C**) β-glucans bind to enzymes to inhibit enzyme–substrate interactions.

6. Conclusions

β-glucans are sustainable polymers that widely exist in natural resources. These biomacromolecules mainly contain β-(1,3)-linked, β-(1,4)-linked and β-(1,6)-linked glycosidic bonds, and adopt several different conformations in solutions, such as a helical conformation, which seems to be the origin of their versatile biofunctions and thus furnish a target for targeting efficacy. For a detailed characterization of β-glucans, both chemical and instrumental methods should be combined to give accurate structural and conformational features, which can then be linked to an understanding of their functionalities. Although there are many reported anti-diabetic mechanisms of β-glucans, only two mechanisms, retardation of macronutrient absorption and inhibition of digestive enzymes, can be well explained through their detailed structures and conformations. However, current research on the anti-diabetic functions of β-glucans is focused on naturally derived β -glucans. It would be worthwhile to explore the underlying anti-diabetic mechanisms using synthetic β-glucans with detailed structural information. With improved understanding of the structure/function relationship of these molecules, precision designs of β-glucans with particular structures and/or conformations could be produced to help control T2D.

Author Contributions: Y.W.: writing—original draft, writing—review & editing. X.X.: writing—review & editing, funding acquisition. M.A.S.: writing—review & editing, supervision, funding acquisition. R.G.G.: writing—review & editing, supervision, project administration, funding acquisition. All authors have read and agreed to the published version of the manuscript.

Funding: Y.W. gratefully acknowledges the support of a University of Queensland Research Training Scholarship. X.X. gratefully acknowledges financial support from National Natural Science Foundation of China (22075213, 21875167, and 21574102), National Key Research and Development Program of China (2016YFD0400202), and Key Research & Development Program of Hubei province (2020BCA079). M.A.S. is supported by an Advance Queensland Industry Research Fellowship, Mater Foundation, Equity Trustees and the L G McCallam Est and George Weaber Trusts. For R.G.G., partial funding was from the National Natural Science Foundation of China, grant number C1304013151101138, and the Priority Academic Program of Jiangsu Higher Education Institutions.

Institutional Review Board Statement: Not applicable.

Informed Consent Statement: Not applicable.

Data Availability Statement: Not applicable.

Conflicts of Interest: The authors declare no conflict of interest.

References

1. Association, A.D. Diagnosis and classification of diabetes mellitus. *Diabetes Care* **2014**, *37*, S81–S90. [CrossRef] [PubMed]
2. Cho, N.; Shaw, J.; Karuranga, S.; Huang, Y.; da Rocha Fernandes, J.; Ohlrogge, A.; Malanda, B. IDF Diabetes Atlas: Global estimates of diabetes prevalence for 2017 and projections for 2045. *Diabetes Res. Clin. Pract.* **2018**, *138*, 271–281. [CrossRef] [PubMed]
3. International Diabetes Federation. *IDF Diabetes Atlas*, 9th ed.; International Diabetes Federation: Brussels, Belgium, 2019.
4. Hu, Y.; Ding, B.; Shen, Y.; Yan, R.-N.; Li, F.-F.; Sun, R.; Jing, T.; Lee, K.-O.; Ma, J.-H. Rapid changes in serum testosterone in men with newly diagnosed type 2 diabetes with intensive insulin and metformin. *Diabetes Care* **2021**, *44*, 1059–1061. [CrossRef] [PubMed]
5. Ling, S.; Zaccardi, F.; Lawson, C.; Seidu, S.I.; Davies, M.J.; Khunti, K. Glucose control, sulfonylureas, and insulin treatment in elderly people with type 2 diabetes and risk of severe hypoglycemia and death: An observational study. *Diabetes Care* **2021**, *44*, 915–924. [CrossRef]
6. Han, S.; Hagan, D.L.; Taylor, J.R.; Xin, L.; Meng, W.; Biller, S.A.; Wetterau, J.R.; Washburn, W.N.; Whaley, J.M. Dapagliflozin, a selective SGLT2 inhibitor, improves glucose homeostasis in normal and diabetic rats. *Diabetes* **2008**, *57*, 1723–1729. [CrossRef] [PubMed]
7. Bae, J.; Huh, J.H.; Lee, M.; Lee, Y.H.; Lee, B.W. Glycaemic control with add-on thiazolidinedione or a sodium-glucose co-transporter-2 inhibitor in patients with type 2 diabetes after the failure of an oral triple antidiabetic regimen: A 24-week, randomized controlled trial. *Diabetes Obes. Metab.* **2021**, *23*, 609–618. [CrossRef]
8. Group, U.P.D.S. Effect of intensive blood-glucose control with metformin on complications in overweight patients with type 2 diabetes (UKPDS 34). *Lancet* **1998**, *352*, 854–865.
9. Bergenstal, R.; Lewin, A.; Bailey, T.; Chang, D.; Gylvin, T.; Roberts, V.; Group, N.M.-v.-E.S. Efficacy and safety of biphasic insulin aspart 70/30 versus exenatide in subjects with type 2 diabetes failing to achieve glycemic control with metformin and a sulfonylurea. *Curr. Med. Res. Opin.* **2009**, *25*, 65–75. [CrossRef]
10. Mafra, D.; Borges, N.A.; Lindholm, B.; Shiels, P.G.; Evenepoel, P.; Stenvinkel, P. Food as medicine: Targeting the uraemic phenotype in chronic kidney disease. *Nat. Rev. Nephrol.* **2020**, *17*, 153–171. [CrossRef]
11. Cui, L.; Chen, L.; Yang, G.; Li, Y.; Qiao, Z.; Liu, Y.; Meng, Y.; Zhou, Y.; Sun, L. Structural characterization and immunomodulatory activity of a heterogalactan from Panax ginseng flowers. *Food Res. Int.* **2021**, *140*, 109859. [CrossRef]
12. Wan, Y.; Shi, H.; Xu, R.; Yin, J.; Nie, S.; Xiong, T.; Xie, M. Origin of hypoglycemic benefits of probiotic-fermented carrot pulp. *J. Agric. Food Chem.* **2019**, *67*, 895–904. [CrossRef]
13. Song, Q.; Wang, Y.; Huang, L.; Shen, M.; Yu, Y.; Yu, Q.; Chen, Y.; Xie, J. Review of the relationships among polysaccharides, gut microbiota, and human health. *Food Res. Int.* **2020**, *140*, 109858. [CrossRef] [PubMed]
14. Chihara, G.; Hamuro, J.; Maeda, Y.Y.; Arai, Y.; Fukuoka, F. Fractionation and purification of the polysaccharides with marked antitumor activity, especially lentinan, from *Lentinus edodes* (Berk.) Sing.(an edible mushroom). *Cancer Res.* **1970**, *30*, 2776–2781.
15. Chihara, G.; Maeda, Y.; Hamuro, J.; Sasaki, T.; Fukuoka, F. Inhibition of mouse sarcoma 180 by polysaccharides from *Lentinus edodes* (Berk.) sing. *Nature* **1969**, *222*, 687–688. [CrossRef]
16. El Enshasy, H.A.; Hatti-Kaul, R. Mushroom immunomodulators: Unique molecules with unlimited applications. *Trends Biotechnol.* **2013**, *31*, 668–677. [CrossRef] [PubMed]
17. Qin, Y.; Xie, J.; Xue, B.; Li, X.; Gan, J.; Zhu, T.; Sun, T. Effect of acid and oxidative degradation on the structural, rheological, and physiological properties of oat β-glucan. *Food Hydrocoll.* **2021**, *112*, 106284. [CrossRef]
18. Sun, T.; Li, J.; Qin, Y.; Xie, J.; Xue, B.; Li, X.; Gan, J.; Bian, X.; Shao, Z. Rheological and functional properties of oat β-glucan with different molecular weight. *J. Mol. Struct.* **2020**, *1209*, 127944. [CrossRef]
19. Xiao, J.B.; Jiang, H. A review on the structure-function relationship aspect of polysaccharides from tea materials. *Crit. Rev. Food Sci. Nutr.* **2015**, *55*, 930–938. [CrossRef]
20. Meng, Y.; Lyu, F.; Xu, X.; Zhang, L. Recent advances in chain conformation and bioactivities of triple-helix polysaccharides. *Biomacromolecules* **2020**, *21*, 1653–1677. [CrossRef]
21. Mejia, S.M.V.; de Francisco, A.; Bohrer, B.M. A comprehensive review on cereal beta-glucan: Extraction, characterization, causes of degradation, and food application. *Crit. Rev. Food Sci. Nutr.* **2020**, *61*, 3693–3704. [CrossRef]
22. Wang, J.T.; Xu, X.J.; Zheng, H.; Li, J.L.; Deng, C.; Xu, Z.H.; Chen, J.H. Structural characterization, chain conformation, and morphology of a beta-(1 – >3)-D-glucan isolated from the fruiting body of dictyophora indusiata. *J. Agric. Food Chem.* **2009**, *57*, 5918–5924. [CrossRef] [PubMed]
23. Deng, C.; Hu, Z.; Fu, H.; Hu, M.; Xu, X.; Chen, J. Chemical analysis and antioxidant activity in vitro of a beta-D-glucan isolated from *Dictyophora indusiata*. *Int. J. Biol. Macromol.* **2012**, *51*, 70–75. [CrossRef] [PubMed]
24. Hua, Y.; Yang, B.; Tang, J.; Ma, Z.; Gao, Q.; Zhao, M. Structural analysis of water-soluble polysaccharides in the fruiting body of *Dictyophora indusiata* and their in vivo antioxidant activities. *Carbohydr. Polym.* **2012**, *87*, 343–347. [CrossRef]
25. Zhang, R.; Edgar, K.J. Properties, chemistry, and applications of the bioactive polysaccharide curdlan. *Biomacromolecules* **2014**, *15*, 1079–1096. [CrossRef]

26. Harada, T.; Misaki, A.; Saito, H. Curdlan—A Bacterial gel-forming beta-1→3-glucan. *Arch. Biochem. Biophys.* **1968**, *124*, 292. [CrossRef]
27. Ahmad, A.; Anjum, F.M.; Zahoor, T.; Nawaz, H.; Ahmed, Z. Extraction and characterization of β-d-glucan from oat for industrial utilization. *Int. J. Biol. Macromol.* **2010**, *46*, 304–309. [CrossRef]
28. Gamel, T.H.; Abdel-Aal, E.-S.M.; Ames, N.P.; Duss, R.; Tosh, S.M. Enzymatic extraction of beta-glucan from oat bran cereals and oat crackers and optimization of viscosity measurement. *J. Cereal Sci.* **2014**, *59*, 33–40. [CrossRef]
29. Okobira, T.; Miyoshi, K.; Uezu, K.; Sakurai, K.; Shinkai, S. Molecular dynamics studies of side chain effect on the β-1, 3-D-glucan triple helix in aqueous solution. *Biomacromolecules* **2008**, *9*, 783–788. [CrossRef]
30. Zhong, K.; Tong, L.; Liu, L.; Zhou, X.; Liu, X.; Zhang, Q.; Zhou, S. Immunoregulatory and antitumor activity of schizophyllan under ultrasonic treatment. *Int. J. Biol. Macromol.* **2015**, *80*, 302–308. [CrossRef] [PubMed]
31. Zhang, Y.; Li, S.; Wang, X.; Zhang, L.; Cheung, P.C.K. Advances in lentinan: Isolation, structure, chain conformation and bioactivities. *Food Hydrocoll.* **2011**, *25*, 196–206. [CrossRef]
32. Meng, Y.; Shi, X.; Cai, L.; Zhang, S.; Ding, K.; Nie, S.; Luo, C.; Xu, X.; Zhang, L. Triple-helix conformation of a polysaccharide determined with light scattering, afm, and molecular dynamics simulation. *Macromolecules* **2018**, *51*, 10150–10159. [CrossRef]
33. Xu, S.; Lin, Y.; Huang, J.; Li, Z.; Xu, X.; Zhang, L. Construction of high strength hollow fibers by self-assembly of a stiff polysaccharide with short branches in water. *J. Mater. Chem. A* **2013**, *1*, 4198–4206. [CrossRef]
34. Ping, Z.H.; Xu, H.; Liu, T.; Huang, J.C.; Meng, Y.; Xu, X.J.; Li, W.H.; Zhang, L.N. Anti-hepatoma activity of the stiff branched beta-D-glucan and effects of molecular weight. *J. Mat. Chem. B* **2016**, *4*, 4565–4573. [CrossRef] [PubMed]
35. Dong, Q.; Jia, L.M.; Fang, J.N. A beta-D-glucan isolated from the fruiting bodies of *Hericium erinaceus* and its aqueous conformation. *Carbohydr. Res.* **2006**, *341*, 791–795. [CrossRef]
36. Wu, D.; Tang, C.; Liu, Y.; Li, Q.; Wang, W.; Zhou, S.; Zhang, Z.; Cui, F.; Yang, Y. Structural elucidation and immunomodulatory activity of a beta-D-glucan prepared by freeze-thawing from *Hericium erinaceus*. *Carbohydr. Polym.* **2019**, *222*, 114996. [CrossRef]
37. Cui, H.; Zhu, X.; Huo, Z.; Liao, B.; Huang, J.; Wang, Z.; Song, C.; Hu, X.; Fang, J. A beta-glucan from *Grifola frondosa* effectively delivers therapeutic oligonucleotide into cells via dectin-1 receptor and attenuates TNFalpha gene expression. *Int. J. Biol. Macromol.* **2020**, *149*, 801–808. [CrossRef] [PubMed]
38. Iino, K.; Ohno, N.; Suzuki, I.; Miyazaki, T.; Yadomae, T.; Oikawa, S.; Sato, K. Structural characterisation of a neutral antitumour β-d-glucan extracted with hot sodium hydroxide from cultured fruit bodies of *Grifola frondosa*. *Carbohydr. Res.* **1985**, *141*, 111–119. [CrossRef]
39. Fang, J.; Wang, Y.; Lv, X.; Shen, X.; Ni, X.; Ding, K. Structure of a β-glucan from *Grifola frondosa* and its antitumor effect by activating Dectin-1/Syk/NF-κB signaling. *Glycoconj. J.* **2012**, *29*, 365–377. [CrossRef] [PubMed]
40. Kim, Y.-R.; Hwang, J.; Koh, H.-J.; Jang, K.; Lee, J.-D.; Choi, J.; Yang, C.-S. The targeted delivery of the c-Src peptide complexed with schizophyllan to macrophages inhibits polymicrobial sepsis and ulcerative colitis in mice. *Biomaterials* **2016**, *89*, 1–13. [CrossRef]
41. Kadam, S.U.; Tiwari, B.K.; O'Donnell, C.P. Extraction, structure and biofunctional activities of laminarin from brown algae. *Int. J. Food Sci. Technol.* **2015**, *50*, 24–31. [CrossRef]
42. Shin, H.J.; Oh, S.J.; Kim, S.I.; Won Kim, H.; Son, J.-H. Conformational characteristics of β-glucan in laminarin probed by terahertz spectroscopy. *Appl. Phys. Lett.* **2009**, *94*, 111911. [CrossRef]
43. Zhang, Y.; Xu, X.; Zhang, L. Gel formation and low-temperature intramolecular conformation transition of a triple-helical polysaccharide lentinan in water. *Biopolymers* **2008**, *89*, 852–861. [CrossRef]
44. Kao, P.F.; Wang, S.H.; Hung, W.T.; Liao, Y.H.; Lin, C.M.; Yang, W.B. Structural characterization and antioxidative activity of low-molecular-weights beta-1,3-glucan from the residue of extracted ganoderma lucidum fruiting bodies. *J. Biomed. Biotechnol.* **2012**, *2012*, 673764. [CrossRef] [PubMed]
45. Liu, Y.; Zhang, J.; Tang, Q.; Yang, Y.; Guo, Q.; Wang, Q.; Wu, D.; Cui, S.W. Physicochemical characterization of a high molecular weight bioactive beta-D-glucan from the fruiting bodies of *Ganoderma lucidum*. *Carbohydr. Polym.* **2014**, *101*, 968–974. [CrossRef]
46. Aimanianda, V.; Clavaud, C.; Simenel, C.; Fontaine, T.; Delepierre, M.; Latgé, J.-P. Cell wall β-(1, 6)-glucan of *Saccharomyces cerevisiae*: Structural characterization and in situ synthesis. *J. Biol. Chem.* **2009**, *284*, 13401–13412. [CrossRef] [PubMed]
47. Wang, J.; Chen, S.; Nie, S.; Cui, S.W.; Wang, Q.; Phillips, A.O.; Phillips, G.O.; Xie, M. Structural characterization and chain conformation of water-soluble β-glucan from wild cordyceps sinensis. *J. Agric. Food Chem.* **2019**, *67*, 12520–12527. [CrossRef] [PubMed]
48. Cui, W.; Wood, P. Relationships Between Structural Features, Molecular Weight and Rheological Properties of Cereal β-D-Glucans. In *Hydrocolloids*; Elsevier: Amstardam, The Netherlands, 2000; pp. 159–168.
49. De Arcangelis, E.; Djurle, S.; Andersson, A.A.M.; Marconi, E.; Messia, M.C.; Andersson, R. Structure analysis of β-glucan in barley and effects of wheat β-glucanase. *J. Cereal Sci.* **2019**, *85*, 175–181. [CrossRef]
50. Roubroeks, J.; Andersson, R.; Åman, P. Structural features of (1→3),(1→4)-β-D-glucan and arabinoxylan fractions isolated from rye bran. *Carbohydr. Polym.* **2000**, *42*, 3–11. [CrossRef]
51. Wood, P.; Weisz, J.; Blackwell, B. Molecular characterization of cereal β-D-glucans. Structural analysis of oat β-D-glucan and rapid structural evaluation of β-D-glucans from different sources by high-performance liquid chromatography of oligosaccharides released by lichenase. *Cereal Chem.* **1991**, *68*, 31–39.

52. Wood, P.; Weisz, J.; Blackwell, B. Structural studies of (1→3),(1→4)-β-D-glucans by 13C-nuclear magnetic resonance spectroscopy and by rapid analysis of cellulose-like regions using high-performance anion-exchange chromatography of oligosaccharides released by lichenase. *Cereal Chem.* **1994**, *71*, 301–307.
53. Beer, M.U.; Wood, P.J.; Weisz, J.; Fillion, N. Effect of cooking and storage on the amount and molecular weight of (1→3)(1→4)-β-D-glucan extracted from oat products by an in vitro digestion system. *Cereal Chem.* **1997**, *74*, 705–709. [CrossRef]
54. Yuan, H.J.; Lan, P.; He, Y.; Li, C.L.; Ma, X. Effect of the modifications on the physicochemical and biological properties of beta-glucan—A critical review. *Molecules* **2020**, *25*, 57. [CrossRef]
55. Zheng, X.; Zhou, F.; Xu, X.; Zhang, L. Uptake of intraperitoneally administrated triple helical β-glucan for antitumor activity in murine tumor models. *J. Mat. Chem. B* **2017**, *5*, 9337–9345. [CrossRef] [PubMed]
56. Cai, Z.; Zhang, H. Recent progress on curdlan provided by functionalization strategies. *Food Hydrocoll.* **2017**, *68*, 128–135. [CrossRef]
57. Geller, A.; Shrestha, R.; Yan, J. Yeast-derived beta-glucan in cancer: Novel uses of a traditional therapeutic. *Int. J. Mol. Sci.* **2019**, *20*, 618. [CrossRef] [PubMed]
58. Zheng, X.; Lu, F.Z.; Xu, X.J.; Zhang, L.N. Extended chain conformation of beta-glucan and its effect on antitumor activity. *J. Mat. Chem. B* **2017**, *5*, 5623–5631. [CrossRef]
59. Yanaki, T.; Norisuye, T.; Fujita, H. Triple helix of *Schizophyllum commune* polysaccharide in dilute solution. 3. Hydrodynamic properties in water. *Macromolecules* **1980**, *13*, 1462–1466. [CrossRef]
60. Xu, S.; Xu, X.; Zhang, L. Effect of heating on chain conformation of branched β-glucan in water. *J. Phys. Chem. B* **2013**, *117*, 8370–8377. [CrossRef]
61. Zhang, X.; Zhang, L.; Xu, X. Morphologies and conformation transition of lentinan in aqueous NaOH solution. *Biopolym. Orig. Res. Biomol.* **2004**, *75*, 187–195.
62. Xu, X.; Zhang, X.; Zhang, L.; Wu, C. Collapse and association of denatured lentinan in water/dimethlysulfoxide solutions. *Biomacromolecules* **2004**, *5*, 1893–1898. [CrossRef]
63. Edge, C.; Parekh, R.; Rademacher, T.; Wormald, M.; Dwek, R. Fast sequencing of oligosaccharides using arrays of enzymes. *Nature* **1992**, *358*, 693–694. [CrossRef]
64. Venkataraman, G.; Shriver, Z.; Raman, R.; Sasisekharan, R. Sequencing complex polysaccharides. *Science* **1999**, *286*, 537–542. [CrossRef] [PubMed]
65. Hofmann, J.; Hahm, H.; Seeberger, P.H.; Pagel, K. Identification of carbohydrate anomers using ion mobility–mass spectrometry. *Nature* **2015**, *526*, 241–244. [CrossRef] [PubMed]
66. Rief, M.; Oesterhelt, F.; Heymann, B.; Gaub, H.E. Single molecule force spectroscopy on polysaccharides by atomic force microscopy. *Science* **1997**, *275*, 1295–1297. [CrossRef] [PubMed]
67. Li, Y.-Q.; Bruun, L.; Pierson, E.S.; Cresti, M. Periodic deposition of arabinogalactan epitopes in the cell wall of pollen tubes of *Nicotiana tabacum* L. *Planta* **1992**, *188*, 532–538. [CrossRef]
68. Wu, X.; Delbianco, M.; Anggara, K.; Michnowicz, T.; Pardo-Vargas, A.; Bharate, P.; Sen, S.; Pristl, M.; Rauschenbach, S.; Schlickum, U.; et al. Imaging single glycans. *Nature* **2020**, *582*, 375–378. [CrossRef]
69. Li, B.; Zhang, N.; Feng, Q.; Li, H.; Wang, D.; Ma, L.; Liu, S.; Chen, C.; Wu, W.; Jiao, L. The core structure characterization and of ginseng neutral polysaccharide with the immune-enhancing activity. *Int. J. Biol. Macromol.* **2019**, *123*, 713–722. [CrossRef]
70. Lee, Y.C. Carbohydrate analyses with high-performance anion-exchange chromatography. *J. Chromatogr. A* **1996**, *720*, 137–149. [CrossRef]
71. Kono, H.; Kondo, N.; Isono, T.; Ogata, M.; Hirabayashi, K. Characterization of the secondary structure and order–disorder transition of a β-(1→3, 1→6)-glucan from *Aureobasidium pullulans*. *Int. J. Biol. Macromol.* **2020**, *154*, 1382–1391. [CrossRef]
72. Roden, M.; Shulman, G.I. The integrative biology of type 2 diabetes. *Nature* **2019**, *576*, 51–60. [CrossRef]
73. Mahajan, A.; Wessel, J.; Willems, S.M.; Zhao, W.; Robertson, N.R.; Chu, A.Y.; Gan, W.; Kitajima, H.; Taliun, D.; Rayner, N.W. Refining the accuracy of validated target identification through coding variant fine-mapping in type 2 diabetes. *Nat. Genet.* **2018**, *50*, 559–571. [CrossRef] [PubMed]
74. Varanko, A.K.; Chilkoti, A. Molecular and materials engineering for delivery of peptide drugs to treat type 2 diabetes. *Adv. Healthc. Mater.* **2019**, *8*, 1801509. [CrossRef] [PubMed]
75. Cabré, C.; Colungo, C.; Vinagre, I.; Jansà, M.; Conget, I. Frequency and awareness of hypoglycemia in patients with Type 2 Diabetes treated with two or more insulin injections in primary care outpatient clinics. *Prim. Care Diabetes* **2020**, *14*, 168–172. [CrossRef] [PubMed]
76. Wu, H.; Esteve, E.; Tremaroli, V.; Khan, M.T.; Caesar, R.; Mannerås-Holm, L.; Ståhlman, M.; Olsson, L.M.; Serino, M.; Planas-Fèlix, M. Metformin alters the gut microbiome of individuals with treatment-naive type 2 diabetes, contributing to the therapeutic effects of the drug. *Nat. Med.* **2017**, *23*, 850. [CrossRef]
77. Cabreiro, F.; Au, C.; Leung, K.-Y.; Vergara-Irigaray, N.; Cochemé, H.M.; Noori, T.; Weinkove, D.; Schuster, E.; Greene, N.D.; Gems, D. Metformin retards aging in C. elegans by altering microbial folate and methionine metabolism. *Cell* **2013**, *153*, 228–239. [CrossRef]

78. Nauck, M.A.; Meininger, G.; Sheng, D.; Terranella, L.; Stein, P.P.; Sitagliptin Study Group. Efficacy and safety of the dipeptidyl peptidase-4 inhibitor, sitagliptin, compared with the sulfonylurea, glipizide, in patients with type 2 diabetes inadequately controlled on metformin alone: A randomized, double-blind, non-inferiority trial. *Diabetes Obes. Metab.* **2007**, *9*, 194–205. [CrossRef]
79. Proks, P.; Reimann, F.; Green, N.; Gribble, F.; Ashcroft, F. Sulfonylurea stimulation of insulin secretion. *Diabetes* **2002**, *51*, S368–S376. [CrossRef]
80. Pajvani, U.B.; Hawkins, M.; Combs, T.P.; Rajala, M.W.; Doebber, T.; Berger, J.P.; Wagner, J.A.; Wu, M.; Knopps, A.; Xiang, A.H. Complex distribution, not absolute amount of adiponectin, correlates with thiazolidinedione-mediated improvement in insulin sensitivity. *J. Biol. Chem.* **2004**, *279*, 12152–12162. [CrossRef]
81. Zhang, H.; Zhang, A.; Kohan, D.E.; Nelson, R.D.; Gonzalez, F.J.; Yang, T. Collecting duct-specific deletion of peroxisome proliferator-activated receptor γ blocks thiazolidinedione-induced fluid retention. *Proc. Natl. Acad. Sci. USA* **2005**, *102*, 9406–9411. [CrossRef]
82. Drucker, D.J. Biological actions and therapeutic potential of the glucagon-like peptides. *Gastroenterology* **2002**, *122*, 531–544. [CrossRef]
83. Bettge, K.; Kahle, M.; Abd El Aziz, M.S.; Meier, J.J.; Nauck, M.A. Occurrence of nausea, vomiting and diarrhoea reported as adverse events in clinical trials studying glucagon-like peptide-1 receptor agonists: A systematic analysis of published clinical trials. *Diabetes Obes. Metab.* **2017**, *19*, 336–347. [CrossRef] [PubMed]
84. Herman, G.A.; Stevens, C.; van Dyck, K.; Bergman, A.; Yi, B.; de Smet, M.; Snyder, K.; Hilliard, D.; Tanen, M.; Tanaka, W. Pharmacokinetics and pharmacodynamics of sitagliptin, an inhibitor of dipeptidyl peptidase IV, in healthy subjects: Results from two randomized, double-blind, placebo-controlled studies with single oral doses. *Clin. Pharmacol. Ther.* **2005**, *78*, 675–688. [CrossRef] [PubMed]
85. Fadini, G.P.; Avogaro, A. Cardiovascular effects of DPP-4 inhibition: Beyond GLP-1. *Vasc. Pharmacol.* **2011**, *55*, 10–16. [CrossRef] [PubMed]
86. Lambeir, A.-M.; Scharpé, S.; de Meester, I. DPP4 inhibitors for diabetes—What next? *Biochem. Pharmacol.* **2008**, *76*, 1637–1643. [CrossRef]
87. Taylor, S.I.; Blau, J.E.; Rother, K.I. SGLT2 inhibitors may predispose to ketoacidosis. *J. Clin. Endocrinol. Metab.* **2015**, *100*, 2849–2852. [CrossRef]
88. Tundis, R.; Loizzo, M.; Menichini, F. Natural products as α-amylase and α-glucosidase inhibitors and their hypoglycaemic potential in the treatment of diabetes: An update. *Mini Rev. Med. Chem.* **2010**, *10*, 315–331. [CrossRef] [PubMed]
89. Pino, J.L.; Mujica, V.; Arredondo, M. Effect of dietary supplementation with oat β-glucan for 3 months in subjects with type 2 diabetes: A randomized, double-blind, controlled clinical trial. *J. Funct. Foods* **2021**, *77*, 104311. [CrossRef]
90. Maki, K.; Galant, R.; Samuel, P.; Tesser, J.; Witchger, M.; Ribaya-Mercado, J.; Blumberg, J.; Geohas, J. Effects of consuming foods containing oat β-glucan on blood pressure, carbohydrate metabolism and biomarkers of oxidative stress in men and women with elevated blood pressure. *Eur. J. Clin. Nutr.* **2007**, *61*, 786–795. [CrossRef]
91. Zhao, C.; Yang, C.; Liu, B.; Lin, L.; Sarker, S.D.; Nahar, L.; Yu, H.; Cao, H.; Xiao, J. Bioactive compounds from marine macroalgae and their hypoglycemic benefits. *Trends Food Sci. Technol.* **2018**, *72*, 1–12. [CrossRef]
92. Nishinari, K.; Takemasa, M.; Fang, Y.; Hossain, K.S.; Tsumura, Y.; Sone, Y.; Fujiwara, M.; Habu, D.; Emoto, M. Effects of xyloglucan with different molar masses on glucose in blood. *Food Hydrocoll.* **2020**, *108*, 105727. [CrossRef]
93. Wood, P.; Beer, M.; Butler, G. Evaluation of role of concentration and molecular weight of oat β-glucan in determining effect of viscosity on plasma glucose and insulin following an oral glucose load. *Br. J. Nutr.* **2000**, *84*, 19–23. [CrossRef]
94. Tosh, S.M.; Bordenave, N. Emerging science on benefits of whole grain oat and barley and their soluble dietary fibers for heart health, glycemic response, and gut microbiota. *Nutr. Rev.* **2020**, *78*, 13–20. [CrossRef]
95. Liu, Y.; Wang, Y.; Zhou, S.; Yan, M.; Tang, Q.; Zhang, J. Structure and chain conformation of bioactive β-D-glucan purified from water extracts of *Ganoderma lucidum* unbroken spores. *Int. J. Biol. Macromol.* **2021**, *180*, 484–493. [CrossRef] [PubMed]
96. Bai, Y.; Atluri, S.; Zhang, Z.; Gidley, M.J.; Li, E.; Gilbert, R.G. Structural reasons for inhibitory effects of pectin on α-amylase enzyme activity and in-vitro digestibility of starch. *Food Hydrocoll.* **2021**, *114*, 106581. [CrossRef]
97. Gong, X.; Sui, L.; Morton, J.; Brennan, M.A.; Brennan, C.S. Investigation of nutritional and functional effects of rice bran protein hydrolysates by using Preferred Reporting Items for Systematic Reviews and Meta-Analysis (PRISMA) guidelines: A review. *Trends Food Sci. Technol.* **2021**, *110*, 798–811. [CrossRef]
98. Ma, B.; Feng, T.; Zhang, S.; Zhuang, H.; Chen, D.; Yao, L. The inhibitory effects of *Hericium erinaceus* β-glucan on in vitro starch digestion. *Front. Nutr.* **2020**, *7*, 348. [CrossRef] [PubMed]
99. Singh, B.P.; Aluko, R.E.; Hati, S.; Solanki, D. Bioactive peptides in the management of lifestyle-related diseases: Current trends and future perspectives. *Crit. Rev. Food Sci. Nutr.* **2021**, 1–14. [CrossRef] [PubMed]
100. Zhai, H.; Gunness, P.; Gidley, M.J. Barley β-glucan effects on emulsification and in vitro lipolysis of canola oil are modulated by molecular size, mixing method, and emulsifier type. *Food Hydrocoll.* **2020**, *103*, 105643. [CrossRef]
101. Zhang, H.; Li, Z.; Tian, Y.; Song, Z.; Ai, L. Interaction between barley β-glucan and corn starch and its effects on the in vitro digestion of starch. *Int. J. Biol. Macromol.* **2019**, *141*, 240–246. [CrossRef]
102. Gong, L.; Feng, D.; Wang, T.; Ren, Y.; Liu, Y.; Wang, J. Inhibitors of α-amylase and α-glucosidase: Potential linkage for whole cereal foods on prevention of hyperglycemia. *Food Sci. Nutr.* **2020**, *8*, 6320–6337. [CrossRef]

103. Fukuhara, G. Analytical supramolecular chemistry: Colorimetric and fluorimetric chemosensors. *J. Photochem. Photobiol. C Photochem. Rev.* **2020**, *42*, 100340. [CrossRef]
104. Tan, H.; Nie, S. Functional hydrocolloids, gut microbiota and health: Picking food additives for personalized nutrition. *FEMS Microbiol. Rev.* **2021**, *45*, fuaa065. [CrossRef]
105. Velikonja, A.; Lipoglavšek, L.; Zorec, M.; Avguštin, G. Alterations in gut microbiota composition and metabolic parameters after dietary intervention with barley beta glucans in patients with high risk for metabolic syndrome development. *Anaerobe* **2019**, *55*, 67–77. [CrossRef] [PubMed]
106. Wang, Y.; Ames, N.P.; Tun, H.M.; Tosh, S.M.; Jones, P.J.; Khafipour, E. High molecular weight barley β-glucan alters gut microbiota toward reduced cardiovascular disease risk. *Front. Microbiol.* **2016**, *7*, 129. [CrossRef] [PubMed]
107. Liu, P.; Wang, Y.; Yang, G.; Zhang, Q.; Meng, L.; Xin, Y.; Jiang, X. The role of short-chain fatty acids in intestinal barrier function, inflammation, oxidative stress, and colonic carcinogenesis. *Pharmacol. Res.* **2021**, *165*, 105420. [CrossRef] [PubMed]
108. Zhao, Q.; Hu, X.; Guo, Q.; Cui, S.W.; Xian, Y.; You, S.; Chen, X.; Xu, C.; Gao, X. Physicochemical properties and regulatory effects on db/db diabetic mice of β-glucans extracted from oat, wheat and barley. *Food Hydrocoll.* **2014**, *37*, 60–68. [CrossRef]

foods

Article

In Vitro Immuno-Modulatory Potentials of Purslane (*Portulaca oleracea* L.) Polysaccharides with a Chemical Selenylation

Ya-Ru Lin [1], Qing-Yun Guan [1], Ling-Yu Li [1], Zhi-Mei Tang [1,2], Qiang Zhang [1,2] and Xin-Huai Zhao [1,2,3,*]

1 School of Biology and Food Engineering, Guangdong University of Petrochemical Technology, Maoming 525000, China; lin17853509810@163.com (Y.-R.L.); mguan4826@163.com (Q.-Y.G.); 17852026615@163.com (L.-Y.L.); tzm1103@gdupt.edu.cn (Z.-M.T.); zhangqiang@gdupt.edu.cn (Q.Z.)
2 Research Centre of Food Nutrition and Human Healthcare, Guangdong University of Petrochemical Technology, Maoming 525000, China
3 Maoming Branch, Guangdong Laboratory for Lingnan Modern Agriculture, Guangdong University of Petrochemical Technology, Maoming 525000, China
* Correspondence: zhaoxh@gdupt.edu.cn

Abstract: The soluble polysaccharides from a non-conventional and edible plant purslane (*Portulaca oleracea* L.), namely PSPO, were prepared by the water extraction and ethanol precipitation methods in this study. The obtained PSPO were selenylated using the Na_2SeO_3-HNO_3 method to successfully prepare two selenylated products, namely SePSPO-1 and SePSPO-2, with different selenylation extents. The assay results confirmed that SePSPO-1 and SePSPO-2 had respective Se contents of 753.8 and 1325.1 mg/kg, while PSPO only contained Se element about 80.6 mg/kg. The results demonstrated that SePSPO-1 and SePSPO-2 had higher immune modulation than PSPO ($p < 0.05$), when using the two immune cells (murine splenocytes and RAW 264.7 macrophages) as two cell models. Specifically, SePSPO-1 and SePSPO-2 were more active than PSPO in the macrophages, resulting in higher cell proliferation, greater macrophage phagocytosis, and higher secretion of the immune-related three cytokines, including tumor necrosis factor-α (TNF-α), interleukin-6 (IL-6), and IL-1β. Meanwhile, SePSPO-1 and SePSPO-2 were more potent than PSPO in the concanavalin A- or lipopolysaccharide-stimulated splenocytes in cell proliferation, or more able than PSPO in the splenocytes to promote interferon-γ secretion but suppress IL-4 secretion, or more capable of enhancing the ratio of T-helper ($CD4^+$) cells to T-cytotoxic ($CD8^+$) cells for the T lymphocytes than PSPO. Overall, the higher selenylation extent of the selenylated PSPO mostly caused higher immune modulation in the model cells, while a higher polysaccharide dose consistently led to the greater regulation effect. Thus, it is concluded that the employed chemical selenylation could be used in the chemical modification of purslane or other plant polysaccharides, when aiming to endow the polysaccharides with higher immuno-modulatory effect on the two immune cells.

Keywords: purslane polysaccharides; selenylation; RAW 264.7 macrophages; splenocytes; immune modulation

Citation: Lin, Y.-R.; Guan, Q.-Y.; Li, L.-Y.; Tang, Z.-M.; Zhang, Q.; Zhao, X.-H. In Vitro Immuno-Modulatory Potentials of Purslane (*Portulaca oleracea* L.) Polysaccharides with a Chemical Selenylation. *Foods* **2022**, *11*, 14. https://doi.org/10.3390/foods11010014

Academic Editors: Jianhua Xie, Yanjun Zhang and Hansong Yu

Received: 18 November 2021
Accepted: 18 December 2021
Published: 21 December 2021

1. Introduction

Purslane (*Portulaca oleracea* L.) is a wild plant belonging to the *Portulacaceae* family. Purslane is widely spread and popular in most areas, including China, Europe, and Mediterranean countries, and is edible but usually regarded as one of these non-conventional plants. More importantly, purslane is regarded as having many biofunctions in both medicine and food fields. It was first recorded in the Compendium of Material Medica that purslane leaves had the ability to clear evil heat and remove toxins [1]. Moreover, recent research results have indicated the emerging functional properties of purslane in the intestine, skin, nerve, respiratory, and other systems. For example, the extract of purslane leaves showed an ability to reduce the severity of colitis through regulating the immune mechanism involved in the pathogenesis of colitis [2]. Besides, it was also found that the purslane juice could protect the rat brain from the rotenone-caused neurotoxicity, as well as apoptosis, by inhibiting excessive oxidative stress [3]. Overall, purslane is considered to contain these bioactive components, including alkaloids, polysaccharides, unsaturated fatty acids, flavonoids, proteins, and others, thus being regarded with various beneficial functions, such as anti-bacterial, anti-fungal, anti-inflammatory, analgesic, muscle relaxant, and wound healing effects. In addition, purslane is unusually rich in aliphatic acids (e.g., the α-linolenic acid) that are important in cholesterol reduction and blood lipid-lowering, as well as anti-thrombotic or anti-cardiovascular effect [4], while the flavone compounds in purslane also are of importance for the vital hypoglycemic and anti-oxidative functions by inhibiting the Akt phosphorylation to enhance the consumption of glucose or by scavenging free radicals and reducing metal ions, such as Fe^{3+} [5]. Overall, the potential health benefits of purslane are still insufficiently investigated so far.

Polysaccharides, a kind of carbohydrates, are made up of more than ten monosaccharide units that are joined through the glycosidic bonds in the branched or unbranched chains. For purslane, it was reported the polysaccharides extracted by water had a molecular mass about 7.3 kDa, with arabinose, galactose, glucose, mannose, rhamnose, and xylose as main saccharide elements [6]. It is worth mentioning that natural polysaccharides have various pharmacological effects, such as anti-cancer, anti-inflammation, anti-oxidation, modulation of gut microbiota, and immune function [7]. For example, the natural polysaccharides might exert hepatoprotective effect by regulating the pathways of inflammation and apoptosis, lipid metabolism, and cytochrome P450 enzymes [8], while the acidic polysaccharides from *Schisandra chinensis*, through reducing the oxidative stress, could protect the acute liver injury induced by ethanol [9]. Additionally, the combined fungal polysaccharides could suppress the hepatotoxicity induced by cyclophosphamide through reducing toxicity markers and preventing inflammatory responses [10]. As is reported, anti-oxidation and hypoglycemic effect are two important biological functions of plant polysaccharides. It was reported that the polysaccharides from garlic (*Allium sativum* L.) bolt and green walnut (*Juglandaceae*) husk possessed anti-oxidant activity to scavenge three radicals or reduce the multi-valent metal ions, such as Fe_3^+ [11,12], while those from bluish dogbane (*Apocynum venetum*) leaves had anti-hypoglycemic effect in type 2 diabetes mice by regulating intestinal flora, along with reducing glucose absorption [13]. Today, cancer is one of the most fatal diseases in the world; thus, the anti-tumor activities of polysaccharides by inhibiting tumor growth and enhancing immunological functions have attracted a special attention in recent [14,15]. For example, the polysaccharides from shiitake mushrooms (*Lentinus edodes*) could exert anti-tumor activities to the colon cancer HT-29 cells via cell proliferation suppression and apoptosis induction, through an internal pathway mediated by reactive oxygen species (ROS) and external pathway engaged with TNF-α [16]. In referring purslane polysaccharides, they were reported to have anti-diabetic effect in diabetic rats and could enhance the immune state of the rats with gastric cancer [6,17].

Immune modulation of natural plant polysaccharides and other components are also sufficiently studied. The immune system is the body's defensive system that performs the immune responses, immune function, and self-protection, and is composed of these elements, such as immune organs, cells, and molecules. The immunological responses consist of the innate and adaptive immune responses, including humoral and cellular immunity, while the responses to external stimulation are considered as one of the body's key defending strategies to prevent and combat external infections, inflammation, and cancers [18]. Some natural substances derived from natural foods have immuno-modulatory effects. Tea polyphenols could increase the immunity of tilapia via promoting the activity and expression of immunoglobulin, enhancing the lysozyme activity, and regulating the NF-κB signaling pathway [19]. The peptides derived from whey and casein proteins had a terrific immuno-modulatory function because the peptides could increase the macrophage phagocytosis, promote splenocyte proliferation, and enhance cytokine secretion [20,21]. In addition, a flavonoid compound, quercetin, also might improve the immunity of *Arbor Acre* broilers [22]. Overall, it was revealed that plant polysaccharides could play an effective immuno-modulatory role in immune systems through activating the macrophages, splenocytes, and other immune cells, promoting the release of cytokines, increasing the growth of immune organs and the secretion of immunoglobulins, and inhibiting the over-activation of the complement system [23]. However, whether a chemical modification of natural polysaccharides will cause positive or negative effects on the immune modulation of the modified polysaccharides is less studied in the present. Thus, such an investigation using the soluble purslane polysaccharides as a target for a chemical selenylation deserves our consideration.

In this study, the soluble polysaccharides from purslane (namely PSPO) were extracted by water at a neutral condition, and then selenylated chemically using the Na_2SeO_3-HNO_3 system for two selenylation extents to prepare two selenylated PSPO products (SePSPO), namely SePSPO-1 and SePSPO-2, respectively. Both SePSPO-1 and SePSPO-2 were assessed for their in vitro immuno-modulatory activities using two immune cells (i.e., the RAW 264.7 macrophages and murine splenocytes) as cell models and the unmodified PSPO as a control. Several indices, such as growth proliferation, phagocytic activity, cellular secretion of five cytokines, and T lymphocyte subpopulations, were measured and compared to reflect the target immune modulation. The purpose of this study was to disclose whether the performed chemical selenylation could cause bioactivity changes for the soluble PSPO in their important immune potential.

2. Materials and Methods

2.1. Materials and Reagents

The RPMI-1640 medium, Dulbecco's modified essential medium with high glucose (DMEM) were purchased from HyClone Co. (Logan, UT, USA), while the fetal bovine serum (FBS) was provided by Thermo Fisher Scientific Inc. (Cleveland, OH, USA). Both neutral red and trypan blue were purchased from Amresco Inc. (Los Angeles, CA, USA), and 3-(4,5-dimethyl-2-thiazolyl)-2,5-diphenyl tetrazolium bromide (MTT), concanavalin A (ConA), and lipopolysaccharide (LPS) were provided by Sigma-Aldrich Chemical Co. (St. Louis, MO, USA). The phosphate-buffered saline (PBS) was the product of Solarbio Science and Technology Co. Ltd. (Beijing, China), while the Hanks' balanced salt solution (HBSS) and the red blood cell lysis buffer were obtained from Beyotime Biotechnology (Beijing, China). The cell counting kit-8 (CCK-8) was the product of Dojindo Laboratories (Kyushu, Japan). Ultrapure water generated from Milli-Q Plus (Millipore Corporation, New York, NY, USA) was used in this study. Other chemicals used in the present work were of analytical grade.

The phycoerythrin (PE)-conjugated anti-mouse $CD8a^+$ antibodies and fluorescein isothiocyanate (FITC) anti-mouse $CD4^+$ antibodies were bought from Miltenyi Biological Technology Co. Ltd. (Bergisch Gladbach, Cologne, Germany), while the enzyme-linked immunosorbent assay (ELISA) kits [mouse interferon-γ (IFN-γ), interleukin-1β (IL-1β), IL-4, IL-6, and tumor necrosis factor-α (TNF-α)] were bought from Boster Biological Engineering Co. Ltd. (Wuhan, China).

2.2. Animal and Cells

The used mice (female BALB/c, 6–8 weeks old) in this study were provided by professional institution Beijing Vital River Experimental Animal Technical Co. Ltd. (Beijing, China). As usual, the mice were maintained for at least 7 d before the experiments performed at Northeast Agricultural University (Harbin, China). In addition, all animal procedures were approved and instructed by Animal Care and Use Committee of Northeast Agricultural University.

The used RAW 264.7 macrophages, provided by Shanghai Branch of Chinese Academy of Sciences (Shanghai, China), were incubated in the DMEM medium supplemented with 10% fetal bovine serum (FBS) and 100 U/mL streptomycin/penicillin. The cells were cultured in an incubator of 37 °C and 5% CO_2 referring to the recommendation of cell supplier.

2.3. Polysaccharide Extraction

The water-extraction and ethanol-precipitation protocols were used to extract PSPO as previously described with minor modification [24,25]. In detail, the dried purslane materials were smashed into powder, blended with water at a 1:20 ratio (*w/v*), added with the thermostable α-amylase of 20 U/mL, and kept at 90 °C for 4 h, followed by a centrifugation at 8000× *g* for 15 min after cooling. Afterwards, the separated supernatant was filtered and mixed with an alkaline protease (100 U/mL) at 55 °C for 8 h to degrade the extracted proteins, and then centrifuged once again at 8000× *g* for 15 min after cooling. The obtained supernatant was concentrated to 1/10 of the original volume by heating and then precipitated by using anhydrous ethanol of three-fold volume at 4 °C for 12 h. The precipitates (i.e., PSPO) were separated by using an 8000× *g* centrifugation for 15 min, washed three times using anhydrous ethanol, and soaked in anhydrous ethyl ether to remove fats and pigments. After that, the PSPO was dialyzed against water for 2 d to remove the small-molecule impurities and salt ions, freeze-dried, and then kept at −20 °C for future usage.

2.4. PSPO Selenylation and Se Detection

The Na_2SeO_3-HNO_3 method was applied to prepare selenylated PSPO according to the previous study [26]. Briefly, 300 mg PSPO powder was dissolved in 20 mL 5% HNO_3, mixed with 30 or 45 mg Na_2SeO_3 and reacted at 75 °C for 8 h. Three-fold volume of anhydrous ethanol was added into the reaction mixture after cooling, while the final system was kept at 4 °C for 12 h. Thus, the precipitates were collected, soaked in anhydrous ethanol five times to remove the unreacted H_2SeO_3, and then freeze-dried to obtain two selenylated products, namely SePSPO-1 and SePSPO-2, respectively. Besides, the same amount of PSPO was mixed with Na_2SeO_3 without HNO_3 and subjected to the same treatments. The yielded PSPO were regarded as the unmodified PSPO and used as a control in this study.

Se contents of the target samples were assessed by the method reported in a previous study [27], using an inductively coupled plasma-mass spectrometer (Agilent Technologies, Santa Clara, CA, USA).

2.5. Assays of Cell Viability and Phagocytic Activity of the Macrophages

The possible effects of PSPO, SePSPO-1, and SePSPO-2 on macrophage viability were measured by a MTT assay, as previously described [28]. Specifically, 100 μL cells (2×10^5 cells/mL) were inoculated into a 96-well plate, followed by 4 h culture. After medium discarding, the adherent cells were exposed to the target samples at 5–80 μg/mL for 24 or 48 h. After removing the supernatants, 100 μL MTT solution was added, and the macrophages were cultured for 4 h. After discarding the supernatants, DMSO of 100 μL was added into each well, while the optical density (OD) value of each well was determined at 450 nm with a microplate reader (Bio-Rad Laboratories, Hercules, CA, USA). The value of macrophage viability was calculated accordingly [29]. Meanwhile, the macrophages of control group without sample treatment were considered with 100% cell viability.

Macrophage phagocytosis was measured by a neutral red assay as a previous study did [30]. In detail, 200 μL cells (2×10^5 cells/mL) were seeded into 96-well plates and cultured for 4 h. The target samples at doses of 5–20 μg/mL were added, while the macrophages were cultured for another 24 h. The samples were replaced by 1% neutral red solution of 100 μL, while the macrophages were incubated for 1.5 h and washed for five times by the PBS (0.1 mmol/L, pH 7.2). The cell lysing solution (ethanol: 1% acetic acid = 1:1, *v/v*) of 200 μL was added into each well, and the macrophages were incubated for another 2 h. The OD values were measured at 540 nm after this incubation at the same microplate reader. Phagocytic index (PI) reflecting the target macrophage phagocytosis was, thus, calculated accordingly [31].

2.6. Assays of Proliferation of the Polysaccharide Samples on Murine Splenocytes

The target splenocytes were obtained from the mice, based on the procedure reported in a previous study [32]. The mice were cervical dislocated and followed by a soak (5 min) in 75% ethanol solution. Thus, the spleens were taken out under an aseptic condition, ground into small pieces in a container with the cold HBSS, and then sifted through a sieve of 200-mesh to obtain the murine splenocytes. The collected splenocytes were cleaned by the PBS, centrifuged at 170× *g* for 5 min, suspended in 5 mL lysis buffer for 3 min, centrifuged and cleaned twice by the PBS, and resuspended in the RPMI-1640 medium fortified with 100 U/mL streptomycin/penicillin and 10% FBS. The splenocytes with a measured viability value more than 98%, adjusted into a fixed cell density of 1×10^6 cells/mL, and then used in this study.

Proliferative effects of the samples on the LPS- or Con A-induced splenocytes were assayed using the CCK-8 method [33]. The splenocytes of 100 μL were plated in 96-well plates and cultured with the target samples at doses of 5–80 μg/mL for 48 h, together with LPS (10 μg/mL) or Con A (5 μg/mL), while the splenocytes incubated with the medium alone and mitogen in the medium were regarded as respective blank and control groups. After discarding the supernatants, the splenocytes were incubated with CCK-8 solutions of 100 μL for 4 h, while the OD values were detected at 450 nm using the same microplate reader.

2.7. Assays of Cytokine Secretion in the Macrophages and Murine Splenocytes

In brief, 2 mL macrophages (2×10^5 cells/mL) were cultured in a 6-well plate for 4 h. After discarding the medium, the target samples at doses of 5–20 μg/mL were added to treat the cells for 24 h. In addition, then, a centrifugation at 500× *g* of 20 min was conducted to collect the supernatants, while the levels of TNF-α, IL-1β, and IL-6 in the supernatants were assayed in accordance with the instructions of the corresponding ELISA kits.

Secretion levels of IL-4 and IL-1β in the splenocytes were assayed using the respective the ELISA kits. Briefly, the splenocytes of 1 mL were incubated with the polysaccharide samples (doses of 5–20 μg/mL) of 0.5 mL in 12-well plates for 48 h. After that, the supernatants were collected using the centrifugation as above, while IL-4 and IL-1β levels were detected by the respective ELISA kits and the suggested protocols.

2.8. Assays of T Lymphocyte Subpopulations

The T lymphocyte subpopulations were analyzed by a flow cytometry protocol according to a previous study [34]. Briefly, the splenocytes (1.5 mL) were plated in 12-well plates, co-cultured with the three polysaccharide samples of 5–20 μg/mL and Con A of 5 μg/mL for 48 h. The cells were collected via a centrifugation at 170× *g* for 5 min, washed twice by the PBS, resuspended in 500 μL PBS, and then adjusted to 1×10^6 cells. The prepared cell suspensions were mixed with FITC-conjugated anti-mouse $CD4^+$ antibody (or PE-conjugated anti-mouse $CD8a^+$ antibody) of 10 μL, kept at 4 °C for 30 min, passed through a sieve of 300-mesh, and then detected at a flow cytometry (Type BD FACS Aria II, BD Bioscience, Franklin Lakes, NJ, USA).

2.9. Statistical Analysis

All data reported in this study were obtained from three independent experiments or assays and expressed as means values ± standard deviations. The one-way ANOVA analysis with Duncan's multiple range tests was used to measure the differences among the mean values, while the $p < 0.05$ was deemed to significant difference. The statistical analysis was conducted using the software SPSS version 16.0 (SPSS, Inc., Chicago, IL, USA).

3. Results

3.1. Macrophage Proliferation and Phagocytosis as Affected by the PSPO and SePSPO

Determined by the classic phenol-H_2SO_4 assaying method, the obtained PSPO were detected with a total saccharide content of 855.4 g/kg and ash content of 91.7 g/kg (dry basis). After the performed chemical selenylation, the obtained SePSPO-1 and SePSPO-2 were detected with Se contents of 753.8 and 1325.1 mg/kg (dry basis), respectively, while the unmodified (i.e., control) PSPO only contained Se of 80.6 mg/kg (dry basis). Compared with the control PSPO, both SePSPO-1 and SePSPO-2 had near 8-fold and 15-fold increases in Se contents. These data confirmed that both SePSPO-1 and SePSPO-2 successfully obtained a chemical selenylation, and Se element (in the status of H_2SeO_3) was covalently bound into the molecules of PSPO. Because PSPO, SePSPO-1, and SePSPO-2 were obviously different in Se contents, this study thereby assessed whether the performed chemical selenylation, as well as the selenylation extent, could affect the immune activity of the target PSPO.

When the three polysaccharide samples were used to treat the macrophages for 24 and 48 h using the five doses (5–80 μg/mL), the data indicated that the target samples all had no cytotoxicity on the cells ($p > 0.05$) (Figure 1a,b) because the treated macrophages showed viability values larger than 100%. With the cell treatment of 24 h, the cells exposed to PSPO, SePSPO-1, and SePSPO-2 showed viability values of 114.7–131.8%, 119.3–133.8%, and 121.1–148.4%, respectively. Using a longer time (48 h) to treat the macrophages, the cells exposed to PSPO, SePSPO-1, and SePSPO-2 showed corresponding viability values of 110.4–129.9%, 118.1–130.7%, and 118.9–130.9%. These data suggested that the samples could promote cell growth. In addition, it was observed that SePSPO-1, and especially SePSPO-2, in all cases could cause higher viability values than PSPO did, suggesting that the performed chemical selenylation resulted in higher bioactivity for PSPO, while higher selenylation extent consistently induced more bioactivity increase. Moreover, it was estimated that the SePSPO-2 dose at 20 μg/mL equaled to a Se intake near 170 μg in the body, considering a well-accepted body fluid volume of 6.4 L. This estimated Se intake is in the range of the recommended daily intake (RDI) value of Se element (50–200 μg) for humans [35]. Thus, two polysaccharide doses of 5 and 20 μg/mL were employed in the later experimental assays to obtain the secretion levels of several cytokines in the macrophages or splenocytes.

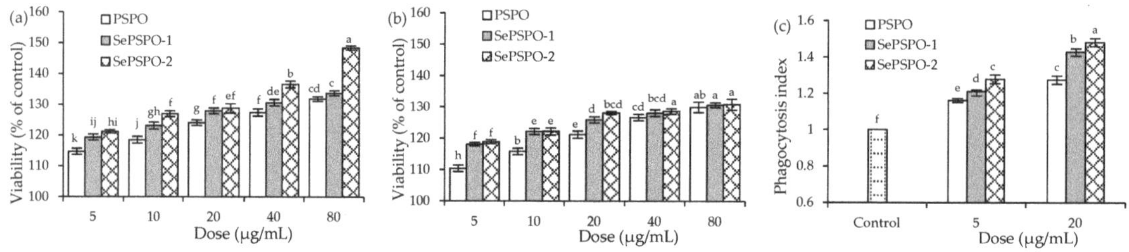

Figure 1. The measured cell viability values (%) of the macrophages exposed to PSPO, SePSPO-1, and SePSPO-2 for 24 (**a**) and 48 h (**b**), as well as the phagocytosis (**c**) of the macrophages exposed to PSPO, SePSPO-1, and SePSPO-2 for 24 h. Different letters lowercase above the columns suggest that the one-way ANOVA of the mean values is significantly different ($p < 0.05$).

The detected phagocytic activities of the macrophages treated by the target samples were reflected in Figure 1c. The control cells had PI value of 1.00. When the cells were exposed to the samples at the doses of 5 and 20 μg/mL for 24 h, they were detected with increased PI values ranging from 1.16 to 1.48, demonstrating a fact that the samples all could enhance macrophage phagocytosis dose-dependently. Overall, PSPO and SePSPO-2 showed the respective lowest and highest potentials to promote macrophage phagocytosis, indicating again that the conducted chemical selenylation caused higher bioactivity for the PSPO, while higher selenylation extent consistently led to more activity enhancement.

3.2. Cytokine Secretion of the Macrophages as Affected by the PSPO and SePSPO

The secretion levels of three cytokines in the macrophages with or without the sample treatments of 24 h are shown in Table 1, using the three cytokines, including IL-6, IL-1β, and

TNF-α, as three evaluation indices. In general, each sample had clear immune promotion on the cells by enhancing the secretion of the cytokines dose-dependently. The control cells had IL-6 level of 10.69 pg/mL, together with respective IL-1β and TNF-α levels of 1.55 and 1.39 pg/mL. Meanwhile, the cells treated by the samples showed much enhanced secretion in IL-6 (13.85–47.58 pg/mL), IL-1β (2.87–11.71 pg/mL), and TNF-α (68.78–144.16 pg/mL) ($p < 0.05$), while a higher polysaccharide dose consistently caused higher cytokine secretion ($p < 0.05$). Data comparison results also demonstrated that PSPO and SePSPO-2 had the respective lowest and highest capacity in the cells to elevate cytokine secretion. That is, the performed chemical selenylation endowed PSPO with a higher ability to elevate cytokine secretions in the treated cells, while higher selenylation extent also caused higher activity for the selenylated PSPO.

Table 1. The measured cytokine levels in the macrophages incubated with PSPO, SePSPO-1, and SePSPO-2 for 24 h.

Cell Group	Dose (μg/mL)	IL-6 (pg/mL)	IL-1β (pg/mL)	TNF-α (pg/mL)
Control	None	10.69 ± 1.24^{f}	1.55 ± 0.40^{f}	61.39 ± 1.70^{g}
PSPO	5	13.85 ± 1.26^{e}	2.87 ± 0.61^{f}	68.78 ± 1.20^{f}
	20	33.81 ± 0.69^{c}	5.48 ± 1.05^{e}	112.98 ± 1.78^{c}
SePSPO-1	5	22.30 ± 1.40^{d}	6.29 ± 0.60^{c}	76.76 ± 1.40^{e}
	20	42.52 ± 0.24^{b}	7.73 ± 0.45^{bc}	128.96 ± 1.55^{b}
SePSPO-2	5	32.81 ± 1.86^{c}	8.53 ± 0.99^{b}	86.90 ± 0.97^{d}
	20	47.58 ± 1.61^{a}	11.71 ± 0.77^{a}	144.16 ± 3.49^{a}

Different lowercase letters as the superscripts after the data in the same column suggest that the one-way ANOVA of the mean values is significantly different ($p < 0.05$).

3.3. Splenocyte Proliferation as Affected by the PSPO and SePSPO

When the target samples were used at the five doses (5–80 μg/mL) with the mitogen to treat the murine splenocytes for 48 h, the treated cells all showed viability values larger than 100% (Figure 2), indicating that the samples had none cytotoxicity on the splenocytes. When the cells were stimulated by Con A, the cells exposed to PSPO, SePSPO-1, and SePSPO-2 showed viability values of 116.1–128.8%, 116.3–130.9%, and 117.4–148.4%, respectively, while the cells stimulated with Con A only had viability value of 112.3% (Figure 2a). Thus, the data suggested that higher polysaccharide doses mostly lead to larger viability values; however, the conducted chemical selenylation and yielded selenylation extent only had a minor effect on the measured viability values, although SePSPO-2 (or SePSPO-1) was slightly active than SePSPO-1 (or PSPO) to increase cell viability. When LPS was also used to stimulate the cells, a similar phenomenon was observed for the measured viability values (Figure 2b). Thus, the target samples were regarded to have immune modulation by promoting the proliferation of the mitogen-stimulated splenocytes, while the used selenylation was not sufficient to enhance PSPO bioactivity towards the stimulated cells.

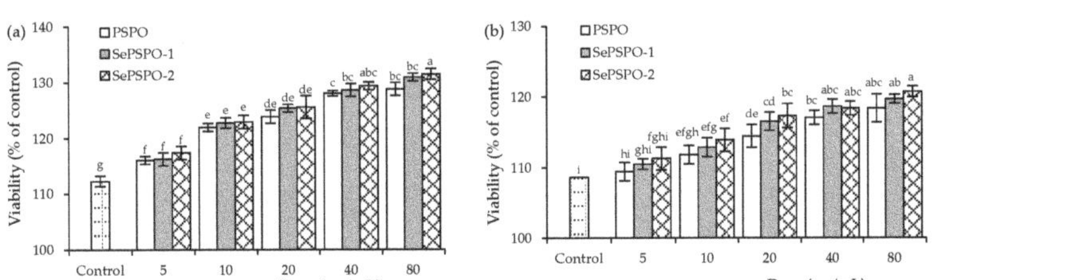

Figure 2. The measured viability values (%) of the splenocytes incubated with PSPO, SePSPO-1, and SePSPO-2 for 48 h and stimulated by Con A (**a**) or LPS (**b**). Different letters lowercase above the columns suggest that the one-way ANOVA of the mean values is significantly different ($p < 0.05$).

3.4. Cytokine Secretion of the Murine Splenocytes as Affected by the PSPO and SePSPO

The secretion levels of two cytokines IFN-γ and IL-4 in the splenocytes responding to the sample treatment of 48 h are reflected in Figure 3. Overall, each sample, in most cases, exerted a significant immune promotion in the cells by increasing the secretion of IFN-γ but clearly decreasing the secretion of IL-4. In detail, the control cells without sample treatment had IFN-γ and IL-4 levels of 3.16 and 61.91 pg/mL, respectively. When the cells were treated by PSPO, SePSPO-1, and SePSPO-2, they were measured with increased IFN-γ secretion levels (4.38–9.24 pg/mL) but distinctly decreased IL-4 secretion levels (56.88–31.90 pg/mL) ($p < 0.05$). Higher sample doses consistently caused higher secretion of IFN-γ but lower secretion of IL-4, while SePSPO-1 and SePSPO-2 were more able than PSPO to modulate the secretion of IFN-γ and IL-4. Although SePSPO-1 and SePSPO-2 had similar ability to enhance IFN-γ section, SePSPO-2 was more able than SePSPO-1 to suppress IL-4 secretion. This fact demonstrated that SePSPO-2 had higher activity in the cells than SePSPO-1. Thus, the conducted chemical selenylation conferred the PSPO with higher immune modulation in the splenocytes via regulating cytokine secretion, while higher selenylation extent generally endowed the selenylated PSPO with higher immune activity.

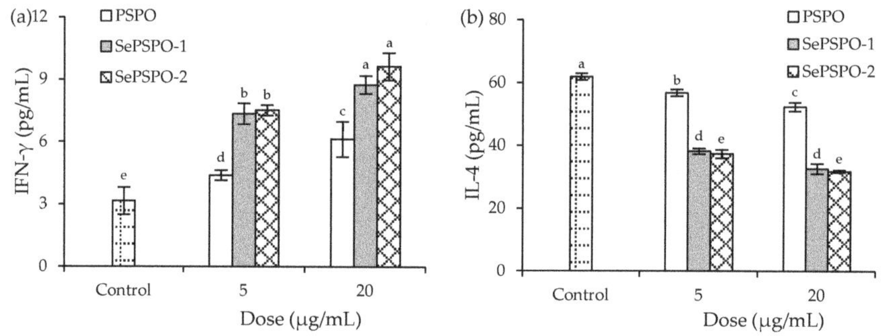

Figure 3. The detected levels of IFN-γ (**a**) and IL-4 (**b**) in the murine splenocytes incubated with PSPO, SePSPO-1, and SePSPO-2 for 48 h. Different letters lowercase above the columns suggest that the one-way ANOVA of the mean values is significantly different ($p < 0.05$).

3.5. T lymphocyte Subpopulations as Affected by the PSPO and SePSPO

When the splenocytes were stimulated by Con A and then treated with the target samples at the two doses of 5 and 20 μg/mL for 48 h, cell proportions of T-helper ($CD4^+$) and T-cytotoxic ($CD8^+$) cells were assayed using the flow cytometry technique (Figure 4). The final results are listed in Table 2, while the calculated $CD4^+/CD8^+$ ratios are also given in this table. The results showed that the control cells had a $CD4^+/CD8^+$ ratio of 2.06, while those treated by the target samples possessed enhanced $CD4^+/CD8^+$ ratios ranging from 2.08 to 2.41. Clearly, PSPO and SePSPO-2 showed the respective lowest and highest capacity to enhance the target $CD4^+/CD8^+$ ratio in the T lymphocytes. Thus, it was confirmed again that the conducted selenylation caused higher bioactivity for the selenylated PSPO, while a higher selenylation extent would make a contribution to activity enhancement.

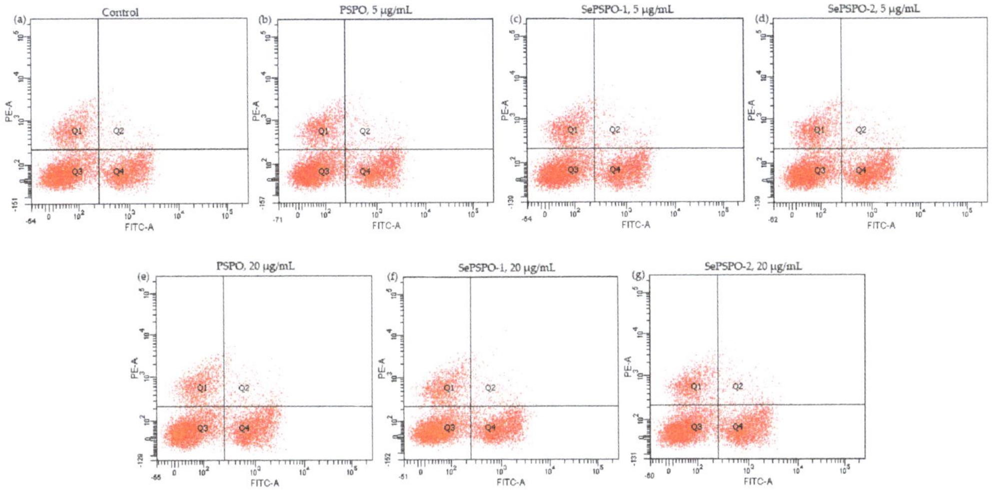

Figure 4. The obtained flow cytometry pictures for the control cells (**a**), and those cells incubated with PSPO (**b**), (**e**), SePSPO-1 (**c**), (**f**), and SePSPO-2 (**d**), (**g**) at the doses of 5 and 20 μg/mL, respectively.

Table 2. The measured T lymphocyte subpopulations in the Con A-treated splenocytes that were incubated with PSPO, SePSPO-1, and SePSPO-2 for 48 h.

Cell Group	Dose (μg/mL)	$CD4^+$ (%)	$CD8^+$ (%)	$CD4^+/CD8^+$ Ratio
Control	None	29.7 ± 1.2	14.4 ± 0.6	2.06 ± 0.01
PSPO	5	30.8 ± 2.0	14.8 ± 1.1	2.08 ± 0.02
	20	31.9 ± 2.8	14.5 ± 1.3	2.20 ± 0.59
SePSPO-1	5	31.9 ± 1.8	14.6 ± 0.8	2.18 ± 0.01
	20	32.3 ± 1.9	14.2 ± 0.7	2.28 ± 0.06
SePSPO-2	5	30.8 ± 0.9	13.6 ± 0.7	2.27 ± 0.08
	20	34.1 ± 3.3	14.1 ± 1.1	2.41 ± 0.09

4. Discussion

Conventional food or agricultural products rich in bioactive substances have been scientifically investigated for their health benefits to the body. It was reported that the acid-soluble pectin isolated from edible okra (*Abelmoschus esculentus*) had both anti-inflammatory and anti-oxidant potentials, and thereby showed a capacity to reduce NO formation and inflammatory cytokines secretion in the LPS-treated RAW 264.7 macrophages [36]. Meanwhile, the pectin from okra stem had the anti-fatigue activity by increasing both blood glucose and glycogen levels in the body [37]. In addition, the okra polysaccharides had anti-inflammation via inhibiting the phosphorylation of IκB and p65 proteins and reducing the secretion of inflammatory cytokines in the LPS-treated RAW 264.7 macrophages [38,39]. Mulberry (*Morus atropurpurea*) fruits were regarded to have beneficial biofunctions [40]. For example, the polysaccharides from black mulberry could reduce ROS formation, improve mitochondrial function, and activate the Nrf2 signaling pathway [41], and the extracts from mulberry fruits showed the anti-cancer effect on the HepG2 cells because they were capable of suppressing cell growth and inducing cell apoptosis [42]. Yam (*Dioscorea Opposita Thunb.*), cultivated widely as food and medicinal materials in China, was also considered to possess various bioactivities [43]. It was revealed in the past studies that yam polysaccharides could ameliorate insulin resistance, decrease the contents of both low-density lipoproteins and total cholesterol [44], enhance the immune functions of the lymphocytes and macrophages [45], or exert anti-tumor effect on a B16 mouse melanoma model or anti-oxidant effect on the diabetic mice [46]. At the same time, special attention has been paid to some non-conventional and edible plants to reveal their health benefits. Hawthorn (*Crataegus pinnatifida*) is known as traditional Chinese medicine, and is also a non-conventional and edible plant. These bioactive components in hawthorn, such as flavonoids and terpenoids, were revealed to have physiological functions in the cardiovascular, digestive, and endocrine systems [47]. It was also found that hawthorn polyphenols could reduce the risks of diabetes [48], or display an immuno-modulatory effect on the lymphocytes [49]. In this study, the purslane polysaccharides and two selenylated products were clarified with a clear immune modulation on the two model cells. Thus, the obtained results in this study were consistent with those results mentioned in these studies, demonstrating a beneficial function of purslane or its components to humans.

Immune modulation is one of the most fundamental biological activities of food components, considering that foodborne toxins and pathogens have various adverse effects in the body [50]. Several bioactive substances, especially those from food materials, such as polysaccharides, curcumin, and bovine lactoferrin, are considered capable of regulating or activating the immune system. Thus, various indices are used to clarify the potential immune modulation of various bioactive substances. For example, ROS formation was applied as an evaluation index [51], when aiming to identify the immune activities of polyphenol curcumin. When assessing the immuno-modulatory potentials of the target lily polysaccharides, both macrophage phagocytosis and NO content were evaluated [29]. Cytokines are the proteins essential to the cellular immunity and, thus, used widely in immune evaluation as the critical indices. For example, the lily polysaccharides and curcumin were found able to increase the secretion levels of IL-6, IL-1β, and TNF-α in the RAW 264.7 macrophages, thereby being suggested as having immuno-modulatory function [29,52]. Moreover, due to their ability to promote splenocyte proliferation, enhance the secretion levels of IL-2 and IFN-γ, and regulate T lymphocyte subpopulations, the polysaccharides from raspberry (*Rubus chingii* Hu) fruits were revealed with immune modulation in the murine splenocytes [53]. In addition, to assess the immune potential of a Zn-fortified bovine lactoferrin, the ratio of $CD4^+/CD8^+$ in the lymphocytes was used as an index [33]. In consistence with these mentioned studies, this study thereby employed these indices, such as growth proliferation, phagocytic activity, cytokine secretion, and T lymphocyte subpopulations, to evaluate the different immuno-modulatory activities of the purslane polysaccharides and two selenylated products, and, subsequently, to reveal whether the performed chemical selenylation, as well as the obtained selenylation extent, had impact on the immune function of the purslane polysaccharides. Additionally, it was found that the selenylated polysaccharides from *Artemisia sphaerocephala* exerted an increased immuno-modulation on the macrophages by promoting the expression of p-p38, p-JNK1/2, and p-ERK1/2 significantly [54], while those from *Hericium erinaceus* showed higher immune activity to the bone marrow-derived dendritic cells via activating both MAPK and NF-κB signaling pathways [55]. Thereby, it is suggested that how the target samples exerted the immuno-modulation in the two immune cells should be investigated in future study, to clarify possible signaling pathways involved in the mentioned biofunction.

Polysaccharides also can be modified to induce corresponding changes in their structures, properties, and more important biological functions [56]. For example, when roundhead wormwood (*Artemisia sphaerocephala*) polysaccharides were sulfated using the chlorosulfuric acid/pyridine system, the resultant sulfated polysaccharides showed higher antitumor activity against the A549, HepG2, and Hela cells via exerting anti-proliferation effect [57]. When the hydroxyl groups of the polysaccharides isolated from the wild pepper (*Morchella angusticeps* peck) were converted to acetyl groups by a chemical esterification, it was found that the used polysaccharide acetylation induced improved immune effect and anti-inflammatory capacity in the macrophages [58]. If the polysaccharides from bast willow (*Cyclocarya paliurus*) were converted into corresponding carboxymethyl products using the chloroacetic acid and ethyl alcohol, their solubility and anti-oxidant properties were promoted [59]. In addition, a chemical phosphorylation of the asparagus (*Radix Cyathulae officinalis*) polysaccharides could yield higher anti-viral activity against the canine parvovirus [60]. In general, the target chemical selenylation is a covalent combination of polysaccharide substrates and H_2SeO_3 (generated from the Na_2SeO_3-HNO_3 system) [61]. When the polysaccharides from alfalfa roots (*Medicago Sativa* L.) were selenylated using the Na_2SeO_3-HNO_3 system, Se content of the selenylated polysaccharides was about 320 mg/kg, and the selenylated polysaccharides in the HepG2 cells showed enhanced anti-tumor and anti-oxidant effects [62]. Moreover, the obtained infrared spectrometry results showed that two new absorption peaks at respective 925 cm^{-1} and 840 cm^{-1} were observed, which indicated the corresponding C-O-Se stretching vibration and Se-O asymmetric stretching [62]. Using the Na_2SeO_3-HNO_3 system, the selenylated lily polysaccharides had a very high level of Se element (39.78 g/kg) and enhanced immune function in the lymphocytes; meanwhile, the critical Se-O-C and O-Se-O bonds were detected [63]. Sharing conclusion consistence with these mentioned studies, the results of this study also confirmed that purslane polysaccharides could be selenylated by the Na_2SeO_3-HNO_3 system to two relative lower selenylation extents (Se contents of 753.8 and 1325.1 mg/kg) and, subsequently, were endowed with improved immune potentials in the two immune cells. Thus, the recent results provided extra evidence for the potential application of this chemical selenylation for the interested polysaccharide modification. However, whether this selenylation also caused a change in other bioactivities for the modified polysaccharides was not investigated in this study. Thus, it is suggested that other bioactivity changes should be an interesting topic in future studies.

5. Conclusions

The present results highlighted that the used chemical selenylation of soluble purslane polysaccharides to the two selenylation extents could induce an enhanced immuno-modulatory effect on both macrophages and splenocytes efficiently, although the unreacted purslane polysaccharides themselves also had immune modulation in the two immune cells. Overall, the selenylated polysaccharides were more active in the macrophages to promote cell growth, phagocytic activity, and the secretion of three cytokines, including TNF-α, IL-1β, and IL-6, or more able in the mitogen-stimulated or non-stimulated splenocytes to enhance cell proliferation, as well as to elevate IFN-γ secretion but suppress IL-4 secretion. The selenylated polysaccharides also possessed a higher capacity of increasing $CD4^+/CD8^+$ ratio in the T lymphocytes. Generally, higher selenylation extent could endow the selenylated polysaccharides with higher bioactivity to the two immune cells, while higher polysaccharide doses also caused greater immune modulation in the cells. Thus, this chemical selenylation might be an applicable technique to enhance the bioactivities, such as immune modulation of natural polysaccharides. It is also suggested that other bioactivity changes of the polysaccharides arisen from the performed chemical selenylation need future investigation.

Author Contributions: Y.-R.L. conducted the experiment, analyzed, and organized data, and wrote the manuscript draft. Q.-Y.G., L.-Y.L. and Z.-M.T. analyzed data. Q.Z. wrote the manuscript draft. X.-H.Z. provided funds, put forward the idea, designed the study, and revised the manuscript. All authors have read and agreed to the published version of the manuscript.

Funding: This study was funded by the Scientific Research Foundation of Guangdong University of Petrochemical Technology (Project No. 2020rc026).

Institutional Review Board Statement: The study was conducted according to the guidelines of the Declaration of Helsinki, and approved by the Animal Ethics Committee of Northeast Agricultural University (NEAUEC20190316, 2019).

Informed Consent Statement: Not applicable.

Data Availability Statement: All data are contained within the article.

Acknowledgments: The authors thank the anonymous reviewers for their valuable advice.

Conflicts of Interest: The authors declare no conflict of interest.

References

1. Nemzer, B.; Al-Taher, F.; Abshiru, N. Phytochemical composition and nutritional value of different plant parts in two cultivated and wild purslane (*Portulaca oleracea* L.) genotypes. *Food Chem.* **2020**, *320*, e126621. [CrossRef] [PubMed]
2. Alfwuaires, M.A.; Algefare, A.I.; Afkar, E.; Salam, S.A.; Abd El-Moaty, H.I.; Badr, G.M. Immunomodulatory assessment of *Portulaca oleracea* L. extract in a mouse model of colitis. *Biomed. Pharmacother.* **2021**, *143*, e112148. [CrossRef] [PubMed]
3. Moneim, A.E.A.; Dkhil, M.A.; Al-Quraishy, S. The potential role of *Portulaca oleracea* as a neuroprotective agent in rotenone-induced neurotoxicity and apoptosis in the brain of rats. *Pestic. Biochem. Physiol.* **2013**, *105*, 203–212. [CrossRef]
4. Uddin, M.K.; Juraimi, A.S.; Ali, M.E.; Ismail, M.R. Evaluation of antioxidant properties and mineral composition of purslane (*Portulaca oleracea* L.) at different growth stages. *Int. J. Mol. Sci.* **2012**, *13*, 10257–10267. [CrossRef] [PubMed]
5. Erkan, N. Antioxidant activity and phenolic compounds of fractions from *Portulaca oleracea* L. *Food Chem.* **2012**, *133*, 775–781. [CrossRef]
6. Bai, Y.; Zang, X.L.; Ma, J.S.; Xu, G.Y. Anti-diabetic effect of *portulaca oleracea* L. polysaccharide and its mechanism in diabetic rats. *Int. J. Mol. Sci.* **2016**, *17*, 1201. [CrossRef] [PubMed]
7. Ren, Y.; Bai, Y.P.; Zhang, Z.; Cai, W.L.; Flores, A.D. The preparation and structure analysis methods of natural polysaccharides of plants and fungi: A review of recent development. *Molecules* **2019**, *24*, 3122. [CrossRef]
8. Qu, J.L.; Huang, P.; Zhang, L.; Qiu, Y.; Qi, H.; Leng, A.J.; Shang, D. Hepatoprotective effect of plant polysaccharides from natural resources: A review of the mechanisms and structure-activity relationship. *Int. J. Biol. Macromol.* **2020**, *161*, 24–34. [CrossRef]
9. Yuan, R.S.; Tao, X.; Liang, S.; Pan, Y.; Li, H.; Sun, J.H.; Ju, W.B.; Li, X.Y.; Chen, J.G.; Wang, C.M. Protective effect of acidic polysaccharide from *Schisandra chinensis* on acute ethanol-induced liver injury through reducing CYP2E1-dependent oxidative stress. *Biomed. Pharmacother.* **2018**, *99*, 537–542. [CrossRef] [PubMed]
10. Fan, S.T.; Huang, X.J.; Wang, S.N.; Li, C.; Zhang, Z.H.; Xie, M.Y.; Nie, S.P. Combinatorial usage of fungal polysaccharides from *Cordyceps sinensis* and *Ganoderma atrum* ameliorate drug-induced liver injury in mice. *Food Chem. Toxicol.* **2018**, *119*, 66–72. [CrossRef]
11. He, N.W.; Zhai, X.C.; Zhang, X.B.; Zhang, X.W.; Wang, X.J. Extraction, purification and characterization of water-soluble polysaccharides from green walnut husk with anti-oxidant and anti-proliferative capacities. *Process Biochem.* **2020**, *88*, 170–179. [CrossRef]
12. Li, G.Q.; Chen, P.F.; Zhao, Y.T.; Zeng, Q.H.; Ou, S.Y.; Zhang, Y.H.; Wang, P.C.; Chen, N.H.; Ou, J.Y. Isolation, structural characterization and anti-oxidant activity of a novel polysaccharide from garlic bolt. *Carbohydr. Polym.* **2021**, *267*, e118194. [CrossRef]
13. Yuan, Y.; Zhou, J.H.; Zheng, Y.F.; Xu, Z.C.; Li, Y.Q.; Zhou, S.; Zhang, C.S. Beneficial effects of polysaccharide-rich extracts from *Apocynum venetum* leaves on hypoglycemic and gut microbiota in type 2 diabetic mice. *Biomed. Pharmacother.* **2020**, *127*, e110182. [CrossRef]
14. Lopez-Legarda, X.; Arboleda-Echavarria, C.; Parra-Saldivar, R.; Rostro-Alanis, M.; Alzate, J.F.; Villa-Pulgarin, J.A.; Segura-Sanchez, F. Biotechnological production, characterization and in vitro antitumor activity of polysaccharides from a native strain of *Lentinus crinitus*. *Int. J. Biol. Macromol.* **2020**, *164*, 3133–3144. [CrossRef]
15. Zeng, Y.J.; Xiang, Y.F.; Sheng, R.L.; Tomas, H.; Rodrigues, J.; Gu, Z.W.; Zhang, H.; Gong, Q.Y.; Luo, K. Polysaccharide-based nanomedicines for cancer immunotherapy: A review. *Bioact. Mater.* **2021**, *6*, 3358–3382. [CrossRef]
16. Wang, J.L.; Li, W.Y.; Huang, X.; Liu, Y.; Li, Q.; Zheng, Z.M.; Wang, K.P. A polysaccharide from *Lentinus edodes* inhibits human colon cancer cell proliferation and suppresses tumor growth in athymic nude mice. *Oncotarget* **2017**, *8*, 610–623. [CrossRef]
17. Li, Y.Q.; Hu, Y.K.; Shi, S.J.; Jiang, L. Evaluation of antioxidant and immuno-enhancing activities of purslane polysaccharides in gastric cancer rats. *Int. J. Biol. Macromol.* **2014**, *68*, 113–116. [CrossRef] [PubMed]
18. Zheng, D.P.; Liwinski, T.; Elinav, E. Interaction between microbiota and immunity in health and disease. *Cell Res.* **2020**, *30*, 492–506. [CrossRef]
19. Guo, H.J.; Chen, C.Y.; Yan, X.; Li, Y.Y.; Wen, X.B.; You, C.H.; Monroig, O.; Tocher, D.R.; Wang, S.Q. Effects of different dietary oil sources on growth performance, antioxidant capacity and lipid deposition of juvenile golden pompano *Trachinotus ovatus*. *Aquaculture* **2021**, *530*, e735923. [CrossRef]
20. Cai, J.Y.; Li, X.; Du, H.M.; Jiang, C.F.; Xu, S.L.; Cao, Y. Immunomodulatory significance of natural peptides in mammalians: Promising agents for medical application. *Immunobiology* **2020**, *225*, e151936. [CrossRef] [PubMed]
21. Reyes-Diaz, A.; Gonzalez-Cordova, A.F.; Hernandez-Mendoza, A.; Reyes-Diaz, R.; Vallejo-Cordoba, B. Immunomodulation by hydrolysates and peptides derived from milk proteins. *Int. J. Dairy Technol.* **2018**, *71*, 1–9. [CrossRef]
22. Yang, J.X.; Maria, T.C.; Zhou, B.; Xiao, F.L.; Li, Y. Quercetin improves immune function in *Arbor Acre* broilers through activation of NF-κB signaling pathway. *Poult. Sci.* **2020**, *99*, 906–913. [CrossRef]
23. Loftus, R.M.; Finlay, D.K. Immunometabolism: Cellular metabolism turns immune regulator. *J. Biol. Chem.* **2016**, *291*, 1–10. [CrossRef]
24. Gong, F.Y.; Li, F.L.; Zhang, L.L.; Li, J.; Zhang, Z.; Wang, G.Y. Hypoglycemic effects of crude polysaccharide from purslane. *Int. J. Mol. Sci.* **2009**, *10*, 880–888. [CrossRef] [PubMed]
25. Chen, Y.; Shen, Z.; Chen, X. Evaluation of free radicals scavenging and immunity-modulatory activities of purslane polysaccharides. *Int. J. Biol. Macromol.* **2009**, *45*, 448–452. [CrossRef]

26. Xu, Y.; Wu, Y.J.; Sun, P.L.; Zhang, F.M.; Linhardt, R.J.; Zhang, A.Q. Chemically modified polysaccharides: Synthesis, characterization, structure activity relationships of action. *Int. J. Biol. Macromol.* **2019**, *132*, 970–977. [CrossRef]
27. Jagtap, R.; Maher, W. Determination of selenium species in biota with an emphasis on animal tissues by HPLC-ICP-MS. *Microchem. J.* **2016**, *124*, 422–529. [CrossRef]
28. Li, G.R.; Xiang, Y.; Zhao, J.; Chang, J.M. *Saccharum Alhagi* polysaccharide-1 and -2 promote the immunocompetence of RAW 264.7 macrophages In Vitro. *Exp. Ther. Med.* **2018**, *15*, 3556–3562. [CrossRef] [PubMed]
29. Pan, G.F.; Xie, Z.W.; Huang, S.X.; Tai, Y.L.; Cai, Q.S.; Jiang, W.; Sun, J.M.; Yuan, Y. Immune-enhancing effects of polysaccharides extracted from *Lilium lancifolium* Thunb. *Int. Immunopharmacol.* **2017**, *52*, 119–126. [CrossRef] [PubMed]
30. Cui, H.Y.; Wang, C.L.; Wang, Y.R.; Li, Z.J.; Zhang, Y.N. The polysaccharide isolated from *Pleurotus nebrodensis* (PN-S) shows immune-stimulating activity in RAW 264.7 macrophages. *Chin. J. Nat. Med.* **2015**, *13*, 355–360. [CrossRef]
31. Wang, C.L.; Cui, H.Y.; Wang, Y.R.; Wang, Z.F.; Li, Z.J.; Chen, M.H.; Li, F.J. Bidirectional immunomodulatory activities of polysaccharides purified from *Pleurotus nebrodensis*. *Inflammation* **2014**, *37*, 83–93. [CrossRef]
32. Silva, M.S.D.; Santos, J.D.; Alves, A.J.; da Silva, R.M.F.; Santos, B.S.; de Lorena, V.M.B.; de Oliveira, G.G.; de Melo, C.M.L.; Goes, A.J.D. Evaluation of the immunomodulatory effect against splenocytes of Balb/c mice of biflorin obtained from *Capraria biflora* by a new isolation method. *Rev. Bras. Farmacogn.-Braz. J. Pharmacogn.* **2019**, *29*, 464–469. [CrossRef]
33. Zhao, H.J.; Zhao, X.H. Effect of the Zn supplementation on immuno-modulatory activities of bovine lactoferrin in the murine splenocytes and RAW 264.7 macrophages. *Biol. Trace Elem. Res.* **2019**, *192*, 287–296. [CrossRef]
34. Wang, Y.J.; Wan, D.L.; Li, Q.M.; Zha, X.Q.; Luo, J.P. Structural characteristics and immunostimulatory activities of a new polysaccharide from *Dendrobium fimbriatum* Hook. *Food Funct.* **2021**, *12*, 3057–3068. [CrossRef] [PubMed]
35. FAO; WHO. *Vitamin and Mineral Requirements in Human Nutrition*, 2nd ed.; World Health Organization: Bangkok, Thailand, 2004; pp. 194–216.
36. Li, Y.; Deng, Y.; Li, Z.; Liu, Z.Q.; Piao, M.Z.; Cui, X.Q. Composition, physicochemical properties, and anti-fatigue activity of water-soluble okra (*Abelmoschus esculentus*) stem pectins. *Int. J. Biol. Macromol.* **2020**, *165*, 2630–2639. [CrossRef]
37. Xiong, B.Y.; Zhang, W.C.; Wu, Z.Y.; Liu, R.; Yang, C.Y.; Hui, A.L.; Huang, X.S.; Xian, Z.J. Preparation, characterization, anti-oxidant and anti-inflammatory activities of acid-soluble pectin from okra (*Abelmoschus esculentus* L.). *Int. J. Biol. Macromol.* **2021**, *181*, 824–834. [CrossRef]
38. Liu, Y.; Ye, Y.F.; Hu, X.B.; Wang, J.H. Structural characterization and anti-inflammatory activity of a polysaccharide from the lignified okra. *Carbohydr. Polym.* **2021**, *265*, e118081. [CrossRef] [PubMed]
39. Olawuyi, I.F.; Lee, W.Y. Structural characterization, functional properties and antioxidant activities of polysaccharide extract obtained from okra leaves (*Abelmoschus esculentus*). *Food Chem.* **2021**, *354*, e129437. [CrossRef] [PubMed]
40. Bhattacharjya, D.; Sadat, A.; Dam, P.; Buccini, D.F.; Mandal, A.K. Current concepts and prospects of mulberry fruits for nutraceutical and medicinal benefits. *Curr. Opin. Food Sci.* **2021**, *40*, 121–135. [CrossRef]
41. Chen, W.; Lu, Y.; Hu, D.W.; Mo, J.L.; Ni, J.D. Black mulberry (*Morus nigra* L.) polysaccharide ameliorates palmitate-induced lipotoxicity in hepatocytes by activating Nrf2 signaling pathway. *Int. J. Biol. Macromol.* **2021**, *172*, 394–407. [CrossRef]
42. Cheng, K.C.; Wang, C.J.; Chang, Y.C.; Hung, T.W.; Lai, C.J.; Kuo, C.W.; Huang, H.P. Mulberry fruits extracts induce apoptosis and autophagy of liver cancer cell and prevent hepatocarcinogenesis in vivo. *J. Food Drug Anal.* **2020**, *28*, 84–93. [CrossRef]
43. Wu, Z.G.; Jiang, W.; Nitin, M.; Bao, X.Q.; Chen, S.L.; Tao, Z.M. Characterizing diversity based on nutritional and bioactive compositions of yam germplasm (*Dioscorea* spp.) commonly cultivated in China. *J. Food Drug Anal.* **2016**, *24*, 367–375. [CrossRef]
44. Cheng, Z.Y.; Hu, M.; Tao, J.; Yang, H.; Yan, P.J.; An, G.P.; Wang, H.L. The protective effects of Chinese yam polysaccharide against obesity-induced insulin resistance. *J. Funct. Foods* **2019**, *55*, 238–247. [CrossRef]
45. Hao, L.X.; Zhao, X.H. Immunomodulatory potentials of the water-soluble yam (*Dioscorea opposita* Thunb) polysaccharides for the normal and cyclophosphamide-suppressed mice. *Food Agric. Immunol.* **2016**, *27*, 667–677. [CrossRef]
46. Huang, R.; Xie, J.H.; Yu, Y.; Shen, M.Y. Recent progress in the research of yam mucilage polysaccharides: Isolation, structure and bioactivities. *Int. J. Biol. Macromol.* **2020**, *155*, 1262–1269. [CrossRef]
47. Wu, J.Q.; Peng, W.; Qin, R.X.; Zhou, H. *Crataegus pinnatifida*: Chemical constituents, pharmacology, and potential applications. *Molecules* **2014**, *19*, 1685–1712. [CrossRef] [PubMed]
48. Liu, S.W.; Yu, J.C.; Fu, M.F.; Wang, X.F.; Chang, X.D. Regulatory effects of hawthorn polyphenols on hyperglycemic, inflammatory, insulin resistance responses, and alleviation of aortic injury in type 2 diabetic rats. *Food Res. Int.* **2021**, *142*, e110239. [CrossRef] [PubMed]
49. Lis, M.; Szczypka, M.; Suszko-Pawlowska, A.; Sokol-Letowska, A.; Kucharska, A.; Obminska-Mrukowicz, B. Hawthorn (*Crataegus monogyna*) phenolic extract modulates lymphocyte subsets and humoral immune response in mice. *Planta Med.* **2020**, *86*, 160–168. [CrossRef] [PubMed]
50. Schepetkin, I.A.; Quinn, M.T. Botanical polysaccharides: Macrophage immunomodulation and therapeutic potential. *Int. Immunopharmacol.* **2006**, *6*, 317–333. [CrossRef] [PubMed]
51. Ye, S.Z.; Li, S.W.; Ma, Y.; Hu, D.; Xiao, F. Curcumin hinders PBDE-47-induced neutrophil extracellular traps release via Nrf2-associated ROS inhibition. *Ecotoxicol. Environ. Saf.* **2021**, *225*, e112779. [CrossRef]
52. Jagetia, G.C.; Aggarwal, B.B. "Spicing up" of the immune system by curcumin. *J. Clin. Immunol.* **2007**, *27*, 19–35. [CrossRef]
53. Su, X.; Zhao, M.; Fu, X.; Ma, X.; Xu, W.; Hu, S. Immunomodulatory activity of purified polysaccharides from *Rubus chingii* Hu fruits in lymphocytes and its molecular mechanisms. *J. Funct. Foods* **2021**, *87*, e104785. [CrossRef]

54. Li, R.; Qin, X.J.; Liu, S.; Zhang, X.Y.; Zeng, X.R.; Guo, H.Y.; Wang, T.; Zhang, Y.D.; Zhang, J.P.; Zhang, J.; et al. HNMP HSO_4 catalyzed synthesis of selenized polysaccharide and its immunomodulatory effect on RAW 264.7 cells via MAPKs pathway. *Int. J. Biol. Macromol.* **2020**, *160*, 1066–1077. [CrossRef]
55. Qin, T.; Ren, Z.; Huang, Y.; Song, Y.; Lin, D.; Li, J.; Ma, Y.; Wu, X.; Qiu, F.; Xiao, Q. Selenizing *Hericium erinaceus* polysaccharides induces dendritic cells maturation through MAPK and NF-kappa B signaling pathways. *Int. J. Biol. Macromol.* **2017**, *97*, 287–298. [CrossRef]
56. Li, S.J.; Xiong, Q.P.; Lai, X.P.; Li, X.; Wan, M.; Zhang, J.N.; Yan, Y.J.; Cao, M.; Lu, L.; Guan, J.M.; et al. Molecular modification of polysaccharides and resulting bioactivities. *Compr. Rev. Food Sci. Food Saf.* **2016**, *15*, 237–250. [CrossRef]
57. Wang, J.L.; Bao, A.J.; Meng, X.H.; Guo, H.Y.; Zhang, Y.D.; Zhao, Y.L.; Kong, W.B.; Liang, J.Y.; Yao, J.; Zhang, J. An efficient approach to prepare sulfated polysaccharide and evaluation of anti-tumor activities in vitro. *Carbohydr. Polym.* **2018**, *184*, 366–375. [CrossRef] [PubMed]
58. Yang, Y.X.; Chen, J.L.; Lei, L.; Li, F.H.; Tang, Y.; Yuan, Y.; Zhang, Y.Q.; Wu, S.R.; Yin, R.; Ming, J. Acetylation of polysaccharide from *Morchella angusticeps* peck enhances its immune activation and anti-inflammatory activities in macrophage RAW 264.7 cells. *Food Chem. Toxicol.* **2019**, *125*, 38–45. [CrossRef]
59. Wang, Z.J.; Xie, J.H.; Shen, M.Y.; Tang, W.; Wang, H.; Nie, S.P.; Xie, M.Y. Carboxymethylation of polysaccharide from *Cyclocarya paliurus* and their characterization and antioxidant properties evaluation. *Carbohydr. Polym.* **2016**, *136*, 988–994. [CrossRef] [PubMed]
60. Feng, H.B.; Fan, J.; Yang, S.P.; Zhao, X.L.; Yi, X. Antiviral activity of phosphorylated *Radix Cyathulae officinalis* polysaccharide against canine parvovirus in vitro. *Int. J. Biol. Macromol.* **2017**, *99*, 511–518. [CrossRef]
61. Cheng, L.Z.; Wang, Y.F.; He, X.X.; Wei, X.L. Preparation, structural characterization and bioactivities of Se-containing polysaccharide: A review. *Int. J. Biol. Macromol.* **2018**, *120*, 82–92. [CrossRef]
62. Gao, P.Y.; Bian, J.; Xu, S.S.; Liu, C.F.; Sun, Y.Q.; Zhang, G.L.; Li, D.Q.; Liu, X.G. Structural features, selenization modification, antioxidant and anti-tumor effects of polysaccharides from alfalfa roots. *Int. J. Biol. Macromol.* **2020**, *149*, 207–214. [CrossRef] [PubMed]
63. Hou, R.; Chen, J.; Yue, C.; Li, X.; Liu, J.; Gao, Z.; Liu, C.; Lu, Y.; Wang, D.; Li, H.; et al. Modification of lily polysaccharide by selenylation and the immune-enhancing activity. *Carbohydr. Polym.* **2016**, *142*, 73–81. [CrossRef] [PubMed]

foods

MDPI

Article

Changes in Gel Structure and Chemical Interactions of *Hypophthalmichthys molitrix* Surimi Gels: Effect of Setting Process and Different Starch Addition

Xin Jiang [1], Qing Chen [1], Naiyong Xiao [1], Yufan Du [1], Qian Feng [1] and Wenzheng Shi [1,2,*]

[1] College of Food Sciences & Technology, Shanghai Ocean University, Shanghai 201306, China; 13053523375@163.com (X.J.); talina_chen@126.com (Q.C.); Xny931215@163.com (N.X.); yfdu122433@163.com (Y.D.); 18361206059@163.com (Q.F.)

[2] National Research and Development Center for Processing Technology of Freshwater Aquatic Products (Shanghai), Shanghai 201306, China

* Correspondence: wzshi@shou.edu.cn; Tel.: +86-156-9216-5859

Abstract: The modifications of histological properties and chemical forces on heated surimi gels with starch addition (0–12 g/100 g surimi) were investigated. Two types of heating processes (direct heating and two-step heating) were carried out on surimi gels in order to reveal the effect of setting on mixed matrices. The results of transverse relaxation time showed less immobile water and free water converted into bound water in a matrix subjected to the setting process. Scanning electron microscope and light microscopy images revealed inefficient starch-swelling in two-step heated gels. Chemical interactions and forces in direct cooking gels were more vulnerable to starch addition, resulting in significant decreases in hydrophobic interaction and sulfhydryl content ($p < 0.05$). With the increment of starch, the disulfide stretching vibrations of the gauche–gauche–gauche conformation were reduced in both gel matrices. The structural variations of different components collectively resulted in changes in texture profile analysis and water holding capacity. Overall, the results demonstrated that starch addition had a great and positive effect on the weak gel matrix by direct heating.

Keywords: surimi; starch; setting; water migration; microstructure; chemical interactions

Citation: Jiang, X.; Chen, Q.; Xiao, N.; Du, Y.; Feng, Q.; Shi, W. Changes in Gel Structure and Chemical Interactions of *Hypophthalmichthys molitrix* Surimi Gels: Effect of Setting Process and Different Starch Addition. *Foods* **2022**, *11*, 9. https://doi.org/10.3390/foods11010009

Academic Editors: Jianhua Xie, Yanjun Zhang and Hansong Yu

Received: 22 November 2021
Accepted: 15 December 2021
Published: 21 December 2021

1. Introduction

Surimi-based products have become an increasingly consumed food with prominent characteristics of convenience, special flavor, and unique texture [1]. In addition, the deep processing technology gave it the qualities of excellent nutritional value and high digestibility [2]. Due to the continuous overexploitation of marine fishing resources, low-value or cultured fish has been paid more attention as a potential alternative raw material for surimi production [3]. Silver carp (*Hypophthalmichthys molitrix*), widely cultured in China, is considered a low-commercial value fish due to its muddy flavor and higher by-product content [4]. However, due to its rapid growth, high yield, and low price, silver carp could replace sea fish as raw material for surimi [5]. Through the processes of rinsing, dehydration, and defatting, the myofibrillar protein could be well retained, and more fishy compounds were removed, which improved the flavor characteristics to a certain extent. Raw surimi could be further processed into prepared foods or ready-to-eat foods in order to satisfy different taste demands for surimi-based products, which realize the value addition of silver carp [6].

Throughout the entire process of thermal gel formation, heat treatment is an important step in determining the quality of the gel matrix [7]. Before heating, free myosin molecules and actomyosin complexes were dispersed, while the network structure complex reinforced by actomyosin was completed as the temperature rose [8]. Based on the unique property of fish protein, heating processes can be divided into direct heating and two-step heating, with significant differences in protein networks between these two treatments [9]. Direct heating treatment provided the gel with soft structure but might cause the formation of protein aggregates in different sizes [10]. In case of two-step heating, an extension of setting time at 4–40 °C prior to heating at 90 °C could strengthen gel properties, which was widely used in making kamaboko [6]. The first setting step, in which more available cross-linking sites become accessible, ensured a gradual sol-gel transition so that an orderly initial protein network formed [11,12]. Subsequent secondary heating at high temperature is used for the final production of thermal gels [13]. Compared with direct heated gels, the two-step heated gels enabled the gel structure to remain more compact even when suffering from physical forces [10]. Hence, the setting process played a vital role in gel processing. However, the addition of some exogenous substances during the actual processing was also necessarily for helping improve gel texture.

Starch, as a popular food additive for improving gels, plays an essential role in reducing the factory processing costs of surimi-based products and meeting consumer taste needs [14,15]. According to the "packing effect" raised by Kong et al. [16], the starch granules swelled and then exerted more pressure on the gel matrix, forming a firm and cohesive gel. Starch swelling in the surimi matrix increased the hardness of the gels, thereby improving overall gel properties. It also found that starch did not crosslink with the surimi protein, whereas it could change the chemical interaction in the surimi matrix [17]. Thus far, studies have mainly emphasized the effects of different starches in improving gel strength [17–19]. Nevertheless, the surimi gel network also affected the filling effect of starch in the mixture, which was rarely reported in research. It was hypothesized that the characteristics of the starch-containing surimi matrix could be affected by the setting processing, showing different change trends in various properties. Accordingly, two heating processes (direct heating process and two-step heating process) were set up to detect the different properties of the mixed matrix and to compare the systematic effect of the setting process on the heated starch–surimi matrix.

Therefore, the research aimed to elucidate water migration, microstructure, chemical interactions, and surimi protein structure in the starch–surimi matrix comprehensively. Physical properties, such as texture profile analysis, whiteness, and water holding capacity, were also detected, providing insight into the combined effect of thermal processing and exogenous additives on texture characteristics.

2. Materials and Methods

2.1. Materials and Reagents

Silver carp frozen surimi (AAA grade, cryoprotectants were 6% sucrose and 0.25% polyphosphate) obtained from Jinli Fishery Food Co., Ltd. (Honghu, China) was cut into pieces weighing about 200 g and stored at −20 °C after vacuum sealing. The moisture and crude protein contents were 75.20% and 14.37%, respectively. Native potato starch was purchased from Hangzhou Starpro Starch Co., Ltd. (Hangzhou, China). The P0006C Detergent Compatible Bradford Protein Assay Kit was obtained from Shanghai Beyotime Biotechnology Co., Ltd. (Shanghai, China). The other chemicals were analytical grade, with the exception of KBr (spectrography), and purchased from Sinopharm Chemical Reagent Co., Ltd. (Shanghai, China).

2.2. Sample Preparation

Frozen surimi was thawed at 4 °C overnight and cut into small pieces. An amount of 200 g of semi-thawed silver carp surimi was blended for 2 min and then mixed with 5 g of NaCl for 4 min. During the blending process, moisture content was adjusted to 80% with iced water that was also used to keep the temperature of mixtures below 10 °C. Native potato starch (0 g, 3 g, 6 g, 9 g, or 12 g/100 g surimi) was added to the surimi and mixed in blender (AM-CG108-1, Appliance Co. of America, Zhuhai, China) for 6 min. Then, starch–surimi combinations were filled into the plastic tubes with a diameter of 25 mm and heated in two different processes. In the direct heating process, the tubes were heated in a water bath at 90 °C for 30 min, obtaining cooking gel (CG). In the two-step heating process, a combination of preheating at 40 °C for 1 h and cooking at 90 °C for 30 min was carried out to obtain setting-cooking gel (SCG) [6]. After heating, all samples were stored at 4 °C for 12 h.

2.3. Low Field Nuclear Magnetic Resonance (LF-NMR)

The relaxation time and moisture distribution were measured by a Niumag Pulsed NMR analyzer (MesoMR23-060H-I, Niumag Electric Co., Shanghai, China). Gel samples were cut into the cylinders with a height of 20 mm, and a CPMG (Carr-Purcell-Meiboom-Gill) pulse sequence was carried out [20,21]. The CPMG parameters are listed in Table 1.

Table 1. CPMG parameters.

CPMG	SF	SW	RFD	Tw	RG1	DRG1	PRG	P1	P2	TE	NECH	NS
Parameter	21	200	0.08	2500	10	3	1	20	40	0.3	8000	8

SF: magnet frequency, MHz; SW: spectral width, kHz; RFD: radio frequency delay, ms; Tw: recycle delay, ms; RG1: regulate analog gain 1, db; DRG1: regulate digital gain 1; PRG: pre-amplified receiver gain; P1: 90° pulse lengths, μs; P2: 180° pulse lengths, μs; TE: pulse gaps between π and π, ms; NECH: echo number; NS: scanning number.

2.4. Scanning Electron Microscope (SEM)

The surimi gels were cut into 3 mm × 3 mm × 1.5 mm pieces and fixed with glutaraldehyde (2.5%, v/v) for 14 h at 4 °C. The fixed samples were rinsed with 0.1 M phosphoric acid buffer (pH 7.2–7.4) three times. After that, the samples were dehydrated with a serious of ethanol solution (30%, 50%, 70%, 80%, 90%, and 100%) and then replaced with tert-butanol solution (absolute ethanol: tert-butanol = 3:1, 1:1, 1:3, 0:1). The dehydrated samples were dried by using a freeze dryer (SCIENTZ-10N, Ningbo Scientz Biotechnology Co., Ltd., Ningbo, China) and sputter-coated with gold. The microstructures were analyzed by an SEM instrument (Hitachi SU5000, Hitachi High-Tech Co., Ltd., Shanghai, China) at an acceleration voltage of 5 kV.

2.5. Light Microscopy (LM)

Surimi gels were cut into 5 mm cubes and immersed with 10% formalin fixture solution (1:10, w/v) for more than 24 h. After being dehydrated with gradient alcohol (75%, 85%, 90%, 95%, and 100%), the samples were placed in xylene followed by wax leaching using an embedding machine (JB-P5, Wuhan Junjie Electronics Co., Ltd., Wuhan, China). After cooling at −20 °C, the paraffin slice was cut with a thickness of 4 μm carried out by a tissue spreader (KD-P, Zhejiang Kehua Instrument Co., Ltd., Jinhua, China). The dewaxed slices were stained using the periodic acid-Schiff (PAS) method according to the method of Jia et al. [22]. The samples were observed using a light microscope (MS500W, Shanghai Meizs Precision Instrument Co., Ltd., Shanghai, China).

2.6. Determination of Non-Covalent Bonds

The chemical interactions or bonds were determined according to the method of Yan et al. [23] with slight modifications. Briefly, 1 g of surimi gels was added 10 mL 0.05 M NaCl (A), 0.6 M NaCl (B), 0.6 M NaCl + 1.5 M urea (C), and 0.6 M NaCl + 8 M urea (D) and homogenized for 2 min by homogenizer (Shanghai Fokker Equipment Co. Ltd., Shanghai, China). The mixtures were placed at 4 °C for 1 h and subsequently centrifuged at 15,000× g for 10 min by refrigerated centrifuge (CR21GIII, Hitachi High-Tech Co., Ltd., Shanghai, China).

The protein concentrations of supernatant collected were determined by using the Biuret method [24]. The contents of nonspecific associations (A), ionic bonds (B), hydrogen bonds (C), and hydrophobic interactions (D) were determined by the concentrations of soluble protein in supernatants.

2.7. Determination of Total Sulfhydryl Groups

The measurement of total SH content was described by Zhang et al. [25]. According to the following equation, the content of SH was calculated.

$$\text{SH groups } \mu\text{mol/ g protein} = \frac{A \times D}{\varepsilon \times d \times c} \times 10^6 \quad (1)$$

In the equation, A and D represent absorbance at 412 nm and the dilution ratio, respectively; ε refers to molar extinction coefficient, 13,600 L/(mol·cm); d is the thickness of cuvette, 1 cm; and c is the protein concentration, 4 mg/mL.

2.8. Fourier Transform Infrared (FT-IR) Spectroscopy

The surimi gels were freeze-dried and ground into powder. An amount of 1.5 mg of powder was mixed with 150 mg KBr and then pressed into a transparent sheet. The spectrums of the matrix were observed by a Nicolet iS5 FT-IR spectrometer (Thermo Fisher Instruments Co., Ltd., Shanghai, China). Samples were then scanned 32 times from 400 to 4000 cm^{-1} with a resolution of 4 cm^{-1} [26]. Relative intensity was analyzed by OMNIC 9.2 software (Thermo Fisher Scientific Inc., Waltham, MA, USA) and PeakFit 4.12 software (Systat Software Inc., San Jose, CA, USA).

2.9. Texture Profile Analysis (TPA)

Equilibrium temperature to 25 °C, the gel matrix was cut into cylinders (20 mm × 20 mm), and TPA was determined by using a TA-XT Plus texture analyzer (Stable Micro Systems Ltd., Godalming, UK) with P/50 cylindrical probe. The pre-test speed and test speed were both 1 $mm \cdot s^{-1}$, the compression percentage was 40%, and the trigger force was 5 g.

2.10. Determination of Whiteness

The surimi gels were cut with a thickness of 10 mm. The L* (lightness), a* (redness/greenness), and b* (yellowness/blueness) values of gels were determined by colorimeter (CR-400, Konica Minolta Japan, Inc., Tokyo, Japan). Whiteness was calculated as follows.

$$\text{Whiteness} = 100 - \left[(100 - L^*)^2 + a^{*2} + b^{*2}\right]^{1/2} \quad (2)$$

2.11. Determination of Water Holding Capacity (WHC)

All samples were cut into 3 g with a thickness of about 5 mm, weighed (W_1) and then placed into three layers of filter paper. After centrifugation of 15 min (5000× *g*, 4 °C), the gels were quickly weighed again (W_2). The equation of water holding capacity is defined as follows:

$$\text{Centrifugal loss } (\%) = \frac{W_2}{W_1} \times 100 \quad (3)$$

$$\text{Water holding capacity } (\%) = \frac{M - CL}{M} \times 100 \quad (4)$$

where CL is the centrifugal loss of surimi gel after heating, and M is the moisture content determined by drying the sample at 105 °C until constant weight.

2.12. Statistical Analysis

All experiments were performed at least three times, and the results were presented as the mean ± standard deviation (SD). Statistical Package for Social Science (SPSS 26, SPSS Inc., Chicago, IL, USA) was used to carry out all statistical analysis with the significance level set at 0.05 based on one-way analyses of variance (ANOVA). Significant differences were evaluated by Duncan's multiple range test ($p < 0.05$).

3. Results and Discussion

3.1. Changes of Relaxation Time and Water Distribution of Starch–Surimi Gels with Non-Setting or Setting Effect

The results of LF-NMR can effectively determine water mobility and distribution in gel matrix [27]. T_{2b} (1–20 ms), T_{21} (20–300 ms), and T_{22} (400–2000 ms) represented the transverse relaxation time of bound water, immobilized water, and free water (Figure 1a, b), respectively. It was obvious the relaxation time of different water migrated. As shown in Figure 1c, d, the peak relaxation time of bound water (T_{P2b}) in gels without starch was 2.39 ms (CG) and 1.94 ms (SCG), between which there was a significant difference ($p < 0.05$) that might be related to the gel structure heated by dissimilar processes. T_{P2b} increased by starch addition and showed a positive correlation with starch content. The migration of T_{P2b} was relevant to different water mobility bound by the protein and starch. The peak relaxation time of immobile water (T_{P21}) dropped from 101.71 ms to 71.71 ms in CG, and from 94.84 ms to 65.36 ms in SCG. The shorter the relaxation time (T_2), the stronger the binding force of water molecules to matrix [12]. Thus, the water mobility of bound water significantly increased with starch incorporation ($p < 0.05$), while the mobility of non-mobile and free water decreased in both matrices. This was consistent with the investigation by Li et al. [28], who found that starch showed a better restriction capacity on the free motion of water molecules. However, the relaxation time between the two types of gels had a significant difference ($p < 0.05$), indicating that starch-swelling differed in gel matrices with different heating treatments.

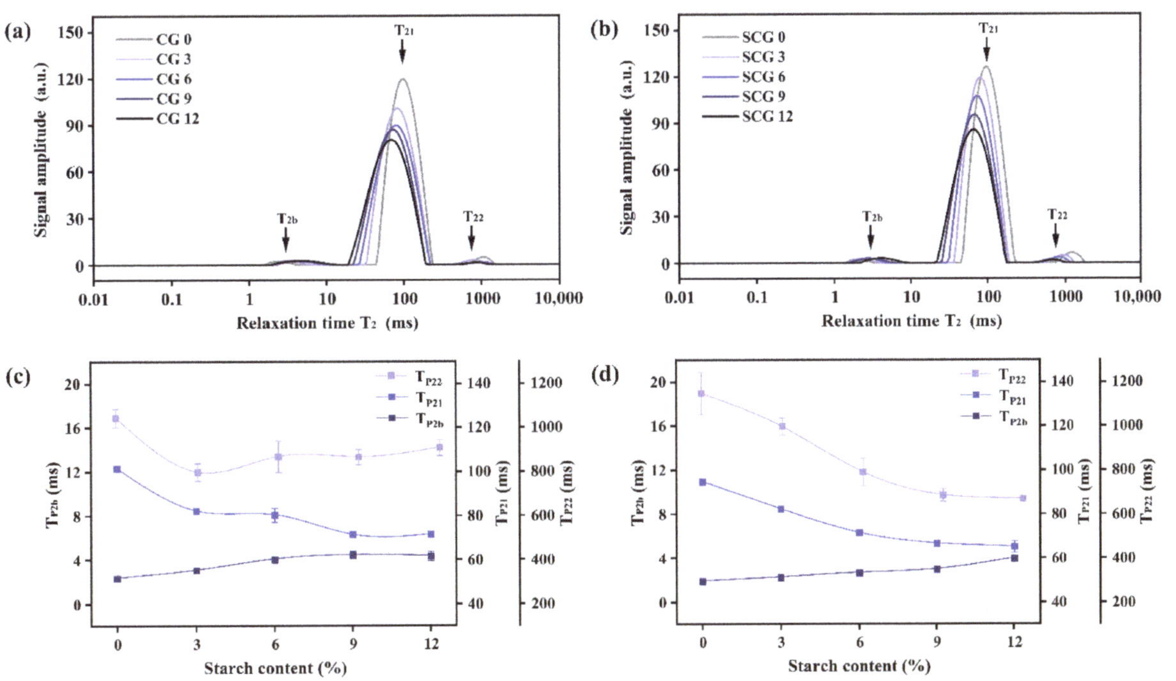

Figure 1. T_2 relaxation time (**a**,**b**) and water mobility (**c**,**d**) of starch–surimi matrix in direct heating process and two-step heating process. CG: gel obtained by direct heating; SCG: gel obtained by two-step heating. 0: without starch; 3, 6, 9, and 12: incorporation with 3%, 6%, 9%, and 12% starch.

The relative moisture content was observed by peak area proportions (P_{T2b}, P_{T21}, and P_{T22}) in Figure 2. It was discovered that immobile water made up the majority of the primary proportion in surimi gel, followed by bound water and free water. P_{T21} showed the relative content of immobile water that increased initially with the addition of starch, resulting in a reduction in free water in both CG and SCG. The P_{T21} of CG and SCG reached a maximum at 6% and 9% starch content, respectively. Subsequently, it decreased and was possibly associated with the dehydration of surimi gels suffered by the compression from starch swelling. Moreover, starch incorporation reduced the relative surimi content in the entire matrix, causing a decrease in immobile water content held by surimi gel network. Notably, in the non-starch containing gels, P_{T21} of CG and SCG was 97.20% and 96.27%, showing that the gel matrix had higher P_{T21} without preincubation at 40 °C. The changes might presumably contribute to overheated time and the squeeze by gel formation in SCG, resulting in a transition from immobile water to free water. Although direct heating formed a poor gel, the juiciness mouthfeel of surimi products improved [29]. Since the hydroxyl groups in starch bound more water, the matrix with starch incorporation significantly differed in P_{T22} ($p < 0.05$). However, P_{T2b} significantly increased with the continuous addition of starch ($p < 0.05$), especially in CG. With 12% starch, P_{T2b} of CG increased by 173.15% compared to 113.89% in SCG. Among the above changes, it indicated that water absorbed by starch existed in the form of immobile water and bound water. Moreover, it was inferred that the starch in direct cooking gels showed better swelling states, which resulted in the phenomena where the internal starch structure tended to easily combine with water.

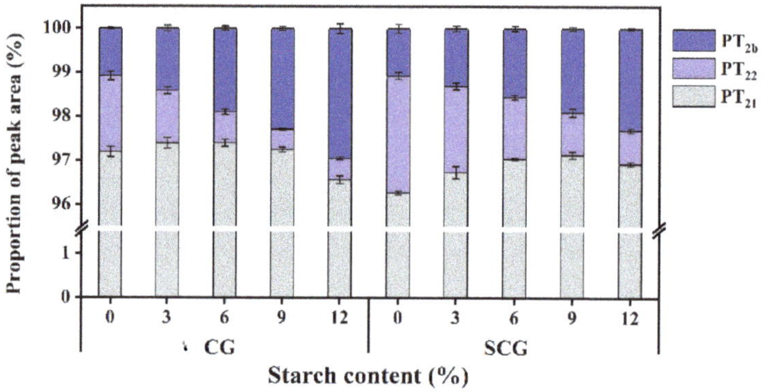

Figure 2. Peak area proportion (P_{T2b}, P_{T21}, and P_{T22}) of different water in starch–surimi matrix subjected to direct heating and two-step heating. Caption: see Figure 1.

3.2. *Changes in Tissue Histology of Starch–Surimi Gels with Non-Setting or Setting Effect*

3.2.1. Light Micrograph

The morphology of potato starch differed appreciably between direct heated gels and two-step heated gels (Figure 3). It was observed that a nearly spherical morphology of starch was shown in SCG, whereas a relatively irregular state was observed in CG. Starch showed non-obstructive expansion in CG with 3% or 6% starch content. The striation structures inside of starch granules were intact with an increase in starch content, showing that the swelling degree of starch decreased in CG. In comparison, it was observed that potato starch granules in SCG were almost circular, especially in the low starch-containing matrices. In addition, large granule starch could not realize valuable swelling in two-step heating. However, regardless of non-setting or setting samples, small starch granules showed extended states, which was more conducive for applying starch in the surimi matrix.

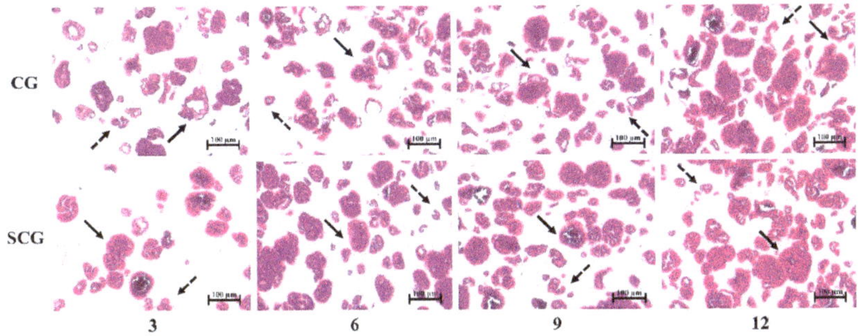

Figure 3. Light microscopy of cooking gels and setting-cooking gels at four starch contents (3%, 6%, 9%, and 12%). ↗ indicates large granule starch; ⇢ indicates small granule starch. Caption: see Figure 1.

3.2.2. Scanning Electron Micrograph

Overall microstructures of starch–surimi gels were observed at 300× to reflect the morphology of starch and surimi matrix after cooking (Figure 4). The gelatinization temperature of potato starch ranged from 56 to 66 °C, and the amorphous region extended after reaching this temperature range [30]. For CG, as the temperature rose above 45–50 °C, the suwari was partially disrupted, resulting in modori formation and unhindered starch swelling [31]. Moreover, the loose gel network might be a key factor of the irregular swelling of starch. In contrast with non-setting gels, SCG showed more substantial resistance to starch swelling that avoided water loss in the gel network structure. It might contribute to the setting step, which resulted in the formation of a stable surimi gel network in SCG. Thus, it could be determined that the setting effect promoted cross-linking in surimi, hindered water absorption, and broke hydrogen bonds in the striations of starch granules.

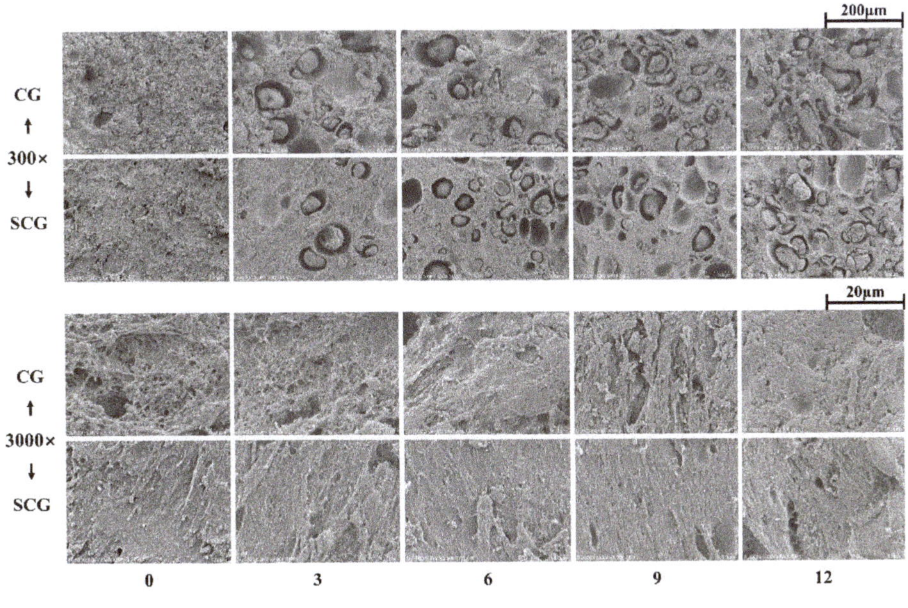

Figure 4. Microstructure micrographs (magnification ×300, ×3000) of starch–surimi matrix subjected with different heating processes: 0, 3, 6, 9, and 12 represented 0%, 3%, 6%, 9%, and 12% starch content, respectively. Caption: See Figure 1.

Viewed at large magnifications, non-starch containing CG exhibited a coarse matrix with large cavities, as shown in Figure 4. However, the addition of the starch made cavities disappear, resulting in more compact surimi. According to the results from Kong et al. [16], these shrunk surimi networks were considered to be responsible for the "packing effect." Meanwhile, for SCG, no significant changes were observed, and smooth structures were shown throughout the addition of starch.

3.3. *Changes in Chemical Interactions of Starch–Surimi Gels with Non-Setting or Setting Effect*

3.3.1. Non-Covalent Bonds

Non-covalent bonds play essential roles in supporting the three-dimensional structure and enhancing gel strength [32]. All of the non-covalent bonds (including the non-specific associations, ionic bonds, hydrogen bonds, and hydrophobic interactions) differed between the two types of heating samples (Table 2). The non-specific associations had significant increases in both CG and SCG ($p < 0.05$) with the increase in starch content, which resulted from a weak link with low molecular proteins in the gel [33]. Li et al. [28] discovered that the non-specific associations of myofibrillar protein were affected by starch with a larger particle size. Hence, the increase in non-specific associations might be attributed to the expansion of potato starch. The ionic bonds generated by electrostatic interactions between peptides also increased [34]. Hydrogen bonds could enhance the rigidity of the gel, but it might be easily destroyed under high temperature. However, the reduction in hydrogen bonds, in turn, allowed the hydration of exposed peptide backbones and, thus, was important in stabilizing bound water [2]. Overall, these weak forces uniformly increased with the increment of starch content, containing non-specific associations, ionic bonds, and hydrogen bonds ($p < 0.05$).

Table 2. Non-covalent bonds and total sulfhydryl groups of starch–surimi mixtures subjected with different heating processes.

Samples	Starch Content (%)	Non-Covalent Bonds (g/L)				Total Sulfhydryl Groups (μmol/g Protein)
		Non-Specific Associations	Ionic Bonds	Hydrogen Bonds	Hydrophobic Interactions	
CG	0	0.99 ± 0.03 ^{Be}	0.85 ± 0.02 ^{Bd}	0.39 ± 0.02 ^{Bd}	7.71 ± 0.35 ^{Aa}	25.17 ± 0.96 ^{Aa}
	3	1.22 ± 0.01 ^{Bd}	1.33 ± 0.02 ^{Ac}	0.63 ± 0.03 ^{Bc}	7.75 ± 0.27 ^{Aa}	20.07 ± 1.75 ^{Ab}
	6	1.30 ± 0.02 ^{Bc}	1.90 ± 0.03 ^{Ab}	1.21 ± 0.01 ^{Bb}	5.64 ± 0.10 ^{Ab}	19.44 ± 2.60 ^{Ab}
	9	2.00 ± 0.01 ^{Bb}	1.84 ± 0.03 ^{Ab}	2.19 ± 0.04 ^{Aa}	4.84 ± 0.10 ^{Ac}	18.29 ± 1.34 ^{Bb}
	12	2.60 ± 0.06 ^{Ba}	2.16 ± 0.06 ^{Aa}	2.21 ± 0.02 ^{Ba}	3.94 ± 0.10 ^{Ad}	12.55 ± 1.10 ^{Bc}
SCG	0	1.58 ± 0.05 ^{Ae}	1.03 ± 0.03 ^{Ad}	1.16 ± 0.02 ^{Ae}	2.95 ± 0.09 ^{Be}	25.62 ± 1.91 ^{Aa}
	3	1.97 ± 0.03 ^{Ad}	1.21 ± 0.02 ^{Bc}	1.82 ± 0.01 ^{Ad}	4.06 ± 0.04 ^{Ba}	21.03 ± 2.02 ^{Abc}
	6	2.43 ± 0.02 ^{Ac}	1.25 ± 0.03 ^{Bc}	2.69 ± 0.04 ^{Aa}	3.47 ± 0.02 ^{Bb}	17.97 ± 1.75 ^{Ac}
	9	2.96 ± 0.04 ^{Ab}	1.33 ± 0.02 ^{Bb}	2.19 ± 0.01 ^{Ac}	3.41 ± 0.03 ^{Bc}	23.20 ± 3.09 ^{Aab}
	12	4.07 ± 0.05 ^{Aa}	1.56 ± 0.01 ^{Ba}	2.36 ± 0.03 ^{Ab}	3.17 ± 0.03 ^{Bd}	22.43 ± 1.89 ^{Aab}

Uppercase letters indicate significant difference ($p < 0.05$) between different heating processes, lowercase letters indicate the difference between gels with different starch content ($p < 0.05$), and the values are expressed as mean ± SD.

Hydrophobic interactions predominate in the gel matrix compared with other non-covalent bonds [17], which are produced by the unfolding action of the protein (above 60 °C) and the exposure of the hydrophobic core. With an increase in starch content, hydrophobic interactions declined in CG (Table 2). The unfolding action of the surimi protein structure during heating contributed to the formation of hydrophobic interactions. It could be presumed that the hydrophilic groups absorbed water, resulting in starch swelling as the temperature rose. Subjected to setting treatment, the hydrophobic interactions tended to increase with relatively minor fluctuations. Compared with CG, the effect of starch addition on hydrophobic interactions in SCG weakened. This was related to the elastic SCG containing disulfide bonds, which had a high extrusion resistance to starch [35].

3.3.2. Total Sulfhydryl Groups

Sulfhydryl groups buried in protein are exposed during heating process and subsequently generate disulfide cross-linking [23]. The disulfide bonds display a type of rheological behavior known as rubber elasticity, and they are critical to maintaining network stability [34]. The concentration of total sulfhydryl groups decreased ceaselessly in CG with increased starch content, as shown in Table 2. It was ascribed to internal changes of protein aggregates in which more peptide chains were unfolded and sulfhydryl groups were exposed. Subsequently, the cross-linking of -SH occurred followed by the formation of more disulfide bonds [36]. Nevertheless, faced with stress from starch, SCG showed minor changes in total sulfhydryl groups, contributing to a more stable structure formed by low-temperature preincubation [37].

3.4. FT-IR Spectroscopy Analysis of Starch–Surimi Gels with Non-Setting or Setting Effect

3.4.1. Amide Bands of Protein

The gel matrix can be analyzed by using FT-IR spectroscopy to detect functional groups associated with intramolecular and intermolecular structures [38]. The amide bands of proteins have several distinct vibrational modes, including amide I, II, and III. Amide I (1600–1700 cm^{-1}) resulted primarily from ν (C=O) and δ(N–H), whereas amide II and III (1480–1580 cm^{-1}; 1200–1350 cm^{-1}) originated from ν(C–N) and δ(N–H) [39]. Among them, amide I was the most useful in reflecting secondary and tertiary structures [9]. Generally, the α-helix, random coil, β-sheet, and β-turn structures correspond to 1650–1660 cm^{-1}, 1660–1665 cm^{-1}, 1665–1680 cm^{-1}, and –1680 cm^{-1} ranges of the amide I band, respectively [40]. Li et al. [25] discovered that the starch did not cause significant shifts in amide bands (Figure 5a, b, Table 3). Non-setting gels and setting gels both showed peak values of amide I at 1654 cm^{-1}, suggesting that α-helix dominated the secondary structure of the protein in the starch–surimi matrix. Although starch could increase the density of the gel matrix and influence chemical interactions, it had little effect on the three-dimensional structure of proteins. In contrast, slight changes in α-helix and random coil occurred between CG and SCG, which was conducive to increasing hydrogen bonds. The α-helix of native and partially denatured proteins and β structures that formed during heating and cooling are both stabilized by hydrogen bonds [2]. Therefore, the secondary structure of surimi protein had no significant change caused by the external physical forces of starch.

Table 3. Secondary structure (amide I) of protein in different starch–surimi matrices.

Samples	Starch Content (%)	α-Helix (%)	Random Coil (%)	β-Sheet (%)	β-Turn (%)
CG	0	26.16 ± 0.75 Aa	26.49 ± 0.26 Aa	25.49 ± 0.18 Aa	21.85 ± 0.73 Aa
	3	26.89 ± 1.38 Ba	25.80 ± 1.13 Aa	25.20 ± 0.96 Aa	22.11 ± 2.84 Aa
	6	26.47 ± 1.07 Aa	26.03 ± 0.53 Aa	25.63 ± 0.61 Aa	21.88 ± 1.02 Aa
	9	27.20 ± 1.01 Ba	26.00 ± 0.23 Aa	25.05 ± 0.47 Aa	21.75 ± 0.35 Aa
	12	27.69 ± 1.03 Ba	25.97 ± 0.21 Aa	25.34 ± 0.16 Aa	21.00 ± 1.06 Aa
SCG	0	28.14 ± 2.39 Aa	25.03 ± 0.75 Ba	24.47 ± 1.32 Ab	22.36 ± 2.34 Aa
	3	28.98 ± 0.55 Aa	25.25 ± 0.32 Aa	24.92 ± 0.29 Aab	20.85 ± 0.26 Aa
	6	27.74 ± 1.19 Aa	24.82 ± 0.31 Ba	25.89 ± 0.15 Aa	21.55 ± 0.99 Aa
	9	29.12 ± 0.62 Aa	24.66 ± 0.48 Ba	25.65 ± 0.47 Aa	20.56 ± 0.49 Ba
	12	29.03 ± 0.38 Aa	25.17 ± 0.40 Ba	25.13 ± 0.31 Aab	20.67 ± 0.52 Aa

Uppercase letters indicate significant difference ($p < 0.05$) between different heating processes, lowercase letters indicate the difference between gels with different starch content ($p < 0.05$), and values are expressed as mean ± SD.

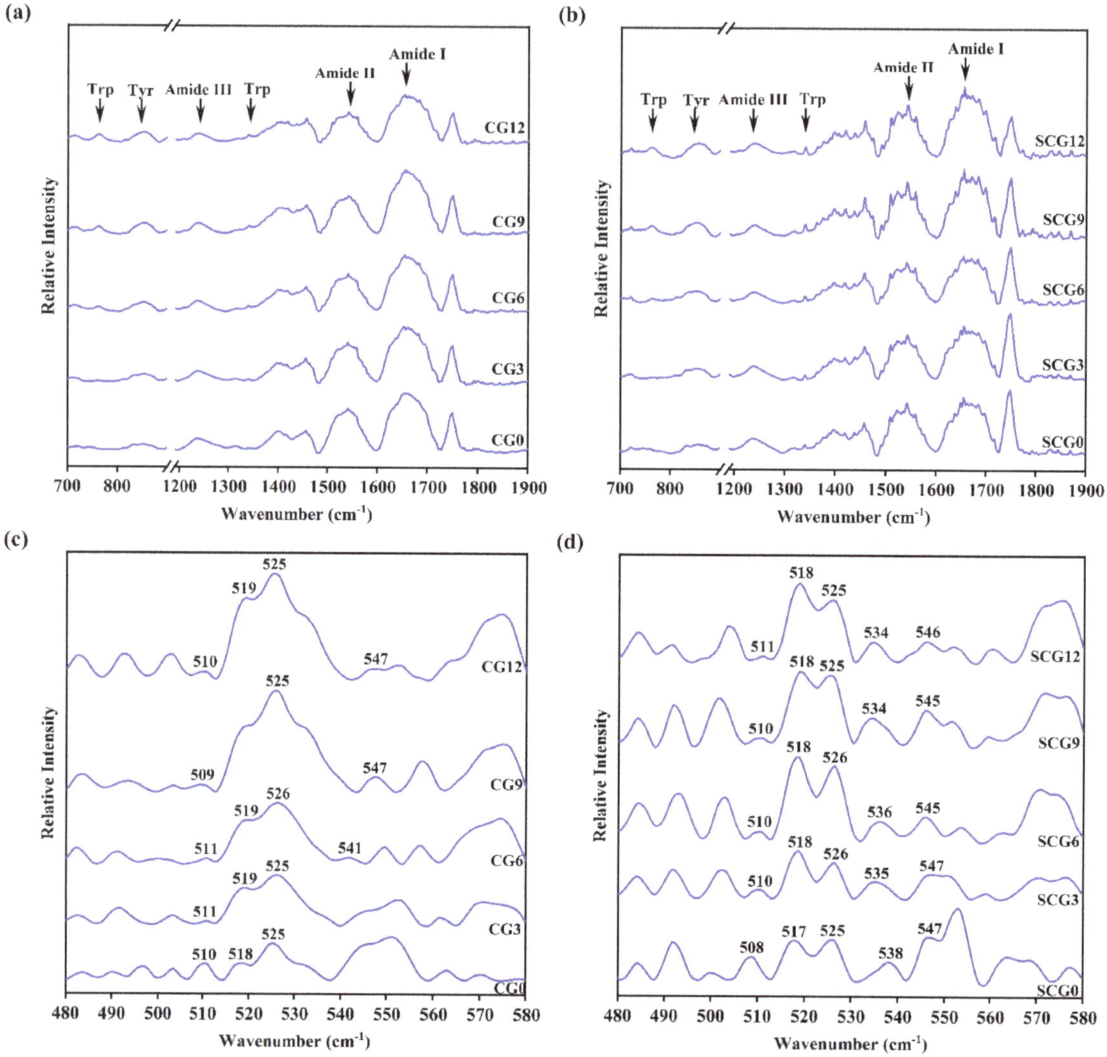

Figure 5. FT-IR spectroscopy (**a**,**b**) and sulfhydryl and disulfides (**c**,**d**) regions of starch–surimi matrix with different treatment. CG0–CG12: cooking gels with starch content (0–12%); SCG0–SCG12: setting-cooking gels with starch content (0–12%).

3.4.2. Tryptophan (Trp) Residue Bands and Tyrosine (Tyr) Doublet Bands

The vibrations near 760 and 1340 cm^{-1} present the microenvironment of Trp residues. Once the Trp residues buried in the hydrophobic environment were exposed to a polar environment, the intensity of bands showed upward trends [41]. The intensity of tryptophan residue bands increased slightly with the increment of starch content in Figure 4 possibly contributing to the exposure of Trp residues [42]. The starch granules occupied the matrix space, which promoted the unfolding of protein structure and then provided impetus to the exposure of the hydrophobic core. A similar result was found in the Tyr doublet bands, which were proposed as a means for determining whether the tyrosine residue was solvent-exposed or buried [43]. If the intensity at 850 cm^{-1} (I_{850}) was higher than I_{830}, this indicated that the Tyr residues at this time changed from the "buried" to the "exposed" state [44]. The distinct vibrations were exhibited differently at 830 and 850 cm^{-1} in both gels (Figure 5a, b). A decreased I_{830} and a concurrent increased I_{850} indicated that there were weak hydrophobic interactions among tyrosine residues. It could be demonstrated that protein solubility of CG increased due to reduced hydrophobic interactions. However, the change of hydrophobic amino acids between CG and SCG was not obvious. Therefore, the setting treatment of surimi did not hinder the effect of starch addition on the spatial structure of hydrophobic amino acids in surimi proteins.

3.4.3. S-S Stretching

The formation and interchange reactions of disulfide are important for non-reversible heat gelation [9]. Moreover, spectral features in the 500–550 cm^{-1} region were correlated with the structural parameters of the disulfide bands (Figure 5c, d). Stretching vibrations located at 510 cm^{-1}, 516–530 cm^{-1}, and 535–545 cm^{-1} have been assigned to gauche–gauche–gauche, gauche–gauche–trans, and trans–gauche–trans, where the –C–S–S–C– torsion angle of 90° was represented as gauche form and that of 180° was represented as trans form [45]. Due to its low potential energy, the gauche–gauche–gauche form was presumably the most stable among the others [46]. Quantitative analysis of FT-IR spectroscopy showed stable bands located at 510 cm^{-1} in CG and SCG and disappeared with the increment of starch addition. The weak bands vibrations of CG significantly increased at 516–530 cm^{-1}. It meant that not only disulfide bond conformation changed from gauche–gauche–gauche to gauche–gauche–trans but also new bonds emerged. The variation was consistent with the decrease in total sulfhydryl groups, indicating that more disulfide bonds formed in CG. Therefore, the conformational shift occurred by a combination with excessive starch, which was conducive to the breaking and reformation of disulfide bonds during the direct heating process. The vibrations rose near 518 cm^{-1} and 525 cm^{-1} in SCG, accompanied by a decrease in bonds near 510 cm^{-1}. Thus, the setting effect resulted in micro-changes in the conformation of the disulfide bonds in SCG. In brief, the results showed that the matrix after the setting process had a specific compressive capacity, causing fewer vibrations (S–S) changes.

3.5. *Changes in Gel Properties of Starch–Surimi Gels with Non-Setting or Setting Effect*

3.5.1. Texture Profile Analysis

Texture profile analysis simulates chewing twice to obtain the characteristic parameters of the gel matrix (Table 4) [47]. A significant increase in hardness was observed along with the elevated starch content due to reduced moisture content and starch swelling. The starch-absorbed water and swelled during the thermal processing, which increased pressure on the surimi gel [22]. No significant difference in springiness occurred in CG with the addition of starch ($p > 0.05$). However, springiness decreased along with the addition of starch in SCG. It is speculated that if the surimi gel had better gelling properties, then it would play a dominant role in the mixed matrix. Thus, the reduction in springiness in SCG was attributed to decreased surimi concentration. The improvement of cohesiveness and resilience in CG presumably contributed to granule-swelling of starch. The gelation and swelling of starch granules might cause an increase in hydration, which enabled better compatibility of protein networks and starch [14]. SCG showed reduced cohesiveness and resilience with excessive starch, which could be explained by how excessive expansion in starch granules attenuated the stability of the surimi matrix. Thus, potato starch filling could ameliorate the weak gel matrices by increasing texture properties such as hardness and cohesiveness. Nevertheless, the addition of potato starch increased the hardness of the gels but at the same time reduced springiness and cohesiveness. However, compared with a large content of starch, CG still showed lower textural properties, which was connected with the limited filling capacity of natural potato starch and the poor texture of unsetting surimi gel.

Table 4. Textural properties of starch–surimi mixtures subjected with different heating processes.

Samples	Starch Content (%)	Hardness (g)	Springiness	Cohesiveness	Resilience	Chewiness (g)
CG	0	476.41 ± 8.31 Be	0.94 ± 0.00 Ba	0.73 ± 0.00 Be	0.40 ± 0.00 Be	325.83 ± 4.29 Be
	3	757.80 ± 20.71 Bd	0.95 ± 0.01 Aa	0.73 ± 0.00 Bd	0.42 ± 0.00 Bd	528.43 ± 20.04 Bd
	6	1097.31 ± 11.53 Bc	0.96 ± 0.01 Aa	0.74 ± 0.00 Bc	0.43 ± 0.00 Bc	780.30 ± 2.10 Bc
	9	1382.00 ± 17.42 Bb	0.94 ± 0.02 Aa	0.75 ± 0.00 Bd	0.44 ± 0.00 Bb	983.86 ± 13.48 Bb
	12	1559.28 ± 17.46 Ba	0.91 ± 0.01 Ab	0.76 ± 0.00 Ba	0.45 ± 0.00 Ba	1090.18 ± 20.44 Ba
SCG	0	868.25 ± 20.61 Ae	0.97 ± 0.00 Aa	0.85 ± 0.01 Aa	0.59 ± 0.00 Aa	716.50 ± 18.45 Ae
	3	1023.91 ± 17.23 Ad	0.96 ± 0.02 Aab	0.84 ± 0.00 Ab	0.57 ± 0.00 Ab	826.50 ± 30.43 Ad
	6	1301.55 ± 27.10 Ac	0.94 ± 0.00 Abc	0.82 ± 0.00 Ac	0.54 ± 0.01 Ac	1005.24 ± 20.65 Ac
	9	1595.79 ± 4.69 Ab	0.93 ± 0.01 Ac	0.81 ± 0.00 Ad	0.52 ± 0.00 Ad	1200.29 ± 5.83 Ab
	12	1874.12 ± 26.36 Aa	0.92 ± 0.01 Ac	0.80 ± 0.00 Ae	0.51 ± 0.00 Ae	1383.62 ± 26.91 Aa

Uppercase letters indicate significant difference ($p < 0.05$) between different heating processes, lowercase letters indicate the difference between gels with different starch content ($p < 0.05$), and values are expressed as mean ± SD.

3.5.2. Whiteness

Whiteness is one of the most essential indications of surimi quality. The variation in whiteness is found to be correlated not only to the conformation of the gel network but also to the color of additives [48,49]. As presented in Table 5, significant increases in whiteness were found with starch contained in both two types of gels ($p < 0.05$). Moreover, two-step heated gels showed higher whiteness in different starch content compared with direct heating gels. The differences in two types of gels might be attributed to a restrictive effect on starch swelling. L*, a*, and b* values decreased as potato starch content increased and presented significant differences between CG and SCG.

Table 5. Whiteness of starch–surimi mixtures subjected with different heating processes.

Samples	Starch Content (%)	L *	a *	b *	Whiteness
CG	0	77.45 ± 0.17 Aa	−0.54 ± 0.12 Aa	6.06 ± 0.35 Aa	76.64 ± 0.14 Ba
	3	75.51 ± 0.31 Ab	−0.60 ± 0.08 Aa	5.77 ± 0.15 Ab	74.83 ± 0.28 Bb
	6	73.95 ± 0.39 Bc	−0.73 ± 0.05 Ab	5.63 ± 0.17 Ab	73.33 ± 0.36 Bc
	9	71.65 ± 0.17 Bd	−1.02 ± 0.08 Ac	4.96 ± 0.18 Ac	71.20 ± 0.19 Bd
	12	71.06 ± 0.28 Be	−1.06 ± 0.05 Ac	4.67 ± 0.06 Ad	70.67 ± 0.27 Be
SCG	0	77.46 ± 0.24 Aa	−0.83 ± 0.06 Ba	5.19 ± 0.16 Ba	76.86 ± 0.25 Aa
	3	75.77 ± 0.35 Ab	−0.88 ± 0.04 Bb	4.92 ± 0.22 Bb	75.26 ± 0.31 Ab
	6	74.87 ± 0.33 Ac	−0.97 ± 0.05 Bc	4.79 ± 0.14 Bb	74.40 ± 0.34 Ac
	9	73.32 ± 0.24 Ad	−1.07 ± 0.05 Bd	4.46 ± 0.20 Bc	72.93 ± 0.21 Ad
	12	71.84 ± 0.21 Ae	−1.28 ± 0.04 Be	4.07 ± 0.06 Bd	71.51 ± 0.21 Ae

Uppercase letters indicate significant difference ($p < 0.05$) between different heating processes, lowercase letters indicate the difference between gels with different starch content ($p < 0.05$), and values are expressed as mean ± SD. L *: lightness; a *: red-green value; b *: yellow-bule value.

3.5.3. Water Holding Capacity

WHC represents the stability of the mixed matrix suffering external forces, which indirectly reflects the compactness of surimi gels and water absorption capacity of starch [50]. As shown in Figure 6, the gel matrix presented a significant increase in WHC accompanied by adding starch in both heating processes ($p < 0.05$). This was in line with previous research by Mi et al. [17] and Luo et al. [14]. It has been revealed that starch had strong hydrophilicity and showed an effective swelling effect with larger particles, which increases the WHC of the gel matrix. The setting effect promoted the formation of the myofibrillar protein network, which could also reduce water loss caused by an external force, thereby increasing WHC. The matrix treated by direct heating showed better WHC with a continuous increase in starch content. It was related to the irregular expansion of starch confirmed by the microstructure (Figure 4). In addition, WHC was markedly correlated with the variations in peak area proportion and peak relaxation time [51]. Some possible assumptions could be introduced to describe the changes of WHC: (i) the increase in WHC was associated with the migration of free water by adding starch; (ii) starch showed better swelling states in CG on account of the higher T_{P2b} and lower T_{P22}; and (iii) the setting effect improved the stability of immobile water, whereas it limited starch swelling.

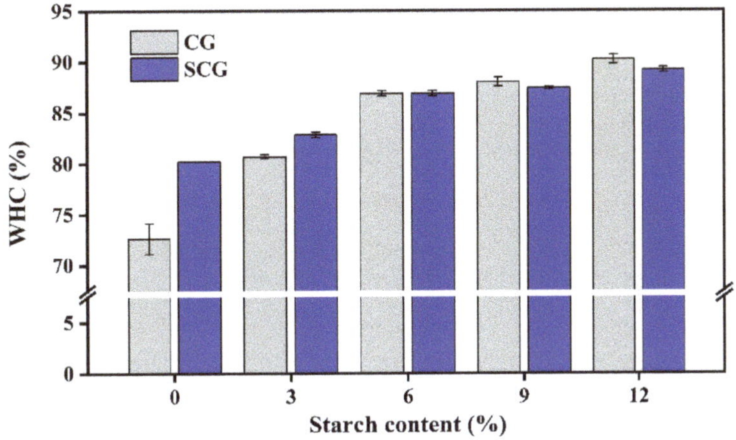

Figure 6. Water holding capacity of different water in starch–surimi matrix subjected to direct heating and two-step heating. Caption: see Figure 1.

4. Conclusions

Changes in the water migration and distribution, tissue microstructures, chemical interactions, protein spatial structure, and physical properties were studied. The addition of starch expanded and caused moisture migration, which was ascribed to the combination of free water and hydrophilic groups in starch. The competition between surimi and starch for water affected the chemical interactions and forces in surimi gel components, thereby indirectly resulting in changes in the microstructure and water holding properties of the gel matrix. For the direct cooking matrix, the porous surimi gel lost water when it was squeezed by starch, resulting in a reduction in pore size and immobile water inside. Moreover, starch was not conducive to the interactions of hydrophobic groups, whereas weak disulfide bonds were generated after adding starch. The enhancement of textural properties might be related to changes in protein structure. Subjected to the setting process, the matrix showed an ascendant trend in non-specific associations, ionic bonds, and hydrogen bonds, which was similar to non-setting gels. Setting-cooking gels with a compact structure were less affected by starch-swelling. However, adding excessive starch destroyed the integrity of SCG and decreased cohesiveness and resilience. Therefore, this study found that the addition of starch content had a better supporting effect on weak gels, which provided some insights for improving the quality of surimi-based products. Moreover, further studies are needed to improve the compatibility and active filling of starch in the two-step heating process.

Author Contributions: Conceptualization, X.J.; methodology, X.J. and N.X.; validation, N.X.; formal analysis, X.J., Q.C., and N.X.; investigation, X.J., Y.D., and Q.F.; resources, X.J., Q.C., and Y.D.; data curation, X.J. and Q.C.; writing—original draft, X.J.; writing—review and editing, Q.C., N.X., Y.D., and Q.F.; visualization, X.J.; supervision, W.S.; project administration, W.S.; funding acquisition, W.S. All authors have read and agreed to the published version of the manuscript.

Funding: This research was funded by the National Key R&D Program of China, grant number 2019YFD0902003.

Institutional Review Board Statement: Not applicable.

Informed Consent Statement: Informed consent was obtained from all subjects involved in the study.

Data Availability Statement: Not applicable.

Conflicts of Interest: The authors declare no conflict of interest.

References

1. Hou, S.-W.; Wei, W.; Wang, Y.; Gan, J.-H.; Lu, Y.; Tao, N.-P.; Wang, X.-C.; Liu, Y.; Xu, C.-H. Integrated recognition and quantitative detection of starch in surimi by infrared spectroscopy and spectroscopic imaging. *Spectrochim. Acta Part A Mol. Biomol. Spectrosc.* **2019**, *215*, 1–8. [CrossRef] [PubMed]
2. Park, J.W.; Lin, T.M.; Yongsawatdigul, J. New developments in manufacturing of Surimi and Surimi seafood. *Food Rev. Int.* **1997**, *13*, 577–610. [CrossRef]
3. Arfat, Y.A.; Benjakul, S. Gelling characteristics of surimi from yellow stripe trevally (*Selaroides leptolepis*). *Int. Aquat. Res.* **2012**, *4*, 5. [CrossRef]
4. Weng, W.; Zheng, W. Silver Carp (*Hypophthalmichthys molitrix*) Surimi Acid-Induced Gel Extract Characteristics: A Comparison with Heat-Induced Gel. *Int. J. Food Prop.* **2015**, *18*, 821–832. [CrossRef]
5. Alipour, H.J.; Rezaei, M.; Shabanpour, B.; Tabarsa, M. Effects of sulfated polysaccharides from green alga Ulva intestinalis on physicochemical properties and microstructure of silver carp surimi. *Food Hydrocoll.* **2018**, *74*, 87–96. [CrossRef]
6. Fang, M.; Xiong, S.; Hu, Y.; Yin, T.; You, J. In vitro pepsin digestion of silver carp (*Hypophthalmichthys molitrix*) surimi gels after cross-linking by Microbial Transglutaminase (MTGase). *Food Hydrocoll.* **2019**, *95*, 152–160. [CrossRef]
7. Matsuoka, Y.; Wan, J.; Ushio, H.; Watabe, S. Thermal gelation properties of white croaker, walleye pollack and deepsea bonefish surimi after suwari treatment at various temperatures. *Fish. Sci.* **2013**, *79*, 715–724. [CrossRef]
8. Yasui, T.; Ishioroshi, M.; Samejima, K. Effect of Actomyosin on Heat-induced Gelation of Myosin. *Agric. Biol. Chem.* **1982**, *46*, 1049–1059. [CrossRef]
9. Liu, R.; Zhao, S.-M.; Xie, B.-J.; Xiong, S.-B. Contribution of protein conformation and intermolecular bonds to fish and pork gelation properties. *Food Hydrocoll.* **2011**, *25*, 898–906. [CrossRef]
10. Jia, R.; Jiang, Q.; Kanda, M.; Tokiwa, J.; Nakazawa, N.; Osako, K.; Okazaki, E. Effects of heating processes on changes in ice crystal formation, water holding capacity, and physical properties of surimi gels during frozen storage. *Food Hydrocoll.* **2019**, *90*, 254–265. [CrossRef]
11. Luo, Y.; Shen, H.; Pan, D.; Bu, G. Gel properties of surimi from silver carp (*Hypophthalmichthys molitrix*) as affected by heat treatment and soy protein isolate. *Food Hydrocoll.* **2008**, *22*, 1513–1519. [CrossRef]
12. Chen, H.; Zou, Y.; Zhou, A.; Xiao, J.; Benjakul, S. Insight into the Effect of Ice Addition on the Gel Properties of *Nemipterus virgatus* Surimi Gel Combined with Water Migration. *Foods* **2021**, *10*, 1815. [CrossRef]
13. Watabe, S.; Ikeda, D.; Mashiro, T.; Kagetakubo, Y.; Takahashi, Y.; Uemura, M.; Mizusawa, N.; Koyama, H.; Yasumoto, K.; Jimbo, M.; et al. Suitability of Japanese codling as a raw material for surimi-based products revealed by primary sequence analysis of myosin heavy chain and thermal gel properties. *Fish. Sci.* **2020**, *86*, 711–719. [CrossRef]
14. Luo, H.; Guo, C.; Lin, L.; Si, Y.; Gao, X.; Xu, D.; Jia, R.; Yang, W. Combined Use of Rheology, LF-NMR, and MRI for Characterizing the Gel Properties of Hairtail Surimi with Potato Starch. *Food Bioprocess Technol.* **2020**, *13*, 637–647. [CrossRef]
15. Jia, R.; Katano, T.; Yoshimoto, Y.; Gao, Y.; Watanabe, Y.; Nakazawa, N.; Osako, K.; Okazaki, E. Sweet potato starch with low pasting temperature to improve the gelling quality of surimi gels after freezing. *Food Hydrocoll.* **2018**, *81*, 467–473. [CrossRef]
16. Kong, C.; Ogawa, H.; Iso, N. Temperature Dependency of Compression Properties of Fish-meat Gel as Affected by Added Starch. *J. Food Sci.* **1999**, *64*, 1048–1051. [CrossRef]
17. Mi, H.; Wang, C.; Su, Q.; Li, X.; Yi, S.; Li, J. The effect of modified starches on the gel properties and protein conformation of Nemipterus virgatus surimi. *J. Texture Stud.* **2019**, *50*, 571–581. [CrossRef] [PubMed]
18. Sun, F.; Huang, Q.; Hu, T.; Xiong, S.; Zhao, S. Effects and mechanism of modified starches on the gel properties of myofibrillar protein from grass carp. *Int. J. Biol. Macromol.* **2014**, *64*, 17–24. [CrossRef] [PubMed]
19. Yang, Z.; Wang, W.; Wang, H.; Ye, Q. Effects of a highly resistant rice starch and pre-incubation temperatures on the physicochemical properties of surimi gel from grass carp (*Ctenopharyn Odon Idellus*). *Food Chem.* **2014**, *145*, 212–219. [CrossRef]
20. Pan, T.; Guo, H.; Li, Y.; Song, J.; Ren, F. The effects of calcium chloride on the gel properties of porcine myosin–κ-carrageenan mixtures. *Food Hydrocoll.* **2017**, *63*, 467–477. [CrossRef]
21. Zhang, H.; She, W.; Li, L.; Wang, W. Effect of temperature rising inhibitor on autogenous shrinkage of cement pastes. *Constr. Build. Mater.* **2019**, *220*, 329–339. [CrossRef]
22. Jia, R.; Katano, T.; Yoshimoto, Y.; Gao, Y.; Nakazawa, N.; Osako, K.; Okazaki, E. Effect of small granules in potato starch and wheat starch on quality changes of direct heated surimi gels after freezing. *Food Hydrocoll.* **2020**, *104*, 105732. [CrossRef]

23. Yan, B.; Jiao, X.; Zhu, H.; Wang, Q.; Huang, J.; Zhao, J.; Cao, H.; Zhou, W.; Zhang, W.; Ye, W.; et al. Chemical interactions involved in microwave heat-induced surimi gel fortified with fish oil and its formation mechanism. *Food Hydrocoll.* **2020**, *105*, 105779. [CrossRef]
24. Reichardt, W.; Eckert, B. Zur Bestimmung des Proteingehaltes von Milch, Käse und Fleisch mit Hilfe der Biuret-Reaktion. *Food/Nahrung* **1991**, *35*, 731–738. [CrossRef] [PubMed]
25. Zhang, L.; Li, Q.; Shi, J.; Zhu, B.; Luo, Y. Changes in chemical interactions and gel properties of heat-induced surimi gels from silver carp (*Hypophthalmichthys molitrix*) fillets during setting and heating: Effects of different washing solutions. *Food Hydrocoll.* **2018**, *75*, 116–124. [CrossRef]
26. Singh, A.; Benjakul, S.; Prodpran, T.; Nuthong, P. Effect of Psyllium (*Plantago ovata Forks*) Husk on Characteristics, Rheological and Textural Properties of Threadfin Bream Surimi Gel. *Foods* **2021**, *10*, 1181. [CrossRef] [PubMed]
27. Zhang, Z.; Regenstein, J.M.; Zhou, P.; Yang, Y. Effects of high intensity ultrasound modification on physicochemical property and water in myofibrillar protein gel. *Ultrason. Sonochem.* **2017**, *34*, 960–967. [CrossRef] [PubMed]
28. Li, X.; Fan, M.; Huang, Q.; Zhao, S.; Xiong, S.; Yin, T.; Zhang, B. Effect of micro- and nano-starch on the gel properties, microstructure and water mobility of myofibrillar protein from grass carp. *Food Chem.* **2022**, *366*, 130579. [CrossRef] [PubMed]
29. Fang, Q.; Shi, L.; Ren, Z.; Hao, G.; Chen, J.; Weng, W. Effects of emulsified lard and TGase on gel properties of threadfin bream (*Nemipterus virgatus*) surimi. *LWT* **2021**, *146*, 111513. [CrossRef]
30. Kim, J.M.; Lee, C.M.; Hufnagel, L.A. Textural properties and structure of starch-reinforced surimi gels as affected by heat-setting. *Food Struct.* **1987**, *6*, 81–89.
31. Stone, A.; Stanley, D. Mechanisms of fish muscle gelation. *Food Res. Int.* **1992**, *25*, 381–388. [CrossRef]
32. Liu, Y.; Sun, Q.; Wei, S.; Xia, Q.; Pan, Y.; Ji, H.; Deng, C.; Hao, J.; Liu, S. Insight into the correlations among rheological behaviour, protein molecular structure and 3D printability during the processing of surimi from golden pompano (*Trachinotus ovatus*). *Food Chem.* **2022**, *371*, 131046. [CrossRef]
33. Jia, D.; Huang, Q.; Xiong, S. Chemical interactions and gel properties of black carp actomyosin affected by MTGase and their relationships. *Food Chem.* **2016**, *196*, 1180–1187. [CrossRef]
34. Li, Z.; Wang, J.; Zheng, B.; Guo, Z. Effects of high pressure processing on gelation properties and molecular forces of myosin containing deacetylated konjac glucomannan. *Food Chem.* **2019**, *291*, 117–125. [CrossRef] [PubMed]
35. Liu, H.; Gao, L.; Ren, Y.; Zhao, Q. Chemical Interactions and Protein Conformation Changes During Silver Carp (*Hypophthalmichthys molitrix*) Surimi Gel Formation. *Int. J. Food Prop.* **2014**, *17*, 1702–1713. [CrossRef]
36. Cao, H.; Fan, D.; Jiao, X.; Huang, J.; Zhao, J.; Yan, B.; Zhou, W.; Zhang, W.; Zhang, H. Effects of microwave combined with conduction heating on surimi quality and morphology. *J. Food Eng.* **2018**, *228*, 1–11. [CrossRef]
37. Chen, Y.; Xu, A.; Yang, R.; Jia, R.; Zhang, J.; Xu, D.; Yang, W. Chemical interactions and rheological properties of hairtail (*Trichiurus haumela*) surimi: Effects of chopping and pressure. *Food Biosci.* **2020**, *38*, 100781. [CrossRef]
38. Yu, W.; Wang, Z.; Pan, Y.; Jiang, P.; Pan, J.; Yu, C.; Dong, X. Effect of κ-carrageenan on quality improvement of 3D printed Hypophthalmichthys molitrix-sea cucumber compound surimi product. *LWT* **2021**, *154*, 112279. [CrossRef]
39. Fevzioglu, M.; Ozturk, O.K.; Hamaker, B.R.; Campanella, O.H. Quantitative approach to study secondary structure of proteins by FT-IR spectroscopy, using a model wheat gluten system. *Int. J. Biol. Macromol.* **2020**, *164*, 2753–2760. [CrossRef] [PubMed]
40. Zhuang, X.; Jiang, X.; Zhou, H.; Chen, Y.; Zhao, Y.; Yang, H.; Zhou, G. Insight into the mechanism of physicochemical influence by three polysaccharides on myofibrillar protein gelation. *Carbohydr. Polym.* **2020**, *229*, 115449. [CrossRef] [PubMed]
41. Wang, L.; Xia, M.; Zhou, Y.; Wang, X.; Ma, J.; Xiong, G.; Wang, L.; Wang, S.; Sun, W. Gel properties of grass carp myofibrillar protein modified by low-frequency magnetic field during two-stage water bath heating. *Food Hydrocoll.* **2020**, *107*, 105920. [CrossRef]
42. Sun, W.; Zhao, Q.; Zhao, M.; Yang, B.; Cui, C.; Ren, J. Structural Evaluation of Myofibrillar Proteins during Processing of Cantonese Sausage by Raman Spectroscopy. *J. Agric. Food Chem.* **2011**, *59*, 11070–11077. [CrossRef]
43. Herrero, A.; Carmona, P.; Cofrades, S.; Jimenez-Colmenero, F. Raman spectroscopic determination of structural changes in meat batters upon soy protein addition and heat treatment. *Food Res. Int.* **2008**, *41*, 765–772. [CrossRef]
44. Yu, D.; Zhang, X.; Zou, W.; Tang, H.; Yang, F.; Wang, L.; Elfalleh, W. Raman spectroscopy analysis of the effect of electrolysis treatment on the structure of soy protein isolate. *J. Food Meas. Charact.* **2021**, *15*, 1294–1300. [CrossRef]
45. Podstawka, E.; Ozaki, Y.; Proniewicz, L.M. Adsorption of S—S Containing Proteins on a Colloidal Silver Surface Studied by Surface-Enhanced Raman Spectroscopy. *Appl. Spectrosc.* **2004**, *58*, 1147–1156. [CrossRef]
46. Pfeuti, G.; Brown, L.S.; Longstaffe, J.G.; Peyronel, F.; Bureau, D.P.; Kiarie, E.G. Predicting the standardized ileal digestibility of crude protein in feather meal fed to broiler chickens using a pH-stat and a FT-Raman method. *Anim. Feed. Sci. Technol.* **2020**, *261*, 114340. [CrossRef]
47. Bourne, M.C. *Food Texture and Viscosity: Concept and Measurement*, 2nd ed.; Academic Press: Salt Lake City, UT, USA, 2002; pp. 182–186.
48. Petcharat, T.; Benjakul, S. Effect of gellan incorporation on gel properties of bigeye snapper surimi. *Food Hydrocoll.* **2018**, *77*, 746–753. [CrossRef]
49. Uresti, R.M.; Velazquez, G.; Ramírez, J.A.; Vázquez, M.; Torres, J.A. Effect of high-pressure treatments on mechanical and functional properties of restructured products from arrowtooth flounder (*Atheresthes stomias*). *J. Sci. Food Agric.* **2004**, *84*, 1741–1749. [CrossRef]

50. Dong, X.; Huang, Y.; Pan, Y.; Wang, K.; Prakash, S.; Zhu, B. Investigation of sweet potato starch as a structural enhancer for three-dimensional printing of Scomberomorus niphonius surimi. *J. Texture Stud.* **2019**, *50*, 316–324. [CrossRef]
51. Duflot, M.; Sánchez-Alonso, I.; Duflos, G.; Careche, M. LF 1H NMR T2 relaxation rate as affected by water addition, NaCl and pH in fresh, frozen and cooked minced hake. *Food Chem.* **2019**, *277*, 229–237. [CrossRef] [PubMed]

foods

MDPI

Article

A Comparative Functional Analysis of Pea Protein and Grass Carp Protein Mixture via Blending and Co-Precipitation

Xiaohu Zhou [1,2,3,4], Chaohua Zhang [1,3,5,*], Wenhong Cao [1,3,5], Chunxia Zhou [1,3,5], Huina Zheng [1,3,5] and Liangzhong Zhao [2,4]

1 College of Food Science and Technology, Guangdong Ocean University, Zhanjiang 524088, China; foodxiaohu@foxmail.com (X.Z.); cchunlin@163.com (W.C.); chunxia.zhou@163.com (C.Z.); zhenghn@gdou.edu.cn (H.Z.)
2 College of Food and Chemical Engineering, Shaoyang University, Shaoyang 422000, China; sys169@163.com
3 Guangdong Provincial Key Laboratory of Aquatic Products Processing and Safety, Zhanjiang 524088, China
4 Hunan Provincial Key Laboratory of Soybean Products Processing and Safety Control, Shaoyang 422000, China
5 Collaborative Innovation Center of Seafood Deep Processing, Dalian Polytechnic University, Dalian 116034, China
* Correspondence: zhangch@gdou.edu.cn

Abstract: Currently, the application of protein mixture derived from plants and animals is of great interest to the food industry. However, the synergistic effects of isolated protein blends (BL) are not well established. Herein, the development of a more effective method (co-precipitation) for the production of protein mixtures from pea and grass carp is reported. Pea protein isolate (PPI), grass carp protein isolate (CPI), and pea–carp protein co-precipitates (Co) were prepared via isoelectric solubilization/precipitation using peas and grass carp as raw materials. Meanwhile, the BL was obtained by blending PPI with CPI. In addition, the subunit composition and functional properties of Co and BL were investigated. The results show that the ratios of vicilin to legumin $\alpha + \beta$ and the soluble aggregates of Co were 2.82- and 1.69-fold higher than that of BL. The surface hydrophobicity of Co was less than that of BL, PPI, and CPI ($p < 0.05$). The solubility of Co was greater than that of BL, PPI, and CPI ($p < 0.05$), and the foaming activity was higher than that of BL and CPI ($p < 0.05$) but slightly lower than that of PPI. In addition, based on the emulsifying activity index, particle size, microstructure, and viscosity, Co had better emulsifying properties than BL, PPI, and CPI. The study not only confirmed that co-precipitation was more effective than blending for the preparation of mixed protein using PPI and CPI but also provided a standard of reference for obtaining a mixture of plant and animal proteins.

Keywords: pea; protein co-precipitates; protein blends; surface hydrophobicity; emulsifying properties

Citation: Zhou, X.; Zhang, C.; Cao, W.; Zhou, C.; Zheng, H.; Zhao, L. A Comparative Functional Analysis of Pea Protein and Grass Carp Protein Mixture via Blending and Co-Precipitation. *Foods* **2021**, *10*, 3037. https://doi.org/10.3390/foods10123037

Academic Editor: Stefania Masci

Received: 31 October 2021
Accepted: 4 December 2021
Published: 7 December 2021

1. Introduction

The increasing demand for high-quality food-grade proteins exhibiting acceptable nutritional and functional properties represents a challenge. Animal proteins are considered to be high-quality proteins but are expensive to produce, with an energy input to protein ratio of 14:1 [1]. Plant proteins may be an economical option to meet the increasing scarcity of protein resources [2]. However, the exclusive use of plant-derived protein is not optimal from a nutritional perspective, as essential amino acid requirements are not met. Nonetheless, mixing plant protein with animal protein can sustain the nutritional requirements of humans. Increasingly, the combination of plant and animal proteins is found to result in a synergistic benefit, which makes the design of foods with both plant and animal proteins a research hotspot [3].

Mixed proteins can be classified into two forms, protein blends (BL) and protein co-precipitates (Co) [4]. The differences in the preparation methods used to generate Co and BL are presented in Figure 1. BL refers to the blending of isolated proteins. Mixtures of pea protein isolate (PPI) and cod protein isolate as well as the combination of soy protein isolate and whey protein isolate exhibit synergistic emulsifying effects; however, when the proportion of plant protein replacing animal protein is increased, the synergistic effect is weakened, indicating that BL application has certain limitations [5,6]. Some studies suggest that the emulsifying gelation of directly mixed plant and animal proteins has an obvious antagonistic effect [7,8]. Concerning the preparation of Co, heterologous proteins are processed via isoelectric solubilization/precipitation (ISP) [9]. Driven by pH, proteins from different sources are simultaneously dissolved and precipitated to promote interaction between heteroproteins; generate additional disulfide bonds; change the subunit composition [10]; and alter the surface charge, solubility, and surface hydrophobicity of the proteins, thus effectively improving their functional properties [11]. Studies have found that legume–rapeseed Co has better nutritional and functional properties than single protein [12], and soy–whey protein Co has high water-holding and gel formation capacity [13]. Previously, we investigated the functional properties of soy–tilapia Co and found that the solubility and emulsifying properties of the Co were better than those of the single protein [14–16]. However, there are few comparative studies of BL and Co.

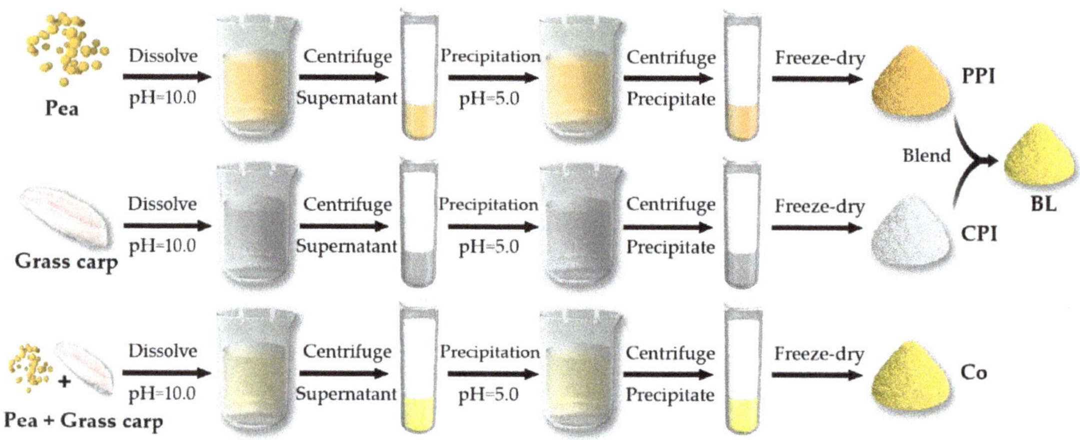

Figure 1. Schematic diagram of the preparation of the pea protein isolate (PPI), grass carp protein isolate (CPI), protein blends (BL), and protein co-precipitates (Co).

Peas (*Pisum sativum* L.) are one of the main legumes grown worldwide, with an annual global output of approximately 14.2 million tons [17]. Peas containing 20–30% protein are rich in lysine but deficient in methionine. Moreover, compared with soybean, pea protein is not associated with a risk of sensitization or transgenic origin. PPI has better emulsifying properties than soybean protein isolate under most pH values and protein concentrations [18]. Grass carp (*Ctenopharyngodon idellus*) is one of the four major carps in China; it is widely consumed in Asian countries and is rich in methionine. Grass carp protein isolate (CPI) can inhibit the flocculation of oil droplets and oxidization of lipids. Therefore, CPI is a potential emulsifier. However, it has poor stability and must be modified or mixed with other proteins for various industrial applications [19,20]. In order to conduct a broader study, pea and grass carp mixed proteins were selected, based on the cost of raw material and the functional properties of the proteins.

The properties of proteins are easily influenced by environmental conditions. pH is the primary factor affecting the properties of proteins, and it regulates electrostatic interactions. It drives the adsorption, reorientation, alignment, and formation of viscoelastic films on the interface, thus changing the stability of the interface [21]. This study evaluated the effects of blending and co-precipitation on the functional properties of pea and grass carp mixed protein by analyzing the solubility, foaming, and emulsifying properties of BL and Co at different pH values. The goal was to provide a basis for the subsequent development of products containing plant and animal proteins.

2. Materials and Methods

2.1. Materials

Split peas were purchased from Foshan Jinnuoyi Processing Factory of Agricultural Products (Foshan, China); the protein content was 21.12% (6.25 of N, wet weight). Grass carp were purchased from the local fish market in Zhanjiang City, Guangdong Province. The gills, heads, tails, bones, and blood were removed, leaving only the white muscles on the back, which were retained and frozen at a low temperature. The fish muscles were brought back to the laboratory and stored at −20 °C; the protein content was 18.27% (6.25 of N, wet weight). Electrophoresis reagents were purchased from Beyotime Biotechnology Co., Ltd. (Shanghai, China). The ANS fluorescence probe was purchased from Shanghai Aladdin Biochemical Technology Co., Ltd. (Shanghai, China). Other reagents were of analytical grade and were ordered from the Guangzhou Chemical Reagent Factory (Guangzhou, China).

2.2. Preparation of PPI, CPI, Co, and BL

Protein samples were prepared using the ISP method [14]. Peas were powdered and sieved through a 100-mesh sieve (0.15 mm). The pea powder was mixed with deionized water at 4 °C at a ratio of 1:9 (*w/v*). The pH was adjusted to 10.0 with 1 M NaOH and dissolved for 30 min, followed by centrifugation at 10,000× *g* and 4 °C for 20 min. The precipitate was removed and 1 M HCl was slowly added to the supernatant. The mixture was fully stirred, and the pH was adjusted to 5.0. At this time, the precipitation was washed with deionized water, and the pH value was adjusted to 7.0, followed by dialysis. PPI was obtained after freeze-drying for 48 h. The fish muscles were defrosted and chopped, and CPI was prepared according to the steps indicated above for PPI. The pea powder was mixed with fish muscle, and the Co was prepared following the steps described above. BL was obtained by directly mixing the powdered PPI and CPI. To ensure similar comparison, the protein ratio of pea and grass carp in BL and Co was maintained at 1:1 (W:W) in this study.

2.3. Sodium Dodecyl Sulfate Polyacrylamide Gel Electrophoresis (SDS-PAGE)

Similar concentrations of BL and Co solutions were prepared. Reducing and non-reducing SDS-PAGE gels were used with 5% spacer gel (upper gel) and 12% separation gel (lower gel) on a vertical gel electrophoresis apparatus at 15 mA. The difference between reducing and non-reducing electrophoresis depends on whether or not the reducing agent dithiothreitol (DTT) is added to the loading buffer. Standard protein markers (16–270 kDa) were used for qualitative analysis of the protein bands in the samples. The protein bands were dyed with Coomassie brilliant blue (CBB) R-250. Later, 40% methyl alcohol and 40% acetic acid were used for decoloration. The relative strength of the dyed protein bands was measured using ImageLab (Bio-Rad Laboratories, Hercules, CA, USA).

2.4. Solubility

According to Kristinsson's method [22], a 0.5% protein solution was prepared. The pH was adjusted to 3.0, 5.0, 7.0, 9.0, and 11.0 and then stirred for 30 min after centrifuging at 10,000× *g* for 20 min. Bovine serum albumin (BSA) was used as the standard and the protein content in the supernatant was measured using the Lowry method at 750 nm. The solubility was calculated as follows:

$$\text{Solubility (\%)} = C/C_0 \times 100\%, \quad (1)$$

where C is the protein content in the sample after centrifugation (mg) and C_0 denotes the protein content in the sample before centrifugation (mg).

2.5. Surface Hydrophobicity (H_0)

The protein surface hydrophobicity was measured using the ANS fluorescence probe as described by Haskard and Li-Chan [23]. A 50 mL aliquot of 0.5% protein solution was added to 10 mmol/L phosphate buffer with pH values of 3.0, 5.0, 7.0, 9.0, and 11.0. Then, the mixtures were stirred for 1 h at room temperature, followed by centrifugation for 30 min at the rate 7500 rpm. The protein concentrations of the supernatant were measured using the Lowry method. After dilution with the same phosphate buffer solution (PBS) (concentration ranging from 0.005 to 0.5 mol/mL), different concentrations of sample solutions were collected (4 mL), to which 40 μL of 8 mmol/L ANS solution (10 mmol/L, prepared with pH 7.0 PBS) was added. After oscillation, the mixtures were held static for 3 min, and the fluorescence intensity of the samples was tested using a G9800A fluorescence spectrophotometer (Cary Eclipse, Agilent, Santa Clara, CA, USA). In this test, the excitation and emission wavelengths were λex = 370 nm and λem = 490 nm, respectively, and the crack was 5 nm. Graphs of fluorescence intensity relative to the protein concentration were plotted. The slope of the initial section is equal to the surface hydrophobicity (H_0) of the protein molecules.

2.6. Foaming Properties and Foaming Stability

With reference to the method of Fekria et al. [24], 1.0% protein solutions were prepared, and the pH was adjusted to 3.0, 5.0, 7.0, 9.0, and 11.0. The protein solutions were stirred at room temperature for 30 min, and specific volumes of the protein solution (V_0) were collected and emulsified with the homogeneous shear machine at a rate of 10,000 r/min. The protein solutions were transferred immediately to the tube to measure the foam volume (V_1). After resting for 30 min, the foam volume (V_{30}) was again measured. The foaming capacity (FC) and foaming stability (FS) were as follows:

$$\mathrm{FC} = \frac{V_1 - V_0}{V_0} \times 100\%\,, \tag{2}$$

$$\mathrm{FS} = \frac{V_{30}}{V_1} \times 100\%\,, \tag{3}$$

2.7. Emulsion Preparation

A 1.0% protein solution was prepared, and the pH was adjusted to 3.0, 5.0, 7.0, 9.0, and 11.0. The solutions were stirred and dissolved for 30 min and then mixed with soybean oil at a ratio of 9:1 (*v/v*). The mixed proteins were blended with a T18 homogeneous shear machine (ULTRA-TURRAX, IKA, GER) for 1 min at a rate of 9000 rpm, followed by additional emulsification for 1 min at a rate of 15,000 rpm. The emulsions were stored under 4 °C.

2.8. Emulsifying Activity Index (EAI) and Emulsifying Stability Index (ESI)

The EAI and ESI of the proteins were measured via turbidimetry [25]. A 0.5% protein solution was prepared, and the pH was adjusted to 3.0, 5.0, 7.0, 9.0, and 11.0. The solutions were stirred and dissolved for 30 min. The protein solutions (6 mL) were mixed with soybean oil (2 mL) and emulsified with the homogeneous shear machine for 1 min (12,000 rpm). Subsequently, 20 μL of each emulsion was collected from the bottom container and dispersed in 4 mL of 0.1% SDS. The absorbance (A_0) was measured at a wavelength of 500 nm. After 10 min, 20 μL of the emulsion was again collected from the bottom container and was dispersed in 4 mL of 0.1% SDS. The absorbance (A_{10}) was tested at a wavelength of 500 nm. The blank control was 0.1% SDS. The EAI and ESI were calculated as follows:

$$\mathrm{EAI} = \frac{2 \times 2.303 \times A_0 \times D}{C \times \varphi \times 10^4} (\mathrm{m^2/g}) \tag{4}$$

$$\mathrm{ESI} = \frac{A_0}{A_0 - A_{10}} \times \Delta t\ (\mathrm{min}) \tag{5}$$

where D is the dilution ratio (200), C denotes the protein concentration (5 mg/mL), φ represents the volume fraction of oil in the emulsion (0.25), L refers to the optical cuvette path length (1 cm), and Δt indicates the measurement interval (10 min).

2.9. Particle Size and Zeta Potential

The particle size distribution of the emulsion, volume-weighted average particle size ($d_{4,3}$), surface-area-weighted average particle size ($d_{3,2}$), and zeta potential under different pH values were measured with a Malvern Zetasizer Nano ZS (Malvern Panalytical, Malvern, UK). The relative refractive index of the emulsion was 1.107, while the refractive indices of soybean oil and water were 1.472 and 1.330, respectively. The average particle size distribution of the emulsification was characterized by $d_{4,3}$. Further, $d_{4,3}$ and $d_{3,2}$ values were measured using deionized water as the dispersing agent. The values of $d'_{4,3}$ and $d'_{3,2}$ were measured using 1.0% SDS as the dispersing agent.

2.10. Interface Protein Absorption (AP) and Interface Protein Content (Γ)

AP and Γ were measured according to the methods of Liang and Tang [26]. The fresh emulsion (1 mL) was centrifuged for 30 min at a rate of 10,000 r/min, resulting in separation into an upper layer of greasy emulsion and a lower layer of protein supernatant. The upper layer was eliminated carefully, and the lower supernatant was collected with an injector and filtered through a 0.22 μm needle-shaped filter. BSA was used as the standard protein, and the protein concentration of the lower supernatant was tested with the Lowry method. The AP and Γ were calculated as follows:

$$AP = \frac{C_0 - C_s}{C_0} \times 100\%, \tag{6}$$

$$\Gamma = C_0 - C_s \frac{1-\varphi}{6\varphi} d'_{3,2}, \tag{7}$$

where C_0 is the protein concentration in the protein solution used to prepare the emulsion (10 mg/mL); C_s denotes the protein concentration of the lower supernatant layer after centrifuging the emulsion (mg/mL); Γ represents the interface protein content (mg/m^2); φ stands for the oil volume fraction (0.1), and $d'_{3,2}$ refers to the surface area average particle size of the emulsion, which was measured using 1.0% SDS as the dispersing agent (μm).

2.11. Confocal Laser Scanning Microscopy

The microstructure of the emulsion was analyzed via confocal laser scanning microscopy (CLSM; FV1200, Olympus, Japan) according to the method of Liu [27]. The emulsion (20 μL) was collected and added to a centrifuge. Simultaneously, 10 μL of 0.1% (*w/v*) Nile red methanol solution and 10 μL 1.0% (*w/v*) Nile blue aqueous solution were added; the reaction was performed in the dark for 15 min. The mixture (10 μL) was collected to produce films and observed with a 100× oil immersion lens. The resolution of the scanning images was 1024 × 1024. Proteins were stained with Nile blue, and laser excitation was performed at an excitation wavelength of 633 nm and a detection wavelength of 655–755 nm. Oil phase dyeing was performed with Nile red, followed by laser excitation at an excitation wavelength of 559 nm and a detection wavelength of 570–625 nm.

2.12. Rheological Properties

The rheological properties of the emulsion were measured with a rheometer (Discovery DHR-2 Dynamic Shear Rheometer, TA, Milford, MA, USA). A 40 mm plate was used to test the apparent viscosity of the emulsion within the shear rate range of 0.1–200 s^{-1}.

2.13. Statistical Analysis

All experimental measurements were performed at least three times. SPSS 25.0 was used for all statistical analyses. Statistical differences between two groups were determined with one-way analysis of variance (ANOVA) and Duncan's multiple range test ($p < 0.05$). The results are expressed as the mean ± standard deviation (SD).

3. Results and Discussion

3.1. Protein Composition Analysis

The protein compositions of BL and Co were analyzed via reducing and non-reducing SDS-PAGE. As can be seen in Figure 2 and Table 1, both BL and Co contained myosin heavy chain (MHC, ~200 kDa), convicilin (~70 kDa), legumin (~62 kDa), vicilin (~50 kDa), actins (AC, ~42 kDa), legumin α (~38 kDa), tropomyosin (TM, ~34 kDa), legumin β (~22 kDa), and aggregates consistent with the literature [28,29]. To sum up, both BL and Co effectively combined pea and CPI, and thus, both are suitable for further testing.

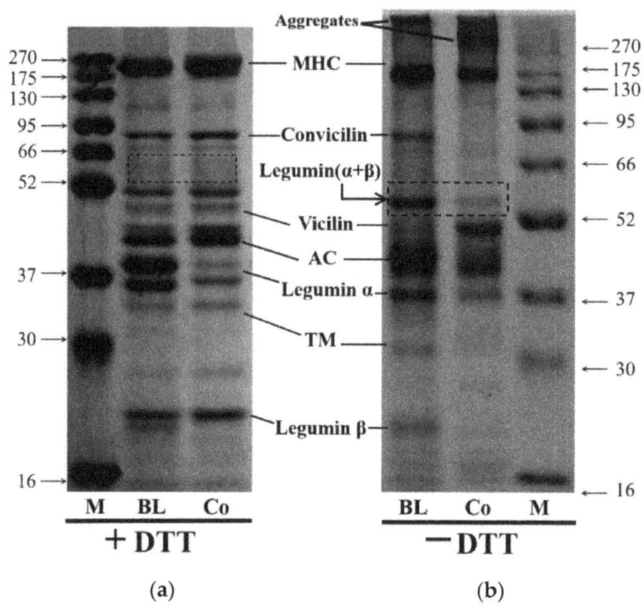

Figure 2. Reducing (**a**) and non-reducing (**b**) electrophoretograms of protein blends (BL) and protein co-precipitates (Co).

Table 1. Relative band optical density (%) of BL and Co (non-reducing electrophoretogram).

Samples	Aggregates	MHC	Convicilin	Leg α + β	Vicilin	AC	Leg α	TM	Leg β	Others	Vicilin/ Leg α + β
BL	11.61	15.90	4.60	9.98	7.22	19.31	10.63	5.31	9.98	5.45	0.72
Co	19.62	15.27	0.11	6.55	13.36	16.46	8.06	5.36	4.63	10.57	2.04

Reducing SDS-PAGE reflects the specific subunit composition of proteins. BL and Co were both composed of pea and carp (1:1). However, the MHC, legumin α, and legumin β protein bands differed significantly in reducing SDS-PAGE, indicating that Co and BL showed different degrees of crosslinking or degeneration with subunits of CPI. Similar studies have suggested that such differences might be attributed to different electrostatic and hydrophobic interactions and the effects of disulfide bond in proteins during preparation of Co by the ISP method [10]. Co contains a higher number of MHCs, indicating that its functional properties might have some advantages [30]. The reduced MHC retention of Co in non-reducing electrophoresis indicates that some of the MHC components of Co entered soluble aggregates. Legumin is composed of six pairs of subunits and each pair contains an α-polypeptide (acid subunit) and a β-polypeptide chain (alkaline subunit) linked via a disulfide bond. The disulfide bond can only be opened under reducing conditions, thus dissociating the protein subunits [31]. SDS can decompose hexamer legumin to 60 kDa dimer legumin α + β under heating conditions. However, DTT destroys the non-covalent bonds of the legumin hexamer as well as the disulfide bonds between the α- and β-polypeptide chains. Hence, no legumin α + β band is detected at 60 kDa in the reducing SDS-PAGE. In contrast, the disulfide bond between the α- and β-polypeptide chains remains intact in the absence of DTT [32]. Therefore, legumin α + β bands appeared at the position of 60 kDa following non-reducing electrophoresis. The optical density of Co was significantly less than that of BL, indicating that legumin α + β components of Co also partially entered the aggregate.

Non-reducing electrophoresis cannot disrupt the disulfide bonds. Therefore, it reflects the original composition of the various proteins in a mixed protein. Compared with proteins separated via reducing SDS-PAGE, many soluble aggregates were blocked at the top of the separated gel and the concentration of Co (19.62%) was 1.69-fold higher than that of BL (11.61%) (Table 1), indicating the presence of additional bonds in Co. Simultaneously, as mentioned above, some of pea and grass carp proteins entered the aggregates, indicating that co-precipitation leads to the formation of disulfide bonds between pea protein and grass carp protein, resulting in altered subunit composition. Current studies indicate that the formation of disulfide bonds between globulin molecules enhances the flexibility of the protein structure and improves its foaming and emulsifying capacity as well as its stability. When proteins are adsorbed onto the O/W interface, the stability of the emulsion is improved [33]. Compared with legumin α + β, vicilin is more flexible and exhibits better interfacial activity [34]. Many studies have confirmed that the vicilin/legumin α + β ratio is positively related to a protein's functional properties, including solubility, and foaming and emulsifying properties [35,36]. Thus, the ratio of vicilin to legumin α + β is very important. As shown in Table 1, the proportion of Co (203.99%) is 2.82-fold larger than that of BL (72.36%). Therefore, the analysis of protein composition indicates that the functional properties of Co might be better than those of BL.

3.2. Solubility and Surface Hydrophobicity (H_0)

Solubility is a key functional characteristic that can be used to evaluate the potential use of protein in food and determine its foaming, emulsification, and gel formation ability [37]. As can be seen in Figure 3, the solubility of Co, BL, PPI, and CPI was significantly pH dependent, with solubility exhibiting a U-shaped curve. The minimum solubility was observed at pH 5.0 because the isoelectric point (PI) of globulin is 4.5 and the isoelectric point of myofibrillar protein is 5.3, which is consistent with other studies of PPI [38], CPI [39], and soy–fish mixed protein [14]. When the pH deviates from the isoelectric point, solubility significantly increases because an increase in net charges on the protein surface reinforces the intermolecular electrostatic repulsive force, which promotes ion–dipole interaction between proteins and water molecules [40].

ANS is a fluorescence probe that is sensitive to external polar groups and interacts with hydrophobic groups of protein molecules to generate fluorescence spectra, which can be used to characterize the exposure of hydrophobic groups in proteins. As shown in Figure 4 under different conditions, H_0 varied inversely with solubility. Thus, a higher solubility was associated with fewer hydrophobic groups on the protein molecular surface [41].

Comprehensive analysis of the solubility and surface hydrophobicity of the mixed proteins indicated that the solubility of Co was the highest under all pH values except at pH 5.0 ($p < 0.05$), which is similar to the result of our previous study of soy–tilapia Co [14]. Co prepared by ISP was precipitated at pH 5.0. Under this circumstance, pea and CPI molecules carry opposite charges and the intermolecular electrostatic interaction is strengthened to generate soluble Co-aggregates of proteins, triggering the reconstruction of hydrophobic and disulfide bonds [42]. Moreover, some subunits, such as vicilin, will be concentrated "selectively", thus changing the protein properties [43]. BL is formed by mixing PPI and CPI powders, and the molecular interaction is not as strong as that of

Co. The molecular interaction of BL involved PPI and CPI at pH 3.0, 7.0, and 9.0, but the solubility (80.77%) at pH 11.0 was a little higher than that of PPI (77.36%) and CPI (79.07%). Under strong alkaline conditions, protein molecules degenerate, and hydrophobic groups are exposed. Protein molecules cluster with each other via hydrophobic bonds, inducing hydrophobic nonlinear superposition [44]. As a result, the H_0 (231.53) of BL was lower than that of PPI (302.36) and CPI (234.41). Due to this hydrophobic synergistic effect, the solubility of BL was higher than that of PPI and CPI. Thus, Co has a better synergistic effect on solubility than BL.

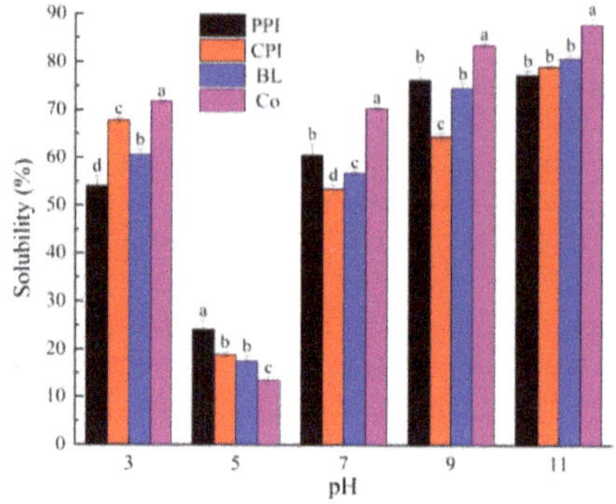

Figure 3. Solubility of proteins under different pH values. Different letters under the same pH value indicate significant differences ($p < 0.05$).

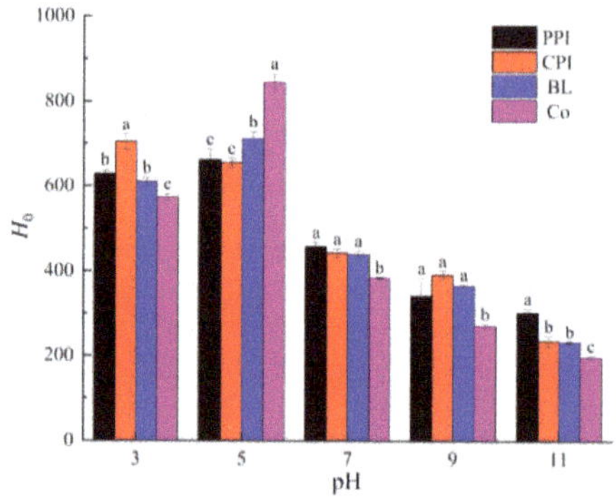

Figure 4. Surface hydrophobicity (H_0) of proteins under different pH values. Different letters under the same pH value indicate significant differences ($p < 0.05$).

3.3. Foaming Properties

Foaming plays a critical role in the texture and structure of foods, such as ice cream, cakes, breads, and meringue. Currently, foaming is achieved by replacing a surfactant with protein. Foaming capacity (FC) and foaming stability (FS) are the most common indices used to describe the foaming properties of protein [45]. The effects of pH on the FC and FS of the proteins are shown in Figures 5 and 6. Overall, FC and FS exhibited a U-shaped variation, similar to those observed for solubility. The foaming properties were the weakest at pH 5.0, which is consistent with other studies [46]. When the pH deviated from the PI, the solubility of the protein was improved and the degree of dispersion of the protein molecules in water was increased. Under these circumstances, the surface charges on the protein molecules are increased and the protein structures are opened to expose the hydrophobic amino acid side chains. Hence, the proteins quickly locate at the air–water interface, and the surface tension is decreased, thus improving the FC and FS of the proteins.

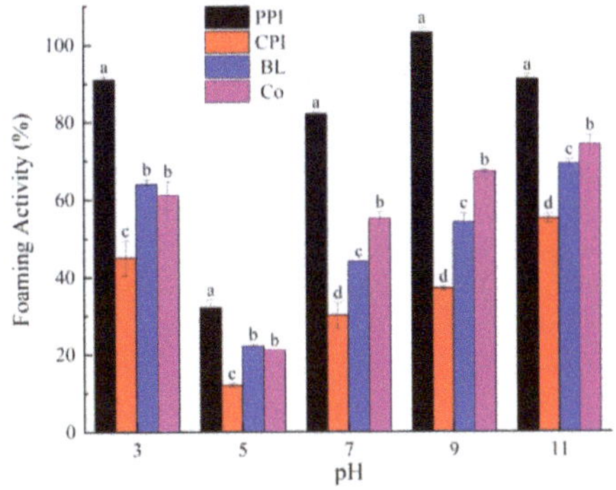

Figure 5. Foaming capacity of proteins under different pH values. Different letters under the same pH value indicate significant differences ($p < 0.05$).

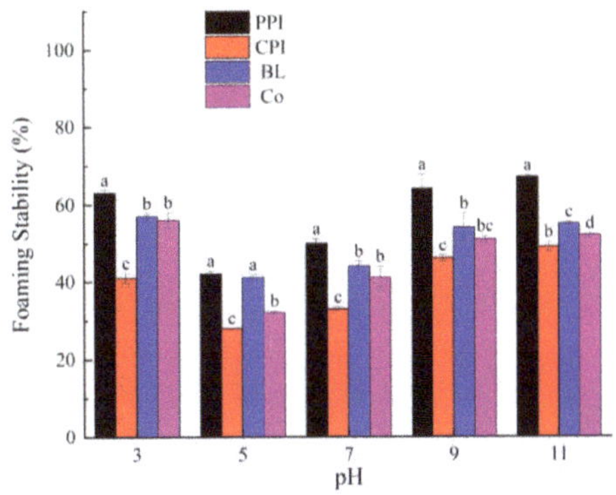

Figure 6. Foaming stability of proteins under different pH values. Different letters under the same pH value indicate significant differences ($p < 0.05$).

Zhao et al. argued that the FC of PPI was similar to that of soy protein isolate, but the FS was slightly higher. PPI is a plant foaming agent with potential application in various food preparations [47]. In this study, the FC and FS of PPI were the highest under all pH values ($p < 0.05$). The maximum value of FC was 103 $\pm$ 1.25 at pH 9.0, which is related to the small molecular weights of PPI, small volume, and weaker intermolecular interaction [48]. CPI exhibited the opposite trend. The molecular weights of fibrillin are relatively high and present a linear structure, thus causing significant steric hindrance. The air–water interfacial absorption capacity of fibrillin is significantly lower than that of globulin, and the compactness of the viscous layer is insufficient. Therefore, the FC and FS of fibrillin are relatively low [49]. The foaming performance of the mixed protein was better than that of CPI, but still inferior to that of PPI at all pH values, i.e., there was no synergistic effect, similar to solubility. The FC of Co was lower than that of BL at pH 3.0 and 5.0, and the FS of Co was lower than that of BL under all pH values. This might be because Co contains additional macromolecular soluble aggregates, making it difficult to form a dense viscous layer on the water–air interface. In brief, the addition of PPI is a potential strategy to improve the foaming properties of CPI.

3.4. *Emulsifying Properties*

3.4.1. EAI and ESI

EAI refers to the oil–water interface area for stabilization of the unit mass of the emulsifier; it reflects the resistance of a protein to emulsion stratification caused by flocculation and aggregation. Thus, the EAI can be used to evaluate the emulsifying properties of proteins. ESI refers to the ability of proteins to remain in a stable state within a certain period, without phase layering or separation [25]. EAI and ESI are key indices that reflect the functional properties of proteins. As shown in Figures 7 and 8, the EAI and ESI of all proteins first decreased and then increased under all pH values. This phenomenon is similar to the trend of the solubility and foaming properties, but contrary to the H_0 value. The emulsifying properties and emulsifying stability near PI were the lowest because, at PI, the protein surface carries few charges and lacks electrostatic repulsion. Accordingly, proteins aggregate, and flocculation occurs [50]. Increasing pH can lead to the degeneration of proteins. The protein molecular conformation in the "melting state" is soft and relaxed. The residual groups of the exposed hydrophobic amino acids point to the oil phase, while the hydrophilic groups point to the water phase, thus enhancing the interaction with oil phases [51]. As a result, the emulsifying properties are enhanced. It can be seen that the pH changes the ability of proteins to stabilize oil droplets via expansion and adsorption on interfaces by regulating electrostatic interactions and H_0 for stability.

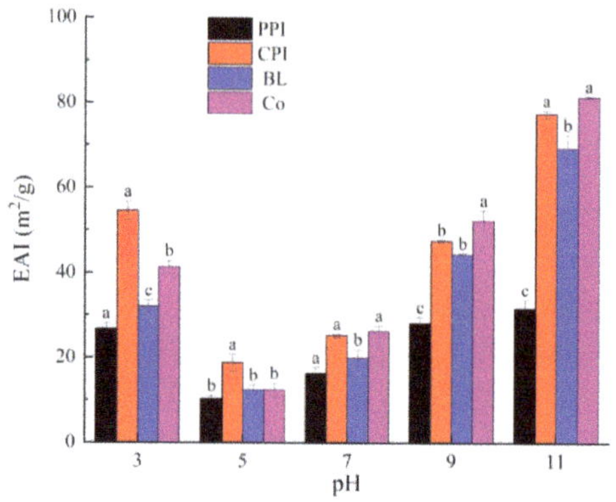

Figure 7. Emulsifying activity index of proteins under different pH values. Different letters under the same pH value indicate significant differences ($p < 0.05$).

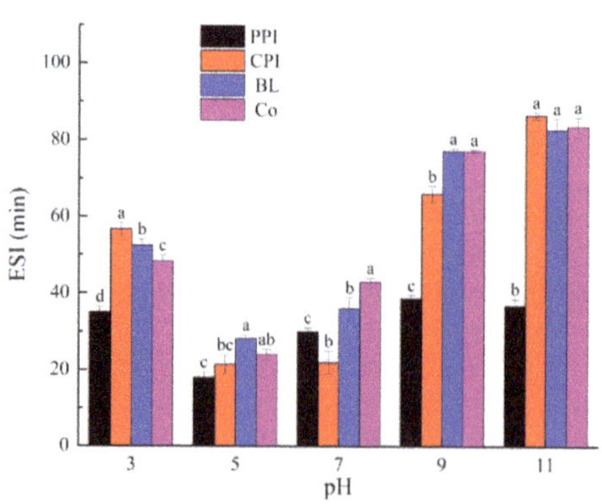

Figure 8. Emulsifying stability index of proteins under different pH values. Different letters under the same pH value indicate significant differences ($p < 0.05$).

The ESI values of BL and Co did not exhibit any clear pattern. At pH 3.0, the ESI of BL was higher than that of Co ($p < 0.05$). At pH 7.0, the ESI of BL was lower than that of Co ($p < 0.05$). There were no significant differences at pH 5.0, 9.0, and 11.0 ($p > 0.05$). However, testing the particle size of the emulsions and analysis under CLSM revealed that Co stabilized smaller oil droplets (discussed later), indicating improved stability. The testing time of ESI was only 10 min, which was not sufficient to complete the maturation process and allow adsorption of the protein on the interface.

3.4.2. Particle Size and Zeta-Potential Analysis

Particle size is an important index used to measure the stability of an emulsion [52]. The particle size distributions of protein emulsions were evaluated using the volume-weighted mean diameter ($d_{4,3}$) (Figure 9), supplemented with the surface-area-weighted mean diameter ($d_{3,2}$), and d'_{43} and d'_{32} values of the emulsion with 1% SDS (Table 2). The effect of the pH value on the particle size was analyzed. The results indicated that the $d_{4,3}$ values of the emulsions at pH 3.0 and 5.0 were relatively high (3.29 ± 0.27 μm and 22.79 ± 2.75 μm for Co) because the electrostatic repulsion among proteins is insufficient to overcome attractions (Figure 10), thus resulting in particle clusters and increasing the particle size. According to Laplace theory, the droplet size of an emulsion decreases with a reduction in two-phase interface tension under constant homogeneity [53]. The $d_{4,3}$ values of the emulsions at pH 9.0 and 11.0 were significantly lower than at other pH values (0.94 ± 0.05 μm and 0.54 ± 0.01 μm for Co), which is consistent with a previous study reporting a significant decrease in the $d_{4,3}$ of soy protein isolate under alkaline pH (pH 9.0–12.0) [54]. Small particle size is conducive to the formation of a stable emulsion [55]. Therefore, the most stable emulsion systems were achieved at pH 11.0. Moreover, as shown in Figures 9 and 10, all proteins exhibited single-peak particle size distributions at pH 3.0 and 11.0, which is related to the high net charges. At pH 5.0 and 7.0, relatively large particle size peaks (10–100 μm) were observed. Under these conditions, oil droplets develop flocculence, which is attributed to bridging protein aggregates between the droplets. Alternatively, it is difficult to achieve ripening and saturation on the oil–water interface due to the low protein–oil ratio, thus leading to bridging flocculation [56]. The particle size of CPI under all pH values was smaller than that of PPI, which is consistent with the results of EAI, because the molecular flexibility of myofibrillar protein (MP) is higher than that of globulin and, thus, leads to directional expansion on the oil–water interface. Therefore, CPI can stabilize smaller oil droplets. The particle size of Co was smaller than that of BL under all pH values, except for pH 5.0 ($p < 0.05$), which demonstrates the higher emulsifying capacity of Co. The trends in $d_{3,2}$, $d'_{4,3}$, and $d'_{3,2}$ variation were similar to that of $d_{4,3}$.

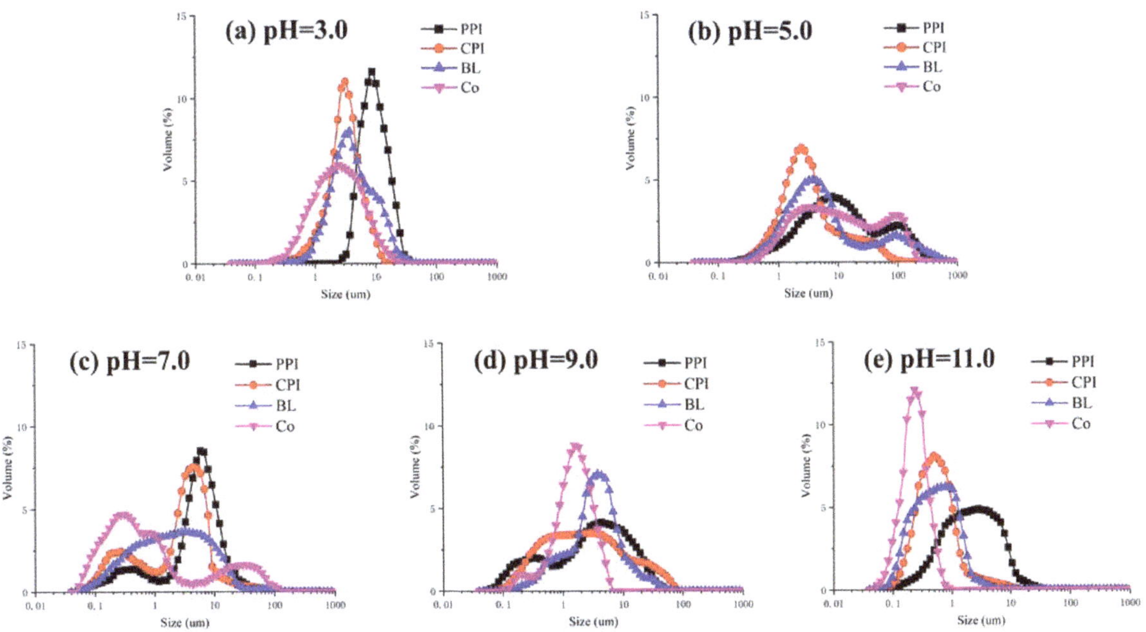

Figure 9. Droplet size distribution of proteins under different pH values. (**a**) pH = 3.0; (**b**) pH = 5.0; (**c**) pH = 7.0; (**d**) pH = 9.0; (**e**) pH = 11.0.

Table 2. Average particle size of proteins at different pH values.

pH	Samples	$d_{4,3}$ (μm)	$d_{3,2}$ (μm)	$d'_{4,3}$ (μm)	$d'_{3,2}$ (μm)
3.0	PPI	10.66 ± 0.56 a	6.75 ± 0.18 a	3.68 ± 0.10 a	2.63 ± 0.05 a
	CPI	3.53 ± 0.06 c	2.58 ± 0.10 b	1.08 ± 0.10 c	1.29 ± 0.07 c
	BL	6.91 ± 0.06 b	3.23 ± 1.56 b	1.84 ± 0.07 b	1.63 ± 0.06 b
	Co	3.29 ± 0.27 c	2.10 ± 0.07 b	0.90 ± 0.01 c	0.87 ± 0.01 d
5.0	PPI	31.91 ± 3.23 a	10.41 ± 1.01 a	4.17 ± 0.13 b	3.77 ± 0.19 a
	CPI	9.85 ± 0.10 d	4.28 ± 0.06 c	2.69 ± 0.22 c	2.01 ± 0.24 b
	BL	17.12 ± 1.46 c	8.21 ± 0.36 b	4.16 ± 0.13 b	3.96 ± 0.34 a
	Co	22.79 ± 2.75 b	9.33 ± 0.23 ab	7.76 ± 0.18 a	3.73 ± 0.10 a
7.0	PPI	4.88 ± 0.19 a	2.73 ± 0.05 a	1.68 ± 0.03 a	1.14 ± 0.07 a
	CPI	2.57 ± 0.18 c	1.56 ± 0.09 c	0.77 ± 0.08 c	0.63 ± 0.02 d
	BL	3.96 ± 0.09 b	2.46 ± 0.17 b	1.45 ± 0.07 b	1.00 ± 0.02 b
	Co	1.17 ± 0.07 d	0.84 ± 0.07 d	0.60 ± 0.03 d	0.85 ± 0.00 c
9.0	PPI	3.20 ± 0.06 a	2.89 ± 0.09 a	1.28 ± 0.05 a	1.16 ± 0.10 a
	CPI	2.01 ± 0.06 b	1.85 ± 0.11 b	0.89 ± 0.02 c	0.72 ± 0.06 b
	BL	3.02 ± 0.14 a	2.75 ± 0.10 a	1.12 ± 0.01 b	1.02 ± 0.05 a
	Co	0.95 ± 0.01 d	0.86 ± 0.00 c	0.68 ± 0.01 d	0.39 ± 0.01 d
11.0	PPI	2.49 ± 0.05 a	2.24 ± 0.19 a	0.98 ± 0.03 a	1.10 ± 0.07 a
	CPI	0.93 ± 0.02 c	0.87 ± 0.02 b	0.47 ± 0.00 b	0.35 ± 0.00 c
	BL	1.15 ± 0.05 b	1.09 ± 0.03 b	0.57 ± 0.02 c	0.47 ± 0.02 b
	Co	0.55 ± 0.01 d	0.53 ± 0.02 c	0.33 ± 0.01 d	0.20 ± 0.00 d

Note: Each value represents the mean ± SD (n = 3). Different letters under the same pH indicate significant differences ($p < 0.05$).

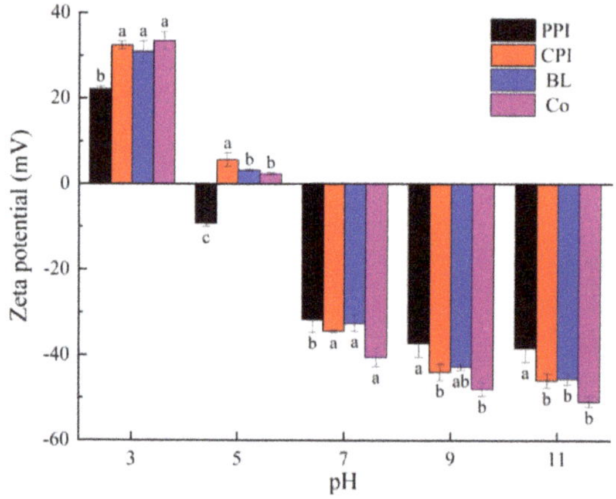

Figure 10. Zeta potential of proteins under different pH values. Different letters under the same pH value indicate significant differences ($p < 0.05$).

Zeta potential is an index of particle repulsion or attraction intensity. The higher the absolute zeta-potential value, the higher the stability of an emulsion [57]. The zeta potentials of droplets under different pH values were evaluated (Figure 10). For all samples, the zeta potential changed from positive to negative when the pH increased from 3.0 to 11.0. This trend can be attributed to the PI values of PPI and CPI, which were 4.5 and 5.2, respectively. At pH 5.0, the zeta potentials of the emulsions were the lowest (2.32 ± 0.31 mV for Co), and the net charges were the lowest, which is related to the negatively charged aggregates on the interfaces. The zeta potentials of the emulsions increased significantly when the pH increased from 7.0 to 11.0 because when the pH is higher than the PI of a protein, electrostatic repulsion increases, and flocculation of droplets is inhibited. Consistent with the particle size results, the zeta potential of CPI was higher than that of PPI ($p < 0.05$).

The zeta potential of Co was higher than that of BL under all pH values, except at pH 5.0, which is attributed to the better dispersing capacity of Co and the higher sensitivity of charging properties to pH. This is related to AP and Γ.

3.4.3. Interface Protein Adsorption (AP) and Interface Protein Content (Γ)

AP refers to the percentage ratio between the protein content adsorbed on the oil droplet surface and the protein content in the continuous phase. Γ refers to the protein content on the unit area of a droplet [43]. The AP (%) and Γ values of PPI, CPI, BL, and Co under different pH values are listed in Table 3. At pH 5.0, the AP (%) of all four protein emulsions was the lowest (2.96–11.42%), indicating the presence of abundant protein isolates in the emulsions. The protein solubility decreased due to the electrostatic shielding effect at pH 5.0. When the pH deviated from the PI, the AP increased gradually, which is consistent with the study of Liang and Tang [43]. The Γ value of an emulsion reflects the thickness of the stable oil–droplet interface membrane. Although it is not directly correlated with the particle size of emulsion droplets or AP, it is still influenced by solubility. The overall trends of variation were also similar to those of AP. Nevertheless, it is interesting that the AP of BL at pH 5.0 was higher than at pH 7.0, and the AP of Co at pH 5.0 was higher than at pH 3.0 and 7.0. This indicates that, during emulsification via high-speed shearing, insoluble proteins can also adsorb onto the interface and an emulsion with a similar stable Pickering effect is formed [58]. The results indicated that the AP% and Γ of CPI were higher than those of PPI, and the AP% and Γ of Co were higher than those of BL. These results further demonstrate that surface hydrophobicity is a major and decisive factor influencing the emulsifying properties of a protein under high solubility.

Table 3. Interface protein adsorption and interface protein content of proteins at different pH values.

Samples	pH = 3.0		pH = 5		pH = 7		pH = 9		pH = 11	
	AP/%	Γ/ (mg/m²)	AP%	Γ/ (mg/m²)	AP%	Γ/ (mg/m²)	AP%	Γ/ (mg/m²)	AP%	Γ/ (mg/m²)
PPI	33.43 ± 1.64 ^{b}	13.61 ± 0.07 ^{a}	2.96 ± 0.72 ^{b}	2.73 ± 0.01 ^{d}	22.94 ± 3.66 ^{c}	4.04 ± 0.10 ^{c}	38.39 ± 2.69 ^{b}	6.60 ± 0.03 ^{a}	48.11 ± 1.57 ^{b}	7.75 ± 0.02 ^{a}
CPI	38.91 ± 4.37 ^{ab}	8.13 ± 0.23 ^{c}	11.42 ± 1.12 ^{a}	3.05 ± 0.02 ^{c}	33.35 ± 0.88 ^{b}	3.11 ± 0.00 ^{d}	36.21 ± 2.87 ^{b}	3.95 ± 0.02 ^{c}	54.94 ± 4.08 ^{b}	2.91 ± 0.09 ^{c}
BL	36.19 ± 2.01 ^{b}	9.18 ± 0.06 ^{b}	9.42 ± 2.26 ^{a}	5.30 ± 0.04 ^{a}	32.50 ± 3.64 ^{b}	4.56 ± 0.12 ^{b}	37.33 ± 1.45 ^{b}	5.66 ± 0.02 ^{b}	54.29 ± 3.70 ^{b}	3.81 ± 0.05 ^{b}
Co	44.15 ± 0.87 ^{a}	5.65 ± 0.03 ^{d}	8.63 ± 0.51 ^{a}	6.57 ± 0.00 ^{b}	48.91 ± 4.31 ^{a}	6.06 ± 0.26 ^{a}	51.71 ± 4.89 ^{a}	3.13 ± 0.11 ^{d}	65.61 ± 4.49 ^{a}	2.03 ± 0.21 ^{d}

Note: Each value represents the mean ± SD (n = 3). Different letters under the same pH value indicate significant differences ($p < 0.05$).

3.4.4. Confocal Laser Scanning Microscopy (CLSM)

CLSM is often used to analyze the microstructure of an emulsion; it reflects the original particle size distribution, dispersion, and stability of emulsion particles. The red and green zones represent oil (Ar laser 559 nm) and protein (He/Ne laser 633 nm), respectively. The microstructures of the protein emulsions under different pH values are shown in Figure 11. The overall results were consistent with the observed particle size results. Almost all oil droplets were spherical and wrapped in an outer protein layer. At pH 5.0, all four protein emulsions showed serious flocculation. The emulsion particles were large and irregular and were mostly distributed in clusters. Although the oil droplets stabilized by Co were small, evident protein aggregation was observed. A pH value approaching PI is the main cause for the bridging flocculation of oil droplets and proteins [6]. When the pH was far from the PI, electrostatic repulsion increased with the increase in the solubility of the protein, while the particle size of the emulsion droplets decreased rapidly. Similarly, the emulsion particle sizes stabilized by BL and Co were smaller and distributed more uniformly under the same pH (e.g., pH 9.0 and 11.0), which is consistent with the results obtained with soy–whey mixed protein at pH 2.0–11.0 [6]. The variation trend of Co was more obvious. CLSM directly validated the EAI/ESI, AP/Γ, and the results of particle size distribution, indicating that Co has a better emulsifying effect than BL.

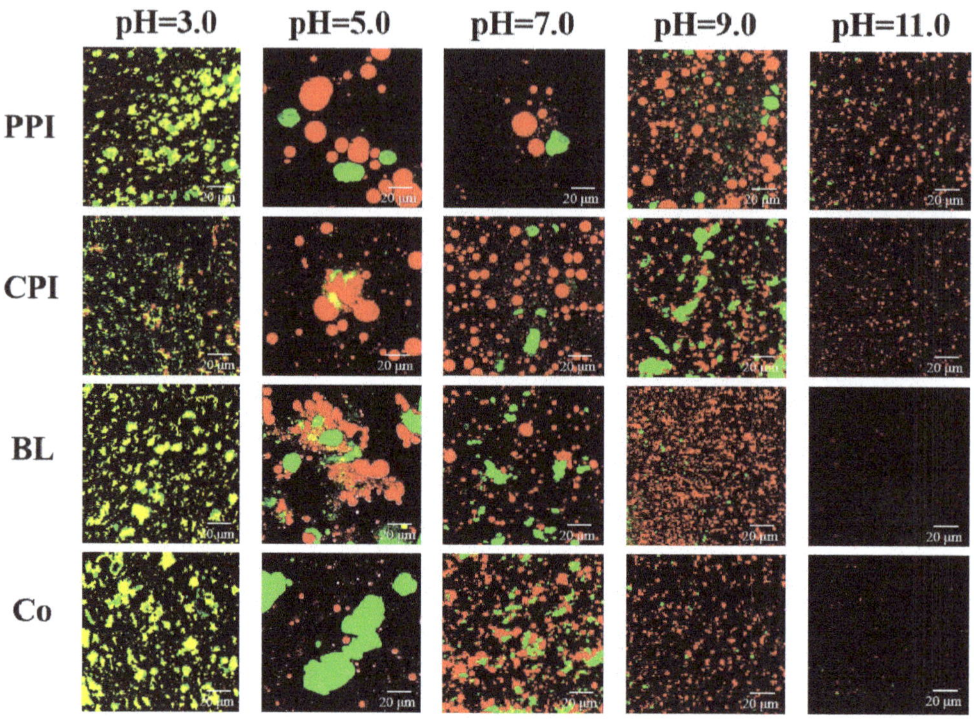

Figure 11. Micromorphology of protein emulsions under different pH values.

3.4.5. Rheological Properties

The processing and storage stability of an emulsion can be predicted based on the apparent rheological properties of the emulsion [59]. The shearing rheological properties of the protein emulsions under different pH values were analyzed by testing the apparent viscosity (Figure 12). With an increase in the shearing rate, the viscosity of the emulsions declined partially, and obvious shearing dilution was observed. This demonstrates that these emulsions are non-Newtonian fluids, and the flocculent structures of the emulsion droplets were destroyed in the shearing process. When the pH was close to the PI, the emulsions exhibited the highest viscosity and shearing dilution behavior due to the high flocculation of droplets, which is consistent with previous results [60]. For spherical oil droplets wrapped by proteins, the electrostatic repulsion among oil droplets decreases as the pH approaches the PI of the adsorbed protein, thus resulting in droplet clustering [61]. According to Stokes' law, the degree of flocculation of an emulsion is positively related to the viscosity and degree of shearing dilution [62]. This is because the aggregate maintains some continuous phases in the structure, thus increasing the viscosity. Due to the external shearing effect, flocculation structural damage causes shearing dilution. Therefore, the viscosity of the Co emulsion was lower than that of PPI, CPI, and BL emulsions when the pH was higher than the PI (pH 7.0–11.0), indicating a low degree of flocculation of the droplets.

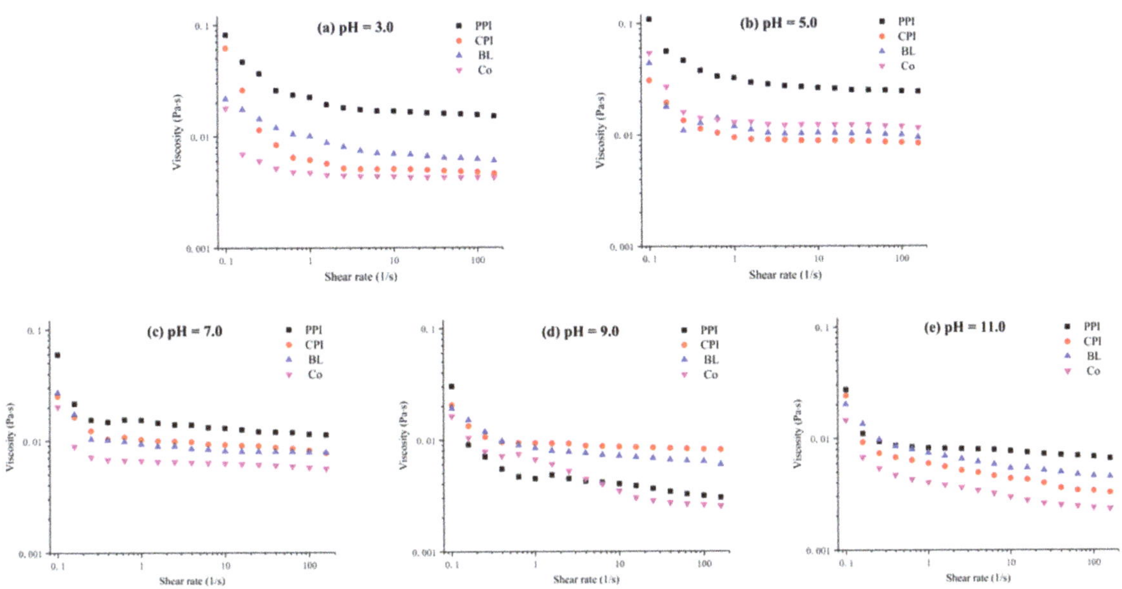

Figure 12. Apparent viscosity of protein emulsions under different pH values. (**a**) pH = 3.0; (**b**) pH = 5.0; (**c**) pH = 7.0; (**d**) pH = 9.0; (**e**) pH = 11.0.

4. Conclusions

The comprehensive analysis shows that co-precipitation changes the subunit composition and hydrophobicity of mixed proteins, and the functional properties of Co and BL under deviating isoelectric points were synergistic. Additionally, the Co prepared by the ISP method exhibited better solubility, foamability, and emulsification than BL comprising a direct blend of isolated proteins, thereby indicating the advantages of the co-precipitation method. Further studies will be performed to investigate the functional properties of Co and protein modification under different temperatures and ion concentrations. This study provides a theoretical basis for further improvement of mixed protein processing methodology.

Author Contributions: Conceptualization: C.Z. (Chaohua Zhang); Methodology: H.Z. and C.Z. (Chunxia Zhou); investigation: X.Z. and L.Z.; resources: C.Z. (Chaohua Zhang) and C.Z. (Chunxia Zhou); supervision: C.Z. (Chaohua Zhang); writing—original draft: X.Z.; writing—review and editing: W.C. and C.Z. (Chaohua Zhang) All authors have read and agreed to the published version of the manuscript.

Funding: This research was funded by the Guangdong Higher Education Institutions Processing and Utilization Innovation Team of High Value of Aquatic Products (GDOU2016030503), the Industrial Development Special Fund Project of Dapeng New District, Shenzhen, China (KY20180111), the Science and Technology Innovation Program of Hunan Province (2019TP1028, 2019SK2122, 2019NK4229, S2021GCZDYF0535), and the Open Project of Hunan Provincial Key Laboratory of Soybean Products Processing and Safety Control (DZPJG202009).

Data Availability Statement: Research data are not shared.

Conflicts of Interest: The authors declare that they have no conflict of interest.

References

1. Pimentel, D.; Pimentel, M. Sustainability of meat-based and plant-based diets and the environment. *Am. J. Clin. Nutr.* **2003**, *78*, 660S–663S. [CrossRef]
2. Pang, X.H.; Yang, Y.; Bian, X.; Wang, B.; Ren, L.K.; Liu, L.L.; Yu, D.H.; Yang, J.; Guo, J.C.; Wang, L.; et al. Hemp (*Cannabis sativa* L.) Seed Protein–EGCG Conjugates: Covalent Bonding and Functional Research. *Foods* **2021**, *10*, 1618. [CrossRef]

3. Alves, A.C.; Tavares, G.M. Mixing animal and plant proteins: Is this a way to improve protein techno-functionalities? *Food Hydrocoll.* **2019**, *97*, 105171. [CrossRef]
4. Wu, C.; Wang, T.; Ren, C.; Ma, W.; Wu, D.; Xu, X.; Wang, L.S.; Du, M. Advancement of food-derived mixed protein systems: Interactions, aggregations, and functional properties. *Compr. Rev. Food Sci. Food Saf.* **2021**, *20*, 627–651. [CrossRef]
5. Tomé, A.S.; Pires, C.; Batista, I.; Sousa, I.; Raymundo, A. Protein gels and emulsions from mixtures of Cape hake and pea proteins. *J. Sci. Food Agric.* **2015**, *95*, 289–298. [CrossRef]
6. Zhang, X.Y.; Qi, B.K.; Xie, F.Y.; Hu, M.; Sun, Y.F.; Hu, L.; Li, L.; Zhang, S.; Li, Y. Emulsion stability and dilatational rheological properties of soy/whey protein isolate complexes at the oil-water interface: Influence of pH. *Food Hydrocoll.* **2021**, *113*, 106391. [CrossRef]
7. Ho, K.; Schroën, K.; Martín-González, M.; Berton-Carabin, C. Synergistic and antagonistic effects of plant and dairy protein blends on the physicochemical stability of lycopene-loaded emulsions. *Food Hydrocoll.* **2018**, *81*, 180–190. [CrossRef]
8. Wu, C.; Yan, X.; Wang, T.; Ma, W.; Xu, X.; Du, M. A self-sorted gel network formed by heating a mixture of soy and cod proteins. *Food Funct.* **2019**, *10*, 5140–5151. [CrossRef]
9. Alu'datt, M.H.; Al-Rabadi, G.J.; Alli, I.; Ereifej, K.; Rababah, T.; Alhamad, M.N.; Torley, P.J. Protein co-precipitates: A review of their preparation and functional properties. *Food Bioprod. Process.* **2013**, *91*, 327–335. [CrossRef]
10. Kristensen, H.T.; Christensen, M.; Hansen, M.S.; Hammershj, M.; Dalsgaard, T.K. Protein–protein interactions of a whey-pea protein co-precipitate. *Int. J. Food Sci. Technol.* **2021**, *56*, 5777–5790. [CrossRef]
11. Bouhallab, S.; Croguennec, T. Spontaneous Assembly and Induced Aggregation of Food Proteins. *Adv. Polym. Sci.* **2013**, *256*, 67–101.
12. Youssef, A.M.; Abu-Foul, N.S.; Moharram, Y.G. Preparation and characteristics of co-precipitate proteins from oil seeds and legumes seeds. *Food Nahrung.* **1995**, *39*, 475–482. [CrossRef]
13. Alu'datt, M.H.; Alli, I.; Nagadi, M. Preparation, characterization and properties of whey-soy proteins co-precipitates. *Food Chem.* **2012**, *134*, 294–300. [CrossRef]
14. Tan, L.; Hong, P.; Yang, P.; Zhou, C.; Xiao, D.; Zhong, T. Correlation between the Water Solubility and Secondary Structure of Tilapia-Soybean Protein Co-Precipitates. *Molecules* **2019**, *24*, 4337. [CrossRef]
15. Dai, Z.Q.; Zhang, M.X.; Zheng, H.N.; Hong, P.Z.; Zhou, C.X. Emulsifying Properties Of Tilapia-Soybean Protein Co-Precipitates. *Food Ferment. Ind.* **2015**, *41*, 63–69.
16. Qi, H.H.; Zhou, C.X.; Zhu, P.H.; Li, T.; Hong, P.Z. Co-precipitated protein from tilapia muscle and soybean meal and its solubility and amino acid composition analysis. *Food Ferment. Ind.* **2018**, *44*, 166–172.
17. FAOSTAT. Production and prices statistics of the Food and Agriculture Organization of the United Nations. 2019. Available online: http://faostat.fao.org (accessed on 26 October 2021).
18. Ge, J.; Sun, C.X.; Corke, H.; Gul, K.; Gan, R.Y.; Fang, Y.P. The health benefits, functional properties, modifications, and applications of pea (*Pisum sativum* L.) protein: Current status, challenges, and perspectives. *Compr. Rev. Food Sci. Food Saf.* **2020**, *19*, 1835–1876. [CrossRef]
19. Tahergorabi, R.; Matak, K.E.; Jaczynski, J. Fish protein isolate: Development of functional foods with nutraceutical ingredients. *J. Funct. Foods* **2015**, *18*, 746–756. [CrossRef]
20. Yang, Y.L.; Peng, X.B.; Dong, Z.; You, Y.; Xu, H.P. Study on emulsifying properties of salt-soluble proteins from grass carp. *Sci. Technol. Food Ind.* **2013**, *34*, 121–125.
21. Chang, C.; Tu, S.; Ghosh, S.; Nickerson, M.T. Effect of pH on the inter-relationships between the physicochemical, interfacial and emulsifying properties for pea, soy, lentil and canola protein isolates. *Food Res. Int.* **2015**, *77*, 360–367. [CrossRef]
22. Kristinsson, H.G.; Hultin, H.O. Changes in Trout Hemoglobin Conformations and Solubility after Exposure to Acid and Alkali pH. *J. Agric. Food Chem.* **2004**, *52*, 3633–3643. [CrossRef]
23. Haskard, C.A.; Li-Chan, E.C. Hydrophobicity of bovine serum albumin and ovalbumin determined using uncharged (PRODAN) and anionic (ANS-) fluorescent probes. *J. Agric. Food Chem.* **1998**, *46*, 2671–2677. [CrossRef]
24. Fekria, A.M.; Ahmed, I.; Ahmed, S.O.; Babiker, E.E. Nutritional and functional characterization of defatted seed cake of two Sudanese groundnut (*Arachis Hypogaea*) cultivars. *Int. Food Res. J.* **2012**, *19*, 629–637.
25. Pearce, K.N.; Kinsella, J.E. Emulsifying properties of proteins: Evaluation of a turbidimetric technique. *J. Agric. Food Chem.* **1978**, *26*, 716–723. [CrossRef]
26. Liang, H.N.; Tang, C.H. Pea protein exhibits a novel Pickering stabilization for oil-in-water emulsions at pH 3.0. *Lwt-Food Sci. Technol.* **2014**, *58*, 463–469. [CrossRef]
27. Liu, C.H.; Wang, R.C.; He, S.H.; Chen, C.L.; Ma, Y. The stability and gastro-intestinal digestion of curcumin emulsion stabilized with soybean oil bodies. *Lwt-Food Sci. Technol.* **2020**, *131*, 109663. [CrossRef]
28. Chen, M.; Lu, J.; Fei, L.; Nsor-Atindana, J.; Fang, Z. Study on the emulsifying stability and interfacial adsorption of pea proteins. *Food Hydrocoll.* **2019**, *88*, 247–255. [CrossRef]
29. Xu, Y.; Wang, R.; Zhao, H.L.; Zhao, J.M.; Li, X.P.; Yi, S.M.; Li, J.R.; Sun, X.T. Binding of aldehydes to myofibrillar proteins as affected by two-step heat treatments. *J. Sci. Food Agric.* **2020**, *100*, 1195–1203. [CrossRef]
30. Kristinsson, H.; Hultin, H. Changes in conformation and subunit assembly of cod myosin at low and high pH and after subsequent refolding. *J. Agric. Food Chem.* **2003**, *51*, 7187–7196. [CrossRef]

31. Casey, R. Immunoaffinity chromatography as a means of purifying legumin from *Pisum*(pea) seeds. *J. Biochem.* **1979**, *177*, 509–520. [CrossRef]
32. Barac, M.; Cabrilo, S.; Pesic, M.; Stanojevic, S.; Zilic, S.; Macej, O.; Ristic, N. Profile and functional properties of seed proteins from six pea (*Pisum sativum*) genotypes. *Int. J. Mol. Sci.* **2010**, *11*, 4973–4990. [CrossRef]
33. Joshi, M.; Adhikari, B.; Aldred, P.; Panozzo, J.F.; Kasapis, S.; Barrow, C.J. Interfacial and emulsifying properties of lentil protein isolate. *Food Chem.* **2012**, *134*, 1343–1353. [CrossRef]
34. Lam, A.C.Y.; Can Karaca, A.; Tyler, R.T.; Nickerson, M.T. Pea protein isolates: Structure, extraction, and functionality. *Food Rev. Int.* **2018**, *34*, 126–147. [CrossRef]
35. Lu, B.Y.; Quillien, L.; Popineau, Y. Foaming and emulsifying properties of pea albumin fractions and partial characterisation of surface-active components. *J. Sci. Food Agric.* **2000**, *80*, 1964–1972. [CrossRef]
36. Kimura, A.; Fukuda, T.; Zhang, M.; Motoyama, S.; Maruyama, N.; Utsumi, S. Comparison of physicochemical properties of 7S and 11S globulins from pea, fava bean, cowpea, and French bean with those of soybean-French bean 7S globulin exhibits excellent properties. *J. Agric. Food Chem.* **2008**, *56*, 10273–10279. [CrossRef]
37. Du, M.; Xie, J.; Gong, B.; Xin, X.; Wei, T.; Xuan, L.; Chang, L.; Xie, M. Extraction, physicochemical characteristics and functional properties of Mung bean protein. *Food Hydrocoll.* **2017**, *76*, 131–140. [CrossRef]
38. Cui, L.; Bandillo, N.; Wang, Y.; Ohm, J.B.; Rao, J. Functionality and structure of yellow pea protein isolate as affected by cultivars and extraction pH. *Food Hydrocoll.* **2020**, *108*, 106008. [CrossRef]
39. Chen, D.; Song, J.; Yang, H.; Xiong, S.B.; Liu, Y.M.; Liu, R. Effects of Acid and Alkali Treatment on the Properties of Proteins Recovered from Whole Gutted Grass Carp (*Ctenopharyngodon idellus*) Using Isoelectric Solubilization/Precipitation. *J. Food Qual.* **2016**, *39*, 707–713. [CrossRef]
40. Wihodo, M.; Moraru, C.I. Physical and chemical methods used to enhance the structure and mechanical properties of protein films: A review. *J. Food Eng.* **2013**, *114*, 292–302. [CrossRef]
41. Kato, A.; Nakai, S. Hydrophobicity determination by a fluorescence probe method and its correlation with surface properties of proteins. *Biochim. Biophys. Acta* **1980**, *624*, 13–20. [CrossRef]
42. Roesch, R.; Juneja, M.; Monagle, C.; Corredig, M. Aggregation of soy/milk mixes during acidification. *Food Res. Int.* **2004**, *37*, 209–215. [CrossRef]
43. Liang, H.N.; Tang, C.H. pH-dependent emulsifying properties of pea [*Pisum sativum* (L.)] proteins. *Food Hydrocoll.* **2013**, *33*, 309–319. [CrossRef]
44. Chi, E.Y.; Krishnan, S.; Randolph, T.W.; Carpenter, J.F. Physical stability of proteins in aqueous solution: Mechanism and driving forces in nonnative protein aggregation. *Pharm. Res.* **2003**, *20*, 1325–1336. [CrossRef]
45. Amagliani, L.; Silva, J.V.C.; Saffon, M.; Dombrowski, J. On the foaming properties of plant proteins: Current status and future opportunities. *Trends Food Sci. Technol.* **2021**, *118*, 261–272. [CrossRef]
46. Adebiyi, A.P.; Aluko, R.E. Functional properties of protein fractions obtained from commercial yellow field pea (*Pisum sativum* L.) seed protein isolate. *Food Chem.* **2011**, *128*, 902–908. [CrossRef]
47. Zhao, H.F.; Shen, C.; Wu, Z.J.; Zhang, Z.; Xu, C.M. Comparison of wheat, soybean, rice, and pea protein properties for effective applications in food products. *J. Food Biochem.* **2020**, *44*, e13157. [CrossRef] [PubMed]
48. Morales, R.; Martínez, K.D.; Ruiz-Henestrosa, V.M.; Pilosof, A.M.R. Modification of foaming properties of soy protein isolate by high ultrasound intensity: Particle size effect. *Ultrason. Sonochem.* **2015**, *26*, 48–55. [CrossRef]
49. Liu, S.; Zhao, P.; Zhang, J.; Xu, Q.; Ding, Y.; Liu, J. Physicochemical and functional properties of silver carp (*Hypophthalmichthys molitrix*) myofibrillar protein glycated with konjac oligo-glucomannan. *Food Hydrocoll.* **2017**, *67*, 216–223. [CrossRef]
50. Lawal, O.S.; Adebowale, K.O.; Ogunsanwo, B.M.; Sosanwo, O.A.; Bankole, S.A. On the fuctional properties of globulin and albumin protein fractions and flours of African locust bean (*Parkia biglobossa*). *Food Chem.* **2005**, *92*, 681–691. [CrossRef]
51. Jiang, J.; Chen, J.; Xiong, Y.L. Structural and emulsifying properties of soy protein isolate subjected to acid and alka line pH-shifting processes. *J. Agric. Food Chem.* **2009**, *57*, 7576–7583. [CrossRef]
52. Zeeb, B.; Herz, E.; McClements, D.J.; Weiss, J. Impact of alcohols on the formation and stability of protein-stabilized nanoemulsions. *J. Colloid Interf. Sci.* **2014**, *433*, 196–203. [CrossRef]
53. Silva, M.; Bui, T.H.; Dharmadana, D.; Zisu, B.; Chandrapala, J. Ultrasoundassisted formation of double emulsions stabilized by casein-whey protein mixtures. *Food Hydrocoll.* **2020**, *109*, 106143. [CrossRef]
54. Gao, H.; Ma, L.; Li, T.Q.; Sun, D.X.; Hou, J.C.; Li, A.L.; Jiang, Z. Impact of ultrasonic power on the structure and emulsifying properties of whey protein isolate under various pH conditions. *Process. Biochem.* **2019**, *81*, 113–122. [CrossRef]
55. Mcclements, D.J. Critical review of techniques and methodologies for characterization of emulsion stability. *Crit. Rev. Food Sci.* **2007**, *47*, 611–649. [CrossRef]
56. Fernandez-Avila, C.; Arranz, E.; Guri, A.; Trujillo, A.J.; Corredig, M. Vegetable protein isolate-stabilized emulsions for enhanced delivery of conjugated linoleic acid in Caco-2 cells. *Food Hydrocoll.* **2016**, *55*, 144–154. [CrossRef]
57. Li, Y.; Wu, C.L.; Liu, J.; Zhu, Y.; Zhang, X.Y.; Jiang, L.Z.; Qi, B.K.; Zhang, X.N.; Wang, Z.J.; Teng, F. Soy protein isolate-phosphatidylcholine nanoemulsions prepared using high-pressure homogenization. *Nanomaterials* **2018**, *8*, 307. [CrossRef]
58. Liu, F.; Tang, C.H. Soy protein nanoparticle aggregates as pickering stabilizers for oil-in-water emulsions. *J. Agric. Food Chem.* **2013**, *61*, 8888–8898. [CrossRef]

59. Romero, A.; Cordobés, F.; Guerrero, A. Influence of pH on linear viscoelasticity and droplet size distribution of highly concentrated O/W crayfish flour-based emulsions. *Food Hydrocoll.* **2009**, *23*, 244–252. [CrossRef]
60. Sriprablom, J.; Luangpituksa, P.; Wongkongkatep, J.; Pongtharangkul, T.; Suphantharika, M. Influence of pH and ionic strength on the physical and rheological properties and stability of whey protein stabilized o/w emulsions containing xanthan gum. *J. Food Eng.* **2019**, *242*, 141–152. [CrossRef]
61. Dickinson, E. Flocculation of protein-stabilized oil-in-water emulsions. *Colloids Surf. B Biointerfaces* **2010**, *81*, 130–140. [CrossRef]
62. Malhotra, A.; Coupland, J.N. The effect of surfactants on the solubility, zeta potential, and viscosity of soy protein isolates. *Food Hydrocoll.* **2004**, *18*, 101–108. [CrossRef]

foods

MDPI

Article

Changes of High-Purity Insoluble Fiber from Soybean Dregs (Okara) after Being Fermented by Colonic Flora and Its Adsorption Capacity

Bo Lyu [1,2,†], Yi Wang [2,3,†], Xin Zhang [1,2], Yuxi Chen [2,3], Hongling Fu [2,3], Tong Liu [4], Jianyu Hao [5], Yang Li [1,2], Hansong Yu [2,3,*] and Lianzhou Jiang [1,2,*]

1 College of Food Science, Northeast Agricultural University, Harbin 150030, China; michael_lvbo@163.com (B.L.); 18249032236@163.com (X.Z.); yangli@neau.edu.cn (Y.L.)
2 Soybean Research & Development Center, Division of Soybean Processing, Chinese Agricultural Research System, Changchun 130118, China; Wangyi284419@163.com (Y.W.); chenyuxi19970624@163.com (Y.C.); 15764381475@163.com (H.F.)
3 College of Food Science and Engineering, Jilin Agricultural University, Changchun 130118, China
4 College of Food Science and Engineering, Changchun University, Changchun 130118, China; liut@ccu.edu.cn
5 School of Food and Biotechnology, Changchun Vocational Institute of Technology, Changchun 130118, China; haojianyu55@126.com
* Correspondence: yuhansong@jlau.edu.cn (H.Y.); jlzname@neau.edu.cn (L.J.); Tel.: +86-133-3176-0468 (H.Y.); +86-139-0465-1669 (L.J.)
† These authors contributed equally to this work.

Abstract: In order to explore the changes and properties of high-purity insoluble dietary fiber from okara (HPIDF) after entering the colon and be fermented by colonic flora, fermented high-purity insoluble dietary fiber (F-HPIDF) was obtained by simulated fermentation in vitro by HPIDF and colonic flora from C57BL/6 mice. For exploring the differences of HPIDF and F-HPIDF, the changes of structure (SEM. FTIR, XRD, particle size, specific surface area, monosaccharide composition) and adsorption properties (water, oil, heavy metal irons, harmful substances) of HPIDF/F-HPIDF were explored. The results showed that F-HPIDF had a higher water-holding capacity (19.17 g/g), water-swelling capacity (24.83 mL/g), heavy metals-adsorption capacity (Cd^{2+}: 1.82 μmol/g; Pb^{2+}: 1.91 μmol/g; Zn^{2+}: 1.30 μmol/g; Cu^{2+}: 0.68 μmol/g), and harmful substances-adsorption capacity (GAC: 0.23 g/g; CAC: 14.80 mg/g; SCAC: 0.49 g/g) than HPIDF due to the changes of structure caused by fermentation. In addition, with the fermentation of HPIDF, some beneficial substances were produced, which might be potential intestinal prebiotics. The study of F-HPIDF strengthens the speculation that HPIDF may have potential bioactivities after entering the colon, which proved that okara-HPIDF may have potential functionality.

Keywords: soybean residue; insoluble dietary fiber; gut microbiota; fermentation in vitro; colon health

Citation: Lyu, B.; Wang, Y.; Zhang, X.; Chen, Y.; Fu, H.; Liu, T.; Hao, J.; Li, Y.; Yu, H.; Jiang, L. Changes of High-Purity Insoluble Fiber from Soybean Dregs (Okara) after Being Fermented by Colonic Flora and Its Adsorption Capacity. *Foods* **2021**, *10*, 2485. https://doi.org/10.3390/foods10102485

Academic Editor: Adriana S. Franca

Received: 10 September 2021
Accepted: 14 October 2021
Published: 17 October 2021

1. Introduction

The functionality of insoluble dietary fiber (IDF) has been gradually accepted by consumers [1–3]. Especially for intestinal health, IDF may have unique physiological activities [4–6]. As a kind of material that can improve intestinal health, IDF had been gradually developed as a kind of functional food material. However, recent studies have shown that some kinds of IDF are not necessarily beneficial to intestinal health [7–9], which may be due to the complex fermentation mode of IDF after entering the colon. It had been proved that roughly ten subtypes of dietary fibers described to date, categorized as soluble or insoluble, with varying chemical structures, and large differences in their fermentation profiles [10]. This means that the changes of gut microbiota and colonic environment are complex. Therefore, the changes of different kinds of IDF in the colon become worthy to study.

Okara (Soybean dregs), the main by-product during traditional soybean products processing and the production of soybean protein isolate (SPI), had caused huge environmental pollution and waste of resources [11], which includes a large amount of dietary fiber (DF). Okara-DF had been proved to be functional, such as decreased content of blood glucose [12], protected the intestine [13], used as a food raw material ingredient [14], etc. However, different from soluble dietary fiber (SDF), the application of Okara-IDF had a core problem—IDF was difficult to purify. The traditional processing method of IDF is often accompanied by a large amount of protein and soluble fiber as remains. It is difficult to evaluate the functionality of one component in the mixture, as while as caused a disturbance, In addition, okara was difficult to dry because of its unique structure [15]. For these reasons, the usage or value-increment of Okara-IDF as a material of functional food was restricted.

To solve the problem, we prepared high purity insoluble dietary fiber from okara (Okara-HPIDF) by enzymatic method, which purity could exceed 90% [16], and proved that it had potential bioactivities. Meanwhile, we proved that Okara-HPIDF had certain functions, such as improving lipid metabolism and so on [17,18]. In the process of studying its relationship with the changes of the colonic environment, we found that Okara-HPIDF could significantly change the composition of colonic flora [18]. This means that Okara-HPIDF was fully fermented by microorganisms in the colon, which may lead to changes in structure, composition, functionality, and even bioactivities of Okara-HPIDF. The above changes had not been studied, because we used germ-free simulated digestive fluid (SGF) in the previous study [16].

In this way, this study aimed to explore the changes after Okara-HPIDF enters the colon, while being fermented by colonic microorganisms. In this work, we were committed to treating Okara-HPIDF by simulated intestinal fluid (SIF) with colonic microorganisms and digestive enzymes, to look for the changes of Okara-HPIDF after fermentation. The structure, monosaccharide composition, and physicochemical properties before and after fermentation were studied, and a more accurate model of simulated fermentation in vitro of Okara-HPIDF was established. This study is a supplement to our team's previous research, to explore the changes and properties of HPIDF after entering the colon and be fermented by colonic flora, which will improve our evaluation system for the potential biological activity of HPIDF and provide a theoretical basis for improving the biological activity of HPIDF. This work will also be the basic theory to functional research of Okara-HPIDF, which is also the core data of the relationship between the intake of Okara-HPIDF and colon health.

2. Materials and Methods

2.1. Preparation of Okara-HPIDF

Low purity dietary fiber from okara (okara-LPDF) was purchased from Shandong Sinoglory Health Food Co., Ltd., Liaocheng, China. After identification, its main components include: dietary fiber $\geq$ 60%, protein $\leq$ 26%, moisture $\leq$ 10%, ash $\leq$ 6%. HPIDF from okara was prepared by complex enzymatic method of LPDF [17]. The purity of HPIDF was 91.25% (according to the enzymatic-gravimetric method described in Chinese national standard GB 5009.88-2014).

2.2. The Fermentation of Okara-HPIDF

The preparation method of simulated intestinal fluid (SIF) was used following the United States Pharmacopeia (USP) [19]. Enzyme-free SIF was prepared in a sterile environment and sterilized at 121 °C for 20 min.

The collection method of feces was similar to the previous study [16] with minor changes: 10 g of feces was collected from C57BL/6 female mice (Beijing Vital River Laboratory Animal Technology Co. Ltd., Beijing, China), which was dissolved in 50 mL phosphate buffer solution (0.1 mol/L, pH = 6.5). After complete mixing, the residue was removed via filtration with four layers of gauze. The suspension was stored at −20 °C in a sterile centrifuge (finished in 1 h).

After the preparation of SIF and the extraction of colonic microorganisms, they were mixed in a ratio of 7:2 (*v*/*v*). The simulated intestinal fluid with a colonic microorganism (SIF-CM) was used as the fermentation environment for HPIDF. The SIF-CM solution was prepared just before use.

2.3. Bacterial 16S rDNA Sequencing

Feces collected in 2.2 were used for 16S rDNA sequencing. DNA extraction from feces was carried out using the QIAamp® DNA Stool Mini Kit (QIAGEN, Germany) according to the manufacturer's instructions. The quality of isolated DNA was evaluated on agarose gel electrophoresis and then DNA concentration was precisely measured using a NanoDrop NC2000 spectrophotometer (Thermo Fisher, Waltham, MA, USA).

The V3-V4 region of the 16S rRNA gene was amplified from the purified DNA of each sample. The primers used for the PCR amplification are as follows: the forward primer 515F (5′-ACTCCTACGGGAGGCAGCA-3′) and the reverse primer 907R (5′-GGACTACHVGGGTWTCTAAT-3′). The concentrations of PCR products were measured by the Quant-iT PicoGreen dsDNA Assay Kit (Invitrogen, Carlsbad, CA, USA). Afterward, all PCR products were pooled in equal amounts, and paired-end sequencing was performed using the Illumina MiSeq platform operated by Shanghai Personal Biotechnology Co., Ltd. (Shanghai, China).

2.4. Simulated Fermentation of Okara-HPIDF In Vitro

After mixing SIF-CM and HPIDF (9:1, *v*/*m*) under sterile conditions, the samples were cultured under anaerobic conditions at 37 °C for 16 h. After fermentation, the samples were filtered, the IDF was taken to determine the contents of the composition of monosaccharide after sterilization. The fermented-HPIDF (F-HPIDF) was used for the follow-up experiments after sterilization and freeze-drying.

2.5. Structure of Okara-HPIDF before/after Fermentation

2.5.1. Scanning Electron Microscopy (SEM)

Merlin Compact scanning electron microscope (SEM; Carl Zeiss Jena GmbH, Jena, Germany) was used to observed the microstructure of HPIDF/F-HPIDF after the spray gold treatment. The scanning images were captured at accelerating voltages of 5.00 kV. All images were recorded at magnifications of 300× (low magnification) and 3000× (high magnification).

2.5.2. Fourier Infrared Spectrum (FT-IR)

The polysaccharide functional group composition of HPIDF/F-HPIDF were performed using a Nicolet iS5 FT-IR spectrometer (Thermo Fisher, Waltham, MA, USA). The spectra were read over the range of 4000–400 cm^{-1} with a resolution of 4 cm^{-1} after the samples were mixed with KBr (1:100, *w*/*w*)

2.5.3. X-ray Diffraction (XRD)

X-ray diffraction (XRD) analysis of the HPIDF/F-HPIDF was conducted according to the previous method [20], with some slight modifications. The diffraction patterns were recorded using a Rigaku Ultima IV (X-ray, Rigaku Corporation, Tokyo, Japan) operating at a voltage of 40 kV, an incident current of 40 mA, an anti-diffusion slit of 2/3, with Cu-Ka radiation (1 μ = 0.154 nm), and a scan speed of 1π/min in the range of diffraction angle from 10 to 80°.

2.5.4. Particle Size and Specific Surface Area

The BT-9300HT laser particle sizer (Bettersize Instruments Ltd., Dandong, China) was used to analyze the particle size and specific surface area of HPIDF/F-HPIDF in which the concentration of the suspension is 4% (*m*/*v*).

2.6. Monosaccharide Composition of Okara-HPIDF before/after Fermentation and the Hydrolyzed

The monosaccharide compositions of HPIDF/F-HPIDF referred to the methods of the preliminary study [16]. The monosaccharide composition of the hydrolyzed was analyzed after freeze-drying by PMP pre-column derivatization and LC-10ATvp & SPD-10AVD HPLC system (Shimadzu, Tokyo, Japan) [21].

2.7. Adsorption Capacity in the Colon of Okara-HPIDF before/after Fermentation

2.7.1. Basic Characteristics of HPIDF/F-HPIDF

We chose water-holding capacity (WHC), water-swelling capacity (WSC), and oil-holding capacity (OHC) as the basic characteristics of HPIDF/F-HPIDF. The determination was described by Zhang et al. [22] with minor changes. Briefly, water-holding capacity (WHC): 1.00 g sample (M1) and 20 mL of water were mixed in a dry centrifuge tube (M0). The sample was kept at RT for 24 h and centrifuged at 4000 rpm for 20 min. The supernatant was removed, and the weight (M2) was measured.

$$\text{WHC (g/g)} = (\text{M2} - \text{M0})/\text{M1} * 100\%$$

Water-swelling capacity (WSC): 0.10 g (M0) sample was taken into the calibration tube (V0), 10 mL water was added and let to stand for 24 h, and the volume (V1) was recorded.

$$\text{WSC (mL/g)} = (\text{V1} - \text{V0})/\text{M0}$$

Oil-holding capacity (OHC): 1.00 g sample (M1) and 20 mL of oil (soybean oil/lard) were mixed in a dry centrifuge tube (M0). The sample was kept at RT for 18 h and centrifuged at 4000 rpm for 20 min. The supernatant was removed, and the weight (M2) was measured.

$$\text{OHC (g/g)} = (\text{M2} - \text{M0})/\text{M1} * 100\%$$

Soybean oil represents unsaturated oil, and lard represents saturated oil.

2.7.2. Heavy Metals-Adsorption Capacity (HMAC)

HMAC was measured by the methods of the previous study [16] with minor changes. The 50 μg/mL standard solution of Pb^{2+}($Pb(NO_3)_2$), Zn^{2+}($Zn(NO_3)_2$), Cu^{2+}($Cu(NO_3)_2$), and Cd^{2+}($Cd(NO_3)_2$) were prepared as the initial concentration to determine the HMAC of HPIDF/F-HPIDF. Total of 10 mL of four kinds of standard solutions were mixed with 0.1 g HPIDF (500 μg of metal ions in the system) separately, and stirred at 37 °C for 14 h before centrifuging at 4000 rpm for 20 min. The 240 Duo GFAAS (Agilent, Lexington, MA, USA) was used to detect the content of Pb^{2+}, Zn^{2+}, Cu^{2+}, and Cd^{2+} in the supernatant according to the preinstalled methods of the workstation.

2.7.3. Potentially Harmful Substances-Adsorption Capacity

Glucose-adsorption capacity (GAC), cholesterol-adsorption capacity (CAC), sodium cholate-adsorption capacity (SCAC), acrylamide-adsorption capacity (AAC), and nitrite-adsorption capacity (NAC) were selected as the measurement indicators of potentially harmful substances-adsorption capacity. The details are as follows:

1. GAC 0.1 g HPIDF/F-HPIDF, mixed with 10 mL 100 mmol/L glucose solution at 37 °C for 16 h, centrifuged at 4000 rpm for 20 min, and the supernatant was taken to determine the glucose concentration using a glucose kit (hexokinase method, A154-2-1, Nanjing Jiancheng Bioengineering Institute).

2. CAC 1 mg/mL cholesterol in ethanol solution was prepared. About 0.2 g HPIDF/F-HPIDF was mixed in 10 mL cholesterol solution. The adsorption and centrifugation conditions were the same as the above. Total cholesterol assay kit (A111-1-1, Nanjing Jiancheng Bioengineering Institute) was used to determine the cholesterol content in the supernatant.
3. SCAC 0.1 g HPIDF/F-HPIDF was mixed with 10 mL sodium cholate standard solution (0.2 g sodium cholate + 15 mmol/L NaCl aq 100 mL). The adsorption and centrifugation conditions were the same as 2.7.3-1. Furfural-sulfuric acid process [23] was used to determine the content of sodium cholate in the supernatant.
4. AAC 0.5 g HPIDF/F-HPIDF was mixed with 50 mL 15 mmol/L acrylamide solution. The adsorption and centrifugation conditions were the same as 2.7.3-1. HPLC was used to determine the concentration of acrylamide in the supernatant according to the method described in GB 5009.204-2014 by 1260 HPLC (Agilent, Lexington, MA, USA).
5. NAC 0.5 g HPIDF/F-HPIDF was mixed with 50 mL 100 mmol/L nitrite (Sodium nitrite) solution. The adsorption and centrifugation conditions were the same as the above. Ultraviolet spectrophotometry method (described in GB 5009.23-2010) was used to determine the concentration of sodium nitrite in the supernatant.

2.8. Statistical Analysis

The measurements of all samples should be repeated for at least three times and the results were expressed as the mean ± standard deviations ($\bar{x} \pm$ SD). ANOVA with Duncan's test and Tukey's test was used to compare the data for the differences, $p < 0.05$ represented significant differences.MS Word and GraphPad Prism 6 (GraphPad Software Inc., San Diego, CA, USA) were used to organize the data, draw the graphs and tables.

3. Results and Discussion

3.1. The Gut Microbiota Structure in the Feces from C57BL/6 Mice

The microbiota structure in the colon was analyzed by high-throughput sequencing spanning the 16S rDNA V3-V4 hypervariable region and shown in Figure 1. At the phylum level, *Firmicutes* (55.17%) and *Bacteroidetes* (43.62%) were abundant. As for the genus level, *Lactobacillus* (14.71%) accounted for the highest abundance of recognizable conventional microorganisms, which was consistent with the microbial composition of normal C57BL/6 mice feces [24]. C57BL/6 mice were used as a suitable animal model to evaluate dietary fiber and colon health. In other studies of HPIDF, we also chose this animal model. Therefore, we selected their feces for gut microbiota isolation and fermented HPIDF in vitro.

3.2. The Structural Changes of HPIDF after Being Fermented

3.2.1. Scanning Electron Micrograph (SEM)

The microstructure of HPIDF/F-HPIDF at 300 times and 3000 times was shown in Figure 2. It can be observed that whether at low magnification or high, F-HPIDF showed a few differences compared to HPIDF. At low magnification (Figure 2a,c), the state of a single dietary fiber particle of F-HPIDF was richer in spatial structure and less smooth than that of HPIDF. In addition, the edge of F-HPIDF's particle was smoother and the tearing feeling was lower than that of HPIDF. At high magnification (Figure 2b,d), the flatness of HPIDF was much higher than that of F-HPIDF, whose structure was chaotic and uneven. This might indicate that F-HPIDF had a higher specific surface area than HPIDF and may be attributed to the removal of some saccharide structure (monosaccharide, etc.,) and the fracture of fiber structure of HPIDF during fermentation. Therefore, we can speculate that some adsorption properties of F-HPIDF should be better than HPIDF because its specific surface area increased (Premise: no functional groups with adsorption properties were removed). This assumption needs to be verified by the following experiments.

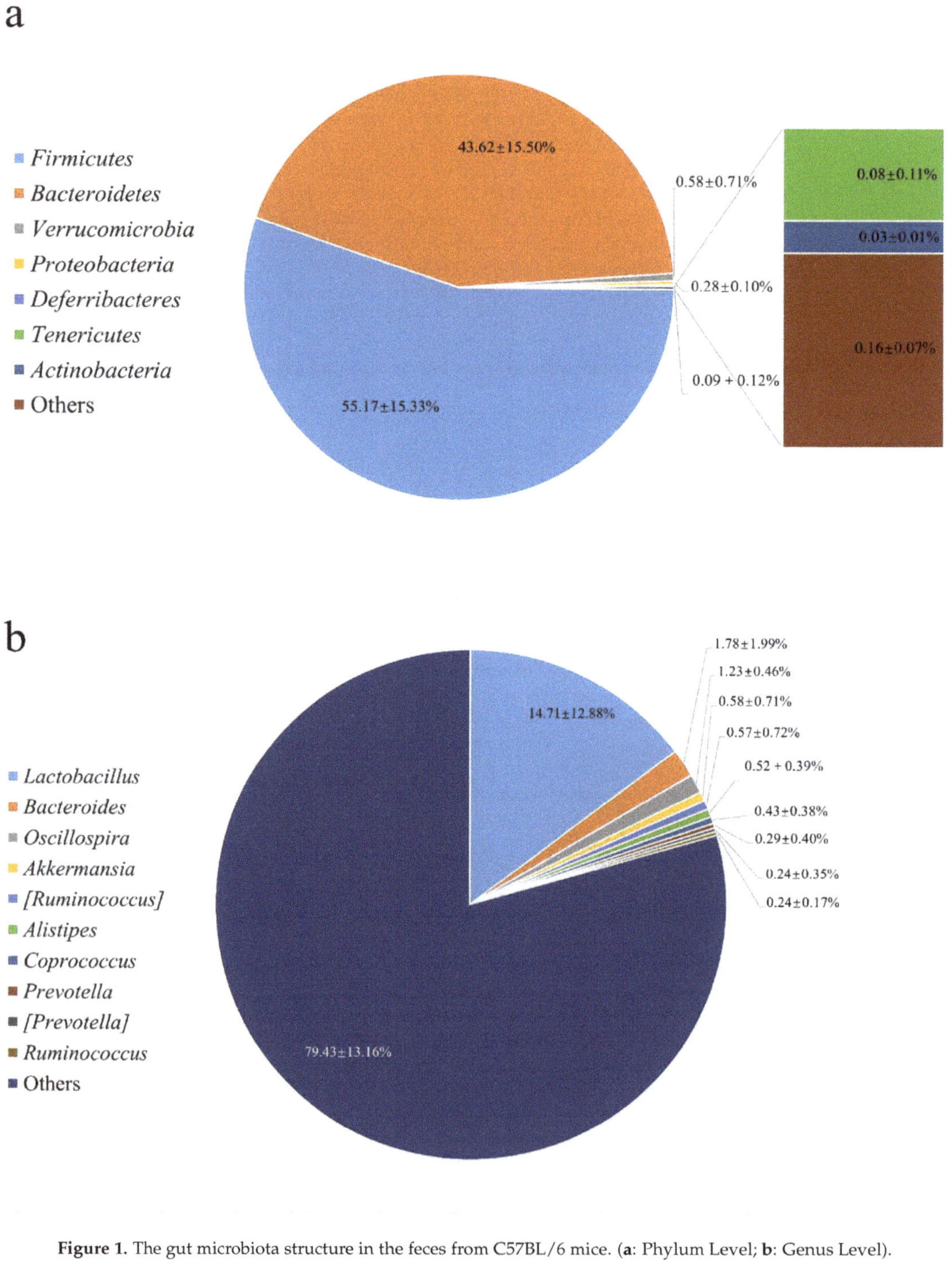

Figure 1. The gut microbiota structure in the feces from C57BL/6 mice. (**a**: Phylum Level; **b**: Genus Level).

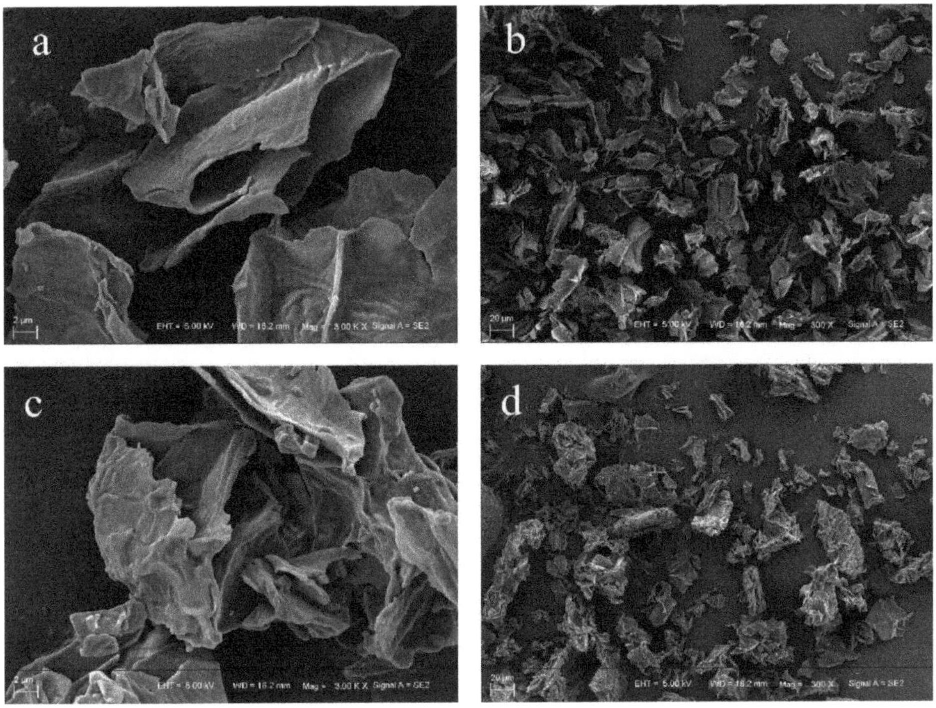

Figure 2. Scanning electron micrograph (SEM) of HPIDF/F-HPIDF. (**a**: HPIDF at high magnification; **b**: HPIDF at low magnification; **c**: F-HPIDF at high magnification; **d**: F-HPIDF at low magnification).

3.2.2. X-ray Diffraction (XRD)

The XRD analysis of HPIDF/F-HPIDF is shown in Figure 3. It could be seen from the XRD spectra that, whether HPIDF or F-HPIDF, there was just one obvious crystal diffraction peak at $2\theta = 22°$, which was a characteristic XRD curve of hemicellulose [25]. Meanwhile, there was little difference between HPIDF and F-HPIDF, which proved our previous speculation that the cellulose content in HPIDF is less [16]. XRD spectra showed the same results. In short, the fermentation process did not change the crystal structure of HPIDF.

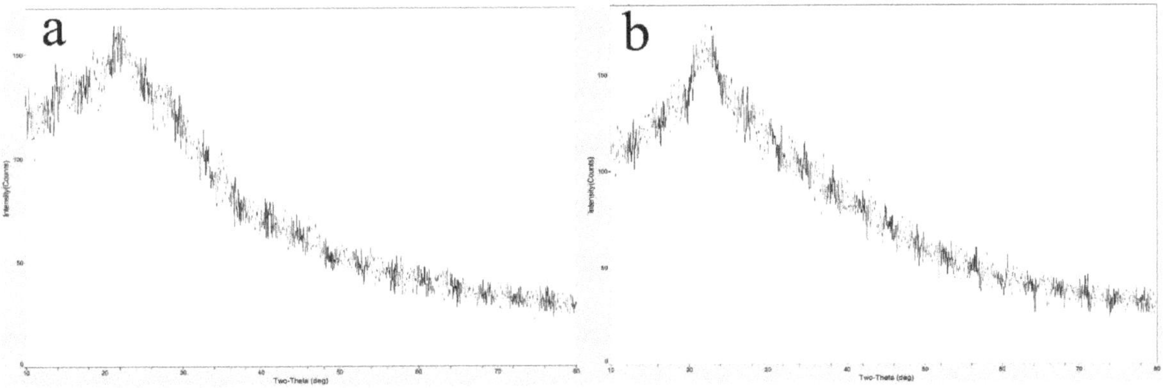

Figure 3. X-ray diffraction (XRD) of HPIDF/F-HPIDF. (**a**: HPIDF; **b**: F-HPIDF).

3.2.3. Fourier Transform Infrared Spectroscopy (FT-IR)

The FR-IR spectrum of HPIDF/F-HPIDF is shown in Figure 4. At 3353 cm^{-1}, 2932 cm^{-1}, 1737 cm^{-1}, 1626 cm^{-1}, 1418 cm^{-1}, 1374 cm^{-1},1248 cm^{-1}, 1057 cm^{-1}, 706 cm^{-1}, 623 cm^{-1}, and 532 cm^{-1}, both HPIDF/F-HPIDF showed significant absorption peak, which represented the composition of their polysaccharide functional groups. Actually, although HPIDF and F-HPIDF showed some differences in peak height, no differences in peak location were observed. In contrast with other studies [26–28], these absorption peaks represented O-H group in cellulose or hemicellulose, C-H stretching of $-CH_3$ or $=CH_2$ on carboxymethyl and methylene, oxygen (CO-OR) stretching vibration in hemicelluloses, C-O stretching vibration in the guaiacyl unit of lignin, glycuronic acid, etc. The other small peaks were not different from the previous study [16].

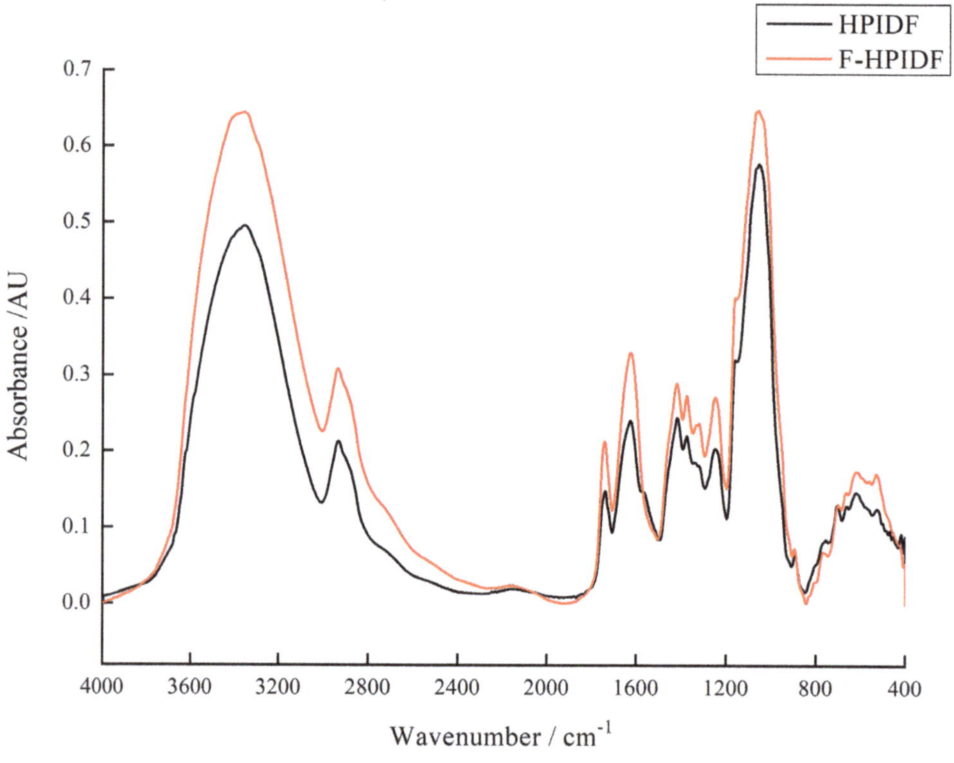

Figure 4. Fourier transform infrared spectrum (FTIR) of HPIDF/F-HPIDF. (Black: HPIDF; Red: F-HPIDF).

Although no differences were found in peak location, F-HPIDF had a markable characteristic in the FT-IR spectrum analysis. All peaks in F-HPIDF are higher than that in HPIDF, which means the fermentation increased the exposure of all functional groups. This result was consistent with the SEM results. The more complex the spatial structure, the greater the exposure of functional groups. This inference needs to be supported by the results of particle size and the specific surface area below.

To sum up, the fermentation might change the structure of HPIDF but did not change the composition, which would make the adsorption of F-HPIDF better than that of HPIDF, so that it may have better potential bioactivity in the colon.

3.2.4. Particle Size and Specific Surface Area

Particle size and specific surface area of HPIDF/F-HPIDF are shown in Table 1,which both showed significant differences ($p < 0.05$). The smaller average particle size and the increase of specific surface area might lead to a higher adsorption capacity of harmful substances and metal ions. This result was in high agreement with the results of SEM and FTIR. These showed that the fermentation had made a great change in the structure of HPIDF. As for the composition, the analysis of monosaccharide composition of HPIDF, F-HPIDF, and hydrolysate was necessary.

Table 1. Particle size and specific surface area of HPIDF/F-HPIDF.

	D_{10} (μm)	D_{25} (μm)	D_{50} (μm)	D_{75} (μm)	Specific Surface Area (m^2/kg)
HPIDF	8.52 ± 0.08 [a]	24.59 ± 0.58 [a]	53.33 ± 1.45 [a]	102.24 ± 5.69 [a]	90.24 ± 1.27 [a]
F-HPIDF	6.02 ± 0.05 [b]	11.25 ± 0.12 [b]	30.00 ± 0.39 [b]	64.60 ± 1.48 [b]	135.75 ± 1.36 [b]

Different lowercase letters indicate a significant difference ($p < 0.05$).

3.3. The Monosaccharide Composition of HPIDF, F-HPIDF, and Hydrolysate

The monosaccharide composition of HPIDF, F-HPIDF, and hydrolysate is shown in Figure 5. The main constituents of HPIDF are shown in Figure 5a, which are same as the previous result [16]. With the progress of colonic fermentation, the monosaccharide composition of HPIDF changed significantly. F-HPIDF (Figure 5b) consisted of a less content of galactose (34.40%) and galacturonic acid (11.47%), meanwhile a higher content of arabinose (27.49%), rhamnose (5.32%), glucose (6.86%), and fucose (3.91%) than HPIDF. This showed that galactose and galacturonic acid were used as the carbon source of bacteria in the colon during fermentation.

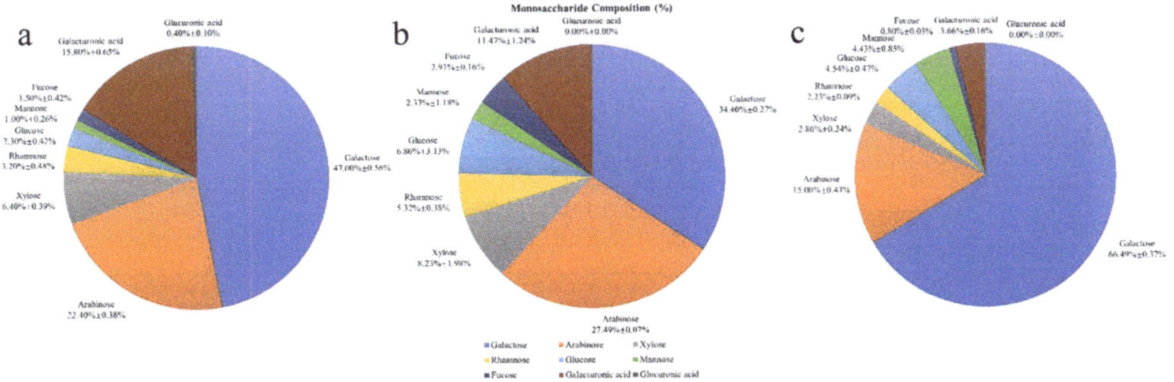

Figure 5. The monosaccharide composition of HPIDF, F-HPIDF, and hydrolysate. (**a**: HPIDF; **b**: F-HPIDF; **c**: Hydrolysate).

As for the hydrolysate (Figure 5c), galactose (66.49%) and arabinose (15.00%) were the main components. The largest component in the hydrolysate should be considered as the monosaccharide that constituted the main component of HPIDF. As know, monosaccharides that make up hemicelluloses are mainly glucose, xylose, mannose, arabinose, and galactose [29]. This result proved that HPIDF is a dietary fiber with hemicellulose as the core. Then, during colonic fermentation, a large amount of hemicellulose was decomposed to free monosaccharides, which is a typical fermentation mode of hemicellulose [30].

The colonic fermentation model of HPIDF with hemicellulose as the main part should be regarded as one of the important findings of this study, which could help us to carry out more accurate research on the metabolic model of colonic microorganisms using HPIDF. On this basis, the metabolic pathway of free monosaccharides in the colon can be studied to make sure if they had any potential functional value. Even the modification of HPIDF could be carried out based on the result, at least we can add modifiers to the functional groups of hemicellulose so that to make these components easier to release in the colon.

In this experiment, the microorganism used to ferment HPIDF was mainly *Lactobacillus* (genus level), and the main components of HPIDF were hemicellulose and lignin. The fermentation mode of hemicellulose by *Lactobacillus* is relatively clear. In short, it is a process of using hemicellulose as the carbon source to produce lactic acid in the colon [31,32]. An increase in lactic acid, as a precursor of propionic acid [33], results in an increase of short-chain fatty acid (SCFA). This phenomenon is beneficial to colon health and may even prevent colon cancer [34,35]. In contrast, the high level of lactic acid in the colon may also have adverse effects on intestinal health [36]. However, we believed that the lactic acid produced by dietary fiber fermentation should not pose this risk because they are ingested non-exogenously.

As the above process occurred, many new monosaccharides would appear in the colon, in which the composition was mainly galactose and arabinose. Galactose is a widely recognized functional monosaccharide with biological activities, such as being conducive to the appreciation of *Bifidobacterium* [37]. Galactose binding to lectin [38], a process that occurs in the colon before cancerization, proved that galactose plays an important role in intestinal health. Arabinose is also an intestinal prebiotic, which can prevent or even treat colitis by affecting the composition of intestinal flora directly [39]. In short, the monosaccharides produced by the fermentation of colonic flora would have a positive effect on intestinal health.

3.4. The Adsorption Capacity of HPIDF/F-HPIDF

3.4.1. WHC, WSC, and OHC of HPIDF/F-HPIDF

The WHC and WSC of HPIDF/F-HPIDF are shown in Figure 6a. Although the WHC and WSC of HPIDF are much higher than other IDFs [40,41], those of F-HPIDF (19.17 g/g, 24.83 mL/g) were still higher than HPIDF ($p < 0.01$). Carboxyl (-COOH) and hydroxyl (-OH) are the common polar functional groups in IDFs. The larger specific surface area and smaller particle size represented more polar groups in contact with the environment. The excellent water absorption of F-HPIDF was consistent with the previous results of the structure changes.

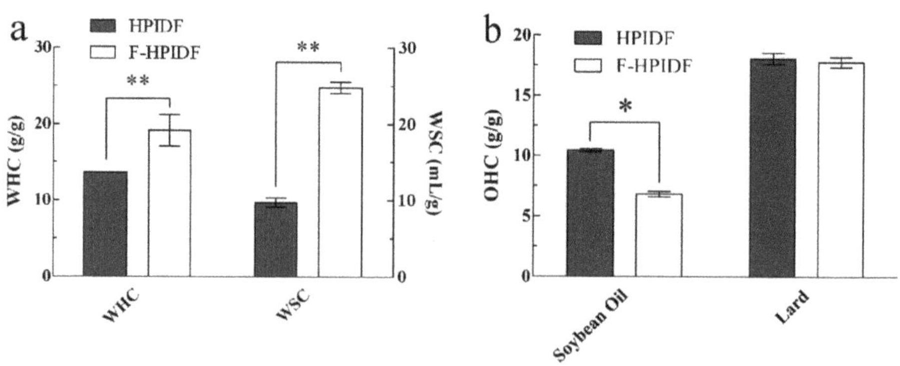

Figure 6. WHC, WSC and OHC of HPIDF/F-HPIDF. (**a**: WHC & WSC; **b**: OHC; *: $p < 0.05$; **: $p < 0.01$).

The OHC of HPIDF/F-HPIDF is shown in Figure 6b. The mechanisms of dietary fiber to adsorb saturated and unsaturated oils are different. So, we chose soybean oil and lard to determine the different OHCs of HPIDF/F-HPIDF. The result shows that the OHC of F-HPIDF for unsaturated fat (6.82 g/g) was lower than that of HPIDF ($p < 0.05$), however not different (17.77 g/g) from that for saturated fat ($p > 0.05$). Common hydrophobic groups include C-H stretching (-CH_3 or =CH_2), ethers (C-O), etc. This just showed that the hydrophobic groups in HPIDF were exposed less with the progress of fermentation, which might be attributed to the decomposition of hemicellulose during the progress described in 3.3. Acetyl should be considered as the characteristic functional group of hemicellulose [42], the decomposition of hemicellulose might lead to the loss of acetyl, which could be the result of the decrease of OHC. In contrast, saturated fat (such as lard) does not contain unsaturated bonds, which leads to a weaker dependence on acetyl than saturated fat. To prevent excessive absorption of fat by the intestine, a higher OHC is desirable. According to the fermentation mode of HPIDF in this study, we may try to modify the non-hemicellulose components so that they will not be decomposed by the gut flora while maintaining a high content of hydrophobic groups.

3.4.2. HMAC of HPIDF/F-HPIDF

The adsorption capacity of HPIDF/F-HPIDF for metal ions is shown in Figure 7. On the whole, the adsorption capacities of F-HPIDF to four kinds of heavy metal were higher than that of HPIDF. The adsorption capacity of F-HPIDF for Cd^{2+} (1.82 μmol/g) and Pb^{2+} (1.91 μmol/g) was similar; meanwhile, the adsorption capacity of Cu^{2+} (0.68 μmol/g) was still the lowest. The HMAC of dietary fiber was directly related to its particle size and specific surface area [43], which led to the phenomenon in this study. This result was consistent with the structural analysis results.

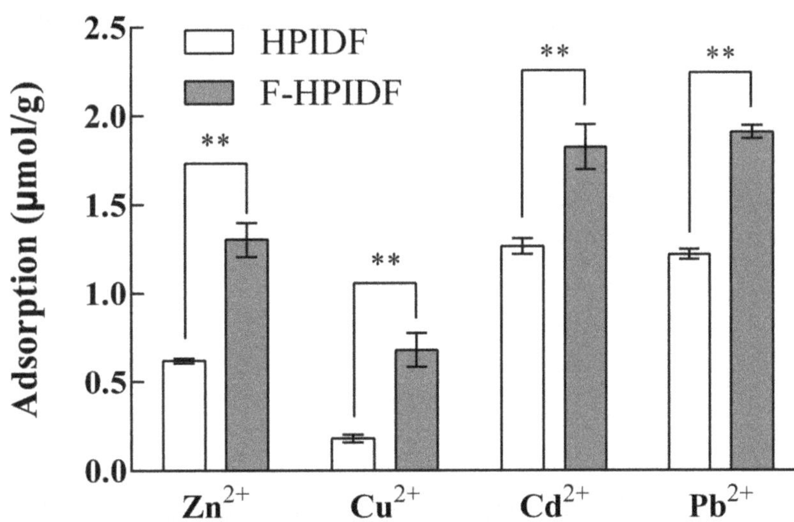

Figure 7. Heavy metal-adsorption capacity of HPIDF/F-HPIDF. (**: $p < 0.01$).

The excessive intake of heavy metals was not only harmful to intestinal health but also might damage multiple organs [44,45]. With the progress of fermentation, the HMAC of HPIDF gradually became stronger, which was a very good situation for the digestive system. Heavy metal ions were difficult to enter the blood and would be excreted via feces. In short, the fermentation process of HPIDF in the colon may prevent the body from ingesting excessive heavy metals.

3.4.3. Potentially Harmful Substances-Adsorption Capacity of HPIDF/F-HPIDF

We used glucose-adsorption capacity, cholesterol-adsorption capacity, sodium cholate-adsorption capacity, acrylamide-adsorption capacity, and nitrite-adsorption capacity to determine the adsorption function to potentially harmful substances of HPIDF and F-HPIDF. The results are shown in Figure 8.

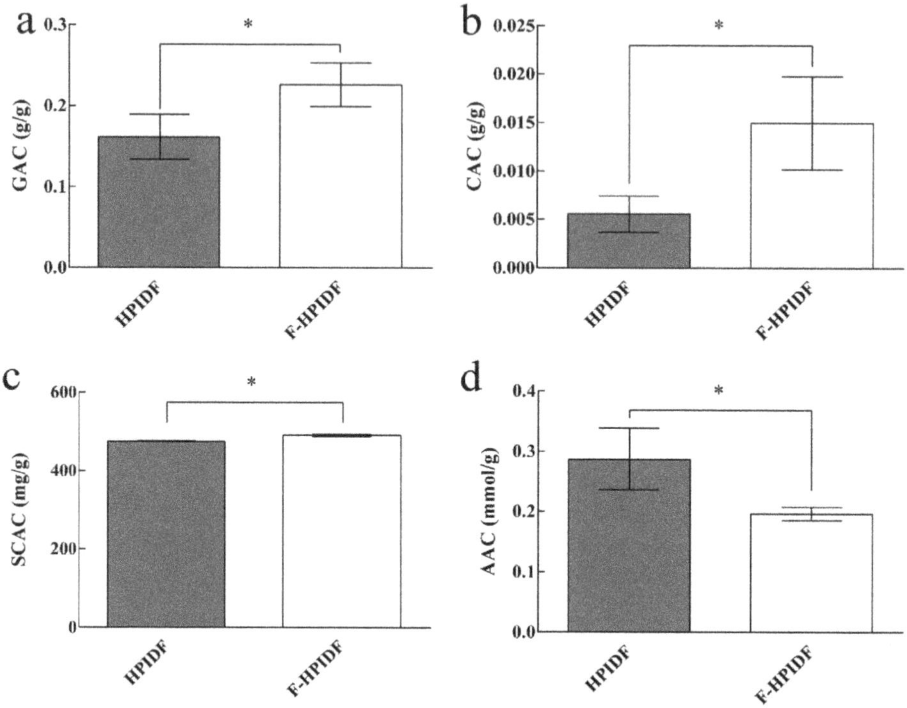

Figure 8. Potentially harmful substances-adsorption capacity of HPIDF/F-HPIDF. (**a**: GAC; **b**: CAC; **c**: SCAC; **d**: AAC; *: $p < 0.05$).

First, the adsorption capacity of HPIDF/F-HPIDF to nitrite was not shown in the results. According to previous research, HPIDF showed nitrite-adsorption capacity only in the gastric juice [16]. The same situation occurred in this study, we even did not find that F-HPIDF had a detectable nitrite-adsorption capacity. Therefore, the result was not shown in the figure. In short, whether it was fermented or not, HPIDF had no nitrite-adsorption capacity in the colon.

Normal amount of glucose is an important energy source for the body, but excessive glucose will cause a burden, which should be called a substance that may affect the health of the body. The GAC of F-HPIDF (Figure 8a) portrayed a better trend (0.23 g/g), suggesting that F-HPIDF may prevent glucose from being over absorbed in the intestine. The GAC of HPIDF is the result of multiple adsorption mechanisms [46]. Therefore, the higher available surface, smaller particle size, and more functional group exposure of F-HPIDF lead to better glucose adsorption than HPIDF. As one of the main places for the body to absorb glucose, the intestine could be prevented from absorbing due to the presence of F-HPIDF, especially the glucose produced by the decomposition of digestive juice.

As shown in Figure 8b, F-HPIDF (14.80 mg/g) exhibited a high cholesterol-adsorption capacity than HPIDF (5.58 mg/g). After F-HPIDF was mixed with cholesterol, the absorption in the colon of cholesterol could be controlled by reducing the solubility of cholesterol [47]. This mode of action was similar to GAC.

The SCAC of HPIDF/F-HPIDF is shown in Figure 8c. F-HPIDF (0.49 g/g) exhibited a small increase in SCAC than HPIDF (0.47 g/g). As a substance produced by bile acids, sodium cholate may lead to intestinal inflammation and even apoptosis [48,49]. A good sodium cholate-adsorption capacity may be a potentially beneficial effect on intestine health. Compared with our previous research [16], the sodium cholate-adsorption capacity in this experiment had a great improvement. We speculated that the reason may be the difference between the adsorption time. In the early stage, when we carried out an adsorption test in the digestive solution, 4 h was used for adsorption, which was much shorter than 14 h. In this experiment, to distinguish the differences in adsorption capacity before and after fermentation, we selected a long time to simulate the retention of dietary fiber in the intestine [50]. This showed that the adsorption of sodium cholate was still going on after 4 h. Then, the cholate adsorption kinetics of F-HPIDF may be also an object worthy of further study.

The acrylamide-adsorption capacity of F-HPIDF (0.20 mmol/g) showed a little decrease than HPIDF (Figure 8d). We had emphasized that the AAC of okara-HPIDF did not change with the gastroenteric environments. Therefore, AAC should only be related to the structure of the dietary fiber. The acrylamide-adsorption capacity on plant DF is generally weak, meanwhile mainly depended on physical adsorption [51]. Then we speculated that the decrease of AAC of F-HPIDF still came from the degradation of hemicellulose, although the specific surface area of F-HPIDF increased with the progress of fermentation. Hemicellulose was used as an adsorption material or matrix of acrylamide in previous studies [52,53].

In summary, the adsorption capacity of HPIDF to potentially harmful substances changed significantly after fermentation by colonic flora. The differences of F-HPIDF and HPIDF in properties were due to the changes in their structure during fermentation in vivo. Therefore, this process depended on the changes caused by the specific metabolic mechanism of colonic microorganisms.

4. Conclusions

In this study, the changes in the structure (SEM, FT-IR, XRD, particle size, specific area, and monosaccharide composition) and the adsorption capacity (WHC, WSC, OHC, heavy metal-adsorption capacity, and harmful substances-adsorption capacity) were analyzed to measure the changes between F-HPIDF and HPIDF. The results suggested that after being fermented by colonic microorganisms, the structure and properties of okara-HPIDF changed greatly. These findings provide a more accurate analysis of HPIDF after entering the digestive system. Meanwhile, the excellent adsorption and physicochemical properties caused by structural changes are beneficial to colon health. The intake of HPIDF might increase the number of beneficial monosaccharides in the colon, which might improve the composition of SCFA and have a beneficial impact on the body. This study also speculated that HPIDF may be modified to enhance its biological activity after its fermentation mode had been identified. The above all proved that HPIDF may have more utilization value.

Author Contributions: B.L.: conceptualization, software, writing—original draft; Y.W.: visualization, software, writing—original draft; X.Z.: writing—review & editing; Y.C.: methodology; H.F.: investigation; T.L.: methodology; J.H.: visualization; Y.L.: investigation; H.Y.: funding acquisition, project administration; L.J.: funding acquisition, project administration. All authors have read and agreed to the published version of the manuscript.

Funding: This research was funded by China Agriculture Research System of MOF and MARA, grant number CARS-04; The Excellent Youth Project of Natural Science Foundation of Heilongjiang Province, grant number YQ2021C023; Major Science and Technology Innovation Projects in Shandong Province, grant number 2019JZZY010722.

Institutional Review Board Statement: The study was conducted according to the guidelines of the Declaration of Helsinki, and approved by the Ethics Committee of Laboratory Animal Welfare and Ethics Committee of Northeast Agricultural University.

Informed Consent Statement: Not applicable.

Data Availability Statement: Raw data can be provided by the corresponding author on request.

Acknowledgments: The authors acknowledge Shanghai Personal Biotechnology Co., Ltd. for their support to the 16S rDNA Sequencing.

Conflicts of Interest: No conflict of interest.

References

1. Hua, M.; Lu, J.; Qu, D.; Liu, C.; Zhang, L.; Li, S.; Chen, J.; Sun, Y. Structure, physicochemical properties and adsorption function of insoluble dietary fiber from ginseng residue: A potential functional ingredient. *Food Chem.* **2019**, *286*, 522–529. [CrossRef]
2. Zhao, G.; Zhang, R.; Dong, L.; Huang, F.; Tang, X.; Wei, Z.; Zhang, M. Particle size of insoluble dietary fiber from rice bran affects its phenolic profile, bioaccessibility and functional properties. *LWT* **2018**, *87*, 450–456. [CrossRef]
3. Meng, X.; Liu, F.; Xiao, Y.; Cao, J.; Wang, M.; Duan, X. Alterations in physicochemical and functional properties of buckwheat straw insoluble dietary fiber by alkaline hydrogen peroxide treatment. *Food Chem. X* **2019**, *3*, 100029. [CrossRef]
4. Hino, S.; Takemura, N.; Sonoyama, K.; Morita, A.; Kawagishi, H.; Aoe, S.; Morita, T. Small intestinal goblet cell proliferation induced by ingestion of soluble and insoluble dietary fiber is characterized by an increase in sialylated mucins in rats. *J. Nutr.* **2012**, *142*, 1429–1436. [CrossRef]
5. McIntosh, G.; Jorgensen, L.; Royle, P. The potential of an insoluble dietary fiber-rich source from barley to protect from DMH-induced intestinal tumors in rats. *Nutr. Cancer* **1993**, *19*, 213–221. [CrossRef] [PubMed]
6. Chen, T.; Chen, D.; Tian, G.; Zheng, P.; Mao, X.; Yu, J.; He, J.; Huang, Z.; Luo, Y.; Luo, J. Effects of soluble and insoluble dietary fiber supplementation on growth performance, nutrient digestibility, intestinal microbe and barrier function in weaning piglet. *Anim. Feed. Sci. Technol.* **2020**, *260*, 114335. [CrossRef]
7. Wilberts, B.L.; Arruda, P.H.; Kinyon, J.M.; Frana, T.S.; Wang, C.; Magstadt, D.R.; Madson, D.M.; Patience, J.F.; Burrough, E.R. Investigation of the impact of increased dietary insoluble fiber through the feeding of distillers dried grains with solubles (DDGS) on the incidence and severity of Brachyspira-associated colitis in pigs. *PLoS ONE* **2014**, *9*, e114741. [CrossRef] [PubMed]
8. Miles, J.P.; Zou, J.; Kumar, M.-V.; Pellizzon, M.; Ulman, E.; Ricci, M.; Gewirtz, A.T.; Chassaing, B. Supplementation of low-and high-fat diets with fermentable fiber exacerbates severity of DSS-induced acute colitis. *Inflamm. Bowel Dis.* **2017**, *23*, 1133–1143. [CrossRef] [PubMed]
9. Singh, V.; San Yeoh, B.; Walker, R.E.; Xiao, X.; Saha, P.; Golonka, R.M.; Cai, J.; Bretin, A.C.A.; Cheng, X.; Liu, Q. Microbiota fermentation-NLRP3 axis shapes the impact of dietary fibres on intestinal inflammation. *Gut* **2019**, *68*, 1801–1812. [CrossRef] [PubMed]
10. Armstrong, H.; Mander, I.; Zhang, Z.; Armstrong, D.; Wine, E. Not All Fibers Are Born Equal; Variable Response to Dietary Fiber Subtypes in IBD. *Front. Pediatr.* **2021**, *8*, 924. [CrossRef]
11. Vong, W.C.; Liu, S.-Q. Biovalorisation of okara (soybean residue) for food and nutrition. *Trends Food Sci. Technol.* **2016**, *52*, 139–147. [CrossRef]
12. Lu, F.; Liu, Y.; Li, B. Okara dietary fiber and hypoglycemic effect of okara foods. *Bioact. Carbohydr. Diet. Fibre* **2013**, *2*, 126–132. [CrossRef]
13. Jiménez-Escrig, A.; Tenorio, M.D.; Espinosa-Martos, I.; Rupérez, P. Health-promoting effects of a dietary fiber concentrate from the soybean byproduct okara in rats. *J. Agric. Food Chem.* **2008**, *56*, 7495–7501. [CrossRef]
14. Fan, X.; Chang, H.; Lin, Y.; Zhao, X.; Zhang, A.; Li, S.; Feng, Z.; Chen, X. Effects of ultrasound-assisted enzyme hydrolysis on the microstructure and physicochemical properties of okara fibers. *Ultrason. Sonochem.* **2020**, *69*, 105247. [CrossRef] [PubMed]
15. Lazarin, R.A.; Falcão, H.G.; Ida, E.I.; Berteli, M.N.; Kurozawa, L.E. Rotating-Pulsed Fluidized Bed Drying of Okara: Evaluation of Process Kinetic and Nutritive Properties of Dried Product. *Food Bioprocess Technol.* **2020**, *13*, 1611–1620. [CrossRef]
16. Bo, L.; Huan, W.; Swallah, M.S.; Hongling, F.; Yue, S.; Zengwang, G.; Xiaohong, T.; Yang, L.; Hansong, Y.; Lianzhou, J. Structure, Properties and Potential Bioactivities of High-purity Insoluble Fibre from Soybean Dregs (Okara). *Food Chem.* **2021**, *364*, 130402.
17. Wang, S.; Sun, W.; Swallah, M.S.; Amin, K.; Lyu, B.; Fan, H.; Zhang, Z.; Yu, H. Preparation and characterization of soybean insoluble dietary fiber and its prebiotic effect on dyslipidemia and hepatic steatosis in high fat-fed C57BL/6J mice. *Food Funct.* **2021**, *12*, 8760–8773. [CrossRef] [PubMed]
18. Wang, B.; Yu, H.; He, Y.; Wen, L.; Gu, J.; Wang, X.; Miao, X.; Qiu, G.; Wang, H. Effect of soybean insoluble dietary fiber on prevention of obesity in high-fat diet fed mice via regulation of the gut microbiota. *Food Funct.* **2021**, *12*, 7923–7937. [CrossRef]
19. Gaddum, J.H. *Hawley's Condensed Chemical Dictionary*; United States Pharmacopeia: Rockville, MD, USA, 2007.
20. Wu, C.; Teng, F.; McClements, D.J.; Zhang, S.; Li, Y.; Wang, Z. Effect of cavitation jet processing on the physicochemical properties and structural characteristics of okara dietary fiber. *Food Res. Int.* **2020**, *134*, 109251. [CrossRef] [PubMed]
21. Zhang, X.; Yu, L.; Bi, H.; Li, X.; Ni, W.; Han, H.; Li, N.; Wang, B.; Zhou, Y.; Tai, G. Total fractionation and characterization of the water-soluble polysaccharides isolated from Panax ginseng CA Meyer. *Carbohydr. Polym.* **2009**, *77*, 544–552. [CrossRef]
22. Zhang, W.; Zeng, G.; Pan, Y.; Chen, W.; Huang, W.; Chen, H.; Li, Y. Properties of soluble dietary fiber-polysaccharide from papaya peel obtained through alkaline or ultrasound-assisted alkaline extraction. *Carbohydr. Polym.* **2017**, *172*, 102–112. [CrossRef] [PubMed]
23. Xie, F.; Wang, Y.; Wu, J.; Wang, Z. Functional properties and morphological characters of soluble dietary fibers in different edible parts of Angelica keiskei. *J. Food Sci.* **2016**, *81*, C2189–C2198. [CrossRef] [PubMed]
24. Lkhagva, E.; Chung, H.-J.; Hong, J.; Tang, W.H.W.; Lee, S.-I.; Hong, S.-T.; Lee, S. The regional diversity of gut microbiome along the GI tract of male C57BL/6 mice. *BMC Microbiol.* **2021**, *21*, 44. [CrossRef] [PubMed]
25. Mohtar, S.S.; Busu, T.N.Z.T.M.; Noor, A.M.M.; Shaari, N.; Mat, H. An ionic liquid treatment and fractionation of cellulose, hemicellulose and lignin from oil palm empty fruit bunch. *Carbohydr. Polym.* **2017**, *166*, 291–299. [CrossRef] [PubMed]
26. Alba, K.; Macnaughtan, W.; Laws, A.; Foster, T.J.; Campbell, G.; Kontogiorgos, V. Fractionation and characterisation of dietary fibre from blackcurrant pomace. *Food Hydrocoll.* **2018**, *81*, 398–408. [CrossRef]

27. Ullah, I.; Yin, T.; Xiong, S.; Zhang, J.; Din, Z.-u.; Zhang, M. Structural characteristics and physicochemical properties of okara (soybean residue) insoluble dietary fiber modified by high-energy wet media milling. *LWT-Food Sci. Technol.* **2017**, *82*, 15–22. [CrossRef]
28. Sharma, V.; Smolin, J.; Nayak, J.; Ayala, J.E.; Scott, D.A.; Peterson, S.N.; Freeze, H.H. Mannose alters gut microbiome, prevents diet-induced obesity, and improves host metabolism. *Cell Rep.* **2018**, *24*, 3087–3098. [CrossRef]
29. Mudgil, D.; Barak, S. Composition, properties and health benefits of indigestible carbohydrate polymers as dietary fiber: A review. *Int. J. Biol. Macromol.* **2013**, *61*, 1–6. [CrossRef]
30. Reddy, N.R.; Palmer, J.K.; Pierson, M.D.; Bothast, R.J. Wheat straw hemicelluloses: Composition and fermentation by human colon Bacteroides. *J. Agric. Food Chem.* **1983**, *31*, 1308–1313. [CrossRef]
31. Garde, A.; Jonsson, G.; Schmidt, A.S.; Ahring, B.K. Lactic acid production from wheat straw hemicellulose hydrolysate by Lactobacillus pentosus and Lactobacillus brevis. *Bioresour. Technol.* **2002**, *81*, 217–223. [CrossRef]
32. Moldes, A.; Torrado, A.; Converti, A.; Dominguez, J. Complete bioconversion of hemicellulosic sugars from agricultural residues into lactic acid by Lactobacillus pentosus. *Appl. Biochem. Biotechnol.* **2006**, *135*, 219–227. [CrossRef]
33. Duncan, S.H.; Barcenilla, A.; Stewart, C.S.; Pryde, S.E.; Flint, H.J. Acetate utilization and butyryl coenzyme A (CoA): Acetate-CoA transferase in butyrate-producing bacteria from the human large intestine. *Appl. Environ. Microbiol.* **2002**, *68*, 5186–5190. [CrossRef]
34. Rafter, J.J. The role of lactic acid bacteria in colon cancer prevention. *Scand. J. Gastroenterol.* **1995**, *30*, 497–502. [CrossRef] [PubMed]
35. Singh, J.; Rivenson, A.; Tomita, M.; Shimamura, S.; Ishibashi, N.; Reddy, B.S. Bifidobacterium longum, a lactic acid-producing intestinal bacterium inhibits colon cancer and modulates the intermediate biomarkers of colon carcinogenesis. *Carcinogenesis* **1997**, *18*, 833–841. [CrossRef]
36. Lin, J.; Nafday, S.M.; Chauvin, S.N.; Magid, M.S.; Pabbatireddy, S.; Holzman, I.R.; Babyatsky, M.W. Variable effects of short chain fatty acids and lactic acid in inducing intestinal mucosal injury in newborn rats. *J. Pediatric Gastroenterol. Nutr.* **2002**, *35*, 545–550. [CrossRef] [PubMed]
37. Torres, D.P.; Gonçalves, M.d.P.F.; Teixeira, J.A.; Rodrigues, L.R. Galacto-oligosaccharides: Production, properties, applications, and significance as prebiotics. *Compr. Rev. Food Sci. Food Saf.* **2010**, *9*, 438–454. [CrossRef]
38. Evans, R.C.; Fear, S.; Ashby, D.; Hackett, A.; Williams, E.; Van Der Vliet, M.; Dunstan, F.D.; Rhodes, J.M. Diet and colorectal cancer: An investigation of the lectin/galactose hypothesis. *Gastroenterology* **2002**, *122*, 1784–1792. [CrossRef]
39. Li, Y.; Pan, H.; Liu, J.; Li, T.; Liu, S.; Shi, W.; Sun, C.; Fan, M.; Xue, L.; Wang, Y. l-Arabinose inhibits colitis by modulating gut microbiota in mice. *J. Agric. Food Chem.* **2019**, *67*, 13299–13306. [CrossRef]
40. Luo, X.; Wang, Q.; Zheng, B.; Lin, L.; Chen, B.; Zheng, Y.; Xiao, J. Hydration properties and binding capacities of dietary fibers from bamboo shoot shell and its hypolipidemic effects in mice. *Food Chem. Toxicol.* **2017**, *109*, 1003–1009. [CrossRef] [PubMed]
41. Wen, Y.; Niu, M.; Zhang, B.; Zhao, S.; Xiong, S. Structural characteristics and functional properties of rice bran dietary fiber modified by enzymatic and enzyme-micronization treatments. *LWT* **2017**, *75*, 344–351. [CrossRef]
42. Wang, S.; Ru, B.; Lin, H.; Sun, W. Pyrolysis behaviors of four O-acetyl-preserved hemicelluloses isolated from hardwoods and softwoods. *Fuel* **2015**, *150*, 243–251. [CrossRef]
43. Kachenpukdee, N.; Santerre, C.R.; Ferruzzi, M.G.; Oonsivilai, R. Modified Dietary Fiber from Cassava Pulp and Assessment of Mercury Bioaccessibility and Intestinal Uptake Using an In Vitro Digestion/Caco-2 Model System. *J. Food Sci.* **2016**, *81*, T1854–T1863. [CrossRef] [PubMed]
44. Zhang, Y.; Huo, X.; Lu, X.; Zeng, Z.; Faas, M.M.; Xu, X. Exposure to multiple heavy metals associate with aberrant immune homeostasis and inflammatory activation in preschool children. *Chemosphere* **2020**, *257*, 127257. [CrossRef] [PubMed]
45. Ma, S.; Ren, B.; Diao, Z.; Chen, Y.; Qiao, Q.; Liu, X. Physicochemical properties and intestinal protective effect of ultra-micro ground insoluble dietary fibre from carrot pomace. *Food Funct.* **2016**, *7*, 3902–3909. [CrossRef]
46. Chen, J.; Zhao, Q.; Wang, L.; Zha, S.; Zhang, L.; Zhao, B. Physicochemical and functional properties of dietary fiber from maca (*Lepidium meyenii* Walp.) liquor residue. *Carbohydr. Polym.* **2015**, *132*, 509–512. [CrossRef]
47. Yao, S.-L.; Xu, Y.; Zhang, Y.-Y.; Lu, Y.-H. Black rice and anthocyanins induce inhibition of cholesterol absorption in vitro. *Food Funct.* **2013**, *4*, 1602–1608. [CrossRef]
48. Morita, S.; Ikeda, Y.; Tsuji, T.; Terada, T. Molecular mechanisms for protection of hepatocytes against bile salt cytotoxicity. *Chem. Pharm. Bull.* **2019**, *67*, 333–340. [CrossRef]
49. Labbé, A.; Ganopolsky, J.G.; Martoni, C.J.; Prakash, S.; Jones, M.L. Bacterial bile metabolising gene abundance in Crohn's, ulcerative colitis and type 2 diabetes metagenomes. *PLoS ONE* **2014**, *9*, e115175. [CrossRef]
50. Tabernero, M.; Venema, K.; Maathuis, A.J.; Saura-Calixto, F.D. Metabolite production during in vitro colonic fermentation of dietary fiber: Analysis and comparison of two European diets. *J. Agric. Food Chem.* **2011**, *59*, 8968–8975. [CrossRef]
51. Zhao, B.; Jiang, H.; Lin, Z.; Xu, S.; Xie, J.; Zhang, A. Preparation of acrylamide/acrylic acid cellulose hydrogels for the adsorption of heavy metal ions. *Carbohydr. Polym.* **2019**, *224*, 115022. [CrossRef] [PubMed]
52. Ferrari, E.; Ranucci, E.; Edlund, U.; Albertsson, A.C. Design of renewable poly (amidoamine)/hemicellulose hydrogels for heavy metal adsorption. *J. Appl. Polym. Sci.* **2015**, *132*. [CrossRef]
53. Kundu, D.; Mondal, S.K.; Banerjee, T. Development of β-cyclodextrin-cellulose/hemicellulose-based hydrogels for the removal of Cd (II) and Ni (II): Synthesis, kinetics, and adsorption aspects. *J. Chem. Eng. Data* **2019**, *64*, 2601–2617. [CrossRef]

foods

MDPI

Article

The Role of Ultrasound in the Preparation of Zein Nanoparticles/Flaxseed Gum Complexes for the Stabilization of Pickering Emulsion

Yinghao Li, Ge Xu, Weiwei Li, Lishuang Lv and Qiuting Zhang *

Department of Food Science and Technology, School of Food Science and Pharmaceutical Engineering, Nanjing Normal University, Nanjing 210023, China; yinghaoli96@163.com (Y.L.); xuge813@163.com (G.X.); liweiwei@njnu.edu.cn (W.L.); lishuanglv@126.com (L.L.)

* Correspondence: qiuting.zhang@njnu.edu.cn; Tel.: +86-152-6182-3407

Abstract: Ultrasound is one of the most commonly used methods to prepare Pickering emulsions. In the study, zein nanoparticles-flaxseed gum (ZNP-FSG) complexes were fabricated through various preparation routes. Firstly, the ZNP-FSG complexes were prepared either through direct homogenization/ultrasonication of the zein and flaxseed gum mixture or through pretreatment of zein and/or flaxseed gum solutions by ultrasonication before homogenization. The Pickering emulsions were then produced with the various ZNP-FSG complexes prepared. ZNP-FSG complexes and the final emulsions were then characterized. We found that the complex prepared by ultrasonication of zein as pretreatment followed by homogenization of the ZNP with FSG ((ZNP$_U$-FSG)H) exhibited the smallest turbidity, highest absolute potential value, relatively small particle size, and formed the most stable complex particles. Meanwhile, complex prepared through direct ultrasonication plus homogenization on the mixture ((ZNP-FSG)HU) showed significantly decreased emulsifying properties and stability. Compared with the complex without ultrasonic treatment, the complex and emulsion, which prepared by ultrasonicated FSG were extremely unstable, and the phase separation phenomenon of the emulsion was observed 30 min after preparation. The above conclusions are also in line with the findings obtained from the properties of the corresponding emulsions, such as the droplets size, microstructure, freeze-thaw stability, and storage stability. It is, therefore, clear that to produce stable Pickering emulsion, ultrasonication should be avoided to apply together at the end of ZNP-FGS complex preparation. It is worth noticing that the emulsions prepared by complex with ultrasonicated zein (ZNP$_U$-FSG)H are smaller, distributed more uniformly, and are able to encapsulate oil droplets well. It was found that the emulsions prepared with ZNP$_U$-FSG remained stable without serum phase for 14 days and exhibited improved stability at low-temperature storage. The current study will provide guidance for the preparation of protein–polysaccharide complexes and Pickering emulsions for future work.

Keywords: zein; flaxseed gum; ultrasound; complex particle; Pickering emulsion

Citation: Li, Y.; Xu, G.; Li, W.; Lv, L.; Zhang, Q. The Role of Ultrasound in the Preparation of Zein Nanoparticles/Flaxseed Gum Complexes for the Stabilization of Pickering Emulsion. *Foods* **2021**, *10*, 1990. https://doi.org/10.3390/foods10091990

Academic Editor: Jayani Chandrapala

Received: 17 July 2021
Accepted: 21 August 2021
Published: 25 August 2021

1. Introduction

The emulsions stabilized by aqueous solid particles are termed Pickering emulsions, which is different from those stabilized by small-molecule surface emulsifiers. Solutions of many organic ingredients, such as protein, glue, flour, milk, starch, saponin, could act as emulsifiers for Pickering emulsion [1]. Pickering emulsions prepared using food-grade colloidal particles can be designed to have high physical stability and overcome the problems of low biocompatibility [2] and environmental pollution [3] caused by traditional synthetic surfactants. With the increasing awareness and pursuit of health and safety from consumers, formulating food with natural materials has been increasingly expected from academia to the food industry. Protein-based colloidal particles are particularly suitable since they are widely available, inexpensive, and show great nutritional benefits [4]. Recently, there are many pieces of research on Pickering emulsions using different colloidal particles, such as starch [5], polysaccharides [6], protein [7], and phospholipids, to form combinations to pursue the emulsification effect closer to that of synthetic chemical surfactants [8]. However, research focusing on the influence of the emulsion preparation process on emulsion stability is still far from enough.

Many techniques have been applied for emulsion preparation, such as high shear mixers (homogenization), ultrasound [9] micro-fluidization, and high-pressure homogenization [10]. Homogenization and ultrasound are the two most commonly used methods to prepare Pickering emulsions. Ultrasound has the advantages of high efficiency, good reproducibility, low cost, simple operation, no pollution, and being able to maintain the internal nutrients of food, which makes the ultrasonic technology widely used. Many reports show ultrasonic treatment could improve the emulsification property of the emulsion stabilized by soy protein isolate [11], chitosan [12], and whey protein-pectin complexes [13]. Moreover, high-intensity ultrasound is more energy-efficient than high-pressure homogenization and microfluidizer [14]. However, the effects of ultrasound on the stability of emulsion were rarely characterized in detail, e.g., effects of ultrasonic parameters, substrates, different processing procedures. In addition, there are few studies on the interaction between different emulsion preparation methods.

In our previous study, we found ultrasonic pretreatment could improve the stability of zein stabilized Pickering emulsion. But if ultrasound is directly applied to zein nanoparticles (ZNP), which is prepared by anti-solvent precipitation method, the stability of ZNP would not have been improved after ultrasonic treatment [15]. Julius W. J. de Folter et al. demonstrated that ZNP system synthesized using anti-solvent precipitation procedure is very stable, specifically at a pH of 3–4. However, Pickering emulsions stabilized by ZNP failed to form stable emulsions due to extreme environmental stresses (pH, heating, and salt) [16]. To improve the stability of the ZNP emulsion, forming a complex with other hydrophilic molecules, such as hydrophilic protein, polysaccharides, or polyphenols, is a common method. Among them, polysaccharides, including sodium alginate, low-acyl gellan gum, gum arabic, chitosan, and carboxymethyl dextrin have been reported to be effective [17–20]. Flaxseed gum (FSG), with good gelling, emulsifying, and rheological properties, was used to form complexes with ZNP [21–24]. Although our previous research showed that flaxseed gum could improve the stability of ZNP emulsion, the influences of different preparation methods and different combinations of ultrasound and homogenization on emulsion have not been studied. Therefore, in the current study, we studied the preparation of zein nanoparticles-flaxseed gum (ZNP-FSG) complexes according to different procedures, including either applying ultrasonication and/or homogenization of the mixed solution without any previous treatment or pretreatment using ultrasonic of one or both ingredients. Both the ZNP-FSG complexes and the final emulsions were characterized. The aim is to devise the preparation route of ZNP-FSG complexes that are best suited for the stabilization of Pickering emulsions [25].

2. Materials and Methods

2.1. Materials

Zein was purchased from Adamas Reagent Co., Ltd. (Shanghai, China). Flaxseed gum was provided by Yuanye Biotechnology Co., Ltd. (Shanghai, China). Corn oil was obtained from a local supermarket (Nanjing, China). Other reagents were analytical grade.

2.2. Preparation and Pretreatment of Samples

In order to analyze the role of ultrasound on Pickering emulsion preparation, different combinations of ultrasound and homogenization and different ultrasound ingredients were compared, 1. zein, 2. FSG, 3. zein-FSG mixtures.

2.2.1. Preparation Individual Solution

Zein stock solution, which was zein dissolved in 70% (v/v) ethanol at 25 mg/mL followed by continuous shaking for 1 h at 25 °C prior to ultrasound treatment.

FSG stock solution was completely dissolved in water at 5 mg/mL by magnetic stirring at 95 °C for 1 h, the pH was adjusted to 4.0 with 0.1 M HCl or NaOH solutions and measured by a pH meter.

2.2.2. Preparation of Zein Nanoparticles (ZNP)

Zein nanoparticles (ZNP) were prepared by anti-solvent precipitation method according to Dai, L. [26]. Briefly, 20 mL of zein stock solution was added dropwise into 60 mL deionized water in an ultrasonic water bath (KH-300DE, Kunshan Hechuang Ultrasonic Instrument Co., Ltd., Kunshan, Jiangsu, China). After magnetic stirring for 30 min, the remained ethanol in the particle dispersions was removed by rotary evaporation (IKA RV 8, IKA Works Guangzhou, China). The final concentration of ZNP was 10 mg/mL. Finally, the pH value of ZNP dispersions was adjusted to 4.0 with 0.1 M HCl or NaOH solutions using a pH meter and stored at 4 °C until use. The nanoparticles are made from natural or ultrasonic zein separately.

2.2.3. Ultrasound Treatment

The sample was subjected to ultrasonic treatment with a frequency of 20 kHz sonicator (Model JY92-IIN, NingBo Xinzhi Technology Co., Ningbo, China), which was equipped with a titanium probe (6 mm of diameter) in an ice bath.

A volume of 20 mL of zein solution (25 mg/mL, w/v) and 20 mL of FSG (5 mg/mL, w/v) was treated, respectively, by ultrasound at 52.95 W·cm^{-2} for 10 min, with pulsed ratios was 5 s on-time/5 s off-time (U). Zein or FSG solution ultrasonicated were termed ZNP$_U$ or FSG$_U$, respectively.

2.3. Preparation of Complex Particles

The complex particles were prepared using different ZNP and FSG by combination with homogenization and/or ultrasonication, and detailed procedures are shown in Figure 1 and Table 1. The differently treated samples were stored at 4 °C until use.

Table 1. The process of complex particles formation.

Sample No.	Complexes	Materials		Process	
		Protein	Polysaccharides	Homogenizing (rpm/3 min)	Ultrasound (W·cm^{-2})
1	ZNP	ZNP	-	-	-
2	ZNP$_U$	ZNP$_U$	-	-	52.95
3	FSG	-	FSG	-	-
4	FSG$_U$	-	FSG$_U$	-	52.95
5	(ZNP-FSG)H	ZNP	FSG	15,000	-
6	(ZNP$_U$-FSG)H	ZNP$_U$	FSG	15,000	52.95
7	(ZNP-FSG$_U$)H	ZNP	FSG$_U$	15,000	52.95
8	(ZNP-FSG)HU	ZNP	FSG	15,000	52.95
9	(ZNP-FSG)U	ZNP	FSG	-	52.95

"-" in Materials indicates without the addition of a certain ingredient; "-" in Process indicates the sample without treatment by ultrasonication or homogenization.

A volume of 20 mL of ZNP-FSG complexes was also treated by ultrasound at 52.95 W·cm^{-2} for 10 min, with pulsed ratios of 5 s on-time/5 s off-time (U). ZNP-FSG solution ultrasonicated was termed (ZNP-FSG)U.

Samples needed to be treated by a high-speed dispersion homogenizer (FJ200-SH, Shanghai specimen model factory, Shanghai, China), the homogenizing conditions were fixed at 15,000 rpm for 3 min, marked as H.

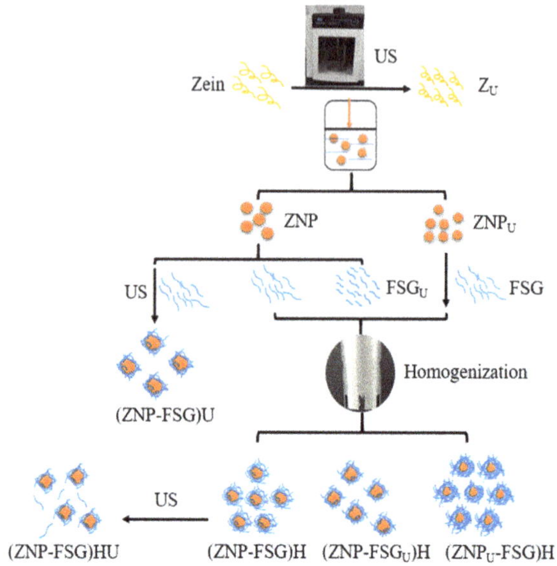

Figure 1. Overall schematic diagram of 11 various preparation routes, such as ultrasonicated zein (Z_U), zein nanoparticles (ZNP), ultrasonicated zein nanoparticles (ZNP_U), flaxseed gum (FSG), ultrasonicated flaxseed gum (FSG_U), zein nanoparticles-flaxseed gum complexes by homogenization (ZNP-FSG)H, zein nanoparticles-flaxseed gum complexes by ultrasonication (ZNP-FSG)U, ultrasonicated zein nanoparticles and flaxseed gum complex by homogenization (ZNP_U-FSG)H, zein nanoparticles and ultrasonicated flaxseed gum complex by homogenization (ZNP-FSG_U)H, ZNP-FSG by homogenization and ultrasonication (ZNP-FSG)HU.

2.4. Preparation of Pickering Emulsions

Pickering emulsions were prepared as described by Yang et al. [27] with slight modifications. The emulsions were prepared by mixing 10 mL of complex and 10 mL corn oil and then homogenized by a high-speed homogenization at 18,000 rpm for 2 min. The resultant emulsions were stored at 4 °C for further characterization.

2.5. Characterization of Complexes

2.5.1. Particle Size and Zeta (ζ)-Potential of Complexes

According to the method of Lv et al. [28], the particle size of the complexes was analyzed by dynamic light scattering using a Zetasizer (Nano-ZS90, Malvern Instruments Co., Ltd., Worcestershire, UK) at a fixed detector angle of 90°. Samples were diluted 50 times by the distilled water at pH 4 adjusted by 0.1 M HCl or NaOH solutions. They were equilibrated for 60 s before detection. The instrument reports the mean particle diameter (z-average) and the polydispersity index (PDI) ranging from 0 to 1. PDI measures the broadness of size distribution, and it provides information on the deviation from mean particle size [29].

Similar to particle size measurement, the ζ-potentials (mV) of the complexes were determined by measuring the electrophoretic mobility of each solution using a Malvern Zetasizer at 25 °C in triplicate.

2.5.2. Turbidity Measurement

Turbidity was determined according to Hosseini [30]. The transmittance of samples in distilled water at pH 4.0 with 0.1 M HCl or NaOH solutions was measured at 600 nm by a UV-Vis spectrophotometer (TU-1810PC, Beijing Purkinje General Instrument Co., Ltd.,

Beijing, China). A turbidity of 0% corresponds to a totally clear solution (transmittance in comparison to the blank is 100%).

2.5.3. Salt Ionic Stability Measurement

The effect of ionic strength on the formation of complexes was determined as described by Lv et al. [28]. The absorbance value of samples with different salt ion concentrations (50, 100, 200, 300, 400, 500, and 1000 mmol/L) was measured at 600 nm to study the stability changes.

2.5.4. Surface Hydrophobicity (H_0)

Surface hydrophobicity of the complex particles was conducted on the basis of the description of Kato et al. [31]. The samples were diluted into a series of concentrations (0.2, 0.4, 0.6, 0.8, and 1 mg/mL) with deionized water with pH adjusted to 4.0 using 0.1 M HCl or NaOH solutions. Then, 50 μL of 1-anilino-8-naphthalene-sulfonate (ANS) solution (8 mM in deionized water of pH 4.0) was mixed with 5 mL of the diluted sample incubated in the darkroom (25 °C) for 15 min. The relative fluorescence intensity of the sample was determined by a fluorescence spectrophotometer (F-380, Tianjin GangDong scientific and technology Co., Ltd., China) at 390 nm (excitation wavelength, slit width 5 nm) and 468 nm (emission wavelength, slit width 5 nm). The initial slope of fluorescence intensity versus protein concentration (%, *w/v*) was calculated by linear regression analysis and used as an index of H_0.

2.5.5. Determination of Emulsifying Properties

The determination of emulsifying property of complex particles was carried out according to the method of Li et al. [32], slightly changed. A total of 30 mL of samples were homogenized at room temperature at 20,000 r/min for 1 min after adding 10 mL of oil. After homogenization, 50 μL of the sample was taken from the bottom immediately and added into 5 mL 0.1% sodium dodecyl sulfate (SDS). The absorbance of the diluted solution was measured at 500 nm, and 0.1% SDS was used as blank. The absorbance measured immediately (A_0) and standing for 10 min (A_{10}) after emulsion formation were used to calculate the emulsifying activity index (EAI) and emulsion stability index (ESI) as follows.

$$\mathrm{EAI} = 2 \times 2.303 \times \frac{A_0 \times \mathrm{DF}}{C \times \varphi \times \theta \times 10000} \quad (1)$$

$$\mathrm{ESI} = \frac{A_0}{A_0 - A_{10}} \times \Delta t \quad (2)$$

where A_0 is the absorbance at 0 min; DF is the dilution factor (100); θ is the proportion of the oil phase (0.25); φ is the optical path of the cuvette (1 cm); C is the concentration of complex particles (g/mL); A_{10} is the absorbance at 10 min; Δt is 10 (min).

2.6. Determination of Stability

2.6.1. Storage Stability of Pickering Emulsions

The stability against the creaming of the emulsions was analyzed via creaming index (CI). The method of Ahmed et al. [11] with slight modifications was used to assess the storage stability of Pickering emulsions. Fresh emulsions (15 mL) were kept in 25 mL-antifreeze glass bottles, then stored at 4 °C for 14 days. And samples were withdrawn for analysis of emulsion properties at regular intervals. The results were calculated using the following equation:

$$\mathrm{CI}\ (\%) = \frac{H_c}{H_0} \times 100 \quad (3)$$

where H_c represents the height of the serum phase, and H_0 represents the total height of emulsion.

2.6.2. Freeze-Thaw Stability of Pickering Emulsions

The freeze-thaw stability of Pickering emulsions was measured by the previous method of Zhu [33]. A total of 15 mL of the emulsion was shifted into freeze-resistant glass bottles and then placed in a freezer at −18 °C for 22 h. Then the emulsions were naturally defrosted at room temperature for 2 h. The process was repeated thrice, the creaming index and oiling off index of the emulsions were recorded after each cycle.

2.7. Droplets Size of Pickering Emulsions

Droplet size distributions of emulsions were measured by a Mastersizer 3000 (Malvern Instruments Co., Ltd., Worcestershire, UK). The Mie theory was applied by considering the following optical properties for corn oil droplets: a refractive index of 1.46 and absorption of 0.001, and for the deionized water, a refractive index of 1.33. Add the sample drop by drop to the deionized water with automatic mixing until an obscuration rate of between 10% and 20% [34]. Prior to the measurements, the samples were thermostated at 25 °C for 5 min and all measurements were carried out over at least 3 times. All emulsions were analyzed 24 h after the emulsification process. The volume-weighted mean diameter ($D_{4,3}$) was reported.

2.8. Microstructure Analysis

The size, shape, and aggregate state of emulsion droplets were quantified by the optical microscopic method described by Yang et al. [27] with slight modifications. Approximately 20 μL of emulsions were deposited on the microscope slide and covered with a coverslip. The morphology of the Pickering emulsion droplet was observed using an inverted optical microscope (CKX41, Olympus, Tokyo, Japan) at 200× magnification.

Confocal Laser Scanning Microscopy (CLSM) were determined by using a Nikon Ti-E-A1R confocal microscope (Nikon Instruments Inc, Tokyo, Japan) to observe the microstructure of the emulsions. The CLSM was operated in the fluorescence mode. Nile red was used to stain the oil phase. All of the stain solutions were prepared at room temperature and stored in a dark place. Approximately 1.0 mL of the samples were completely mixed with a 10 μL aliquot of the Nile red (0.1%, *w/v*). The mixtures were equilibrated for 30 min before observation. Approximately 20 μL of the stained samples was placed on slide and carefully covered with a coverslip, ensuring that there was no trapped air gap or bubbles between the mixture and the cover slip. The images were collected using 488 nm excitation wavelengths for Nile red. Each image contained 1024 × 1024 pixels.

2.9. Statistical Analysis

All experiments were performed in triplicate, and the value was expressed as mean value ± standard deviation. The statistical differences among means were evaluated using a one-way analysis of variance and Duncan's test ($p < 0.05$).

3. Results and Discussion

3.1. Effects of Ultrasound Process on the Preparation of ZNP-FSG Complexes

3.1.1. Emulsifying Properties

The emulsifying activity index (EAI) refers to the ability of proteins absorbed into the oil/water interface. The emulsifying stability index (ESI) is an indicator of the ability of a protein to maintain emulsion stability [35]. The EAI and ESI of the complexes subjected to various ultrasound and homogenization treatments are presented in Figure 2. The results showed that the EAI and ESI of untreated control sample ZNP were 6.09 ± 0.62 m^2/g and 12.64 ± 0.43 min, respectively. In comparison with the ZNP, the EAI and ESI of homogenization treated complex (ZNP-FSG)H were significantly increased, up to 10.47 ± 1.20 m^2/g and 42.47 ± 4.10 min, respectively. Interestingly, no significant difference was observed between the EAI and ESI of (ZNP-FSG)H and (ZNP_U-FSG)H, indicating the emulsification property of the homogenization-treated complex was independent from the ultrasonic treatment of zein. Moreover, (ZNP-FSG)H and (ZNP_U-FSG)H exhibited higher EAI and

ESI than that of the (ZNP-FSG$_U$)H, implying that pretreatment of FSG with ultrasonication was not favorable to improve the emulsifying properties of the complex. More importantly, we found that the (ZNP-FSG)HU and (ZNP-FSG)U complex showed significantly lower EAI and ESI than that of (ZNP-FSG)H. This result indicates that further treatment with ultrasonication could dramatically destroy the emulsifying capability of the complex. In fact, the (ZNP-FSG)H and (ZNP$_U$-FSG)H complex had the highest EAI and ESI among all the samples in the current study. Finally, there was no obvious difference among the EAI and ESI between ZNP, (ZNP-FSG)HU, and (ZNP-FSG)U. That is to say, adding FSG did not necessarily improve the emulsification property of ZNP, which also depends on the pretreatment and emulsifying process. The improved emulsifying properties of ZNP-FSG complexes may be correlated with the solubility, structural changes of zein, electrostatic interaction, and hydrophobic interaction between ZNP and FSG [32,36,37].

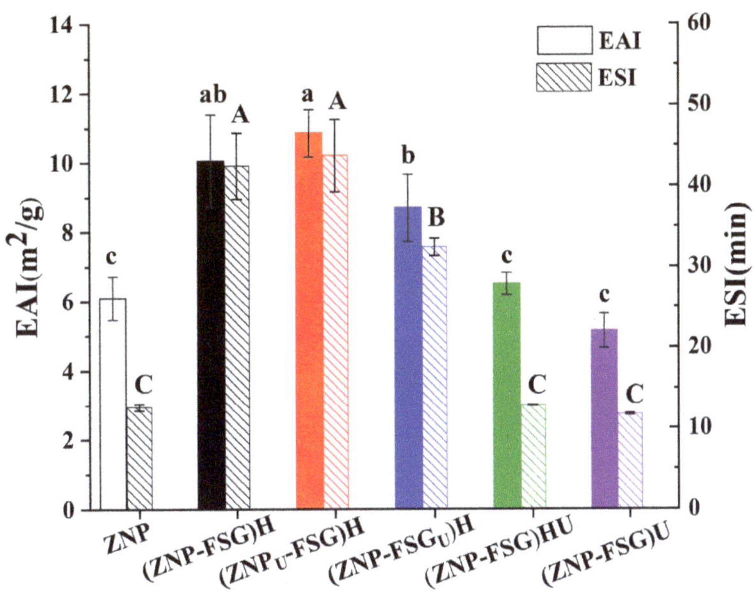

Figure 2. Emulsifying activity index (EAI) and emulsifying stability index (ESI) of emulsions stabilized by ZNP-FSG complexes subjected to various ultrasound and homogenization treatments.

3.1.2. Turbidity Changes

Dispersion turbidity indicates the number and size of dispersed particles to some content [38], which reflects the dispersion state and aggregation degree. Figure 3 illustrates the turbidity of ZNP and ZNP-FSG complexes. Turbidity of most complexes decreased after ultrasound pretreatments, except (ZNP-FSG)HU and (ZNP-FSG)U, which increased. It may be because the stable balances of original complexes were destroyed by extra ultrasound. Contrary to our results, Ma et al. [37] found the ultrasound treatment significantly decreased the turbidity of the soy protein isolate (SPI)—citrus pectin (CP) complex. The type of polysaccharide is probably one of the reasons for the different phenomenon. The turbidity of (ZNP$_U$-FSG) is lowest, and the turbidity of (ZNP-FSG)H is lower than that of (ZNP-FSG$_U$)H. From this, we hypothesized that the particle size changes of proteins induced by ultrasonic treatment decreased the turbidity of the solution most [39], which was related to the smaller protein particle size obtained by ultrasonic treatment. (ZNP-FSG$_U$)H might have had higher turbidity because of the degradation of FSG [36].

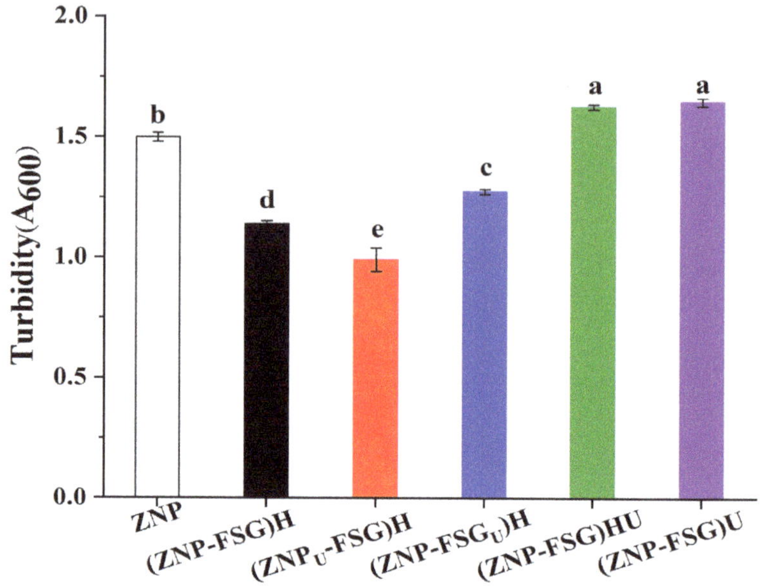

Figure 3. Turbidity of ZNP-FSG complexes subjected to various ultrasound and homogenization treatments.

3.1.3. Salt Ionic Stability

The stability of the complexes at different ionic strengths (50 mM, 100 mM, 200 mM, 300 mM, 400 mM, 500 mM, and 1000 mM) was evaluated by visual observation and turbidity in Figure 4. In Figure 4a, ZNP started to form flocculation and precipitation at a 50 mM NaCl due to electrostatic screening effects [40]. The flocculation extent kept increasing with increased salt concentration. Different from ZNP, it can be seen from the picture that the stability of the ZNP-FSG complex at different salt ion concentrations improvement was attributed to the addition of FSG, even though differences were noticed among complexes with different treatments. The result indicates that FSG is effective in protecting ZNP from salt-induced aggregation, which is probably due to the weakened electrostatic screening effects of ZNP brought by the electrostatic interaction between ZNP and FSG [40]. However, the tolerance of (ZNP-FSG)HU to salt was relatively poor, and flocculation can be clearly observed, as shown in Figure 4a.

The turbidity results, as shown from Figure 4b, is in agreement with the visual observation. The turbidity of (ZNP_U-FSG)H and (ZNP-FSG)H decreased as the NaCl concentration increased. The turbidity of (ZNP-FSG_U)H, (ZNP-FSG)U, and (ZNP-FSG)HU did not show a great relationship with the increasing salt ion concentration but fluctuated to different degrees. As we can see, the turbidity of (ZNP-FSG)HU and (ZNP-FSG)U showed an obvious upward trend at the salt ion concentration of 1000 mM. This is probably caused by the decreased interaction between ZNP and FSG, which resulted from the electrostatic shielding effect at high ion concentrations. At all the tested ion concentrations, (ZNP_U-FSG)H complex showed lower turbidity than the others, which indicated that (ZNP_U-FSG)H had the best tolerance of salt. Overall, the complexes with different treatments had a good stability at the salt ion concentration of 1000 mM except for (ZNP-FSG)HU. Increases of (ZNP-FSG)U and (ZNP-FSG)HU in turbidity illustrate the continued aggregation of soluble complexes into insoluble protein/polysaccharide complexes due to charge neutralization.

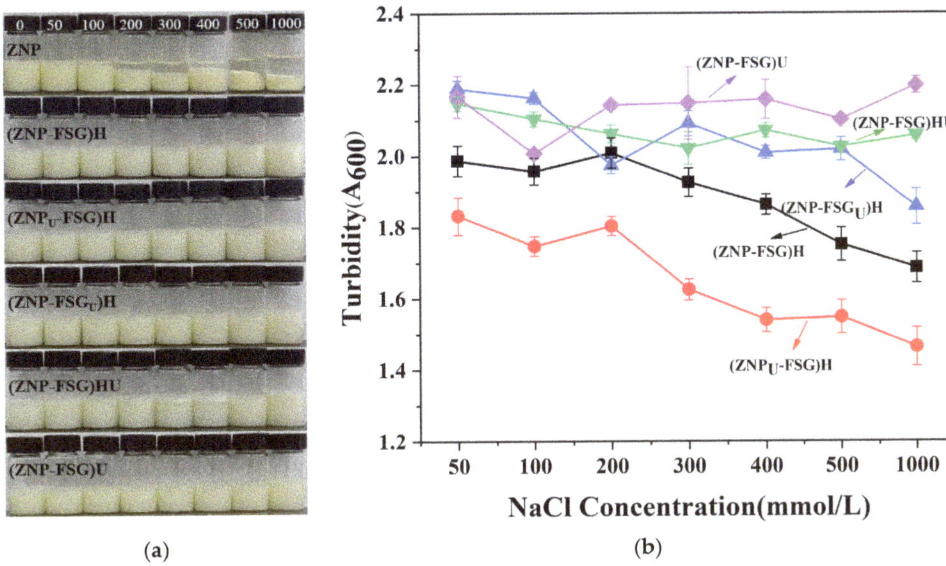

Figure 4. Visual appearance (**a**) and turbidity (**b**) of various ZNP-FSG complexes prepared in different NaCl concentration, i.e., black for (ZNP-FSG)H, red for (ZNP$_U$-FSG)H, blue for (ZNP-FSG$_U$)H, green for (ZNP-FSG)HU, and purple for (ZNP-FSG)U.

3.1.4. Surface Hydrophobicity

Surface hydrophobicity (H_0) is closely related to the emulsifying properties, stability, and function of the complex/particles [41], and a balanced surface hydrophobicity has a beneficial effect on surface activity and interfacial performance in emulsions. The surface hydrophobicity of the complexes is shown in Figure 5. We concluded that the H_0 of ZNP could be increased by ultrasonic pretreatment on zein. Ultrasonic cavitation can effectively expose the buried hydrophobic regions of zein from the interior of molecules to the surface [42]. Similar results were obtained by Gao et al. [14], investigating the effect of ultrasound on whey isolated protein. The H_0 of (ZNP-FSG)H complexes significantly decreased after adding FSG, which was ascribed to FSG blocking the binding sites between the fluorescent probe and hydrophobic groups of protein. Wang et.al [43] also reported that the hydrophobicity of chitosan-zein complexes decreased due to the addition of chitosan. The surface hydrophobicity of (ZNP$_U$-FSG)H was higher than that of (ZNP-FSG)H, which indicated that the increase in H_0 of ZNP after ultrasound improved the H_0 of (ZNP$_U$-FSG)H. The H_0 value of (ZNP-FSG)HU and (ZNP-FSG)U showed an extreme decrease after sonication, which was independent of whether it had been homogenized or not. This was probably due to the contact between the ultrasound and protein increasing when FSG separates from ZNP after sonication, and hydrophobic groups were reburied. A similar result was reported in the effects of dynamic high-pressure microfluidization treatment on the functional and structural properties of potato protein isolate and its complex with chitosan by Hu et al. [44]. The changing trend of surface hydrophobicity is consistent with that of emulsification capability. Previous research [45] has shown that the adsorption rate was shown to scale with the relative hydrophobicity, which is considered to be helpful to improve emulsifying properties.

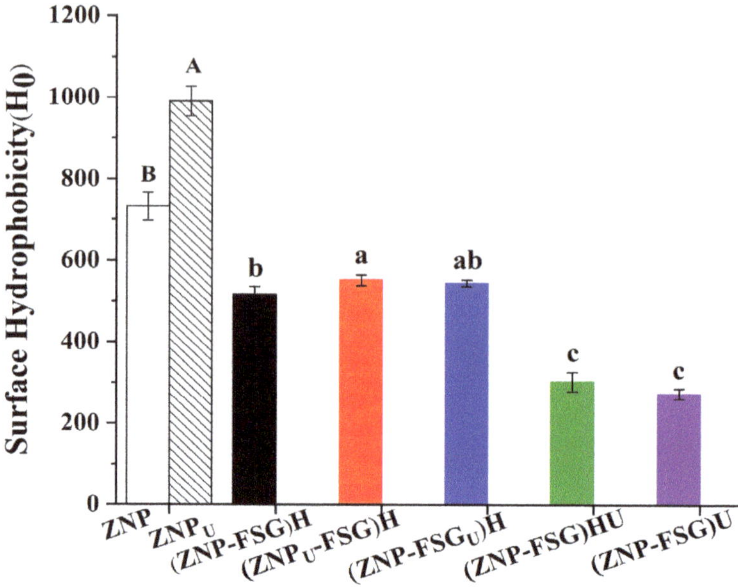

Figure 5. Surface Hydrophobicity (H_0) of ZNP-FSG complexes subjected to various ultrasound and homogenization treatments.

3.1.5. Particle Size

In order to study the effects of the pretreatment methods and operating conditions on different samples, the particle size of complexes was characterized, as shown in Figure 6a. Ultrasound pretreatment was able to significantly decrease the particle size of ZNP from 226.1 nm to 185.4 nm, which was possibly related to the structural changes of protein upon ultrasonication [46]. The particle size of FSG decreased from 767.0 nm to 332.7 nm after ultrasound treatment, presumed to be caused by the degradation of FSG after ultrasound [47]. Similar results were reported for ultrasonication-treated Persian gum and gum tragacanth [48]. The particle size of both (ZNP-FSG)U and (ZNP-FSG)HU was the lowest among all complexes, and there was no significant difference between them. We found out that if the complex eventually requires ultrasonic treatment, there was no difference in particle whether the intermediate process was homogenized or not. There was no significant difference in particle size of (ZNP-FSG$_U$)H and (ZNP-FSG)H, revealing the ultrasound on FSG playing no effect on the particle size of the complex. However, the pre-ultrasonication on zein significantly decreased the particle size of the complex (ZNP$_U$-FSG)H]. Since the particle size of (ZNP-FSG)HU was lower than that of (ZNP-FSG)H, it can be concluded that the extra ultrasonication is effective to further decrease the particle size of the homogenized complex. In conclusion, ultrasonic treatment of different ingredients has a significant effect on the particle size of the complex prepared. However, homogenization on complexes does not have significant effects on the particle size.

PDI is usually used to represent the particle size distribution of dispersion [37]. As shown in Figure 6b, the PDI values of FSG and FSG$_U$ are significantly higher than other samples, indicating a broad particle size distribution [29]. The PDI of (ZNP-FSG$_U$)H, (ZNP-FSG)HU and (ZNP-FSG)U were higher than that of other complexes, which were all around 0.30. The PDI of (ZNP$_U$-FSG)H was the lowest (0.21), showing the uniformity of particle size distribution was increased due to the pre-ultrasonication on zein. The particle size distribution of various ingredients or ZNP -FSG complexes subjected to various ultrasound and homogenization treatments are showing in Figure S1. As is known to all, PDI is not the only reason that describes emulsion stability. We should continue to analyze together with other properties of the complexes.

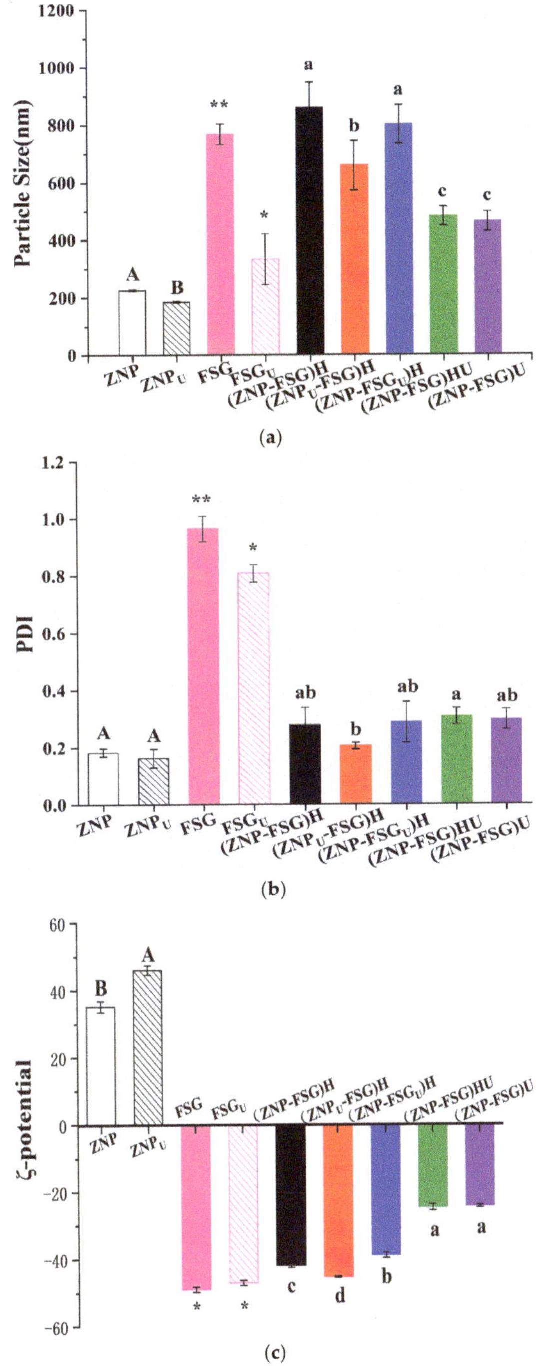

Figure 6. Particle size (**a**), PDI (**b**), and ζ-potential (**c**) of ZNP-FSG complexes subjected to various ultrasound and homogenization treatments. Different lowercase letters (a, b), different capital letters (A, B) and asterisk (*, **) indicated significant difference ($p < 0.05$).

3.1.6. ζ-Potential

Figure 6c also shows the ζ-potential of ZNP-FSG complexes subjected to various ultrasound and homogenization processes. The stability of complexes could be studied by measuring the electrostatic stability of the colloidal particles [49]. Particles with ζ-potentials more positive than +30 mV or more negative than −30 mV are normally considered stable [50]. In the absence of FSG, the ζ-potential of the ZNP suspension was around +35 mV, which was due to the fact that the pH was below its isoelectric point (IP) and, therefore, ZNP had a net positive charge. The ζ-potential value of ZNP was increased from +35.2 mV to +45.9 mV after ultrasound pretreatment, suggesting that pretreatment on zein could improve the stability of ZNP. This was consistent with previous results that the ZNP system was stable below pH 6.0 [16]. In fact, ZNP remained stable for several days at 4 °C. The ζ-potential value of FSG was increased from −49.7 mV to −47.1 mV after ultrasound treatment, indicating that ultrasonic treatment slightly lowered the stability of FSG. According to Figure 5, the ζ-potential of complexes were negative, which is because the negatively charged FSG molecules interacted with the ZNP through electrostatic force and aggregates on the surface of positively charged ZNP, resulting in charge reversal. Similar results were found in the study on complexes of WPI gel-chitosan and zein-carboxymethyl dextrin [28,51].

The ζ-potential values of ZNP-FSG complexes subjected to various ultrasound and homogenization treatments were obviously different from each other, suggesting that the pretreatment and operating process significantly affected the stability of the complexes. The absolute zeta potentials of complexes ranking by increasing order is (ZNP-FSG)U ≈ (ZNP-FSG)HU < (ZNP-FSG_U)H < (ZNP-FSG)H < (ZNP_U-FSG)H. First, the absolute zeta potential of (ZNP-FSG)U was equal to that of (ZNP-FSG)HU and was the smallest among all. Thus, it can be seen that homogenization of the complex prior to ultrasonic treatment had no effect on their stability. Second, the absolute ζ-potential of (ZNP-FSG)HU was lower than that of (ZNP-FSG)H, indicating the stability of the complex was significantly decreased by ultrasonic treatment after homogenization, regardless of the ingredients of ultrasound. Based on the above results, we concluded that in order to obtain ZNP-FSG complexes with better stability, ultrasound and homogenization cannot be superimposed on the complex. Maybe the mechanical force and cavitation induced by ultrasound increase the energy and entropy of ZNP-FSG system and break the balance made by homogenization, even the particle size or type of interaction force may be affected as well [14]. Ultrasonic treatment has the capability of decreasing the size of protein [52] and reduces the viscosity of polysaccharides by degrading chains of them [48]. The ζ-potential of (ZNP-FSG_U)H reduced in comparison with ZNP, which was ascribed to the decrease of amounts and distributions of charges on the ZNP surface because of the depolymerizing of FSG during the ultrasonic process [30].

However, when applying ultrasonic pretreatment on a single ingredient before homogenization, the complex had better stability. Sample (ZNP_U-FSG)H showed the maximum absolute ζ-potential value and the best stability, and indicated its better stability than (ZNP-FSG)H or (ZNP-FSG_U)H. We found that the selection of ultrasonic ingredients significantly affected the stability of the complex. When the ultrasonic pretreatment ingredient was protein, the ultrasonic had a positive effect on the stability of the complex, whereas if the ultrasonic pretreatment ingredient was FSG, a negative effect on the stability of the complex was observed. Some possible reasons for the results were speculated from different mechanisms of ultrasonic treatment on different ingredients.

In addition, we tried to discuss the relationship between particle size and the potential of the ZNP-FSG complex. There was no direct correlation between the stability and particle size of the complex nor an obvious correlation between ζ-potential and particle size changes of the ZNP-FSG complex. Although ultrasound is widely used in the pretreatment of various samples in preparing a Pickering emulsion for emulsification and homogenization, its effect on homogeneity was revealed for the first time. Above all, the selection of ingredients for treatment and combination of process technology is very important for obtaining a good effect on the stability of protein–polysaccharides complex during the emulsification and homogenization.

3.2. *Effects of Ultrasound Process on the Emulsion of ZNP-FSG Complexes*

3.2.1. Storage Stability of Emulsions

Effects of ultrasonic pretreatment on the physicochemical changes of ZNP-FSG complex and the characteristics of subsequent oil-in-water emulsions were investigated. The creaming index is the ratio of the height of the serum phase to the total height of the emulsion. The creaming index of the emulsions stored at 4 °C for 14 days is shown in Table 2. The creaming index was used as a measure of the stability of Pickering emulsions. The occurrence of the phase separation phenomenon proves that the emulsion has poor storage stability [53]. The appearance of the emulsions prepared with ZNP showed slight creaming as early as 30 min after the emulsions were prepared. The creaming index value of ZNP-E exhibited the highest of 25% after 30 min and then increased to 30%, indicating that ZNP-E possessed the worst stability among all the emulsions. The creaming index of (ZNP-FSG$_U$)-E, (ZNP-FSG)HU-E, and (ZNP-FSG)U-E in turn increase. In other words, their stability decreases in turn. However, (ZNP-FSG)-E and (ZNP$_U$-FSG)-E remained physically stable for 14 days at 4 °C. This was consistent with the result that their corresponding complexes had the highest EAI and ESI values.

Table 2. The creaming index (%) of emulsions stabilized by ZNP-FSG complexes subjected to various ultrasound and homogenization treatments.

Creaming Index (%)	Time								
	0.5 h	**2 h**	**1 d**	**2 d**	**3 d**	**4 d**	**7 d**	**10 d**	**14 d**
ZNP-E	25	25	25	27.5	30	30	30	30	30
(ZNP-FSG)H-E	0	0	0	0	0	0	0	0	0
(ZNP$_U$-FSG)H-E	0	0	0	0	0	0	0	0	0
(ZNP-FSG$_U$)H-E	7.5	7.5	8	10	10	10	25	25	25
(ZNP-FSG)HU-E	17.5	22.5	25	25	25	25	25	25	25
(ZNP-FSG)U-E	17.5	25	25	25	25	25	25	25	25

3.2.2. Freeze-Thaw Stability of Emulsions

The emulsions might undergo the freeze-thaw process. Many emulsions were extremely unstable after three cycles of freeze-thaw treatment, some phenomena such as creaming, flocculation, and coalescence occurred [54]. The freeze-thaw stability of all emulsions stabilized by complexes subjected to various ultrasound and homogenization treatments were evaluated by the creaming index and oiling off and are shown in Figure 7.

It could be seen from Figure 7 that both the oiling off and creaming index of emulsions increased with the number of freeze-thaw treatment cycles. Obviously, ZNP-E had the highest oiling off and creaming index. Compared with ZNP-E, the oiling off and creaming index of ZNP-FSG complexes subjected to various ultrasound and homogenization treatments all decreased sharply, indicating that the addition of FSG improved the freeze-thaw stability of the emulsion. However, different treatments play important roles in the process. As expected, the oiling off and the creaming index of (ZNP-FSG)U-E and (ZNP-FSG)HU-E were higher than other emulsions except for ZNP-E. Surprisingly, (ZNP-FSG$_U$)H-E showed the lowest oiling off, followed by (ZNP$_U$-FSG)H-E, and then the (ZNP-FSG)H-E, the sequence of which was different from the order of the creaming index. The creaming index of (ZNP$_U$-FSG)-E was the lowest after three freeze-thaw cycles, followed by (ZNP-FSG$_U$)H-E and (ZNP-FSG)H-E.

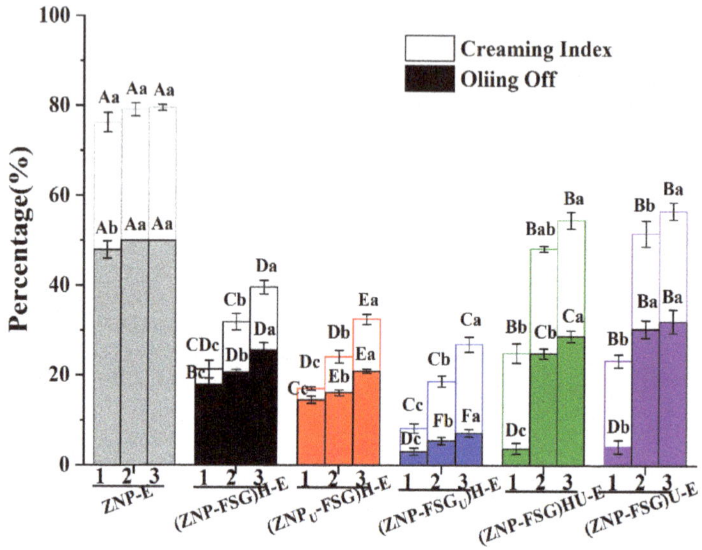

Figure 7. The oiling off and creaming index of the Pickering emulsions stabilized by ZNP-FSG complexes subjected to various ultrasound and homogenization treatments. 1, 2, 3 represent freeze-thaw cycles, respectively.

After the emulsion is frozen, the droplets can undergo coalescence because of the oil phase and water phase crystallization-induced destruction of the emulsions interface film [55]. Then, the addition of FSG could increase the thickness of the interfacial film and provide enough repulsion to prevent droplet aggregation [56]. The improvement of the freeze-thaw stability of the emulsion is closely related to the emulsification and surface hydrophobicity of the ZNP-FSG complexes [57]. As discussed above, ZNP$_U$-FSG)H exhibited superior emulsifying properties mainly attributed to the great ability of decreasing the interfacial tension. Thus, a decrease in freeze-thawing stability of (ZNP-FSG)U-E and (ZNP-FSG)HU-E would be directly associated with a combination of reduced emulsification ability and surface hydrophobicity of complexes. Next, the structure of emulsions was further studied.

3.2.3. Emulsion Droplets Size Distribution

The appearance and droplets size of emulsions (micron scale) stabilized by complexes subjected to various ultrasound and homogenization treatments were presented in Figure 8. The Pickering emulsion of ZNP-FSG can be infinitely diluted in water, indicating they were oil/water emulsions. As shown in Figure 8, all the emulsions exhibited unimodal distributions. It can be observed that emulsion stabilized by individual ZNP tended to separate into two layers. The droplet size of the emulsion decreased from 60.03 $\pm$ 5.72 μm to 48.43 $\pm$ 3.34 μm with the addition of FSG, and no serum phase was observed. The droplets size of (ZNP-FSG)HU-E and (ZNP-FSG)U-E were 60.03 $\pm$ 1.35 μm and 59.97 $\pm$ 3.16 μm, respectively. However, the emulsions broke after 30 min at room temperature.

Emulsion droplets size is one of the key parameters to characterize the properties of the emulsion. The droplets size distribution of emulsion depends on the breakup and coalescence of droplets. The high shear force (ultrasound and homogenization) can cause breakup of droplets, and the droplets coalescence may be attributed to surfactants [35]. The presence of FSG decreased the interfacial wettability and promoted complexes to absorb irreversibly on the oil/water interface, which is beneficial to the stability of the emulsion. However, the changing trend of particle size of emulsion was similar to that of complexes, (ZNP$_U$-FSG)H-E obtained the lowest droplets size of emulsions (46.9 $\pm$ 0.30 μm) because of the small particle size of the complex. According to the results of storage stability and freeze-thaw stability above, (ZNP$_U$-FSG)H-E showed relatively good storage stability and freeze-thaw stability compared with other groups of emulsions.

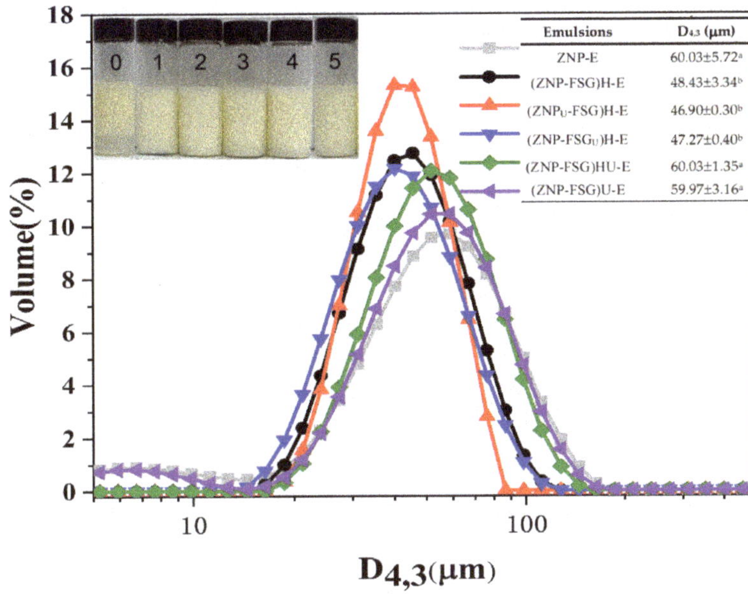

Figure 8. The particle size distribution of emulsions stabilized by ZNP-FSG complexes subjected to various ultrasound and homogenization treatments. Insets are photographs of fresh emulsions (0. ZNP-E, 1. (ZNP-FSG)H-E, 2. (ZNP$_U$-FSG)H-E, 3. (ZNP-FSG$_U$)H-E, 4. (ZNP-FSG)HU-E, 5. (ZNP-FSG)U-E). The emulsion droplets sizes are shown in the table.

3.2.4. Microstructure of Emulsions

Figure 9 shows the optical microscope photographs and CLSM of Pickering emulsions prepared by complexes subjected to various ultrasound and homogenization treatments. The droplets size distribution is consistent with the result of the droplets microstructure of the emulsion. The droplets of emulsion stabilized by ZNP alone has large droplets size and irregular shape. The creaming index of ZNP-E reached 25% in 30 min and eventually reached the maximum of 30% in 4 days. Pickering emulsions stabilized by ZNP-FSG had no aqueous phase, manifesting that the addition of FSG increased the stability of Pickering emulsions. (ZNP$_U$-FSG)H-E showed significantly more droplets and more regular shapes compared to other emulsions, which is consistent with the observations for emulsions stabilized by pea protein with ultrasound [52]. Compared with (ZNP$_U$-FSG)H-E, the droplets size of (ZNP-FSG$_U$)H-E showed obviously thicker walls and irregular shapes. In the course of observation, we found that the droplet size of (ZNP-FSG)U-E and (ZNP-FSG)HU-E increased gradually over time, and the phase separation phenomenon was observed in 30 min, suggesting their stability were poor. In addition, the presence of oil droplets can be observed clearly on the emulsion droplet surface, which could be explained by two reasons. Firstly, ultrasound increased the frequency of contact collisions between droplets, leading to droplet coalescence. Secondly, over-processing of complexes was caused by ultrasonic and high-speed homogenization.

Figure 9b shows CLSM images of Pickering emulsions after 24 h storage at 4 °C. Corn oil was dyed with Nile red. All the above results showed that homogenization had no significant effect on the complexes and emulsion, thus (ZNP-FSG)U-E was not shown in the CLSM diagram. (ZNP$_U$-FSG)-E can prevent collisions and coalescence of droplets and contained a fairly uniform dispersion of spherical oil droplets. The bigger oil droplet size was observed in (ZNP-FSG$_U$)-E and (ZNP-FSG)HU-E. The oil droplet size variation observed by CLSM was in agreement with the storage stability.

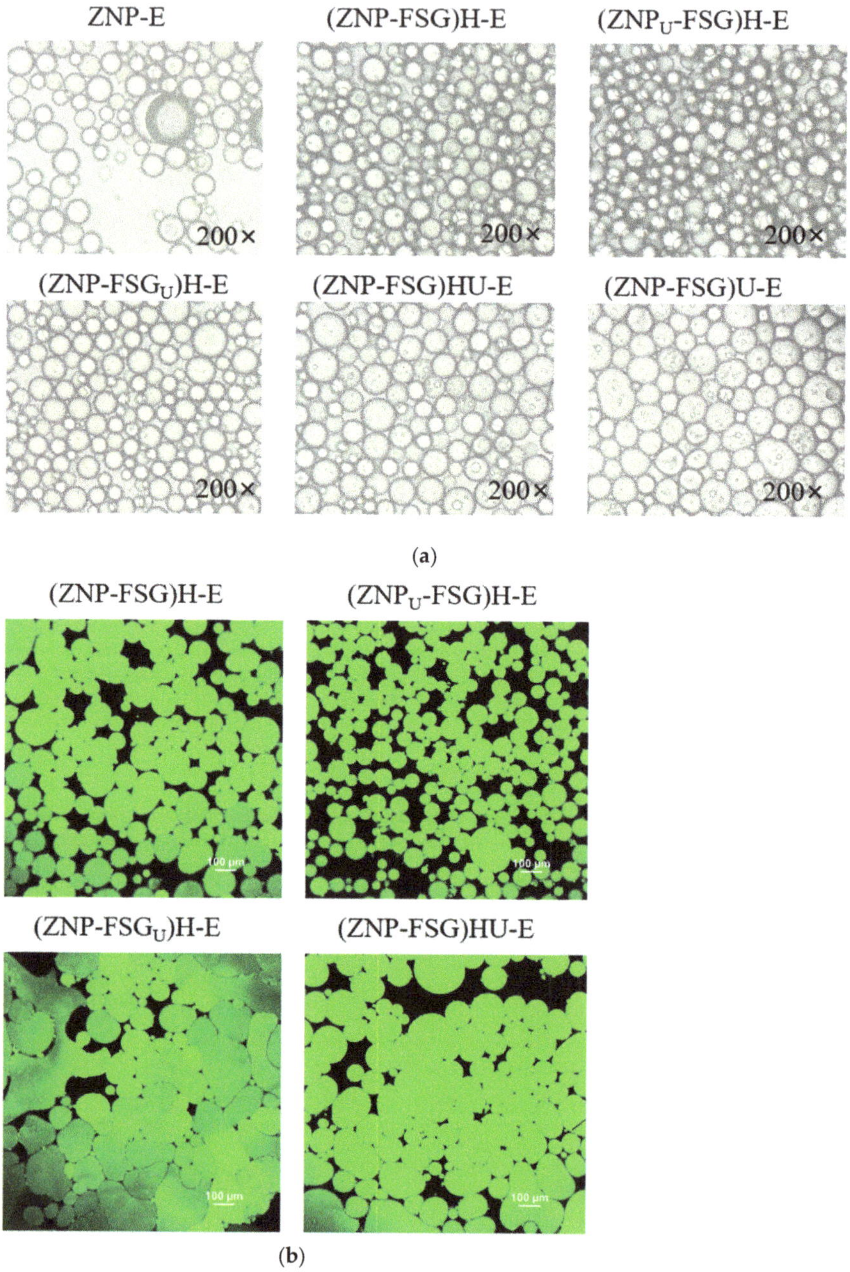

Figure 9. Optical micrographics (**a**) and Confocal laser scanning microscopy (CLSM) (**b**) of emulsions stabilized by ZNP and FSG complexes subjected to various ultrasound and homogenization treatments.

4. Conclusions

In summary, different ultrasonication and homogenization treatments combination on zein and FSG can significantly influence the properties of the complexes and the emulsion formed by the complexes. Ultrasonic treatment affected the interaction between ZNP and FSG. The complex formed by ZNP_U and FSG has the highest absolute zeta potential value, better emulsification properties (higher EAI and ESI), and the highest surface hydrophobicity among all, which indicates that the complex was stable and can be used as a good Pickering emulsions stabilizer among those complexes subjected to various ultrasound and homogenization treatments. However, the properties of (ZNP-FSG)U and (ZNP-FSG)HU are very different from those of (ZNP_U-FSG)H. (ZNP_U-FSG)H-E with good storage stability and freeze-thaw stability can be observed in the microscope with a clear, regular, and uniform spherical structure. In conclusion, these results indicate that the timing of using ultrasound in the preparation of complexes and emulsions is very important. These results provide new ideas and a theoretical basis for improving the properties of protein–polysaccharide complexes and emulsions.

Supplementary Materials: The following are available online at https://www.mdpi.com/article/10.3390/foods10091990/s1, Figure S1: The particle size distribution of 9 various ingredients or ZNP-FSG complexes subjected to various ultrasound and homogenization treatments.

Author Contributions: Conceptualization, Y.L.; data curation, Y.L.; methodology, Y.L., G.X., W.L. and L.L.; investigation, Y.L. and Q.Z.; software, Q.Z., W.L. and L.L.; writing—original draft preparation, Y.L. and Q.Z.; writing—review and editing, Q.Z.; visualization, Y.L. and G.X.; All authors have read and agreed to the published version of the manuscript.

Funding: This study was financially supported by National Natural Science Foundation of China (grant numbers 31701608) and the Natural Science Research Projects of Jiangsu Higher Education Institution (grant numbers 17KJD550002).

Institutional Review Board Statement: Not applicable.

Informed Consent Statement: Not applicable.

Data Availability Statement: The datasets generated for this study are available on request to the corresponding author.

Acknowledgments: The authors are grateful for the financial supports from National Natural Science Foundation of China (No. 31701608). The authors would like to thank Huijuan Zheng for her assistance.

Conflicts of Interest: The authors declare no conflict of interest.

References

1. Mwangi, W.W.; Lim, H.P.; Low, L.E.; Tey, B.T.; Chan, E.S. Food-grade Pickering emulsions for encapsulation and delivery of bioactives. *Trends Food Sci. Technol.* **2020**, *100*, 320–332. [CrossRef]
2. Xiao, J.; Li, Y.Q.; Huang, Q.R. Recent advances on food-grade particles stabilized Pickering emulsions: Fabrication, characterization and research trends. *Trends Food Sci. Technol.* **2016**, *55*, 48–60. [CrossRef]
3. Ying, G.-G. Fate, behavior and effects of surfactants and their degradation products in the environment. *Environ. Int.* **2006**, *32*, 417–431. [CrossRef] [PubMed]
4. Shi, A.-M.; Feng, X.-Y.; Wang, Q.; Adhikari, B. Pickering and high internal phase Pickering emulsions stabilized by protein-based particles: A review of synthesis, application and prospective. *Food Hydrocoll.* **2020**, *109*, 106117. [CrossRef]
5. Zhu, F. Starch based Pickering emulsions: Fabrication, properties, and applications. *Trends Food Sci. Technol.* **2019**, *85*, 129–137. [CrossRef]
6. Wei, Z.-J.; Wang, C.-Y.; Zou, S.-W.; Liu, H.; Tong, Z. Chitosan nanoparticles as particular emulsifier for preparation of novel pH-responsive Pickering emulsions and PLGA microcapsules. *Polymer* **2012**, *53*, 1229–1235. [CrossRef]
7. Ning, F.-J.; Ge, Z.-Z.; Qiu, L.; Wang, X.-Q.; Luo, L.-P.; Xiong, H.; Huang, Q.-R. Double-induced se-enriched peanut protein nanoparticles preparation, characterization and stabilized food-grade pickering emulsions. *Food Hydrocoll.* **2020**, *99*, 105308. [CrossRef]
8. Zhang, X.; Liang, H.S.; Li, J.; Wei, X.L.; Li, B. Improving the emulsifying property of gliadin nanoparticles as stabilizer of Pickering emulsions: Modification with sodium carboxymethyl cellulose. *Food Hydrocoll.* **2020**, *107*, 105936. [CrossRef]
9. Ashokkumar, M. Applications of ultrasound in food and bioprocessing. *Ultrason. Sonochem.* **2015**, *25*, 17–23. [CrossRef]

10. Li, Y.; Wu, C.-L.; Liu, J.; Zhu, Y.; Zhang, X.-Y.; Jiang, L.-Z.; Qi, B.-K.; Zhang, X.-N.; Wang, Z.-J.; Teng, F. Soy Protein Isolate-Phosphatidylcholine Nanoemulsions Prepared Using High-Pressure Homogenization. *Nanomaterials* **2018**, *8*, 307. [CrossRef]
11. Taha, A.; Hu, T.; Zhang, Z.; Bakry, A.M.; Khalifa, I.; Pan, S.-Y.; Hu, H. Effect of different oils and ultrasound emulsification conditions on the physicochemical properties of emulsions stabilized by soy protein isolate. *Ultrason. Sonochem.* **2018**, *49*, 283–293. [CrossRef]
12. Zhang, K.-M.; Mao, Z.-J.; Huang, Y.-C.; Xu, Y.; Huang, C.-D.; Guo, Y.; Ren, X.-e.; Liu, C.-Y. Ultrasonic assisted water-in-oil emulsions encapsulating macro-molecular polysaccharide chitosan: Influence of molecular properties, emulsion viscosity and their stability. *Ultrason. Sonochem.* **2020**, *64*, 105018. [CrossRef]
13. Albano, K.M.; Nicoletti, V.R. Ultrasound impact on whey protein concentrate-pectin complexes and in the O/W emulsions with low oil soybean content stabilization. *Ultrason. Sonochem.* **2018**, *41*, 562–571. [CrossRef] [PubMed]
14. Taha, A.; Ahmed, E.; Ismaiel, A.; Ashokkumar, M.; Xu, X.-Y.; Pan, S.-Y.; Hu, H. Ultrasonic emulsification: An overview on the preparation of different emulsifiers-stabilized emulsions. *Trends Food Sci. Technol.* **2020**, *105*, 363–377. [CrossRef]
15. Sun, Y.; Li, Y.-H.; WULANDARI; Lv, L.-S.; Zhang, Q.-T. Ultrasonic pretreatment improves the stability of zein and Pickering emulsion. *Food Ferment Ind.* **2021**, *47*, 97–106.
16. de Folter, J.W.J.; van Ruijven, M.W.M.; Velikov, K.P. Oil-in-water Pickering emulsions stabilized by colloidal particles from the water-insoluble protein zein. *Soft Matter.* **2012**, *8*, 6807–6815. [CrossRef]
17. Yan, X.-J.; Ma, C.-C.; Cui, F.-Z.; McClements, D.J.; Liu, X.-B.; Liu, F.-G. Protein-stabilized Pickering emulsions: Formation, stability, properties, and applications in foods. *Trends Food Sci. Technol.* **2020**, *103*, 293–303. [CrossRef]
18. Yan, J.; Liang, X.; Ma, C.; McClements, D.J.; Liu, X.; Liu, F. Design and characterization of double-cross-linked emulsion gels using mixed biopolymers: Zein and sodium alginate. *Food Hydrocoll.* **2021**, *113*, 106473. [CrossRef]
19. Jiang, Y.; Zhang, C.; Yuan, J.; Wu, Y.; Li, F.; Waterhouse, G.I.N.; Li, D.; Huang, Q. Exploiting the robust network structure of zein/low-acyl gellan gum nanocomplexes to create Pickering emulsion gels with favorable properties. *Food Chem.* **2021**, *349*, 129112. [CrossRef]
20. Wang, Y.; Li, D.; Wang, L.-J.; Adhikari, B. The effect of addition of flaxseed gum on the emulsion properties of soybean protein isolate (SPI). *J. Food Eng.* **2011**, *104*, 56–62. [CrossRef]
21. Liu, J.; Shim, Y.Y.; Shen, J.-H.; Wang, Y.; Reaney, M.J.T. Whey protein isolate and flaxseed (*Linum usitatissimum* L.) gum electrostatic coacervates: Turbidity and rheology. *Food Hydrocoll.* **2017**, *64*, 18–27. [CrossRef]
22. Liu, J.; Shim, Y.Y.; Tse, T.J.; Wang, Y.; Reaney, M.J.T. Flaxseed gum a versatile natural hydrocolloid for food and non-food applications. *Trends Food Sci. Technol.* **2018**, *75*, 146–157. [CrossRef]
23. Wang, Y.; Wang, L.-J.; Li, D.; Özkan, N.; Chen, X.D.; Mao, Z.-H. Effect of flaxseed gum addition on rheological properties of native maize starch. *J. Food Eng.* **2008**, *89*, 87–92. [CrossRef]
24. Pham, L.B.; Wang, B.; Zisu, B.; Truong, T.; Adhikari, B. Microencapsulation of flaxseed oil using polyphenol-adducted flaxseed protein isolate-flaxseed gum complex coacervates. *Food Hydrocoll.* **2020**, *107*, 105944. [CrossRef]
25. Zeng, T.; Wu, Z.-L.; Zhu, J.-Y.; Yin, S.-W.; Tang, C.-H.; Wu, L.-Y.; Yang, X.-Q. Development of antioxidant Pickering high internal phase emulsions (HIPEs) stabilized by protein/polysaccharide hybrid particles as potential alternative for PHOs. *Food Chem.* **2017**, *231*, 122–130. [CrossRef]
26. Dai, L.; Sun, C.-X.; Wei, Y.; Mao, L.; Gao, Y.-X. Characterization of Pickering emulsion gels stabilized by zein/gum arabic complex colloidal nanoparticles. *Food Hydrocoll.* **2018**, *74*, 239–248. [CrossRef]
27. Yang, H.; Su, Z.-W.; Meng, X.-H.; Zhang, X.-Y.; Kennedy, J.F.; Liu, B.-J. Fabrication and characterization of Pickering emulsion stabilized by soy protein isolate-chitosan nanoparticles. *Carbohydr. Polym.* **2020**, *247*, 116712. [CrossRef]
28. Lv, P.-F.; Wang, D.; Chen, Y.-L.; Zhu, S.-X.; Zhang, J.-B.; Mao, L.-K.; Gao, Y.-X.; Yuan, F. Pickering emulsion gels stabilized by novel complex particles of high-pressure-induced WPI gel and chitosan: Fabrication, characterization and encapsulation. *Food Hydrocoll.* **2020**, *108*, 105992. [CrossRef]
29. Gibis, M.; Rahn, N.; Weiss, J. Physical and oxidative stability of uncoated and chitosan-coated liposomes containing grape seed extract. *Pharmaceutics* **2013**, *5*, 421–433. [CrossRef]
30. Hosseini, S.M.H.; Emamdjomeh, Z.; Razavi, S.H.; Moosavimovahedi, A.A.; Saboury, A.A.; Atri, M.S.; Paul, V.D.M. β-Lactoglobulin–sodium alginate interaction as affected by polysaccharide depolymerization using high intensity ultrasound. *Food Hydrocoll.* **2013**, *32*, 235–244. [CrossRef]
31. Kato, A.; Nakai, S. Hydrophobicity determined by a fluorescence probe method and its correlation with surface properties of proteins. *Biochim. Biophys. Acta* **1980**, *624*, 13–20. [CrossRef]
32. Li, C.; Huang, X.-J.; Peng, Q.; Shan, Y.-Y.; Xue, F. Physicochemical properties of peanut protein isolate–glucomannan conjugates prepared by ultrasonic treatment. *Ultrason. Sonochem.* **2014**, *21*, 1722–1727. [CrossRef]
33. Zhu, Y.-Q.; McClements, D.J.; Zhou, W.; Peng, S.-F.; Zhou, L.; Zou, L.-Q.; Liu, W. Influence of ionic strength and thermal pretreatment on the freeze-thaw stability of Pickering emulsion gels. *Food Chem.* **2020**, *303*, 125401. [CrossRef]
34. Du, Q.; Tang, J.; Xu, M.; Lyu, F.; Zhang, J.; Qiu, Y.; Liu, J.; Ding, Y. Whey protein and maltodextrin-stabilized oil-in-water emulsions: Effects of dextrose equivalent. *Food Chem.* **2021**, *339*, 128094. [CrossRef] [PubMed]
35. Sui, X.-N.; Bi, S.; Qi, B.-K.; Wang, Z.-J.; Zhang, M.; Li, Y.; Jiang, L.-Z. Impact of ultrasonic treatment on an emulsion system stabilized with soybean protein isolate and lecithin: Its emulsifying property and emulsion stability. *Food Hydrocoll.* **2017**, *63*, 727–734. [CrossRef]

36. Cui, R.-B.; Zhu, F. Ultrasound modified polysaccharides: A review of structure, physicochemical properties, biological activities and food applications—ScienceDirect. *Trends Food Sci. Technol.* **2020**, *107*, 491–508. [CrossRef]
37. Ma, X.-B.; Yan, T.-Y.; Hou, F.-R.; Chen, W.-J.; Miao, S.; Liu, D.-H. Formation of soy protein isolate (SPI)-citrus pectin (CP) electrostatic complexes under a high-intensity ultrasonic field: Linking the enhanced emulsifying properties to physicochemical and structural properties. *Ultrason. Sonochem.* **2019**, *59*, 104748. [CrossRef]
38. Gharibzahedi, S.M.T.; Smith, B. The functional modification of legume proteins by ultrasonication: A review. *Trends Food Sci. Technol.* **2020**, *98*, 107–116. [CrossRef]
39. Hu, H.; Cheung, I.W.Y.; Pan, S.Y.; Li-Chan, E.C.Y. Effect of high intensity ultrasound on physicochemical and functional properties of aggregated soybean β-conglycinin and glycinin. *Food Hydrocoll.* **2015**, *45*, 102–110. [CrossRef]
40. Wu, W.-H.; Kong, X.-Z.; Zhang, C.-M.; Hua, Y.-F.; Chen, Y.-M. Improving the stability of wheat gliadin nanoparticles—Effect of gum arabic addition. *Food Hydrocoll.* **2018**, *80*, 78–87. [CrossRef]
41. Yan, S.; Xu, J.; Zhang, S.; Li, Y. Effects of flexibility and surface hydrophobicity on emulsifying properties: Ultrasound-treated soybean protein isolate. *LWT* **2021**, *142*, 110881. [CrossRef]
42. Jin, J.; Ma, H.-L.; Wang, K.; Yagoub, A.E.-G.A.; Owusu, J.; Qu, W.-J.; He, R.-H.; Zhou, C.-S.; Ye, X.-F. Effects of multi-frequency power ultrasound on the enzymolysis and structural characteristics of corn gluten meal. *Ultrason. Sonochem.* **2015**, *24*, 55–64. [CrossRef]
43. Wang, L.-J.; Yin, S.-W.; Wu, L.-Y.; Qi, J.-R.; Guo, J.; Yang, X.-Q. Fabrication and characterization of Pickering emulsions and oil gels stabilized by highly charged zein/chitosan complex particles (ZCCPs). *Food Chem.* **2016**, *213*, 462–469. [CrossRef]
44. Hu, C.; Xiong, Z.-Y.; Xiong, H.-G.; Chen, L.; Zhang, Z.-L. Effects of dynamic high-pressure microfluidization treatment on the functional and structural properties of potato protein isolate and its complex with chitosan. *Food Res. Int.* **2021**, *140*, 109868. [CrossRef]
45. Delahaije, R.J.; Gruppen, H.; Giuseppin, M.L.; Wierenga, P.A. Quantitative description of the parameters affecting the adsorption behaviour of globular proteins. *Colloids Surf. B* **2014**, *123*, 199–206. [CrossRef]
46. O'Sullivan, J.J.; Park, M.; Beevers, J.; Greenwood, R.W.; Norton, I.T. Applications of ultrasound for the functional modification of proteins and nanoemulsion formation: A review. *Food Hydrocoll.* **2016**, *71*, 299–310. [CrossRef]
47. Jambrak, A.R.E.; Herceg, Z.; Ubari, D.; Babi, J.; Brni, M.; Brni, S.R.; Bosiljkov, T.; ?Vek, D.; Tripalo, B.; Gelo, J. Ultrasound effect on physical properties of corn starch. *Carbohydr. Polym.* **2010**, *79*, 91–100. [CrossRef]
48. Raoufi, N.; Kadkhodaee, R.; Fang, Y.; Phillips, G.O. Ultrasonic degradation of Persian gum and gum tragacanth: Effect on chain conformation and molecular properties. *Ultrason. Sonochem.* **2019**, *52*, 311–317. [CrossRef]
49. Wang, C.-N.; Wang, H.; Sun, X.-M.; Sun, Y.-X.; Guo, M.-R. Heat-Induced Interactions between Whey Protein and Inulin and Changes in Physicochemical and Antioxidative Properties of the Complexes. *Int. J. Mol. Sci.* **2019**, *20*, 4089. [CrossRef]
50. Mounsey, J.S.; O'Kennedy, B.T.; Fenelon, M.A.; Brodkorb, A. The effect of heating on β-lactoglobulin–chitosan mixtures as influenced by pH and ionic strength. *Food Hydrocoll.* **2008**, *22*, 65–73. [CrossRef]
51. Meng, R.; Wu, Z.-Z.; Xie, Q.-T.; Cheng, J.-S.; Zhang, B. Preparation and characterization of zein/carboxymethyl dextrin nanoparticles to encapsulate curcumin: Physicochemical stability, antioxidant activity and controlled release properties. *Food Chem.* **2021**, *340*, 127893. [CrossRef] [PubMed]
52. Sha, L.; Koosis, A.O.; Wang, Q.L.; True, A.D.; Xiong, Y.L.L. Interfacial dilatational and emulsifying properties of ultrasound-treated pea protein. *Food Chem.* **2021**, *350*, 129271. [CrossRef] [PubMed]
53. Taha, A.; Ahmed, E.; Hu, T.; Xu, X.-Y.; Pan, S.-Y.; Hu, H. Effects of different ionic strengths on the physicochemical properties of plant and animal proteins-stabilized emulsions fabricated using ultrasound emulsification. *Ultrason. Sonochem.* **2019**, *58*, 104627. [CrossRef] [PubMed]
54. Chen, Y.-B.; Zhu, X.-F.; Liu, T.-X.; Lin, W.-F.; Tang, C.-H.; Liu, R. Improving freeze-thaw stability of soy nanoparticle-stabilized emulsions through increasing particle size and surface hydrophobicity. *Food Hydrocoll.* **2019**, *87*, 404–412. [CrossRef]
55. Zhang, A.-Q.; Cui, Q.; Zhou, M.; Wang, X.-B.; Zhao, X.-H. Improving freeze–thaw stability of soy protein isolate-glucosamine emulsion by transglutaminase glycosylation. *Food Bioprod. Process.* **2021**, *128*, 77–83. [CrossRef]
56. Wong, B.T.; Day, L.; Augustin, M.A. Deamidated wheat protein–dextran Maillard conjugates: Effect of size and location of polysaccharide conjugated on steric stabilization of emulsions at acidic pH. *Food Hydrocoll.* **2011**, *25*, 1424–1432. [CrossRef]
57. Wang, Y.-Y.; Zhang, A.-Q.; Wang, X.-B.; Xu, N.; Jiang, L.-Z. The radiation assisted-Maillard reaction comprehensively improves the freeze-thaw stability of soy protein-stabilized oil-in-water emulsions. *Food Hydrocoll.* **2020**, *103*, 105684. [CrossRef]

foods

MDPI

Article

Beneficial Effects of *Holothuria leucospilota* Polysaccharides on Fermentability In Vivo and In Vitro

Wanting Wang [1,2,3], Yiqiong Yuan [1,2,3], Jun Cao [1,2,3], Xuanri Shen [1,2,3,4] and Chuan Li [1,2,3,4,*]

1 Engineering Research Center of Utilization of Tropical Polysaccharide Resources of Ministry of Education, Hainan University, Haikou 570228, China; wangwanting97@163.com (W.W.); yiqiongyuan@126.com (Y.Y.); juncaoyd2007@126.com (J.C.); shenxuanri2009@163.com (X.S.)
2 Hainan Provincial Engineering Research Centre of Aquatic Resources Efficient Utilization in the South China Sea, Hainan University, Haikou 570228, China
3 School of Food Science and Engineering, Hainan University, Haikou 570228, China
4 Collaborative Innovation Center of Provincial and Ministerial Co-Construction for Marine Food Deep Processing, Dalian Polytechnic University, Dalian 116034, China
* Correspondence: lichuan@hainanu.edu.cn

Abstract: This work aimed to investigate the in-vitro and in-vivo fermentation behaviors of *Holothuria leucospilota Polysaccharides* (HLP) and the impact on mouse liver antioxidant activity. HLP showed excellent fermentability during in vitro experiments, which was characterized by increased levels of total sugar consumption and short-chain fatty acids (SCFAs). During in vitro fecal fermentation, the fucose contents in the HLP fermentation products (0.174 mg/mL) were higher than those of xylose and galactosamine during the first three hours, and fucose disappeared after 24 h. The concentrations of the generated SCFAs increased to 111.13 mmol/mL after in-vitro fermentation at 48 h. After 28 days of oral administration, the SCFA contents that were detected in the feces of mice treated with high HLP doses were significantly higher than those in the feces of mice treated with lower doses and the normal group. In addition, histological observations demonstrated that HLP increased the number of goblet cells without causing hepatocellular injury. Moreover, the increased glutathione peroxidase (GSH-Px) and superoxidase dismutase (SOD) activities and decreased malondialdehyde (MDA) contents in the mouse livers treated with HLP suggested the good performance of HLP with respect to liver antioxidants.

Keywords: *Holothuria leucospilota*; polysaccharides; fermentation behaviors; antioxidant activity

Citation: Wang, W.; Yuan, Y.; Cao, J.; Shen, X.; Li, C. Beneficial Effects of *Holothuria leucospilota* Polysaccharides on Fermentability In Vivo and In Vitro. *Foods* **2021**, *10*, 1884. https://doi.org/10.3390/foods10081884

Academic Editors: Fani Mantzouridou, Jianhua Xie, Yanjun Zhang and Hansong Yu

Received: 4 July 2021
Accepted: 12 August 2021
Published: 15 August 2021

1. Introduction

Polysaccharides are inextricably associated with human health and exist in our foods, which are polymeric carbohydrate macromolecules consisting of multiple monosaccharide units linked by glycosides [1,2]. Recently, developing more bio-based or green chemical food ingredients has become a global trend to meet the increasingly diversified and healthy diet [3]. Exploring new polysaccharide material sources with useable characteristics has attracted the interest of researchers, as polysaccharides with different functions have been isolated from various plants and animals, bacteria, and fungi [4,5]. Despite the lack of large-scale industrial applications, it still makes great sense to understand the available natural resources worldwide [6]. Nondigestible polysaccharides cannot be absorbed immediately by the gastrointestinal digestive system [7]. Trillions of microorganisms coexist in the human intestine and are collectively referred to as the gut microbiota and play a vital role in human diets and host health [8]. However, indigestible polysaccharides can pass through the gastrointestinal tract to the distal gut and then be fermented by the gut microbiota [9] and produce some beneficial metabolites, such as short-chain fatty acids (SCFAs) [10]. SCFAs greatly influence the host physiology and energy homeostasis, especially in maintaining epithelial barrier function and preventing colorectal cancer [11].

As an edible marine resource and traditional medicine, sea cucumber has been used in many Asian countries to treat diverse diseases based on its antioxidant, antithrombotic, and antitumor activities. *Holothuria leucospilota* (Echinodermata: Holothuroidea, *H. leucospilota*) is a tropical sea cucumber species present in the Indo-Pacific region [12]. Among the varieties of active substances in its body wall, polysaccharides are the most noticeable ingredient, accounting for up to 31% of the total organic matter of dried sea cucumbers [13]. The HLP has been shown to be made of typical acidic polysaccharides with a backbone of intercalated 1,4-GlcA and 1,3-GalNAc, branched with fucose with different sulfate patterns and is composed of galactosamine, fucose, and glucuronic acids with ratios of 39.08%, 35.72%, and 10.72%, respectively [14,15]. Our previous works reported that HLP alleviated the symptoms of type 2 diabetes mellitus in rats by improving the gut microbiome [16]. In addition, Yuan et al. showed its excellent antioxidant activity after gastric and intestinal digestion in vitro [17]. However, the information regarding how HLP affects host health is still insufficient. The in-vitro and in-vivo fermentation characteristics and antioxidant activities of HLP have not been comparably explored.

Accordingly, the present study was designed to investigate the in-vitro and in-vivo fermentation behaviors of HLP by characterizing the pH, total sugar, reducing sugar, free monosaccharide, and SCFAs. In addition, the liver malondialdehyde (MDA), glutathione peroxidase (GSH-Px), and superoxidase dismutase (SOD) activities in mice were determined after 28 days of oral administration. This study is expected to promote the application of functional HLP in the food and pharmacological industries.

2. Materials and Methods

2.1. Preparation of the Polysaccharides

The dried sea cucumbers, *H. leucospilota* (80–130 mm, 22–47 g), were purchased from the Market Property Development Co., Ltd. (Haikou, China). Professor Yongqin Feng of Hainan University confirmed the species of the sea cucumber sample.

The polysaccharides were extracted from the *H. leucospilota* sample according to a previously reported method with slight modifications [15]. The dried sea cucumber body wall powder was mixed with sodium acetate buffer, papain, EDTA solution, and cysteine. The mixture was centrifuged (4500× *g*, 15 min, 4 °C), and a chlorohexadecyl pyridinium (CPC) solution was added to the clear supernatant. After 12 h, the mixture was centrifuged (4500× *g*, 15 min, 4 °C), and the precipitated polysaccharides were redissolved in NaCl: ethanol (100:15, *v*/*v*) solution with 95% ethanol. The sediments were collected after centrifugation (3500× *g*, 15 min, 4 °C) and were reconstituted with distilled water. After decolorization and deproteinization, the polysaccharides were dialyzed with distilled water for 72 h. The HLP with extremely low protein content (0.26%) was obtained by concentration and lyophilization.

2.2. In-Vitro Fermentation of HLP

2.2.1. Preparation of Human Intestinal Microbiota

Fecal inocula were collected from eight healthy volunteers who had not taken any antibiotics within three months (male: female = 1:1, average age 20–30 years). Equal amounts of feces from each person were pooled. Samples of 50 g were diluted with 200 mL of Dulbecco's phosphate-buffered saline (10%, D-PBS), which was followed by homogenization. The mixture was filtered through four-layer nylon gauze, and the obtained filtrate was collected.

2.2.2. In-Vitro Fermentation Procedure

The polysaccharides solution, fermentation medium solution, and fecal solution were added to each ampoule tube. The sealed ampoule tubes were incubated in a shaker (250 rpm, 37 °C) for 6, 12, 24, and 48 h. After incubation, the anaerobic tubes were placed in an ice water bath for 10 min for termination. After centrifugation (8000× *g*, 10 min), the supernatant was transferred and stored at −20 °C before analysis.

2.2.3. Determination of Total Sugar, Reducing Sugar, and Free Monosaccharides in the Fermented Products

The total sugar contents were determined by the phenol sulfuric acid method with glucose as the standard. The dinitrosalicylic acid (DNS) method was utilized to determine the reducing sugar contents [18]. The free monosaccharides were determined by gas chromatography (GC) according to the methods of a previous study [19].

2.2.4. Determination of pH and SCFA Content

The fermentation supernatant pH levels were measured with an MP511 Lab pH Meter (Sanxin Apparatus, Shanghai, China).

Anhydrous ethanol and the fermentation supernatants (or SCFAs standard solutions) were added to tubes and mixed. The mixtures were supplemented with concentrated sulfuric acid and n-hexane. The upper organic phase was used to determine the concentrations of SCFAs. SCFAs analysis was performed by an Agilent 6890A GC equipped with a flame ionization detector (FID) and HP-FFAP column (30 m × 0.32 mm × 0.25 μm, Agilent Technologies, Santa Clara, CA, USA).

2.3. *In Vivo Fermentation of HLP*

Thirty-two male Kunming mice were purchased from the Tianqin Biotechnology Co., Ltd. (Changsha, China), Certificate Number: SCXK (Xiang) 2014−0011. The mice were housed under controlled environmental conditions at a temperature of 23 ± 1 °C and a 12/12 h light/dark cycle with ad libitum water and food. All procedures were performed according to the Guidelines of the Care and Use of Laboratory Animals. The experiments were approved by the Animal Experimentation Ethics Committee of Hainan Medical University.

The mice were randomly divided into four groups and received intragastric administrations once every day: low-dose (HLP-L, 50 mg/kg body weight) group; medium-dose (HLP-M, 100 mg/kg body weight) group; high-dose (HLP-H, 200 mg/kg body weight) group; and an equal volume of distilled water as the normal group (ND). The health of the mice was recorded daily for 28 days. Fecal samples from each group were collected after gavage every fortnight. Finally, the mice were sacrificed by cervical dislocation after anesthesia with chloral hydrate. The liver and small intestine tissues were taken under sterile conditions and transferred to a −80 °C refrigerator for preservation.

2.4. *Liver Antioxidant Activity Assay*

The mouse liver tissues were added to ice-cold physiological saline, homogenized, and centrifuged (4 °C, 3000× *g* and 10 min). The malondialdehyde (MDA) contents and superoxide dismutase (SOD) and glutathione peroxidase (GSH-Px) activities in the liver were measured with the corresponding kits (Jiancheng, Nanjing, China).

2.5. *Histological Examinations*

The ileum and liver samples were fixed in 10% neutral buffered formalin solution for histopathological processing. After dehydration, the fixed tissues were embedded in paraffin and were sectioned and dyed with hematoxylin-eosin (H&E). The stained areas were viewed with an optical microscope (Nikon Eclipse, Tokyo, Japan).

2.6. *Statistical Analysis*

The results were presented as the means ± standard deviation (SD). The data were analyzed using one-way analysis of variance (ANOVA) using SPSS 25.0 software (SPSS Inc., Chicago, IL, USA) and compared using Duncan's test. $p < 0.05$ was considered to be significantly different.

3. Results and Discussion

3.1. Changes in Total Sugar, Reducing Sugar, and Free Monosaccharide Contents during In-Vitro Fermentation

The enzymes that are encoded by the human microbiota cleave the glycosidic bonds of carbohydrates to fermentable saccharides [20]. The breakdown of glycosidic bonds leads to the release of some terminal, reducing sugars of the polysaccharides. Hence, the total sugar and reducing sugar contents are useful to evaluate the fermentation extent of polysaccharides [21]. As illustrated in Table 1, the initial total sugar concentration was 1.099 mg/mL. After six hours of fermentation, the total sugar concentration decreased by 56.77% to 0.475 mg/mL, and the rate of decrease was faster than that afterward. The total sugar concentration was 0.218 mg/mL at the end of fermentation, which decreased by 0.881 mg/mL from the initial concentration. However, despite being widely used in many reported studies, the phenol-sulfuric method is not specific enough to determine the residual carbohydrate contents from in-vitro fermentation. Consequently, the total sugar contents that were determined at 0 h were not consistent with the original amount of HLP in the culture medium.

Table 1. Changes in total sugar, reducing sugar and free monosaccharide contents for in-vitro fermentation.

Time (h)	Total Sugar (mg/mL)	Reducing Sugar (mg/mL)	Fucose (mg/mL)	Xylose (mg/mL)	Galactosamine (mg/mL)
0	1.099 ± 0.177^{d}	0.432 ± 0.014^{c}	ND	ND	ND
3	0.928 ± 0.051^{c}	0.833 ± 0.025^{d}	0.174 ± 0.004^{c}	0.063 ± 0.009^{b}	0.017 ± 0.002^{a}
6	0.475 ± 0.066^{b}	0.274 ± 0.034^{b}	0.134 ± 0.001^{b}	0.086 ± 0.004^{c}	0.016 ± 0.001^{a}
12	0.265 ± 0.018^{a}	0.104 ± 0.004^{a}	0.119 ± 0.009^{a}	0.029 ± 0.001^{a}	0.029 ± 0.004^{b}
24	0.234 ± 0.019^{a}	0.102 ± 0.004^{a}	ND	0.036 ± 0.001^{a}	0.016 ± 0.007^{a}
48	0.218 ± 0.009^{a}	0.097 ± 0.002^{a}	ND	ND	ND

Values are the means ± SD of three replicates. Data that are followed by different letters in the same column are significantly different ($p < 0.05$).

Meanwhile, the concentration of reducing sugar was initially 0.432 mg/mL, and it increased to 0.833 mg/mL after three hours, which was followed by a sharp drop to 0.274 mg/mL at six hours. These changes occurred because the generation rate of reducing sugar was faster than its consumption rate at the beginning of fermentation. Thus, the total sugar concentrations showed a declining trend throughout the fermentation process, while the reducing sugar concentrations increased and then followed by a decrease. Similar changes were observed in a study of the in-vitro fermentation of polysaccharides from *Sargassum thunbergii* [21]. The free monosaccharide contents in the fermented products reflect the hierarchy of polysaccharide utilization [22]. As shown in Table 1, free monosaccharides were detected after three hours of fermentation (the fucose, xylose, and galactosamine concentrations were 0.174, 0.063, and 0.017 mg/mL, respectively). The results showed that HLP was degraded into free monosaccharides by the gut microbiota combined with reducing sugar variations. The degradation rate was lower than the monosaccharide utilization rate in the first three hours. Furthermore, fucose was not detected after 24 h of fermentation, while xylose and galactosamine were not detected after 48 h of fermentation. The disappearance of free monosaccharides occurred because the gut microbiota rapidly utilized them. This result was in accordance with the changing trend reported in another study of *Oudemansiella radicata* polysaccharides [19]. Interestingly, the fucose concentrations were higher than those of xylose and galactosamine from 3 h to 12 h, and fucose disappeared earlier, which was different from xylose and galactosamine. It has been reported that the constituent monosaccharides and spatial structures significantly influence the fermentability of polysaccharides [23]. Fucose was consumed on the branches with a residual content of xylose, and then, galactosamine in the backbone was consumed [14]. In addition, diverse fecal bacteria prefer to utilize different monosaccharides that are liberated from polysaccharides [24]. Thus, it was speculated that fucose might be linked to the side chain in the HLP structure and be more prone to be utilized or degraded by the microbiota [14,21].

3.2. Changes in pH Values and SCFA Contents during In-Vitro Fermentation

The variations in pH values of the fermented products reflected the fermentation surroundings and are summarized in Table 2. The pH value in the human colon is approximately 5–7, and this slight acidity is due to some acidic metabolites (such as SCFAs) that are produced by intestinal microorganisms. The initial pH value of the HLP fermentation system exhibited slight acidity at 6.87 (0 h). Thereafter, a rapid and significant pH decrease (to 4.49) was observed during HLP fermentation from 0 h to 12 h ($p < 0.05$), which indicated that some acid metabolites were produced during fermentation. Many other studies have also reported that the pH values of fermentation solutions declined with the extension of fermentation time [25]. However, the pH values of the fermented products exhibited an increasing trend after 24 h, which were similar to the results reported by Gao et al. [26]. Presumably, this might be due to the production of some alkaline metabolites in the culture medium. A weakly acidic environment has been proven to promote the proliferation of beneficial bacteria in the intestinal tract and inhibit the growth of harmful bacterial [27]. Thus, HLP promotes gut health by maintaining acidic environments.

Table 2. Changes in pH and SCFAs in fermented cultures that were supplemented with HLP at different times during in vitro fermentation.

Time (h)	0	6	12	24	48
pH	6.87 ± 0.01 ^{e}	4.59 ± 0.02 ^{f}	4.49 ± 0.00 ^{a}	5.55 ± 0.01 ^{c}	5.91 ± 0.02 ^{d}
Acetic acid (mmol/mL)	9.65 ± 0.26 ^{a}	39.00 ± 5.90 ^{b,c}	70.71 ± 1.48 ^{d}	34.99 ± 1.3 ^{b}	45.53 ± 9.86 ^{b}
Propionic acid (mmol/mL)	2.93 ± 0.05 ^{a}	7.04 ± 1.01 ^{b}	16.00 ± 0.92 ^{c}	16.26 ± 0.92 ^{c}	18.26 ± 3.93 ^{c}
Isobutyric acid (mmol/mL)	0.26 ± 0.01^{a}	0.36 ± 0.04 ^{a}	0.15 ± 0.02 ^{a}	2.32 ± 0.13 ^{c}	3.15 ± 0.58 ^{d}
Butyric acid (mmol/mL)	2.70 ± 0.31 ^{a}	3.09 ± 0.84 ^{a}	3.53 ± 0.43 ^{a}	28.30 ± 1.61 ^{b}	33.76 ± 7.20 ^{c}
Isovaleric acid (mmol/mL)	0.40 ± 0.04 ^{a}	0.63 ± 0.02 ^{a}	0.32 ± 0.03 ^{a}	2.99 ± 0.18 ^{b}	3.93 ± 0.68 ^{c}
Pentanoic acid (mmol/mL)	0.64 ± 0.08 ^{a}	0.42 ± 0.02 ^{a}	0.41 ± 0.07 ^{a}	4.62 ± 0.18 ^{b}	6.51 ± 1.34 ^{c}
Total (mmol/mL)	16.58 ± 0.74 ^{a}	50.54 ± 5.99 ^{b}	91.12 ± 2.88 ^{c}	89.47 ± 4.30 ^{c}	111.13 ± 23.19 ^{d}

Values are the means ± SD of three replicates. Data that are followed by different letters on the same line are significantly different ($p < 0.05$).

SCFAs, such as acetic acid, propionic acid, and butyric acid, which are produced during fermentation, play an indispensable role in sustaining human health [28]. Nevertheless, it is difficult to determine the SCFA contents that are generated in vivo, as over 95% of SCFAs are rapidly absorbed and metabolized by the host. The SCFA concentrations in fecal samples are greatly affected by the intestinal transport times and cannot accurately reflect the proximal colon environment [29]. Hence, combining in-vitro and in-vivo fermentation methods more accurately describes the actual fermentation behavior of polysaccharides. The total SCFAs concentration increased with fermentation time (Table 2). After 48 h of fermentation, the total SCFAs content increased by 111.13 mmol/L. The SCFA amounts are related to the microbial compositions, substrate sources, and intestinal action times [30]. Therefore, the variety of SCFAs and their concentrations during fermentation are generally used to determine polysaccharide fermentability.

Acetic acid is usually converted from pyruvate and is utilized by the liver, muscles, and peripheral tissues through blood circulation and is the most abundant SCFA in the colon [31]. As histone deacetylase inhibitors, propionic and butyric acids participate in the immune regulation process [32]. Among the various SCFAs, the acetic acid, propionic acid, and butyric acid contents were comparably higher and changed more distinctly. The acetic acid concentrations showed a continuously increasing trend from the initial 12 h period and decreased gradually from 12 to 24 h. Nevertheless, the propionic acid and butyric acid contents increased gradually throughout the entire process. This result was probably explained by the acetate CoA-transferase route, namely the gut bacteria further metabolized acetate to butyrate via butyryl-CoA [33]. It is noteworthy that the acetic acid concentration reached its peak value at 12 h, which was consistent with the trend of pH values. The fact that the pH of acetic acid is lower than that of other SCFAs at the same concentrations contributes to this phenomenon. Additionally, the butyric acid concentration was evidently high because the weakly acidic environment was conducive to the metabolism and reproduction of butyrate-producing bacteria. The growth of pH-sensitive pathogens and certain carcinogenic saprophytes (such as *Clostridium* and *Enterobacteriaceae*) was effectively inhibited [29]. In addition, Chen et al. also found that the isobutyric acid and isovaleric acid levels were relatively lower, which resulted from few intestinal bacteria producing isobutyrate and isovalerate [4,19]. HLP may be degraded and utilized by the gut microbiota during in-vitro fermentation, which produced abundant SCFAs.

3.3. Changes in pH Values and SCFA Contents in Mouse Feces

The intestinal microenvironment is a complex ecological and reaction system. The appropriately decreased pH values in the intestine contributed to the growth of probiotics and inhibition of pathogens [34]. The fecal pH value is generally used as an indicator to respond to intestinal health. Based on seven days of oral administration, significant differences in pH values appeared among the different groups (Figure 1). In particular, the fecal pH values of the HLP-H and HLP-M groups were obviously lower than those of the HLP-L and ND groups. This result suggested that the intake of HLP decreased pH in the mouse intestine and that a high-dose intake had a greater influence on pH. The colon and fecal SCFA contents are closely associated with human health.

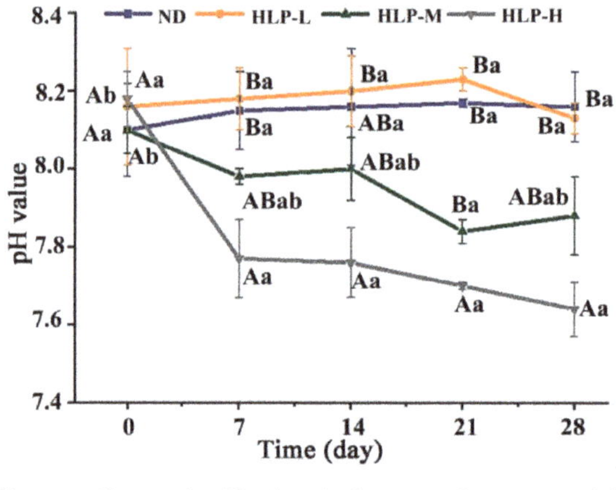

Figure 1. Changes in pH values in the mouse feces among different groups. The different uppercase and lowercase letters around the points represent significant differences in various groups and times (day), respectively ($p < 0.05$).

When compared to the ND group, after HLP treatments for two and four weeks, the total SCFAs concentrations in the HLP-M and HLP-H groups increased significantly ($p < 0.05$, Figure 2). The HLP-H group increased most notably among all groups after four weeks, from 13.4 mmol/L to 20.67 mmol/L. The total SCFAs contents after four weeks were significantly higher than those after two weeks. This result was consistent with the changing trend of pH values in mouse feces and indicated the close relationship between the mouse fecal pH values and SCFA contents in the mouse colon. After four weeks of continuous gavage, both the HLP-M and HLP-H doses caused significant increases in the acetic acid and butyric acid concentrations in feces compared to the ND group ($p < 0.05$). In addition, the fecal isobutyric acid concentrations in the HLP-H group were significantly higher than those in the ND group after four weeks ($p < 0.05$). This result suggested that the intestinal environment of the medium-dose and high-dose HLP-treated mice was affected for four weeks.

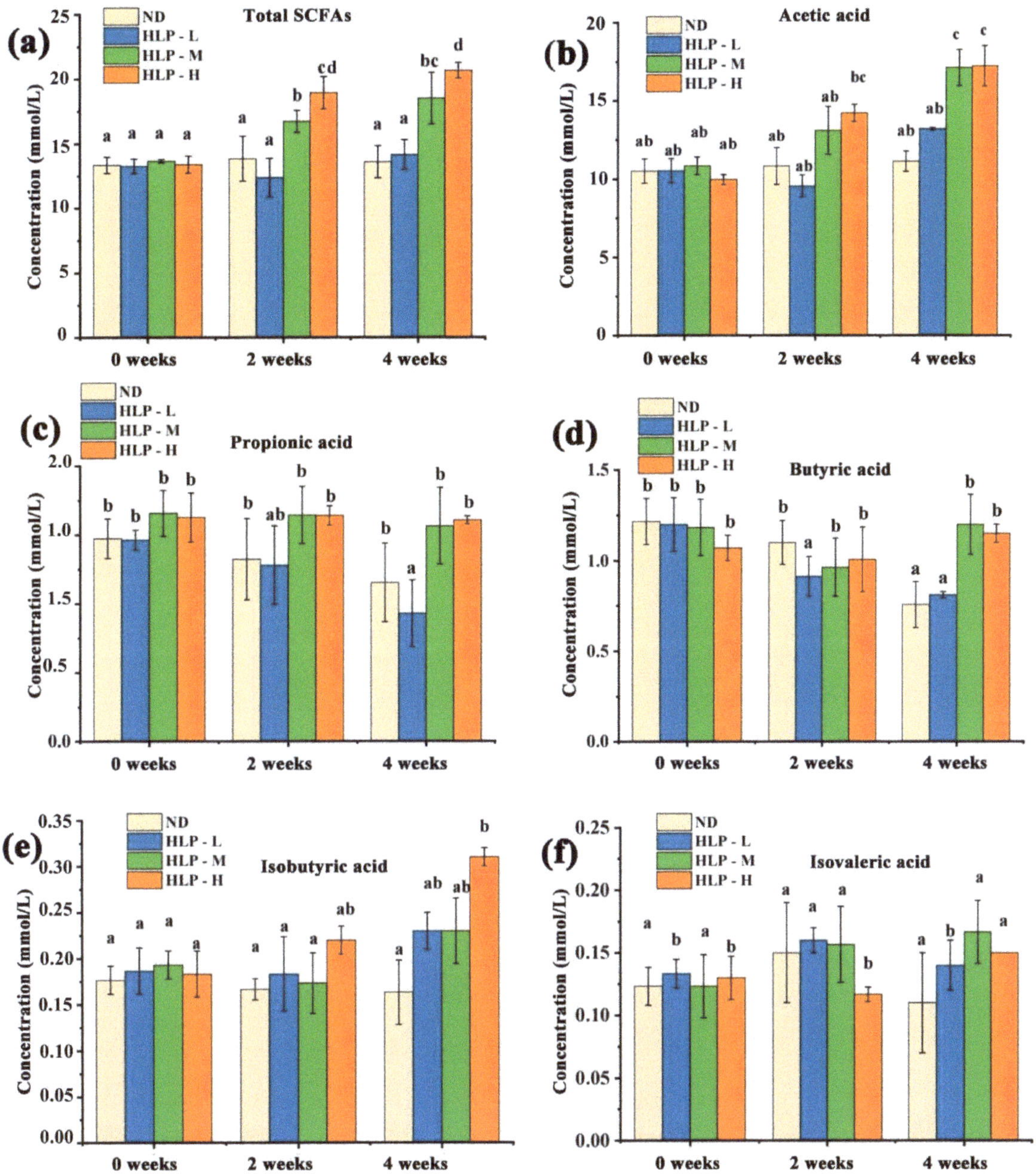

Figure 2. Changes in total SCFAs (**a**), acetic acid (**b**), propionic acid (**c**), butyric acid (**d**), isobutyric acid (**e**), and isovaleric acid (**f**) contents in mouse feces. Values with different superscripts represent significant differences ($p < 0.05$).

3.4. Histological Analysis of the Ileum and Liver

The ileum in mice of the HLP-treated groups exhibited normal histological features (Figure 3a). The ileal villi lengths and crypt depths of all mouse groups were similar. However, the goblet cell numbers among the villi clearly increased in the ileum compared to those of the ND group. Goblet cells are one of the most distinct cell types in the epithelium for intrinsic mucosal immunity and synthesize and secrete mucin, which forms a mucosal barrier to protect epithelial cells [35]. As shown in Figure 3b, the hepatocytes in the ND group were arranged in an orderly manner and were distributed radially around the hepatic cords. The liver tissue structures of the HLP-treated groups were normal and clearly showed the nucleolus, central veins, and abundant cytoplasm. This result implied that, from a pathological perspective, HLP administration would not cause hepatocellular injury.

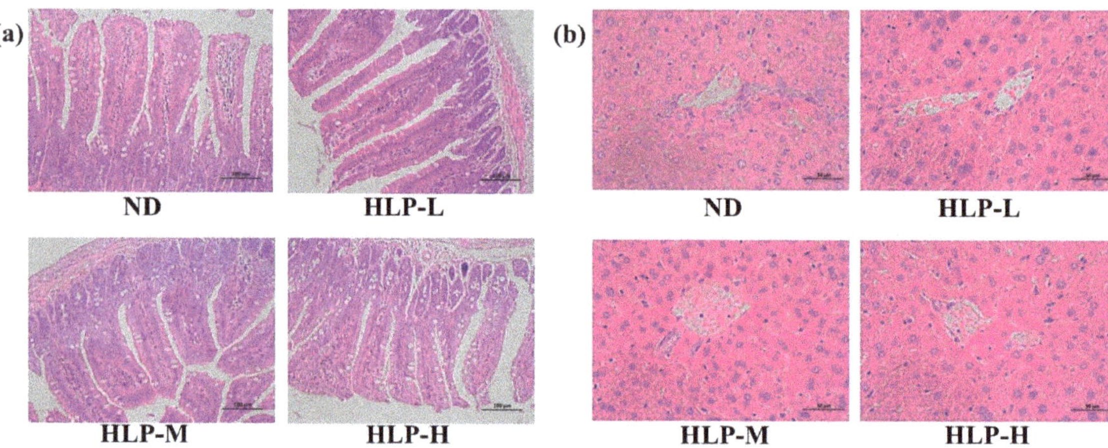

Figure 3. Histological examination of the ileum (**a**) and liver (**b**) by H&E staining (200×).

3.5. Effects of HLP on Oxidative Stress Parameters in Mouse Livers

Oxidative stress, with excessive formation of reactive oxygen species (ROS), is a significant participant in diabetes and cardiovascular disease development and progression. High ROS levels damage cell structures and induce lipid peroxidation, which causes further injury to the surrounding tissues. Meanwhile, ROS also reduces antioxidant enzyme activities in the body, which results in further oxidative damage [36]. As important antioxidant enzymes in the defense system, both GSH-Px and SOD play crucial roles in protecting tissues against oxidative damage that is induced by superoxide anions [37]. GSH-Px catalyzes the formation of glutathione and reacts with ROS to decrease oxidant levels and lipid oxidation. SOD is the most strongly antioxidant enzyme in the first line of defense against ROS.

The GSH-Px activities in the livers of the HLP-H group were significantly higher ($p < 0.05$) than those of the ND group (Figure 4a). Although there were no differences ($p > 0.05$) in SOD activity among the various groups, the SOD activities in the livers of the HLP-M and HLP-H groups were slightly higher than those of the ND group (Figure 4b). The increasing GSH-Px and SOD levels in the liver indicated that high-dose HLP enhanced the antioxidant defense system's reaction to oxidative stress. MDA, a lipid peroxidation secondary product, is involved in forming lipid radicals and reflects the degree of cytotoxicity and cellular damage [38]. The liver MDA levels in the HLP-treated groups significantly declined compared with those in the ND group (Figure 4c). However, no significant differences were observed among the different HLP-treated groups in reducing the MDA levels ($p > 0.05$). This result indicated that HLP exhibited antioxidant activity by activating endogenous antioxidative enzymes.

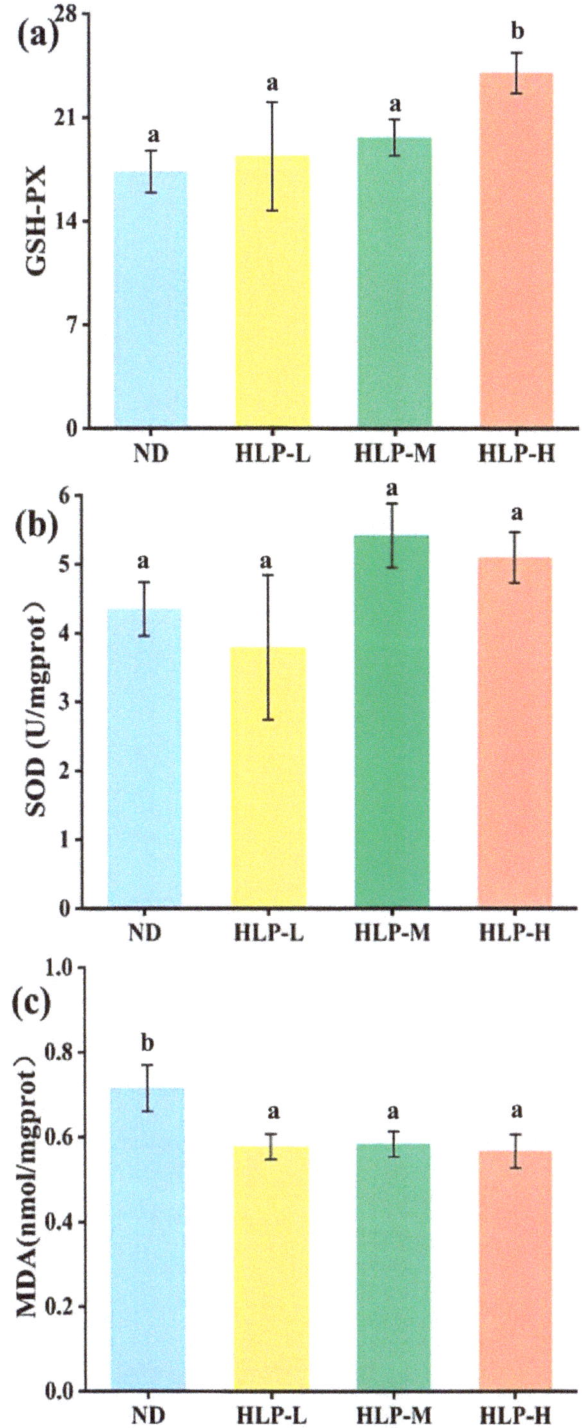

Figure 4. Effects of HLP on GSH-Px activity (**a**), SOD activity (**b**), and MDA content (**c**) in the mouse livers. Values with different letters are significantly different ($p < 0.05$).

4. Conclusions

The polysaccharides from *H. leucospilota* were utilized and decomposed into free monosaccharides by the intestinal microbiota. During in-vitro human fecal fermentation, the pH values and total sugar contents decreased, while various SCFA contents increased. After 28 days of high-dose treatment, the acetic and isobutyric acid levels in mouse feces increased significantly. Additionally, HLP exhibited antioxidant activity by enhancing the GSH-Px and SOD levels and decreasing the MDA contents in mouse livers. This study suggested that HLP plays an important role in potential health benefits by promoting intestinal health and antioxidant activity.

Author Contributions: Conceptualization, C.L.; methodology, W.W. and J.C.; software, Y.Y. and J.C.; formal analysis, Y.Y.; investigation, Y.Y.; resources, C.L.; writing—original draft preparation, W.W.; writing—review and editing, C.L.; supervision, J.C. and C.L.; project administration, X.S. All authors have read and agreed to the published version of the manuscript.

Funding: This research was funded by the Hainan Provincial Key Research and Development Program, grant number ZDYF2020170.

Institutional Review Board Statement: The study was conducted according to the guidelines of the Care and Use of Laboratory Animals published by the U.S. National Institutes of Health. All methods and materials in the study were granted Ethical approval from Animal Experimentation Ethics Committee of Hainan Medical University. (No. HYLL-2019-038; 21/09/2019).

Data Availability Statement: All raw data supporting reported results is available from authors upon request.

Conflicts of Interest: The authors declare that they have no conflict of interest.

References

1. Lu, X.; Chen, J.; Guo, Z.; Zheng, Y.; Rea, M.C.; Su, H.; Zheng, X.; Zheng, B.; Miao, S. Using polysaccharides for the enhancement of functionality of foods: A review. *Trends Food Sci. Technol.* **2019**, *86*, 311–327. [CrossRef]
2. Xie, J.-H.; Jin, M.-L.; Morris, G.; Zha, X.-Q.; Chen, H.-Q.; Yi, Y.; Li, J.-E.; Wang, Z.-J.; Gao, J.; Nie, S.-P.; et al. Advances on Bioactive Polysaccharides from Medicinal Plants. *Crit. Rev. Food Sci. Nutr.* **2015**, *56*, S60–S84. [CrossRef] [PubMed]
3. Rakariyatham, K.; Zhou, D.; Rakariyatham, N.; Shahidi, F. Sapindaceae (*Dimocarpus longan* and *Nephelium lappaceum*) seed and peel by-products: Potential sources for phenolic compounds and use as functional ingredients in food and health applications. *J. Funct. Foods* **2020**, *67*, 103846. [CrossRef]
4. Van Dam, J.E.G.; Van Den Broek, L.A.; Boeriu, C.G. Polysaccharides in Human Health Care. *Nat. Prod. Commun.* **2017**, *12*. [CrossRef]
5. Zhang, X.; Aweya, J.J.; Huang, Z.-X.; Kang, Z.-Y.; Bai, Z.-H.; Li, K.-H.; He, X.-T.; Liu, Y.; Chen, X.-Q.; Cheong, K.-L. In vitro fermentation of *Gracilaria lemaneiformis* sulfated polysaccharides and its agaro-oligosaccharides by human fecal inocula and its impact on microbiota. *Carbohydr. Polym.* **2020**, *234*, 115894. [CrossRef] [PubMed]
6. Wang, Z.; Erasmus, S.W.; Liu, X.; Van Ruth, S.M. Study on the Relations between Hyperspectral Images of Bananas (*Musa* spp.) from Different Countries, their Compositional Traits and Growing Conditions. *Sensors* **2020**, *20*, 5793. [CrossRef]
7. Li, Y.; Shabani, K.I.; Liu, H.; Guo, Q.; Liu, X. Structural, physicochemical and rheological properties of a novel native starch obtained from *Rhizoma Gastrodiae*. *Food Struct.* **2020**, *25*, 100148. [CrossRef]
8. Shang, Q.; Jiang, H.; Cai, C.; Hao, J.; Li, G.; Yu, G. Gut microbiota fermentation of marine polysaccharides and its effects on intestinal ecology: An overview. *Carbohydr. Polym.* **2018**, *179*, 173–185. [CrossRef]
9. Huang, F.; Liu, Y.; Zhang, R.; Bai, Y.; Dong, L.; Liu, L.; Jia, X.; Wang, G.; Zhang, M. Structural characterization and in vitro gastrointestinal digestion and fermentation of litchi polysaccharide. *Int. J. Biol. Macromol.* **2019**, *140*, 965–972. [CrossRef]
10. Mou, J.; Li, Q.; Shi, W.; Qi, X.; Song, W.; Yang, J. Chain conformation, physicochemical properties of fucosylated chondroitin sulfate from sea cucumber *Stichopus chloronotus* and its in vitro fermentation by human gut microbiota. *Carbohydr. Polym.* **2019**, *228*, 115359. [CrossRef]
11. Tian, Y.; Xu, Q.; Sun, L.; Ye, Y.; Ji, G. Short-chain fatty acids administration is protective in colitis-associated colorectal cancer development. *J. Nutr. Biochem.* **2018**, *57*, 103–109. [CrossRef] [PubMed]
12. Huang, W.; Huo, D.; Yu, Z.; Ren, C.; Jiang, X.; Luo, P.; Chen, T.; Hu, C. Spawning, larval development and juvenile growth of the tropical sea cucumber *Holothuria leucospilota*. *Aquaculture* **2018**, *488*, 22–29. [CrossRef]
13. Wang, G. *Studies on chemical constituents from Holothuria leucospilota*; Hainan University: Haikou, China, 2018.
14. Qiu, P.; Wu, F.; Yi, L.; Chen, L.; Jin, Y.; Ding, X.; Ouyang, Y.; Yao, Y.; Jiang, Y.; Zhang, Z. Structure characterization of a heavily fucosylated chondroitin sulfate from sea cucumber (*H. leucospilota*) with bottom-up strategies. *Carbohydr. Polym.* **2020**, *240*, 116337. [CrossRef] [PubMed]

15. Yuan, Y.; Li, C.; Zheng, Q.; Wu, J.; Zhu, K.; Shen, X.; Cao, J. Effect of simulated gastrointestinal digestion in vitro on the antioxidant activity, molecular weight and microstructure of polysaccharides from a tropical sea cucumber (*Holothuria leucospilota*). *Food Hydrocoll.* **2018**, *89*, 735–741. [CrossRef]
16. Zhao, F.; Liu, Q.; Cao, J.; Xu, Y.; Pei, Z.; Fan, H.; Yuan, Y.; Shen, X.; Li, C. A sea cucumber (*Holothuria leucospilota*) polysaccharide improves the gut microbiome to alleviate the symptoms of type 2 diabetes mellitus in Goto-Kakizaki rats. *Food Chem. Toxicol.* **2019**, *135*, 110886. [CrossRef] [PubMed]
17. Zhang, Y.; Wang, H.; Zhang, L.; Yuan, Y.; Yu, D. *Codonopsis lanceolata* polysaccharide CLPS alleviates high fat/high sucrose diet-induced insulin resistance via anti-oxidative stress. *Int. J. Biol. Macromol.* **2019**, *145*, 944–949. [CrossRef]
18. Su, A.; Ma, G.; Xie, M.; Ji, Y.; Li, X.; Zhao, L.; Hu, Q. Characteristic of polysaccharides from *Flammulina velutipes* in vitro digestion under salivary, simulated gastric and small intestinal conditions and fermentation by human gut microbiota. *Int. J. Food Sci. Technol.* **2019**, *54*, 2277–2287. [CrossRef]
19. Liu, Y.; Li, Y.; Ke, Y.; Li, C.; Zhang, Z.; Wu, Y.; Hu, B.; Liu, A.; Luo, Q.; Wu, W. In vitro saliva-gastrointestinal digestion and fecal fermentation of *Oudemansiella radicata* polysaccharides reveal its digestion profile and effect on the modulation of the gut microbiota. *Carbohydr. Polym.* **2020**, *251*, 117041. [CrossRef]
20. Chen, C.; Huang, Q.; Fu, X.; Liu, R.H. In vitro fermentation of mulberry fruit polysaccharides by human fecal inocula and impact on microbiota. *Food Funct.* **2016**, *7*, 4637–4643. [CrossRef]
21. Fu, X.; Cao, C.; Ren, B.; Zhang, B.; Huang, Q.; Li, C. Structural characterization and in vitro fermentation of a novel polysaccharide from *Sargassum thunbergii* and its impact on gut microbiota. *Carbohydr. Polym.* **2017**, *183*, 230–239. [CrossRef]
22. Xie, J.-H.; Tang, W.; Jin, M.-L.; Li, J.-E.; Xie, M.-Y. Recent advances in bioactive polysaccharides from *Lycium barbarum* L., *Zizyphus jujuba* Mill, *Plantago* spp., and *Morus* spp.: Structures and functionalities. *Food Hydrocoll.* **2016**, *60*, 148–160. [CrossRef]
23. Salvador, V.; Cherbut, C.; Barry, J.-L.; Bertrand, D.; Bonnet, C.; Delort-Laval, J. Sugar composition of dietary fibre and short-chain fatty acid production during in vitro fermentation by human bacteria. *Br. J. Nutr.* **1993**, *70*, 189–197. [CrossRef]
24. Cockburn, D.; Koropatkin, N.M. Polysaccharide Degradation by the Intestinal Microbiota and Its Influence on Human Health and Disease. *J. Mol. Biol.* **2016**, *428*, 3230–3252. [CrossRef]
25. Fu, Y.; Zhang, J.; Chen, K.; Xiao, C.; Fan, L.; Zhang, B.; Ren, J.; Fang, B. An in vitro fermentation study on the effects of Dendrobium officinale polysaccharides on human intestinal microbiota from fecal microbiota transplantation donors. *J. Funct. Foods* **2018**, *53*, 44–53. [CrossRef]
26. Gao, Y.; Guo, Q.; Zhang, K.; Wang, N.; Li, C.; Li, Z.; Zhang, A.; Wang, C. Polysaccharide from *Pleurotus nebrodensis*: Physicochemical, structural characterization and in vitro fermentation characteristics. *Int. J. Biol. Macromol.* **2020**, *165*, 1960–1969. [CrossRef] [PubMed]
27. Slavin, J. Fiber and Prebiotics: Mechanisms and Health Benefits. *Nutrients* **2013**, *5*, 1417–1435. [CrossRef] [PubMed]
28. Ding, Y.; Yan, Y.; Peng, Y.; Chen, D.; Mi, J.; Lu, L.; Luo, Q.; Li, X.; Zeng, X.; Cao, Y. In vitro digestion under simulated saliva, gastric and small intestinal conditions and fermentation by human gut microbiota of polysaccharides from the fruits of *Lycium barbarum*. *Int. J. Biol. Macromol.* **2018**, *125*, 751–760. [CrossRef]
29. Besten, G.D.; van Eunen, K.; Groen, A.K.; Venema, K.; Reijngoud, D.-J.; Bakker, B.M. The role of short-chain fatty acids in the interplay between diet, gut microbiota, and host energy metabolism. *J. Lipid Res.* **2013**, *54*, 2325–2340. [CrossRef]
30. Wong, J.M.W.; de Souza, R.; Kendall, C.W.C.; Emam, A.M.; Jenkins, D.J.A. Colonic Health: Fermentation and Short Chain Fatty Acids. *J. Clin. Gastroenterol.* **2006**, *40*, 235–243. [CrossRef]
31. Cook, S.I.; Sellin, J.H. Review article: Short chain fatty acids in health and disease. *Aliment. Pharmacol. Ther.* **1998**, *12*, 499–507. [CrossRef]
32. Donohoe, D.R.; Collins, L.B.; Wali, A.; Bigler, R.; Sun, W.; Bultman, S.J. The Warburg Effect Dictates the Mechanism of Butyrate-Mediated Histone Acetylation and Cell Proliferation. *Mol. Cell* **2012**, *48*, 612–626. [CrossRef]
33. Duncan, S.; Holtrop, G.; Lobley, G.E.; Calder, A.G.; Stewart, C.S.; Flint, H.J. Contribution of acetate to butyrate formation by human faecal bacteria. *Br. J. Nutr.* **2004**, *91*, 915–923. [CrossRef] [PubMed]
34. Walker, A.; Duncan, S.; Leitch, C.; Child, M.W.; Flint, H.J. pH and Peptide Supply Can Radically Alter Bacterial Populations and Short-Chain Fatty Acid Ratios within Microbial Communities from the Human Colon. *Appl. Environ. Microbiol.* **2005**, *71*, 3692–3700. [CrossRef]
35. Kim, Y.S.; Ho, S.B. Intestinal Goblet Cells and Mucins in Health and Disease: Recent Insights and Progress. *Curr. Gastroenterol. Rep.* **2010**, *12*, 319–330. [CrossRef]
36. Jiao, Y.; Kuang, H.; Hu, J.; Chen, Q. Structural characterization and anti-hypoxia activities of polysaccharides from the sporocarp, fermentation broth and cultured mycelium of *Agaricus bitorquis* (Quél.) Sacc. Chaidam in mice. *J. Funct. Foods* **2018**, *51*, 75–85. [CrossRef]
37. Cui, J.-J.; Yuan, J.-F.; Zhang, Z.-Q. Anti-oxidation activity of the crude polysaccharides isolated from Polygonum Cillinerve (Nakai) Ohwi in immunosuppressed mice. *J. Ethnopharmacol.* **2010**, *132*, 512–517. [CrossRef]
38. Li, Q.; Chen, G.; Chen, H.; Zhang, W.; Ding, Y.; Yu, P.; Zhao, T.; Mao, G.; Feng, W.; Yang, L.; et al. Se-enriched *G. frondosa* polysaccharide protects against immunosuppression in cyclophosphamide-induced mice via MAPKs signal transduction pathway. *Carbohydr. Polym.* **2018**, *196*, 445–456. [CrossRef] [PubMed]

foods

MDPI

Review

Comprehensive Review of Polysaccharide-Based Materials in Edible Packaging: A Sustainable Approach

Yuan Zhao [1,2], Bo Li [1,3], Cuicui Li [1], Yangfan Xu [1], Yi Luo [1], Dongwu Liang [1,*] and Chongxing Huang [1,2]

1 School of Light Industry & Food Engineering, Guangxi University, 100 Daxue Road, Nanning 530004, China; zy199113@163.com (Y.Z.); 13654529477@163.com (B.L.); licuicui081@163.com (C.L.); xyflower23@163.com (Y.X.); ly956662@163.com (Y.L.); huangcx@gxu.edu.cn (C.H.)
2 Guangxi Key Laboratory of Clean Pulp & Papermaking and Pollution Control, College of Light Industry and Food Engineering, Guangxi University, Nanning 530004, China
3 Key Laboratory of Processing Suitability and Quality Control of the Special Tropical Crops of Hainan Province, Wanning 571533, China
* Correspondence: liangdongwu@gxu.edu.cn; Tel.: +86-139-7885-1223

Abstract: Edible packaging is a sustainable product and technology that uses one kind of "food" (an edible material) to package another kind of food (a packaged product), and organically integrates food with packaging through ingenious material design. Polysaccharides are a reliable source of edible packaging materials with excellent renewable, biodegradable, and biocompatible properties, as well as antioxidant and antimicrobial activities. Using polysaccharide-based materials effectively reduces the dependence on petroleum resources, decreases the carbon footprint of the "product-packaging" system, and provides a "zero-emission" scheme. To date, they have been commercialized and developed rapidly in the food (e.g., fruits and vegetables, meat, nuts, confectioneries, and delicatessens, etc.) packaging industry. However, compared with petroleum-based polymers and plastics, polysaccharides still have limitations in film-forming, mechanical, barrier, and protective properties. Therefore, they need to be improved by reasonable material modifications (chemical or physical modification). This article comprehensively reviews recent research advances, hot issues, and trends of polysaccharide-based materials in edible packaging. Emphasis is given to fundamental compositions and properties, functional modifications, food-packaging applications, and safety risk assessment of polysaccharides (including cellulose, hemicellulose, starch, chitosan, and polysaccharide gums). Therefore, to provide a reference for the development of modern edible packaging.

Keywords: polysaccharide-based materials; edible packaging; cellulose; hemicellulose; starch; chitosan; polysaccharide gums

Citation: Zhao, Y.; Li, B.; Li, C.; Xu, Y.; Luo, Y.; Liang, D.; Huang, C. Comprehensive Review of Polysaccharide-Based Materials in Edible Packaging: A Sustainable Approach. *Foods* **2021**, *10*, 1845. https://doi.org/10.3390/foods10081845

Academic Editor: Pascal Degraeve

Received: 8 June 2021
Accepted: 8 August 2021
Published: 10 August 2021

1. Introduction

Since the 19th century, petroleum-based polymers and plastics have occupied a major position in food packaging, but most are non-renewable, non-biodegradable, difficult to recycle, and carelessly discarded as garbage after use, thereby contributing to ecological environmental deterioration and possible health hazards [1]. Under various natural and anthropogenic forces, plastic fragments (from waste plastic containers, sheets, and films) break down into small particle sizes, further generating microplastics with a diameter smaller than 5 mm [1–3]. According to Lebreton et al. [4], over 79,000 tons of plastic waste float on the Great Pacific Garbage Patch, and the content of marine microplastics has increased rapidly from 0.4 kg/km^2 in the 1970s to 1.23 kg/km^2 in 2015. Then Barrett et al. [5] estimated that there could be as much as 14.4 million tonnes of microplastics in the top 9 cm of sediment throughout the global ocean, which was 34–57 times more than that at the ocean surface. Moreover, microplastics have been ubiquitously detected in oceans (from the continental shelf to deep-sea waters [6], from the eastern North Pacific Ocean [3] to the Indian Ocean [7], and from coral reef to whales [8]), freshwater systems [9], airborne [10], plants, animals, and even humans [11,12]. Unfortunately, the plastic (including microplastics) pollution is posing a serious threat to the global environment and human health. Therefore, it is of great significance for packaging to develop a series of renewable environment-friendly materials to replace the traditional petrochemical-based materials, among which the edible material is one of the most promising materials.

Edible packaging material is a kind of sustainable material that takes natural, edible and digestible "food" as raw material and is processed by modern material forming technology. It has excellent biocompatibility and biodegradability and can be consumed by animals or humans along with the food, while satisfying the basic functions of packaging (e.g., protection and transport), thus avoiding packaging waste pollution [13,14]. The design of edible packaging was originally inspired by the "peel/skin" of fruits and vegetables, and now edible packaging has been widely applied to various forms of food packaging (e.g., films, coatings, sheets, bags, cups, trays, and lids), as shown in Figure 1. In addition, edible packaging materials are non-toxic harmless, can be in direct contact with food, and even can be used as carriers of some antioxidative, antibacterial and/or nutritional factors to improve the sensory quality and nutritional value of foods [14,15].

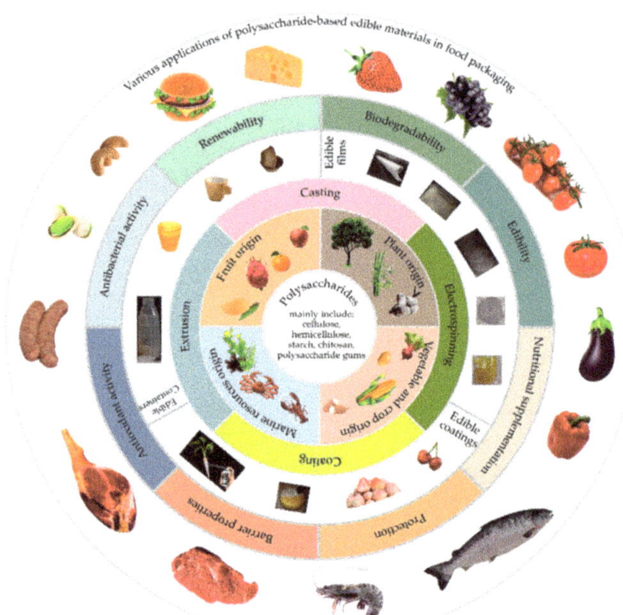

Figure 1. The major sources, types, processing methods, product forms, and food preservation applications of polysaccharide-based edible packaging.

To date, edible packaging materials include three natural biopolymers: polysaccharides, proteins, and lipids, among which polysaccharides (the most abundant natural macromolecules in nature, low processing cost and special function) occupy the most important position [13]. Polysaccharides are complex carbohydrates with varying degrees of polymerization and are composed of monosaccharides linked by α-1,4-, β-1,4-, or α-1,6-glycosidic bonds [16]. The polysaccharides commonly applied in edible packaging are cellulose, hemicellulose, starch, chitosan, and polysaccharide gums, which are used as the main matrix of packaging materials, and processed into polysaccharide-based edible films or layers by casting, coating, electrospinning, or extrusion technologies (Figure 1) [15–18].

The main value of polysaccharide-based edible packaging materials is to protect the quality of food, prolong their shelf life, and improve the functional characteristics, economic benefits, and sustainability of the packaging [15,19]. Compared with traditional packaging materials (such as paper, plastic, metal, and glass), polysaccharide-based materials have two significant advantages: Edibility and environmentally friendly performance. Compared with protein- and lipid-based packaging materials, polysaccharides have better chemical stability and processing adaptability, a greater range of sources, and lower cost. According to relevant studies, polysaccharide-based materials have good gases, aromas, and lipids barrier properties [20–24]; and even some polysaccharides and their derivatives have antioxidant and antimicrobial activities, which can effectively protect foods (e.g., fruits, vegetables, meat, aquatic products, nuts, confectioneries, and delicatessens), and extend their shelf life [15,19]. Furthermore, developing polysaccharide-based materials effectively reduces the dependence on petroleum resources, decreases the carbon footprint of the "product-packaging" system, and meets the strategic requirements of global sustainable packaging.

This article reviews the latest advances in the major polysaccharide-based edible packaging materials (cellulose, hemicellulose, starch, chitosan, and polysaccharide gums) from the viewpoints of fundamental compositions, properties, functional modification, application, and safety, highlights the potential of polysaccharides in food packaging, and provides the trends of these materials in modern packaging technology.

2. Fundamental Compositions and Properties of Various Polysaccharides

The functional characteristics of food packaging are not only related to the properties and main deterioration modes of packaged foods, but also depend on the compositions and properties of the packaging materials. Therefore, the relevant discussion of various polysaccharides has important guiding significance for analyzing the applicability of different polysaccharides in food packaging, as well as the selection of corresponding modification and application schemes. The major and minor sources, similarities and differences in compositions and structures of five polysaccharides, as well as their outstanding advantages as edible packaging are shown in Table 1. Meanwhile, the molecular structure models of different polysaccharides are shown in Figure 2.

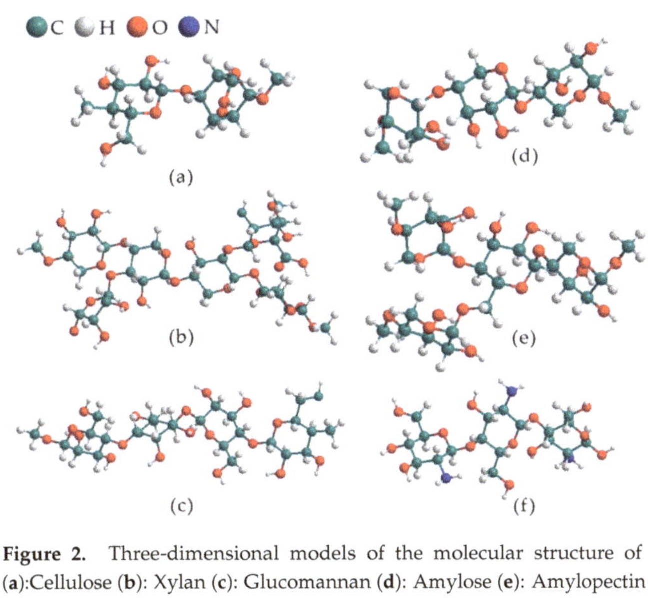

Figure 2. Three-dimensional models of the molecular structure of various polysaccharides. (**a**):Cellulose (**b**): Xylan (**c**): Glucomannan (**d**): Amylose (**e**): Amylopectin (**f**): Chitosan.

Table 1. Sources, compositions, structures, and outstanding characteristics of five polysaccharides for edible packaging application.

Polysaccharides		Sources	Molecular Structure Characteristics	Functional Advantages
Cellulose		• Major: wood and cotton • Minor: certain peels, husks, bagasse, algae, vegetables, tunicates fungi, invertebrates, and bacteria [16,24,25]	• Comprise anhydroglucose units connected by β-glycosidic bonds • Contains numerous hydroxyl groups [26]	• Good chemical stability, gelation, and film-forming properties • Good mechanical properties, and barrier capacities to oxygen and lipids • Renewable, biodegradable, biocompatibility • Soluble dietary fiber and food additive [13,16,27] □ Compared with ordinary cellulose, nanocellulose has a higher elastic modulus, tensile strength, crystallinity, lower coefficient of thermal expansion, large specific surface area, high reactivity, and small size effects [28]
Hemicellulose	Xylan	• Major: hardwoods, gramineous plants • Minor: certain crops and their processing by-products [29–32]	• Composed of (1→4) bonds connected to the main chain of β-D-pyranose units and different side groups connected by (1→2) and/or (1→3) bonds • Contains numerous hydroxyl groups [29,30,32]	• Good gelation, and film-forming properties • Good mechanical and gas barrier properties (But these properties are slightly worse than those of cellulose) • Renewable, biodegradable, biocompatibility • Soluble dietary fiber and food additive [16,29,30,32]
	Glucomannan	Softwoods, tubers and seeds of *Amorphophallus konjac* plants [30,31,33]	• Composed of D-glucopyranosyl and D-mannopyranosyl connected by β-(1→4) bonds • Contains numerous hydroxyl groups [16,30,31]	

Table 1. *Cont.*

Polysaccharides		Sources	Molecular Structure Characteristics	Functional Advantages
Starch	Amylose	• Major: corn, rice, wheat, cassava, and potatoes [34,35] • Minor: banana, mango, breadfruit [34,35], oca [36], jackfruit and lotus seeds [37], and pineapple stems [38]	• Composed of α-D glucose connected by α-(1→4) glycosidic bonds; has no branched structure or only a small amount of branched structures connected by α-(1→6) glycosidic bonds • Only hydrophobic hydrogen atoms inside the helix structure, and hydrophilic hydroxyl groups outside it [16,39–41]	• Good mechanical properties, oxygen barrier property, and processability • Renewable, biodegradable, recyclable, biocompatibility • Low processing cost • Food additive [15,42,43] ☐ Gelatinize, regenerate, swell, and a certain proportion of starch aqueous solution behaves as non-Newtonian fluid (The above characteristics are not available in polysaccharides such as cellulose, hemicellulose, chitosan and alginate); and semi-permeable to carbon dioxide [41,44–46] ☐ Worse gelation, film-forming properties, and transparency if compared to cellulose, hemicellulose, chitosan, and polysaccharide gums [15,16]
	Amylopectin		• The main chain is composed of α-D-(1→4) glycosidic bonds, and the side chain is composed of α-(1→6) glycosidic bonds; the structure is more complex and arranged radially in a concentric form • Contains numerous hydroxyl groups [44,47,48]	

Table 1. *Cont.*

Polysaccharides	Sources	Molecular Structure Characteristics	Functional Advantages
Chitosan	• Major: the shells of crustaceans such as shrimps, crabs, insects • Minor: the cell walls of lower plants, bacteria, and fungi [49]	• Composed of N-acetyl-D-glucosamine and D-glucosamine (occupies a larger proportion, generally > 55%) connected by β-(1→4) glycosidic bonds • Contains numerous amino and hydroxyl groups, and a few acetylamino [16,50,51]	• Good gelation, film-forming properties and processing suitability • Good mechanical, oxygen and lipids barrier, and adsorptive properties (The tensile strength and swelling power of chitosan films prepared at higher drying temperatures and solute concentrations improved relatively) • Renewable, biodegradable, biocompatibility • Food additive [51–54] □ The high specific surface area, large aspect ratio, and small size effect of nano-chitosan can further improve the biological activity, biocompatibility, and adsorption properties of ordinary chitosan [55,56] □ Good antioxidant activity; and excellent antimicrobial activity, with effective inhibition of most gram-negative and -positive bacteria and fungi (These properties differ from those of cellulose, hemicellulose and starch) [57,58]

Table 1. *Cont.*

Polysaccharides		Sources	Molecular Structure Characteristics	Functional Advantages
Polysaccharide gums	Pectin	• Major: fruit and vegetable processing residues such as citrus peel, apple peel, sweet potato residue, and beet residue [59] • Minor: the peels of passion fruit [60], lime [61], dragon fruit [62], fig [63], grapefruit [64], pomegranate [39], lemon [65], and hawthorn [66]; and sunflower heads without seeds [67], *Premna microphylla* Turcz leaves, and Creeping fig seeds [68]	• An acidic heteropolysaccharide composed of D-galacturonic acid and other neutral sugars; the fine structure of the domain has not yet been fully clarified [15] • Contains numerous hydroxyl and carboxyl groups [63,66] • Complex metal ions such as Fe^{2+} or Cu^{2+}; enhance the activity of antioxidant enzymes such as superoxide dismutase and catalase [63,66]	• Better gelling and film-forming properties than cellulose and starch • Better water-retaining properties and transparency than cellulose and starch • Good oxygen and lipids barrier properties (Their products have oil-, grease-, and odor-proofing capabilities and can effectively slow down the oxidation of food lipids) • Renewable, biodegradable, biocompatibility • Soluble dietary fiber and food additive (e.g., water sacrificial agent) [15,16,60,69–71] □ Pectin and its derivatives extracted from the peel of certain fruits (e.g., fig, lemon, apple, and hawthorn) have antioxidant and antimicrobial activities; furthermore, pectin has a weak antibacterial effect, but its degradation products (especially pectin enzymatic hydrolysis products) have an obvious inhibitory effect on common foodborne pathogens such as *Staphylococcus aureus*, *Escherichia coli*, and *Vibrio parahaemolyticus* [63,65,66,72,73] □ Alginate exhibit polyanion behavior in an aqueous solution and have a certain amount adhesion [16,74] □ The commonly used agar for packaging is agarose, and its molecules can interact through hydrogen bonds to form a continuous and firm network structure [71]

Table 1. *Cont.*

Polysaccharides	Sources	Molecular Structure Characteristics	Functional Advantages
Alginate	• Major: cell wall and intercellular mucilage of brown algae such as *Laminaria*, *Kelp*, *Durvillaea potatorum*, and *Sargassum* • Minor: some *Pseudomonas*, nitrogen-fixing bacteria, and other bacteria that can produce mucous capsules [15,16,74–76]	• A long-chain linear copolymer connected by β-D-mannuronic acid and α-guluronic acid, according to (1→4) bonds • Contains numerous -COO- groups • Its products usually include sodium alginate, potassium alginate, calcium alginate, zinc alginate, and magnesium alginate [16,74,76–78]	
Carrageenan	Cell walls of marine red algae, such as *Eucheuma*, *Chondrus*, *Gigartina*, *Gelidium*, and *Hypnea* [15,79]	• A linear galactosan composed of sulfated or non-sulfated galactose and 3,6-dehydrated galactose alternately connected by α-(1→3) and β-(1→4) glycosidic bonds • Divided into seven types (e.g., κ, ι, λ, γ, ν, ξ, and μ-type) according to the different binding forms of sulfate esters • Contains numerous hydroxyl groups [15,79]	
Agar	Marine red algae, such as ferns, asparagus, laver, *Gelidium*, and *Gracilaria* [80]	• A galactose polymer composed of agarose and agaropectin • Agarose is a non-ionic polysaccharide without sulfate (salt) and comprises 3,6-dehydration-α-L-galactose and β-D-galactose residues alternately connected by (1→3) glycosidic bonds • Contains numerous hydroxyl groups [15,69,80]	

Although the reported polysaccharides differ in source, composition, structure, and characteristics, they generally have good gelation, film-forming, mechanical, and barrier properties, and are abundant, renewable, edible and biodegradable. In particular, there are many kinds of hemicellulose and polysaccharide gums, but the ones commonly used in packaging are xylan, glucomannan, pectin, alginate, carrageenan, and agar. These polysaccharides can be processed into different forms of packaging (including films, coatings, containers, sponges, and gels) through various material technologies, and have tremendous potential in the development and application of edible packaging in the future.

However, compared with traditional petroleum-based polymers and plastics, polysaccharide-based materials still have many disadvantages, mainly including the following:

(1) The chemical and thermal stability of polysaccharides are poor, which is not conducive to their subsequent molding processing. In particular, the materials formed by only one kind of polysaccharide are often brittle, easy to crack or wrinkle, have high shrinkage after molding, and have poor mechanical properties.
(2) Polysaccharide-based materials contain many hydroxyl, amino or carboxyl groups, which result in high hydrophilicity, easy swelling by moisture, and poor water vapor barrier and moisture resistance. Moreover, they are sensitive to water, and their hydrogen bonding actions, microstructures and internal stress would change after moisture absorption; which resulted in a significant decrease in the mechanical strength of polysaccharide-based materials at high relative humidity [81–83].
(3) Cellulose, hemicellulose, starch, agar, and other polysaccharides (except chitosan, pectin, and their derivatives) would provide nutrients and facilitate the growth and reproduction of microorganisms, which is not conducive to food storage.

3. Modifications of Various Polysaccharide-Based Materials for Edible Packaging

Given the above limitations, polysaccharide-based materials should be modified based on the actual application requirements to optimize their functional properties and promote their application in edible packaging. The existence of functional groups such as hydroxyl, amino, acetylamino, and carboxyl groups in polysaccharides creates conditions for their material modification. Currently, the commonly used modification techniques are chemical and physical modifications.

3.1. Chemical Modifications of Polysaccharide-Based Materials

Common methods of polysaccharide chemical modification include functional group modification, graft copolymerization, and cross-linking (Table 2). Functional group modification refers to the modification of some functional groups on the main chain and/or side chain of polysaccharides to obtain modified polysaccharides with improved physical and chemical properties through etherification (e.g., carboxymethylation and hydroxypropylation), esterification (e.g., organic acid and anhydride esterification), quaternization, and acylation [32,84–86]. Graft copolymerization refers to the process by which the polysaccharide active groups (e.g., hydroxyl, amino, and carboxyl) react with other monomers to obtain target polysaccharides under the action of an initiator or radiation [32,87]. Cross-linking refers to the process in which polysaccharides are polymerized within themselves or with macromolecules of other materials under the action of cross-linking agents (which can improve the cross-linking degree between substances) to obtain cross-linked polysaccharides with a network structure, thus enhancing the stability and physical properties of polysaccharides [85]. For example, polysaccharides are linked with proteins (whose mechanical properties are often better than polysaccharides) to obtain polysaccharide-protein complexes with optimized properties based on reducing the electrostatic free energy of the system by electrostatic interaction. In particular, during cross-linking, the thermal and mechanical properties of polysaccharide-based materials can be further improved by using carboxylic acid or calcium ions as cross-linking agents [88].

Table 2. Chemical modification methods and effects of various polysaccharides.

Polysaccharides	Modification Methods	Edible Packaging Materials	Th	MC/%	WS/%	TS	EB/%	WVP	Functional Characteristics
Cellulose	Methylation (etherification)	Methylcellulose (MC) [89]	0.041	27.3	100	55	36	2.78×10^{-10}	Better water solubility and mechanical properties than native cellulose
		MC [90]	0.048		98.9	31.4	16.2	7.95×10^{-11}	
		MC [91]	0.062		100	15.78	15.4	1.19×10^{-4}	
	Carboxymethylation	Carboxymethyl cellulose (CMC) [92]	0.142	16.55		10.48	42.37	1.198×10^{-3}	Improve transparency, thermal stability, salt tolerance and acid resistant properties
		CMC [93]	0.097		21.19	0.23	60.21	7.41×10^{-7}	
		CMC [94]	0.05			56	6.5	11.18×10^{-11}	
		CMC [95]	0.070	22.71	75.08	14.18	10.54	3.36×10^{-10}	
	Hydroxyethylation (etherification)	Hydroxyethyl cellulose (HEC) [96]	0.07		93.26			WVTR: 18.94	• The water retention capacity of HEC is higher than that of MC, and it has a good thickening effect • The dispersion of HEC is worse than that of MC and HPMC
	Hydroxypropylation (etherification)	Hydroxypropylated cellulose (HPC) [87]	0.04			7.0	7.5	7.52×10^{-5}	Enhance mechanical and barrier properties
		Hydroxypropyl methyl cellulose (HPMC) [92]	0.110	24.54		19.25	37.56	95.66×10^{-5}	
		HPMC [97]			100	33.0	13.4	1.34×10^{-10}	
		Average methoxyl content/hydroxypropyl content (M/HP): 3.05 [83]	0.025			30.83	6.06	2.036×10^{-4}	
		M/HP: 2.26 [83]	0.044			52.13	11.89	4.136×10^{-4}	
		M/HP: 3.05 [83]	0.030			67.28	17.37	2.566×10^{-4}	
		HPMC [98]	0.079			53.02	10.32	6.85×10^{-5}	
	Acetic acid esterification	Acetylated cellulose (DS 0.54) [82]	0.04–0.12						• Surface -OH are replaced by non-polar $-COCH_3$ • Better hydrophobic property (SWCA 73° > native cellulose 48°) and thermal stability

Table 2. *Cont.*

Polysaccharides	Modification Methods	Edible Packaging Materials	Th	MC/%	WS/%	TS	EB/%	WVP	Functional Characteristics
	Cross-linking	Tea catechins-cross-linked MC [90]	0.060		25.5	73.7	2.8	2.84×10^{-11}	Light barrier, antioxidant and antibacterial properties
		Dialdehyde carboxymethyl cellulose (DCMC) crosslinkedfeather keratin (FK) [99]	0.09–0.15	17.9	43.8	2.1	26.8	3.3×10^{-10}	• Covalent bonds and hydrogen bonds occurred between FK and DCMC • DCMC could slightly improve the water resistance, water vapor barrier property and flexibility, whereas reducing tensile strength
	Graft copolymerization	MC-g-2-hydroxyethyl methacrylate [100]	0.025					8.6×10^{-5}	• Puncture strength was 282 N/mm; Puncture deformation was ≈5.501 mm • Grafted MC-based films' surface appeared better smoothness
Hemicellulose	Carboxymethylation	Carboxymethyl xylan (DS 0.30) [101]	0.052			28.0	1.6	1.6×10^{-5}	• OP: 47×10^{-9}
	Hydroxypropylation	Hydroxypropylated birch xylan [87]	0.04			39.0	4.5	1.5×10^{-5}	• Hydroxypropyl groups acted as inner plasticizers • Better barrier and mechanical properties; OP: 6.5×10^{-9}
	Esterification	2-dodecenyl succinic anhydride-modified xylan (DS 0.31) [101]	0.057			29.3	6.0	0.69×10^{-5}	• Better barrier and mechanical properties; OP: 42 $\times 10^{-9}$
	Acetylation	Acetylated bleached hemicellulose (DS 1.8) [82]	0.04–0.12			44.1	5.7		• Better hydrophobic (SWCA 72° > Unmodified 57°), thermal stability and mechanical properties
	Cross-linking	Add citric acid into wheat straw hemicelluloses matrix containing cellulose nanocrystals [102]		7.0–7.5	47.41	9.76	3.94	4.09×10^{-4}	• Citric acid worked as crosslinker and plasticizer • Enhanced modulus, elongation, water resistance, and water vapor barrier property

Table 2. *Cont.*

Polysaccharides	Modification Methods	Edible Packaging Materials	Th	MC/%	WS/%	TS	EB/%	WVP	Functional Characteristics
Starch	Carboxymethylation	Carboxymethyl starch (as functional master batch or raw material) [103]							• Does not tend to retrogradation (recrystallization) • Higher thermal stability and water solubility
	Hydroxypropylation	Hydroxypropylated rice starch (Molar substitution: 0.022–0.033) [104]			4.46–5.97	3.88–5.53	79.57–132.58	4.19–5.75 × 10^{-5}	• Improved swelling capacity, viscosity and paste clarity • Higher elongation at break, water vapor permeability, film solubility, and transparency
	Acetylation	Acetylated cassava starch [96]	0.04		28.73			WVTR: 12.84	• Reduce the gelatinization temperature • Increase the hydrophobicity and tensile strength
	Esterification	Thermoplastic/succinated cassava starch (as functional master batch or raw material) [84]							• The starch developed B- and V-type structures • Smoother, continuous, and homogeneous starch matrix as the percentage of 2-Octen-1-ylsuccinic anhydride (OSA) increased, including numerous partially gelatinized granules • The incorporation of OS groups via reactive extrusion imparts better thermal stability (improved by 20%)
		Starch-laurate esters (DS 0.45–2.92) [105,106]							• Lauric acid (C_{12}) replaces -OH groups, and the modified starch shows melting thermoplastic behavior and hydrophobicity • Better thermal stability, clarity, mechanical properties (the elastic storage modulus could reach 226 MPa at room temperature) • With DS increasing, glass transition temperature and tensile strength increase while melting temperature decreases

Table 2. *Cont.*

Polysaccharides	Modification Methods	Edible Packaging Materials	Th	MC/%	WS/%	TS	EB/%	WVP	Functional Characteristics
	Cross-linking	Add citric acid (as crosslinker and plasticizer) into carboxymethyl potato starch (DS 0.5) matrix [107]	0.2–0.3		58	0.16	26		• Best mechanical performance and thermal stability containing 30% (*w*/*w*) citric acid; E: 0.65 MPa; Glass-transition temperature: 58 °C • An excess of citric acid could lead to carboxymethyl starch hydrolysis
		Add citric acid into corn starch (DS ≈ 0.98), and then blended with grape juice [108]	0.17		59	0.24	63.68	4.7×10^{-4}	• Citric acid acted as crosslinker and plasticizer • Better transparency, water barrier property and elongation at break than native film • OP: 6.2×10^{-9}
		Add sodium trimetaphosphate into corn starch (DS 0.95), and then the modified starch was blended with grape juice [108]	0.17		55	0.38	16.47	3.84×10^{-4}	• OP: 2.51×10^{-9} • Sodium trimetaphosphate acted as crosslinker • Better barrier properties and tensile strength
	Click chemistry: Cu(I) catalyzed azide-alkyne [3 + 2] cycloaddition (CuAAC)	Amphiprotic starch derivatives linked 1,2,3-triazole (as antibacterial raw material) [109]							• Enhanced antibacterial activities for *E. coli* and *S. aureus*, and inhibitory activity decreased in the order: CBTST > CMTST > BMTST > HMTST > starch; the inhibitory index of CBTST attained 97% above at 1.0 mg/mL • 1,2,3-triazole substituted groups with stronger electron-withdrawing ability relatively possessed greater antibacterial activity
Chitosan (CS)	Carboxymethylation	Carboxymethyl chitosan (DS 0.49) [110]	0.159			21.25	42	6.65×10^{-11}	• Biodegradable and soluble over a wide range of pH • High viscosity
	Graft copolymerization	Ascorbic acid was chemically grafted into CS backbones to form chitosan ascorbate (DS 0.88) [111]	0.067	10.7	43.0	22	12.0	6.3×10^{-10}	• Improved light barrier, water solubility, and water vapor barrier • Antioxidant activity (EC_{50} < 0.025)
		Chitosan acetate (DS 0.60) [111]	0.070	24.3	20.4	43	31	8.6×10^{-10}	• Better thermal stability and mechanical properties; EC_{50} > 1.60

Table 2. *Cont.*

Polysaccharides	Modification Methods	Edible Packaging Materials	Th	MC/%	WS/%	TS	EB/%	WVP	Functional Characteristics
	Cross-linking	Add fulvic acid (as crosslinker) into konjac glucomannan/chitosan matrix [33]				57.79	21.04	5.25	• Better thermal stability, optical, water vapor barrier properties, and tensile strength • Improved antimicrobial activity (when the addition of fulvic • acid ≤ 0.01% *w*/*w*)
Polysaccharide gums	Carboxymethylation	Carboxymethyl agar (CMA) [112]							• Gel skeleton microstructure of CMA was porous network structure, and the pore size of CMA became smaller and denser with the increase of DS • Hygroscopicity increased, but thermal stability decreased
	Cross-linking	Add calcium chloride (as crosslinker) into citrus pectin/CMC composite matrix [113]				468	10.6	4.45×10^{-11}	• Carboxyl group from pectin are mainly involved in interactions with CMC, whereas -OH groups are mainly involved in self-associated hydrogen bonding of biopolymers • Better thermal stability and mechanical properties (E: 4.4 ± 0.66 GPa), but worse water vapor barrier

Note: DS: Degree of substitution; Th: Thickness, mm; MC: Moisture content; WS: Water solubility; TS: Tensile strength, MPa; EB: Elongation at break; WVP: Water vapor permeability, $g \cdot m^{-1} \cdot s^{-1} \cdot Pa^{-1}$; WVTR: Water vapor transmission rate, $g \cdot h^{-1} \cdot m^{-2}$; OP: Oxygen permeability, $cm^3 \cdot m^{-1} \cdot d^{-1} \cdot Pa^{-1}$; SWCA: Static water contact angle; EC_{50}: Antioxidant value against the DPPH radical (namely, the mass concentration of antioxidants produced a 50% scavenging effect against active free radicals), mg/mL; E: Young's modulus; HMTST: 6-hydroxymethyltriazole-6-deoxy starch; BMTST: 6-bromomethyltriazole-6-deoxy starch; CMTST: 6-chloromethyltriazole-6-deoxy starch; CBTST: 6-carboxyltriazole-6-deoxy starch.

For cellulose, the goal of chemical modification is to reduce the hydrogen bond strength and improve the processing adaptability of the materials. Various properties of cellulose-based materials (e.g., permeability, solubility, mechanical properties, barrier properties, and thermoplastic behavior) can be adjusted by changing the degree of substitution, type of chemicals, and polymer chain length [114]. Methylation, carboxymethylation, hydroxypropylation, and acetic acid esterification are often used to replace the hydroxyl groups of cellulose (Table 2). For instance, the mechanical and water vapor barrier properties of edible films, prepared using modified hydroxypropyl methylcellulose (HPMC), obtained by increasing the degree of hydroxyl substitution and relative molecular weight, were significantly improved [83]. The modified methylcellulose (MC) has high solubility and efficient oxygen and lipid barrier properties. A water-soluble edible packaging bag made of MC/HPMC composites has better mechanical and barrier properties, which are suitable for packaging dry food ingredients [81]. Furthermore, compared with other polymers in the previous literature, the tensile strength of MC films (15.78 MPa) was better than that of collagen and whey protein films [91], and even was higher than that of low-density polyethylene films (0.9–14 MPa) and poly(ε-caprolactone) (14 MPa). Moreover, the corresponding elongation at break (15.4%) was superior to polystyrene (2–3%), poly (3-hydroxybutyrate) (5–8%), and poly(L-lactic acid) (9%) [115,116]. Besides, the cross-linking [90] and graft copolymerization [100] could give cellulose-based materials better surface morphology and mechanical properties, even light resistance, antioxidant, and/or antimicrobial activities.

Chemical modification of hemicellulose is often conducted through carboxymethylation, hydroxypropylation, esterification, acetylation, and cross-linking (Table 2) [32,87,101,102]. Ramos et al. [101] prepared two kinds of functional xylans, carboxymethyl xylan (CMX) and 2-dodecenyl succinic anhydride-modified xylan (X-2-DSA), using beech xylan as the raw material, and then prepared different films. The results showed that X-2-DSA film possessed similar tensile strength and oxygen permeability to CMX film. Whereas the elongation at break of X-2-DSA film was almost 3.75 times that of the latter one, and the water vapor permeability of CMX film was about 2.3 times that of the former. These phenomena might be due to the replacement of some hydroxyl groups by non-polar long aliphatic carbon chains of dodecenyl succinic anhydride, which obtains plasticizing effect and makes the xylan less polar. Additionally, Mikkone et al. [87] modified xylan by hydroxypropylation (playing an internal plasticization role) and sorbitol was added as an external plasticizer to prepare a xylan-based barrier film via the casting method. The results indicated that the combination of xylan and sorbitol with a certain degree of hydroxypropyl substitution (from low to medium is 0.3 to 1.1) improved the film formability, flexibility, thermal stability, and barrier properties of the composite films. In particular, the composite film with the lowest hydroxypropyl substitution degree (0.3) had the best comprehensive properties (e.g., the highest tensile strength and the lowest oxygen and water vapor permeabilities), and the best biomass use and biodegradability.

The objective of chemical modification of the original starch is to reduce its moisture absorption and water sensitivity, heighten the compatibility of starch with other hydrophobic materials, and improve its processing adaptability [117]. Therefore, researchers often use highly hydrophobic groups to replace hydrophilic -OH groups through chemical modification methods, such as carboxymethylation, acetylation, esterification [84,106], polymer grafting, cross-linking, and "click chemistry", which reduce the polarity of starch-based materials and improve their mechanical properties (Table 2). Liu et al. [118] first prepared carboxylated starch (which has higher hydrophilicity and polarity than that of native starch, but lower gelatinization temperature and enthalpy) by bio-α-amylase catalysis, and then introduced CMC into the modified starch matrix to enhance the hydrophobicity, thermal stability and mechanical strength of starch-based materials. In particular, the tensile strength of carboxylated starch composite films reached a maximum value of 44.8 MPa at 15% CMC addition, the hydrophobic property was effectively improved when CMC > 10%, and the static water contact angle was 66.8° at 35% CMC addition. Similarly, other researchers have modified starch by chemical methods first, but then combined it with the unmodified natural starch to produce better composites [17,119,120]. Notably, the FDA has limitations on the reagents and reactions, which are used for the manufacturing of food-grade modified starch [46], so we should follow the applicable regulations and standards when preparing starch-based edible packaging, as well as other polysaccharides edible packaging.

The purpose of the chemical modification of chitosan is to increase its water solubility, thermal stability, mechanical properties, barrier properties, and antibacterial activity, and the main chemical methods include carboxymethylation, acylation, quaternization, graft copolymerization, and cross-linking (Table 2) [58,88,121]. For example, carboxymethyl chitosan was formed by introducing carboxymethyl into N or O atoms of the chitosan skeleton through reactions of halogenated acetic acid or glyoxylic acid, thus enhancing the water solubility and adhesion [58]. This modification could also improve the antibacterial properties, with a wide range of carboxymethylation degrees. *O*-carboxymethyl and *N,O*-carboxymethyl chitosans showed better antibacterial activity than ordinary chitosan, and with the increase in carboxymethylation, the antibacterial activity of *O*-carboxymethyl chitosan increased first, then decreased [122]. Likewise, the water solubility and antimicrobial activity of the original chitosan also improved by grafting glycidyltrimethylammonium chloride [123] or nisin [124] onto the chitosan chain. Furthermore, Li et al. [125] introduced monophenol and ortho-diphenol to chitosan to obtain functionalized chitosan derivatives owned high antioxidant activity, which the EC_{50} of inhibition of DPPH, hydroxyl (·OH), and superoxide ($O_2\cdot^-$) radical-scavenging was 0.041–0.172, 0.010–0.089, and 0.014–0.038 mg/mL, respectively. Tan et al. [126] synthesized amino- and acylhydrazine-functionalized chitosan derivatives via 1,2,3-triazole and 1,2,3-triazolium by Cuprous-catalyzed azide-alkyne cycloaddition and *N*-methylation, which displayed stronger antioxidant capacity (especially against superoxide anion radical) than pristine and hydroxyl-modified chitosan. Besides, *N*-methylation of 1,2,3-triazoles further strengthened their antioxidant action. These chitosan derivatives had no cytotoxicity on L929 (at 0.0625 mg/mL) or HaCaT (at 0.1 mg/mL) cells, showing bright prospect in novel antioxidant edible packaging. In addition, the reaction of amino and hydroxyl groups in chitosan with polyaldehydes, polyesters, or polyethers can lead to cross-linking in the composite system, forming a three-dimensional network structure, thus, enhancing the thermal stability, mechanical properties, and barrier properties of chitosan [33,88]. Notably, the introduction of a cross-linking agent can further improve the properties of chitosan-based materials [33,127]. Chen et al. [33] added fulvic acid as a cross-linking agent to a konjac glucomannan/chitosan matrix to improve the thermostability, optical properties, and tensile strength (57.79 MPa, increased by 41.16%) of the composite film, while reduced its WVP (as low as 5.25 $g\cdot Pa^{-1}\cdot s^{-1}\cdot m^{-1}$, decreased by 39.31%).

In addition, the chemical modifications of polysaccharide gums (e.g., pectin, alginate, carrageenan, and agar) are mainly carboxymethylation, hydroxylation, acylation, esterification, graft copolymerization, and cross-linking (Table 2) [86,88,113,128]. For instance, Cao et al. [112] modified the original agar via carboxymethylation, while decreasing the dissolving temperature, gelling temperature, gel strength, hardness, fragility, adhesiveness, gumminess, and chewiness of carboxymethyl agar (CMA) by increasing carboxymethyl groups, conversely improving the springiness and cohesiveness of CMA, and enhancing the compactness of CMA skeleton structures. Based on polysaccharide gums and carboxymethyl cellulose being rich in active groups (-COOH and -OH) and have polyanion properties, Šešlija et al. [113] modified pectin with carboxymethyl cellulose and added glycerol and calcium chloride (which promote cross-linking through calcium ions), thus improving the thermal stability and mechanical strength of the composite film.

3.2. Physical Modifications of Polysaccharide-Based Materials

The most common and simple method for physical modification of polysaccharides is blending, namely blending one kind of polysaccharide with another or more edible materials (e.g., another polysaccharide, protein, and lipid), while supplementing with edible plasticizers, compatibilizers, antioxidants or antibacterial agents, and other small molecular additives (e.g., glycerin, essential oil, and other plant extracts). Therefore, complementary advantages of different materials are achieved while optimizing their comprehensive functions (Table 3) [81]. For example, proteins and polysaccharides are blended to form edible composites, in which positively charged proteins and anionic polysaccharides are attracted to each other to form highly structured compounds, and the water solubility, interfacial properties, adsorption, mechanical properties, and barrier properties of the composites are better than those of a single material [129,130]. Furthermore, when adding lipids into the polysaccharide/protein matrix, polysaccharides or proteins with high surface activity reduces the surface tension in the lipid emulsion, forms a space layer around the lipid droplets to enhance the emulsifying ability, promotes the stability of the emulsion, and ensure the mechanical strength and structural integrity of the composites. However, hydrophobic lipids reduce water migration and enhance the water resistance and water vapor barrier properties of the composites [131,132]. Overall, the water resistance, barrier properties, mechanical properties, heat sealing properties, and transparency of the polysaccharide-based composites could be further optimized, and even new functional activities could be developed by adjusting the composition and proportion of raw materials during blending. In general, the cohesion of a complex material increases with an increase in the length and polarity of the polymer chain, thus, improving the strength and abrasion resistance of its products, as well as the barrier properties to gas, water vapor, and solute. However, the enhancement of structural cohesion would lead to a decrease in the flexibility, porosity, and transparency of materials. Therefore, the types, proportions, and processing methods of raw materials should be explored according to the application requirements of polysaccharide-based edible packaging.

Table 3. Physical modification methods and effects of various polysaccharides.

Polysaccharides	Modification Methods	Th	MC/%	WS/%	TS	EB/%	WVP	Functional Characteristics
Cellulose	Blend CMC with gelatin and add *Dianthus barbatus* essential Oil [93]	0.100		9.86	0.16	68.37	2.19×10^{-7}	• More flexible • Better antioxidant and antimicrobial activities
	Add dipalmitoyl lecithin liposomes loaded with quercetin and rutin to CMC matrix [133]	0.035–0.045						• Antioxidant activity • Sustained-release function (preserve poly- • phenols and control their release)
	Add α-tocopherol and a mixture of polysorbate 80 and lecithin to CMC matrix [94]				44	18.5	12.45×10^{-11}	• More flexible • Antioxidant activity and sustained-release function
	Add spent coffee grounds polysaccharides to CMC matrix [95]	0.070	21.63	50.52	26.04	6.84	3.36×10^{-10}	Light barrier, antioxidant and antimicrobial properties
	Add cypress (*Cupressus sempervirens*) cone seeds extracts to HPMC matrix [98]	0.084			61.04	7.67	5.16×10^{-5}	Light barrier and antioxidant properties
Hemicellulose	Add cellulose nanocrystals into wheat straw hemicelluloses matrix [102]		7.0–7.5	93.75	11.25	3.13	8.376×10^{-4}	Improved tensile strength, modulus, water resistance, and water vapor barrier property
	Blend acetylated hemicellulose (DS 1.7) with acetylated nanocellulose (DS 2.34) [134]	0.250		17.67	10.59	15.49		Increasing DS and loading of acetylated nanocellulose, increased hydrophobicity (SWCA 68.29°) of composite and reduced its solubility in food simulants
	Blend konjac glucomannan (KGM) with microcrystalline cellulose [135]	40.53			5.12		WVTR: 3.38	Improved thermal stability, barrier and mechanical properties compared pure KGM film
	Add polydopamine functionalized microcrystalline cellulose into KGM matrix [135]	43.01			8.51		WVTR: 1.67	• Better thermal stability, barrier and mechanical properties
	Add CS/gallic acid nanoparticles into KGM matrix [127]				42.50	26.61	11.25×10^{-11}	• Better thermal stability, water vapor barrier and tensile strength • Obtain UV barrier and antimicrobial activity (*S. aureus* and *E. coli* O157:H7)

Table 3. *Cont.*

Polysaccharides	Modification Methods	Th	MC/%	WS/%	TS	EB/%	WVP	Functional Characteristics
	Blend KGM with zein and add curcumin [136]				7.34			• Better hydrophobic (SWCA: 32.6–57.5°), thermal stability and mechanical properties • Good antioxidant (DPPH value: 42.6–51.48%) and antimicrobial activities
	Blend KGM with pectin [137]	0.048	17.91		15.75	22	1.76×10^{-10}	• Improved mechanical properties compared pure KGM or pectin film • SWCA: 69.50°; DPPH value: 10.50%
	Add tea polyphenol into KGM/pectin matrix [137]	0.061	16.13		21.03	16.94	1.37×10^{-10}	• Better thermal stability, hydrophobicity, water vapor barrier, and tensile strength • Improved antioxidant and antimicrobial activities (e.g., *E. coli* and *S. aureus*) • SWCA: 88.43°; DPPH value: 50.46%
	Blend KGM with shellac [138]	0.106			13.8	20.5	11.28×10^{-5}	• Improved thermal stability, water resistance (SWCA 63.3°) and mechanical properties
Starch	Blend acetylated cassava starch with hydroxyethyl cellulose [96]	0.06		61.24			WVTR: 16.27	• Films with higher concentrations of hydroxyethyl cellulose were thicker, more transparent and hygroscopic • -OH groups in hydroxyethyl cellulose might have strongly bonded to the -COOH from acetylated starch, increasing the WS
	Blend carboxymethyl potato starch (DS 0.8) with carboxymethyl cellulose (DS 2.6) and add citric acid and glycerol [139]	0.2–0.3			3.4	29		• E: 4.9 MPa • Improve thermal, mechanical and hydrophilic properties
	Blend octenylsuccinated- (DS 0.0425) with native- sweet potato starch and add glycerol [120]	0.091	13.41	15.25	0.72	260	5.69×10^{-11}	• SWCA: 91.59°; Oil permeability: 0.149 ± 0.010 $g \cdot mm \cdot d^{-1} \cdot m^{-2}$ • Enhance moisture-proof property, elongation at break and transparency • Damage tensile strength and surface morphology

Table 3. *Cont.*

Polysaccharides	Modification Methods	Th	MC/%	WS/%	TS	EB/%	WVP	Functional Characteristics
	Blend acetylated- with native-corn starches and add glycerol to form thermoplastic corn starch [17]	0.129	9.26		23.99	6.14	1.20×10^{-10}	• Better barrier properties; OP: 2.57×10^{-5}, CO_2 Permeability: 3.32×10^{-5} $cm^3 \cdot d^{-1} \cdot m^{-1} \cdot Pa^{-1}$ • Maintain mechanical properties
	Add CS into thermoplastic corn starch [140]	0.138			12.5	1.64	0.87×10^{-9}	• Higher UV absorption and opacity • Better barrier and mechanical properties • Antimicrobial property (e.g., *S. aureus* and *E. coli*)
	Add chitin into thermoplastic corn starch [140]	0.121			12.6	1.86	0.59×10^{-9}	
	Blend rice starch with carboxymethyl chitosan (DS 0.49) [110]	0.143			18.5	35	4.70×10^{-11}	• Better transparency, thermal stability, and mechanical properties • Delayed biodegradation
	Blend hydroxypropyl high-amylose starch with pomegranate peel [141]	0.11			24.32	9.39		• Good antibacterial properties (*S. aureus* and *Salmonella*) • Better mechanical properties (e.g., stiffness, modulus, tensile strength and drop impact strength); E: 611.79 ± 72.11 MPa, Energy at peak load: 3.69 ± 0.43 J
Chitosan	Blend CS ascorbate (DS 0.80) with MC [89]	0.044	21.9	61	35	24.4	2.93×10^{-10}	• Better water solubility, barrier and mechanical properties • Maintain antioxidant activity (EC_{50}: 1.30)
	Blend CS with carboxymethyl chitosan and add nisin [142]	0.048		45.4	9.2	19.8	7.65×10^{-10}	• Carboxymethyl chitosan possessed plasticizing effect, led to higher EB, lower TS, and thermal stability • Nisin reduces transparency and mechanical properties, but improves antimicrobial activity for *Listeria monocytogenes* and water solubility • Combination of CS with CMCS improves the antimicrobial activity
	Blend CS with carboxymethyl chitosan [142]	0.021		15.4	25.4	58.4	3.43×10^{-10}	
	Add nisin into CS matrix [142]	0.043		37.5	11.4	15.3	6.35×10^{-10}	
	Blend CS with gelatin and add thymol [143]	0.104					WVTR: 2.18	• Antioxidant and antifungal properties
	Blend CS with starch and add thymol [143]	0.108					WVTR: 1.32	
	Blend CS with propolis extract [144]				17.5	12.1	0.578×10^{-8} OP: 0.21×10^{-8}	• Better gas barrier and mechanical properties • Antimicrobial (e.g., *S. aureus*, *Salmonella Enteritidis*, *E. coli*, and *Pseudomonas aeruginosa*) and antioxidant activities

Table 3. *Cont.*

Polysaccharides	Modification Methods	Th	MC/%	WS/%	TS	EB/%	WVP	Functional Characteristics
	Blend agar with acid hydrolyzed cotton linter cellulose nanocrystals (which neutralized with NaOH) [145]	0.052			33.7	30.7	$1.9 5 \times 10^{-9}$	• E: 0.72 ± 0.01 GPa; SWCA: 39.1° • Better optical, thermal stability, mechanical, and water vapor barrier properties (When the addition of cellulose nanocrystals ≤5%)
	Blend agarose with CS [146]	0.013			42.35	16	6.95×10^{-11}	• Better hydrophobicity (SWCA: 97.7°) and mechanical properties, but slightly higher WVP
	Blend pectin (75–80% degree of esterification) with corn flour [147]	0.06	21.2	70.7	7.47		0.022×10^{-7}	• Improved mechanical, structural, thermal, and water vapor barrier properties • Antioxidant activity, DPPH value: 13.97 ± 3.08%
Polysaccharide gums	Blend CS (prepared from *Callinectes sapidus*) with (high methoxyl pectin (prepared from *Citruis sinensis Osbeck* peel) [148]	0.082	16.9		17.5	35	0.97×10^{-15}	• Better water vapor barrier and mechanical properties
	Blend gum tragacanth with locust bean gum [149]	0.047	13.07		20.28	1.10	0.83×10^{-4}	• Improved transparency, water barrier, and mechanical properties • Decreased surface tension (53.97 ± 0.28 mN/m) could enhance the spreadability and coating integrity when applied to foods
	Blend low methoxyl with pectin sodium caseinate at pH 3 [150] and pH 7 [151]	0.040	14.5		15.64	9.35		• Better E (182.97 ± 6.48 MPa) and TS • Exist attractive interactions between the two negatively charged biopolymers • Charge neutrality occurred for a sodium caseinate/low methoxyl pectin ratio corresponding to the maximal coacervation
	Add *Origanum vulgare* subsp. *viride* essential oil into basil seed gum [152]	0.060	17.92				3.69×10^{-11}	• Improved water vapor barrier • Antioxidant and antimicrobial activities

Table 3. *Cont.*

Polysaccharides	Modification Methods	Th	MC/%	WS/%	TS	EB/%	WVP	Functional Characteristics
	Add fish protein hydrolysate into agar matrix [153]	0.044		48.86	19.89	42.70	10.04×10^{-11}	• Higher mechanical properties, WVP, solubility, and yellowness • Alcalase hydrolysate exhibited antimicrobial effect against five tested microorganisms (e.g., *Staphylococcus aureus*, *Yersinia enterocolitica*, *Aeromonas hydrophila*, *Debaryomyces hansenii* and *Listeria innocua*)
	Add clove essential oil into agar matrix [153]	0.061		20.86	10.16	3.93	9.37×10^{-11}	Better hydrophobicity, antioxidant and antimicrobial activities

DS: Degree of substitution; Th: Thickness, mm; MC: Moisture content; WS: Water solubility; TS: Tensile strength, MPa; EB: Elongation at break; WVP: Water vapor permeability, $g \cdot m^{-1} \cdot s^{-1} \cdot Pa^{-1}$; WVTR: Water vapor transmission rate, $g \cdot h^{-1} \cdot m^{-2}$; OP: Oxygen permeability, $cm^3 \cdot m^{-1} \cdot d^{-1} \cdot Pa^{-1}$; SWCA: Static water contact angle; EC_{50}: Antioxidant value against the DPPH radical (namely, the mass concentration of antioxidants produced a 50% scavenging effect against active free radicals), mg/mL; DPPH value: 2,2-diphenyl-1-picrylhydrazyl radical scavenging activity; E: Young's modulus.

For cellulose, the functional properties of cellulose-based packaging materials can be further optimized through physical blending reinforcers, barrier factors, antioxidants, or antimicrobials into the cellulose matrix (Table 3). Esther et al. [154] significantly improved the antioxidant, antibacterial, and barrier properties of carboxymethyl cellulose-based edible films by adding concentrated bay leaf essential oil. The results showed that when the content of essential oil was 15% (*w*/*w*), compared with the unmodified carboxymethyl cellulose film, the antioxidant activity of the composite film was improved (as high as 99%), which slowed down lipid oxidation in food and effectively inhibited the growth of *Escherichia coli* and *Candida glabrata*, the water vapor barrier property was increased by 50%, and almost 100% ultraviolet light was blocked. Other studies have found that the antioxidant and antibacterial activities of cellulose-based materials can also be improved by adding dipalmitoyl lecithin liposomes (loaded with quercetin and rutin) [133], α-tocopherol [94], spent coffee grounds' polysaccharides [95] and bacteriocin (from *Bacillus methylotrophicus* BM47) [27].

Functional hemicellulose-based edible materials can be obtained by the physical blending of hemicellulose with other polysaccharides, proteins, lipids, or other animal and plant extracts (Table 3) [127,135,136]. Along with konjac glucomannan (KGM), Wang et al. [135] improved the thermal stability, mechanical and water vapor barrier properties of KGM-based edible films by introducing microcrystalline cellulose loaded with polydopamine. Wu et al. [127] integrated chitosan/gallic acid nanoparticles with a KGM matrix to reduce the free volume of this blending system, significantly improving the mechanical and barrier properties of the edible composite film, while endowing the films with good antibacterial activity (for *Staphylococcus aureus* and *E. coli* O157:H7). Likewise, electrospun KGM/zein edible nanofiber films loaded with curcumin were prepared by Wang et al. [136] for application in food packaging. The addition of zein caused an increase in the thermal properties and hydrophobicity based on the interactions of hydrogen bonds between KGM and zein, whereas curcumin functioned as an antioxidant [2,2-diphenyl-1-picrylhydrazyl (DPPH) radical scavenging activity increased by about 15%] and antibacterial (the bacteriostatic zone for *E. coli* and *S. aureus* was 12–20 mm).

Various extracts or processing residues of animals and plants, such as cellulose, chitosan, propolis, protein, gallic acid, resveratrol, curcumin, and essential oils are often used in the blending modification of starch (Table 3) [108,141,155,156]. They have a wide range of sources and low cost, which could enhance the stability, mechanical, and barrier properties of starch-based materials, and even give them antioxidant, antibacterial, or ultraviolet light-shielding performances. For instance, Ali et al. [141] increased the mechanical properties (e.g., Young's modulus, tensile strength, stiffness, and drop impact strength) of hydroxypropyl high-amylose starch-based films by adding pomegranate peel ground powder, and endowed the films with an inhibitory effect on the growth of *S. aureus* and *Salmonella*.

Chitosan is often uniformly blended with small molecular additives (e.g., glycerol, essential oils, and other plant extracts), or with other natural polymers (e.g., other polysaccharides, proteins, and lipids) to improve the comprehensive properties of chitosan-based composites (Table 3) [51,144,157,158]. Siripatrawan et al. [144] improved the functional properties of chitosan-based edible films by incorporating propolis containing high polyphenols, specifically enhancing the tensile strength, elongation at break, total phenol content, and antioxidant and antibacterial activities of the composite films, while reducing their oxygen and WVP. Likewise, Rambabu et al. [157] added mango leaf extract (MLE) to chitosan to significantly improve the tensile strength and surface hydrophobicity of the chitosan-based composite film and reduce its WVP, water solubility, and elongation at break. Moreover, the antioxidant activity of the composite film was higher than both the original chitosan and commercial PA/PE films (in which the antioxidant activity of the edible composite film containing 5% extracts was 56% higher than the PA/PE film).

In addition, polysaccharide-gum based composites with improved performance can be obtained by uniformly blending cellulose [113,145,159], starch [147,160,161], chitosan [146,148,162], another polysaccharide gum [149,163], as well as proteins [150,151], lipids [164–167], essential oils, and probiotics [64,152,153,168,169] with the original polysaccharide-gum matrix (Table 3). By adding nanocellulose (usually $\leq$ 5% w/w) to the agar matrix, Oun [145] and Shankar [159] et al. significantly improved the tensile strength, water vapor barrier and thermal stability of the agar-based edible films. Likewise, the addition of starch to agar by Phan [160] and Fekete [161] also enhanced the water resistance and water vapor barrier properties of the composite films and reduced the overall cost of the composites. When the amount of cassava starch was 20% (w/w), the WVP at 57–22% relative humidity differential of the composite film was reduced to about 3.33×10^{-11} $g \cdot m^{-1} \cdot s^{-1} \cdot Pa^{-1}$, which is 53.8% less than pure agar film. When the added amount was 50% (w/w), the WVP was about 2.99 $\times 10^{-11}$ $g \cdot m^{-1} \cdot s^{-1} \cdot Pa^{-1}$, which was 58.5% less than pure agar film [160]. Furthermore, the blending of chitosan (an alkali-soluble polysaccharide) and acidic polysaccharide gums produces electrostatic interactions, which makes the structure of the composite compact and without phase separation, thus, leading to better mechanical and barrier properties than a single material, and even improves the antibacterial and ultraviolet light-shielding properties of the composite [146,148,162]. Sodium caseinate was introduced into low methoxyl pectin by Eghbal et al. [150,151] to adjust the water content and absorption, as well as the mechanical and optical properties of the composites. The results indicated that the protein content affected the properties of the composites; the highest amount of complex coacervates of blending liquids was formed at a sodium caseinate/low methoxy pectin ratio of 2, at which the ζ-potential value was zero and the turbidity reached the highest value. While the ratio was 0.05, the Young's modulus (182.97 $\pm$ 6.48 MPa) and tensile strength (15.64 $\pm$ 1.74 MPa) of the composite films were the highest, which were all higher than those of the pure pectin film. In addition, lipids (e.g., beeswax, shortening, and shellac) are the most effective natural substances to enhance the water and moisture resistances of polysaccharide gums [165–167]. Active extracts (e.g., various plant essential oils) could not only strengthen the thermal stability, and mechanical and barrier properties of polysaccharide gum-based materials, but also improve their antioxidative, antibacterial, and other functional characteristics [64,152,153,168,169].

4. Applications of Various Polysaccharide-Based Materials in Edible Packaging

Original polysaccharides can form self-assembled films, coatings, or microcapsules under the action of hydrogen bonding, van der Waals, or electrostatic forces, form hydrogels with a three-dimensional network structure through gelation, and form composites by combining them with other edible materials (e.g., proteins, lipids, probiotics, and other natural active small molecule substances), which can apply to different food packaging (Figure 1). The predominant use of polysaccharide-based edible materials is to serve as an auxiliary means of packaging. They can effectively delay the migration of water, gas, oil, and solute by providing a selective barrier, retain volatile flavor compounds and mechanical integrities of foods, improve treatment properties of foods, or even be used as non-toxic carriers of food additives (e.g., antioxidants, anti-browning, and antimicrobial agents) integrated into the packaging to improve the sensory properties of foods and extend their shelf lives.

In the following section, the potential application of five polysaccharides (cellulose, hemicellulose, starch, chitosan, and polysaccharide gums) in targeting the edible packaging sector are briefly described. Whereas, Table 4 includes the main preparation methods, packaging forms, packaged objects, and packaging effects of the different polysaccharide-based materials; the main preparation methods are shown in Figure 3.

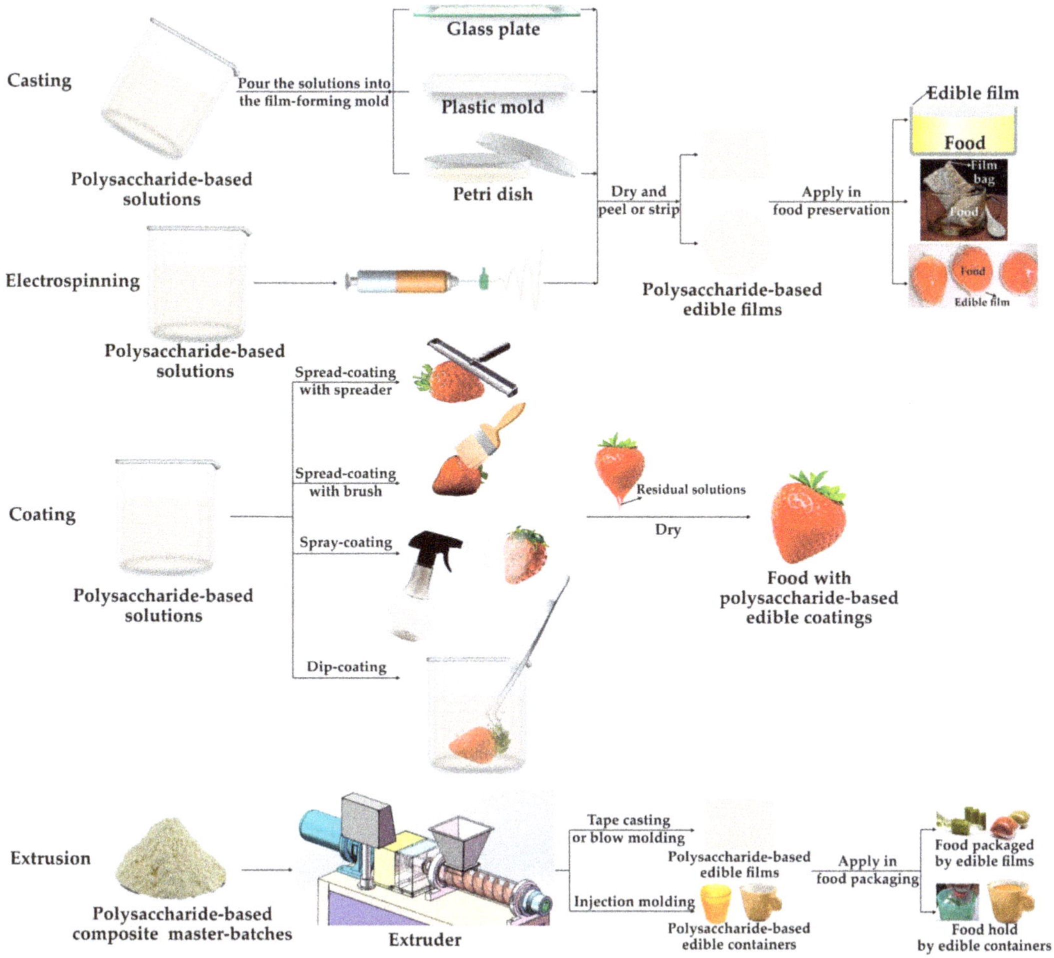

Figure 3. Different manufacture methods of polysaccharide-based edible packaging.

Table 4. Applications of five kinds of polysaccharide-based edible materials in food packaging.

Food		Edible Packaging & Preparation Method	St	Mass Loss/%	Dp/%	TSS/%	TA/%	pH	Vc Mass	TSP	Packaging Effects
Fruit	Strawberry	CMC/bacteriocin from *Bacillus methylotrophicus* BM47 coating; Dip-coating [27]	4–16 °C	10.5; 12 d	0; 12 d	8.6; 12 d	1.09; 12 d	3.34; 12 d	24.5; 12 d	9–10; 12 d	• Reduce the weight loss and decay percentage of strawberries • Inhibit the decrease of antioxidant activity and the propagation of the fungus • Extend the shelf life by 4 or more days
		KGM/pullulan film; Casting [170]	4–14 °C	25		6	0.55		0.015 μg/mL		• Decrease weight loss; Slow down fruit aging • When the concentration of KGM was 1% with the mass ratio of KGM/pullulan 2:1, films exhibited the best preservation effect • Extend the shelf life to 14 days
		CS/gelatin/thymol coating; Dip-coating [143]	4–7 °C	1–2	1.67	7.16				6.71	• Both coatings protect strawberries against fungal (Botrytis cinerea) decay, improve the physiochemical parameters and shelf lives (extend by 2–3 days) • The composite coating containing starch possesses higher TSP, antioxidant activity and catalase activity, and lower mass loss, Dp, TSS, guaiacol peroxidase, polyphenol oxidase, total anthocyanins, polygalacturonase and pectin-lyase than that containing gelatin; especially the antioxidant activity value/(mmoleq ascorbic acid/g food) of the former (26.34) was higher than the latter (25.31) • The preservation effect of CS/starch/thymol coating was better than that of CS/gelatin/thymol coating
		CS/starch/thymol coating; Dip-coating [143]	4–7 °C	0.61	0	6.95				7.06	
	Grape	CS/*Mentha* (piperita L. or x villosa Huds) essential oil coating; Dip-coating [171]	25–12 °C; 12–24 °C			11.2–12.6° Brix	42.9–47.3 mmol H^+/100 g food				• Delay and even inhibit the appearance of postharvest mold (e.g., *Aspergillus niger, Botrytis cinerea, Penicillium expansum, and Rhizopus stolonifer*) infection in table grapes • Reduce respiration and transpiration across the fruit surface, thus delaying senescence and extend shelf lives
	Banana	Rice starch/ι-carrageenan/sucrose fatty acid esters coating; Spray-coating [172]	20–14 °C	4.5		20.5°Brix	0.25				• Reduce the weight loss, firmness (6.89 N), chlorophyll degradation, and respiration rate of Cavendish banana • Delay the ethylene production and starch degradation rate during storage • Extend the postharvest life for 12 days (40% extension) in the absence of refrigerated storage

Table 4. *Cont.*

	Food	Edible Packaging & Preparation Method	St	Mass Loss/%	Dp/%	TSS/%	TA/%	pH	Vc Mass	TSP	Packaging Effects
	Guava	Acetylated cassava starch/hydroxyethyl cellulose coating; Dip-coating [96]	25–13 °C	13.15		8.0	0.66		20.5		• Allow the guava respiration but still delayed the ripening process; • Reduce mass loss, increase firmness, and maintain green skin color • Extend the shelf life of guava
	Apricot	Basil seed gum/*Origanum vulgare* subsp. *viride* essential oil coating; Dip-coating [152]	4–8 °C	6.9		15				230	• Kept quality and increased shelf-life of cut apricots • Good antioxidant (EC_{50}: 31.2 μg/mL; DPPH value: 22.7 g/kg) and antimicrobial properties (e.g., aerobic mesophilic, yeasts, and molds)
Vegetable	Cherry tomato	KGM/nisin coating; Spread-coating [173]	25–16 °C	9.5	Decay index: 0.133	6.22					• Reduce the rotting index, weight loss rate, soluble solids content, and hardness of cherry tomato (Firmness of coated fruit was 47.02% higher than that of the control group) • Induce peroxidase activity of cherry tomato • Maintain sensory quality and extend shelf life
	Cucumber	Konjac glucomannan/saffron petal extract coating; Spread-coating [174]	4–5 °C			17.56				0.17	• Reduce mesophilic bacteria and fungi populations; especially when the concentration of the extracts was 4%, the antimicrobial effect was most effective • Improve the soluble solids, antioxidant activity, and soluble phenols (DPPH value could reach 20 mg/g) • Decrease spoilage • Keep quality features and prolong the shelf life
	Tomato/Chilly/Brinjal	CS nanoparticles coating; Dip-coating [56]	25–5 °C	0.21/3.3/0.53							• Inhibit the growth of *Rhizoctonia solani, Fusarium oxysporum, Collectotrichum acutatum, and Phytophthora infestans* during storage • Significant antioxidant activity; reduce the weight loss of vegetables, and prolong the shelf lives
	Food	**Edible Packaging & Preparation Method**	**St**	**PV**	**TBARS**	**TVB-N**	**DPPH/%**	**ABTS/%**	**pH**	**TSP**	**Packaging Effects**
Nut	Pistachio	CMC/gelatin/*Dianthus barbatus* essential oil coating; Dip-coating [93]	25 °C–6 months	0.1–3.5							• Slow down the lipid oxidation of pistachios • Inhibit the growth of three aflatoxin-producing molds on pistachios, including *Aspergillus flavus* (PTCC-5004), *Aspergillus parasiticus* (PTCC-5286), and *Aspergillus parasiticus* (PTCC-5018) during storage • Extend the shelf life

Table 4. *Cont.*

Food		Edible Packaging & Preparation Method	St	Mass Loss/%	Dp/%	TSS/%	TA/%	pH	Vc Mass	TSP	Packaging Effects
	Cashew nut	CS/mango leaf extract film; Casting [157]	30–28 °C	2.88							• Better oxidation resistance than the commercial PA/PE and pure chitosan films • Inhibit the lipid oxidation and remain the sensory quality of cashew nuts during storage
Meat	Chicken breast	Rye starch/Rosehip extract film; Casting [175]	4–9 °C		0.59		80.22	96.87			• Reduce the generations of peroxide and TBARS; DPPH value of films was 25.62 mg GAE/g films • Inhibit the lipid oxidation in chicken breast • Prolong the shelf life
		Corn starch/gelatin/N-α-lauroyl-l-arginine ethyl ester monohydrochloride film; Casting [155]	4–19 °C		0.2				5.88		• Good antimicrobial activity (The microbiological limit of acceptability for total viable counts was reached after 16 d) • Composite films with non-oxidized starch better preserved the quality attributes of chicken than oxidized starch-based coating • Extended the shelf-life of chicken to 16 d
		Oxidized corn starch/gelatin/N-α-lauroyl-l-arginine ethyl ester monohydrochloride coating; Spread-coating [155]	4–19 °C		1.52; 9 d				5.72; 9 d		• Oxidized starch increased the antibacterial effectiveness, but enhanced lipid oxidation (the limit of acceptability in terms of TBARS was reached after 9 d) • Extended the shelf-life of chicken to 9 d
	Pork	Cassava starch/*Lycium ruthenicum* Murr anthocyanins film; Casting [176]	25 °C–48 h			10.89–16 h; 17.21–24 h				6.15 –16 h; 6.49 –24 h	• Delay the lipid oxidation of pork • Achieve real-time and visual monitor for the pork freshness
	Beef loin	CS/cumin essential oil-loaded nanoemulsion film; Casting [177]	3–21 °C		1.39	12			5.4		• Withstand low-dose gamma irradiation (GI) at 2.5 kGy • Inhibited the growth of *L. monocytogenes*, *E. coli* O157:H7 and *Salmonella typhimurium* • Slow down the increasing of TVB-N and pH • Shelf life was extended at least 14 days combined with GI and refrigerated storage
	Ham	Iota-carrageenan/rosemary extract coating; Dip-coating [178]	5–15 °C								• Inhibit the growth of aerobic mesophilic microorganisms, coliforms, lactic acid bacteria, and yeasts • Remain the moisture, hardness (3779 g), and color of hams over the 15/days of storage

Table 4. *Cont.*

	Food	Edible Packaging & Preparation Method	St	Mass Loss/%	Dp/%	TSS/%	TA/%	pH	Vc Mass	TSP	Packaging Effects
	Goat meat sausage	Maltodextrin/calcium alginate/*Tinospora cordifolia* extracts film; Casting [179]	−18–21 °C		0.54				6.79		• Reduce the production of TBARS and free fatty acid (FFA); FFA reached 0.352% Oleic acid at 21 d • Inhibit the reproduction of microorganisms (total plate, psychrophilic, yeast, and mold) • Maintain the sensory quality of goat meat sausages
Oil	Olive oil	HPMC/cypress seed extract film; Casting [98]	23–23 °C	<20 (legal limit)							• Slow down the oxidation of olive oil during 23 days of accelerated storage • Shelf life could reach at least 7 days
Oil	Soybean oil	Pomelo peel flours/tea polyphenol film; Casting [64]	23–15 °C	31.58			74.39				• Significantly decrease peroxide value to delay oil oxidation during storage • Inhibit the growth of *E. coli*, *S. aureus* and other bacteria, especially the inhibition of Gram-positive bacteria is stronger than Gram-negative bacteria
Oil	Soybean oil	Lime peel pectin/coconut water/lime peel extract film; Casting [180]	27–30 °C		3.39						• Total phenolic content, DPPH value and ABTS value of composite films were 81.01 mg GAE/g film, 43.50 μM Trolox/g film, and 543.14 μM Trolox/g film, respectively • Retarded soybean oil oxidation during storage by delaying hydroperoxide (primary lipid oxidation products) degradation
Aquatic prod-uct	Salmon	Cowpea starch/maqui berry extract film; Casting [181]	4–6 °C	1	0.63		42.39	88.46			• Delay the lipid oxidation of salmon and extend its shelf life
Aquatic prod-uct	Hake	Agar/green tea extract/probiotic bacteria film; Casting [169]	4–15 °C			25			7.01		• Inhibit the growth of spoilage microorganisms, especially H_2S-producing bacteria • Decrease the TVB-N, TMA-N, and pH value • Increased the beneficial lactic acid bacteria in hake • Extend the shelf life of hake at least for a week
Aquatic prod-uct	Flounder fillets	Agar/fish protein hydrolysate film; Casting [153]	5–15 °C			29.80			7.05		• Decrease TVB-N and pH values; Delay the growth of bacteria groups, especially H_2S-producing microorganisms • Extend the shelf life of flounder fillets from 10 days to 15 days by improving biochemical and microbiological parameters in the last stages of the chilled storage • Film with protein hydrolysate had no sensory limitation of essential oil, but its preservation effect was slightly lower
Aquatic prod-uct	Flounder fillets	Agar/clove essential oil film; Casting [153]	5–15 °C			25.83			6.76		

Table 4. *Cont.*

Food	Edible Packaging & Preparation Method	St	Mass Loss/%	Dp/%	TSS/%	TA/%	pH	Vc Mass	TSP	Packaging Effects
Beluga sturgeon fillets	Jujube gum/nettle oil-loaded nanoemulsions coating; Coating [182]	4–15 °C	2.64	1.22	16.42 mg N/100 g			6.42		• Warner–Bratzler shear force: 18.74 N; FAA: 0.94 • Reduce the weight and cooking losses, and pH changes; Delay the textural and color deterioration • Inhibit the lipid oxidation and foodborne bacteria growth; Prolong the shelf life
Shrimp	Sweet potato starch/thyme essential oil coating; Dip-coating [183]	4–8 °C		0.3–0.5					8	• Maintain the sensory properties (e.g., textural, hardness and color) and freshness • Reduce pH value, lipid oxidation, bacteria count, and melanosis; Extend the shelf life

Note: St: Storage time, day; Dp: Decay percentage; TSS: Total soluble solids; TA: Titratable acidity; TSP: Total soluble phenolic, mg Gallic acid equivalent (GAE)/100 g food; Vitamin C mass: Vc mass, mg/100 g food; PV: Peroxide value, meq (peroxides or O_2)/kg food; TBARS: Thiobarbituric acid reactive substances, mg malondialdehyde (MDA)/kg food. TVB-N: Total volatile basic nitrogen, mg/100 g food; TMA-N: Trimethylamine nitrogen, mg/100 g food; ABTS value: 2,2′-azino-bis (3-ethylbenzothiazoline-6-sulphonic acid) radical scavenging activity; DPPH value: 2,2-diphenyl-1-picrylhydrazyl radical scavenging activity.

4.1. Applications of Cellulose

Cellulose is commonly applied in food packaging (e.g., fruits, vegetables, and oils) as edible films, coatings, and emulsions to protect the sensory qualities of foods and extend their shelf lives (Table 4). Rhimi et al. [98] added cypress seed extract to an HPMC matrix and prepared edible composite films using the casting method, and then applied them in olive oil packaging (as shown in Figure 3). The results indicated that compared with pure HPMC films, the tensile strength of the composite films was significantly improved (up to 15.13%) and the WVP was reduced (24.66% at most), which slowed down the oxidation of olive oil during 23 days storage. The lowest WVP, greatest opacity, and highest antioxidant capacity of the composite films were obtained with the highest extract concentration. Therefore, the peroxide value of olive oil sealed with composite films (containing 2% *w*/*v* extract) after accelerated storage for 11 days was 10 times lower than when sealed with pure HPMC films.

It is also noteworthy that cellulose is usually added to other edible materials as a reinforcing or toughening agent to improve the properties of composites. In the blends with collagen and whey protein, methylcellulose was responsible for the increase in tensile strength, water vapor barrier, and thermal properties. While, the prepared methylcellulose-based edible materials (Figure 4) could maintain their integrity for months, be completely biodegraded in 10 days in soil (Figure 5), and when immersed in hot or cold water showed total solubilization in around 30 s upon manual shaking [91]. The edible packaging has immense potential applications in soluble sachets for powdered foods, as well as oil containers and capsules for instant foods (Figure 6). Furthermore, the addition of cellulose nanocrystals to soybean protein could improve the tensile strength and barrier properties (the static water contact angle increased, and the moisture content, WVP, and reduced oxygen permeability) of the edible composite film, and enable the film to obtain ultraviolet light-shielding performance on the premise of appropriate transparency [184]. In addition, the creaming stability and ability to form an elastic gel-like network of beeswax-in-water (O/W) Pickering emulsions could be improved by blending with cellulose nanofibrils/carboxymethyl chitosan. Meanwhile, the complex edible films cast by modified emulsions had good tensile strength (5.0 MPa at a strain of 2.2%) and low WVP ($<2 \times 10^{-7}$ $g \cdot h^{-1} \cdot m^{-1} \cdot Pa^{-1}$), and could inhibit the growth of *S. aureus* and *E. coli*, a promising application for antiseptic and fresh-keeping packaging for berry fruits [185].

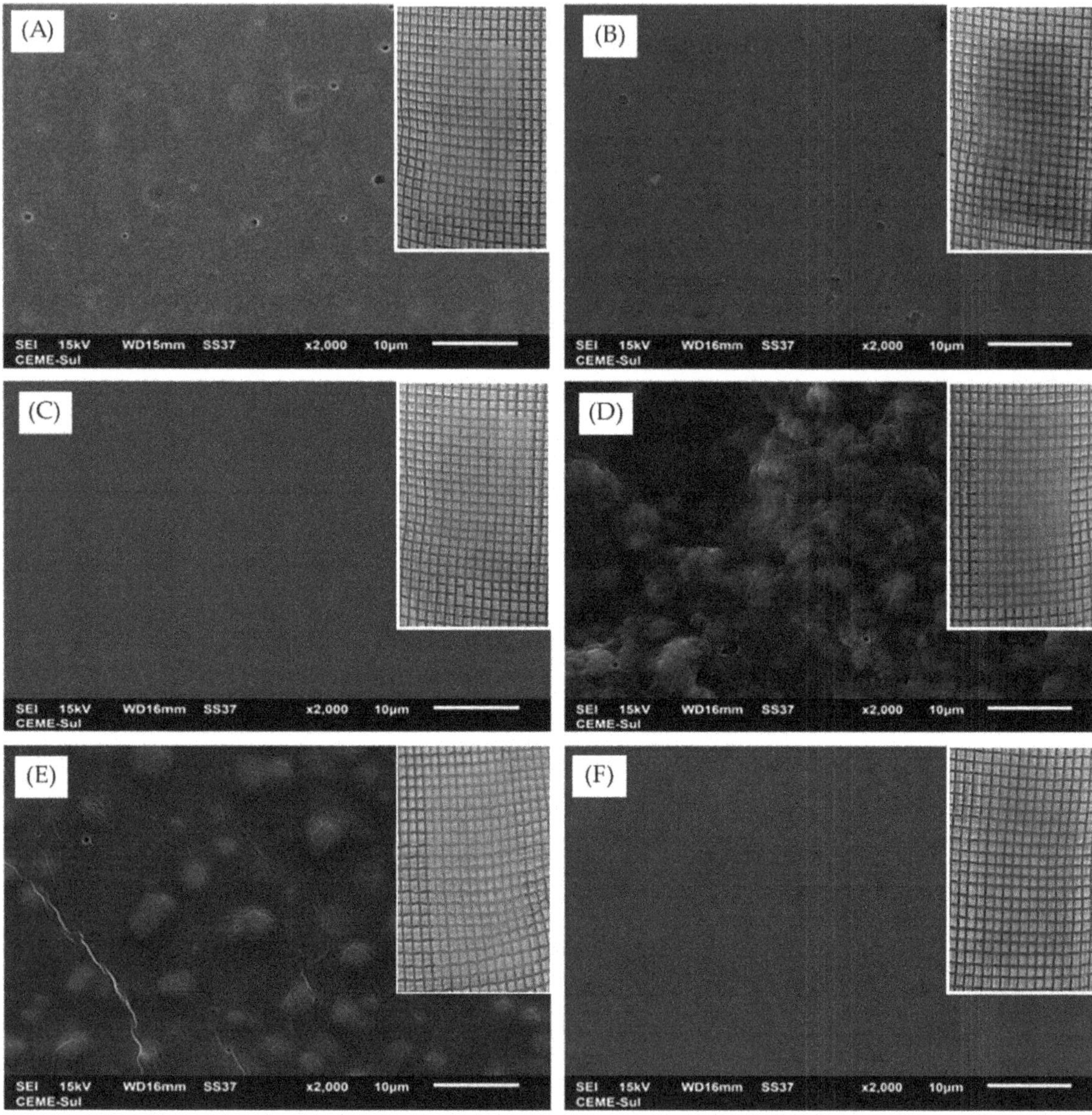

Figure 4. Scanning electron microscopy images (×2000) and physical photos of different edible films. (**A**) Collagen film; (**B**) Whey protein film; (**C**) Methylcellulose film; (**D**) Collagen/whey protein blend film; (**E**) Collagen/methylcellulose blend film; (**F**) Whey protein/methylcellulose blend film. (Adapted with permission from Filipini [91]; published by John Wiley and Sons, 2020).

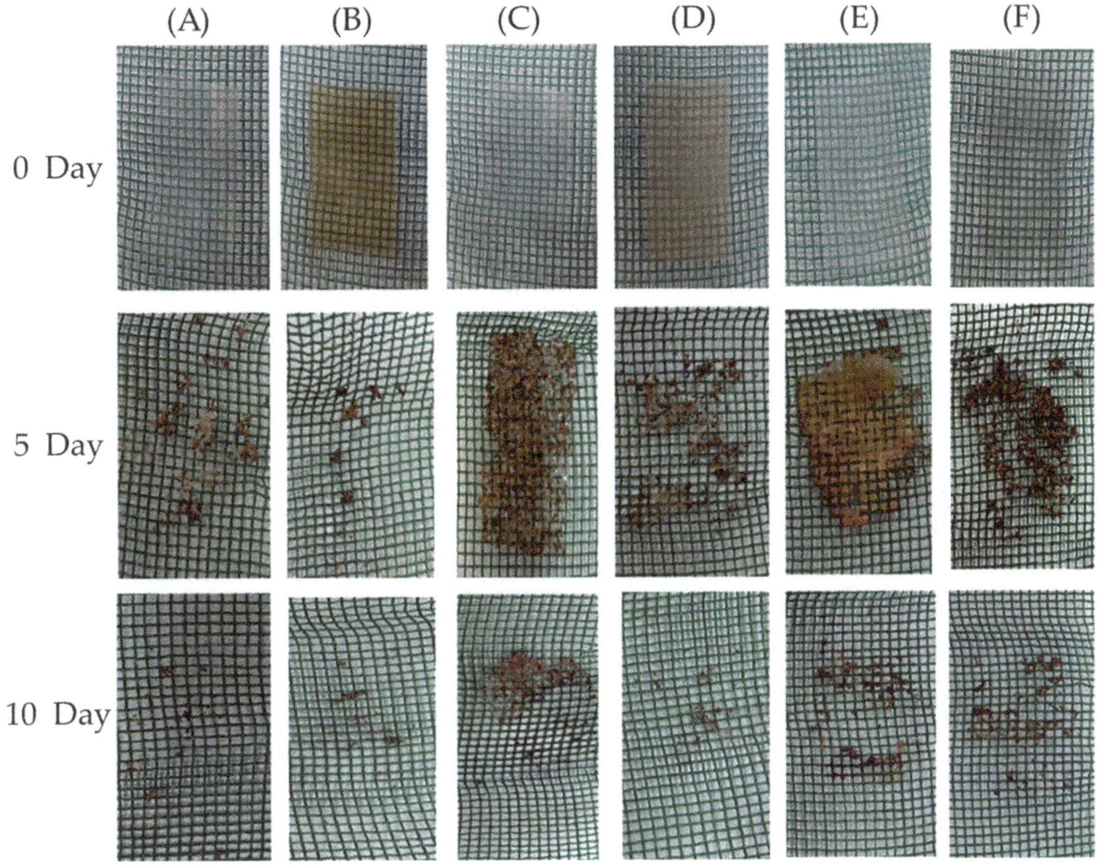

Figure 5. Biodegradability in the soil of different edible films. (**A**) Collagen film; (**B**) Whey protein film; (**C**) Methylcellulose film; (**D**) Collagen/whey protein blend film; (**E**) Collagen/methylcellulose blend film; (**F**) Whey protein/methylcellulose blend film. (Adapted with permission from Filipini [91]; published by John Wiley and Sons, 2020).

Figure 6. Prototype photos of different edible packaging. From (**A–G**) are methylcellulose sachets containing soybean oil, salt, whey protein, powdered coffee, powdered juice, rice, and cookies, respectively; (**H**) Whey protein/methylcellulose edible sachet containing oil; (**I**) Whey protein edible film for the coffee capsule. (Reproduced with permission from Filipini [91]; published by John Wiley and Sons, 2020).

4.2. Applications of Hemicellulose

Hemicellulose is usually used in edible packaging as films, coatings, or modifying additives, which is like cellulose (Table 4). Taking KGM as an example, Yan et al. [170] introduced pullulan into the KGM matrix to cast edible composite films for strawberry preservation. They showed that the mechanical and barrier properties of the composite films were markedly enhanced because of the intermolecular interaction between KGM and pullulan; 1% (*w*/*v*) KGM/pullulan (with a mass ratio of 2:1) composite film significantly decreased the weight loss and maintained the titratable acidity, soluble solids, ascorbic acid, and skin color on strawberry preservation, thus slowing fruit aging, improving the quality during storage, and extending their shelf life to 14 days. Hashemi et al. [174] blended saffron petal extract with a KGM matrix to cast edible complex films, while coating fresh-cut cucumbers (as shown in Figure 3). The results indicated that saffron petal extracts markedly improved the transparency and moisture content of the complex films, reduced their WVP, and even endowed them with promising antioxidant and antimicrobial properties. Furthermore, this composite coating reduced mesophilic bacterial and fungal populations during cucumber storage (in which 4% extracts were considered as the most effective additives), improved the soluble solids content, antioxidant activity, and soluble phenols of coated sliced cucumbers, thus decreasing their spoilage, maintaining their quality features, and prolonging their shelf lives. Wang et al. [18] introduced zein into KGM matrix by electrospinning to form stable homogeneous nanofibril films, which the hydrophobicity was improved (SWCA of the composite film increased from 7.5° to 57.5°). Furthermore, they added curcumin into the above nanofibers to form a functional nanofilm with advanced antioxidant (scavenging activity increased about 15%) and antibacterial (a large inhibitory zone of 12–20 mm for *E. coli* and *S. aureus*) activities, as well as better thermal stability, water resistance and tensile strength.

4.3. Applications of Starch

Starch is often compounded with other edible materials to fabricate edible films or coatings, which are widely used in different food packagings, such as fruits, vegetables, meat, seafood, confectioneries, cakes, and pastries to block the migration of oxygen and grease and help improve the appearance, texture, and processing performance of foods (Table 4). Go et al. [175] added rosehip extracts to rye starch matrix to cast edible composite films and applied them in chicken breast packaging. The flexibility, optical properties, and antioxidant activity of the composite films were improved, and the highest ABTS and DPPH radical scavenging activities were observed in films containing 1.0% extracts (96.87% and 80.22%, respectively). Moreover, chicken breasts packaged with these films had lower peroxide and thiobarbituric acid reactive substance values than those packaged with original rye starch film, as well as the non-packaged control, suggesting that the edible composite films could effectively inhibit lipid oxidation and prolong its shelf life. Likewise, incorporating maqui berry extract [181], carvacrol, and chitosan [186] in starch-based edible composites (e.g., edible films and coatings) retarded lipid oxidation in fish, ham, and other foods, inhibited the growth of foodborne pathogens, and extended the shelf life of foods. Qin et al. [176] added *Lycium ruthenicum* Murr anthocyanins to cassava starch to manufacture a freshness indicator film with both intelligent pH sensitivity and edibility for pork packaging. The results showed that the barrier ability, tensile strength, and antioxidant activity of the composite film were improved by hydrogen bond interactions between anthocyanins and starch chains. Moreover, this composite film achieved real-time and visual monitoring of pork freshness based on its color change with pork quality during storage.

Furthermore, a significant difference from other polysaccharides is that original starch exposed to shear and high temperature (supplemented with water and processing aids) could be converted into thermoplastic starch-based materials, and then various starch-based edible packaging containers (e.g., film, cup, tray, and plate) can be obtained through extrusion, compression, or injection molding (Figure 3) [17,140,187,188].

4.4. Applications of Chitosan

Currently, chitosan-based edible packaging (as a film and coating) has been widely used in the packaging of fruits (e.g., strawberries, apples, kiwi, and grapes), vegetables (e.g., tomato, pepper, and eggplant), meats, and nuts to retain food quality and prolong their shelf life (Table 4) [56,171,177,189–192]. These edible packages mainly achieve food preservation by reducing the transpiration rate, delaying browning or lipid oxidation, and inhibiting the growth of spoilage microorganisms.

Divya et al. [56] coated chitosan nanoparticle solutions on the surfaces of tomatoes, chilies, and brinjals using the dip-coating method (Figure 3). The edible coatings had a good inhibitory effect on *Rhizoctonia solani*, *Fusarium oxysporum*, *Collectotrichum acutatum*, and *Phytophthora infestans* during 5 days of storage, had significant antioxidant activity, reduced the weight loss of these vegetables, and prolonged their shelf lives. Perdones et al. [189] applied chitosan-lemon essential oil dip-coatings to strawberry preservation. The results indicated that these edible coatings could control strawberry fungal decay during storage and affect the metabolic pathways and volatile profile by promoting the formation of esters and dimethyl furfural and incorporating terpenes into the fruit volatiles in a short time. Likewise, Dini et al. [177] packaged beef loins in chitosan-based edible films containing cumin essential oil nanoemulsions supplemented with irradiation treatment. The results showed that the edible composite films could withstand low-dose gamma irradiation at 2.5 kGy, while inhibiting the growth of *L. monocytogenes*, *E. coli* O157:H7, and *Salmonella typhimurium* in beef loins during the 21/days refrigerated storage, and slowed down the increasing level of total volatile basic nitrogen and pH value of beef, thus effectively enhancing the microbiological safety, quality, and storage life.

4.5. Applications of Polysaccharide Gums

Polysaccharide gums (e.g., pectin, alginate, carrageenan, and agar) are commonly used in edible packaging as gels, films, and coatings for food preservation of fruits (e.g., apple, peach, cherry), vegetables (e.g., tomato, papaya, and lettuce), meats, and seafood (Table 4), and even have commoditized packaging of pure water and other beverages. These polysaccharide-gum based edible materials could effectively reduce the dryness degree of food surfaces, prevent food from water loss and atrophy, and are beneficial to slow down lipid oxidation and surface discoloration of foods, as well as inhibit the reproduction of spoilage microorganisms, thereby extending the shelf life of foods [19,70].

López et al. [169] added green tea extract to an agar solution containing glycerin and glucose to prepare the substrates by casting, and then coated the substrates with probiotic strains (*Lactobacillus paracasei* L26 and *Bifidobacterium lactis* B94) to acquire the edible composite films and further apply in hake packaging. The results showed that during 15 days of storage, the edible composite films effectively inhibited the growth of spoilage microorganisms, especially the H_2S-producing bacteria, causing a decrease in TVB-N, trimethylamine nitrogen (TMA-N), and pH value of the hake, and increased the beneficial lactic acid bacteria, thus leading to its shelf life extension for at least a week. Additionally, maltodextrin/calcium alginate edible casting films containing Tinospora cordifolia extracts were fabricated by Kalem et al. [179] and then used as casings substitutes for goat meat sausages. It was found that edible films with antibacterial and antioxidant properties could significantly reduce the production of thiobarbituric acid reacting substances and free fatty acids in sausages during storage, inhibit the reproduction of microorganisms (total plate, psychrophilic, and yeast and mold), and maintain the sensory quality of goat meat sausages. Similarly, the cooked ham portions were dipped in iota-carrageenan-based coating solutions containing rosemary extract, ascorbic acid, calcium chloride, α-tocopherol, and glycerol by Carocho et al. [178] for food preservation. The results showed the edible coating based on the above solutions inhibited the growth of microorganisms and retained the sensory quality of hams over the 15-days of storage.

5. Safety Risk Assessment of Polysaccharide-Based Edible Packaging

Edible packaging serves to protect food and act as a ready-to-eat "food", which provides valuable nutrients and energy [193]. In theory, food-grade polysaccharides made from natural edible constituents used in most studies are non-toxic, and edible packaging prepared from these polysaccharides could be consumed by animals or humans without health risk [15]. However, to be edible actually, the materials (including substrates and additives) used in the formulations should be green, non-toxic, safe and meet applicable regulations or standards (e.g., GRAS—Generally Recognized as Safe by the FDA-U.S. Food and Drug Administration).

Uncertainties and knowledge gaps on the possible health effects and long-term safety of polysaccharides and their modifying additives, when used in edible packaging, are still the most important concern. To date, very few studies have been published regarding the effects of polysaccharides-based edible packaging upon ingestion, and the absorption, distribution, metabolism, and excretion after oral exposure, and the potential interactions of polysaccharides with packaged food components [194]. Most edible films and coatings, discussed in this review, focus on the preparation and characterization of materials, with little follow-up food safety risk assessment.

Therefore, polysaccharides-based edible packaging must be exhaustively studied, they are easier to transfer constituents into foods than petroleum-based polymers. The first step in assessing the potential hazard of polysaccharides-based packaging for a comprehensive risk assessment, in terms of consumer safety, is to evaluate their potential migration into food (usually according to Regulation (EU) No. 10/2011 on plastic materials and articles) [195]. In particular, the solubility of polysaccharides that migrate in the food matrix and/or upon gastrointestinal passage is a crucial factor.

In addition, toxicological risk and dietary exposure assessment are important for polysaccharides edible packaging. Barreto et al. [196] prepared two kinds of onion (*Allium cepa L.)* puree-based edible films by casting, namely unwashed hydrothermally treated pulp (HTP) and washed hydrothermally treated pulp (W-HTP), and then assessed their genotoxicological safety. The cellular viability demonstrated that HTP films showed greater cytotoxicity than W-HTP films; and the mutagenic activity indicated that both HTP and W-HTP films were not able to statistically increase the frequencies of the biomarkers for chromosome damage (micronucleus test) at the tested concentrations. However, the HTP films showed signs of mutagenicity in the Ames test (gene mutations), suggesting caution in their use. Therefore, W-HTP onion-based edible films are harmless and possess safety potential application in food packaging, supporting the first level of evidence. For the additives, Sohrabi et al. [197] evaluated the potential cyto-genotoxicity of ascorbyl palmitate (AP, a widely used food additive) on Human Umbilical Vein Endothelial Cells (HUVECs). The results indicated that the growth of HUVECs was decreased upon treatment with AP in dose-and time-dependent manner, and AP induced apoptosis by up-regulation of caspase-3, 9 and down-regulation of Bcl-2 ratio. Therefore, AP application in the edible packaging industry should be carefully considered.

Zheng et al. [198] prepared hydroxypropylated-Phosphated-modified glutinous rice starch and evaluated its safety through acute and 28-day repeated oral toxicity tests. The results showed that the modified starch possessed more than 10,000 mg/kg LD_{50} value, was belong to non-toxic. Moreover, its acceptable daily intake for a normal person (70 kg) should be less than 38,900 mg, which means that the recommended intake (RNI) is no more than 38,900 mg/d. Asmar et al. [199] dipped the potato sticks into chitosan or pectin hydrocolloid coating solutions before frying to reduce the acrylamide and oil content of French fries. Then, the Daily Intake (DI) (Table 5) and Margin of Exposure (MOE) (Figure 7) were further calculated by considering the six following age groups (as stated from EFSA) to estimate variations in risk assessment by applying coating solutions. The results showed that, compared with the control sample (reached highest acrylamide concentration 2089 $\mu g \cdot kg^{-1}$), the edible polysaccharides coating reduced the acrylamide content by 48% for pectin and >38% for chitosan, respectively. Moreover, the increasing MOE value indicated that recurring coatings could provide advantages to consumers, especially for the ones from 1 to 65 years old, and the pectin coating was the most effective.

Table 5. Dietary intake of acrylamide consumption [$ng \cdot (kg \cdot body \cdot weight)^{-1} \cdot day^{-1}$] based on the median of the estimated consumption of fried potatoes treated with the coating solutions. (Adapted with permission from Al-Asmar [199]; published by MDPI, 2018).

Age Groups	Control	Chitosan Coating	Pectin Coating
Toddlers	1387	858	720
Other children	1521	941	790
Adolescents	1072	663	557
Adults	719	445	374
Elderly	536	332	279
Very elderly	417	258	217

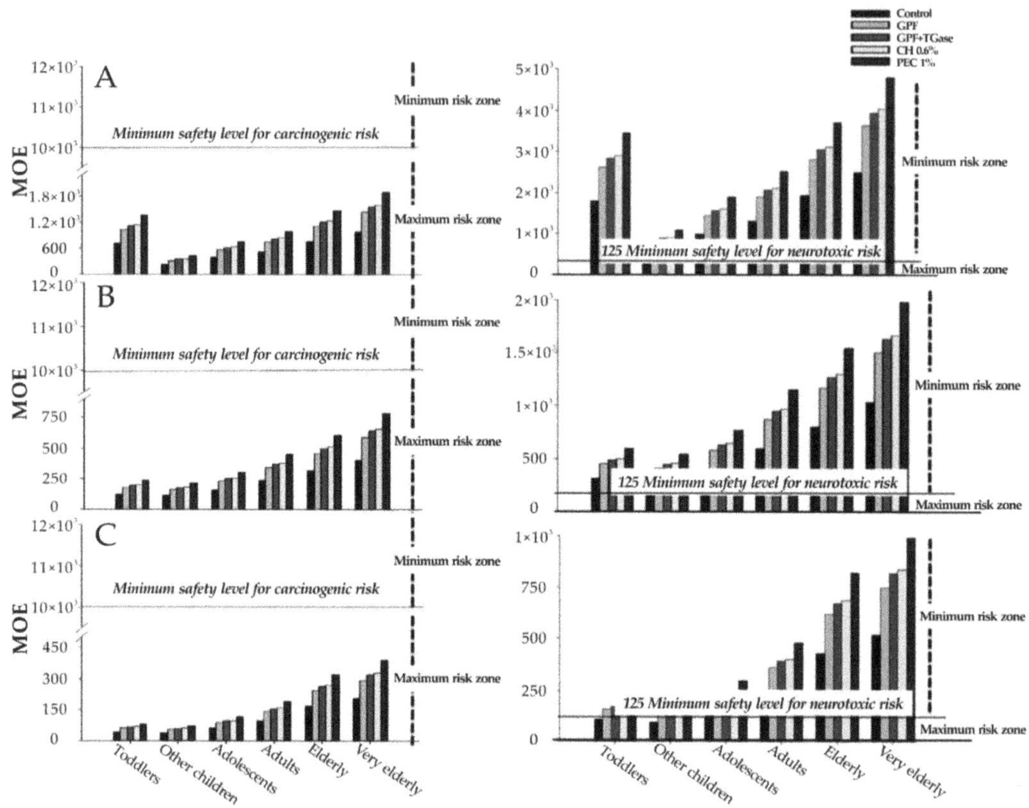

Figure 7. MOE values for carcinogenicity (**left panel**) and neurotoxic (**right panel**) of acrylamide through the consumption of French fries that were both uncoated and coated with hydrocolloid coating solutions. Samples were coated with different polysaccharides-based coatings made of PEC, pectin; and CH, chitosan. "Uncoated" represents the control sample dipped in distilled water, across different consumer age groups: (**A**) minimum, (**B**) median, and (**C**) maximum of consumption levels estimated from the 2015 EFSA report. (Adapted with permission from Al-Asmar [199]; published by MDPI, 2018).

Overall, a series of safety studies can be conducted on edible materials based on relevant regulations and standards (e.g., FDA for Preparation of Food Contact Notifications for Food Contact Substances-Toxicology Recommendations), such as composition analysis (including nutritional composition and possible natural toxic substances), hygienic tests (heavy metals, pesticide residues), and toxicological tests [including acute oral toxicity test, three genetic toxicity tests (Ames test, mammalian red blood cell micronucleus test and mouse spermatocyte chromosome aberration test), 90 d oral toxicity test and teratogenicity test], and further combined with the population, history of consumption, and the survey results of adverse reactions to assess the safety of polysaccharides-based edible packaging comprehensively.

6. Conclusions and Prospects

Edible packaging is a vital component of sustainable packaging. It significantly expands the source of packaging materials, reduces the dependence on non-renewable petroleum resources, and efficiently uses food processing waste. Polysaccharides are the major study objects of edible packaging materials. Considering the advantages and limitations of polysaccharides, researchers currently use various modifications to optimize the material's comprehensive properties, such as film-forming, mechanical and barrier properties, and antioxidant and antibacterial activities. They have successfully developed a variety of polysaccharide-based edible packaging materials such as ink, microcapsules, coatings, films, and sheets, which are applied to food packaging. These materials can provide selective barriers to prevent the migration of water, gas, and lipid in the food-packaging system, effectively retain the flavor and nutrition of food, and extend its shelf life (e.g., fruits, vegetables, meat, aquatic products, nuts, confectioneries, and delicatessens, etc.).

In general, polysaccharide-based edible packaging plays a key role in the environmental protection of food packaging and the high value of food processing waste, and it is one of the best alternative non-renewable resources. Although numerous studies on polysaccharide-based edible packaging have been reported in the past 10 years, they are still mainly on the laboratory scale and are less industrialized. Herein, trends of research and application of polysaccharide-based edible packaging will mainly focus on the following four aspects:

(1) Development of more edible polysaccharide-based materials: To date, the primary sources of polysaccharide-based edible packaging are plants and animals. Microorganisms, such as bacteria and fungi are also a great potential source, especially for the development of marine microorganisms.
(2) Multi-functional modification of polysaccharide-based materials: Modification of materials based on practical application requirements is a hot topic in material science. Defects in mechanical properties and water resistance are difficulties in the application of polysaccharide-based materials. Whereas, hydrophilicity is a key factor influencing these performances. Therefore, selecting the most reasonable modification technology (e.g., high stability, low cost, safety, convenience, and easy industrialization) for new polysaccharide-based materials to improve their properties would be an important topic for future research on polysaccharide-based edible packaging. Therefore, the design of the binding mode between polymer chains, the design of monomer molecular group structures, and the realization of more functional effects according to different food requirements (e.g., flavor regulation, acid and alkali resistance, amphiphobicity, and controlled release of functional factors) are the development trend of the future modification of polysaccharide-based materials.
(3) Application expansion and comprehensive evaluation of polysaccharide-based edible packaging: To conduct engineering research on cost reduction and large-scale production, and to evaluate the economic, environmental, and social benefits of polysaccharide-based edible packaging (the life cycle sustainability assessment theoretical model is recommended here), is the only way for the application and promotion of polysaccharides in packaging and food fields in the future.
(4) A deeper knowledge and practice of safety risk assessment for polysaccharides: Understand the potential exposure of polysaccharides through migration into food, the interaction of polysaccharides with food constituent, and their effects upon ingestion, which could verify polysaccharides-based edible packaging safety for commercial purposes and provide a reference for dietary reference intake of residents.

Author Contributions: Conceptualization and writing—original draft preparation, Y.Z.; investigation, B.L., C.L., Y.X. and Y.L.; writing—review and editing, Y.Z., B.L. and D.L.; supervision and funding acquisition, D.L. and C.H. All authors have read and agreed to the published version of the manuscript.

Funding: This research was funded by the Natural Science Foundation of China (Grant number 31960484) and the Natural Science Foundation of Guangxi Province (Grant number 2019JJD120012).

Institutional Review Board Statement: Not applicable.

Informed Consent Statement: Not applicable.

Data Availability Statement: Not applicable.

Conflicts of Interest: The authors declare no conflict of interest.

References

1. Jang, M.; Shim, W.J.; Cho, Y.; Han, G.M.; Song, Y.K.; Hong, S.H. A close relationship between microplastic contamination and coastal area use pattern. *Water Res.* **2020**, *171*, 115400. [CrossRef] [PubMed]
2. Eckert, E.M.; Di Cesare, A.; Kettner, M.T.; Arias-Andres, M.; Fontaneto, D.; Grossart, H.-P.; Corno, G. Microplastics increase impact of treated wastewater on freshwater microbial community. *Environ. Pollut.* **2018**, *234*, 495–502. [CrossRef] [PubMed]
3. Egger, M.; Nijhof, R.; Quiros, L.; Leone, G.; Royer, S.-J.; McWhirter, A.C.; Kantakov, G.A.; Radchenko, V.I.; Pakhomov, E.A.; Hunt, B.P.V.; et al. A spatially variable scarcity of floating microplastics in the eastern North Pacific Ocean. *Environ. Res. Lett.* **2020**, *15*, 114056. [CrossRef]
4. Lebreton, L.; Slat, B.; Ferrari, F.; Sainte-Rose, B.; Aitken, J.; Marthouse, R.; Hajbane, S.; Cunsolo, S.; Schwarz, A.; Levivier, A.; et al. Evidence that the Great Pacific Garbage Patch is rapidly accumulating plastic. *Sci. Rep.* **2018**, *8*, 4666. [CrossRef]
5. Barrett, J.; Chase, Z.; Zhang, J.; Holl, M.M.B.; Willis, K.; Williams, A.; Hardesty, B.D.; Wilcox, C. Microplastic Pollution in Deep-Sea Sediments From the Great Australian Bight. *Front. Mar. Sci.* **2020**, *7*, 808. [CrossRef]
6. Eo, S.; Hong, S.H.; Song, Y.K.; Han, G.M.; Seo, S.; Shim, W.J. Prevalence of small high-density microplastics in the continental shelf and deep sea waters of East Asia. *Water Res.* **2021**, *200*, 117238. [CrossRef]
7. Connan, M.; Perold, V.; Dilley, B.J.; Barbraud, C.; Cherel, Y.; Ryan, P.G. The Indian Ocean 'garbage patch': Empirical evidence from floating macro-litter. *Mar. Pollut. Bull.* **2021**, *169*, 112559. [CrossRef]
8. Auta, H.S.; Emenike, C.U.; Fauziah, S.H. Distribution and importance of microplastics in the marine environment: A review of the sources, fate, effects, and potential solutions. *Environ. Int.* **2017**, *102*, 165–176. [CrossRef]
9. Yan, Z.; Chen, Y.; Bao, X.; Zhang, X.; Ling, X.; Lu, G.; Liu, J.; Nie, Y. Microplastic pollution in an urbanized river affected by water diversion: Combining with active biomonitoring. *J. Hazard. Mater.* **2021**, *417*, 126058. [CrossRef]
10. Chen, G.; Feng, Q.; Wang, J. Mini-review of microplastics in the atmosphere and their risks to humans. *Sci. Total. Environ.* **2020**, *703*, 135504. [CrossRef]
11. Amato-Lourenço, L.F.; Carvalho-Oliveira, R.; Júnior, G.R.; dos Santos Galvão, L.; Ando, R.A.; Mauad, T. Presence of airborne microplastics in human lung tissue. *J. Hazard. Mater.* **2021**, *416*, 126124. [CrossRef]
12. Ragusa, A.; Svelato, A.; Santacroce, C.; Catalano, P.; Notarstefano, V.; Carnevali, O.; Papa, F.; Rongioletti, M.C.A.; Baiocco, F.; Draghi, S.; et al. Plasticenta: First evidence of microplastics in human placenta. *Environ. Int.* **2021**, *146*, 106274. [CrossRef] [PubMed]
13. Hassan, B.; Chatha, S.A.S.; Hussain, A.I.; Zia, K.M.; Akhtar, N. Recent advances on polysaccharides, lipids and protein based edible films and coatings: A review. *Int. J. Biol. Macromol.* **2018**, *109*, 1095–1107. [CrossRef] [PubMed]
14. Garcia, M.P.M.; GómezGuillén, M.C.; LópezCaballero, M.E.; BarbosaCánovas, G.V. *Edible Films and Coatings: Fundamentals and Applications*, 1st ed.; CRC Press: Boca Raton, FL, USA, 2017.
15. Mohamed, S.A.A.; El-Sakhawy, M.; El-Sakhawy, M.A.-M. Polysaccharides, Protein and Lipid -Based Natural Edible Films in Food Packaging: A Review. *Carbohydr. Polym.* **2020**, *238*, 116178. [CrossRef]
16. Nešić, A.; Cabrera-Barjas, G.; Dimitrijević-Branković, S.; Davidović, S.; Radovanović, N.; Delattre, C. Prospect of Polysaccharide-Based Materials as Advanced Food Packaging. *Molecules* **2020**, *25*, 135. [CrossRef]
17. López, O.V.; Zaritzky, N.E.; Grossmann, M.V.E.; García, M.A. Acetylated and native corn starch blend films produced by blown extrusion. *J. Food Eng.* **2013**, *116*, 286–297. [CrossRef]
18. Wang, P.; Li, Y.; Zhang, C.; Feng, F.; Zhang, H. Sequential electrospinning of multilayer ethylcellulose/gelatin/ethylcellulose nanofibrous film for sustained release of curcumin. *Food Chem.* **2020**, *308*, 125599. [CrossRef] [PubMed]
19. Dehghani, S.; ValiHosseini, S.M. Regenstein, J. Edible films and coatings in seafood preservation: A review. *Food Chem.* **2018**, *240*, 505–513. [CrossRef]
20. Benbettaïeb, N.; Gay, J.-P.; Karbowiak, T.; Debeaufort, F. Tuning the Functional Properties of Polysaccharide–Protein Bio-Based Edible Films by Chemical, Enzymatic, and Physical Cross-Linking. *Compr. Rev. Food Sci. Food Saf.* **2016**, *15*, 739–752. [CrossRef]
21. Embuscado, M.E.; Huber, K.C. *Edible Films and Coatings for Food Applications*, 1st ed.; Springer: New York, NY, USA, 2009. [CrossRef]
22. Nechita, P.; Roman, M. Review on Polysaccharides Used in Coatings for Food Packaging Papers. *Coatings* **2020**, *10*, 566. [CrossRef]
23. Rastogi, V.K.; Samyn, P. Bio-Based Coatings for Paper Applications. *Coatings* **2015**, *5*, 887–930. [CrossRef]
24. Lavoine, N.; Desloges, I.; Dufresne, A.; Bras, J. Microfibrillated cellulose—Its barrier properties and applications in cellulosic materials: A review. *Carbohydr. Polym.* **2012**, *90*, 735–764. [CrossRef] [PubMed]
25. Habibi, Y.; Lucia, L.A.; Rojas, O.J. Cellulose Nanocrystals: Chemistry, Self-Assembly, and Applications. *Chem. Rev.* **2010**, *110*, 3479–3500. [CrossRef]
26. Brown, R.M. The Biosynthesis of Cellulose. *J. Macromol. Sci. Part A* **1996**, *33*, 1345–1373. [CrossRef]
27. Tumbarski, Y.; Nikolova, R.; Petkova, N.; Ivanov, I.; Lante, A. Biopreservation of Fresh Strawberries by Carboxymethyl Cellulose Edible Coatings Enriched with a Bacteriocin from Bacillus methylotrophicus BM47. *Food Technol. Biotechnol.* **2019**, *57*, 230–237. [CrossRef]
28. Yu, H.-Y.; Zhang, H.; Song, M.-L.; Zhou, Y.; Yao, J.; Ni, Q.-Q. From Cellulose Nanospheres, Nanorods to Nanofibers: Various Aspect Ratio Induced Nucleation/Reinforcing Effects on Polylactic Acid for Robust-Barrier Food Packaging. *ACS Appl. Mater. Interfaces* **2017**, *9*, 43920–43938. [CrossRef] [PubMed]
29. Ebringerová, A. Structural Diversity and Application Potential of Hemicelluloses. *Macromol. Symp.* **2005**, *232*, 1–12. [CrossRef]

30. Scheller, H.V.; Ulvskov, P. Hemicelluloses. *Annu. Rev. Plant Biol.* **2010**, *61*, 263–289. [CrossRef] [PubMed]
31. Ebringerová, A.; Hromádková, Z.; Heinze, T. Hemicellulose. In *Polysaccharides I: Structure, Characterization and Use*; Heinze, T., Ed.; Springer: Berlin/Heidelberg, Germany, 2005. [CrossRef]
32. Mikkonen, K.S.; Tenkanen, M. Sustainable food-packaging materials based on future biorefinery products: Xylans and mannans. *Trends Food Sci. Technol.* **2012**, *28*, 90–102. [CrossRef]
33. Chen, X.; Pang, J.; Wu, C. Fabrication and characterization of antimicrobial food packaging materials composed of konjac glucomannan, chitosan and fulvic acid. *Food Sci.* **2021**, *42*, 232–239. [CrossRef]
34. Dai, L.; Zhang, J.; Cheng, F. Effects of starches from different botanical sources and modification methods on physicochemical properties of starch-based edible films. *Int. J. Biol. Macromol.* **2019**, *132*, 897–905. [CrossRef] [PubMed]
35. Zhu, F. Underutilized and unconventional starches: Why should we care? *Trends Food Sci. Technol.* **2020**, *100*, 363–373. [CrossRef]
36. Zhu, F.; Cui, R. Comparison of molecular structure of oca (Oxalis tuberosa), potato, and maize starches. *Food Chem.* **2019**, *296*, 116–122. [CrossRef]
37. Wang, Q.; Zheng, Y.; Zhuang, W.; Lu, X.; Luo, X.; Zheng, B. Genome-wide transcriptional changes in type 2 diabetic mice supplemented with lotus seed resistant starch. *Food Chem.* **2018**, *264*, 427–434. [CrossRef] [PubMed]
38. Nakthong, N.; Wongsagonsup, R.; Amornsakchai, T. Characteristics and potential utilizations of starch from pineapple stem waste. *Ind. Crop. Prod.* **2017**, *105*, 74–82. [CrossRef]
39. Zhang, Y.; Li, B.; Xu, F.; He, S.; Zhang, Y.; Sun, L.; Zhu, K.; Li, S.; Wu, G.; Tan, L. Jackfruit starch: Composition, structure, functional properties, modifications and applications. *Trends Food Sci. Technol.* **2021**, *107*, 268–283. [CrossRef]
40. Leloup, V.M.; Colonna, P.; Ring, S.G.; Roberts, K.; Wells, B. Microstructure of amylose gels. *Carbohydr. Polym.* **1992**, *18*, 189–197. [CrossRef]
41. Pérez, S.; Bertoft, E. The molecular structures of starch components and their contribution to the architecture of starch granules: A comprehensive review. *Starch-Stärke* **2010**, *62*, 389–420. [CrossRef]
42. Jiang, Z.; Neetoo, H.; Chen, H. Efficacy of freezing, frozen storage and edible antimicrobial coatings used in combination for control of Listeria monocytogenes on roasted turkey stored at chiller temperatures. *Food Microbiol.* **2011**, *28*, 1394–1401. [CrossRef] [PubMed]
43. Neetoo, H.; Ye, M.; Chen, H. Bioactive alginate coatings to control Listeria monocytogenes on cold-smoked salmon slices and fillets. *Int. J. Food Microbiol.* **2010**, *136*, 326–331. [CrossRef] [PubMed]
44. Zhu, F. Relationships between amylopectin internal molecular structure and physicochemical properties of starch. *Trends Food Sci. Technol.* **2018**, *78*, 234–242. [CrossRef]
45. Singh, S.; Singh, N.; Isono, N.; Noda, T. Relationship of Granule Size Distribution and Amylopectin Structure with Pasting, Thermal, and Retrogradation Properties in Wheat Starch. *J. Agric. Food Chem.* **2010**, *58*, 1180–1188. [CrossRef] [PubMed]
46. Kumar, N. Polysaccharide-based component and their relevance in edible film/coating: A review. *Nutr. Food Sci.* **2019**, *49*, 793–823. [CrossRef]
47. Le Corre, D.; Bras, J.; Dufresne, A. Starch Nanoparticles: A Review. *Biomacromolecules* **2010**, *11*, 1139–1153. [CrossRef] [PubMed]
48. Zhang, Y.; Li, B.; Zhang, Y.; Xu, F.; Zhu, K.; Li, S.; Tan, L.; Wu, G.; Dong, W. Effect of degree of polymerization of amylopectin on the gelatinization properties of jackfruit seed starch. *Food Chem.* **2019**, *289*, 152–159. [CrossRef] [PubMed]
49. Sandford, P. *Chitin and Chitosan: Commercial Uses and Potential Applications in Chitin and Chitosan: Sources, Chemistry, Biochemistry, Physical Properties and Applicaties*; Skjak-braek, G., Anthonsen, T., Sandford, P., Eds.; Elsevier Applied Science: New York, NY, USA, 1989; pp. 51–69.
50. Yen, M.-T.; Yang, J.-H.; Mau, J.-L. Physicochemical characterization of chitin and chitosan from crab shells. *Carbohydr. Polym.* **2009**, *75*, 15–21. [CrossRef]
51. Wang, H.; Qian, J.; Ding, F. Emerging Chitosan-Based Films for Food Packaging Applications. *J. Agric. Food Chem.* **2018**, *66*, 395–413. [CrossRef] [PubMed]
52. Szymańska, E.; Winnicka, K. Stability of Chitosan—A Challenge for Pharmaceutical and Biomedical Applications. *Mar. Drugs* **2015**, *13*, 1819–1846. [CrossRef] [PubMed]
53. Jayakumar, R.; Menon, D.; Manzoor, K.; Nair, S.V.; Tamura, H. Biomedical applications of chitin and chitosan based nanomaterials—A short review. *Carbohydr. Polym.* **2010**, *82*, 227–232. [CrossRef]
54. Homez-Jara, A.; Daza, L.D.; Aguirre, D.M.; Muñoz, J.A.; Solanilla, J.F.; Váquiro, H.A. Characterization of chitosan edible films obtained with various polymer concentrations and drying temperatures. *Int. J. Biol. Macromol.* **2018**, *113*, 1233–1240. [CrossRef] [PubMed]
55. O'Callaghan, K.A.M.; Kerry, J.P. Preparation of low- and medium-molecular weight chitosan nanoparticles and their antimicrobial evaluation against a panel of microorganisms, including cheese-derived cultures. *Food Control.* **2016**, *69*, 256–261. [CrossRef]
56. Divya, K.; Smitha, V.; Jisha, M.S. Antifungal, antioxidant and cytotoxic activities of chitosan nanoparticles and its use as an edible coating on vegetables. *Int. J. Biol. Macromol.* **2018**, *114*, 572–577. [CrossRef]
57. Dutta, P.K.; Tripathi, S.; Mehrotra, G.K.; Dutta, J. Perspectives for chitosan based antimicrobial films in food applications. *Food Chem.* **2009**, *114*, 1173–1182. [CrossRef]
58. Sahariah, P.; Másson, M. Antimicrobial Chitosan and Chitosan Derivatives: A Review of the Structure–Activity Relationship. *Biomacromolecules* **2017**, *18*, 3846–3868. [CrossRef] [PubMed]

59. Pérez-Jiménez, J.; Saura-Calixto, F. Fruit peels as sources of non-extractable polyphenols or macromolecular antioxidants: Analysis and nutritional implications. *Food Res. Int.* **2018**, *111*, 148–152. [CrossRef]
60. Abboud, K.Y.; Iacomini, M.; Simas, F.F.; Cordeiro, L.M.C. High methoxyl pectin from the soluble dietary fiber of passion fruit peel forms weak gel without the requirement of sugar addition. *Carbohydr. Polym.* **2020**, *246*, 116616. [CrossRef] [PubMed]
61. Rodsamran, P.; Sothornvit, R. Microwave heating extraction of pectin from lime peel: Characterization and properties compared with the conventional heating method. *Food Chem.* **2019**, *278*, 364–372. [CrossRef] [PubMed]
62. Nguyen, B.M.N.; Pirak, T. Physicochemical properties and antioxidant activities of white dragon fruit peel pectin extracted with conventional and ultrasound-assisted extraction. *Cogent Food Agric.* **2019**, *5*, 1633076. [CrossRef]
63. Gharibzahedi, S.M.T.; Smith, B.; Guo, Y. Pectin extraction from common fig skin by different methods: The physicochemical, rheological, functional, and structural evaluations. *Int. J. Biol. Macromol.* **2019**, *136*, 275–283. [CrossRef] [PubMed]
64. Wu, H.; Lei, Y.; Zhu, R.; Zhao, M.; Lu, J.; Xiao, D.; Jiao, C.; Zhang, Z.; Shen, G.; Li, S. Preparation and characterization of bioactive edible packaging films based on pomelo peel flours incorporating tea polyphenol. *Food Hydrocoll.* **2019**, *90*, 41–49. [CrossRef]
65. Presentato, A.; Scurria, A.; Albanese, L.; Lino, C.; Sciortino, M.; Pagliaro, M.; Zabini, F.; Meneguzzo, F.; Alduina, R.; Nuzzo, D.; et al. Superior Antibacterial Activity of Integral Lemon Pectin Extracted via Hydrodynamic Cavitation. *ChemistryOpen* **2020**, *9*, 628–630. [CrossRef] [PubMed]
66. Chen, X.; Qi, Y.; Zhu, C.; Wang, Q. Effect of ultrasound on the properties and antioxidant activity of hawthorn pectin. *Int. J. Biol. Macromol.* **2019**, *131*, 273–281. [CrossRef] [PubMed]
67. Muñoz-Almagro, N.; Prodanov, M.; Wilde, P.J.; Villamiel, M.; Montilla, A. Obtainment and characterisation of pectin from sunflower heads purified by membrane separation techniques. *Food Chem.* **2020**, *318*, 126476. [CrossRef] [PubMed]
68. Chen, J. Extraction and Physicochemical Characterization of Pectins from some Special Resources and the Mechanism of Pectin Degradation Induced by Dynamic High Pressure Microfluidization. Ph.D. Thesis, Nanchang University, Nanchang, China, 2013.
69. Mostafavi, F.S.; Zaeim, D. Agar-based edible films for food packaging applications—A review. *Int. J. Biol. Macromol.* **2020**, *159*, 1165–1176. [CrossRef]
70. Tavassoli-Kafrani, E.; Shekarchizadeh, H.; Masoudpour-Behabadi, M. Development of edible films and coatings from alginates and carrageenans. *Carbohydr. Polym.* **2016**, *137*, 360–374. [CrossRef]
71. Nieto, M.B. Structure and Function of Polysaccharide Gum-Based Edible Films and Coatings. In *Edible Films and Coatings for Food Applications*; Huber, K., Embuscado, M., Eds.; Springer: New York, NY, USA, 2009; pp. 57–112.
72. Ciriminna, R.; Fidalgo, A.; Meneguzzo, F.; Presentato, A.; Scurria, A.; Nuzzo, D.; Alduina, R.; Ilharco, L.M.; Pagliaro, M. Pectin: A Long-Neglected Broad-Spectrum Antibacterial. *ChemMedChem* **2020**, *15*, 2228–2235. [CrossRef] [PubMed]
73. Martinov, J.; Krstić, M.; Spasić, S.; Miletić, S.; Stefanović-Kojić, J.; Nikolić-Kokić, A.; Blagojević, D.; Spasojević, I.; Spasić, M.B. Apple pectin-derived oligosaccharides produce carbon dioxide radical anion in Fenton reaction and prevent growth of Escherichia coli and Staphylococcus aureus. *Food Res. Int.* **2017**, *100*, 132–136. [CrossRef]
74. Senturk Parreidt, T.; Müller, K.; Schmid, M. Alginate-Based Edible Films and Coatings for Food Packaging Applications. *Foods* **2018**, *7*, 170. [CrossRef]
75. Kloareg, B.; Quatrano, R.S. Structure of the cell walls of marine algae and ecophysiological functions of the matrix polysaccharides. *Oceanogr. Mar. Biol. Annu. Rev.* **1988**, *26*, 259–315.
76. Abraham, R.E.; Su, P.; Puri, M.; Raston, C.L.; Zhang, W. Optimisation of biorefinery production of alginate, fucoidan and laminarin from brown seaweed Durvillaea potatorum. *Algal Res.* **2019**, *38*, 101389. [CrossRef]
77. Comaposada, J.; Marcos, B.; Bou, R.; Gou, P. Influence of surfactants and proteins on the properties of wet edible calcium alginate meat coatings. *Food Res. Int.* **2018**, *108*, 539–550. [CrossRef]
78. Costa, M.J.; Marques, A.M.; Pastrana, L.M.; Teixeira, J.A.; Sillankorva, S.M.; Cerqueira, M.A. Physicochemical properties of alginate-based films: Effect of ionic crosslinking and mannuronic and guluronic acid ratio. *Food Hydrocoll.* **2018**, *81*, 442–448. [CrossRef]
79. Delattre, C.; Fenoradosoa, T.A.; Michaud, P. Galactans: An overview of their most important sourcing and applications as natural polysaccharides. *Braz. Arch. Biol. Technol.* **2011**, *54*, 1075–1092. [CrossRef]
80. Armisén, R.; Gaiatas, F. 4-Agar. In *Handbook of Hydrocolloids*, 2nd ed.; Phillips, G.O., Williams, P.A., Eds.; Woodhead Publishing: Sawston, UK, 2009. [CrossRef]
81. Janjarasskul, T.; Krochta, J.M. Edible Packaging Materials. *Annu. Rev. Food Sci. Technol.* **2010**, *1*, 415–448. [CrossRef] [PubMed]
82. Gordobil, O.; Egüés, I.; Urruzola, I.; Labidi, J. Xylan–cellulose films: Improvement of hydrophobicity, thermal and mechanical properties. *Carbohydr. Polym.* **2014**, *112*, 56–62. [CrossRef] [PubMed]
83. Otoni, C.G.; Lorevice, M.V.; Moura, M.R.d.; Mattoso, L.H.C. On the effects of hydroxyl substitution degree and molecular weight on mechanical and water barrier properties of hydroxypropyl methylcellulose films. *Carbohydr. Polym.* **2018**, *185*, 105–111. [CrossRef]
84. Fonseca-Florido, H.A.; Soriano-Corral, F.; Yañez-Macías, R.; González-Morones, P.; Hernández-Rodríguez, F.; Aguirre-Zurita, J.; Ávila-Orta, C.; Rodríguez-Velázquez, J. Effects of multiphase transitions and reactive extrusion on in situ thermoplasticization/succination of cassava starch. *Carbohydr. Polym.* **2019**, *225*, 115250. [CrossRef]
85. Li, S.; Xiong, Q.; Lai, X.; Li, X.; Wan, M.; Zhang, J.; Yan, Y.; Cao, M.; Lu, L.; Guan, J.; et al. Molecular Modification of Polysaccharides and Resulting Bioactivities. *Compr. Rev. Food Sci. Food Saf.* **2016**, *15*, 237–250. [CrossRef]
86. Cumpstey, I. Chemical Modification of Polysaccharides. *ISRN Org. Chem.* **2013**, *2013*, 417672. [CrossRef]

87. Mikkonen, K.S.; Laine, C.; Kontro, I.; Talja, R.A.; Serimaa, R.; Tenkanen, M. Combination of internal and external plasticization of hydroxypropylated birch xylan tailors the properties of sustainable barrier films. *Eur. Polym. J.* **2015**, *66*, 307–318. [CrossRef]
88. Garavand, F.; Rouhi, M.; Razavi, S.H.; Cacciotti, I.; Mohammadi, R. Improving the integrity of natural biopolymer films used in food packaging by crosslinking approach: A review. *Int. J. Biol. Macromol.* **2017**, *104*, 687–707. [CrossRef]
89. Tan, W.; Zhang, J.; Zhao, X.; Li, Q.; Dong, F.; Guo, Z. Preparation and physicochemical properties of antioxidant chitosan ascorbate/methylcellulose composite films. *Int. J. Biol. Macromol.* **2020**, *146*, 53–61. [CrossRef] [PubMed]
90. Yu, S.-H.; Tsai, M.-L.; Lin, B.-X.; Lin, C.-W.; Mi, F.-L. Tea catechins-cross-linked methylcellulose active films for inhibition of light irradiation and lipid peroxidation induced β-carotene degradation. *Food Hydrocoll.* **2015**, *44*, 491–505. [CrossRef]
91. da Silva Filipini, G.; Romani, V.P.; Guimarães Martins, V. Blending collagen, methylcellulose, and whey protein in films as a greener alternative for food packaging: Physicochemical and biodegradable properties. *Packag. Technol. Sci.* **2021**, *34*, 91–103. [CrossRef]
92. Zheng, H.; Zhang, G.; Zhuang, C.; Yang, T.; Zheng, Y.; Zheng, C.; Zhong, Y. Effects of Different Substrates on the Properties of Edible Films for Oily Foods. *Food Res. Dev.* **2018**, *39*, 18–23. [CrossRef]
93. Mohammadi, M.; Azizi, M.H.; Zoghi, A. Antimicrobial activity of carboxymethyl cellulose-gelatin film containing Dianthus barbatus essential oil against aflatoxin-producing molds. *Food Sci. Nutr.* **2020**, *8*, 1244–1253. [CrossRef] [PubMed]
94. Martelli, S.M.; Motta, C.; Caon, T.; Alberton, J.; Bellettini, I.C.; do Prado, A.C.P.; Barreto, P.L.M.; Soldi, V. Edible carboxymethyl cellulose films containing natural antioxidant and surfactants: α-tocopherol stability, in vitro release and film properties. *LWT* **2017**, *77*, 21–29. [CrossRef]
95. Ballesteros, L.F.; Cerqueira, M.A.; Teixeira, J.A.; Mussatto, S.I. Production and physicochemical properties of carboxymethyl cellulose films enriched with spent coffee grounds polysaccharides. *Int. J. Biol. Macromol.* **2018**, *106*, 647–655. [CrossRef]
96. Francisco, C.B.; Pellá, M.G.; Silva, O.A.; Raimundo, K.F.; Caetano, J.; Linde, G.A.; Colauto, N.B.; Dragunski, D.C. Shelf-life of guavas coated with biodegradable starch and cellulose-based films. *Int. J. Biol. Macromol.* **2020**, *152*, 272–279. [CrossRef]
97. Zhai, X.; Qin, Y.; Lu, H.; Dai, Y.; Zhang, H.; Wang, W.; Hou, H.; Chen, N. Preparation and Properties of Edible Films of High-Amylose Corn Starch/HPMC (Hydroxypropyl Methylcellulose). *J. Chin. Cereals Oils Assoc.* **2019**, *34*, 33–38.
98. Rhimi, W.; Boulila, A.; Gheribi, R.; Khwaldia, K. Development, characterization and application of hydroxypropylmethylcellulose films enriched with cypress seed extract. *Rsc Adv.* **2018**, *8*, 23615–23622. [CrossRef]
99. Dou, Y.; Zhang, L.; Zhang, B.; He, M.; Shi, W.; Yang, S.; Cui, Y.; Yin, G. Preparation and Characterization of Edible Dialdehyde Carboxymethyl Cellulose Crosslinked Feather Keratin Films for Food Packaging. *Polymers* **2020**, *12*, 158. [CrossRef]
100. Salmieri, S.; Khan, R.A.; Safrany, A.; Lacroix, M. Gamma rays-induced 2-hydroxyethyl methacrylate graft copolymerization on methylcellulose-based films: Structure analysis and physicochemical properties. *Ind. Crop. Prod.* **2015**, *70*, 64–71. [CrossRef]
101. Ramos, A.; Sousa, S.; Evtuguin, D.V.; Gamelas, J.A.F. Functionalized xylans in the production of xylan-coated paper laminates. *React. Funct. Polym.* **2017**, *117*, 89–96. [CrossRef]
102. Pereira, P.H.F.; Waldron, K.W.; Wilson, D.R.; Cunha, A.P.; Brito, E.S.d.; Rodrigues, T.H.S.; Rosa, M.F.; Azeredo, H.M.C. Wheat straw hemicelluloses added with cellulose nanocrystals and citric acid. Effect on film physical properties. *Carbohydr. Polym.* **2017**, *164*, 317–324. [CrossRef]
103. Milotskyi, R.; Bliard, C.; Tusseau, D.; Benoit, C. Starch carboxymethylation by reactive extrusion: Reaction kinetics and structure analysis. *Carbohydr. Polym.* **2018**, *194*, 193–199. [CrossRef] [PubMed]
104. Woggum, T.; Sirivongpaisal, P.; Wittaya, T. Characteristics and properties of hydroxypropylated rice starch based biodegradable films. *Food Hydrocoll.* **2015**, *50*, 54–64. [CrossRef]
105. Ojogbo, E.; Blanchard, R.; Mekonnen, T. Hydrophobic and Melt Processable Starch-Laurate Esters: Synthesis, Structure–Property Correlations. *J. Polym. Sci. Part A: Polym. Chem.* **2018**, *56*, 2611–2622. [CrossRef]
106. Blohm, S.; Heinze, T. Synthesis and properties of thermoplastic starch laurates. *Carbohydr. Res.* **2019**, *486*, 107833. [CrossRef]
107. Wilpiszewska, K.; Antosik, A.K.; Zdanowicz, M. The Effect of Citric Acid on Physicochemical Properties of Hydrophilic Carboxymethyl Starch-Based Films. *J. Polym. Environ.* **2019**, *27*, 1379–1387. [CrossRef]
108. Yıldırım-Yalçın, M.; Şeker, M.; Sadıkoğlu, H. Development and characterization of edible films based on modified corn starch and grape juice. *Food Chem.* **2019**, *292*, 6–13. [CrossRef] [PubMed]
109. Tan, W.; Li, Q.; Wang, H.; Liu, Y.; Zhang, J.; Dong, F.; Guo, Z. Synthesis, characterization, and antibacterial property of novel starch derivatives with 1,2,3-triazole. *Carbohydr. Polym.* **2016**, *142*, 1–7. [CrossRef] [PubMed]
110. Suriyatem, R.; Auras, R.A.; Rachtanapun, P. Improvement of mechanical properties and thermal stability of biodegradable rice starch–based films blended with carboxymethyl chitosan. *Ind. Crop. Prod.* **2018**, *122*, 37–48. [CrossRef]
111. Tan, W.; Dong, F.; Zhang, J.; Zhao, X.; Li, Q.; Guo, Z. Physical and Antioxidant Properties of Edible Chitosan Ascorbate Films. *J. Agric. Food Chem.* **2019**, *67*, 2530–2539. [CrossRef] [PubMed]
112. Cao, M.; Liu, X.; Luan, J.; Zhang, X. Characterization of physicochemical properties of carboxymethyl agar. *Carbohydr. Polym.* **2014**, *111*, 449–455. [CrossRef]
113. Šešlija, S.; Nešić, A.; Škorić, M.L.; Krušić, M.K.; Santagata, G.; Malinconico, M. Pectin/Carboxymethylcellulose Films as a Potential Food Packaging Material. *Macromol. Symp.* **2018**, *378*, 1600163. [CrossRef]
114. Ling, X.; Yang, C.; Ning, J. Research progress in modification of cellulose and application. *J. Text. Sci. Eng.* **2020**, *37*, 60–85.
115. Dias, M.V.; de Medeiros, H.S.; Soares, N.d.F.F.; Melo, N.R.d.; Borges, S.V.; Carneiro, J.d.D.S.; Pereira, J.M.T.d.A.K. Development of low-density polyethylene films with lemon aroma. *LWT Food Sci. Technol.* **2013**, *50*, 167–171. [CrossRef]

116. Bastarrachea, L.; Dhawan, S.; Sablani, S.S. Engineering Properties of Polymeric-Based Antimicrobial Films for Food Packaging: A Review. *Food Eng. Rev.* **2011**, *3*, 79–93. [CrossRef]
117. Shah, U.; Naqash, F.; Gani, A.; Masoodi, F.A. Art and Science behind Modified Starch Edible Films and Coatings: A Review. *Compr. Rev. Food Sci. Food Saf.* **2016**, *15*, 568–580. [CrossRef] [PubMed]
118. Liu, C.; Qin, S.; Xie, J.; Lin, X.; Zheng, Y.; Yang, J.; Kan, H.; Shi, Z. Using Carboxymethyl Cellulose as the Additive With Enzyme-Catalyzed Carboxylated Starch to Prepare the Film With Enhanced Mechanical and Hydrophobic Properties. *Front. Bioeng. Biotechnol.* **2021**, *9*, 638546. [CrossRef]
119. Gurgel de Carvalho, L.G.; do Nascimento Marques, N.; da Silva Fernandes, R.; Villetti, M.A.; Men de Sá Moreira de Souza, F.; de Carvalho Balaban, R. Effect of Starch Laurate Addition on the Properties of Mango Kernel Starch Films. Materials Research. *Mater. Res.* **2021**, *24*, e20200331. [CrossRef]
120. Li, J.; Ye, F.; Liu, J.; Zhao, G. Effects of octenylsuccination on physical, mechanical and moisture-proof properties of stretchable sweet potato starch film. *Food Hydrocoll.* **2015**, *46*, 226–232. [CrossRef]
121. Guo, L.-Y.; Lu, H.-Q.; Rackemann, D.; Shi, C.; Li, W.; Li, K.; Doherty, W.O.S. Quaternary ammonium-functionalized magnetic chitosan microspheres as an effective green adsorbent to remove high-molecular-weight invert sugar alkaline degradation products (HISADPs). *Chem. Eng. J.* **2021**, *416*, 129084. [CrossRef]
122. Anitha, A.; Divya Rani, V.V.; Krishna, R.; Sreeja, V.; Selvamurugan, N.; Nair, S.V.; Tamura, H.; Jayakumar, R. Synthesis, characterization, cytotoxicity and antibacterial studies of chitosan, O-carboxymethyl and N,O-carboxymethyl chitosan nanoparticles. *Carbohydr. Polym.* **2009**, *78*, 672–677. [CrossRef]
123. Tang, R.; Zhang, Y.; Zhang, Y.; Yu, Z. Synthesis and characterization of chitosan based dye containing quaternary ammonium group. *Carbohydr. Polym.* **2016**, *139*, 191–196. [CrossRef] [PubMed]
124. Zhu, X.; Wu, H.; Yang, J.; Tong, J.; Yi, J.; Hu, Z.; Hu, J.; Wang, T.; Fan, L. Antibacterial activity of chitosan grafting nisin: Preparation and characterization. *React. Funct. Polym.* **2015**, *91–92*, 71–76. [CrossRef]
125. Li, Q.; Mi, Y.; Tan, W.; Guo, Z. Highly efficient free radical-scavenging property of phenolic-functionalized chitosan derivatives: Chemical modification and activity assessment. *Int. J. Biol. Macromol.* **2020**, *164*, 4279–4288. [CrossRef]
126. Tan, W.; Zhang, J.; Zhao, X.; Dong, F.; Li, Q.; Guo, Z. Synthesis and antioxidant action of chitosan derivatives with amino-containing groups via azide-alkyne click reaction and N-methylation. *Carbohydr. Polym.* **2018**, *199*, 583–592. [CrossRef] [PubMed]
127. Wu, C.; Li, Y.; Du, Y.; Wang, L.; Tong, C.; Hu, Y.; Pang, J.; Yan, Z. Preparation and characterization of konjac glucomannan-based bionanocomposite film for active food packaging. *Food Hydrocoll.* **2019**, *89*, 682–690. [CrossRef]
128. Chen, J.; Liu, W.; Liu, C.-M.; Li, T.; Liang, R.-H.; Luo, S.-J. Pectin Modifications: A Review. *Crit. Rev. Food Sci. Nutr.* **2015**, *55*, 1684–1698. [CrossRef]
129. Turgeon, S.L.; Beaulieu, M.; Schmitt, C.; Sanchez, C. Protein–polysaccharide interactions: Phase-ordering kinetics, thermodynamic and structural aspects. *Curr. Opin. Colloid Interface Sci.* **2003**, *8*, 401–414. [CrossRef]
130. Ye, A. Complexation between milk proteins and polysaccharides via electrostatic interaction: Principles and applications—A review. *Int. J. Food Sci. Technol.* **2008**, *43*, 406–415. [CrossRef]
131. Wu, Y.; Weller, C.L.; Hamouz, F.; Cuppett, S.L.; Schnepf, M. Development and application of multicomponent edible coatings and films: A review. In *Advances in Food and Nutrition Research*; Academic Press: Cambridge, MA, USA, 2002; Volume 44, pp. 347–394.
132. Alizadeh Sani, M.; Ehsani, A.; Hashemi, M. Whey protein isolate/cellulose nanofibre/TiO2 nanoparticle/rosemary essential oil nanocomposite film: Its effect on microbial and sensory quality of lamb meat and growth of common foodborne pathogenic bacteria during refrigeration. *Int. J. Food Microbiol.* **2017**, *251*, 8–14. [CrossRef]
133. Silva-Weiss, A.; Quilaqueo, M.; Venegas, O.; Ahumada, M.; Silva, W.; Osorio, F.; Giménez, B. Design of dipalmitoyl lecithin liposomes loaded with quercetin and rutin and their release kinetics from carboxymethyl cellulose edible films. *J. Food Eng.* **2018**, *224*, 165–173. [CrossRef]
134. Mugwagwa, L.R.; Chimphango, A.F.A. Enhancing the functional properties of acetylated hemicellulose films for active food packaging using acetylated nanocellulose reinforcement and polycaprolactone coating. *Food Packag. Shelf Life* **2020**, *24*, 100481. [CrossRef]
135. Wang, L.; Lin, L.; Chen, X.; Tong, C.; Pang, J. Synthesis and characteristics of konjac glucomannan films incorporated with functionalized microcrystalline cellulose. *Colloids Surf. A: Physicochem. Eng. Asp.* **2019**, *563*, 237–245. [CrossRef]
136. Wang, L.; Mu, R.-J.; Li, Y.; Lin, L.; Lin, Z.; Pang, J. Characterization and antibacterial activity evaluation of curcumin loaded konjac glucomannan and zein nanofibril films. *LWT* **2019**, *113*, 108293. [CrossRef]
137. Lei, Y.; Wu, H.; Jiao, C.; Jiang, Y.; Liu, R.; Xiao, D.; Lu, J.; Zhang, Z.; Shen, G.; Li, S. Investigation of the structural and physical properties, antioxidant and antimicrobial activity of pectin-konjac glucomannan composite edible films incorporated with tea polyphenol. *Food Hydrocoll.* **2019**, *94*, 128–135. [CrossRef]
138. Du, Y.; Wang, L.; Mu, R.; Wang, Y.; Li, Y.; Wu, D.; Wu, C.; Pang, J. Fabrication of novel Konjac glucomannan/shellac film with advanced functions for food packaging. *Int. J. Biol. Macromol.* **2019**, *131*, 36–42. [CrossRef]
139. Wilpiszewska, K.; Antosik, A.K.; Schmidt, B.; Janik, J.; Rokicka, J. Hydrophilic Films Based on Carboxymethylated Derivatives of Starch and Cellulose. *Polymers* **2020**, *12*, 2447. [CrossRef] [PubMed]
140. Lopez, O.; Garcia, M.A.; Villar, M.A.; Gentili, A.; Rodriguez, M.S.; Albertengo, L. Thermo-compression of biodegradable thermoplastic corn starch films containing chitin and chitosan. *LWT Food Sci. Technol.* **2014**, *57*, 106–115. [CrossRef]

141. Ali, A.; Chen, Y.; Liu, H.; Yu, L.; Baloch, Z.; Khalid, S.; Zhu, J.; Chen, L. Starch-based antimicrobial films functionalized by pomegranate peel. *Int. J. Biol. Macromol.* **2019**, *129*, 1120–1126. [CrossRef] [PubMed]
142. Zimet, P.; Mombrú, Á.W.; Mombrú, D.; Castro, A.; Villanueva, J.P.; Pardo, H.; Rufo, C. Physico-chemical and antilisterial properties of nisin-incorporated chitosan/carboxymethyl chitosan films. *Carbohydr. Polym.* **2019**, *219*, 334–343. [CrossRef]
143. Badawy, M.E.I.; Rabea, E.I.; El-Nouby, M.; Ismail, R.I.A.; Taktak, N.E.M. Strawberry Shelf Life, Composition, and Enzymes Activity in Response to Edible Chitosan Coatings. *Int. J. Fruit Sci.* **2017**, *17*, 117–136. [CrossRef]
144. Siripatrawan, U.; Vitchayakitti, W. Improving functional properties of chitosan films as active food packaging by incorporating with propolis. *Food Hydrocoll.* **2016**, *61*, 695–702. [CrossRef]
145. Oun, A.A.; Rhim, J.-W. Effect of post-treatments and concentration of cotton linter cellulose nanocrystals on the properties of agar-based nanocomposite films. *Carbohydr. Polym.* **2015**, *134*, 20–29. [CrossRef] [PubMed]
146. Cao, Q.; Zhang, Y.; Chen, W.; Meng, X.; Liu, B. Hydrophobicity and physicochemical properties of agarose film as affected by chitosan addition. *Int. J. Biol. Macromol.* **2018**, *106*, 1307–1313. [CrossRef]
147. Rai, S.K.; Chaturvedi, K.; Yadav, S.K. Evaluation of structural integrity and functionality of commercial pectin based edible films incorporated with corn flour, beetroot, orange peel, muesli and rice flour. *Food Hydrocoll.* **2019**, *91*, 127–135. [CrossRef]
148. Baron, R.D.; Pérez, L.L.; Salcedo, J.M.; Córdoba, L.P.; Sobral, P.J.d.A. Production and characterization of films based on blends of chitosan from blue crab (Callinectes sapidus) waste and pectin from Orange (Citrus sinensis Osbeck) peel. *Int. J. Biol. Macromol.* **2017**, *98*, 676–683. [CrossRef]
149. Mostafavi, F.S.; Kadkhodaee, R.; Emadzadeh, B.; Koocheki, A. Preparation and characterization of tragacanth–locust bean gum edible blend films. *Carbohydr. Polym.* **2016**, *139*, 20–27. [CrossRef]
150. Eghbal, N.; Yarmand, M.S.; Mousavi, M.; Degraeve, P.; Oulahal, N.; Gharsallaoui, A. Complex coacervation for the development of composite edible films based on LM pectin and sodium caseinate. *Carbohydr. Polym.* **2016**, *151*, 947–956. [CrossRef]
151. Eghbal, N.; Degraeve, P.; Oulahal, N.; Yarmand, M.S.; Mousavi, M.E.; Gharsallaoui, A. Low methoxyl pectin/sodium caseinate interactions and composite film formation at neutral pH. *Food Hydrocoll.* **2017**, *69*, 132–140. [CrossRef]
152. Hashemi, S.M.B.; Mousavi Khaneghah, A.; Ghaderi Ghahfarrokhi, M.; Eş, I. Basil-seed gum containing Origanum vulgare subsp. viride essential oil as edible coating for fresh cut apricots. *Postharvest Biol. Technol.* **2017**, *125*, 26–34. [CrossRef]
153. da Rocha, M.; Alemán, A.; Romani, V.P.; López-Caballero, M.E.; Gómez-Guillén, M.C.; Montero, P.; Prentice, C. Effects of agar films incorporated with fish protein hydrolysate or clove essential oil on flounder (*Paralichthys orbignyanus*) fillets shelf-life. *Food Hydrocoll.* **2018**, *81*, 351–363. [CrossRef]
154. Rincón, E.; Serrano, L.; Balu, A.M.; Aguilar, J.J.; Luque, R.; García, A. Effect of Bay Leaves Essential Oil Concentration on the Properties of Biodegradable Carboxymethyl Cellulose-Based Edible Films. *Materials* **2019**, *12*, 2356. [CrossRef] [PubMed]
155. Moreno, O.; Atarés, L.; Chiralt, A.; Cruz-Romero, M.C.; Kerry, J. Starch-gelatin antimicrobial packaging materials to extend the shelf life of chicken breast fillets. *LWT* **2018**, *97*, 483–490. [CrossRef]
156. Panrong, T.; Karbowiak, T.; Harnkarnsujarit, N. Thermoplastic starch and green tea blends with LLDPE films for active packaging of meat and oil-based products. *Food Packag. Shelf Life* **2019**, *21*, 100331. [CrossRef]
157. Rambabu, K.; Bharath, G.; Banat, F.; Show, P.L.; Cocoletzi, H.H. Mango leaf extract incorporated chitosan antioxidant film for active food packaging. *Int. J. Biol. Macromol.* **2019**, *126*, 1234–1243. [CrossRef]
158. Souza, V.G.L.; Pires, J.R.A.; Rodrigues, C.; Coelhoso, I.M.; Fernando, A.L. Chitosan Composites in Packaging Industry—Current Trends and Future Challenges. *Polymers* **2020**, *12*, 417. [CrossRef] [PubMed]
159. Shankar, S.; Rhim, J.-W. Preparation of nanocellulose from micro-crystalline cellulose: The effect on the performance and properties of agar-based composite films. *Carbohydr. Polym.* **2016**, *135*, 18–26. [CrossRef]
160. Phan The, D.; Debeaufort, F.; Voilley, A.; Luu, D. Biopolymer interactions affect the functional properties of edible films based on agar, cassava starch and arabinoxylan blends. *J. Food Eng.* **2009**, *90*, 548–558. [CrossRef]
161. Fekete, E.; Bella, É.; Csiszár, E.; Móczó, J. Improving physical properties and retrogradation of thermoplastic starch by incorporating agar. *Int. J. Biol. Macromol.* **2019**, *136*, 1026–1033. [CrossRef] [PubMed]
162. Liu, K.; Lin, X.; Chen, L.; Huang, L.; Cao, S. Dual-functional chitosan–methylisothiazolinone/microfibrillated cellulose biocomposites for enhancing antibacterial and mechanical properties of agar films. *Cellulose* **2014**, *21*, 519–528. [CrossRef]
163. Sousa, A.M.M.; Gonçalves, M.P. Strategies to improve the mechanical strength and water resistance of agar films for food packaging applications. *Carbohydr. Polym.* **2015**, *132*, 196–204. [CrossRef]
164. Luangtana-anan, M.; Soradech, S.; Saengsod, S.; Nunthanid, J.; Limmatvapirat, S. Enhancement of Moisture Protective Properties and Stability of Pectin through Formation of a Composite Film: Effects of Shellac and Plasticizer. *J. Food Sci.* **2017**, *82*, 2915–2925. [CrossRef]
165. Pérez-Gago, M.B.; Rhim, J.-W. Chapter 13—Edible Coating and Film Materials: Lipid Bilayers and Lipid Emulsions. In *Innovations in Food Packaging, 2nd ed*; Han, J.H., Ed.; Academic Press: San Diego, CA, USA, 2014. [CrossRef]
166. Galus, S.; Kadzińska, J. Food applications of emulsion-based edible films and coatings. *Trends Food Sci. Technol.* **2015**, *45*, 273–283. [CrossRef]
167. Zhang, R.; Wang, W.; Zhang, H.; Dai, Y.; Dong, H.; Hou, H. Effects of hydrophobic agents on the physicochemical properties of edible agar/maltodextrin films. *Food Hydrocoll.* **2019**, *88*, 283–290. [CrossRef]

168. Nisar, T.; Wang, Z.-C.; Yang, X.; Tian, Y.; Iqbal, M.; Guo, Y. Characterization of citrus pectin films integrated with clove bud essential oil: Physical, thermal, barrier, antioxidant and antibacterial properties. *Int. J. Biol. Macromol.* **2018**, *106*, 670–680. [CrossRef]
169. López de Lacey, A.M.; López-Caballero, M.E.; Montero, P. Agar films containing green tea extract and probiotic bacteria for extending fish shelf-life. *LWT Food Sci. Technol.* **2014**, *55*, 559–564. [CrossRef]
170. Yan, Y.; Duan, S.; Zhang, H.; Liu, Y.; Li, C.; Hu, B.; Liu, A.; Wu, D.; He, J.; Wu, W. Preparation and characterization of Konjac glucomannan and pullulan composite films for strawberry preservation. *Carbohydr. Polym.* **2020**, *243*, 116446. [CrossRef]
171. Guerra, I.C.D.; de Oliveira, P.D.L.; Santos, M.M.F.; Lúcio, A.S.S.C.; Tavares, J.F.; Barbosa-Filho, J.M.; Madruga, M.S.; de Souza, E.L. The effects of composite coatings containing chitosan and Mentha (piperita L. or x villosa Huds) essential oil on postharvest mold occurrence and quality of table grape cv. Isabella. *Innov. Food Sci. Emerg. Technol.* **2016**, *34*, 112–121. [CrossRef]
172. Thakur, R.; Pristijono, P.; Bowyer, M.; Singh, S.P.; Scarlett, C.J.; Stathopoulos, C.E.; Vuong, Q.V. A starch edible surface coating delays banana fruit ripening. *LWT* **2019**, *100*, 341–347. [CrossRef]
173. Pan, T.; Shi, T.; Wen, D.; Yan, G. Study on Preservation of Cherry Tomato by Konjac Glucomannan Composite Coating Solution. *Food Ind.* **2020**, *41*, 60–63.
174. Hashemi, S.M.B.; Jafarpour, D. The efficacy of edible film from Konjac glucomannan and saffron petal extract to improve shelf life of fresh-cut cucumber. *Food Sci. Nutr.* **2020**, *8*, 3128–3137. [CrossRef] [PubMed]
175. Go, E.-J.; Song, K.B. Antioxidant Properties of Rye Starch Films Containing Rosehip Extract and Their Application in Packaging of Chicken Breast. *Starch-Stärke* **2019**, *71*, 1900116. [CrossRef]
176. Qin, Y.; Liu, Y.; Yong, H.; Liu, J.; Zhang, X.; Liu, J. Preparation and characterization of active and intelligent packaging films based on cassava starch and anthocyanins from Lycium ruthenicum Murr. *Int. J. Biol. Macromol.* **2019**, *134*, 80–90. [CrossRef] [PubMed]
177. Dini, H.; Fallah, A.A.; Bonyadian, M.; Abbasvali, M.; Soleimani, M. Effect of edible composite film based on chitosan and cumin essential oil-loaded nanoemulsion combined with low-dose gamma irradiation on microbiological safety and quality of beef loins during refrigerated storage. *Int. J. Biol. Macromol.* **2020**, *164*, 1501–1509. [CrossRef] [PubMed]
178. Carocho, M.; Heleno, S.; Rodrigues, P.; Barreiro, M.F.; Barros, L.; Ferreira, I.C.F.R. A novel natural coating for food preservation: Effectiveness on microbial growth and physicochemical parameters. *LWT* **2019**, *104*, 76–83. [CrossRef]
179. Kalem, I.K.; Bhat, Z.F.; Kumar, S.; Wang, L.; Mudiyanselage, R.J.; Bhat, H.F. Tinospora cordifolia: A novel bioactive ingredient for edible films for improved lipid oxidative and microbial stability of meat products. *J. Food Process. Preserv.* **2018**, *42*, e13774. [CrossRef]
180. Rodsamran, P.; Sothornvit, R. Lime peel pectin integrated with coconut water and lime peel extract as a new bioactive film sachet to retard soybean oil oxidation. *Food Hydrocoll.* **2019**, *97*, 105173. [CrossRef]
181. Baek, S.-K.; Kim, S.; Song, K.B. Cowpea starch films containing maqui berry extract and their application in salmon packaging. *Food Packag. Shelf Life* **2019**, *22*, 100394. [CrossRef]
182. Gharibzahedi, S.M.T.; Mohammadnabi, S. Effect of novel bioactive edible coatings based on jujube gum and nettle oil-loaded nanoemulsions on the shelf-life of Beluga sturgeon fillets. *Int. J. Biol. Macromol.* **2017**, *95*, 769–777. [CrossRef]
183. Alotaibi, S.; Tahergorabi, R. Development of a sweet potato starch-based coating and its effect on quality attributes of shrimp during refrigerated storage. *LWT* **2018**, *88*, 203–209. [CrossRef]
184. Yu, Z.; Sun, L.; Wang, W.; Zeng, W.; Mustapha, A.; Lin, M. Soy protein-based films incorporated with cellulose nanocrystals and pine needle extract for active packaging. *Ind. Crop. Prod.* **2018**, *112*, 412–419. [CrossRef]
185. Xie, B.; Zhang, X.; Luo, X.; Wang, Y.; Li, Y.; Li, B.; Liu, S. Edible coating based on beeswax-in-water Pickering emulsion stabilized by cellulose nanofibrils and carboxymethyl chitosan. *Food Chem.* **2020**, *331*, 127108. [CrossRef] [PubMed]
186. Zhao, Y.; Teixeira, J.S.; Saldaña, M.D.A.; Gänzle, M.G. Antimicrobial activity of bioactive starch packaging films against Listeria monocytogenes and reconstituted meat microbiota on ham. *Int. J. Food Microbiol.* **2019**, *305*, 108253. [CrossRef]
187. Volpe, V.; De Feo, G.; De Marco, I.; Pantani, R. Use of sunflower seed fried oil as an ecofriendly plasticizer for starch and application of this thermoplastic starch as a filler for PLA. *Ind. Crop. Prod.* **2018**, *122*, 545–552. [CrossRef]
188. Meng, L.; Liu, H.; Yu, L.; Duan, Q.; Chen, L.; Liu, F.; Shao, Z.; Shi, K.; Lin, X. How water acting as both blowing agent and plasticizer affect on starch-based foam. *Ind. Crop. Prod.* **2019**, *134*, 43–49. [CrossRef]
189. Perdones, Á.; Escriche, I.; Chiralt, A.; Vargas, M. Effect of chitosan–lemon essential oil coatings on volatile profile of strawberries during storage. *Food Chem.* **2016**, *197*, 979–986. [CrossRef]
190. Duran, M.; Aday, M.S.; Zorba, N.N.D.; Temizkan, R.; Büyükcan, M.B.; Caner, C. Potential of antimicrobial active packaging 'containing natamycin, nisin, pomegranate and grape seed extract in chitosan coating' to extend shelf life of fresh strawberry. *Food Bioprod. Process.* **2016**, *98*, 354–363. [CrossRef]
191. Sahraei Khosh Gardesh, A.; Badii, F.; Hashemi, M.; Ardakani, A.Y.; Maftoonazad, N.; Gorji, A.M. Effect of nanochitosan based coating on climacteric behavior and postharvest shelf-life extension of apple cv. Golab Kohanz. *LWT* **2016**, *70*, 33–40. [CrossRef]
192. Kaya, M.; Česonienė, L.; Daubaras, R.; Leskauskaitė, D.; Zabulionė, D. Chitosan coating of red kiwifruit (Actinidia melanandra) for extending of the shelf life. *Int. J. Biol. Macromol.* **2016**, *85*, 355–360. [CrossRef]
193. Ciurzyńska, A.; Cieśluk, P.; Barwińska, M.; Marczak, W.; Ordyniak, A.; Lenart, A.; Janowicz, M. Eating Habits and Sustainable Food Production in the Development of Innovative "Healthy" Snacks. *Sustainability* **2019**, *11*, 2800. [CrossRef]
194. Pitkänen, L.; Heinonen, M.; Mikkonen, K.S. Safety considerations of plant polysaccharides for food use: A case study on phenolic-rich softwood galactoglucomannan extract. *Food Funct.* **2018**, *9*, 1931–1943. [CrossRef] [PubMed]

195. Picó, Y. Chapter 9—Safety Assessment and Migration Tests. In *Nanomaterials for Food Packaging*; Cerqueira, M.Â.P.R., Lagaron, J.M., Pastrana Castro, L.M., de Oliveira Soares Vicente, A.A.M., Eds.; Elsevier: Amsterdam, The Netherlands, 2018. [CrossRef]
196. Barreto, M.R.; Aleixo, N.A.; Silvestre, R.B.; Fregonezi, N.F.; Barud, H.d.S.; Dias, D.d.S.; Ribeiro, C.A.; Resende, F.A. Genotoxicological safety assessment of puree-only edible films from onion bulb (*Allium cepa* L.) for use in food packaging-related applications. *J. Food Sci.* **2020**, *85*, 201–208. [CrossRef]
197. Sohrabi, Y.; Mohammadzadeh-Aghdash, H.; Baghbani, E.; Dehghan, P.; Ezzati Nazhad Dolatabadi, J. Cytotoxicity and Genotoxicity Assessment of Ascorbyl Palmitate (AP). *Food Additive Adv. Pharm. Bull.* **2018**, *8*, 341–346. [CrossRef]
198. Zheng, X.; Zhou, Y.; Yang, L. Safty evaluation of hydroxypropylated-phosphorylated modified glutinous rice starch in KM mice. *Food Sci. Technol.* **2017**, *42*, 229–233.
199. Al-Asmar, A.; Naviglio, D.; Giosafatto, C.V.L.; Mariniello, L. Hydrocolloid-Based Coatings are Effective at Reducing Acrylamide and Oil Content of French Fries. *Coatings* **2018**, *8*, 147. [CrossRef]

foods

MDPI

Article

Hemp (*Cannabis sativa* L.) Seed Protein–EGCG Conjugates: Covalent Bonding and Functional Research

Xin-Hui Pang [1], Yang Yang [1], Xin Bian [1], Bing Wang [1], Li-Kun Ren [1], Lin-Lin Liu [1], De-Hui Yu [1], Jing Yang [1], Jing-Chun Guo [2], Lei Wang [2], Xiu-Min Zhang [3], Han-Song Yu [4,5,*] and Na Zhang [1,*]

1 Key Laboratory of Food Science and Engineering of Heilongjiang Province, College of Food Engineering, Harbin University of Commerce, Songbei District, Harbin 150076, China; pxh10010@163.com (X.-H.P.); foodyangyang@163.com (Y.Y.); bianbian1225@163.com (X.B.); iceking85@163.com (B.W.); likun931006@163.com (L.-K.R.); liulinlin721@163.com (L.-L.L.); yudehui1213@163.com (D.-H.Y.); YJYJ6620@163.com (J.Y.)

2 Heilongjiang Academy of Sciences, Harbin 150000, China; guojingchun@163.com (J.-C.G.); wleileiyu@163.com (L.W.)

3 Beijing Academy of Food Sciences, Beijing 100068, China; zxmchnfood@163.com

4 College of Food Science and Engineering, Jilin Agricultural University, Changchun 130118, China

5 Division of Soybean Processing, Soybean Research & Development Center, Chinese Agricultural Research System, Changchun 130118, China

* Correspondence: yuhansongvip@163.com (H.-S.Y.); foodzhangna@163.com (N.Z.)

Abstract: In order to make HPI have a wide application prospect in the food industry, we used EGCG to modify HPI. In this study, we prepared different concentrations (1, 2, 3, 4, and 5 mM) of (−)-epigallocatechin gallate (EGCG) covalently linked to HPI and use methods such as particle size analysis, circular dichroism (CD), and three-dimensional fluorescence spectroscopy to study the changes in the structure and functional properties of HPI after being covalently combined with EGCG. The particle size data indicated that the covalent HPI-EGCG complex was larger than native HPI, and the particle size was mainly distributed at about 200 μm. CD and three-dimensional fluorescence spectroscopy analyses showed that the conformation of the protein was changed by conjugation with EGCG. The β-sheet content decreased from 82.79% to 66.67% after EGCG bound to the protein, and the hydrophobic groups inside the protein were exposed, which increased the hydrophobicity of the protein and changed its conformation. After HPI and 1 mM of EGCG were covalently bonded, the solubility and emulsifying properties of the covalent complex were improved compared with native HPI. These results indicated that HPI-EGCG conjugates can be added in some foods.

Keywords: hemp protein isolate; (−)-epigallocatechin gallate; covalent binding; structural changes; emulsifying properties

Citation: Pang, X.-H.; Yang, Y.; Bian, X.; Wang, B.; Ren, L.-K.; Liu, L.-L.; Yu, D.-H.; Yang, J.; Guo, J.-C.; Wang, L.; et al. Hemp (*Cannabis sativa* L.) Seed Protein–EGCG Conjugates: Covalent Bonding and Functional Research. *Foods* **2021**, *10*, 1618. https://doi.org/10.3390/foods10071618

Academic Editor: Barry J. Parsons

Received: 4 June 2021
Accepted: 1 July 2021
Published: 13 July 2021

1. Introduction

Hemp (*Cannabis sativa* L.) is a widely cultivated plant, and has an especially important role in industry. It is divided into two types according to the 9-tetrahydrocannabinol (THC) content: industrial hemp and drug hemp [1]. A 0.3% THC standard has been established by the European Union for this classification. If the THC is less than 0.3%, industrial hemp is allowed to be grown in China and Canada. Hemp is grown for industrial use and harvested for its fiber, seeds, and oil. Hemp seeds are rich in phytosterols, omega-3, and omega-6 essential fatty acids and protein (approximately 25% of dry weight). Moreover, there are many kinds of essential amino acids, and all of the essential amino acids needed by the human body are contained, with relatively high glutamic acid and arginine content [2,3]. Histidine is an essential amino acid for infants less than eight months old, and it plays an important role in the prevention of cardiovascular diseases in middle-aged and elderly people and in the growth of children. Hemp seeds are considered to be a good source of high-quality protein suitable for the elderly and infants. Therefore, they have been used in the production of various foods with high nutritional value. Importantly, the emergence of hemp varieties with low THC content has increased the utilization of hemp in food production [4]. Hemp seeds contain more than 30% oil and 25% protein. Currently, hemp is mainly used for the extraction of hemp oil because of the high oil content in the seeds [5]. However, compared with other plant proteins, hemp protein has low solubility, which is attributed to the aggregation of edestin (11S globulin) at pH values lower than 7.0. Therefore, hemp protein needs to be modified to increase its range of applications.

In recent years, various chemical modification methods, such as phosphorylation, glycosylation, deamidation, and succinylation, have been proven to be effective in improving the functional properties of proteins [6]. Phosphorylation, deamidation, and succinylation methods all result in chemical residues, which lead to a decrease in the nutritional value of the protein. The Maillard reaction is the main method of glycosylation, but it is difficult to control. Overreaction affects the flavor and quality of the products. Therefore, some researchers have proposed the use of polyphenols to modify proteins. In recent years, many researchers have begun to study the interaction between polyphenols and proteins. Proteins and polyphenols can interact through both non-covalent and covalent bonds. Non-covalent interactions are reversible, whereas covalent interactions are irreversible [7]. There are two types of methods used to form protein–polyphenol conjugates: non-enzymatic (alkaline and free radical reactions) and enzymatic (polyphenol oxidase, laccase, and tyrosinase) [8]. The mechanism of the alkaline reaction involves the oxidation of phenolic compounds to quinone compounds in an alkaline solution [9]. Quinone compounds usually react further with nucleophilic amino acid residues in the protein chain. Wei conducted research on the covalent complex of sodium caseinate, β-lactoglobulin, lactoferrin, and α-whey protein with EGCG and found that the covalent complex significantly improved the stability of the β-carotene emulsion [10]. Some researchers have also covalently combined phenolic compounds with flaxseed protein isolate (FPI). They found that the covalent complex of polyphenols and has a higher emulsifying ability than FPI [11]. When He studied the covalent binding of ovalbumin (OVA) and EGCG, he also found that the emulsification of OVA was improved [12]. The addition of phenol allows the protein to acquire phenolic hydroxyl groups, thereby changing the functional properties of the protein.

(−)-Epigallocatechin gallate (EGCG) is the main component of polyphenols and catechins in green tea [13]. EGCG has beneficial physiological activities, such as protecting against free radical DNA damage, antioxidative effects [14], inhibiting tumor growth, reducing serum low-density lipoprotein and cholesterol levels, improving vascular proliferation, and protecting against cardiovascular diseases [15].

Research on polyphenols and proteins has mainly focused on non-covalent interactions, but these interactions are reversible and unstable. Therefore, we chose to study covalent interactions between polyphenols and proteins. The covalent complexes formed by hemp protein isolate (HPI) and EGCG were used to investigate these interactions. Zeta potential values, CD spectra, and three-dimensional fluorescence spectra were used to explore the structure of the HPI-EGCG complex. The functional characteristics of the covalent complex were determined by analyzing the reactive groups, polyphenol content, the size of the particles, and other relevant indicators. The results of this research provide some theories and provide a research basis for future researchers for the application of the HPI-EGCG complex in the food industry as an emulsifier.

2. Materials and Methods

2.1. Experimental Materials

Hemp was purchased from the Heilongjiang Academy of Sciences (Harbin, China). Hemp protein samples were prepared by alkali-soluble acid precipitation. EGCG, *o*-phthaldialdehyde (OPA), and 5,5′-dithio-2-nitrobenzoic acid (DTNB) were purchased from Solarbio Life Science Co., Ltd. (Beijing, China). Other reagents were of analytical grade and were purchased from Tianjin Kermel Chemical Reagent Co., Ltd. (Tianjin, China).

2.2. Extraction of HPI

HPI was extracted according to Tang's method, with some changes [16]. Degreased powder was dissolved in 20 volumes of distilled water, and the extraction solution was adjusted to a pH of 9.0 generally with 2 mol/L of NaOH, stirred at 50 °C for 1 h, and the centrifugal force was set to 10,000× *g* at 4 °C. The centrifugal time was 20 min to yield a clear liquid. The pH of the clear liquid was adjusted to 4.8 with 2 mol/L of HCl, the centrifugal force was set to 8000× *g* at 4 °C, the centrifugal time was 20 min, and the liquid was cooled overnight. The supernatant was discarded and the remaining precipitate was washed with distilled water three times to remove the salt. The resulting precipitate was then dissolved in distilled water, adjusted to a pH of 7.0, placed in a low-temperature refrigerator for pre-freezing for 24 h, and freeze-dried. The protein content of the freeze-dried powder, as determined by the Kjeldahl method, was 93.7% (N × 6.25).

2.3. Preparation of Covalent HPI-EGCG Complexes

Covalent HPI-EGCG complexes were prepared using a previously described method [17], with slight modifications. HPI (0.5 g) was dissolved in 50 mL of distilled water. Different concentrations of the EGCG mother liquor were prepared, and the HPI and EGCG solution was mixed in a volume ratio of 1:1 to ensure that the concentration of EGCG in the mixed solution was 0, 1, 2, 3, 4, or 5 mM. The solution was adjusted to a pH of 9.0 and magnetically stirred for 24 h under an aerobic condition at 25 °C and then dialyzed. The HPI-EGCG conjugate was obtained by freeze-drying.

2.4. Determination of Sulfhydryl Content

The determination of sulfhydryl content was based on the method of previous researchers [18], with slight modifications. Ellman reagent was prepared with 4 mg of DTNB and 1 mL of Tris-glycine. Tris-glycine (contained 8 M of urea) buffer was used to prepare an HPI solution at a concentration of 3 mg/mL. Fifty microliters of Ellman reagent was then added to 5 mL of the HPI solution and reacted for 1 h in a dark environment. The absorbance value under ultraviolet light was 412 nm, and the sulfhydryl content was calculated using the following formula:

$$\text{sulfhydryl content } (\mu\text{mol/g protein}) = (73.53 \times A_{412} \times D)/C \qquad (1)$$

where A_{412} is the absorbance at 412 nm, D is the dilution coefficient, and C is the protein concentration of the HPI sample (mg/mL).

2.5. Determination of Free Amino Group Content

The free amino group content of HPI and the HPI-EGCG complexes was determined using the *o*-phthaldialdehyde (OPA) method [19]; 80 mg of OPA powder was accurately weighted, added to 4 mL of methanol solution, and mixed to prepare OPA reagent. Then, 50 mL was added to 0.1 M of borax solution, 200 μL of β-mercaptoethanol, and 5 mL of sodium lauryl sulfate solution (20%, *w*/*w*). The resulting solution was transferred to a volumetric flask and made up to 100 mL with distilled water. Two hundred microliters of the sample solution, at a concentration of 1 mg/mL, was mixed with 4 mL of the OPA reagent and vortexed to obtain a homogeneous solution. After protecting from light for 2 h, the spectrophotometer was set to 340 nm and the absorbance of the solution was measured. L-leucine solutions of different concentrations were used to construct a standard curve of the absorbance value vs. the L-leucine concentration. The content of free amino groups was calculated with the obtained absorbance value through the standard curve.

2.6. Determination of EGCG Content

Total phenol equivalents were determined in the HPI and HPI-EGCG complex samples [20]. Folin–Ciocalteu reagent (2.5 mL, 0.2 N) was added to 0.5 mL of HPI-EGCG complex solution and mixed thoroughly using a wandering oscillator. After reacting for

5 min, 2 mL of an Na_2CO_3 solution (7.5%, *w*/*v*) was added, and the mixture was shaken for 30 s and reacted for 2 h in the dark. Absorbance was then measured at 760 nm, usually with distilled water as a blank. EGCG standard solutions of different concentrations were used to construct a standard curve of absorbance vs. EGCG concentration. The EGCG content of the sample was then calculated according to the standard curve, and the final result was expressed as μmol EGCG/g protein.

2.7. Zeta Potential and Particle Size Determination

According to the method used by previous researchers [21], the zeta potential and particle size of the covalent complex were measured, with some modifications.

The average particle size distribution of each sample was determined, and a sample solution was prepared with a concentration of 0.2 mg/mL using a Malvern particle size analyzer (Malvern Panalytical, Malvern, UK). The samples were centrifuged at 4000× *g* for 15 min and passed through a 0.45 μm cellulose acetate filter membrane, and the particle size was then measured at room temperature.

A Nano ZS90 zetasizer was used to determine the zeta potential, according to the method used by previous researchers [22], with slight modifications. Each sample was measured six times, and the average value was calculated as the final result.

2.8. Circular Dichroism Analysis

Far-ultraviolet circular dichroism (CD) can analyze the changes in the secondary structure of the HPI-EGCG complex [23]. A sample solution with a protein concentration of 0.5 mg/mL was prepared in distilled water. Far-ultraviolet CD measurements were performed in a 1 mm diameter quartz cuvette using an MOS-450 spectrophotometer (French Bio-Logic SAS, Seyssinet-Pariset, France). The following parameters were used: spectral scan range, 190–250 nm; scan resolution, 1 nm; scan rate, 100 nm/min; and general band width, 2.0 nm. Samples were scanned nine times and the average value was calculated.

2.9. Three-Dimensional Fluorescence Spectroscopy

The change in fluorescence intensity of the samples was measured using an F-6000 fluorometer spectrophotometer (Hitachi Ltd., Tokyo, Japan) [24]. A 1 mg/mL sample was prepared, and the samples were placed in a quartz cuvette for measurement. The excitation wavelength (λ_{Ex}) was set at 200–350 nm. Continuous scanning of the three-dimensional fluorescence spectrum was performed at an emission wavelength (λ_{Em}) ranging from 200 to 500 nm, with an initial λ_{Ex} of 200 nm, a scanning speed of 3000 nm/min, and a slit width of 5 nm.

2.10. HPI-EGCG Conjugate Properties

2.10.1. Solubility

The total protein content change of the sample solutions was determined using the Garcia method [25], with slight modifications; 20 mL of the 5.0 mg/mL protein sample was stirred for 1 h at a constant temperature of 25 °C. The samples were then centrifuged at 3000× *g* for 15 min, and the amount of soluble protein was determined using the following formula:

$$SI\ (\%) = W_2/W_1 \times 100 \quad (2)$$

where SI is the protein solubility, W_1 is the original protein quantity (g), and W_2 is the soluble protein quantity (g).

2.10.2. Emulsifying Properties

Emulsification and emulsification stability were determined as previously described [26]. Briefly, 5 mL of soybean salad oil were added to 15 mL of a concentration of 3.0 mg/mL HPI-EGCG conjugate solution at a pH of 7.0 and centrifuged at 9000× *g* for 1 min. A 50 μL sample of the emulsion was removed and mixed with 0.1% (*w*/*v*) sodium dodecyl sulfate. The absorbance of the sample at 500 nm at 0 and 30 min was recorded. The following formulas were used to calculate the Emulsification Activity Index (EAI) and Emulsification Stability Index (ESI):

$$\text{EAI } (m^2/g) = 2 \times 2.303 \times A_0 \times \text{dilution factor}/100{,}000 \times \lambda \times C \quad (3)$$

$$\text{ESI } (\%) = A_{30}/A_0 \quad (4)$$

where the dilution factor is 100, C is the protein concentration (g/mL), λ is the volumetric oil fraction, A_0 is the absorbance measured at 0 min, and A_{30} is the absorbance measured at 30 min.

2.11. Cryo-Scanning Electron Microscopy

A cryogenic scanning electron microscope (JSM-7100; Jeol Europe, Zaventem, Belgium) was used to observe the microstructure of the emulsion. A field emission scanning electron microscope was used to capture still images of the microstructure of the emulsions. The emulsion was added dropwise into the short tube of a low-temperature scanning electron microscope located on the holder and frozen in nitrogen. Subsequently, the short tube was transferred to an ultra-vacuum low-temperature chamber, and the emulsion was broken and etched at −95 °C for 60 s. The broken surface was coated in platinum and then transferred to the freezing stage (−190 °C) of the scanning electron microscope. Images were captured using the microscope control software [27].

2.12. Statistical Analysis

Each group of experiments was repeated in triplicate, and the data were analyzed using SPSS 23.0 software (IBM, Armonk, NY, USA). The variance of each treatment group and an independent-sample Student's *t*-test were used to determine whether there were significant differences between the samples ($p < 0.05$). The results are expressed as mean ± standard deviation (SD).

3. Results and Discussion

3.1. Determination of HPI and EGCG Covalent Reactive Groups and EGCG Content

In an alkaline environment, the phenolic hydroxyl group of EGCG is oxidized to produce quinones, which changes the sulfhydryl or amino groups of the protein. C-S or C-N bonds are formed during the covalent bonding of polyphenols and proteins. In this study, an increase or decrease in the content of sulfhydryl and free amino groups concentrations of HPI during the reaction with EGCG and the amount of polyphenol binding in the reaction was measured (Table 1). In the process of covalent bonding of HPI and EGCG, the content of side-chain amino groups in the protein also changed with changes in the amount of EGCG added.

Table 1. The sulfhydryl group, free amino group, and EGCG concentrations in HPI-EGCG conjugates.

Sample	Sulfhydryl Groups (μmol/g Protein)	Free Amino Groups (μmol/g Protein)	EGCG Content (μmol/g Protein)
HPI	28.41 ± 0.06^{a}	225.33 ± 0.7^{a}	—
HPI-1 mM EGCG	17.48 ± 0.13^{b}	202.02 ± 5.1^{b}	23.08 ± 1.1^{c}
HPI-2 mM EGCG	14.58 ± 0.14^{c}	196.02 ± 1.9^{c}	55.79 ± 1.8^{c}
HPI-3 mM EGCG	12.53 ± 0.08^{d}	173.39 ± 1.2^{d}	87.38 ± 3.0^{bc}
HPI-4 mM EGCG	9.14 ± 0.01^{e}	84.39 ± 2.3^{e}	100.15 ± 0.9^{b}
HPI-5 mM EGCG	9.96 ± 0.03^{e}	65.29 ± 2.8^{f}	109.80 ± 3.3^{a}

Values are expressed as the mean ± SD. Different letters in the same column indicate significant differences ($p < 0.05$).

The sulfhydryl contents of HPI and the HPI-EGCG covalent complex were measured within the allowable error range of the method. EGCG was reacted with HPI to form covalent bonds. When the concentration of EGCG continued to rise, the sulfhydryl content of HPI decreased from 28.41 μmol/g HPI to 9.14 μmol/g HPI (Table 1). The process of converting sulfhydryl groups into disulfide bonds can be prevented by urea, and therefore, EGCG can be derived from the decrease in free sulfhydryl content. Covalent bonding occurred with the free sulfhydryl group of HPI; that is, the phenol rings of polyphenols, such as EGCG, can covalently react with the nucleophilic groups of HPI to form C-N and C-S bonds, which illustrates the occurrence of covalent reactions. This is the main reason for the reduction in the number of sulfhydryl groups [28].

The OPA method can sensitively reflect relevant information about the side chain of lysine, the N-terminal residues, and α- or ε-amino groups of proteins, which are the main reactive groups measured by the OPA method. As shown in Table 1, compared with control HPI, there was a downward trend in the content of free amino groups in the covalent complex of HPI and EGCG. When the concentration of EGCG continued to increase, the number of free amino groups in the conjugate greatly decreased. Because free amino groups were quantified in the presence of a 1% concentration sodium dodecyl sulfate, a denaturant that destroys non-covalent protein interactions, we deduced that the decrease in the number of free amino groups may have indicated that the free amino groups of HPI were covalently linked with EGCG. Free amino content groups in HPI were expected to decrease after covalent bonding with EGCG, because the amino acid residues bind to quinone after EGCG oxidation. The ε-amino group contained in lysine was oxidized to form a carbonyl group, and a condensation reaction occurred with the free amino groups, thus reducing the number of amino groups [29].

To further confirm the covalent linkage between HPI and EGCG, the Folin–Ciocalteu method was used to determine the content of EGCG in the solution of the HPI-EGCG conjugate. When determining polyphenol content, HPI samples were pre-measured to have the same protein concentration as the HPI-EGCG samples under the same experimental conditions. Additionally, HPI was used as a blank control to reduce the interference caused by polyphenols. It can be seen from Table 1 that the concentration of EGCG increased and the number of polyphenols increased. However, when the EGCG concentration was 4 mM and 5 mM, the polyphenol content did not significantly change. This implies that, at the concentration of 4 mM, EGCG had reacted with all of the active groups exposed by specific amino acids. It can also be seen from Table 1 that, as more polyphenols were bound, the content of sulfhydryl groups was reduced, as well as the content of free amino groups.

3.2. Particle Size Distribution and Zeta Potential Change Analysis

It can be seen from the particle size distribution and zeta potential changes of the HPI-EGCG complexes and HPI samples that, compared with HPI, the covalent HPI-EGCG complex had larger particles (Figure 1). The particle size of the covalently bound protein was mainly about 200 μm. In addition, with an increase in the amount of EGCG added, the height of the peak representing covalent complexes with a particle size of less than 100 μm decreased. The reason for this phenomenon is related to the covalent bonding of EGCG and the protein, as described in the following.

The zeta potential represents the amount of charge on or near the particle surface. The amount of charge is affected by the liquid environment (pH, ionic strength, or presence of surfactants) [30]. The electric potential is related to the size of the interaction between the charged groups, and plays an important role. It can be seen that, as the EGCG content continued to increase, the absolute zeta potential values of the samples increased (Figure 1). This increase in absolute zeta potential is due to the fact that the EGCG molecule is a negatively charged molecule, and, after binding, the protein is also negatively charged. The dispersion and aggregation of particles are related to their surface charge [31]. The absolute value of the zeta potential becomes larger indicates that the system is relatively stable. In the covalently bound samples, the oxidation and cross-linking of EGCG formed a stable network structure. Therefore, as the content of EGCG continued to increase, so too did both the zeta value and the size of the protein particles.

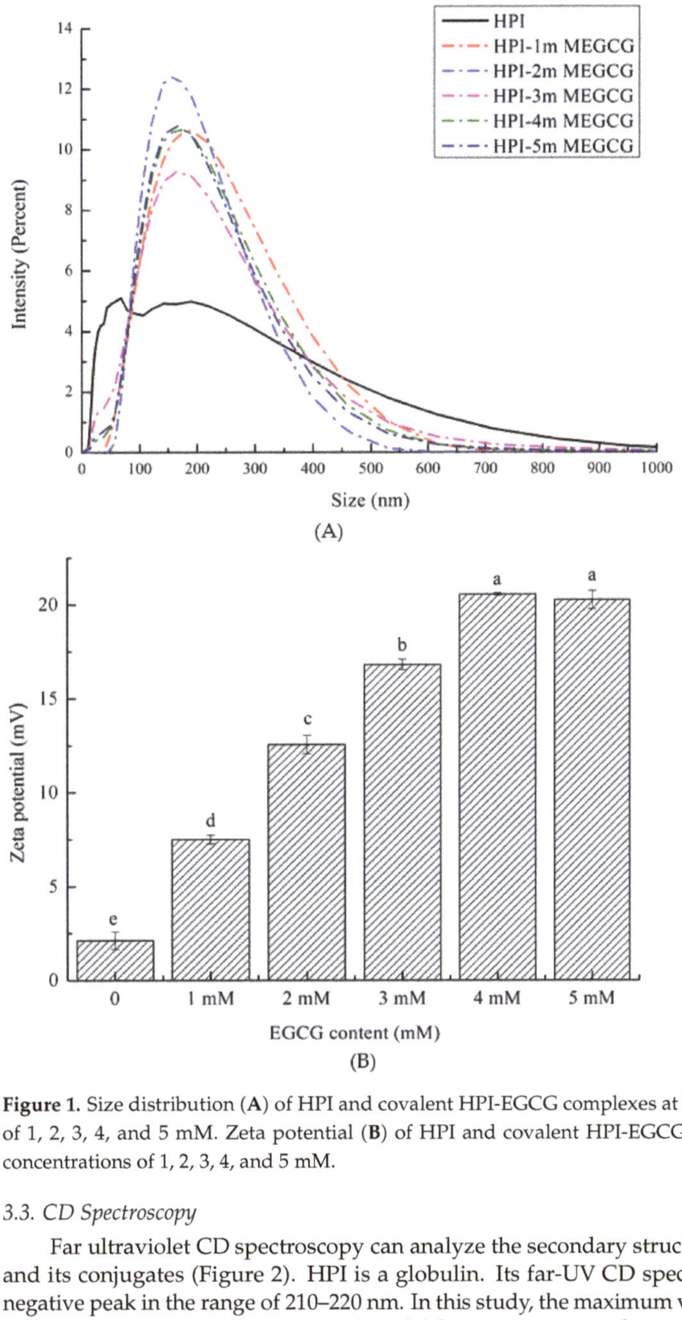

Figure 1. Size distribution (**A**) of HPI and covalent HPI-EGCG complexes at EGCG concentrations of 1, 2, 3, 4, and 5 mM. Zeta potential (**B**) of HPI and covalent HPI-EGCG complexes at EGCG concentrations of 1, 2, 3, 4, and 5 mM.

3.3. CD Spectroscopy

Far ultraviolet CD spectroscopy can analyze the secondary structure changes of HPI and its conjugates (Figure 2). HPI is a globulin. Its far-UV CD spectrum has a specific negative peak in the range of 210–220 nm. In this study, the maximum value of the negative peak shifted from 213 to 208 nm as the EGCG content continued to rise. Table 2 illustrates the changes in the secondary structure of the HPI-EGCG complex. After the protein was covalently bound to EGCG, the maximum negative value showed an increasing trend. That is, the binding of EGCG and HPI reduced the ellipticity of the protein as measured by CD spectroscopy, which indicated that the secondary structure of HPI had changed after EGCG was added. The secondary structural of natural HPI was composed of 1.79% α-helix, 82.79% β-sheet, and 15.50% unordered. After covalent bonding, the β-sheet content decreased, while the α-helix content and random coil content both increased. The decrease in β-sheet content is the structure that determines the flexibility of protein molecules. The increase in their content indicates that the protein chain becomes loose after covalent interaction and the protein is stretched [32]. Liu also observed this change in a study of the covalent interaction between porcine bone protein hydrolysate and rutin, in which the β-sheet and β-turn content decreased and the α-helix and random helix content increased [33]. In addition, the β-sheet content affects the internal hydrophobic groups of the protein. After covalent bonding, the β-sheet content of the complex was reduced, allowing the internal hydrophobic groups to be exposed, which increased the hydrophobicity of the protein. This change is a sign of a change in protein conformation [34]. Furthermore, Damodaran's research also found that α-helix and random coil structures are more flexible than β-sheet structures, which indicates that covalent bonding increases the flexibility of the secondary conformation of HPI [35]. These changes are brought about by the binding of hydroxyl groups at the protein hydroxyl, amino, or sulfhydryl sites in the reaction system, which significantly changes the secondary structure of the protein.

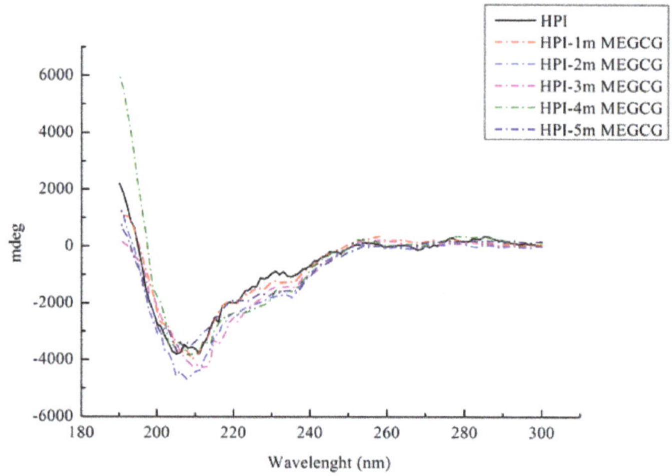

Figure 2. Far-UV CD spectroscopy of HPI and HPI-EGCG conjugates.

Table 2. Secondary structure contents of HPI and HPI–EGCG conjugates.

Sample	α-Helix (%)	β-Sheet (%)	β-Turn (%)	Random Coil (%)
HPI	1.79	82.79	0	15.50
HPI-1mMEGCG	1.87	71.16	0	27.30
HPI-2mMEGCG	3.33	66.86	2.33	27.76
HPI-3mMEGCG	3.90	66.67	3.40	25.53
HPI-4mMEGCG	5.41	69.60	0	24.43
HPI-5mMEGCG	1.94	70.67	1.43	25.67

3.4. Three-Dimensional Fluorescence Spectroscopy

Three-dimensional fluorescence spectroscopy was used to determine the effect of EGCG on the structure of HPI. Peak (a) in the three-dimensional contour map represents Raman scattering. The formation of the complex between EGCG and HPI increased the particle size and enhanced the light scattering effect, thereby increasing the fluorescence intensity of peak (a). Peak 1 shows that the characteristic peaks produced by tryptophan and tyrosine turned blue after the addition of EGCG, which indicated that the fluorescence intensity of the protein decreased after EGCG was combined with HPI [36]. It can be seen from that, as the amount of EGCG increased, the color of the characteristic peak became lighter as the contour line became thinner, and the fluorescence intensity of the protein decreased (Figure 3). The reason for this phenomenon is the covalent interaction between the Trp or Tyr residues in HPI and the quinone formed by the oxidation of polyphenols [37]. Therefore, the fluorescence intensity of HPI decreased after combining with EGCG, which indicated a change in the conformation of HPI. These results confirmed the covalent bonding between EGCG and HPI. Under aerobic conditions, phenol is easily oxidized to quinone, and further oxidation causes a dimerization reaction. Therefore, it is oxidized under aerobic conditions to form *o*-quinone, which can be reduced by the amino groups of the protein side chain, thereby forming a covalent C-N or C-S bond between the HPI and EGCG. These C-N and C-S bonds can form dimers through cross-linking reactions to provide new polymers.

3.5. Functional Properties of SPI-EGCG Conjugates

3.5.1. Solubility

The solubility of HPI refers to the percentage of soluble protein in HPI under a specific environment. The solubility of a protein affects its ability to emulsify. HPI had good solubility at 1 mM in (Figure 4). Compared with native HPI, the solubility of HPI-EGCG increased from 39.4% to 50.6%. Thus, the conjugation of HPI with EGCG markedly altered its solubility. The solubility of the conjugate increased at the optimal polyphenol loading (1 mM). The conjugation of EGCG enhanced the solubility of HPI, probably due to EGCG and HPI molecules combining to react, which changed the net charge of the protein molecules. Negatively charged EGCG bound to the hydrophobic area on the HPI surface. On the one hand, it increased the surface electronegativity of the HPI particles due to electrostatic repulsion between the particles, which promoted the dispersion of the protein particles in the aqueous solution to avoid coagulation. On the other hand, the hydrophilic tail of the phenolic hydroxyl group enhanced the interaction between HPI and water molecules [38]. Solubility showed a downward trend at other concentrations (2 mM, 3 mM, 4 mM, and 5 mM) due to excessive polyphenols, which caused protein molecules and polyphenols to form aggregates. The low solubility of such aggregates increased the turbidity of a solution [39]. The result we got is consistent with the result of Quan's study [40], who found that when the concentration of polyphenols exceeds the concentration of protein binding sites, proteins and polyphenols form aggregates, thereby greatly reducing their solubility. Therefore, the covalent bonding of EGCG and HPI resulted in a substantial change in the solubility of HPI.

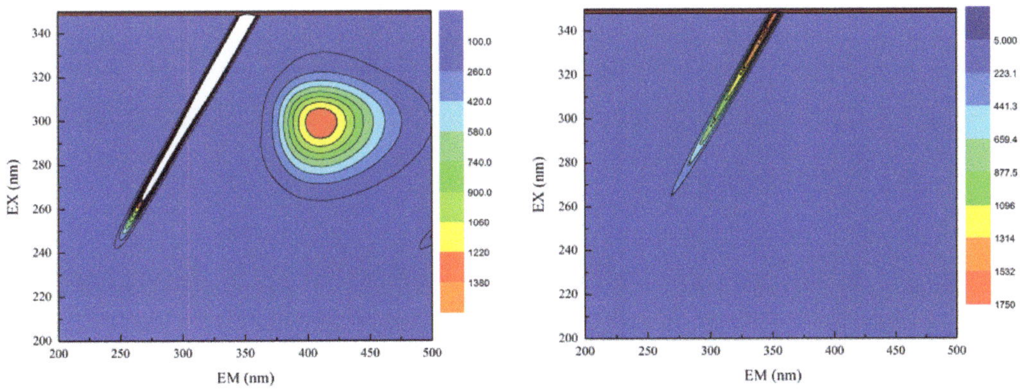

(A) (B)

Figure 3. *Cont.*

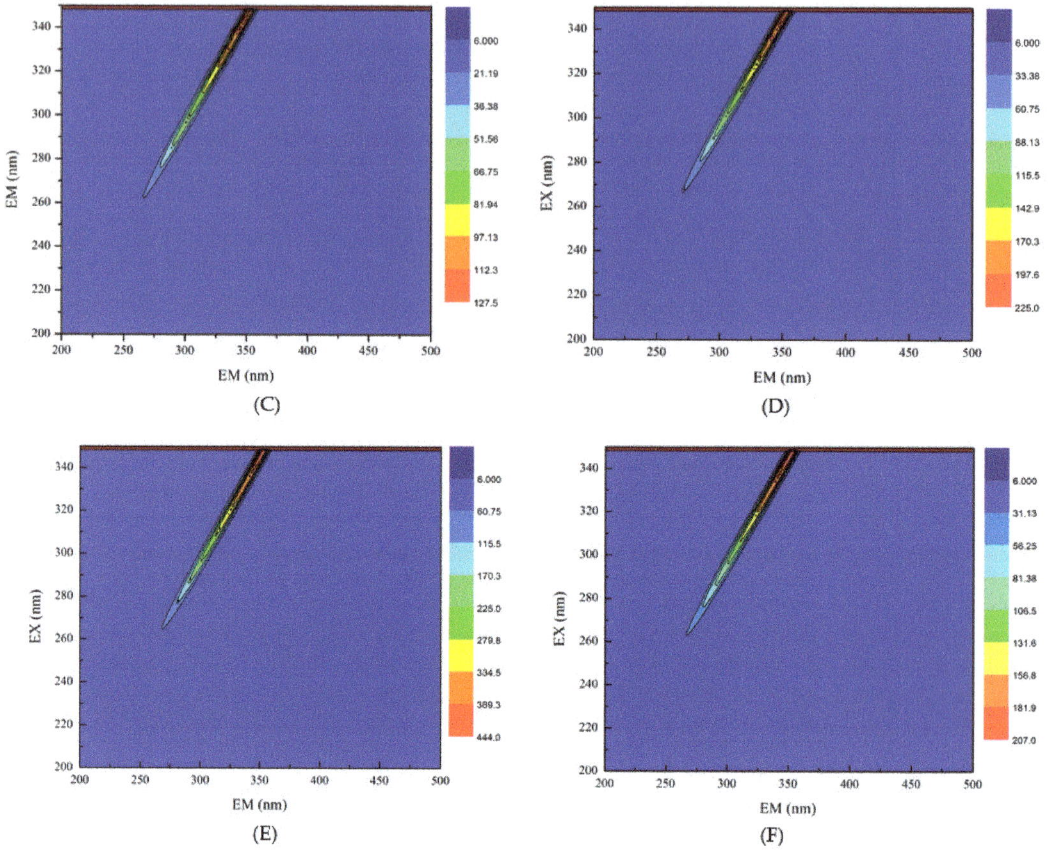

Figure 3. Three-dimensional fluorescence spectra of HPI and covalent HPI-EGCG complexes. (**A**) HPI, (**B**) HPI-1 Mm EGCG, (**C**) HPI-2 mM EGCG, (**D**) HPI-3 mM EGCG, (**E**) HPI-4 mM EGCG, and (**F**) HPI-5 mM EGCG, all suspended in H_2O (Ex is the excitation wavelength and Em is the emission wavelength).

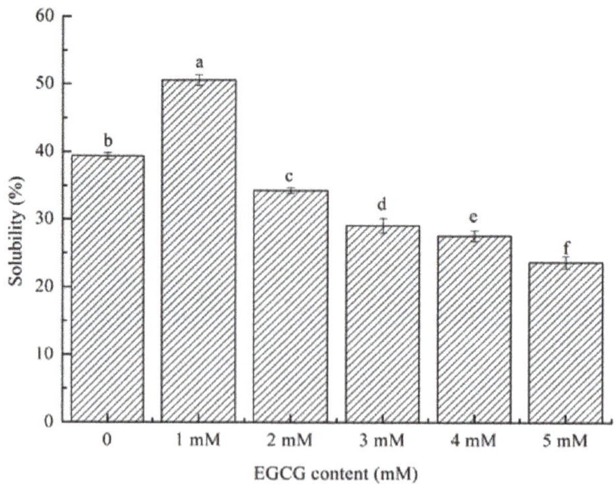

Figure 4. (a-f) Solubility of HPI and covalent HPI-EGCG complexes at EGCG concentrations of 1, 2, 3, 4, and 5 mM.

3.5.2. Emulsifying Properties of HPI-EGCG Conjugates

The EAI reflects the interfacial tension of protein droplets at the oil–water interface and the ability to stabilize the emulsion. It is determined by protein–protein and protein–lipid interactions. The ESI refers to the stable strength of the emulsion in the dispersion [41]. It can be seen that, except for the 1 mM EGCG complex, the EAI of the HPI-EGCG complexes was lower than that of the control in (Figure 5). Specifically, the ESI value first increased and then decreased with increasing EGCG concentration, reaching a maximum when EGCG was added at 1 mM. Protein–protein and protein-lipid interaction affects the emulsifying properties of proteins and plays an important role in the middle [42]. The combination of protein and the appropriate concentration of EGCG may change the protein–protein interactions between molecules, thereby reducing the interfacial tension of the oil–water interface [43]. The improvement in emulsification performance may be due to the change in the flexibility of HPI after combining with EGCG. The solubility of the protein is improved and the surface hydrophobicity is increased, making protein particles more stable at the oil–water interface [44]. The emulsification of lactoferrin has also been shown to improve when covalently linked with EGCG.

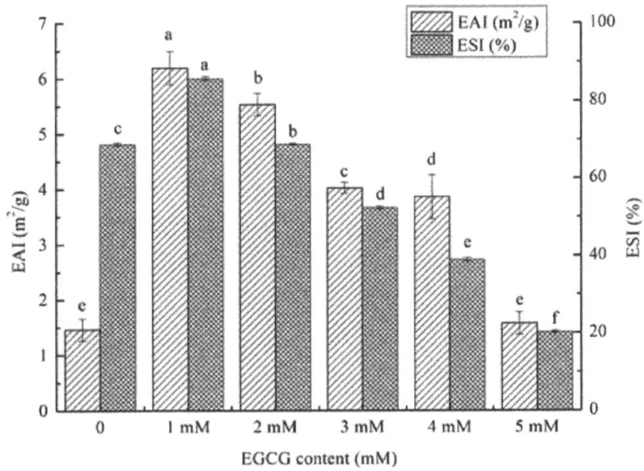

Figure 5. (a–f) Emulsifying properties of HPI and covalent HPI-EGCG complexes at EGCG concentrations of 1, 2, 3, 4, and 5 mM (EAI is the Emulsifying Activity Index and ESI is the Emulsifying Stability Index).

3.6. Cryo-Scanning Electron Microscopy Microstructure

Cryo-scanning electron microscopy images can be seen for HPI and covalent HPI-EGCG complex emulsions in Figure 6. The oil droplets were uniformly attached to the surface of the HPI-EGCG molecules, which had a micro-spherical structure [45]. This result is consistent with the results of a previous study showing that when anthocyanins are covalently bound to HPI, they can disrupt the protein peptide chain and enhance the interactions between droplets to form an emulsion [46]. It can be seen that the native state of the HPI emulsion is flocculated, in which fat globules flocculate without breaking the membrane. The droplets of the emulsion continued to aggregate, and as aggregation increased, flocculation also increased, and emulsification occurred more rapidly [47]. After HPI and EGCG were covalently bound, the flocculation phenomenon in the emulsion was suppressed, with the greatest uniformity seen when EGCG was at 1 mM. This shows that the negative charge of EGCG helped stabilize the emulsion and make it more uniform [48].

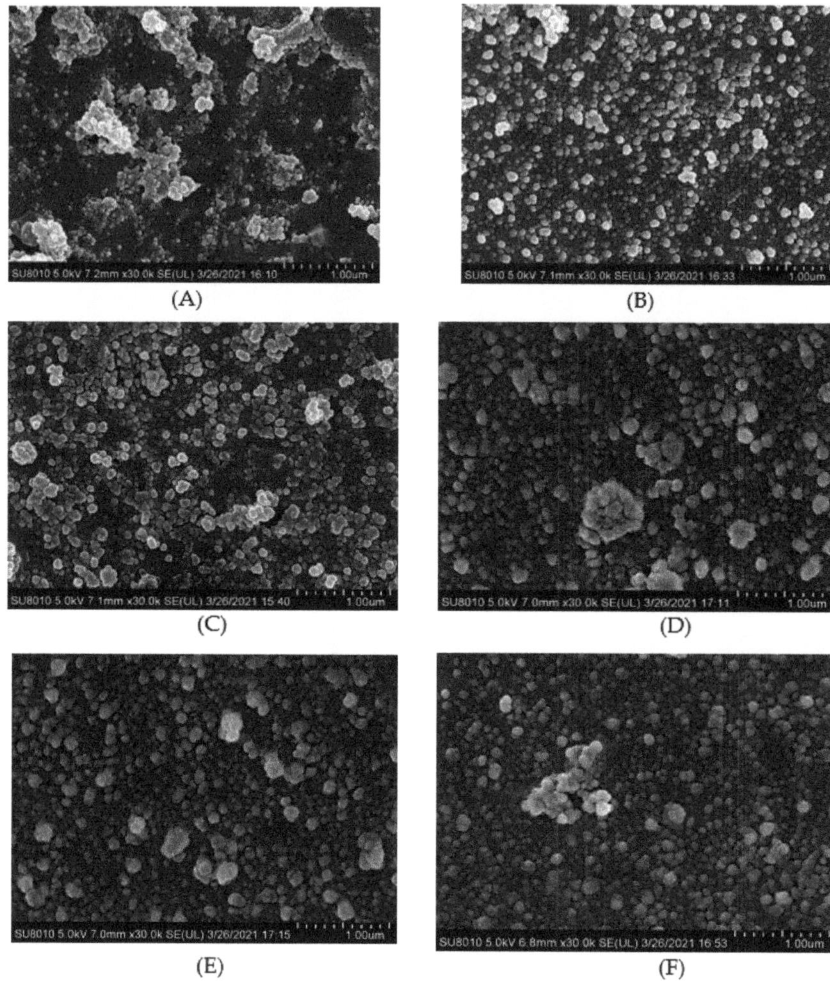

Figure 6. Cryo-scanning electron microscope images of HPI and covalent HPI-EGCG complexes at EGCG concentrations of 1, 2, 3, 4, and 5 mM. (**A**) HPI, (**B**) HPI-1 Mm EGCG, (**C**) HPI-2 mM EGCG, (**D**) HPI-3 mM EGCG, (**E**) HPI-4 mM EGCG, and (**F**) HPI-5 mM EGCG.

4. Conclusions

In this study, a protein–polyphenol covalent complex was formed by combining HPI with different concentrations of a polyphenol. The addition of EGCG caused changes in the structure of HPI and improved functional properties. After HPI is combined with 1 mM of EGCG, the emulsification and solubility of the covalent complex are improved. These data provide the theoretical basis for the application of polyphenols and protein covalent complex food processing emulsifier.

Author Contributions: X.B. and L.-L.L.: methodology, validation, and writing—original draft. Y.Y. and D.-H.Y.: formal analysis. X.-H.P. and L.-K.R.: data curation and writing—review. J.-C.G., L.W. and X.-M.Z.: methodology and validation. B.W. and J.Y.: editing. N.Z. and H.-S.Y.: supervision, project administration. All authors have read and agreed to the published version of the manuscript.

Funding: We appreciate the financial support from the National Natural Science Foundation of China (32072258), the Major Science and Technology Program of Heilongjiang (2019ZX08B02, 2020ZX08B02), Harbin University of Commerce "young innovative talents" support program (2019CX06, 2020CX26,

2019CX34), Heilongjiang Academy of Sciences (High Technology Research Institute) (Horizontal Project) Functional evaluation of hemp oil and protein and development of high-value food and central financial support for the development of local colleges and universities.

Institutional Review Board Statement: This article does not involve any animal experiments.

Informed Consent Statement: This article does not involve any experiments with humans as recipients.

Data Availability Statement: Research data are not shared.

Conflicts of Interest: The authors declare that they have no conflict of interest.

References

1. Blade, S.F. Industrial hemp in Alberta. In *Alberta Hemp Symposia Proceedings*; Blade, S., Ed.; Alberta Agriculture, Food and Rural Development: Red Deer, AB, Canada, 1998; pp. 2–11.
2. Malomo, S.A.; He, R.; Aluko, R.E. Structural and Functional Properties of Hemp Seed Protein Products. *J. Food Sci.* **2014**, *79*, C1512–C1521. [CrossRef] [PubMed]
3. Tang, C.-H.; Wang, X.-S.; Yang, X.-Q. Enzymatic hydrolysis of hemp (*Cannabis sativa* L.) protein isolate by various proteases and antioxidant properties of the resulting hydrolysates. *Food Chem.* **2009**, *114*, 1484–1490. [CrossRef]
4. Lu, R.-R.; Qian, P.; Sun, Z.; Zhou, X.-H.; Chen, T.-P.; He, J.-F.; Zhang, H.; Wu, J. Hempseed protein derived antioxidative peptides: Purification, identification and protection from hydrogen peroxide-induced apoptosis in PC12 cells. *Food Chem.* **2010**, *123*, 1210–1218. [CrossRef]
5. Pojić, M.; Mišan, A.; Sakač, M.; Dapčević Hadnađev, T.; Šarić, B.; Milovanović, I. Characterization of by-products originating from hemp oil processing. *J. Agric. Food Chem.* **2014**, *62*, 12436–12442. [CrossRef]
6. Morand, M.; Guyomarc'H, F.; Legland, D.; Famelart, M.-H. Changing the isoelectric point of the heat-induced whey protein complexes affects the acid gelation of skim milk. *Int. Dairy J.* **2012**, *23*, 9–17. [CrossRef]
7. Jakobek, L. Interactions of polyphenols with carbohydrates, lipids and proteins. *Food Chem.* **2015**, *175*, 556–567. [CrossRef] [PubMed]
8. Liu, F.; Ma, C.; Gao, Y.; McClements, D.J. Food-Grade Covalent Complexes and Their Application as Nutraceutical Delivery Systems: A Review. *Compr. Rev. Food Sci. Food Saf.* **2017**, *16*, 76–95. [CrossRef]
9. Hurell, R.F.; Finot, P.A. Nutritional consequences of the reactions between proteins and oxidized polyphenolic acids. *Adv. Exp. Med. Biol.* **1984**, *177*, 423–435.
10. Wei, Z.; Yang, W.; Fan, R.; Yuan, F.; Gao, Y. Evaluation of structural and functional properties of protein–EGCG complexes and their ability of stabilizing a model β-carotene emulsion. *Food Hydrocoll.* **2015**, *45*, 337–350. [CrossRef]
11. Phama, L.B.; Wang, B.; Zisuc, B.; Adhikaria, B. Covalent modification of flaxseed protein isolate by phenolic compounds and the structure and functional properties of the adducts. *Food Chem.* **2019**, *293*, 463–471. [CrossRef]
12. He, W.Y.; Xu, H.X.; Lu, Y.Q.; Zhang, T.T.; Li, S.M.; Lin, X.; Xu, B.Q.; Wu, X.L. Function, digestibility and allergenicity assessment of ovalbumin–EGCG conjugates. *J. Funct. Foods* **2019**, *61*, 103490. [CrossRef]
13. Pinto, M.D.S. Tea: A new perspective on health benefits. *Food Res. Int.* **2013**, *53*, 558–567. [CrossRef]
14. Gramza, A.; Korczak, J. Tea constituents (*Camellia sinensis* L.) as antioxidants in lipid systems. *Trends Food Sci. Technol.* **2005**, *16*, 351–358. [CrossRef]
15. Wolfram, S. Effects of Green Tea and EGCG on Cardiovascular and Metabolic Health. *J. Am. Coll. Nutr.* **2007**, *26*, 373S–388S. [CrossRef]
16. Tang, C.-H.; Ten, Z.; Wang, X.-S.; Yang, X.-Q. Physicochemical and Functional Properties of Hemp (*Cannabis sativa* L.) Protein Isolate. *J. Agric. Food Chem.* **2006**, *54*, 8945–8950. [CrossRef] [PubMed]
17. Liu, F.; Ma, C.; McClements, D.J.; Gao, Y. A comparative study of covalent and non-covalent interactions between zein and polyphenols in ethanol-water solution. *Food Hydrocoll.* **2017**, *63*, 625–634. [CrossRef]
18. Beveridge, T.; Toma, S.J.; Nakai, S. Determination of SH-and SS-groups in some food proteins using Ellman's reagent. *J. Food Sci.* **1974**, *39*, 49–51. [CrossRef]
19. Chen, N.; Zhao, Q.; Sun, W.; Zhao, M. Effects of Malondialdehyde Modification on the In Vitro Digestibility of Soy Protein Isolate. *J. Agric. Food Chem.* **2013**, *61*, 12139–12145. [CrossRef]
20. Fei, Q.; Gao, Y.; Zhang, X.; Sun, Y.; Hu, B.; Zhou, L.; Jabbar, S.; Zeng, X. Effects of Oolong Tea Polyphenols, EGCG, and EGCG3″Me on Pancreatic α-Amylase Activity In Vitro. *J. Agric. Food Chem.* **2014**, *62*, 9507–9514. [CrossRef] [PubMed]
21. Chen, F.; Ou, S.-Y.; Tang, C.-H. Core–Shell Soy Protein–Soy Polysaccharide Complex (Nano)particles as Carriers for Improved Stability and Sustained Release of Curcumin. *J. Agric. Food Chem.* **2016**, *64*, 5053–5059. [CrossRef]
22. Schneider, M.; Esposito, D.; Lila, M.A.; Foegeding, E.A. Formation of whey protein–polyphenol meso-structures as a natural means of creating functional particles. *Food Funct.* **2016**, *7*, 1306–1318. [CrossRef]
23. Kanakis, C.; Hasni, I.; Bourassa, P.; Tarantilis, P.; Polissiou, M.; Tajmir-Riahi, H.-A. Milk β-lactoglobulin complexes with tea polyphenols. *Food Chem.* **2011**, *127*, 1046–1055. [CrossRef]
24. Li, Y.; Liu, B.; Jiang, L.; Regenstein, J.M.; Jiang, N.; Poias, V.; Zhang, X.; Qi, B.; Li, A.; Wang, Z. Interaction of soybean protein isolate and phosphatidylcholine in nanoemulsions: A fluorescence analysis. *Food Hydrocoll.* **2019**, *87*, 814–829. [CrossRef]

25. Garcia, V.M.; Lopez, A.M.; Hayes, M. Assessment of the functional properties of protein extracted from the brown seaweed *Himanthalia elongata* (Linnaeus) S. F. Gray. *Food Res. Int.* **2017**, *99*, 971–978. [CrossRef]
26. Gao, J.; Li, X.M.; Chen, F.L.; Piekoszewski, W.; Yang, Y.; Wang, B.; Liu, L.L.; Guan, H.N.; Shi, Y.G.; Zhang, N. The effect of extreme acid-induced soybean glycinin molten globules state on foaming ability. *Food Hydrocoll.* **2020**, *105*, 105819–105827. [CrossRef]
27. Rutkevičius, M.; Allred, S.; Velev, O.D.; Velikov, K.P. Stabilization of oil continuous emulsions with colloidal particles from water-insoluble plant proteins. *Food Hydrocoll.* **2018**, *8*, 89–95. [CrossRef]
28. Liu, F.; Wang, D.; Sun, C.; Gao, Y. Influence of polysaccharides on the physicochemical properties of lactoferrin–polyphenol conjugates coated β-carotene emulsions. *Food Hydrocoll.* **2016**, *52*, 661–669. [CrossRef]
29. Furuhashi, M.; Hatasa, Y.; Kawamura, S.; Shibata, T.; Akagawa, M.; Uchida, K. Identification of Polyphenol-Specific Innate Epitopes That Originated from a Resveratrol Analogue. *Biochemistry* **2017**, *56*, 4701–4712. [CrossRef] [PubMed]
30. Malhotra, A.; Coupland, J.N. The effect of surfactants on the solubility, zeta potential, and viscosity of soy protein isolates. *Food Hydrocoll.* **2004**, *18*, 101–108. [CrossRef]
31. Song, X.; Zhou, C.; Fu, F.; Chen, Z.; Wu, Q. Effect of high-pressure homogenization on particle size and film properties of soy protein isolate. *Ind. Crop. Prod.* **2013**, *43*, 538–544. [CrossRef]
32. Zhou, S.-D.; Lin, Y.-F.; Xu, X.; Meng, L.; Dong, M.-S. Effect of non-covalent and covalent complexation of (−)-epigallocatechin gallate with soybean protein isolate on protein structure and in vitro digestion characteristics. *Food Chem.* **2020**, *309*, 125718. [CrossRef]
33. Liu, F.; Wang, D.; Sun, C.; McClements, D.J.; Gao, Y. Utilization of interfacial engineering to improve physicochemical stability of beta-carotene emulsions: Multilayer coatings formed using protein and protein-polyphenol conjugates. *Food Chem.* **2016**, *205*, 129–139. [CrossRef] [PubMed]
34. Yong, Y.H.; Yamaguchi, S.; Matsumura, Y. Effects of Enzymatic Deamidation by Protein-Glutaminase on Structure and Functional Properties of Wheat Gluten. *J. Agric. Food Chem.* **2006**, *54*, 6034–6040. [CrossRef] [PubMed]
35. Zhu, H.; Damodaran, S. Heat-induced conformational changes in whey protein isolate and its relation to foaming properties. *J. Agric. Food Chem.* **1994**, *42*, 846–855. [CrossRef]
36. Sui, X.; Sun, H.; Qi, B.; Zhang, M.; Li, Y.; Jiang, L. Functional and conformational changes to soy proteins accompanying anthocyanins: Focus on covalent and non-covalent interactions. *Food Chem.* **2018**, *245*, 871–878. [CrossRef] [PubMed]
37. Jia, Z.; Zheng, M.; Tao, F.; Chen, W.; Huang, G.; Jiang, J. Effect of covalent modification by (−)-epigallocatechin-3-gallate on physicochemical and functional properties of whey protein isolate. *LWT* **2016**, *66*, 305–310. [CrossRef]
38. Harshadrai, S.R.; Rawel, M.; Kroll, J. Influence of a sugar moiety (rhamnosylgluco-side) at 3-O position on the reactivity of quercetin with whey proteins. *Int. J. Biol. Macromol.* **2003**, *32*, 109–120.
39. Ozdal, T.; Capanoglu, E.; Altay, F. A review on protein–phenolic interactions and associated changes. *Food Res. Int.* **2013**, *51*, 954–970. [CrossRef]
40. Quan, T.H.; Benjakul, S.; Sae-leaw, T.; Balange, A.K.; Maqsood, S. Protein–polyphenol conjugates: Antioxidant property, functionalities and their applications. *Trends Food Sci. Technol.* **2019**, *91*, 507–517. [CrossRef]
41. Liu, C.; Wang, X.; Ma, H.; Zhang, Z.; Gao, W.; Xiao, L. Functional properties of protein isolates from soybeans stored under various conditions. *Food Chem.* **2008**, *111*, 29–37. [CrossRef]
42. Bos, M.A.; van Vliet, T. Interfacial rheological properties of adsorbed protein layers and surfactants: A review. *Adv. Colloid Interface Sci.* **2011**, *91*, 437–471. [CrossRef]
43. Perrin, E.; Bizot, H.; Cathala, B.; Capron, I. Chitin Nanocrystals for Pickering High Internal Phase Emulsions. *Biomacromolecules* **2014**, *15*, 3766–3771. [CrossRef] [PubMed]
44. Li, D.; Zhao, Y.; Wang, X.; Tang, H.; Wu, N.; Wu, F.; Yu, D.; Elfalleh, W. Effects of (+)-catechin on a rice bran protein oil-in-water emulsion: Droplet size, zeta-potential, emulsifying properties, and rheological behavior. *Food Hydrocoll.* **2020**, *98*, 105306. [CrossRef]
45. Burgos Díaz, C.; Wandersleben, T.; Marqués, A.M.; Rubilar, M. Multilayer emulsions stabilized by vegetable proteins and polysaccharides. *Curr. Opin. Colloid Interface Sci.* **2016**, *25*, 51–57. [CrossRef]
46. Ju, M.; Zhu, G.; Huang, G.; Shen, X.; Zhang, Y.; Jiang, L.; Sui, X. A novel pickering emulsion produced using soy protein-anthocyanin complex nanoparticles. *Food Hydrocoll.* **2020**, *99*, 105329. [CrossRef]
47. Sun, C.; Gunasekaran, S. Effects of protein concentration and oil-phase volume fraction on the stability and rheology of menhaden oil-in-water emulsions stabilized by whey protein isolate with xanthan gum. *Food Hydrocoll.* **2009**, *23*, 165–174. [CrossRef]
48. Matsuyama, S.; Kazuhiro, M.; Nakauma, M.; Funami, T.; Nambu, Y.; Matsumiya, K.; Matsumura, Y. Stabilization of whey protein isolate-based emulsions via complexation with xanthan gum under acidic conditions. *Food Hydrocoll.* **2021**, *111*, 106365. [CrossRef]

foods MDPI

Article

Synergistic Effects of the Jackfruit Seed Sourced Resistant Starch and *Bifidobacterium pseudolongum* subsp. *globosum* on Suppression of Hyperlipidemia in Mice

Zeng Zhang [1], Yuanyuan Wang [1], Yanjun Zhang [2], Kaining Chen [3], Haibo Chang [1], Chenchen Ma [1], Shuaiming Jiang [1], Dongxue Huo [1], Wenjun Liu [4], Rajesh Jha [5] and Jiachao Zhang [1,5,*]

1 Key Laboratory of Food Nutrition and Functional Food of Hainan Province, College of Food Science and Engineering, Hainan University, Haikou 570228, China; zzeng66@163.com (Z.Z.); wang2019yuanyuan@163.com (Y.W.); c15364690101@163.com (H.C.); mcc19970104@163.com (C.M.); jsm15501859060@163.com (S.J.); 17784625873@126.com (D.H.)
2 Spice and Beverages Research Institute, Chinese Academy of Tropical Agricultural Science, Wanning 571533, China; zhangyanjun0305@163.com
3 Hainan Provincial People's Hospital, Haikou 570311, China; kainch@sina.com
4 Key Laboratory of Dairy Biotechnology and Engineering, Ministry of Education P.R.C., Key Laboratory of Dairy Products Processing, Ministry of Agriculture and Rural Affairs China, Inner Mongolia Agricultural University, Hohhot 010018, China; wjliu168@163.com
5 Department of Human Nutrition, Food and Animal Sciences, College of Tropical Agriculture and Human Resources, University of Hawaii at Manoa, Honolulu, HI 96822, USA; rjha@hawaii.edu
* Correspondence: jiachao@hainanu.edu.cn

Citation: Zhang, Z.; Wang, Y.; Zhang, Y.; Chen, K.; Chang, H.; Ma, C.; Jiang, S.; Huo, D.; Liu, W.; Jha, R.; et al. Synergistic Effects of the Jackfruit Seed Sourced Resistant Starch and *Bifidobacterium pseudolongum* subsp. *globosum* on Suppression of Hyperlipidemia in Mice. *Foods* **2021**, *10*, 1431. https://doi.org/10.3390/foods10061431

Academic Editor: Robert G. Gilbert

Received: 25 April 2021
Accepted: 30 May 2021
Published: 21 June 2021

Abstract: Approximately 17 million people suffer from cardiovascular diseases caused by hyperlipidemia, making it a serious global health concern. Among others, resistant starch (RS) has been widely used as a prebiotic in managing hyperlipidemia conditions. However, some studies have reported limited effects of RS on body weight and blood lipid profile of the host, suggesting further investigation on the synergistic effects of RS in combination with probiotics as gut microbes plays a role in lipid metabolism. This study evaluated the effects of jackfruit seed sourced resistant starch (JSRS) as a novel RS on mice gut microbes and hyperlipidemia by performing 16s rRNA and shotgun metagenomic sequencing. The results showed that 10% JSRS had a limited preventive effect on bodyweight and serum lipid levels. However, the JSRS promoted the growth of *Bifidobacterium pseudolongum*, which indicated the ability of *B. pseudolongum* for JSRS utilization. In the validation experiment, *B. pseudolongum* interacted with JSRS to significantly reduce bodyweight and serum lipid levels and had a therapeutic effect on hepatic steatosis in mice. Collectively, this study revealed the improvements of hyperlipidemia in mice by the synergistic effects of JSRS and *B. pseudolongum*, which will help in the development of "synbiotics" for the treatment of hyperlipidemia in the future.

Keywords: hyperlipidemia; resistant starch; *Bifidobacterium pseudolongum*; gut microbes; synbiotics

1. Introduction

Hyperlipidemia is a chronic systemic metabolic disease with lower levels of high-density lipoprotein cholesterol (HDL-C) and higher levels of total cholesterol (TC), triglycerides (TG), and low-density lipoprotein cholesterol (LDL-C) due to abnormal fat transport or metabolism [1]. It is considered as one of the risk factors of cardiovascular diseases, including atherosclerosis [2], coronary heart disease [3], and diabetes [4]. Intestinal microbes, the "second genome" of the human body [5], are inextricably linked to these diseases, and the intestinal microbiota of such patients is significantly different from that of healthy people [6]. The gut microbiome in hyperlipidemic subjects is also characterized by low microbial diversity, such as a high abundance of some taxa from the phylum Actinobacteria and lower abundance of many taxa from phyla Proteobacteria and Bacteroidetes [7].

Li et.al [8] found that *Alistipes*, *Intestinibacter*, *Subdoligranulum*, and *unidentified Ruminococcaceae* in the gut were significantly negatively correlated to TG, which was an important indicator of hyperlipidemia.

Resistant starch (RS) is defined as "the sum of starch and starch degradation products not absorbed by the small intestine of a healthy individual" [9], which is naturally found in cereal grains, seeds, heated starch, or starch-containing food [10]. RS has been widely recognized for its beneficial effects, such as improving insulin resistance and glucose homeostasis [11,12], maintaining colon health [13], controlling body weight [14], elevating large-bowel short-chain fatty acids (SCFAs) [15], and especially lowering blood lipid [16,17]. It has been suggested that a high dosage of RS administration in hamsters could increase HDL-C concentration and decrease TG, TC, and LDL-C concentrations to ameliorate hyperlipidemia [18], and *Bifidobacteria* and *Lactobacillus* in the gut were dramatically increased and positively correlated with blood lipid levels. Some of these functions of RS are linked to their fermentation characteristics, thus labeled as prebiotic. However, there are also inconsistent reports. Zhang et al. [19] tested the effect of RS on 19 volunteers and found no significant difference in body weight and HDL-C. However, LDL-C decreased significantly after four weeks of treatment. This was similar to a study reporting that body weight and liver triglycerides of C57BL/6J mice did not change after being fed a 45%-fat diet with 20% high-amylose-maize RS [20]. This might be because different types of RS are affected differently due to their fermentation characteristics, thereby affecting the gut physiology and health differently [21], which is affected by the gut microbial ecology [22]. Thus, these inconsistencies might be due to neglecting the interaction between RS and the key bacteria in the intestine. Therefore, we wanted to explore the crucial role of microbiota during the digestion of RS. Generally, jackfruit seed is a good source of RS [23]. However, there is limited or no information about the effect of jackfruit seed sourced resistant starch (JSRS) on the intestinal microbiota and hyperlipidemia, requiring exploration.

To address the problem, a two-stage experiment was conducted. In Stage I, we revealed the effect of JSRS on hyperlipidemia-related indexes and gut microbiome in mice. Meanwhile, *Bifidobacterium pseudolongum* was identified as the key microorganism that can utilize RS in the gut through 16s rRNA and shotgun metagenomic data. Based on this result, in vitro and in vivo validation experiments in Stage II were conducted to determine the ability of *B. pseudolongum (Bifidobacterium pseudolongum* subsp. *globosum)* to utilize JSRS and to explore the synergistic effects between the two for the prevention and treatment of hyperlipidemia in mice. This study highlights the irreplaceable role of intestinal microbes in the utilization of RS and a new idea that intervention studies of functional macromolecules, including RS, need to consider the involvement of exogenous microorganisms.

2. Methods

2.1. Animal Feeding and Diet Formula

Four-week-old C57BL/6 J male mice (n = 40) were sourced from Hunan SJA Laboratory Animal Co. Ltd. China, which were bred and housed in specific pathogen-free conditions with 12-h day and night light cycles at 22 ± 2 °C temperature and 55 ± 10% relative humidity. The normal-fat diet (NFD) was made up of 41% corn, 26% bran, 29% bean cake, 1% salt, 1% bone meal, 1% lysine, and 1% other. The high-fat diet (HFD) was made up of 1% cholesterol, 10% egg yolk, 10% lard, 0.2% sodium cholate, and 78.8% NFD. The proximate nutrient content of the NFD was as follows: crude protein, 18%; crude fat, 4%; crude fiber, 5%; calcium, 1.0–1.8%; and phosphorus, 0.6–1.2%. The proximate nutrient content of the HFD was as follows: crude protein, 17.6%; crude fat, 19.7%; crude fiber, 4%; calcium, 0.81–1.45%; and phosphorus, 0.56–1.04%. JSRS was provided by the Spice and Beverage Research Institute, Chinese Academy of Tropical Agricultural Sciences, Haikou, China. The *B. pseudolongum* was provided by Inner Mongolia Agricultural University, China.

2.2. Stage I, the Exploration Experiment

After a 7-day adaptation period, mice were randomly divided into 4 groups: NFD (n = 10); NFD plus JSRS (90% NFD plus 10% JSRS, n = 10); HFD (n = 10); HFD plus JSRS (90% HFD plus 10% JSRS, n = 10). Body weight was assessed at weeks 1, 2, 4, and 8. Feces were collected at weeks 2, 4, 8, and stored at −80 °C until further microbiota analysis. After 8 weeks of intervention, all mice were subjected to a 16-h fast and then euthanized and dissected. Blood was collected, and serum was isolated to determine TG as a blood lipid indicator. All fecal samples from week 2 and week 4 were used for high-throughput sequencing of the V3-V4 region of the bacterial 16s rRNA gene [24], and fecal samples from week 8 were processed for deep metagenomic sequencing [25]. The experimental design is presented graphically in Figure 1A.

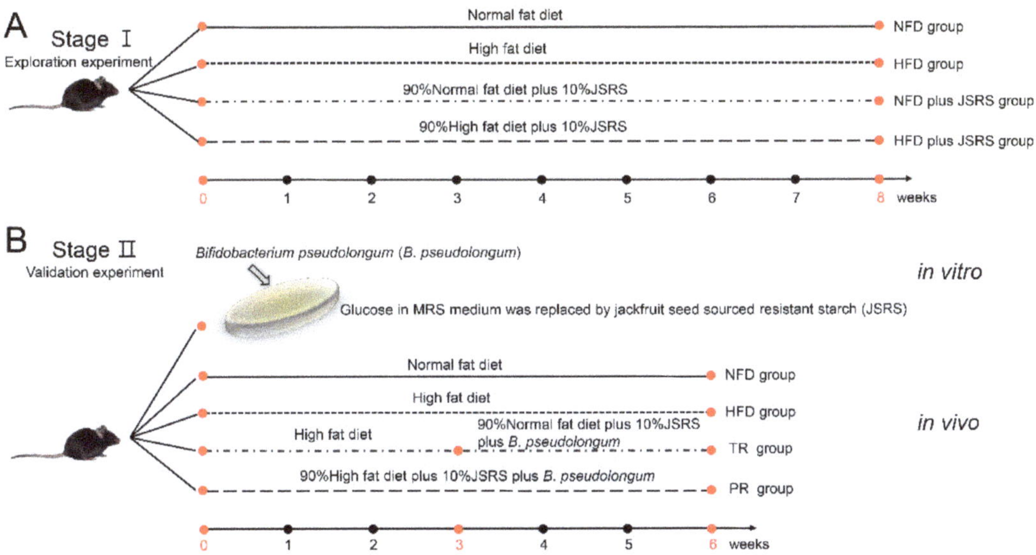

Figure 1. Experimental design. (**A**) In stage I, mice were randomly divided into four groups: NFD (normal-fat diet, n = 5); NFD plus JSRS (90% normal-fat diet plus 10% jackfruit seed sourced resistant starch, n = 6); HFD (high-fat diet, n = 6); HFD plus JSRS (90% high-fat diet plus 10% jackfruit seed sourced resistant starch, n = 6). The mice were kept for eight weeks. (**B**) The 4 groups that were treated for 3 weeks in order to make nutritionally obese mice model: NFD (normal-fat diet, n = 10); TR (HFD was chosen for the first 3 weeks for the development of obese mice model, then, 90% NFD plus 10% JSRSplus 8Log CFU *B. pseudolongum* infusions done later n = 10); HFD (high-fat diet, n = 10); PR (90% HFD plus 10% JSRS plus 8Log *B. pseudolongum* infusions, n = 10). After 3 weeks, three mice in each group were euthanized to observe abdominal fat accumulation. Then a 3-week intervention was continued on respective dietary treatments until the end of the experiment.

2.3. Stage II, the Validation Experiment

In vitro validation. *B. pseudolongum* was inoculated into MRS agar medium (agar 20 g, glucose 20 g, peptone 10 g, beef extract 10 g, yeast extract 5 g, $C_6H_5O_7(NH_4)_3$ 2 g, Tween-801 mL, CH_3COONa 5 g, K_2HPO_4 2 g, $MgSO_4$ 0.58 g, $MnSO_4$ 0.25 g, water 1 L) as the control and JSRS agar medium (agar 20 g, JSRS 20 g, peptone 10 g, beef extract 10 g, yeast extract 5 g, $C_6H_5O_7(NH_4)_3$ 2 g, Tween-80 1 mL, CH_3COONa 5 g, K_2HPO_4 2 g, $MgSO_4$ 0.58 g, $MnSO_4$ 0.25 g, water 1 L) anaerobic culture for 48 h, and the total number of microbial colonies were calculated. To determine the extent of starch utilization by *B. pseudolongum*, the plate was treated with iodine plus potassium iodide solution, and diameter of the colony and transparent circle was calculated.

In vivo validation. After 2 weeks of acclimatization on an NFD to the laboratory environment, mice (n = 40) were randomly subdivided equally into 4 groups [NFD, (n = 10); TR (HFD was fed for the first 3 weeks for making the obese mice model, then, 90% NFD plus 10% JSRS plus 8Log CFU *B. pseudolongum* infusions were done later, n = 10); HFD (n = 10); PR (90% HFD plus 10% JSRS plus 8Log CFU *B. pseudolongum* infusions, n = 10)] and were treated for 3 weeks to make the nutritionally obese mice model. After 3 weeks, three mice from each group were euthanized to determine abdominal fat accumulation. The rest of the mice were continued on the respective dietary treatment for another 3 weeks. The NFD and HFD groups were gavaged with an equal volume of normal saline as controls (Figure 1B). Body weight was recorded weekly. Feces were collected weekly and stored at −80 °C. After 8 weeks of intervention, mice were subjected to a 16-h fast and then were euthanized. Livers and abdominal fat were excised and weighed quickly. Then, livers were washed by phosphate buffer saline and processed for further analysis.

2.4. Scanning Electron Microscopy

The JSRS particle structure (granule form) analysis was done using scanning electron microscopy (SEM) as described by Zhang et al. [26] with some modifications. The samples were fixed on a sample holder with a silver plate and coated with a platinum film. The obtained samples were observed under a scanning electron microscope (Verios G4 UC, Thermo Fisher Scientific, Brno, South Moravia, Czech Republic).

2.5. Serum Lipid Levels Determination

The blood from the eyeball was collected and centrifuged at 571× g for 15 min at 4 °C. Commercial assay kits (Jian Cheng Biotechnology Co., Ltd., Nanjing, China) were used to measure blood lipid indices, including TG, TC, HDL-C, and LDL-C. The results of hepatic lipids were corrected for total protein concentration.

2.6. Liver Histology and Morphometric Assessment

The largest lobe of the fresh liver from each mouse in the different treatment groups was fixed in paraformaldehyde at room temperature and embedded in paraffin. Five-micrometer-thick sections of the liver tissue were cut, dewaxed, stained with hematoxylin-eosin (H&E), dehydrated, and sealed with neutral gum. To determine possible histopathological changes, the slides were then observed under an upright optical microscope (Nikon, Shanghai, China) for routine morphological evaluation and image acquisition (Servicebio Technology Co., Ltd., Wuhan, China).

2.7. DNA Extraction and High throughput Metagenomic Sequencing

Total DNA was extracted from fecal samples by using the QIAamp® DNA Stool Mini Kit (Qiagen, Hilden, Germany) according to the manufacturer's instruction, and purity and integrity of DNA were determined by 0.8% agarose gel electrophoresis. The concentration of DNA was determined by NanoDrop 2000 (Novogene Company, Beijing, China). The V3-V4 region of the 16s ribosomal RNA (rRNA) gene was amplified as previously described [24]. The shotgun metagenomic sequencing was performed by Illumina HiSeq 2500 instrument (Novogene, Beijing, China). About 150 bp paired-end gene fragments were obtained by sequencing, and a sequencing library with the length of the DNA fragments of 300 bp was prepared.

2.8. Bioinformatics for Amplification and Shotgun Metagenomic Sequencing Analysis

According to the quality scores, the sequences were filtered by using the sliding window approach. The low-quality sequences for which the average quality score over a 50-nt sliding window dropped below 30 were removed from the raw sequences. The QIIME (v1.9) platform was used for bioinformatics analysis [27]. The Operational Taxonomic Units (OTU) were selected by QIIME (default settings). OTUs were classified with a 97% threshold identity. An RDP Classifier was used to obtain the taxonomy against the RDP database using the representative [28]. The metagenomic reads were trimmed using Sickle software and subsequently aligned to the mouse genome (GRCm38.p6) to remove the host DNA fragments for subsequent analysis. For metagenomic species annotation, MetaPhlan2 software for taxonomic classification was employed [29]. HUMAnN2 was performed for metagenomic functional features and metabolic-pathway annotation based on the UniRef90 database [29].

2.9. Statistical Analysis and Figure Construction

The statistical analyses were performed using R. Data are presented as mean ± standard error (SE). The differential abundances of the genera were tested using the Wilcoxon rank-sum test and the Kruskal–Wallis test, and $p < 0.05$ were considered to be significantly different genera. Principal coordinate analysis (PCoA) was performed using the "ade4" package. The package "ggplot2" was used for generating line charts, box plots, and bar charts. The heatmap was constructed using the "pheatmap" package. The *p*-value was filtered and corrected by the "DESeq2" package in the bubble diagram. The microbial ecological networks were inferred by Spearman's rank correlation coefficient from the metagenomic sequencing data and visualized in Cytoscape (Version 3.7.1).

3. Results

3.1. The Granules Form of JSRS

The morphology of JSRS granules is shown in Figure S1A. The SEM images of JSRS samples showed that the starch granules were mainly semi-oval or bell-shaped. The particle size of starches was about 10 μm, which was similar to our previous finding [26].

3.2. 10% JSRS Could neither Significantly Restrain the Body Weight Gain nor Maintain Serum Lipid Levels

There was no significant difference in body weight among treatment groups ($p > 0.05$) at the beginning of the study (0 week). After eight weeks, the body weight of the HFD group and HFD plus JSRS group were significantly higher than that of the NFD group. There was no significant difference between the HFD and HFD plus JSRS groups (Figure S1B). Moreover, the TG levels were significantly lower in the NFD and NFD plus JSRS groups as compared with the HFD group, while the HFD plus JSRS group was significantly higher than that of the NFD group (Figure S1C).

3.3. The JSRS Regulated Dysbiosis in Intestinal Microbiota Caused by a High-Fat Diet

The alpha and beta diversity of the mice gut microbes in different groups were compared among the second, fourth, and eighth weeks based on the high-throughput sequencing data of 16s rRNA. In the fourth week's results, the points representing NFD and HFD groups had different clustering trends in the PCoA plots (Figure 2A), which indicated that the HFD could significantly affect the gut microbes. In the eighth week, the alpha diversity index of the JSRS group remained at a high level, and the Shannon index of the NFD plus JSRS group was significantly higher than the NFD group (Figure S2). The gut microbe structure of the HFD plus JSRS group was more similar to that of the NFD group, compared with that of the HFD group at week eight (Figure 2B), which implied the addition of JSRS to the HFD made it possible to correct the HFD-induced intestinal dysbiosis.

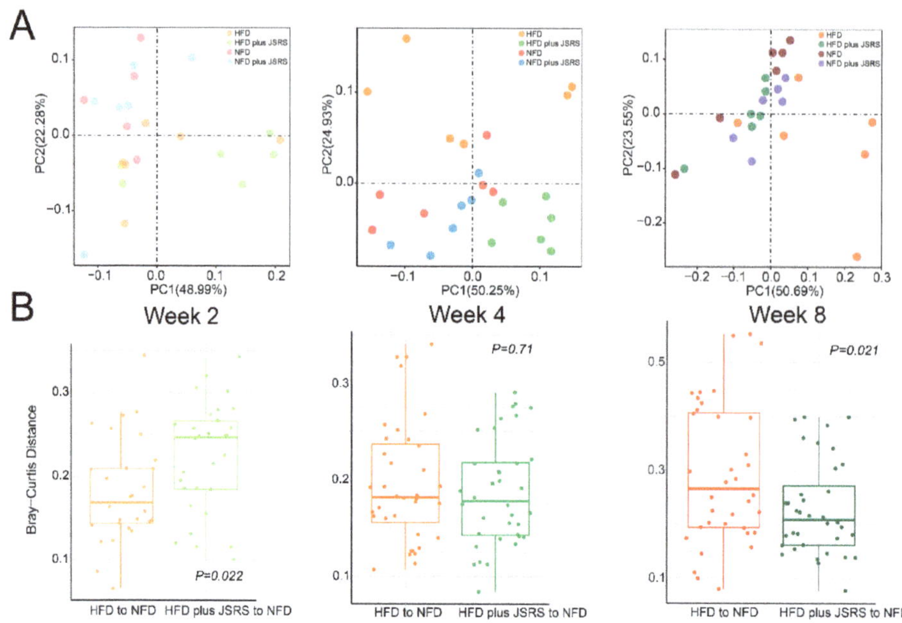

Figure 2. Changes in microbial structure and similarity at different time points. (**A**) Principal coordinate analysis (PCoA) showed the changes in microbial composition based on bray-curtis distance among four groups at weeks 2, 4, and 8 in Stage I. Each point represents the composition of the microbiota of one sample. (**B**) Comparison of the similarity between HFD group to NFD group and HFD plus JSRS group to NFD group based on Bray–Curtis distance at week 2, 4, and 8 in Stage I. Lower values indicate similarity to the NFD group. (JSRS, Jackfruit seed sourced resistant starch).

3.4. Bifidobacterium Pseudolongum Was the Potential Functional Intestinal Microorganism in JSRS Utilization

Since there was a difference in the intestinal microbiota between NFD and HFD groups, the differential microorganisms at the genus levels in JSRS utilization were found on Mann–Whitney tests by combined NFD plus JSRS group and HFD plus JSRS group into the JSRS group. The remaining two groups were combined into the NFD and HFD group (N and H group), based on the high-throughput 16s rRNA sequencing abundance results at week 2 and week 4. Differential microorganisms at the species level in the eighth week were calculated by the same method, as the abundance results were derived from metagenomic sequencing data.

In the fourth week, the genus abundance of *Bifidobacterium, Parabacteroides, Akkermansia, Robinsoniella,* and *Unclassified Beijerinckiaceae* in the JSRS group was significantly higher than that in the N and H group, while the *Bifidobacterium* and *Parabacteroides* were also significantly different in the second week. The abundance of *B. pseudolongum, Akkermansia muciniphila,* and *Subdoligranulum unclassified* were significantly higher in the JSRS group than in the N and H group at week eight (Figure 3A). Overall, the average abundance of *Bifidobacterium* was significantly higher than that of the N and H group at weeks two and four, which highlighted the importance of the species belonging to the *Bifidobacterium* genus. Accordingly, for fecal samples collected at week eight, shotgun metagenomic sequencing was applied to identify the potential specific species of *Bifidobacterium,* which utilized JSRS in mice guts. Metagenomic species annotation results confirmed that *B. pseudolongum* was the dominant species capable of utilizing JSRS for their growth in mice gut (Figure 3A). Using the same grouping method, the top twenty metabolic pathways with significant baseline differences were screened (Figure 3B). Compared with the JSRS group, 6-hydroxymethyl-dihydropterin diphosphate biosynthesis I, anaerobic energy metabolism, anhydro-muropeptides recycling, chorismate biosynthesis from 3-

dehydroquinate, L-lysine biosynthesis, Purine nucleobases degradation, superpathway of N-acetylglucosamine, N-acetylmannosamine, N-acetylneuraminate degradation, and Superpathway of phospholipid biosynthesis were significantly more enriched in the N and H group. The remaining 12 pathways such as L-tryptophan biosynthesis and superpathway of pyrimidine ribonucleotides de novo biosynthesis were enriched in JSRS group. Collectively, the major differential metabolic pathways were more active in the JSRS group. JSRS significantly altered specific metabolic pathways in mice gut microbes.

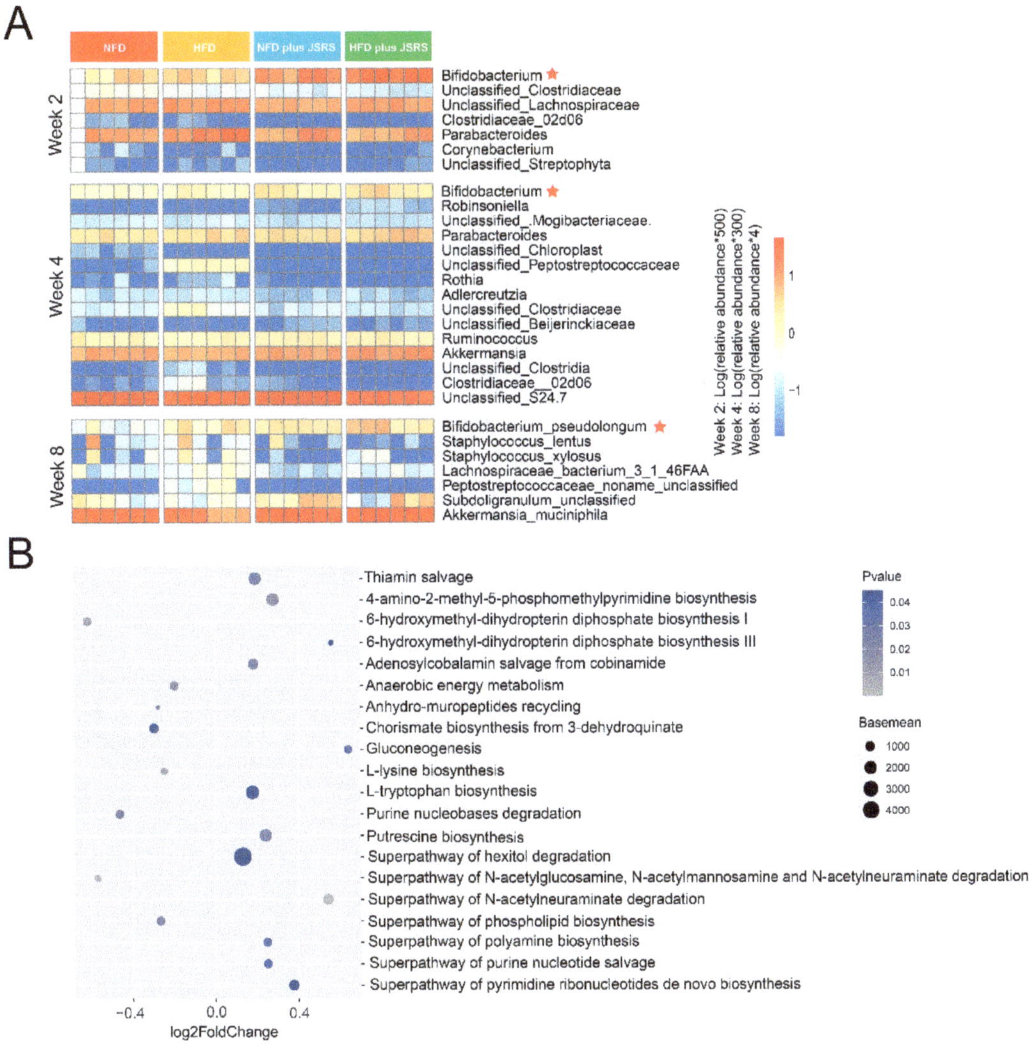

Figure 3. Comparison of differential microbial and metabolic pathways. (**A**) Variations of differential bacteria in different groups at weeks 2, 4, and 8 in Stage I. The significant difference genera (week 2 and week 4) and species (week 8) were screened (Wilcoxon rank-sum test). The depth degree of color represents the relative abundance of the genus or species (blue indicates a small number, and red indicates a large number). (**B**) Metabolic pathway analysis was performed based on the shotgun metagenomic sequencing data in Stage I. Significantly different pathways with baseline ranking in the top 20 were screened out. Log_2 foldchange value greater than 0 indicated that the enrichment abundance in the JSRS group (the combination of the HFD plus JSRS group and the NFD plus JSRS group) is greater than that in the N and H group (the combination of the NFD group and the HFD group). (JSRS, Jackfruit seed sourced resistant starch).

3.5. The Ability of B. pseudolongum to Utilize the JSRS In Vitro

Since we found *B. pseudolongum* in mice gut to be strongly correlated with JSRS intake, to test this hypothesis, we performed an in vitro validation experiment in Stage II. A strain of *B. pseudolongum* from the gut of cattle was used in an in vitro validation experiment. The glucose in the MRS agar medium was replaced entirely by JSRS (JSRS agar medium), and *B. pseudolongum* was inoculated. After 48 h, the average growth of *B. pseudolongum* in the starch agar medium reached 6Log CFU/mL, although the average growth of the control group was 7Log CFU/mL. The ratio of JSRS utilization circle reached 8.3 (Figure S3A). These results proved that *B. pseudolongum* could grow by using JSRS as the primary carbon source in vitro.

3.6. The Synergistic Effect of JSRS and B. pseudolongum Suppressed Hyperlipidemia in Mice

Based on the results above, another hypothesis in the present study is that the synergistic effect of JSRS and *B. pseudolongum* could suppress hyperlipidemia in mice. Therefore, we performed an in vivo validation experiment to test this hypothesis (Figure 1B). At week three, there was no significant difference in body weight between the TR and HFD groups, both of which were significantly higher than the NFD group. The body weight of the TR group was significantly higher than the NFD group (Figure 4A), although there was no significant difference in body weight between the two groups. After another three weeks, the body weight of the TR group not only gradually decreased but also was lower than the NFD group (Figure 4B). Moreover, the body weight of the PR group was consistently lower than the HFD group, even lower than the NFD group. The synergistic effect of JSRS and *B. pseudolongum* prevented high fat diet-induced obesity in mice.

Data on serum lipid levels and abdominal fat of mice further verified that the synergistic effect could reduce abdominal fat and blood lipid in mice. At week six, with the synergistic effect of both, the abdominal white fat weight was effectively reduced and controlled in PR and TR groups, which was significantly lower than that in the HFD group. TR group was slightly higher than the NFD group (Figure 4C), but no significant differences existed. The same trend was found in the anatomy of mice (Figure S3B). Similarly, the level of blood lipids of mice reinforced the conclusion that JSRS and *B. pseudolongum* worked together to reduce body weight and serum lipid levels in mice. The levels of TC, TG, and LDL showed the highest trend in the HFD group (Figure 4D). The levels of these three blood lipid parameters were significantly reduced by the synergistic effect, although there was no significant difference in HDL-C level between the four groups. TC and TG levels were lower in the TR group than in the NFD group, while they were not statistically significantly different in LDL-C levels between the two groups.

The histopathology results showed that the liver structure of the NFD group was normal, and the liver lobules were clearly visible. The liver cells were intact, without rupture or autolysis, and closely arranged. No fatty vacuoles were observed. In contrast, the liver of the HFD group mice had many fatty vacuoles and disturbed hepatic cord distribution. Hepatocytes were loosely arranged and had fuzzy margins. Evidently, the HFD caused severe pathological changes in the liver of mice. The fatty infiltration of hepatocytes was significantly reduced in the TR group, although the arrangement was slightly disturbed. Compared with the HFD group, the number of fatty vacuoles in the liver of the PR group was significantly reduced, and the hepatocytes had a clearer contour and tighter structure (Figure 4E).

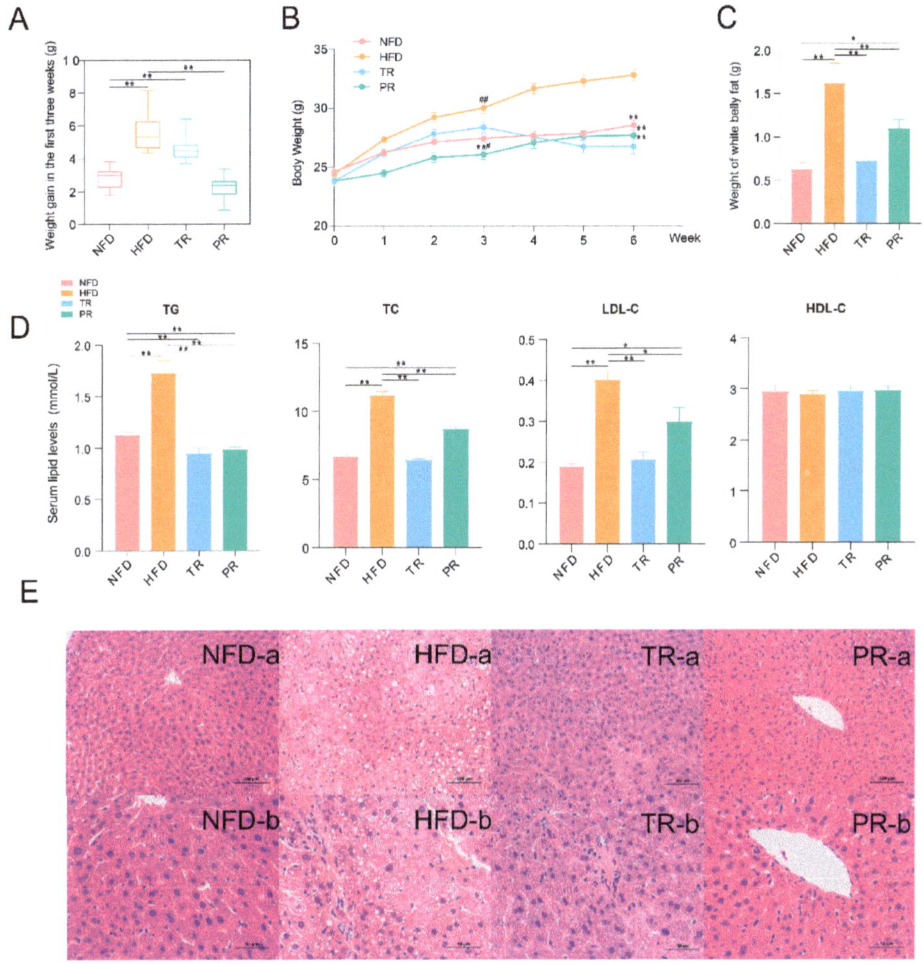

Figure 4. Effect of Jackfruit seed sourced resistant starch (JSRS)plus *Bifidobacterium pseudolongum* on hyperlipidemia in mice. (**A**) Body weight of four groups at each time point in Stage II. (Wilcoxon rank-sum test, * $p < 0.05$, ** $p < 0.01$ compared with HFD; # $p < 0.05$, ## $p < 0.01$ compared with NFD, error bar: mean ± SE) (**B**) Weight gain was compared between the four groups during the first three weeks of the study at each time point in Stage II (Wilcoxon rank-sum test, * $p < 0.05$, ** $p < 0.01$, NS means no significant difference, error bar: mean ± SE) (**C**) Comparison of the weight of white fat in the abdomen of four groups of mice. (Wilcoxon rank-sum test, * $p < 0.05$, ** $p < 0.01$, error bar: mean ± SE) (**D**) Comparison of TC, TG, LDL-C, and HDL-C levels among the four groups in Stage II. (Wilcoxon rank-sum test, * $p < 0.05$, ** $p < 0.01$, error bar: mean ± SE) (**E**) Histopathological analysis of the liver sections of mice in four groups at 200× and 400× magnification in Stage II. ("-a" and "-b" were liver section images at the same position, but with different magnification. "-a" means 200×; "-b" means 400×).

3.7. The Potential Mechanism of the Synergistic Effect of JSRS and B. pseudolongum for Improving the Symptom of Hyperlipidemia in Mice

The previous analysis revealed the differential bacteria and metabolic pathways. To better reveal the correlation of JSRS, metagenomic species, microbial metabolic pathways, and serum lipid indicators, we calculated Spearman's rank correlation coefficients based on the data from Stage I of the experiment and constructed correlation networks for them (Figure 5). *B. pseudolongum*, *Subdoligranulum unclassified*, and *Akkermansia muciniphila* were positively correlated with JSRS, and *B. pseudolongum* had the highest correlation with JSRS compared to the other two bacteria. The adenosylcobalamin salvage from cobinamide I (M) as well as TG was positively correlated with *B. pseudolongum;* conversely, the thiamin salvage (E), putrescine biosynthesis (P), and superpathway of polyamine biosynthesis (S) were negatively correlated. The network diagram showed the interconnections and interactions among JSRS, differential strains, differential metabolic pathways, and TG.

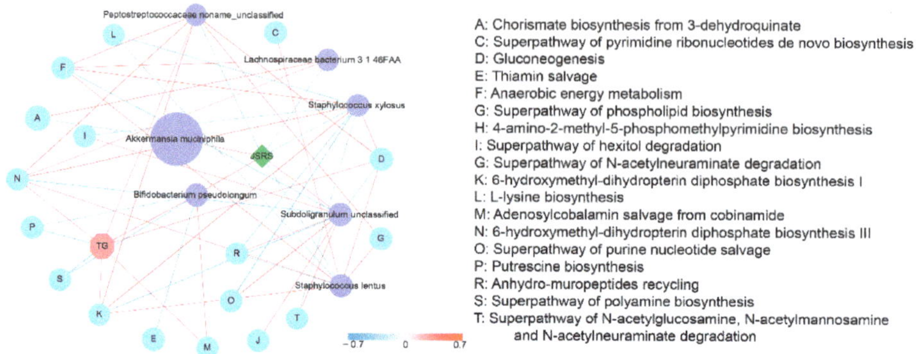

Figure 5. The correlation network among Jackfruit seed sourced resistant starch (JSRS), differential species, metabolic pathways, and TG. The R-value less than −0.4 or greater than 0.4 was selected. The width and color (red indicates a positive correlation, while blue indicates a negative correlation) of the edge are proportional to the correlation intensity. The green node represents JSRS, purple nodes represent species, light blue nodes represent metabolic pathways, and light red nodes represent TG. The node size is proportional to the mean abundance in the respective population.

4. Discussion

The results of this study show that only 10% of JSRS had a limited preventive effect against hyperlipidemia in mice. On the other hand, the combination of JSRS and *B. pseudolongum* improved not only body weight, abdominal white fat, and serum lipid levels, but also had preventive and therapeutic effects on hepatic steatosis in liver cells. In previous studies, JSRS was able to regulate blood glucose levels and lipid metabolism [30], and prevent and treat obesity [31]. However, it has also been shown that the effects of JSRS on body weight and visceral fat were limited [12,32]. Similarly, this study shows that the addition of 10% JSRS to diet did not significantly reduce body weight and serum lipid levels in mice. The possible reason for this result may be that a high-fat diet could minimize the RS fermentation, resulting in lower levels of propionic acid and butyric acid [33]. SCFAs are closely related to the host's health, especially some metabolic diseases [34,35]. This condition led to an attenuated beneficial effect of RS in rats on a 20% high-fat diet, compared to a low-fat diet [33]. Another possible reason may be the structural differences of RS. It has been reported that the RS with lower crystallinity and double helix content had a poor effect on the body weight and serum TC content of mice induced by a high-fat diet. Resistant starch with high crystallinity and double helix content had a more stable structure, resulting in a more stable and slow fermentation in the intestine, allowing the substance to be evenly distributed in the intestine and meeting the needs of the distal colon [17]. Moreover, the variation of RS polymorphism would also lead to ecological changes in the microbial community structure of the colon. Different types of RS produce slightly different proportions of SCFAs and might undergo different fermentation patterns in the gut of obese mice, resulting in different effects on intestinal health [36]. However, our results indicated that JSRS ingestion could significantly elicit a response from *B. pseudolongum*. Furthermore, we observed that the major differential metabolic pathways such as L-tryptophan biosynthesis, gluconeogenesis III and thiamin salvage II were more active in the JSRS group. It has been shown that L-tryptophan can regulate immunity [37] and intestinal homeostasis [38]. Proper nutritional supplementation with tryptophan can prevent or reduce inflammation of the gut [39]. Bacterial fermentation of fiber can improve glycemic control by producing succinate and thus activating intestinal gluconeogenesis [40]. The thiamin, as vitamin B1, is a cofactor for many enzymes indispensable for glucose and energy metabolism [41]. Apparently, JSRS can significantly improve the metabolism of intestinal microorganisms in mice and may be involved in related pathways such as host immunity and energy metabolism. Based on these findings, we hypothesized that the synergistic effect of JSRS and *B. pseudolongum* could prevent hyperlipidemia in mice. The hypothesis was then validated in the in vitro and in vivo experiments in Stage II. JSRS is not only a prebiotic but also a carbon source that promotes the growth of certain beneficial microorganisms in the gut. The pairing of prebiotics and probiotics is also gradually emerging in many studies. Ma et al. [25] found that constant supplementation with low galactose improves the stability of the intestinal microbes. Thus, probiotics may be able to work more reliably or better in the gut if they are supported by their substrates. Numerous studies have shown that the combined effects of probiotics and prebiotics can positively impact a wide range of human diseases, such as enhancing Disease Activity Score (DAS 28) and visual analog scores in rheumatoid patients, modulating plasma nitric oxide, rising glutathione [42], improving some symptoms of type 2 diabetes and quality of life [43], prolonging life, and reducing liver cancer cell proliferation in mice with malignant leukemia [44]. Research of synbiotics is expected to be promoted more with the updated definition of synbiotics [45]. *Bifidobacteria* are major degraders of RS, producing butyrate during the breakdown of RS to improve host health [46,47], and RS is the prebiotic capable of regulating changes in gut microbes [48]. There is no doubt that the combination of JSRS and *B. pseudolongum* as potential "synbiotics" can have a far more positive impact on the gut and health of mice than only JSRS supplementation.

In this research, differential microbes capable of utilizing JSRS in the intestine of mice on HFD and HFD were found. Interestingly, although 10% JSRS did not significantly improve body weight and serum TG level in mice, it was still able to regulate certain critical intestinal microbes, such as *Akkermansia*, *Ruminococcus*, and *Bifidobacterium*, which could promote the production of SCFAs that affect metabolic diseases positively to the host [49,50].

5. Conclusions

JSRS maintained the homeostasis of the intestinal microbes by correcting the damaging effects of a high-fat diet on gut microbes. As expected, *B. pseudolongum*, which are differential microorganisms at weeks two, four, and eight, could use JSRS to grow both in vitro and in vivo, and exerted excellent synergistic effects to treat and prevent hyperlipidemia and had a positive effect on hepatic steatosis in mice. Collectively, this study provided evidence on the suppression of hyperlipidemia in mice by the synergistic effects of JSRS and *B. pseudolongum*, which can support the development of "synbiotics" for the treatment of hyperlipidemia in the future. Collectively, this study provided evidence on the suppression of hyperlipidemia in mice by the synergistic effects of JSRS and *B. pseudolongum*, which can support the development of "synbiotics" for the treatment of hyperlipidemia in the future. By understanding the effects of JSRS on the gut microbiome and hyperlipidemia, it will help to obtain better applications in different fields, and provide a reference for the functional development of new varieties in the starch industry. We suggest the application of JSRS in combination with *B. pseudolongum* (synbiotics) to target the intestinal microbiota to alleviate hyperlipidemia in order to achieve the high value utilization of JSRS as a functional health food.

Supplementary Materials: The following are available online at https://www.mdpi.com/article/10.3390/foods10061431/s1, Figure S1. Effects of Jackfruit seed sourced resistant starch (JSRS) on body weight and serum lipid levels of mice in Stage I. Figure S2. Alpha diversity analysis at different time points among the four groups in Stage I. Figure S3. Verification experiment of *Bifidobacterium pseudolongum* in vitro and the anatomy of mice in Stage II.

Author Contributions: Conceptualization, J.Z. and R.J.; Methodology, J.Z.; Formal Analysis, Z.Z., Y.W., C.M. and S.J.; Investigation, H.C., Resources, Y.Z., J.Z. and W.L.; Data Curation, Z.Z. and Y.W.; Writing—Original Draft Preparation, Z.Z. and Y.W.; Writing—Review and Editing, Z.Z., K.C., J.Z. and R.J.; Visualization, Z.Z., D.H., and Y.W.; Supervision, J.Z.; Project Administration, J.Z., Funding Acquisition, J.Z. All authors have read and agreed to the published version of the manuscript.

Funding: This work was supported by the Natural Science Foundation of Hainan Province (No. 320RC513) and Key R & D programs in Hainan (No. ZDYF2019150 and ZDYF2018111).

Institutional Review Board Statement: The study was conducted according to the guidelines of the Declaration of Helsinki and approved by the Ethics Committee of Hainan University (AWER Issue No. HNU-EC-FS39, 09 May 2019).

Data Availability Statement: The sequence data reported in this paper have been deposited in the NCBI database (metagenomic sequencing data: PRJNA669624, PRJNA669572).

Acknowledgments: We sincerely appreciate Yanjun Zhang from Spice and Beverage Research Institute, Chinese Academy of Tropical Agricultural Sciences, who supplied the jackfruit seed sourced resistant starch for the present study.

Conflicts of Interest: The authors declare no conflict of interest.

Abbreviations

The following abbreviations are used in this manuscript:

RS	Resistant starch
JSRS	Jackfruit seed sourced resistant starch
LDL-C	Low-density lipoprotein cholesterol
HDL-C	High-density lipoprotein cholesterol
TC	Total cholesterol
TG	Triglycerides

References

1. Panahi, Y.; Ahmadi, Y.; Teymouri, M.; Johnston, T.P.; Sahebkar, A. Curcumin as a potential candidate for treating hyper-lipidemia: A review of cellular and metabolic mechanisms. *J. Cell Physiol.* **2018**, *233*, 141–152. [CrossRef]
2. Li, W.; Liu, M.; Wu, B.; Liu, H.; Wang, L.-C.; Tan, S. Serum lipid levels and 3-month prognosis in Chinese patients with acute stroke. *Adv. Ther.* **2008**, *25*, 329–341. [CrossRef] [PubMed]
3. Chiavaroli, L.; Nishi, S.K.; Khan, T.A.; Braunstein, C.R.; Glenn, A.J.; Mejia, S.B.; Rahelic, D.; Kahleova, H.; Salas-Salvado, J.; Jenkins, D.J.A.; et al. Portfolio Dietary Pattern and Cardiovascular Disease: A Systematic Review and Meta-analysis of Con-trolled Trials. *Prog. Cardiovasc. Dis.* **2018**, *61*, 43–53. [CrossRef] [PubMed]
4. Khursheed, R.; Singh, S.K.; Wadhwa, S.; Gulati, M.; Awasthi, A. Therapeutic potential of mushrooms in diabetes mellitus: Role of polysaccharides. *Int. J. Biol. Macromol.* **2020**, *164*, 1194–1205. [CrossRef] [PubMed]
5. Sender, R.; Fuchs, S.; Milo, R. Are We Really Vastly Outnumbered? Revisiting the Ratio of Bacterial to Host Cells in Hu-mans. *Cell* **2016**, *164*, 337–340. [CrossRef] [PubMed]
6. Hwang, I.Y.; Chang, M.W. Engineering commensal bacteria to rewire host–microbiome interactions. *Curr. Opin. Biotechnol.* **2020**, *62*, 116–122. [CrossRef]
7. Wang, Z.; Koonen, D.; Hofker, M.; Fu, J. Gut microbiome and lipid metabolism: From associations to mechanisms. *Curr. Opin. Lipidol.* **2016**, *27*, 216–224. [CrossRef]
8. Li, H.; Liu, B.; Song, J.; An, Z.; Zeng, X.; Li, J.; Jiang, J.; Xie, L.; Wu, W. Characteristics of Gut Microbiota in Patients with Hy-pertension and/or Hyperlipidemia: A Cross-Sectional Study on Rural Residents in Xinxiang County, Henan Province. *Microorganisms* **2019**, *7*, 399. [CrossRef]
9. Charalampopoulos, D.; Wang, R.; Pandiella, S.; Webb, C. Application of cereals and cereal components in functional foods: A review. *Int. J. Food Microbiol.* **2002**, *79*, 131–141. [CrossRef]
10. Fuentes-Zaragoza, E.; Riquelme-Navarrete, M.J.; Sánchez-Zapata, E.; Pérez-Álvarez, J.A. Resistant starch as functional in-gredient: A review. *Food Res. Int.* **2010**, *43*, 931–942. [CrossRef]
11. Bodinham, C.L.; Smith, L.; Thomas, E.L.; Bell, J.D.; Swann, J.R.; Costabile, A.; Russell-Jones, D.; Umpleby, A.M.; Robertson, M.D. Efficacy of increased resistant starch consumption in human type 2 diabetes. *Endocr. Connect.* **2014**, *3*, 75–84. [CrossRef] [PubMed]
12. Johnston, K.L.; Thomas, E.L.; Bell, J.D.; Frost, G.S.; Robertson, M.D. Resistant starch improves insulin sensitivity in metabol-ic syndrome. *Diabet. Med.* **2010**, *27*, 391–397. [CrossRef]
13. Zhao, Y.; Hasjim, J.; Li, L.; Jane, J.-L.; Hendrich, S.; Birt, D.F. Inhibition of Azoxymethane-Induced Preneoplastic Lesions in the Rat Colon by a Cooked Stearic Acid Complexed High-Amylose Cornstarch. *J. Agric. Food Chem.* **2011**, *59*, 9700–9708. [CrossRef]

14. Harazaki, T.; Inoue, S.; Imai, C.; Mochizuki, K.; Goda, T. Resistant starch improves insulin resistance and reduces adipose tissue weight and CD11c expression in rat OLETF adipose tissue. *Nutrients* **2014**, *30*, 590–595. [CrossRef] [PubMed]
15. Regina, A.; Bird, A.; Topping, D.; Bowden, S.; Freeman, J.; Barsby, T.; Kosar-Hashemi, B.; Li, Z.; Rahman, S.; Morell, M. High-amylose wheat generated by RNA interference improves indices of large-bowel health in rats. *Proc. Natl. Acad. Sci. USA* **2006**, *103*, 3546–3551. [CrossRef]
16. Nichenametla, S.N.; Weidauer, L.A.; Wey, H.E.; Beare, T.M.; Specker, B.L.; Dey, M. Resistant starch type 4-enriched diet low-ered blood cholesterols and improved body composition in a double blind controlled cross-over intervention. *Mol. Nutr. Food Res.* **2014**, *58*, 1365–1369. [CrossRef]
17. Xu, J.; Ma, Z.; Li, X.; Liu, L.; Hu, X. A more pronounced effect of type III resistant starch vs. type II resistant starch on ameliorating hyperlipidemia in high fat diet-fed mice is associated with its supramolecular structural characteristics. *Food Funct.* **2020**, *11*, 1982–1995. [CrossRef]
18. Liang, R.; Huang, J.; Wu, X.; Fan, J.; Xu, Y.; Wu, C.; Jin, Y.; Zhou, R. Effect of raw material and starters on the metabolite con-stituents and microbial community diversity of fermented soy sauce. *J. Sci. Food Agric.* **2019**, *99*, 5687–5695. [CrossRef]
19. Zhang, L.; Ouyang, Y.; Li, H.; Shen, L.; Ni, Y.; Fang, Q.; Wu, G.; Qian, L.; Xiao, Y.; Zhang, J.; et al. Metabolic phenotypes and the gut microbiota in response to dietary resistant starch type 2 in normal-weight subjects: A randomized crossover trial. *Sci. Rep.* **2019**, *9*, 1–11. [CrossRef]
20. Kieffer, D.A.; Piccolo, B.D.; Marco, M.L.; Kim, E.B.; Goodson, M.L.; Keenan, M.J.; Dunn, T.N.; Knudsen, K.E.B.; Martin, R.J.; Adams, S.H. Mice Fed a High-Fat Diet Supplemented with Resistant Starch Display Marked Shifts in the Liver Metabolome Concurrent with Altered Gut Bacteria. *J. Nutr.* **2016**, *146*, 2476–2490. [CrossRef]
21. Jha, R.; Zijlstra, R.T. Physico-chemical properties of purified starch affect their in vitro fermentation characteristics and are linked to in vivo fermentation characteristics in pigs. *Anim. Feed. Sci. Technol.* **2019**, *253*, 74–80. [CrossRef]
22. Jha, R.; Fouhse, J.M.; Tiwari, U.P.; Li, L.; Willing, B.P. Dietary Fiber and Intestinal Health of Monogastric Animals. *Front. Vet. Sci.* **2019**, *6*, 48. [CrossRef]
23. Zhang, Y.; Li, B.; Xu, F.; He, S.; Zhang, Y.; Sun, L.; Zhu, K.; Li, S.; Wu, G.; Tan, L. Jackfruit starch: Composition, structure, functional properties, modifications and applications. *Trends Food Sci. Technol.* **2021**, *107*, 268–283. [CrossRef]
24. Ma, C.C.; Huo, D.X.; You, Z.K.; Peng, Q.N.; Jiang, S.M.; Chang, H.B.; Zhang, J.C.; Zhang, H.P. Differential pattern of indige-nous microbiome responses to probiotic Bifidobacterium lactis V9 consumption across subjects. *Food Res. Int.* **2020**, *136*. [CrossRef]
25. Ma, C.C.; Wasti, S.; Huang, S.; Zhang, Z.; Mishra, R.; Jiang, S.M.; You, Z.K.; Wu, Y.X.; Chang, H.B.; Wang, Y.Y.; et al. The gut microbiome stability is altered by probiotic ingestion and improved by the continuous supplementation of galactooligo-saccharide. *Gut Microbes* **2020**, *12*. [CrossRef]
26. Zhang, Y.; Hu, M.; Zhu, K.; Wu, G.; Tan, L. Functional properties and utilization of Artocarpus heterophyllus Lam seed starch from new species in China. *Int. J. Biol. Macromol.* **2018**, *107*, 1395–1405. [CrossRef]
27. Caporaso, J.G.; Kuczynski, J.; Stombaugh, J.; Bittinger, K.; Bushman, F.D.; Costello, E.K.; Fierer, N.; Peña, A.G.; Goodrich, J.K.; Gordon, J.I.; et al. QIIME Allows Analysis of High-Throughput Community Sequencing data. *Nat. Methods* **2010**, *7*, 335–336. [CrossRef]
28. Cole, J.R.; Chai, B.; Farris, R.J.; Wang, Q.; Kulam-Syed-Mohideen, A.S.; McGarrell, D.M.; Bandela, A.M.; Cardenas, E.; Garrity, G.; Tiedje, J.M. The ribosomal database project (RDP-II): Introducing myRDP space and quality controlled public data. *Nucleic Acids Res.* **2006**, *35*, D169–D172. [CrossRef]
29. Franzosa, E.A.; McIver, L.J.; Rahnavard, G.; Thompson, L.R.; Schirmer, M.; Weingart, G.; Lipson, K.S.; Knight, R.; Caporaso, J.G.; Segata, N.; et al. Species-level functional profiling of metagenomes and metatranscriptomes. *Nat. Methods* **2018**, *15*, 962–968. [CrossRef]
30. Zhang, L.; Li, H.T.; Shen, L.; Fang, Q.C.; Qian, L.L.; Jia, W.P. Effect of Dietary Resistant Starch on Prevention and Treatment of Obesity-related Diseases and Its Possible Mechanisms. *Biomed. Environ. Sci.* **2015**, *28*, 291–297.
31. Shen, R.-L.; Zhang, W.-L.; Dong, J.-L.; Ren, G.-X.; Chen, M. Sorghum resistant starch reduces adiposity in high-fat diet-induced overweight and obese rats via mechanisms involving adipokines and intestinal flora. *Food Agric. Immunol.* **2013**, *26*, 120–130. [CrossRef]
32. Fukushima, M.; Ohashi, T.; Kojima, M.; Ohba, K.; Shimizu, H.; Sonoyama, K.; Nakano, M. Low density lipoprotein receptor mRNA in rat liver is affected by resistant starch of beans. *Lipids* **2001**, *36*, 129–134. [CrossRef]
33. Charrier, J.A.; Martin, R.J.; McCutcheon, K.L.; Raggio, A.M.; Goldsmith, F.; Goita, M.; Senevirathne, R.N.; Brown, I.L.; Pelkman, C.; Zhou, J.; et al. High fat diet partially attenuates fermentation responses in rats fed resistant starch from high-amylose maize. *Obesist* **2013**, *21*, 2350–2355. [CrossRef]
34. Fluitman, K.S.; Wijdeveld, M.; Nieuwdorp, M.; Ijzerman, R.G. Potential of butyrate to influence food intake in mice and men. *Gut* **2018**, *67*, 1203–1204. [CrossRef]
35. Canfora, E.; Jocken, J.W.; Blaak, E.E. Short-chain fatty acids in control of body weight and insulin sensitivity. *Nat. Rev. Endocrinol.* **2015**, *11*, 577–591. [CrossRef]
36. Lehmann, U.; Jacobasch, G.; Schmiedl, D. Characterization of Resistant Starch Type III from Banana (Musa acuminata). *J. Agric. Food Chem.* **2002**, *50*, 5236–5240. [CrossRef]
37. Gao, J.; Xu, K.; Liu, H.; Liu, G.; Bai, M.; Peng, C.; Li, T.; Yin, Y. Impact of the Gut Microbiota on Intestinal Immunity Mediat-ed by Tryptophan Metabolism. *Front. Cell Infect. Microbiol.* **2018**, *8*, 13. [CrossRef]

38. Platten, M.; Nollen, E.A.A.; Röhrig, U.F.; Fallarino, F.; Opitz, C.A. Tryptophan metabolism as a common therapeutic target in cancer, neurodegeneration and beyond. *Nat. Rev. Drug Discov.* **2019**, *18*, 379–401. [CrossRef]
39. Krautkramer, K.A.; Fan, J.; Bäckhed, F. Gut microbial metabolites as multi-kingdom intermediates. *Nat. Rev. Genet.* **2021**, *19*, 77–94. [CrossRef] [PubMed]
40. De Vadder, F.; Kovatcheva-Datchary, P.; Zitoun, C.; Duchampt, A.; Bäckhed, F.; Mithieux, G. Microbiota-Produced Succinate Improves Glucose Homeostasis via Intestinal Gluconeogenesis. *Cell Metab.* **2016**, *24*, 151–157. [CrossRef]
41. Hayashi, M.; Nosaka, K. Characterization of Thiamin Phosphate Kinase in the Hyperthermophilic Archaeon Pyrobaculum calidifontis. *J. Nutr. Sci. Vitaminol.* **2015**, *61*, 369–374. [CrossRef]
42. Zamani, B.; Farshbaf, S.; Golkar, H.R.; Bahmani, F.; Asemi, Z. Synbiotic supplementation and the effects on clinical and metabolic responses in patients with rheumatoid arthritis: A randomised, double-blind, placebo-controlled trial. *Br. J. Nutr.* **2017**, *117*, 1095–1102. [CrossRef]
43. Horvath, A.; Leber, B.; Feldbacher, N.; Tripolt, N.; Rainer, F.; Blesl, A.; Trieb, M.; Marsche, G.; Sourij, H.; Stadlbauer, V. Ef-fects of a multispecies synbiotic on glucose metabolism, lipid marker, gut microbiome composition, gut permeability, and quality of life in diabesity: A randomized, double-blind, placebo-controlled pilot study. *Eur. J. Nutr.* **2020**, *59*, 2969–2983. [CrossRef]
44. Bindels, L.B.; Neyrinck, A.M.; Claus, S.P.; Le Roy, C., I; Grangette, C.; Pot, B.; Martinez, I.; Walter, J.; Cani, P.D.; Delzenne, N.M. Synbiotic approach restores intestinal homeostasis and prolongs survival in leukaemic mice with cachexia. *ISME J.* **2015**, *10*, 1456–1470. [CrossRef] [PubMed]
45. Swanson, K.S.; Gibson, G.R.; Hutkins, R.; Reimer, R.A.; Reid, G.; Verbeke, K.; Scott, K.P.; Holscher, H.D.; Azad, M.B.; Delzenne, N.M.; et al. The International Scientific Association for Probiotics and Prebiotics (ISAPP) consensus statement on the definition and scope of synbiotics. *Nat. Rev. Gastroenterol. Hepatol.* **2020**, *17*, 687–701. [CrossRef] [PubMed]
46. Crittenden, R.; Morris, L.; Harvey, M.; Tran, L.; Mitchell, H.; Playne, M. Selection of a Bifidobacterium strain to complement resistant starch in a synbiotic yoghurt. *J. Appl. Microbiol.* **2001**, *90*, 268–278. [CrossRef]
47. Shao, Y.; Huo, D.; Peng, Q.; Pan, Y.; Jiang, S.; Liu, B.; Zhang, J. Lactobacillus plantarum HNU082-derived improvements in the intestinal microbiome prevent the development of hyperlipidaemia. *Food Funct.* **2017**, *8*, 4508–4516. [CrossRef]
48. Cerqueira, F.M.; Photenhauer, A.L.; Pollet, R.M.; Brown, H.A.; Koropatkin, N.M. Starch Digestion by Gut Bacteria: Crowdsourcing for Carbs. *Trends Microbiol.* **2020**, *28*, 95–108. [CrossRef]
49. Cani, P.D. Microbiota and metabolites in metabolic diseases. *Nat. Rev. Endocrinol.* **2019**, *15*, 69–70. [CrossRef]
50. Ze, X.; Le Mougen, F.; Duncan, S.H.; Louis, P.; Flint, H.J. Some are more equal than others: The role of "keystone" species in the degradation of recalcitrant substrates. *Gut Microbes* **2013**, *4*, 236–240. [CrossRef] [PubMed]

foods

MDPI

Article

Effects of Three Types of Polymeric Proanthocyanidins on Physicochemical and In Vitro Digestive Properties of Potato Starch

Jiahui Xu [1], Taotao Dai [2], Jun Chen [1], Xuemei He [2], Xixiang Shuai [1], Chengmei Liu [1] and Ti Li [1,*]

1 State Key Laboratory of Food Science and Technology, Nanchang University, Nanchang 330047, China; xl19961019@126.com (J.X.); chen-jun1986@hotmail.com (J.C.); shuaixixiang1989@163.com (X.S.); liuchengmei@aliyun.com (C.L.)

2 Agro-Products Processing Science and Technology Research Institute, Guangxi Academy of Agricultural Sciences, Nanning 530007, China; daitaotao@gxaas.net (T.D.); xuemeihe1981@126.com (X.H.)

* Correspondence: liti@ncu.edu.cn

Abstract: The effects of three types of polymeric proanthocyanidins (PPC) with different degrees of polymerization (DP), namely PPC1 (DP = 6.39 ± 0.13), PPC2 (DP = 8.21 ± 0.76), and PPC3 (DP = 9.92 ± 0.21), on the physicochemical characteristics and in vitro starch digestibility of potato starch were studied. PPC addition (5%, *w/w*) increased the gelatinization temperature and decreased some viscosity indices of potato starch, including the peak, trough, breakdown, and setback viscosities. Starch-PPC pastes showed reduced thixotropy and improved stability and gelling properties compared to starch paste. The three types of proanthocyanidins all showed evident inhibitory effects on the digestion and retrogradation of potato starch, including short-term and long-term retrogradation. Among the three, PPC with a lower DP had stronger effects on the starch short-term retrogradation and gelling performance, whereas larger PPC molecules exhibited a greater impact on starch recrystallization and digestive characteristics. The research consequences were conducive to explore the application of functional PPC in starch-based food processing.

Keywords: polymeric proanthocyanidin; degree of polymerization; starch; pasting; retrogradation; digestibility

Citation: Xu, J.; Dai, T.; Chen, J.; He, X.; Shuai, X.; Liu, C.; Li, T. Effects of Three Types of Polymeric Proanthocyanidins on Physicochemical and In Vitro Digestive Properties of Potato Starch. *Foods* **2021**, *10*, 1394. https://doi.org/10.3390/foods10061394

Academic Editor: Mario M. Martinez

Received: 20 May 2021
Accepted: 7 June 2021
Published: 16 June 2021

1. Introduction

Potato starch (PoS) has attracted extensive attention in regard to its physicochemical characteristics, such as digestibility, fine textural properties, high swelling power, and transparency. In addition to producing sacchariferous products, potato starch not only plays a unique role in processing flour products, snacks, and chemical starch, but it also serves as a food additive, including gelling agents and film-forming agents used in dairy products, baked foods, beverages, and other foods [1,2]. However, the application of native potato starch has sometimes been limited due to its thermal instability, easy retrogradation, and high content of rapidly digesting starch. Mixing potato starch with polyphenols has been a research hotspot, because polyphenols could play an important role in enhancing starch performance in different applications [3–6].

Polymeric proanthocyanidins (PPC) are a kind of polyphenol widely distributed in the plant realm [7]. These compounds are complicated polymers of flavan-3-ol monomer units, with a degree of polymerization (DP) that is generally more than five [8]. Evidence suggests that they are more efficient in the regulation of starch digestion compared to monomeric polyphenols. Li et al. [9] and Mkandawire et al. [10] indicated a positive relationship between the DP value (DP $\geq$ 5) of proanthocyanidin and the inhibition of α-amylase and α-glucosidase, accordingly modifying starch intestinal glucose release. Barros et al. [11] found that proanthocyanidin-rich sorghum extracts elevated the content of resistant starch in corn starch to nearly three times that of monomer polyphenols. Extensive research has shown that physicochemical properties and biological activities of proanthocyanidins depend on the DP. The physical conformation of high DP proanthocyanidins provides more hydrophobic sites, while lower DP counterparts exhibit a stronger water-soluble capacity due to their considerable external hydroxyl groups [12]. Besides this, Arimboor and Arumughan [13] and Dai et al. [14] discovered that smaller-molecular proanthocyanidins had stronger antioxidant effects. Khadri et al. [15] illustrated that the anti-inflammatory activity of lemon grass proanthocyanidin relied on the DP. Pierini et al. [16] and Kawahara et al. [17] found that proanthocyanidin with a higher DP achieved a stronger growth inhibition influence on some specific cancer cells (e.g., hepatoma carcinoma, esophageal adenocarcinoma, and colon cancer cells).

To date, several papers have focused on the influence of proanthocyanidin on pasting, thermal, rheological, and digestion characteristics of potato starch. Zhang et al. [18] found that grape seed proanthocyanidins increased thermostability, reduced rapid digestion, and decreased the final viscosity and hardness of PoS gel significantly. Gao et al. [19] reported that oligomeric procyanidins (DP $\approx$ 2.88) improved the viscoelasticity and restrained the retrogradation of PoS paste. However, these studies concentrated on the influence of proanthocyanidins of a certain DP on the potato starch properties, while there have been few studies that have investigated the association between the DP value and the effect, especially for polymeric proanthocyanidin with different DPs. Proanthocyanidins with different degrees of polymerization might exert varied effects on the characteristics of potato starch. Thus, this research provides a new perspective for the influences of DP-sensitive polymeric proanthocyanidin with different DPs on the physical and chemical characteristics and in vitro starch digestibility of potato starch. It is likely to supply a theoretical basis for quality improvement of PoS-based foodstuffs.

2. Materials and Methods

2.1. Materials

Polymeric proanthocyanidins (PPC, purity $\geq$ 95%) derived from grape seed were purchased from Solarbio Technology Co., Ltd. (Beijing, China). Potato starch (PoS) was obtained from Rogate Starch Co., Ltd. (Jiangsu, China), and the contents of lipid, water, ash, and amylose were 0.32%, 9.86%, 0.25%, and 26.61% (*w/w*), respectively. All other chemicals were of reagent grade and supplied by Aladdin Co., Ltd. (Shanghai, China).

2.2. Purification and Fractionation of PPC by a Sephadex LH-20 Column

The PPC was fractionated with a Sephadex LH-20 column, according to a means expounded in the literature, with certain modifications [8,19]. A total of 6 g of proanthocyanidin sample was dissolved in a methanol/water solution with the volume ratio of 1:1 and loaded in a Sephadex LH-20 column that was pretreated. Then, the pillar was washed with 50% methanol until the eluent became colorless to clear other phenolic substances, glycosides, and some flavan-3-ol monomers. The column was then developed with the sequence of acetone/methanol/water solutions (respectively, 2:5:3, 4:3:3, and 7:0:3 *v/v/v*), and the volume of all solvent mixtures was 600 mL. The grading elutions were named as fraction 1, 2, and 3, respectively. Then, the organic solvents of these fractions were removed by vacuum rotary evaporation and lyophilized to drying. The obtained components were, respectively, labeled PPC1, PPC2, and PPC3.

2.3. Determination of the Degree of Polymerization of Proanthocyanidin

The average degrees of polymerization of PPC1, PPC2, and PPC3 were determined by the vanillin-hydrochloric acid method [20,21]. Briefly, the standard curves of catechin's mass concentration and molar concentration were drawn in sequence (the standard curve of catechin mass concentration was $y = 2020.6x + 0.6963$, $R^2 = 0.9954$; the standard curve of catechin molar concentration was $y = 0.1756x - 0.0013$, $R^2 = 0.9999$). A total of 10 mg of the proanthocyanidin sample thoroughly dissolved in 20 mL methanol solution, which was then diluted 20 times with methanol and acetic acid and, respectively, named as liquid A and liquid B. One milliliter of liquid A and liquid B was applied to measure the content (m) and the amount of substance (n) of proanthocyanidin according to the methods of standard curves of catechin mass concentration and catechin molar concentration, respectively [22]. In this research, the value of DP was computed by the formula below (1):

$$DP = \frac{m}{M \times n} \quad (1)$$

where m refers to the proanthocyanidin content, n is the amount of substance of proanthocyanidin, and M is the relative molecular mass of catechin.

2.4. Rapid Viscosity Analysis (RVA)

The pasting characteristics of PoS and PoS-PPC blends were determined using RVA (Perten RVA 4500, Stockholm, Sweden), according to preceding papers [18]. Based on our previous research, we chose 20:1 as the mass ratio of starch to PPC [18]. Briefly, 2 g of starch was mixed with proanthocyanidins (PPC1, PPC2, and PPC3) at doses of 0% and 5% (*w/w*), and added with 20 mL of distilled water. The slurries of PoS and PoS-PPC samples were characterized using the RVA, in accordance with the description of Li et al. [23]. The preliminary stirring speed was 960 r/min; then, it was kept at 160 r/min after 10s. The sample was maintained at 50 °C for 90s, heated and maintained at 95 °C for 150s, and decreased and kept at 50 °C for 90s. The pasting curves and parameters of starch pastes were generated by TCW (Thermocline for Windows). Some of the gelatinized samples were transferred for rheological analysis, and others were reserved at 4 °C for a week to prepare the retrograded samples. In addition, the same mass ratio of starch to PPC was used to investigate DSC, XRD, and in vitro digestibility of starch-PPC samples.

2.5. Rheological Measurements

The rheological characteristics of PoS and PoS-PPC compounds were analyzed by an MCR 302 rheometer (Anton-Paar, Graz, Austria) with a parallel plate measuring system, according to Kong et al. [24] and Chen et al. [4]. The gels of starch and starch-PPC mixtures obtained from Section 2.4 were transferred to a rheometer board with a probe type of PP50 and a gap of 0.5 mm. The pastes were equilibrated at room temperature for 5 min before measurement.

2.5.1. Steady Shear Analysis

The changes in shear stress of the samples were determined within the range of increasing shear rate from 0–1000 s^{-1} and then decreasing from 1000–0 s^{-1}. The obtained curve was fitted with the Power Law equation for regression fitting:

$$\sigma = K \cdot \gamma^n \quad (2)$$

where σ is the shear stress (Pa). γ is the shear rate (s^{-1}). K is the consistency coefficient ($Pa \cdot s^n$). n refers to the flow behavior index ($n < 1$ for pseudoplasticity fluid, and $n = 1$ for Newtonian fluid).

2.5.2. Dynamic Rheological Analysis

An oscillatory frequency sweep mode (0.2–20 Hz) was conducted at room temperature with 1% strain. The storage modulus (G') and loss modulus (G'') were obtained.

2.6. Differential Scanning Calorimetry (DSC)

The thermal characteristics of PoS and PoS-PPC samples were measured with DSC (7000X, HITACHI, Japan) on the basis of He et al. [25] and Xu et al. [26]. Some parameters were achieved from the DSC curves, including the onset (T_o), peak (T_p), conclusion temperature (T_c), gelatinization (ΔH_g), and retrogradation enthalpy (ΔH_r). Finally, the retrogradation degree R ($\Delta H_r/\Delta H_g$) was computed on account of previous research [27].

2.7. X-ray Diffraction (XRD)

XRD (D8 Advance, Bruker, Karlsruhe, Germany) was used to determine the crystallinity of PoS and PoS-PPC complexes, based on a previous paper [28]. The scanning range of the diffraction angle (2θ) was 4°–40°. The relative crystallinity was computed by an integral using the Origin software (3):

$$Relative\ crystallinity\ (\%) = \frac{A_c}{A_c + A_a} \times 100 \tag{3}$$

here A_c and A_a stand for the crystal and amorphous areas, respectively.

2.8. In Vitro Digestion

The in vitro digestibility of PoS and PoS-PPC compounds was analyzed in accordance with the preceding literature, with some modifications [29,30]. A total of 200 mg of starch or starch-PPC mixture samples was dissolved in 5 mL water, fully gelatinized, cooled, and added with 15 mL phosphate buffered solution (pH = 5.2). Subsequently, 5 mL of mixed digestive enzyme fluid (13 U/mg porcine pancreatic α-amylase, 260 U/mg glucoamylase) was added to the PoS or PoS-PPC mixtures, and then incubated at 37 °C. Other steps refer to previous studies. The contents of rapidly digested starch (RDS), slowly digested starch (SDS), and resistant starch (RS) in starch fractions were computed with the Equations (4)–(6) below:

$$RDS\ (\%) = \frac{G20 - FG}{TG} \times 0.9 \times 100 \tag{4}$$

$$SDS\ (\%) = \frac{G120 - G20}{TG} \times 0.9 \times 100 \tag{5}$$

$$RS\ (\%) = \frac{TG - (RDS + SDS)}{TG} \times 100 \tag{6}$$

Here, *G20* and *G120* refer to the contents of glucose generated at 20 and 120 min of the hydrolysis process, respectively. *FG* is the content of free glucose in samples, and *TG* is the total amount of starch in the original sample.

2.9. Statistical Analyses

The data were represented as the means ± standard deviation (SD) conducted in triplicate analyses. Duncan's test was carried out in SPSS 24.0 statistical software, and at $p < 0.05$, the significant difference was set.

3. Results and Discussion

3.1. Qualitative Analysis of PPC

On the basis of the difference in the molecular weight, molecular structure, and molecular polarity of proanthocyanidins with different DPs, we reported on the use of such a gradient with acetone/methanol/water solutions (respectively, 2:5:3, 4:3:3, and 7:0:3 *v/v/v*) to preparatively separate the proanthocyanidins [9,31]. With the increase of acetone volume fraction, the average degrees of polymerization of the fractionated proanthocyanidins increased successively, which were 6.39 ± 0.13, 8.21 ± 0.76, and 9.92 ± 0.21, respectively.

3.2. Pasting Properties

Table 1 and Figure 1 exhibit the pasting properties of PoS in the presence of PPC with three DPs. As shown in Figure 1, with the temperature increasing, starch granules began to swell under the action of water, the crystalline region was destroyed, and amylose subsequently diffused and leached out [27]. Visually, all of the viscosity curves increased markedly but showed an obvious difference between PoS and PoS-PPC mixtures, with 3491, 2397, 2736, and 2591 mPa·s for PoS, PoS-PPC1, PoS-PPC2, and PoS-PPC3, respectively. The addition of PPC1, PPC2, and PPC3 led to significant decreases in peak viscosity (PV) and increases of pasting temperature (PT) of PoS from 68.63 to 68.66 and 69.00 and 69.53 °C, respectively, which indicates that PPC delayed the rapid swelling of starch granules. Similar results were reported in previous works [18], which found that the released polyphenols crosslinked starch with water through polyphenolic hydroxyl groups. On the one hand, PPC diminished water usability for swelling of starch particles via competition for water and then decreased the amount of leaching amyloses and overall viscosity. On the other hand, the connection between PPC and starch chains, especially amylose, weakened the direct interaction between starch granules and moisture [32]. The newly formed starch-PPC complex might have wrapped onto the surface of starch particles and further inhibited the hydration of amorphous regions and pasting process of starch. Moreover, the negative charges of phosphate monoesters in PoS repelled each other, which promoted the rapid swelling of PoS granules. However, interactions between phosphate monoesters of starch and hydroxy groups of proanthocyanidin might happen during hydration, thus retarding the blends' gelatinization [33].

Table 1. Gelatinization characteristics of PoS and PoS-PPC complexes.

Samples	PV (mPa·s)	TV (mPa·s)	BD (mPa·s)	FV (mPa·s)	SB (mPa·s)	PT (°C)
PoS	3491 ± 1 ^{a}	1588 ± 12 ^{a}	1906 ± 27 ^{a}	1833 ± 13 ^{c}	397 ± 8 ^{a}	68.63 ± 0.01 ^{b}
PoS-PPC1	2397 ± 7 ^{d}	1495 ± 6 ^{c}	902 ± 1 ^{d}	1778 ± 11 ^{d}	248 ± 3 ^{d}	68.66 ± 0.04 ^{b}
PoS-PPC2	2736 ± 20 ^{b}	1535 ± 6 ^{b}	1202 ± 16 ^{b}	1864 ± 4 ^{b}	283 ± 6 ^{c}	69.00 ± 0.57 ^{a}
PoS-PPC3	2591 ± 2 ^{c}	1524 ± 4 ^{b}	1061 ± 2 ^{c}	1919 ± 1 ^{a}	330 ± 2 ^{b}	69.53 ± 0.04 ^{a}

Different letters in a column signify significant differences ($p < 0.05$). PoS: potato starch; PPC: polymeric proanthocyanidin; PV: peak viscosity; TV: trough viscosity; FV: final viscosity; PT: pasting temperature; BD: breakdown; SB: setback.

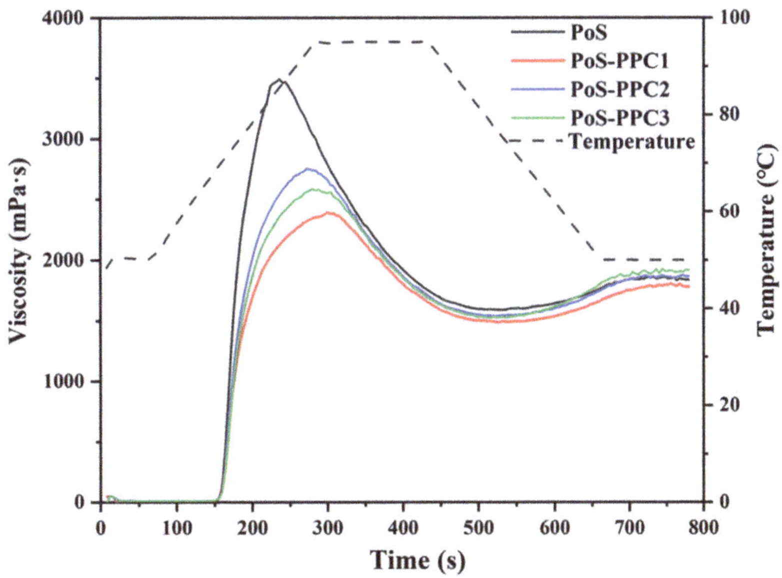

Figure 1. RVA gelatinization curves of native PoS and PoS-PPC pastes. PoS: potato starch; PPC: polymeric proanthocyanidin.

Among the three types of PPCs, PPC1 had the strongest effect on the pasting viscosity. This could be explained by the hygroscopicity of proanthocyanidins varying with their DP. PPCs with a slightly lower DP exhibited stronger moisture-absorption capacity because of a lot of external hydroxyl groups, thus reducing the exposure of starch molecules to water [19]. Presumably, PPCs of a smaller size could hamper the entanglement and formation of the double helix crystalline region of amylose, resulting in a decrease in starch viscosity. As the DP value increases, proanthocyanidin molecules become larger and their spatial structures get more complicated [34]. Vernhet et al. [35] described that highly condensed proanthocyanidins showed a denser shape with ramifications. To maintain the system stability of PPC molecules, intramolecular hydrophobic interactions and hydrogen bonds between the groups tend to be formed, which lead to a less extended structure and lower solubility.

In the case of BD, it reflects the stability and integrality of the starch, and a high BD value suggests that granules were easily damaged [33,36,37]. PPC1, PPC2, and PPC3 addition caused decreases of the BD value from 1906 to 902, 1202, and 1061 mPa·s, respectively. The results showed that PPC could improve the thermal and shearing stability of potato starch. The effect was most pronounced for short-chain PPCs with the smallest molecules. This might be attributed to the increased conformational constraints for higher-DP proanthocyanidins, which impeded the interaction between proanthocyanidins and starch molecules to form a network structure within a relatively short time. The SB value reflects the degree of amylose reaggregation and recrystallization during short-term retrogradation [24]. The three types of PPCs could significantly reduce SB values to varying degrees, for PPC1 the strongest and PPC3 the least, from 397 to 248 and 330 mPa·s, respectively. This phenomenon might be because the connection of PPC and amyloses was stronger than the rearrangement between amylose molecules, resulting in a decrease of SB value [24]. In addition, PPC molecules perhaps generated a water layer around starch granules to limit the rearrangement of amylose during the gelatinization process; low-DP proanthocyanidin had a higher water retention capacity and reduced the fluidity of starch chain, thereby restraining short-term retrogradation of starch.

3.3. Rheological Measurements

3.3.1. Steady Shear Analysis

The flow characteristic curves of the PoS and PoS-PPC complexes are shown in Figure 2, and all the fitting parameters are listed in Table 2. Since the correlation between shear stress and shear rate was nonlinear and all n values were less than 1, the PoS and PoS-PPC blends were non-Newtonian fluids, exhibiting a shear-thinning property (a pseudoplasticity behavior). As for all the samples, stress shear hysteresis loops appeared in the shearing process, indicating a thixotropic effect. A large hysteresis ring area signifies a strong thixotropic property [38]. The K value is related to the starch viscosity, and the larger the K value, the poorer the system mobility and shear stability [39]. The hysteresis ring area of starch and consistency coefficient, K, decreased with the addition of PPC, which suggested that the starch paste thixotropy was reduced and stability was improved by adding PPC, and the effect was more noteworthy for PPC with a lower DP. This finding was in agreement with the BD result of RVA. The same trends of K value and hysteresis ring area of starch added with polysaccharides have been reported before, and these results indicated that the relative stability and uniformity of the starch gel structure was enhanced [4,19].

Table 2. The steady shear rheological parameters of PoS and PoS-PPC pastes modelled by the Power Law equation.

Samples	Hysteresis Loops Area ($Pa \cdot s^{-1}$)	Up Curve			Down Curve		
		$K/Pa \cdot s^n$	n	R^2	$K/Pa \cdot s^n$	n	R^2
PoS	52,972.6	43.55	0.37	0.98	12.11	0.45	0.98
PoS-PPC1	32,590.2	16.81	0.29	0.99	8.93	0.44	0.98
PoS-PPC2	33,943.7	17.82	0.36	0.99	9.30	0.46	0.99
PoS-PPC3	38,943.9	19.56	0.35	0.97	9.05	0.45	0.90

PoS: potato starch; PPC: polymeric proanthocyanidin; K: the consistency coefficient; n: the flow behavior index.

The value of n is inversely correlated with the pseudoplasticity degree of the entire system [31], PPC replacement decreasing the n value, for PPC1 the most. This meant that the presence of PPC made the entangled starch molecular chains have a greater tendency to be straightened or dispersed under the action of the external force, which was conducive to the flow of starch paste, consequently lowering the viscosity more obviously. The result was unanimous to the change of the RVA test.

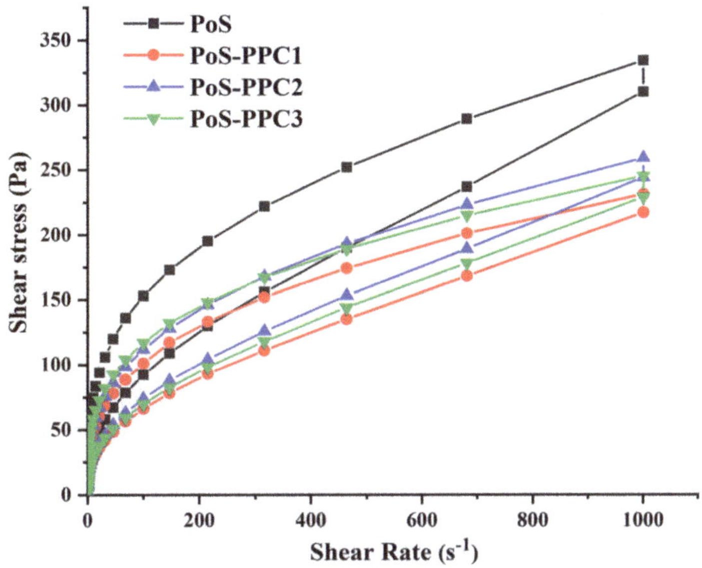

Figure 2. Steady flow curve of PoS and PoS-PPC pastes. PoS: potato starch; PPC: polymeric proanthocyanidin.

3.3.2. Dynamic Rheological Analysis

The viscoelastic properties were determined by dynamic rheological analysis (Figure 3). The storage modulus (G′) represents the starch elastic properties and reflects the ability of the material to restore its original state after deformation, while the loss modulus (G″) represents the viscous nature of starch colloid and characterizes the ability of the material to resist flowing [4]. G′ and G″ of all samples showed mild frequency dependence; in the total frequency range, the value of G′ was markedly higher than the G″ value, indicating that PoS and PoS-PPC blends displayed weak gel-like behavior [40]. Replacement of PoS with PPC increased the mechanical modulus, especially G′. This result indicated that PPC prominently affected the elastic properties instead of the viscous characteristics of the PoS-PPC mixtures. As a higher G′ means greater rigidity and strength of the PoS paste network structure, proancyanidins with a slightly lower DP could strengthen the gel network of starch more significantly [41]. Because amylose is the main polymer that forms the cross-linked network [42], we can speculate that the connection of PPC and starch internal amyloses was enhanced through hydrogen bonds, resulting in the increase of the elastic property of starch paste. The more externally active hydroxyl groups of PPC1 increase the entanglement points between molecular chains, facilitating the formation of a three-dimensional gel network structure [40].

In summary, starch-PPC pastes had better gelling properties than starch paste, and the lower the degree of PPC polymerization, the more obvious the effect. PPC addition could make PoS more suitable candidates as gelling agents for food industry manufacturing, including soft candies, ice cream, and meat foods production, thereby improving these products' nutrition, texture, and stability [43].

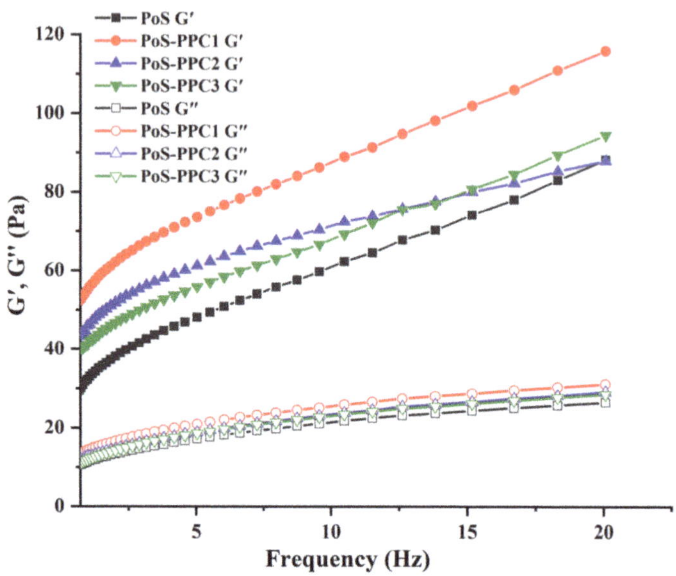

Figure 3. Variation tendency of storage modulus (**G′**) and loss modulus (**G″**) with frequency for PoS and PoS-PPC mixtures. PoS: potato starch; PPC: polymeric proanthocyanidin.

3.4. Thermodynamic Properties

The thermodynamic parameters obtained from DSC are shown in Table 3. The PPC addition increased the pasting transition temperatures (T_o, T_p, and T_c) of PoS within 2 °C, indicating that PPC could delay starch gelatinization [28]. The results were supposedly attributed to the partial undissolved PPC attached to the starch granules, which enlarged steric hindrance and suppressed the swelling and fracture of starch. Similar effects have been observed by Xiao et al. [44] and Zheng et al. [30], who found that black tea extract and proanthocyanidins from Chinese berry leaves discernibly increased the gelatinization temperatures of rice starch, and they ascribed it to the stability of interactions between these polyphenols and starch.

Table 3. Thermal properties of PoS and PoS-PPC blends.

Samples	T_o (°C)	T_p (°C)	T_c (°C)	ΔH_g (J/g)	ΔH_r (J/g)	*R* (%)
PoS	60.51 ± 0.29 ^{c}	62.43 ± 0.21 ^{c}	65.54 ± 0.24 ^{b}	13.50 ± 0.12 ^{a}	4.54 ± 0.03 ^{a}	33.63
PoS-PPC1	61.14 ± 0.01 ^{b}	63.16 ± 0.11 ^{b}	66.61 ± 0.15 ^{a}	12.17 ± 0.13 ^{b}	3.13 ± 0.12 ^{b}	25.68
PoS-PPC2	61.71 ± 0.05 ^{a}	63.63 ± 0.09 ^{a}	67.22 ± 0.21 ^{a}	8.70 ± 0.19 ^{c}	1.93 ± 0.07 ^{c}	22.18
PoS-PPC3	61.45± 0.07 ^{ab}	63.74 ± 0.08 ^{a}	66.84 ± 0.11 ^{a}	8.07 ± 0.09 ^{d}	1.52 ± 0.08 ^{d}	18.83

Different letters in a column signify significant differences ($p < 0.05$). PoS: potato starch; PPC: polymeric proanthocyanidin. T_o: onset temperature; T_p: peak temperature; T_c: conclusion temperature; ΔH_g: gelatinization enthalpy; ΔH_r: retrogradation enthalpy; *R*: the degree of retrogradation ($\Delta H_r/\Delta H_g$) × 100.

The gelatinization enthalpy value ΔH_g represents the energy required for the melting of the starch crystallization zone, especially for the double helix structure of amylopectin crystals [45]. Compared to the control (13.50 J/g), the ΔH_g for starch gelatinized with 5% PPC1, PPC2, and PPC3 decreased to 12.17, 8.70, and 8.07 J/g, respectively. PPC has a polyhydroxyl structure. The OH groups of PPC interacted with the side chain of amylopectin and bound to the amorphous area of starch particles to varying degrees. They facilitated the transition from the water-absorbing swelling in the amorphous zone to the crystalline zone for the starch granules, thereby altering the coupling force between the crystalline and amorphous matrix [46]. The consequent hydration easily led to a decrease in gelatinization energy.

The ΔH_r and R value of starch mainly reflect the recrystallization level of amylopectin and the long-term retrogradation degree of starch [40]. With the addition of PPC, the ΔH_r and R values of starch declined significantly, indicating that the degree of order and crystallinity of starch decreased during the retrogradation process. On the one hand, the strong hydrophilicity of PPC reduced the content of free or available moisture in the starch-PPC system. On the other hand, PPC associated with starch molecules by intense hydrogen bonding and hydrophobicity interactions and, consequently, played a role in steric exclusion, which all limited the activity of the starch chains and retarded retrogradation [47]. Among the three types of PPC, PPC3 had the strongest suppression effect on starch retrogradation, then PPC2, and finally, PPC1. This might be ascribed to the increased binding affinity and specificity for larger proanthocyanidins with a multidentate character [48], which also showed that macromolecular proanthocyanidins possibly had a greater impact on amylopectin and could better inhibit amylopectin recrystallization.

3.5. X-ray Diffraction Patterns

The results of XRD are shown in Figure 4. Native potato starch (NPoS) displayed a typical B-type crystal structure, with representative peaks appearing at $2\theta = 5°, 15°, 17°, 21°$, and 24°. During gelatinization, the crystalline structure of starch granules was destroyed by hydrothermal treatment and converted into an amorphous form [33]. However, during retrogradation, the starch chains tended to recombine into an ordered crystal structure and show a B-type XRD spectra ($2\theta \approx 17°$), which is principally due to rearrangement of the amylopectin part [6,49]. The supplementation of PPC led to the disappearance of starch B-type diffraction peaks, indicating that PPC could effectively restrain the recrystallization of amylopectin. Studies have shown that starch will form V-type crystals when combined with fatty acids, iodine, and other substances, and the V-type inclusion shows representative diffraction peaks at 20.0° and sometimes at 7.0° and 13.0° [50]. The complex had an obvious diffraction peak at $2\theta = 13.1°$, which was primarily due to the formation of amylose–lipid complexes, further hindering amylose rearrangement [28].

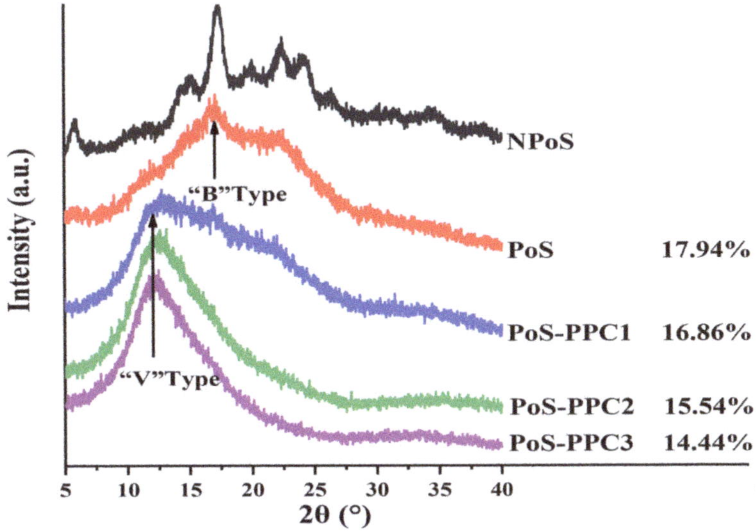

Figure 4. XRD diagrams of PoS and PoS-PPC blends, and the relative crystallinity is shown next to the sample name. NPoS, native potato starch; PoS, potato starch; and PPC, polymeric proanthocyanidin. PoS: potato starch; PPC: polymeric proanthocyanidin; 2θ: Diffraction angle.

After adding PPC1, PPC2, and PPC3, the relative crystallinity of potato starch decreased from 17.94% to 16.86%, 15.54%, and 14.44%, respectively. It is speculated that when embedded in the starch-moisture matrix, PPC could have interfered with the rearrangement of starch chains, especially for amylopectins, and a hydrogen-bond interaction might have been the main cause of their mutual effects [51]. The result also showed that larger proanthocyanidin molecules exerted a more significant influence on starch crystallinity than smaller ones, which indicated that proanthocyanidins with high DP values were more likely to destroy the long-range ordered crystalline structure and, thus, retard starch long-term retrogradation. Due to their large steric hindrance, macromolecular polyphenols are difficult to insert into the cavity of the amylose double helix to form inclusion compounds through hydrophobic interactions [52,53]. In other words, long-chain PPCs affect the overall distribution and arrangement instead of short-range double helical construction of the starch chains [30]. This result was consistent with the DSC.

PoS is often served as a substrate for producing starchy packaging films due to its edibility and biodegradability. However, on account of the shortcomings of PoS in processing, starch film production faces the problems of poor mechanical properties and easy crystallization [54,55]. Through our research, we found that PPC as a functional compound can not only serve as a mechanical reinforcement to form an interpenetrating network with starch and enhance the mechanical characteristics of starch films, but it can also suppress the recrystallization and retrogradation of starch, thus maintaining the preserved starch film as flexible and stretchable.

3.6. *In Vitro Digestibility*

As presented in Figure 5, the influences of three types of PPC on the RDS decrease (by 6.06% of PPC1, 9.93% of PPC2, and 13.68% of PPC3) and SDS increase (by 13.41% of PPC1, 21.90% of PPC2, and 20.18% of PPC3) were all significant ($p < 0.05$). PPC3 also showed a prominent effect in the improvement of RS content, with 16.35%. These results indicated that PPC could effectively hinder the digestibility of potato starch, and the inhibition impact was closely associated with the DP of PPC. In general, the replacement of PoS with PPC3 had the strongest suppression of starch digestibility among the three.

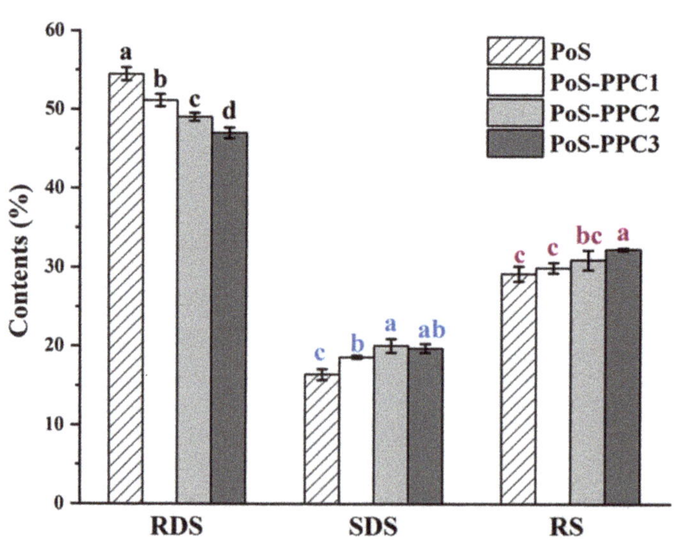

Figure 5. Influence of PPC on in vitro digestibility of potato starch. RDS, rapidly digesting starch; SDS, slowly digesting starch; and RS, resistant starch. PoS, potato starch and PPC, polymeric proanthocyanidin. Different colors of lowercase letters stand for significant differences ($p < 0.05$).

For the following probable reasons, on the one hand, PPC could bind to the amino acid residues and non-competitively inhibit the catalysis of α-amylase and α-glucosidase, mainly through hydrophobic interaction, thus reducing the activity of enzymes [56]. As the DP value of PPC increases, the molecule hydrophobicity tends to be enhanced, which is advantageous to the formation of multiple interaction sites between PPC and amylase [9,11]. Zhou et al. [57] explained that the affinities for α-amylase varies among procyanidins with different DPs, mainly due to their different stabilization on the surface near to the catalytic core of α-amylase. In addition, the low swelling degree was unfavorable to the interaction of enzymes with the starch particles, thereby reducing the content of RDS. On the other hand, much of the literature has showed that starch in the presence of polyphenols might form a nonordered crystal structure with a slow digestion characteristic [5,58]. However, large-molecular-weight proanthocyanidin was likely to provide more hydroxyl groups and hydrophobic domains to favor the interactions with starch molecules. Based on this, by adding candidates of PPC, it was expected to produce natural anti-diabetic drugs or chronically digestive starch that are conductive to postprandial glucose control.

4. Conclusions

In this study, we utilized three types of PPC with different DPs (6.39 ± 0.13, 8.21 ± 0.76, and 9.92 ± 0.21, respectively) to research the influence of PPC on physicochemical and in vitro digestion properties for better application of PPC in starchy food. PPC enhanced the thermal stability and delayed starch gelatinization of PoS. It pronouncedly affected the elastic properties, instead of the viscous properties, of PoS pasting. The reduced thixotropy indicated that the relative stability and uniformity of the starch gel structure was enhanced, showing better gelling properties. Starch digestion and retrogradation could be restrained by PPCs with different DPs in varying degrees. Among the three, PPCs with a lower DP more evidently influenced starch gelling performance, whereas larger PPC molecules exhibited a greater impact on starch long-term retrogradation and digestive properties. This might be attributed to the different type of starch chains combined with PPCs of different DP. The findings of the present study may render a beneficial recommendation for the application of PPC to modify gelatinization, thermodynamic, rheological, and digestive characteristics of PoS-based foods.

Author Contributions: Conceptualization, J.X., T.D., J.C., C.L., and T.L.; methodology, J.X. and T.D.; data curation, J.X., and T.D.; writing–original draft preparation, J.X., T.D., and J.C.; writing–review and editing, J.X., T.D., and J.C.; visualization, X.H., and X.S.; supervision, J.X., C.L., and T.L. All authors have read and agreed to the published version of the manuscript.

Funding: This work was supported by the China National Natural Science Foundation (Grant No. 31760441 and 32060524), the China Postdoctoral Science Foundation (Grant No. 2020M6832211), and the Postdoctoral Foundation of Guangxi Academy of Agricultural Sciences (Grant No. 2020037).

Institutional Review Board Statement: Not applicable.

Informed Consent Statement: Not applicable.

Data Availability Statement: Data are contained within the article.

Conflicts of Interest: The authors declare no conflict of interest.

References

1. Mu, T.; Sun, H.; Zhang, M.; Wang, C. Sweet Potato Starch and its Series Products. *Sweet Potato Process. Technol.* **2017**, 1–48. [CrossRef]
2. Sharma, R.; Dar, B.; Sharma, S.; Singh, B. In Vitro digestibility, cooking quality, bio-functional composition, and sensory properties of pasta incorporated with potato and pigeonpea flour. *Int. J. Gastron. Food Sci.* **2021**, *23*, 100300. [CrossRef]
3. Zhu, F. Interactions between starch and phenolic compound. *Trends Food Sci. Technol.* **2015**, *43*, 129–143. [CrossRef]
4. Chen, N.; Chen, L.; Gao, H.; Zeng, W. Mechanism of bridging and interfering effects of tea polyphenols on starch molecules. *J. Food Process. Preserv.* **2020**, *44*. [CrossRef]
5. Lv, Y.; Zhang, L.; Li, M.; He, X.; Hao, L.; Dai, Y. Physicochemical properties and digestibility of potato starch treated by ball milling with tea polyphenols. *Int. J. Biol. Macromol.* **2019**, *129*, 207–213. [CrossRef]

6. Xiao, H.; Lin, Q.; Liu, G.-Q.; Wu, Y.; Wu, W.; Fu, X. Inhibitory Effects of Green Tea Polyphenols on the Retrogradation of Starches from Different Botanical Sources. *Food Bioprocess Technol.* **2013**, *6*, 2177–2181. [CrossRef]
7. Dai, T.; Li, T.; Li, R.; Zhou, H.; Liu, C.; Chen, J.; McClements, D.J. Utilization of plant-based protein-polyphenol complexes to form and stabilize emulsions: Pea proteins and grape seed proanthocyanidins. *Food Chem.* **2020**, *329*, 127219. [CrossRef]
8. Ito, C.; Oki, T.; Yoshida, T.; Nanba, F.; Yamada, K.; Toda, T. Characterisation of proanthocyanidins from black soybeans: Isolation and characterisation of proanthocyanidin oligomers from black soybean seed coats. *Food Chem.* **2013**, *141*, 2507–2512. [CrossRef]
9. Li, Q.; Chen, J.; Li, T.; Liu, C.; Zhai, Y.; McClements, D.J.; Liu, J. Separation and characterization of polyphenolics from underutilized byproducts of fruit production (*Choerospondias axillaris* peels): Inhibitory activity of proanthocyanidins against glycolysis enzymes. *Food Funct.* **2015**, *6*, 3693–3701. [CrossRef]
10. Mkandawire, N.L.; Kaufman, R.C.; Bean, S.; Weller, C.L.; Jackson, D.S.; Rose, D.J. Effects of Sorghum (*Sorghum bicolor* (L.) Moench) Tannins on α-Amylase Activity and in Vitro Digestibility of Starch in Raw and Processed Flours. *J. Agric. Food Chem.* **2013**, *61*, 4448–4454. [CrossRef]
11. Barros, F.; Awika, J.M.; Rooney, L.W. Interaction of Tannins and Other Sorghum Phenolic Compounds with Starch and Effects on in Vitro Starch Digestibility. *J. Agric. Food Chem.* **2012**, *60*, 11609–11617. [CrossRef]
12. Amoako, D.B.; Awika, J.M. Polymeric tannins significantly alter properties and in vitro digestibility of partially gelatinized intact starch granule. *Food Chem.* **2016**, *208*, 10–17. [CrossRef]
13. Arimboor, R.; Arumughan, C. Effect of Polymerization on Antioxidant and Xanthine Oxidase Inhibitory Potential of Sea Buckthorn (*H. rhamnoides*) Proanthocyanidins. *J. Food Sci.* **2012**, *77*, C1036–C1041. [CrossRef]
14. Dai, T.; Chen, J.; McClements, D.J.; Hu, P.; Ye, X.; Liu, C.; Li, T. Protein–polyphenol interactions enhance the antioxidant capacity of phenolics: Analysis of rice glutelin–procyanidin dimer interactions. *Food Funct.* **2019**, *10*, 765–774. [CrossRef]
15. Khadri, A.; Neffati, M.; Smiti, S.; Falé, P.; Lino, A.R.L.; Serralheiro, M.L.M.; Araújo, M.E.M. Antioxidant, antiacetylcholinesterase and antimicrobial activities of *Cymbopogon schoenanthus* L. Spreng (lemon grass) from Tunisia. *LWT* **2010**, *43*, 331–336. [CrossRef]
16. Pierini, R.; Kroon, P.A.; Guyot, S.; Ivory, K.; Johnson, I.T.; Belshaw, N.J. Procyanidin effects on oesophageal adenocarcinoma cells strongly depend on flavan-3-ol degree of polymerization. *Mol. Nutr. Food Res.* **2008**, *52*, 1399–1407. [CrossRef]
17. Kawahara, S.-I.; Ishihara, C.; Matsumoto, K.; Senga, S.; Kawaguchi, K.; Yamamoto, A.; Suwannachot, J.; Hamauzu, Y.; Makabe, H.; Fujii, H. Identification and characterization of oligomeric proanthocyanidins with significant anti-cancer activity in adzuki beans (*Vigna angularis*). *Heliyon* **2019**, *5*, e02610. [CrossRef]
18. Zhang, Z.; Tian, J.; Fang, H.; Zhang, H.; Kong, X.; Wu, D.; Zheng, J.; Liu, D.; Ye, X.; Chen, S. Physicochemical and Digestion Properties of Potato Starch Were Modified by Complexing with Grape Seed Proanthocyanidins. *Molecules* **2020**, *25*, 1123. [CrossRef]
19. Gao, L.; Zhang, C.; Chen, J.; Liu, C.; Dai, T.; Chen, M.; Li, T. Effects of proanthocyanidins on the pasting, rheological and retrogradation properties of potato starch. *J. Sci. Food Agric.* **2021**. [CrossRef]
20. Yang, H.; Ye, X.; Liu, D.; Chen, J.; Zhang, J.; Shen, Y.; Yu, D. Characterization of Unusual Proanthocyanidins in Leaves of Bayberry (*Myrica rubra* Sieb. et Zucc.). *J. Agric. Food Chem.* **2011**, *59*, 1622–1629. [CrossRef]
21. Butler, L.G.; Price, M.L.; Brotherton, J.E. Vanillin assay for proanthocyanidins (condensed tannins): Modification of the solvent for estimation of the degree of polymerization. *J. Agric. Food Chem.* **1982**, *30*, 1087–1089. [CrossRef]
22. Bordiga, M.; Travaglia, F.; Locatelli, M.; Coïsson, J.D.; Arlorio, M. Characterisation of polymeric skin and seed proanthocyanidins during ripening in six *Vitis vinifera* L. cv. *Food Chem.* **2011**, *127*, 180–187. [CrossRef]
23. Li, Y.-T.; Wang, R.-S.; Liang, R.-H.; Chen, J.; He, X.-H.; Chen, R.-Y.; Liu, W.; Liu, C.-M. Dynamic high-pressure microfluidization assisting octenyl succinic anhydride modification of rice starch. *Carbohydr. Polym.* **2018**, *193*, 336–342. [CrossRef]
24. Kong, X.-R.; Zhu, Z.-Y.; Zhang, X.-J.; Zhu, Y.-M. Effects of Cordyceps polysaccharides on pasting properties and in vitro starch digestibility of wheat starch. *Food Hydrocoll.* **2020**, *102*, 105604. [CrossRef]
25. He, X.-H.; Luo, S.-J.; Chen, M.-S.; Xia, W.; Chen, J.; Liu, C.-M. Effect of industry-scale microfluidization on structural and physicochemical properties of potato starch. *Innov. Food Sci. Emerg. Technol.* **2020**, *60*, 102278. [CrossRef]
26. Xu, J.; Li, X.; Chen, J.; Dai, T.; Liu, C.; Li, T. Effect of polymeric proanthocyanidin on the physicochemical and in vitro digestive properties of different starches. *LWT* **2021**, *148*, 111713. [CrossRef]
27. Wang, R.; Wan, J.; Liu, C.; Xia, X.; Ding, Y. Pasting, thermal, and rheological properties of rice starch partially replaced by inulin with different degrees of polymerization. *Food Hydrocoll.* **2019**, *92*, 228–232. [CrossRef]
28. He, X.-H.; Xia, W.; Chen, R.-Y.; Dai, T.-T.; Luo, S.-J.; Chen, J.; Liu, C.-M. A new pre-gelatinized starch preparing by gelatinization and spray drying of rice starch with hydrocolloids. *Carbohydr. Polym.* **2020**, *229*, 115485. [CrossRef] [PubMed]
29. Englyst, H.N.; Kingman, S.M.; Cummings, J.H. Classification and measurement of nutritionally important starch fractions. *Eur. J. Clin. Nutr.* **1992**, *46*, S33–S50. [PubMed]
30. Zheng, Y.; Tian, J.; Kong, X.; Wu, D.; Chen, S.; Liu, D.; Ye, X. Proanthocyanidins from Chinese berry leaves modified the physicochemical properties and digestive characteristic of rice starch. *Food Chem.* **2021**, *335*, 127666. [CrossRef]
31. Taylor, A.W.; Barofsky, E.; Kennedy, J.A.; Deinzer, M.L. Hop (*Humulus lupulus* L.) Proanthocyanidins Characterized by Mass Spectrometry, Acid Catalysis, and Gel Permeation Chromatography. *J. Agric. Food Chem.* **2003**, *51*, 4101–4110. [CrossRef]
32. Pan, J.; Li, M.; Zhang, S.; Jiang, Y.; Lv, Y.; Liu, J.; Liu, Q.; Zhu, Y.; Zhang, H. Effect of epigallocatechin gallate on the gelatinisation and retrogradation of wheat starch. *Food Chem.* **2019**, *294*, 209–215. [CrossRef]

33. Liu, S.; Shen, M.; Xiao, Y.; Luo, Y.; Xie, J. Effect of maize, potato, and pea starches with *Mesona chinensis* polysaccharide on pasting, gelatinization properties, granular morphology and digestion. *Food Hydrocoll.* **2020**, *108*, 106047. [CrossRef]
34. Saucier, C.; Mirabel, M.; Daviaud, F.; Longieras, A.; Glories, Y. Rapid fractionation of grape seed proanthocyanidins. *J. Agric. Food Chem.* **2001**, *49*, 5732–5735. [CrossRef]
35. Vernhet, A.; Carrillo, S.; Poncet-Legrand, C. Condensed Tannin Changes Induced by Autoxidation: Effect of the Initial Degree of Polymerization and Concentration. *J. Agric. Food Chem.* **2014**, *62*, 7833–7842. [CrossRef]
36. Mahmood, K.; Kamilah, H.; Shang, P.L.; Sulaiman, S.; Ariffin, F.; Alias, A.K. A review: Interaction of starch/non-starch hydrocolloid blending and the recent food applications. *Food Biosci.* **2017**, *19*, 110–120. [CrossRef]
37. Zhang, H.; Sun, B.; Zhang, S.; Zhu, Y.; Tian, Y. Inhibition of wheat starch retrogradation by tea derivatives. *Carbohydr. Polym.* **2015**, *134*, 413–417. [CrossRef]
38. Ma, S.; Zhu, P.; Wang, M. Effects of konjac glucomannan on pasting and rheological properties of corn starch. *Food Hydrocoll.* **2019**, *89*, 234–240. [CrossRef]
39. Lin, L.; Huang, J.; Zhao, L.; Wang, J.; Wang, Z.; Wei, C. Effect of granule size on the properties of lotus rhizome C-type starch. *Carbohydr. Polym.* **2015**, *134*, 448–457. [CrossRef]
40. Sun, Y.; Wang, M.; Ma, S.; Wang, H. Physicochemical characterization of rice, potato, and pea starches, each with different crystalline pattern, when incorporated with Konjac glucomannan. *Food Hydrocoll.* **2020**, *101*, 105499. [CrossRef]
41. Sandhu, K.S.; Siroha, A.K. Relationships between physicochemical, thermal, rheological and in vitro digestibility properties of starches from pearl millet cultivars. *LWT* **2017**, *83*, 213–224. [CrossRef]
42. Yuris, A.; Goh, K.K.T.; Hardacre, A.K.; Matia-Merino, L. Understanding the interaction between wheat starch and *Mesona chinensis* polysaccharide. *LWT* **2017**, *84*, 212–221. [CrossRef]
43. Masina, N.; Choonara, Y.E.; Kumar, P.; du Toit, L.C.; Govender, M.; Indermun, S.; Pillay, V. A review of the chemical modification techniques of starch. *Carbohydr. Polym.* **2017**, *157*, 1226–1236. [CrossRef]
44. Xiao, H.; Lin, Q.; Liu, G.-Q.; Yu, F. Evaluation of Black Tea Polyphenol Extract Against the Retrogradation of Starches from Various Plant Sources. *Molecules* **2012**, *17*, 8147–8158. [CrossRef]
45. Thakur, R.; Pristijono, P.; Golding, J.B.; Stathopoulos, C.E.; Scarlett, C.; Bowyer, M.; Singh, S.P.; Vuong, Q.V. Effect of starch physiology, gelatinization, and retrogradation on the attributes of rice starch-ι-carrageenan film. *Starch Stärke* **2017**, *70*, 1700099. [CrossRef]
46. Xiao, Y.; Zheng, M.; Yang, S.; Li, Z.; Liu, M.; Yang, X.; Lin, N.; Liu, J. Physicochemical properties and in vitro digestibility of proso millet starch after addition of Proanthocyanidins. *Int. J. Biol. Macromol.* **2021**, *168*, 784–791. [CrossRef]
47. Ye, J.; Yang, R.; Liu, C.; Luo, S.; Chen, J.; Hu, X.; Wu, J. Improvement in freeze-thaw stability of rice starch gel by inulin and its mechanism. *Food Chem.* **2018**, *268*, 324–333. [CrossRef]
48. Prigent, S.; Voragen, A.; Van Koningsveld, G.; Baron, A.; Renard, C.M.; Gruppen, H. Interactions between globular proteins and procyanidins of different degrees of polymerization. *J. Dairy Sci.* **2009**, *92*, 5843–5853. [CrossRef]
49. Niu, H.; Zhang, M.; Xia, X.; Liu, Q.; Kong, B. Effect of porcine plasma protein hydrolysates on long-term retrogradation of corn starch. *Food Chem.* **2018**, *239*, 172–179. [CrossRef]
50. Lesmes, U.; Beards, E.J.; Gibson, G.R.; Tuohy, K.; Shimoni, E. Effects of Resistant Starch Type III Polymorphs on Human Colon Microbiota and Short Chain Fatty Acids in Human Gut Models. *J. Agric. Food Chem.* **2008**, *56*, 5415–5421. [CrossRef]
51. Luo, D.; Li, Y.; Xu, B.; Ren, G.; Li, P.; Li, X.; Han, S.; Liu, J. Effects of inulin with different degree of polymerization on gelatinization and retrogradation of wheat starch. *Food Chem.* **2017**, *229*, 35–43. [CrossRef] [PubMed]
52. Igoumenidis, P.E.; Zoumpoulakis, P.; Karathanos, V.T. Physicochemical interactions between rice starch and caffeic acid during boiling. *Food Res. Int.* **2018**, *109*, 589–595. [CrossRef] [PubMed]
53. Yu, M.; Liu, B.; Zhong, F.; Wan, Q.; Zhu, S.; Huang, D.; Li, Y. Interactions between caffeic acid and corn starch with varying amylose content and their effects on starch digestion. *Food Hydrocoll.* **2021**, *114*, 106544. [CrossRef]
54. Han, G.; Guo, R.; Yu, Z. A review on biodegradable starch-based films. *Chin. J. Bioprocess Eng.* **2019**, *17*, 460–465. [CrossRef]
55. Estevez-Areco, S.; Guz, L.; Famá, L.; Candal, R.; Goyanes, S. Bioactive starch nanocomposite films with antioxidant activity and enhanced mechanical properties obtained by extrusion followed by thermo-compression. *Food Hydrocoll.* **2019**, *96*, 518–528. [CrossRef]
56. Barrett, A.H.; Farhadi, N.F.; Smith, T.J. Slowing starch digestion and inhibiting digestive enzyme activity using plant flavanols/tannins—A review of efficacy and mechanisms. *LWT* **2018**, *87*, 394–399. [CrossRef]
57. Zhou, P.; Zhang, L.; Li, W.; Zhang, S.; Luo, L.; Wang, J.; Sun, B. In Vitro evaluation of the anti-digestion and antioxidant effects of grape seed procyanidins according to their degrees of polymerization. *J. Funct. Foods* **2018**, *49*, 85–95. [CrossRef]
58. Sun, L.; Miao, M. Dietary polyphenols modulate starch digestion and glycaemic level: A review. *Crit. Rev. Food Sci. Nutr.* **2020**, *60*, 541–555. [CrossRef]

foods

MDPI

Article

Dietary Fiber Modulates the Fermentation Patterns of Cyanidin-3-*O*-Glucoside in a Fiber-Type Dependent Manner

Zixin Yang [1,†], Ting Huang [1,†], Ping Li [2], Jian Ai [1,3], Jiaxin Liu [1], Weibin Bai [2] and Lingmin Tian [1,*]

1 Department of Food Science and Engineering, Jinan University, Guangzhou 510632, China; yangzixin121@163.com (Z.Y.); huangting081@163.com (T.H.); aijian23@163.com (J.A.); liujiaxin9898@163.com (J.L.)
2 Guangdong Engineering Technology Center of Food Safety Molecular Rapid Detection, Institute of Food Safety and Nutrition, Jinan University, Guangzhou 510632, China; liping5477@163.com (P.L.); baiweibin@163.com (W.B.)
3 Department of Food Science and Engineering, College of Food Science, Shanghai Ocean University, Shanghai 201306, China
* Correspondence: lingmintian@jnu.edu.cn
† These two authors have contributed equally to this work.

Abstract: The interactions between cell-wall polysaccharides and polyphenols in the gastrointestinal tract have attracted extensive attention. We hypothesized that dietary fiber modulates the fermentation patterns of cyanidin-3-*O*-glucoside (C3G) in a fiber-type-dependent manner. In the present study, the effects of four dietary fibers (fructose-oligosaccharides, pectin, β-glucan and arabinoxylan) on the modulation of C3G fermentation patterns were investigated through in vitro fermentation inoculated with human feces. The changes in gas volume, pH, total carbohydrate content, metabolites of C3G, antioxidant activity, and microbial community distribution during in vitro fermentation were analyzed. After 24 h of fermentation, the gas volume and total carbohydrate contents of the four dietary-fiber-supplemented groups respectively increased and decreased to varying degrees. The results showed that the C3G metabolites after in vitro fermentation mainly included cyanidin, protocatechuic acid, 2,4,6-trihydroxybenzoic acid, and 2,4,6-trihydroxybenzaldehyde. Supplementation of dietary fibers changed the proportions of C3G metabolites depending on the structures. Dietary fibers increased the production of short-chain fatty acids and the relative abundance of gut microbiota *Bifidobacterium* and *Lactobacillus*, thus potentially maintaining colonic health to a certain extent. In conclusion, the used dietary fibers modulate the fermentation patterns of C3G in a fiber-type-dependent manner.

Keywords: dietary fiber; fermentation patterns; cyanidin-3-*O*-glucoside; gut microbiota

Citation: Yang, Z.; Huang, T.; Li, P.; Ai, J.; Liu, J.; Bai, W.; Tian, L. Dietary Fiber Modulates the Fermentation Patterns of Cyanidin-3-O-Glucoside in a Fiber-Type Dependent Manner. *Foods* **2021**, *10*, 1386. https://doi.org/10.3390/foods10061386

Academic Editors: Jianhua Xie, Yanjun Zhang and Hansong Yu

Received: 11 May 2021
Accepted: 8 June 2021
Published: 16 June 2021

1. Introduction

Anthocyanins, a class of phenolic compounds belonging to the flavonoid group, are mainly found in the epidermal tissues of plants, such as vegetables, fruits, and flowers [1]. Anthocyanins are responsible for organoleptic colors such as red, purple, and blue [2]. Cyanidin-3-*O*-glucoside (C3G) is one of the most widely distributed anthocyanins in fruits [3], accounting for 94% of the total anthocyanin content in blackberries [4,5]. C3G serves as key antioxidant. Anthocyanins are glycosides of polyhydroxy and polymethoxy derivatives of 2-phenylbenzopyrylium or flavylium salts [2]. These compounds have a high antioxidant capacity [6] and the ability to prevent a variety of chronic diseases associated with inflammation [7,8] and improve the colon environment [9].

Dietary phenols, such as C3G, are substrates for a variety of enzymes in the liver, small intestine, and colon [10,11], mainly enzymes produced by gut microbiota. The colon is therefore an active site for C3G metabolism. When C3G is present in the gastrointestinal tract (GIT), colonic microbiota decompose them into a variety of bioactive substances and phenolic acids, such as protocatechuic acid (PCA) and 2,4,6-trihydroxybenzaldehyde (TB) [12], derived from the parent compound [12–14]. C3G is likely to primarily exert its influence through C3G metabolites [15]. In addition, in our previous study [16], we found that C3G plays a role in regulating the prebiotic activity of the microbial composition in rats. Another study found that the fermentation of C3G stimulated the proliferation of *Bifidobacteria* and *Lactobacillus*; both effects can affect intestinal health [17].

In addition to polyphenols, dietary fiber could modulate gut microbiota in a structure-dependent manner, releasing short-chain fatty acids (SCFAs) in the GIT. In recent years, the interactions between anthocyanins and dietary fibers have attracted extensive attention. A significant increase in plasma antioxidant capacity was found 8 h after the acute ingestion of high dietary fiber and polyphenols (antioxidant capacity determined by $ABTS^{\bullet+}$ and FRAP assays) [18]. The previous study has suggested that certain pectin fractions can increase C3G stability during in vitro digestion [19] in upper GIT, and consequently change the substrates for microbiota in the colon. The interactions between anthocyanins and dietary fibers could benefit gut health. However, how this interaction affects the fermentation of dietary fiber and C3G is still unclear, especially in terms of how different dietary fibers modulate C3G fermentation patterns depending on fiber type.

In this study, we investigated the interplay of C3G with four different dietary fibers during in vitro fermentation. We used in vitro fermentation model to determine the effects of different dietary fibers (β-glucan, arabinoxylan, fructo-oligosaccharide, and pectin) on the C3G metabolites, antioxidant capacity, SCFAs, and the microbiota composition in the colonic environment.

2. Materials and Methods

2.1. Materials

Oat β-glucan (BG) and wheat arabinoxylan (AX) were purchased from Shaanxi Pioneer Biotech Co., Ltd. (Xi'an, Shaanxi, China). Pectin (PEC) with 71% degree of esterification was provided by Herbstreith & Fox (Neuenbürg, Germany). Fructo-oligosaccharide (FOS) was obtained from Shanghai Yuanye Bio-Technology Co., Ltd. (Shanghai, China). Cyanidin-3-*O*-glucoside (C3G) was extracted and purified from black soybean peels (Anhui province, China) according to the method we described before [20,21]. The purity of the C3G was 95%.

2.2. In Vitro Fermentation Model

2.2.1. Medium

The culture medium was based on the modified standard ileal efflux medium (SIEM), as previously described [22]. Per liter, the medium contained the following: 4.5 g NaCl, 2.5 g K_2HPO_4, 0.45 g $CaCl_2{\cdot}2H_2O$, 0.5 g $MgSO_4{\cdot}7H_2O$, 0.005 g $FeSO_4{\cdot}7H_2O$, 0.05 g ox bile, 0.01 g haemin, 0.4 g cystein, 3 g bactopeptone, and 3 g casein. We then added 1 mL of a vitamin mixture, per liter of the SIEM contained: 1 mg menadione, 2 mg D-biotin, 0.5 mg vitamin B-12, 10 mg pantothenate, 5 mg nicotinamide, 5 mg *para*-aminobenzoic acid, and 4 mg thiamine.

2.2.2. Fecal Inoculum

Fecal samples were obtained from four healthy donors aged 22 to 23 years (mean age 23) who did not have any intestinal diseases and had not been treated with antibiotics for the previous three months. The donors' mean Body Mass Index (BMI) was 26.3 kg/m^2 (range: 22.8–35.1). Fecal samples were collected in a centrifuge tube with a sterile phosphatic buffer saline (PBS); then they were swirled and centrifuged (5000× *g*, 10 min). The mixture was diluted with a sterile PBS buffer to 0.1 g/mL to obtain the final fecal inoculum.

2.2.3. In Vitro Fermentation

The FOS, PEC, BG, and AX were each mixed with the C3G (CF, CP, CB, and CA groups), and each of the four mixtures was added to the SIEM culture medium as the substrates. The final fermentation system contained a 10 mg/mL concentration of dietary fiber and a 1 mg/mL concentration of C3G. Another experimental group supplemented only with a 1 mg/mL concentration of C3G (C group) was designed. The medium without any dietary fiber or C3G was used as blank control (N group). The mixture (3.6 mL) of medium and fermentation substrate was placed into an anaerobic fermentation tube; then, under anaerobic conditions, we added 0.4 mL of fecal inoculum. The in vitro fermentation was carried out in a vibrating incubator (37 °C, 100 rpm), and the fermentation fluids were collected and stored in a refrigerator, at −80 °C, after 0, 0.5, 1, 2, 4, 12, and 24 h. Each group was repeated three times in this study.

2.3. Antioxidant Capacity of the Fermentation Samples

The DPPH free-radical scavenging ability of the fermentation samples was determined according to a previous method [23].

The determination of ABTS radical cation scavenging activity was based on the method described before [24].

2.4. Gas Volume and pH

The measurement of gas volume during fermentation was based on a previous study [25]. This was done by emptying the gas in a syringe, and then inserting the syringe needle into the rubber stopper of the anaerobic bottle, allowing the gas produced in the anaerobic bottle to drain into the syringe. After fermentation, the volume of produced gas in each syringe was recorded.

The pH value was performed at 0, 4, 12, and 24 h after starting the colonic incubation, using a pH meter (PHS-3C, Leici, China) with a pH electrode (PY-P22-2S, Sartorius, Germany).

2.5. Determination of Total Carbohydrate Contents

Total carbohydrate content was determined according to the phenol–sulfuric acid method. We mixed 0.45 mL of 72% concentrated sulfuric acid and lyophilized 1 mL fermented sample in a test tube; then we heated the mixture at 30 °C for 1 h. We then added 4.95 mL of distilled water and heated the mixture at 100 °C for 3 h to obtain the hydrolyzed liquid for testing. The hydrolyzed sample was diluted at an appropriate dilution factor. A total of 200 μL of the diluted solution and 800 μL of a 2.5% phenol solution were mixed into the test tube. Then, 2.5 mL of concentrated sulfuric acid was added to the tube, and the mixture was mixed well and cooled to room temperature. The absorption value was read at 490 nm, and the total carbohydrate concentration of the fermentation liquid was calculated with a linear regression equation established for glucose.

2.6. Analysis of C3G and Metabolites

The High-Performance Liquid Chromatography (HPLC) system (Thermo U3000, USA) was used for the chromatographic separation of the C3G and its metabolites in all sample groups during fermentation. Reversed-phase chromatography was performed with a C18 column (4.6 × 250 mm, 5 μm; Thermo, Waltham, MA, USA). The mobile phase was a 2% formic acid aqueous solution (A) and acetonitrile (B). The gradient elution program was as follows: 0–34 min, 6–30% B; 34–35 min, 30–90% B; 35–40 min, 95–95% B; 40–41 min, 95–95% B; and 41–45 min, 6–6% B.

The Ultraviolet (UV) absorption intensity of the eluent was read at 280 and 520 nm. Compared with the peak area of standards for C3G and C3G metabolites, the contents of the C3G and C3G metabolite at different fermentation times were calculated.

2.7. DNA Extraction and 16S rRNA Sequence

Total genomic DNA was extracted by using a MagPure Soil DNA KF Kit (Catalogue No. D6356-02), and the manufacturer's instructions were followed. Quality and quantity of DNA were verified with NanoDrop and agarose gel. Then, the sample was diluted to 1 ng/μL and stored at −20 °C until further processing. Two primer pairs,

343F-5′-TACGGRAGGCAGCAG-3′ and 798R-5′-AGGGTATCTAATCCT-3′, were conducted to amplify the V3/V4 region (343–798) of the 16S rRNA gene. For library construction, a 30-μL PCR mixture was prepared as follows: 10–50 ng of DNA template, 1 μL of the forward and reverse PCR primers (5 pmol/μL each), 15 μL 2× Gflex PCR buffer (Takara), and 0.6 μL of Tks Gflex DNA polymerase (1.25 U/μL, Takara). An appropriate volume of double-distilled water (ddH_2O) was added to bring the volume up to 30 μL. The PCR program was as follows: initial denaturation at 94 °C for 5 min, followed by 26 cycles of denaturation at 94 °C for 30 s, annealing at 56 °C for 30 s, extension at 72 °C for 20 s, and a final extension at 72 °C for 5 min.

2.8. Analyses of SCFAs

SCFA contents, including acetic, propionic, butyric, valeric, isobutyric, and isovaleric acids in the fermentation fluids, were analyzed by using gas chromatography (GC; Shimadzu G2010Plus, Kyoto, Japan) on a DB-FFAP chromatographic capillary column (30 m × 0.530 mm × 1.00 μm; Agilent, Santa Clara, CA, USA), with calculated standard curves based on the method described in Tian et al. [26], with minor modifications. A total of 200 μL of the fermentation samples was mixed with 200 μL of 0.3 mg/mL 2-ethylbutyric acid and 50 μL of 0.15 M oxalic acid. The mixture was filtered at 0.22 μm into a vial for analysis.

2.9. Statistical Analysis

GraphPad Prism 8.0 (San Diego, CA, USA) was used for the graphic presentation. Data other than microbial diversity and composition were expressed as means ± standard deviation (SD), and the differences among all the groups were tested with a one-way analysis of variance (ANOVA) with multiple post hoc comparisons. A value of $p < 0.05$ was considered to be statistically significant. Chao1, Shannon, and Simpson indices were applied to evaluate alpha diversity. Weighted-UniFrac distance was used to evaluate beta diversity in different fermented samples.

3. Results and Discussion

3.1. Antioxidant Capacity

The results for DPPH and $ABTS^{\bullet+}$ scavenging ability over the 24 h fermentation period are shown in Figure 1A,C. The antioxidant activity of each group decreased after 24 h in vitro fermentation; however, the decrease was not significant in every group. This may be due to the lack of metabolic absorption pathways in our simulated colon process, causing the phenolic substances to remain in the fermentation broth after degradation. Compared with 0 h, the DPPH radical scavenging power decreased 23.4% for C, 20.6% for CF, 17.9% for CP, 27.8% for CB ($p < 0.05$), and 23.2% for CA ($p < 0.05$), respectively (Figure 1B). The $ABTS^{\bullet+}$ scavenging power of the fermentation samples showed a similar trend (Figure 1C,D). The antioxidant activity in the samples indicated that the mixtures of C3G with FOS, PEC, or AX could all retain the antioxidant active components in the fermentation broth to varying degrees. C3G and C3G metabolites are considered to be major contributors to antioxidant capacity, due to their chemical structure [2]. The uronic acids–rich fractions in polysaccharides have higher antioxidant activities and are significant for the biological activities of polysaccharides [27]. Pectin contains the greatest amounts of uronic acids among the four dietary fibers, so the CP group had the strongest antioxidant activity in vitro. The mixture of C3G and BG had an antagonistic effect on antioxidant activity, and this might be due to its structure and provided dose, as well as the strong hydrogen bonds between the two metabolites [28].

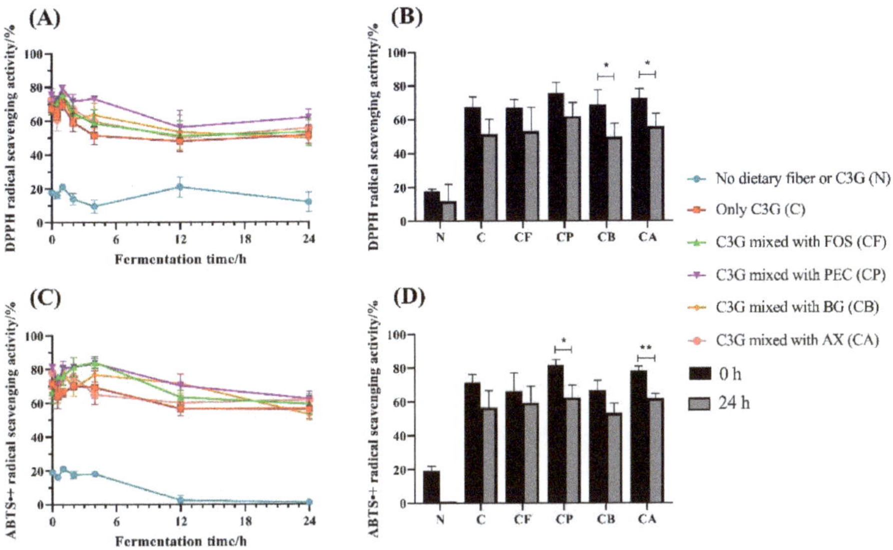

Figure 1. Changes in antioxidant activity, DPPH (**A**), and ABTS$^{\bullet+}$ (**C**) radical scavenging rate for C3G mixed with different polysaccharides during in vitro fermentation; differences in DPPH (**B**) and ABTS$^{\bullet+}$ (**D**) radical scavenging rate in vitro fermentation for 0 and 24 h. Asterisks indicate significant difference between fermentation for 0 and 24 h, according to an ordinary one-way ANOVA: * ($p < 0.05$) and ** ($p < 0.01$). Values are means ± SD for $n = 3$.

3.2. Total Gas Production and pH

The total gas production for C3G and the different polysaccharides during in vitro fermentation is shown in Figure 2A. The cumulative gas production of the groups with added polysaccharides was higher than that of group N and group C. Cumulative gas production increased gradually with increasing fermentation time. The CP, CB, and CA groups had no obvious gas-production changes after 12 h, while the CF group still showed an increasing trend in gas production. By the end of the 24 h fermentation time, the volume of cumulative gas production for the CF, CP, CB, and CA groups had reached 4.36 ± 0.19, 0.55 ± 0.07, 1.19 ± 0.05, and 2.87 ± 0.05 mL, respectively. Fermentation of FOS from using pig fecal microbiota produced a higher volume than that of the other carbohydrates used as substrates [29]. After 24 h in vitro fermentation, the results for gas production with different polysaccharide substrates were similar to those of previous studies [30].

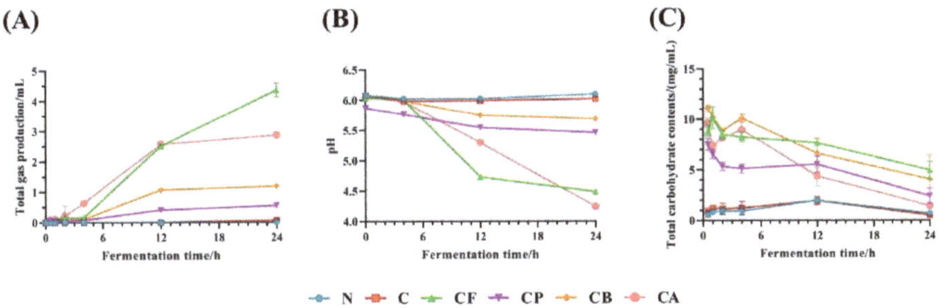

Figure 2. Cumulative gas production (**A**), pH changes (**B**), and total carbohydrate contents (**C**) during in vitro fermentation for C3G and different polysaccharides.

The fermentation of carbohydrates by human gut microbiota usually produces gases. Production of H_2 is often necessary for the cycling of $NAD^+/NADH$ during fermentation, and CO_2 is released whenever decarboxylation occurs [31]. C3G did not have sufficient carbohydrates for the fermentation by gut microbiota; therefore, the gas volume of C group was very low (Figure 2A). The chemical composition of the prebiotics and the composition of the microbiota are related to the amount of produced gas during fermentation [31]. A previous study showed that inulin has a higher potential to produce H_2 than pectin during in vitro fermentation, due to its composition, and its total gas production was significantly higher than pectin [31]. This result is similar to that in this study; the gas production of FOS is significantly higher than that of PEC. In another study comparing the fermentation characteristics of FOS and BG, it was also found that FOS produced more gas than BG when fermented by gut microbiota [32].

Gas volume, pH, and SCFA production are all indices reflecting the degree of fermentation, and they are generally correlated with each other. The data of gas volume and pH (Figure 2B) indicated that CF and CA groups were the two most fully fermented groups. The more full fermentation of CA was also reflected in the SCFA production, while the SCFA production of CF is not consistent with its gas volume and pH, which may be due to the accumulation of SCFA intermediates (lactic acid) [33].

3.3. Total Carbohydrate Contents

Total carbohydrate content in fermentation broth at different fermentation times was determined through phenol–sulfuric acid method. In Figure 2C, the total carbohydrate content of each group with polysaccharide substrate generally decreased with increasing fermentation time. In general, utilization by gut microbiota was the highest for AX, followed by PEC and then BG; the utilization of FOS and PEC was mainly concentrated in the first 4 h of fermentation. Total carbohydrate contents in the N and the C groups were also about 1 mg/mL, which may be related to the carbohydrate from microbiota. FOS, PEC, BG, and AX are polysaccharides that can rapidly fermented in vitro static fermentation models [34]; thus, their degradation time is mainly concentrated in the first 12 h. The type and complexity of glycosidic linkages, monosaccharide molecular arrangement, molecular weight, particle size, solubility, and physical properties of DFs led to different fermentation rates and degrees of DFs in this study [34].

Polysaccharides were consumed in descending order: AX (85.2%) > PEC (68.3%) > BG (63.7%) > FOS (43.5%). Normally, the degradation of AX takes longer [35], but in this study, after 24 h, the degradation of AX was the most complete among the four polysaccharide groups. This may be due to the fact that the different preferences of microbiota in DFs degradation of different colonic locations [36] and different hosts' dietary preferences, and leading to the highest degree of AX degradation.

3.4. Changes of C3G and C3G Metabolites during Fermentation

We used HPLC to analyze the metabolites in fermentation broth of each group at each time point, compared to standard C3G metabolites. The results obtained are shown in Figure 3A. Figure 3A shows that the peak area for C3G decreased with the increasing of fermentation time in all the groups, but that the C3G was degraded and metabolized to varying degrees in the different groups. Notably, C3G was almost completely metabolized in C, CP, and CA groups after 24 h in vitro fermentation. However, there was still a high proportion of C3G in the fermentation broth of the CF group, and only 1/2 to 2/3 of the original C3G had been consumed in the CB group, indicating that the presence of FOS and BG may have a protective effect on C3G to some extent. In addition, C3G in group CA was the earliest to degrade among all the groups; this may be related to the high amount of *Lactobacillus* and *Bifidobacteria* in group CA [37].

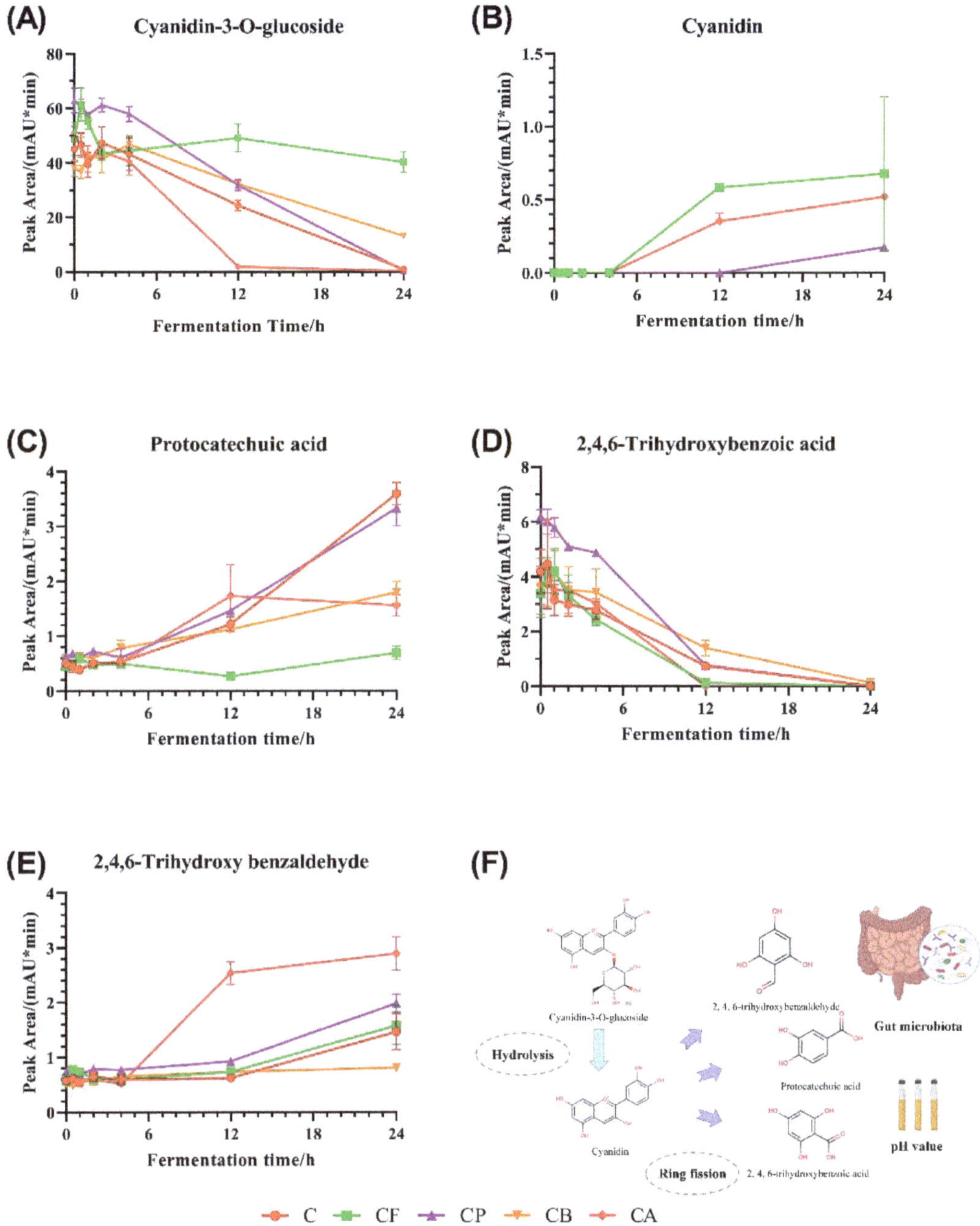

Figure 3. Changes in anthocyanin metabolites (**A**), Cyanidin-3-O-Glucoside (**B**), Cyanidin (**C**), PCA (**D**), 2,4,6-trihydroxybenzoic acid (TBA) (**E**), and (TB) during in vitro fermentation and the structural changes in metabolites (**F**).

Cyanidin is the hydrolyzed product of the C3G glycosidic bond [38] (Figure 3F). The changes in the peak area for cyanidin in six groups of samples at seven time points after integration are shown in Figure 3B. Cyanidin was detected in the fermentation broth in only three of the six groups. Generally speaking, the changes in C3G and cyanidin should be inversely proportional; that is, the more cyanidin breaks away from the C3G glycosidic bond, the more cyanidin will be retained in the fermentation broth. However, the C3G peak area for the CF group was still relatively high after 24 h, and this phenomenon was also observed for the cyanidin peak area in the corresponding integral graph. Therefore, we speculate that the co-fermentation of FOS and C3G may have protective effects on both C3G and cyanidin. Although the C3G in the CB group was not totally consumed by gut microbiota completely within 24 h, cyanidin was not detected in the fermentation broth, indicating that the protective effects of BG on cyanidin was not obvious. Most of the C3G in the CA group had been consumed before 12 h of culture, but cyanidin was still detected at 24 h, indicating that AX protected cyanidin from fermentation by microbiota.

Previous studies have investigated the metabolism of C3G in rats or in vitro fermentation model, and identified PCA, TBA, and TB as the major colonic metabolites [39,40] that were affected by gut microbiota and the pH value of the environment [38] (Figure 3F). Therefore, we integrated the peak area of PCA, TBA, and TB, which were the C3G metabolites in the fermentation broth. Figure 3C–E shows that PCA and TB did not significantly change during the 0–4 h phase, but did gradually increase after 4 h, which is in accordance with the degradation status of C3G after 4 h. The main metabolites in the CF group was cyanidin, and the main metabolites for CA were TB and cyanidin. These acidic metabolites may be absorbed by epithelial cells via monocarboxylic acid transporter [41].

Different polysaccharide substrates have different effects on the degradation of C3G by gut microbiota. FOS slowed down the degradation of C3G and prevented the further degradation of cyanidin. PEC modified the end product of C3G, so that the protocatechuic acid became the mainly metabolites. Similar to FOS, BG had the same inhibitory effect on C3G degradation, but the inhibitory effect is not as obvious as FOS, and the final main production was PCA. CA had the effect of promoting the degradation of C3G, which was almost degraded after 12 h of in vitro fermentation, and the main metabolite was TB. The different ways in which polysaccharides regulate C3G metabolism will help us to target those metabolites that are beneficial to health.

3.5. Alpha- and Beta-Diversity of Gut Microbiota

In order to evaluate the impact of the different C3G-polysaccharide co-cultures on in vitro gut microbiota, the bacterial lineages of the in vitro fermentation broth were characterized. Alpha-diversity analysis was performed to reveal the diversity of species in each individual sample. The constructed dilution curves of diversity index show the difference in species richness. The Chao 1 (Figure 4A) index shows that there was no significant difference in species richness among the six experimental groups. Only the species richness in groups C and CP increased slightly; this was beneficial to the formation of a complete bacterial community structure. Both the Shannon (Figure 4B) and Simpson (Figure 4C) indices account for abundance and distribution of the species present. The addition of the four different polysaccharides to C3G significantly improved the diversity of gut microbiota after in vitro fermentation for 24 h, as indicated by the Shannon and Simpson indices, especially for PEC and BG.

Beta-diversity illustrates the differences between different populations. PCoA was used to further analyze the variation of the four polysaccharides mixed with C3G (Figure 4D). PCoA with Weighted Unifrac method showed that the first and second principal components explained 83.31% and 9.25% of the variation in microbial diversity, respectively. As can be seen in Figure 4D, the microorganisms in CF, CP, CB, and CA exhibited varying degrees of ability to utilize the substrate, indicating fine microbial differentiation due to the substrate chemistry. For PCoA, statistical analyses (ADONIS) showed that the four C3G-polysaccharide mixtures had a significant effect on the microbial community ($p < 0.001$), which further confirmed the important effect of dietary fiber type on microbial community structure [42]. In contrast to CF, CP, CB, and CA groups, the N and C groups clustered closely in the PCoA plot, indicating a significant similarity (low beta-diversity) between the two samples. These results indicate that C3G fermentation alone has little effect on changes in microbial community.

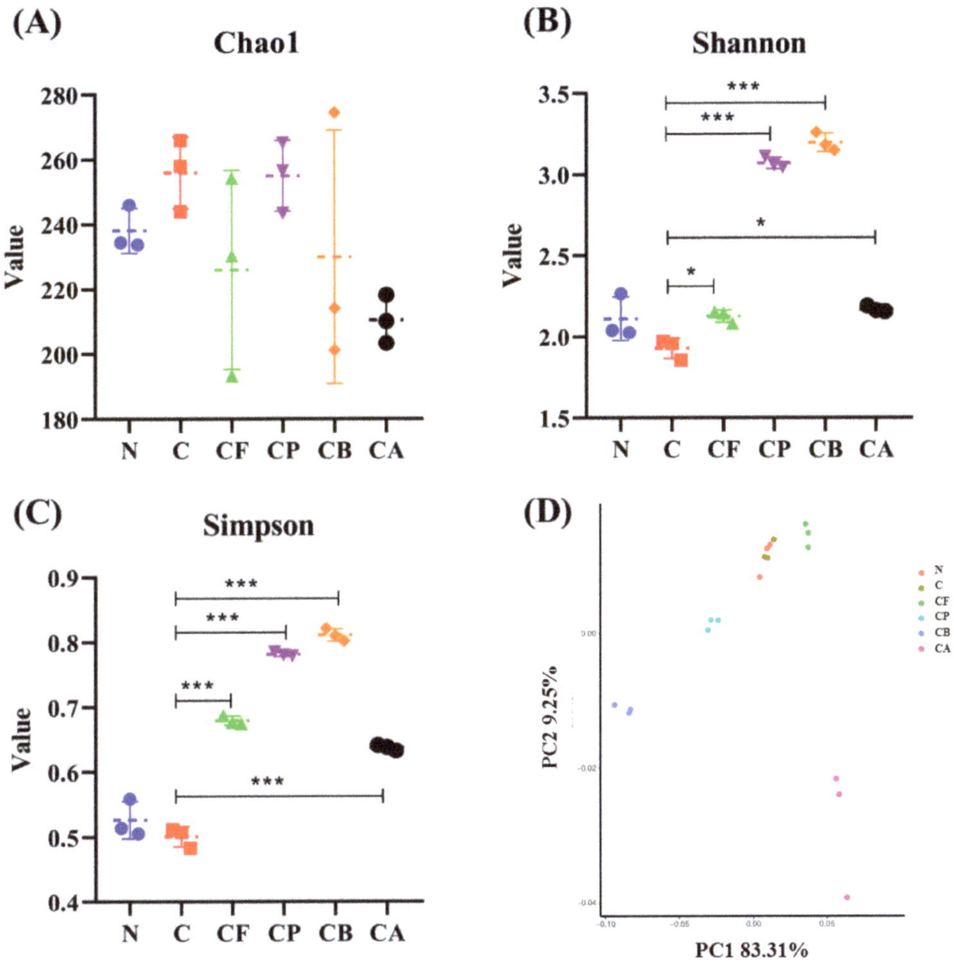

Figure 4. Analysis of different microbial α-diversity and β-diversity indices in the six experimental groups. The Chao1 (**A**) index was used as richness estimators. The Shannon (**B**) and Simpson (**C**) indices were used as diversity estimators. Principal Coordinates Analysis (PCoA) (**D**) was based on the weighted UniFrac distance matrix generated from all the samples in each group. Asterisks indicate significant difference between fermentation with C3G only and with C3G mixed with different polysaccharides according to an ordinary one-way ANOVA: * ($p < 0.05$), and *** ($p < 0.001$). Values are means ± SD for $n = 3$, as below.

3.6. Microbial Community Distribution

Bacterial taxonomy at the genus level, at 24 h of incubation, with each substrate, is shown in Figure 5. We analyzed the sequencing results of each group, and Proteobacteria and Bacteroidetes were the dominant bacteria detected in all samples. The dominant microbiota at the genus level were *Escherichia–Shigella*, *Klebsiella*, *Bacteroides*, and *Lactobacillus*. In order to further determine the difference in microbial community among the six groups at the genus level, statistical analyses were conducted on the dominant species at the genus level.

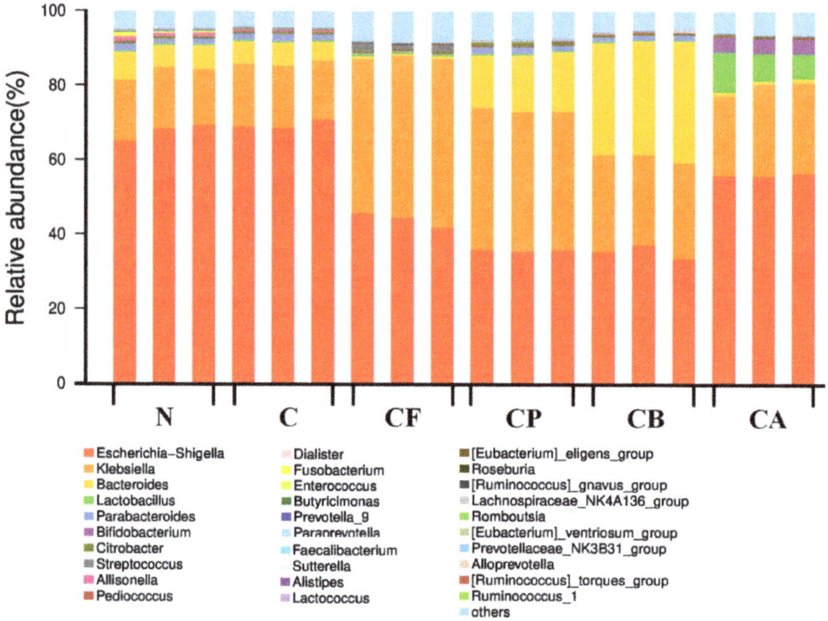

Figure 5. Relative abundance by genus level. Each column represents one sample. Different colors indicate different bacteria in genus.

We performed an analysis of ANOVA and counted six different abundant microbiomes at the genus level (Figure 6). *Escherichia–Shigella* could impact the health of the host, leading to an unstable gut microbiota associated with low-grade inflammation or even cause chronic colitis. Thus, lower levels of fecal *Escherichia–Shigella* are likely to be beneficial. In this study, the *Escherichia–Shigella* group predominated during fermentation, possibly due to the presence of residual amounts of oxygen during inoculum activation. However, at the same time, we noticed that the four C3G-polysaccharide complexes all reduced the abundance of *Escherichia–Shigella* (Figure 6A) in the fermentation samples to varying degrees, which is beneficial to the health of the GIT. This observation could be attributed to the fact that the fermentation of polysaccharides resulted in an acidic environment that was not suitable for the growth of *Escherichia–Shigella*. Both *Bacteroides* (Figure 6C) and *Lactobacillus* (Figure 6D) can induce an acidic environment [43], which could explain why the abundance of *Escherichia–Shigella* in CB and CP groups was lower than that in the other two groups (CF and CA). *Lactobacillus* species have significant effects on intestinal permeability, inflammation, body weight, and obesity [44,45]. A high presence of *Bifidobacterium* is inversely associated with increased fat mass, glucose tolerance, adipose tissue inflammation, and lipopolysaccharide (LPS) levels [46]. In our study, the abundance of *Lactobacillus* (Figure 6D) and *Bifidobacterium* (Figure 6E), two types of beneficial bacteria commonly found in gut microbiota, was improved; this was consistent with previous a study [17]. The four polysaccharides and C3G have some degree of probiotic-growth-promoting effect. In addition, the compound substance C3G–AX significantly increased the abundance of *Lactobacillus* and *Bifidobacterium*, and was apparently higher than that of other groups. Previous studies have demonstrated a more bifidogenic effect for AX than BG [35]. An increase in *Bifidobacterium* and *Lactobacillus* induced by anthocyanins had also been previously reported [47,48]. C3G could be degraded by β-glucosidases secreted by the microbiota to provide energy and more suitable pH for the growth of the probiotics [17]. However, since we did not set up a pure polysaccharide group, it is not clear whether this effect of promoting probiotic growth is synergistic. *Fusobacterium* positively correlated with fasting blood glucose, glycosylated serum protein, LPS, insulin resistance index (HOMA-IR), and three inflammatory factors (TNF-alpha, IL-1 beta, and IL-6) in rats [49]. We compared the abundance of *Fusobacterium* (Figure 6F) in our six groups after 24 h of in vitro fermentation. Our results showed that C3G alone and the different C3G–polysaccharide complexes significantly decreased the abundance of *Fusobacterium*, and that the polysaccharides enhanced the effect of C3G to a similar extent.

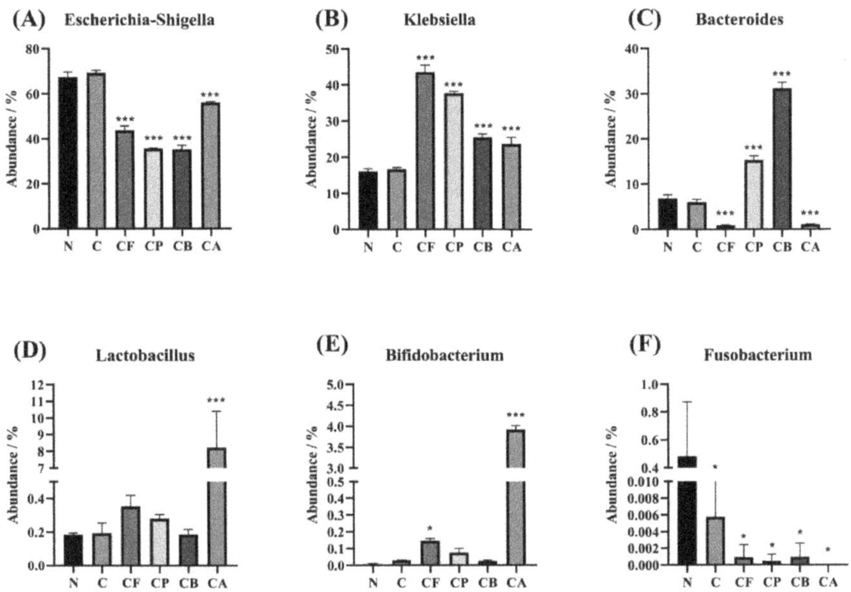

Figure 6. Significant differences among the five substrates and the control at the genus level. Asterisks indicate significant difference between N group and other five groups, respectively: * ($p < 0.05$), and *** ($p < 0.001$). Values are means ± SD for $n = 3$.

3.7. SCFAs Production

The main SCFAs in the gut, including acetic acid, propionic acid, and butyric acid, are generally considered to be beneficial for human health. The SCFA plays a crucial role in maintaining the barrier function of the gut, regulating epithelial proliferation, preventing colorectal cancer, and modulating immune responses [50,51]. Therefore, accurately determining the level of SCFAs is important for evaluating the fermentation behavior of C3G mixed with dietary fibers in the gut. However, in vivo produced SCFAs are difficult to measure, since more than 95% of SCFAs are rapidly absorbed and metabolized by the GIT [52]. For this reason, the concentrations of SCFAs in fecal samples do not truly represent those in the proximal colon. Compared with in vivo methods, in vitro batch fermentation is a more promising approach that can be adapted to determine the actual fermentation behavior of polysaccharides more accurately [34]. Thus, we employed batch fermentation and used gas chromatography to determine SCFAs. To study the effects of in vitro fermentation on SCFA production in C3G mixed with four different polysaccharide treatments, we measured the concentrations of acetic acid, propionic acid, butyric acid, valeric acid, isobutyric acid, and isovaleric acid.

Total SCFA concentrations in the fermentation broth (Figure 7) increased gradually with the extension of fermentation time. Compared with the N group, SCFA concentrations in the fermentation broth for C3G only and C3G mixed with different polysaccharides were higher. This indicates that both C3G alone and C3G mixed with different polysaccharides can promote SCFA production by gut microbiota. SCFAs are derived from microbial fermentation of dietary fibers [53], so the total SCFA production of the four groups with polysaccharide substrates was significantly higher than the other two groups (N and C3G alone). Total SCFA concentrations in the fermentation broth for group C were also higher than for group N; however, the effect was not significant. This could be explained by the breakdown of the glucoside bonds in C3G by bacterial enzymes, resulting in the situation that microbiota used the glucose to produce SCFAs [54].

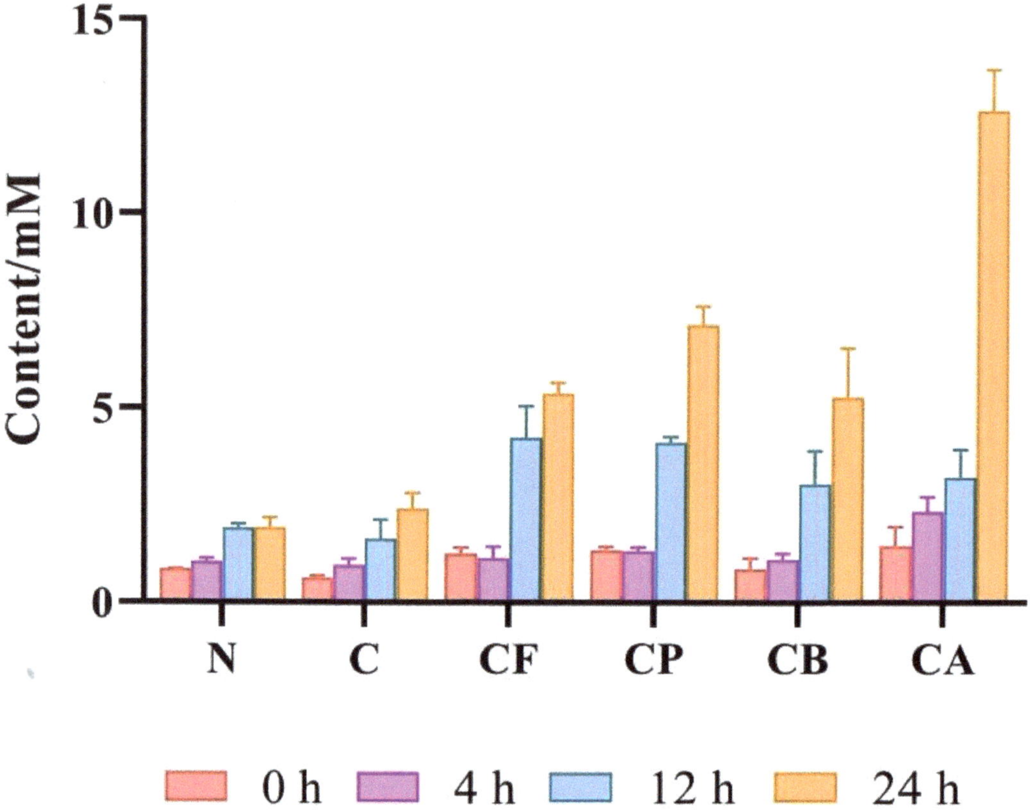

Figure 7. Total C3G SCFAs mixed with different polysaccharides during in vitro fermentation.

In this study, the concentrations of acetic acid (Figure 8A) in the fermentation broth were the highest, and they gradually increased with the extension of the fermentation time. At 24 h of fermentation, the acetic acid concentrations in the fermentation broths with a polysaccharide substrate were significantly higher than for the N and C groups. In particular, acetic acid production in the CA group was significantly higher than that of all the other groups. This may be because the combination of C3G and AX increases the abundance of *Bifidobacterium*, and acetic acid is the final product of *Bifidobacterium*. Acetic acid can enter the blood circulation system and be oxidized by the brain, heart, and peripheral tissues [55]. Acetic acid may be closely related to the regulation of body weight, because high concentration of acetic acid can promote the expression of AMP kinase and inhibit the synthesis of fat [56]. Therefore, the combination of C3G and AX may be more effective than other combinations in controlling weight gain.

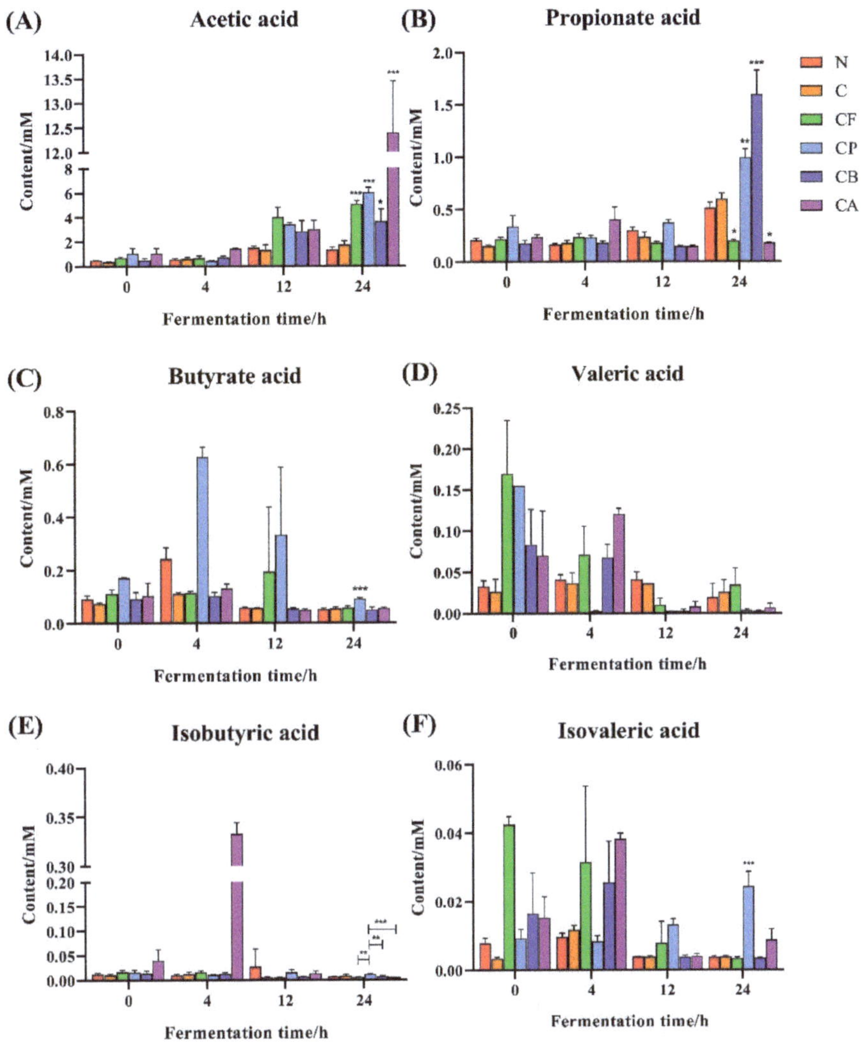

Figure 8. Effects of C3G mixed with different polysaccharides on SCFAs during in vitro fermentation: (**A**) acetic acid, (**B**) propionic acid, (**C**) butyric acid, (**D**) valeric acid, (**E**) isobutyric acid, and (**F**) isovaleric acid. Asterisks indicate significant difference between N group and other five groups at 24 h of fermentation, respectively, except for isobutyric acid: * ($p < 0.05$), ** ($p < 0.01$) and *** ($p < 0.001$). Values are means ± SD for $n = 3$

Propionic acid (Figure 8B) produced the second highest SCFA concentrations in our study. Surprisingly, combining C3G with FOS and AX did not increase propionic acid production. However, mixing C3G with PEC and BG did significantly increase propionic acid production in the fermentation broth, especially for the CB group. Previous studies have similar results to ours [30,43]. A significant increase in the Bacteroides (Figure 5E) from β-glucan fermentation may explain the increase in propionate production [57]. Propionic acid is absorbed by the liver, participates in gluconeogenesis, and inhibits cholesterol synthesis [58], so we hypothesize that the combination of C3G and BG has the potential to lower cholesterol.

Butyric acid (Figure 8C) is an important source of energy for intestinal epithelial cells; can regulate the growth and apoptosis of epithelial cells and immune cells; can inhibit colitis and colon cancer; and can play a crucial role in maintaining the integrity of the intestinal mucosal barrier and in stabilizing the intestinal microenvironment [59]. Mixing C3G and PEC significantly increased butyric acid production during in vitro fermentation, with levels peaking at 4 h. This may be due to the fact that pectin promotes the growth of *Faecalibacterium* and *Ruminococcaceae* during fermentation [60], thus increasing butyric acid production, which is subsequently consumed as an energy source by microbiota. The results of Harris et al. on butyric acid production of different DFs in the mid fermentation were similar to ours [61]. However, in the mid-late fermentation, β-glucan and FOS did not produce significantly higher butyric acid, as shown in a previous study [62], which may due to the changes the metabolic preference of microorganisms affecting by C3G. In addition, the amounts of isobutyric, isovaleric, and valeric acid (Figure 8D–F) were very low. Indeed, few bacteria are known to produce these three acids [63].

4. Conclusions

In conclusion, the used dietary fibers modulate the fermentation patterns of cyanidin-3-*O*-glucoside in a fiber-type-dependent manner. Depending on its structure, each compound provided a specific carbon source for the fermentation of different bacterial populations, thereby influencing the beneficial taxa and producing different amounts of SCFAs and gases in vitro. The present work illustrated the interaction among anthocyanins, polysaccharides, and gut microbiota, which could provide a reference for elucidating the health effects of the supplementation of C3G-polysaccharide diets. For future research, the individual dietary fiber groups will be included to study the potential synergistic effects of C3G and dietary fiber on fermentation patterns.

Author Contributions: All authors participated in the design of the study. Conceptualization, W.B. and L.T.; data curation, Z.Y. and P.L.; formal analysis, Z.Y. and T.H.; funding acquisition, W.B. and L.T.; investigation, Z.Y. and P.L.; methodology, Z.Y., P.L., J.L., and L.T.; supervision, L.T.; visualization, J.A.; writing—original draft, Z.Y. and T.H. The authors thank Vicki Moore for correcting the English and Oebiotech (Shanghai, China) for performing 16S rRNA sequencing. All authors have read and agreed to the published version of the manuscript.

Funding: This study was supported by the Natural Science Foundation of China (NSFC, Nos. 31901706 and No.31911530698), the Natural Science Foundation of Guangdong (2019A1515012237), the Science and Technology Program of Guangzhou (201903010081), the Pearl River Talent Plan (2017GC010387), the Youth Science and Technology Innovation Talent of Guangdong TeZhi Plan (2019TQ05N770), and the International Science and Technology Cooperation Project of Guangdong (2021A0505030043).

Institutional Review Board Statement: This study did not require ethical approval.

Informed Consent Statement: Informed consent was obtained from all subjects involved in the study.

Data Availability Statement: Data available in a publicly accessible repository.

Conflicts of Interest: The authors declare no conflict of interest.

References

1. Vauzour, D. Dietary Polyphenols as Modulators of Brain Functions: Biological Actions and Molecular Mechanisms Underpinning Their Beneficial Effects. *Oxidative Med. Cell. Longev.* **2012**, *2012*, 914273. [CrossRef]
2. Kong, J.M.; Chia, L.S.; Goh, N.K.; Chia, T.F.; Brouillard, R. Analysis and biological activities of anthocyanins. *Phytochemistry* **2003**, *64*, 923–933. [CrossRef]
3. Wu, X.L.; Prior, R.L. Systematic identification and characterization of anthocyanins by HPLC-ESI-MS/MS in common foods in the United States: Fruits and berries. *J. Agric. Food Chem.* **2005**, *53*, 2589–2599. [CrossRef] [PubMed]
4. Lee, S.G.; Vance, T.M.; Nam, T.-G.; Kim, D.-O.; Koo, S.I.; Chun, O.K. Contribution of Anthocyanin Composition to Total Antioxidant Capacity of Berries. *Plant Foods Hum. Nutr.* **2015**, *70*, 427–432. [CrossRef]
5. Tan, J.J.; Li, Y.L.; Hou, D.X.; Wu, S.S. The Effects and Mechanisms of Cyanidin-3-Glucoside and Its Phenolic Metabolites in Maintaining Intestinal Integrity. *Antioxidants* **2019**, *8*, 479. [CrossRef] [PubMed]
6. Tsuda, T.; Shiga, K.; Ohshima, K.; Kawakishi, S.; Osawa, T. Inhibition of lipid peroxidation and the active oxygen radical scavenging effect of anthocyanin pigments isolated from *Phaseolus vulgaris* L. *Biochem. Pharmacol.* **1996**, *52*, 1033–1039. [CrossRef]

7. Esposito, D.; Chen, A.; Grace, M.H.; Komarnytsky, S.; Lila, M.A. Inhibitory Effects of Wild Blueberry Anthocyanins and Other Flavonoids on Biomarkers of Acute and Chronic Inflammation In Vitro. *J. Agric. Food Chem.* **2014**, *62*, 7022–7028. [CrossRef]
8. Lee, S.G.; Kim, B.; Yang, Y.; Pham, T.X.; Park, Y.-K.; Manatou, J.E.; Koo, S.I.; Chun, O.K.; Lee, J.-Y. Berry anthocyanins suppress the expression and secretion of proinflammatory mediators in macrophages by inhibiting nuclear translocation of NF-kappa B independent of NRF2-mediated mechanism. *J. Nutr. Biochem.* **2014**, *25*, 404–411. [CrossRef] [PubMed]
9. Si, X.; Bi, J.F.; Chen, Q.Q.; Cui, H.J.; Bao, Y.W.; Tian, J.L.; Shu, C.; Wang, Y.H.; Tan, H.; Zhang, W.J.; et al. Effect of Blueberry Anthocyanin-Rich Extracts on Peripheral and Hippocampal Antioxidant Defensiveness: The Analysis of the Serum Fatty Acid Species and Gut Microbiota Profile. *J. Agric. Food Chem.* **2021**, *69*, 3658–3666. [CrossRef]
10. Tapiero, H.; Tew, K.D.; Ba, G.N.; Mathe, G. Polyphenols: Do they play a role in the prevention of human pathologies? *Biomed. Pharmacother.* **2002**, *56*, 200–207. [CrossRef]
11. Rechner, A.R.; Kuhnle, G.; Bremner, P.; Hubbard, G.P.; Moore, K.P.; Rice-Evans, C.A. The metabolic fate of dietary polyphenols in humans. *Free Radic. Biol. Med.* **2002**, *33*, 220–235. [CrossRef]
12. Chen, Y.; Chen, H.; Zhang, W.J.; Ding, Y.Y.; Zhao, T.; Zhang, M.; Mao, G.H.; Feng, W.W.; Wu, X.Y.; Yang, L.Q. Bioaccessibility and biotransformation of anthocyanin monomers following in vitro simulated gastric-intestinal digestion and in vivo metabolism in rats. *Food Funct.* **2019**, *10*, 6052–6061. [CrossRef]
13. Ludwig, I.A.; Mena, P.; Calani, L.; Borges, G.; Pereira-Caro, G.; Bresciani, L.; Del Rio, D.; Lean, M.E.J.; Crozier, A. New insights into the bioavailability of red raspberry anthocyanins and ellagitannins. *Free Radic. Biol. Med.* **2015**, *89*, 758–769. [CrossRef] [PubMed]
14. Kay, C.D.; Pereira-Caro, G.; Ludwig, I.A.; Clifford, M.N.; Crozier, A. Anthocyanins and Flavanones Are More Bioavailable than Previously Perceived: A Review of Recent Evidence. *Annu. Rev. Food Sci. Technol.* **2017**, *8*, 155–180. [CrossRef]
15. Bharat, D.; Cavalcanti, R.R.M.; Petersen, C.; Begaye, N.; Cutler, B.R.; Costa, M.M.A.; Ramos, R.; Ferreira, M.R.; Li, Y.Y.; Bharath, L.P.; et al. Blueberry Metabolites Attenuate Lipotoxicity-Induced Endothelial Dysfunction. *Mol. Nutr. Food Res.* **2018**, *62*, 8. [CrossRef]
16. Chen, G.W.; Wang, G.; Zhu, C.J.; Jiang, X.W.; Sun, J.X.; Tian, L.M.; Bai, W.B. Effects of cyanidin-3-O-glucoside on 3-chloro-1,2-propanediol induced intestinal microbiota dysbiosis in rats. *Food Chem. Toxicol.* **2019**, *133*, 9. [CrossRef]
17. Zhu, Y.S.; Sun, H.J.; He, S.D.; Lou, Q.Y.; Yu, M.; Tang, M.M.; Tu, L.J. Metabolism and prebiotics activity of anthocyanins from black rice (*Oryza sativa* L.) In Vitro. *PLoS ONE* **2018**, *13*, e0195754. [CrossRef]
18. Perez-Jimenez, J.; Serrano, J.; Tabernero, M.; Arranz, S.; Diaz-Rubio, M.E.; Garcia-Diz, L.; Goni, I.; Saura-Calixto, F. Bioavailability of Phenolic Antioxidants Associated with Dietary Fiber: Plasma Antioxidant Capacity after Acute and Long-Term Intake in Humans. *Plant Foods Hum. Nutr.* **2009**, *64*, 102–107. [CrossRef] [PubMed]
19. Koh, J.; Xu, Z.M.; Wicker, L. Blueberry pectin and increased anthocyanins stability under in vitro digestion. *Food Chem.* **2020**, *302*, 8. [CrossRef]
20. Li, X.S.; Guo, J.T.; Jiang, X.W.; Sun, J.X.; Tian, L.M.; Jiao, R.; Tang, Y.G.; Bai, W.B. Cyanidin-3-*O*-glucoside protects against cadmium-induced dysfunction of sex hormone secretion via the regulation of hypothalamus-pituitary-gonadal axis in male pubertal mice. *Food Chem. Toxicol.* **2019**, *129*, 13–21. [CrossRef] [PubMed]
21. Jiang, X.W.; Zhu, C.J.; Li, X.S.; Sun, J.X.; Tian, L.M.; Bai, W.B. Cyanidin-3-*O*-glucoside at Low Doses Protected against 3-Chloro-1,2-propanediol Induced Testis Injury and Improved Spermatogenesis in Male Rats. *J. Agric. Food Chem.* **2018**, *66*, 12675–12684. [CrossRef]
22. Minekus, M.; Smeets-Peeters, M.; Bernalier, A.; Marol-Bonnin, S.; Havenaar, R.; Marteau, P.; Alric, M.; Fonty, G.; Huis in't Veld, J.H. A computer-controlled system to simulate conditions of the large intestine with peristaltic mixing, water absorption and absorption of fermentation products. *Appl. Microbiol. Biotechnol.* **1999**, *53*, 108–114. [CrossRef] [PubMed]
23. Zhu, Y.; Jiang, J.; Yue, Y.; Feng, Z.; Chen, J.; Ye, X. Influence of mixed probiotics on the the bioactive composition, antioxidant activity and appearance of fermented red bayberry pomace. *LWT Food Sci. Technol.* **2020**, *133*, 110076. [CrossRef]
24. Xiao, Y.; Rui, X.; Xing, G.L.; Wu, H.; Li, W.; Chen, X.H.; Jiang, M.; Dong, M.S. Solid state fermentation with Cordyceps militaris SN-18 enhanced antioxidant capacity and DNA damage protective effect of oats (*Auena sativa* L.). *J. Funct. Food.* **2015**, *16*, 58–73. [CrossRef]
25. Koecher, K.J.; Noack, J.A.; Timm, D.A.; Klosterbuer, A.S.; Thomas, W.; Slavin, J.L. Estimation and Interpretation of Fermentation in the Gut: Coupling Results from a 24 h Batch in Vitro System with Fecal Measurements from a Human Intervention Feeding Study Using Fructooligosaccharides, Inulin, Gum Acacia, and Pea Fiber. *J. Agric. Food Chem.* **2014**, *62*, 1332–1337. [CrossRef]
26. Tian, L.M.; Scholte, J.; Borewicz, K.; van den Bogert, B.; Smidt, H.; Scheurink, A.J.W.; Gruppen, H.; Schols, H.A. Effects of pectin supplementation on the fermentation patterns of different structural carbohydrates in rats. *Mol. Nutr. Food Res.* **2016**, *60*, 2256–2266. [CrossRef]
27. Jin, M.L.; Zhao, K.; Huang, Q.S.; Xu, C.L.; Shang, P. Isolation, structure and bioactivities of the polysaccharides from Angelica sinensis (Oliv.) Diels: A review. *Carbohydr. Polym.* **2012**, *89*, 713–722. [CrossRef] [PubMed]
28. Wang, J.Q.; Hu, S.Z.; Nie, S.P.; Yu, Q.; Xie, M.Y. Reviews on Mechanisms of In Vitro Antioxidant Activity of Polysaccharides. *Oxidative Med. Cell. Longev.* **2016**, *2016*, 5692852. [CrossRef] [PubMed]
29. Musco, N.; Calabro, S.; Infascelli, F.; Tudisco, R.; Lombardi, P.; Grossi, M.; Addi, L.; Neto, B.P.; Cutrignelli, M.I. In vitro fermentation of structural carbohydrate-rich feeds using faecal inoculum from pigs. *Ital. J. Anim. Sci.* **2015**, *14*, 557–562. [CrossRef]

30. Carlson, J.L.; Erickson, J.M.; Hess, J.M.; Gould, T.J.; Slavin, J.L. Prebiotic Dietary Fiber and Gut Health: Comparing the in Vitro Fermentations of Beta-Glucan, Inulin and Xylooligosaccharide. *Nutrients* **2017**, *9*, 1361. [CrossRef]
31. Yu, X.Q.; Gurry, T.; Nguyen, L.T.T.; Richardson, H.S.; Alm, E.J. Prebiotics and Community Composition Influence Gas Production of the Human Gut Microbiota. *mBio* **2020**, *11*, 13. [CrossRef] [PubMed]
32. Van den Abbeele, P.; Kamil, A.; Fleige, L.; Chung, Y.; De Chavez, P.; Marzorati, M. Different Oat Ingredients Stimulate Specific Microbial Metabolites in the Gut Microbiome of Three Human Individuals In Vitro. *ACS Omega* **2018**, *3*, 12446–12456. [CrossRef] [PubMed]
33. Jonathan, M.C.; van den Borne, J.; van Wiechen, P.; da Silva, C.S.; Schols, H.A.; Gruppen, H. In vitro fermentation of 12 dietary fibres by faecal inoculum from pigs and humans. *Food Chem.* **2012**, *133*, 889–897. [CrossRef]
34. Wang, M.M.; Wichienchot, S.; He, X.W.; Fu, X.; Huang, Q.; Zhang, B. In vitro colonic fermentation of dietary fibers: Fermentation rate, short-chain fatty acid production and changes in microbiota. *Trends Food Sci. Technol.* **2019**, *88*, 1–9. [CrossRef]
35. Harris, S.; Monteagudo-Mera, A.; Kosik, O.; Charalampopoulos, D.; Shewry, P.; Lovegrove, A. Comparative prebiotic activity of mixtures of cereal grain polysaccharides. *AMB Express* **2019**, *9*, 7. [CrossRef] [PubMed]
36. Riva, A.; Kuzyk, O.; Forsberg, E.; Siuzdak, G.; Pfann, C.; Herbold, C.; Daims, H.; Loy, A.; Warth, B.; Berry, D. A fiber-deprived diet disturbs the fine-scale spatial architecture of the murine colon microbiome. *Nat. Commun.* **2019**, *10*, 11. [CrossRef]
37. Hidalgo, M.; Oruna-Concha, M.J.; Kolida, S.; Walton, G.E.; Kallithraka, S.; Spencer, J.P.E.; Gibson, G.R.; de Pascual-Teresa, S. Metabolism of Anthocyanins by Human Gut Microflora and Their Influence on Gut Bacterial Growth. *J. Agric. Food Chem.* **2012**, *60*, 3882–3890. [CrossRef]
38. He, J.; Wallace, T.C.; Keatley, K.E.; Failla, M.L.; Giusti, M.M. Stability of Black Raspberry Anthocyanins in the Digestive Tract Lumen and Transport Efficiency into Gastric and Small Intestinal Tissues in the Rat. *J. Agric. Food Chem.* **2009**, *57*, 3141–3148. [CrossRef]
39. Hanske, L.; Engst, W.; Loh, G.; Sczesny, S.; Blaut, M.; Braune, A. Contribution of gut bacteria to the metabolism of cyanidin 3-glucoside in human microbiota-associated rats. *Br. J. Nutr.* **2013**, *109*, 1433–1441. [CrossRef] [PubMed]
40. Aura, A.M.; Martin-Lopez, P.; O'Leary, K.A.; Williamson, G.; Oksman-Caldentey, K.M.; Poutanen, K.; Santos-Buelga, C. In vitro metabolism of anthocyanins by human gut microflora. *Eur. J. Nutr.* **2005**, *44*, 133–142. [CrossRef]
41. Faria, A.; Fernandes, I.; Norberto, S.; Mateus, N.; Calhau, C. Interplay between Anthocyanins and Gut Microbiota. *J. Agric. Food Chem.* **2014**, *62*, 6898–6902. [CrossRef] [PubMed]
42. Tuncil, Y.E.; Nakatsu, C.H.; Kazem, A.E.; Arioglu-Tuncil, S.; Reuhs, B.; Martens, E.C.; Hamaker, B.R. Delayed utilization of some fast-fermenting soluble dietary fibers by human gut microbiota when presented in a mixture. *J. Funct. Food.* **2017**, *32*, 347–357. [CrossRef]
43. Chen, M.; Fan, B.; Liu, S.J.; Imam, K.; Xie, Y.Y.; Wen, B.T.; Xin, F.J. The in vitro Effect of Fibers with Different Degrees of Polymerization on Human Gut Bacteria. *Front. Microbiol.* **2020**, *11*, 12. [CrossRef]
44. Donato, K.A.; Gareau, M.G.; Wang, Y.J.J.; Sherman, P.M. Lactobacillus rhamnosus GG attenuates interferon-gamma and tumour necrosis factor-alpha-induced barrier dysfunction and pro-inflammatory signalling. *Microbiol. Sgm* **2010**, *156*, 3288–3297. [CrossRef]
45. D'Argenio, G.; Cariello, R.; Tuccillo, C.; Mazzone, G.; Federico, A.; Funaro, A.; De Magistris, L.; Grossi, E.; Callegari, M.L.; Chirico, M.; et al. Symbiotic formulation in experimentally induced liver fibrosis in rats: Intestinal microbiota as a key point to treat liver damage? *Liver Int.* **2013**, *33*, 687–697. [CrossRef] [PubMed]
46. Delzenne, N.M.; Neyrinck, A.M.; Cani, P.D. Modulation of the gut microbiota by nutrients with prebiotic properties: Consequences for host health in the context of obesity and metabolic syndrome. *Microb. Cell Factories* **2011**, *10*, S17. [CrossRef]
47. Yan, Y.M.; Peng, Y.J.; Tang, J.L.; Mi, J.; Lu, L.; Li, X.Y.; Ran, L.W.; Zeng, X.X.; Cao, Y.L. Effects of anthocyanins from the fruit of Lycium ruthenicum Murray on intestinal microbiota. *J. Funct. Food.* **2018**, *48*, 533–541. [CrossRef]
48. Morais, C.A.; de Rosso, V.V.; Estadella, D.; Pisani, L.P. Anthocyanins as inflammatory modulators and the role of the gut microbiota. *J. Nutr. Biochem.* **2016**, *33*, 1–7. [CrossRef]
49. Xu, T.; Ge, Y.; Du, H.; Li, Q.; Xu, X.; Yi, H.; Wu, X.; Kuang, T.; Fan, G.; Zhang, Y. Berberis kansuensis extract alleviates type 2 diabetes in rats by regulating gut microbiota composition. *J. Ethnopharmacol.* **2021**, *273*, 113995. [CrossRef]
50. Hu, J.M.; Lin, S.L.; Zheng, B.D.; Cheung, P.C.K. Short-chain fatty acids in control of energy metabolism. *Crit. Rev. Food Sci. Nutr.* **2018**, *58*, 1243–1249. [CrossRef]
51. Kabat, A.M.; Srinivasan, N.; Maloy, K.J. Modulation of immune development and function by intestinal microbiota. *Trends Immunol.* **2014**, *35*, 507–517. [CrossRef]
52. Topping, D.L.; Clifton, P.M. Short-chain fatty acids and human colonic function: Roles of resistant starch and nonstarch polysaccharides. *Physiol. Rev.* **2001**, *81*, 1031–1064. [CrossRef]
53. Koh, A.; De Vadder, F.; Kovatcheva-Datchary, P.; Backhed, F. From Dietary Fiber to Host Physiology: Short-Chain Fatty Acids as Key Bacterial Metabolites. *Cell* **2016**, *165*, 1332–1345. [CrossRef] [PubMed]
54. Eker, M.E.; Aaby, K.; Budic-Leto, I.; Brncic, S.R.; Nehir, E.; Karakaya, S.; Simsek, S.; Manach, C.; Wiczkowski, W.; de Pascual-Teresa, S. A Review of Factors Affecting Anthocyanin Bioavailability: Possible Implications for the Inter-Individual Variability. *Foods* **2020**, *9*, 2. [CrossRef]
55. Rios-Covian, D.; Ruas-Madiedo, P.; Margolles, A.; Gueimonde, M.; de los Reyes-Gavilan, C.G.; Salazar, N. Intestinal Short Chain Fatty Acids and their Link with Diet and Human Health. *Front. Microbiol.* **2016**, *7*, 9. [CrossRef] [PubMed]

56. Yamashita, H.; Fujisawa, K.; Ito, E.; Idei, S.; Kawaguchi, N.; Kimoto, M.; Hiemori, M.; Tsuji, H. Improvement of obesity and glucose tolerance by acetate in type 2 diabetic Otsuka Long-Evans Tokushima Fatty (OLETF) rats. *Biosci. Biotechnol. Biochem.* **2007**, *71*, 1236–1243. [CrossRef] [PubMed]
57. Hughes, S.A.; Shewry, P.R.; Gibson, G.R.; McCleary, B.V.; Rastall, R.A. In vitro fermentation of oat and barley derived beta-glucans by human faecal microbiota. *Fems Microbiol. Ecol.* **2008**, *64*, 482–493. [CrossRef]
58. den Besten, G.; van Eunen, K.; Groen, A.K.; Venema, K.; Reijngoud, D.J.; Bakker, B.M. The role of short-chain fatty acids in the interplay between diet, gut microbiota, and host energy metabolism. *J. Lipid Res.* **2013**, *54*, 2325–2340. [CrossRef] [PubMed]
59. Han, X.; Song, H.; Wang, Y.; Sheng, Y.; Chen, J. Sodium butyrate protects the intestinal barrier function in peritonitic mice. *Int. J. Clin. Exp. Med.* **2015**, *8*, 4000–4007.
60. Yang, J.Y.; Martinez, I.; Walter, J.; Keshavarzian, A.; Rose, D.J. In vitro characterization of the impact of selected dietary fibers on fecal microbiota composition and short chain fatty acid production. *Anaerobe* **2013**, *23*, 74–81. [CrossRef] [PubMed]
61. Harris, H.C.; Morrison, D.J.; Edwards, C.A. Impact of the source of fermentable carbohydrate on SCFA production by human gut microbiotain vitro—A systematic scoping review and secondary analysis. *Crit. Rev. Food Sci. Nutr.* **2020**, *12*, 1–12. [CrossRef] [PubMed]
62. Ashaolu, T.J.; Ashaolu, J.O.; Adeyeye, S.A.O. Fermentation of prebiotics by human colonic microbiotain vitroand short-chain fatty acids production: A critical review. *J. Appl. Microbiol.* **2021**, *130*, 677–687. [CrossRef] [PubMed]
63. Chen, L.G.; Liu, J.W.; Ge, X.D.; Xu, W.; Chen, Y.; Li, F.W.; Cheng, D.L.; Shao, R. Simulated digestion and fermentation in vitro by human gut microbiota of polysaccharides from *Helicteres angustifolia* L. *Int. J. Biol. Macromol.* **2019**, *141*, 1065–1071. [CrossRef] [PubMed]

foods

MDPI

Article

Thermal Inactivation Kinetics of Kudzu (*Pueraria lobata*) Polyphenol Oxidase and the Influence of Food Constituents

Junping Liu [1], Jiayan Zhang [1], Tao Liao [1], Lei Zhou [1], Liqiang Zou [1], Yafei Liu [1], Li Zhang [2] and Wei Liu [1,3,*]

1 State Key Laboratory of Food Science and Technology, Nanchang University, Nanchang 330047, China; ljpfood@163.com (J.L.); ncuzhangjiayan@163.com (J.Z.); 15180122056@163.com (T.L.); ncuskzhoulei@163.com (L.Z.); zouliqiang2010@163.com (L.Z.); ncuspyliuyafei@163.com (Y.L.)

2 Key Laboratory of Tropical Crop Products Processing of Ministry of Agriculture and Rural Affairs, Agricultural Products Processing Research Institute of Chinese Academy of Tropical Agricultural Sciences, Guangdong 524001, China; zhang1993123321@163.com

3 National R&D Center for Freshwater Fish Processing, Jiangxi Normal University, Nanchang 330022, China

* Correspondence: liuwei@ncu.edu.cn; Tel.: +86-0791-88305872 (ext. 8106)

Abstract: The thermal inactivation kinetics of kudzu (*Pueraria lobata*) polyphenol oxidase (PPO) were investigated in model and food systems. PPO in kudzu tissue (tPPO) showed a higher thermostability than that of PPO in crude extract (cPPO) and purification fractions (pPPO). The PPO inactivation rate constant (k) increased with an increase in temperature, and tPPO showed the lowest k value, followed by that of cPPO and pPPO at the same temperature, indicating that PPO in the food system was more resistant to thermal treatment. Food constituents (pectin, starch, sucrose, and bovine serum albumin) in the food system decreased the activity of PPO but increased the thermostability of PPO, among which pectin exhibited the strongest protective effect against thermal inactivation, and the influence of sucrose was much slighter than that of other macromolecules. Fluorescence emission spectra indicated that pPPO exhibited stronger interactions with pectin than sucrose, and pPPO with pectin showed a more stable conformation under thermal treatment.

Keywords: polyphenol oxidase; kudzu; thermostability; thermal inactivation kinetics; food constituents; conformation

Citation: Liu, J.; Zhang, J.; Liao, T.; Zhou, L.; Zou, L.; Liu, Y.; Zhang, L.; Liu, W. Thermal Inactivation Kinetics of Kudzu (*Pueraria lobata*) Polyphenol Oxidase and the Influence of Food Constituents. *Foods* **2021**, *10*, 1320. https://doi.org/10.3390/foods10061320

Academic Editor: Montserrat Dueñas Paton

Received: 14 May 2021
Accepted: 4 June 2021
Published: 8 June 2021

1. Introduction

Kudzu (*Pueraria lobata*) is a herbaceous, perennial plant native to East Asia and belongs to the Leguminosae family. It is a rich source of starch, dietary fiber, minerals, and isoflavonoids and possesses many nutritional and pharmacological activities [1–3], such as antioxidant, anticancer, anti-inflammation, and neuroprotective properties. Additionally, kudzu is used as a vegetable and edible ingredient in the preparation of various foods [4]. For example, kudzu starch is used in the production of noodles, beverages, oral liquids, and various dishes, and it serves as a sauce thickener [5]. Food products containing components of kudzu have begun to attract interest from both consumers and researchers due to their health benefits. However, enzymatic browning of kudzu during storage, handling, and processing negatively affects its sensory and nutritional properties. Polyphenol oxidase (PPO) is the primary enzyme involved in enzymatic browning [6,7]. Therefore, the inactivation of PPO is important in improving the quality attributes of kudzu products.

Thermal treatment has been widely used as a non-chemical means for enzyme inactivation in food processing. The thermal inactivation of PPO from fruits and vegetables has been well evaluated in mushrooms [8,9], pear [10,11], apple [12,13], chestnut kernel [14], and pineapple [15]. The thermostability of PPO highly depended on the environmental conditions, and an obvious difference was observed in food and model systems. Purified apple PPO was almost completely inactivated after incubation at 65 °C for 20 min [13], while PPO in apple slices still retained about 40% relative activity in the same condition [12].

Compared to crude PPO extract from pear [11], PPO in pear puree [10] showed higher thermostability at a temperature range from 30 to 100 °C. The food system differs from the model system since the complex food constituents may affect the enzymatic properties during food processing [16]. The impact of pectin [17,18], protein [19,20], and sugar [21] on enzymatic properties have been investigated. However, there are few reports comparing the thermostability of PPO in food and model systems in the same study, and the effect of food constituents on the thermal inactivation kinetics remains unclear.

In this study, PPO from kudzu was purified and characterized. The thermal stability and inactivation kinetics of kudzu tissue PPO, crude PPO, and purified PPO were evaluated and compared with each other. To further illustrate the lower thermal inactivation effect in kudzu tissue, the content of food constituents, including pectin, protein, sucrose, and starch, in kudzu was analyzed, and the effect of these constituents on PPO activity, thermal inactivation kinetics, and conformation change was also evaluated.

2. Materials and Methods

2.1. Materials

Kudzu root (*Pueraria lobata*) was harvested from the plantation at Shangrao, Jiangxi province in eastern China, and stored at 4 °C until processing. Triton X-100, polyvinylpolypyrrolidone (PVPP), and *m*-hydroxydiphenyl were obtained from Aladdin Chemicals Co. (Shanghai, China). Catechol was obtained from Macklin Chemicals Co. (Shanghai, China). Pectin was purchased from Sigma-Aldrich Co. (Shanghai, China). Bovine serum albumin (BSA), starch, sucrose, D-Galacturonic acid, D-Glucose ammonium sulfate, tetramethylethylenediamine (TEMED), phenylmethylsulfonyl fluoride (PMSF), SDS-PAGE loading buffer, and standard protein mixture (Marker) were purchased from Solarbio Science and Technology Co. (Beijing, China). All the other chemicals were of analytical grade. Double-distilled water was used for preparing all the solutions.

2.2. PPO Extraction

The extraction and purification of PPO from kudzu were carried out according to the method of Liu et al. [22] with modifications. After washing and peeling, kudzu was sliced and mashed using a blender (HX-PB1053, AUX, Zhejiang, China) with 0.1 M phosphate buffer (pH 6.5) containing 2% (*w*/*v*) polyvinylpolypyrrolidone (PVPP), 0.5 M NaCl, and 1% (*v*/*v*) triton X-100 in a proportion of 1:1 (*v*/*v*) to yield puree. The mixture was stirred in an ice-water bath for 2 h and centrifuged (SORVALL Biofuge primo R, Thermo, Germany) for 20 min at 11,000× *g* at 4 °C. The supernatant was collected as the crude PPO (cPPO) and stored at 4 °C until use.

2.3. PPO Purification

To obtain purified PPO (pPPO), a two-step purification was conducted. Firstly, ammonium sulfate was added to cPPO, and the precipitate obtained between 20% and 30% saturation was collected by centrifugation at 11,000× *g* for 25 min at 4 °C. The precipitate was dissolved in phosphate buffer (pH 6.5) and dialyzed against the same buffer at 4 °C for 24 h. Then, the dialyzed fraction was loaded onto a DEAE Sepharose fast flow column (8 cm × 1 cm, GE, Healthcare Bio-Sciences AB, Sweden), which was pre-equilibrated with 50 mM phosphate buffer (pH 6.5). The column was eluted with a linear gradient of sodium chloride (0–0.5 M) in 50 mM phosphate buffer (pH 6.5) at a flow rate of 0.2 mL/min. SDS-PAGE electrophoresis was performed to determine purity [23].

2.4. Constituent Determination

Four main constituents (pectin, protein, starch, and sucrose) in kudzu tissue, the cPPO solution, and the pPPO solution were analyzed. The content of pectin was determined by *m*-hydroxybiphenyl and expressed as galacturonic acid equivalents [22]. Protein concentration was determined by the Bradford method [24], using bovine serum albumin as standard.

Starch and sucrose contents were determined by acid hydrolysis and expressed as reducing sugar equivalents.

2.5. PPO Activity Assay

PPO activity was determined based on the method of Terefe et al. [25] with slight modifications. The reaction mixture consisted of 2.7 mL of McIlvaine buffer (pH 5.0), 0.2 mL of 0.4 M catechol solution, and 0.1 mL of PPO sample. The absorbance of the assay mixture was measured at 25 °C in the kinetic mode at 420 nm for 1 min using a UV-visible spectrophotometer (UV-1600PC, Mapada, Shanghai, China). The relative activities of PPO were calculated in accordance with Equation (1):

$$\text{Relative activity}(\%) = \frac{\text{Activity of treated enzyme}}{\text{Activity of untreated enzyme}} \times 100\% \quad (1)$$

2.6. Sample Preparation

Before thermal treatment, kudzu slices (20 g) were vacuum packed in polyethylene bags (5 cm × 5 cm), while the cPPO and pPPO (0.2 mL) samples with 2.7 mL of McIlvaine buffer (pH 5.0) were filled in capillary tubes (5 mL). Samples were kept at 4 °C for 2 h to minimize heating lags.

The effects of food constituents on the activity of pPPO were examined, with pectin, BSA, starch, and sucrose in concentrations ranging from 0.5% to 3% (w/v), respectively. The resulting mixtures were labeled as "pectin-pPPO", "BSA-pPPO", "starch-pPPO", and sucrose-pPPO", respectively. They were stirred in an ice-water bath for 1 h and were immediately used in the experiment.

2.7. Thermal Processing

The thermal stability and thermal inactivation kinetics of PPO in kudzu tissue (tPPO), cPPO, pPPO, pectin-pPPO, BSA-pPPO, starch-pPPO, and sucrose-pPPO were tested according to the method of Tan et al. [26]. The thermal stability was measured by incubating samples at temperatures ranging from 45 to 90 °C at 10 min. Thermal inactivation kinetics was carried out at temperatures ranging from 50 to 80 °C for 0–60 min. After thermal treatment, all samples were rapidly cooled in ice-water, then the activity of the sample was determined.

The thermal inactivation kinetics of samples was described by the first-order kinetic reaction (Equation (2)):

$$\ln(A_t/A_0) = -kt \quad (2)$$

where A_0 is the initial activity of the samples, A_t is the residual activity after treatment time t (min), and k is the inactivation rate constant (min^{-1}).

The Arrhenius law was used to describe the temperature dependence of the k-values and to estimate the activation energy by linear regression analysis (Equation (3)):

$$\ln(k) = \ln(k_0) - \frac{E_a}{R \cdot T} \quad (3)$$

where k_0 is the Arrhenius constant (min^{-1}), T is the absolute temperature (K), and k is the inactivation rate constants at T (min^{-1}). E_a is the activation energy (kJ/mol), and R is the universal gas constant (8.314 J/(mol·K)).

The thermodynamic parameters of inactivation (change in enthalpy (ΔH), Gibbs free energy (ΔG), and entropy (ΔS)) were determined using the equations presented below (Equations (4)–(6)):

$$\Delta G = -R \cdot T \cdot \ln\left(\frac{k \cdot h_p}{K_B \cdot T}\right) \quad (4)$$

$$\Delta H = E_a - R \cdot T \quad (5)$$

$$\Delta S = \left(\frac{\Delta H - \Delta G}{T}\right) \tag{6}$$

where R is the universal gas constant (8.314 J/(mol·K)), k is the inactivation rate constant at T (s^{-1}), h_p is the Planck's constant (6.6262×10^{-34} J·s), and K_B is the Boltzmann's constant (1.3806×10^{-23} J/K).

The half-life ($t_{1/2}$) value of inactivation is given by Equation (7):

$$t_{1/2} = \frac{\ln(2)}{k} \tag{7}$$

The decimal reduction time (D value) is the time needed to reduce the initial activity by 90%, given by Equation (8):

$$D = \frac{\ln(10)}{k} \tag{8}$$

The Z value is the temperature needed to reduce the D value by one log cycle (temperature sensitivity parameter), and it is obtained by plotting the log D values versus the corresponding temperatures.

2.8. Fluorescence Spectra Analysis

Fluorescence emission spectra of pPPO, pectin-pPPO, and sucrose-pPPO were measured using a F-4500 spectrophotometer (Hitachi, Tokyo, Japan). The details are as follows: 0.6 mg/mL pPPO was added into quartz cuvettes and treated with 0, 0.2%, 0.4%, 0.6%, 0.8%, and 1.0% (w/v) pectin and sucrose to investigate their impact on the tertiary structure of pPPO. In addition, pPPO, pectin (0.4%)-pPPO, pectin (1.0%)-pPPO, and sucrose (1.0%)-pPPO were subjected to thermal processing at 50–80 °C to further explore the effect of pectin and sucrose on the conformation change against heat. As described by Zhou et al. [27], all the samples were excited at 280 nm and scanned from 420 to 300 nm at room temperature (25 ± 1 °C), with the excitation and emission slit bandwidths of 5 nm.

2.9. Statistical Analysis

All the experiments were performed at least in triplicate. The values are presented as means ± standard deviation (SD). The significant differences were determined based on an analysis of variance (ANOVA) with significance at 95% confidence.

3. Results and Discussion

3.1. Purification and Characterization of PPO from Kudzu

PPO from kudzu was purified by ammonium sulfate fractionation together with DEAE Sepharose fast flow column. PPO was successfully purified 38.4-fold with 32.9% recovery following DEAE Sepharose fast flow column (Supplementary Material Table S1). After conducting the SDS-PAGE electrophoresis (Supplementary Material Figure S1), only a single prominent protein band of approximately 21 kDa was obtained, which confirmed the effectiveness of this purification procedure. The optimum pH of PPO from kudzu ranged from 4.5 to 5.0, and the Michaelis–Menten constant (K_m) and maximum velocity (V_{max}) were estimated to be 23.54 mM and 0.37 ΔOD/min (catechol as substrate), respectively (Supplementary Material Figure S2). Moreover, the general constituents (starch, protein, pectin, and sucrose) in kudzu were determined. Starch showed the highest content (21.42%) in kudzu tissue (Supplementary Material Table S2). After extraction, the cPPO solution still contained a certain amount of starch, pectin, sucrose, and protein, but starch, pectin, and sucrose were no longer detected in the pPPO solution and the protein concentration was significantly reduced.

3.2. Thermal Stability of PPO

Thermal processing from 45 to 90 °C for 10 min resulted in different inactivation effects on kudzu tissue PPO (tPPO), crude PPO (cPPO), and purified PPO (pPPO) (Figure 1).

The activity of tPPO, cPPO, and pPPO decreased gradually with increasing processing temperatures. After 10 min treatment at 45 to 60 °C, only slight inactivation was found, and the activity of tPPO, cPPO, and pPPO remained above 80% at 60 °C, which indicated that kudzu PPO exhibited high resistance to mild thermal processing between 45 and 60 °C. With the continuous increase in temperature, PPO activity declined sharply, and cPPO and pPPO were completely inactivated at 85 °C, while tPPO was completely inactivated at 90 °C. Between 60 and 90 °C, pPPO showed the fastest reduction rate in activity, followed by cPPO and tPPO. After 10 min treatment at 80 °C, the residual activity of pPPO was only 11.0%, whereas the values of cPPO and tPPO were 24.7% and 42.1%, respectively. These results illustrate that tPPO had the strongest thermostability while pPPO had the weakest. PPO in the food system displayed higher thermostability than that in the model system. This may be due to the effect of constituents in kudzu tissue.

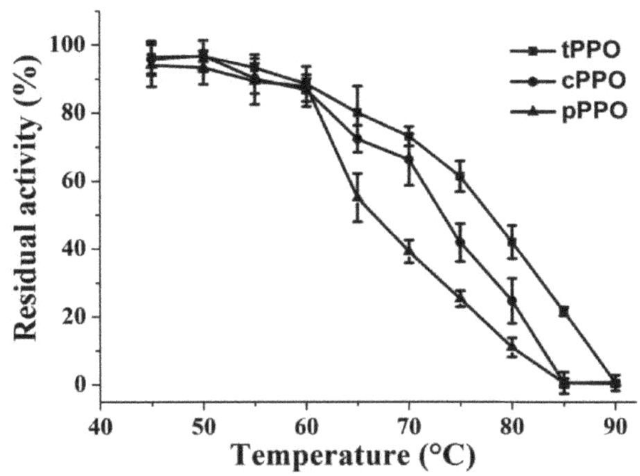

Figure 1. Thermal stability of PPO at temperatures ranging from 45 to 90 °C for 10 min. tPPO (■): PPO in kudzu tissue. cPPO (•): PPO from the crude kudzu extract. pPPO (▲): purified PPO from kudzu.

In our previous studies, it was found that PPO in pear puree [10] showed higher residual activity than that of the partially purified PPO from pear [11] after thermal treatment in the temperature range of 30–100 °C. The results are consistent with the viewpoint in this study that PPO in the food system was more resistant to heat, and a reasonable explanation for this could be that food constituents such as proteins and carbohydrates can protect the zymoprotein from heat [28].

3.3. Thermal Inactivation Kinetics and Thermodynamics of PPO

Based on the results of the thermal stability, the thermal inactivation of PPO was performed at temperatures that ranged from 50 to 80 °C and the incubation time ranged from 0 to 60 min (Figure 2). Lower activity of PPO was observed in all samples as a result of a higher temperature or longer incubation time. As shown in Figure 2a, when the temperature of thermal processing was 50 °C, the thermal inactivation of pPPO was the fastest, followed by that of cPPO, while that of tPPO was the slowest. The residual activities of tPPO, cPPO, and pPPO were 81.8%, 81.6%, and 74.9% after thermal processing at 50 °C for 60 min. It could be clearly seen that the semilog relationship between the residual activity of PPO from kudzu and treatment time followed first-order kinetics well, with the correlation coefficients (R^2) ranging from 0.9708 to 0.9994. The *k* value of tPPO, cPPO, and pPPO at 50 °C increased successively and were 0.32×10^{-2}, 0.36×10^{-2}, and 0.52×10^{-2} min^{-1}, respectively (Table 1). Similar results were observed in PPOs treated at 60, 70, and 80 °C (Figure 2b–d). With the increase in temperature, the difference during samples gradually became more obvious, and at 80 °C, the *k* value of tPPO, cPPO, and pPPO were 8.57 × 10^{-2}, 14.63 × 10^{-2}, and 23.67 × 10^{-2} min^{-1}, respectively. It was observed that for tPPO, cPPO, and pPPO, the k value increased with the increase in processing temperature, indicating that kudzu PPO was more sensitive at higher temperatures. The highest D value (729.18 min) and $t_{1/2}$ value (219.50 min) were found in tPPO at 50 °C, which also supports the viewpoints that higher temperature leads to lower stability, and tPPO was the most stable. In comparison with cPPO from other sources, the D values and $t_{1/2}$ values were 73.99 and 22.29 min for bayberry [29], 39.0 and 11.8 min for mushroom [8], and 17.3 and 5.2 min for peach [30] at 60 °C, respectively, which are lower than 158.20 and 47.62 min in this study, indicating that PPO from kudzu was more thermostable than the above-mentioned PPO. The Z values of tPPO, cPPO, and pPPO were 21.25, 19.21, and 17.47 °C, and the corresponding E_a were 102.68, 113.67, and 125.05 kJ/mol, respectively, the lower Z value and higher E_a suggests an increased susceptibility to heat [8].

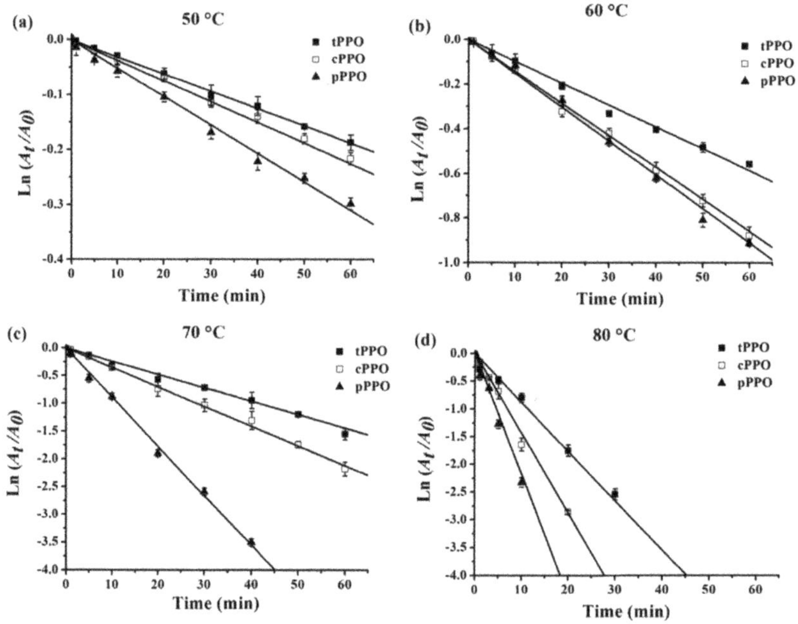

Figure 2. Thermal inactivation kinetics curves of PPO at 50 °C (**a**), 60 °C (**b**), 70 °C (**c**), and 80 °C (**d**) for 0–60 min. tPPO (■): PPO in kudzu tissue. cPPO (□): PPO from crude extract of kudzu. pPPO (▲): purified PPO from kudzu.

Table 1. Thermal inactivation parameters of PPO in kudzu tissue (tPPO), crude extract (cPPO), purification fractions (pPPO), and pPPO in the presence of food constituents [1].

Enzyme	Temperature (°C)	k (10^{-2} Min^{-1})	D min	$t_{1/2}$ min	E_a (kJ mol^{-1})	Z (°C)	ΔH (kJ mol^{-1})	ΔG (kJ mol^{-1})	ΔS (J mol^{-1} K^{-1})
tPPO	50	0.32 ± 0.02	729.18 ± 46.66	219.50 ± 14.05	102.68 ± 2.18	21.25 ± 0.46	99.99 ± 2.18 ^{c}	83.83 ± 0.17 ^{abc}	50.02 ± 6.28 ^{g}
	60	0.98 ± 0.02	235.02 ± 4.80	70.75 ± 1.44			99.91 ± 2.18 ^{c}	83.37 ± 0.06 ^{de}	49.63 ± 6.38 ^{g}
	70	2.50 ± 0.04	92.12 ± 1.34	27.73 ± 0.40			99.82 ± 2.18 ^{c}	83.29 ± 0.04 ^{def}	48.19 ± 6.25 ^{g}
	80	8.57 ± 0.34	26.90 ± 1.10	8.10 ± 0.33			99.74 ± 2.18 ^{c}	82.18 ± 0.12 ^{ij}	49.71 ± 6.32 ^{g}
cPPO	50	0.36 ± 0.01	639.94 ± 17.78	192.64 ± 5.35	113.67 ± 0.86	19.21 ± 0.15	110.98 ± 0.86 ^{b}	83.48 ± 0.07 ^{cde}	77.88 ± 2.41 ^{cde}
	60	1.46 ± 0.05	158.20 ± 5.43	47.62 ± 1.64			110.90 ± 0.86 ^{b}	82.28 ± 0.10 ^{ij}	81.05 ± 2.67 ^{cd}
	70	3.50 ± 0.22	65.90 ± 4.15	19.84 ± 1.25			110.82 ± 0.86 ^{b}	82.33 ± 0.18 ^{ij}	80.66 ± 1.99 ^{cd}
	80	14.63 ± 0.96	15.78 ± 1.05	4.75 ± 0.31			110.73 ± 0.86 ^{b}	80.62 ± 0.19 ^{l}	85.28 ± 2.91 ^{c}
pPPO	50	0.52 ± 0.02	440.54 ± 18.96	132.62 ±5.71	125.05 ± 0.47	17.47 ± 0.09	122.36 ± 0.47 ^{a}	82.48 ± 0.12 ^{hi}	112.94 ± 1.52 ^{b}
	60	1.54 ± 0.03	149.24 ± 3.08	44.93 ± 0.93			122.28 ± 0.47 ^{a}	82.12 ± 0.06 ^{ij}	113.73 ± 1.18 ^{b}
	70	8.88 ± 0.65	26.03 ± 1.85	7.84 ± 0.56			122.20 ± 0.47 ^{a}	79.68 ± 0.21 ^{m}	120.3 9± 0.79 ^{ab}
	80	23.67 ± 1.53	9.76 ± 0.63	2.94 ± 0.19			122.11 ± 0.47 ^{a}	79.20 ± 0.19 ^{n}	121.50 ± 1.83 ^{ab}
Pectin-pPPO	50	0.27 ± 0.01	842.66 ± 17.58	253.67± 5.29	107.53 ± 1.78	20.29 ± 0.33	104.85 ± 1.78 ^{bc}	84.22± 0.06 ^{a}	58.41 ± 5.02 ^{fg}
	60	0.78 ± 0.041	297.02 ±15.93	89.41 ± 4.80			104.76 ± 1.78 ^{bc}	84.02 ± 0.15 ^{ab}	58.74 ± 5.15 ^{fg}
	70	2.35 ± 0.06	98.16 ± 2.31	29.55 ± 0.70			104.68 ± 1.78 ^{bc}	83.47 ± 0.07 ^{cde}	60.06 ± 5.17 ^{efg}
	80	8.32 ± 0.49	27.73 ±1.63	8.35 ± 0.49			104.60 ± 1.78 ^{bc}	82.27 ± 0.17 ^{ij}	63.22 ± 5.53 ^{defg}
BSA-pPPO	50	0.34 ± 0.01	677.62 ± 19.94	203.98 ± 6.00	109.79 ± 2.60	19.90 ± 0.48	107.10 ± 2.60 ^{b}	83.63 ± 0.08 ^{bcd}	66.45 ± 7.16 ^{defg}
	60	0.89 ± 0.04	258.11 ± 11.82	77.70 ± 3.56			107.02 ± 2.60 ^{b}	83.63 ± 0.13 ^{bcd}	66.23 ± 7.29 ^{defg}
	70	3.63 ± 0.24	63.55 ±4.09	19.13 ±1.23			106.94 ± 2.60 ^{b}	82.22 ± 0.19 ^{ij}	69.97 ± 7.79 ^{cdef}
	80	10.12 ± 0.46	22.79 ± 1.06	6.86 ± 0.32			106.85 ± 2.60 ^{b}	81.70 ± 0.14 ^{k}	71.23 ± 7.72 ^{cdef}
Starch-pPPO	50	0.44 ± 0.01	519.62 ± 13.74	156.42 ± 4.14	110.94 ± 1.23	19.64 ± 0.22	108.25 ± 1.23 ^{b}	82.92 ± 0.07 ^{fg}	71.73 ± 3.29 ^{cdef}
	60	1.05 ± 0.05	218.94 ± 10.49	65.91 ± 3.16			108.17 ± 1.23 ^{b}	83.17 ± 0.13 ^{ef}	70.77 ± 3.27 ^{cdef}
	70	3.84 ± 0.06	59.92 ± 0.94	18.04 ± 0.28			108.09 ± 1.23 ^{b}	82.06 ± 0.04 ^{jk}	73.69 ± 3.58 ^{cdef}
	80	14.34 ± 0.31	16.06 ± 0.35	4.84 ± 0.10			108.00 ± 1.23 ^{b}	80.67 ± 0.06 ^{l}	77.39 ± 3.53 ^{cde}
Sucrose-pPPO	50	0.48 ± 0.01	483.11 ± 5.89	145.43 ± 1.77	129.34 ± 3.31	17.26 ± 0.25	126.65 ± 3.31 ^{a}	82.72 ± 0.03 ^{gh}	124.39 ± 9.45 ^{ab}
	60	1.35 ± 0.02	171.00 ± 1.95	51.48 ± 0.59			126.57 ± 3.31 ^{a}	82.49 ± 0.03 ^{hi}	124.81 ± 9.43 ^{ab}
	70	8.59 ± 0.24	26.83 ± 0.75	8.08 ± 0.23			126.49 ± 3.31 ^{a}	79.77 ± 0.08 ^{m}	132.29 ± 9.25 ^{a}
	80	21.97 ±1.26	10.50 ± 0.59	3.16 ± 0.18			126.40 ± 3.31 ^{a}	79.42 ± 0.17 ^{mn}	133.04 ± 9.16 ^{a}

[a–n] Different lower-case letters within a row show statistically significant differences among values ($p < 0.05$). [1] Pectin, bovine serum albumin, starch and sucrose at the concentration of 1.0% (w/v) was added to pPPO solution (pectin-pPPO, BSA-pPPO, starch-pPPO, and sucrose-pPPO). k is the inactivation rate constant (min^{-1}). D is the decimal reduction time. $t_{1/2}$ is the half-life value. E_a is the activation energy (kJ/mol). Z is the temperature needed to reduce the D value by one log cycle. ΔH is change in enthalpy. ΔG is Gibbs free energy. ΔS is entropy.

Thermodynamic parameters provide information on the enzyme stability in the thermal inactivation process, which was helpful for evaluating the possible effects of the stabilization or destabilization of the structure of an enzyme molecule and interactions between the enzyme and media [31]. ΔH is an indicator of the number of non-covalent bonds destroyed during the inactivation process [8]. The higher ΔH value of pPPO compared with that of tPPO and cPPO at the same temperature indicates that more non-covalent bonds were broken during the formation of transition states. When increasing the processing temperature, no significant changes of ΔH were noted for the three PPO samples, which suggests that the heat capacity of PPO did not change. ΔG is another important parameter related to the stability of the enzyme; a lower ΔG represents the lower stability of the enzyme [8] and a more spontaneous inactivation reaction [31]. ΔS is a physical quantity that reflects the degree of disorder of microscopic particles in the system. From 50 to 80 °C, the ΔG of tPPO, cPPO, and pPPO decreased, and the corresponding ΔS increased (with the exception of that of the tPPO at 65 and 70 °C and that of cPPO at 70 °C), which proves that the ordered structure and thermal stability of kudzu PPO decreased with the increasing temperature. Overall, the thermodynamics were closely related to the kinetics and unanimously showed that increasing the temperature led to a decrease in the stability of different PPO samples; at the same temperature, tPPO had the greatest resistance against heat followed by cPPO; pPPO was the most sensitive to heat.

These variations in inactivation kinetics and thermodynamic parameters of different PPO samples may be attributed to environmental conditions and food matrixes [26]. Ji et al. [31] observed that the $t_{1/2}$ value of purified β-galactosidase (8.25 and 0.58 min at 55 and 60 °C, respectively) was significantly smaller than that of the crude enzyme (10.13 and 0.75 min) in the temperature range of 55–65 °C. Furthermore, Zhou et al. [11] found that pear PPO showed higher resistance to thermal inactivation in the food systems than in model systems. Food constituents in the food system or crude extract were rich and various, which might affect the thermal inactivation of the enzyme.

3.4. Effect of Food Constituents on the Activity of PPO

In order to further investigate the effect of food constituents on the activity of PPO, exogenous sucrose, starch, pectin, and BSA were added to pPPO (Figure 3). The results showed that different food constituents have different effects on PPO. BSA showed the greatest impact on PPO activity: the activity of PPO was significantly decreased to 70.2% as the concentration of BSA was 1.0% (w/v), and it was only 13.6% when the concentration of BSA increased to 3.0% (w/v). Compared with BSA, the effect of sucrose, starch, and pectin on pPPO activity was much smaller. For sucrose, starch, and pectin, PPO activity changed slightly when the concentration of constituents was lower than 1.0% (w/v). As the concentration increased, the effects of sucrose, starch, and pectin on PPO activity became distinct. The residual activities of sucrose-pPPO, starch-pPPO, and pectin-pPPO were 90.7%, 82.7%, and 75.1% when the additive amount reached 2.0%, and 86.6%, 63.2%, and 45.1% when the additive amount reached 3.0% (w/v). In general, all four food constituents displayed an inhibition effect on PPO activity, and BSA showed the strongest inhibition effect, followed by pectin and starch, and sucrose exhibited the weakest inhibition effect. Similarly, some previous reports studied the effects of food constituents on enzyme activity and found that different additives result in different influences on different enzymes. For instance, Isleroglu and Turker [32] reported saccharides (gum arabic and inulin) as coating materials to protect the microbial transglutaminase. Bayarri et al. [33] reported that the addition of 0–0.2% (w/v) pectin induced a decrease in lysozyme activity, and the minimum activity was found as the additive concentration of pectin reached 0.02%. Chisari et al. [34] found that 72.1% *D*-glucose and *D*-fructose treatment resulted in a decrease in the PPO activity of "Madame Moutot" strawberry fruit by ~50% and ~60%, respectively, while the addition of 1.8% *D*-glucose and *D*-fructose led to a slight increase in PPO activity. In this study, only the limited inhibition effect on PPO activity was observed with the addition of low molecular compound sucrose, while macromolecular compounds, such as BSA, pectin, and starch, showed a stronger inhibition effect on PPO.

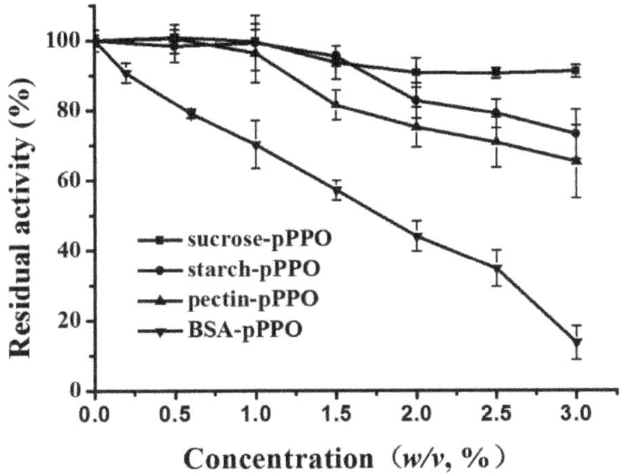

Figure 3. Effect of pectin, bovine serum albumin, starch and sucrose on the activity of purified PPO (pectin-pPPO, BSA-pPPO, starch-pPPO, and sucrose-pPPO) in concentrations ranging from 0.5% to 3% (w/v).

According to the study by Matsue and Miyawaki [35], the catalytic ability of the enzyme may be largely related to water activity and solvent ordering, and Chakraborty et al. [21] reported that sugars could change water activity, which may be a reasonable explanation for the decrease in PPO activity caused by sucrose, starch, and pectin. In addition, the abundant hydroxyl groups in starch and pectin were conducive to the formation of hydrogen bonds, which may disrupt the equilibrium of the enzyme's native structure in solvents and thus exhibit an inhibition effect [21]. Moreover, PPO can form a complex with pectin through a bridging mechanism, and the formation of electrostatic bridges may induce aggregation of PPO [36], which will hide the active site of the enzyme and lead to a lower substrate affinity and PPO activity. BSA exhibited the strongest inhibitory ability among the four food constituents. BSA molecule contains a free sulfhydryl group and 17 disulfide bonds that can be converted into sulfhydryl groups. The formed sulfhydryl groups can chelate and even remove the essential copper on the active sites of PPO, which was the factor that directly resulted in a decrease in PPO activity [37].

3.5. Effect of Food Constituents on the Thermal Inactivation Kinetics and Thermodynamics of PPO

Figure 4 presents the role of pectin, BSA, starch, and sucrose in the thermal inactivation of pPPO, and the thermal inactivation of pPPO, pectin-pPPO, BSA-pPPO, starch-pPPO, and sucrose-pPPO, fitting a linear relationship (R^2 = 0.9721–0.9992). As shown in Figure 4a, pectin, BSA, starch, and sucrose reduced the thermal inactivation rate of pPPO, and the residual activities of pPPO, pectin-pPPO, BSA-pPPO, starch-pPPO, and sucrose-pPPO were 74.9%, 83.6%, 82.0%, 77.1%, and 75.1%, respectively, after thermal processing at 50 °C for 60 min, and the corresponding k values were 0.52×10^{-2}, 0.27×10^{-2}, 0.34×10^{-2}, 0.44×10^{-2}, and 0.48×10^{-2} min^{-1}, respectively (Table 1). Similar results are shown in Figure 4b–d, and the corresponding k values of pectin-pPPO, BSA-pPPO, starch-pPPO, and sucrose-pPPO were 0.78×10^{-2}, 0.89×10^{-2}, 1.05×10^{-2}, and 1.35×10^{-2} min^{-1} at 60 °C, 2.35×10^{-2}, 3.63×10^{-2}, 3.84×10^{-2} and 8.59×10^{-2} min^{-1} at 70 °C; and 8.32×10^{-2}, 10.12×10^{-2}, 14.34×10^{-2}, and 21.97×10^{-2} min^{-1} at 80 °C, respectively. It can be noted that all four constituents exhibited protective effects on pPPO at all temperatures, and pectin showed the strongest protective effect, followed by BSA and starch, while sucrose showed the worst protection effect. These conclusions are also supported by gradually decreasing D values, $t_{1/2}$ values, and Z values and the gradually increasing E_a in the order of pectin-pPPO, BSA-pPPO, starch-pPPO, and sucrose-pPPO. As the thermal processing temperature increased, ΔG gradually decreased, ΔS gradually increased, and ΔH did not change significantly. These changes indicated that the stability of pectin-pPPO, BSA-pPPO, starch-pPPO, and sucrose-pPPO decreased with the increase in processing temperature. These results are also consistent with the results presented in Section 3.3, wherein cPPO and tPPO, which contain food constituents including pectin, protein, starch, and sucrose, had higher thermostability than that of pPPO. Similarly, Liu et al. [17] reported that pectin had a protective effect on the activity of peach PPO during thermal processing, and the protective effect was closely related to the concentration of pectin. Li et al. [19] found that pectin, BSA, sucrose, and trehalose could protect soluble acid invertase against HPP inactivation and the protective effect of sucrose and trehalose was weaker than that of pectin and BSA.

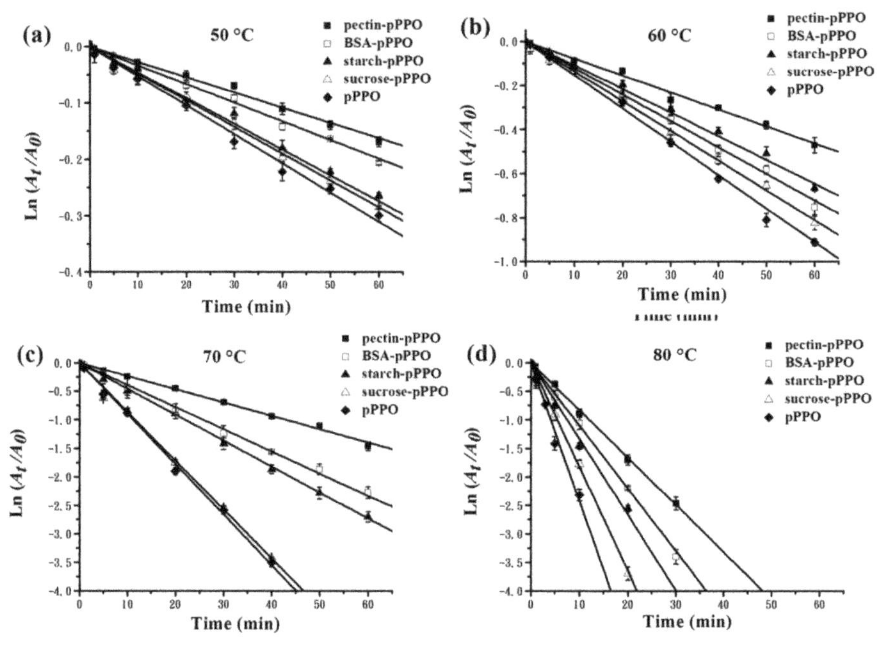

Figure 4. Thermal inactivation kinetics curves of purified PPO (pPPO) in the present of pectin (pectin-pPPO), bovine serum albumin (BSA-pPPO), starch(starch-pPPO), and sucrose (sucrose-pPPO) at a concentration of 1.0% (w/v) at 50 °C (**a**), 60 °C (**b**), 70 °C (**c**), and 80 °C (**d**) for 0–60 min.

Generally, the stability of an enzyme is affected by hydrogen bonds, hydrophobic bonds, metal bonds, ionic interactions, and disulfide bridges [38]. PPO tends to form complexes with proteins or carbohydrates, and the combination of PPO hydrophobic clusters with protein or carbohydrate molecules can shield the hydrophobic area on its surface, thereby protecting the structure of PPO from unfolding and enhancing its resistance against thermal inactivation [17,19]. In addition, the hydrophobic interactions and hydrogen bonds established between PPO and these food constituents, as well as the increase in solution viscosity caused by the additives, may also improve the thermal stability of PPO.

3.6. Effect of Food Constituents on Tertiary Structure of PPO

The effects of pectin and sucrose on PPO tertiary structure were significantly different (Figure 5). Native pPPO displayed a 340.5 nm peak wavelength with a corresponding fluorescence intensity of 287.6, which is similar to values of mushroom PPO reported by Zhou et al. [27]. Sucrose with a concentration of 0.2–1.0% (w/v) did not significantly change the shape of pPPO fluorescence spectra (Figure 5a). Meanwhile 0.4%, 0.6%, 0.8%, and 1.0% (*w*/*v*) pectin caused a 0.5, 1.0, 1.5, and 2.0 nm redshift in the peak wavelength and a 47.3%, 58.1%, 68.7%, and 73.5% decrease, respectively, in the fluorescence intensity of pPPO (Figure 5b). The shape change in the fluorescence spectra was correlated with the conformational change in pPPO as well as the polarity change in the environment around tyrosine and tryptophan residues [39]. The more dramatic changes in pPPO in the presence of pectin rather than sucrose indicated that pectin had stronger interactions with pPPO, and the driving force may be the hydrogen bond, hydrophobic interaction, and electrostatic interaction. Moreover, the reduction in the environmental pH and hence proton transfer induced by pectin could be another reasonable explanation, according to Li et al. [19].

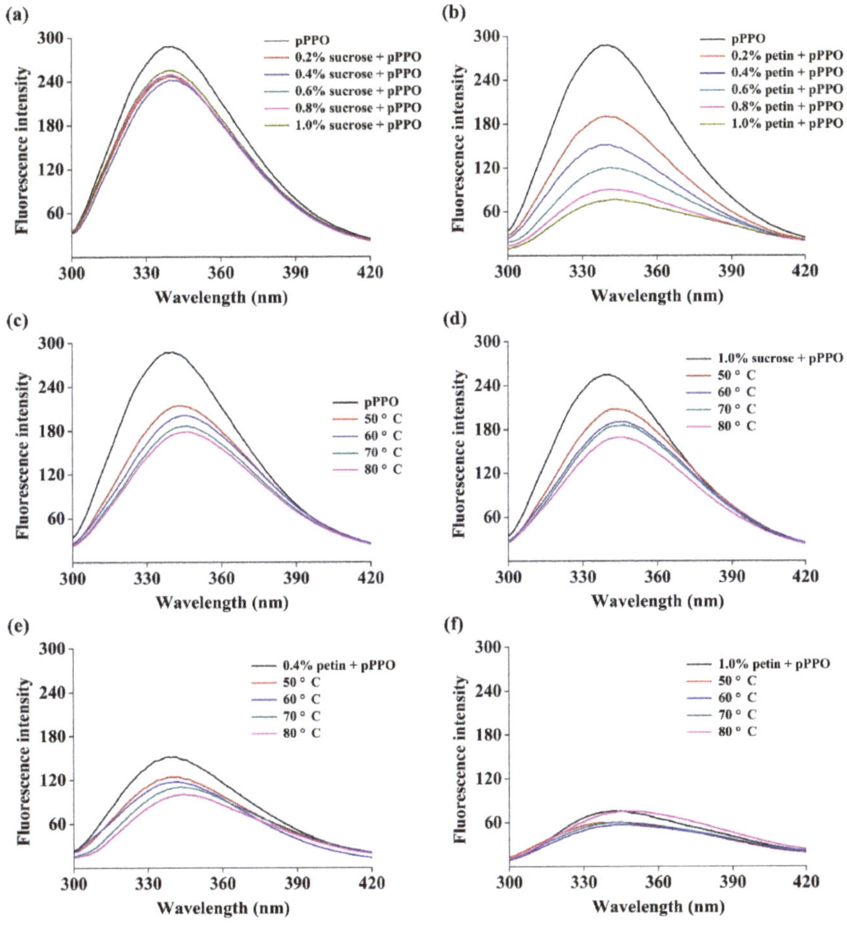

Figure 5. Fluorescence emission spectra of pPPO with sucrose (**a**) and pectin (**b**) at room temperature. From top to down, control sample (pPPO), pPPO with sucrose or pectin at the concentration of 0.2%, 0.4%, 0.6%, 0.8%, and 1.0% (*w*/*v*), respectively. Fluorescence emission spectra of pPPO (**c**), pPPO with sucrose (1.0%) (**d**), pPPO with pectin (0.4%) (**e**), and pPPO with pectin (1.0%) (**f**) during thermal treatment from room temperature (25 °C) to 80 °C.

Thermal processing led to decreases in fluorescence intensity and a redshift in the peak wavelength of pPPO. The fluorescence intensities decreased to 74.5%, 69.9%, 64.8%, and 62.2%, with thermal processing at 50, 60, 70, and 80 °C, respectively, and the corresponding peak wavelengths were redshifted by 2.0, 4.0, 5.5, and 6.0 nm, respectively (Figure 5c). With the addition of 1.0% (*w*/*v*) sucrose, the redshifts of the pPPO fluorescence spectra were 2.5, 5.0, 6.0, and 5.5 nm at 50, 60, 70, and 80 °C (Figure 5d), which are very close to the values obtained in the absence of sucrose. In comparison, the peak wavelengths of pPPO were redshifted by 1.0, 1.0, 2.5, and 3.5 nm with 0.4% (*w*/*v*) pectin and 1.5, 2.0, 2.5, and 4.0 nm with 1.0% (*w*/*v*) pectin at 50, 60, 70, and 80 °C, respectively (Figure 5e,f), which are smaller than the values obtained in the case without pectin. These results illustrate that thermal processing could disrupt the tertiary structure of pPPO; however, the presence of pectin provided protection for the pPPO structure against thermal processing. This supports the results presented in Section 3.5, wherein pectin provided pPPO with stronger protection than of sucrose at all temperatures. Heating could unfold the pPPO molecules and expose them to amino acid residues [40], which may be conducive to the association of pPPO with pectin and hence improve its structural and thermal stability.

4. Conclusions

In this study, kudzu (*Pueraria lobata*) PPO with different constituent contents was successfully obtained by different purification procedures. It was noted that the thermal stability and inactivation rate constant of tPPO and cPPO containing food constituents was significantly higher than that of pPPO, confirming that food constituents were important factors leading to the differential thermal stability of PPO in the model and food systems. Food constituents pectin, BSA, starch, and sucrose exhibited a protective effect on pPPO against thermal processing, and the protective effect decreased orderly. Fluorescence spectrum results demonstrated that pectin could form stronger interactions with pPPO than sucrose and thus improve the structural and thermal stability of pPPO. In conclusion, the enzymatic reaction of PPO in the real food system is complex. This study provides a scientific and technological basis for the further evaluation of browning control and contributes to its practical application in foods.

Supplementary Materials: The following are available online at https://www.mdpi.com/article/10.3390/foods10061320/s1. Table S1: Purification of PPO from kudzu; Table S2: Change of constituent content in kudzu tissue, crude PPO (cPPO) solution, and purified PPO (pPPO) solution; Figure S1: SDS-PAGE electrophoresis of the purified PPO; Figure S2: Effect of pH on PPO activity (a) and Lineweaver–Burk plots of PPO (b).

Author Contributions: Conceptualization, J.L., L.Z. (Lei Zhou), and W.L.; methodology, T.L.; validation, J.L., J.Z., and Y.L.; formal analysis, J.L. and L.Z. (Lei Zhou); investigation, T.L.; resources, L.Z. (Lei Zhou), and L.Z. (Liqiang Zou).; data curation, J.L., J.Z., and Y.L.; writing—original draft preparation, J.L.; writing—review and editing, J.L. and T.L.; visualization, J.Z. and T.L.; supervision, W.L.; project administration, L.Z. (Lei Zhou); funding acquisition, J.L., L.Z. (Lei Zhou), L.Z. (Liqiang Zou), L.Z. (Li Zhang), and W.L. All authors have read and agreed to the published version of the manuscript.

Funding: This research was funded by the National Natural Science Foundation of China (grant numbers 31860452), the China Postdoctoral Science Foundation (grant numbers 2019M652288, 2020T130276), the Free exploration project of the State Key Laboratory of Food Science and Technology, Nanchang University (grant numbers SKLF-ZZB-201919), the Foundation of Key Laboratory of Tropical Crop Products Processing of Ministry of Agriculture and Rural Affairs, China (grant numbers KFKT202005), the Youth Science Foundation of Jiangxi Province (grant numbers 20181BAB214019), and the Jiangxi Province Education Department Science Foundation (grant numbers GJJ160190).

Institutional Review Board Statement: Not applicable.

Informed Consent Statement: Not applicable.

Data Availability Statement: The data presented in this study are available in this article.

Conflicts of Interest: The authors declare no conflict of interest.

References

1. Huang, Q.; Zhang, H.; Xue, D. Enhancement of antioxidant activity of radix puerariae and red yeast rice by mixed fermentation with monascus purpureus. *Food Chem.* **2017**, *226*, 89–94. [CrossRef]
2. Wang, S.; Yang, Z.; Peng, N.; Zhou, J.; Yong, X.; Yuan, H.; Zheng, T. Optimization of ionic liquids-based microwave-assisted hydrolysis of puerarin and daidzein derivatives from radix puerariae lobatae extract. *Food Chem.* **2018**, *256*, 149–155. [CrossRef] [PubMed]
3. Liu, J.; Shi, Y.C.; Lee, D.Y. Applications of *pueraria lobata* in treating diabetics and reducing alcohol drinking. *Chin. Herb. Med.* **2019**, *11*, 141–149. [CrossRef]
4. Hung, P.V.; Morita, N. Chemical compositions, fine structure and physicochemical properties of kudzu (*Pueraria lobata*) starches from different regions. *Food Chem.* **2007**, *105*, 749–755. [CrossRef]
5. Xu, L.; Shi, W.; Cai, C.; Zhong, W.; Tu, K. Rapid and nondestructive detection of multiple adulterants in kudzu starch by near infrared (NIR) spectroscopy and chemometrics. *LWT-Food Sci. Technol.* **2015**, *61*, 590–595. [CrossRef]
6. Mayer, A.M. Polyphenol oxidases in plants and fungi: Going places? a review. *Phytochemistry* **2006**, *67*, 2318–2331. [CrossRef] [PubMed]
7. Siddiq, M.; Dolan, K.D. Characterization of polyphenol oxidase from blueberry (*Vaccinium corymbosum* L.). *Food Chem.* **2017**, *218*, 216–220. [CrossRef]
8. Gouzi, H.; Depagne, C.; Coradin, T. Kinetics and thermodynamics of the thermal inactivation of polyphenol oxidase in an aqueous extract from Agaricus bisporus. *J. Agric. Food Chem.* **2012**, *60*, 500–506. [CrossRef]
9. Xiong, Z.; Liu, W.; Zhou, L.; Zou, L.; Chen, J. Mushroom (*Agaricus bisporus*) polyphenoloxidase inhibited by apigenin: Multi-spectroscopic analyses and computational docking simulation. *Food Chem.* **2016**, *203*, 430–439. [CrossRef] [PubMed]
10. Zhou, L.; Liu, W.; Stockmann, R.; Terefe, N.S. Effect of citric acid and high pressure thermal processing on enzyme activity and related quality attributes of pear puree. *Innov. Food Sci. Emerg.* **2018**, *45*, 196–207. [CrossRef]
11. Zhou, L.; Liu, W.; Terefe, N.S. The inactivation kinetics of soluble and membrane-bound polyphenol oxidase in pear during thermal and high-pressure processing. *Food Bioprocess Technol.* **2018**, *11*, 1039–1049. [CrossRef]
12. Niu, S.; Xu, Z.; Fang, Y.; Zhang, L.; Yang, Y.; Liao, X.; Hu, X. Comparative study on cloudy apple juice qualities from apple slices treated by high pressure carbon dioxide and mild heat. *Innov. Food Sci. Emerg.* **2010**, *11*, 91–97. [CrossRef]
13. Han, Q.Y.; Liu, F.; Li, M.; Wang, K.L.; Ni, Y.Y. Comparison of biochemical properties of membrane-bound and soluble polyphenol oxidase from Granny Smith apple (*Malus* × *domestica Borkh*). *Food Chem.* **2019**, *289*, 657–663. [CrossRef]
14. Gong, Z.; Li, D.; Liu, C.; Cheng, A.; Wang, W. Partial purification and characterization of polyphenol oxidase and peroxidase from chestnut kernel. *LWT-Food Sci. Technol.* **2015**, *60*, 1095–1099. [CrossRef]
15. Chutintrasri, B.; Noomhorm, A. Thermal inactivation of polyphenoloxidase in pineapple puree. *LWT-Food Sci. Technol.* **2006**, *39*, 492–495. [CrossRef]
16. Zhou, L.; Liao, T.; Liu, W.; Zou, L.; Liu, C.; Terefe, N.S. Inhibitory effects of organic acids on polyphenol oxidase: From model systems to food systems. *Crit. Rev. Food Sci.* **2019**, *6*, 1–28. [CrossRef]
17. Liu, L.; Cao, S.; Yang, H.; Qi, X. Pectin plays an important role on the kinetics properties of polyphenol oxidase from honeydew peach. *Food Chem.* **2015**, *168*, 14–20. [CrossRef]
18. Ormus, S.; Oulahal, N.; Noel, C.; Degraeve, P.; Gharsallaoui, A. Effect of low methoxyl (LM) pectin complexation on the thermal and proteolytic inactivation of lysozyme: A kinetic study. *Food Hydrocoll.* **2015**, *43*, 812–818. [CrossRef]
19. Li, R.; Wang, Y.; Ling, J.; Liao, X. Effects of high pressure processing on activity and structure of soluble acid invertase in mango pulp, crude extract, purified form and model systems. *Food Chem.* **2017**, *231*, 96–104. [CrossRef]
20. Huang, N.; Cheng, X.; Hu, W.; Pan, S. Inactivation, aggregation, secondary and tertiary structural changes of germin-like protein in satsuma mandarine with high polyphenol oxidase activity induced by ultrasonic processing. *Biophys. Chem.* **2015**, *197*, 18–24. [CrossRef]
21. Chakraborty, S.; Baier, D.; Knorr, D.; Mishra, H.N. High pressure inactivation of polygalacturonase, pectinmethylesterase and polyphenoloxidase in strawberry puree mixed with sugar. *Food Bioprod. Process.* **2015**, *95*, 281–291. [CrossRef]
22. Liu, F.; Zhao, J.H.; Wen, X.; Ni, Y.Y. Purification and structural analysis of membrane-bound polyphenol oxidase from fuji apple. *Food Chem.* **2015**, *183*, 72–77. [CrossRef] [PubMed]
23. Laemli, U.K. Cleavage of structural proteins during the assembly of the head of bacteriophage T4. *Nature* **1970**, *227*, 680–685. [CrossRef]
24. Bradford, M.M. A rapid and sensitive method for the quantization of microgram quantities of prote in utilizing the principle of protein-dye binding. *Anal. Biochem.* **1976**, *72*, 248–254. [CrossRef]
25. Terefe, N.S.; Delon, A.; Buckow, R.; Versteeg, C. Blueberry polyphenol oxidase: Characterization and the kinetics of thermal and high pressure activation and inactivation. *Food Chem.* **2015**, *188*, 193–200. [CrossRef] [PubMed]
26. Tan, T.C.; Cheng, L.H.; Bhat, R.; Rusul, G.; Easa, A.M. Composition, physicochemical properties and thermal inactivation kinetics of polyphenol oxidase and peroxidase from coconut (*Cocos nucifera*) water obtained from immature, mature and overly-mature coconut. *Food Chem.* **2014**, *142*, 121–128. [CrossRef]
27. Zhou, L.; Liao, T.; Liu, J.P.; Zou, L.Q.; Liu, C.M.; Liu, W. Unfolding and inhibition of polyphenoloxidase induced by acidic pH and mild thermal treatment. *Food Bioprocess Tech.* **2019**, *12*, 1907–1916. [CrossRef]

28. Terefe, N.S.; Yang, Y.H.; Knoerzer, K.; Buckow, R.; Versteeg, C. High pressure and thermal inactivation kinetics of polyphenol oxidase and peroxidase in strawberry puree. *Innov. Food Sci. Emerg.* **2010**, *11*, 52–60. [CrossRef]
29. Chourio, A.M.; Fabiola, S.F.; Zahid, M.; Martinez-Monteagudo, S.I.; Saldaña, M.D.A. Inactivation of peroxidase and polyphenoloxidase in coconut water using pressure-assisted thermal processing. *Innov. Food Sci. Emerg.* **2018**, *49*, 41–50. [CrossRef]
30. Lopes, A.M.; Toralles, R.P.; Rombaldi, C.V. Thermal inactivation of polyphenoloxidase and peroxidase in jubileu clingstone peach and yeast isolated from its spoiled puree. *Food Sci. Technol.* **2014**, *34*, 150–156. [CrossRef]
31. Ji, D.; Oey, I.; Agyei, D. Purification, characterization and thermal inactivation kinetics of β-galactosidase from Lactobacillus leichmannii 313. *LWT-Food Sci. Technol.* **2019**, *116*, 108545. [CrossRef]
32. Isleroglu, H.; Turker, I. Thermal inactivation kinetics of microencapsulated microbial transglutaminase by ultrasonic spray-freeze drying. *LWT-Food Sci. Technol.* **2019**, *101*, 653–662. [CrossRef]
33. Bayarri, M.; Oulahal, N.; Degraeve, P.; Gharsallaoui, A. Properties of lysozyme/low methoxyl (LM) pectin complexes for antimicrobial edible food packaging. *J. Food Eng.* **2014**, *131*, 18–25. [CrossRef]
34. Chisari, M.; Barbagallo, R.N.; Spagna, G. Characterization of polyphenol oxidase and peroxidase and influence on browning of cold stored strawberry fruit. *J. Agric. Food Chem.* **2007**, *55*, 3469–3476. [CrossRef] [PubMed]
35. Matsue, S.; Miyawaki, O. Influence of water activity and aqueous solvent ordering on enzyme kinetics of alcohol dehydrogenase, lysozyme, and β-galactosidase. *Enzyme Microb. Technol.* **2000**, *26*, 342–347. [CrossRef]
36. Jones, O.G.; Decker, E.A.; Mcclements, D.J. Formation of biopolymer particles by thermal treatmentof β-lactoglobulin-pectin complexes. *Food Hydrocoll.* **2009**, *23*, 1312–1321. [CrossRef]
37. Friedman, M.; Bautista, F.F. Inhibition of polyphenol oxidase by thiols in the absence and presence of potato tissue suspensions. *J. Agric. Food Chem.* **1995**, *43*, 69–76. [CrossRef]
38. Kishore, V.; Gowda, S.; Krishna, S.; Sharma, K.; Rashmi, M.; Nishita, K.P. Bovine serum albumin a potential thermostabilizer: A study on α-amylase. *J. Appl. Microbiol.* **2014**, *2*, 37–41. [CrossRef]
39. Zhou, L.; Wu, J.; Hu, X.; Zhi, X.; Liao, X. Alterations in the activity and structure of pectin methylesterase treated by high pressure carbon dioxide. *J. Agric. Food Chem.* **2009**, *57*, 1890–1895. [CrossRef]
40. Zhou, L.; Liu, W.; Zou, L.; Xiong, Z.; Hu, X.; Chen, J. Aggregation and conformational change of mushroom (*Agaricus bisporus*) polyphenoloxidase subjected to thermal treatment. *Food Chem.* **2017**, *214*, 423–431. [CrossRef]

foods

MDPI

Review

Natural Food Polysaccharides Ameliorate Inflammatory Bowel Disease and Its Mechanisms

Yikun Wang [1], Haibin Zhu [2], Xiaoji Wang [3], Yue Yu [2] and Jianhua Xie [1,2,*]

[1] Queen Mary School, Jiangxi Medical College, Nanchang University, Nanchang 330006, China; yiqueenwang@163.com
[2] State Key Laboratory of Food Science and Technology, Nanchang University, Nanchang 330047, China; zhuhaibin163@163.com (H.Z.); zongyu_li@email.ncu.edu.cn (Y.Y.)
[3] School of Chemical Engineering and Energy Technology, Dongguan University of Technology, Dongguan 523808, China; wangxj@dgut.edu.cn
* Correspondence: jhxie@ncu.edu.cn; Tel.: +86-0791-8830-4347

Abstract: Natural polysaccharides and their metabolites' short chain fatty acids (SCFAs) have attracted much attention. Recently, they have shown great potential in attenuating systemic inflammation activities, especially in inflammatory bowel disease (IBD). IBD is a complex pathological process and is related to epithelial damage and microbiota imbalance in the gut. Recent studies have indicated that natural polysaccharides could improve IBD recovery by different mechanisms. They could not only influence the ratio of intestine microbiota, but also regulate the secretion levels of immunity cytokines through multiple pathways, the latter including modulation of the TLR/MAPK/NF-κB signaling pathways and stimulation of G-protein-coupled receptors. Moreover, they could increase intestinal integrity and modulate oxidative stress. In this review, recent research about how natural polysaccharides impact the pathogenesis of IBD are summarized to prove the association between polysaccharides and disease recovery, which might contribute to the secretion of inflammatory cytokines, improve intestine epithelial damage, reduce oxidative stress, sustain the balanced microenvironment of the intestines, and finally lower the risk of IBD.

Keywords: polysaccharides; inflammatory bowel disease; mechanism

Citation: Wang, Y.; Zhu, H.; Wang, X.; Yu, Y.; Xie, J. Natural Food Polysaccharides Ameliorate Inflammatory Bowel Disease and Its Mechanisms. *Foods* **2021**, *10*, 1288. https://doi.org/10.3390/foods10061288

Academic Editor: Barry J. Parsons

Received: 4 May 2021
Accepted: 2 June 2021
Published: 4 June 2021

1. Introduction

Inflammatory bowel disease (IBD) is represented by ulcerative colitis (UC) and Crohn's disease (CD), which manifest as chronic inflammation of the gut [1]. It has a higher prevalence and a rising incidence in developing countries in recent years [2]. The clinical features of IBD include pathological damage, ulceration of intestinal mucosa, diarrhea, as well as bloody stool [3]. Various factors have been involved in the pathogenesis of IBD, such as environmental exposure, diet shortage, innate and required immunity deficiency, or gut microbiota dysregulation [4,5]. Until now, the conventional treatment of IBD has still used drugs such as methylprednisolone, infliximab, and cyclosporine. However, the continued refractory as well as the serious adverse effects are strongly troubling to the patients [6]. Recently, it has been indicated that polysaccharides from the natural source could alleviate IBD symptoms, which might be an alternative element in the treatment of IBD [7].

Natural polysaccharides can improve intestine damage recovery, through their structural characteristics, molecular weight, and bioactivities. The diversity of microbiota is significantly reduced comparing the IBD patient with normal people. The microbiota dysbiosis in the gut is recognized as the increase of *Proteobacteria* and *Bacteroidetes*, and simultaneously a decrease in the abundance of *Firmicutes*, *Lactobacillis* and *Clostridia* [8,9]. Polysaccharides can be metabolized by gut microbiota and finally fermented into SCFAs, then SCFAs modulate the gut microbiota ratio via increase the abundance of probiotics and decrease the ratio of bacteria harmful to the gut mucosa.

Recently, more and more research has attached great importance to the immunity effect in the IBD development [10]. On the one hand, the supplement of natural polysaccharides up-regulates the tight junction proteins, which not only decreases the neutrophil infiltration through intestine mucosa, but also promotes the recovery of the epithelial barrier of the colon. On the other hand, administrating polysaccharides can influence the production of inflammatory cytokines in the IBD patients' gut. As the indispensable component of the immune responses, inflammatory cytokines send signals to the numerous immune cells, which are critical to cellular immunity regulation. Excessive pro-inflammatory cytokines, like interleukin (IL)-6, IL-1β, tumor necrosis factor (TNF)-α, induce diseases associated with immunity reaction, while anti-inflammatory cytokines ameliorate the inflammation and promote recovery [9]. Thus, recovering the balance of inflammatory cytokine production via natural polysaccharides is an essential mechanism to the gut defensive response.

The excretion of inflammatory cytokines is regulated by the inflammatory signaling pathways including toll-like receptors (TLR), mitogen-activated protein kinase (MAPK), nuclear factor-κB (NF-κB), and G-protein coupled receptors, which participate in regulating the phosphorylation level of TLR2/4/6, ERK, IκB, GPR41, and GPR43 [11]. By regulating these signal transduction pathways, natural polysaccharides and their metabolites ameliorate IBD symptoms in the damaged tissues. Polysaccharides alleviate oxidative stress through down-regulating the production level of oxygen free radicals including myeloperoxidase (MPO), nitric oxide (NO), malondialdehyde (MDA), which subsequently facilitate IBD symptoms [12].

Most of the reviews related to natural polysaccharides have paid attention to their structures and anti-inflammatory applications in pharmaceuticals. However, few discussed the effects of natural polysaccharides and their mechanism involved in colitis in detail and in its entirety. This current review summarizes the latest linkages of natural polysaccharides and IBD, including the types of natural polysaccharides that alleviate the IBD symptoms and the mechanisms of moderation of this disease. The main objective is to discuss the relationship between natural polysaccharides and IBD treatment.

2. Polysaccharides and Their Impact on Inflammatory Bowel Diseases

2.1. Polysaccharides from Hericium Erinaceus

A purified polysaccharide (EP-1) isolated from *Hericium Erinaceus* has been studied for its anti-gastric/colitis ulcer activity [11]. The molecular weight of EP-1 is around 3.1 kDa and the structure is constituted of glucose, mannose, and galactose, whose backbone is α-d-Glc(1→3) and β-d-Glc(1→3) and both of them are branches with C-4 position [13].

Recently, EP-1 has been confirmed to relieve the symptoms of ulcerative colitis induced by acetic acid [14]. As it enters the lower gastrointestinal tract, the polysaccharide would be fermented by the gut microbiota to form SCFAs. The carbon from different polysaccharides gets used as the unique energy source for particular microbes in intestine [15]. Therefore, the polysaccharides could change the composition of the gut microbiota, which is companied with changes of the SCFAs. EP-1 changes the bacteria community through recovering the levels of *Firmicutes, Bacteroidetes, Proteobacteria*, and *Actinobacteria* with the increase of SCFAs, including ethanoic acid and butanoic acid, in the colon. SCFAs including acetate, propionate, and butyrate are the last degraded product metabolized from polysaccharides by the gut microbiota [16], which influence the diversity of intestinal microbiota, and contribute vitally to immune response regulation in the gastrointestinal system [17].

Another polysaccharide was extracted from *Hericium erinaceus*, whose molecular weight is 86.67 kDa. It was also reported to down-regulate the expression of IL-6, IL-1β, TNF-α, and cyclooxygenase-2 (COX-2); induce iNOS; and decrease the expression of related mRNA in dextran sulfate sodium (DSS)-induced mice [18]. iNOS belongs to the systems of reactive oxygen/nitrogen species (ROS/RNS) generation that stimulate oxidative stress. Oxidative stress is tightly connected with the production of inflammatory cytokines and inflammatory cell penetration [19]. As the polysaccharides reduce the expression of iNOS, there is the amelioration of oxidative stress. Otherwise, iNOS stimulates the manufacture of the NO and promotes the expression of COX-2; the latter could catalyze the prostaglandin production finally induced the inflammation responses. All of these implementations depend upon the NF-κB signaling pathway [20].

2.2. *Sarcodon aspratus Polysaccharides*

The water-soluble polysaccharide (HCP) extracted from *Sarcodon aspratus* with a molecular weight of 6.7×10^3 kDa was reported having immunomodulatory function in the intestine. HCP has a backbone structure of (1→6)-linked-α-d-glucopyranosyl residues. HCP facilitates the immunity function of macrophages cells through the TLR4-MAPK-NFκB signaling pathway, which enhances the phagocytic activity and promotes the expression of NO, iNOS. HCP works toward regulating the immunity in the intestine through moderating inflammatory cytokines secreted from immunity cells [21].

2.3. *Dictyophora indusiata Polysaccharide*

Dictyophora indusiata polysaccharide (DIP) is isolated from mushrooms with a molecular weight of 1132 kDa. The main linkage-type of DIP including (1→3)-linked α-L-Man, (1→2, 6)-linked α-d-Glc, (1→6)-linked β-d-Glc, (1→6)-linked β-d-Gal, and (1→6)-linked β-d-Man. DIP could markedly facilitate the excretion of NO, TNF-α, and IL-6 from macrophages involved in complement receptor 3 (CR3) [22]. DIP has a latent treatment impact against colitis by increasing the level of *Bacteroidaceae* and *Enterobacteriaceae* in the colitis microenvironment. The administration of DSS-induced mice with DIP significantly reduced the pro-inflammatory cytokines (TNF-α, IL-1β, IL-6, IFN-γ) and enhanced anti-inflammatory cytokines (IL-4, IL-10) [23]. The treatment of DIP with the DSS-induced mice restrains the biomarker of M1 macrophages CD86. It changed the macrophage subset distribution by reducing the production of M1 subsets from splenic macrophages and made deviate polarization to M2 subsets [24]. The macrophages have a tight connection to the inflammatory cytokines' generation. The M1 macrophages are connected to IL-6 as well as TNF-α, and M2 macrophages are associated with the generation of IL-10 [25]. DIP also can down-regulate the MPO level dramatically in a dose-dependent manner. As an inflammatory marker, the association between MPO and IBD is that it identifies the infiltration of inflammatory cells and tissue damage. Simultaneously, DIP also decreases NO activity and increases T-SOD levels. The latter would give rise to the production of NO [26], which triggers tissue damage and induces inflammatory responses during the occurring of IBD [18].

Recent studies have found that DIP not only decreased endotoxins, including lipopolysaccharides, but also maintained the normal level of tight junction (TJ) proteins expression including claudin-1, occludin, and zonula occludens (ZO)-1 in the antibiotic-induced mice [27]. Broad spectrum antibiotics have been demonstrated to cause gut dysbiosis [28] and destroy epithelial barrier integrity [29], thus raising the hazard of IBD occurrence in the neonates [30]. DIP could restore the function of the gut barrier. The gut mucosal barrier prevents the pathogens and toxins from invasion into the intestine, which has an essential function involved in maintaining the intestine in healthy condition. Additionally, DIP also inhibited the colonic tissue apoptosis in the ulcerative colitis mice by improving the expression of Bcl-2 and Bax [24].

2.4. *Flammuliana velutipes Polysaccharids*

Flammuliana velutipes polysaccharides (FVP) are comprised of glucans, galactan, xyloglucan, and xylomannan, whose backbone is α-(1→4)-d-glucose residues with branches consisting of (1→6)-linked α-d-glucopyranosyl [31]. FVP has various bioactivities including immunity regulation [32], anti-oxidant, anti-microbial, anti-tumor properties, and so on [33]. FVP could recover colon injury induced by DSS, and moderate the dysbiosis in the gut microbiota, such as by boosting the *Firmicutes/Bacteroidetes* ratio and enhancing the *Lachnospiraceae* abundance [34]. *Lachnospiraceae* is able to restrain intestinal immune homeostasis via promoting the SCFA production [35], which might explain the phenomenon that FVP increased the production of SCFAs in the gut.

2.5. *Pleurotus eryngii* Polysaccharides

Treatment of *Pleurotus eryngii* polysaccharides (PEP) could increase the abundance of *Porphyromonadaceae, Rikenellaceae, Bacteroidaceae,* and *Lactobacillaceae* on C57BL/6 mice. More importantly, PEP increased the SCFA production in the mice colon. SCFAs exert numerous functions in the mice including regulating gut immunity conditions, changing the mucus layer composition and restraining the homeostasis of the intestine [36]. At the same time, PEP inhibited the expression of ROS and NO in RAW264.7 macrophages, and sulphonated PEP showed more powerful anti-inflammatory activity [37].

A water-soluble polysaccharide (EPA-1) isolated from *Pleurotus eryngii* (99.7 kDa) consisting of mannose, glucose, and galactose with the ratio of 2.2:1.0:3.2 EPA-1 not only increased the manufacturing of NO and pro-inflammatory cytokines including TNF-α, IL-1, as well as IL-6, but also exhibited the improved expression level of phosphorylated p38, ERK, JNK, and NF-κB. It reflects the EPA-1 influence inflammatory response via the effect on MAPK and the NF-κB signal mediated pathway [38]. EPA-1 exerts intestine immunity to enhance its protective effect. The immunity cytokines including IL-6 motivated signaling pathways could promote the various leukocytes recruiting and activation [39], which is subsequently followed by possessing invading microorganisms and further diminishment of epithelial damage. Therefore, EPA-1 ameliorates IBD through moderate inflammatory cytokines. Above all these results have indicated that polysaccharides from *Pleurotus eryngii* have the function to alleviate IBD. Therefore, it is thought that PEP reduces the risk of IBD and benefits colitis recovery [40].

2.6. *Gracilaria* Polysaccharides

Gracilaria caudata polysaccharide consists of 1→3-linked-β-d-galactose and 1→4-linked-3,6-anhydro-α-L-galactose [41] with a molecular weight of 116 kDa [42]. Administrated *G. caudata* polysaccharide could significantly reduce neutrophil infiltration in acetic acid-induced mice via decreasing the expression of MPO levels together with restoring the colon from acetic acid injuries by reducing oxidative stress, including the diminishment of GSH consumption and iNOS expression in the gut [43].

Another species from genus of *Gracilaria* is *Graciliaria lemaneiformis*, the polysaccharides of *Graciliaria lemaneiformis* (SP), which also has sulfated groups. In the mice model developed by Ren et al. [18], SP was orally administrated 200, 400, 600 mg/kg/day for 3 weeks and the arresting weight loss following subsequently by the improved conditions of mice. Histological analysis showed that the colon structure, including the crypt number and muscular structure, was improved by raising the SP dosage, suggesting that a high dose of SP protects the goblet cells and crypt structure in the gut. SP also could repress the intestinal endotoxin and lipopolysaccharide-binding protein production, together with the reduction of TNF-α, IL-6, and IL-1β expression [44]. All of these markers indicate that SP alleviates the symptoms of IBD.

2.7. *Blidingia minima* Polysaccharides

Blidingia minima often grow in Subei Shoal in China and have similar morphology characteristics as *Enteromorpha* species. Polysaccharides extracted from *Enteromorpha* species have revealed various beneficial functions such as, for instance, anti-inflammatory and antioxidant activities [45]. Recently, Song et al. administrated *Blidingia minima* polysaccharides (BMP) to DSS-induced mice, and found a decrease in the expression of anti-inflammatory cytokine IL-10 and an increase in the production of pro-inflammatory cytokines including IL-1β and TNF-α. The expression of NF-κB, IκB-α, and AKT was also increased, thus, BMP might modulate the motivation of NF-κB and AKT signaling pathway to ameliorate IBD [46].

2.8. *Arctium lappa* L. Polysaccharides

Arctium lappa L. is one of the well-studied traditional medicines. The polysaccharide (ALP-1) isolated from *Arctium lappa* L. has the molecular weight of 5.12 kDa and is constituted of a (2→1)-β-d-fructofuranose backbone linked to a terminal of (2→1)-α-d-glucopyranose at the non-reducing end and a (2→6)-β-d-fructofuranose branching. The anti-inflammatory impression of this alkaline extraction polysaccharide is thought to relate to its specific structure feature the (2→1)-linked type in fructan [47]. ALP-1 caused a higher relative abundance of Lactobacillaceae, Lachnospiraceae, as well as Ruminococcaceae, and a lower relative abundance of Bacteroides plus *Staphylococcus* in DSS-induced male ICR mice. It also led to a remarkable decline of pro-inflammatory cytokines such as IL-1β, IL-6, TNF-α, and improved levels of anti-inflammatory cytokines including IL-10 and immunoglobulin A (Ig A) in DSS-induced male ICR mice [48]. The pathogenic microorganisms are easier to be phagocytosed through phagocytic cells [49]. Similar effects were noted in another study conducted by Zhang and his coworkers where the same inflammatory cytokine results were observed when treating with ALP-1 in lipopolysaccharide (LPS)-induced RAW264.7 macrophage [50].

An alkali-soluble polysaccharide (ASPP) has proved to have a stronger anti-inflammatory and immunomodulatory effect [47,51]. Female ICR mice were administrated with ASPP 400 mg/kg body weight for 28 days. They displayed a higher abundance of probiotics such as *Firmicutes, Alistipes, Odoribacter, Lachnospiraceae,* as well as *Lactobacillus* in mice [52]. *Lactobacillus* not only has a reverse relationship with IL-1β but also prevents epithelial barrier disruptions by blocking TNF-α [53]. There is also an increment in the *Firmicutes/Bacteroides* ratio, which has a positive connection with a protective effect on IBD patients [54]. Except for elevating this ratio in the colon, ASPP simultaneously decreases the level of the harmful bacteria *Proteobacteria,* which boosts the excessive pro-inflammatory cytokines expression [55]. Otherwise, ASPP inhibits the expression of pro-inflammatory cytokines including IL-6, IL-1β, TNF-α, and IL-10 and decreases the overproduction of NO through down-regulating iNOS expression [56]. As one of the pro-inflammatory molecules, NO can stimulate the immune system [57] and is mediated by NF-κB and MAPKs pathways [58].

2.9. *Morinda citrifolia* L. Polysaccharides

Morinda citrifolia (Noni) has been utilized as medicine for a hundred years. Polysaccharides extracted from noni fruit (NFP) have been shown to exert numerous benefits such as being anti-diabetic, anti-inflammatory, anti-oxidant, and anti-cancer [59]. NFP is comprised of galacturonic acid, rhamnogalacturonan-I, arabinogalactan, and arabinan with a molecular weight of nearly 456 kDa. NFP could improve the tight junction protein production, including ZO-1 and occludins, and thus enhance the gut barrier integrity and decrease the possibility of endotoxemia in DSS-induced IBD mice [60].

2.10. *Astragalus membranaceus* Polysaccharides

Astragalus membranaceus has been used for curing numerous conditions such as wounds, diabetes, leukemia, eye disease, and nephritis for several centuries [61]. Recent studies have proved that the *Astragalus* polysaccharides (APS) could exert immunomodulation and anti-oxidant effects. In DSS-induced male C57BL/6 mice, APS was administrated 200 mg/kg/day for the first 3 days. It promoted the colitis-related clinical indices, and noticeably, when compared with the DSS-induced colitis group, a histopathology assessment revealed that all of the ulcerations and mucosal, edema, and neutrophil infiltration in the epithelium of intestine were significantly reduced. APS also down-regulated production of MPO and pro-inflammatory cytokines including TNF-α, IL-1β, and IL-6 as well as the expression of NF-κB in DSS-induced mice. Therefore, APS could relieve symptoms of colitis through modulating the NF-κB signaling pathway [62].

2.11. *Dendrobium officinale* Polysaccharides

Dendrobium officinale has been used as a Chinese medicine for thousands of years. Previous studies have displayed that the polysaccharides from *Dendrobium officinale* (DOP) make a significant contribution against inflammation and immunity modulating [63]. Recently, based on the specific endo-β-1,4-mannanase linear structure of DOP, Zhang et al. have separated the core domain of Dendrobium officinale polysaccharides (EDOP) after enzymatic. It consists of glucose and mannose with the molar ratio of 1.00:4.76 and is composed of (1→4)-β-d-Glcp and (1→4)-β-d-Manp with attached 2-O-acetylated groups [64]. The effects of EDOP on colitis were evaluated in DSS-induced colitis male Balb/c mice. Both DOP and EDOP could reduce the overexpression of pro-inflammatory cytokines including IL-1β, IL-6, and TNF-α together with their mRNA expression levels simultaneously, which could moderate the IBD to some extent. DOP and EDOP treatment increased the abundance of *Bacteroides, Lactobacillus,* plus *Ruminococcaceae*, and down-regulated the level of *Proteobacteria* and *Akkermansia* in the meantime. *Ruminococcaceae* is one of the common probiotics that could promote the degradation of complex polysaccharides followed by the increased production of SCFAs [65]. *Akkermansia* improves pro-inflammatory cytokine manufacturing and diminishes the SCFAs expression, as a result of a positive relationship with IBD [66]. DOP and EDOP also restored the mRNA expression level of GPR41/43, which reactivated the metabolism of SCFAs in the gut. These studies might explain that DOP and EDOP treatment recovered the level of acetic acid in the colon, thus promoting SCFA production in the gut. DOP can ameliorate IBD through exhibiting depression of NLRP3 inflammasome and β-arrestin1, following by subsequently regulating the expression of TLR-2,4,6,9 [67].

2.12. *Lycium barbarum* Polysaccharides

Lycium barbarum (Goji berry) has been utilized in medicine and diet for thousands of years. The major backbone of the polysaccharide isolated from *Lycium barbarum* (LBP) is comprised of (1→3)-β-d-galactopyranosyl, (1→6)-β-d-galactopyranosyl, and (1→4)-α-d-galactopyranosyluronic acid residues, and the molecular weight cover from $2.1–6.5 \times 10^3$ kDa [68]. The supplementation of LBP promoted growth of intestinal probiotics such as *Akkermansia, Lactobacillus,* and Prevotellaceae. It also significantly increased the expression level of TGF-β and IL-6 in serum and sIgA in colon tissues [69]. Furthermore, treatment with LBP ameliorated the colitis symptoms in DSS-induced colitis C57BL/6 mice, which includes alleviating the damage of mucus and reducing the neutrophil infiltration in colonic tissues [68]. LBP could inhibit the production of chemokine ligand 1 (CXCL1), and monocyte chemoattractant protein 1 (MCP-1), together with the down-regulation of COX-2 [70]. CXCL1 is one of the potential chemo-attractants that could trigger neutrophils recruitment and activation in inflammation [71]. MCP-1 can attract immunity cells including monocytes, macrophages, and lymphocytes [72]. Both of them play an essential role in the occurring of colitis. The results have suggested that LBP might alleviate the IBD symptoms by restoring the normal level in the gut and decreasing the neutrophil infiltration.

2.13. *Codonopsis pilosula* Polysaccharides

Codonopssi pilosula polysaccharides (CREP) were used as an effective medicine for ulcerative colitis. CREP is comprised of mannose, rhamnose, galacturonic acid, glucose, galactose, and arabinose at the molar ratio of 1.00:3.26:27.87:10.87:7.70:9.94. Tang et al. found that CREP could ameliorate colitis symptoms in the DSS-induced mice combined with APS. In addition to the administrative effect, APS decreases the expression of pro-inflammatory cytokines and supplies CREP in the treatment, causing down-regulation of TJs proteins including occludin, ZO-1, claudins, and MUC-2 [73].

Especially, administration of CREP in the DSS-induced mice leads to the up-regulation of IL-22. IL-22 is one of the cytokines that belongs to the IL-10 family and can protect the gastrointestinal tract in the colitis models. A previous study has confirmed that the blocking IL-22 pathway prolongs recovery in the DSS-induced colitis model. IL-22 makes a crucial protective contribution to intestine integrity. On the one hand, IL-22 could facilitate the goblet cells' secret mucins, which form the stabilized external barrier to sequester the direct interaction between epithelial cells and microorganisms [74]. On the other hand, it also could promote trauma healing through activating the STAT3 pathway in epithelial cells [75]. Therefore, CREP have potential effects to ameliorate IBD symptoms.

2.14. The Polysaccharides from Purple Sweet Potato

Purple sweet potato contains three polysaccharides consisting of water-soluble polysaccharide (WPSPP), dilute alkali-soluble polysaccharide, and concentrated alkali-soluble polysaccharide [76]. The backbone of the alkali-soluble polysaccharide (ASPP) obtained from purple sweet potato is composed of 1,4-linked Glcp with side chains attached to the O-6 position [77]. Sun et al. found that female ICR mice had higher levels of acetate and propionate when administrated with ASPP 400 mg/kg body-weight once a day for 30 days. ASPP inhibited the pro-inflammatory cytokines IL-1β, IL-6, and TNF-α, and the anti-inflammatory property of ASPP was by up-regulating the level of SCFAs [78].

A purified polysaccharide (WPSPP-1) from purple sweet potato was constituted of 1,6-α-d-Glcp, 1,4-α-d-Glcp, 1,2-α-d-Manp, and 1,4,6-α-d-Glcp with β-d-Glcp on the branches, and exhibited a molecular weight of 10^3 kDa [79]. The administration of WPSPP-1 can promote SCFA production including acetate, propionate, and butyrate. It recovered inflammatory lesions through improving IL-10, SOD, and T-AOC levels and reducing IL-1β, TNF-α, IL-6 together with MDA levels. Meanwhile, it increased the ratio of Firmicutes/Bacteroides and decreased the proportion of Proteobacteria, exerting a beneficial effect on the colonic microbiota modulation [54].

2.15. Others

Polysaccharides (TFPS) from the flowers of *Camellia sinensis* L, which are comprised of two kinds of polysaccharides with a molecular weight of 4.4 kDa and 31 kDa [80] exhibit immunostimulant activity. TFPS have a probiotic-promoting effect which works through increasing the ratio of probiotics in IBD, which significantly changes the composition of the metabolite SCFAs. In IBD feces, the relative abundances of *Escherichia/Shigella, Enterococcus, Collinsella, Lactobacillus,* and *Bifidobacterium* were increased, while the relative abundances of *Enterobacter, Ptococcus, Bacteroides, Clostridium XlVa, Megasphaera, Roseburia, Granulicatella, Akkermansia* and *Fusobacterium* decreased [81].

The polysaccharides from *Ficus carica* (FCPS) were considered to be used in nutritional therapy due to their anti-oxidant and anti-inflammatory function. FCPS has a molecular weight of 98.9 kDa and is comprised of rhamnose, arabinose, glucose, galactose as well as galacturonic acid with ratios of 5.8:18.1:3.3:19.3:53.5. The administration of FCPS in DSS-treated mice could down-regulate the mRNA expression of IL-1β, TLR4, and iNOS; the high-dose FCPS essentially decreased the mRNA expression of TNF-α, IL-6, MCP1, and COX-2, which resulted in the inhibition of colon inflammatory response [82].

FCPS treatment could decrease the abundance of *Esherichia* and *Clostridium. Esherichia* always multiplies the danger of colitis and *Clostridium* is related to the production of SCFAs. The administration of FCPS can improve the level of *Prevotella, Bacteroides, Butyricicoccus,* and Coprococcus. All of them have been reported to be beneficial in preventing IBD through down-regulating the expression of SCFAs [83]. The effects of some natural polysaccharides on inflammatory bowel disease are shown in Table 1.

Table 1. Summary of the effects of some natural polysaccharides on inflammatory bowel disease amelioration.

Name	Source	Origin	Molecular Weight (kDa)	Model	Targets	Reference
EP-1	*Hericium erinaceus*	Mushroom	3.1	Acetic acid-induced mice	Recover the level of *Firmicutes, Bacteroidetes, Proteobacteria,* and *Actinobacteria*	[17]
HECP	*Hericium erinaceus*	Mushroom	87	DSS-induced colitis mice	Decrease expression of pro-inflammatory cytokines (IL-6, IL-1β, TNF-α), COX-2, iNOS	[18]
HCP	*Sarcodon aspratus*	Mushroom	670	RAW264.7 cells	Increase the expression of NO, NOS, cytokines and phagocytic activity Stimulate TLR4-MAPK-NFkB signaling pathway	[21]
DP1	*Dictyophora indusiata*	Mushroom	1132	RAW264.7 cells	Increase the level of *Bacteroidaceae* and *Enterobacteriaceae* Reduce the pro-inflammatory cytokines (TNF-α, IL-1β, IL-6, IFN-γ), MPO, NO and enhance anti-inflammatory cytokines (IL-4, IL-10), T-SOD; Inhibit CD86 Increase tight junction proteins including claudin-1, occludin, and ZO-1; and expression of Bcl-2 and Bax	[22–24]
FVP	*Flammuliana velutipes*	Mushroom	dozens to hundreds	DSS-induced colitis mice	Decrease the level of MPO, NO, and increase SOD Raise the ratio of *Firmicutes/Bacteroidetes* and enhance the abundance of *Lachnospiraceae*	[34]
PEP	*Pleurotus eryngii*	[34]	426	C57BL/6 mice	Increase the abundance of *Porphyromonadaceae, Rikenellaceae, Bacteroidaceae,* and *Lactobacillaceae* Decrease SCFA production	[41]
EPA-1	*Pleurotus eryngii*	Mushroom	99.7	RAW 264.7 cells	Increase the production of pro-inflammatory cytokines TNF-α, IL-1, and IL-6 Increase expression level of phosphorylated p38, ERK, JNK, and NF-κB	[43]
SP	*Graciliaria lemaneiformis*	Seaweed		DSS-induced colitis mice	Inhibit activation of MPO Repress intestinal endotoxin and lipopolysaccharide-binding protein production Decrease the expression of TNF-α, IL-6, IL-1β	[44]
BMP	*Blidingia minima*	Seaweed		DSS-induced mice	Decrease the expression of MPO and EPO Increase the expression of IL-10 and decrease the production L-1β and TNF-α, NF-κB, IκB-α, and AKT	[46]
ALP-1	*Arctium lappa* L.	Chinese herbs	5.12	Systemic inflammatory mice	Decrease the IL-1β, IL-6, TNF-α; increase IL-10 and Ig A; Increase the abundance of *Lactobacillaceae, Lachnospiraceae,* and *Ruminococcaceae* and inhibit the abundance of *Bacteroides* and *Staphylococcus*	[45,46]

Table 1. *Cont.*

Name	Source	Origin	Molecular Weight (kDa)	Model	Targets	Reference
ASPP	*Arctium lappa L.*	Chinese herbs	120	LPS-induced inflammatory cell	Increase the expression of IL-10 and decrease MPO Increase the abundance of *Firmicutes, Alistipes, Odoribacter,* and the *Firmicutes/Bacteroides* ratio	[52]
NFP	*Morinda citrifolia*	Chinese herbs	456	DSS-induced mice	Improve the tight junction proteins production including zonula, occludens-1, and occludins	[60]
APS	*Astragalus membranaceus*	Chinese herbs		DSS-induced mice	Down-regulating production of MPO, NF-κB, and pro-inflammatory cytokines (TNF-α, IL-1β, IL-6)	[62]
DOP/EDOP	*Dendrobium officinale*	Chinese herbs		DSS-induced mice	Inhibit NLRP3 inflammasome and β-arrestin1 Decrease the expression of pro-inflammatory cytokines including IL-1β, IL-6, and TNF-α; Activate GPR41/43 signaling pathway Increase the abundance of *Bacteroides, Lactobacillus* plus *Ruminococcaceae*, but down-regulate the levels of *Proteobacteria* and *Akkermansia*	[64]
LBP	*Goji berry*	Chinese herbs	2.1–6.5 × 10^3		Inhibit the expression of CXCL1, MCP-1, COX-2, and IL-6 Increase the abundance of *Akkermansia, Lactobacillus,* and *Prevotellaceae* Increase the expression level of TGF-β and sIgA	[70]
CREP	*Codonopssi pilosula*	Chinese herbs	2.0 × 10^3/7.3	DSS-induced mice	Down-regulation of TJ proteins including Occludin, ZO-1, claudins, and MUC-2 Up-regulation of IL-22	[73]
WPSPP-1	*Purple sweet potato*	Plant	10^3	DSS-induced mice	Increase the expression of pro-inflammatory cytokines IL-10 and decrease the expression of IL-1β, TNF-α, IL-6. Promote the level of SOD and T-AOC, and inhibit the production of MDA	[54]
ASPP	*Purple sweet potato*	Plant		DSS-induced mice	Inhibit the pro-inflammatory cytokines including IL-1β, IL-6, and TNF-α Increase the production of SCFAs	[78]
TFPS	*Camellia sinensis* L.	Plant	4.4/31	Human stool samples	Increase the abundance of *Escherichia/Shigella, Enterococcus, Collinsella, Lactobacillus,* and *Bifidobacterium* Decrease the abundance of *Enterobacter, Streptococcus, Bacteroides, Clostridium XIVa, Megasphaera, Roseburia, Granulicatella, Akkermansia,* and *Fusobacterium*	[81]
FCPS	*Ficus carica*	Plant	98.9	DSS-treated mice	Decrease expression of IL-1β, iNOS, TNF-α, MCP1, IL-6, and COX-2 Decrease the abundance of *Esherichia* and *Clostridium* Increase the abundance of *Prevotella, Bacteroides, Butyricicoccus,* and *Coprococcus*	[83]

3. Mechanisms

Natural polysaccharides regulate the expression of cytokines engaged in the inflammatory responses of IBD through multiple signaling pathways (Figure 1). Polysaccharides could ameliorate IBD symptoms via decreasing the neutrophil infiltration and regulation of oxidative stress. The following is a discussion of the potential mechanism of the effect of polysaccharides on therapy for IBD in Table 2.

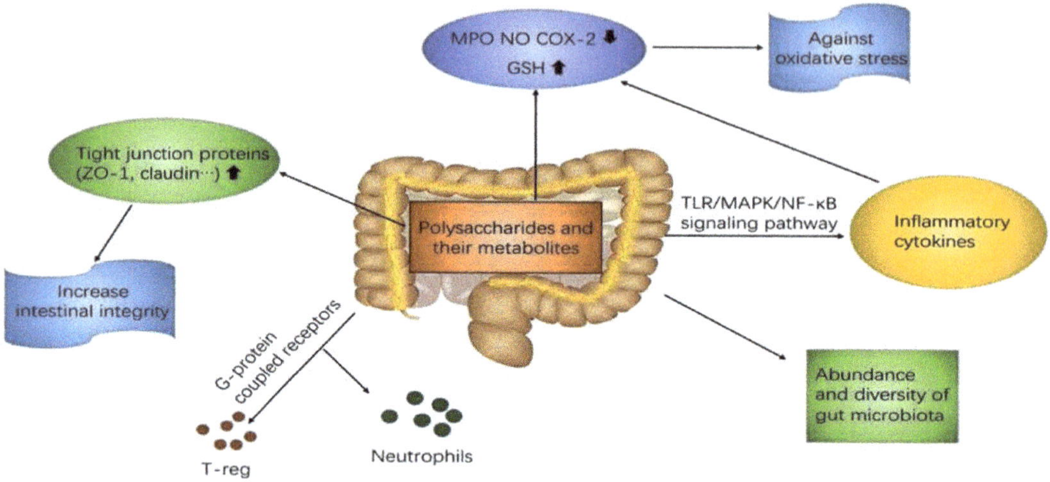

Figure 1. The effects of natural polysaccharides on IBD.

Table 2. The mechanisms of natural polysaccharides involved in colon inflammation amelioration.

Mechanism Involved	Source of Polysaccharides	Results	Reference
MAPK transduction signaling pathway	*Ganoderma lucidum and G. sinense/Pleurotus eryngii*	Increase the phosphorylation level of ERK, JNK, and p38 in macrophage cells	[38,84]
	Dictyophora indusiata	Decrease the phosphorylation level of ERK and pro-inflammatory cytokines in DSS-induced mice	[27]
NF-κB signaling pathway	*Blidingia minima/H. erinaceus/Arctium lappa L./Purple sweet potato*	Restore phosphorylation level of NF-κB, IκB-α, and AKT in colitis mice model	[18,46,53]
G-protein-coupled receptors	*Hericium Erinaceus*	Increase the expression of GPR41/43 in acetic acid-induced colitis mice	[85]
	Dendrobium officinaleon	Inhibit β-arrestin1and NLRP3 inflammasome signaling pathway in DSS-induced mice	[86]
Increase intestinal integrity via upregulate tight junction proteins	*Ficus carica*	Increase the expression of light junction protein claudin-1 and decrease the expression of TNF-α and IL-1β in DSS-induced mice	[82]
	Dictyophora indusiata	Increase the expression of claudin-1, occludin, and ZO-1 in DSS-induced mice	[27]
Regulation of oxidative stress	*Graciliaria lemaneiformis/Flammuliana velutipes*	Decrease the expression of MPO and NO in DSS-induced mice	[34,44]
	Hericium Erinaceus	Decrease the expression of COX-2 in acetic acid-induced mice	[14]
	Graciliaria caudata	Restoration of GSH and decrease of MDA in acetic acid-induced mice	[43]

3.1. TLRs-MAPK/NF-κB Mediated Signal Transduction Pathway

TLRs belong to the pattern recognition receptors and they are widely found on the external sides of immune cells such as macrophages, neutrophils, and lymphocytes. In the immunity signaling pathway, the process of TLR-recruiting adaptors enhances the production of numerous pro-inflammatory cytokines, which is followed by the activation of pro-inflammatory signaling pathways and host defense pathogens by participating in the signal transducer and activator [87]. After binding to polysaccharide ligands, TLRs motivate TNF receptor-associated factor 6 (TRAF6), which consequently activates MAPK and NF-κB. The anti-inflammatory impact of polysaccharide from *Sarcodon aspratus* was checked into LPS induced-RAW264.7 macrophage cells and it turned out to increase the production level of TLR4 as well as the activation of macrophage cells. It revealed that SAP exerts its protective effect on the intestine by enhancing the immunity reaction. Otherwise, when treating DSS-induced colitis mice with FCPS, it not only decreased the TLR4 expression but also down-regulated the manufacturing of pro-inflammatory cytokines including IL-1β, TNF-α, and IL-6 [82]. The outcome explains that the mechanism of FCPS alleviates IBD by inhibiting the immunity reaction.

MAPK is a family of serine/threonine protein kinases that are comprised of c-jun N-terminal kinase (JNK), p38, and extracellular signal-regulated kinase 1/2 (ERK 1/2) [88]. NF-κB is a transcription factor that is involved in the management of the expression of critical immunity and inflammation regulatory genes [89]. NF-κB always keeps an inactive complex form binding with inhibitory protein IκB in the normal condition [38]. As the cells are stimulated, NF-κB is activated and IκB is phosphorylated by IKK. Then, the activated complex form is transferred into the nucleus and regulates the expression of inflammatory cytokines and proteins. The underlying mechanism of polysaccharide- regulated intestine protective effects is considerably connected with the governed expression of MAPK and NF-κB. For example, when either of the polysaccharides obtained from *Ganoderma lucidum* and *G. sinense* or *Pleurotus eryngii* was administrated with LPS-induced RAW264.7 macrophage cells, the results revealed that all of the phosphorylation level of ERK, JNK, p38, and NF-κB were up-regulated [38,84], which indicated that all of these polysaccharides enhance the immune regulation effects. In another study, treating a high dose of DIP could significantly reduce the phosphorylation level of ERK and NF-κB and pro-inflammatory cytokines in DSS-induced mice [23]. These results are consistent with the previous results that polysaccharides have immunomodulatory activities on the immune cells, and further confirm that polysaccharides could ameliorate IBD via modulating the MAPK and NF-κB signaling pathway.

3.2. G-Protein-Coupled Receptors

G-protein-coupled receptors (GPRs) are a family of receptors for different second messengers. It has been demonstrated that GPRs highly expressed in the gastrointestinal tract could ameliorate IBD by promoting the colon epithelium repair [90] and improving the glucagon-like peptide-2 production [91]. In the previous study, GPR43 has been found in the neutrophils and colon [92]. GPR43 is vital to neutrophil recruitment in intestinal inflammation and their deficiency induces severe inflammation. SCFAs would bind to GPR41/43 and influence anti-inflammation activity in the cells [85]. SCFAs metabolized from natural polysaccharides also stimulate the GPR43 on colonic T cells followed by up-regulating the expression of Foxp3 [90] and facilitating peripheral T-reg cells, which finally inhibits the inflammatory response in the intestinal mucosa [35]. Otherwise, activated GPR43 involves the NF-κB pathway, which inhibits the expression of pro-inflammatory cytokines such as IL-6 and IL-1β. It also indicates that GPR43 is beneficial to attenuate the inflammation responses in the gut [93]. β-arrestin1, one of the important regulators of the GPR signaling pathway [94], makes a crucial contribution in the activation of nucleotide leucine repeat pyrin 3 (NLRP3) inflammasome [95]. NLRP3 inflammasome exerts importance in the inflammation and immunity responses via the caspase-1 activation and IL-1β maturation. The motivation of NLRP3 inflammasome could aggravate gut inflammation [96] and the variant of NLRP3 in the cells would trigger severe IBD [86]. Therefore, the pro-inflammatory effects of β-arrestin1 and NLRP3 were mediated through the cytokines in the inflammatory intestine like the DSS-induced colitis models. However, the researchers have offered evidence that administration of *Dendrobium officinaleon* polysaccharides (DOPS) ameliorated the injury of the colon and moderated the symptoms of colitis including the rapid body weight loss and the somatotype decreased in DSS-induced mice. The inhibition effect of DOPS to the NLRP3 inflammasome pathway was showed by down-regulation of the IL-1β, IL-18, and caspase-1 expression. The results also displayed that DOPS could suppress the β-arrestin1 signaling pathway. Moreover, in vitro experiments indicated that DOPS significantly inhibit their expressions in the LPS-stimulated NCM 460 cells, the results were the same as before in vivo [97]. Above all the results have indicated that natural polysaccharides could regulate the expression of GPRs on behalf of alleviating the IBD.

3.3. *Increase Intestinal Integrity via Up-regulating Tight Junction Proteins*

Gut integrity is vital to maintaining the host's innate immunity, which could prevent intestine damage from lipopolysaccharides (LPS) and toxins [98]. The tight junction proteins such as the ZO family, claudins, and occludins are responsible for the diffusion of epithelium and among cells. Numerous inflammatory responses would affect the tight junction proteins expressions and since colitis makes the levels change, the expression of tight junction proteins is always decreased in IBD [99]. TNF-α plays an important role in the mediation of occludin internalization, and the occludin has a positive relationship with gut permeability. Although the cytokines increase the gut permeability, overexpression of occludin could ameliorate it [100]. Moreover, the other mechanism involved in gut integrity regulation is TNF-α activation of the NF-κB signal transduction pathway. The inhibition of NF-κB could prevent colitis mice from having diarrhea and serious loss of water, which demonstrated the modulatory function of NF-κB in barrier capability [101]. In the colitis patient, TNF-α and IL-1β not only increase the infiltration of neutrophils, but also cause damage to the intestinal barrier, which may induce diarrhea symptoms in patients [102]. Otherwise, TNF-α attracts the myosin light-chain phosphorylation, which damages the tight junction protein. Then the risky factors and microbes like LPS, the signals from gram-negative bacteria, could enter the intestine through the mucous membrane leak, which would bind to toll-like receptors and stimulate the TLR-mediated signaling pathway [103], further provoking the inflammatory response. IFN-γ diminishes the ZO-1 and occluding expression through the adenosine monophosphate-activated protein kinase dependent pathway without regard to the energy level of the cell. Therefore, it also could increase barrier permeability [104]. FCPS treatment could ameliorate colitis symptoms by restoring the expression of light junction protein claudin-1 and down-regulating the level of TNF-α and IL-1β in the DSS-induced colitis mice [82]. The animal models' results proved that FCPS could facilitate the repair of damaged intestinal barriers, thus blocking the penetration of endangering molecules and microbes through the mucous membrane [105]. In addition, the gastrointestinal protective effect of DIP has been reported to recover the expression level of claudin-1, occludin, and ZO-1 in the DSS-induced colitis mice [27]. These studies indicated that despite the fact that these inflammatory cytokines are always presented in the intestine at the same time during inflammation and harmful to the patient due to the disconnection of the tight junction proteins [106], the natural polysaccharides could restrain the response of colon inflammation and promote IBD recovery via increasing the tight junction protein expression in the gut and thereby protect the colon structure of the IBD patient (Figure 2).

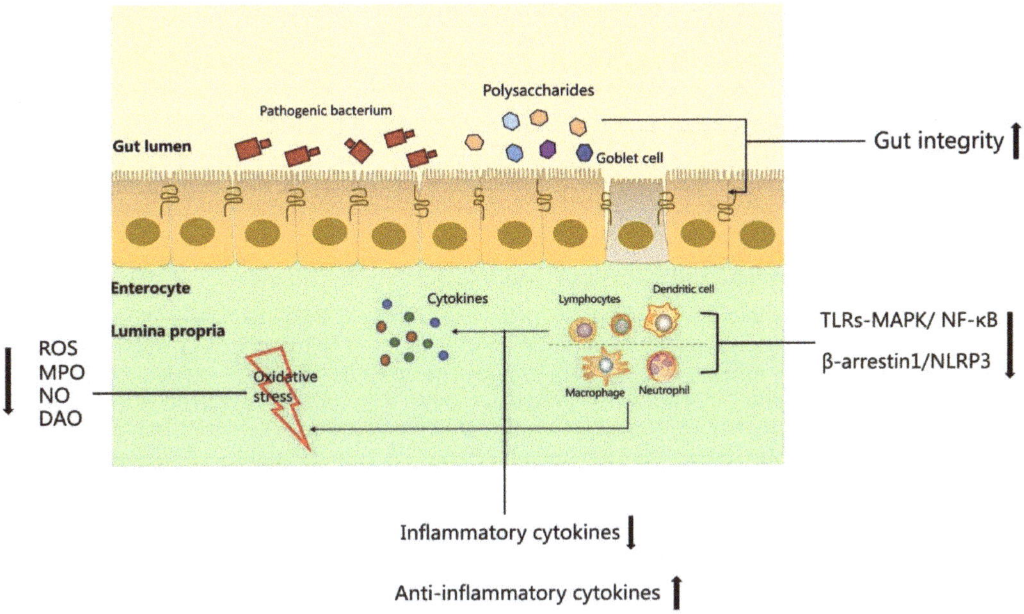

Figure 2. The summary of mechanisms about polysaccharides involved in IBD patient intestine.

3.4. Regulation of Oxidative Stress

In the intestine, immunity plays a critical role in the pathogenesis of IBD. As the inflammation induces the T-cell response through the secretion of inflammatory cytokines, the other activated immune cells, including macrophages and neutrophils [107], induce significantly increased production of oxygen free radicals, which cause further damage in colon tissue. The synthesis of oxygen free radicals and the antioxidase in the intestine, including ROS, myeloperoxidase (MPO), nitric oxide (NO), malondialdehyde (MDA), plasmic diamine oxidase (DAO), and glutathione (GSH). ROS have various physiological functions but under the condition of oxidative stress, overproduction of it is harmful to intestine cell membrane lipids [108], DNA, and proteins. ROS also are responsible for causing diarrhea, one of the most common symptoms in IBD, due to the excessive secretion of water and electrolytes [109]. MPO, as one of the reactive oxygen species, is always regarded as the inflammatory marker of neutrophil infiltration. However, polysaccharides might ameliorate damage through inhibiting neutrophil infiltration in the gut. Han et al. have treated DSS-induced mice with SP and found the expression level of MPO significantly decreased in the gut [44].

After being stimulated by oxidative stress, iNOS belonging to reactive oxygen/nitrogen species (ROS/RNS)-generating systems catalyzes the production of NO. As one of the markers of inflammation, NO is released by the neutrophils in damaged tissues and induces an immunity inflammatory response, whose generation has a positive relationship with the level of oxygen free radicals [26]. Prior research has shown that administrated FVP would lead to a decline in NO content in the gut of DSS-induced colitis models [34]. Except for promoting the production of NO, iNOS also improves the expression of COX-2 in the meanwhile. COX-2 could catalyze prostaglandin production and finally induce the inflammation response. Shao et al. administrated acetic acid-induced mice with EP-1 and found both the expression of COX-2 and related mRNA were down-regulated, which might prove that polysaccharides could balance the immunity in the intestinal system via regulating the oxidative stress, following subsequently moderate IBD [14]. The above results have indicated that polysaccharides make a critical contribution to the down-regulation of reactive oxygen species and therefore exert an essentially protective effect on the colon of patients.

The different concentration of MDA and GSH in the tissues has revealed that diverse levels of oxidants caused by the epithelial cell rupture. Dutra et al. have confirmed that *G. caudata* PLS suppresses the acetic acid-induced UC mice [43], which induces the restoration of the GSH and down-regulation of MDA level in the cells. In the pathogenesis of IBD, above all the results have proved that polysaccharides could decrease the production of reactive oxygen species and their metabolites as well as increase the level of anti-oxidant molecules meanwhile to ameliorate IBD that is caused by prooxidant substances.

4. Future Outlooks and Conclusions

Natural polysaccharides have the potential ability to be used in IBD treatment. Many types of polysaccharides from mushroom, seaweed, herbs, and plants can be fermented in the colon, which not only changes the diversity of gut microbiota, but also recovers gut health via stimulating various types of immune cells and motivating numerous immunity-related signaling pathways. The main mechanism of polysaccharides on inflammatory bowel disease relies on immune regulation, anti-oxidation, and regulation of probiotics in the intestine.

Natural polysaccharides have attracted considerable attention because of their low side effects, nontoxicity to individuals, and easy availability in diet. However, further examination is necessary to explore the link between natural polysaccharides and immune regulation in diseases, together with analyzing the variety of microbiota in the intestine. Systemic studies would help create better understanding of the specific mechanism that the function of polysaccharides exert on diseases. Thus, complementary alternative treatments could be provided to IBD patients. Additionally, for the purpose of better applying polysaccharides to normal immunity recovery, more clinical work is needed on the related foods and drugs to ensure the safety of dosage and half-life time. More innovative polysaccharide-based foods and medicines for IBD can be expected under a comprehensive understanding of the structure, biological activities, and underlying mechanism of polysaccharides.

Author Contributions: Conceptualization, Y.W.; Writing—original draft preparation, Y.W.; Validation, H.Z.; Investigation, Y.Y.; Visualization, Writing—review, X.W.; Writing—review, supervision and editing, J.X. All authors have read and agreed to the published version of the manuscript.

Funding: This research received no external funding.

Institutional Review Board Statement: Studies not involving humans or animals.

Informed Consent Statement: Patient consent was waived.

Data Availability Statement: Data is contained within the article.

Conflicts of Interest: The authors declare no conflict of interest.

References

1. Ramos, G.P.; Papadakis, K.A. Mechanisms of Disease: Inflammatory Bowel Diseases. *Mayo Clin. Proc.* **2019**, *94*, 155–165. [CrossRef]
2. Ng, S.C.; Shi, H.Y.; Hamidi, N.; Underwood, F.E.; Tang, W.; Benchimol, E.I.; Panaccione, R.; Ghosh, S.; Wu, J.C.Y.; Chan, F.K.L.; et al. Worldwide incidence and prevalence of inflammatory bowel disease in the 21st century: A systematic review of population-based studies. *Lancet* **2017**, *390*, 2769–2778. [CrossRef]
3. Popivanova, B.K.; Kitamura, K.; Wu, Y.; Kondo, T.; Kagaya, T.; Kaneko, S.; Oshima, M.; Fujii, C.; Mukaida, N. Blocking TNF-alpha in mice reduces colorectal carcinogenesis associated with chronic colitis. *J. Clin. Investig.* **2008**, *118*, 560–570. [CrossRef] [PubMed]
4. Elinav, E.; Thaiss, C.A.; Flavell, R.A. Analysis of microbiota alterations in inflammasome-deficient mice. *Methods Mol. Biol.* **2013**, *1040*, 185–194. [CrossRef]
5. Mar, J.S.; Nagalingam, N.A.; Song, Y.; Onizawa, M.; Lee, J.W.; Lynch, S.V. Amelioration of DSS-induced murine colitis by VSL#3 supplementation is primarily associated with changes in ileal microbiota composition. *Gut Microbes* **2014**, *5*, 494–503. [CrossRef]
6. Kaur, M.; Dalal, R.L.; Shaffer, S.; Schwartz, D.A.; Rubin, D.T. Inpatient Management of Inflammatory Bowel Disease-Related Complications. *Clin. Gastroenterol. Hepatol.* **2020**, *18*, 1346–1355. [CrossRef] [PubMed]

7. Niu, W.; Chen, X.; Xu, R.; Dong, H.; Yang, F.; Wang, Y.; Zhang, Z.; Ju, J. Polysaccharides from natural resources exhibit great potential in the treatment of ulcerative colitis: A review. *Carbohydr. Polym.* **2021**, *254*, 117189. [CrossRef]
8. Kostic, A.D.; Xavier, R.J.; Gevers, D. The microbiome in inflammatory bowel disease: Current status and the future ahead. *Gastroenterology* **2014**, *146*, 1489–1499. [CrossRef] [PubMed]
9. Shao, P.; Chen, M.; Pei, Y.; Sun, P. In intro antioxidant activities of different sulfated polysaccharides from chlorophytan seaweeds *Ulva fasciata*. *Int. J. Biol. Macromol.* **2013**, *59*, 295–300. [CrossRef] [PubMed]
10. Kaser, A.; Zeissig, S.; Blumberg, R.S. Inflammatory bowel disease. *Annu. Rev. Immunol.* **2010**, *28*, 573–621. [CrossRef]
11. Kany, S.; Vollrath, J.T.; Relja, B. Cytokines in Inflammatory Disease. *Int. J. Mol. Sci.* **2019**, *20*, 6008. [CrossRef] [PubMed]
12. Tian, T.; Wang, Z.; Zhang, J. Pathomechanisms of Oxidative Stress in Inflammatory Bowel Disease and Potential Antioxidant Therapies. *Oxid. Med. Cell. Longev.* **2017**, *2017*, 1–18. [CrossRef]
13. Wang, M.; Gao, Y.; Xu, D.; Gao, Q. A polysaccharide from cultured mycelium of *Hericium erinaceus* and its anti-chronic atrophic gastritis activity. *Int. J. Biol. Macromol.* **2015**, *81*, 656–661. [CrossRef]
14. Shao, S.; Wang, D.; Zheng, W.; Li, X.; Zhang, H.; Zhao, D.; Wang, M. A unique polysaccharide from *Hericium erinaceus* mycelium ameliorates acetic acid-induced ulcerative colitis rats by modulating the composition of the gut microbiota, short chain fatty acids levels and GPR41/43 respectors. *Int. Immunopharmacol.* **2019**, *71*, 411–422. [CrossRef]
15. Song, Q.; Wang, Y.; Huang, L.; Shen, M.; Yu, Y.; Yu, Q.; Chen, Y.; Xie, J. Review of the relationships among polysaccharides, gut microbiota, and human health. *Food Res. Int.* **2021**, *140*, 109858. [CrossRef] [PubMed]
16. Wang, X.; Wang, X.; Jiang, H.; Cai, C.; Li, G.; Hao, J.; Yu, G. Marine polysaccharides attenuate metabolic syndrome by fermentation products and altering gut microbiota: An overview. *Carbohydr. Polym.* **2018**, *195*, 601–612. [CrossRef]
17. Pryde, S.E.; Duncan, S.H.; Hold, G.L.; Stewart, C.S.; Flint, H.J. The microbiology of butyrate formation in the human colon. *FEMS Microbiol. Lett.* **2002**, *217*, 133–139. [CrossRef] [PubMed]
18. Ren, Y.; Geng, Y.; Du, Y.; Li, W.; Lu, Z.M.; Xu, H.Y.; Xu, G.H.; Shi, J.S.; Xu, Z.H. Polysaccharide of *Hericium erinaceus* attenuates colitis in C57BL/6 mice via regulation of oxidative stress, inflammation-related signaling pathways and modulating the composition of the gut microbiota. *J. Nutr. Biochem.* **2018**, *57*, 67–76. [CrossRef] [PubMed]
19. Zhu, H.; Li, Y.R. Oxidative stress and redox signaling mechanisms of inflammatory bowel disease: Updated experimental and clinical evidence. *Exp. Biol. Med.* **2012**, *237*, 474–480. [CrossRef]
20. Chen, Y.C.; Yang, L.L.; Lee, T.J.F. Oroxylin A inhibition of lipopolysaccharide-induced iNOS and COX-2 gene expression via suppression of nuclear factor-κB activation. *Biochem. Pharmacol.* **2000**, *59*, 1445–1457. [CrossRef]
21. Chen, J.; Liu, J.; Yan, C.; Zhang, C.; Pan, W.; Zhang, W.; Lu, Y.; Chen, L.; Chen, Y. Sarcodon aspratus polysaccharides ameliorated obesity-induced metabolic disorders and modulated gut microbiota dysbiosis in mice fed a high-fat diet. *Food Funct.* **2020**, *11*, 2588–2602. [CrossRef]
22. Liao, W.; Luo, Z.; Liu, D.; Ning, Z.; Yang, J.; Ren, J. Structure characterization of a novel polysaccharide from *Dictyophora indusiata* and its macrophage immunomodulatory activities. *J. Agric. Food Chem.* **2015**, *63*, 535–544. [CrossRef] [PubMed]
23. Kanwal, S.; Joseph, T.P.; Aliya, S.; Song, S.; Saleem, M.Z.; Nisar, M.A.; Wang, Y.; Meyiah, A.; Ma, Y.; Xin, Y. Attenuation of DSS induced colitis by *Dictyophora indusiata* polysaccharide (DIP) via modulation of gut microbiota and inflammatory related signaling pathways. *J. Funct. Foods* **2020**, *64*, 103641. [CrossRef]
24. Wang, Y.; Ji, X.; Yan, M.; Chen, X.; Kang, M.; Teng, L.; Wu, X.; Chen, J.; Deng, C. Protective effect and mechanism of polysaccharide from *Dictyophora indusiata* on dextran sodium sulfate-induced colitis in C57BL/6 mice. *Int. J. Biol. Macromol.* **2019**, *140*, 973–984. [CrossRef]
25. Mills, C.D. Anatomy of a discovery: m1 and m2 macrophages. *Front. Immunol.* **2015**, *6*, 212. [CrossRef] [PubMed]
26. Li, R.; Chen, Y.; Shi, M.; Xu, X.; Zhao, Y.; Wu, X.; Zhang, Y. Gegen Qinlian decoction alleviates experimental colitis via suppressing TLR4/NF-κB signaling and enhancing antioxidant effect. *Phytomedicine* **2016**, *23*, 1012–1020. [CrossRef]
27. Kanwal, S.; Joseph, T.P.; Owusu, L.; Xiaomeng, R.; Meiqi, L.; Yi, X. A Polysaccharide Isolated from *Dictyophora indusiata* Promotes Recovery from Antibiotic-Driven Intestinal Dysbiosis and Improves Gut Epithelial Barrier Function in a Mouse Model. *Nutrients* **2018**, *10*, 1003. [CrossRef]
28. Antonopoulos, D.A.; Huse, S.M.; Morrison, H.G.; Schmidt, T.M.; Sogin, M.L.; Young, V.B. Reproducible community dynamics of the gastrointestinal microbiota following antibiotic perturbation. *Infect. Immun.* **2009**, *77*, 2367–2375. [CrossRef]
29. Shao, P.; Chen, X.; Sun, P. Chemical characterization, antioxidant and antitumor activity of sulfated polysaccharide from *Sargassum horneri*. *Carbohydr. Polym.* **2014**, *105*, 260–269. [CrossRef]
30. Kronman, M.P.; Zaoutis, T.E.; Haynes, K.; Feng, R.; Coffin, S.E. Antibiotic exposure and IBD development among children: A population-based cohort study. *Pediatrics* **2012**, *130*, e794–e803. [CrossRef]
31. Besednova, N.N.; Zaporozhets, T.S.; Kuznetsova, T.A.; Makarenkova, I.D.; Kryzhanovsky, S.P.; Fedyanina, L.N.; Ermakova, S.P. Extracts and Marine Algae Polysaccharides in Therapy and Prevention of Inflammatory Diseases of the Intestine. *Mar. Drugs* **2020**, *18*, 289. [CrossRef]
32. Kodama, N.; Murata, Y.; Nanba, H. Administration of a Polysaccharide from *Grifola frondosa* Stimulates Immune Function of Normal Mice. *J. Med. Food* **2004**, *7*, 141–145. [CrossRef] [PubMed]
33. Tang, C.; Hoo, P.C.; Tan, L.T.; Pusparajah, P.; Khan, T.M.; Lee, L.H.; Goh, B.H.; Chan, K.G. Golden Needle Mushroom: A Culinary Medicine with Evidenced-Based Biological Activities and Health Promoting Properties. *Front. Pharmacol.* **2016**, *7*, 474. [CrossRef] [PubMed]

34. Zhang, R.; Yuan, S.; Ye, J.; Wang, X.; Zhang, X.; Shen, J.; Yuan, M.; Liao, W. Polysaccharide from *Flammuliana velutipes* improves colitis via regulation of colonic microbial dysbiosis and inflammatory responses. *Int. J. Biol. Macromol.* **2020**, *149*, 1252–1261. [CrossRef]
35. Furusawa, Y.; Obata, Y.; Fukuda, S.; Endo, T.A.; Nakato, G.; Takahashi, D.; Nakanishi, Y.; Uetake, C.; Kato, K.; Kato, T.; et al. Commensal microbe-derived butyrate induces the differentiation of colonic regulatory T cells. *Nature* **2013**, *504*, 446–450. [CrossRef]
36. Morampudi, V.; Dalwadi, U.; Bhinder, G.; Sham, H.P.; Gill, S.K.; Chan, J.; Bergstrom, K.S.; Huang, T.; Ma, C.; Jacobson, K.; et al. The goblet cell-derived mediator RELM-beta drives spontaneous colitis in Muc2-deficient mice by promoting commensal microbial dysbiosis. *Mucosal Immunol.* **2016**, *9*, 1218–1233. [CrossRef]
37. Li, S.; Shah, N.P. Anti-inflammatory and anti-proliferative activities of natural and sulphonated polysaccharides from *Pleurotus eryngii*. *J. Funct. Foods* **2016**, *23*, 80–86. [CrossRef]
38. Xu, D.; Wang, H.; Zheng, W.; Gao, Y.; Wang, M.; Zhang, Y.; Gao, Q. Charaterization and immunomodulatory activities of polysaccharide isolated from *Pleurotus eryngii*. *Int. J. Biol. Macromol.* **2016**, *92*, 30–36. [CrossRef]
39. Brand, S.; Beigel, F.; Olszak, T.; Zitzmann, K.; Eichhorst, S.T.; Otte, J.M.; Diepolder, H.; Marquardt, A.; Jagla, W.; Popp, A.; et al. IL-22 is increased in active Crohn's disease and promotes proinflammatory gene expression and intestinal epithelial cell migration. *Am. J. Physiol. Gastrointest. Liver Physiol.* **2006**, *290*, G827–G838. [CrossRef]
40. Ma, G.; Kimatu, B.M.; Zhao, L.; Yang, W.; Pei, F.; Hu, Q. In vivo fermentation of a *Pleurotus eryngii* polysaccharide and its effects on fecal microbiota composition and immune response. *Food Funct.* **2017**, *8*, 1810–1821. [CrossRef]
41. Gericke, M.; Heinze, T. Homogeneous tosylation of agarose as an approach toward novel functional polysaccharide materials. *Carbohydr. Polym.* **2015**, *127*, 236–245. [CrossRef]
42. Alencar, P.O.C.; Lima, G.C.; Barros, F.C.N.; Costa, L.E.C.; Ribeiro, C.V.P.E.; Sousa, W.M.; Sombra, V.G.; Abreu, C.M.W.S.; Abreu, E.S.; Pontes, E.O.B.; et al. A novel antioxidant sulfated polysaccharide from the algae *Gracilaria caudata*: In vitro and in vivo activities. *Food Hydrocoll.* **2019**, *90*, 28–34. [CrossRef]
43. Dutra, N.L.S.; de Brito, T.V.; de Aguiar Magalhães, D.; Sousa, S.G.; Batista, J.A.; Pereira, C.M.C.; dos Santos Ferreira, J.; da Rocha Rodrigues, L.; do Nascimento Lima, J.V.; de Albuquerque, I.F.; et al. Sulfated polysaccharide extracted from seaweed Gracilaria caudata attenuates acetic acid-induced ulcerative colitis. *Food Hydrocoll.* **2021**, *111*, 106221. [CrossRef]
44. Han, R.; Wang, L.; Zhao, Z.; You, L.; Pedisić, S.; Kulikouskaya, V.; Lin, Z. Polysaccharide from *Gracilaria Lemaneiformis* prevents colitis in Balb/c mice via enhancing intestinal barrier function and attenuating intestinal inflammation. *Food Hydrocoll.* **2020**, *109*, 106048. [CrossRef]
45. Jiao, L.; Li, X.; Li, T.; Jiang, P.; Zhang, L.; Wu, M.; Zhang, L. Characterization and anti-tumor activity of alkali-extracted polysaccharide from *Enteromorpha intestinalis*. *Int. Immunopharmacol.* **2009**, *9*, 324–329. [CrossRef] [PubMed]
46. Song, W.; Li, Y.; Zhang, X.; Wang, Z. Potent anti-inflammatory activity of polysaccharides extracted from *Blidingia minima* and their effect in a mouse model of inflammatory bowel disease. *J. Funct. Foods* **2019**, *61*, 103494. [CrossRef]
47. Shi, L. Bioactivities, isolation and purification methods of polysaccharides from natural products: A review. *Int. J. Biol. Macromol.* **2016**, *92*, 37–48. [CrossRef] [PubMed]
48. Wang, Y.; Zhang, N.; Kan, J.; Zhang, X.; Wu, X.; Sun, R.; Tang, S.; Liu, J.; Qian, C.; Jin, C. Structural characterization of water-soluble polysaccharide from *Arctium lappa* and its effects on colitis mice. *Carbohydr. Polym.* **2019**, *213*, 89–99. [CrossRef]
49. Zhang, Y.; Wang, J.-H.; Luo, J.-P.; Zhang, Q.; Lu, J. The Structure-Activity Relationship and Molecular Mechanism of Anti-tumor Polysaccharide Isolated from *Dendrobium Nobile* Lindl. *Curr. Top. Nutraceutical Res.* **2019**, *17*, 153–163.
50. Zhang, N.; Wang, Y.; Kan, J.; Wu, X.; Zhang, X.; Tang, S.; Sun, R.; Liu, J.; Qian, C.; Jin, C. In vivo and in vitro anti-inflammatory effects of water-soluble polysaccharide from *Arctium lappa*. *Int. J. Biol. Macromol.* **2019**, *135*, 717–724. [CrossRef]
51. Zou, M.; Chen, Y.; Sun-Waterhouse, D.; Zhang, Y.; Li, F. Immunomodulatory acidic polysaccharides from *Zizyphus jujuba* cv. Huizao: Insights into their chemical characteristics and modes of action. *Food Chem.* **2018**, *258*, 35–42. [CrossRef] [PubMed]
52. Zhang, X.; Zhang, N.; Kan, J.; Sun, R.; Tang, S.; Wang, Z.; Chen, M.; Liu, J.; Jin, C. Anti-inflammatory activity of alkali-soluble polysaccharides from *Arctium lappa* L. and its effect on gut microbiota of mice with inflammation. *Int. J. Biol. Macromol.* **2020**, *154*, 773–787. [CrossRef] [PubMed]
53. Hsieh, C.Y.; Osaka, T.; Moriyama, E.; Date, Y.; Kikuchi, J.; Tsuneda, S. Strengthening of the intestinal epithelial tight junction by *Bifidobacterium bifidum*. *Physiol. Rep.* **2015**, *3*, e12327. [CrossRef] [PubMed]
54. Gou, Y.; Sun, J.; Liu, J.; Chen, H.; Kan, J.; Qian, C.; Zhang, N.; Jin, C. Structural characterization of a water-soluble purple sweet potato polysaccharide and its effect on intestinal inflammation in mice. *J. Funct. Foods* **2019**, *61*, 103502. [CrossRef]
55. Powell, N.; Walker, A.W.; Stolarczyk, E.; Canavan, J.B.; Gokmen, M.R.; Marks, E.; Jackson, I.; Hashim, A.; Curtis, M.A.; Jenner, R.G.; et al. The transcription factor T-bet regulates intestinal inflammation mediated by interleukin-7 receptor+ innate lymphoid cells. *Immunity* **2012**, *37*, 674–684. [CrossRef]
56. Rahat, M.A.; Hemmerlein, B. Macrophage-tumor cell interactions regulate the function of nitric oxide. *Front. Physiol.* **2013**, *4*, 144. [CrossRef]
57. Tham, C.S.; Whitaker, J.; Luo, L.; Webb, M. Inhibition of microglial fatty acid amide hydrolase modulates LPS stimulated release of inflammatory mediators. *FEBS Lett.* **2007**, *581*, 2899–2904. [CrossRef]

58. Meng, Z.; Yan, C.; Deng, Q.; Gao, D.F.; Niu, X.L. Curcumin inhibits LPS-induced inflammation in rat vascular smooth muscle cells in vitro via ROS-relative TLR4-MAPK/NF-κB pathways. *Acta Pharmacol. Sin.* **2013**, *34*, 901–911. [CrossRef]
59. Ahmad, A.N.; Mat Daud, Z.A.; Ismail, A. Review on potential therapeutic effect of *Morinda citrifolia* L. *Curr. Opin. Food Sci.* **2016**, *8*, 62–67. [CrossRef]
60. Jin, M.; Wang, Y.; Yang, X.; Yin, H.; Nie, S.; Wu, X. Structure characterization of a polysaccharide extracted from noni (*Morinda citrifolia* L.) and its protective effect against DSS-induced bowel disease in mice. *Food Hydrocoll.* **2019**, *90*, 189–197. [CrossRef]
61. Han, Y.; Son, S.J.; Akhalaia, M.; Platonov, A.; Son, H.J.; Lee, K.H.; Yun, Y.S.; Song, J.Y. Modulation of radiation-induced disturbances of antioxidant defense systems by ginsan. *Evid. Based Complementary Altern. Med.* **2005**, *2*, 529–536. [CrossRef]
62. Lv, J.; Zhang, Y.; Tian, Z.; Liu, F.; Shi, Y.; Liu, Y.; Xia, P. *Astragalus* polysaccharides protect against dextran sulfate sodium-induced colitis by inhibiting NF-kB activation. *Int. J. Biol. Macromol.* **2017**, *98*, 723–729. [CrossRef]
63. Guo, L.; Qi, J.; Du, D.; Liu, Y.; Jiang, X. Current advances of *Dendrobium officinale* polysaccharides in dermatology: A literature review. *Pharm. Biol.* **2020**, *58*, 664–673. [CrossRef]
64. Zhang, Y.; Wu, Z.; Liu, J.; Zheng, Z.; Li, Q.; Wang, H.; Chen, Z.; Wang, K. Identification of the core active structure of a *Dendrobium officinale* polysaccharide and its protective effect against dextran sulfate sodium-induced colitis via alleviating gut microbiota dysbiosis. *Food Res. Int.* **2020**, *137*, 109641. [CrossRef]
65. Yilmaz, B.; Juillerat, P.; Oyas, O.; Ramon, C.; Bravo, F.D.; Franc, Y.; Fournier, N.; Michetti, P.; Mueller, C.; Geuking, M.; et al. Microbial network disturbances in relapsing refractory Crohn's disease. *Nat. Med.* **2019**, *25*, 323–336. [CrossRef] [PubMed]
66. Han, Y.; Song, M.; Gu, M.; Ren, D.; Zhu, X.; Cao, X.; Li, F.; Wang, W.; Cai, X.; Yuan, B.; et al. Dietary Intake of Whole Strawberry Inhibited Colonic Inflammation in Dextran-Sulfate-Sodium-Treated Mice via Restoring Immune Homeostasis and Alleviating Gut Microbiota Dysbiosis. *J. Agric. Food Chem.* **2019**, *67*, 9168–9177. [CrossRef] [PubMed]
67. Zhang, Y.; Pan, X.; Ran, S.; Wang, K. Purification, structural elucidation and anti-inflammatory activity in vitro of polysaccharides from *Smilax china* L. *Int. J. Biol. Macromol.* **2019**, *139*, 233–243. [CrossRef] [PubMed]
68. Wu, D.-T.; Guo, H.; Lin, S.; Lam, S.-C.; Zhao, L.; Lin, D.-R.; Qin, W. Review of the structural characterization, quality evaluation, and industrial application of *Lycium barbarum* polysaccharides. *Trends Food Sci. Technol.* **2018**, *79*, 171–183. [CrossRef]
69. Xie, J.H.; Jin, M.L.; Morris, G.A.; Zha, X.Q.; Chen, H.Q.; Yi, Y.; Li, J.E.; Wang, Z.J.; Gao, J.; Nie, S.P.; et al. Advances on Bioactive Polysaccharides from Medicinal Plants. *Crit. Rev. Food Sci. Nutr.* **2016**, *56* (Suppl. 1), S60–S84. [CrossRef]
70. Kang, Y.; Xue, Y.; Du, M.; Zhu, M.J. Preventive effects of Goji berry on dextran-sulfate-sodium-induced colitis in mice. *J. Nutr. Biochem.* **2017**, *40*, 70–76. [CrossRef] [PubMed]
71. Cao, J.J.; Lv, Q.Q.; Zhang, B.; Chen, H.Q. Structural characterization and hepatoprotective activities of polysaccharides from the leaves of *Toona sinensis* (A. Juss) Roem. *Carbohydr. Polym.* **2019**, *212*, 89–101. [CrossRef] [PubMed]
72. Metzemaekers, M.; Gouwy, M.; Proost, P. Neutrophil chemoattractant receptors in health and disease: Double-edged swords. *Cell. Mol. Immunol.* **2020**, *17*, 433–450. [CrossRef]
73. Tang, S.; Liu, W.; Zhao, Q.; Li, K.; Zhu, J.; Yao, W.; Gao, X. Combination of polysaccharides from *Astragalus membranaceus* and *Codonopsis pilosula* ameliorated mice colitis and underlying mechanisms. *J. Ethnopharmacol.* **2021**, *264*, 113280. [CrossRef]
74. Sugimoto, K.; Ogawa, A.; Mizoguchi, E.; Shimomura, Y.; Andoh, A.; Bhan, A.K.; Blumberg, R.S.; Xavier, R.J.; Mizoguchi, A. IL-22 ameliorates intestinal inflammation in a mouse model of ulcerative colitis. *J. Clin. Investig.* **2008**, *118*, 534–544. [CrossRef] [PubMed]
75. Neufert, C.; Pickert, G.; Zheng, Y.; Wittkopf, N.; Warntjen, M.; Nikolaev, A.; Ouyang, W.; Neurath, M.F.; Becker, C. Activation of epithelial STAT3 regulates intestinal homeostasis. *Cell Cycle* **2010**, *9*, 652–655. [CrossRef]
76. Sun, J.; Zhou, B.; Tang, C.; Gou, Y.; Chen, H.; Wang, Y.; Jin, C.; Liu, J.; Niu, F.; Kan, J.; et al. Characterization, antioxidant activity and hepatoprotective effect of purple sweet potato polysaccharides. *Int. J. Biol. Macromol.* **2018**, *115*, 69–76. [CrossRef]
77. Chen, H.; Sun, J.; Liu, J.; Gou, Y.; Zhang, X.; Wu, X.; Sun, R.; Tang, S.; Kan, J.; Qian, C.; et al. Structural characterization and anti-inflammatory activity of alkali-soluble polysaccharides from purple sweet potato. *Int. J. Biol. Macromol.* **2019**, *131*, 484–494. [CrossRef]
78. Sun, J.; Chen, H.; Kan, J.; Gou, Y.; Liu, J.; Zhang, X.; Wu, X.; Tang, S.; Sun, R.; Qian, C.; et al. Anti-inflammatory properties and gut microbiota modulation of an alkali-soluble polysaccharide from purple sweet potato in DSS-induced colitis mice. *Int. J. Biol. Macromol.* **2020**, *153*, 708–722. [CrossRef]
79. Wu, Q.; Qu, H.; Jia, J.; Kuang, C.; Wen, Y.; Yan, H.; Gui, Z. Characterization, antioxidant and antitumor activities of polysaccharides from purple sweet potato. *Carbohydr. Polym.* **2015**, *132*, 31–40. [CrossRef] [PubMed]
80. Xiao, J.B.; Jiang, H. A review on the structure-function relationship aspect of polysaccharides from tea materials. *Crit. Rev. Food Sci. Nutr.* **2015**, *55*, 930–938. [CrossRef] [PubMed]
81. Chen, D.; Chen, G.; Chen, C.; Zeng, X.; Ye, H. Prebiotics effects in vitro of polysaccharides from tea flowers on gut microbiota of healthy persons and patients with inflammatory bowel disease. *Int. J. Biol. Macromol.* **2020**, *158*, 968–976. [CrossRef] [PubMed]
82. Zou, Q.; Zhang, X.; Liu, X.; Li, Y.; Tan, Q.; Dan, Q.; Yuan, T.; Liu, X.; Liu, R.H.; Liu, Z. *Ficus carica* polysaccharide attenuates DSS-induced ulcerative colitis in C57BL/6 mice. *Food Funct.* **2020**, *11*, 6666–6679. [CrossRef]
83. Meng, X.L.; Li, S.; Qin, C.B.; Zhu, Z.X.; Hu, W.P.; Yang, L.P.; Lu, R.H.; Li, W.J.; Nie, G.X. Intestinal microbiota and lipid metabolism responses in the common carp (*Cyprinus carpio* L.) following copper exposure. *Ecotoxicol. Environ. Saf.* **2018**, *160*, 257–264. [CrossRef]

84. Li, L.F.; Liu, H.B.; Zhang, Q.W.; Li, Z.P.; Wong, T.L.; Fung, H.Y.; Zhang, J.X.; Bai, S.P.; Lu, A.P.; Han, Q.B. Comprehensive comparison of polysaccharides from *Ganoderma lucidum* and *G. sinense*: Chemical, antitumor, immunomodulating and gut-microbiota modulatory properties. *Sci. Rep.* **2018**, *8*, 6172. [CrossRef]
85. Kobayashi, M.; Mikami, D.; Kimura, H.; Kamiyama, K.; Morikawa, Y.; Yokoi, S.; Kasuno, K.; Takahashi, N.; Taniguchi, T.; Iwano, M. Short-chain fatty acids, GPR41 and GPR43 ligands, inhibit TNF-alpha-induced MCP-1 expression by modulating p38 and JNK signaling pathways in human renal cortical epithelial cells. *Biochem. Biophys. Res. Commun.* **2017**, *486*, 499–505. [CrossRef]
86. Zhou, L.; Liu, T.; Huang, B.; Luo, M.; Chen, Z.; Zhao, Z.; Wang, J.; Leung, D.; Yang, X.; Chan, K.W.; et al. Excessive deubiquitination of NLRP3-R779C variant contributes to very-early-onset inflammatory bowel disease development. *J. Allergy Clin. Immunol.* **2021**, *147*, 267–279. [CrossRef]
87. Huang, L.; Shen, M.; Morris, G.A.; Xie, J. Sulfated polysaccharides: Immunomodulation and signaling mechanisms. *Trends Food Sci. Technol.* **2019**, *92*, 1–11. [CrossRef]
88. Phull, A.R.; Majid, M.; Haq, I.U.; Khan, M.R.; Kim, S.J. In vitro and in vivo evaluation of anti-arthritic, antioxidant efficacy of fucoidan from *Undaria pinnatifida* (Harvey) Suringar. *Int. J. Biol. Macromol.* **2017**, *97*, 468–480. [CrossRef] [PubMed]
89. Yamamoto, Y.; Gaynor, R.B. Therapeutic potential of inhibition of the NF-κB pathway in the treatment of inflammation and cancer. *J. Clin. Investig.* **2001**, *107*, 135–142. [CrossRef] [PubMed]
90. Tsukahara, T.; Hamouda, N.; Utsumi, D.; Matsumoto, K.; Amagase, K.; Kato, S. G protein-coupled receptor 35 contributes to mucosal repair in mice via migration of colonic epithelial cells. *Pharmacol. Res.* **2017**, *123*, 27–39. [CrossRef] [PubMed]
91. Kato, S.; Utsumi, D.; Matsumoto, K. G protein-coupled receptor 40 activation ameliorates dextran sulfate sodium-induced colitis in mice via the upregulation of glucagon-likepeptide-2. *J. Pharmacol. Sci.* **2019**, *140*, 144–152. [CrossRef]
92. Tedelind, S.; Westberg, F.; Kjerrulf, M.; Vidal, A. Anti-inflammatory properties of the short-chain fatty acids acetate and propionate: A study with relevance to inflammatory bowel disease. *World J. Gastroenterol.* **2007**, *13*, 2826–2832. [CrossRef]
93. Lee, H.A.; Lee, D.Y.; Cho, H.M.; Kim, S.Y.; Iwasaki, Y.; Kim, I.K. Histone deacetylase inhibition attenuates transcriptional activity of mineralocorticoid receptor through its acetylation and prevents development of hypertension. *Circ. Res.* **2013**, *112*, 1004–1012. [CrossRef]
94. DeWire, S.M.; Ahn, S.; Lefkowitz, R.J.; Shenoy, S.K. β-Arrestins and cell signaling. *Annu. Rev. Physiol.* **2007**, *69*, 483–510. [CrossRef] [PubMed]
95. Mao, K.; Chen, S.; Wang, Y.; Zeng, Y.; Ma, Y.; Hu, Y.; Zhang, H.; Sun, S.; Wu, X.; Meng, G.; et al. β-arrestin1 is critical for the full activation of NLRP3 and NLRC4 inflammasomes. *J. Immunol.* **2015**, *194*, 1867–1873. [CrossRef] [PubMed]
96. Higashimura, Y.; Tanaka, Y.; Takagi, T.; Uchiyama, K.; Mizushima, K.; Niki, E.; Naito, Y. Trans-unsaturated fatty acid activates NLRP3 inflammasome in macrophages and exacerbates intestinal inflammation in mice. *Biochem. Biophys. Res. Commun.* **2020**, *529*, 243–250. [CrossRef] [PubMed]
97. Liang, J.; Chen, S.; Chen, J.; Lin, J.; Xiong, Q.; Yang, Y.; Yuan, J.; Zhou, L.; He, L.; Hou, S.; et al. Therapeutic roles of polysaccharides from *Dendrobium officinale* on colitis and its underlying mechanisms. *Carbohydr. Polym.* **2018**, *185*, 159–168. [CrossRef]
98. Johansson, M.E.; Larsson, J.M.; Hansson, G.C. The two mucus layers of colon are organized by the MUC2 mucin, whereas the outer layer is a legislator of host-microbial interactions. *Proc. Natl. Acad. Sci. USA* **2011**, *108* (Suppl. 1), 4659–4665. [CrossRef]
99. Mees, S.T.; Mennigen, R.; Spieker, T.; Rijcken, E.; Senninger, N.; Haier, J.; Bruewer, M. Expression of tight and adherens junction proteins in ulcerative colitis associated colorectal carcinoma: Upregulation of claudin-1, claudin-3, claudin-4, and β-catenin. *Int. J. Colorectal Dis.* **2009**, *24*, 361–368. [CrossRef]
100. Marchiando, A.M.; Shen, L.; Graham, W.V.; Weber, C.R.; Schwarz, B.T.; Austin, J.R.; Raleigh, D.R.; Guan, Y.; Watson, A.J.M.; Montrose, M.H.; et al. Caveolin-1–dependent occludin endocytosis is required for TNF-induced tight junction regulation in vivo. *J. Cell Biol.* **2010**, *189*, 111–126. [CrossRef]
101. Wang, F.; Graham, W.V.; Wang, Y.; Witkowski, E.D.; Schwarz, B.T.; Turner, J.R. Interferon-γ and Tumor Necrosis Factor-α Synergize to Induce Intestinal Epithelial Barrier Dysfunction by Up-Regulating Myosin Light Chain Kinase Expression. *Am. J. Pathol.* **2005**, *166*, 409–419. [CrossRef]
102. Cao, W.; Vrees, M.D.; Potenti, F.M.; Harnett, K.M.; Fiocchi, C.; Pricolo, V.E. Interleukin 1β-induced production of H2O2 contributes to reduced sigmoid colonic circular smooth muscle contractility in ulcerative colitis. *J. Pharmacol. Exp. Ther.* **2004**, *311*, 60–70. [CrossRef]
103. Masaki, T.; Kishiki, T.; Kojima, K.; Asou, N.; Beniya, A.; Matsuoka, H. Recent trends (2016–2017) in the treatment of inflammatory bowel disease. *Ann. Gastroenterol. Surg.* **2018**, *2*, 282–288. [CrossRef] [PubMed]
104. Scharl, M.; Paul, G.; Barrett, K.E.; McCole, D.F. AMP-activated protein kinase mediates the interferon-gamma-induced decrease in intestinal epithelial barrier function. *J. Biol. Chem.* **2009**, *284*, 27952–27963. [CrossRef]
105. Kucharzik, T.; Walsh, S.V.; Chen, J.; Parkos, C.A.; Nusrat, A. Neutrophil Transmigration in Inflammatory Bowel Disease Is Associated with Differential Expression of Epithelial Intercellular Junction Proteins. *Am. J. Pathol.* **2001**, *159*, 2001–2009. [CrossRef]
106. Ye, D.; Ma, I.; Ma, T.Y. Molecular mechanism of tumor necrosis factor-alpha modulation of intestinal epithelial tight junction barrier. *Am. J. Physiol. Gastrointest. Liver Physiol.* **2006**, *290*, G496–G504. [CrossRef]
107. Knutson, C.G.; Mangerich, A.; Zeng, Y.; Raczynski, A.R.; Liberman, R.G.; Kang, P.; Ye, W.; Prestwich, E.G.; Lu, K.; Wishnok, J.S.; et al. Chemical and cytokine features of innate immunity characterize serum and tissue profiles in inflammatory bowel disease. *Proc. Natl. Acad. Sci. USA* **2013**, *110*, E2332–E2341. [CrossRef]

108. Pizzino, G.; Irrera, N.; Cucinotta, M.; Pallio, G.; Mannino, F.; Arcoraci, V.; Squadrito, F.; Altavilla, D.; Bitto, A. Oxidative Stress: Harms and Benefits for Human Health. *Oxidative Med. Cell. Longev.* **2017**, *2017*, 8416763. [CrossRef]
109. Alemany-Cosme, E.; Saez-Gonzalez, E.; Moret, I.; Mateos, B.; Iborra, M.; Nos, P.; Sandoval, J.; Beltran, B. Oxidative Stress in the Pathogenesis of Crohn's Disease and the Interconnection with Immunological Response, Microbiota, External Environmental Factors, and Epigenetics. *Antioxidants* **2021**, *10*, 64. [CrossRef]

foods

MDPI

Article

Physicochemical and Structural Characterization of Potato Starch with Different Degrees of Gelatinization

Fen Xu [1,2], Liang Zhang [1], Wei Liu [1], Qiannan Liu [1], Feng Wang [1], Hong Zhang [1], Honghai Hu [1,*] and Christophe Blecker [2]

1 Institute of Food Science and Technology, Chinese Academy of Agricultural Sciences, Key Laboratory of Agro-Products Processing, Ministry of Agriculture and Rural Affairs, Beijing 100193, China; xufen@caas.cn (F.X.); zhangliang@caas.cn (L.Z.); liuwei@caas.cn (W.L.); liuqiannan@caas.cn (Q.L.); wangfeng@caas.cn (F.W.); zhanghong03@caas.cn (H.Z.)

2 Department of Food Science and Formulation, Gembloux Agro-Bio Tech, Université de Liège, Passage des Déportés 2, B-5030 Gembloux, Belgium; christophe.blecker@uliege.be

* Correspondence: huhonghai@caas.cn; Tel.: +86-10-62816503

Abstract: Starch gelatinization has been widely studied previously, but there is still a lack of systematical research on the relationship between the degree of starch gelatinization (DSG) and its physicochemical and structural properties. In this study, potato starch samples with DSG ranging from 39.41% to 90.56% were obtained by hydrothermal treatment. The thermal, rheological, and structural properties, as well as the water-binding capacity of samples were investigated. A starch solution with a DSG of 39.41% was partially sedimented at room temperature, while starch with a DSG of 56.11% can form a stable paste with a fine shear-thinning property, as well as samples with a DSG larger than 56.11%. The endothermic enthalpy, gelatinization range, and short-range ordered structure of starch were negatively correlated with DSG, whereas onset gelatinization temperature, apparent viscosity, and water-binding capacity were positively correlated. The viscoelasticity of starch gels was negatively correlated with the DSG after full gelatinization (DSG > 39.41%). Starch granules gradually lose their typical shape and less birefringence can be observed with increasing DSG. Hydrothermal treatment has a more significant effect on the amount of exposed hydroxyl groups than the ordered and amorphous structures of partially gelatinized starch. This study built linear correlations between starch physicochemical properties and the DSG and provided comprehensive insight into the characteristics of partially gelatinized potato starch.

Keywords: pre-gelatinization; thermal properties; rheology; structural properties; microstructure; starch gel

Citation: Xu, F.; Zhang, L.; Liu, W.; Liu, Q.; Wang, F.; Zhang, H.; Hu, H.; Blecker, C. Physicochemical and Structural Characterization of Potato Starch with Different Degrees of Gelatinization. *Foods* **2021**, *10*, 1104. https://doi.org/10.3390/foods10051104

Academic Editors: Jianhua Xie, Yanjun Zhang, Hansong Yu and Joana S. Amaral

Received: 22 March 2021
Accepted: 12 May 2021
Published: 17 May 2021

1. Introduction

Potato starch is widely used in the food industry owing to its low cost, abundant availability, and more importantly, its unique physicochemical properties compared to other commercially available starches [1,2]. Potato starch can hydrate quickly and forms a paste with higher viscosity when heated, and a clearer gel when cooled, which differ from those of other starches such as corn, wheat, and rice [1,3,4]. The natural properties of potato starch are exploited by the food industry to provide products the required texture, appearance, density, and storage stability. However, native potato starch has some drawbacks, such as poor solubility and poor stability against heat and shear during pasting. These drawbacks have limited its application in industry. Consequently, various physical (heat-moisture treatment, annealing, pre-gelatinization, and high pressure treatment), chemical (cross-linking, substitution, acid hydrolysis, and oxidation), and genetic modification techniques have been used to modify potato starch and enhance its physicochemical properties to meet the demands of consumers [5–7]. Pre-gelatinization is a method widely used to modify potato starch, and pre-gelatinized starch has been extensively applied in the food industry [8,9].

Potato starch can be gelatinized in hot water by breaking the intermolecular hydrogen bond and destroying the arrangement of micelle structure in starch granules. After being fully gelatinized, the starch granules are porous with broken hydrogen bonds, birefringence disappears, and the crystalline order is lost through rapid drying at high temperature [10]. The degree of starch gelatinization (DSG) plays a very important role in imparting desirable product textures. Many starch-based food formulations do not require fully gelatinized starch for the best product characteristics. So far, most studies have been conducted using either native or fully gelatinized potato starch, and there has been a lack of attention on incomplete gelatinized potato starch.

Nevertheless, there have been studies on partial starch gelatinization induced by pre-heating for different times/temperatures or ball milling for different times. Studies reported that the pre-gelatinization treatment significantly changed the granule size distribution and rheological properties of starch dispersions [11–14]. Recent studies also showed that partially gelatinized starch with a lower DSG resulted in increased pasting viscosities and gel textural parameters, while the opposite was noted at higher DSG and native starch [15]. The partially gelatinized potato starch with different DSGs can be used as a thickener or stabilizer in many food products that receive a minimal heating process, for example, instant soups, desserts, powders in beverages, frozen noodles, and baking products [14]. The thermal, rheological, and structural characteristics of partially gelatinized starch can provide essential information for the selection of optimum processing conditions of starch and starchy foods [16]. However, systematical studies involving the physicochemical and structural properties of partially gelatinized potato starch with different DSGs and their correlations are still limited.

The objective of this study is to understand the relationship between the DSG and its physicochemical and structural properties. Thus, we prepared 14 partially gelatinized potato starch samples with relatively evenly distributed DSG data points by hydrothermal treatment. The water-binding capacity and thermal, rheological, and structural properties were analyzed, and the correlations between these physicochemical properties of starch and its DSG were established.

2. Materials and Methods

2.1. Raw Materials

Potato starch (PS) containing 89.32% starch, 0.27% protein, and 0.42% ash, based on the dry matters of samples, was purchased from Aladdin Bio-Chem Technology Co., Ltd. (Shanghai, China). Moisture content was determined by drying samples to a constant weight in a 105 °C oven (AOAC 925.09). Crude protein was determined by the Kjeldahl method, with the total nitrogen content being converted to protein content by a conversion factor of 6.25 (AOAC 979.09). Ash content was determined by dry mineralization of samples in a muffle oven at 550 °C for 8 h (AOAC 923.03), and starch content was determined by a total starch assay kit (K-TSTA; Megazyme, Ireland).

2.2. Sample Preparation

Deionized water was stirred in beakers at 300 rpm using a thermostatic water bath maintained at 59 ± 1 °C and 60 ± 1 °C, separately. Native starch powder was added after the deionized water reached the fixed temperature. Each batch of dispersion at a concentration of 10% (*w/w*) was thoroughly stirred for 1, 3, 6, 9, 12, 15, or 18 min, separately. Heating conditions were determined based on pre-experiments and the onset gelatinization temperature of native potato starch. The selected heating temperature provides starch samples with different DSGs in a convenient time period. Starch samples with a higher DSG are difficult to obtain when treated with a lower temperature in a relatively short time period, while starch is gelatinized too quickly when treated with a higher temperature. We therefore chose 59 °C and 60 °C as the temperatures and obtained a series of starch samples with different DSGs. After hydrothermal treatment, the beakers were quickly moved into ice water and cooled to room temperature before being put in the freezer. These partially gelatinized starch dispersions were freeze-dried for around 48 h using a vacuum freeze dryer (GenesisTM SQ, Virtis, Warminster, PA, USA) [15,17]. Starch samples were ground into powders with a mortar after drying and the resulting powder samples were sealed in plastic bags and stored in a silica-gel desiccator at 25 °C.

2.3. Differential Scanning Calorimetry (DSC)

The thermal properties of starch samples were investigated using a differential scanning calorimeter (DSC8000, PerkinElmer, Waltham, MA, USA). Powder samples (3 mg) were weighed into aluminum pans, and 10 μL of distilled water was added before the pans were hermetically sealed. The pans were equilibrated at room temperature for 2 h before heating from 20 °C to 100 °C at a rate of 5 °C/min [18,19]. A sealed empty aluminum pan was used as a reference. Onset temperature (T_0), peak temperature (T_p), conclusion temperature (T_c), and endothermic enthalpy (ΔH) were calculated from the DSC curves using the equipment software. Gelatinization range (R) was computed as ($T_c - T_0$), and the peak height index (PHI) was calculated by the ratio $\Delta H/(T_p - T_0)$, as previously described [20]. The degree of starch gelatinization (DSG) was determined using Equation (1) given below [21,22].

$$\text{DSG}\ (\%) = (1 - \Delta H_{sample}/\Delta H_{native}) \times 100\% \tag{1}$$

where ΔH_{native} and ΔH_{sample} are the enthalpy change due to gelatinization of native and modified starches, respectively.

2.4. Water-Binding Capacity (WBC)

The WBC of starch samples was measured as previously described [23] with slight modifications. A suspension of 5 g of starch (dry wt. basis) in 75 mL of distilled water was agitated for 1 h at 150 rpm at 20 °C, 40 °C, 60 °C, and 80 °C, separately, and then centrifuged (3000× *g*) for 10 min. Free water was removed from the wet starch which was then continuously drained for 10 min at room temperature (20 °C). The wet starch was calculated using the difference between wet starch with tubes and empty tubes. The WBC of starch granules was calculated using the following formula:

$$\text{WBC}\ (\%) = (W_2 - W_1)/W_1 \times 100\% \tag{2}$$

where W_1 and W_2 are the dry weight of original starch and wet weight of starch, respectively.

2.5. Fourier Transform Infrared (FTIR) Spectroscopy

The short-range molecular order of starch samples was determined using a FTIR spectrometer (Tensor 27, Bruker Opticals Company, Rheinstetten, Germany). FTIR spectra were recorded from 4000 to 600 cm^{-1} at a resolution of 4 cm^{-1} with an accumulation of 64 scans using the attenuated total reflectance (ATR) accessory. All spectra were baseline corrected automatically by OMNIC 8.2 and deconvoluted from 1200 to 800 cm^{-1} with a half-bandwidth of 19 cm^{-1} and an enhancement factor of 1.9. Intensity measurements were performed on the deconvoluted spectra by recording the peak height of absorbance bands from the baseline. The ratio of absorbance at 1047/1022 cm^{-1} and 1022/995 cm^{-1} was used to estimate the short-range ordered structure of starch [6,24].

2.6. Scanning Electron Microscopy (SEM)

The microstructures of the starch samples were observed using field-emission environmental scanning electron microscopy (SU8010, Hitachi, Tokyo, Japan) with a 10-KV acceleration voltage. Powder samples were added to double-sided adhesive tape mounted on an aluminum stub and sprayed with gold. Samples were photographed at 500× and 1000× magnification.

2.7. Polarized Light Microscopy (PLM)

The birefringence of starch samples was observed under polarized light with a binocular microscope (BA310 Pol, Motic, Xiamen, China) at a magnification of 100×. The starch samples were transferred onto a slide glass and a drop of water was then added to each sample before observation.

2.8. Rheological Measurement

The rheological properties of the starch samples were determined using a rheometer (Physica MCR301; Anton Paar, Graz, Austria). Starch suspensions at 10% (*w/w*) concentration after hydrothermal treatment (as described in the Section 2.2) were loaded onto the measuring apparatus after rapid cooling to room temperature. The 5 cm in diameter parallel-plate geometry with a 1 mm gap was used for all the measurements. The outer edge of the sample was coated with silicone oil to minimize water loss during measurements [23]. Temperature was controlled using a water bath system connected to a Peltier system in the bottom plate to accurately control temperature during rapid heating and cooling.

2.8.1. Steady Shear Viscosity Measurement

Steady shear tests were programmed to increase the shear rate from 0.01 s^{-1} to 100 s^{-1} with six points per decade. Apparent viscosity values were obtained as a function of shear rate. The flow behaviors of mixed starch samples were analyzed using a power law equation [25,26].

$$\sigma = K \cdot \gamma^{n} \tag{3}$$

where σ (Pa) is the shear stress, K (Pa s^n) is the consistency index, γ (s^{-1}) is the shear rate, and n is the flow index [26].

2.8.2. Temperature Sweep

In the temperature sweep test, the strain and frequency were set at 0.5% and 1 Hz, respectively (within the linear viscoelastic region). The temperature was set from 25 °C to 75 °C at a heating rate of 2 °C/min and cooled from 75 °C to 25 °C at a rate of 5 °C/min. The storage modulus (G′), loss modulus (G″), and loss tangent (tanδ = G″/G′) were obtained as a function of temperature [27].

2.8.3. Frequency Sweep

Frequency sweep tests were conducted in situ after the temperature sweep. Mechanical spectra of starch gel samples were recorded in the range of 0.1–100 rad/s with 10 points per decade, and the controlled variable was set at 5% strain (within the linear viscoelastic region) [26,28]. All frequency sweep tests were performed at 25 °C. G′, G″, and tanδ were obtained by changing the frequency. The power law models represented in Equations (4) and (5) were applied to describe the frequency dependence of G′ and G″, respectively.

$$G' = K' \cdot \omega^{n'} \tag{4}$$

$$G'' = K'' \cdot \omega^{n''} \tag{5}$$

In Equations (4) and (5), K′, K″, n′, and n″ are the corresponding fitting parameters, respectively, and ω is the angular frequency (Hz). K′ and K″ reflect the elastic and viscous properties, respectively. n′ and n″ are referred to as the frequency exponents and can provide useful information regarding the viscoelastic nature of materials [25].

2.9. Correlation Analysis

The physicochemical properties of potato starch that depend on its DSG were analyzed. Scatter diagrams were plotted, and the linear variation regression equation and correlation coefficient were calculated using Excel. SPSS was used to conduct the bivariate correlation test of the two datasets; the P value was obtained to determine the significance level of linear correlation between the starch properties and the DSG.

2.10. Statistical Analysis

Data are reported as averages of triplicate observations and expressed as the mean ± standard deviation. ANOVA followed by Tukey's test at the 0.05 significance level was used to evaluate the differences among sample means. ANOVA was conducted using SPSS 18.0 (SPSS Inc., Chicago, IL, USA). All graphs were generated using Excel and Origin 9.

3. Results and Discussion

3.1. Thermal Properties

The DSC thermograms in Figure 1A,B clearly show the phase transition peaks of starch samples, and as heating time increased, there was an obvious decrease in the areas under the thermal transition peak above the extrapolation, and the curves tend to be flat. The DSC parameters including T_0, ΔH, R, and PHI are shown in Table 1. The T_P and T_c of the starch samples showed no significant differences (data not shown). The T_0 and ΔH of the native potato starch samples were 56.94 °C and 12.73 J/g, respectively, which were values close to previous reports [29,30]. The T_0 of the partially gelatinized starch samples was higher than that of native potato starch, and it increased as the heating time and temperature increased. The gelatinization temperature of the partially gelatinized starch modified by spray drying or high hydrostatic pressure was also reported to be higher than that of the native starch [13,31]. The increasing DSC thermal transition temperature of the heat-gelatinized starch could be due to the colloidal molecular structure of the starch granules, the amylopectin chain length, and the reordering of the crystalline structure after hydrolysis [31]. Specifically, a previous study reported that the increasing gelatinization temperature of pre-heating starch was interpreted as being due to the disruption of less stable crystallites in the first instance, followed by the melting of the remaining and more stable crystallites at a higher temperature [15,32]. However, partially gelatinized starch modified by ball milling showed the opposite variation tendency [12]. The difference might be due to the different residual crystalline structure of partially gelatinized starch after various modifications. On the contrary, ΔH, R, and PHI decreased as the heating time and temperature increased. Table 1 also shows that the calculated DSG of partially gelatinized potato starch varied from 39.41% to 90.56%. As the heating time and temperature increased, the DSG of the starch samples increased. Although the DSG of the starch sample heated at 59 °C for 9 min was slightly higher than the sample heated for 12 min, no significant difference was observed between them. The thermal parameters are known to depend on the stability of the amorphous and crystalline regions of starch [33]. Specifically, thermal parameters depend on the thickness of crystals, their polymorphic structure, and the free energy of the surface of the face side [34]. Samples with higher heating temperatures and longer heating times decreased the crystalline regions and the crystal strength within a starch granule, thus requiring less energy for full gelatinization. This may explain the decrease of R, PHI, ΔH, and the increase of DSG of potato starch as the heating time and temperature increased.

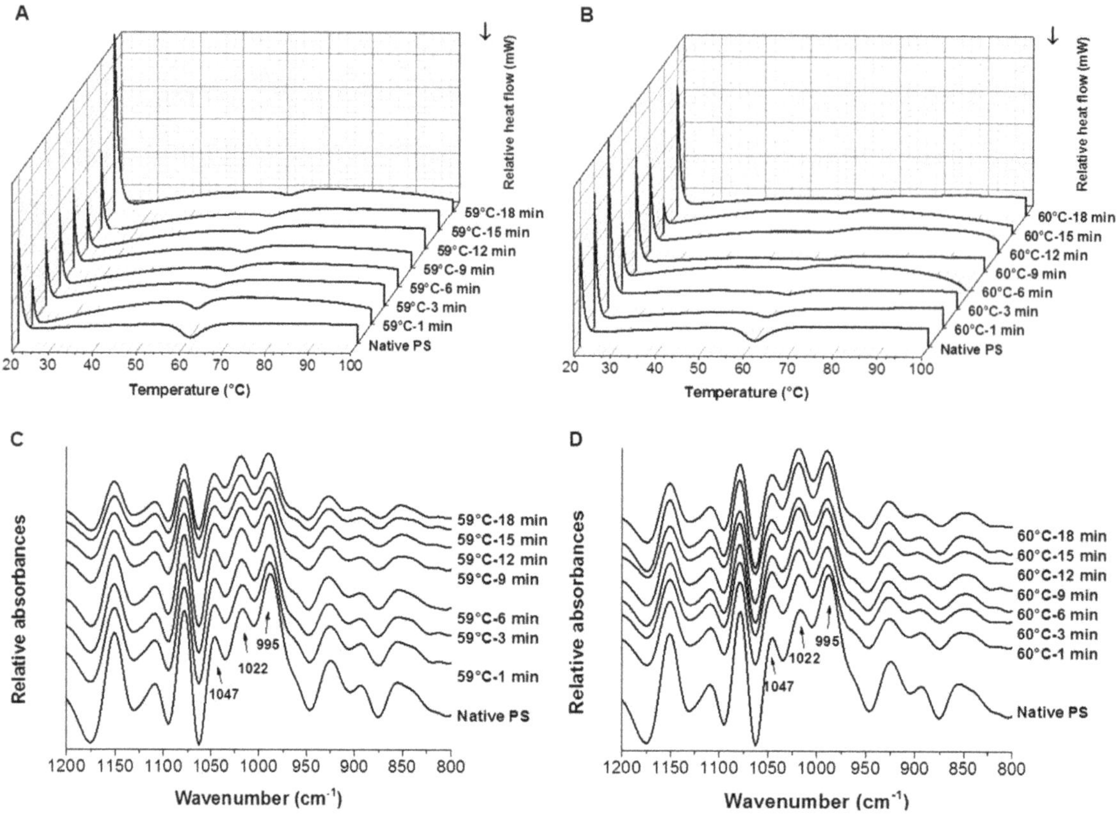

Figure 1. DSC thermograms (**A**,**B**) and FTIR spectra (**C**,**D**) of native and partially gelatinized potato starch samples (**A**,**C**: pre-heated at 59 °C; **B**,**D**: pre-heated at 60 °C).

3.2. Water-Binding Capacity (WBC)

Table 2 shows the WBC of starch samples under 20 °C, 40 °C, 60 °C, and 80 °C. Results indicated that the WBC of all partially gelatinized starch samples was higher than that of the native potato starch. With increased pre-heating time, the WBC of starch samples significantly increased. Moreover, a significant increase in the WBC value can be observed when the reheating temperature increased from 20 °C to 60 °C. However, the WBC did not significantly increase when the reheating temperature further increased from 60 °C to 80 °C. Similarly, a previous study on pre-gelatinized rice starch showed that the water absorption index of modified samples was significantly higher than that of native starch due to the disruption of crystalline structure and the gelatinization of starch [17]. The differences in the WBC of potato starch samples may be attributed to variation in their granular structures [27]. The loose association of amylose and amylopectin molecules in starch granules is responsible for high WBC [35]. The formation of hydrogen and covalent bonds between starch chains engaged by hydroxyl groups could reduce the WBC, and differences in the availability of water-binding sites may also lead to changes in the WBC [36]. In this case, the crystalline structure of samples may be (partially) disrupted due to the breakage of intra- and intermolecular hydrogen bonds when the starch is pre-heated in excess water. The hydroxyl groups of amylose and amylopectin were exposed to a certain extent, potentially forming hydrogen bonds with water molecules and, thus, causing the increased WBC of starch samples with higher DSGs.

Table 1. DSC parameters of native and partially gelatinized starch samples.

Starch	T_0 (°C)	ΔH (J/g)	R	PHI	DSG (%)
Native PS	56.94 ± 0.06 ^{BCde}	12.73 ± 0.09 ^{Aa}	8.46 ± 0.06 ^{Dbc}	3.2 ± 0.01 ^{Aa}	0
59 °C—1 min	56.31 ± 0.06 ^{C}	6.56 ± 0.43 ^{B}	9.49 ± 0.42 ^{C}	2.32 ± 0.11 ^{B}	39.41 ± 3.98 ^{C}
59 °C—3 min	56.86 ± 0.45 ^{BC}	4.66 ± 0.81 ^{C}	11.02 ± 0.33 ^{AB}	1.39 ± 0.42 ^{C}	56.93 ± 7.50 ^{B}
59 °C—6 min	57.40 ± 0.37 ^{B}	4.97 ± 0.33 ^{C}	11.41 ± 0.04 ^{A}	1.54 ± 0.12 ^{C}	60.99 ± 2.59 ^{B}
59 °C—9 min	57.61 ± 0.58 ^{B}	4.26 ± 0.02 ^{C}	10.88 ± 0.58 ^{AB}	1.31 ± 0.23 ^{C}	66.56 ± 0.12 ^{B}
59 °C—12 min	56.99 ± 0.41 ^{BC}	4.46 ± 0.37 ^{C}	10.31 ± 0.54 ^{BC}	1.24 ± 0.12 ^{CD}	64.98 ± 2.88 ^{B}
59 °C—15 min	57.42 ± 0.40 ^{B}	4.40 ± 0.43 ^{C}	11.08 ± 0.30 ^{AB}	1.37 ± 0.14 ^{C}	65.43 ± 3.39 ^{B}
59 °C—18 min	59.05 ± 0.25 ^{A}	2.38 ± 0.84 ^{D}	9.60 ± 0.34 ^{C}	0.79 ± 0.19 ^{D}	81.28 ± 6.60 ^{A}
60 °C—1 min	56.50 ± 0.18 ^{e}	5.59 ± 0.19 ^{b}	12.46 ± 0.03 ^{a}	1.38 ± 0.09 ^{b}	56.11 ± 1.52 ^{d}
60 °C—3 min	58.40 ± 0.25 ^{c}	3.50 ± 0.16 ^{c}	10.82 ± 1.77 ^{ab}	0.96 ± 0.10 ^{c}	72.55 ± 1.24 ^{c}
60 °C—6 min	57.38 ± 0.23 ^{d}	3.10 ± 0.13 ^{cd}	11.45 ± 0.08 ^{a}	0.86 ± 0.02 ^{cd}	75.63 ± 1.01 ^{bc}
60 °C—9 min	60.26 ± 0.15 ^{a}	2.03 ± 0.02 ^{de}	11.24 ± 0.98 ^{a}	0.55 ± 0.03 ^{ef}	84.06 ± 0.15 ^{ab}
60 °C—12 min	57.36 ± 0.19 ^{d}	2.25 ± 1.32 ^{de}	10.59 ± 0.98 ^{ab}	0.65 ± 0.28 ^{de}	82.37 ± 10.38 ^{abc}
60 °C—15 min	59.15 ± 0.01 ^{b}	1.92 ± 0.39 ^{de}	6.84 ± 1.15 ^{c}	0.65 ± 0.12 ^{de}	84.94 ± 3.04 ^{ab}
60 °C—18 min	60.54 ± 0.58 ^{a}	1.20 ± 0.03 ^{e}	9.83 ± 1.71 ^{ab}	0.33 ± 0.03 ^{f}	90.56 ± 0.20 ^{a}

Data are means ± SD. $^{A, B, C, D.}$ represent the significant difference of starch samples in column by heating at 59 °C ($p < 0.05$); $^{a, b, c, d, e, f}$ represent the significant difference of starch samples in column by heating at 60 °C ($p < 0.05$).

Table 2. The WBC of native and partially gelatinized starch samples (10^{-2}%).

Starch	20 °C	40 °C	60 °C	80 °C
Native PS	0.85 ± 0.02 $^{Efb'}$	1.34 ± 0.17 $^{Adb'}$	7.74 ± 0.60 $^{Dca'}$	7.11 ± 0.21 $^{Dca'}$
59 °C—1 min	2.88 ± 0.03 $^{Dd'}$	4.14 ± 0.34 $^{Bc'}$	10.45 ± 0.02 $^{Ca'}$	9.43 ± 0.61 $^{Cb'}$
59 °C—3 min	3.15 ± 0.02 $^{Dc'}$	4.74 ± 0.05 $^{Bb'}$	13.70 ± 0.27 $^{Aba'}$	13.52 ± 0.04 $^{Ba'}$
59 °C—6 min	4.05 ± 0.13 $^{Cc'}$	6.14 ± 0.32 $^{Ab'}$	13.76 ± 0.63 $^{Aba'}$	14.77 ± 0.36 $^{Aa'}$
59 °C—9 min	5.67 ± 0.28 $^{Ac'}$	6.57 ± 1.24 $^{Ab'}$	14.42 ± 0.47 $^{Aa'}$	15.02 ± 0.14 $^{Aa'}$
59 °C—12 min	5.05 ± 0.15 $^{Bb'}$	6.25 ± 0.01 $^{Ab'}$	13.19 ± 0.63 $^{Ba'}$	14.30 ± 0.58 $^{Aba'}$
59 °C—15 min	5.29 ± 0.45 $^{ABc'}$	6.52 ± 0.59 $^{Ab'}$	14.80 ± 0.07 $^{Aa'}$	14.93 ± 0.08 $^{Aa'}$
59 °C—18 min	5.81 ± 0.40 $^{Ab'}$	6.58 ± 0.14 $^{Ab'}$	14.14 ± 0.60 $^{Aba'}$	14.76 ± 0.41 $^{Aa'}$
60 °C—1 min	3.01 ± 0.05 $^{ed'}$	4.33 ± 0.10 $^{cc'}$	13.63 ± 0.06 $^{ba'}$	12.83 ± 0.42 $^{bb'}$
60 °C—3 min	3.85 ± 0.20 $^{dd'}$	5.23 ± 0.02 $^{bcc'}$	13.66 ± 0.78 $^{bb'}$	15.25 ± 0.19 $^{aa'}$
60 °C—6 min	4.67 ± 0.12 $^{cc'}$	6.26 ± 0.24 $^{bb'}$	14.24 ± 0.34 $^{aba'}$	14.87 ± 0.19 $^{aa'}$
60 °C—9 min	6.08 ± 0.18 $^{bc'}$	8.13 ± 0.33 $^{ab'}$	14.78 ± 0.01 $^{aa'}$	15.09 ± 0.24 $^{aa'}$
60 °C—12 min	5.76 ± 0.89 $^{bc'}$	7.88 ± 0.50 $^{ab'}$	14.73 ± 0.07 $^{aa'}$	14.90 ± 0.17 $^{aa'}$
60 °C—15 min	6.02 ± 0.30 $^{ab'}$	8.84 ± 1.16 $^{ab'}$	14.90 ± 0.06 $^{aa'}$	15.02 ± 0.23 $^{aa'}$
60 °C—18 min	8.42 ± 0.09 $^{bc'}$	8.91 ± 0.13 $^{ab'}$	14.92 ± 0.07 $^{aa'}$	15.00 ± 0.01 $^{aa'}$

Data are means ± SD. $^{A, B, C, D, E.}$ represent the significant difference of starch samples in column by heating at 59 °C ($p < 0.05$); $^{a, b, c, d, e, f}$ represent the significant difference of starch samples in column by heating at 60 °C ($p < 0.05$); $^{a', b', c', d'}$ represent the significant difference of starch samples heated at a different temperature ($p < 0.05$).

3.3. FTIR Spectra of Native and Partially Gelatinized Starch

The deconvoluted FTIR spectra of the native and partially gelatinized potato starch samples are shown in Figure 1C,D. FTIR spectroscopy was suggested to be sensitive to the so-called short-range order, defined as the double helical order, rather than the long-range order which is related to the packing of double helices [37]. Bands in the spectral region of 800–1200 cm^{-1} reflected changes in polymer conformation and the hydration of processed starches [38]. Bands at 1047 and 1022 cm^{-1} were associated with the ordered and amorphous structures of starch, respectively. The band at 995 cm^{-1} was mainly caused by the bending vibration of C-OH, corresponding to the hydrogen bond structure formed between the hydroxyl groups of starch macromolecules. The FTIR spectra of potato starch samples clearly show that, as pre-heating time increased, the intensities of the peak around 1022 cm^{-1} gradually increased, the peak around 1047 cm^{-1} became flatter, and the peak around 995 cm^{-1} significantly decreased. The ratio 1022/995 cm^{-1} significantly increased, while the ratio 1047/1022 cm^{-1} slightly decreased as heating temperature and time increased (Supplementary Data Figure S1). These data clearly show that higher DSGs of starch lead to the loss of the short-range ordered structure. Sevenou et al. [37] found that the ratio 1047/1022 cm^{-1} of native and fully gelatinized potato starch decreased from 1.37 cm^{-1} to 0.30 cm^{-1}, and the ratio 1022/995 cm^{-1} increased from 0.40 cm^{-1} to 2.50 cm^{-1}. Moreover, the variation of maximum absorbance at 995 cm^{-1} changed more significantly as a function of heating time than at 1047 cm^{-1}, indicating that the hydrothermal treatment has a more significant effect on the amount of exposed hydroxyl groups than the ordered and amorphous structures of starch. The significant decrease of the peak around 995 cm^{-1} also explains the increase of the WBC of samples as the pre-heating time increased, since the band at 995 cm^{-1} mainly represents the hydrogen bond structure formed between the hydroxyl groups of starch macromolecules.

3.4. Morphological Properties of Native and Partially Gelatinized Starch

The microstructure of the native and partially gelatinized potato starch granules is presented in Figure 2. The image clearly shows that the native potato starch granules are characterized by an oval or elliptical shape with a smooth surface (Figure 2A). Starch samples heated at 59 °C and 60 °C for 1 min showed most starch granules retaining their original shape. However, with increased heating time (Figure 2B–G), the starch granules started to break and collapsed into flakes or blocks, and gradually lost their typical oval shape. The oval or elliptical shape of starch granules is largely not observed after extensive heating. The structure changes of the granules under hydrothermal treatment are mainly caused by starch gelatinization. In this process, water penetrates the starch granules and initiates irreversible swelling and microcrystal melt [9]. The polarized light micrographs showed that native potato starch granules have strong birefringence patterns (Figure 2a), reflecting a high degree of ordered molecular orientation and average radial orientation of helical structures in native potato starch [39]. The partial gelatinization treatment resulted in the disappearance of birefringence in potato starch granules to varying degrees. Compared with native potato starch sample, the partially gelatinized starch samples showed larger voids at the granule center. The loss of radial orientation in partially gelatinized potato starch granules is mainly caused by increased mobility of starch chains at the granule center following hydrothermal treatment. More thermal energy was imparted to starch chains during heating and resulted in the destruction of more molecules double helix structure; thus, less birefringence was observed.

3.5. Rheological Properties of Native and Partially Gelatinized Starch

3.5.1. Apparent Viscosity

The flow curves of starch solutions/pastes are presented in Figure 3A,B. The apparent viscosity curves decreased as the shear rate increased, and the steady shear curves were almost parallel to one another, except for the native starch solution and starch sample pre-heated at 59 °C for 1 min. This result indicated that these samples exhibited shear-thinning behavior, albeit at different levels. The shear-thinning behavior of starch samples is attributed to the destroying of the molecular network in starch pastes due to applied shear strain [9]. Similarly, shear-thinning behavior was also observed in pre-heated Peruvian carrot starch and cross-linked waxy maize starch dispersions [25,40]. Table 3 shows the power-law fitting parameters of the steady flows. Data for the native starch solution and starch sample pre-heated at 59 °C for 1 min (DSG = 39.41%) are not shown because these two samples failed to form uniform pastes, were unstable, and sedimentation occurred during the rheology measurements; thus, they do not fit the power law model. Partially gelatinized potato starch with a DSG of 56.11% can form a stable paste with a fine shear-thinning property, as well as starch samples with a DSG larger than 56.11%. Equation (3) can nicely fit the rest of the potato starch flow curves with R^2 between 0.997 and 0.999. The consistency index (K) of the starch pastes increased significantly with increased heating time and temperature. This indicated that starch pastes with higher DSGs showed higher apparent viscosity. All flow behavior indexes (n) were less than 1.0, further indicating the pseudoplastic and shear-thinning behavior of the samples [41]. The n value increased with the heating time and temperature, indicating that starch samples with higher DSGs showed reduced pseudoplasticity. Pseudoplasticity can be attributed to the progressive orientation and alignment of the starch molecules, and the breaking of hydrogen bonds formed among amylose molecules under the influence of the shear field [42]. A high degree of macromolecular disorganization enhanced the solubility of gelatinized potato starches, leading to the formation of viscous pseudoplastics [43].

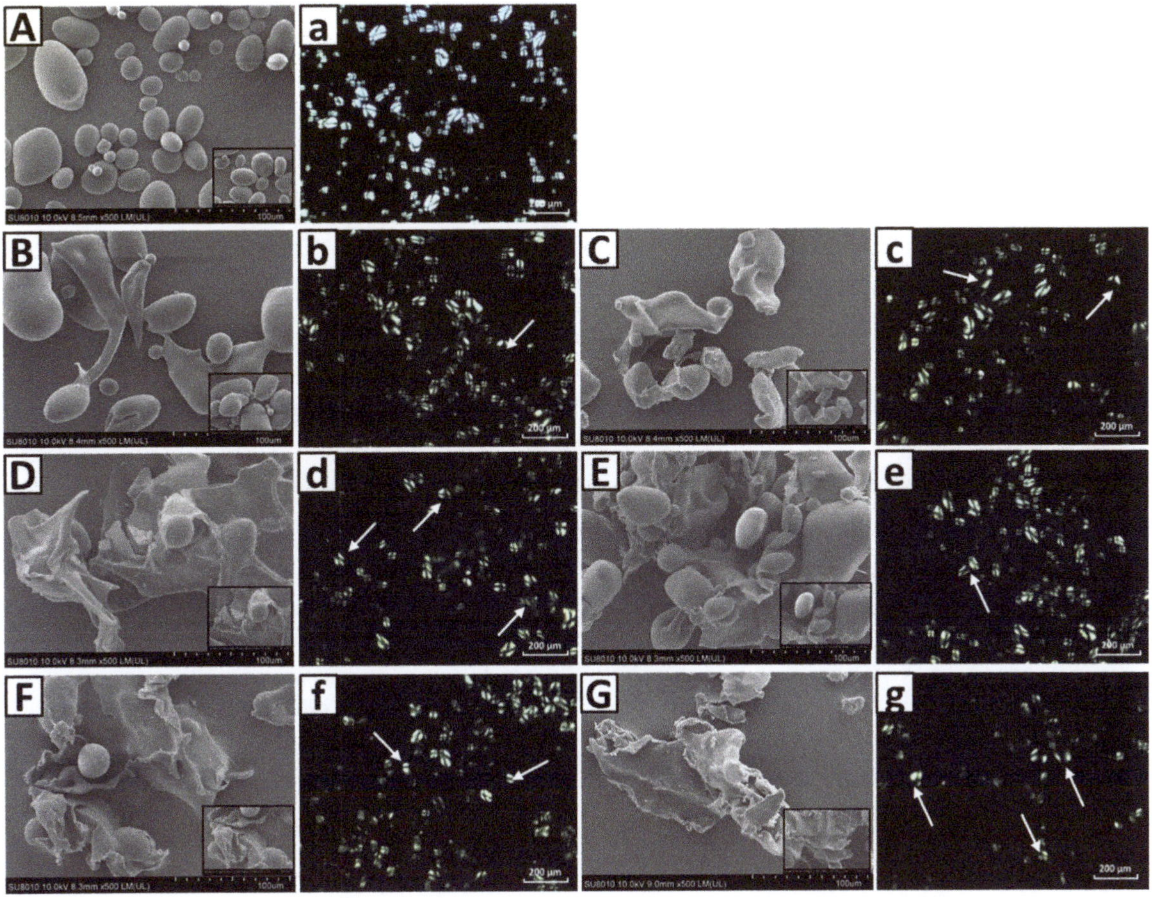

Figure 2. Scanning electron micrographs (upper case: **A–G**) and polarized light micrographs (lower case: **a–g**) of the native (**A**,**a**) and partially gelatinized potato starch granules (**B–G**,**b–g**) heated at 59 °C (**B–D**,**b–d**) and 60 °C (**E–G**,**e–g**) for different times (**B**,**b**,**E**, textbfe: 1 min; **C**,**c**,**F**,**f**: 6 min; **D**,**d**,**G**,**g**: 18 min).

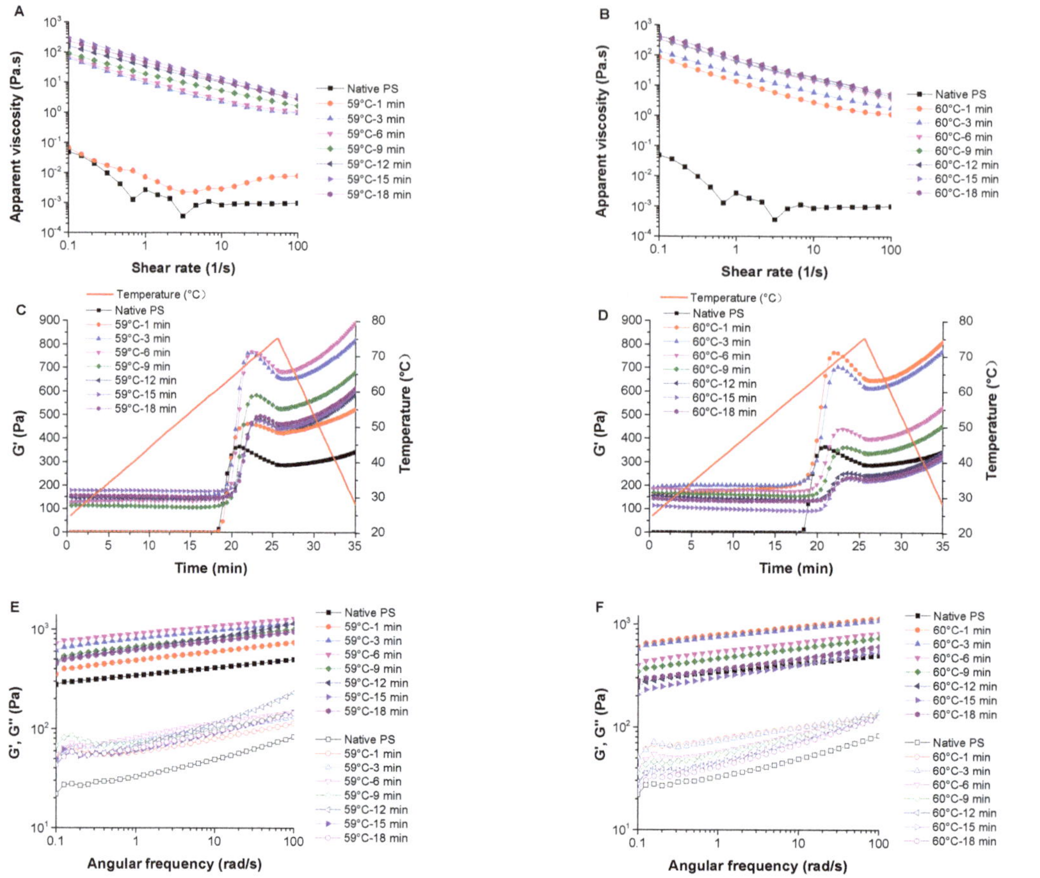

Figure 3. The apparent viscosity flow (**A**,**B**), temperature dependence of G′ (**C**,**D**), and frequency dependence of G′ (solid symbols) and G″ (open symbols) (**E**,**F**) of native and partially gelatinized potato starch samples (**A**,**C**,**E**: pre-heated at 59 °C; **B**,**D**,**F**: pre-heated at 60 °C).

Table 3. Effect of shear rate and frequency on the consistency index (K, K′, and K″), flow behavior index (n, n′, and n″), and determination coefficient (R^2, R'^2, and R''^2) of starch pastes and gels pre-heated at 59 °C and 60 °C.

Starch	K (Pa s^n)	n	R^2	K′ (Pa s^n)	n′	R'^2	K″ (Pa s^n)	n″	R''^2
Native PS	—	—	—	343.44 ± 38.54^{Ec}	0.08 ± 0.01^{Be}	0.996	33.2 ± 4.08^{Cd}	0.19 ± 0.03^{Abc}	0.981
59 °C—1 min	—	—	—	485.71 ± 25.19^{DE}	0.09 ± 0.00^{AB}	0.994	63.14 ± 7.54^{B}	0.12 ± 0.01^{A}	0.914
59 °C—3 min	9.74 ± 1.22^{E}	-0.83 ± 0.06^{C}	0.998	803.67 ± 0.92^{AB}	0.08 ± 0.00^{B}	0.995	78.15 ± 1.54^{A}	0.11 ± 0.00^{A}	0.988
59 °C—6 min	11.61 ± 0.72^{E}	-0.81 ± 0.01^{BC}	0.998	907.70 ± 43.09^{A}	0.08 ± 0.01^{B}	0.996	83.18 ± 4.51^{A}	0.13 ± 0.01^{A}	0.992
59 °C—9 min	19.89 ± 1.51^{D}	-0.66 ± 0.07^{A}	0.999	665.12 ± 11.21^{BC}	0.09 ± 0.00^{AB}	0.992	78.76 ± 1.74^{A}	0.12 ± 0.00^{A}	0.862
59 °C—12 min	37.04 ± 1.62^{C}	-0.64 ± 0.02^{A}	0.999	636.78 ± 160.52^{CD}	0.12 ± 0.03^{A}	0.990	66.98 ± 5.35^{B}	0.23 ± 0.12^{A}	0.980
59 °C—15 min	60.78 ± 2.83^{A}	-0.71 ± 0.02^{AB}	0.999	620.32 ± 63.77^{CD}	0.10 ± 0.00^{AB}	0.997	64.64 ± 0.88^{B}	0.17 ± 0.01^{A}	0.963
59 °C—18 min	48.66 ± 1.60^{B}	-0.69 ± 0.03^{A}	0.999	613.45 ± 24.59^{CD}	0.10 ± 0.00^{AB}	0.996	65.56 ± 0.67^{B}	0.17 ± 0.01^{A}	0.986
60 °C—1 min	13.35 ± 0.86^{b}	-0.80 ± 0.03^{b}	0.999	784.36 ± 48.37^{a}	0.08 ± 0.00^{e}	0.996	76.96 ± 5.33^{a}	0.11 ± 0.00^{d}	0.990
60 °C—3 min	24.45 ± 0.75^{b}	-0.75 ± 0.01^{ab}	0.999	747.29 ± 10.63^{a}	0.09 ± 0.00^{e}	0.996	75.34 ± 1.86^{a}	0.12 ± 0.00^{d}	0.982
60 °C—6 min	62.90 ± 1.20^{a}	-0.72 ± 0.01^{ab}	0.999	536.33 ± 32.29^{b}	0.09 ± 0.00^{de}	0.996	59.40 ± 1.97^{b}	0.17 ± 0.01^{c}	0.981
60 °C—9 min	67.92 ± 1.07^{a}	-0.70 ± 0.01^{a}	0.999	459.54 ± 29.51^{b}	0.10 ± 0.01^{cd}	0.998	55.44 ± 1.01^{b}	0.19 ± 0.00^{bc}	0.992
60 °C—12 min	88.19 ± 4.34^{a}	-0.70 ± 0.01^{a}	0.999	353.52 ± 62.90^{c}	0.12 ± 0.01^{ab}	0.997	46.16 ± 1.93^{c}	0.21 ± 0.01^{ab}	0.987
60 °C—15 min	76.04 ± 34.53^{a}	-0.70 ± 0.11^{a}	0.999	302.18 ± 25.68^{c}	0.13 ± 0.01^{a}	0.998	46.51 ± 0.52^{c}	0.21 ± 0.01^{ab}	0.993
60 °C—18 min	89.90 ± 6.80^{a}	-0.68 ± 0.01^{a}	0.998	361.52 ± 51.89^{c}	0.11 ± 0.01^{bc}	0.997	41.66 ± 2.72^{c}	0.23 ± 0.01^{a}	0.992

Data are means ± SD. $^{A, B, C, D, E}$ represent the significant difference of starch samples in column by heating at 59 °C ($p < 0.05$); $^{a, b, c, d, e}$ represent the significant difference of starch samples in column by heating at 60 °C ($p < 0.05$).

3.5.2. Temperature Sweep

Figure 3C,D shows the temperature dependence of G′ of starch samples. Apart from the native starch sample and partially gelatinized starch samples pre-heated for 1 min at 59 °C, the peak G′ of other starch samples decreased as the heating time increased. This indicated that the elasticity modulus of partially gelatinized starch decreased at the re-gelatinization process with its DSG. The hydrogen bonds between starch molecules of the partially gelatinized starch samples were broken to some extent in the pre-gelatinization process, thus decreasing its modulus in the reheating process. For the native potato starch samples, starch granules first gradually swelled and had a relatively high interaction between starch molecules, requiring more energy to break the bond. This may primarily explain why the peak G′ decreased with the increased DSG. Tanδ against temperature is used to identify the gelatinization onset in oscillatory tests by the temperature at the maximum point of Tanδ [44]. As the heating time increased, the gelatinization onset temperature of the samples pre-heated at 59 °C increased from 58.4 °C to 65.9 °C, and that of the samples pre-heated at 60 °C increased from 60.13 °C to 67.90 °C (Supplementary Data Figure S2 and Table S1). This phenomenon indicated that higher degrees of starch gelatinization delayed the re-gelatinization process. This might be due to the disruption of less stable crystallites in the first instance, followed by the melting of the remaining more stable crystallites at the higher temperature [15,32]. The retrogradation may also influence the gelatinization temperature of partially gelatinized starch by forming different crystalline structures. The results are consistent with the onset temperature (T_0) obtained from the DSC measurements.

3.5.3. Frequency Sweep

Frequency sweep tests were performed on the starch gels formed in situ after the temperature sweep. The G′ and G″ of the samples were dependent on the oscillation frequency, indicating a typical viscoelastic nature of the starch gels [42] (Figure 3E,F). G′ and G″ increased as the frequency increased; therefore, the overall chain mobility within the network was relatively high. G′ was higher than G″ at the same angular frequency throughout the entire tested angular frequency range (0.1 to 100 rad/s), indicating that these starch gels exhibited dominant elastic behavior compared to viscous behavior [45]. The power-law fitting parameters affected by the frequency are shown in Table 3. The power law model represents the experimental data of G′ and G″ since the determination coefficients (R'^2 and R''^2) of G′ and G″ were all above 0.86. The K′ of all starch gel samples was higher than K″ at the same angular frequency, consistent with the changes of G′ and G″. K′ and K″ decreased with the increase of pre-heating time, indicating that the viscoelastic moduli of starch gels decreased with increasing DSG after full gelatinization. The tanδ of starch gel samples ranged from 0.06 to 0.24 (Supplementary Figure S2). Similarly, a previous study reported starch films exhibiting a tanδ from 0.05 to 0.25 [45]. Moreover, as a dimensionless parameter, tanδ increased with increased pre-heating time, indicating the viscous nature was dominant over the elastic nature. The pre-gelatinization treatment under certain conditions (forming uniform starch pastes with a low DSG) can provide the starch gels relatively high viscoelasticity after full gelatinization. The n′ and n″ values of all starch gels were very similar, indicating that these samples have similar frequency sensitivity.

3.6. Correlation between the DSG and Its Physicochemical Properties

The physicochemical properties of potato starch as a function of the DSG are shown in Figure 4. Results showed a linear dependent relation between the starch properties and DSG, with their corresponding coefficients of determination (r^2) differing. The prominence detection of the two datasets analyzed was less than 0.01, indicating that these starch properties were significantly correlated with the DSG when the significance level was below 0.01. Both the T_0 from the DSC test and the temperature at peak tanδ from the temperature sweep of the rheometer were positively correlated with the DSG (Figure 4A,B), indicating that the gelatinization temperature of partially gelatinized potato starch increased in the reheating process as the DSG increased. The WBC and K values were also positively correlated with the DSG (Figure 4C,G). The ratios of absorbance 1047/1022 cm^{-1} were negatively correlated with the DSG, while the ratios 1022/995cm^{-1} were positively correlated with the DSG (Figure 4D,E), confirming the loss of the short-range ordered structure of the starch samples with a higher DSG. Moreover, the r^2 of 1022/995cm^{-1} with DSG was 0.844, which is much higher than that of the 1047/1022cm^{-1} with DSG (0.363). Thus, the value of 1022/995 cm^{-1} from the FTIR better represents the DSG of partially gelatinized potato starch after hydrothermal treatment. The maximum point of G′ in the reheating process, K′, and K″ of potato starch samples with DSGs higher than 39.41% were also negatively correlated with their DSG, indicating that the viscoelasticity of starch gels decreased with increasing DSG (Figure 4F,H,I).

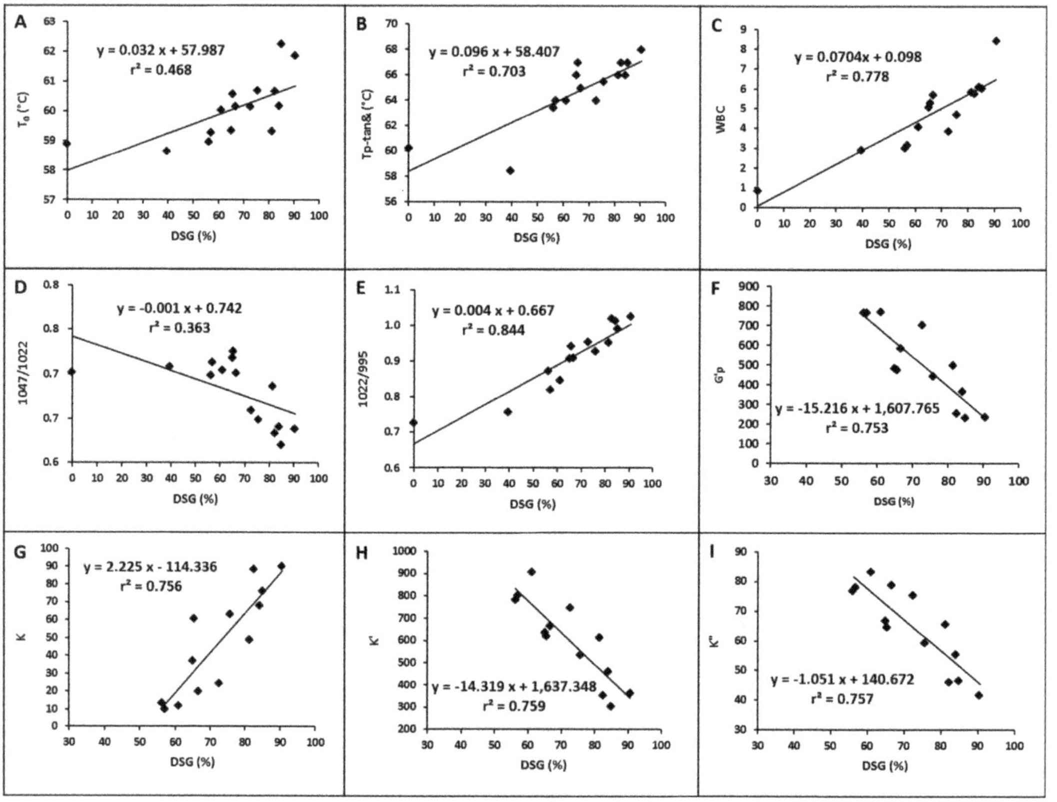

Figure 4. Relationship between the physicochemical properties of potato starch and its degree of gelatinization (**A**: T_0 from DSC; **B**: temperature at the peak tanδ from temperature sweep; **C**: WBC at 20 °C; **D**: ratios of absorbance 1047/1022 cm^{-1} from FTIR; **E**: ratios of absorbance 1022/995 cm^{-1} from FTIR; **F**: peak G′ from temperature sweep; **G**: K value from steady shear test; **H**: K′ value from frequency sweep; **I**: K″ value from frequency sweep).

4. Conclusions

Native and partially gelatinized potato starch (DSG = 39.41%) cannot form uniform pastes in cold water (10%, *w*/*w*). However, partially gelatinized potato starch with a DSG of 56.11% can form stable paste with a fine shear-thinning property, as well as starch samples with a DSG larger than 56.11%. The apparent viscosity, WBC, and gelatinization onset of partially gelatinized potato starch increased as the DSG increased. Higher starch DSG led to the loss of the short-range ordered structure, and we also found that hydrothermal treatment has a more significant effect on the amount of exposed hydroxyl groups than the ordered and amorphous structures of starch. Partially gelatinized potato starch prepared under certain conditions (form uniform starch pastes in cold water with a low DSG) could form starch gels with relatively high viscoelasticity after reheating. The loss of oval or elliptical structures and the disappearance of birefringence of the starch granules can be clearly observed with increasing DSG. A correlation analysis between the starch physicochemical properties and the DSG further confirmed their linear relationships. The findings are expected help researchers better design the processing conditions for partially gelatinized potato starch and to provide information to better formulate food systems with added potato starch. However, since a heating process is normally required in food processing, further study is still needed to investigate the gelatinization and retrogradation of partially gelatinized potato starch with different DSGs during the reheating process. The molecular mechanism leading to the corresponding gelling behaviors still needs to be revealed.

Supplementary Materials: The following are available online at https://www.mdpi.com/article/10.3390/foods10051104/s1, Table S1: The gelatinization onset temperatures of native and partially gelatinized potato starch samples pre-heated at 59 °C and 60 °C for different times (min) by calculating temperatures at the maximum points of Tanδ, Figure S1: Ratios of absorbance 1047/1022 cm^{-1} (open symbols) and 1022/995 cm^{-1} (solid symbols) of native and partially gelatinized potato starch samples as a function of heating time, Figure S2: Temperature (A, B) and frequency (C, D) dependence of tanδ of native and partially gelatinized potato starch samples (A, C: pre-heated at 59 °C; B, D: pre-heated at 60 °C).

Author Contributions: Conceptualization, F.X. and H.H.; methodology, F.X.; software, F.X.; validation, L.Z., W.L. and Q.L.; formal analysis, F.X.; investigation, W.L.; resources, L.Z.; data curation, F.X.; writing—original draft preparation, F.X.; writing—review and editing, F.W.; visualization, H.Z.; supervision, C.B.; project administration, H.H.; funding acquisition, F.X. and H.H. All authors have read and agreed to the published version of the manuscript.

Funding: This research was funded by the National Natural Science Foundation of China, grant number: 31601507; the Agricultural Science and Technology Innovation Program, grant number: CAAS-ASTIP-IFST; the China Agricultural Research System, grant number: CARS-09-P27; and the Special Fund for Agro-scientific Research in the Public Interest of China, grant number: 201503001-2.

Data Availability Statement: The data presented in this study are available on request from the corresponding author.

Conflicts of Interest: The authors declare no conflict of interest.

References

1. Yusuph, M.; Tester, R.F.; Ansell, R.; Snape, C.E. Composition and properties of starches extracted from tubers of different potato varieties grown under the same environmental conditions. *Food Chem.* **2003**, *82*, 283–289. [CrossRef]
2. Hong, J.; Chen, R.; Zeng, X.A.; Han, Z. Effect of pulsed electric fields assisted acetylation on morphological, structural and functional characteristics of potato starch. *Food Chem.* **2016**, *192*, 15–24. [CrossRef] [PubMed]
3. Zhang, C.; Lim, S.-T.; Chung, H.-J. Physical modification of potato starch using mild heating and freezing with minor addition of gums. *Food Hydrocoll.* **2019**, *94*, 294–303. [CrossRef]
4. Alvani, K.; Qi, X.; Tester, R.F.; Snape, C.E. Physico-chemical properties of potato starches. *Food Chem.* **2011**, *125*, 958–965. [CrossRef]
5. Wootton, M.; Bamunuarachch, K.A. Water Binding Capacity of Commercial Produced Native and Modified Starches. *Starch Stärke* **1978**, *30*, 306–309. [CrossRef]
6. Wang, J.; Zhu, H.; Li, S.; Wang, S.; Wang, S.; Copeland, L. Insights into structure and function of high pressure-modified starches with different crystalline polymorphs. *Int. J. Biol. Macromol.* **2017**, *102*, 414–424. [CrossRef]
7. Muneer, F.; Andersson, M.; Koch, K.; Menzel, C.; Hedenqvist, M.S.; Gallstedt, M.; Plivelic, T.S.; Kuktaite, R. Nanostructural morphology of plasticized wheat gluten and modified potato starch composites: Relationship to mechanical and barrier properties. *Biomacromolecules* **2015**, *16*, 695–705. [CrossRef] [PubMed]
8. Wang, H.; Xiao, N.; Wang, X.; Zhao, X.; Zhang, H. Effect of pregelatinized starch on the characteristics, microstructures, and quality attributes of glutinous rice flour and dumplings. *Food Chem.* **2019**, *283*, 248–256. [CrossRef] [PubMed]
9. Fu, Z.; Che, L.; Li, D.; Wang, L.; Adhikari, B. Effect of partially gelatinized corn starch on the rheological properties of wheat dough. *LWT Food Sci. Technol.* **2016**, *66*, 324–331. [CrossRef]

10. Han, H.; Hou, J.; Yang, N.; Zhang, Y.; Chen, H.; Zhang, Z.; Shen, Y.; Huang, S.; Guo, S. Insight on the changes of cassava and potato starch granules during gelatinization. *Int. J. Biol. Macromol.* **2019**, *126*, 37–43. [CrossRef]
11. Juszczak, L.; Witczak, M.; Ziêba, T.; Fortuna, T. Rheological behaviour of heated potato starch dispersions. *Int. Agrophys.* **2012**, *26*, 381–386. [CrossRef]
12. Huang, Z.; Xie, X.; Chen, Y.; Lu, J.; Tong, Z. Ball-milling treatment effect on physicochemical properties and features for cassava and maize starches. *C. R. Chim.* **2008**, *11*, 73–79. [CrossRef]
13. Yu, S.-X.; Mu, T.-H.; Zhang, M.; Zhao, Z.-K. Effects of inorganic salts on the structural and physicochemical properties of high-hydrostatic-pressure-gelatinized sweet potato starch. *Starch Stärke* **2016**, *68*, 980–988. [CrossRef]
14. Dos Santos, T.P.R.; Franco, C.M.L.; Leonel, M. Gelatinized sweet potato starches obtained at different preheating temperatures in a spray dryer. *Int. J. Biol. Macromol.* **2020**, *149*, 1339–1346. [CrossRef] [PubMed]
15. Liu, Y.; Chao, C.; Yu, J.; Wang, S.; Wang, S.; Copeland, L. New insights into starch gelatinization by high pressure: Comparison with heat-gelatinization. *Food Chem.* **2020**, *318*, 126493. [CrossRef] [PubMed]
16. Torres, M.D.; Chenlo, F.; Moreira, R. Rheological Effect of Gelatinisation Using Different Temperature-Time Conditions on Potato Starch Dispersions: Mechanical Characterisation of the Obtained Gels. *Food Bioproc. Tech.* **2018**, *11*, 132–140. [CrossRef]
17. Liu, Y.; Chen, J.; Luo, S.; Li, C.; Ye, J.; Liu, C.; Gilbert, R.G. Physicochemical and structural properties of pregelatinized starch prepared by improved extrusion cooking technology. *Carbohydr. Polym.* **2017**, *175*, 265–272. [CrossRef] [PubMed]
18. Sui, Z.; Yao, T.; Zhao, Y.; Ye, X.; Kong, X.; Ai, L. Effects of heat-moisture treatment reaction conditions on the physicochemical and structural properties of maize starch: Moisture and length of heating. *Food Chem.* **2015**, *173*, 1125–1132. [CrossRef]
19. Sui, Z.; Shah, A.; BeMiller, J.N. Crosslinked and stabilized in-kernel heat-moisture-treated and temperature-cycled normal maize starch and effects of reaction conditions on starch properties. *Carbohydr. Polym.* **2011**, *86*, 1461–1467. [CrossRef]
20. Krueger, B.R.; Knutson, C.A.; Inglett, G.E.; Walker, C.E. A differential scanning calorimetry study on the effect of annealing on gelatinization behaviour of corn starch. *J. Food Sci.* **1987**, *52*, 715–718. [CrossRef]
21. Fu, Z.; Wang, L.; Zou, H.; Li, D.; Adhikari, B. Studies on the starch-water interactions between partially gelatinized corn starch and water during gelatinization. *Carbohydr. Polym.* **2014**, *101*, 727–732. [CrossRef]
22. Taghinezhad, E.; Khoshtaghaza, M.H.; Minaei, S.; Suzuki, T.; Brenner, T. Relationship between Degree of Starch Gelatinization and Quality Attributes of Parboiled Rice during Steaming. *Rice Sci.* **2016**, *23*, 339–344. [CrossRef]
23. Kaur, L.; Singh, N.; Sodhi, N.S. Some properties of potatoes and their starches II. Morphological, thermal and rheological properties of starches. *Food Chem.* **2002**, *79*, 183–192. [CrossRef]
24. Wang, S.; Li, P.; Yu, J.; Guo, P.; Wang, S. Multi-scale structures and functional properties of starches from Indica hybrid, Japonica and waxy rice. *Int. J. Biol. Macromol.* **2017**, *102*, 136–143. [CrossRef] [PubMed]
25. Albano, K.M.; Franco, C.M.L.; Telis, V.R.N. Rheological behavior of Peruvian carrot starch gels as affected by temperature and concentration. *Food Hydrocoll.* **2014**, *40*, 30–43. [CrossRef]
26. Witczak, T.; Witczak, M.; Ziobro, R. Effect of inulin and pectin on rheological and thermal properties of potato starch paste and gel. *J. Food Eng.* **2014**, *124*, 72–79. [CrossRef]
27. Singh, J.; Singh, N. Studies on the morphological, thermal and rheological properties of starch separated from some Indian potato cultivars. *Food Chem.* **2001**, *75*, 67–77. [CrossRef]
28. Sun, X.D.; Arntfield, S.D. Gelation properties of myofibrillar/pea protein mixtures induced by transglutaminase crosslinking. *Food Hydrocoll.* **2012**, *27*, 394–400. [CrossRef]
29. Przetaczek-Roznowska, I.; Fortuna, T.; Wodniak, M.; Labanowska, M.; Pajak, P.; Krolikowska, K. Properties of potato starch treated with microwave radiation and enriched with mineral additives. *Int. J. Biol. Macromol.* **2019**, *124*, 229–234. [CrossRef] [PubMed]
30. Singh, J.; Singh, N. Studies on the morphological and rheological properties of granular cold water soluble corn and potato starches. *Food Hydrocoll.* **2003**, *17*, 63–72. [CrossRef]
31. Fu, Z.; Wang, L.; Li, D.; Adhikari, B. Effects of partial gelatinization on structure and thermal properties of corn starch after spray drying. *Carbohydr. Polym.* **2012**, *88*, 1319–1325. [CrossRef]
32. Wang, S.; Zhang, X.; Wang, S.; Copeland, L. Changes of multi-scale structure during mimicked DSC heating reveal the nature of starch gelatinization. *Sci. Rep.* **2016**, *6*, 28271. [CrossRef]
33. Chung, H.-J.; Li, X.-Q.; Kalinga, D.; Lim, S.-T.; Yada, R.; Liu, Q. Physicochemical properties of dry matter and isolated starch from potatoes grown in different locations in Canada. *Food Res. Int.* **2014**, *57*, 89–94. [CrossRef]
34. Błaszczak, W.; Fornal, J.; Kiseleva, V.I.; Yuryev, V.P.; Sergeev, A.I.; Sadowska, J. Effect of high pressure on thermal, structural and osmotic properties of waxy maize and Hylon VII starch blends. *Carbohydr. Polym.* **2007**, *68*, 387–396. [CrossRef]
35. Soni, P.L.; Sharma, H.W.; Bisen, S.S.; Srivastava, H.C.; Gharia, M.M. Unique Physico-chemical Properties of Sal (Shorea robusta) Starch. *Starch Stärke* **1987**, *39*, 411–413. [CrossRef]
36. Hoover, O.R.; Sosulski, S.F. Effect of Cross-Linking on Functional Properties of Legume Starches. *Starch Stärke* **1986**, *38*, 149–155. [CrossRef]
37. Sevenou, O.; Hill, S.E.; Farhat, I.A.; Mitchell, J.R. Organisation of the external region of the starch granule as determined by infrared spectroscopy. *Int. J. Biol. Macromol.* **2002**, *31*, 79–85. [CrossRef]

38. Wei, C.; Qin, F.; Zhou, W.; Xu, B.; Chen, C.; Chen, Y.; Wang, Y.; Gu, M.; Liu, Q. Comparison of the crystalline properties and structural changes of starches from high-amylose transgenic rice and its wild type during heating. *Food Chem.* **2011**, *128*, 645–652. [CrossRef]
39. Van Riemsdijk, L.E.; Sprakel, J.H.B.; van der Goot, A.J.; Hamer, R.J. Elastic Networks of Protein Particles. *Food Biophys.* **2009**, *5*, 41–48. [CrossRef]
40. Ravindra, P.; Genovese, D.B.; Foegeding, E.A.; Rao, M.A. Rheology of heated mixed whey protein isolate/cross-linked waxy maize starch dispersions. *Food Hydrocoll.* **2004**, *18*, 775–781. [CrossRef]
41. Zhang, X.; Tong, Q.; Zhu, W.; Ren, F. Pasting, rheological properties and gelatinization kinetics of tapioca starch with sucrose or glucose. *J. Food Eng.* **2013**, *114*, 255–261. [CrossRef]
42. Meng, Y.C.; Sun, M.H.; Fang, S.; Chen, J.; Li, Y.H. Effect of sucrose fatty acid esters on pasting, rheological properties and freeze–thaw stability of rice flour. *Food Hydrocoll.* **2014**, *40*, 64–70. [CrossRef]
43. Gryszkin, A.; Zięba, T.; Kapelko, M.; Buczek, A. Effect of thermal modifications of potato starch on its selected properties. *Food Hydrocoll.* **2014**, *40*, 122–127. [CrossRef]
44. Jekle, M.; Mühlberger, K.; Becker, T. Starch–gluten interactions during gelatinization and its functionality in dough like model systems. *Food Hydrocoll.* **2016**, *54*, 196–201. [CrossRef]
45. Shi, A.M.; Wang, L.J.; Li, D.; Adhikari, B. characterization of starch films containing starch nanoparticles. Part 2: Viscoelasticity and creep properties. *Carbohydr. Polym.* **2013**, *96*, 602–610. [CrossRef] [PubMed]

foods MDPI

Article

Effect of an Antibacterial Polysaccharide Produced by *Chaetomium globosum* CGMCC 6882 on the Gut Microbiota of Mice

Xincheng Sun [1,2], Zichao Wang [3,*], Xuyang Hu [1,2], Chengxin Zhao [1,2], Xiaogen Zhang [1,2] and Huiru Zhang [3]

1 Henan Key Laboratory of Cold Chain Food Quality and Safety Control, Zhengzhou University of Light Industry, Zhengzhou 450001, China; sunxinch@zzuli.edu.cn (X.S.); xyhu@zzuli.edu.cn (X.H.); chxzhao@zzuli.edu.cn (C.Z.); xgzhang@zzuli.edu.cn (X.Z.)
2 Collaborative Innovation Center of Food Production and Safety, Zhengzhou 450001, China
3 College of Biological Engineering, Henan University of Technology, Zhengzhou 450001, China; zhr@haut.edu.cn
* Correspondence: zcwang@haut.edu.cn

Abstract: Previously, a polysaccharide produced by *Chaetomium globosum* CGMCC 6882 was found to have antibacterial activity, but its toxic effects on body health and gut microbiota were concealed. Recent results showed that this polysaccharide was safe to Caco-2 cells and mice, while it reduced the body weight gain of mice from 10.5 ± 1.21 g to 8.4 ± 1.17 g after 28 days administration. Acetate, propionate, butyrate and total short-chain fatty acids concentrations increased from 23.85 ± 1.37 μmol/g, 10.23 ± 0.78 μmol/g, 7.15 ± 0.35 μmol/g and 41.23 ± 0.86 μmol/g to 42.77 ± 1.29 μmol/g, 20.03 ± 1.44 μmol/g, 12.06 ± 0.51 μmol/g and 74.86 ± 2.07 μmol/g, respectively. Furthermore, this polysaccharide enriched the abundance of gut microbiota and the *Firmicutes*/*Bacteroidetes* ratio was increased from 0.5172 to 0.7238. Overall, this study provides good guidance for the promising application of polysaccharides as preservatives in foods and in other fields in the future.

Keywords: polysaccharide; antibacterial activity; toxicity; body health; gut microbiota

Citation: Sun, X.; Wang, Z.; Hu, X.; Zhao, C.; Zhang, X.; Zhang, H. Effect of an Antibacterial Polysaccharide Produced by *Chaetomium globosum* CGMCC 6882 on the Gut Microbiota of Mice. *Foods* **2021**, *10*, 1084. https://doi.org/10.3390/foods10051084

Academic Editors: Jianhua Xie, Yanjun Zhang and Hansong Yu

Received: 29 March 2021
Accepted: 11 May 2021
Published: 13 May 2021

1. Introduction

The intestinal tract is home to a large number of complex and diverse gut microbiota [1], such as different types of bacteria, viruses and fungi [2]. As a bridge between diet and host health, gut microbiota not only affects the digestion and absorption of nutrients in the diet, but also regulates the normal physiological functions and the occurrence of diseases in the host [3]. Recently, with the help of next-generation high-throughput sequencing technology, bioinformatics and metagenomics [4], researchers have verified that gut microbiota is vital to host health, and the disruption of gut microbiota has been shown to be associated with multiple diseases, including metabolic syndrome [5], obesity [6], tumor [7], diabetes [8], HIV [9], flu [10], fatigue [11], brain health [12], etc. At the same time, intestinal flora transplantation has shown promising application prospects in the treatment of diseases [13].

With their antibacterial and growth-promoting properties, antibiotics are widely used in disease treatment and daily production. However, abuse of antibiotics not only increases the antibiotic residue in foods and the resistance of disease-fighting microorganisms [14], but it also leads to drug-resistant genes being transmitted from livestock and microorganisms to humans [15]. Meanwhile, antibiotics can directly damage body health via disrupting the homeostasis of gut microbiota in the intestinal tract. For example, Cox et al. [16] found that mice treated with continuous low doses of penicillin could develop a higher body weight due to the disruption of the gut microbiota. Zhang et al. [2] reported that the gene expression and metabolic homeostasis of mice were affected by the administration of perfluorooctane sulfonate. Xu et al. [17] demonstrated that antibiotics could promote tumor initiation in mice by inducing gut microbiota dysbiosis. Therefore, looking for a new generation of safe, high-efficient, widely applicable and non-toxic antibiotics has been growing increasingly important.

As a kind of macromolecule connected by more than ten monosaccharides through a glycoside bond, the gut microbiota could convert polysaccharides into short-chain fatty acids (SCFAs), such as acetic, propionic and butyric acid, thus having a positive effect on gut microbiota and body health [18]. However, the effects of antibacterial polysaccharides on body health, especially the gut microbiota in the intestinal tract, are concealed and poorly understood. Herein, this work assayed the toxicity of an antibacterial polysaccharide (GCP) produced by *Chaetomium globosum* CGMCC 6882 [19] to Caco-2 cells. Secondly, the effects of GCP on the body weight and serum biochemistry of normal mice were detected. Finally, the influence of GCP on the gut microbiota of normal mice was assessed. We hope that this work could provide some help and guidance for the application of bacteriostatic polysaccharides.

2. Materials and Methods

2.1. Preparation of GCP

The preparation of GCP produced from *C. globosum* CGMCC 6882 was based on the methods reported in our previous work [20]. Briefly, fermentation liquid was filtered and centrifuged at 12,000× *g* for 30 min to remove mycelium and cells. The supernatant was de-proteinized by adding three volumes Sevag solution, then three volumes cold alcohol were added and it was kept at 4 °C overnight to precipitate GCP. The crude GCP re-dissolved in distilled water was de-pigmented with AB-8 macroporous resin (Beijing NuoqiYa Biotechnology Co., Ltd., Beijing, China) and then dialyzed for 48 h in distilled water. After this, GCP solution was filtered with a 0.22 μm filter and applied to a Sepharose CL-6B column (2.5 cm × 60 cm) for further purification, eluted with 0.1 mol/L NaCl solution at a flow rate of 0.6 mL/min, and the fraction was then collected. In the end, the purified GCP was lyophilized for further experiments.

2.2. Cell Viability Assay

The toxicity of GCP to Caco-2 cells (American Type Culture Collection, ATCC, HTB037) was measured by the 3-(4,5-Dimethylthiazol-2-yl)-2,5-bromo diphenyltetrazolium (MTT) assay reported in our previous work [21] with some modifications. Dimethyl sulfoximine (DMSO), Dulbecco's modified Eagle medium (DMEM) and MTT were brought from Sigma-Aldrich (Shanghai, China). Meanwhile, fetal bovine serum, penicillin and streptomycin were brought from Sangon Biotech (Shanghai, China). The Caco-2 cells were cultured in DMEM containing 10% (*v*/*v*) fetal bovine serum, 100 U/mL penicillin and 100 μg/mL of streptomycin at 37 °C in a humidified 5% CO_2 incubator (Series 8000 WJ, Thermo Fisher Seientific, Waltham, MA, USA). Before experiment, the dried GCP powder was dissolved in different concentrations of DMEM solution (100–600 μg/mL) and DMEM solution without GCP was used as the control. Briefly, Caco-2 cells were seeded into 96-well plates at a concentration of 2×10^4 cells/mL and incubated at 37 °C in 5% CO_2 for 24 h before treatment.

Then, 100 μL GCP at different concentrations was added into wells and cultured for another 24 h. Afterwards, 20 μL of 5 mg/mL MTT was added. After 4 h of incubation, cell supernatant was discarded and 150 μL DMSO was added to dissolve the insoluble crystals in the cell. In the end, the absorbance of each well was recorded by a microplate reader (Bio-Rad Laboratories, Inc., Pleasanton, CA, USA) at 490 nm.

2.3. Experimental Design and Samples Collection

Specific pathogen free-male mice (20 ± 1 g) were purchased from the Laboratory Animal Center of Henan province (SCXK: 2017-0002; Zhengzhou, China). All mice were held in independent cages and kept in specific pathogen-free conditions at temperatures of 24 ± 1 °C, humidity of 60 ± 5%, and with a light to dark cycle of 12 h/12 h. During experiments, all mice were monitored every day, and the experiments were performed strictly according to the guidelines for the care and use of laboratory animals (Henan University of Technology, Zhengzhou, China). Forty mice were randomly divided into four groups (n = 10) after adaption for 7 days. One group was used as the normal control group (NC), and another three groups were designed as the experimental group and treated with 100 μg/mL GCP (low-dose group), 200 μg/mL GCP (middle-dose group) and 400 μg/mL GCP (high-dose group), respectively. Then, mice in the experimental groups were orally administered 0.5 mL GCP once a day, and mice in the normal control group were administered equal distilled water. The animal experiments lasted for 28 days and all mice were weighted weekly. During the whole experiment, all mice had free access to a basic diet and distilled water. At the end of experiment, all mice were killed after fasting for 12 h. Blood samples were collected from the orbit and centrifuged at 3000 r/min for 10 min to collect the serum. Meanwhile, the contents of the cecum were immediately collected in plastic tubes (1.5 mL) and stored at −80 °C for further analyses.

2.4. Serum Biochemical Index Detection

The levels of aspartate transaminase, alanine aminotransferase, total protein, albumin, globulin, urea, high-density lipoprotein and low-density lipoprotein, and the glucose concentration in the serum, were tested using the serum analyzer (BS-420, Shenzhen Mindray Biomedical Electronics Co., Ltd., Wuhan, China).

2.5. Measurement of SCFAs

The concentration of SCFAs in cecum contents was analyzed according to the method of Wu et al. [22] with some modifications. Briefly, 50 mg of cecum contents was suspended in 500 μL of saturated sodium chloride solution and vortexed uniformly for 30 min. Then, the solutions were acidified with 20 μL of 10% H_2SO_4 and extracted with 1 mL ethyl ether. After this, the mixtures were centrifuged at 12,000× g and 4 °C for 10 min, and the obtained organic layer of supernatant was mixed with 0.25 g of anhydrous sodium sulfate for 5 min to remove water. In the end, the supernatant was filtered with 0.22 μm organic-based filter membrane and the SCFAs in the organic layer were analyzed by a 7890A GC system (Agilent Technologies Inc., Santa Clara, CA, USA) equipped with a flame ionization detector; the carrier gas was N_2, the shunt ratio was 20:1 and the flow rate was 1.5 mL/min. The chromatographic column was HP-INNOWAX (Agilent, 30 m × 0.25 mm × 0.25 μm), and the temperature procedure was as follows: temperature was increased from 60 °C to 190 °C at 20 °C/min and maintained for 4 min. The injection temperature was 200 °C, the ionization temperature was 250 °C and the injection volume was 5 μL. The standard curve was made by the external standard method and the concentrations of SCFAs were calculated according to the standard curve.

2.6. DNA Extraction of Cecum Contents and High-Throughput Sequencing

The cecum contents were sent to Majorbio Co., Ltd., China (Shanghai, China) for DNA extraction and sequencing of 16S rRNA gene. Briefly, the total microbial DNA was extracted from cecum contents (n = 10) with the DNA extraction kit and tested by agarose gel electrophoresis. The V3-V4 hypervariable region of 16S rRNA was amplified by polymerase chain reaction (PCR), and the primer sequences were 338F (5′-ACTCCTACGGGAGGCAGCAG-3′) and 806R (5′-CTCCTACGGGAGG CAGCAG-3′). Then, the PCR products of equimolar concentrations were sequenced using Illumina MiSeq platforms according to the operation manual.

2.7. Statistical Analysis

Data were expressed as the mean ± standard deviation (SD). Data were subjected to one-way ANOVA, and significance differences were analyzed using SPSS version 19.0 (IBM Company, Armonk, NY, USA).

3. Results and Discussion

3.1. Cell Viability Assay

Antibiotics easily cause some side effects in the treatment of diseases, such as antibiotic-associated diarrhea [23]. Even though it is the bacteriocin approved by JECFA, nisin could also bring some adverse effects to the gut microbiota of the body [24]. As shown in Figure 1, with the experimental concentrations of GCP from 100 μg/mL to 600 μg/mL, cell viability increased from 101 ± 0.8% to 115 ± 1.3% ($p < 0.05$), suggesting that GCP was not toxic to Caco-2 cells. This toxicity result was similar to that of *Ganoderma lucidum* polysaccharide reported in our earlier work [21] and in works by other researchers. For example, Zhang et al. [25] found that the alkali-soluble polysaccharides from *Arctium lappa* L. had no toxicity to RAW264.7 cells. Caillot et al. [26] reported that the blackberry wine polysaccharides had no toxicity to RAW 264.7 macrophages. Meanwhile, during the whole experiment, there were no signs of disease or death in mice, indicating the security of GCP to mice. Furthermore, He et al. [27] demonstrated that a novel polysaccharide produced by *Streptomyces Virginia* H03 was safe to mice when administered at doses of 500 mg/kg/day. Therefore, the concentrations of GCP used in the following tests were 100 μg/mL, 200 μg/mL and 400 μg/mL, respectively.

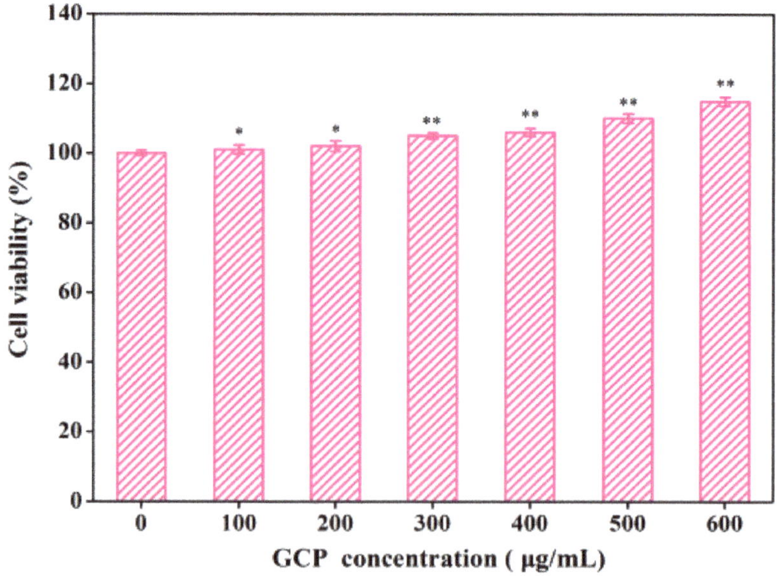

Figure 1. Cell viability assay of GCP to Caco-2 cells ($n = 5$). Significance was determined through ANOVA, * $p < 0.05$, ** $p < 0.01$.

3.2. Effect of GCP on the Body Weight of Normal Mice

As can be seen from Table 1, the administration of GCP decreased the average body weight of mice, and the body weight gain of mice significantly ($p < 0.05$) decreased from 10.5 ± 1.21 g to 8.4 ± 1.17 g. Chen et al. [23] reported that *Pueraria lobata* polysaccharide could effectively reduce the average body weight of mice. Meanwhile, Tian et al. [28] and Yin et al. [6] demonstrated that *Lycium ruthenicum* and resveratrol had similar weight loss effects as those seen in this work. However, Yang et al. [29] and Wei et al. [30] found that flaxseed polysaccharides and *Musa basjoo* had almost no effects on the weight loss of mice, which was inconsistent with the results of this work. The weight gain control of GCP might relate to the metabolism of acetate, alanine aminotransferase and aspartate aminotransferase [6,28], and this will be analyzed in the following work.

Table 1. Effect of GCP on the body weight of mice ($n = 10$).

Mice Weight	Control Group	GCP Concentration (μg/mL)		
		100	200	400
0 day (g)	20.2 ± 1.57 [a]	20.4 ± 1.38 [a]	20.0 ± 2.23 [a]	20.8 ± 2.42 [a]
7 day (g)	23.7 ± 2.27 [b]	23.5 ± 2.09 [b]	22.9 ± 3.37 [b]	22.5 ± 2.45 [b]
14 day (g)	26.5 ± 2.34 [c]	26.1 ± 1.99 [c]	25.7 ± 2.54 [c]	25.4 ± 2.08 [c]
21 day (g)	28.1 ± 2.14 [d]	27.9 ± 2.23 [d]	27.4 ± 2.15 [d]	26.9 ± 2.64 [d]
28 day (g)	30.7 ± 2.61 [e]	30.1 ± 2.45 [e]	29.1 ± 2.38 [e]	29.2 ± 3.17 [e]
Weight gain (g)	10.5 ± 1.21 [a]	9.7 ± 1.09 [b]	9.1 ± 0.88 [c]	8.4 ± 1.17 [d]

Different letters (a–e) represent significant differences, $p < 0.05$.

3.3. Effect of GCP on the Serum Biochemistry of Normal Mice

The effects of GCP on physiological parameters in the serum of mice are shown in Table 2. There was a decreasing trend ($p < 0.05$) in the concentration of aspartate transaminase and alanine aminotransferase in the mice serum after the administration of GCP; the aspartate transaminase concentration decreased from 38.5 ± 3.37 U/L to 35.4 ± 1.87 U/L, and the alanine aminotransferase concentration reduced from 121.5 ± 6.43 U/L to 105.4 ± 10.91 U/L. Meanwhile, Guo et al. [31] and Yin et al. [6] reported that the weight loss effects of nanobubble water and resveratrol were partly related to the reduction inaspartate transaminase and alanine aminotransferase in the body, which could partly explain why the GCP reduced the weight of the mice in Table 1. However, the activities of total protein, albumin, globulin, urea, high-density lipoprotein, low-density lipoprotein and the glucose in the mice serum among the four groups underwent little change, suggesting that GCP had no toxicity to mice.

Table 2. Effect of GCP on the serum biochemistry of mice ($n = 10$).

Serum Biochemistry	Control Group	GCP Concentration (μg/mL)		
		100	200	400
Aspartate transaminase (U/L)	38.5 ± 3.37 [a]	37.3 ± 2.98 [a]	36.2 ± 3.29 [b]	35.4 ± 1.87 [b]
Alanine aminotransferase (U/L)	121.5 ± 6.43 [a]	115.3 ± 9.31 [b]	111.9 ± 11.34 [c]	105.4 ± 10.91 [d]
Total protein (g/L)	53.5 ± 3.04 [a]	54.5 ± 5.09 [a]	54.8 ± 4.10 [a]	53.9 ± 2.08 [a]
Albumin (g/L)	31.7 ± 1.63 [a]	32.1 ± 3.14 [a]	31.9 ± 2.68 [a]	31.2 ± 2.10 [a]
Globulin (g/L)	20.8 ± 1.41 [a]	21.1 ± 2.15 [a]	20.9 ± 3.32 [a]	20.7 ± 2.47 [a]
Urea (mmol/L)	11.62 ± 0.83 [a]	11.75 ± 1.22 [a]	11.69 ± 0.91 [a]	11.66 ± 1.13 [a]
High density lipoprotein (mmol/L)	1.66 ± 0.03 [a]	1.68 ± 0.02 [a]	1.63 ± 0.03 [a]	1.65 ± 0.06 [a]
Low density lipoprotein (mmol/L)	0.11 ± 0.008 [a]	0.09 ± 0.004 [a]	0.12 ± 0.012 [a]	0.10 ± 0.007 [a]
Glucose (mmol/L)	5.74 ± 0.41 [a]	5.84 ± 0.12 [a]	5.76 ± 0.51 [a]	5.79 ± 0.32 [a]

Different letters (a–d) represent significant differences, $p < 0.05$.

3.4. Effect of GCP on the SCFAs of Normal Mice

Due to the glycoside bonds between monosaccharides and their complex structure, most polysaccharides are resistant to saliva and gastric and small intestinal juices, which are further utilized by gut microbiota to produce SCFAs [18,32]. As illustrated in Figure 2, the concentrations of SCFAs in the control group were significantly lower than in experimental groups ($p < 0.01$). After administration of 400 μg/mL GCP for 28 days, the acetate concentration in the cecum contents of mice increased from 23.85 ± 1.37 μmol/g to 42.77 ± 1.29 μmol/g, the propionate concentration increased from 10.23 ± 0.78 μmol/g to 20.03 ± 1.44 μmol/g, the butyrate concentration increased from 7.15 ± 0.35 μmol/g to 12.06 ± 0.51 μmol/g, and the total SCFAs increased from 41.23 ± 0.86 μmol/g to 74.86 ± 2.07 μmol/g. SCFAs could inhibit the growth and reproduction of pathogenic bacteria by reducing the acidity of the intestinal environment, but could also produce a positive systematic physiological effect on the host via regulating the innate and adaptive immune systems and intestinal permeability [33]. Furthermore, SCFAs may also be conducive to losing weight by promoting satiety [34], especially acetate, thus explaining the reduction in body weight gain caused by GCP in Table 1. Many factors affect the utilization of polysaccharides by gut microbiota, such as linkage mode, chain type, molecular weight, sulfate content, etc. [35]. In future work, we will investigate the effect of the GCP's structure on the gut microbiota utilization.

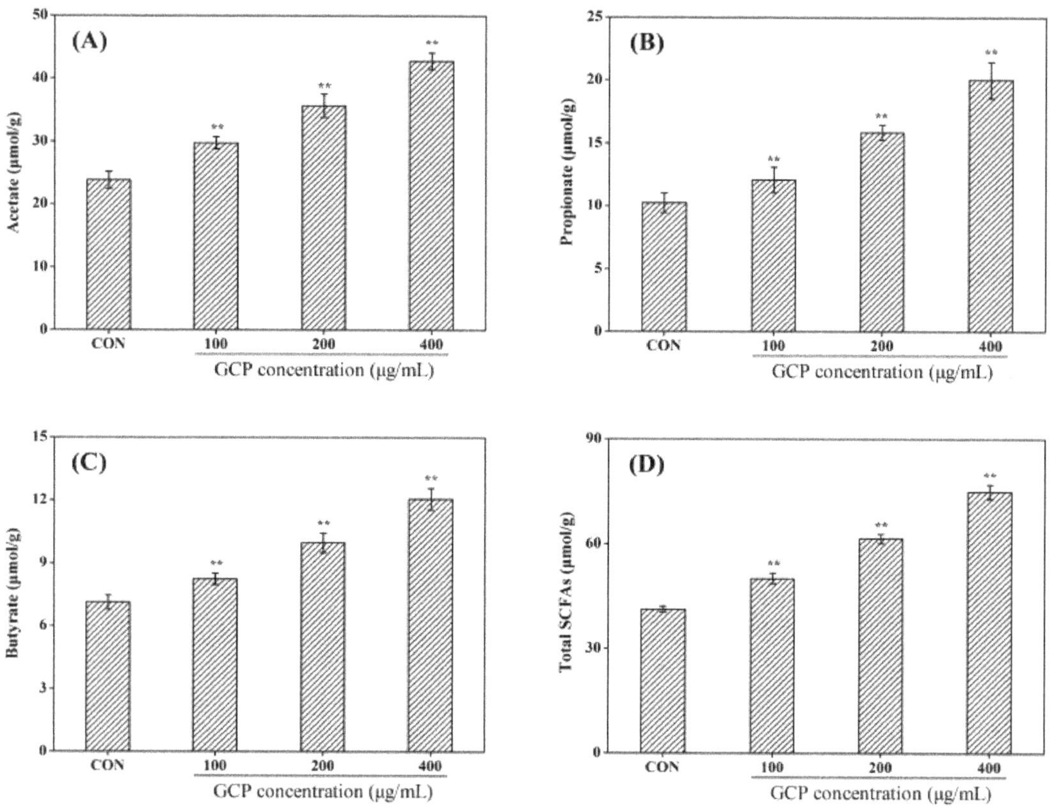

Figure 2. Effect of GCP on the concentrations of acetate, propionate, butyrate and total SCFAs in the cecum of mice ($n = 10$). (**A**) Acetate, (**B**) propionate, (**C**) butyrate, (**D**) total SCFAs. CON: control group. Significance was determined through ANOVA, ** $p < 0.01$.

3.5. Effect of GCP on the Composition and Diversity of Gut Microbiota

3.5.1. Diversity Analysis of the Structure of Gut Microbiota

Through α-diversity analysis, a series of statistical analysis indexes was used to estimate and reflect the abundance and diversity of microbial communities. Sobs, ACE and Chao index reflected the species richness of operational taxonomic units (OTUs) in the gut microbiota, while the Simpson and Shannon indexes reflected the differences in species diversity in the gut microbiota [22]. As shown in Table 3, the Sobs, ACE, Chao and Shannon index of the gut microbiota in the experimental groups showed an increasing trend ($p < 0.05$) in a concentration-dependent manner compared to the control group, and the Simpson index showed a downward trend ($p < 0.05$). This indicates that GCP increased the species richness and diversity of the gut microbiota in the cecum contents of mice.

Table 3. Effect of GCP on the α-diversity of gut microbiota ($n = 10$).

Diversity Index	Control Group	GCP Concentration (μg/mL)		
		100	200	400
Sobs	360.66 ± 18.64 ^{a}	368.66 ± 15.21 ^{b}	382.00 ± 21.35 ^{c}	379.50 ± 26.33 ^{c}
ACE	401.41 ± 29.32 ^{a}	408.41 ± 35.03 ^{b}	416.06 ± 27.06 ^{c}	414.54 ± 38.09 ^{c}
Chao1	404.96 ± 25.54 ^{a}	418.25 ± 34.17 ^{b}	426.88 ± 29.13 ^{c}	424.96 ± 34.10 ^{c}
Simpson	0.074 ± 0.09 ^{a}	0.057 ± 0.01 ^{b}	0.049 ± 0.05 ^{c}	0.043 ± 0.09 ^{d}
Shannon	3.84 ± 0.19 ^{a}	3.94 ± 0.27 ^{b}	4.19 ± 0.31 ^{c}	4.30 ± 0.39 ^{d}

Different letters (a–d) represent significant differences, $p < 0.05$.

The different numbers of OTUs are illustrated among the four groups by the Venn diagram in (Figure 3). Among all the OTUs in this work, 371 were shared by all groups. Meanwhile, the numbers of OTUs shared by experimental groups and control group were 414 (low-dose group), 441 (middle-dose group) and 425 (high-dose group). Furthermore, 88 OTUs were detected in the low-dose group but not in the control group, and 77 and 88 OTUs were detected separately in the middle-dose group and high-dose group. Furthermore, a different group had its own separate set of OTUs: 2 in the control group, 8 in the low-dose group, 7 in the middle-dose group and 12 in the high-dose group. However, the total numbers of OTUs in the control group, low-dose group, middle-dose group and high-dose group were 472, 502, 518 and 513, respectively. This suggested that GCP increased the species richness of the gut microbiota in the cecum of mice.

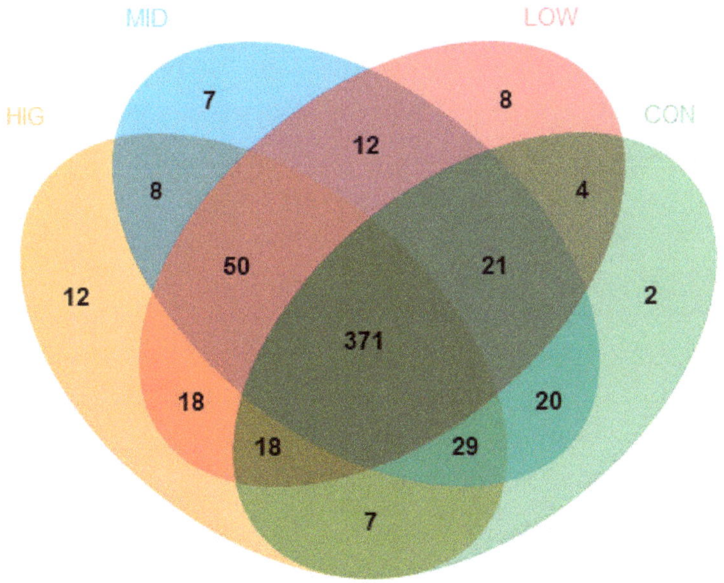

Figure 3. Venn diagram of colon gut microbiota ($n = 10$). CON: control group; LOW: 100 μg/mL GCP; MID: 200 μg/mL GCP; HIG: 400 μg/mL GCP.

3.5.2. Composition Analysis of the Gut Microbiota

The relative abundance of cecum gut microbiota composition of each group is shown in Figure 4. At the phylum level, the gut microbiota in the four groups mainly consisted of *Bacteroidetes*, *Firmicutes*, *Proteobacteria*, *Spirochaetae* and *Verrucomicrobia*, and these five phyla in all groups represented approximately ≥95% of the sequences. The relative abundances of *Bacteroidetes* in the four groups were 62.41%, 55.38%, 54.05% and 58.97%, and those of *Firmicutes* were 32.28%, 33.81%, 39.12% and 38.36%. Notably, th *Firmicutes/Bacteroidetes* (F/B) ratio increased from 0.5172 in the control group to 0.6105 (low-dose group), 0.7238 (middle-dose group) and 0.6505 (high-dose group), showing an increasing trend. The increased dose of GCP might have an effect on the change in ratio of F/B in mice gut microbiota, but as the two main communities that affect energy metabolism homeostasis [36], researchers have different opinions on the functions of *Firmicutes* and *Bacteroidetes*. Wu et al. [22] found that *Cyclocarya paliurus* polysaccharides alleviated the liver inflammation of mice by increasing the F/B ratio in their gut microbiota. However, Yang et al. [29] demonstrated that *Linum usitatissimum* L. polysaccharides mitigated the high-fat diet-induced metabolic syndrome in mice, which did not affect the relative abundance of *Bacteroidetes*, but reduced the relative abundance of *Firmicutes*. At the same time, a few researchers have suggested that obesity is not associated with the ratio of F/B [37,38]. The utilization and digestion of polysaccharides by gut microbiota are affected by many factors, including monosaccharide composition, linkage mode, chain type, molecular weight, sulfate content, etc. [4,29]. In future work, we will investigate the effect of high-dose GCP with a relatively low F/B ratio (0.6505), compared with the 0.7238 of the middle-dose group.

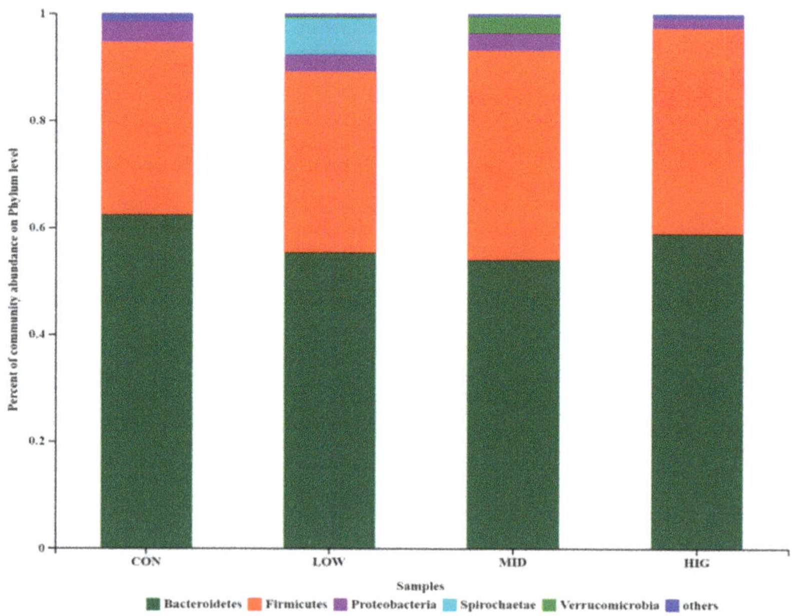

Figure 4. Relative abundance of the gut microbiota at the phylum level (n = 10). CON: control group; LOW: 100 µg/mL GCP; MID: 200 µg/mL GCP; HIG: 400 µg/mL GCP.

4. Conclusions

The toxicity of antibacterial polysaccharides to the body and gut microbiota is poorly understood. The results in the present work show that the antibacterial polysaccharide of GCP was safe for Caco-2 cells and mice. Meanwhile, GCP reduced the body weight gain of mice and increased the SCFAs concentration in the colon. Furthermore, GCP increased the diversity of gut microbiota and the *Firmicutes/Bacteroidetes* ratio. In future work, the specific bacteria in the microbiota of a gut affected by GCP, especially *Firmicutes* and *Bacteroidetes*, will be investigated.

Author Contributions: Conceptualization and methodology, X.S.; software and validation, X.H.; investigation, resources and data curation, C.Z.; writing—original draft preparation, X.Z.; writing—review and editing, Z.W.; visualization and supervision, H.Z. All authors have read and agreed to the published version of the manuscript.

Funding: This work is supported by Natural Science Foundation of Henan Province (212300410131), Henan major public welfare project (201300110100), Henan Province Human Resources and Social Security department Science and Sechnology Activities Programs for overseas students, Henan Key Laboratory of Cold Chain Food Quality and Safety Control (CCFQ2020-YB-18), Henan major public welfare project (201300110100), and The major breeding project of Zhengzhou university of light industry (2020ZDPY0102).

Institutional Review Board Statement: The experiments were performed strictly according to the guidelines for the care and use of laboratory animals of Henan University of Technology.

Informed Consent Statement: Not applicable.

Data Availability Statement: Not applicable.

Conflicts of Interest: The authors declare that there are no conflict of interest.

References

1. Nicholson, J.K.; Holmes, E.; Kinross, J.; Burcelin, R.; Gibson, G.; Jia, W.; Pettersson, S. Host-gut microbiota metabolic interactions. *Science* **2012**, *336*, 1262. [CrossRef] [PubMed]
2. Zhang, L.; Rimal, B.; Nichols, R.G.; Tian, Y.; Smith, P.B.; Hatzakis, E.; Chang, S.-C.; Butenhoff, J.L.; Peters, J.M.; Patterson, A.D. Perfluorooctane sulfonate alters gut microbiota-host metabolic homeostasis in mice. *Toxicology* **2020**, *431*, 152365. [CrossRef]
3. Li, Z.-T.; Zhu, L.; Zhang, W.-L.; Zhan, X.-B.; Gao, M.-J. New dynamic digestion model reactor that mimics gastrointestinal function. *Biochem. Eng. J.* **2020**, *154*, 107431. [CrossRef]
4. Shang, Q.; Jiang, H.; Cai, C.; Hao, J.; Li, G.; Yu, G. Gut microbiota fermentation of marine polysaccharides and its effects on intestinal ecology: An overview. *Carbohydr. Polym.* **2018**, *179*, 173–185. [CrossRef]
5. Hu, B.; Ye, C.; Leung, E.L.-H.; Zhu, L.; Hu, H.; Zhang, Z.; Zheng, J.; Liu, H. *Bletilla striata* oligosaccharides improve metabolic syndrome through modulation of gut microbiota and intestinal metabolites in high fat diet-fed mice. *Pharmacol. Res.* **2020**, *159*, 104942. [CrossRef] [PubMed]
6. Yin, X.; Liao, W.; Li, Q.; Zhang, H.; Liu, Z.; Zheng, X.; Zheng, L.; Feng, X. Interactions between resveratrol and gut microbiota affect the development of hepatic steatosis: A fecal microbiota transplantation study in high-fat diet mice. *J. Funct. Foods* **2020**, *67*, 103883. [CrossRef]
7. Liu, L.; Li, M.; Yu, M.; Shen, M.; Wang, Q.; Yu, Y.; Xie, J. Natural polysaccharides exhibit anti-tumor activity by targeting gut microbiota. *Int. J. Biol. Macromol.* **2019**, *121*, 743–751. [CrossRef]
8. Zhao, L.; Zhang, F.; Ding, X.; Wu, G.; Lam, Y.Y.; Wang, X.; Fu, H.; Xue, X.; Lu, C.; Ma, J.; et al. Gut bacteria selectively promoted by dietary fibers alleviate type 2 diabetes. *Science* **2018**, *359*, 1151–1156. [CrossRef]
9. Vujkovic-Cvijin, I.; Dunham, R.M.; Iwai, S.; Maher, M.C.; Albright, R.G.; Broadhurst, M.J.; Hernandez, R.D.; Lederman, M.M.; Huang, Y.; Somsouk, M.; et al. Dysbiosis of the Gut Microbiota Is Associated with HIV Disease Progression and Tryptophan Catabolism. *Sci. Transl. Med.* **2013**, *5*, 193ra91. [CrossRef]
10. Trompette, A.; Gollwitzer, E.S.; Pattaroni, C.; Lopez-Mejia, I.C.; Riva, E.; Pernot, J.; Ubags, N.; Fajas, L.; Nicod, L.P.; Marsland, B.J. Dietary Fiber Confers Protection against Flu by Shaping Ly6c− Patrolling Monocyte Hematopoiesis and CD8+ T Cell Metabolism. *Immunity* **2018**, *48*, 992–1005.e8. [CrossRef]
11. Zhong, L.; Ma, N.; Zheng, H.; Ma, G.; Zhao, L.; Hu, Q. Tuber indicum polysaccharide relieves fatigue by regulating gut microbiota in mice. *J. Funct. Foods* **2019**, *63*, 103580. [CrossRef]
12. Al-Khafaji, A.H.; Jepsen, S.D.; Christensen, K.R.; Vigsnæs, L.K. The potential of human milk oligosaccharides to impact the microbiota-gut-brain axis through modulation of the gut microbiota. *J. Funct. Foods* **2020**, *74*, 104176. [CrossRef]
13. Chen, D.; Wu, J.; Jin, D.; Wang, B.; Cao, H. Fecal microbiota transplantation in cancer management: Current status and per-spectives. *Int. J. Cancer* **2019**, *145*, 2021–2031. [CrossRef]
14. van Boeckel, T.P.; Glennon, E.E.; Chen, D.; Gilbert, M.; Robinson, T.P.; Grenfell, B.B.; Levin, S.A.S.; Bonhoeffer, S.; Laxminarayan, R.R. Reducing antimicrobial use in food animals. *Science* **2017**, *357*, 1350–1352. [CrossRef]
15. Wang, Y.; Zhang, R.; Li, J.; Wu, Z.; Yin, W.; Schwarz, S.; Tyrrell, J.M.; Zheng, Y.; Wang, S.; Shen, Z.; et al. Comprehensive resistome analysis reveals the prevalence of NDM and MCR-1 in Chinese poultry production. *Nat. Microbiol.* **2017**, *2*, 16260. [CrossRef] [PubMed]
16. Cox, L.M.; Yamanishi, S.; Sohn, J.; Alekseyenko, A.V.; Leung, J.M.; Cho, I.; Kim, S.G.; Li, H.; Gao, Z.; Mahana, D.; et al. Altering the Intestinal Microbiota during a Critical Developmental Window Has Lasting Metabolic Consequences. *Cell* **2014**, *158*, 705–721. [CrossRef] [PubMed]

17. Xu, C.; Ruan, B.; Jiang, Y.; Xue, T.; Wang, Z.; Lu, H.; Wei, M.; Wang, S.; Ye, Z.; Zhai, D.; et al. Antibiotics-induced gut microbiota dysbiosis promotes tumor initiation via affecting APC-Th1 development in mice. *Biochem. Biophys. Res. Commun.* **2017**, *488*, 418–424. [CrossRef]
18. Fang, Q.; Hu, J.; Nie, Q.; Nie, S. Effects of polysaccharides on glycometabolism based on gut microbiota alteration. *Trends Food Sci. Technol.* **2019**, *92*, 65–70. [CrossRef]
19. Wang, Z.; Zhu, J.; Li, W.; Li, R.; Wang, X.; Qiao, H.; Sun, Q.; Zhang, H. Antibacterial mechanism of the polysaccharide produced by Chaetomium globosum CGMCC 6882 against Staphylococcus aureus. *Int. J. Biol. Macromol.* **2020**, *159*, 231–235. [CrossRef] [PubMed]
20. Wang, Z.; Xue, R.; Cui, J.; Wang, J.; Fan, W.; Zhang, H.; Zhan, X. Antibacterial activity of a polysaccharide produced from Chaetomium globosum CGMCC 6882. *Int. J. Biol. Macromol.* **2019**, *125*, 376–382. [CrossRef]
21. Wang, Z.; Zhang, H.; Shen, Y.; Zhao, X.; Wang, X.; Wang, J.; Fan, K.; Zhan, X. Characterization of a novel polysaccharide from *Ganoderma lucidum* and its absorption mechanism in Caco-2 cells and mice model. *Int. J. Biol. Macromol.* **2018**, *118*, 320–326. [CrossRef]
22. Wu, T.; Shen, M.; Guo, X.; Huang, L.; Yang, J.; Yu, Q.; Chen, Y.; Xie, J. *Cyclocaryapaliurus* polysaccharide alleviates liver inflammation in mice via beneficial regulation of gut microbiota and TLR4/MAPK signaling pathways. *Int. J. Biol. Macromol.* **2020**, *160*, 164–174. [CrossRef] [PubMed]
23. Chen, R.; Liu, B.; Wang, X.; Chen, K.; Zhang, K.; Zhang, L.; Fei, C.; Wang, C.; Liu, Y.; Xue, F.; et al. Effects of polysaccharide from *Pueraria lobata* on gut microbiota in mice. *Int. J. Biol. Macromol.* **2020**, *158*, 740–749. [CrossRef] [PubMed]
24. Cao, Y.; Liu, H.; Qin, N.; Ren, X.; Zhu, B.; Xia, X. Impact of food additives on the composition and function of gut microbiota: A review. *Trends Food Sci. Technol.* **2020**, *99*, 295–310. [CrossRef]
25. Zhang, X.; Zhang, N.; Kan, J.; Sun, R.; Tang, S.; Wang, Z.; Chen, M.; Liu, J.; Jin, C. Anti-inflammatory activity of alkali-soluble polysaccharides from *Arctium lappa* L. and its effect on gut microbiota of mice with inflammation. *Int. J. Biol. Macromol.* **2020**, *154*, 773–787. [CrossRef]
26. Caillot, A.R.C.; Bezerra, I.D.L.; Palhares, L.C.G.F.; Santana-Filho, A.P.; Chavante, S.F.; Sassaki, G.L. Structural characterization of blackberry wine polysaccharides and immunomodulatory effects on LPS-activated RAW 264.7 macrophages. *Food Chem.* **2018**, *257*, 143–149. [CrossRef]
27. He, F.; Yang, Y.; Yang, G.; Yu, L. Studies on antibacterial activity and antibacterial mechanism of a novel polysaccharide from *Streptomyces virginia* H03. *Food Control.* **2010**, *21*, 1257–1262. [CrossRef]
28. Tian, B.; Zhao, J.; An, W.; Zhang, J.; Cao, X.; Mi, J.; Zhao, J.; Zhang, Y.; Li, J. *Lyciumruthenicum* diet alters the gut microbiota and partially enhances gut barrier function in male C57BL/6 mice. *J. Funct. Foods* **2019**, *52*, 516–528. [CrossRef]
29. Yang, C.; Xu, Z.; Deng, Q.; Huang, Q.; Wang, X.; Huang, F. Beneficial effects of flaxseed polysaccharides on metabolic syndrome via gut microbiota in high-fat diet fed mice. *Food Res. Int.* **2020**, *131*, 108994. [CrossRef]
30. Wei, T.; Bao, J.-Y.; Yang, H.-H.; Lin, J.-F.; Zheng, Q.-W.; Ye, Z.-W.; Zou, Y.; Li, X.; Jiang, Z.-L.; Guo, L.-Q. *Musa basjoo* regulates the gut microbiota in mice by rebalancing the abundance of probiotic and pathogen. *Microb. Pathog.* **2019**, *131*, 205–211. [CrossRef]
31. Guo, Z.; Hu, B.; Wang, H.; Kong, L.; Han, H.; Li, K.; Sun, S.; Lei, Z.; Shimizu, K.; Zhang, Z. Supplementation with nanobubble water alleviates obesity-associated markers through modulation of gut microbiota in high-fat diet fed mice. *J. Funct. Foods* **2020**, *67*, 103820. [CrossRef]
32. Pan, L.; Han, Y.; Zhou, Z. In Vitro prebiotic activities of exopolysaccharide from *Leuconostocpseudomesenteroides* XG5 and its effect on the gut microbiota of mice. *J. Funct. Foods* **2020**, *67*, 103853. [CrossRef]
33. Rizzetto, L.; Fava, F.; Tuohy, K.M.; Selmi, C. Connecting the immune system, systemic chronic inflammation and the gut mi-crobiome: The role of sex. *J. Autoimm.* **2018**, *92*, 12–34. [CrossRef] [PubMed]
34. Canfora, E.E.; Jocken, J.W.; Blaak, E.E. Short-chain fatty acids in control of body weight and insulin sensitivity. *Nat. Rev. Endocrinol.* **2015**, *11*, 577–591. [CrossRef]
35. Fu, Z.; Han, L.; Zhang, P.; Mao, H.; Zhang, H.; Wang, Y.; Gao, X.; Liu, E. *Cistanche* polysaccharides enhance echinacoside absorption In Vivo and affect the gut microbiota. *Int. J. Biol. Macromol.* **2020**, *149*, 732–740. [CrossRef]
36. Goodrich, J.K.; Waters, J.L.; Poole, A.C.; Sutter, J.L.; Koren, O.; Blekhman, R.; Beaumont, M.; van Treuren, W.; Knight, R.; Bell, J.T.; et al. Human Genetics Shape the Gut Microbiome. *Cell* **2014**, *159*, 789–799. [CrossRef]
37. Singh, A.; Zapata, R.C.; Pezeshki, A.; Reidelberger, R.D.; Chelikani, P.K. Inulin fiber dose-dependently modulates energy balance, glucose tolerance, gut microbiota, hormones and diet preference in high-fat-fed male rats. *J. Nutr. Biochem.* **2018**, *59*, 142–152. [CrossRef]
38. Gao, R.; Zhu, C.; Li, H.; Yin, M.; Pan, C.; Huang, L.; Kong, C.; Wang, X.; Zhang, Y.; Qu, S.; et al. Dysbiosis Signatures of Gut Microbiota Along the Sequence from Healthy, Young Patients to Those with Overweight and Obesity. *Obesity* **2018**, *26*, 351–361. [CrossRef]

MDPI
St. Alban-Anlage 66
4052 Basel
Switzerland
www.mdpi.com

Foods Editorial Office
E-mail: foods@mdpi.com
www.mdpi.com/journal/foods

www.ingramcontent.com/pod-product-compliance
Lightning Source LLC
Chambersburg PA
CBHW061454220926
51850CB00004B/86

* 9 7 8 3 0 3 6 5 9 6 7 4 7 *